Worldwide Evaluations of Quinoa—Biodiversity and Food Security under Climate Change Pressures

Worldwide Evaluations of Quinoa—Biodiversity and Food Security under Climate Change Pressures

Editors

Cataldo Pulvento
Didier Bazile

MDPI • Basel • Beijing • Wuhan • Barcelona • Belgrade • Manchester • Tokyo • Cluj • Tianjin

Editors
Cataldo Pulvento
Department of Soil, Plant and Food Sciences (DISSPA)
University of Bari
Bari
Italy

Didier Bazile
UMR Savoirs, Environnement et Sociétés (SENS)
CIRAD
Montpellier
France

Editorial Office
MDPI
St. Alban-Anlage 66
4052 Basel, Switzerland

This is a reprint of articles from the Special Issue published online in the open access journal *Plants* (ISSN 2223-7747) (available at: www.mdpi.com/journal/plants/special_issues/quinoa).

For citation purposes, cite each article independently as indicated on the article page online and as indicated below:

LastName, A.A.; LastName, B.B.; LastName, C.C. Article Title. *Journal Name* **Year**, *Volume Number*, Page Range.

ISBN 978-3-0365-7043-3 (Hbk)
ISBN 978-3-0365-7042-6 (PDF)

Cover image courtesy of Didier Bazile

Contents

About the Editors . . . . **ix**

Preface to "Worldwide Evaluations of Quinoa—Biodiversity and Food Security under Climate Change Pressures" . . . . **xi**

Cataldo Pulvento and Didier Bazile
Worldwide Evaluations of Quinoa—Biodiversity and Food Security under Climate Change Pressures: Advances and Perspectives
Reprinted from: *Plants* **2023**, *12*, 868, doi:10.3390/plants12040868 . . . . **1**

Mohamed Houssemeddine Sellami, Cataldo Pulvento and Antonella Lavini
Agronomic Practices and Performances of Quinoa under Field Conditions: A Systematic Review
Reprinted from: *Plants* **2020**, *10*, 72, doi:10.3390/plants10010072 . . . . **5**

Maryam Bahrami, Rezvan Talebnejad, Ali Reza Sepaskhah and Didier Bazile
Irrigation Regimes and Nitrogen Rates as the Contributing Factors in Quinoa Yield to Increase Water and Nitrogen Efficiencies
Reprinted from: *Plants* **2022**, *11*, 2048, doi:10.3390/plants11152048 . . . . **25**

Irfan Afzal, Shahzad Maqsood Ahmed Basra, Hafeez Ur Rehman, Shahid Iqbal and Didier Bazile
Trends and Limits for Quinoa Production and Promotion in Pakistan
Reprinted from: *Plants* **2022**, *11*, 1603, doi:10.3390/plants11121603 . . . . **45**

Muhammad Bilal Hafeez, Shahid Iqbal, Yuanyuan Li, Muhammad Sohail Saddiq, Shahzad M. A. Basra and Hui Zhang et al.
Assessment of Phenotypic Diversity in the USDA Collection of Quinoa Links Genotypic Adaptation to Germplasm Origin
Reprinted from: *Plants* **2022**, *11*, 738, doi:10.3390/plants11060738 . . . . **65**

Hafeez ur Rehman, Hesham F. Alharby, Hassan S. Al-Zahrani, Atif A. Bamagoos, Nadiah B. Alsulami and Nadiyah M. Alabdallah et al.
Enriching Urea with Nitrogen Inhibitors Improves Growth, N Uptake and Seed Yield in Quinoa (*Chenopodium quinoa* Willd) Affecting Photochemical Efficiency and Nitrate Reductase Activity
Reprinted from: *Plants* **2022**, *11*, 371, doi:10.3390/plants11030371 . . . . **79**

Kathryn Dumschott, Nathalie Wuyts, Christian Alfaro, Dalma Castillo, Fabio Fiorani and Andrés Zurita-Silva
Morphological and Physiological Traits Associated with Yield under Reduced Irrigation in Chilean Coastal Lowland Quinoa
Reprinted from: *Plants* **2022**, *11*, 323, doi:10.3390/plants11030323 . . . . **93**

Phara De Bock, Gerda Cnops, Hilde Muylle, Paul Quataert, Mia Eeckhout and Filip Van Bockstaele
Physicochemical Characterization of Thirteen Quinoa (*Chenopodium quinoa* Willd.) Varieties Grown in North-West Europe—Part II
Reprinted from: *Plants* **2022**, *11*, 265, doi:10.3390/plants11030265 . . . . **119**

Rodrigo A. Chorbadjian, María I. Ahumada, Francisco Urra, Mario Elgueta and Todd M. Gilligan
Biogeographical Patterns of Herbivore Arthropods Associated with *Chenopodium quinoa* Grown along the Latitudinal Gradient of Chile
Reprinted from: *Plants* **2021**, *10*, 2811, doi:10.3390/plants10122811 . . . . **137**

Phara De Bock, Filip Van Bockstaele, Hilde Muylle, Paul Quataert, Pieter Vermeir and Mia Eeckhout et al.
Yield and Nutritional Characterization of Thirteen Quinoa (*Chenopodium quinoa* Willd.) Varieties Grown in North-West Europe—Part I
Reprinted from: *Plants* **2021**, *10*, 2689, doi:10.3390/plants10122689 . . . . **151**

Jesús E. Cárdenas-Castillo, José Delatorre-Herrera, Luisa Bascuñán-Godoy and Juan Pablo Rodriguez
Quinoa (*Chenopodium quinoa* Wild.) Seed Yield and Efficiency in Soils Deficient of Nitrogen in the Bolivian Altiplano: An Analytical Review
Reprinted from: *Plants* **2021**, *10*, 2479, doi:10.3390/plants10112479 . . . . **177**

M. Iftikhar Hussain, Muhammad Farooq, Qamar Abbas Syed, Anum Ishaq, Abdullah Ahmed Al-Ghamdi and Ashraf A. Hatamleh
Botany, Nutritional Value, Phytochemical Composition and Biological Activities of Quinoa
Reprinted from: *Plants* **2021**, *10*, 2258, doi:10.3390/plants10112258 . . . . **197**

Miguel García-Parra, Diego Roa-Acosta, Víctor García-Londoño, Brigitte Moreno-Medina and Jesús Bravo-Gomez
Structural Characterization and Antioxidant Capacity of Quinoa Cultivars Using Techniques of FT-MIR and UHPLC/ESI-Orbitrap MS Spectroscopy
Reprinted from: *Plants* **2021**, *10*, 2159, doi:10.3390/plants10102159 . . . . **215**

Sara Granado-Rodríguez, Susana Vilariño-Rodríguez, Isaac Maestro-Gaitán, Javier Matías, María José Rodríguez and Patricia Calvo et al.
Genotype-Dependent Variation of Nutritional Quality-Related Traits in Quinoa Seeds
Reprinted from: *Plants* **2021**, *10*, 2128, doi:10.3390/plants10102128 . . . . **231**

Giovanni Beccari, Mara Quaglia, Francesco Tini, Euro Pannacci and Lorenzo Covarelli
Phytopathological Threats Associated with Quinoa (*Chenopodium quinoa* Willd.) Cultivation and Seed Production in an Area of Central Italy
Reprinted from: *Plants* **2021**, *10*, 1933, doi:10.3390/plants10091933 . . . . **253**

Luis Cruces, Eduardo de la Peña and Patrick De Clercq
Field Evaluation of Cypermethrin, Imidacloprid, Teflubenzuron and Emamectin Benzoate against Pests of Quinoa (*Chenopodium quinoa* Willd.) and Their Side Effects on Non-Target Species
Reprinted from: *Plants* **2021**, *10*, 1788, doi:10.3390/plants10091788 . . . . **273**

Clara S. Stanschewski, Elodie Rey, Gabriele Fiene, Evan B. Craine, Gordon Wellman and Vanessa J. Melino et al.
Quinoa Phenotyping Methodologies: An International Consensus
Reprinted from: *Plants* **2021**, *10*, 1759, doi:10.3390/plants10091759 . . . . **291**

Patrícia Carvalho da Silva, Walter Quadros Ribeiro Junior, Maria Lucrecia Gerosa Ramos, Sonia Maria Costa Celestino, Alberto do Nascimento Silva and Raphael Augusto das Chagas Noqueli Casari et al.
Quinoa for the Brazilian Cerrado: Agronomic Characteristics of Elite Genotypes under Different Water Regimes
Reprinted from: *Plants* **2021**, *10*, 1591, doi:10.3390/plants10081591 . . . . **343**

Elsa Helena Manjarres-Hernández, Diana Marcela Arias-Moreno, Ana Cruz Morillo-Coronado, Zaida Zarely Ojeda-Pérez and Agobardo Cárdenas-Chaparro
Phenotypic Characterization of Quinoa (*Chenopodium quinoa* Willd.) for the Selection of Promising Materials for Breeding Programs
Reprinted from: *Plants* **2021**, *10*, 1339, doi:10.3390/plants10071339 . . . . **361**

Carla Colque-Little, Daniel Buchvaldt Amby and Christian Andreasen
A Review of *Chenopodium quinoa* (Willd.) Diseases—An Updated Perspective
Reprinted from: *Plants* **2021**, *10*, 1228, doi:10.3390/plants10061228 . . . . **377**

Sifeddine Rafik, Mohamed Rahmani, Juan Pablo Rodriguez, Said Andam, Amine Ezzariai and Mohamed El Gharous et al.
How Does Mechanical Pearling Affect Quinoa Nutrients and Saponin Contents?
Reprinted from: *Plants* **2021**, *10*, 1133, doi:10.3390/plants10061133 . . . . **417**

Fabiana Antognoni, Giulia Potente, Stefania Biondi, Roberto Mandrioli, Lorenzo Marincich and Karina B. Ruiz
Free and Conjugated Phenolic Profiles and Antioxidant Activity in Quinoa Seeds and Their Relationship with Genotype and Environment
Reprinted from: *Plants* **2021**, *10*, 1046, doi:10.3390/plants10061046 . . . . **439**

Javier Matías, Verónica Cruz and María Reguera
Heat Stress Impact on Yield and Composition of Quinoa Straw under Mediterranean Field Conditions
Reprinted from: *Plants* **2021**, *10*, 955, doi:10.3390/plants10050955 . . . . **455**

José Delatorre-Herrera, Karina B. Ruiz and Manuel Pinto
The Importance of Non-Diffusional Factors in Determining Photosynthesis of Two Contrasting Quinoa Ecotypes (*Chenopodium quinoa* Willd.) Subjected to Salinity Conditions
Reprinted from: *Plants* **2021**, *10*, 927, doi:10.3390/plants10050927 . . . . **467**

Elhadji Thiam, Asmaa Allaoui and Ouafae Benlhabib
Quinoa Productivity and Stability Evaluation through Varietal and Environmental Interaction
Reprinted from: *Plants* **2021**, *10*, 714, doi:10.3390/plants10040714 . . . . **489**

Emma M. McGinty, Kevin M. Murphy and Amber L. Hauvermale
Seed Dormancy and Preharvest Sprouting in Quinoa (*Chenopodium quinoa* Willd)
Reprinted from: *Plants* **2021**, *10*, 458, doi:10.3390/plants10030458 . . . . **503**

Juan A. González, Leonardo Hinojosa, María I. Mercado, José-Luis Fernández-Turiel, Didier Bazile and Graciela I. Ponessa et al.
A Long Journey of CICA-17 Quinoa Variety to Salinity Conditions in Egypt: Mineral Concentration in the Seeds
Reprinted from: *Plants* **2021**, *10*, 407, doi:10.3390/plants10020407 . . . . **517**

Abdelaziz Hirich, Sifeddine Rafik, Mohamed Rahmani, Amira Fetouab, Fatima Azaykou and Kaoutar Filali et al.
Development of Quinoa Value Chain to Improve Food and Nutritional Security in Rural Communities in Rehamna, Morocco: Lessons Learned and Perspectives
Reprinted from: *Plants* **2021**, *10*, 301, doi:10.3390/plants10020301 . . . . **531**

Leonardo Hinojosa, Alex Leguizamo, Carlos Carpio, Diego Muñoz, Camilo Mestanza and José Ochoa et al.
Quinoa in Ecuador: Recent Advances under Global Expansion
Reprinted from: *Plants* **2021**, *10*, 298, doi:10.3390/plants10020298 . . . . **557**

About the Editors

Cataldo Pulvento

Cataldo Pulvento is an Associate Professor of Agronomy and Crop Science at the University of Bari. Prior to joining the University of Bari in March 2021, Pulvento was a researcher at the Italian National Council of Research from 2005 to 2021. Pulvento received his Ph.D. in Mediterranean Agronomy from the University of Bari in 2011. His research is focused on herbaceous crops, with a particular focus on quinoa and amaranth, soil–water plant interactions, best agronomic practices, and crop tolerance to abiotic stresses. Since 2005, he has wrote and participated in regional, national, and international research projects, e.g., EU-funded TRUSTFARM, SWUP-MED, and Protein2Food, Quinoa Felix, etc., and he has also coordinated technology transfer projects for the development of the quinoa supply chain in Italy. He has cooperated with the FAO as a quinoa expert for technical cooperation projects on quinoa in central Asia and Arab Countries.

Didier Bazile

Dr Didier Bazile is a senior scientist based at the French Agricultural Research Centre for International Development, an organization working towards the sustainable development of tropical and Mediterranean regions (CIRAD). Didier is an active researcher in agroecology, with a focus on agrobiodiversity and plant genetic resources. He is an expert in the area of Neglected and Underutilized Crops production and the In Situ Conservation of Biodiversity. He received his Ph.D. in Rural Geography (Toulouse, 1998) and graduated as the Director of Research (HdR) (Montpellier, 2014), working towards Biodiversity and Natural Resources Management considering both conservation and the sustainable use of biodiversity. After different international experiences (Africa, Asia, and South America) for the European Union, the World Bank Group, and French Ministries, he joined the CIRAD in 2001 as a Principal Scientist where he has since developed an ecosystem approach and promoted participatory research. From 2008 to 2012, he adopted the position of Associated Professor of the Institute of Geography (PUCV, Valparaiso, Chile). He worked in close collaboration with Andean national research institutions for characterizing quinoa genetic resources and traditional knowledge. During 2014–2016, he was invited to FAO-HQ as a Visiting Expert; he supported the preparation and animation of an informal stakeholder dialogue on a global networking mechanism on in situ conservation and on-farm management of plant genetic resources for food and agriculture (PGRFA). After three years as the Cirad Regional Director for the Mediterranean, the Middle East, and the Balkans regions, he was appointed as CIRAD Biodiversity Advisor. Bazile wishes to integrate his experience on multi-level and multiple-actor dialogues for facilitating new partnerships, with a particular focus on gender dimensions.

Preface to "Worldwide Evaluations of Quinoa—Biodiversity and Food Security under Climate Change Pressures"

Quinoa (Chenopodium quinoa Willd.) is a herbaceous plant which has been domesticated for more than 5000 years BP in the Andean region. The crop is characterized by a very high biodiversity, which allows it to adapt easily considering the very different pedoclimatic conditions it faces, as well as makes it resistant to abiotic stresses and climate change. Moreover, quinoa is distinguished by its exceptional nutritional characteristics, such as the content and quality of its proteins, minerals, lipids, and tocopherols. These features have determined, since the 1990s, the growing interest in quinoa crop by the scientific community and international organizations. In 2013, the United Nations Organization for Food and Agriculture (FAO) celebrated the "International Year of Quinoa" to valorize its biodiversity for fighting against food insecurity. Several experiments around the world have taken place in order to study the quinoa plant, evaluating the adaptability of different genotypes in new environments and its response to various laboratory stimuli and cultivation best practices. This volume collates the most recent developments from studies on quinoa worldwide.

Cataldo Pulvento and Didier Bazile
Editors

plants

Editorial

Worldwide Evaluations of Quinoa—Biodiversity and Food Security under Climate Change Pressures: Advances and Perspectives

Cataldo Pulvento [1,*] and Didier Bazile [2,3,*]

1 Department of Soil, Plant and Food Science (DISSPA) University of Bari, 70121 Bari, Italy
2 CIRAD, UMR SENS, F-34398 Montpellier, France
3 SENS, Univ Montpellier, CIRAD, F-34398 Montpellier, France
* Correspondence: cataldo.pulvento@uniba.it (C.P.); didier.bazile@cirad.fr (D.B.)

1. Introduction

Quinoa (*Chenopodium quinoa* Willd.) is an Andean herbaceous crop that has attracted increasing interest in recent years thanks to its ecophysiological behavior and the nutritional characteristics of its seeds. The quinoa boom followed the celebration of the International Year of Quinoa in 2013 by the United Nations (FAO), when numerous initiatives were implemented to spread the positive characteristics that make quinoa a suitable crop with which to fight world hunger. In this Special Issue, we wanted to summarize the state of the art and the main research activities that are currently underway in different parts of the world.

2. Ecophysiological Traits and Adaptability

A bibliographic analysis of selected papers published from 2000 to 2020 highlighted that the number of studies on the best agronomic practices for quinoa strongly increased after 2013, when the FAO celebrated the International Year of Quinoa and disseminated the importance of quinoa as a high-quality protein crop resistant to unfavorable environments. Experimentation activity especially increased in countries characterized by a hot, arid climate and water scarcity (Morocco, Egypt, Burkina Faso, and the UAE), as well as in countries at risk of water and salt stress due to climate change (Italy, Greece, Turkey, Pakistan, and the USA), with trials focused on the effect of deficit irrigation and the use of saline water on quinoa yield and quality [1]. The same theme was also taken up by the papers published in this Special Issue; quinoa confirmed its adaptability to arid environments such as the Brazilian Cerrado, where water regimes between 309 and 389 mm do not reduce grain yield with respect to higher irrigation volumes [2].

In the same way, a field experiment in the southern Atacama Desert in Chile to investigate the responses to reduced irrigation of nine previously selected coastal lowland self-pollinated (CLS) lines and the commercial cultivar Regalona showed that several lines performed best when faced with a 50% reduction in irrigation [3].

Bharami et al. [4] studied the yield response of quinoa cv. Titicaca under field conditions in Iran and showed that the application of 75% of full irrigation requirements led to NO_3-N accumulation in upper soil layers, thus facilitating nitrogen uptake and reduced nitrate loss to deeper layers of the soil and allowing for a reduction in the optimal nitrogen fertilization level for the study area.

Quinoa responds positively to fertilization in the Bolivian Altiplano [5], with differences among irrigated and rainfed conditions; quinoa can produce 1850 kg grains ha^{-1} with 50 kg N ha^{-1} under irrigated conditions and 670 kg grains ha^{-1} with 15 kg N ha^{-1} in rainfed conditions.

Rehman et al. [6] demonstrated, in Pakistan, that urea enriched with urease and nitrification inhibitors simultaneously can be used to improve the N uptake, seed yield, and grain protein contents in quinoa.

Citation: Pulvento, C.; Bazile, D. Worldwide Evaluations of Quinoa—Biodiversity and Food Security under Climate Change Pressures: Advances and Perspectives. *Plants* **2023**, *12*, 868. https://doi.org/10.3390/plants12040868

Received: 30 November 2022
Revised: 7 February 2023
Accepted: 9 February 2023
Published: 15 February 2023

Quinoa was confirmed as maybe being a complementary crop in the marginal lands of high salinity in Egypt and the Mediterranean region [7]. Delatorre-Herrera et al. [8] demonstrated that the salinity tolerance of salares ecotypes is due to their ability to control non-diffusional components, indicating their superior photosynthetic capacity under salt stress conditions. Quinoa has also been proven to be a promising crop in cases of heat stress, with increased values of crude protein, ash, phosphorus, calcium, and relative feed [9].

Many papers from the literature are focused on the study of the best time and density for sowing, which represent the main agronomic practices for the introduction of a crop to a new environment [1]. In this Special Issue, new field trials evaluated the adaptability of quinoa to new environments in terms of yield, quality, and physiochemical characteristics in Belgium [10,11], Morocco [12], Pakistan [13], and Spain [14], in addition to selecting promising materials for breeding programs under greenhouse conditions [15]. The cultivation of quinoa was also reviewed in Pakistan [16] and Ecuador [17]. A large group of researchers from universities and research institutes from all over the world have proposed standard methodology guidelines [18] to be used for the phenotypic characterization of quinoa in order to improve comparability among field trials across the globe and to facilitate collaborations with the Global Collaborative Network on Quinoa (gcn-quinoa.org). Aspects related to quinoa diseases were reviewed by Colque-Little et al. [19], who summarized existing information on symptoms and causal agents. In Central Italy, the presence of *P. variabilis* and *F. equiseti* was monitored on *C. quinoa* [20]. Seed dormancy and breeding as well as nonbreeding strategies for enhancing resistance to preharvest sprouting in quinoa were reviewed by McGynti et al. [21].

Other ecological aspects, such as the geographical distribution of herbivore arthropods that affect the production of quinoa [22] and the impact of insecticides on insect pests of quinoa, as well as their side effects on the arthropod community [23], were analyzed in Chile and Peru.

3. Quinoa Seed Quality and Post-Harvest Activities

Hussain et al. [24] summarized recent findings regarding the nutritional and phytochemical properties of quinoa grains. A spectroscopy analysis of different quinoa cultivars grown under greenhouse conditions was conducted by García-Parra et al. [25] to evaluate the structural characterization and antioxidant capacity of quinoa.

The profiles of bioactive compounds in seeds of two quinoa varieties, Regalona-Baer and Titicaca, grown in Northern Italy, compared to that of seeds of those varieties grown in Chile and Denmark, were respectively assessed in order to establish the best conditions (genotype/geographical cultivation zone) leading to seed enrichment in functional compounds [26].

The pearling of quinoa seeds, nutrients, and saponin contents was evaluated to determine a correct standard for postharvest seed processing [27]; the description of a project aimed at the development of a potential quinoa value chain in order to improve food and nutritional security in rural communities in Rehamna, Morocco, was also reported [28].

Author Contributions: Writing—original draft preparation, C.P.; writing—review and editing, C.P. and D.B. All authors have read and agreed to the published version of the manuscript.

Conflicts of Interest: The authors declare no conflict of interest.

References

1. Sellami, M.H.; Pulvento, C.; Lavini, A. Agronomic Practices and Performances of Quinoa under Field Conditions: A Systematic Review. *Plants* **2021**, *10*, 72. [CrossRef]
2. da Silva, P.C.; Ribeiro Junior, W.Q.; Ramos, M.L.G.; Celestino, S.M.C.; Silva, A.d.N.; Casari, R.A.d.C.N.; Santana, C.C.; de Lima, C.A.; Williams, T.C.R.; Vinson, C.C. Quinoa for the Brazilian Cerrado: Agronomic Characteristics of Elite Genotypes under Different Water Regimes. *Plants* **2021**, *10*, 1591. [CrossRef]
3. Dumschott, K.; Wuyts, N.; Alfaro, C.; Castillo, D.; Fiorani, F.; Zurita-Silva, A. Morphological and Physiological Traits Associated with Yield under Reduced Irrigation in Chilean Coastal Lowland Quinoa. *Plants* **2022**, *11*, 323. [CrossRef]

4. Bahrami, M.; Talebnejad, R.; Sepaskhah, A.R.; Bazile, D. Irrigation Regimes and Nitrogen Rates as the Contributing Factors in Quinoa Yield to Increase Water and Nitrogen Efficiencies. *Plants* **2022**, *11*, 2048. [CrossRef]
5. Cárdenas-Castillo, J.E.; Delatorre-Herrera, J.; Bascuñán-Godoy, L.; Rodriguez, J.P. Quinoa (*Chenopodium quinoa* Wild.) Seed Yield and Efficiency in Soils Deficient of Nitrogen in the Bolivian Altiplano: An Analytical Review. *Plants* **2021**, *10*, 2479. [CrossRef]
6. Rehman, H.u.; Alharby, H.F.; Al-Zahrani, H.S.; Bamagoos, A.A.; Alsulami, N.B.; Alabdallah, N.M.; Iqbal, T.; Wakeel, A. Enriching Urea with Nitrogen Inhibitors Improves Growth, N Uptake and Seed Yield in Quinoa (*Chenopodium quinoa* Willd) Affecting Photochemical Efficiency and Nitrate Reductase Activity. *Plants* **2022**, *11*, 371. [CrossRef]
7. González, J.A.; Hinojosa, L.; Mercado, M.I.; Fernández-Turiel, J.-L.; Bazile, D.; Ponessa, G.I.; Eisa, S.; González, D.A.; Rejas, M.; Hussin, S.; et al. A Long Journey of CICA-17 Quinoa Variety to Salinity Conditions in Egypt: Mineral Concentration in the Seeds. *Plants* **2021**, *10*, 407. [CrossRef]
8. Delatorre-Herrera, J.; Ruiz, K.B.; Pinto, M. The Importance of Non-Diffusional Factors in Determining Photosynthesis of Two Contrasting Quinoa Ecotypes (*Chenopodium quinoa* Willd.) Subjected to Salinity Conditions. *Plants* **2021**, *10*, 927. [CrossRef]
9. Matías, J.; Cruz, V.; Reguera, M. Heat Stress Impact on Yield and Composition of Quinoa Straw under Mediterranean Field Conditions. *Plants* **2021**, *10*, 955. [CrossRef]
10. De Bock, P.; Van Bockstaele, F.; Muylle, H.; Quataert, P.; Vermeir, P.; Eeckhout, M.; Cnops, G. Yield and Nutritional Characterization of Thirteen Quinoa (*Chenopodium quinoa* Willd.) Varieties Grown in North-West Europe—Part I. *Plants* **2021**, *10*, 2689. [CrossRef]
11. De Bock, P.; Cnops, G.; Muylle, H.; Quataert, P.; Eeckhout, M.; Van Bockstaele, F. Physicochemical Characterization of Thirteen Quinoa (*Chenopodium quinoa* Willd.) Varieties Grown in North-West Europe—Part II. *Plants* **2022**, *11*, 265. [CrossRef]
12. Thiam, E.; Allaoui, A.; Benlhabib, O. Quinoa Productivity and Stability Evaluation through Varietal and Environmental Interaction. *Plants* **2021**, *10*, 714. [CrossRef]
13. Hafeez, M.B.; Iqbal, S.; Li, Y.; Saddiq, M.S.; Basra, S.M.A.; Zhang, H.; Zahra, N.; Akram, M.Z.; Bertero, D.; Curti, R.N. Assessment of Phenotypic Diversity in the USDA Collection of Quinoa Links Genotypic Adaptation to Germplasm Origin. *Plants* **2022**, *11*, 738. [CrossRef]
14. Granado-Rodríguez, S.; Vilariño-Rodríguez, S.; Maestro-Gaitán, I.; Matías, J.; Rodríguez, M.J.; Calvo, P.; Cruz, V.; Bolaños, L.; Reguera, M. Genotype-Dependent Variation of Nutritional Quality-Related Traits in Quinoa Seeds. *Plants* **2021**, *10*, 2128. [CrossRef]
15. Manjarres-Hernández, E.H.; Arias-Moreno, D.M.; Morillo-Coronado, A.C.; Ojeda-Pérez, Z.Z.; Cárdenas-Chaparro, A. Phenotypic Characterization of Quinoa (*Chenopodium quinoa* Willd.) for the Selection of Promising Materials for Breeding Programs. *Plants* **2021**, *10*, 1339. [CrossRef]
16. Afzal, I.; Basra, S.M.A.; Rehman, H.U.; Iqbal, S.; Bazile, D. Trends and Limits for Quinoa Production and Promotion in Pakistan. *Plants* **2022**, *11*, 1603. [CrossRef]
17. Hinojosa, L.; Leguizamo, A.; Carpio, C.; Muñoz, D.; Mestanza, C.; Ochoa, J.; Castillo, C.; Murillo, A.; Villacréz, E.; Monar, C.; et al. Quinoa in Ecuador: Recent Advances under Global Expansion. *Plants* **2021**, *10*, 298. [CrossRef]
18. Stanschewski, C.S.; Rey, E.; Fiene, G.; Craine, E.B.; Wellman, G.; Melino, V.J.; S. R. Patiranage, D.; Johansen, K.; Schmöckel, S.M.; Bertero, D.; et al. Quinoa Phenotyping Methodologies: An International Consensus. *Plants* **2021**, *10*, 1759. [CrossRef]
19. Colque-Little, C.; Amby, D.B.; Andreasen, C. A Review of *Chenopodium quinoa* (Willd.) Diseases—An Updated Perspective. *Plants* **2021**, *10*, 1228. [CrossRef]
20. Beccari, G.; Quaglia, M.; Tini, F.; Pannacci, E.; Covarelli, L. Phytopathological Threats Associated with Quinoa (*Chenopodium quinoa* Willd.) Cultivation and Seed Production in an Area of Central Italy. *Plants* **2021**, *10*, 1933. [CrossRef]
21. McGinty, E.M.; Murphy, K.M.; Hauvermale, A.L. Seed Dormancy and Preharvest Sprouting in Quinoa (*Chenopodium quinoa* Willd). *Plants* **2021**, *10*, 458. [CrossRef]
22. Chorbadjian, R.A.; Ahumada, M.I.; Urra, F.; Elgueta, M.; Gilligan, T.M. Biogeographical Patterns of Herbivore Arthropods Associated with *Chenopodium quinoa* Grown along the Latitudinal Gradient of Chile. *Plants* **2021**, *10*, 2811. [CrossRef]
23. Cruces, L.; de la Peña, E.; De Clercq, P. Field Evaluation of Cypermethrin, Imidacloprid, Teflubenzuron and Emamectin Benzoate against Pests of Quinoa (*Chenopodium quinoa* Willd.) and Their Side Effects on Non-Target Species. *Plants* **2021**, *10*, 1788. [CrossRef]
24. Hussain, M.I.; Farooq, M.; Syed, Q.A.; Ishaq, A.; Al-Ghamdi, A.A.; Hatamleh, A.A. Botany, Nutritional Value, Phytochemical Composition and Biological Activities of Quinoa. *Plants* **2021**, *10*, 2258. [CrossRef]
25. García-Parra, M.; Roa-Acosta, D.; García-Londoño, V.; Moreno-Medina, B.; Bravo-Gomez, J. Structural Characterization and Antioxidant Capacity of Quinoa Cultivars Using Techniques of FT-MIR and UHPLC/ESI-Orbitrap MS Spectroscopy. *Plants* **2021**, *10*, 2159. [CrossRef]
26. Antognoni, F.; Potente, G.; Biondi, S.; Mandrioli, R.; Marincich, L.; Ruiz, K.B. Free and Conjugated Phenolic Profiles and Antioxidant Activity in Quinoa Seeds and Their Relationship with Genotype and Environment. *Plants* **2021**, *10*, 1046. [CrossRef]
27. Rafik, S.; Rahmani, M.; Rodriguez, J.P.; Andam, S.; Ezzariai, A.; El Gharous, M.; Karboune, S.; Choukr-Allah, R.; Hirich, A. How Does Mechanical Pearling Affect Quinoa Nutrients and Saponin Contents? *Plants* **2021**, *10*, 1133. [CrossRef]

28. Hirich, A.; Rafik, S.; Rahmani, M.; Fetouab, A.; Azaykou, F.; Filali, K.; Ahmadzai, H.; Jnaoui, Y.; Soulaimani, A.; Moussafir, M.; et al. Development of Quinoa Value Chain to Improve Food and Nutritional Security in Rural Communities in Rehamna, Morocco: Lessons Learned and Perspectives. *Plants* **2021**, *10*, 301. [CrossRef]

plants

MDPI

Review

Agronomic Practices and Performances of Quinoa under Field Conditions: A Systematic Review

Mohamed Houssemeddine Sellami [1], Cataldo Pulvento [2,*] and Antonella Lavini [1]

1 Institute for Agricultural and Forestry Systems in the Mediterranean (ISAFOM), National Research Council of Italy (CNR), 80055 Portici (NA), Italy; mohamed.sellami@isafom.cnr.it (M.H.S.); antonella.lavini@isafom.cnr.it (A.L.)

2 Institute for Biosciences and Bioresources (IBBR), National Research Council of Italy (CNR), Via Amendola, 165/A, 70126 Bari, Italy

* Correspondence: cataldo.pulvento@ibbr.cnr.it

Abstract: Quinoa (*Chenopodium quinoa* Willd.) is one of the most popular emerging food crops in the Andean region. It is tolerant to environmental stresses and characterized by interesting nutritional traits. Thus, it has the potential to contribute to food and nutrition security in marginal environments. In this study, we conducted a systematic review integrated with a bibliometric analysis of cropping practices of quinoa under field conditions. The analysis is based on published data from the literature relating to the period 2000–2020. A total of 33 publications were identified, revealing that scientific research on the agronomic practices and performances of quinoa under field conditions is still limited. Africa, Asia, and Europe were the leading research production sites in this field and together provided over 81% of the total scientific production. There were no papers from the Australian continent. The number of papers screened dealing with tillage and weed control management was very limited. The keyword co-occurrence network analyses revealed that the main topics addressed in the scientific literature related to the effect of "variety" and "deficit irrigation", followed by "water quality", "fertilization", and "sowing date" on seed yield. Results from this study will permit us to identify knowledge gaps and limited collaboration among authors and institutions from different countries. Salinity, sowing density, and sowing date were the agronomic interventions affecting productive response the most.

Keywords: quinoa; systematic review; bibliometric analysis; concept network analysis; agronomic practices

Citation: Sellami, M.H.; Pulvento, C.; Lavini, A. Agronomic Practices and Performances of Quinoa under Field Conditions: A Systematic Review. *Plants* **2021**, *10*, 72. https://doi.org/10.3390/plants10010072

Received: 9 December 2020
Accepted: 28 December 2020
Published: 31 December 2020

1. Introduction

In recent years, quinoa (*Chenopodium quinoa* Willd.) cultivation has expanded to several countries beyond its area of origin due to increasing interest, market development, research, and promotion [1]. Thanks to its high-quality protein content, quinoa is considered a promising candidate for enhancing high-quality plant protein food production in the world [2]. It is recognized as a crop of great value in terms of tolerance to abiotic stresses and it is one of the most nutritious food crops currently known [3]. However, there are still many issues, including a lack of knowledge of best management practices, which need to be addressed in order to introduce quinoa crop in marginal areas [4].

Till the early 1900s, the cultivation of quinoa remained limited to its countries of origin. In the following century, quinoa arrived in Africa, Asia, Europe, and North America [1]. Until 1999, few studies were published in peer-review journals concerning agronomic practices in the open field. Risi and Galewey [5] evaluated the effect of sowing density and sowing date on different genotypes of quinoa grown in the UK. Jacobsen et al. [6] analyzed the effects of varying the nitrogen fertilization rate, seed rate, row spacing, harvesting method, and harvest date on quinoa in Denmark. In 1998, Vacher [7] analyzed the effect of drought on quinoa in Bolivia. Besides these few examples, most of the studies carried out before the new millennium on quinoa were reported in the so-called gray literature; many of the experimental trials were conducted in the countries of origin, but not available in international databases.

Despite the importance of quinoa in marginal areas, its adaptability to unfavorable environments, high protein content, and the interesting nutritional properties of the seed, few studies have been conducted on its yield responses to different strategies of agronomic management under field conditions; this represents a gap in research in this field.

Therefore, there is a need for a systematic review integrated with a bibliometric analysis to answer the question: what is the research gap in agronomic management and performance of quinoa under field conditions?

A systematic review (SR) is defined as a review of the evidence on a formulated question that uses systematic and explicit methods to identify, select, and critically appraise relevant primary research; it is used to extract and analyze data from the studies included in the review [8]. Bibliometric or research impact is the quantitative method of analyzing citations and content for academic journals, books, and researchers. The quantitative impact of a given publication is appraised by measuring the number of times a certain work is cited by other resources [9].

This study aims to apply a systematic review integrated with a bibliometric analysis to evaluate the research trend of the last twenty years (2000–2020) on the subject of cropping practices of quinoa production under field conditions.

2. Results

2.1. Screening Process

In total 520 sources of literature were identified within academic databases (after the removal of 252 duplicates or non-journal papers), of which 33 were selected and analyzed (Table 1), to provide 513 observations. The screening process is described in Figure 1.

2.2. Evolution Articles over the Years

Figure 2 shows the annual scientific production dealing with the agronomic practices and performances of quinoa under field conditions in the world. The research on the effect of agronomic practices on quinoa under field conditions is considered relatively young and started less than twenty years ago. In fact, the first paper in this research area was published in 2003. The number of published research papers has fluctuated over the last two decades, reaching a peak of eight during 2019. This leads us to infer that this rise in the number of articles over the years represents an increasing interest in this research area.

2.3. Geographical Distribution of Articles

Our screening process reveals that accessible published research on the agronomic practices and performances of quinoa under field conditions in the world, with high reporting standard suitable for this systematic review, is concentrated in Europe (13 articles, 39%), followed by Africa (8 articles, 24%), and Asia (6 articles, 18%) (Table 2, Figure 3). These three continents together represent more than 80% of the research papers published in the past two decades. Such research is lacking in the Australian continent, a large part of which is arid.

Table 1. Description of the experimental studies in the selected references.

N°	References	No. of Observations	Location	Climatic Zone	Soil Texture	Year	No. of Genotypes	Yield Range (t ha^{-1})	Agronomic Management
1	Jacobsen et al. [10]	18	Tastrup (DK)	Cfb	Coarse	2004; 2005	1	1.17–1.99	Weed Control
2	Pulvento et al. [11]	12	Vitulazio (IT)	Csa	Medium	2009; 2010	1	1.96–3.1	[Deficit irrigation; Salinity]
3	Jacobsen and Christiansen [12]	6	Tastrup (DK)	Cfb	Coarse	2007–2009	1	1.74–2.27	Fertilization
4	Pulvento et al. [13]	2	Vitulazio (IT)	Csa	Medium	2006	1	1.5–3.28	Sowing date
5	Geren [14]	14	Bornova (TR)	Csa	Medium	2013; 2014	1	0.87–3.31	Fertilization
6	Ince Kaya and Yazar [15]	14	Adana (TR)	Csa	Fine	2010; 2012	1	1.51–2.99	Deficit irrigation; Salinity
7	Yazar et al. [16]	31	Adana (TR)	Csa	Fine	2010–2012	1	1.28–3.17	[Deficit irrigation; Salinity]
8	Alvar-Beltrán et al. [17]	13	Bobo Dioulasso (BF)	Aw	Coarse	2017; 2018	1	0.23–1.36	[Deficit irrigation; Fertilization]
9	Alvar-Beltrán et al. [18]	36	Bobo Dioulasso (BF)	Aw	Coarse	2017	2	0.01–1.91	[Deficit irrigation; Fertilization]; Sowing date
10	Asher et al. [19]	24	Avnei Eitan (IL)	Csa	Medium	2016; 2017	6	1.52–6.34	Sowing date
11	Basra et al. [20]	10	Faisalabad (PK)	BWh	Coarse	2010	2	0.01–0.02	Fertilization
12	Fghire et al. [21]	8	Marrakech (MA)	BSh	Coarse	2011; 2012	1	1.10–4.09	Deficit irrigation
13	Hirich et al. [22]	10	Agadir (MA)	BSh	Medium	2012	1	0.13–3.07	Sowing date
14	Hirich et al. [23]	6	Agadir (MA)	BSh	Medium	2011	1	1.70–3.30	[Deficit irrigation; Fertilization]
15	Filali et al. [24]	20	Agadir (MA)	BSh	Medium	2012	5	1.60–4.60	Deficit irrigation
16	Bertero and Ruiz [25]	8	Pergamino (AR)	Csc	Medium	2003	4	2.08–4.94	Sowing density
17	Bilalis et al. [26]	12	Agrinio (GR)	Csa	Medium	2010; 2011	1	2.18–2.64	[Tillage; Fertilization]
18	Çolak et al. [27]	22	Adana (TR)	Csa	Fine	2016; 2017	1	1.86–3.02	Deficit irrigation
19	Eisa et al. [28]	2	Sinai (EG)	BWh	Coarse; Medium	2014	1	0.44–1.15	Salinity
20	Fischer et al. [29]	12	Nuble (CL)	Csb	Medium	2010	3	3.61–5.84	Deficit irrigation
21	Geerts et al. [30]	21	Patacamaya; Condoriri (BO)	BSk	Medium	2005; 2006	1	0.79–2.10	[Deficit irrigation; Fertilization]; [Deficit irrigation.; Sowing date]; Deficit irrigation

Table 1. *Cont.*

N°	References	No. of Observations	Location	Climatic Zone	Soil Texture	Year	No. of Genotypes	Yield Range (t ha^{-1})	Agronomic Management
22	Hinojosa et al. [31]	12	Pullman (US)	Csa	Coarse	2016	6	0.00–0.96	Deficit irrigation
23	Hinojosa et al. [32]	32	Pullman; Chimacum; Mount Vernon (US)	Csa; Csb	Medium; Coarse	2016; 2017	6	0.01–2.94	Deficit irrigation
24	Hirich et al. [33]	6	Agadir (MA)	BSh	Medium	2011	1	1.70–3.30	[Deficit irrigation; Fertilization]
25	Iqbal et al. [34]	8	Faisalabad; Pindi Bhattian (PK)	BWh	Coarse; Medium	2013	2	1.59–3.06	Salinity
26	Kakabouki et al. [35]	16	Agrinio (GR)	Csa	Medium	2012; 2013	1	1.90–2.63	[Tillage; Fertilization]
27	Kakabouki et al. [36]	24	Agrinio (GR)	Csa	Medium	2011–2013	1	1.90–2.66	[Tillage; Fertilization]
28	Karyotis et al. [37]	16	Larissa (GR)	BSk	Fine	2001	8	0.52–2.30	Salinity
29	Martınez et al. [38]	12	Coquimbo; Ovalle (CH)	BWk	Coarse	2004; 2005	2	0.55–7.80	Deficit irrigation
30	Noulas et al. [39]	24	Larissa (GR)	BSk	Fine	1996; 1998; 2001	10	0.52–2.30	Sowing density; Salinity; Sowing date
31	Präger et al. [40]	8	Stuttgart (DE)	Cfb	Medium	2016	2	2.89–3.93	Sowing date
32	Razzaghi et al. [41]	16	Shiraz (IR)	Csa	Medium	2016; 2017	1	0.26–0.86	Deficit irrigation
33	Rezzouk et al. [42]	38	Dubai (UAE)	BWh	Coarse	2016	19	0.53–6.35	Salinity

Agronomic management in square brackets represents multiple interventions. Köppen–Geiger climate zone: Aw: tropical wet-dry climate; BSh: hot semi-arid climate; BSk: cold semi-arid climate; BWh: hot desert climate; BWk: cold desert climate; Cfa: humid subtropical; Cfb: maritime; Csa: interior Mediterranean; Csb: coastal Mediterranean; and Csc: cold-summer Mediterranean climate. DK: Denmark; IT: Italy; TR: Turkey; BF: Burkina Faso; IL: Israel; PK: Pakistan; MA: Morocco; AR: Argentina; GR: Greece; EG: Egypt; CL: Chile; BO: Bolivia; US: United States of America; CH: Switzerland; DE: Germany; IR: Iran; and UAE: United Arab Emirates. Where two years are separated by a semicolon this indicates that an experience is not repeated each year; where two years are separated by a hyphen this indicates that an experience repeated over time.

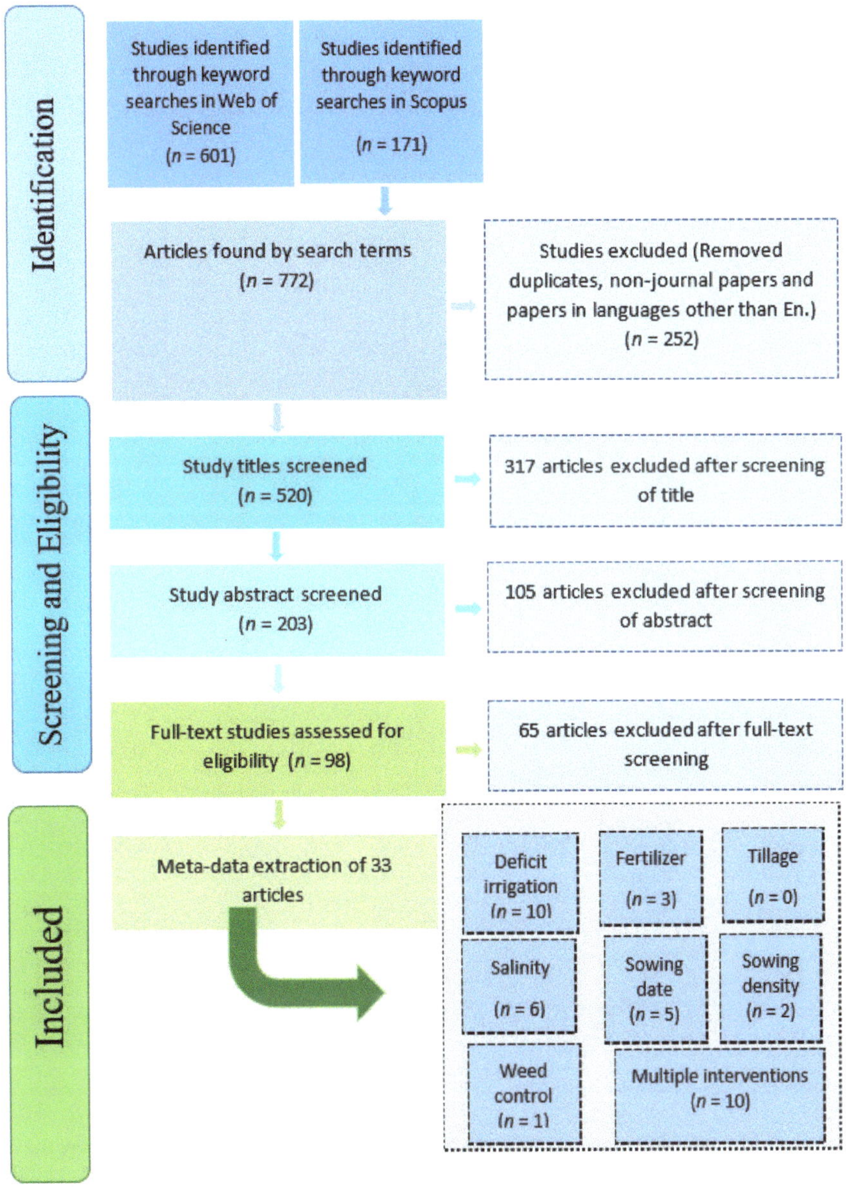

Figure 1. Selection of studies for inclusion in the systematic review (*n*—the number of studies).

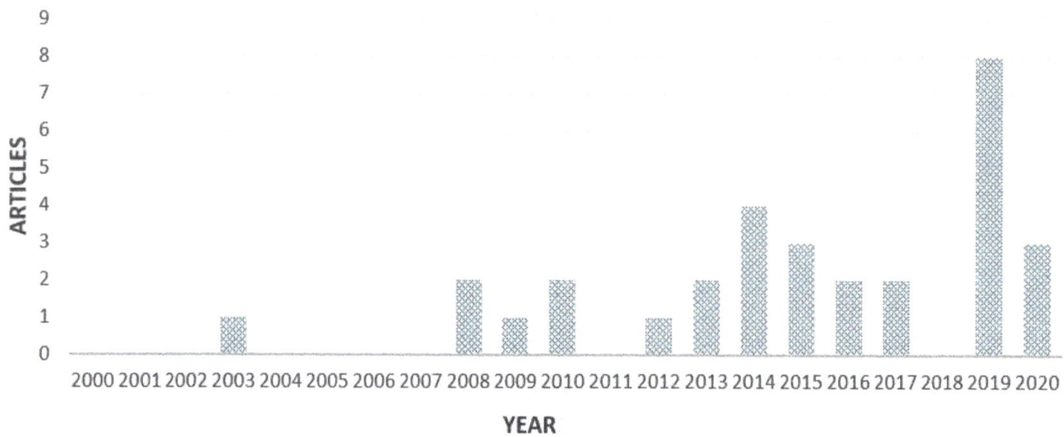

Figure 2. Annual scientific production.

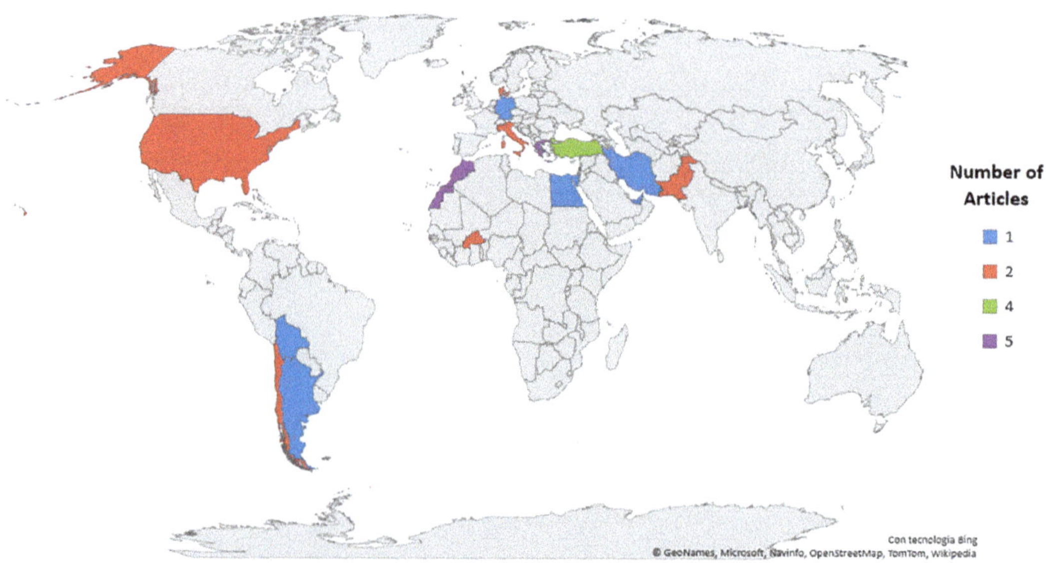

Figure 3. Number of published research articles worldwide.

In the following text, numbers in brackets indicate the number of research articles published in the categories described. Greece and Morocco ($n = 5$) were the most frequently studied countries, followed by Turkey ($n = 4$) (Figure 3).

Table 2. Continent-wise descriptive statistical parameters (minimum, maximum, mean, median, SD, and CV) relative to yield (t ha^{-1}) for all agronomic managements.

Continent	No. of Countries	List of Countries	No. of Cases	No. of Observations	No. of Genotypes	Yield (t ha^{-1})					C.V. [2] (%)
						Minimum	Maximum	Mean	Median	S.D. [1]	
Africa	3	Burkina Faso, Egypt, Morocco	8	101	10	0.01	4.60	1.42	1.08	1.27	89
Asia	5	Iran, Israel, Pakistan, Turkey, United Arab Emirates	6	118	29	0.01	6.35	2.55	2.45	1.40	64
Europe	4	Denmark, Germany, Greece, Italy	13	197	14	0.52	3.93	2.02	2.05	0.69	33
North America	1	US	2	44	6	0.002	2.94	0.65	0.20	0.87	134
South America	3	Argentina, Bolivia, Chile	4	53	11	0.55	7.80	2.65	2.10	1.45	58

[1] Standard deviation; [2] Coefficient of variation.

The most commonly studied climatic zones were Csa (n = 13) with a hot-summer Mediterranean climate (Table 3).

Table 3. Number of observations included in the meta-dataset as per the Köppen–Geiger climate zone.

Main Climate Groups	Köppen–Geiger Climate Zone	Name of the Climate Zone	No. of Articles	No. of Observations
Tropical Climates	Aw	Aw: tropical wet-dry climate	2	49
Dry Climates	BSh	BSh: hot semi-arid climate	6	54
	BSk	BSk: cold semi-arid climate	3	58
	BWh	BWh: hot desert climate	4	54
	BWk	BWk: cold desert climate	1	12
Temperate Climates	Cfa	Cfa: humid subtropical	1	3
	Cfb	Cfb: maritime	3	32
	Csa	Csa: interior Mediterranean	13	232
	Csb	Csb: coastal Mediterranean	2	30
	Csc	Csc: cold-summer Mediterranean climate	1	8

2.4. Management and Duration of Trials

Seven main groups of agronomic management were identified during the screening: deficit irrigation (n = 10), salinity (n = 6), fertilization (n = 3), sowing density (n = 2), sowing date (n = 5), weed control (n = 1), and multiple interventions (n = 10). The number of articles and observations reporting investigations of each group of treatments is shown in Figures 3 and 4, and Table 3.

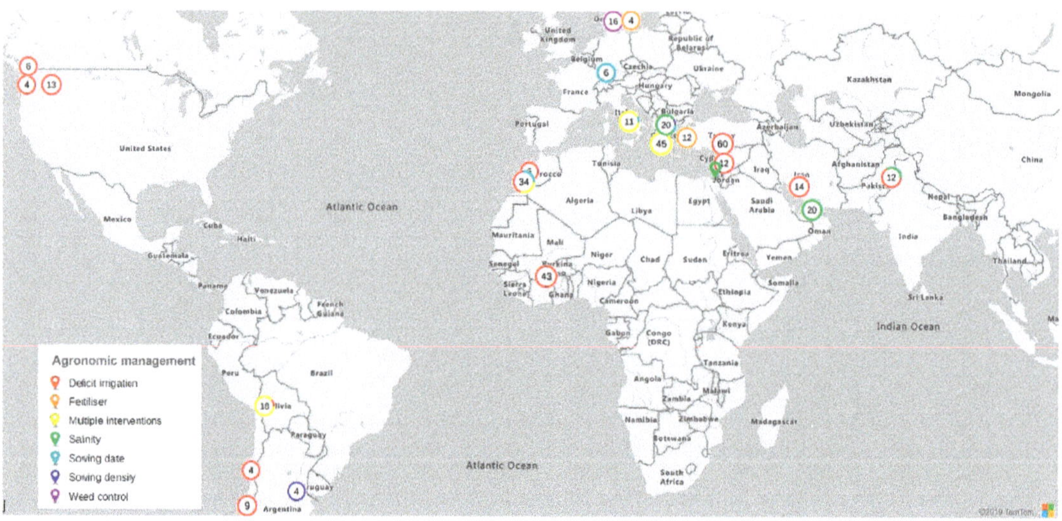

Figure 4. World map showing the number of observations per country. The writing and color in each Doughnut chart represent the total number of observations for each study area and interventions (deficit irrigation, salinity, fertilizer, sowing date, sowing density, weed control, and multiple interventions), respectively.

In the continent of Europe we found all types of agronomic management, with salinity, sowing date, and multiple interventions (e.g., tillage and fertilization) being the most frequently studied treatments (3, 3, and 5 articles, respectively), while in North America, only deficit irrigation was found (2 articles) (Figure 5).

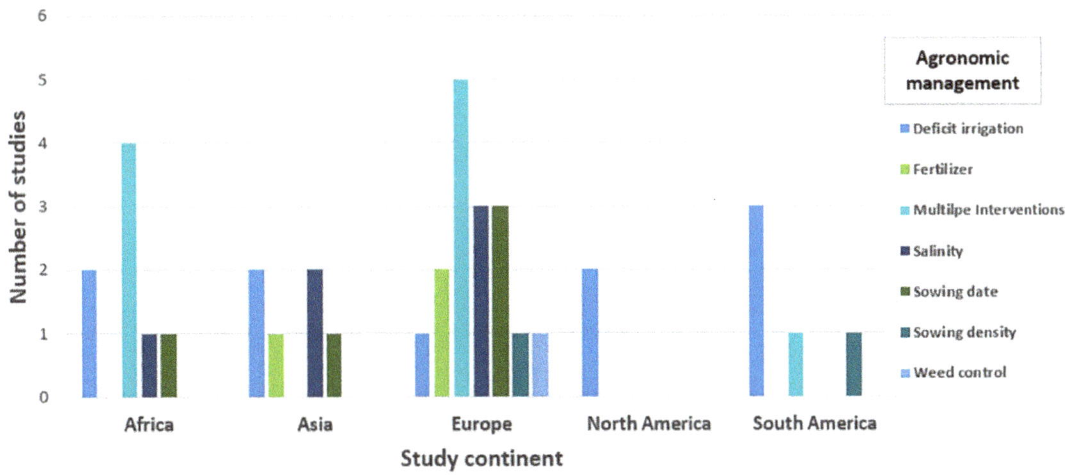

Figure 5. The number of studies undertaken across continents. The numbers are separated by the agronomic management group investigated in each study. Studies may be present in more than one agronomic management category.

A major proportion of the studies were carried out in temperate zones (59% of the total observations), of which 35% were deficit irrigation and 37% multiple interventions (e.g., tillage/fertilization; deficit irrigation/salinity). Salinity treatment represented 50% of the total observations in the arid zones, while 100% of the total observations in the tropical zones were represented by multiple interventions (e.g., deficit irrigation and fertilization) (Table 4).

Table 4. Summary of agronomic managements factors used in the systematic review.

Agronomic Management	Articles	Obs.	Countries	Tropical Obs.	Arid Obs.	Temperate Obs.	Yield (t ha^{-1})
Overall	33	513	16	49	160	304	2.00 ± 1.35
Sowing Date	5	48	5	0	14	34	2.94 ± 1.63
Sowing Density	2	12	2	0	4	8	2.81 ± 1.15
Salinity	6	86	5	0	80	6	2.32 ± 1.43
Deficit Irrigation	10	145	7	0	40	105	1.97 ± 1.57
Fertilization	3	30	3	0	10	20	1.40 ± 1.12
Weed Control	1	18	1	0	0	18	1.69 ± 0.22
Multiple Interventions	10	174	6	49	12	113	1.67 ± 0.89

Obs.: Observations.

A large number of studies were carried out over one year (21) and less commonly over two years (12), with only one study carried out over three years (Figure 6). The longest studies occurred in Europe (three years) with tillage and fertilization treatments in the Agrinio area, Greece (Table 1).

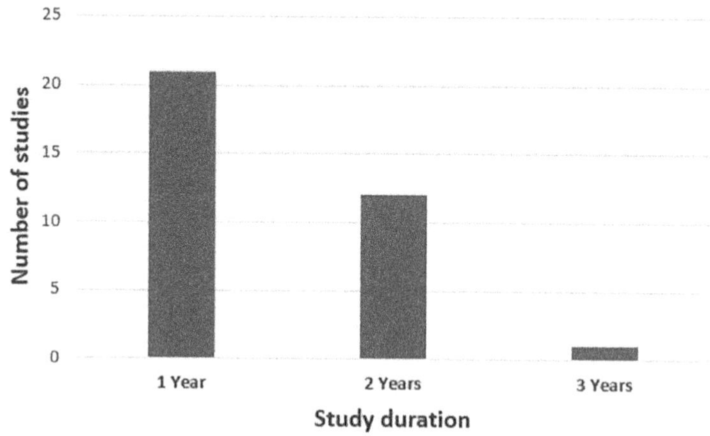

Figure 6. Duration of trials included in the systematic review.

2.5. Most Productive Institutions and Analysis of Source Publications

Figure 7 shows the top twenty most productive institutions. According to the bibliometric analysis, the most productive institutions were the University of Copenhagen, Denmark with eight articles, followed by the Agricultural University of Athens, Greece; Çukurova University, Turkey; University of Agriculture Faisalabad, Pakistan; and the University of Concepción, Chile, who produced three articles each.

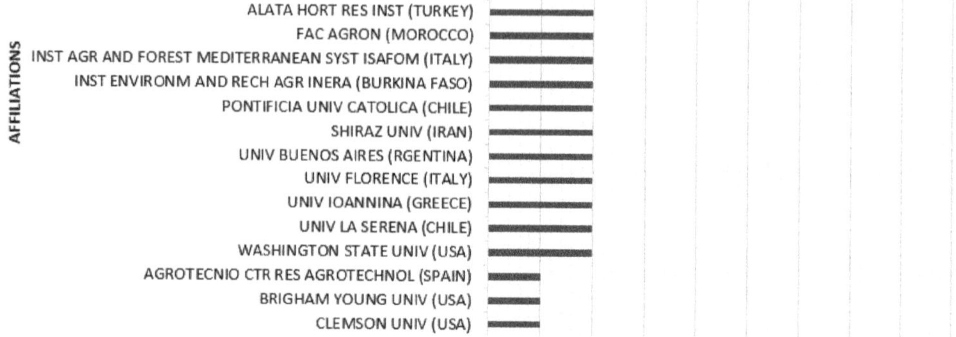

Figure 7. The top 20 most productive institutions in terms of publications.

Figure 8 shows the collaboration network among the leading institutions. The network was drawn from the institution × institution adjacency matrix counting the co-authored publications. In the open-source R package bibliometrix [43], we considered only the first 20 institutions, with a threshold of at least one co-authored publications. The institutions were classified into six clusters, with the first cluster formed by the University of Florence, Italy, and Institut de l'Environnement et de Recherches Agricoles, Burkina Fasso, who are closely connected to each other. A similar situation was observed in respect of two institutions from Turkey, the Alata Horticultural Research Institute and Çukurova University. The third cluster represented a strong connection between two US universities, Washington State University and Brigham Young University, with the National Research Centre in Egypt as well as Durham University in the UK. Another cluster was represented by the University of Copenhagen, Denmark, which is closely connected to the Hassan II Institute of Agronomy and Veterinary Medicine, Morocco, and weakly connected to the Pakistan universities. The fifth cluster is represented by the different Greek universities connected to the Pontifical Catholic University of Chile. Finally, the Agrotecnio Centre for Research in Agrotechnology and the University of Barcelona in Spain were connected with the International Center for Biosaline Agriculture in the UAE.

This analysis is useful for identifying potential partners and cooperative organizations and opening up prospects for research cooperation in this field.

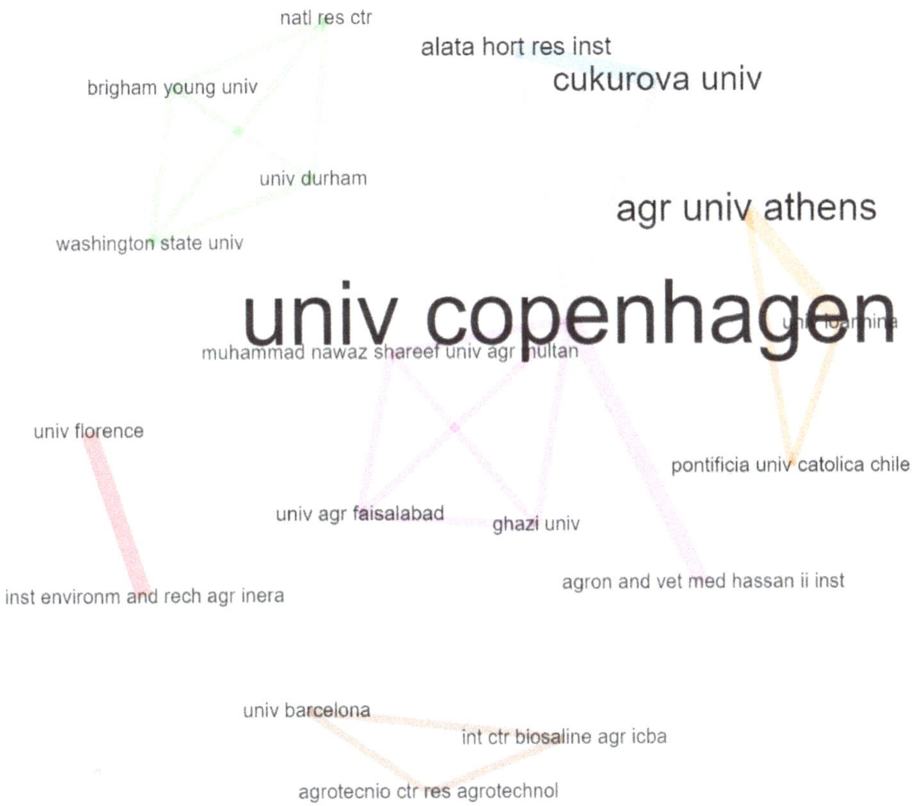

Figure 8. Institution collaboration network. The total number of papers related to each institution proportionally sizing the corresponding label. The thicker the line, the closer the connection between the two institutions.

The bibliometric analysis showed that between 2000 and 2020, the 33 papers included in the systematic review were published in 20 journals. Figure 9 shows the top 10 journals that published articles related to the agronomic practices and performances of quinoa under field conditions. According to the analysis, the journals mostly selected by authors were the Journal of Agronomy and Crop Science followed by Irrigation and Drainage.

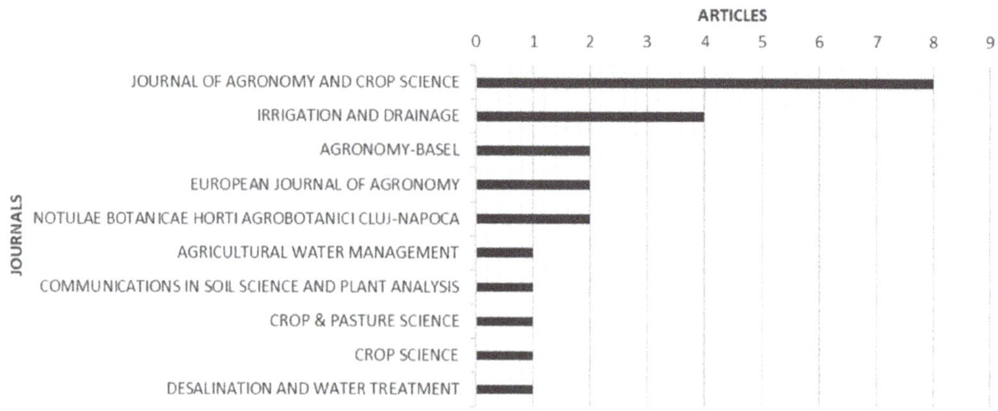

Figure 9. Top 10 journals that published articles related to the agronomic practices and performances of quinoa under field conditions.

2.6. Concept Network Analysis

2.6.1. Terms Analysis

A concept network analysis was performed to extract the terms most used in the title and abstract fields of the publications selected for final review in this research. Our bibliometric analysis revealed the presence of 1206 terms used in the 33 articles. The minimum number of occurrences of a term used in VOSviewer software [44] was set to 6. Accordingly, of the 1206 terms, 17 met the threshold. Figure 10 shows the concept network map for the titles of the 33 articles included in the systematic review and the corresponding abstracts. The 17 terms are classified into three different clusters, in which these terms are based on the same topics (co-occurrence). The higher the number of co-occurrences of two terms, the closer will they be located on the map. The mapped data revealed that the quinoa research fields are related to the effect of the agronomic management (water quality, deficit irrigation, fertilization, sowing date), variety, and soil on seed yield of quinoa. The terms with the highest occurrences are seed yield, variety, deficit irrigation, full irrigation, and fertilization, with a number of occurrences of 23, 19, 25, 20, and 18, respectively.

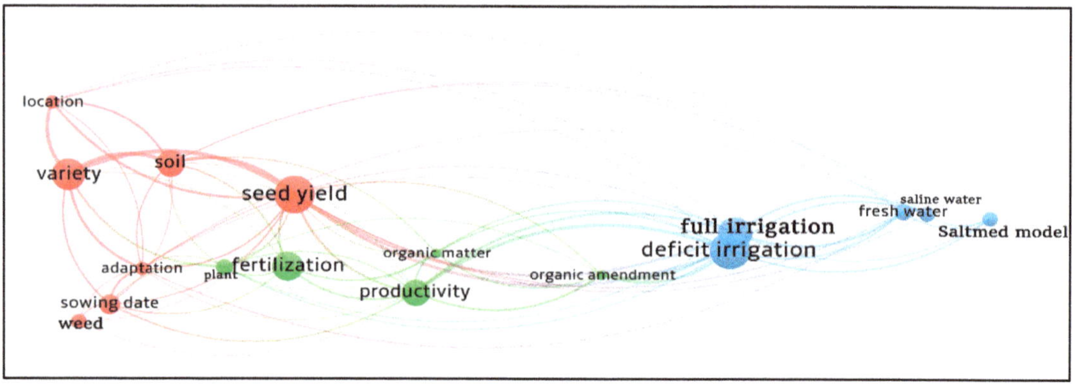

Figure 10. Concept network map for the titles of the 33 articles included in the systematic review and the corresponding abstracts. Map produced by VOSviewer.

2.6.2. Authors Keywords Analysis

Figure 11a shows the author's keywords map. This analysis was performed by VOSviewer. The minimum number of occurrences of a keyword was set to two. Consequently, of the 110 identified keywords, 17 met the threshold. This analysis identified four clusters: quinoa (modeling), seed yield, crop (*Chenopodium quinoa*) and abiotic stress (Figure 11a).

Figure 11b shows the thematic evolution of the author's keywords based on the average times they appeared in our collection of articles. We find that the keywords related to water use efficiency, harvest index, and drought stress appeared early while those related to deficit irrigation, abiotic stress, and growth analysis appeared later.

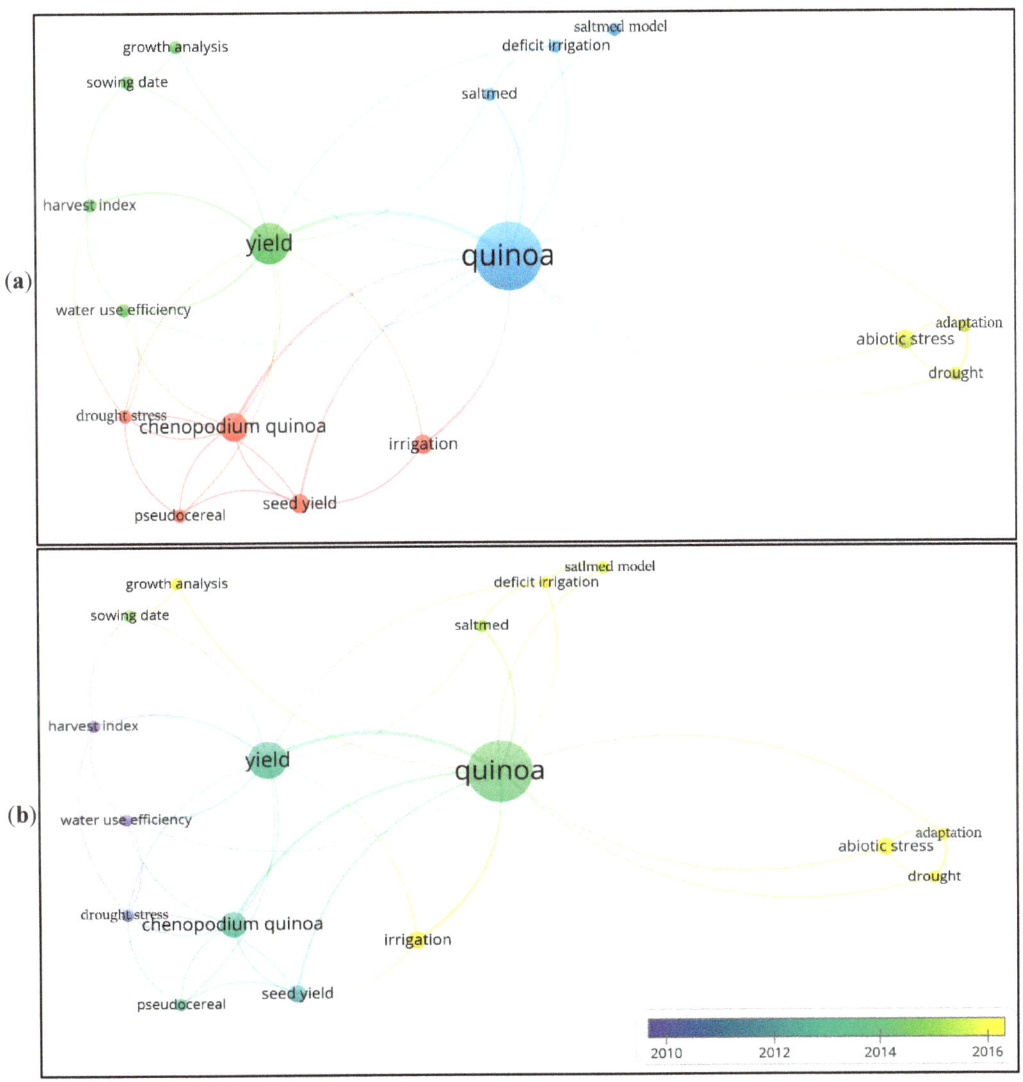

Figure 11. Authors keywords maps. (**a**) Network visualization of authors keywords co-occurrence. (**b**) Thematic evolution of authors keywords in the field of research on the agronomic practices and performances of quinoa under field conditions 2000–2020. Map produced by VOSviewer.

2.7. Overall Yield Across Factors of Variation

Figure 12 shows that the variation of yield due to environment, agronomic management, and soil factors was quite large. Yield response varied across the ten climatic zones with the lowest value of 0 t ha^{-1} obtained in the tropical climates zone (Aw) and the highest value of 7.80 t ha^{-1} observed in the cold desert climate zone (BWk). It can be seen from Figure 12a that the highest seeds yield was recorded in the arid climate zone and the lowest in the tropical climate zone. The humid subtropical zone (Cfa) was the most homogeneous class, followed by the Aw zone. The hot desert climate zone (BWh) and the coastal Mediterranean zone (Csb) were the least homogeneous classes. The yield values for zones Aw, BSk, and Cfa were clearly behind those of BWh, BWk, and Csb.

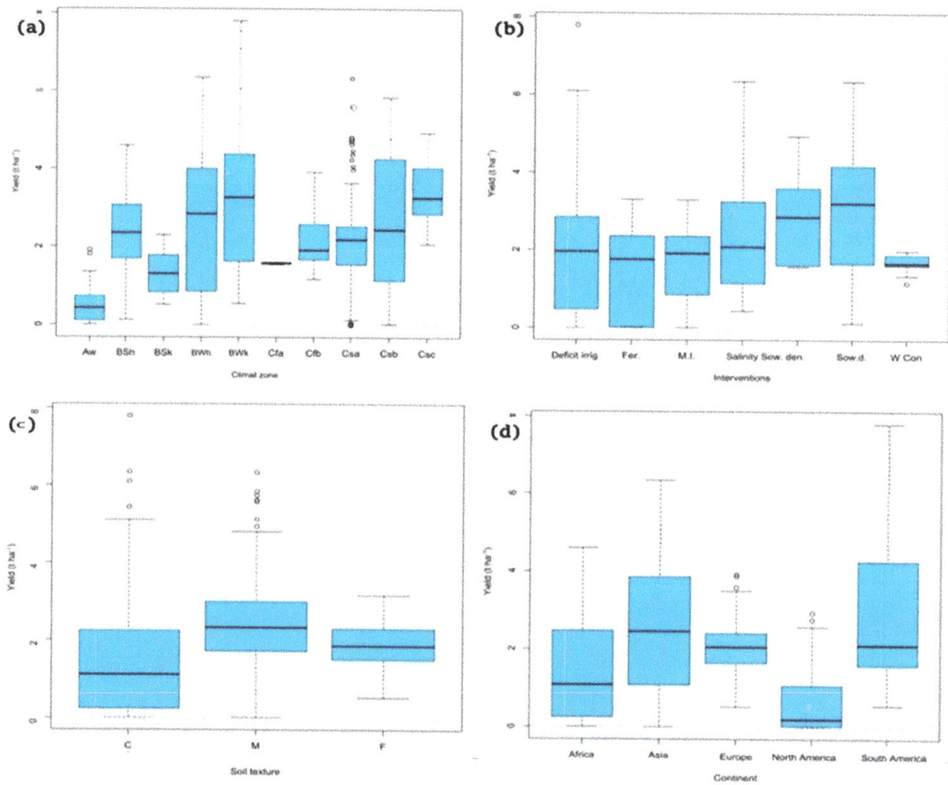

Figure 12. Box plots of patterns of yield (t ha^{-1}) for all articles (n = 33) across: (**a**) different groups of climatic zones, Aw: tropical wet-dry climate; BSh: hot semi-arid climate; BSk: cold semi-arid climate; BWh: hot desert climate; BWk: cold desert climate; Cfa: humid subtropical; Cfb: maritime; Csa: interior Mediterranean; Csb: coastal Mediterranean; and Csc: cold-summer Mediterranean climate, (**b**) different groups of agronomic management, A: deficit irrigation; AB: deficit irrigation and salinity; AD: deficit irrigation and fertilizer; B: salinity; C: tillage; CD: tillage and fertilizer; D: fertilizer; E: sowing density; EF: sowing density and sowing date; F: sowing date; and G: weed control, (**c**) different groups of soil texture, C: coarse soil; M: medium soil; F: fine soil, and (**d**) different groups of the continent. Box edges represent the upper and lower quantile with the median value shown in the middle of the box. The small circles on the box plot relate to outliers.

Table 4 and Figure 12b show the yield variation between different agronomic managements. The highest value of 6.35 t ha^{-1} was obtained in the salinity and sowing date treatment. Salinity, sowing density, and sowing date treatments were the agronomic interventions most influential to productive response; fertilization and multiple interventions were less impactful, with average yield values ranging from 1.40 and 1.67 t ha^{-1}, respectively.

The variation of yield between soil texture showed that medium soil (M) and fine soil (F) were the most productive, with average yield values ranging from 2.36 to 1.83 t ha^{-1}, respectively. The highest value of 7.80 t ha^{-1} was obtained in the coarse soil (C), with fine soil (F) being the most homogeneous class (Figure 12c).

Studies conducted in South America and Asia showed the highest yields, ranging from 0.55 to 7.80 t ha^{-1}, and 0.01 to 6.35 t ha^{-1}, respectively (Figure 12d). In contrast, the yield values of the North American continent were clearly behind other continents. The European continent was the most homogeneous class, with yield values ranging from 0.52 to 3.93 t ha^{-1}.

3. Discussion

The systematic review integrated with bibliometric analysis allows the identification of a research gap in the cropping practices of quinoa production under field conditions, highlighting the necessity to develop research and to establish global research networks in order to include different scientists worldwide, especially from arid regions.

The analysis of co-occurrence terms and the author's keywords identified the main topics of actual research. Over the last 20 years, of the top 17 terms in the 1206 used in 33 articles, five (seed yield, variety, deficit irrigation, full irrigation, and fertilization) registered the highest co-occurrence frequency, indicating that from 2000 to 2020, research was primarily focused on these topics. The high occurrence of the terms seed yield and deficit irrigation in the titles and abstracts of the analyzed papers may indicate the focus of most of the papers on the effects of deficit irrigation on seed yield. According to Radhakrishnan et al. [45], the keyword co-occurrence network analyses can be performed quickly to explore a wide range of literature and can provide a knowledge map and insights before conducting a rigorous conventional systematic review. In the early stages of research on the cropping practices of quinoa production under field conditions, the studies focused on topics related to water use efficiency, harvest index, and drought stress. Later on, the focus was on more recent topics related to deficit irrigation and abiotic stress. In the last twenty years, drought signals in the field have been confirmed by a large number of field studies [46]. The major agricultural use of water is for irrigation, which is thus affected by decreased supply. In recent years, significant effort was made to increase the efficiency of water use through the use of deficit irrigation strategies [47].

The bibliographic analysis of the selected studies highlighted the beginning of the study of best agronomic practices for quinoa production in 2003, probably due to the impact of specific research projects conducted from 1990 to 2000 [48]. The number of studies strongly increased after 2013, when the FAO celebrated the International Year of Quinoa; this activity of disseminating the importance of quinoa as a crop resistant to unfavorable environments and a high-quality protein source focused world attention on it [49].

The highlighting of the resistance of quinoa to abiotic stresses and its adaptability to different environments led to an increase in studies, especially in countries characterized by a hot, arid climate and water scarcity (Morocco, Egypt, Burkina Faso, and the UAE) and countries at risk of water and salt stress due to climate change (Italy, Greece, Turkey, Pakistan, and the US). Many of the studies in these countries were related to the evaluation of the effect of deficit irrigation and the use of saline water on quinoa. Much importance has also been bestowed on the study of the best time and sowing density, which represent the main agronomic practices for the introduction of a crop in a new environment.

The analyzed papers showed that quinoa is able to guarantee seed yields in line with those of the countries of origin even in different climatic conditions and different soils texture; in fact, even if quinoa prefers well-drained soils and warm beds [50], it has been shown to adapt well and guarantee high yields even in clayey and less-drained soils.

The geographical distribution of the studies confirmed that quinoa adapts to different pedo-climatic regions.

The papers analyzed have also shown that agronomic practices such as sowing date and sowing density are very important parameters that affect seed yield; the evaluation of the best sowing date is fundamental in the case of quinoa introduction in new environments.

Studies on the effect of irrigation management, such as deficit irrigation or use of saline water, have confirmed that these are sustainable practices conducive to the optimization of production by reducing the use of water resources; moreover, quinoa can be grown in marginal environments [51] characterized by scarcity and poor-quality of water.

The total number of papers found in peer-reviewed journals appears still limited compared to the potential of quinoa to adapt to different environments and the great genetic diversity that distinguishes this crop. Today, it is estimated that there are over 6000 varieties cultivated by farmers [52], each of these accessions with distinctive genetic characteristics requiring specific study.

4. Materials and Methods

4.1. Literature Research

A systematic review (SR) was conducted across two bibliographic databases (ISI Web of Science™ and Scopus™), to identify studies related to the agronomic practices of quinoa (*Chenopodium quinoa*) production in the World. The studies were published between 2000 and 2020 in peer-reviewed journals written in English. The searches of academic databases were performed on 5 October 2020. In bibliographic databases the following search strings were used to search "topic words" combined with Boolean operators: ((field OR cultivar* OR genotyp* OR crop* OR farm* OR cultivat* OR accessions) AND (yield OR grain OR product* OR seed*) AND (quinoa or (Chenopodium and quinoa))). The wildcards * represent any number of characters.

4.2. Inclusion and Exclusion Criteria

We used a highly robust and rational systematic review methodology to synthesize the evidence from a wide range of sources. In this study, we constrained the SR by defining boundaries to include: (I) studies conducted only under field conditions, but not under glasshouse conditions and pots; and (II) studies that focused on crop productivity, omitting forestry, fisheries, livestock, and other non-food crop agricultural sectors. Following the SR convention, the search terms were based on the four PICO components (population, intervention, comparator, and outcome) (Table 5). A list of references included in the SR meta-database is provided in Table 1.

Table 5. Defining the PICO terms for the research "question" used in this study.

PICO	Description
Population	Agriculture: food crops under field conditions
	World: study included all the countries in all the continents
Intervention	Management included sowing date, sowing density, fertilizer, tillage, salinity, deficit irrigation, and weed control
Comparator	Impacts and/or benefits
Outcomes	Yield, yield gap, potential yield, farmer yield, and attainable yield

4.3. Screening

Following the removal of duplicates, in order to extract yield information data from accepted papers were entered into Endnote (online bibliographic management software) (version basic; Clarivate Analytics, https://access.clarivate.com/#/login?app=endnote); all the references were retrieved and screened for relevance using the following inclusion criteria: every selected study was screened through three stages: title, abstract, and full text.

At each level, records containing or likely to contain relevant information were identified and taken to the next stage.

4.4. Coding and Data Extraction

Meta-data (descriptive categorical information regarding citations, study setting, design, and methods) were extracted from included studies following full-text assessment.

The investigated treatments (agronomic management) were recorded for each study as categorical variables where possible; in this case, a complete disjunctive coding of our variables (treatments investigated) was carried out. This means that variables are dichotomous, assuming the value "1" should the keyword be associated with the paper, and "0" if not. This coding was done according to methods described by Cuccurullo et al. [53].

Data for different years or experimental conditions (i.e., cultivars or other experimental factors) within each publication were treated as independent observations. Data were obtained directly from tables and if data were provided in graphical form, means were extracted using WebPlotDigitizer [54].

4.5. Bibliometric Analysis and Concept Network Analysis

The meta-data from SR was analyzed for the year of publication, the journal, and the frequency of terms and keywords used by authors. This analysis was carried out with the aid of a software package of comprehensive science mapping analysis, bibliometrix [43] in R studio software [55]. The package is available through the Comprehensive R Archive Network (CRAN, https://cran.r-project.org). We used the VOSviewer software developed by scientometricians [44] (http://www.vosviewer.com) for concept network mapping to generate keywords and term maps.

5. Conclusions

This study analyzed the scientific production in the last 20 years on the adoption of agronomic practices for the cultivation of quinoa under field conditions. The results showed that since 2003 there has been a fair amount of scientific production; however, only after 2013, the International Year of Quinoa celebrated by FAO, was there a significant increase in the number of papers.

In many cases, the results of previous experiments mostly carried out in South America have been published in journals and/or volumes not indexed.

The analysis revealed a greater interest in studying the date and density of sowing, and irrigation (deficit irrigation and with saline waters) as sustainable practices to ensure good yields in different environments. The studies analyzed have emphasized that the best agronomic practices can guarantee good production of quinoa even in marginal environments or those characterized by abiotic stress (drought and salinity). Studies have shown that quinoa can also be grown on fine, less-drained soils.

Data reported in the screened paper need to be analyzed more deeply; a useful approach to this could be a meta-analysis using the relative yield as an effect-size estimator. The meta-analysis would allow investigation of the interaction between different continental regions, environmental and agronomic management, and the effect of these factors on yield response.

Scientific production still appears limited and no publications relating to experiments were found for the Australian continent, which is characterized by large arid areas and marginal environments.

Several agronomic practices should be explored such as weed control (only one article has been detected), use of wastewater, soil processing, etc.

This systematic review can be used by researchers to identify deficiencies and best practices in research methodology, fostering collaboration, especially between arid country researchers, increase the field research, and exploit research results at the maximum extent. It would make a significant contribution to the expansion of quinoa in different environments.

Author Contributions: Conceptualization, M.H.S.; Methodology, M.H.S.; Formal analysis, M.H.S.; Investigation, M.H.S.; Resources, M.H.S.; Data curation, M.H.S.; Writing—original draft preparation, M.H.S., and C.P.; Writing—review and editing, M.H.S., C.P., and A.L.; Visualization, M.H.S.; Supervision, C.P. All authors have read and agreed to the published version of the manuscript.

Funding: This research received no external funding.

Institutional Review Board Statement: Not applicable.

Informed Consent Statement: Not applicable.

Data Availability Statement: Data sharing not applicable. No new data were created or analyzed in this study. Data sharing is not applicable to this article.

Conflicts of Interest: The authors declare no conflict of interest.

References

1. Alandia, G.; Rodriguez, J.P.; Jacobsen, S.E.; Bazile, D.; Condori, B. Global expansion of quinoa and challenges for the Andean region. *Glob. Food Sec.* **2020**, *26*, 100429. [CrossRef]
2. Pulvento, C.; Jacobsen, S.-E.; Alandia, G.; Prins, U.; d'Andria, R.; Sellami, M.H.; Grimberg, A.; Carlsson, A.S.; Capannini, S.; Lavini, A. Evaluation of quinoa adaptability under European conditions to enhance high quality food protein production. In Proceedings of the Quinoa for Future Food and Nutrition Security in Marginal Environments Conference, Dubai, United Arab Emirates, 6–8 December 2016; p. 28.
3. Abdelhamid, M.T. Quinoa (*Chenopodium quinoa* Willd.), a potential new crop for Egypt. In Proceedings of the Quinoa for Future Food and Nutrition Security in Marginal Environments Conference, Dubai, United Arab Emirates, 6–8 December 2016; p. 65.
4. Elouafi, I. Foreword. In Proceedings of the Quinoa for Future Food and Nutrition Security in Marginal Environments Conference, Dubai, United Arab Emirates, 6–8 December 2016; p. 10.
5. Risi, J.; Galwey, N.W. Effects of sowing date and sowing rate on plant development and grain yield of quinoa (*Chenopodium quinoa*) in a temperate environment. *J. Agric. Sci.* **1991**, *117*, 325–332. [CrossRef]
6. Jacobsen, S.E.; Jørgensen, I.; Stølen, O. Cultivation of quinoa (*Chenopodium Quinoa*) under temperate climatic conditions in Denmark. *J. Agric. Sci.* **1994**, *122*, 47–52. [CrossRef]
7. Vacher, J.J. Responses of two main Andean crops, Quinoa (*Chenopodium quinoa* Willd) and Papa Amarga (*Solanum juzepczukii* Buk.) to drought on the Bolivian Altiplano: Significance of local adaptation. *Agric. Ecosyst. Environ.* **1998**, *68*, 99–108. [CrossRef]
8. NHS Centre for Reviews and Dissemination. *Undertaking Systematic Reviews of Research on Effectiveness: CRD's Guidance for Those Carrying Out or Commissioning Reviews*, 2nd ed.; CRD Report 4; University of York: York, UK, 2001.
9. UMD. Research Guides. University of Maryland. 2020. Available online: https://lib.guides.umd.edu/bibliometrics/bibliometrics (accessed on 9 November 2020).
10. Jacobsen, S.E.; Christiansen, J.L.; Rasmussen, J. Weed harrowing and inter-row hoeing in organic grown quinoa (*Chenopodium quinoa* Willd.). *Outlook Agric.* **2010**, *39*, 223–227. [CrossRef]
11. Pulvento, C.; Riccardi, M.; Lavini, A.; D'Andria, R.; Ragab, R. Saltmed model to simulate yield and dry matter for quinoa crop and soil moisture content under different irrigation strategies in South Italy. *Irrig. Drain.* **2013**, *62*, 229–238. [CrossRef]
12. Jacobsen, S.E.; Christiansen, J.L. Some agronomic strategies for organic quinoa (*Chenopodium quinoa* Willd.). *J. Agron. Crop Sci.* **2016**, *202*, 454–463. [CrossRef]
13. Pulvento, C.; Riccardi, M.; Lavini, A.; d'Andria, R.; Iafelice, G.; Marconi, E. Field trial evaluation of two *Chenopodium quinoa* genotypes grown under rain-fed conditions in a typical Mediterranean environment in South Italy. *J. Agron. Crop Sci.* **2010**, *196*, 407–411. [CrossRef]
14. Geren, H. Effects of different nitrogen levels on the grain yield and some yield components of Quinoa (*Chenopodium quinoa* Willd.) under Mediterranean climatic conditions. *Turk. J. Field Crop.* **2015**, *20*, 59–64. [CrossRef]
15. Kaya, Ç.I.; Yazar, A. Saltmed model performance for quinoa irrigated with fresh and saline water in a Mediterranean environment. *Irrig. Drain.* **2016**, *65*, 29–37. [CrossRef]
16. Yazar, A.; Incekaya, Ç.; Sezen, S.M.; Jacobsen, S.E. Saline water irrigation of quinoa (*Chenopodium quinoa*) under Mediterranean conditions. *Crop Pasture Sci.* **2015**, *66*, 993–1002. [CrossRef]
17. Alvar-Beltrán, J.; Dao, A.; Marta, A.D.; Saturnin, C.; Casini, P.; Sanou, J.; Orlandini, S. Effect of drought, nitrogen fertilization, temperature, and photoperiodicity on quinoa plant growth and development in the Sahel. *Agronomy* **2019**, *9*, 607. [CrossRef]
18. Alvar-Beltran, J.; Saturnin, C.; Dao, A.; Marta, A.D.; Sanou, J.; Orlandini, S. Effect of drought and nitrogen fertilization on quinoa (*Chenopodium quinoa* Willd.) under field conditions in Burkina Faso. *Ital. J. Agrometeorol.* **2019**, *24*, 33–43.
19. Asher, A.; Galili, S.; Whitney, T.; Rubinovich, L. The potential of Quinoa (*Chenopodium quinoa*) cultivation in Israel as a dual-purpose crop for grain production and livestock feed. *Sci. Hortic.* **2020**, *272*, 109534. [CrossRef]
20. Basra, S.M.A.; Iqbal, S.; Afzal, I. Evaluating the response of nitrogen application on growth, development and yield of quinoa genotypes. *Int. J. Agric. Biol.* **2014**, *16*, 886–892.
21. Fghire, R.; Wahbi, S.; Anaya, F.; Ali, O.I.; Benlhabib, O.; Ragab, R. Response of Quinoa to different water management strategies: Field experiments and Saltmed model application results. *Irrig. Drain.* **2015**, *64*, 29–40. [CrossRef]

22. Hirich, A.; Choukr-Allah, R.; Jacobsen, S.E. Quinoa in Morocco–Effect of sowing dates on development and yield. *J. Agron. Crop Sci.* **2014**, *200*, 371–377. [CrossRef]
23. Hirich, A.; Choukr-Allah, R.; Jacobsen, S.E. Deficit irrigation and organic compost improve growth and yield of quinoa and pea. *J. Agron. Crop Sci.* **2014**, *200*, 390–398. [CrossRef]
24. Filali, K.; Hirich, A.; Ouafae, B.; Choukr-Allah, R.; Ragab, R. Yield and dry matter simulation using the Saltmed model for five quinoa (*Chenopodium quinoa*) accessions under deficit irrigation in South Morocco. *Irrig. Drain.* **2017**, *66*, 340–350.
25. Bertero, H.D.; Ruiz, R.A. Determination of seed number in sea level quinoa (*Chenopodium quinoa* Willd.) cultivars. *Eur. J. Agron.* **2008**, *28*, 186–194. [CrossRef]
26. Bilalis, D.; Kakabouki, I.; Karkanis, A.; Travlos, I.; Triantafyllidis, V.; Hela, D. Seed and saponin production of organic quinoa (*Chenopodium quinoa* Willd.) for different tillage and fertilization. *Not. Bot. Horti Agrobot. Cluj-Napoca* **2012**, *40*, 42–46. [CrossRef]
27. Çolak, B.Y.; Yazar, A.; Alghory, A.; Tekin, S. Evaluation of crop water stress index and leaf water potential for differentially irrigated quinoa with surface and subsurface drip systems. *Irrig. Sci.* **2020**, *39*, 1–20.
28. Eisa, S.S.; Eid, M.A.; Abd El-Samad, E.H.; Hussin, S.A.; Abdel-Ati, A.A.; El-Bordeny, N.E.; Ali, S.H.; Al-Sayed, H.M.A.; Lotfy, M.E.; Masoud, A.M.; et al. *Chenopodium quinoa* Willd. A new cash crop halophyte for saline regions of Egypt. *Aust. J. Crop Sci.* **2017**, *11*, 343–351. [CrossRef]
29. Fischer, S.; Wilckens, R.; Jara, J.; Aranda, M. Variation in antioxidant capacity of quinoa (*Chenopodium quinoa* Willd) subjected to drought stress. *Ind. Crop Prod.* **2013**, *46*, 341–349. [CrossRef]
30. Geerts, S.; Raes, D.; Garcia, M.; Vacher, J.; Mamani, R.; Mendoza, J.; Huanca, R.; Morales, B.; Miranda, R.; Cusicanqui, J.; et al. Introducing deficit irrigation to stabilize yields of quinoa (*Chenopodium quinoa* Willd.). *Eur. J. Agron.* **2008**, *28*, 427–436. [CrossRef]
31. Hinojosa, L.; Sanad, M.N.M.E.; Jarvis, D.E.; Steel, P.; Murphy, K.; Smertenko, A. Impact of heat and drought stress on peroxisome proliferation in quinoa. *Plant J.* **2019**, *99*, 1144–1158. [CrossRef] [PubMed]
32. Hinojosa, L.; Kumar, N.; Gill, K.S.; Murphy, K.M. Spectral reflectance indices and physiological parameters in quinoa under contrasting irrigation regimes. *Crop Sci.* **2019**, *59*, 1927–1944. [CrossRef]
33. Hirich, A.; Choukr-Allah, R.; Jacobsen, S.E. The combined effect of deficit irrigation by treated wastewater and organic amendment on quinoa (*Chenopodium quinoa* Willd.) productivity. *Desalin. Water Treat.* **2014**, *52*, 2208–2213. [CrossRef]
34. Iqbal, S.; Basra, S.M.A.; Afzal, I.; Wahid, A.; Saddiq, M.S.; Hafeez, M.B.; Jacobsen, S.E. Yield potential and salt tolerance of quinoa on salt-degraded soils of Pakistan. *J. Agron. Crop Sci.* **2019**, *205*, 13–21. [CrossRef]
35. Kakabouki, I.P.; Roussis, I.E.; Papastylianou, P.; Kanatas, P.; Hela, D.; Katsenios, N.; Fuentes, F. Growth analysis of quinoa (*Chenopodium quinoa* Willd.) in response to fertilization and soil tillage. *Not. Bot. Horti Agrobot.* **2019**, *47*, 1025–1036. [CrossRef]
36. Kakabouki, I.P.; Roussis, I.; Hela, D.; Papastylianou, P.; Folina, A.; Bilalis, D. Root growth dynamics and productivity of quinoa (*Chenopodium quinoa* Willd.) in response to fertilization and soil tillage. *Folia Hortic.* **2019**, *31*, 285–299. [CrossRef]
37. Karyotis, T.; Iliadis, C.; Noulas, C.; Mitsibonas, T. Preliminary research on seed production and nutrient content for certain quinoa varieties in a saline-sodic soil. *J. Agron. Crop Sci.* **2003**, *189*, 402–408. [CrossRef]
38. Martínez, E.A.; Veas, E.; Jorquera, C.; San Martín, R.; Jara, P. Re-introduction of quínoa into arid Chile: Cultivation of two lowland races under extremely low irrigation. *J. Agron. Crop Sci.* **2009**, *195*, 1–10. [CrossRef]
39. Noulas, C.; Tziouvalekas, M.; Vlachostergios, D.; Baxevanos, D.; Karyotis, T.; Iliadis, C. Adaptation, agronomic potential, and current perspectives of quinoa under Mediterranean conditions: Case studies from the lowlands of central Greece. *Commun. Soil Sci. Plant Anal.* **2017**, *48*, 2612–2629.
40. Präger, A.; Boote, K.J.; Munz, S.; Graeff-Hönninger, S. Simulating growth and development processes of quinoa (*Chenopodium quinoa* Willd.): Adaptation and evaluation of the CSM-CROPGRO model. *Agronomy* **2019**, *9*, 832.
41. Razzaghi, F.; Bahadori-Ghasroldashti, M.R.; Henriksen, S.; Sepaskhah, A.R.; Jacobsen, S.E. Physiological characteristics and irrigation water productivity of quinoa (*Chenopodium quinoa* Willd.) in response to deficit irrigation imposed at different growing stages- a field study from southern Iran. *J. Agron. Crop Sci.* **2020**, *206*, 390–404. [CrossRef]
42. Rezzouk, F.Z.; Shahid, M.A.; Elouafi, I.A.; Zhou, B.; Araus, J.L.; Serret, M.D. Agronomic performance of irrigated quinoa in desert areas: Comparing different approaches for early assessment of salinity stress. *Agric. Water Manag.* **2020**, *240*, 106205. [CrossRef]
43. Aria, M.; Cuccurullo, C. Bibliometrix: An R-tool for comprehensive science mapping analysis. *J. Informetr.* **2017**, *11*, 959–975. [CrossRef]
44. Van Eck, N.J.; Waltman, L. Software survey: VOSviewer, a computer program for bibliometric mapping. *Scientometrics* **2010**, *84*, 523–538. [CrossRef]
45. Radhakrishnan, S.; Erbis, S.; Isaacs, J.A.; Kamarthi, S. Novel keyword co-occurrence network-based methods to foster systematic reviews of scientific literature. *PLoS ONE* **2017**, *12*, e0172778.
46. Stikic, R.; Savíc, S.; Jovanovic, Z.; Jacobsen, S.E.; Liu, F.; Jensen, C.R. Deficit irrigation strategies: Use of stress physiology knowledge to increase water use efficency in tomato and potato. In *Horticulture in 21st Century*; Sampson, A.N., Ed.; Series: Botanical Research and Practices; Nova Science Publishers: New York, NY, USA, 2010; pp. 161–178.
47. Kirda, C. *Deficit Irrigation Scheduling Based on Plant Growth Stages Showing Water Stress Tolerance*; Deficit Irrigation Practices, Water Reports; Food and Agricultural Organization of the United Nations: Rome, Italy, 2002; p. 22.
48. Bazile, D.; Baudron, F. Dinámica de expansión mundial del cultivo de la quinua respecto a su alta biodiversidad. In *Estado Del Arte De La Quinua En El Mundo En 2013*; Bazile, D., Bertero, H.D., Nieto, C., Eds.; FAO y CIRAD: Santiago de Chile, Chile; Montpellier, Francia, 2014; pp. 49–64.

49. Bazile, D.; Jacobsen, S.E.; Verniau, A. The global expansion of quinoa: Trends and limits. *Front. Plant Sci.* **2016**, *7*, 622. [CrossRef] [PubMed]
50. Plants for a Future. *Chenopodium quinoa*–Willd. 2007. Available online: https://pfaf.org/user/Plant.aspx?LatinName=Chenopodium+quinoa (accessed on 6 December 2020).
51. Choukr-Allah, R.; Rao, N.K.; Hirich, A.; Shahid, M.; Alshankiti, A.; Toderich, K.; Gill, S.; Butt, K.U.R. Quinoa for marginal environments: Toward future food and nutritional security in MENA and Central Asia regions. *Front. Plant Sci.* **2016**, *7*, 346. [CrossRef] [PubMed]
52. Rojas, W.; Pinto, M.; Alanoca, C.; Pando, L.G.; Lobos, P.L.; Alercia, A.; Diulgheroff, S.; Padulosi, S.; Bazile, D. Quinoa genetic resources and ex situ conservation. In *State of the Art Report on Quinoa around the World 2013*; Bazile, D., Bertero, H.D., Nieto, C., Eds.; FAO y CIRAD: Santiago de Chile, Chile; Montpellier, Francia, 2015; pp. 56–82.
53. Cuccurullo, C.; Aria, M.; Sarto, F. Foundations and trends in performance management. A twenty-five years bibliometric analysis in business and public administration domains. *Scientometrics* **2016**, *108*, 595–611. [CrossRef]
54. Rohatgi, A. WebPlotDigitizer 2011. Available online: https://apps.automeris.io/wpd/ (accessed on 10 October 2020).
55. R Core Team. *R: A Language and Environment for Statistical Computing*; R Foundation for Statistical Computing: Vienna, Austria, 2013.

plants

Article

Irrigation Regimes and Nitrogen Rates as the Contributing Factors in Quinoa Yield to Increase Water and Nitrogen Efficiencies

Maryam Bahrami [1,2], Rezvan Talebnejad [1,2,*], Ali Reza Sepaskhah [1,2] and Didier Bazile [3,4,*]

1 Water Engineering Department, Shiraz University, Shiraz 7144113131, Iran
2 Drought Research Center, Shiraz University, Shiraz 7144113131, Iran
3 CIRAD, UMR Savoirs, Environnement et Sociétés (SENS), CEDEX, 34398 Montpellier, France
4 SENS, CIRAD, IRD, Université Paul Valery Montpellier 3, Univ Montpellier, CEDEX, 34199 Montpellier, France
* Correspondence: rtalebnejad@shirazu.ac.ir (R.T.); didier.bazile@cirad.fr (D.B.)

Abstract: Sustainable field crop management has been considered to reach the food security issue due to global warming and water scarcity. The effect of deficit irrigation and nitrogen rates on quinoa yield is a challenging issue in those areas. In this regard, the interaction effects of different N rates (0, 125, 250, and 375 kg N ha^{-1}) and irrigation regimes [full irrigation (FI) and deficit irrigation at 0.75 FI and 0.5 FI] on quinoa yield and water and nitrogen efficiencies were evaluated with a two-year field experiment. Increasing nitrogen fertilizer application levels from 250 to 375 kg N ha^{-1} under FI and deficit irrigation did not cause a significant difference in seed yield and the total dry matter of quinoa. Furthermore, 20% and 34% reductions were observed for nitrogen use efficiency (NUE) and nitrogen yield efficiency with the application of 375 kg N ha^{-1} compared with that obtained in 250 kg N ha^{-1} nitrogen fertilizer, respectively. Therefore, a Nitrogen application rate of 250 kg ha^{-1} and applying 0.75 FI is suggested as the optimum rate to reach the highest seed water use efficiency (0.7 kg m^{-3}) and NUE (0.28 kg m^{-3}) to gain 4.12 Mg ha^{-1} quinoa seed yield. Under non-limited water resource conditions, an FI and N application rate of 375 kg ha^{-1} could be used for higher seed yield; however, under water-deficit regimes, an N application rate of 250 kg ha^{-1} could be adequate. However, questions about which environmental factors impressively restricted the quinoa growth for optimizing the potential yield need further investigation.

Keywords: *Chenopodium quinoa* Willd.; deficit irrigation; nitrogen fertilizer rate; water use efficiency; nitrogen uptake; residual soil NO_3-N

Citation: Bahrami, M.; Talebnejad, R.; Sepaskhah, A.R.; Bazile, D. Irrigation Regimes and Nitrogen Rates as the Contributing Factors in Quinoa Yield to Increase Water and Nitrogen Efficiencies. *Plants* **2022**, *11*, 2048. https://doi.org/10.3390/plants11152048

Academic Editor: Jim Moir

Received: 6 July 2022
Accepted: 1 August 2022
Published: 5 August 2022

1. Introduction

Recent studies revealed that crop production must be increased by around 25–70% to meet the worldwide food demands in 2050 [1,2] due to rapid population growth. Therefore, it is expected to use sustainable field management [3] and use resilient crops to increase crop products and reach the food security mission; whereas environmental changes [4] such as global climate change [5] and scarce water availability have put negative pressure on agricultural production [6–8].

Quinoa crop (*Chenopodium quinoa* Willd.) is an alternative option for farmers due to its adaptability to various agroecosystems [6,9,10], which has been considered widely in the last decades [7,9–11]. The adaptation of quinoa to some environmental changes such as drought or water stress [12,13], salinity [14–16], and frost [17] has made it favorable amongst scientists for securing food production [18–21]. A systematic review showed that the mean value of quinoa seed yield amongst the Asian countries (Iran, Israel, Pakistan, Turkey, United Arab Emirates) was about 2.55 Mg ha^{-1} (in the range of 0.01–6.35 Mg ha^{-1}) while it was about 2.65 Mg ha^{-1} in South America (ranging from 0.55 to 7.8 Mg ha^{-1}). Africa, Europe, and North America (with the lowest mean seed yield, 0.65 Mg ha^{-1}), respectively, were placed after them [16]. Quinoa seed is rich in proteins, fats, fiber, minerals, vitamins, and essential amino acids [22–24], which extensively enhanced quinoa's popularity.

Deficit irrigation has been introduced as a proper irrigation regime to increase water use efficiency in quinoa production [14,25], which has been an acceptable strategy, especially in arid and semi-arid areas that suffer from water scarcity [9,26]. Therefore, it could be a valuable option to reach the optimum quinoa seed yield in areas with water scarcity [27–30]. In this regard, findings by Geerts et al. [25] based on some field experiments in Bolivia and crop modeling, indicated that a deficit in irrigation to 55% of full irrigation had no significant effect on dry matter, while it enhanced seed yield and water use efficiency [25]. Nevertheless, another study in Bolivia reported the flowering and seed filling stages as the most sensitive stages to the water stress conditions [31]. It is correlated with exposure to high temperatures at this stage that have cumulative effects [6,32,33]. Therefore, deficit irrigation and high temperatures at the flowering and seed filling stages play an important role in quinoa seed yield.

Furthermore, nitrogen plays a vital role in increasing crop productivity as it increases crop yield for the unit of applied water [34,35] and seed quality [32,33]. Geren [33] investigated the effects of different N rates (0, 50, 75, 100, 125, 150, and 175 kg N ha^{-1}) on quinoa seed yield and its components, which indicated a positive increase in seed yield by increasing N rates. This investigation indicated that for the Mediterranean conditions, a rate of 150 kg N ha^{-1} was appropriate to earn a 2.95 Mg ha^{-1} seed yield with a 16% protein concentration. However, Oelke et al. [36] suggested the possibility of a 4.5 Mg ha^{-1} seed yield in Colorado by increasing the N rate to 170 and 200 kg N ha^{-1} [36]. Likewise, Kaul et al. reported a 94% rise in quinoa seed yield by increasing the N rate to 120 kg N ha^{-1} [37].

Despite the influence of the increasing N rate on quinoa seed yield and its components in different soil types [38], the interaction effects between the N rates and other field practices such as irrigation could be important in different climate conditions [39,40]. Limited research on nitrogen fertilizer and irrigation water management is available in European humid weather conditions [38] or arid weather conditions in quinoa native regions [32]. For example, the results of Alandia et al. [35] showed that N might confer a certain degree of drought tolerance to quinoa as seed quality and yield of N-fertilized plants were not affected by drought stress. Their results under controlled conditions serve as a basis to elucidate drought tolerance mechanisms activated with N fertilization.

Soil organic matter content is lower in soil with high temperatures in arid and semi-arid weather conditions. Moreover, microbial activity is affected by soil water content [41]. Therefore, the nitrogen fertilizer application rate in arid climates is higher than those in humid climates. However, documented research on quinoa nitrogen fertilizer management in arid and semi-arid weather conditions in non-native regions (out of the Andes area) is not well documented. The complex cycle of nitrogen in the environment under different soil and weather conditions associated with nitrogen degradation and assimilation mechanism is an essential aspect of nitrogen fertilizer application [42]. Quinoa cultivation is becoming popular in non-native arid and semi-arid regions; however, there is still a gap in its yield response to different field management practices, especially in these climate conditions. Moreover, the rate of the optimum nitrogen fertilizer is still unknown under field conditions in non-native quinoa cultivation areas with water scarcity. Therefore, investigating the effect of different irrigation regimes and different nitrogen rates on quinoa yield, irrigation water use efficiency, and N use efficiency was the objective of this study.

2. Results

2.1. Water Use

Figure 1 illustrates the irrigation depth in three irrigation regimes along with reference evapotranspiration, mm, during the two growing seasons. Irrigation depths before applying irrigation regimes treatments were 326 mm and 256 mm with a 6-day interval in the first and second year, respectively. Irrigation treatments were initiated at the vegetative with bud formation stage. The highest value of irrigation depth belonged to full irrigation in both years (850 mm and 714 mm in the first and second years, respectively). Data showed that irrigation depths in 0.75 FI were 719 mm and 600 mm in both years, with a similar reduction of 15% and 15.7% compared to full irrigation. Likely, in 0.5 FI, irrigation depths were 588 mm and 485 mm, which decreased by 30% and 31.5% in the first and second years, respectively, compared to full irrigation (Figure 1).

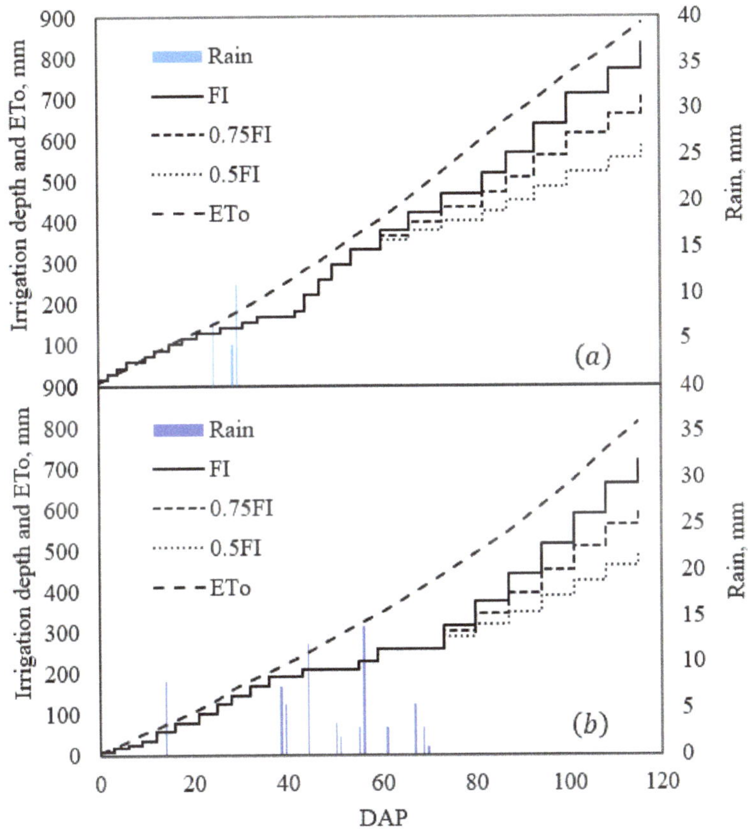

Figure 1. Irrigation depth, ET_o, and rainfall during the quinoa growing season; (**a**) first year; (**b**) second year. DAP: days after planting.

There was no significant interaction between the effect of irrigation regimes and nitrogen application rate on the soil water contents before irrigation, according to the analysis of variance ($p < 0.05$). Furthermore, there was no significant difference in soil water contents between 0.75 FI and 0.5 FI, while they decreased by 8.2% compared to FI (Table 1). Furthermore, the mean soil water content in the control N treatment (0 kg N ha^{-1}) was statistically higher than that in the 250 kg N ha^{-1}. Nitrogen application at the rate of 250 kg N ha^{-1} noticeably decreased soil water contents before irrigation events by 4.5% compared to non-fertilized treatment due to higher crop growth and higher water uptake.

Table 1. Seasonal mean soil water content, seasonal evapotranspiration, evaporation, transpiration, seed, and total dry matter water use efficiencies (SWUE, DWUE) for two nitrogen application rates and irrigation regimes averaged in two growing seasons.

Parameters	Irrigation Treatment	Nitrogen Application Rate (kg N ha^{-1})	
		0	250
Mean volumetric soil water content before irrigation, %	FI	23.8 a *	22.5 ab
	0.75 FI	22.2 ab	21.0 b
	0.5 FI	21.2 b	20.8 b
Seasonal evapotranspiration, mm	FI	802.50 a	782.1 a
	0.75 FI	676.5 b	655.9 b
	0.5 FI	542.2 c	528.0 c
Seasonal transpiration, mm	FI	377.4 b	479.3 a
	0.75 FI	295.7 c	396.2 b
	0.5 FI	236.2 c	363.8 b
Seasonal evaporation, mm	FI	425.1 a	302.8 b
	0.75 FI	380.9 a	259.7 b
	0.5 FI	306.0 b	164.2 c
SWUE, kg m^{-3}	FI	0.43 c	0.61 b
	0.75 FI	0.42 c	0.70 a
	0.5 FI	0.29 d	0.77 a
DWUE, kg m^{-3}	FI	1.04 c	1.48 b
	0.75 FI	1.08 c	1.54 b
	0.5 FI	0.93 d	1.80 a

* Means followed by the same letter in each trait are not significantly different at a 5% level of probability.

2.2. Seasonal Evapotranspiration

Actual crop evapotranspiration was estimated using water balance and averaged in two growing seasons, which is presented in Table 1. Overall, in the first year, reference potential evapotranspiration was higher than that in the second year due to higher temperature during the growing season (Figure 2); however, the difference was not statistically significant according to the analysis of variance ($p < 0.05$). The results indicated a significant difference amongst the irrigation strategies, in which FI had the highest ET (802.5 mm) followed by 0.5 FI with the lowest ET (528.0 mm) (Table 1). In non-fertilized treatment, ET decreased by 15% and 32% in 0.75 FI and 0.5 FI, respectively, compared to FI. Likewise, there was a reduction of 19% and 48% of ET in 0.75 FI and 0.5 FI in 250 kg N ha^{-1}, respectively, compared to FI. Furthermore, nitrogen application did not show a significant impact on ET in both full and deficit irrigation regimes (Table 1).

2.3. Seasonal Evaporation and Transpiration

The mean value of soil surface seasonal evaporation over two growing seasons (Table 1) indicated that there was a significant difference in evaporation between the two nitrogen treatments (0 and 250 kg N ha^{-1}) at all irrigation regimes. The nitrogen application rate of 250 kg N ha^{-1} remarkably decreased soil evaporation by 4.5%, on average, compared to non-fertilized treatment. Therefore, nitrogen application statistically reduced the soil evaporation during the growing season due to higher crop canopy development and higher coverage of the soil surface so that about 53–56.4% and 32–39% of ET was related to evaporation in non-fertilized and fertilized treatments, respectively. On the other hand, the evaporation did not significantly differ between FI and 0.75 FI (Table 1) in two nitrogen application rates, though both were different from the 0.5 FI. On average, deficit irrigation regimes decreased evaporation by 12% and 35% in 0.75 FI and 0.5 FI, respectively, compared to FI (Table 1). The two-year average of seasonal crop transpiration over the growing seasons is presented in Table 1. The highest value (479.3 mm) of crop transpiration was observed in the FI and 250 kg N ha^{-1}, which was significantly higher than the other treatments. The application of 250 kg N ha^{-1} enhanced crop transpiration by 27, 34, and 54% compared to non-fertilized treatment in FI, 0.75, and 0.5 FI treatment, respectively (Table 1). Therefore, nitrogen application increased crop transpiration due to improving crop canopy growth, especially in treatments with water stress conditions. Additionally, deficit irrigation regimes as 0.75 FI and 0.5 FI remarkably decreased seasonal crop transpiration compared to FI, though they were not statistically different (Table 1). Generally, quinoa transpiration dropped by 19% and 30% in 0.75 FI and 0.5 FI, respectively, in comparison with that obtained in FI (Table 1).

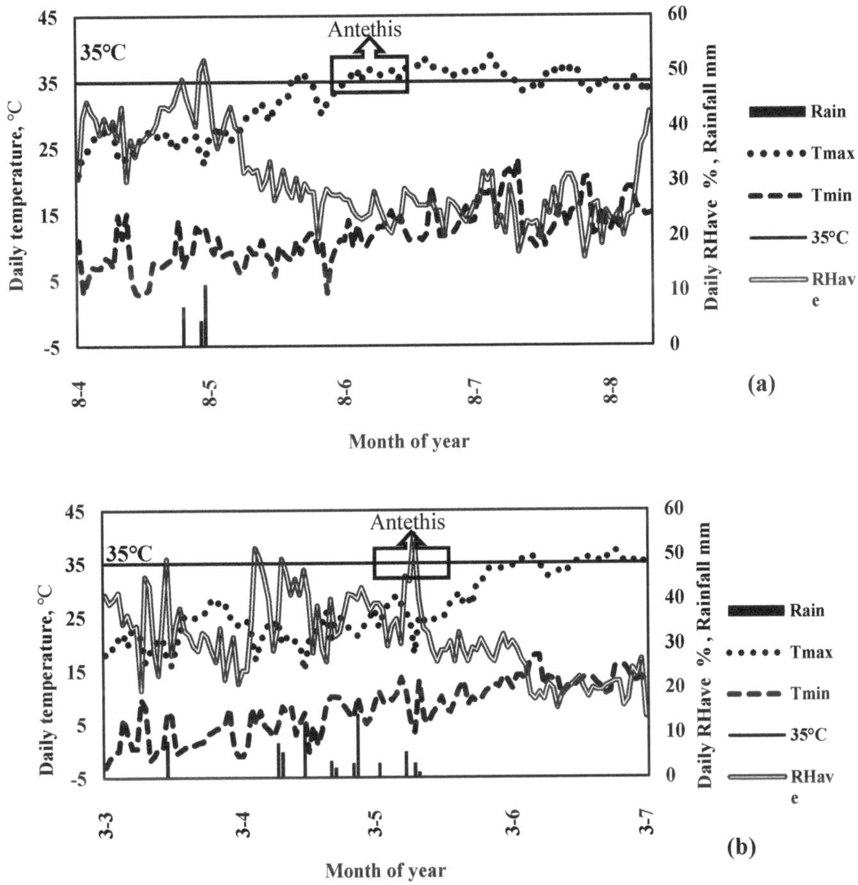

Figure 2. Daily maximum and minimum air temperature (T_{max}, T_{min}), relative humidity (RHavg), and rainfall during both growing periods. (**a**) 2017; (**b**) 2018.

2.4. Yield and Yield Components

There was a significant interaction between the effect of irrigation regimes and nitrogen application rate on quinoa seed yield, total dry matter, and harvest index ($p < 0.05$). Therefore, the interactive effects are presented in Table 2.

Table 2. Seed yield (Mg ha^{-1}), the two-year mean of total dry matter (Mg ha^{-1}), harvest index and estimated seed yield in the first year', hypothetically discarding the pest and unfavorable air temperature (Mg ha^{-1}), total N uptake, (kg N ha^{-1}), and residual soil NO_3-N, (kg N ha^{-1}) in different N application rates and irrigation regimes.

Parameters	Irrigation Treatment	Nitrogen Application Rate (kg N ha^{-1}) 0	125	250	375
Seed yield, Mg ha^{-1} (2017)	FI	0.77 cde *	0.84 bc	0.89 ab	0.96 a
	0.75 FI	0.55 gh	0.68 ef	0.85 bc	0.8 bcd
	0.5 FI	0.47 h	0.54 h	0.73 def	0.64 fg
Seed yield, Mg ha^{-1} (2018)	FI	2.96 gf	3.91 bcd	4.27 b	4.71 a
	0.75 FI	2.63 g	3.41 def	4.12 bc	3.76 cd
	0.5 FI	1.39 h	2.63 g	3.58 de	3.17 ef
Total dry matter, Mg ha^{-1}	FI	6.37 bc	7.43 abc	8.86 a	9.03 a
	0.75 FI	5.79 c	6.81 bc	7.87 ab	7.40 abc
	0.5 FI	4.23 d	6.31 bc	7.19 bc	6.91 bc
Harvest index	FI	0.40 bc	0.45 a	0.42 ab	0.45 a
	0.75 FI	0.40 bc	0.44 ab	0.45 a	0.44 ab
	0.5 FI	0.31 d	0.37 c	0.43 ab	0.40 bc
Estimated seed yield in the first year (2017), Mg ha^{-1}	FI	2.10 gh	2.83 cd	3.13 b	3.44 a
	0.75 FI	1.98 h	2.51 ef	3.08 bc	2.68 de
	0.5 FI	1.22 i	2.01 h	2.59 def	2.35 fg
Total N uptake, kg N ha^{-1}	FI	96.56 de	129.61 c	159.15 b	186.49 a
	0.75 FI	83.53 e	117.85 c	147.68 b	148.57 b
	0.5 FI	50.49 f	98.90 d	130.82 c	129.71 c
Residual soil NO_3-N, kg N ha^{-1}	FI	49.98 h	92.59 f	135.61 d	155.59 c
	0.75 FI	81.01 g	96.81 f	127.05 d	150.56 c
	0.5 FI	111.94 e	117.22 e	202.96 b	216.26 a

* Means followed by the same letter in each trait are not significantly different at the 5% level of probability.

2.4.1. Seed Yield

Quinoa seed yield was significantly different between the two years (2017–2018) according to the analysis of variance ($p < 0.05$); therefore, the measured data were separately presented and discussed (Table 2). In the first year, seed yield dramatically decreased by 80% compared to that obtained in the second year. The highest seed yield was 0.96 Mg ha^{-1} in the first year, which was 80% lower than that obtained in the second year. This dramatic decrease in the first year occurred due to facing air temperatures higher than 35 °C in the flowering stage (Figure 2).

In the first year, the effects of different irrigation regimes and nitrogen rates on quinoa seed yield were significant (Table 2). Nitrogen application increased seed yield in FI, and the highest value was obtained in 375 kg N ha^{-1} while there was no significant difference in seed yield between 375 kg N ha^{-1} and 250 kg N ha^{-1} in the first year with pest damage and unfavorable air temperature. However, in the second year with favorable air temperature, the highest yield with a significant increase was obtained in 375 kg N ha^{-1} and FI regime. In the deficit irrigations, on the other hand, the highest value of seed yield was achieved at 250 kg N ha^{-1}, and adding more N declined the quinoa seed yield (Table 2). Deficit irrigation as 0.75 FI and 0.5 FI reduced seed yield by 15% and 30%, respectively, compared to FI. In comparison with FI, applying the 0.5 FI regime decreased quinoa seed yield by 38.9%, 35.7%, 17.9%, and 33% in the nitrogen application rates at 0, 125, 250, and 375 kg N ha^{-1}, respectively. Even though seed yield was much higher in the second year than that in the first year, the same trend in the results was obtained in the second year, which indicated the significant effect of different N application rates and irrigation strategies on quinoa seed yield (Table 2). Despite the first year, the interaction effect of N application rate and irrigation regimes was significant in the second year ($p < 0.05$). In FI, the highest seed yield was obtained in 375 kg N ha^{-1} (4.71 Mg ha^{-1}), followed by 250 kg N ha^{-1} (4.27 Mg ha^{-1}), whereas the maximum seed yield was obtained in 250 kg N ha^{-1} in 0.75 FI and 0.5 FI as 4.12 Mg ha^{-1} and 3.58 Mg ha^{-1}, respectively. Although various irrigation regimes showed a significant difference in seed yield, N application rates of 250 and 375 kg N ha^{-1} did show any significant difference in seed yield. The application of 125 kg N ha^{-1} enhanced quinoa seed yield by 42.5% compared to non-fertilized treatment, whereas applying 375 kg N ha^{-1} increased seed yield by 69%. A reduction of 53% and 16% in seed yield was observed in 0.5 FI with 0 kg N ha^{-1} and 250 kg N ha^{-1} application rates compared to 375 kg N ha^{-1}, respectively. Likely, 0.5 FI with 125 kg N ha^{-1} and 375 kg N ha^{-1} decreased seed yield by 32.7% compared to FI (Table 2). Generally, increasing N fertilizer to 250 kg N ha^{-1} noticeably enhanced the quinoa seed yield; therefore, statistically, it could be the proper rate of N application in rising seed yield for deficit irrigation regimes. However, under non–limited water resource conditions, an FI and N application rate of 375 kg ha^{-1} could be used for higher seed yield.

2.4.2. Total Dry Matter

Quinoa total dry matter (TDM) was not significantly different in the two years ($p < 0.05$). Therefore, the mean values of TDM over two years are presented in Table 2. The interaction effect between N application rates and irrigation regimes was significant ($p < 0.05$). The highest value of TDM was obtained in FI at the rate of 375 kg N ha^{-1} (9.03 Mg ha^{-1}), while in deficit irrigation regimes, 250 kg N ha^{-1} had the highest TDM values (Table 2). However, there was no significant difference between 250 kg N ha^{-1} and 375 kg N ha^{-1} in TDM in FI for all irrigation regimes. When the N application rate exceeded 250 kg N ha^{-1}, it caused a rise of 40.5% in TDM compared to non-fertilized treatment. In deficit irrigations (0.75 FI and 0.5 FI), TDM matter did not show a significant difference when the N rate exceeded 125 kg N ha^{-1}. However, considering the seed yield increase, the N rate higher than 250 kg N ha^{-1} did not show a significant effect on seed production. Furthermore, a linear relationship between quinoa TDM and total transpiration during the growing season, indicated that increasing transpiration directly increased total dry matter (TDM = 0.0188 T, $R^2 = 0.92$, $p < 0.001$, SE = 0.45, n = 18). This relationship can be used for comparing quinoa transpiration efficiency with other cereals.

2.4.3. Harvest Index

A reduction of 80% in quinoa seed yield occurred in the first year as compared with the second year. This was due to high air temperature in the flowering stage and pests damage in the first year compared to the second year. Therefore, the harvest index was determined only for the second year (Table 2). The results indicated that the interaction effect between N application rate and irrigation regimes was significant, as well as the main effects ($p < 0.05$). The harvest index varied between 0.31 and 0.45 for quinoa in the second year. Although 0.75 FI in all fertilized treatments did not show a significant difference in harvest index with those in FI, harvest index significantly decreased in 0.5 FI and N rates by 25.5%, 17.8%, and 11% in non-fertilized, 125 kg N ha^{-1}, and 375 kg N ha^{-1}, respectively, compared to FI. However, the harvest index in 0.5 FI and 250 kg N ha^{-1} were statistically the same as that in FI (Table 2). Overall, the N application effectively increased harvest index in comparison with that obtained in non-fertilized treatment, especially in 0.5 FI. In 0.75 FI, N application of 250 kg N ha^{-1} and 375 kg N ha^{-1} raised harvest index by 12.5% and 10%, respectively, compared to non-fertilized treatment. Likewise, 0.5 FI increased the harvest index by 19.4%, 38.7, and 29% in N application rates of 125, 250, and 375, respectively. Taking the second-year harvest index into account, quinoa seed yield for the first year was estimated by multiplying HI and first-year dry matter (Table 2). Therefore, seed yield was estimated by omitting pest and heat damage to seed yield in the first year. According to this result, the highest seed yield (3.44 Mg ha^{-1}) was obtained in FI with 375 kg N ha^{-1} and the lowest value (1.22 Mg ha^{-1}) in 0.5 FI with non-fertilized treatment (Table 2).

2.5. Total Nitrogen Uptake

The total N uptake is estimated by seed N uptake plus straw N uptake (Table 2). The analysis of variance showed that the interaction effects between irrigation regimes and N application rates on total N uptake were significant ($p < 0.05$). Generally, a high-water deficit (0.5 FI) resulted in a high reduction in total N uptake (Table 2). In FI, increasing the N rate from 250 to 375 kg N ha^{-1} significantly raised total N uptake by 17%, whereas this increase was not significant in 0.75 FI and 0.5 FI. In non-fertilized treatment, irrigation application of 0.75 FI and 0.5 FI dropped total N uptake by 13.5% and 48% compared to FI. Likewise, in 375 kg N ha^{-1}, the application of 0.75 FI and 0.5 FI decreased total N uptake compared to FI by 20.3% and 30.5%, respectively. Furthermore, increasing N application rates significantly increased total N uptake (Table 2). For example, the application of 125 kg N ha^{-1} compared to non-fertilized treatment significantly enhanced total N uptake by 34%, 41%, and 95.9% in FI, 0.75 FI, and 0.5 FI, respectively. Therefore, the highest value of total N uptake was obtained in 375 kg N ha^{-1} and FI, while the lowest value was obtained in non-fertilized treatment and 0.5 FI.

2.6. Residual Soil NO_3-N

The residual soil NO_3-N in three different depths, including, 0–30 cm, 30–60 cm, and 60–90 cm, was measured before sowing and after harvest in both years, and the total residual NO_3-N, in the soil profile is presented in Table 2. The results indicated that the interaction effect of irrigation regimes and N application rates on soil NO_3-N was statistically significant ($p < 0.05$). In non-fertilized conditions, an irrigation regime of 0.75 FI noticeably enhanced the residual soil NO_3-N by 62% compared to FI, while, in fertilized treatments, there was no significant difference between 0.75 FI and FI in residual soil NO_3-N (Table 2). By contrast, 0.5 FI significantly increased residual soil NO_3-N in all N rates compared to FI and 0.75 FI. In this study, the highest residual soil NO_3-N was obtained in 0.5 FI and 375 kg N ha^{-1} when the FI and non-fertilized treatment had the lowest value (Table 2). Furthermore, measured residual soil NO_3-N at different soil depths after harvest averaged in both years is presented in Figure 2. This figure illustrates the variation of soil NO_3-N in all treatments after harvest compared to that before planting. In FI, residual soil NO_3-N in all N application rates at the topsoil layer (0–30 cm) decreased at the end of the growing season (Figure 3). Although, it increased at deep layers (60–90 cm), especially at higher N rates (250 kg N ha^{-1} and 375 kg N ha^{-1}) due to N leaching to the lower soil layers. Application of 0.75 FI led to NO_3-N accumulation at upper soil layers compared to that in FI. A dramatic increase was observed at the 30–60 cm soil layer by applying 375 kg N ha^{-1}. When the irrigation water decreased to a lower level as 0.5 FI, NO_3-N accumulation in the middle of the soil profile (30–60 cm) increased, which remarkably exceeded the initial soil NO_3-N at topsoil (0–30 cm) and 60–90 cm in two application N rates of 250 kg N ha^{-1} and 375 kg N ha^{-1}. Therefore, a higher reduction in irrigation water led to a much higher increase in residual soil NO_3-N, almost the same as increasing N application rate effects on residual soil NO_3-N.

2.7. The Relationship between Total N Uptake and Transpiration

Due to the influential role of seasonal transpiration (T) in N uptake (TNU), the relationship between these parameters was obtained for two application N rates of 0 kg N ha^{-1} and 250 kg N ha^{-1} as follows:

$$\begin{gathered}\text{TNU} = 0.195\ \text{T for } 0\ \text{N ha}^{-1}\ \text{nitrogen rate} \\ R^2 = 0.76,\ p < 0.001,\ \text{SE} = 0.35,\ n = 9\end{gathered} \tag{1}$$

$$\begin{gathered}\text{TNU} = 0.16\ \text{T} + 41.4\ \text{for } 250\ \text{N ha}^{-1}\ \text{nitrogen rate} \\ R^2 = 0.75,\ p < 0.001,\ \text{SE} = 0.40,\ n = 9\end{gathered} \tag{2}$$

where TNU and T are total nitrogen uptake (kg ha^{-1}) and seasonal transpiration (mm), respectively. It is indicated that passive absorption of N in the plant is regulated by transpiration. Therefore, transpiration is an effective force in N uptake in two N application rates of 0 kg N ha^{-1} and 250 kg N ha^{-1}. However, the slope of T in the regression is higher in the N application rate of 0 kg N ha^{-1}. According to Equation (2), at an N application rate of 250 kg ha^{-1} about 41.4 kg N ha^{-1} [intercept of Equation (2)] is absorbed by active absorption from the root zone. It is indicated that at high N application rates, both N absorption mechanism is effective [Equation (2)].

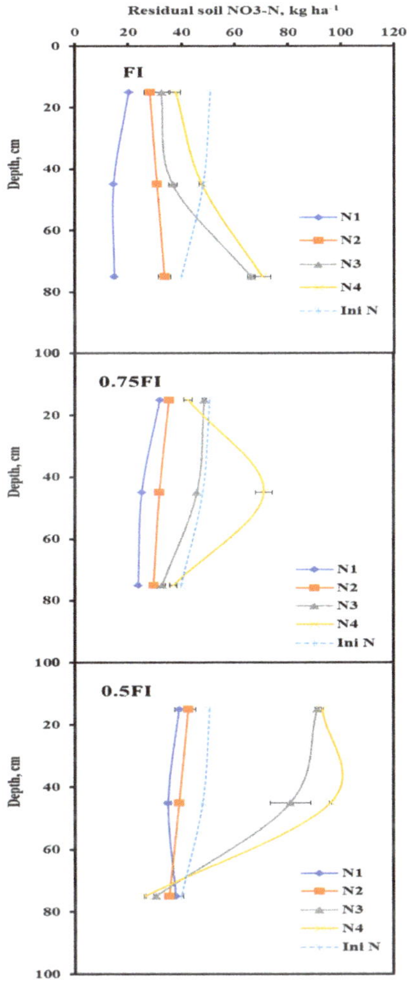

Figure 3. Mean residual soil NO_3-N at different soil depths after harvest over two years.

2.8. The Relationship between Soil Total Available N and Yield

Due to the influential role of soil total available N in producing crop yield and N uptake, the relationships between these parameters were obtained (Figure 3). Soil total available N (STAN) included soil N, applied N by fertilizer, and N applied by irrigation water. Soil N was considered to be residual N at planting plus the N mineralization that was determined considering 3% of the soil organic N mineralized to inorganic N [43].The relationship between seed yield and soil total available N indicated that in all irrigation regimes, seed yield increased gradually by increasing the soil total available N and reached a maximum value at a specific value of soil total available N, after which it dropped (Figure 4a).

A similar trend was found for the total dry matter and crop N uptake (Figure 4b,c). The maximum value of the soil total available N was different in various irrigation regimes. Therefore, the maximum values to reach the highest seed yield were obtained by numerical differentiation of the equation for each irrigation regime. These values are 822, 639, and 634 kg N ha^{-1} in FI, 0.75 FI, and 0.5 FI, respectively. By subtracting applied N by irrigation water, mineralized N, and soil N before planting, the optimum amount of N fertilizer was individually estimated for different irrigation strategies. In FI, 0.75 FI, and 0.5 FI, the maximum N fertilizer rate in the study area is estimated as 467, 287, and 285 kg N ha^{-1}, respectively. Overall, seed yield, total dry matter, and N uptake decreased when applied water dropped from full irrigation to 0.5 FI. Considering the total available soil N and irrigation water depth, a relationship was found to predict quinoa seed yield as follows:

$$SY = -1.758 \times 10^{-5}\,N^2 - 1.143 \times 10^{-5}\,I^2 + 23.22 \times 10^{-3}\,N + 0.0192\,I - 10.96 \quad R^2 = 0.88,\ p < 0.001,\ SE = 0.22,\ n = 9 \tag{3}$$

where SY, N, and I are the seed yield (Mg ha^{-1}), soil total available N (kg ha^{-1}), and irrigation water depth (mm), respectively. This relationship could be used in modeling to predict seed yield based on soil total available N and applied water. Further, contour (iso-quant) plots were developed in (Figure 5) to show the combined effect of soil total available N and irrigation water depth on seed yield for practical use by farm managers. The quadratic equation implied that the optimum N application rate would be different in various irrigation regimes to reach the maximum seed yield.

2.9. *Water Use Efficiency*

The analysis of variance showed a significant effect of the interaction between irrigation regimes and the N application rates on the seed water use efficiency (SWUE) ($p < 0.05$). Table 1 presents the comparison between means of SWUE for two N application rates at different irrigation regimes. Crop evapotranspiration was calculated from Equation (4) to determine WUE. As the soil water content was measured by neutron probe tubes located at two N application rates (0 and 250 Kg N ha^{-1}), WUE is presented for these two treatments. Generally, the N application significantly raised SWUE and DWUE. Adding 250 kg N ha^{-1} noticeably increased SWUE by 42%, 67%, and 165.5% in FI, 0.75 FI, and 0.5 FI, respectively. There was a significant difference between FI and deficit irrigation regimes in SWUE when the 250 kg N ha^{-1} rate was applied, whereas no significant difference was observed between 0.75 FI and 0.5 FI. By contrast, non-fertilized treatment and 0.5 FI significantly decreased SWUE by 31% compared to 0.75 FI (Table 1). Furthermore, total dry matter water use efficiency (DWUE) was determined at different irrigation regimes and two N application rates (Table 1). The highest DWUE of 1.8 kg m^{-3} was obtained in 0.5 FI with the application of 250 kg N ha^{-1}, whereas it was 19% higher than those obtained in FI and 0.75 FI. In non-fertilized treatment, DWUE decreased by 12% in 0.5 FI compared to those obtained in FI and 0.75 FI. Accordingly, increasing the N application rate to 250 kg N ha^{-1} enhanced DWUE by 42%, 43%, and 94% in FI, 0.75 FI, and 0.5 FI, respectively. Generally, deficit irrigation could not increase SWUE and DWUE in non-fertilized treatment.

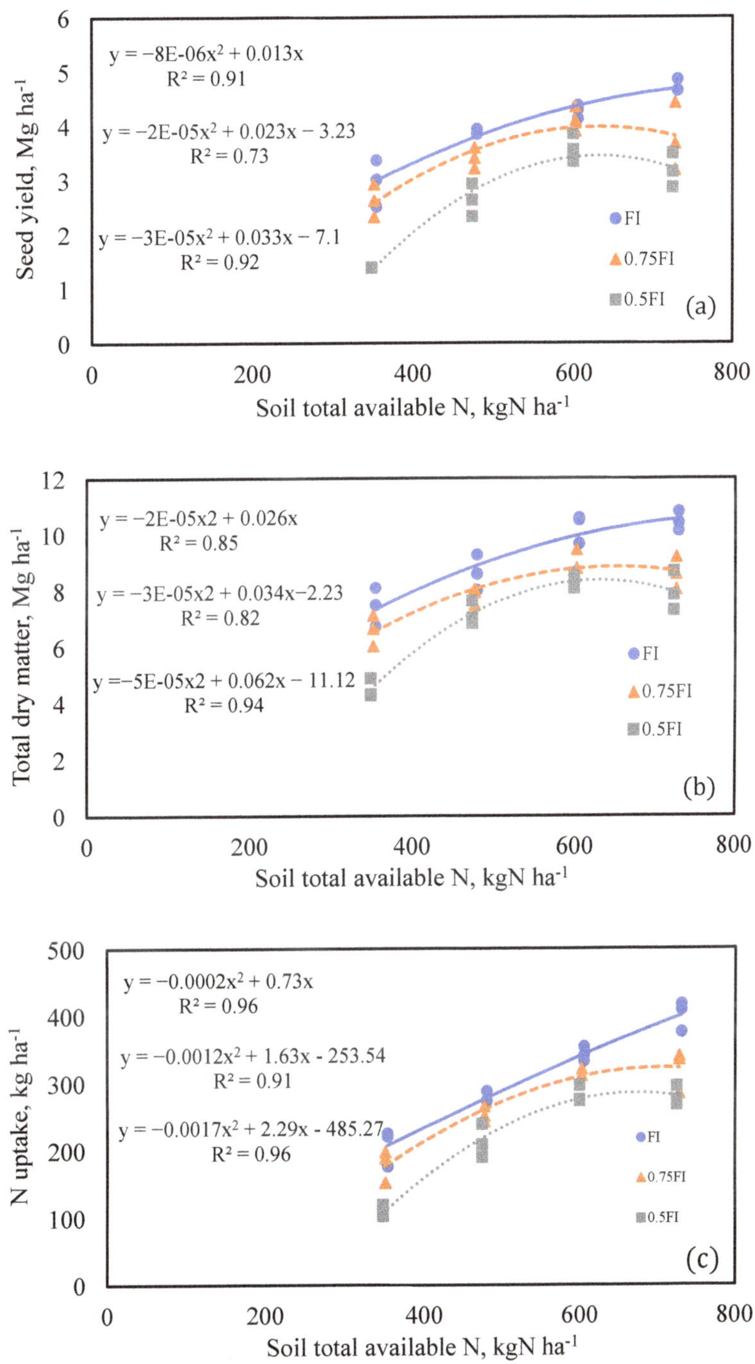

Figure 4. The relationship between soil total available N and quinoa seed yield, dry matter, and crop N uptake.

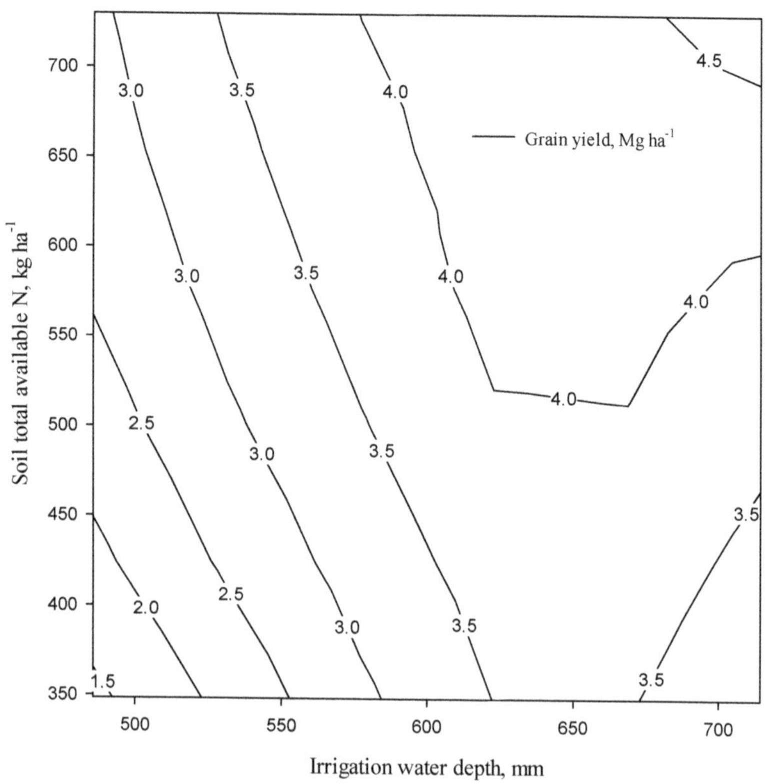

Figure 5. Nomo graph showing the relationship between irrigation water depth and soil total available N for obtaining different quinoa seed yields.

2.10. Nitrogen Efficiencies

The results of the estimated nitrogen use efficiency (NUE) and nitrogen yield efficiency (NYE) at different irrigation regimes and N application rates in the second year are presented in Table 3. There is no significant interaction between the effect of irrigation regimes and nitrogen application rate on nitrogen use efficiencies ($p < 0.05$). However, the main effects of the irrigation regimes and nitrogen rates had a significant effect on both NUE and NYE ($p < 0.05$). In general, the increasing N application rate decreased NUE though there was no significant difference between 125 kg N ha^{-1} and 250 kg N ha^{-1}. Increasing the N application rate from 125 kg N ha^{-1} and 250 kg N ha^{-1} to 375 kg N ha^{-1} noticeably decreased NUE by 29% and 21%, respectively (Table 3). Reducing water application to half (0.50 FI) significantly increased NUE, whereas 0.5 FI increased NUE by 19% and 29% compared to FI and 0.75 FI, respectively. Same as the NUE, the increasing N rate decreased NYE with no significant difference between 125 kg N ha^{-1} and 250 kg N ha^{-1}. The nitrogen application rate of 375 kg N ha^{-1} dropped NYE by 44.3% and 34% compared to 125 kg N ha^{-1} and 250 kg N ha^{-1}, respectively (Table 3). In addition, no significant difference was found between 0.75 FI and FI in NYE, while the application of 0.5 FI increased NYE by 54% compared to FI. Table 3 presented the mean physiological N efficiency (NPE) at different irrigation regimes and N application rates. Analysis of variance showed that the effect of irrigation regimes was the only significant factor in NPE ($p < 0.05$). Compared to FI, deficit irrigation of 0.75 FI significantly increased NPE by 26.7%, whereas 0.5 FI raised NPE by 17%, which was not statistically different from FI.

Table 3. Nitrogen use efficiency (NUE) and nitrogen yield efficiency (NYE) and physiological N efficiency (NPE) at different irrigation regimes and N application rates in the second year.

	NUE, kg m^{-3}	NYE, kg kg^{-1}	NPE, kg kg^{-1}
Irrigation treatment			
FI	0.26 b *	5.94 b	26.98 b
0.75 FI	0.24 b	5.12 b	34.19 a
0.5 FI	0.31 a	7.87 a	31.55 ab
Nitrogen rate (kg N ha^{-1})			
125	0.31 a	7.90 a	29.77 a
250	0.28 a	6.64 a	34.08 a
375	o.22 b	4.40 b	28.87 a

* Means followed by the same letter in each trait are not significantly different at the 5% level of probability.

3. Discussion

3.1. Quinoa Yield Is Affected by N Application Rate and Irrigation Regimes

Comparing the results of quinoa seed yield in two growing seasons, apart from the irrigation and N application rates, showed an unexpected drop in seed yield (0.96 Mg ha^{-1}) in the first year. However, TDM was not different in the two growing seasons. Therefore, the results from these experiments could be generalized for field management decisions in arid and semi-arid weather conditions. The dramatic decrease in the first year occurred due to the late planting dates in the first year and air temperatures higher than 35 °C in the flowering stage (Figure 2). The value of 35 °C for air temperature is considered the threshold value for quinoa in Algeria, Lebanon, Yemen, and Iraq during the flowering stage [6,44–47]. This could be due to the strong dependence of pollen moisture on-air vapor pressure deficit [48]. Therefore, a slight temperature increase higher than 35 °C would greatly reduce seed yield [49]. The increasing N application rate remarkably raised seed yield, and the highest seed yield in the second year was 4.7 Mg ha^{-1} in 375 kg N ha^{-1} with FI (Table 2), which is inconsistent with findings by Shams [50], and Kakabouki et al. [2] who reported the rise of quinoa seed yield with increasing N application rate to 120, 200, and 360 kg N ha^{-1}. Additionally, in the weather conditions of Turkey, the highest seed yield, ranged between 4.1 Mg ha^{-1} and 8.7 Mg ha^{-1} [39], whereas it was 3.3 Mg ha^{-1} in the south of Morocco [51]. Deficit irrigation (0.5 FI) significantly reduced quinoa seed yield, while there was no significant difference between FI and 0.75 FI. Therefore, an irrigation regime of 0.75 FI is suggested in the case of scarce water. Similar findings were reported by Yazar et al. [52]. According to our findings, in water scarcity conditions, 0.75 an FI and N application rate of 250 kg N ha^{-1} is recommended.

Considering the total dry matter, increasing the N application rate higher than 125 kg N ha^{-1} did not show a significant increase in the dry matter; however, the highest value of DM (9.03 Mg ha^{-1}) was obtained in 375 kg N ha^{-1} (Table 2). Similarly, Kakabouki et al. [2] reported no significant difference in the dry matter between two N application rates of 200 kg N ha^{-1} and 345 kg N ha^{-1}. In the Mediterranean conditions, the highest average DM was reported as 8.5 kg ha^{-1} [51]. The linear relationship between quinoa TDM and total transpiration during the growing season indicated that increasing transpiration directly increased total dry matter (TDM = 0.0188 T). The comparison between the slope of the linear relationship between quinoa total dry matter and seasonal transpiration with that reported by Azizian and Sepaskhah [53] for maize as a C_4 crop (0.028 Mg ha^{-1}) and Bahari-Sadi for saffron as a C_3 crop (0.010 Mg ha^{-1}) indicated that maize could produce higher TDM in given seasonal transpiration as compared with quinoa as a C_3 crop. However, quinoa can produce higher TDM in given seasonal transpiration as compared with saffron in similar weather conditions. In general, based on our findings, it is recommended to use 375 kg N ha^{-1} for reaching the optimum quinoa yield in full irrigation (no water scarcity) with an average of 24% soil water content before irrigation, and 250 kg N ha^{-1} is required for 0.75 FI and 0.5 FI (scare water conditions) with an average of 21.0 and 20.8% soil water content before irrigation, respectively.

3.2. Water and N Efficiencies as Influenced by N Application Rate

In the current study, the highest water use efficiency obtained in 0.5 FI with 250 kg N ha^{-1} was 0.77 kg m^{-3}, which was lower than 1.2 kg m^{-3} obtained in Italy [54]. Increasing N rates resulted in higher water use efficiency; however, the inverse impact of exceeding the N application rate from 250 kg N ha^{-1} was observed (Table 1). Similarly, increasing the N application rate reduced all the N efficiencies (NUE, NYE, and NPE), which is in agreement with the findings of Kakabouki et al. [2] for quinoa and Mehrabi and Sepaskhah [55] for winter wheat. Therefore, N application at a rate of 250 kg N ha^{-1} could be suggested as the optimum rate, reaching the highest water use efficiency and N use efficiencies in the study region. Although an increase was found in the water and N use efficiencies in the application of 0.5 FI with 250 kg N ha^{-1}, it is not recommended to produce quinoa crops because it might not be economic, as it remarkably decreases seed yield.

3.3. Nitrogen Uptake Mechanism and N Application Rate

Nutrient ion's movement from the soil solution to the vascular center of the root cell membrane may be passive or active. Passive absorption is the movement across a membrane from higher to lower concentrations. A similar gradient leads to crop transpiration. Active absorption requires metabolical energy. As with the uptake of other nutrients, N uptake activities are both strongly regulated by a high plant N status [56]. A different nitrogen uptake mechanism is discussed by [57–59] in the physiological and molecular mechanism. According to our findings, passive absorption is an effective force in N uptake in two N application rates of 0 kg N ha^{-1} and 250 kg N ha^{-1}. However, at high N application rates, both passive and active N absorption mechanism is effective.

4. Materials and Methods

4.1. Experimental Site and Design

The experimental research was conducted in the Experimental Research Station of the Agricultural College, Shiraz University, Shiraz, Iran, during two growing seasons (2017 and 2018). The station is located at 29°56′ N, 52°02′ E and is 1810 m above sea level, with a semi-arid climate. Maximum and minimum temperatures (T_{max}, T_{min}), average relative humidity, and rainfall during both growing seasons are presented in Figure 6. T_{max} and T_{min} during the first growing season were, on average, 32.4 °C and 12 °C, respectively. In contrast, the mean daily T_{max} and T_{min} during the second growing season were 27 °C and 7 °C, respectively. The rainfall depths during the two growing seasons were 29 and 64 mm, respectively.

The experimental design was a factorial arrangement with randomized complete blocks in three replications. The treatments consisted of four nitrogen application rates of 0, 125, 250, and 375 kg N ha^{-1} (N1, N2, N3, and N4, respectively) and three levels of irrigation water regimes. The irrigation regimes included full irrigation (FI), 75% of full irrigation (0.75 FI), and 50% of full irrigation (0.5 FI). Therefore, 36 plots were constructed with the dimensions of 2 m × 2 m and placed at a distance of 1.5 m from each other (Figure 6a). The physical and chemical properties of soil adapted from Yarami and Sepaskhah (2015) are presented in Table 4 [60].

Figure 6. Experimental plot design at planting date (**a**), quinoa at seed filling stage (**b**) and the micro-lysimeter (**c**) for soil water evaporation measurement.

Table 4. Soil and water properties.

Physical and Chemical Properties	Soil Depth (cm)			Irrigation Water
	0–30	30–60	60–90	
Field capacity (cm cm^{-3})	0.32	0.32	0.32	
Permanent wilting point (cm cm^{-3})	0.17	0.19	0.19	
Bulk density (g cm^{-3})	1.4	1.47	1.51	
Sand%	11	10	16	
Silt%	56	51	50	
Clay%	33	39	34	
Texture	SCL *	SCL	SCL	
EC (dS m^{-1})	0.74	0.51	0.49	0.58
Cl^- (meq L^{-1})	5.31	3.05	2.90	0.50
Na^+ (meq L^{-1})	3.29	1.97	1.91	0.48
Ca^{2+} (meq L^{-1})	5.43	4.16	4.07	1.80
Mg^{2+} (meq L^{-1})	3.50	2.88	2.84	2.0

* Silty clay loam.

After a deep plowing (30 cm) during the land preparation, triple superphosphate (CaH_2PO_4, including 46% P_2O_5) at a rate of 30 kg ha^{-1} was mixed with the soil surface layer. The quinoa seeds (Titicaca cultivar, developed by University of Copenhagen, Denmark) were planted in 6 rows with 0.33 m spacing. The seeds were placed in 0.01–0.02 m depth with a 0.15 m distance on the row. The sowing dates were 6 April and 3 March in the first and second growing seasons, respectively, with the average air temperature of 10 to 12 °C. Two weeks after germination, the crops were hand-thinned to obtain a uniform density of

20 plants per plot. Physical weed control was also conducted every two weeks during the growing seasons. Nitrogen as urea (46% N) was added to the soil surface at the vegetative and seed filling stages (Figure 6b), which were 60, and 80 days after sowing, respectively.

4.2. Irrigation Requirement

The irrigation water depth was determined using the daily potential evapotranspiration and crop coefficient for quinoa. The potential evapotranspiration was estimated using the modified FAO-Penman–Monteith method [61]. The crop coefficients at initial, mid-season, and late season growth stages for quinoa were also determined by Talebnejad and Sepaskhah as 0.58, 1.2, and 0.8, respectively [14]. Therefore, irrigation depth as crop evapotranspiration (ET_c) was determined by multiplying crop coefficient and potential evapotranspiration during the irrigation intervals. The irrigation treatments were initiated at the vegetative with bud formation stage with a 7-day interval as surface basin irrigation. The total irrigation water depths were 850 and 714 mm for the two growing seasons, respectively.

4.3. Crop Actual Evapotranspiration

Crop actual evapotranspiration (ET) during each growing season is estimated by following the soil water balance equation considering negligible deep percolation:

$$ET = I + P \pm \Delta S \quad (4)$$

where I and P are the irrigation and precipitation depth (cm), respectively. ΔS implies soil water variations (cm) in each interval.

The soil water content was measured by a neutron probe three times during the growing season at four soil depths (0–30 cm, 30–60 cm, 60–90 cm, and 90–120 cm) for two nitration application rates (0 and 250 kg N ha^{-1}). Soil evaporation was measured using micro-lysimeters [55], which were installed in the plots with different irrigation regimes and two N treatments (N1 and N3) with three replications. Micro-lysimeters were made of a small cylinder with 10 cm diameter and 30 cm height filled up with the same field soil (Figure 6c). The micro-lysimeters were weighted between irrigation intervals. Then, decreasing in micro-lysimeter weight was divided per the micro-lysimeter area to determine the evaporation from the soil surface.

4.4. Field Measurements

Characterizing the plants in the fields needs to be managed with attention considering the last international consensus on quinoa phenotyping methodologies [62]. Ten to fourteen days after the last irrigation event, four rows in the middle of each plot were harvested [134 and 122 days after planting (17 August and 3 July) in the first and second year, respectively]. The panicles were separated, and the seeds and shoots were oven-dried at 72 °C for about 48–72 h to determine seed yield, shoot dry weight, and total dry matter. The harvest index was calculated as seed yield divided by the total top dry matter. The seed and shoot nitrogen concentrations were measured by the Kjeldahl method then nitrogen uptake was determined by multiplying seed and shoot N concentrations by their relevant dry weight. Before N application and after harvesting, soil samples were taken from three depths, 0–30 cm, 30–60 cm, and 60–90 cm, and air-dried to determine the residual soil nitrate spectrophotometrically.

4.5. Water and Nitrogen Use Efficiencies

Seed water use efficiency (SWUE) and dry matter water use efficiency (DWUE) are determined as the ratio between seed yield (SY) or total dry matter (TDM) and crop water use (ET). Furthermore, nitrogen use efficiency as a vital component in obtaining net yield with high quality has different descriptions such as nitrogen use efficiency (NUE), nitrogen yield efficiency (NYE), and physiological N efficiency (NPE) [63] as follows:

$$\text{NUE} = \frac{NU_a - NU_c}{NR_a - NR_c} \tag{5}$$

$$\text{NYE} = \frac{Y_a - Y_c}{NR_a - NR_c} \tag{6}$$

$$\text{NPE} = \frac{Y_a - Y_c}{NU_a - NU_c} \tag{7}$$

where NU_a, Y_a, and NR_a are the quinoa N uptake, quinoa seed yield, and nitrogen application rate, kg ha^{-1}, respectively, in different treatments. While NU_c, Y_c, and NR_c are the quinoa N uptake, seed yield, and nitrogen application rate, kg ha^{-1}, respectively, in the control treatment (N application rate of 0 kg ha^{-1}).

4.6. Statistical Analysis

Statistical analysis was carried out using software SAS 9.4 software (SAS Institute Inc., Carry, NC, USA). The interaction effects between irrigation regimes and N application rates were evaluated using the analysis of variance (ANOVA), and the year was incorporated into the model. The means were compared at a 5% level of probability using Duncan's multiple range tests.

5. Conclusions

The nitrogen fertilizer application rate in arid and semi-arid areas is higher due to warmer climates. However, deficit irrigation improves NUE. Therefore, the effect of different irrigation regimes and nitrogen rates on quinoa yield is a challenging issue in those areas. Increasing nitrogen fertilizer application levels from 250 to 375 kg N ha^{-1} under FI and deficit irrigation (0.75 FI and 0.5 FI) did not cause a significant difference in grain yield and the total dry matter of quinoa. Furthermore, the application of 0.75 FI led to NO_3-N accumulation in upper soil layers compared to that in FI, which facilitated nitrogen uptake and reduced nitrate loss to deeper layers of the soil. Therefore, an appropriate nitrogen application rate of 250 kg N ha^{-1} and deficit irrigation of 0.75 FI is suggested as the optimum management in the study area, with challenging water scarcity based on SWUE, NUE, and total nitrogen uptake. However, in the area with ample water resources, FI with 375 kg, N ha^{-1} could be recommended based on seed yield and total nitrogen uptake. These factors should be considered as struggling with the potential quinoa seed yield and considering fertilizer application environmental impacts. Questions about which environmental factors impressively restricted the quinoa growth for optimizing the potential yield need further investigations.

Author Contributions: Conceptualization, R.T. and A.R.S.; methodology, R.T. and M.B.; software, M.B.; investigation and data curation, M.B.; writing—original draft preparation, M.B. and R.T.; writing—review and editing, R.T. and A.R.S. and D.B.; supervision, R.T. and A.R.S.; project administration, R.T.; funding acquisition, R.T. and D.B. All authors have read and agreed to the published version of the manuscript.

Funding: This research was funded by Grant #97GCU1M222614 and #98GRC1M222614 of Shiraz University Research Council, Drought Research Center, the Center of Excellent for On-Farm Water Management, and Iran National Science Foundation (NSF).

Institutional Review Board Statement: Not applicable.

Informed Consent Statement: Not applicable.

Data Availability Statement: The data presented in this study are available in the article.

Acknowledgments: Authors gratefully acknowledge Shiraz University Research Council, Drought Research Center, the Center of Excellent for On-Farm Water Management, and Iran National Science Foundation (NSF) for funding.

Conflicts of Interest: The authors declare no conflict of interest.

References

1. Hunter, M.C.; Smith, R.G.; Schipanski, M.E.; Atwood, L.W.; Mortensen, D.A. Agriculture in 2050: Recalibrating Targets for Sustainable Intensification. *Bioscience* **2017**, *67*, 386–391. [CrossRef]
2. Kakabouki, I.; Bilalis, D.; Karkanis, A.; Zervas, G.; Hela, D. Effects of Fertilization and Tillage System on Growth and Crude Protein Content of Quinoa (*Chenopodium quinoa* Willd.): An Alternative Forage Crop. *Emirates J. Food Agric.* **2014**, *26*, 18–24. [CrossRef]
3. Leakey, R.; Kranjac-Berisavljevic, G.; Caron, P.; Craufurd, P.; Martin, A.M.; McDonald, A.; Abedini, W.; Afiff, S.; Bakurin, N.; Bass, S. *Impacts of AKST on Development and Sustainability Goals*; Island Press: New York, NY, USA, 2009.
4. Fathi, F.; Sheikhzeinoddin, A.; Talebnejad, R. Environmental and Economic Risk Management of Seed Maize Production in Iran. *J. Clean. Prod.* **2020**, *258*, 120772. [CrossRef]
5. Ruiz, K.B.; Biondi, S.; Oses, R.; Acuña-Rodríguez, I.S.; Antognoni, F.; Martinez-Mosqueira, E.A.; Coulibaly, A.; Canahua-Murillo, A.; Pinto, M.; Zurita, A.; et al. Quinoa Biodiversity and Sustainability for Food Security under Climate Change. A Review. *Agron. Sustain. Dev.* **2014**, *34*, 349–359. [CrossRef]
6. García-Parra, M.A.; Roa-Acosta, D.F.; Stechauner-Rohringer, R.; García-Molano, F.; Bazile, D.; Plazas-Leguizamón, N. Effect of Temperature on the Growth and Development of Quinoa Plants (*Chenopodium quinoa* Willd.): A Review on a Global Scale. *Sylwan* **2020**, *164*, 411–423.
7. Ruiz-Carrasco, K.; Antognoni, F.; Coulibaly, A.K.; Lizardi, S.; Covarrubias, A.; Martínez, E.A.; Molina-Montenegro, M.A.; Biondi, S.; Zurita-Silva, A. Variation in Salinity Tolerance of Four Lowland Genotypes of Quinoa (*Chenopodium quinoa* Willd.) as Assessed by Growth, Physiological Traits, and Sodium Transporter Gene Expression. *Plant Physiol. Biochem.* **2011**, *49*, 1333–1341. [CrossRef]
8. Hatfield, J.L.; Prueger, J.H. Temperature Extremes: Effect on Plant Growth and Development. *Weather Clim. Extrem.* **2015**, *10*, 4–10. [CrossRef]
9. Kaya, Ç.I.; Yazar, A.; Sezen, S.M. SALTMED Model Performance on Simulation of Soil Moisture and Crop Yield for Quinoa Irrigated Using Different Irrigation Systems, Irrigation Strategies and Water Qualities in Turkey. *Agric. Agric. Sci. Procedia* **2015**, *4*, 108–118. [CrossRef]
10. Choukr-Allah, R.; Rao, N.K.; Hirich, A.; Shahid, M.; Alshankiti, A.; Toderich, K.; Gill, S.; Butt, K.U.R. Quinoa for Marginal Environments: Toward Future Food and Nutritional Security in MENA and Central Asia Regions. *Front. Plant Sci.* **2016**, *7*, 346. [CrossRef]
11. Bazile, D.; Bertero, H.D.; Nieto, C. *State of the Art Report on Quinoa around the World in 2013*; FAO & CIRAD: Rome, Italy, 2015.
12. Talebnejad, R.; Sepaskhah, A.R. Effect of deficit irrigation and different saline groundwater depths on yield and water productivity of quinoa in lysimiter. *Agric. Water Manag.* **2015**, *159*, 225–238. [CrossRef]
13. Talebnejad, R.; Sepaskhah, A.R. Modification of Transient State Analytical Model under Different Saline Groundwater Depths, Irrigation Water Salinities and Deficit Irrigation for Quinoa. *Int. J. Plant Prod.* **2016**, *10*, 365–390. [CrossRef]
14. Talebnejad, R.; Sepaskhah, A.R. Effect of Different Saline Groundwater Depths and Irrigation Water Salinities on Yield and Water Use of Quinoa in Lysimeter. *Agric. Water Manag.* **2015**, *148*, 177–188. [CrossRef]
15. Ruiz, K.B.; Biondi, S.; Martínez, E.A.; Orsini, F.; Antognoni, F.; Jacobsen, S.-E. Quinoa–a Model Crop for Understanding Salt-Tolerance Mechanisms in Halophytes. *Plant Biosyst. Int. J. Deal. All Asp. Plant Biol.* **2016**, *150*, 357–371. [CrossRef]
16. Sellami, M.H.; Pulvento, C.; Lavini, A. Agronomic Practices and Performances of Quinoa under Field Conditions: A Systematic Review. *Plants* **2021**, *10*, 72. [CrossRef]
17. Jacobsen, S.E.; Monteros, C.; Christiansen, J.L.; Bravo, L.A.; Corcuera, L.J.; Mujica, A. Plant Responses of Quinoa (*Chenopodium quinoa* Willd.) to Frost at Various Phenological Stages. *Eur. J. Agron.* **2005**, *22*, 131–139. [CrossRef]
18. Bazile, D.; Pulvento, C.; Verniau, A.; Al-Nusairi, M.S.; Ba, D.; Breidy, J.; Hassan, L.; Mohammed, M.I.; Mambetov, O.; Otambekova, M. Worldwide Evaluations of Quinoa: Preliminary Results from Post International Year of Quinoa FAO Projects in Nine Countries. *Front. Plant Sci.* **2016**, *7*, 850. [CrossRef]
19. De Ron, A.M.; Sparvoli, F.; Pueyo, J.J.; Bazile, D. *The Challenge of Protein Crops as a Sustainable Source of Food and Feed for the Future*; Frontiers Media SA: Lausanne, Switzerland, 2017; ISBN 2889451623.
20. Andreotti, F.; Bazile, D.; Biaggi, C.; Callo-Concha, D.; Jacquet, J.; Jemal, O.M.; King, O.I.; Mbosso, C.; Padulosi, S.; Speelman, E.N. When Neglected Species Gain Global Interest: Lessons Learned from Quinoa's Boom and Bust for Teff and Minor Millet. *Glob. Food Sec.* **2022**, *32*, 100613. [CrossRef]
21. Maliro, M.F.A.; Abang, M.; Mukankusi, C.; Lung'aho, M.; Fenta, B.; Wanderi, S.; Kapa, R.; Okiro, O.A.; Kooma, E.; Mwaba, C.; et al. *Prospects for Quinoa Adaptation and Utilization in Eastern and Southern Africa: Technological, Institutional and Policy Considerations*; Food & Agriculture Org: Rome, Italy, 2021; ISBN 9251336873.

22. Nowak, V.; Du, J.; Charrondière, U.R. Assessment of the Nutritional Composition of Quinoa (*Chenopodium quinoa* Willd.). *Food Chem.* **2016**, *193*, 47–54. [CrossRef]
23. Bilalis, D.; Roussis, I.; Fuentes, F.; Kakabouki, I.; Travlos, I. Organic Agriculture and Innovative Crops under Mediterranean Conditions. *Not. Bot. Horti Agrobot. Cluj-Napoca* **2017**, *45*, 323–331. [CrossRef]
24. Comai, S.; Bertazzo, A.; Bailoni, L.; Zancato, M.; Costa, C.V.L.; Allegri, G. The Content of Proteic and Nonproteic (Free and Protein-Bound) Tryptophan in Quinoa and Cereal Flours. *Food Chem.* **2007**, *100*, 1350–1355. [CrossRef]
25. Geerts, S.; Raes, D.; Garcia, M.; Taboada, C.; Miranda, R.; Cusicanqui, J.; Mhizha, T.; Vacher, J. Modeling the Potential for Closing Quinoa Yield Gaps under Varying Water Availability in the Bolivian Altiplano. *Agric. Water Manag.* **2009**, *96*, 1652–1658. [CrossRef]
26. Razzaghi, F.; Bahadori-Ghasroldashti, M.R.; Henriksen, S.; Sepaskhah, A.R.; Jacobsen, S. Physiological Characteristics and Irrigation Water Productivity of Quinoa (*Chenopodium quinoa* Willd.) in Response to Deficit Irrigation Imposed at Different Growing Stages—A Field Study from Southern Iran. *J. Agron. Crop Sci.* **2020**, *206*, 390–404. [CrossRef]
27. Geerts, S.; Raes, D.; Garcia, M.; Mendoza, J.; Huanca, R. Crop Water Use Indicators to Quantify the Flexible Phenology of Quinoa (*Chenopodium quinoa* Willd.) in Response to Drought Stress. *Field Crop. Res.* **2008**, *108*, 150–156. [CrossRef]
28. Garcia, M.; Raes, D.; Jacobsen, S.E. Evapotranspiration Analysis and Irrigation Requirements of Quinoa (*Chenopodium quinoa*) in the Bolivian Highlands. *Agric. Water Manag.* **2003**, *60*, 119–134. [CrossRef]
29. Geerts, S.; Raes, D.; Garcia, M.; Del Castillo, C.; Buytaert, W. Agro-Climatic Suitability Mapping for Crop Production in the Bolivian Altiplano: A Case Study for Quinoa. *Agric. For. Meteorol.* **2006**, *139*, 399–412. [CrossRef]
30. Talebnejad, R.; Sepaskhah, A.R.; Bahrami, M. Quinoa, a Promising Halophyte with Modified Planting Date, and Minimum Water and Pesticide Requirements for Fars Province, Iran. In *Future of Sustainable Agriculture in Saline Environments*; CRC Press: Boca Raton, FL, USA, 2021; pp. 413–426. ISBN 1003112323.
31. Geerts, S.; Raes, D.; Garcia, M.; Vacher, J.; Mamani, R.; Mendoza, J.; Huanca, R.; Morales, B.; Miranda, R.; Cusicanqui, J.; et al. Introducing Deficit Irrigation to Stabilize Yields of Quinoa (*Chenopodium quinoa* Willd.). *Eur. J. Agron.* **2008**, *28*, 427–436. [CrossRef]
32. Repo-Carrasco, R.; Espinoza, C.; Jacobsen, S.-E.-E. Nutritional Value and Use of the Andean Crops Quinoa (*Chenopodium quinoa*) and Kañiwa (Chenopodium Pallidicaule). *Food Rev. Int.* **2003**, *19*, 179–189. [CrossRef]
33. Geren, H. Effects of Different Nitrogen Levels on the Grain Yield and Some Yield Components of Quinoa (*Chenopodium quinoa* Willd.) under Mediterranean Climatic Conditions. *Turk. J. Field Crop.* **2015**, *20*, 59–64. [CrossRef]
34. Alandia, G.; Jacobsen, S.; Kyvsgaard, N.C.; Condori, B.; Liu, F. Nitrogen Sustains Seed Yield of Quinoa under Intermediate Drought. *J. Agron. Crop Sci.* **2016**, *202*, 281–291. [CrossRef]
35. Mahbod, M.; Zand-Parsa, S.; Sepaskhah, A.R. Modification of Maize Simulation Model for Predicting Growth and Yield of Winter Wheat under Different Applied Water and Nitrogen. *Agric. Water Manag.* **2015**, *150*, 18–34. [CrossRef]
36. Oelke, E.A.; Oplinger, E.S.; Teynor, T.M.; Putnam, D.H.; Doll, J.D.; Kelling, K.A.; Durgan, B.R.; Noetzel, D.M. Safflower. In *Alternative Field Crops Manual*; Purdue University: St. Paul, MN, USA, 1992; pp. 1–8.
37. Kaul, H.-P.; Kruse, M.; Aufhammer, W. Yield and Nitrogen Utilization Efficiency of the Pseudocereals Amaranth, Quinoa, and Buckwheat under Differing Nitrogen Fertilization. *Eur. J. Agron.* **2005**, *22*, 95–100.
38. Razzaghi, F.; Plauborg, F.; Jacobsen, S.E.; Jensen, C.R.; Andersen, M.N. Effect of Nitrogen and Water Availability of Three Soil Types on Yield, Radiation Use Efficiency and Evapotranspiration in Field-Grown Quinoa. *Agric. Water Manag.* **2012**, *109*, 20–29. [CrossRef]
39. Alvar-Beltrán, J.; Dao, A.; Dalla Marta, A.; Saturnin, C.; Casini, P.; Sanou, J.; Orlandini, S. Effect of Drought, Nitrogen Fertilization, Temperature and Photoperiodicity on Quinoa Plant Growth and Development in the Sahel. *Agronomy* **2019**, *9*, 607. [CrossRef]
40. Wang, N.; Wang, F.-X.; Shock, C.C.; Meng, C.-B.; Huang, Z.-J.; Gao, L.; Zhao, J.-Y. Evaluating Quinoa Stem Lodging Susceptibility by a Mathematical Model and the Finite Element Method under Different Agronomic Practices. *Field Crop. Res.* **2021**, *271*, 108241. [CrossRef]
41. Marschner, H. *Marschner's Mineral Nutrition of Higher Plants*; Academic Press: Cambridge, MA, USA, 2011; ISBN 0123849063.
42. Below, F.E. Nitrogen Metabolism and Crop Productivity. *Handb. plant Crop Physiol.* **2002**, 2, 385–406.
43. Mehrabi, F.; Sepaskhah, A.R. Partial Root Zone Drying Irrigation, Planting Methods and Nitrogen Fertilization Influence on Physiologic and Agronomic Parameters of Winter Wheat. *Agric. Water Manag.* **2019**, *223*, 105688. [CrossRef]
44. Breidy, J. *Final Report on Quinoa Evaluation Trials in Lebanon*; Rome FAO: Rome, Italy, 2015.
45. Djamal, S. *Technical Assistance for the Introduction of Quinoa and Appropriation/Institutionalization of Its Production in Algeria-Second Evaluation Report*; Rome FAO: Rome, Italy, 2015.
46. Hassan, L. *Iraq Final Evaluation Report on Quinoa*; Rome FAO: Rome, Italy, 2015.
47. Saeed, A.L. *Yemen Progress Report on Quinoa*; Rome FAO: Rome, Italy, 2015.
48. Fonseca, A.E.; Westgate, M.E. Relationship between Desiccation and Viability of Maize Pollen. *Field Crop. Res.* **2005**, *94*, 114–125. [CrossRef]
49. Curti, R.N.; De la Vega, A.J.; Andrade, A.J.; Bramardi, S.J.; Bertero, H.D. Adaptive Responses of Quinoa to Diverse Agro-Ecological Environments along an Altitudinal Gradient in North West Argentina. *Field Crop. Res.* **2016**, *189*, 10–18. [CrossRef]
50. Shams, A.S. Response of Quinoa to Nitrogen Fertilizer Rates under Sandy Soil Conditions. In Proceedings of the 13th International Conference Agriculture, Faculty of Agriculture, Benha University, Benha, Egypt, 10 September 2012; pp. 9–10.

51. Hirich, A.; Choukr-Allah, R.; Jacobsen, S. Deficit Irrigation and Organic Compost Improve Growth and Yield of Quinoa and Pea. *J. Agron. Crop Sci.* **2014**, *200*, 390–398. [CrossRef]
52. Yazar, A.; Incekaya, Ç.; Sezen, S.M.; Jacobsen, S.-E. Saline Water Irrigation of Quinoa (*Chenopodium quinoa*) under Mediterranean Conditions. *Crop Pasture Sci.* **2015**, *66*, 993–1002. [CrossRef]
53. Azizian, A.; Sepaskhah, A.R. Maize Response to Different Water, Salinity and Nitrogen Levels: Agronomic Behavior. *Int. J. Plant Prod.* **2014**, *8*, 107–130.
54. Riccardi, M.; Pulvento, C.; Lavini, A.; d'Andria, R.; Jacobsen, S.E. Growth and Ionic Content of Quinoa under Saline Irrigation. *J. Agron. Crop Sci.* **2014**, *200*, 246–260. [CrossRef]
55. Mehrabi, F.; Sepaskhah, A.R. Interaction Effects of Planting Method, Irrigation Regimes, and Nitrogen Application Rates on Yield, Water and Nitrogen Use Efficiencies of Winter Wheat (Triticum Aestivum). *Int. J. Plant Prod.* **2018**, *12*, 265–283. [CrossRef]
56. Reid, R.; Hayes, J. Mechanisms and Control of Nutrient Uptake in Plants. *Int. Rev. Cytol.* **2003**, *229*, 73–115. [PubMed]
57. Glass, A.D.M. Nitrate Uptake by Plant Roots. *Botany* **2009**, *87*, 659–667. [CrossRef]
58. Antonopoulos, V.Z. Modelling of Water and Nitrogen Balances in the Ponded Water and Soil Profile of Rice Fields in Northern Greece. *Agric. Water Manag.* **2010**, *98*, 321–330. [CrossRef]
59. Hanson, B.R.; Šimůnek, J.; Hopmans, J.W. Evaluation of Urea–Ammonium–Nitrate Fertigation with Drip Irrigation Using Numerical Modeling. *Agric. Water Manag.* **2006**, *86*, 102–113. [CrossRef]
60. Yarami, N.; Sepaskhah, A.R. Saffron Response to Irrigation Water Salinity, Cow Manure and Planting Method. *Agric. Water Manag.* **2015**, *150*, 57–66. [CrossRef]
61. Razzaghi, F.; Sepaskhah, A.R. Calibration and Validation of Four Common ET0 Estimation Equations by Lysimeter Data in a Semi-Arid Environment. *Arch. Agron. Soil Sci.* **2012**, *58*, 303–319. [CrossRef]
62. Jacobson, D.A.; Streich, J.C.; Stanschewski, C.S.; Rey, E.; Fiene, G.; Craine, E.; Wellman, G.; Melino, V.; Patiranage, D.S.R.; Johansen, K. Quinoa Phenotyping Methodologies: An International Consensus. *Plants* **2021**, *10*, 1759.
63. Singh, R.J.; Ahlawat, I.P.S. Dry Matter, Nitrogen, Phosphorous, and Potassium Partitioning, Accumulation, and Use Efficiency in Transgenic Cotton-Based Cropping Systems. *Commun. Soil Sci. Plant Anal.* **2012**, *43*, 2633–2650. [CrossRef]

plants

MDPI

Review

Trends and Limits for Quinoa Production and Promotion in Pakistan

Irfan Afzal [1,*], Shahzad Maqsood Ahmed Basra [1], Hafeez Ur Rehman [1], Shahid Iqbal [2] and Didier Bazile [3,4,*]

1 Department of Agronomy, University of Agriculture, Faisalabad 38040, Pakistan; shehzadcp@gmail.com (S.M.A.B.); hafeezcp@gmail.com (H.U.R.)
2 Department of Agronomy, MNS-University of Agriculture, Multan 66000, Pakistan; shahid.iqbal@mnsuam.edu.pk
3 CIRAD, UMR SENS, 34398 Montpellier, France
4 SENS, CIRAD, IRD, University Paul Valery Montpellier 3, 34090 Montpellier, France
* Correspondence: iafzal@uaf.edu.pk (I.A.); didier.bazile@cirad.fr (D.B.)

Abstract: Quinoa is known as a super food due to its extraordinary nutritional qualities and has the potential to ensure future global food and nutritional security. As a model plant with halophytic behavior, quinoa has potential to meet the challenges of climate change and salinization due to its capabilities for survival in harsh climatic conditions. The quinoa crop has received worldwide attention due to its adoption and production expanded in countries out of the native Andean region. Quinoa was introduced to Pakistan in 2009 and it is still a new crop in Pakistan. The first quinoa variety was registered in 2019, then afterward, its cultivation started on a larger scale. Weed pressure, terminal heat stress, stem lodging, bold grain size, and an unstructured market are the major challenges in the production and promotion of the crop. The potential of superior features of quinoa has not been fully explored and utilized. Hence, there is a need to acquire more diverse quinoa germplasm and to establish a strong breeding program to develop new lines with higher productivity and improved crop features for the Pakistan market. Mechanized production, processing practices, and a structured market are needed for further scaling of quinoa production in Pakistan. To achieve these objectives, there is a dire need to create an enabling environment for quinoa production and promotion through the involvement of policymakers, research institutions, farmers associations, and the private sector.

Keywords: Andean regions; abiotic stresses; nutrition profile; value chain; developing countries; germplasm diversity

Citation: Afzal, I.; Basra, S.M.A.; Rehman, H.U.; Iqbal, S.; Bazile, D. Trends and Limits for Quinoa Production and Promotion in Pakistan. *Plants* **2022**, *11*, 1603. https://doi.org/10.3390/plants11121603

Academic Editor: Andreas W. Ebert

Received: 18 April 2022
Accepted: 15 June 2022
Published: 18 June 2022

1. Introduction

Climate change, water shortage, and increasing salinization including malnourishment and chronic dietary problems are the major challenges for sustainable agriculture as well as for food and nutritional security of the burgeoning population. It is the right time to diversify cropping systems by introducing new crops to achieve sustainable development goals [1]. Quinoa is an ideal candidate crop which may contribute to environmental and food sustainability owing to its high adaptability to a wide range of growing conditions [2]. Quinoa is gaining popularity due to its functional and nutritional characteristics [3]. It can achieve higher productivity and maintain nutritional quality in different environments where conventional crops cannot perform well. Moreover, quinoa has potential for climate resistance to different stresses such as salinity, drought, and frostlike conditions [4–6]. It is an annual, mainly self-pollinated, dicotyledonous, and C_3 crop for CO_2 fixation during photosynthesis [4].

Quinoa has a high nutritional profile with 10–18% seed proteins [7,8] and 4.1–8.8% fats [9]. It is ideal for celiac patients because it is gluten free. The whole plant can be used as feed for both humans and animals. Its leaves are also used as a salad because they have the same nutritional value as spinach and mustard [10]. Quinoa grain is rich in all amino-acids, vitamins (A, E, B2), carbohydrates, minerals (K, Fe, Ca, Mn), and healthy supportive fatty acids (Omega-3) [9]. Its grains are ground into flour as wheat and used for further purposes such as bread formation, beer formation, and fermented drinks [11].

Quinoa has been cultivated in more than 120 countries worldwide with major producers including Peru, Bolivia, Ecuador, USA, Columbia, Chile, and Brazil [12].

The quinoa plant life cycle is divided into vegetative and reproductive stages. Each phase is dependent on day length and temperature [13] due to which it has wide adaptability [9]. Quinoa plants take about 40–89 days for bud appearance, 7–50 days for the anthesis stage, and 66–135 days for maturity after anthesis [14]. However, the crop reaches maturity within 109–182 days in Europe [15,16].

In Pakistan, quinoa was introduced for the first time in central Punjab by the University of Agriculture, Faisalabad (UAF), to increase diversity in the cropping system and environmental sustainability [17,18]. Now it is well adapted and grown in all provinces of Pakistan.

Over 7 million hectares in Pakistan are affected by soil salinity. Research indicates that quinoa can be grown on salt affected soils with electrical conductivity (ECe) 10 to 15 dS cm^{-1} in South Punjab [19,20]. It can even tolerate salinity and arsenic stress due to less uptake of toxic ions and higher activities of antioxidant enzymes [21]. Therefore, quinoa crop has potential for salt affected soils. Despite huge potential and wide adaptability, lack of awareness about nutritional and health benefits and unstructured markets are major challenges in upscaling quinoa crop in Pakistan. This review highlights the current trends in quinoa research, its cultivation and future challenges in quinoa production, and value chain development in Pakistan.

2. Germplasm Collection and Evaluation

Only a few quinoa varieties have been commercialized out of more than 3000 landraces identified in the Andean countries [22]. Cultivated quinoa has plentiful seed colors (>10), but the marketable grain is usually white, red, and black. During 2009, quinoa was introduced successfully in Pakistan based on a collection of 170 accessions from the USDA, USA, and Denmark [17]. Out of 170 quinoa lines tested, only four accessions were found to be widely adapted to the local climatic conditions of Pakistan and valuable for domestic production. Basic farming practices have been developed by optimizing sowing time; sowing method; and nitrogen, phosphorus, and potassium (75:60:50 kg ha^{-1}) requirements under Faisalabad conditions. The preliminary trials have shown that quinoa is also well acclimatized to the different agro-ecological conditions of Punjab. Yields obtained (3.2 tonnes/ha) and nutritional profiles investigated in these environments are equivalent to native regions of quinoa production [16]. Likely, adaptability trials across different parts of the country including KPK and Sindh are in progress.

The University of Agriculture, Faisalabad, has conducted trials on genetic variability for a wide range of quinoa types under agro-climatic conditions of Faisalabad, Punjab-Pakistan, in collaboration with King Abdullah University of Science and Technology, Saudi Arabia. About 370 accessions have been phenotyped for morphological, phenological, and yield traits under field and for postharvest management of quinoa seed during the years 2019–2021 [unpublished data].

The UAF-Q7 is the first approved variety of quinoa in Pakistan [23]. The basic production practices for this variety have been optimized [17]. The UAF-Q7 variety has a hollow stem with a tap root system, and its leaf shape resembles the goose foot type. Its plant height ranges from 110 to 150 cm. The panicle shape is an intermediate type with a green color at flowering that turns brown at maturity. It matures in 130–140 days and has an average yield potential of 3.2 t ha^{-1}.

3. Developments in Quinoa Research

Due to high grain yield, biomass, and nutritional quality, quinoa is regarded as a dual-purpose crop both for grain production and livestock feed [24]. After seed harvesting, there is potential for quinoa growers to market straw a forage crop [24–26]. Generally, genotypes with thick stems, more branches, and moderate plant height producing higher biomass are ideal for fodder and genotypes with compact inflorescence are ideal for grain purpose [27]. Presently, some quinoa lines with high nutritional profile, biomass, and low saponin contents have been evaluated for fodder purposes [unpublished data]. The assessment of the digestibility and palatability potential of these quinoa lines for ruminants is in progress.

3.1. Germplasm Diversity

Quinoa is cultivated from sea level to 4000 m with a broad agroecological adaptation to different types of soils [28]. It is an Andean crop that originated around Lake Titicaca in Peru and Bolivia, the area with greatest diversity and genetic variation. Currently, quinoa is grown in countries spanning five continents, including North America, Europe, Asia, Africa, and Oceania. The center of quinoa diversity is the southern Andean highlands viz. Bolivia and Peru have huge variability and Bolivia's gene bank center has more than 5000 accessions [29]. Quinoa varieties are genetically grouped into two main groups: lowland and highland. Fuentes et al. [30] mapped quinoa's genetic structure by matching it with natural geographical edaphic climatic constraints and the social linguistic context of ancient people inhabiting the Andes region [30].

Huge diversity also exists in the quinoa germplasm based on morphological and physiological adaptability to various climatic conditions [31]. Yield mainly depends upon the phenological and seed related attributes of a crop and duration between each stage. Under local conditions of Faisalabad, Pakistan, exotic accessions with medium crop duration and more plant biomass produced higher grain yield as compared to long duration genotypes. The number of lateral branches in quinoa plants vary according to the genotypes and the crop condition. Accessions with more branches and inflorescence express more plant biomass and yields as compared to accessions with a single panicle per plant [16].

According to Sosa-Zuniga et al. [32], 15 panicle colors and 3 types of panicle shapes (Glomerulate, intermediate, and amaranthiform) are reported in quinoa at physiological maturity. The large grain size in quinoa is preferred [3]. Apart from phenological and grain characteristics, quinoa genotypes also diversified in terms of nutritional quality as protein contents ranged from 11 to 16% in selected genotypes adapted in Pakistan [33]. Fewer studies have reported on the role of phytates in quinoa as it is known as an anti-nutritional factor.

3.2. Production Practices

Quinoa can grow in a range of soils from clayey to sandy including marginal soils with a pH of 4.5–9.5 [13]. For quinoa cultivation in a new environment, sandy loam soils with good drainage, appreciable organic matter, and nutrients should be preferred. In Pakistan, quinoa crop has been preliminarily tested on sandy loam and clay loam soils with a pH range of 7.4–8.8, medium in fertility and low in organic matter contents (0.77%) under semi-arid regions of Punjab (elevation 184 m above sea level 31.4187° N, 73.0791° E; elevation 190 m above sea level 31.8950° N, 73.2706° E) [17,34,35] and Sindh (20 m above sea level) [36]. Quinoa is grown during rabi season as a spring crop in most parts of the country except for northern areas. The window of plasticity for planting ranges from 15 October to 15 December and favorable time for its growth and yield potential is during November under irrigated conditions [17,34,36]. A delay in planting the crop usually prolongs growth, reduces grain filling, and delays crop maturity with a substantial reduction in seed yield [37] and response may be genotype specific [38–41].

The crop sowing requires fine textured, well drained, and levelled seedbed with optimal moisture for its germination; but it is important to know that quinoa is sensitive to high moisture due to its small seed [4]. The planting method and geometry are critical in crop establishment of quinoa because of slow growth rate until the bud formation stage, otherwise weed-crop competition becomes greater to affect yield. Timely sowing of quinoa can provide a head start over weeds as crop may obtain good growth during this period.

Experimentally, quinoa has been cultivated on ridges manually or by hand drill on normal and salt affected soils to sowing depth of 2–3 cm at field capacity level [17,34]. Ridge cultivation is usually practiced by many growers with plant distance of 15 cm on 75 cm spaced ridges [17]. Weeds are cumbersome to control; hence optimum plant density is important to reduce weed competition. Experimentation is in progress by planting quinoa at 30 cm inter-row distance and a plant distance of 11 cm using the drill method. Seed rate in quinoa depends on the method of sowing viz. 5–7 kg ha^{-1} for the drill method and 4–5 kg ha^{-1} for ridge cultivation. Nonetheless, high biomass, growth, and yield have also been reported in quinoa sown on beds with 75 cm width and 15 cm plant distance of a furrow on both sides for water flow under irrigated conditions [35]. This method has an advantage of planting quinoa on both sides of the beds compared to ridge planting with a single row [35]. Though, further studies on resource use efficiency in terms of water, fertilizer, and radiation including stem breaking under high wind and thunderstorms are required.

Quinoa is produced in marginal lands of its native regions. Although, the crop is very fertilized and irrigation input is responsive under irrigated conditions. In Pakistan, a recommendation for nitrogen (N), phosphorus (P), and potassium (K) (N:P:K) using 75:60:50 kg ha^{-1} for quinoa cultivation is being followed. A full dose of the phosphorus and potassium and 1/2 dose of nitrogen are applied as basal and the remaining at the flowering stage [17,34]. Usually, high N application has been reported to delay maturity, increase plant height, and the crop may be susceptible to lodging [17]. Alandia et al. [42] discovered that increase in N rate 80–160 kg ha^{-1} resulted in a 10–15% rise in seed yield, while enhancing N rate up to 240 kg ha^{-1} resulted in negligible seed output. Furthermore, extensive research concerning nutrition in relation to soil type should be conducted before recommending farming practices for any specific location. As Pakistan soils are of alkaline nature and low in organic matter and micronutrients (Z, Fe, Mn, B), these essential micronutrients should be included in the basic fertility plan to harvest high quality quinoa grains.

Quinoa is a drought-tolerant crop and has a low water requirement, though yield is significantly affected by irrigation [2]. Between three and four irrigations are required by a quinoa crop during its growing cycle; however, crop stages critical for its irrigation during the vegetative and grain formation period remain to define for its successful adaptation to semi-arid condition of country. Heavy watering throughout the panicle development phases has been reported to extend crop maturity and increase plant height, suggesting that the crop might be prone to lodging [Personal observation].

Various narrow and broad leaf weeds occur in quinoa fields and are mainly influenced by the type of sowing method, planting geometry, and plant density. Quinoa plants resemble its wild relatives *C. album* and *C. murale*, during the early growth period. Therefore, quinoa seedlings must be differentiated for proper identification of weeds and their control. As there is no chemical weed control yet established due to sensitivity of Chenopodium to herbicides, weeds are controlled manually. In research trials, weeds are controlled usually at 2–4 true leaf and bud formation stages to achieve optimum plant density [17,35]. Studies are much needed to establish the critical crop weed competition period in quinoa and combined application of different pre- and post-emergence herbicide formulations without detrimental effects on soil and plant foliage including their residual effects on the environment. Nonetheless, an integrated approach which involves mechanical, cultural, chemical, and biological control is called sustainable weed management in quinoa organic production.

3.3. Abiotic Stresses

Abiotic stresses are becoming the most devastating threat that limits agricultural productivity for most of the crops [43]. One of the possible solutions to cope with abiotic stresses is the cultivation of stress resistant crops to abridge the food requirement [44]. Quinoa is cultivated due to its abiotic stress tolerance behavior [45]. Due to this potential and unique nutritional profile, FAO termed quinoa as "Future Smart Food" and advocated for its promotion, especially in salt affected and drought prone areas [22,44].

Quinoa genotypes well adapted to local conditions had been evaluated for salt tolerance, heat, and phytoremediation potential [34,46].

Quinoa has been identified as a facultative halophyte with better salt tolerance [47] and a high variability in salinity tolerance among quinoa genotypes has been reported [48–50]. Saleem et al. [51] investigated the salt tolerance behavior of various quinoa lines grown hydroponically at 100 mM NaCl salinity level and found that Q7 and Q9 lines had better chlorophyll content index, free proline, ascorbic acid, and carotenoids contents but gaseous exchange traits decreased in Q7 plants under saline environment. In another study, Iqbal et al. [34] found an improvement in water relations, leaf photosynthetic rate, K^+ contents in leaf, proline, phenolics, morphological and yield related attributes and ultimately increased grain yield at 10 dS m^{-1}. Quinoa performance decreased drastically at 30 dS m^{-1}. Iqbal et al. [19] also found that under natural salt affected conditions (9.8 and 13.9 dS m^{-1}), leaf antioxidants, K^+, total phenolics, and proline contents increased compared to control conditions while 1000-seed weight, grain protein, Cu^{+2}, Ca^{+2}, and Zn^{+2} contents were not affected [19]. However, seed and biological yields diminished under high salinity (>13.9 dS m^{-1}) might be ascribed to poor seedling emergence caused by dispersion effects in sodic soil [52]. Yet, seed yield reported by Iqbal et al. [19] was higher ($\approx$1 t ha^{-1}) than world average yield under salt affected conditions [3]. Abbas et al. [20] reported that quinoa significantly improved plant biomass, grain number and weight, antioxidants, total chlorophyll, and relative water contents at 10.5 dS m^{-1}.

High temperature is one of the limitations to widespread cultivation of quinoa. Under the climate change scenario, the high temperature causes drastic effects on plant functions [53]. Rashid et al. [54] reported that quinoa plants under terminal heat stress induced 76 days after sowing produced less chlorophyll contents and decreased gaseous exchange parameters, seed yield and its nutrients. Contrastingly, plant height, antioxidants, seed Mg^{+2}, K^+, and Na^+ contents were increased in heated plants as compared to controlled conditions. In another study, Rashid et al. [55] observed lower gaseous exchange, panicle length, 1000 seed weight, seed yield, seed Ca^{+2}, K^+, and chlorophyll content during anthesis when exposed to terminal heat stress. Control quinoa seeds, on the other hand, showed more antioxidant enzymes activity [54,55]. Quinoa performance was negatively affected when it was planted late in Pakistan conditions. At temperatures above 35 °C, quinoa performance suffers due to phenological changes which promote more vegetative growth than reproductive growth [3,17]. Quinoa is a cool season crop and sensitive to high temperature stress for grain production.

Heavy metal toxicity hinders the physiological, biochemical, and morphological responses which ultimately limits the yield of crops [56]. On the other hand, tolerance and plasticity in quinoa against heavy metals have been reported [57–59]. Parvez et al. [21] reported that at 150 μM arsenic (As) stress, seedling biomass, and chlorophyll contents were decreased while antioxidant enzymes increased. Under lead (Pb) 100 mg kg^{-1} and 60 mg kg^{-1} cadmium (Cd) stress, quinoa seedling biomass, and membrane stability index decreased, while tissue Pb, Cd, and antioxidant enzymes were increased [60].

Recently, Naeem et al. [61] found a decrease in seedling vigor and membrane stability index and a concomitant increase in root/shoot growth, SOD, POD and CAT activity including grain Cd contents at 75 mg kg^{-1} cadmium (Cd) stress. Haseeb et al. [46] found a decrease in morphological, yield related attributes and final grain yield and an increase in soluble phenolics, root, stem, leaf, and seed Pb contents at 100 mg kg^{-1} Pb stress. More importantly, Pb contents in quinoa grain were within the permissible limits (0.3 mg kg^{-1} DW) as per FAO/WHO guidelines [61]. This depicts the phyto-extraction capacity of quinoa against industrial effluents, mainly heavy metals.

3.4. Phenotyping Approaches

Phenotyping is a foundation of plant breeding and grain yield is the most reliable phenotypic trait in the breeding programs [62]. Secondary traits also contribute to crop improvement depending upon genotype by environment interaction under various environmental conditions [63]. Three main classes of phenotyping are identified in the literature: handy, high-throughput, and precision phenotyping traits to tackle current bottlenecks to yield improvement [64]. Many useful phenotypes were established with the publication of *Descriptors for Quinoa and Wild Relatives* [64] and the guidelines for distinctness, uniformity, and stability testing of quinoa under CPVO system [65]. However, a detailed explanation of the important traits was lacking and needs further investigation. Regarding precision and high-throughput phenotyping through remote sensing, no work has been reported on quinoa in Pakistan. Studies for a consensus on phenotyping methods for 400 quinoa accessions in the field with the international collaboration are in progress, during which the phenotyping protocols at different phenological stages, maturity time, harvest and postharvest phases throughout the growing season have been established.

Quinoa genotypes show different behavior in phenological stages and duration to complete their lifecycle according to the latitude, altitude, and environmental conditions especially photoperiod and temperature of a region [65]. Sosa-Zuniga et al. [32] presented the most recent and comprehensive description of phenological stages of quinoa in accordance with the BBCH criteria. For reliable and stable phenotyping, defined phenological phases are critical. Researchers defined eight major phases of quinoa crop development. However, stage five, inflorescence, is the crucial bordering phase between vegetative and reproductive growth stages. Additionally, stage six, flowering, is highly associated with yield related traits. Moreover, sowing and harvest dates are also important to record according to the local conditions.

The duration of each quinoa stage is highly dependent upon temperature and photoperiod which is different for each quinoa variety [13]. In Pakistani conditions, exotic quinoa accessions along with UAF-Q7 reached the inflorescence emergence stage within 45–71 days after sowing and completed anthesis at 70–108 days after sowing. The accessions having emergence with UAF-Q7 completed the physiological maturity stage within 101–144 days after sowing (unpublished data; Table 1). In South America, days to flowering varies from 71 to 101, days to maturity varies from 117 to 157 days after emergence and seed yield (t/ha) varies from 0.32 to 9.33 [66]. In European region conditions, the total growth duration of quinoa crop varies from 109 to 182 days. In England, the appearance of true leaves to the visible floral bud initiation stage varies from 41 to 89 days, the visible floral bud stage to anthesis stage ranges from 7 to 53 days and maturity is reached from 65 to 135 days after anthesis [67].

Table 1. Description of phenological stages of quinoa accession under agro-climatic conditions of Faisalabad-Pakistan during 2020–2021.

Sr #	Description of Stage	Days after Sowing	Image
1	Emergence of Cotyledons	4–5	
2	Emergence of true leaves	17–19	
3	Visible bud appearance	65–68	
4	Anthesis	83–85	
5	Physiological maturity	125–126	
6	Harvest maturity	153–168	

3.5. Grain Nutritional Profile

Pakistan has the world's sixth highest population by human index, which has a drastic impact on the world food program. A decrease in food security and safety has led to a child stunting rate of 45% in Pakistan, ranking 8th among 132 nations [68]. Such conditions increase the healthcare costs of this lower-middle-income country. Quinoa's diverse nutritional profile can offset prevalent nutrient deficiencies related to the lack of nutrient-dense or biofortified crops. Nasir et al. [33] investigated the nutritional profiles of grains obtained from Pakistan's well adapted quinoa genotypes (Q1, Q2, Q7, and Q9). Genotypes of quinoa were evaluated with special emphasis on functional properties and digestibility of its proteins. Proteins of all genotypes had good functional properties, i.e., water absorption capacity (2.81–3.82%), oil absorption capacity (2.72–3.03%), and foaming capacity (9.09–10.05%). Proteins also exhibited outstanding in vitro digestibility (75.95–78.11%), protein efficiency ratio (3.5–3.78%), net protein ratio (3.9–4.69%), net protein utilization (70.75–73.78%), biological value (79.15–81.74%), and true digestibility (87.66–90.57%). Fats were also studied, and various fatty acids were found including oleic acid (26.28–31.62%), palmitic acid (11.39–13.25%), α-Linoleic acid (4.45–7.71%), and Linoleic acid (47.73–52.02%).

Iqbal et al. [19] estimated the nutritional profile of quinoa grains obtained from crops grown on fertile and salt affected soils. Highly significant results showed the resilient nutritional profile of quinoa grains via depicting no change in the quality of grain protein contents. Astonishingly, seeds harvested from salt-affected soils were rich in potassium, magnesium, and manganese. Mineral profiles of quinoa grains adapted to Pakistani soils are given in Tables 2 and 3.

Table 2. Comparative proximate analysis of UAF-Q7 quinoa cultivar based on published reports [7,33,69–71].

%	[33]	[7]	[69]	[70]	[71]
Ash	2.44	3.80	3.20	3.00	3.70
Protein	13.47	16.50	16.70	15.60	12.50
Fat	5.59	6.30	5.50	7.40	8.50
Fiber	2.71	3.80	10.50	2.90	1.90

Note: [33] values are average of four genotypes.

Table 3. Comparative mineral analysis of quinoa grains based on published reports [7,8,33,72].

Minerals (mg/kg)	[33]	[8]	[7]	[72]
Ca	691.00	940.00	1487.00	1020.00
Copper	4.49	37.00	51.00	ND
Iron	64.47	168.00	132.00	105.00
Potassium	8877.98	ND	9267.00	8225.00
Magnesium	2115.70	2700.00	2496.00	ND
Manganese	32.72	ND	ND	ND
Sodium	48.14	ND	ND	ND
Phosphorous	4523.55	1400.00	3837.00	1400.00
Sulphur	1549.06	ND	ND	ND
Zinc	28.67	48.00	44.00	ND

Ref. [33] values are average of four genotypes; ND = Not detected.

Vega-Gálvez et al. [73] studied detailed characterization of the nutritional composition of six quinoa varieties grown in Southern Europe. High contents of potassium, phosphorus, and magnesium along with low saponin contents were reported in these quinoa varieties. Nonetheless, further studies are required to explore amino acid profile, antioxidants, and identification of bioactive compounds such as kaempferol and quercetin.

3.6. Seed Storage

Quinoa seed quality depends on environmental conditions at the time of harvesting and storage [74]. Proper handling and safe storage ensure seed quality at the time of sowing. Temperature, moisture contents, and oxygen are important factors that influence seed longevity [75] but elevated seed moisture is the most critical factor responsible for loss of seed quality during storage [76,77]. Poor storage enhances the attack of storage insect pests, which promotes deterioration, and eventually death of seeds [78]. Due to inadequate storage, both natural and economic resources are spoiled if poor quality seeds are sown in the field [79]. So, the quality of seed should be maintained during production, harvesting and storage to ensure the availability of highly viable seed at the time of planting.

Quinoa seed is spherical and consists of a peripherally curved embryo surrounded by a large central perisperm, a two-layered pericarp, and a seed coat. A micropylar endosperm in the form of a cone surrounds the radicle tip [80]. Quinoa seeds lose viability more rapidly than cereals because of the porosity in the integument, which allows a seed to easily gain or lose moisture and may initiate germination in the panicle [74]. Initial quality of seed, temperature, and humidity during storage and rate of aging process influence seed longevity [81]. This aging process varies among quinoa accessions [82]. Quinoa seed deteriorates with an inadequate storage environment, particularly at high relative humidity and temperature [74]. Recently, Kibar et al. [83] reported loss in viability of quinoa seed packed in traditional bags during storage at ambient conditions. Conversely, if a seed is dried properly and packed in hermetically sealed storage bags, the quality of quinoa seed could be maintained as reported in other cereals [76]. If seed loses its viability under ambient storage conditions, then it would be very difficult to obtain optimum plant population in the field as quinoa seed is very sensitive at the seedling development stage. Furthermore, environmental factors such as high temperature and moisture during production can also influence seed quality of quinoa.

During storage, seed moisture contents, relative humidity, and storage temperature are the main factors that determine the viability of quinoa and rate of deterioration [84,85]. Dry storage of seed for a short-term period preserves its biological value. For long-term and reliable storage, specific cold storage conditions have been used [86]. Despite that, seeds still deteriorate at a reduced rate in the dry state due to very low levels of metabolism [87,88]. Decline in seed quality is initially seen as a decrease in rapidity and synchronicity of germination. An increasing delay to germination is also accompanied by an increased frequency of abnormal seedlings in quinoa seeds and eventually demonstrates a loss of viability. Quinoa seed is orthodox and hygroscopic in nature so it can gain moisture from atmosphere and become susceptible in storage. Seed moisture determines the total life span of vigorous seed so drying is performed after harvesting for reducing moisture contents and to increase storage duration. For quinoa, approximately 10% moisture contents are best for prolonged storage [89]. At 18–20% moisture content and 70–80% RH, the respiration rate increases, and metabolic reactions start. Temperature increases the rate of deterioration in the presence of moisture contents and humidity. High temperature along with high moisture content promotes dormancy as well as ageing in quinoa.

4. Quinoa Consumption and Product Development

The grain composition of quinoa shows health benefits concerning contents of fatty acids, minerals, good quality protein, and bioactive compounds. For these reasons, its consumption is adopted by health-conscious citizens. Quinoa is consumed as a significant ingredient in meatballs and salads and is used to prepare cookies as gluten-free products. Several other products that include quinoa ingredients are multigrain flour such as Maxgrain product to supplement nutrition for people consuming monotonous single grain flour especially for diabetic and celiac patients. A recent development is the launch of CERELAC with oat and quinoa by Nestle-Pakistan for nourishing infants and children. Several other local quinoa-based recipes are being branded to increase quinoa consumption as "Kheer", milkshakes, fruit salads, chapati, kebabs, and vegetable salads. Quinoa may be added to bread flour after evaluating the functional properties and digestibility of protein of available quinoa genotypes [33]. In another study, Mahmood et al. [90] evaluated the rheological properties of quinoa, buckwheat, and wheat doughs and sensory properties of cookies made from their flours. They found good nutritional benefits and high sensory acceptability from composite flour having 10% quinoa and 10% buckwheat. It is proven that quinoa genotypes grown in Pakistan have a strong nutritional profile, especially better protein quality [33,90]. Thus, it can be utilized in cereal-based products for achieving higher quality and value addition. Almost 21 food companies have introduced quinoa products in the country and most of these companies are also involved in the export of quinoa in UAE and European countries (Table 4).

Table 4. Companies marketing quinoa products in Pakistan. (Website accessed date is 8 June 2022).

Sr #	Company Name	Product	Price (USD/kg)	Website
1	Khalis Things	Whole grain washed quinoa	8.33	https://khalisthings.com
2	The Soul Food Company	Prewashed quinoa	9.61	https://getsoulfood.com
3	Amna's	Organic quinoa	7.77	https://amnasorganics.com
4	Virsa agri farms	Quinoa grain, multigrain flour	5.50	https://virsaproducts.com.pk
5	Shazday	White quinoa flour	8.88	https://freshbasket.com.pk
6	Farm Fresh	White quinoa, multigrain flour	6.00	https://farmfresh.com.pk
7	Syed Flour Mills	Quinoa ka Dalya	10.00	www.tradekey.com.pk
8	Gold Tree Millers	White quinoa	5.50	https://goldtreemillers.com
9	Hunter Foods	White and tri-color quinoa	13.30	https://www.hunterfoods.com
10	Family Foods	Organic White Quinoa	4.44	info@familyfoodproducts.com.pk
11	One Organics	Whole grain quinoa	4.22	https://www.daraz.pk
12	Natures Hug	Tri-color quinoa	20.20	https://www.alfatah.pk
13	Morganic	White quinoa grain, multigrain flour	9.40	https://www.morganic.com.
14	Quill (Bin Hashim Pharmacy)	Quinoa grain	9.08	https://binhashimonline.pk
15	Nutricles	Multigrain flour	8.05	https://nutricles.com
16	Healthhut	Quinoa grain	8.80	https://www.healthhut.pk
17	Daali Earth Foods	Quinoa grain	8.50	https://www.daaliearthfoods.com.pk
18	Ashley Foods	Quinoa grain	8.05	https://www.ashleyfoodsinc.com
19	Natural Foods	Quinoa grain	9.30	https://naturals.pk
20	Meadows Organic	Organic white quinoa	9.00	http://meadows-glutenfree.com
21	Sarang Herbs and Food	Quinoa grain and flour	8.33	https://sarang.com.pk

5. Challenges in Quinoa Production and Promotion

Quinoa expansion and production started across the globe after its recognition by the United Nations in 2013 [22,91]. During the year 2007–2008, it was agreed in West France to grow quinoa "d'Anjou" in the Loire area. Trials in Italy indicated that quinoa can be grown in southern regions, and it thrives even in harsh natural conditions. Positive studies have also been performed in Morocco, Greece, and the Indian Subcontinent (India and Pakistan). India is particularly interested in establishing its own quinoa markets [3].

A common issue in these countries is a system with a small or non-existent market, farmers who are risk averse and severe lack of information and technical diffusion. Finally, there is a growing trend to test this crop under local conditions to expand national markets. As a result of the rise in demand, price hikes were observed which have tripled between 2006 and 2013 [3]. Carimentrand et al. [92] concentrated on the various approaches of selling mixed quinoa grain in local markets by responding to international and domestic demand for standardized quinoa products. Agro-industrial enterprises and exporters have encouraged farmers in Peru and Bolivia to plant improved quinoa varieties to meet market demand for uniform and large grains. Community resilience and socio-economic challenges of the quinoa market must be taken into account concerning environmental challenges in quinoa value chain [93]. It is emphasized that rising worldwide market prices have resulted in a drop in consumption, specifically in quinoa growing regions [94].

Despite the growing worldwide recognition of the health benefits of quinoa, the barriers to its widespread adoption remain significant. Institutions and farmers are facing a lack of knack in terms of planting, harvesting, distribution, and overall management. Furthermore, rural residents are unaware of the crop's nutritional benefits; they are not used to the taste and lack recipes to incorporate this product into traditional dishes for consumption [95]. Lack of factors such as information, training change in agronomic, and plant protection practices are the constraints in adoption of quinoa cultivation [96].

5.1. Mechanization

For sustainable cultivation of quinoa crop, proper management of chemical fertilizers and farm machinery are the key factors [97]. Additional characteristics include modifying land use and mechanization of agricultural practices [92]. When compared to a manual production system under rain fed, using mechanized production and processing practices combined with irrigation and organic amendment can reduce processing costs from 2.8 to 1.2 USD kg^{-1} [98].

Sowing methods have a great influence on growth, morphology, yield, and biomass accumulation. The raised bed planting technique is superior for obtaining high grain yield under the irrigated conditions of Pakistan [35]. Quinoa seed sowing by hand is being practiced in developing countries such as Pakistan which is labor intensive and high seed rate demanding. Similarly, harvesting is also performed by hands so mechanization at sowing and harvesting times is a big challenge for the quinoa growers in the developing countries.

The industrial processing of quinoa is crucial to ensure the consumer or supplier is provided with clean quinoa, free of impurities and saponins. Since 2009, quinoa was introduced in Pakistan, but its cultivation is limited because of bitterness in approved varieties which is attributed to its high saponin contents. Farmers are practicing a traditional method of washing and drying for its removal which is a labor-intensive process. The timely introduction of mechanized system at harvest and postharvest stages has various advantages over traditional practices. In Morocco, mechanical pearling, on the other hand decreased saponin content by 68%, compared to 57% using both conventional abrasion and cleaning [98].

5.2. Weed Control

Weed control is an important crop husbandry practice since quinoa grows in a season when its wild relatives, such as *Chenopodium album* and *Chenopodium murale*, compete for light, water, nutrients, and space. It is difficult for common farmers to distinguish among all these at early growth stages. The only alternatives for weed management are cultural methods such as uprooting or interculture between rows, which raises production costs. No chemical control for broadleaf weeds is currently available, although chemical control for narrow leaf weeds is available in the form of selective weedicides.

5.3. Photoperiod Sensitivity and Heat Tolerance

The quinoa is cultivated as spring crop during November and harvested in April and May. Early crop growth stages usually enjoy low temperature (12–22 °C) which later increases during the reproductive period. The delayed sown quinoa often experiences high temperature (>30 °C) during flowering and anthesis termed as "Terminal Heat Stress" [99]. High temperature above 35 °C during flowering and seed filling stages causes significant reductions in seed yields of quinoa [53] associated with reduced pollen viability and empty inflorescence. The delayed cultivation of quinoa has been shown to reduce shoot and root growth traits, seed and biological yields including harvest index [18,56]. In addition, late sowing crop takes more days to complete the true leaf, four leaves, multiple leaves, and bud formation stages [18]. Under open door plexiglass fitted canopies, with a light transmission index of about 0.8, quinoa plants exposed to terminal heat (±7 °C) during the anthesis stage reduced the panicle length and weight, 100-seed weight and seed yield per plant including above ground dry matter in quinoa genotype UAF-Q7 [53,54]. These reduced yields were attributed to a decrease in gas exchange attributes, photosynthetic pigments, and decline in enzyme activities of antioxidants' defense system under terminal heat [55]. Delayed sowing of quinoa with terminal heat had also reduced seed nutritional quality [55]. Nonetheless, high temperature stress during flowering has been found to produce longer panicles and more branches with delayed maturity showing quinoa adapt avoidance mechanisms to heat (Personal observation). As quinoa is a photoperiod sensitive crop, its cultivation in new regions is influenced by day length [36]. This crop is also genotype specific, which may affect crop growth duration [13,14]. Therefore, to reduce the negative impact of terminal heat, good yielding cultivars with early to medium duration should be identified in Pakistan's irrigated conditions.

5.4. Control of Plant Height and Lodging Resistance

The crop has been ignored for decades and only rudimentary genetic changes have been made until now. To achieve maximum potential of quinoa as a fully domesticated crop, attempts to develop the plant by breeding have been limited [100]. Cereal crops lodging resistance is mostly determined by plant height [101]. Quinoa plants can grow up to 3 m tall in South America, posing a threat for lodging [102]. Additionally, environmental conditions influence plant height in quinoa and several experiments have found a negative link between plant height and seed yields for certain cultivars [102]. Under Faisalabad, Punjab, Pakistan condition, quinoa gain undesirable height (more than 120 cm) when day length and temperature start increasing after mid-February [17]. This result might be due to the amaranth form nature of adaptable genotypes that leads to lodging and stem breakage if a storm prevails. It may also be due to the hollow nature of the stem in quinoa [3]. Studies are in progress to use gibberellic acid inhibitor to control height and to avoid lodging and stem breakage issue in quinoa cultivars. Besides that, phenotyping studies are in progress to identify short stature genotypes from germplasm collection obtained from various countries.

To avoid lodging without detrimental effects on quinoa yields, efforts should focus on genes that influence plant height. The quinoa genome includes two homologues of wheat Rht-B1/Rht-D1 (AUR62039523 and AUR62014191), which are both homologues of Arabidopsis RGA1 and encode a transcription factor involved in gibberellin signal transduction [103]. In comparison, no direct homologue of the GA20ox2 gene has been discovered [100].

5.5. Grain Number/Size and Yield Stability

Grain size is a desirable trait of crop improvement and consumer preference. Quinoa grains from southern highlands of Bolivia are of larger size than other ecotypes and affected by temperature during grain filling in this region due to photoperiod sensitivity of these ecotypes [104]. However, it should be noted that grain weight is more strongly affected by the grain filling rate [105,106] than by the duration of grain filling [107]. On the other hand, grain growth rate is negatively affected by high temperature and longer photoperiods; therefore, it is possible to select larger grains through breeding without affecting the duration of the grain growth under the irrigated conditions of Pakistan for genotypes of early to medium duration.

Similarly, selection for seed yield and seed sizes can be achieved simultaneously as both traits are independent of genetics and environment (G $\times$ E) interaction among quinoa cultivars [104]. As yield from a farm scale is rarely reported, yield data obtained across different environments of approved cultivar can be used to establish a reference point or baseline to begin improvement for yield. The average yield potential (1500–2000 kg ha^{-1}) of recently introduced quinoa cultivar UAF-Q7 in Pakistan can be utilized as a baseline for further quinoa genotype selection and yield enhancement under irrigated conditions. Even so, gains in quinoa yield should not be achieved at the expense of decreased nutritional and end-use quality.

Several other traits such as leaf area, total chlorophyll, number of branches, dry weight, and inflorescence per plant including harvest index have a positive association with seed yield. Hence, they can also be used as indirect selection traits in the yield improvement of quinoa crop [67,106].

5.6. Molecular Breeding and Genetic Approaches for Traits Improvement

Several breeding methods such as hybridization, interspecific crosses including simple, and reciprocal and passive crossing are carried out in quinoa to recombine desirable traits found in different species to next generation and for significant variation under abiotic conditions [108]. Individual and mass selection are applied for seed multiplication of quinoa cultivars developed from landraces to preserve their identity and composition of established cultivars while mutagenesis has been employed for improvement in plant type for vigor, yield potential, and decrease in saponin contents in quinoa [108,109]. In Pakistan, currently, information on quinoa breeding is scanty.; However, the selection of genotypes based on their adaptability, yield performance and low saponin contents is in progress. Conversely, considering the challenges of early vigor, seed size, yield stability, lodging resistance, heat tolerance, and low saponin grain contents, individual and mass selection and mutagenesis breeding techniques can be of potential application to develop a sustainable breeding program in Pakistan.

Recently, Jarvis et al. [110] published high-quality genome data for quinoa which has opened new avenues for using targeted genome editing for evaluating adoption of this crop into new geographical areas different from its origin such as Pakistan and improving its agronomic performance.

As an allotetraploid species, novel genome-editing technologies, such as CRISPR can be used efficiently to develop new varieties with reduced plant height to improve lodging resistance and knock out genes of saponin contents to produce sweet quinoa. However, this would require regulation through GM legislation before commercialization. Alternatively, technologies such as high-end TILLING as molecular breeding tool can be applied to speed up the varietal development program [100].

Marker assisted selection (MAS) following the identification of quantitative traits loci (QTLs) for increasing seed size and grain number and combining them in cultivar with similar genetic background can be a potential target for improving seed yield in quinoa. For this, two close homologues AtCKX5 and another two of AtCKX3 have been mapped in quinoa genome [100,110]. Likely, a two-gene sequence associated with saponin production has been identified that needs to be repressed in advanced generations to produce saponin free quinoa varieties. This will reduce 30% of the costs associated with quinoa production [98,110].

5.7. Socio-Economic Constraints and Adaptability in Existing Cropping Patterns

Quinoa adapts easily to existing technology practices followed in a region and responds by expressing all traits for better agronomic performance of a highly productive crop. Since the declaration of International Year of Quinoa in 2013 by the United Nations, with increasing demand, crop cultivation has been expanded in more than 120 countries and commercially produced outside Andean regions including US, Canada, India and China [22,111,112]. Even in the Middle East and North Africa Region (MENA) and the European Mediterranean regions, the crop has been successfully cultivated in marginal environments on salt affected soils and when using salt water for irrigation [22,113–116]. Furthermore, its adaptation to these environments is based on experimental results and several genotypes have been in the process of selection and approval [113,117–119].

In Pakistan, since the release of the first quinoa cultivar UAF-Q7 in 2019 for commercial cultivation, the crop has been cultivated on more than 200 ha. Several progressive growers, food companies, and retailers are involved in selling to local and international markets as quinoa grain and value-added products.

If quinoa cultivation sees a further expansion in Pakistan, it will become a major crop. On the other hand, it may also stagnate if market demand decreases, or consumer demand fluctuates. Pakistan needs to adjust its quinoa production to market-driven demand of both local and international markets and has to meet internationally agreed quality standards to be able to compete with other stakeholders in the region including MENA, China, and India. The ball of contention between small to medium quinoa growers is the market access and quality maintenance for sustaining a profitable business. Now it is not merely a nutritional concern, as daily requirement can be fulfilled via quinoa's replacement to wheat and rice. However, studies on quantitative comparisons are missing. Additionally, information on phytoremediation potential is missing in case of this notorious halophyte. There is still a considerable lack of research concerning biomass production of quinoa and its value as forage. Further, growing quinoa in existing cropping systems will compete with major crops for cultivated areas such as wheat or oilseeds or we have to cultivate the crop in small areas in rotation with other crops, such as rice–wheat or cotton–wheat. On other hand, we are also facing challenges of urbanization by bringing more cultivated areas under housing schemes. Pakistan spends millions of USD on oilseed import to meet vegetable oil requirements. It is still a burning question whether producing quinoa will reduce the burden of imports or not. Furthermore, growing quinoa on cultivated land under irrigated conditions as a low input crop with less fertilizers may degrade the soil even more. Nonetheless, increased quinoa production in Pakistan will raise concerns about its long-term sustainability as compared to the Andean area, where average yields of 600 kg ha^{-1} may lag if prices increase or decrease in the long run [22].

6. Conclusions and Future Prospects

Quinoa is famous due to its extraordinary nutritional profile, climate resilience, and extreme adaptability to adverse climates. Thus, it is the most potential crop that can ensure future global food and nutritional security in the developing countries. Despite quinoa expansion in more than 120 countries worldwide, the quinoa cultivation in Pakistan is still in experimentation since its introduction in 2009. The first commercial variety UAF-Q7 was released in 2019, and it is being cultivated throughout the country. There is lack of a breeding program for germplasm improvement regarding superior features of quinoa such as high yield and adaptability in different agro-ecological conditions. Access to more quinoa germplasm for maximizing genetic diversity is needed. There is lack of awareness about the nutritional and health benefits of quinoa among consumers and the unstructured market for farmers are major challenges in the promotion of crop. The relatively low productivity of existing quinoa variety, lack of quality seed, undesirable traits, and high market prices compared to other crops restricts its further scaling in Pakistan.

Quinoa requires continued promotion until it becomes a part of the main food chain of common people. The role of policy makers, research institutions, farmers, and supply chain are important for its production and consumption. It is very important to vitalize the market and promote its consumption which in turn will trigger the increasing demand for quinoa production among smallholder farmers. Unfortunately, there is limited development in quinoa related products. In addition to the local market, the international market should be explored for export of high-quality quinoa grains matching the consumer demand with good branding.

Candidate lines with low saponin content and grains of bold size could be helpful for marketing purposes and reducing production costs [95]. The high yielding quinoa varieties with wide adaptability under various agroecological zones are required. Pakistan needs to develop organic certification bodies for achieving maximum returns from this crop in the global market. Postharvest operations for saponin removal are complicated and need investment in mechanization to reduce the procedure. Mechanization in quinoa cultivation due to troublesome weed pressure need identification of cultivars with herbicide tolerance and of early vigor to reduce crop-weed competition. For this purpose, it is desired to introduce low-cost machinery for production and processing of quinoa among the growers and industrialists. As a spring crop, heat stress during the reproductive period due to increasing temperature is challenging to reduce detrimental effects on yield. Given the high protein content in the vegetative parts of quinoa, varieties with high biomass and productivity can be of particular interest as a nutritious fodder for livestock. Germplasm enhancement efforts through pre-breeding, quantitative and participatory breeding, as well as marker assisted selection for potential traits such as grain yield, high biomass, less saponin, and pollen viability need to be explored in adaptable quinoa germplasm. The successful development of quinoa value chains in Morocco offers a perspective to improve food and nutritional diversity of quinoa in Pakistan in a similar way [98].

Author Contributions: Conceptualization, I.A. and S.M.A.B.; writing—original draft preparation, I.A., H.U.R. and S.I.; writing—review and editing, S.M.A.B., H.U.R. and D.B.; supervision, I.A. and D.B.; funding acquisition, D.B. All authors have read and agreed to the published version of the manuscript.

Funding: This research received no external funding.

Institutional Review Board Statement: Not applicable.

Informed Consent Statement: Not applicable.

Data Availability Statement: Not applicable.

Conflicts of Interest: The authors declare no conflict of interest.

References

1. IPCC. *Climate Change and Land: An IPCC Special Report on Climate Change, Desertification, Land Degradation, Sustainable Land Management, Food Security, and Greenhouse Gas Fluxes in Terrestrial Ecosystems, CH5*; Shukla, P.R., Skea, J., Calvo Buendia, E., Masson-Delmotte, V., Pörtner, H.-O., Roberts, D.C., Zhai, P., Slade, R., Connors, S., van Diemen, R., et al., Eds.; IPCC: Geneva, Switzerland, 2019.
2. Saddiq, M.S.; Wang, X.; Iqbal, S.; Hafeez, M.B.; Khan, S.; Raza, A.; Iqbal, J.; Maqbool, M.M.; Fiaz, S.; Qazi, M.A. Effect of Water Stress on Grain Yield and Physiological Characters of Quinoa Genotypes. *Agronomy* **2021**, *11*, 1934. [CrossRef]
3. Bazile, D.; Bertero, H.D.; Nieto, C. *State of the Art Report on Quinoa around the World in 2013*; FAO: Rome, Italy, 2015.
4. Jacobsen, S.-E.; Mujica, A.; Jensen, C. The resistance of quinoa (*Chenopodium quinoa* Willd.) to adverse abiotic factors. *Food Rev. Int.* **2003**, *19*, 99–109. [CrossRef]
5. Gómez-Pando, L.R.; Álvarez-Castro, R.; Eguiluz-De La Barra, A. Effect of salt stress on Peruvian germplasm of *Chenopodium quinoa* Willd.: A promising crop. *J. Agron. Crop Sci.* **2010**, *196*, 391–396. [CrossRef]
6. González, J.A.; Hinojosa, L.; Mercado, M.I.; Fernández-Turiel, J.-L.; Bazile, D.; Ponessa, G.I.; Eisa, S.; González, D.A.; Rejas, M.; Hussin, S. A long journey of CICA-17 quinoa variety to salinity conditions in Egypt: Mineral concentration in the seeds. *Plants* **2021**, *10*, 407. [CrossRef] [PubMed]
7. Kozioł, M.J. Chemical composition and nutritional evaluation of quinoa (*Chenopodium quinoa* Willd.). *J. Food Compos. Anal.* **1992**, *5*, 35–68. [CrossRef]

8. Repo-Carrasco, R.; Espinoza, C.; Jacobsen, S.-E. Nutritional value and use of the Andean crops quinoa (*Chenopodium quinoa*) and kañiwa (*Chenopodium pallidicaule*). *Food Rev. Int.* **2003**, *19*, 179–189. [CrossRef]
9. Stikic, R.; Glamoclija, D.; Demin, M.; Vucelic-Radovic, B.; Jovanovic, Z.; Milojkovic-Opsenica, D.; Jacobsen, S.-E.; Milovanovic, M. Agronomical and nutritional evaluation of quinoa seeds (*Chenopodium quinoa* Willd.) as an ingredient in bread formulations. *J. Cereal Sci.* **2012**, *55*, 132–138. [CrossRef]
10. Galwey, N. The potential of quinoa as a multi-purpose crop for agricultural diversification: A review. *Ind. Crops Prod.* **1992**, *1*, 101–106. [CrossRef]
11. Bhargava, A.; Shukla, S.; Ohri, D. *Chenopodium quinoa*—An Indian perspective. *Ind. Crops Prod.* **2006**, *23*, 73–87. [CrossRef]
12. Alandia, G.; Rodriguez, J.; Jacobsen, S.-E.; Bazile, D.; Condori, B. Global expansion of quinoa and challenges for the Andean region. *Glob. Food Secur.* **2020**, *26*, 100429. [CrossRef]
13. Jacobsen, S.-E.; Stølen, O. Quinoa-morphology, phenology and prospects for its production as a new crop in Europe. *Eur. J. Agron.* **1993**, *2*, 19–29. [CrossRef]
14. Risi, J.; Galwey, N. The pattern of genetic diversity in the Andean grain crop quinoa (*Chenopodium quinoa* Willd.). I. Associations between characteristics. *Euphytica* **1989**, *41*, 147–162.
15. Alandia, G.; Odone, A.; Rodriguez, J.P.; Bazile, D.; Condori, B. Quinoa—Evolution and future perspectives. In *The Quinoa Genome*; Springer: Berlin/Heidelberg, Germany, 2021; pp. 179–195.
16. Hafeez, M.B.; Iqbal, S.; Li, Y.; Saddiq, M.S.; Basra, S.M.; Zhang, H.; Zahra, N.; Akram, M.Z.; Bertero, D.; Curti, R.N. Assessment of Phenotypic Diversity in the USDA Collection of Quinoa Links Genotypic Adaptation to Germplasm Origin. *Plants* **2022**, *11*, 738. [CrossRef] [PubMed]
17. Basra, S.; Iqbal, S.; Afzal, I. Evaluating the response of nitrogen application on growth development and yield of quinoa genotypes. *Int. J. Agric. Biol.* **2014**, *16*, 886–892.
18. Sajjad, A.; Munir, H.; Ahmed Anjum, S.; Tanveer, M.; Rehman, A. Growth and development of *Chenopodium quinoa* genotypes at different sowing dates. *J. Agric. Res.* **2014**, *52*, 535–546.
19. Iqbal, S.; Basra, S.M.; Afzal, I.; Wahid, A.; Saddiq, M.S.; Hafeez, M.B.; Jacobsen, S.E. Yield potential and salt tolerance of quinoa on salt-degraded soils of Pakistan. *J. Agron. Crop Sci.* **2019**, *205*, 13–21. [CrossRef]
20. Parvez, S.; Abbas, G.; Shahid, M.; Amjad, M.; Hussain, M.; Asad, S.A.; Imran, M.; Naeem, M.A. Effect of salinity on physiological, biochemical and photostabilizing attributes of two genotypes of quinoa (*Chenopodium quinoa* Willd.) exposed to arsenic stress. *Ecotoxicol. Environ. Saf.* **2020**, *187*, 109814. [CrossRef]
21. Abbas, G.; Amjad, M.; Saqib, M.; Murtaza, B.; Asif Naeem, M.; Shabbir, A.; Murtaza, G. Soil sodicity is more detrimental than salinity for quinoa (*Chenopodium quinoa* Willd.): A multivariate comparison of physiological, biochemical and nutritional quality attributes. *J. Agron. Crop Sci.* **2021**, *207*, 59–73. [CrossRef]
22. Bazile, D.; Jacobsen, S.-E.; Verniau, A. The global expansion of quinoa: Trends and limits. *Front. Plant Sci.* **2016**, *7*, 622. [CrossRef]
23. Akram, M.Z.; Basra, S.M.A.; Hafeez, M.B.; Khan, S.; Nazeer, S.; Iqbal, S.; Saddiq, M.S.; Zahra, N. Adaptability and yield potential of new quinoa lines under agro-ecological conditions of Faisalabad-Pakistan. *Asian J. Agric. Biol.* **2021**, *2*, 202005301. [CrossRef]
24. Asher, A.; Galili, S.; Whitney, T.; Rubinovich, L. The potential of quinoa (*Chenopodium quinoa*) cultivation in Israel as a dual-purpose crop for grain production and livestock feed. *Sci. Hortic.* **2020**, *272*, 109534. [CrossRef]
25. Ramos, N.; Cruz, A. Evaluation of seven seasonal crops for forage production during the dry season in Cuba. *Cuba. J. Agric. Sci.* **2002**, *36*, 271–276.
26. Peiretti, P.; Gai, F.; Tassone, S. Fatty acid profile and nutritive value of quinoa (*Chenopodium quinoa* Willd.) seeds and plants at different growth stages. *Anim. Feed. Sci. Technol.* **2013**, *183*, 56–61. [CrossRef]
27. Shah, S.S.; Shi, L.; Li, Z.; Ren, G.; Zhou, B.; Qin, P. Yield, agronomic and forage quality traits of different quinoa (*Chenopodium quinoa* Willd.) genotypes in northeast China. *Agronomy* **2020**, *10*, 1908. [CrossRef]
28. Hong, S.-Y.; Cheon, K.-S.; Yoo, K.-O.; Lee, H.-O.; Cho, K.-S.; Suh, J.-T.; Kim, S.-J.; Nam, J.-H.; Sohn, H.-B.; Kim, Y.-H. Complete chloroplast genome sequences and comparative analysis of *Chenopodium quinoa* and *C. album*. *Front. Plant Sci.* **2017**, *8*, 1696. [CrossRef]
29. Rojas, W.; Pinto, M.; Alanoca, C.; Gomez Pando, L.; Leon-Lobos, P.; Alercia, A.; Diulgheroff, S.; Padulosi, S.; Bazile, D. Quinoa genetic resources and ex situ conservation. In *State of the Art Report on Quinoa around the World in 2013*; FAO: Rome, Italy, 2015.
30. Fuentes, F.; Bazile, D.; Bhargava, A.; Martinez, E.A. Implications of farmers' seed exchanges for on-farm conservation of quinoa, as revealed by its genetic diversity in Chile. *J. Agric. Sci.* **2012**, *150*, 702–716. [CrossRef]
31. Bertero, H.D. Effects of photoperiod, temperature and radiation on the rate of leaf appearance in quinoa (*Chenopodium quinoa* Willd.) under field conditions. *Ann. Bot.* **2001**, *87*, 495–502. [CrossRef]
32. Sosa-Zuniga, V.; Brito, V.; Fuentes, F.; Steinfort, U. Phenological growth stages of quinoa (*Chenopodium quinoa*) based on the BBCH scale. *Ann. Appl. Biol.* **2017**, *171*, 117–124. [CrossRef]
33. Nasir, M.A.; Pasha, I.; Butt, M.S.; Nawaz, H. Biochemical characterization of quinoa with special reference to its protein quality. *Pak. J. Agric. Sci.* **2015**, *52*, 731–737.
34. Iqbal, S.; Maqsood Ahmed Basra, S.; Afzal, I.; Wahid, A. Exploring Potential of Well Adapted Quinoa Lines for Salt Tolerance. *Int. J. Agric. Biol.* **2017**, *19*, 933–940. [CrossRef]

35. Ali, S.; Chattha, M.U.; Hassan, M.U.; Khan, I.; Chattha, M.B.; Iqbal, B.; Rehman, M.; Nawaz, M.; Amin, M.Z. Growth, biomass production, and yield potential of quinoa (*Chenopodium quinoa* Willd.) as affected by planting techniques under irrigated conditions. *Int. J. Plant Prod.* **2020**, *14*, 427–441. [CrossRef]
36. Rizvi, M.A.; Ali, S.A.; Munir, I.; Yasmeen, K.; Abid, R.; Ahmed, S. Cultivation of new emerging agro-nutritional crop of quinoa at Madinat al-Hikmah Karachi, Sindh, Pakistan. *Open Plant Sci. J.* **2017**, *10*, 70–81. [CrossRef]
37. Rashid, N.; Abdul, W.; Basra, S.; Muhammad, A. Foliar spray of moringa leaf extract, sorgaab, hydrogen peroxide and ascorbic acid improve leaf physiological and seed quality traits of quinoa (*Chenopodium quinoa*) under terminal heat stress. *Int. J. Agric. Biol.* **2020**, *23*, 811–819.
38. Casini, P. Seed yield of two new quinoa (*Chenopodium quinoa* Willd.) breeding lines as affected by sowing date in Central Italy. *Acta Agric. Slov.* **2019**, *113*, 51–62. [CrossRef]
39. Nanduri, K.R.; Hirich, A.; Salehi, M.; Saadat, S.; Jacobsen, S.E. Quinoa: A new crop for harsh environments. In *Sabkha Ecosystems*; Springer: Berlin/Heidelberg, Germany, 2019; pp. 301–333.
40. Murphy, K.S.; Matanguihan, J. *Quinoa: Improvement and Sustainable Production*; John Wiley & Sons: Hoboken, NJ, USA, 2015.
41. Rehman, H.U.; Alharby, H.F.; Al-Zahrani, H.S.; Bamagoos, A.A.; Alsulami, N.B.; Alabdallah, N.M.; Iqbal, T.; Wakeel, A. Enriching Urea with Nitrogen Inhibitors Improves Growth, N Uptake and Seed Yield in Quinoa (*Chenopodium quinoa* Willd.) Affecting Photochemical Efficiency and Nitrate Reductase Activity. *Plants* **2022**, *11*, 371. [CrossRef]
42. Alandia, G.; Jacobsen, S.E.; Kyvsgaard, N.; Condori, B.; Liu, F. Nitrogen sustains seed yield of quinoa under intermediate drought. *J. Agron. Crop Sci.* **2016**, *202*, 281–291. [CrossRef]
43. Shahzad, A.; Ullah, S.; Dar, A.A.; Sardar, M.F.; Mehmood, T.; Tufail, M.A.; Shakoor, A.; Haris, M. Nexus on climate change: Agriculture and possible solution to cope future climate change stresses. *Environ. Sci. Pollut. Res.* **2021**, *28*, 14211–14232. [CrossRef]
44. Li, H.; Rasheed, A.; Hickey, L.T.; He, Z. Fast-forwarding genetic gain. *Trends Plant Sci.* **2018**, *23*, 184–186. [CrossRef]
45. Grenfell-Shaw, L.; Tester, M. Abiotic stress tolerance in quinoa. In *The Quinoa Genome*; Springer: Berlin/Heidelberg, Germany, 2021; pp. 139–167.
46. Haseeb, M.; Basra, S.M.; Irfan, A.; Abdul, W. Quinoa response to lead: Growth and lead partitioning. *Int. J. Agric. Biol.* **2018**, *20*, 338–344. [CrossRef]
47. Waqas, M.; Yaning, C.; Iqbal, H.; Shareef, M.; ur Rehman, H.; Bilal, H.M. Synergistic consequences of salinity and potassium deficiency in quinoa: Linking with stomatal patterning, ionic relations and oxidative metabolism. *Plant Physiol. Biochem.* **2021**, *159*, 17–27. [CrossRef]
48. Roman, V.J.; den Toom, L.A.; Gamiz, C.C.; van der Pijl, N.; Visser, R.G.; van Loo, E.N.; van der Linden, C.G. Differential responses to salt stress in ion dynamics, growth and seed yield of European quinoa varieties. *Environ. Exp. Bot.* **2020**, *177*, 104146. [CrossRef]
49. Qureshi, A.S.; Daba, A.W. Evaluating Growth and Yield Parameters of Five Quinoa (*Chenopodium quinoa* W.) Genotypes under Different Salt Stress Conditions. *J. Agric. Sci.* **2020**, *12*, 128. [CrossRef]
50. Nguyen, L.V.; Bertero, D.; Nguyen, L.V. Genetic variation in root development responses to salt stresses of quinoa. *J. Agron. Crop Sci.* **2020**, *206*, 538–547. [CrossRef]
51. Saleem, M.A.; Ahmed Basra, S.M.; Afzal, I.; Iqbal, S.; Saddiq, M.S.; Naz, S. Exploring the potential of quinoa accessions for salt tolerance in soilless culture. *Int. J. Agric. Biol.* **2017**, *19*, 233–240. [CrossRef]
52. Alcívar, M.; Zurita-Silva, A.; Sandoval, M.; Muñoz, C.; Schoebitz, M. Reclamation of saline–sodic soils with combined amendments: Impact on quinoa performance and biological soil quality. *Sustainability* **2018**, *10*, 3083. [CrossRef]
53. Hinojosa, L.; Matanguihan, J.B.; Murphy, K.M. Effect of high temperature on pollen morphology, plant growth and seed yield in quinoa (*Chenopodium quinoa* Willd.). *J. Agron. Crop Sci.* **2019**, *205*, 33–45. [CrossRef]
54. Rashid, N.; Basra, S.M.; Shahbaz, M.; Iqbal, S.; Hafeez, M.B. Foliar applied moringa leaf extract induces terminal heat tolerance in quinoa. *Int. J. Agric. Biol.* **2018**, *20*, 157–164.
55. Rashid, N.; Khan, S.; Wahid, A.; Basra, S.M.A.; Alwahibi, M.S.; Jacobsen, S.E. Impact of natural and synthetic growth enhancers on the productivity and yield of quinoa (*Chenopodium quinoa* willd.) cultivated under normal and late sown circumstances. *J. Agron. Crop Sci.* **2021**. [CrossRef]
56. Zahra, N.; Hafeez, M.B.; Shaukat, K.; Wahid, A.; Hasanuzzaman, M. Fe toxicity in plants: Impacts and remediation. *Physiol. Plant.* **2021**, *173*, 201–222. [CrossRef]
57. Ramzan, A.; Muhammad, A.; Muhammad, S.; Basra, S. Hyperaccumulation response of quinoa to soil-applied cadmium: A better choice as cadmium phytoremediating plant. *Int. J. Agric. Biol.* **2020**, *24*, 603–609.
58. Ramzani, P.M.A.; Shan, L.; Anjum, S.; Ronggui, H.; Iqbal, M.; Virk, Z.A.; Kausar, S. Improved quinoa growth, physiological response, and seed nutritional quality in three soils having different stresses by the application of acidified biochar and compost. *Plant Physiol. Biochem.* **2017**, *116*, 127–138. [CrossRef] [PubMed]
59. Hinojosa, L.; González, J.A.; Barrios-Masias, F.H.; Fuentes, F.; Murphy, K.M. Quinoa abiotic stress responses: A review. *Plants* **2018**, *7*, 106. [CrossRef] [PubMed]
60. Amjad, M.; Iqbal, M.M.; Abbas, G.; Farooq, A.B.U.; Naeem, M.A.; Imran, M.; Murtaza, B.; Nadeem, M.; Jacobsen, S.-E. Assessment of cadmium and lead tolerance potential of quinoa (*Chenopodium quinoa* Willd.) and its implications for phytoremediation and human health. *Environ. Geochem. Health* **2022**, *44*, 1487–1500. [CrossRef] [PubMed]

61. Naeem, M.A.; Shabbir, A.; Amjad, M.; Abbas, G.; Imran, M.; Murtaza, B.; Tahir, M.; Ahmad, A. Acid treated biochar enhances cadmium tolerance by restricting its uptake and improving physio-chemical attributes in quinoa (*Chenopodium quinoa* Willd.). *Ecotoxicol. Environ. Saf.* **2020**, *191*, 110218. [CrossRef]
62. Stanschewski, C.S.; Rey, E.; Fiene, G.; Craine, E.B.; Wellman, G.; Melino, V.J.; Patiranage, D.S.; Johansen, K.; Schmöckel, S.M.; Bertero, D. Quinoa phenotyping methodologies: An international consensus. *Plants* **2021**, *10*, 1759. [CrossRef] [PubMed]
63. Reynolds, M.; Chapman, S.; Crespo-Herrera, L.; Molero, G.; Mondal, S.; Pequeno, D.N.; Pinto, F.; Pinera-Chavez, F.J.; Poland, J.; Rivera-Amado, C. Breeder friendly phenotyping. *Plant Sci.* **2020**, *295*, 110396. [CrossRef]
64. Bioversity International; FAO; PROINPA; INIAF; IFAD. *Descriptors for Quinoa (Chenopodium quinoa Willd.) and Wild Relatives*; Bioversity International: Rome, Italy; Fundación PROINPA: La Paz, Bolivia; Instituto Nacional de Innovación Agropecuaria y: La Paz, Bolivia, 2013; Volume 1.
65. CPVO. Protocol for Tests on Distinctness, Uniformity and Stability *Chenopodium quinoa* Willd. 2021. Available online: https://cpvo.europa.eu/sites/default/files/documents/chenopodium.pdf (accessed on 15 August 2021).
66. Bhargava, A.; Shukla, S.; Ohri, D. Genetic variability and interrelationship among various morphological and quality traits in quinoa (*Chenopodium quinoa* Willd.). *Field Crops Res.* **2007**, *101*, 104–116. [CrossRef]
67. Jacobsen, S.-E. Adaptation of quinoa (*Chenopodium quinoa*) to Northern European agriculture: Studies on developmental pattern. *Euphytica* **1997**, *96*, 41–48. [CrossRef]
68. Institute, I.F.P.R. *Global Nutrition Report 2016: From Promise to Impact: Ending Malnutrition by 2030*; International Food Policy Research Institute: Washington, DC, USA, 2016.
69. Wright, K.; Huber, K.; Fairbanks, D.J.; Huber, C. Isolation and characterization of *Atriplex hortensis* and sweet *Chenopodium quinoa* starches. *Cereal Chem.* **2002**, *79*, 715–719. [CrossRef]
70. Ruales, J.; Nair, B.M. Nutritional quality of the protein in quinoa (*Chenopodium quinoa* Willd.) seeds. *Plant Foods Hum. Nutr.* **1992**, *42*, 1–11. [CrossRef]
71. Dini, A.; Rastrelli, L.; Saturnino, P.; Schettino, O. A compositional study of *Chenopodium quinoa* seeds. *Food/Nahrung* **1992**, *36*, 400–404. [CrossRef]
72. Gonzalez, J.; Roldan, A.; Gallardo, M.; Escudero, T.; Prado, F. Quantitative determinations of chemical compounds with nutritional value from Inca crops: *Chenopodium quinoa* ('quinoa'). *Plant Foods Hum. Nutr.* **1989**, *39*, 331–337. [CrossRef]
73. Vega-Gálvez, A.; Miranda, M.; Vergara, J.; Uribe, E.; Puente, L.; Martínez, E.A. Nutrition facts and functional potential of quinoa (*Chenopodium quinoa* willd.), an ancient Andean grain: A review. *J. Sci. Food Agric.* **2010**, *90*, 2541–2547. [CrossRef] [PubMed]
74. Romero, G.; Heredia, A.; Chaparro-Zambrano, H.N. Germinative potential in quinoa (*Chenopodium quinoa* Willd.) seeds stored under cool conditions. *Rev. UDCA Actual. Divulg. Científica* **2018**, *21*, 341–350.
75. Afzal, I.; Kamran, M.; Basra, S.M.A.; Khan, S.H.U.; Mahmood, A.; Farooq, M.; Tan, D.K. Harvesting and post-harvest management approaches for preserving cottonseed quality. *Ind. Crops Prod.* **2020**, *155*, 112842. [CrossRef]
76. Afzal, I.; Jaffar, I.; Zahid, S.; Rehman, H.U.; Basra, S.M.A. Physiological and biochemical changes during hermetic storage of *Moringa oleifera* seeds. *S. Afr. J. Bot.* **2020**, *129*, 435–441. [CrossRef]
77. Bradford, K.J.; Dahal, P.; Van Asbrouck, J.; Kunusoth, K.; Bello, P.; Thompson, J.; Wu, F. The dry chain: Reducing postharvest losses and improving food safety in humid climates. In *Food Industry Wastes*; Elsevier: Amsterdam, The Netherlands, 2020; pp. 375–389.
78. Devaiah, S.P.; Pan, X.; Hong, Y.; Roth, M.; Welti, R.; Wang, X. Enhancing seed quality and viability by suppressing phospholipase D in Arabidopsis. *Plant J.* **2007**, *50*, 950–957. [CrossRef]
79. De Vitis, M.; Hay, F.R.; Dickie, J.B.; Trivedi, C.; Choi, J.; Fiegener, R. Seed storage: Maintaining seed viability and vigor for restoration use. *Restor. Ecol.* **2020**, *28*, S249–S255. [CrossRef]
80. Prego, I.; Maldonado, S.; Otegui, M. Seed structure and localization of reserves in *Chenopodium quinoa*. *Ann. Bot.* **1998**, *82*, 481–488. [CrossRef]
81. Shelar, V.; Shaikh, R.; Nikam, A. Soybean seed quality during storage: A review. *Agric. Rev.* **2008**, *29*, 125–131.
82. Ceccato, D.; Delatorre-Herrera, J.; Burrieza, H.; Bertero, H.D.; Martinez, E.A.; Delfino, I.; Moncada, S.; Bazile, D.; Castellión, M. Seed physiology and response to germination conditions. In *State of the Art Report on Quinoa around the World in 2013*; FAO-CIRAD: Santiago, Chile, 2015; pp. 131–142.
83. Kibar, H.; Temel, S.; Yücesan, B. Kinetic modeling and multivariate analysis on germination parameters of quinoa varieties: Effects of storage temperatures and durations. *J. Stored Prod. Res.* **2021**, *94*, 101880. [CrossRef]
84. McDonald, M.B. Seed moisture and the equilibrium seed moisture content curve. *Seed Technol.* **2007**, *29*, 7–18.
85. Bakhtavar, M.A.; Afzal, I. Climate smart Dry Chain Technology for safe storage of quinoa seeds. *Sci. Rep.* **2020**, *10*, 12554. [CrossRef] [PubMed]
86. Balešević-Tubić, S.; Tatić, M.; Đorđević, V.; Nikolić, Z.; Subić, J.; Đukić, V. Changes in soybean seeds as affected by accelerated and natural aging. *Rom. Biotechnol. Lett.* **2011**, *6*, 6740–6747.
87. Bewley, J.D.; Black, M. Seeds. In *Seeds*; Springer: Berlin/Heidelberg, Germany, 1994; pp. 1–33.
88. Bewley, J.D. Seed germination and dormancy. *Plant Cell* **1997**, *9*, 1055. [CrossRef]
89. Bewley, J.D.; Black, M. *Seeds: Physiology of Development and Germination*; Springer Science & Business Media: Berlin/Heidelberg, Germany, 2013.

90. Mahmood, S.; Pasha, I.; Iqbal, M.W.; Riaz, T.; Adnan, M.; Chitrakar, B.; Azam, M. Rheological and sensory attributes of wheat, quinoa and buckwheat composite flour and their use in bakery products. *J. Food Sci. Technol. Nepal* **2019**, *11*, 25–31. [CrossRef]
91. Angeli, V.; Miguel Silva, P.; Crispim Massuela, D.; Khan, M.W.; Hamar, A.; Khajehei, F.; Graeff-Hönninger, S.; Piatti, C. Quinoa (*Chenopodium quinoa* Willd.): An overview of the potentials of the "golden grain" and socio-economic and environmental aspects of its cultivation and marketization. *Foods* **2020**, *9*, 216. [CrossRef]
92. Carimentrand, A.; Ballet, J. When Fair Trade increases unfairness: The case of quinoa from Bolivia. *Cah. FREE* **2010**, *5*, 17.
93. Walsh-Dilley, M. Negotiating hybridity in highland Bolivia: Indigenous moral economy and the expanding market for quinoa. *J. Peasant. Stud.* **2013**, *40*, 659–682. [CrossRef]
94. Ayala Macedo, G. Consumption of quinoa in Peru. *Food Rev. Int.* **2003**, *19*, 221–227. [CrossRef]
95. Ahmad, R.; Malik, K.A. *Prospects for Saline Agriculture*; Springer: Berlin/Heidelberg, Germany, 2002.
96. Ashraf, E.; Babar, R.; Yaseen, M.; Shurjeel, H.K.; Fatima, N. Assessing the Impact of Quinoa Cultivation Adopted to Produce a Secure Food Crop and Poverty Reduction by Farmers in Rural Pakistan. *Int. J. Agric. Biosyst. Eng.* **2017**, *11*, 465–469.
97. Lotfalian Dehkordi, A.; Forootan, M. Estimation of energy flow and environmental impacts of quinoa cultivation through life cycle assessment methodology. *Environ. Sci. Pollut. Res.* **2020**, *27*, 21836–21846. [CrossRef] [PubMed]
98. Hirich, A.; Rafik, S.; Rahmani, M.; Fetouab, A.; Azaykou, F.; Filali, K.; Ahmadzai, H.; Jnaoui, Y.; Soulaimani, A.; Moussafir, M. Development of quinoa value chain to improve food and nutritional security in rural communities in rehamna, morocco: Lessons learned and perspectives. *Plants* **2021**, *10*, 301. [CrossRef] [PubMed]
99. Tariq, A.; Ashraf, I.; Ahmed, M.; Muscolo, A.; Basra, S.; Reynolds, M. Evaluation of physiological and morphological traits for improving spring wheat adaptation to terminal heat stress. *Plants* **2021**, *10*, 455.
100. López-Marqués, R.L.; Nørrevang, A.F.; Ache, P.; Moog, M.; Visintainer, D.; Wendt, T.; Østerberg, J.T.; Dockter, C.; Jørgensen, M.E.; Salvador, A.T. Prospects for the accelerated improvement of the resilient crop quinoa. *J. Exp. Bot.* **2020**, *71*, 5333–5347. [CrossRef]
101. Weng, F.; Zhang, W.; Wu, X.; Xu, X.; Ding, Y.; Li, G.; Liu, Z.; Wang, S. Impact of low-temperature, overcast and rainy weather during the reproductive growth stage on lodging resistance of rice. *Sci. Rep.* **2017**, *7*, 46596. [CrossRef]
102. Maliro, M.F.; Guwela, V.F.; Nyaika, J.; Murphy, K.M. Preliminary studies of the performance of quinoa (*Chenopodium quinoa* Willd.) genotypes under irrigated and rainfed conditions of central Malawi. *Front. Plant Sci.* **2017**, *8*, 227. [CrossRef]
103. Silverstone, A.L.; Ciampaglio, C.N.; Sun, T.-P. The Arabidopsis RGA gene encodes a transcriptional regulator repressing the gibberellin signal transduction pathway. *Plant Cell* **1998**, *10*, 155–169. [CrossRef]
104. Bertero, H.D.; De la Vega, A.; Correa, G.; Jacobsen, S.; Mujica, A. Genotype and genotype-by-environment interaction effects for grain yield and grain size of quinoa (*Chenopodium quinoa* Willd.) as revealed by pattern analysis of international multi-environment trials. *Field Crops Res.* **2004**, *89*, 299–318. [CrossRef]
105. Mignone, C.; Bertero, H. Identificación del período crítico de determinación del rendimiento en quínoas de nivel del mar. In Proceedings of the Congreso Internacional de la Quinua, Iquiqu, Chile, 23–27 October 2007; pp. 23–26.
106. Bertero, H.D.; Ruiz, R.A. Determination of seed number in sea level quinoa (*Chenopodium quinoa* Willd.) cultivars. *Eur. J. Agron.* **2008**, *28*, 186–194. [CrossRef]
107. López, L.M.; Capparelli, A.; Nielsen, A.E. Traditional post-harvest processing to make quinoa grains (*Chenopodium quinoa* var. quinoa) apt for consumption in Northern Lipez (Potosí, Bolivia): Ethnoarchaeological and archaeobotanical analyses. *Archaeol. Anthropol. Sci.* **2011**, *3*, 49–70. [CrossRef]
108. Murphy, K.M.; Matanguihan, J.B.; Fuentes, F.F.; Gómez-Pando, L.R.; Jellen, E.N.; Maughan, P.J.; Jarvis, D.E. Quinoa breeding and genomics. *Plant Breed. Rev.* **2018**, *42*, 257–320.
109. Gomez-Pando, L.R.; Eguiluz-de la Barra, A. Developing genetic variability of quinoa (*Chenopodium quinoa* Willd.) with gamma radiation for use in breeding programs. *Sci. Res.* **2013**, *4*, 28368. [CrossRef]
110. Jarvis, D.E.; Ho, Y.S.; Lightfoot, D.J.; Schmöckel, S.M.; Li, B.; Borm, T.J.; Ohyanagi, H.; Mineta, K.; Michell, C.T.; Saber, N. The genome of *Chenopodium quinoa*. *Nature* **2017**, *542*, 307–312. [CrossRef] [PubMed]
111. Yang, X.; Qin, P.; Guo, H.; Ren, G. Quinoa industry development in China. *Cienc. Investig. Agrar. Rev. Latinoam. Cienc. Agric.* **2019**, *46*, 208–219.
112. Ramesh, K.; Devi, K.S.; Gopinath, K.; Devi, M.U. Physiological Indices, yield and yield attributes of quinoa (*Chenopodium quinoa* Willd.) as influenced by dates of sowing and varied crop geometry. *Int. J. Curr. Microbiol. App. Sci.* **2017**, *6*, 1023–1034. [CrossRef]
113. Noulas, C.; Tziouvalekas, M.; Vlachostergios, D.; Baxevanos, D.; Karyotis, T.; Iliadis, C. Adaptation, agronomic potential, and current perspectives of quinoa under mediterranean conditions: Case studies from the lowlands of central Greece. *Commun. Soil Sci. Plant Anal.* **2017**, *48*, 2612–2629.
114. Mahmoud, A.H. Production of quinoa (*Chenopodium quinoa*) in the marginal environments of South Mediterranean region: Nile Delta, Egypt. *Egypt. J. Soil Sci.* **2017**, *57*, 329–337. [CrossRef]
115. Eisa, S.S.; Eid, M.; Abd El-Samad, E.; Hussin, S.; Abdel-Ati, A.; El-Bordeny, N.; Ali, S.; Al-Sayed, H.M.; Lotfy, M.; Masoud, A. '*Chenopodium quinoa*' Willd. A new cash crop halophyte for saline regions of Egypt. *Aust. J. Crop Sci.* **2017**, *11*, 343–351.
116. Choukr-Allah, R.; Rao, N.K.; Hirich, A.; Shahid, M.; Alshankiti, A.; Toderich, K.; Gill, S.; Butt, K.U.R. Quinoa for marginal environments: Toward future food and nutritional security in MENA and Central Asia regions. *Front. Plant Sci.* **2016**, *7*, 346. [CrossRef]
117. Shams, A.S. Preliminary evaluation of new quinoa genotypes under sandy soil conditions in Egypt. *Agric. Sci.* **2018**, *9*, 1444. [CrossRef]

118. Mustafa, T.; Temel, S. Studies on the adaptation of quinoa (*Chenopodium quinoa* Willd.) to Eastern Anatolia Region of Turkey. *AGROFOR* **2017**, *2*. [CrossRef]
119. Gardner, M.; Maliro, M.F.; Goldberger, J.R.; Murphy, K.M. Assessing the potential adoption of quinoa for human consumption in central Malawi. *Front. Sustain. Food Syst.* **2019**, *3*, 52. [CrossRef]

plants

MDPI

Article

Assessment of Phenotypic Diversity in the USDA Collection of Quinoa Links Genotypic Adaptation to Germplasm Origin

Muhammad Bilal Hafeez [1], Shahid Iqbal [2,3], Yuanyuan Li [4,*], Muhammad Sohail Saddiq [5], Shahzad M. A. Basra [1], Hui Zhang [4], Noreen Zahra [6], Muhammad Z. Akram [1], Daniel Bertero [7] and Ramiro N. Curti [8,*]

1 Department of Agronomy, University of Agriculture, Faisalabad 38040, Pakistan; bilalhafeez32@gmail.com (M.B.H.); shehzadbasra@gmail.com (S.M.A.B.); zakram97@yahoo.com (M.Z.A.)
2 Department of Agronomy, Muhammad Nawaz Shareef, University of Agriculture, Multan 66000, Pakistan; sahu.shehdi@yahoo.com
3 Instititute of Plant Breeding and Biotechnology, Muhammad Nawaz Shareef, University of Agriculture, Multan 66000, Pakistan
4 Shandong Provincial Key Laboratory of Plant Stress Research, College of Life Science, Shandong Normal University, Jinan 250014, China; lyyzhh@gmail.com
5 Department of Agronomy, Ghazi University, Dera Ghazi Khan 32200, Pakistan; sohail1540@gmail.com
6 Department of Botany, University of Agriculture, Faisalabad 38040, Pakistan; noreenzahra59@gmail.com
7 Cátedra de Producción Vegetal and Instituto de Investigaciones Fisiológicas y Ecológicas Vinculadas a la Agricultura (IFEVA)—CONICET, Facultad de Agronomía, Universidad de Buenos Aires, Buenos Aires C1417DSE, Argentina; bertero@agro.uba.ar
8 Laboratorio de Investigaciones Botánicas (LABIBO), Facultad de Ciencias Naturales and Sede Regional Sur, Universidad Nacional de Salta—CCT-CONICET, Salta 4400, Argentina
* Correspondence: yuanyuan_li@outlook.com (Y.L.); rcurti@agro.uba.ar (R.N.C.)

Citation: Hafeez, M.B.; Iqbal, S.; Li, Y.; Saddiq, M.S.; Basra, S.M.A.; Zhang, H.; Zahra, N.; Akram, M.Z.; Bertero, D.; Curti, R.N. Assessment of Phenotypic Diversity in the USDA Collection of Quinoa Links Genotypic Adaptation to Germplasm Origin. *Plants* **2022**, *11*, 738. https://doi.org/10.3390/plants11060738

Academic Editors: Cataldo Pulvento and Didier Bazile

Received: 28 January 2022
Accepted: 3 March 2022
Published: 10 March 2022

Abstract: Quinoa's germplasm evaluation is the first step towards determining its suitability under new environmental conditions. The aim of this study was to introduce suitable germplasm to the lowland areas of the Faisalabad Plain that could then be used to introduce quinoa more effectively to that region. A set of 117 quinoa genotypes belonging to the USDA quinoa collection was evaluated for 11 phenotypic quantitative traits (grain yield (Y), its biological and numerical components plus phenological variables) in a RCBD during two consecutive growing seasons at the University of Agriculture, Faisalabad, Pakistan under mid-autumn sowings. Genotypic performance changed across the years, however most phenotypic traits showed high heritability, from 0.75 for Harvest Index (HI) to 0.97 for aerial biomass (B) and Y. Ordination and cluster analyses differentiated four groups dominated by genotypes from: Peru and the Bolivian Highlands (G1); the Bolivian Highlands (G2); the Ballón collection (regarded as a cross between Bolivian and Sea Level (Chilean) genotypes) plus Bolivian Highlands (G3); and Ballón plus Sea Level (G4), this latter group being the most differentiated one. This genetic structure shared similarities with previous groups identified using SSR markers and $G \times E$ data from an international quinoa test. G4 genotypes showed the highest Y associated with higher B and seed numbers (SN), while HI made a significant contribution to yield determination in G2 and seed weight (SW) in G3. G1 and G2 showed the lowest Y associated with a lower B and SN. Moreover, SW showed a strongly negative association with SN in G2. Accordingly, G4 followed by G3 are better suited to the lowland areas of Faisalabad plain and the physiological traits underlying yield determination among genotypic groups should be considered in future breeding programs.

Keywords: breeding; *Chenopodium quinoa* Willd.; genetic structure; germplasm; heritability; effects of genotype by environment interaction

1. Introduction

The growing popularity of quinoa in the recent years made the crop require little introduction [1,2]. This is the consequence of sustained demand for its seed in the international market, particularly in the Health Food sector in developed countries [3]. Quinoa prices increased in a sustained way in the last few decades with a peak in 2014, reaching ~7000 US$ per ton [4]. This was accompanied by a parallel increased interest in quinoa evaluation and production in many countries outside the Andes, its traditional growing region [5,6]. It was not only its good nutritional balance and high prices, but also the capacity of the crop to face many limitations like water deficits, salinity, frost or poor soils that contributed to this interest [7,8]. Quinoa is currently being evaluated in all continents and commercial production is underway in a significant number of them [5,9,10]. Germplasm from different countries were evaluated and new breeding programs started in the US, Canada, the E.U., Australia, Israel, China, India and the Middle East, among others.

Germplasm evaluation, ideally involving contrasting origins when performed for the first time, is central to guide breeding and agronomic management [11,12]. For quinoa, the biggest undertaking was that organized by the FAO two decades ago [13] in which 24 genotypes with origins ranging from Colombia to Chile plus Europe were evaluated in locations in South America, Europe, Africa and Asia. This evaluation used pattern analysis and identified four genetic and four environmental groups based on yield performance. A strong degree of genotype by environment (G $\times$ E) interaction was detected demanding a clear characterization of the target environment and the identification of genotypes specifically suited to them. A similar analysis, this time including a more detailed analysis of the numerical (seed number and seed weight) and physiological (crop biomass and harvest index) components of yield was performed for local germplasm in N.W. Argentina by [14]. In addition to these analyses, where genotypic, environmental and G $\times$ E components of variation were identified, many more evaluations were performed with varying degree of detail in the analysis, such as in India [15], the Middle East and North Africa [16–18], Central Africa [19,20] and China [9]. More detailed descriptions of quinoa evaluation at national levels can be found in [2], in their report on the State of the Art of Quinoa in the World in 2013.

Besides these agronomic evaluations, quinoa genetic studies were approached using molecular markers. The first researcher to complete this work, [21], using allozyme markers, identified two distinct groups, from Central and Southern Chile and from the Central Andes, with a less clear distinction between germplasm from the Northern Andes (Colombia to Peru) and Southern Andes (Peru to Northwest Argentina and Northern Chile). A more recent evaluation [22] used microsatellite markers to characterize the USDA's (United States Department of Agriculture) quinoa collection and was able to distinguish between Sea Level (quinoa originating or selected from accessions from low altitude environments in Central and Southern Chile [21]) and Andean types, plus a third group, named the Ballón collection according to the name of the germplasm donor to the USDA quinoa collection, which overlapped with both groups. This latter group originated in spontaneous crosses between Andean and Central Chile germplasm when multiplied at high field temperatures in New Mexico (E. Ballón pers. comm.). Using SNPs, Mizuno et al. [23] confirmed the structure of these three sub-groups (Northern and Southern Highlands plus Lowlands (central and southern Chile)).

Given the current legal restrictions to access germplasm from the Andean Region [24], the USDA germplasm collection (including more than 140 accessions) has been the basis of several adaptation and breeding programs internationally [15–18,20]. This collection includes accessions from most countries of traditional cultivation and provided sources of traits like male sterility [25] adaptations to hot environments [17,26,27], resistance to pre-harvest sprouting [28,29] or prolonged seed viability [30]. Based on this information, a better understanding of its agronomic performance is central and there are no published reports of it for this collection. Furthermore, as the Middle East is an area of recent fast expansion of quinoa cultivation, understanding quinoa agronomic behavior in these Mediterranean climates is also central in its own right.

The aim of this study was to screen a collection of quinoa genotypes for adaptation to yield under the low altitude environments of the Faisalabad region, Northern Pakistan. Specifically, we measured yield, its biological determinants and components plus phenology in 117 quinoa genotypes from the USDA germplasm collection to determine: (i) the relative contribution of genotype, environment and genotype-by-environment interaction effect to the phenotypic variation of those traits; (ii) the role of these traits in capturing the patterns of genotypic adaptation; (iii) the association between these patterns and the origin of materials; and (iv) the physiological trait combinations underlying yield-formation among genetic groups. Our working hypothesis is that the structure of phenotypic variation expressed in the patterns of performance of quinoa in Pakistan reflects the geographical origin of the materials.

2. Material and Methods

2.1. Plant Genetic Materials and Field Experiments

One hundred-seventeen quinoa genotypes belonging to the USDA-NPGS (United States Department of Agriculture) collection were utilized in this study (Table S1). A two-years field study was conducted through a randomized complete block design (RCBD) with three replications per genotype and year in order to evaluate the performance of this collection. The crop was sown on 21 November and 20 November in 2016 and 2017, respectively, at the Agronomy Farm (73.88° E, 31.88° N; elevation 184 masl) of the University of Agriculture of Faisalabad (UAF), Pakistan. Daily recorded values of air temperature and rainfall events at the experimental site were obtained from the Agricultural Meteorology Cell, UAF. For each experiment, the site soil was ploughed to 30 cm depth and subsequently harrowed. Ridges, 30 cm in height, were prepared at 75 cm spacing. Seeds were sown on the top of each ridge at 15 cm spacing with a hand dibble method. Each experimental plot comprised two ridges with a length of 3 m. Two terminal quinoa plants at each end of the ridge were used as guard plants, giving a net plot size of two ridges of 2.75 m in length. For fertilization, nitrogen (N), phosphorus (P) and potassium (K) were applied at 75, 60, and 60 kg ha^{-1}, respectively, during both years. Fertilizers used were urea (46% N), diammonium phosphate (18% N, 46% P_2O_5) and sulfate of potassium (50% K_2O). A half dose of N and a full dose of P and K were applied as a basal dose at sowing and the remaining N was added 75 days after sowing. During the entire crop period, four irrigations of 330 mm total were applied including the pre-sowing irrigation, recommended as optimal for the cultivation of quinoa, for the conditions of Pakistan [31]. The field was kept free from weeds by hand hoeing, whereas no insecticide/pesticide and herbicide were used throughout both experimental years.

2.2. Measurements

A set of 11 phenotypic traits encompassing grain yield-related attributes and crop phenological stages were obtained for each plot. Grain yield (Y, g m^{-2}) and its biological determinants aerial biomass (B, g m^{-2}), harvest index (HI) plus terminal panicle grain yield (Pan Y, g m^{-2}), and the numerical components seed number (SN, # m^{-2}) and weight (SW, mg) were obtained from five uniform tagged plants from the middle of each row, avoiding border plants. Plants were harvested with a sickle, sundried for three to five days and later on threshed manually to measure yield attributes using a digital scale. HI was estimated as the Y/B ratio. SN was estimated considering the final harvest data as the ratio of Y to the average individual SW. Individual SW was estimated by manually counting and weighing 1000 grains in each replicate plot. In addition, for the five plants tagged in each plot, the plant height (Ht, cm) and terminal panicle length (Pan Ht, cm) were measured with a ruler. Ht was measured from the soil to the top of the terminal panicle, whereas the Pan Ht was measured between the basal and the last node below the main panicle.

Crop development stages (recorded when at least 50% of the plants in each plot had reached the stage) were determined as emergence, first anthesis (extrusion of anthers completed), and physiological maturity (when grain could not be nail dented). Thus, we defined the main phenological phases as the emergence–anthesis (E–Ant) and anthesis-physiological maturity (Ant–PM) periods. Besides that, the crop duration (Cycle) period was recorded as the time between emergence and harvest.

2.3. Data Analysis

Linear mixed models were set up to examine the relative contribution of genotype and genotype-by-environment interaction effects to the phenotypic variation of all phenotypic traits across experiments. The phenotypic observation y_{ijk} on genotype i in block k of environment j was modelled following the expression:

$$y_{ijk} = \mu + G_i + E_j + (GE)_{ij} + B_{k(j)} + \varepsilon_{ijk}$$

with μ designating the general intercept, $G_i \sim N(0, \sigma^2{}_g)$ is the random main effect of the i-th genotype, E_j is the fixed main effect of the j-th environment (years), $GE_{ij} \sim N(0, \sigma^2{}_{ge})$ is the random interaction effect of the i-th genotype and the j-th environment, $B_{k(j)} \sim N(0, \sigma^2{}_b)$> is the random effect of the k-th block nested within the environment j-th, and ε_{ijk} is the residual plot error associated with the observation y_{ijk}. All linear mixed models were fitted with the function gamen_met of the R package metan (multi-environment trial analysis) [32]. This function estimates the variance components of random effects by Restricted Maximum Likelihood (REML), whereas their significance by a likelihood ratio (LRT) test, comparing a full model with all random terms with each other, without one of the random terms (reduced model). In addition, broad-sense heritability (H) was computed for all traits.

To determine the role of phenotypic traits in capturing similarities in genotypic response, Pattern Analysis (PA), defined as the combined application of ordination and classification multivariate analyses (principal component analysis (PCA) and Cluster Analysis (CA)), was used. This set of analyses was based on predictions computed from random terms obtained from REML analyses (see above). First, we computed the Best Linear Unbiased Predictors (BLUPs) for genotype, and genotype-by environment interaction terms, and then the predictions adjusted by the BLUPs' effects were used to build an array of 117 genotypes × 11 traits. Previous to PA, the array was trait-standardized by removing the traits' grand mean and dividing the remainder by the within-trait standard deviation. For classification (CA), the hierarchical agglomerative method, with the incremental sum of squares as fusion criteria [33], was chosen, using the Euclidean distance. A dendrogram of genotypes was produced in order to investigate the grouping of genotypes according to the evaluated phenotypic traits. The optimal number of clusters was defined according to 30 indices computed from the NbClust package [34]. Then, the results of the dendrogram obtained were compared with the genetic diversity study of [22], to investigate the relatedness between the patterns of genotypic responses and the origin of materials. In addition, we tested the differences among groups for all agronomic traits by means of analyses of variance. Mean comparisons were based on Tukey's HSD test.

For ordination, a PCA was performed using the singular value decomposition algorithm on the Euclidean standardized distance of the two-way array of genotypes × traits, using the FactoMineR package [35]. In order to investigate the interrelations between genotypes and traits, a Biplot of the first and second axis was obtained. In the Biplot the symbols (genotypes) were depicted according to the dendrogram results obtained from CA. To study the physiological traits combinations underlying the yield-formation among genetic groups, we approximated the correlations between grain yield components by inspecting the angles formed between vectors (traits) in the Biplot. Rules of interpretation according to Biplot properties are: angles below 90° approximate a positive association between vectors; angles above 90° approximate a negative association; and angles at 90°

indicate no association. All statistical analyses were performed within the R environment version 4.0.5 (2021).

3. Results

3.1. Effects of Growing Conditions on Phenotypic Variation

Growing conditions in Faisalabad, Pakistan were similar among the experimental years, but with a lower minimum temperature and rainfall during the second season (Table 1). Climatic conditions of these two particular years matched the general climate patterns for the same locality as corroborated by comparison with weather data from the years 1981–2015 in the NASA website (https://power.larc.nasa.gov/data-access-viewer/, accessed on 15 October 2021). The seasonal photoperiod varied from 10.4 to 12.9 h day^{-1} for the crop cycle, with most genotypes flowering during spring and maturing in summer (i.e., under high photo-thermal conditions). The environmental means for phenotypic traits, i.e., the average across genotypes, differed between growing seasons for Pan Ht, B, Pan Y, SW, E–Ant and Cycle, with a trend towards higher values in the second growing season (Table 1).

Table 1. Growing conditions in Faisalabad and their effect on measured phenotypic traits. Maximum, minimum and mean temperature are ranges of monthly values whereas rainfall are totals for the growing season (November–April). For phenotypic traits the values are shown as means ± SE.

	Season		
	2016	2017	
Max. temp. (°C)	17.6–37.7	21.5–36.8	(27.3–35.4) [a]
Min. temp. (°C)	8.2–20.9	5.5–20.8	(2.2–13.0) [a]
Mean temp. (°C)	12.9–29.3	13.5–28.8	(12.5–27.7) [a]
Rainfall (mm)	60	36	(24.4) [a]
Ht (cm) [b]	94.3 ± 1.6	97.6 ± 1.8 **	
Pan Ht (cm)	29.3 ± 0.5	29.4 ± 0.4	
B (g m^{-2})	67.9 ± 3.2	71.0 ± 3.1 *	
HI	35.4 ± 0.6	35.7 ± 0.6	
Pan Y (g m^{-2})	17.0 ± 0.9	18.3 ± 0.9 **	
Y (g m^{-2})	24.6 ± 1.3	25.6 ± 1.3	
SN (# m^{-2})	$7.7 \times 10^6 \pm 4 \times 10^5$	$7.9 \times 10^6 \pm 3.9 \times 10^5$	
SW (mg)	$3.1 \times 10^{-3} \pm 3.6 \times 10^{-5}$	$3.2 \times 10^{-3} \pm 4 \times 10^{-5}$ **	
E—Ant (days)	64.7 ± 0.3	67.9 ± 0.3 **	
Ant–PM (days)	51.3 ± 0.5	50.6 ± 0.5	
Cycle (days)	129.6 ± 0.5	134.9 ± 0.4 **	

** significant at $p < 0.01$, * significant at $p < 0.05$. [a] Range values correspond to the historical weather records during the last 35 years (from 1981 to 2015) downloaded from the NASA website (https://power.larc.nasa.gov/data-access-viewer/, accessed on 15 October 2021); [b] Abbreviations: Ht: plant height, Pan Ht: terminal panicle length, B: aerial biomass, HI: harvest index, Pan Y: terminal panicle grain yield: Y: yield, SN: seed number, SW: seed weight, E–Ant: emergence-anthesis period: Ant–PM: anthesis–physiological maturity period, and Cycle: crop duration.

3.2. Variance Components and Heritability

Estimated components of variance and their relative contribution to phenotypic variation are shown in Table 2. The G effect contributed significantly to phenotypic variation in all traits, whereas the G × E effect also contributed significantly to phenotypic variation save for Pan Ht, B, Pan Y and Y (Table 2). The G term accounted for a larger proportion of variation than the G × E term for phenotypic traits, in which both sources of variation were significant (with G/G × E ranges from 5.1 (SW) to 25.3 (SN)). Consequently, high estimates of broad-sense heritability were observed (ranging from 0.82 for SW to 0.97 for B and Y, respectively; Table 2).

Table 2. Relative contribution of estimated variance components and broad-sense heritability (*H*) for 11 phenotypic traits measured in 117 genotypes of quinoa across two years in Faisalabad.

Trait	σ^2_g	σ^2_{ge}	σ^2_g/σ^2_{ge}	*H*
Ht	942.0 **	46.9 **	20.1	0.90
Pan Ht	60.1 **	0.15 ns		0.87
B	3364.0 **	7.8 ns		0.97
HI	89.5 **	11.0 **	8.1	0.75
Pan Y	17.7 *	15.9 ns		0.96
Y	57.3 **	32.2 ns		0.97
SN	5.2×10^{13} **	2.0×10^{12} **	25.3	0.93
SW	4.3×10^{-7} **	8.4×10^{-8} **	5.1	0.82
E–Ant	35.1 **	4.0 **	8.8	0.87
Ant–PM	78.8 **	7.2 **	10.9	0.90
Cycle	63.8 **	5.2 **	12.3	0.92

** significant at $p < 0.01$, * significant at $p < 0.05$, ^{ns} non-significant.

3.3. Phenotypic Variation Patterns in the USDA Germplasm Collection

The 117 quinoa genotypes clustered into four clearly distinct groups (Figure 1). Group 1 (G1) consisted of 35 entries dominated by genotypes from the Bolivian (17) and Peruvian (12) Highlands plus one accession from the Peruvian Inter-Andean Valleys (Rosa de Junín, genotype 68), three from the Ballón collection (genotypes 12, 19 and 44), one from Sea Level (genotype 108), and one from an unknown origin (genotype 110) (Figure 1 and Table S1). Most of these genotypes belong to the Andean group according to [22] and were classified in the northern highland subgroup based on microsatellite's markers (Table S1). Genotypes from this group showed the lowest values for Y, B, HI, SN and SW, but the highest values for traits such as Ant–PM and Cycle (Table 3). Group 2 (G2) consisted of 22 entries with most genotypes from the Bolivian Highlands (16) plus three accessions from Sea Level (genotypes 74, 104 and 107), two from Peru (genotypes 54 and 61) and one from the Ballón collection (genotype 40) (Figure 1 and Table S1). Most of these genotypes belong to the Andean group and were classified in the southern highland subgroup based on microsatellites' markers (Table S1). Genotypes from this group showed the lowest values for most phenotypic traits, but higher values for Y, B, HI, SN and SW compared with G1 (Table 3). Group 3 (G3) consisted of 31 entries dominated by genotypes from the Ballón collection (21) and the Bolivian Highlands (5) plus three accessions from Northwest Argentina (genotypes 67, 71 and 72) and two from Sea Level (genotypes 73 and 109) (Figure 1 and Table S1). Most of these genotypes were classified in the lowland and southern highland subgroup based on microsatellite's markers (Table S1). Genotypes from this group showed higher values for most phenotypic traits compared with G1–G2 and the highest SW within the evaluated collection (Table 3). Group 4 (G4) was composed of 29 entries dominated by genotypes from the Ballón collection (17) and Sea Level (9) plus the Isluga (genotype 23), Cochabamba (genotype 69) and Plant Virus (genotype 43) (Figure 1 and Table S1). Most of these genotypes were classified in the lowland group based on microsatellites' markers (Table S1). This group showed the highest values for most phenotypic traits, except for traits such as SW, Ant–PM and Cycle (Table 3).

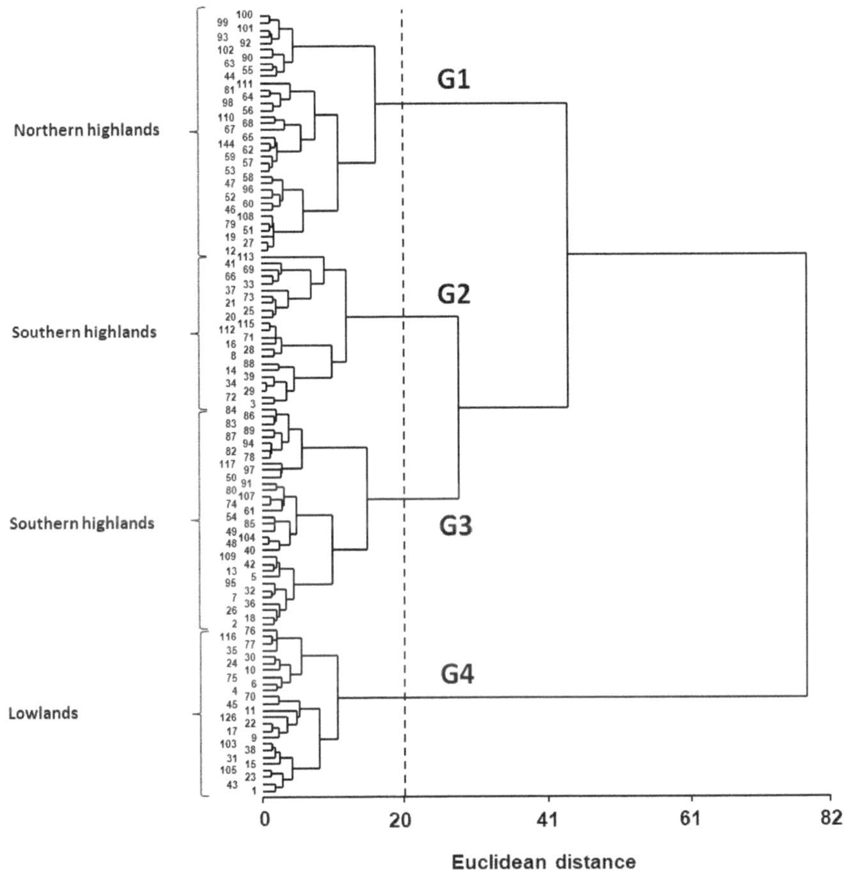

Figure 1. Dendrogram showing 117 genotypes of quinoa grouped according to Ward cluster analysis using 11 phenotypic traits. To the left of the dendrogram the groups are named according to the ecotype classification based on microsatellite markers by [22]. To the right of the dendrogram the groups are named according to the classification obtained in the present study; G1: genotypes from Peruvian and Bolivian highlands, G2: genotypes from Bolivian highlands, G3: genotypes from Ballón collection and Bolivian highlands and G4: genotypes from Ballón collection and Sea Level. See Table S1 for more information about genotypes codes and both classifications.

Table 3. Agronomic traits of the four genotype groups resulting from a hierarchical agglomerative clustering method.

Group	Ht [a] (cm)	Pan Ht (cm)	B (g m^{-2})	HI	Pan Y (g m^{-2})	Y (g m^{-2})	SN (# m^{-2})	SW (mg)	E–Ant (days)	Ant–PM (days)	Cycle (days)
G1	79.7 a	29.8 b	26.2 a	0.31 a	5.0 a	6.6 a	2.8×10^6 a	2.6 a	63 a	60 b	138 a
G2	67.9 a	25.0 a	26.0 a	0.37 bc	7.7 a	9.9 a	3.2×10^6 a	3.3 b	63 a	45 a	123 c
G3	110.5 b	29.3 b	73.4 b	0.35 b	17.0 b	23.8 b	6.6×10^6 b	3.6 b	69 b	46 a	131 b
G4	121.4 b	32.3 b	150.5 c	0.40 c	41.2 c	60.4 c	1.9×10^7 c	3.3 b	69 b	49 a	133 b

[a] Abbreviations: Ht: plant height; Pan Ht: terminal panicle length; B: aerial biomass; HI: harvest index (%); Pan Y: terminal panicle grain yield; Y: yield; SN: seed number; SW: seed weight; E–Ant: emergence–anthesis period; Ant–PM: anthesis–physiological maturity period; Cycle: crop duration. Different letters following agronomic values indicate significant differences based on Tukey's HSD test.

Results of the ordination analyses are displayed in the Biplot of the 1st and 2nd principal components, which, when combined, accounted for ~65% of total variation (Figure 2). The trait vectors covered a wide range of Euclidean space suggesting a strong contrast among the phenotypic traits evaluated. The angle between grain yield (Y) and its biological determinants and components (B, SN, Pan Y, HI and SW) plus E–Ant, is smaller than 90° (Figure 2), which suggests that most of these traits are positively associated within the collection evaluated. Traits such as HI and SW are positively correlated, but both lack association with Ht as their angles are close to 90° (Figure 2); whereas strong and negative associations (angles larger than 90°) were found between Y plus its related traits (B, SN, Pan Y, HI and SW) and Ant–PM plus Cycle (Figure 2). In turn, the E–Ant phase lacks association with Ant–PM duration as their angles are close to 90° (Figure 2).

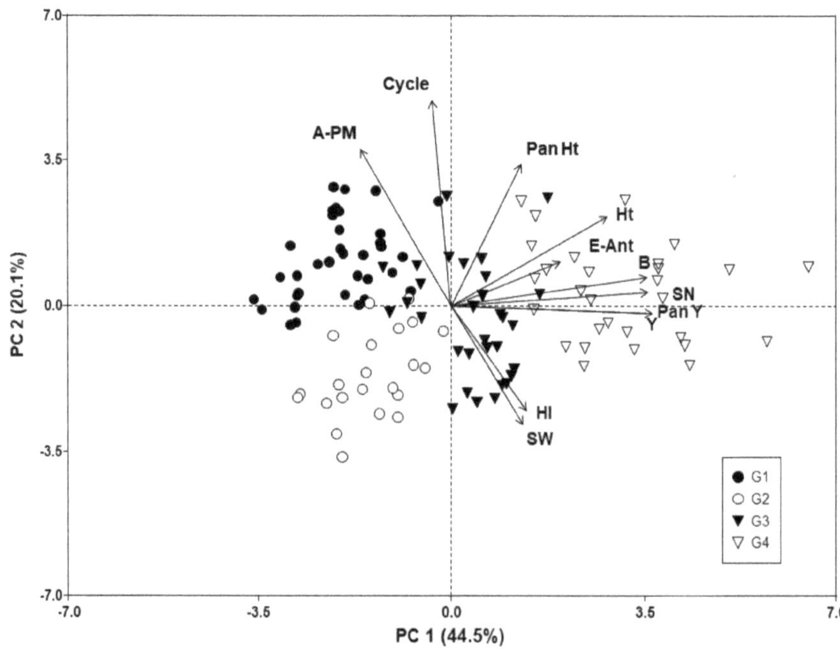

Figure 2. Biplot of the 1st and 2nd principal components for 117 accessions of quinoa described by 11 phenotypic traits. Genotypes are represented by symbols and traits by vectors. Same color symbols indicate genotype groups with members of a similar response pattern. G1: Peruvian and Bolivian Highlands; G2: Bolivian highlands; G3: Ballón collection and Bolivian highlands; and G4: Ballón collection and Sea Level.

The 1st principal component (PC1) explained 44.5% of the total variation and ordered the genotypes according to Y, its related traits (B, SN, Pan Y, HI, and SW) and time to anthesis. As indicated by Figure 2, genotypes with higher Y, B, HI, SN, Pan Y, Ht, and E–Ant duration were placed to the right of the Biplot. Most of these genotypes are from G4 (Ballón collection plus Sea Level) and represent the most differentiated group (Figures 1 and 2). Genotypes to the left side of the PC1 showed lower values for Y and its related traits plus time to anthesis (Table 3 and Figure 2). Most of them are from G1 and G2 (Peru and Bolivian Highlands) (Figure 2 and Table S1). Genotypes from G3 (Ballón collection plus Bolivian Highlands) are located in an intermediate place on PC1 (Figure 2). They are characterized by the most Y and its related traits, similar to those from G4 and to some extent to G2 (mainly by HI, SW, and Cycle), but by Pan Ht (terminal panicle length) to those from G1 (Table 3 and Figure 2).

The second principal component (PC2) explained 20.1% of the total variation and accounted for the effects of the contrasting traits Ant–PM, Pan Ht, Cycle, and HI plus SW, emphasizing the differences between groups G1 and G2 (Figure 2). G1 tended to be at the top left-hand quadrant of the Biplot, which indicated that it had high values for Pan Ht, Ant–PM, and Cycle, but low values for HI plus SW and included the genotypes with the longest duration within the collection (Figure 2 and Table 3). The genotypes from G2 showed contrasting values for most of those phenotypic traits and are placed toward the bottom left-hand quadrant of the Biplot (Figure 2), showing high values for HI and SW, but lower values for Pan Ht, Ant–PM, and Cycle, and could be considered as the genotypic set that was the most precocious within the collection (Table 3).

4. Discussion

The results of this study show the first phenotyping assessment carried out on a comprehensive set of quinoa accessions from the USDA germplasm collection accounting for yield and its components. This collection was the basis of several adaptations and breeding programs internationally and, as such, made a highly significant contribution to the global expansion of this crop [17,36–42]. According to our results, the environment contributed to phenotypic variation in most traits (Table 1); however, their magnitude was relatively low compared with contributions of G and G × E effects. This is surprising as usually the E component accounts for the highest proportion of variation, also for quinoa [13–15] and is explained by its similarity between both evaluation years (which reflected general climate patterns for the region as mentioned). Consequently, the G effects, accounting for a large proportion of variation and its contribution, was relatively high compared to G × E effects for most traits (Table 2). These results match those found under Tropical and Mediterranean conditions involving subsets of genotypes originating from the USDA collection [15,41]. Accordingly, the overall picture arising from this comparison highlights the major role of G effects in determining phenotypic values in a large set of representative genotypes from that collection.

The high heritability observed for all traits implies that phenotypic variation reflects the patterns of genotypic adaptation. The genotypic groups found in the present study clearly showed differences in their performance (Table 3). Phenotypic trait combinations observed in genotypes from G3 and G4 determined their higher suitability for cultivation under the Mediterranean conditions of Faisalabad. This set of genotypes with the highest values for yield and its related traits showed intermediate crop cycle durations (Table 3). Conversely, genotypes from G1 and G2 with the lowest values for most traits showed either a longer (G1) or shorter (G2) crop cycle, respectively (Table 3). This pattern of response was observed in other evaluations conducted at high latitudes or under tropical and Mediterranean conditions [15,43,44]. Early or late-maturing genotypes, in general, have been shown to perform better within a narrow range of environments, whereas genotypes with an intermediate crop cycle are better adapted to a broader range of conditions [13,14,45,46]. Genotypic variations in sensitivity to temperature and photoperiod conditions are the main factors controlling phenology and explains this contrast in genotypic adaptation in quinoa [47,48].

The hierarchical agglomerative groupings identified four genotypic groups associated with the genotype's environments of origin (Figure 1). This grouping shows a close correspondence with the proposed quinoa genetic groups based on molecular studies [22,23], which distinguished three groups of accessions, namely Lowland, Northern Highlands and Southern Highlands, corresponding to G4, G1 and G2–G3 of this study, respectively (Table S1). Moreover, the four genotypic groups found here partially resemble those found in early evaluations conducted using a large set of quinoa cultivars representing all environments of origin of the crop [13] or on a local basis with germplasm from N.W. Argentina [14]. The difference between this classification and those obtained on the basis of yield performance was observed regarding the genotypic composition among groups. Within G1, accessions from the northern highlands dominated, however southern highlands' accessions were also represented (Table S1). Besides, while accessions from the southern highlands were distributed among G2 and G3, they clearly dominated in G2 (Table S1). These results contrast with the clear distinction between northern- and southern highlands' types found by [13]. The highland accessions from the N.W Argentina germplasm were grouped with accessions belonging to the Ballón collection in G3 (Table S1), while it was expected that they would be grouped with accessions from G2. On the other hand, a correspondence was observed between accessions included in G4 and the sea-level type as in [13,47].

The genotypes originating from the Ballón collection have been little evaluated in other studies [2,15,17,49], and were represented with at least one accession in all groups, and clearly dominated in G3 and G4 (Figure 1 and Table S1). This grouping pattern reinforces the notion of an inter-regional origin for the Ballón genotypes arising from spontaneous crosses between genotypes from southern Andean Highlands (G3) and Sea Level (G4) groups (E. Ballón pers. com.) [22]. Genotypic adaptation in quinoa has largely been related to the environment of the origin of the materials. Thus, several early and recent introduction programs outside the Andean region were frequently based on accessions originating from Sea Level environments or from genotypes of that origin [43,44,46]. The results presented here are in line with the above statements, as genotypes originating from Sea Level were the best performers in the Faisalabad environments (Table 3 and Figure 2). However, genotypes from the Ballón collection also showed phenotypic traits' combinations that were desirable for cultivation in those sites and deserve further exploration in future studies. An example of this is accession AMES 13,737 (2 Want) which exhibited good performance in Central Argentina [50,51] and the heat-stressed conditions of UAE [17].

The physiological traits' combinations underlying yield-formation differed among the four genetic groups and resembled results from other studies [13,14]. High aerial biomass and a long time to anthesis were suitable traits' combinations for determining higher grain yield associated with a higher seed number (Figure 2). This set of phenotypic traits was observed in genotypes from G4 and to lesser extent from G3 (Figure 2). In addition, both genetic groups showed medium crop cycle duration (Table 3). A longer or shorter time for seed-filling duration determines poor performance as the genotypes with the longest time (G1) or the earliest (G2) showed a lower grain yield (Figure 2 and Table 3). However, early-maturing genotypes from G2 and G3 improved grain yield and seed weight associated with increases in harvest index (Figure 2). These different combinations of underlying physiological traits determining cultivars with similar grain yield performance have been observed in several major crops [52–55], and also in quinoa [14]. In addition, the lack of association found between the developmental phases of E–Ant and Ant–PM (Figure 2), implies that there is scope for manipulating the plant's developmental duration through breeding targeted to obtain genotypes with different duration combinations by crossing accessions from groups showing contrasting developmental phase durations as proposed by [56], and also observed in crops like soybean [57]. In this sense, future breeding programs aimed to increase grain yield in quinoa under Faisalabad cultivation conditions and similar environments under Mediterranean Climates could be targeted by combining phenotypic traits found in genotypes from groups G3 and G4, or by exploiting indirect selection for phenotypic traits determining yield across both groups, respectively. Finally, more testing sites are needed across countries from North Africa and East Asia, which share similarities in the Mediterranean conditions with Faisalabad, to evaluate the degree of repeatability in the genotypic response patterns observed in the present study.

5. Conclusions

The performance (average yields) of quinoa under Pakistan's Mediterranean conditions was poorer than evaluations carried out in locations with similar climate regimes [17,27,58,59], with only G4 reaching agronomic significance. However, the patterns of genotypic adaptation were reflected by the strong genotype effect found in all yields and its related traits, whereas the patterns of variation matched the environments of origin of materials. Furthermore, the physiological traits combinations underlying yield-formation varied among the genotypic groups, determining differences in adaptation to yield. The genotypes originating from Sea Level, the Ballón collection and the Bolivian Highlands (G3 and G4) show a suitable set of phenotypic traits able to expand quinoa cultivation under the low altitude environments of the Faisalabad region. In addition, seminal breeding programs established with the prospect to advance the adaptation of quinoa to these conditions could benefit from exploiting indirect selection for traits contributing to yield generation among genotypes from those groups.

Supplementary Materials: The following supporting information can be downloaded at: https://www.mdpi.com/article/10.3390/plants11060738/s1, Table S1: Passport data on quinoa germplasm evaluated under Faisalabad environmental conditions.

Author Contributions: M.B.H., S.I. and S.M.A.B. drafted the experimental design and M.B.H. performed the experiments; N.Z., M.S.S. and M.Z.A. helped in data collection; R.N.C., Y.L., H.Z. and D.B. did data analysis, figures, and initial draft of manuscript text. All authors have read and agreed to the published version of the manuscript.

Funding: This work was partially supported by Shandong Provincial "Bohai Granary" Science and Technology Demonstration Project (2019BHLC001).

Institutional Review Board Statement: Not applicable.

Informed Consent Statement: Not applicable.

Data Availability Statement: The data presented in this study are available on request from the corresponding author. The data are not publicly available due to data protection for varietal development purposes.

Acknowledgments: We gratefully acknowledge the U.S. National Plant Germplasm System for seeds' contribution.

Conflicts of Interest: The authors declare no conflict of interest.

References

1. Murphy, K.S.; Matanguihan, J. *Quinoa: Improvement and Sustainable Production*; John Wiley & Sons: Hoboken, NJ, USA, 2015.
2. Bazile, D.; Bertero, H.D.; Nieto, C. *State of the Art Report on Quinoa around the World in 2013*; Oficina Regional de la FAO para América Latina y el Caribe; FAO: Rome, Italy, 2015.
3. Haros, C.M.; Schoenlechner, R. *Pseudocereals: Chemistry and Technology*; John Wiley & Sons: Hoboken, NJ, USA, 2017.
4. Economist, T. Big Agribusiness Wants to Make Quinoa More Mainstream. Traders Hope That Reliable Domestic Supply Will Entice Foodmakers to Use More of the Crop as an Ingredient in Processed Snacks. 2019. Available online: https://www.economist.com/business/2019/05/25/big-agribusiness-wants-to-make-quinoa-more-mainstream (accessed on 10 January 2019).
5. Alandia, G.; Rodriguez, J.; Jacobsen, S.-E.; Bazile, D.; Condori, B. Global expansion of quinoa and challenges for the Andean region. *Glob. Food Secur.* **2020**, *26*, 100429. [CrossRef]
6. Tapia, M.E. History of the Quinuas in South America. In *The Quinoa Genome*; Springer: Gewerbestrasse, Switzerland; Cham, Switzerland, 2021; pp. 1–12.
7. Hinojosa, L.; González, J.A.; Barrios-Masias, F.H.; Fuentes, F.; Murphy, K.M. Quinoa abiotic stress responses: A review. *Plants* **2018**, *7*, 106. [CrossRef] [PubMed]
8. Grenfell-Shaw, L.; Tester, M. Abiotic Stress Tolerance in Quinoa. In *The Quinoa Genome*; Springer: Gewerbestrasse, Switzerland; Cham, Switzerland, 2021; pp. 139–167.
9. Yang, X.; Qin, P.; Guo, H.; Ren, G. Quinoa industry development in China. *Cienc. Investig. Agrar. Rev. Latinoam. Cienc. Agric.* **2019**, *46*, 208–219.
10. Angeli, V.; Miguel Silva, P.; Crispim Massuela, D.; Khan, M.W.; Hamar, A.; Khajehei, F.; Graeff-Hönninger, S.; Piatti, C. Quinoa (*Chenopodium quinoa* Willd.): An overview of the potentials of the "Golden Grain" and socio-economic and environmental aspects of its cultivation and marketization. *Foods* **2020**, *9*, 216. [CrossRef] [PubMed]
11. Ortiz, R. Leveraging Agricultural Biodiversity for Crop Improvement and Food Security. In *Routledge Handbook of Agricultural Biodiversity*; Routledge: London, UK, 2017; pp. 285–297.
12. Kilian, B.; Dempewolf, H.; Guarino, L.; Werner, P.; Coyne, C.; Warburton, M.L. Crop Science special issue: Adapting agriculture to climate change: A walk on the wild side. *Crop Sci.* **2021**, *61*, 32–36. [CrossRef]
13. Bertero, H.; De la Vega, A.; Correa, G.; Jacobsen, S.; Mujica, A. Genotype and genotype-by-environment interaction effects for grain yield and grain size of quinoa (*Chenopodium quinoa* Willd.) as revealed by pattern analysis of international multi-environment trials. *Field Crops Res.* **2004**, *89*, 299–318. [CrossRef]

14. Curti, R.N.; De la Vega, A.; Andrade, A.J.; Bramardi, S.J.; Bertero, H.D. Multi-environmental evaluation for grain yield and its physiological determinants of quinoa genotypes across Northwest Argentina. *Field Crops Res.* **2014**, *166*, 46–57. [CrossRef]
15. Bhargava, A.; Shukla, S.; Ohri, D. Genetic variability and interrelationship among various morphological and quality traits in quinoa (*Chenopodium quinoa* Willd.). *Field Crops Res.* **2007**, *101*, 104–116. [CrossRef]
16. Bazile, D.; Pulvento, C.; Verniau, A.; Al-Nusairi, M.S.; Ba, D.; Breidy, J.; Hassan, L.; Mohammed, M.I.; Mambetov, O.; Otambekova, M. Worldwide evaluations of quinoa: Preliminary results from post international year of quinoa FAO projects in nine countries. *Front. Plant Sci.* **2016**, *7*, 850. [CrossRef]
17. Choukr-Allah, R.; Rao, N.K.; Hirich, A.; Shahid, M.; Alshankiti, A.; Toderich, K.; Gill, S.; Butt, K.U.R. Quinoa for marginal environments: Toward future food and nutritional security in MENA and Central Asia regions. *Front. Plant Sci.* **2016**, *7*, 346. [CrossRef]
18. Rodriguez, J.P.; Ono, E.; Choukr-Allah, R.; Abdelaziz, H. Cultivation of quinoa (*Chenopodium quinoa*) in desert ecoregion. In *Emerging Research in Alternative Crops*; Springer: Gewerbestrasse, Switzerland; Cham, Switzerland, 2020; pp. 145–161.
19. Maliro, M.F.; Guwela, V.F.; Nyaika, J.; Murphy, K.M. Preliminary studies of the performance of quinoa (*Chenopodium quinoa* Willd.) genotypes under irrigated and rainfed conditions of central Malawi. *Front. Plant Sci.* **2017**, *8*, 227. [CrossRef]
20. Maliro, M.; Njala, A.L. Agronomic performance and strategies of promoting Quinoa (*Chenopodium quinoa* Willd) in Malawi. *Cienc. Investig. Agrar. Rev. Latinoam. Cienc. Agric.* **2019**, *46*, 82–99. [CrossRef]
21. Wilson, H.D. Quinua and relatives (Chenopodium sect. Chenopodium subsect. Celluloid). *Econ. Bot.* **1990**, *44*, 92–110. [CrossRef]
22. Christensen, S.; Pratt, D.B.; Pratt, C.; Nelson, P.; Stevens, M.; Jellen, E.N.; Coleman, C.E.; Fairbanks, D.J.; Bonifacio, A.; Maughan, P.J. Assessment of genetic diversity in the USDA and CIP-FAO international nursery collections of quinoa (*Chenopodium quinoa* Willd.) using microsatellite markers. *Plant Genet. Resour.* **2007**, *5*, 82–95. [CrossRef]
23. Mizuno, N.; Toyoshima, M.; Fujita, M.; Fukuda, S.; Kobayashi, Y.; Ueno, M.; Tanaka, K.; Tanaka, T.; Nishihara, E.; Mizukoshi, H. The genotype-dependent phenotypic landscape of quinoa in salt tolerance and key growth traits. *DNA Res.* **2020**, *27*, dsaa022. [CrossRef]
24. Bazile, D.; Jacobsen, S.-E.; Verniau, A. The global expansion of quinoa: Trends and limits. *Front. Plant Sci.* **2016**, *7*, 622. [CrossRef]
25. Ward, S.M. Response to selection for reduced grain saponin content in quinoa (*Chenopodium quinoa* Willd.). *Field Crops Res.* **2000**, *68*, 157–163. [CrossRef]
26. Jarvis, D.; Shwen, H.Y.; Lightfoot, D.J.; Schmöckel, S.M.; Bo, L.; Borm, T.J.; Hajime, O.; Katsuhiko, M.; Michell, C.T.; Noha, S.; et al. The genome of *Chenopodium quinoa*. *Nature* **2017**, *542*, 307–312. [CrossRef]
27. Ahmadi, S.H.; Solgi, S.; Sepaskhah, A.R. Quinoa: A super or pseudo-super crop? Evidences from evapotranspiration, root growth, crop coefficients, and water productivity in a hot and semi-arid area under three planting densities. *Agric. Water Manag.* **2019**, *225*, 105784. [CrossRef]
28. Ceccato, D.V.; Bertero, H.D.; Batlla, D. Environmental control of dormancy in quinoa (*Chenopodium quinoa*) seeds: Two potential genetic resources for pre-harvest sprouting tolerance. *Seed Sci. Res.* **2011**, *21*, 133–141. [CrossRef]
29. Ceccato, D.; Bertero, D.; Batlla, D.; Galati, B. Structural aspects of dormancy in quinoa (*Chenopodium quinoa*): Importance and possible action mechanisms of the seed coat. *Seed Sci. Res.* **2015**, *25*, 267–275. [CrossRef]
30. Castellión, M.; Matiacevich, S.; Buera, P.; Maldonado, S. Protein deterioration and longevity of quinoa seeds during long-term storage. *Food Chem.* **2010**, *121*, 952–958. [CrossRef]
31. Akram, M.Z.; Basra, S.M.A.; Hafeez, M.B.; Khan, S.; Nazeer, S.; Iqbal, S.; Saddiq, M.S.; Zahra, N. Adaptability and yield potential of new quinoa lines under agro-ecological conditions of Faisalabad-Pakistan. *Asian J. Agric. Biol.* **2021**, *2*, 202005301. [CrossRef]
32. Olivoto, T.; Lúcio, A.D.C. metan: An R package for multi-environment trial analysis. *Methods Ecol. Evol.* **2020**, *11*, 783–789. [CrossRef]
33. Ward, J., Jr. Hierarchical grouping to optimize an objective function. *J. Am. Stat. Assoc.* **1963**, *58*, 236–244. [CrossRef]
34. Charrad, M.; Ghazzali, N.; Boiteau, V.; Niknafs, A. NbClust: An R package for determining the relevant number of clusters in a data set. *J. Stat. Softw.* **2014**, *61*, 1–36. [CrossRef]
35. Lê, S.; Josse, J.; Mazet, F. Package 'FactoMineR'. *J. Stat. Softw.* **2008**, *25*, 1–18.
36. Afiah, S.A.; Hassan, W.A.; Al Kady, A. Assessment of six quinoa (*Chenopodium quinoa* Willd.) genotypes for seed yield and its attributes under Toshka conditions. *Zagazig J. Agric. Res.* **2018**, *45*, 2281–2294. [CrossRef]
37. Saddiq, M.S.; Wang, X.; Iqbal, S.; Hafeez, M.B.; Khan, S.; Raza, A.; Iqbal, J.; Maqbool, M.M.; Fiaz, S.; Qazi, M.A. Effect of Water Stress on Grain Yield and Physiological Characters of Quinoa Genotypes. *Agronomy* **2021**, *11*, 1934. [CrossRef]
38. Isobe, K.; Sugiyama, H.; Okuda, D.; Murase, Y.; Harada, H.; Miyamoto, M.; Koide, S.; Higo, M.; Torigoe, Y. Effects of sowing time on the seed yield of quinoa (*Chenopodium quinoa* Willd) in South Kanto, Japan. *Agric. Sci.* **2016**, *7*, 146.
39. Iqbal, S.; Basra, S.M.; Afzal, I.; Wahid, A.; Saddiq, M.S.; Hafeez, M.B.; Jacobsen, S.E. Yield potential and salt tolerance of quinoa on salt-degraded soils of Pakistan. *J. Agron. Crop Sci.* **2019**, *205*, 13–21. [CrossRef]
40. Hussain, M.I.; Muscolo, A.; Ahmed, M.; Asghar, M.A.; Al-Dakheel, A.J. Agro-Morphological, Yield and Quality Traits and Interrelationship with Yield Stability in Quinoa (*Chenopodium quinoa* Willd.) Genotypes under Saline Marginal Environment. *Plants* **2020**, *9*, 1763. [CrossRef]
41. De Santis, G.; Ronga, D.; Caradonia, F.; Ambrosio, T.D.; Troisi, J.; Rascio, A.; Fragasso, M.; Pecchioni, N.; Rinaldi, M. Evaluation of two groups of quinoa (*Chenopodium quinoa* Willd.) accessions with different seed colours for adaptation to the Mediterranean environment. *Crop Pasture Sci.* **2018**, *69*, 1264–1275. [CrossRef]

42. Colque-Little, C.; Abondano, M.C.; Lund, O.S.; Amby, D.B.; Piepho, H.-P.; Andreasen, C.; Schmöckel, S.; Schmid, K. Genetic variation for tolerance to the downy mildew pathogen Peronospora variabilis in genetic resources of quinoa (*Chenopodium quinoa*). *BMC Plant Biol.* **2021**, *21*, 41. [CrossRef]
43. De Santis, G.; D'Ambrosio, T.; Rinaldi, M.; Rascio, A. Heritabilities of morphological and quality traits and interrelationships with yield in quinoa (*Chenopodium quinoa* Willd.) genotypes in the Mediterranean environment. *J. Cereal Sci.* **2016**, *70*, 177–185. [CrossRef]
44. Jacobsen, S.E. The scope for adaptation of quinoa in Northern Latitudes of Europe. *J. Agron. Crop Sci.* **2017**, *203*, 603–613. [CrossRef]
45. Christiansen, J.; Jacobsen, S.-E.; Jørgensen, S. Photoperiodic effect on flowering and seed development in quinoa (*Chenopodium quinoa* Willd.). *Acta Agric. Scand. Sect. B-Soil Plant Sci.* **2010**, *60*, 539–544.
46. Thiam, E.; Allaoui, A.; Benlhabib, O. Quinoa Productivity and Stability Evaluation through Varietal and Environmental Interaction. *Plants* **2021**, *10*, 714. [CrossRef]
47. Curti, R.N.; De la Vega, A.; Andrade, A.J.; Bramardi, S.J.; Bertero, H.D. Adaptive responses of quinoa to diverse agro-ecological environments along an altitudinal gradient in North West Argentina. *Field Crops Res.* **2016**, *189*, 10–18. [CrossRef]
48. Bertero, H.; King, R.; Hall, A. Modelling photoperiod and temperature responses of flowering in quinoa (*Chenopodium quinoa* Willd.). *Field Crops Res.* **1999**, *63*, 19–34. [CrossRef]
49. EL-Harty, E.H.; Ghazy, A.; Alateeq, T.K.; Al-Faifi, S.A.; Khan, M.A.; Afzal, M.; Alghamdi, S.S.; Migdadi, H.M. Morphological and Molecular Characterization of Quinoa Genotypes. *Agriculture* **2021**, *11*, 286. [CrossRef]
50. Gómez, M.B.; Curti, R.N.; Bertero, H.D. Seed weight determination in quinoa (*Chenopodium quinoa* Willd.). *J. Agron. Crop Sci.* **2022**, *208*, 243–254. [CrossRef]
51. Gómez, M.B.; Castro, P.A.; Mignone, C.; Bertero, H.D. Can yield potential be increased by manipulation of reproductive partitioning in quinoa (*Chenopodium quinoa*)? Evidence from gibberellic acid synthesis inhibition using Paclobutrazol. *Funct. Plant Biol.* **2011**, *38*, 420–430. [CrossRef] [PubMed]
52. Rotundo, J.L.; Borrás, L.; De Bruin, J.; Pedersen, P. Physiological strategies for seed number determination in soybean: Biomass accumulation, partitioning and seed set efficiency. *Field Crops Res.* **2012**, *135*, 58–66. [CrossRef]
53. de la Vega, A.J.; Hall, A.J.; Kroonenberg, P.M. Investigating the physiological bases of predictable and unpredictable genotype by environment interactions using three-mode pattern analysis. *Field Crops Res.* **2002**, *78*, 165–183. [CrossRef]
54. De La Vega, A.J.; Chapman, S.C.; Hall, A.J. Genotype by environment interaction and indirect selection for yield in sunflower: I. Two-mode pattern analysis of oil and biomass yield across environments in Argentina. *Field Crops Res.* **2001**, *72*, 17–38. [CrossRef]
55. Chapman, S.C.; Crossa, J.; Edmeades, G.O. Genotype by environment effects and selection for drought tolerance in tropical maize. I. Two mode pattern analysis of yield. *Euphytica* **1997**, *95*, 1–9. [CrossRef]
56. Risi, J.; Galwey, N. The pattern of genetic diversity in the Andean grain crop quinoa (*Chenopodium quinoa* Willd). I. Associations between characteristics. *Euphytica* **1989**, *41*, 147–162. [CrossRef]
57. Nico, M.; Miralles, D.J.; Kantolic, A.G. Natural post-flowering photoperiod and photoperiod sensitivity: Roles in yield-determining processes in soybean. *Field Crops Res.* **2019**, *231*, 141–152. [CrossRef]
58. Naneli, I.; Tanrikulu, A.; Dokuyucu, T. Response of the quinoa genotypes to different locations by grain yield and yield components. *Int. J. Agric. Innov. Res.* **2017**, *6*, 446–451.
59. Khaitov, B.; Karimov, A.A.; Toderich, K.; Sultanova, Z.; Mamadrahimov, A.; Allanov, K.; Islamov, S. Adaptation, grain yield and nutritional characteristics of quinoa (Chenopodium quinoa) genotypes in marginal environments of the Aral Sea basin. *J. Plant Nutr.* **2020**, *44*, 1365–1379. [CrossRef]

plants

MDPI

Article

Enriching Urea with Nitrogen Inhibitors Improves Growth, N Uptake and Seed Yield in Quinoa (*Chenopodium quinoa* Willd) Affecting Photochemical Efficiency and Nitrate Reductase Activity

Hafeez ur Rehman [1,*], Hesham F. Alharby [2], Hassan S. Al-Zahrani [2], Atif A. Bamagoos [2], Nadiah B. Alsulami [2], Nadiyah M. Alabdallah [3], Tahir Iqbal [1,4] and Abdul Wakeel [5]

1 Department of Agronomy, University of Agriculture, Faisalabad 38040, Pakistan; tahiriqbaldgk29@gmail.com
2 Department of Biological Sciences, Faculty of Science, King Abdulaziz University, Jeddah 21589, Saudi Arabia; halharby@kau.edu.sa (H.F.A.); hsalzahrani@kau.edu.sa (H.S.A.-Z.); abamagoos@kau.edu.sa (A.A.B.); nbalsulami@kau.edu.sa (N.B.A.)
3 Department of Biology, College of Science, Imam Abdulrahman Bin Fasial University, P.O. Box 1982, Dammam 31441, Saudi Arabia; nmalabdallah@iau.edu.sa
4 Department of Botany, University of Agriculture, Faisalabad 38040, Pakistan
5 Institute of Soil and Environmental Sciences, University of Agriculture, Faisalabad 38040, Pakistan; abdulwakeel77@gmail.com
* Correspondence: hafeezcp@gmail.com or h.rehman@uaf.edu.pk

Citation: Rehman, H.u.; Alharby, H.F.; Al-Zahrani, H.S.; Bamagoos, A.A.; Alsulami, N.B.; Alabdallah, N.M.; Iqbal, T.; Wakeel, A. Enriching Urea with Nitrogen Inhibitors Improves Growth, N Uptake and Seed Yield in Quinoa (*Chenopodium quinoa* Willd) Affecting Photochemical Efficiency and Nitrate Reductase Activity. *Plants* **2022**, *11*, 371. https://doi.org/10.3390/plants11030371

Academic Editors: Cataldo Pulvento and Didier Bazile

Received: 27 December 2021
Accepted: 20 January 2022
Published: 29 January 2022

Abstract: Quinoa is a climate resilience potential crop for food security due to high nutritive value. However, crop variable response to nitrogen (N) use efficiency may lead to affect grain quality and yield. This study compared the performance of contrasting quinoa genotypes (UAF Q-7, EMS-line and JQH1) to fertilizer urea enriched with urease and nitrification inhibitors (NIs; 1% (*w*/*w*) thiourea + boric acid + sodium thiosulphate), ordinary urea and with no N as control. Application of NIs-enriched urea improved plant growth, N uptake and chlorophyll values in quinoa genotype UAF-Q7 and JHQ1, however, highest nitrate reductase (NR) activity was observed in EMS-line. Quinoa plants supplied with NIs-enriched urea also completed true and multiple leaf stage, bud formation, flowering, and maturity stages earlier than ordinary urea and control, nevertheless, all quinoa genotypes reached true and multiple leaf stage, flowering and maturity stages at same time. Among photosynthetic efficiency traits, application of NIs-enriched urea expressed highest photosynthetic active radiations (PAR), electron transport rate (ETR), current fluorescence (Ft) and reduced quantum yield (Y) in EMS line. Nitrogen treatments had no significant difference for panicle length, however, among genotypes, UAF-Q7 showed highest length of panicle followed by others. Among yield attributes, NIs-enriched urea expressed maximum 1000-seed weight and seed yield per plant in JQH-1 hybrid and EMS-line. Likely, an increase in quinoa grain protein contents was observed in JQH-1 hybrid for NIs-enriched urea. In conclusion, NIs-enriched urea with urease and nitrification inhibitors simultaneously can be used to improve the N uptake, seed yield and grain protein contents in quinoa, however, better crop response was attributed to enhanced plant growth and photosynthetic efficiency.

Keywords: nitrogen use efficiency; inhibitors; chlorophyll fluorescence; grain protein

1. Introduction

Quinoa has exceptional nutritional grain value containing high protein contents and balanced amino acids while its enduring potential for abiotic stress tolerance makes it future potential crop both for nutritional and food security [1–3]. Since last decade, quinoa cultivation has spread into non-native geographical areas of world due to wide diversity of its ecotypes especially photoperiod response adaptation to specific agro-climatic conditions [1,2]. Nonetheless, quinoa growth and development are affected by environmental and genetic variations. For example, among environmental factors, nutrients especially nitrogen (N) improve vegetative growth in quinoa by affecting crop leaf area and growth rate, photosynthesis and N metabolism enzymes thereby increasing grain weight and yield [4–6]. These growth and yield responses in quinoa are variable to N supply and genotypic specific. The nitrogen use efficiency (NUE) is tightly linked to N uptake and its utilization efficiency (NUtE) that varies among genotypes, application method and time including rate in quinoa [6–10]. For instance, high total plant N uptake, its apparent recovery efficiency and harvest index including biomass and seed N contents was observed for 200 kg N ha^{-1}, however, N utilization efficiency (NUtE) and remobilization from vegetative tissues to seed was found low in quinoa [9]. Improved growth and seed yield has also been reported in response to N application [10]. Likely, delayed flowering, extended seed filling period and improved photosynthetic pigments including seed yield was observed for 150 kg N ha^{-1} applied in two splits at 6–8 leaves and anthesis stages, respectively, compared to three splits and control with no N [11]. Nonetheless, variable response to N fertilization for growth including relative and crop growth rates, and seed yield was observed at low N rate in two quinoa cultivars [7]. Bascuñán-Godoy et al. [6] compared three quinoa cultivars similar in phenology for NUE traits under low and high N conditions. Photosynthetic rate, protein contents and leaf dry mass correlated positively with seed yield while proline contents, NH_4^+ assimilation and glutamine synthetase activity were correlated negatively under both N regimes. Nonetheless, high yields were correlated positively with seed weight under low N condition. Total N uptake in quinoa also vary with duration of crop growth cycle. Quinoa genotypes with shorter (NL-6) and longer growth duration (2-Want) accumulated similar total N before anthesis while differed after anthesis. Genotype 2-Want of longer growth duration accumulated 250 kg N ha^{-1} compared to NL-6 genotype with shorter growth behavior had reduced total uptake of 164 kg N ha^{-1} [8,12]. This variable response to N uptake after anthesis was associated with its remobilization towards reproductive structures [8].

Quinoa take up N as nitrates and application of N fertilizers had been found to improve its yield, water and NUE [13,14]. Among different sources, urea is mostly commonly used N source, most of which volatilizes as NH_3, lost as NO_2 into atmosphere and leaches as NO_3 into the soil with low NUE which rarely exceeds 33% [15].

Among several strategies, application of urea enriched with urease and nitrification inhibitors can reduce N losses and improve crop productivity, hence NUE. Urease inhibitors reduce urea hydrolysis into NH_3 by slowing down the urease enzyme activity [16,17]. This inhibited activity of urease decreases pH around urea molecules and NH_3 concentration in soil, thus reducing its volatilization and increases the retention of applied N in soil to improve its plant availability for longer period [18]. On other hand, enhanced retention of NH_4 ions into soil for longer time induces nitrification and dentification losses associated with NO_3 leaching and N_2O emission [19]. These leaching and denitrification losses can be minimized by nitrification inhibitors which limit oxidation of NH_4^+ to NO_3^- by reducing the activity of nitrifying bacteria [20,21], however, retention of NH_4^+ by application of these inhibitors may further increase risk of volatilization [22,23]. Likely, boric acid has the potential to inhibit urease activity in soil [24]. While ammonium thiosulphate (ATS) as a good source of N and sulphur (S) for plants has the potential to inhibit both hydrolysis and nitrification without harming the soil microbial pool [25]. Therefore, rather than individual application, effective approach is to utilize these inhibitors in combination. The combined application of NIs reduces multiple losses associated with volatilization and denitrification [26,27] had been found to improve yield 5.7 and 8.0% of N uptake in rice [28], 22–36% increase in biomass and 23–32% of N uptake in pasture [29]. The combined application of different inhibitors, for instance, boric acid and 3, 4-Dimethylpyrazole phosphate (DMPP), urease and nitrification have also the potential to inhibit the N transformation synergistically [24], increase yield by 7.5% and NUE by 12.9% especially at low N dose [30,31], respectively. Nonetheless, these increases in N uptake and yield were associated with reduced losses of N and environmental footprints [24,28–31].

As uptake of N by quinoa varies with crop stage, response may be genotype specific and N application level, and no information is available for application of urea enriched with urease and nitrification inhibitors on growth, plant N uptake, photosynthetic efficiency, and yield response. Genotypes used in present study have contrasting behavior for growth including a cultivated variety, a hybrid and EMS line. Present study hypothesized that urea enriched with N inhibitors (NIs; boric acid, thiourea and sodium thiosulphate) improves its availability to quinoa plant at reduced N dose, thereby affecting crop performance and NUE.

2. Results

2.1. Plant Growth and Photosynthetic Pigments

Urea enriched with nitrogen inhibitors (35 kg N ha^{-1} + NI's) improved the growth, chlorophyll values and plant N uptake in quinoa genotypes compared to control (0 kg N ha^{-1}) and ordinary urea (70 kg N ha^{-1}) (Table 1). There was an increase in shoot, root fresh and their dry weights, shoot and root lengths of quinoa genotypes UAF-Q7 and JHQ1 compared to control. Nonetheless, these increases in shoot dry weight, root fresh and dry weights were significantly similar to ordinary urea. Maximum and similar SPAD-chlorophyll values was found for enriched and ordinary urea application compared to control, while among genotypes, highest and significantly similar chlorophyll values were found between EMS-line and UAF-Q7 (Table 1).

2.2. Plant N Uptake and Nitrate Reductase Activity (NR)

Application of NIs-enriched urea significantly improved the plant N uptake and NR activity in quinoa genotypes compared to control and vice versa response was observed for nitrogen utilization efficiency (NUtE). Urea enriched with NI's showed highest plant N uptake in quinoa genotypes UAF-Q7 and EMS-line compared to control with minimum uptake. NIs-enriched urea application also expressed highest nitrate reductase (NR) activity in quinoa plants as compared to control and ordinary urea showing similar NR activity (Figure 1b). Among genotypes, EMS-line exhibited highest NR activity that was significantly similar to JQH-1 hybrid (Figure 1a).

2.3. Crop Phenology

Quinoa plants applied with NIs-enriched urea completed true and multiple leaf stage, bud formation, flowering, and maturity stages earlier than ordinary urea and control treatments. However, quinoa plants applied with enriched and ordinary urea attained panicle emergence, flowering and maturity at similar time compared to control with delayed in these attributes. Among genotypes, UAF-Q7 observed delayed bud formation and panicle emergence compared to other genotypes while all quinoa genotypes exhibited true and multiple leaf stage, flowering, and maturity at same time (Table 2).

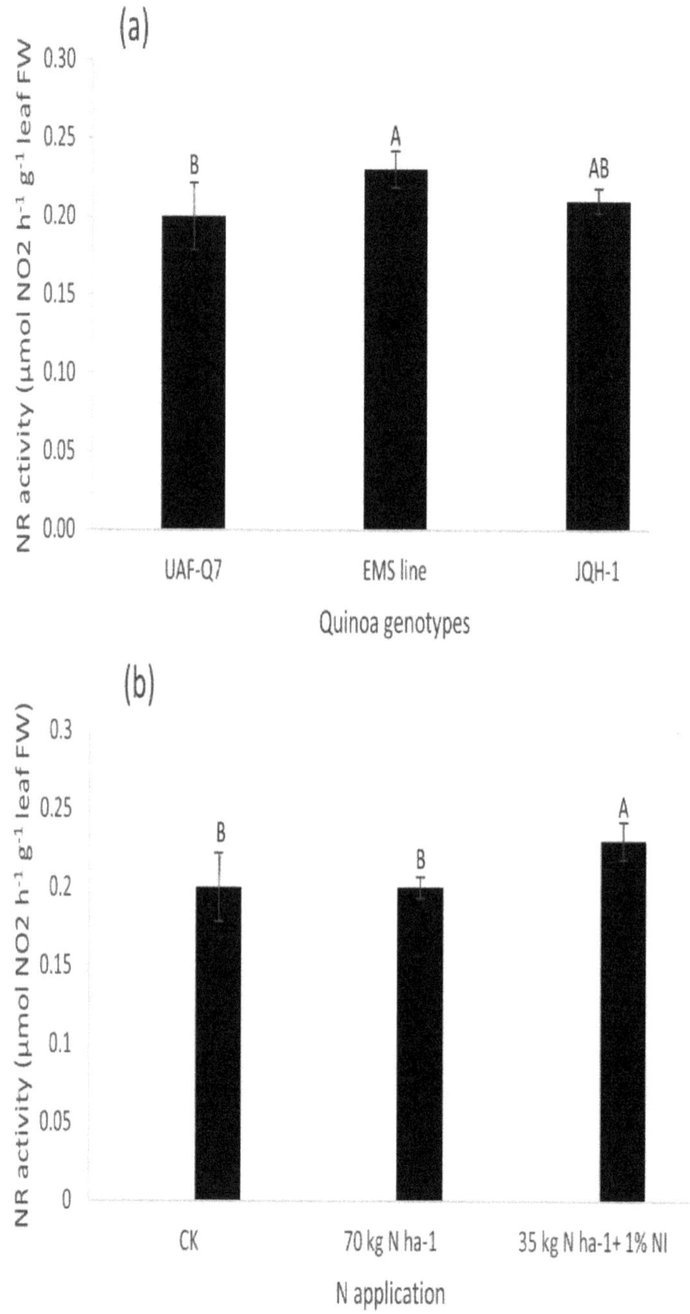

Figure 1. Nitrate reductase activities in three quinoa genotypes (**a**) and three fertilizer treatments (**b**) at bud formation stage Columns show mean of three replicates, whereas bar shows standard error. Means sharing same alphabets are not significantly different at $p \leq 0.5$.

Table 1. NIs-enriched urea effects on growth, plant N uptake, nitrogen utilization efficiency (NUtE) and SPAD-chlorophyll in three quinoa genotypes at bud formation stage.

Quinoa Genotypes	Shoot Fresh Weight (g)				Root Fresh Weight (g)			
	CK (0 kg N ha^{-1})	70 kg N ha^{-1}	35 kg N ha^{-1} + 1% NI	Means Genotypes	CK (0 kg N ha^{-1})	70 kg N ha^{-1}	35 kg N ha^{-1} + 1% NI	Means Genotypes
UAF-Q7	6.10 d	12.35 b	18.10 a	12.18 A	0.30 bc	0.40 ab	0.53 a	0.41 A
EMS line	9.27 c	4.73 d	11.50 bc	8.50 B	0.33 bc	0.10 d	0.40 ab	0.28 B
JQH-1	4.20 d	12.77 b	16.20 a	11.06 A	0.17 cd	0.33 bc	0.33 bc	0.28 B
Means N	6.52 A	9.95 B	15.27 A		0.27 B	0.28 B	0.42 A	
HSD	G = 1.36, N = 1.36, G × N = 2.36				G = 0.10, N = 0.10, G × N = 0.18			

Quinoa Genotypes	Shoot Dry Weight (g)				Root Dry Weight (g)			
	CK (0 kg N ha^{-1})	70 kg N ha^{-1}	35 kg N ha^{-1} + 1% NI	Means Genotypes	CK (0 kg N ha^{-1})	70 kg N ha^{-1}	35 kg N ha^{-1} + 1%NI	Means Genotypes
UAF-Q7	0.82 cde	1.17 abc	1.48 a	1.15	0.06 b	0.08 ab	0.11 a	0.08 A
EMS line	0.10 bcd	0.49 e	1.24 ab	0.90	0.06 b	0.02 c	0.07 b	0.05 B
JQH-1	0.60 de	1.33 ab	1.51 a	1.15	0.02 b	0.06 b	0.08 ab	0.06 B
Means	0.80 B	0.10 B	1.41 A		0.05 B	0.05 B	0.09 A	
HSD	G = n.s., N = 0.24, G × N = 0.42				G = 0.02, N = 0.02, G × N = 0.03			

Quinoa Genotypes	Shoot Length (cm)				Root Length (cm)			
	CK (0 kg N ha^{-1})	70 kg N ha^{-1}	35 kg N ha^{-1} + 1% NI	Means Genotypes	CK (0 kg N ha^{-1})	70 kg N ha^{-1}	35 kg N ha^{-1} + 1% NI	Means Genotypes
UAF-Q7	16.00 e	23.80 b	24.15 b	21.31 B	5.50 e	6.80 de	11.50 a	7.93
EMS line	21.77 bc	6.67 de	23.50 b	20.64 B	8.30 c	7.05 cd	7.40 cd	7.58
JQH-1	19.60 cd	28.00 a	29.00 a	25.53 A	9.80 b	7.85 cd	8.10 cd	8.58
Means	19.12 c	22.8 B	25.55 A		7.88 B	7.23 B	9.00 A	
HSD	G = 1.81, N = 1.80, G × N = 3.12				G = n.s., N = 0.86, G ×N = 1.49			

Quinoa Genotypes	SPAD-Chlorophyll Values				Plant N Uptake (mg N Plant^{-1})			
	CK (0 kg N ha^{-1})	70 kg N ha^{-1}	35 kg N ha^{-1} + 1% NI	Means Genotypes	CK (0 kg N ha^{-1})	70 kg N ha^{-1}	35 kg N ha^{-1} +1%NI	Means Genotypes
UAF-Q7	40.60	44.40	45.10	43.37 AB	1.71 d	3.31 c	6.84 a	3.95
EMS line	46.00	44.55	45.73	45.43 A	3.00 c	1.67 d	4.99 b	3.22
JQH-1	35.90	44.03	43.47	41.13 B	1.12 d	3.58 c	5.92 ab	3.53
Means	40.83 B	44.33 A	44.77 A		1.94 C	2.82 B	5.92 A	
HSD	G = 2.96, N = 2.96, G × N = n.s.				G = n.s., N = 0.66, G × N = 1.15			

Quinoa Genotypes	NUtE (g DW per mg N)			
	CK (0 kg N ha^{-1})	70 kg N ha^{-1}	35 kg N ha^{-1} + 1% NI	Means Genotypes
UAF-Q7	0.51 a	0.38 bc	0.26 e	0.42 A
EMS line	0.35 c	0.31 d	0.26 e	0.36 B
JQH-1	0.39 b	0.39 bc	0.27 de	0.26 C
Means	0.39 A	0.31 C	0.35 B	
HSD	G = 0.02, N = 0.02, G × N = 0.04			

Letters among and within columns denote significant differences in means for nitrogen and between cultivars at $p \leq 0.05$.

2.4. Photochemical Efficiency Traits and SPAD-Chlorophyll Values

The photochemical efficiency traits and chlorophyll were affected significantly by application of NIs-enriched urea in quinoa genotypes. Quinoa plants applied with NIs-enriched urea showed highest photosynthetic active radiations (PAR), electron transport rate (ETR), current fluorescence (Ft) and reduced quantum yield (Y) in EMS line followed by two other genotypes. Lowes values for these attributes were observed in ordi nary urea compared to control. The Ft values and SPAD-chlorophyll values were similar when ordinary and NIs-enriched urea applied in UAF-Q7 and JQH-1 genotypes, respectively (Table 3).

Table 2. Influence of NIs-enriched urea on phenological development in three quinoa genotypes.

Quinoa Genotypes	Days to True Leaf				Days to Multiple Leaf			
	CK (0 kg N ha^{-1})	70 kg N ha^{-1}	35 kg N ha^{-1} + 1% NI	Means Genotypes	CK (0 kg N ha^{-1})	70 kg N ha^{-1}	35 kg N ha^{-1} + 1% NI	Means Genotypes
UAF-Q7	19.67 a	13.33 c	14.00 c	15.67	28.33 ab	26.33 cde	25.33e	26.67
EMS line	18.33 ab	19.00 a	15.67 bc	17.67	29.33 a	28.00 abc	26.00 de	27.78
JQH-1	18.33 ab	18.00 ab	13.33 c	16.56	26.67 bcde	27.00 bcde	27.67 abcd	27.11
Means N	18.78 A	16.78 B	14.33 C		28.11 A	27.11 AB	26.33 C	
HSD	G = n.s., N = 1.59, G × N = 2.75				G = n.s., N = 1.08, G × N = 1.87			
Quinoa Genotypes	**Days to Bud Formation**				**Days to Panicle Emergence**			
	CK (0 kg N ha^{-1})	70 kg N ha^{-1}	35 kg N ha^{-1} + 1% NI	Means Genotypes	CK (0 kg N ha^{-1})	70 kg N ha^{-1}	35 kg N ha^{-1} + 1%NI	Means Genotypes
UAF-Q7	40.33	40.00	40.33	40.22 A	59.67	59.00	56.67	58.44 A
EMS line	40.00	39.67	36.67	38.78 B	55.67	57.33	55.33	55.44 B
JQH-1	37.33	38.67	36.33	37.44 C	59.00	54.67	53.33	55.66 B
Means	39.22 A	39.44 A	37.78 B		58.11 A	55.44 B	55.67 B	
HSD	G = 1.15, N = 1.15, G × N = n.s.				G = 1.82, N = 1.82, G ×N = n.s.			
Quinoa Genotypes	**Days to Flowering**				**Days to Maturity**			
	CK (0 kg N ha^{-1})	70 kg N ha^{-1}	35 kg N ha^{-1} + 1% NI	Means Genotypes	CK (0 kg N ha^{-1})	70 kg N ha^{-1}	35 kg N ha^{-1} + 1% NI	Means Genotypes
UAF-Q7	79.33	77.67	77.00	78.00	127.67	125.33	121.67	124.89
EMS line	79.33	78.00	76.00	77.78	128.00	124.00	122.67	124.89
JQH-1	78.00	77.00	75.33	76.78	125.33	126.00	122.67	124.67
Means	78.88 A	77.56 AB	76.11 B		127.00 A	125.11 AB	122.33 B	
HSD	G = n.s., N = 1.97, G × N = n.s.				G = n.s., N = 3.55, G × N = n.s.			

Letters among and within columns denote significant differences in means for nitrogen and between cultivars at $p \leq 0.05$.

2.5. Seed Yield and Its Attributes

Application of NIs-enriched urea produced tallest plants in genotype UAF-Q7 that was significantly similar to ordinary urea in same genotype. Minimum plant height was found in control plants with no supplemental N. There was no difference observed for panicle length in N treatments compared to control, however, among genotypes, UAF-Q7 expressed maximum panicle length followed by JQH1 hybrid.

Application of NIs-enriched urea showed maximum 1000-seed weight in quinoa genotypes UAF-Q7 and EMS-line. However, this increase was similar to ordinary urea in JQH-1 and EMS-line. Likely, highest seed yield per plant was found for NIs-enriched urea compared to ordinary urea while minimum seed yield per plant in control plants without supplemental N. Among genotypes, highest seed yield was expressed by JQH-1 hybrid followed by EMS-line while minimum in UAF-Q7 genotype.

2.6. Seed Protein Contents

Application of NIs-enriched urea showed highest seed protein contents in harvested grains of JHQ-1 hybrid that was similar to ordinary urea in same genotype. However, minimum seed protein contents were found in UAF-Q7 genotype in control treatment (Table 4).

Table 3. Effects of NI-enriched urea on photochemical efficiency and SPAD-chlorophyll values in three quinoa genotypes at panicle emergence stage.

Quinoa Genotypes	SPAD-Chlorophyll				Photosynthetic Active Radiation (PAR)			
	CK (0 kg N ha^{-1})	70 kg N ha^{-1}	35 kg N ha^{-1} + 1% NI	Means Genotypes	CK (0 kg N ha^{-1})	70 kg N ha^{-1}	35 kg N ha^{-1} + 1% NI	Means Genotypes
UAF-Q7	43.77 bc	46.93 ab	48.07 a	46.26 A	887.00 d	584.70 ef	1268.70 b	913.44 A
EMS line	42.33 c	34.90 d	48.30 a	41.84 B	486.00 f	584.70 ef	1445.70 a	838.78 B
JQH-1	35.65 d	47.97 a	47.80 ab	43.81 B	657.00 e	649.70 e	1074.30 c	793.67 B
Means N	40.58 C	43.27 B	48.06 A		676.70 B	606.30 C	1262.90 A	
HSD	G = 2.42, N = 2.42, G × N = 4.19				G = 67.743, N = 67.743, G × N = 117.33			

Quinoa Genotypes	Electron Transport Rate (ETR)				Current Fluorescence Value (Ft)			
	CK (0 kg N ha^{-1})	70 kg N ha^{-1}	35 kg N ha^{-1} + 1% NI	Means Genotypes	CK (0 kg N ha^{-1})	70 kg N ha^{-1}	35 kg N ha^{-1} + 1%NI	Means Genotypes
UAF-Q7	214.07 b	169.07 bc	176.97 bc	186.70 AB	471.67 bc	607.33 a	450.00 bc	509.67
EMS line	164.47 bc	139.83	294.43 a	199.58 A	422.00 c	423.33 c	635.00 a	493.44
JQH-1	126.87 c	148.23 c	184.13 bc	153.08 B	467.00 bc	386.00 c	556.33 ab	469.78
Means	168.47 B	152.38 B	218.51 A		453.56 B	472.22 B	547.11 A	
HSD	G = 36.703, N = 36.703, G × N = 63.57				G = n.s., N = 71.709, G × N = 124.20			

Quinoa Genotypes	Quantum Yield (Y)			
	CK (0 kg N ha^{-1})	70 kg N ha^{-1}	35 kg N ha^{-1} + 1% NI	Means Genotypes
UAF-Q7	0.62 a	0.57 ab	0.42 c	0.54
EMS line	0.57 ab	0.63 a	0.50 bc	0.57
JQH-1	0.47 c	0.60 a	0.44 c	0.51
Means	0.55 A	0.6 A	0.46 B	
HSD	G = n.s., N = 0.0470, G × N = 0.0814			

Letters among and within columns denote significant differences for nitrogen and between cultivars at $p \leq 0.05$.

Table 4. Effects of NI-enriched urea on yield components and seed protein contents in three quinoa genotypes at maturity.

Quinoa Genotypes	Plant Height (cm)				Panicle Length (cm)			
	CK (0 kg N ha^{-1})	70 kg N ha^{-1}	35 kg N ha^{-1} + 1% NI	Means Genotypes	CK (0 kg N ha^{-1})	70 kg N ha^{-1}	35 kg N ha^{-1} + 1% NI	Means Genotypes
UAF-Q7	70 b	81 a	88 a	80 A	27	23	26	25 A
EMS line	41 c	49 c	43 c	44 B	15	15	13	14 C
JQH-1	51 c	49 c	49 c	50 B	18	19	20	19 B
Means N	54	60	60		20	19	19	
HSD	G = 5.61, N = n.s., G × N = 9.72				G = 2.59, N = n.s., G × N = n.s.			

Table 4. *Cont.*

Quinoa Genotypes	1000 Seed Yield (g)				Seed Yield Per Plant (g)			
	CK (0 kg N ha^{-1})	70 kg N ha^{-1}	35 kg N ha^{-1} + 1% NI	Means Genotypes	CK (0 kg N ha^{-1})	70 kg N ha^{-1}	35 kg N ha^{-1} + 1%NI	Means Genotypes
UAF-Q7	0.28 d	0.44 bc	0.53 a	0.42 B	1.16	2.32	2.92	2.13 C
EMS line	0.42 c	0.47 abc	0.54 a	0.48 A	2.46	3.17	3.93	3.19 B
JQH-1	0.48 abc	0.52 ab	0.52 ab	0.51 A	3.40	3.54	5.33	4.09 A
Means	0.39 C	0.48 B	0.53 A		2.34 C	3.01 B	4.06 A	
HSD	G = 0.05, N = 0.05, G × N = 0.08				G = 0.60, N = 0.59, G × N = n.s.			

Quinoa Genotypes	Seed Protein Contents (%)			
	CK (0 kg N ha^{-1})	70 kg N ha^{-1}	35 kg N ha^{-1} + 1% NI	Means Genotypes
UAF-Q7	10.16 d	13.22 c	15.66 b	13.01 B
EMS line	14.66 bc	14.83 bc	15.66 bc	15.05 A
JQH-1	14.32 bc	16.44 ab	18.8 a	16.53 A
Means	13.05 A	14.83 B	16.7 A	
HSD	G = 1.5758, N = 1.5758, G × N = 2.72			

Letters among and within columns denote significant differences in means for nitrogen and between cultivars at $p \leq 0.05$.

3. Discussion

Urea fertilizers are often used as nitrogen (N) source worldwide due to high N contents. However, it is rapidly hydrolyzed to ammonia (NH_3) and carbon dioxide (CO_2) in soils. Nonetheless, application of urea enriched with urease and nitrification inhibitors are well known to synchronize N supply with crop demand to increase N use efficiency [32,33]. The present study evaluated the potential of urea enriched with N inhibitors (NI's; 1% boric acid, thiourea and sodium thiosulphate + 35 kg N ha^{-1}) to improve crop growth, photochemical efficiency, N uptake, seed yield and protein contents in contrasting quinoa genotypes compared to ordinary urea (70 kg N ha^{-1}) and no N as control (0 kg N ha^{-1}). The NIs enriched urea improved the growth, chlorophyll values and N uptake in quinoa genotypes [Table 1] are associated with its increased soil availability and is also an important component of chlorophyll structure [34]. Urea enriched with N-(n-butyl) thiophosphoric triamide (NBPT) inhibitor had been reported to improve photosynthetic pigments, plant growth and seed yield in cotton owing to increased N uptake. Likely, positive relationship of leaf N with chlorophyll contents has also been reported [34,35]. Increased N uptake and dry matter with application of urea enriched with NBPT in cotton [36] also reflected in plant fresh and dry biomass of quinoa when NIs-enriched urea was applied in this study [Table 1]. Increased NR activity in quinoa of present study [Table 1] was associated with beneficial effects of thiourea in enhancing N metabolism [36]. Thiourea application is reported to improve chlorophyll contents, photosynthetic activity, starch, and soluble protein levels in plants [37] confirms diverse functions of SH group in thiourea molecule. However, variable response of quinoa genotypes to NIs-enriched urea for total N uptake seems effect of NI's to inhibit urease and nitrification [38]. Curti et al. [8] and Gomez et al. [12] reported that quinoa genotypes had no difference in total N content before anthesis independent of their growth duration which is also evident in present study where JHQ-1 hybrid showed low N uptake [Table 1] because of dilution effect associated with higher crop growth rate of quinoa hybrid not matching with N uptake [8,12]. Usually, application of N delays the crop phenological development, however, application of NIs-enriched urea helped quinoa plants to complete different stages of crop development earlier than ordinary urea and control treatments [Table 2]. There are no evidence showing the effects of NIs-enriched urea with NI's on crop maturity, nevertheless, delay in development stages for N applied in splits had been reported in quinoa validated in present study findings [11]. Plant height and panicle length are genetic and stable characters, however, strongly affected by environmental factors including N. No significant difference observed between these traits for enriched and ordinary urea could be attributed to similar gains in photosynthetic efficiency of these genotypes [Table 4]. Nonetheless, increase in 1000-seed weight and seed yield of quinoa plant could be attributed to total N uptake which consistently contributed to enhance photosynthetic carbon fixation by affecting sink capacity [36,39]. The combined use of urease and nitrification inhibitors in NIs-enriched urea increased plant N uptake improving its efficiency associated with reduced losses, thereby increased yield and seed proteins contents [40]. The increase in seed protein contents in quinoa of present study by application of NIs-enriched urea [Table 4] were associated with slow-release effect to enhance retention and bioavailability of NH_4-N for direct uptake and its remobilization to affect protein contents in grains [41,42]. Increase in protein contents with application of NH_4-N [43] and inhibitors was associated with improved plant N uptake and its remobilization had also been reported in wheat and some other crops [44–46]. Gupta et al. [47] suggested that plants remobilize and constitute mechanism of re-uptake of resources such as NH_4^+ under N limited condition to ameliorate the increased NH_4^+ levels derived from different physiological process after anthesis. Nonetheless, present study results are in consistent with Bascuñán-Godoy et al. [48] where positive relationship between seed N% and seed amino acid was reported under low N supply and support the hypothesis to remobilize resources under limited vs. sufficient conditions [49].

4. Materials and Methods

4.1. Experimental Details

The seed of three quinoa genotypes, EMS-line and were collected from alternative crops lab, Department of Agronomy, University of Agriculture Faisalabad, Pakistan. These three genotypes varied in growth behavior and include a cultivated variety (UAF Q-7), a hybrid (JQH-1) and EMS line. The experiment was comprised of two factors as quinoa genotypes and fertilizer treatments including no N (0 kg N ha^{-1}) (control), 75 kg N ha^{-1} (recommended) [50] and 35 kg N ha^{-1} enriched with 1% NI (Thiourea + Boric acid + Sodium thiosulphate). Experimental design used was completely randomized with two factors factorial. The tenth seed of each genotype were sown in earthen pots each filled with 5 kg soil at field capacity level under the wire house condition with exposure to natural growing condition. After seedling establishment, three plants per experimental unit were maintained for further growth and assays. The experimental treatments were randomized completely with factorial arrangement in three replications. Other fertilizers including phosphorus and potassium were thoroughly mixed in soil before sowing using 50 kg ha^{-1} of single super phosphate and 120 kg ha^{-1} of sulphate of potash, respectively. While half of nitrogen (N) was applied at sowing and other half at flowering stage. The pots were irrigated 10 and 15 days after sowing (DAS) for optimum growth of quinoa crop and later when required.

4.2. Determination of Plant Growth, Nitrogen Uptake, Utilization Efficiency and Seed Protein Contents

After 30–40 days of sowing at bud formation stage, only one plant was randomly uprooted for measuring plant growth traits including shoot and root lengths with measuring scale. The quinoa plant shoots fresh and root fresh weights was determined and oven-dried at 98 °C for 48 h for dry weights. The SPAD chlorophyll value of upper most leaf of quinoa plants was measured at the growth stage of BBCH 18 with the help of SPAD-501, Minolta Japan. For the determination of seedling N and grain protein, 0.5 g of plant and grain sample separately was taken into the Pyrex tube and added with 10 mL of concentrated H_2SO_4 for digestion and allowed to stand for overnight. After which, distillation was performed by Kjeldahl apparatus and then titrated against 0.1% H_2SO_4 [51]. Flasks were placed on the hot plate and heated at the temperature ranging between 100–150 °C for 30 min, then 2 mL of 30% H_2O_2 was added upon cooling, and heated again at increasing temperature of up to 300 °C. The processes were repeated till the solution became transparent and the final volume kept 50 mLby addition of distilled water. Distillation was performed by extracting 10 mLof the digested sample and 10 mLof 40% NaOH solution was added in the tube. The 5 mLboric acid solution (4%) was added to the receiver flask with 2–3 drops of mixed indicator and 40 mLof total solution was obtained by distillation and titrated against 0.01 N H_2SO_4 until the original color of methyl red appeared and the values were noted. For protein contents, nitrogen values were multiplied with a factor 5.95. Plant N uptake (mg N plant^{-1}) and its utilization efficiency (NUtE; g DW per mg N) were measured as suggested by Merigout et al. [52] and Wang et al. [53], respectively.

4.3. Measurement of Nitrate Reductase Activity

At bud formation stage, leaf samples harvested at bud formation stage were shifted to icebox and stored at −30 °C until use [54]. For enzyme assay, leaf harvest (0.25 g) after grinding was extracted with 1 mL digestion buffer (1 mM EDTA + 10 mM cysteine) using ice-cold pestle mortar. The leaf extract was filtered through cheese cloth and homogenates was transferred to 2 mL falcon tubes, centrifuged at 10,000 rpm for 10 min to collect supernatant. Enzyme extract (1 mL) was added with 0.25 mL phosphate buffer (pH 7.5), 0.1 mL KNO_3, 2 mM NADH and 0.35 mL distilled water to initiate reaction. After incubation at 30 °C for 15 min, reaction was terminated by the addition of 1% sulphanilamide and 0.02% naphthyl ethylenediamine reagent of 0.5 mL of each and kept it for 30 min to settle down. The absorbance of extract was measured on 540 nm using spectrophotometer (UV 4000, ORI Germany). The NR activity was calculated following Kaiser and Lewis [55].

4.4. Determination of Phenological Traits

All the crop growth stage of quinoa were recorded at regular intervals according to the BBCH scale [56]. The first true leaves (BBCH stage 11) developed 14–15 days after sowing and multiple leaf stage (BBCH 18) was recorded 25–26 days after sowing. The bud formation stage after 35–40 days of sowing was coded as BBCH 50. The panicle emergence (BBCH 59) started after the 50–55 days of sowing and flowering (BBCH 67) in the inflorescence of quinoa plant after the 70–75 days after sowing. After flowering, the grains of quinoa start to ripened and milky stage recorded after 90–95 after sowing was coded as BBCH 81 and the maturity stage after 120–130 days after sowing as BBCH 90.

4.5. Measurement of Photosynthetic Efficiency and SPAD-Chlorophyll Value

Photochemical efficiency traits were measured at 12:00–1:00 pm in a bright sunny-day using photosynthetic efficiency analyzer (MINI-PAM-II) and chlorophyll values by SPAD-chlorophyll meter (SPAD-501, Minolta, Osaka, Japan) at panicle emergence stage (BBCH 59–60) from the upper most leaf. The data regarding current fluorescence value (Ft), electron transport rate (ETR), photosynthetically active radiations (PAR) and effective photochemical yield (Fv/Fm) was measured.

4.6. Seed Yield and Its Related Traits

At physiological maturity, height of plant and main panicle length was measured with a measuring scale. At harvesting, grains number per plant and 1000-grain yield were recorded. The grains number per plant was counted by threshing the panicles and 1000 grain weight was measured by using seed counter.

4.7. Statistical Analysis

Analysis of variance technique was performed to analyze the data statistically using Statistix 8.1 software (Hamburg, Germany) and differences among treatment means were computed by least significance difference (LSD) test at 5% probability.

5. Conclusions

The present study showed that enriching urea with urease and nitrification inhibitors (1% boric acid, thiourea and sodium thiosulphate) simultaneously have potential to reduce crop N requirements and its application can improve the N use efficiency, seed yield and grain protein contents in quinoa irrespective of genotypes. Nonetheless, better crop response at (NI; 1% boric acid, thiourea and sodium thiosulphate + 35 kg N ha^{-1}) was attributed to photosynthetic efficiency and increased N uptake in quinoa.

Author Contributions: Conceptualization, H.u.R. and H.F.A.; methodology, H.u.R. and A.W.; software, A.A.B.; validation, H.u.R., H.F.A., N.B.A., N.M.A., H.S.A.-Z., A.A.B. and T.I.; formal analysis, H.u.R.; investigation, H.u.R. and T.I.; resources, H.F.A. and A.W.; data curation, H.S.A.-Z.; writing—original draft preparation, H.u.R.; writing—review and editing, H.u.R., A.W. and N.B.A.; visualization, N.M.A.; supervision, H.u.R.; project administration, H.F.A.; funding acquisition, H.F.A. All authors have read and agreed to the published version of the manuscript.

Funding: This project was funded by Deputyship for Research & Innovation, Ministry of Education in Saudi Arabia for funding this research work through the project number (IFPIP: 539-130-1442).

Institutional Review Board Statement: Not applicable.

Data Availability Statement: Data is contained within the article.

Acknowledgments: The authors extend their appreciation to the Deputyship for Research & Innovation, Ministry of Education in Saudi Arabia for funding this research work through the project number (IFPIP: 539-130-1442), and King Abdulaziz University, DSR, Jeddah, Saudi Arabia.

Conflicts of Interest: The authors declare no conflict of interest.

References

1. Martínez, E.A.; Veas, E.; Jorquera, C.; San Martín, R.; Jara, P. Re-introduction of quínoa into arid Chile: Cultivation of two lowland races under extremely low irrigation. *J. Agron. Crop. Sci.* **2009**, *195*, 1–10. [CrossRef]
2. Bazile, D.; Jacobsen, S.-E.; Verniau, A. The global expansion of quinoa: Trends and limits. *Front. Plant Sci.* **2016**, *7*, 622. [CrossRef]
3. Lutz, M.; Bascuñán-Godoy, L. The Revival of Quinoa: A Crop for Health. In *Superfood and Functional Food—An Overview of Their Processing and Utilization*; Elsevier: Amsterdam, The Netherlands, 2017.
4. Schulte, E.G.; Kaul, H.-P.; Kruse, M.; Aufammer, W. Yield and nitrogen utilization efficiency of the pseudocereals amaranth, quinoa, and buckwheat under differing nitrogen fertilization. *Eur. J. Agron.* **2005**, *22*, 95–100. [CrossRef]
5. Razzaghi, F.; Plauborg, F.; Jacobsen, S.; Jense, C.; Andersen, M. Effect of nitrogen and water availability of three soil types on yield, radiation use efficiency and evapotranspiration in field-grown quinoa. *Agric. Water Manag.* **2012**, *109*, 20–29. [CrossRef]
6. Bascuñán-Godoy, L.; Reguera, M.; Abdel-Tawab, Y.M.; Blumwald, E. Water deficit stress-induced changes in carbon and nitrogen partitioning in Chenopodium quinoa willd. *Planta* **2016**, *243*, 591–603. [CrossRef] [PubMed]
7. Kansomjet, P.; Thobunluepop, P.; Lertmongkol, S.; Sarobol, E.; Kaewsuwan, P.; Junhaeng, P.; Pipattanawong, N.; Iván, M.T. Response of physiological characteristics, seed yield and seed quality of quinoa under difference of nitrogen fertilizer management. *Am. J. Plant Physiol.* **2017**, *12*, 20–27. [CrossRef]
8. Curti, R.N.; Sanahuja, M.D.C.; Vidueiros, S.M.; Pallaro, A.N.; Bertero, H.D. Trade-off between seed yield components and seed composition traits in sea level quinoa in response to sowing dates. *Cereal Chem.* **2018**, *95*, 734–741. [CrossRef]
9. Kakabouki, I.P.; Dimitra, H.; Ioannis, R.; Papastylianou, P.; Sestras, A.F.; Bilalis, D.J. Influence of fertilization and soil tillage on nitrogen uptake and utilization efficiency of quinoa crop (*Chenopodium quinoa* Willd.). *J. Soil Sci. Plant Nutr.* **2018**, *18*, 220–235. [CrossRef]
10. Kakabouki, I.; Roussis, I.E.; Papastylianou, P.; Kanatas, P.; Hela, D.; Katsenios, N.; Fuentus, F. Growth analysis of quinoa (*Chenopodium quinoa* Willd.) in response to fertilization and soil tillage. *Not. Bot. Horti Agrobot.* **2019**, *47*, 1025–1036. [CrossRef]
11. Ebrahimikia, M.; Moeini, M.J.; Marvi, H.; Hasheminejhad, Y.; Ganjehie, M.G. Agro-physiological response of quinoa (*Chenopodium quinoa* Willd.) to the nitrogen application rate and split application method. *J. Soil Sci. Plant Nutr.* **2021**, *21*, 3437–3450. [CrossRef]
12. Gómez, M.B.; Aguirre Castro, P.; Mignone, C.; Bertero, H.D. Can yield potential be increased by manipulation of reproductive partitioning in quinoa? Evidence from giberellic acid synthesis inhibition using Paclobutrazol. *Funct. Plant Biol.* **2011**, *38*, 420–430. [CrossRef] [PubMed]
13. Geren, H.; Kavut, Y.; Fakültesi, M. Altınbaş—Eg. effect of different row spacings on the grain yield and some yield characteristics of quinoa (Chenopodium quinoa Wild.) under Bornova ecological conditions. *Turk. J. F. Cr.* **2015**, *20*, 59–64.
14. Shams, A.S. Response of quinoa to nitrogen fertilizer rates under sandy soil conditions. In Proceedings of the 13th International Conference of Agronomy, Benha, Egypt, 9–10 September 2012; pp. 195–205.
15. Nosengo, N. Fertilized to death. *Nature* **2003**, *425*, 894–895. [CrossRef] [PubMed]

16. Malhi, S.S.; Oliver, E.; Mayerle, G.; Kruger, G.; Gill, K.S. Improving effectiveness of seed row-placed urea with urease inhibitor and polymer coating for durum wheat and canola. *Commun. Soil Sci. Plant Anal.* **2003**, *34*, 1709–1727. [CrossRef]
17. Sanz-Cobena, A.; Misselbrook, T.H.; Arce, A.; Mingot, J.I.; Diez, J.A.; Vallejo, A. An inhibitor of urease activity effectively reduces ammonia emissions from soil treated with urea under mediterranean conditions. *Agric. Ecosyst. Environ.* **2008**, *126*, 243–249. [CrossRef]
18. Schwab, G.J.; Murdock, L.W. *Nitrogen Transformation Inhibitors and Controlled Release Urea*; Department of Plant and Soil Sciences, Cooperative extension service, University of Kentucky, College of Agriculture: Lexington, KY, USA, 2010.
19. Dougherty, W.; Collins, D.; Zwieten, L.V.; Rowlings, D. Nitrification (DMPP) and urease (NBPT) inhibitors had no effect on pasture yield, nitrous oxide emissions, or nitrate leaching under irrigation in a hot-dry climate. *Soil Res.* **2016**, *54*, 675–683. [CrossRef]
20. Subbarao, G.V.; Ito, O.; Sahrawat, K.L.; Berry, W.L.; Nakahara, K.; Ishikawa, T.; Watanabe, T.; Suenaga, K.; Rondon, M.; Rao, I.M. Scope and strategies for regulation of nitrification in agricultural systems—Challenges and opportunities. *Rev. Plant Sci.* **2007**, *2*, 303–335. [CrossRef]
21. Zaman, M.; Nguyen, M.L.; Blennerhassett, J.D.; Quin, B.F. Reducing NH_3, N_2O and NO_3^--N losses from a pasture soil with urease or nitrification inhibitors and elemental S-amended nitrogenous fertilizers. *Biol. Fertil. Soils* **2008**, *44*, 693–705. [CrossRef]
22. Kim, D.G.; Saggar, S.; Roudier, P. The effect of nitrification inhibitors on soil ammonia emissions in nitrogen managed soils: A meta-analysis. *Nutr. Cycl. Agroecosyst.* **2012**, *93*, 51–64. [CrossRef]
23. Zhang, M.; Fan, C.H.; Li, Q.L.; Li, B.; Zhu, Y.Y.; Xiong, Z.Q. A 2-yr field assessment of the effects of chemical and biological nitrification inhibitors on nitrous oxide emissions and nitrogen use efficiency in an intensively managed vegetable cropping system. *Agric. Ecosyst. Environ.* **2015**, *201*, 43–50. [CrossRef]
24. Gao, J.; Luo, J.; Lindsey, S.; Shi, Y.; Wei, Z.; Wang, L.; Zhang, L. Effects of boric acid on urea-N transformation and DMPP efficiency. *J. Sci. Food Agric.* **2020**, *101*, 1091–1099. [CrossRef] [PubMed]
25. Margon, A.; Parente, G.; Piantanida, M.; Cantone, P.; Leita, L. Novel investigation on ammonium thiosulphate (ATS) as an inhibitor of soil urease and nitrification. *Agric. Sci.* **2015**, *6*, 1502–1512. [CrossRef]
26. Chien, S.H.; Prochnow, L.I.; Cantarella, H. Recent developments of fertilizer production and use to improve nutrient efficiency and minimize environmental impacts. *Adv. Agron.* **2009**, *102*, 267–322. [CrossRef]
27. Kong, X.; Duan, Y.; Schramm, A.; Eriksen, J.; Petersen, S.O. 3,4-Dimethylpyrazole phosphate (DMPP) reduces activity of ammonia oxidizers without adverse effects on nontarget soil microorganisms and functions. *Appl. Soil Ecol.* **2016**, *105*, 67–75. [CrossRef]
28. Linquist, B.A.; Lijun, L.; van Kessel, C.; van Groenigen, K.J. Enhanced efficiency nitrogen fertilizers for rice systems: Meta-analysis of yield and nitrogen uptake. *Field Cr. Res.* **2013**, *154*, 246–254. [CrossRef]
29. Lam, S.K.; Suter, H.; Bai, M.; Walker, C.; Davies, R.; Mosier, A.R.; Chen, D. Using urease and nitrification inhibitors to decrease ammonia and nitrous oxide emissions and improve productivity in a subtropical pasture. *Sci. Total Environ.* **2018**, *644*, 1531–1535. [CrossRef]
30. Abalos, D.; Jeffery, S.; Sanz-Cobena, A.; Guardia, G.; Vallejo, A. Meta-analysis of the effect of urease and nitrification inhibitors on crop productivity and nitrogen use efficiency. *Agric. Ecosyst. Environ.* **2014**, *189*, 136–144. [CrossRef]
31. Khan, M.A.; Shah, Z.; Rab, A.; Arif, M.; Shah, T. Effect of urease and nitrification inhibitors on wheat yield. *Sarhad J. Agric.* **2013**, *29*, 371–378.
32. Glibert, P.M.; Harrison, J.; Heil, C.; Seitzinger, S. Escalating worldwide use ofurea–a global change contributing to coastal eutrophication. *Biogeochemestry* **2006**, *77*, 441–463. [CrossRef]
33. Rehman, H.; Basra, S.M.A.; Wahid, A. Optimizing nitrogen-split application time to improve dry matter accumulation and yield in dry direct seeded rice. *Int. J. Agric. Biol.* **2013**, *15*, 1560–8530.
34. Kawakami, E.M.; Oosterhuis, D.M.; Snider, J.L.; Mozaffari, M. Physiological and yield responses of field-grown cotton to application of urea with the urease inhibitor NBPT and the nitrification inhibitor DCD. *Eur. J. Agron.* **2012**, *43*, 147–154. [CrossRef]
35. Buscaglia, H.; Varco, J. Early detection of cotton leaf nitrogen status using leaf reflectance. *J. Plant Nutr.* **2002**, *25*, 2067. [CrossRef]
36. Bondada, B.R.; Oosterhuis, D.M. Canopy photosynthesis, specific leaf weight, and yield components of cotton under varying nitrogen supply. *J. Plant Nutr.* **2001**, *24*, 469–477. [CrossRef]
37. Garg, B.K.; Burman, U.; Kathju, S. Influence of thiourea on photosynthesis, nitrogen metabolism and yield of cluster bean (*Cyamopsis tetragonoloba* (L.) Taub under rainfed conditions of Indian arid zone. *Plant Growth Regul.* **2006**, *48*, 237–245. [CrossRef]
38. Zaman, M.; Zaman, S.; Nguyen, M.L.; Smith, T.J.; Nawaz, S. The effect of urease and nitrification inhibitors on ammonia and nitrous oxide emissions from simulated urine patches in pastoral system: A two-year study. *Sci. Total Environ.* **2013**, *465*, 97–106. [CrossRef]
39. Pettigrew, W. Environmental effects on cotton fiber carbohydrate concentration and quality. *Crop Sci.* **2001**, *41*, 1108–1111. [CrossRef]
40. Guardia, G.; Sanz-Cobena, A.; Sanchez-Martín, L.; Fuertes-Mendizábal, T.; González-Murua, C.; Álvarez, J.M.; Chadwick, D.; Vallejo, A. Urea-based fertilization strategies to reduce yield-scaled N oxides and enhance bread-making quality in a rainfed Mediterranean wheat crop. *Agric. Ecosys. Environ.* **2018**, *265*, 421–431. [CrossRef]
41. Artola, E.; Cruchaga, S.; Ariz, I.; Moran, J.F.; Garnica, M.; Houdusse, F.; Garcia Mina, J.M.; Irigoyen, I.; Lasa, B.; Aparicio-Tejo, P.M. Effect of N-(n-butyl) thiophosphoric triamide on urea metabolism and the assimilation of ammonium by *Triticum aestivum* L. *Plant Growth Regul.* **2011**, *63*, 73–79. [CrossRef]

42. Cruchaga, S.; Lasa, B.; Jauregui, I.; González-Murua, C.; Aparicio-Tejo, P.M.; Ariz, I. Inhibition of endogenous urease activity by NBPT application reveals differential N metabolism responses to ammonium or nitrate nutrition in pea plants: A physiological study. *Plant Soil* **2013**, *373*, 813–827. [CrossRef]
43. Fuertes-Mendizábal, T.; González-Torralba, J.; Arregui, L.M.; González-Murua, C.; González-Moro, M.B.; Estavillo, J.M. Ammonium as sole N source improves grain quality in wheat. *J. Sci. Food Agric.* **2013**, *93*, 2162–2171. [CrossRef]
44. Qiao, C.; Liu, L.; Hu, S.; Compton, J.E.; Greaver, T.L.; Li, Q. How inhibiting nitrification affects nitrogen cycle and reduces environmental impacts of anthropogenic nitrogen input. *Glob. Chang. Biol.* **2015**, *21*, 1249–1257. [CrossRef] [PubMed]
45. Thapa, R.; Chatterjee, A.; Awale, R.; McGranahan, D.A.; Daigh, A. Effect of enhanced efficiency fertilizers on nitrous oxide emissions and crop yields: A meta-analysis. *Soil Sci. Soc. Am. J.* **2016**, *80*, 1121–1134. [CrossRef]
46. Xue, C.; Rossmann, A.; Schuster, R.; Koehler, P.; Mühling, K.H. Split nitrogen application improves wheat baking quality by influencing protein composition rather than concentration. *Front. Plant Sci.* **2016**, *7*, 738. [CrossRef] [PubMed]
47. Gupta, N.; Gupta, A.; Gaur, V.; Kumar, A. Relationship of nitrogen use efficiency with the activities of enzymes involved in nitrogen uptake and assimilation of finger millet genotypes grown under different nitrogen inputs. *Sci. World J.* **2012**, *2012*, 625731. [CrossRef] [PubMed]
48. Bascuñán-Godoy, L.; Sanhueza, C.; Pinto, K.; Cifuentes, L.; Reguera, M.; Briones, V.; Zurita-Silva, A.; Álvarez, R.; Morales, A.; Silva, H. Nitrogen physiology of contrasting genotypes of *Chenopodium quinoa* Willd. (Amaranthaceae). *Sci. Rep.* **2018**, *8*, 17524. [CrossRef] [PubMed]
49. Masclaux-Daubresse, C.; Françoise, D.-V.; Dechorgnat, J.; Chardon, F.; Gaufichon, L.; Suzuki, A. Nitrogen uptake, assimilation and remobilization in plants: Challenges for sustainable and productive agriculture. *Ann. Bot.* **2010**, *105*, 1141–1157. [CrossRef] [PubMed]
50. Basra, S.M.A.; Iqbal, S.; Afzal, I. Evaluating the response of nitrogen application on growth, development and yield of quinoa genotypes. *Int. J. Agric. Biol.* **2014**, *16*, 886–892.
51. Ryan, J.; Estefan, G.; Rashid, A. *Soil and Plant Analysis Laboratory Manual*; ICARDA: Beirut, Lebanon, 2001.
52. Mérigout, P.; Gaudon, V.; Quilleré, I.; Briand, X.; Vedele, F.D. Urea use efficiency of hydroponically grown maize and wheat. *J. Plant Nutr.* **2008**, *31*, 427–443. [CrossRef]
53. Wang, J.; Dun, X.; Shi, J.; Wang, X.; Liu, G.; Wang, H. Genetic dissection of root morphological traits related to nitrogen use efficiency in *Brassica napus* L. under two contrasting nitrogen conditions. *Front. Plant Sci.* **2017**, *8*, 1709. [CrossRef]
54. Abd-El Baki, G.K.; Siefritz, F.; Man, H.M.; Weiner, H.; Kaldenhoff, R.; Kaiser, W.M. Nitrate reductase in Zea mays L. under salinity. *Plant Cell Environ.* **2000**, *23*, 515–521. [CrossRef]
55. Kaiser, J.J.; Lewis, O.A.M. Nitrate reductase and glutamine synthetase activity in leaves and roots of nitrate-fed *Helianthus annuus* L. *Plant Soil.* **1984**, *77*, 127–130. [CrossRef]
56. Sosa-Zuniga, V.; Brito, V.; Fuentes, F.; Steinfort, U. Phenological growth stages of quinoa (*Chenopodium quinoa*) based on the BBCH scale. *Ann. Appl. Biol.* **2017**, *171*, 117–124. [CrossRef]

plants

Article

Morphological and Physiological Traits Associated with Yield under Reduced Irrigation in Chilean Coastal Lowland Quinoa

Kathryn Dumschott [1,2,†], Nathalie Wuyts [3,†], Christian Alfaro [4,5,6], Dalma Castillo [4,5,6], Fabio Fiorani [3,*] and Andrés Zurita-Silva [4,5,6,*]

1 Institute for Biology I, BioSC, RWTH Aachen University, 52056 Aachen, Germany; k.dumschott@fz-juelich.de
2 Institute of Bio- and Geosciences, Bioinformatics (IBG-4), Forschungszentrum Jülich GmbH, 52425 Jülich, Germany
3 Institute of Bio- and Geosciences, Plant Sciences (IBG-2), Forschungszentrum Jülich GmbH, 52425 Jülich, Germany; n.wuyts@fz-juelich.de
4 Centro de Investigación Intihuasi (AZS), Instituto de Investigaciones Agropecuarias, La Serena 1722093, Chile; calfaro@inia.cl (C.A.); dalma.castillo@inia.cl (D.C.)
5 Centro de Investigación Rayentué (CA), Instituto de Investigaciones Agropecuarias, Rengo 2940000, Chile
6 Centro de Investigación Quilamapu, Instituto de Investigaciones Agropecuarias, Chillán 3780000, Chile
* Correspondence: f.fiorani@fz-juelich.de (F.F.); andres.zurita@inia.cl (A.Z.-S.); Tel.: +49-2461-61-8642 (F.F.); +56-512-222-390 (A.Z.-S.)
† These authors contributed equally to this work.

Abstract: Quinoa (*Chenopodium quinoa* Willd.) is a genetically diverse crop that has gained popularity in recent years due to its high nutritional content and ability to tolerate abiotic stresses such as salinity and drought. Varieties from the coastal lowland ecotype are of particular interest due to their insensitivity to photoperiod and their potential to be cultivated in higher latitudes. We performed a field experiment in the southern Atacama Desert in Chile to investigate the responses to reduced irrigation of nine previously selected coastal lowland self-pollinated (CLS) lines and the commercial cultivar Regalona. We found that several lines exhibited a yield and seed size superior to Regalona, also under reduced irrigation. Plant productivity data were analyzed together with morphological and physiological traits measured at the visible inflorescence stage to estimate the contribution of these traits to differences between the CLS lines and Regalona under full and reduced irrigation. We applied proximal sensing methods and found that thermal imaging provided a promising means to estimate variation in plant water use relating to yield, whereas hyperspectral imaging separated lines in a different way, potentially related to photosynthesis as well as water use.

Keywords: *Chenopodium quinoa* Willd.; field trial; hyperspectral imaging; phenotyping; quinoa; reduced irrigation; thermal imaging; yield

Citation: Dumschott, K.; Wuyts, N.; Alfaro, C.; Castillo, D.; Fiorani, F.; Zurita-Silva, A. Morphological and Physiological Traits Associated with Yield under Reduced Irrigation in Chilean Coastal Lowland Quinoa. *Plants* **2022**, *11*, 323. https://doi.org/10.3390/plants11030323

Academic Editors: Cataldo Pulvento and Didier Bazile

Received: 21 December 2021
Accepted: 11 January 2022
Published: 26 January 2022

1. Introduction

Quinoa (*Chenopodium quinoa* Willd.) is a highly nutritious member of the Amaranthaceae, originating from the Andean region of Central and South America. Originally cultivated by the Incas during pre-Colombian times [1], quinoa became a staple food of the Incan Empire and is now considered an important food crop in many South American countries [2]. Quinoa grains are gluten free and highly nutritious, containing high-quality protein and all essential amino acids, vitamins, minerals and antioxidants (flavonoids and polyphenols) that contribute to the health-promoting effects of this food crop [3–9]. Additionally, the seeds exhibit a high content of unsaturated fatty acids (oleic, linoleic, and α-linolenic acids) and a close to optimal omega-6/omega-3 ratio, which support the oil quality of this crop [10]. On the other hand, the seeds accumulate saponins, commonly considered anti-nutritional factors due to their hemolytic, membranolytic, and fungitoxic activities [11]. In recent decades, quinoa has become a target of research worldwide due to its potential contribution to food security [2,12]. Quinoa is a genetically diverse crop that thrives in the heterogeneous environments of the Andean region from which it originated [13]. Quinoa genotypes, landraces and cultivars can be classified into five ecotypes, which exhibit wide ranges of adaptations to elevation, annual rainfall, soil fertility, temperature and photoperiod [2].

Quinoa is also known for its ability to grow in marginal environments and tolerate a range of adverse growth conditions, such as high salinity [14–17], heat [18] and drought [15,19,20]. Detailed studies of quinoa subjected to drought stress have been conducted, both in the field and in greenhouse experiments, providing insight into the key physiological adaptations of quinoa. Increased water use efficiency associated with abscisic acid (ABA)-induced stomatal closure [21–24] is a common strategy implemented by quinoa in response to drought stress. Leaf ABA concentration has been observed to accumulate in response to increased stress in the field, suggesting it is an important mechanism for drought tolerance in quinoa [21]. In addition to ABA signaling, hydraulic regulation through changes in turgor may play an equally important role in stomatal closure for certain quinoa varieties subject to water stress [25]. Another strategy is the induction of metabolic adaptations that improve tolerance to osmotic or water stress, and which involve an increased synthesis of osmoprotectants such as free amino acids, proline, and soluble sugars (glucose, trehalose), to enhance scavenging of reactive oxygen species and to protect plants from destructive oxidative reactions [26,27]. Reductions in photosynthetic rates and the efficiency of photosystem II have also been observed in quinoa plants in response to water deficit [24,28]. Strategies used to mitigate water deficit stress have been shown to differ between quinoa varieties depending on their geographical origin. In a study involving two quinoa varieties, Sun et al. [23] observed that the relatively slower growth rates and smaller leaf areas of varieties originating from adverse environments were more effective at tolerating drought stress compared to fast-growing varieties originating from nutrient-rich environments, due to reduced overall transpiration and water loss [23].

To date, most drought stress studies involving quinoa have focused on only a few bred varieties and seldom compare ecotypes or ecotype-derived self-pollinated progenies. The diversity of quinoa and the establishment of new breeding and improvement programs to develop new varieties better adapted to different environmental conditions remain largely unexplored [13,29–31]. To this end, unique germplasm collections representing local and regional biodiversity are of particular interest as a source of variation. One major consideration when selecting quinoa material for this study was the sensitivity of quinoa to photoperiod, as this trait can significantly limit quinoa adaptation and breeding efforts in high-latitude regions, such as Europe [29,32]. Disruption to seed filling and maturation, resulting in continued vegetative growth and flowering, has been observed in photoperiod-sensitive lines grown in photoperiods of longer than 12 h [29,32]. Quinoa cultivars from the coastal lowland ecotype show an insensitivity to photoperiod, as they are adapted to the coastal conditions of southern Chile (latitudes up to ~40° S) and have already been used for European-bred cultivars from Denmark and the Netherlands [29,33,34]. Therefore, we focused on the coastal lowland ecotype and selected lines from the INIA SeedBank collection (Chile), based on morphology and yield observed in the field and further developed in a breeding program run by INIA Chile.

The physiological characterization and classification of lines according to their tolerance to drought stress customarily involves destructive or laborious measurements of traits such as leaf water potential or photosynthesis and stomatal conductance. These methods are often time-consuming and difficult to apply on a large scale. Screening of genotypes for pre-breeding can be facilitated by using more high-throughput phenotyping methods to measure photosynthetic status, spectral reflectance and canopy temperature [35,36]. These methods, in particular vegetation indices based on spectral reflectance measurements, have recently begun to be applied in quinoa research on drought stress [37,38].

In this study, we implemented a combination of standard and more recent approaches to measure productivity and underlying morphological and physiological traits of nine novel Chilean coastal lowland (CLS) lines and one commercial cultivar, Regalona Baer (hereafter referred to as cv Regalona), grown in a field experiment subject to a full and reduced irrigation regime starting from the branching stage (extended BBCH 20 [39]). Standard approaches included the non-invasive measurement of chlorophyll fluorescence and destructive measurements of water potential, relative water content, and shoot morphology. These were tested alongside recent approaches which can be scaled to high throughput, including hyperspectral and thermal imaging. All measurements were taken at the visible inflorescence stage of development (extended BBCH 59 [39]) and compared to seed yield and seed size determined at physiological maturity, as well as plant development over the course of the trial. We aimed to determine differences in yield and seed size among the selected CLS lines compared to cv Regalona, and to detect the underlying morphological and physiological traits that may contribute to yield and seed size determination under full and reduced irrigation. A better understanding of how well-established and more recently developed phenotyping methods can be deployed to determine traits contributing to yield under adverse conditions will help improve the effectiveness of future quinoa breeding programs and crop management.

2. Results

2.1. Agronomical Traits at Harvest

A significant treatment effect was observed for plant yield ($p < 0.01$), with plants receiving reduced irrigation (RI) yielding less than fully irrigated (FI) plants (Figure 1). There was no significant genotype or genotype by treatment interaction effect. However, lines CLS-1 and CLS-2 sustained a comparable yield despite a reduced water supply, which was similar to the response observed for the commercial cv Regalona. In the other CLS lines, the yield was decreased by water deficit (Figure 1). When considering seed weight per plant, treatment was again the only significant factor ($p < 0.01$), with average seed weights of 2.42 g and 1.92 g for FI and RI plants, respectively. Interestingly, lines CLS-1 and CLS-2 increased their average seed weight per plant in response to RI (1.88 g vs. 2.01 g for CLS-1 FI vs. RI samples, and 2.45 g vs. 2.94 g for CLS-2 FI vs. RI samples) (Figure 1).

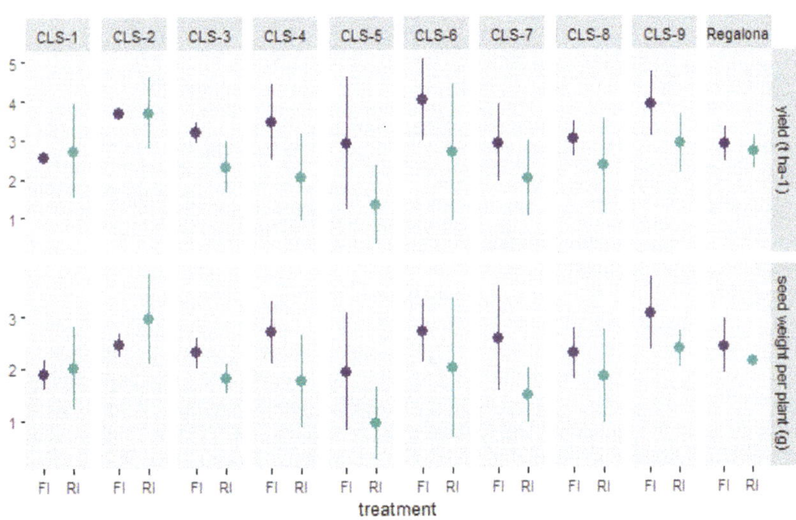

Figure 1. Productivity of nine coastal lowland self-pollinated lines (CLS) and commercial cv Regalona in a field experiment under full irrigation (FI, purple) and reduced irrigation (RI, green). Productivity is presented as yield expressed in t ha^{-1} (**top** panel) and the seed weight per plant (**bottom** panel) collected in 1 m of two central rows per plot. Data are presented as the mean (dot) and the standard deviation around the mean (vertical bar).

For the 1000 seed weight, both the treatment ($p < 0.0001$) and genotype effects ($p < 0.0001$) were significant, but not the genotype by treatment interaction effect. The average 1000 seed weight was 3.08 g for FI plants and 2.86 g for RI plants (Table 1). The highest weights corresponded to lines CLS-3 and CLS-7, whereas the lowest weights were recorded in lines CLS-8, CLS-6 and cv Regalona. It is noteworthy that among the highest 1000 seed weights, CLS-7 was not affected by the RI treatment (Figure S1).

Table 1. One-thousand seed weight and proportional weight of 1.7 mm caliber seeds measured for nine coastal lowland self-pollinated lines (CLS) and commercial cv Regalona in a field experiment under full irrigation (FI) and reduced irrigation (RI). Values are means across treatment ± standard deviation.

Genotype	1000 Seed Weight (g)	1.7 mm Caliber Seed Weight (g) [1]
CLS-1	3.22 ± 0.25 ab	5.93 ± 1.10 a
CLS-2	3.22 ± 0.19 ab	4.35 ± 0.92 b
CLS-3	3.28 ± 0.17 a	5.83 ± 0.51 a
CLS-4	3.02 ± 0.08 bc	4.23 ± 0.54 bde
CLS-5	2.82 ± 0.26 cde	4.80 ± 1.13 bc
CLS-6	2.68 ± 0.19 de	3.78 ± 0.66 def
CLS-7	3.25 ± 0.10 a	5.67 ± 0.55 ac
CLS-8	2.60 ± 0.26 e	3.35 ± 0.72 ef
CLS-9	2.90 ± 0.40 cd	3.05 ± 1.59 fg
Regalona	2.73 ± 0.16 de	2.35 ± 0.86 g

[1] Out of a 10 g sample of seeds, the total weight of seeds that are larger than 1.7 mm. $n = 6$. Letters indicate significant differences between lines ($p < 0.05$).

Significant differences at both the treatment ($p < 0.05$) and genotype ($p < 0.0001$) level were found for the 1.7 mm-caliber seed weight. The average weight for the FI treatment was 4.82 g, whereas the RI average was significantly lower (3.85 g). The highest 1.7 mm-caliber seed size weight was observed for line CLS-1, followed by CLS-3 and CLS-7 (Table 1). Interestingly, these three lines did not show similar decreases in the RI treatment, but for CLS-3 and CLS-7 the average difference between treatments was comparatively small (Figure S1). When also considering lower caliber seeds (1.4 mm and 1.18 mm), seed size displayed an inverse effect. A decrease in the proportion of higher caliber seeds occurred simultaneously with an increase in the proportion of lower caliber seeds (Figure S1).

For panicle length, we found a significant effect of genotype ($p < 0.0001$) and genotype by treatment interaction ($p < 0.0001$). Within the FI treatment, CLS-2 and cv Regalona had the shortest panicles, followed by CLS-7 and CLS-9, whereas lines CLS-1, CLS-5, CLS-6 and CLS-8 had the longest panicles. In the RI treatment, CLS-5 still had the largest panicle length, followed by CLS-1, CLS-6 and CLS-8. The other CLS lines and cv Regalona all had shorter panicles. Panicle width showed a significant treatment ($p < 0.01$), genotype ($p < 0.0001$) and genotype by treatment interaction effect ($p < 0.0001$). The panicle width was reduced in the RI treatment for lines CLS-2, CLS-3, CLS-4 and CLS-9 (Figure S2b). Within the FI treatment, panicles were smallest in width in CLS-7 and largest in CLS-6, CLS-8 and CLS-9. The widest panicles in the RI treatment were found for lines CLS-6, CLS-8 and CLS-5. The lines CLS-6 and CLS-8 had the overall largest panicles, together with CLS-5 whose panicles even increased in size under RI.

The plant height at harvest ranged from 99 to 127 cm (with an average height of 115 cm) in the FI treatment, and from 87 to 108 cm (with an average height of 97 cm) in the RI treatment (Figure S3a). For all genotypes except CLS-5 and CLS-8, plant height was significantly ($p < 0.01$) reduced under RI conditions (Figure S3b). The largest effect of RI on plant height was observed for CLS-2 and CLS-6, whereas the effect was small for CLS-8 and CLS-5 (Figure 2). Plant height recorded for line CLS-8 was relatively short in both treatments, while CLS-5 plants were considered medium height when compared to the other lines (Figure S3). The tallest lines at harvest were CLS-1, CLS-3, CLS-4, CLS-6 and CLS-7, whereas plants of CLS-9 and cv Regalona were the shortest in both treatments.

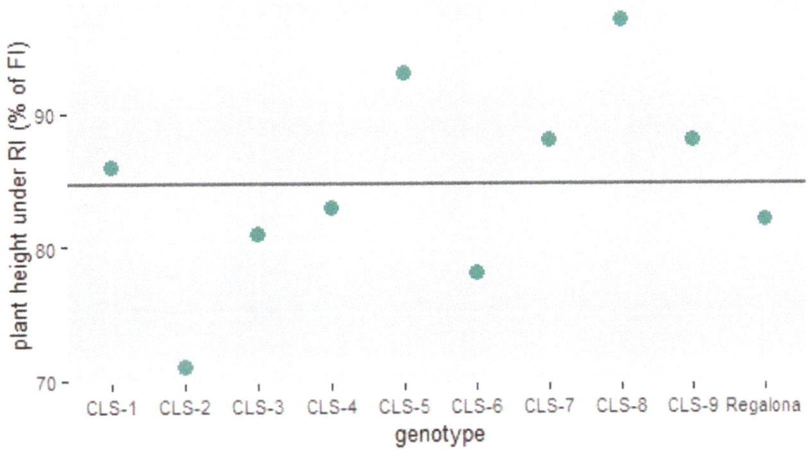

Figure 2. Plant height at harvest under reduced irrigation (RI) expressed as a percentage of plant height under full irrigation (FI) for nine coastal lowland self-pollinated lines (CLS) and commercial cv Regalona. The mean plant height under RI as a % of FI is indicated by a horizontal line.

Yield responses to reduced water availability in a genotype panel can also be assessed by means of drought and yield tolerance indices. These are based on the overall effect on yield or by comparing genotypes of interest to the yield reduction in a set of reference genotypes [40]. The drought tolerance index (DTI) was 0.76 in our field trial involving nine CLS lines and the commercial cv Regalona as a reference. Therefore, an overall 24% yield loss occurred under RI conditions. The DTI of CLS-1 and CLS-2 was the highest (1.41 and 1.32, respectively), as no yield loss was recorded under RI, while cv Regalona had a DTI of 1.23 and was also relatively unaffected by drought (Table 2). Among the other lines, CLS-5 was the most affected by drought with a DTI of 0.60. In this field experiment, all CLS lines, except CLS-1 and CLS-2, performed worse than cv Regalona. As recommended by Ober et al. and del Pozo et al. [40,41], we also calculated the yield tolerance index (YTI) because it may differentiate between genotypes that perform better under drought stress due to an inherently high yield potential, and those that intrinsically have greater drought tolerance. This difference was observed in our field trial for CLS-1 and CLS-2, which had YTI's of 0.66 and 1.29, respectively (Table 2). CLS-2 was therefore a high yielding line despite the applied treatment, whereas CLS-1 was in itself more drought tolerant. The line CLS-8 was also found to be drought tolerant per se, most likely due to its overall smaller size, which resulted in reduced total transpiration. CLS-9 had higher yields under drought due to its inherently higher yield potential, albeit this effect was still less pronounced when compared to CLS-2. Finally, CLS-6 performed better than cv Regalona, despite being more sensitive to drought (Table 2).

Table 2. Drought and yield tolerance indices according to Ober et al. [40] and del Pozo et al. [41] calculated for nine coastal lowland self-pollinated lines (CLS) and commercial cv Regalona grown in a field experiment under full irrigation and reduced irrigation. The drought tolerance index (DTI) is based on the drought intensity index across all genotypes in the experiment (DII, 0.76), whereas the DTI-R is based on the DII of Regalona (0.93) as a reference. The yield tolerance index (YIT) is calculated across all genotypes in the experiment.

Genotype	DTI	DTI-R	YTI
CLS-1	1.41	1.15	0.66
CLS-2	1.32	1.08	1.29
CLS-3	0.96	0.78	0.69
CLS-4	0.78	0.63	0.67
CLS-5	0.60	0.49	0.36
CLS-6	0.88	0.71	1.01
CLS-7	0.90	0.74	0.55
CLS-8	1.02	0.83	0.66
CLS-9	0.98	0.80	1.07
Regalona	1.23	1	0.72

2.2. *Thermal Index for Plant Responses to Drought and Irrigation*

Thermal index 1 (TI1, °C) was obtained at 46 and 47 days after sowing (DAS), i.e., at the visible inflorescence stage for all plots in both treatments. A TI1 close to zero means cool leaves corresponding to high transpiration, whereas more negative values indicate warmer leaves, suggesting stomatal closure. We found a significant effect of treatment and measurement day, as well as their interaction ($p < 0.001$), but no significant effect of genotype. The measurement day effect originated from the timing of irrigation. At 46 DAS, plants had not been irrigated for almost three days and loss of turgor in leaves was observed in both the FI and RI treatments. All plots were irrigated at noon of 46 DAS, but in the afternoon, TI1 was still very negative in both treatments, indicating that the leaves had not fully recovered and transpiration was still low (Table 3). At 47 DAS, 24 h after irrigation, all values became less negative, suggesting recovery from acute drought, although the treatment effect was noticeable with values closer to zero for FI plants. This was most pronounced for CLS-2, CLS-9 and cv Regalona, and may indicate higher transpiration rates and/or reduced stomatal closure under FI conditions compared to the other lines. For CLS-4 and CLS-8, TI1 values under FI conditions were still more negative, potentially indicating faster stomatal closure and reduced transpiration under high vapor pressure deficit conditions in the afternoon. On the other hand, lines CLS-3 and CLS-4 had the least negative TI1 under RI conditions at 47 DAS, suggesting that transpiration was reduced less, compared to the other lines and cv Regalona.

2.3. *Hyperspectral Indices as Proxies for Plant Trait Measurements*

Hyperspectral imaging data acquired at the visible inflorescence stage were processed to obtain vegetation indices. These included published vegetation indices (VIs, Table S1, [42–64]), calculated based on wavelengths selected for other species and mostly obtained by remote sensing. In addition, we compared the relative reflectance of quinoa across the complete spectral region (File S1).

Table 3. The thermal index 1 (TI1, °C) calculated based on thermal infrared data for nine coastal lowland self-pollinated lines (CLS) and commercial cv Regalona in a field experiment under full irrigation (FI) and reduced irrigation (RI) conditions. Images were acquired at the visible inflorescence stage on two consecutive measurement days, expressed in days after sowing (DAS). Values are means ± standard deviation.

Genotype	46 DAS		47 DAS	
	FI	RI	FI	RI
CLS-1	−5.55 ± 0.65	−5.39 ± 0.38	−1.10 ± 0.85 †	−2.72 ± 0.61 †
CLS-2	−5.00 ± 1.72	−4.17 ± 0.79	−0.02 ± 1.23 †	−2.68 ± 1.04 *
CLS-3	−5.21 ± 0.74	−3.85 ± 0.78	−0.37 ± 1.20 †	−1.69 ± 0.65 †
CLS-4	−6.13 ± 1.62	−4.38 ± 0.20	−1.70 ± 0.27 †	−1.87 ± 0.88 †
CLS-5	−5.07 ± 1.38	−5.96 ± 1.37	−1.10 ± 1.28 †	−2.27 ± 0.79 †
CLS-6	−5.03 ± 1.42	−4.91 ± 1.96	−0.81 ± 1.12 †	−2.38 ± 1.24
CLS-7	−5.48 ± 0.60	−5.29 ± 0.67	−1.30 ± 1.09 †	−2.57 ± 1.33 †
CLS-8	−5.22 ± 0.88	−5.19 ± 0.54	−2.29 ± 1.06 †	−3.23 ± 0.77 †
CLS-9	−4.24 ± 1.22	−5.06 ± 0.81	−0.12 ± 0.90 †	−3.37 ± 0.64 †*
Regalona	−4.36 ± 1.23	−4.55 ± 1.90	−0.17 ± 0.48 †	−2.75 ± 0.11 *

$n = 3$. * indicates significant differences between treatments per genotype and measurement day ($p < 0.05$). † indicates significant differences between measurement days 46 DAS and 47 DAS within genotype and treatment ($p < 0.05$).

We performed a three-way ANOVA to find out whether VIs could detect differences between genotypes and treatments across repetitions. All main effects (genotype, treatment and repetition) were significant, as well as all interactions. We therefore investigated the effect sizes of the main effects and their interactions (Figure S4). Effect sizes were the smallest for treatment in all VIs, which was further confirmed by the results of the two-way ANOVA per repetition. Either genotype or repetition, and their interaction, had the largest effect size depending on the VI. All NDVI-related VIs had very small effect sizes, whereas the RGRI and MCARI2 of the same group of structure-related VIs showed the overall largest effect sizes, together with the RGI and G. Both calculations include wavelengths in the red and green region of the spectrum. RGI and G were followed in effect size by the WCI (red, green, NIR), CRI2 (blue-green, red-edge) and BRI (blue, green). We further analyzed the effect sizes for pairwise contrasts between genotypes within each treatment. Figure 3 shows the VIs for which an effect size larger than one standard deviation across the three repetitions was found. The VIs MCARI2, RGI, RGRI and G showed the largest effect sizes and were associated with a larger number of contrasts between genotypes in both treatments. CLS-7 differed from six other genotypes in the FI treatment, followed by CLS-3 and cv Regalona (five genotypes), CLS-5 (four genotypes), and CLS-9 (three genotypes). In the RI treatment, CLS-2 and CLS-4 appeared at the forefront, with large effect size differences with five and four other genotypes, respectively. CLS-7 and CLS-3 still differed from each other and three other genotypes, whereas cv Regalona only differed from two genotypes.

The WBI, which represents the extent of water-sensitive depression between 900 nm and 970 nm, showed consistent differences in mean values across repetitions between FI and RI treatments for CLS-1, CLS-6 and CLS-7. The effect sizes, on the other hand, were small. The WBI was not different between treatments for CLS-2, CLS-5 and CLS-8. For the remaining lines, no consistent treatment effects were detected. For the other water content-related index WCI, consistent differences in mean values between FI and RI were found for CLS-3, CLS-7, CLS-8 and CLS-9, with effect sizes close to two and three standard deviations for CLS-3 and CLS-9 respectively, and one standard deviation for CLS-7 and CLS-8.

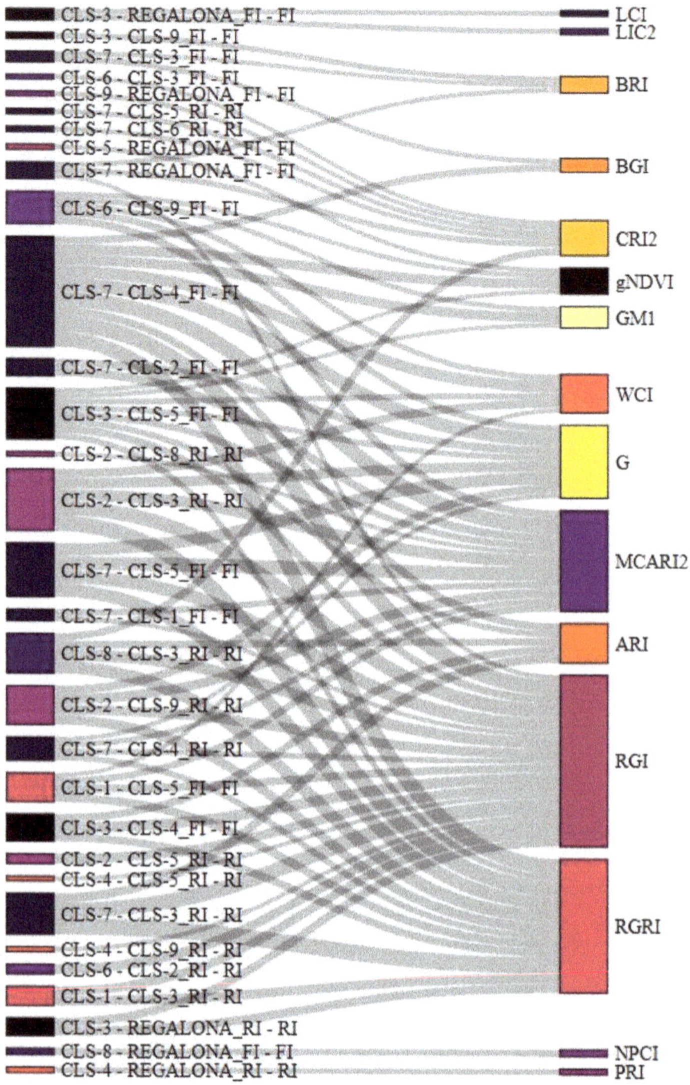

Figure 3. Quinoa genotype pairwise contrasts for the full irrigation (FI) and reduced irrigation (RI) treatments and vegetation indices (VIs) with consistently high effect sizes (larger than one standard deviation) across the three repetitions. The size of the nodes' rectangle is proportional to the number of connected VIs per genotype contrast (left) or the number of genotype contrasts per VI (right). The online figure is interactive and highlights interactions when pointing the cursor to genotype contrasts and VIs.

2.4. Relationship between Agronomical Traits at Harvest, Plant Phenology, and Morphological and Physiological Traits at the Onset of Flowering

Observations on the progress of phenological stages were made throughout the growing season, but differences between CLS lines and cv Regalona were only found in the duration of the seed maturation stages (File S3). At the visible inflorescence stage (extended BBCH 59 [39]), we measured plant morphological and physiological traits by well-established methods, details of which and measurement results are provided in File S4.

Traits included plant height, stem diameter, stem and leaf biomass, stem and leaf water potential, leaf relative water content, and the quantum efficiency of photosystem II.

Morphological traits measured at the visible inflorescence stage showed a significant positive correlation with yield and yield components (Figure 4). Plant height measured at harvest correlated positively with seed size. Among the physiological traits measured at the visible inflorescence stage, the leaf relative water content and stem water potential showed a positive correlation with yield (Figure 4).

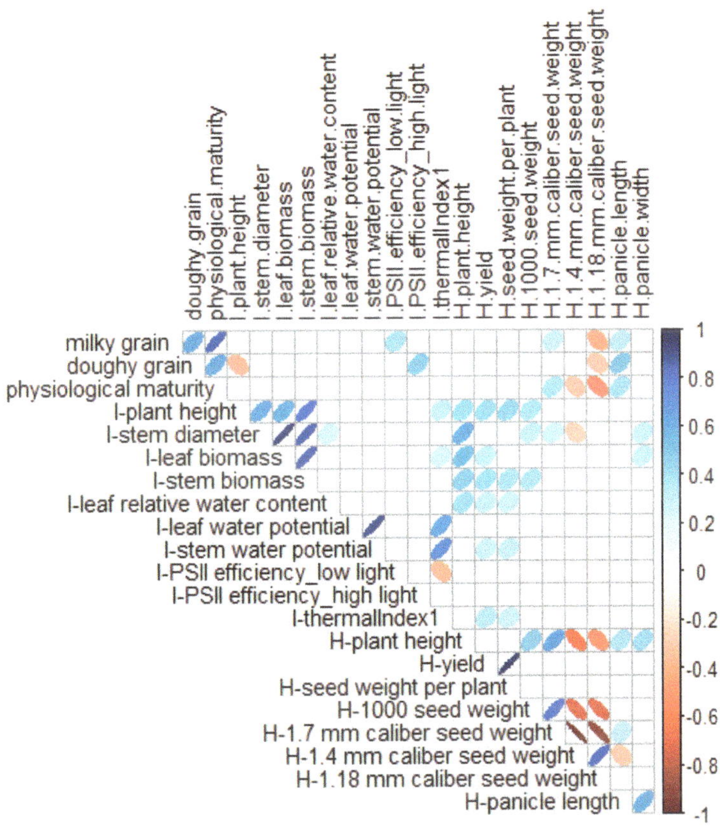

Figure 4. Correlation of phenological, morphological, physiological, and agronomical traits measured across development, at the visible inflorescence stage (traits preceded by 'I-'), or at harvest (traits preceded by 'H-') in nine coastal lowland self-pollinated lines and commercial cv Regalona grown in a field experiment under full irrigation (FI) and reduced irrigation (RI). The color of the ellipses represents the correlation coefficient (R) which can be positive (blue) or negative (red). The slimness of the ellipses represents the coefficient of determination (R^2). Only significant correlations ($p < 0.05$) are shown.

A principal component analysis (PCA) across treatments explained 59.3% of the variability and separated the treatments along principal component 1 (PC1, 34.1%), with the exception of CLS-3 (Figure 5a). The traits that most contributed to PC1 consisted of morphological traits measured at the visible inflorescence stage, including stem and leaf biomass, stem diameter and plant height (Figure 5b). The latter observation explains why CLS-3-RI did not separate well from CLS-3-FI, because for this line, we did not detect a reduction in shoot biomass under RI conditions at the visible inflorescence stage (File S4). Stem water potential, leaf water content and thermal index 1, measured at the visible inflorescence stage, also contributed to PC1. Treatments were further separated by yield, the 1000 seed weight and plant height at harvest. Principle component 2 (PC2), which explained a further 25.2% of the variability, distinguished CLS lines from cv Regalona, with CLS-1 and CLS-5 being the most distant from cv Regalona as well as the other CLS lines under FI conditions (Figure 5a). Under RI conditions, CLS-5 remained the most distant, whereas CLS-9 was similar to cv Regalona. The traits that most contributed to PC2 were the timing of the milky and doughy grain stages, physiological maturity of the grains, seed size, panicle length, seed weight per plant and the efficiency of PSII under high-light conditions (Figure 5b). According to the PCA, plant height, stem biomass, stem water potential and thermal index 1, all measured at the visible inflorescence stage, were most closely related to yield, whereas panicle length and efficiency of PSII were associated with seed maturation stages. Plant height at harvest was related to seed size (Figure 5b). We did not find any similarity between the classification of the lines according to the DTI and YTI and the position of the lines in the PCA plot.

In a PCA of vegetation indices in combination with agronomical, morphological and physiological traits, PC1, PC2 and PC3 explained 88% of variability (Figure S5). A majority of traits contributing the most to the PC1-3 consisted of VIs developed as proxies for chlorophyll content. Only one water-related VI was in the top ten traits of PC1. The top-scoring VIs corresponded to the VIs for which a consistently large effect size across the three repetitions was found as well as a large number of contrasts between genotypes (Figure 3). For CLS-3 and CLS-7, treatments were most separated along PC1 and PC2 (Figure S5), whereas CLS-4 and CLS-5 (FI and RI), and CLS-6, CLS-9 and CLS-3 (RI) grouped together along PC1. For lines CLS-1 and CLS-4 and cv Regalona, treatments did not separate along PC1 or PC2. No consistent correspondence between DTI and YTI and clustering of lines was detected based on VIs. Also, no significant correlation with yield was found for any of the VIs, and only NPCI and SIPI showed a weakly significant correlation with 1000 seed weight (Table S2). The WBI correlated weakly but significantly with the 1.7 mm and 1.18 mm caliber seed weights. A majority of significant correlations were observed for morphological and physiological traits, with VIs including blue and red wavelengths correlating with morphological traits and stem and leaf water potential, and those calculated using green wavelengths correlating with leaf relative water content (Table S2). The strongest correlations overall were detected for MCARI and PSII efficiency, and PRI and stem and leaf water potential.

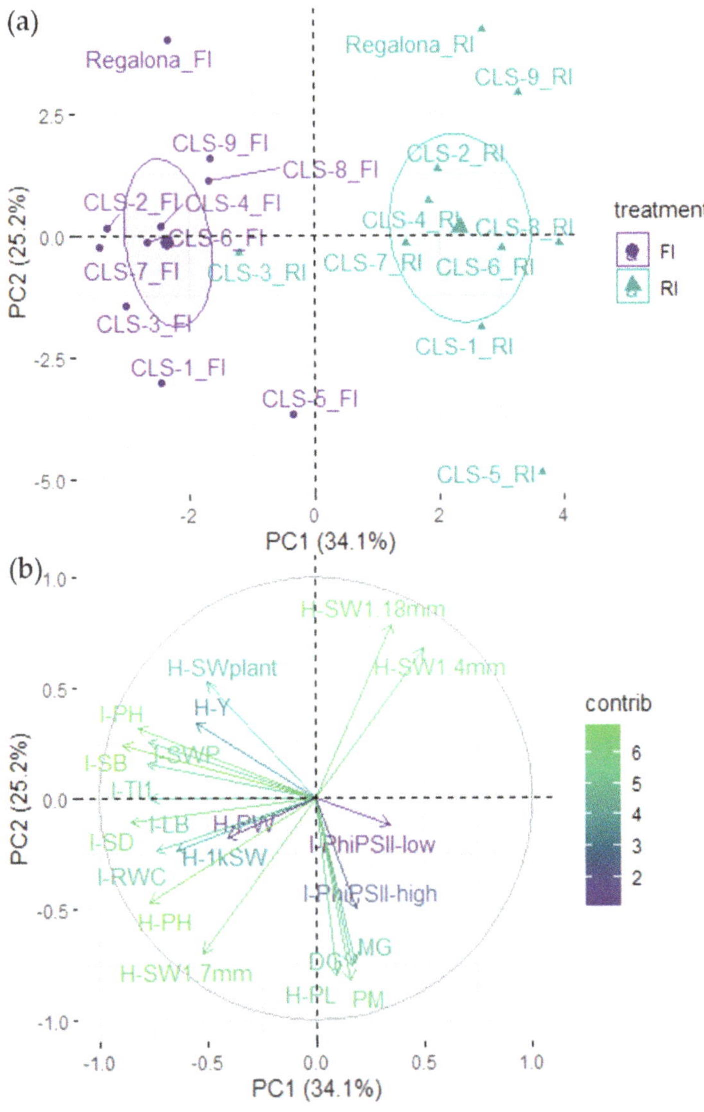

Figure 5. Principal component analysis of agronomical, morphological and physiological traits measured in nine coastal lowland self-pollinated lines (CLS) and commercial cv Regalona grown in a field experiment under full irrigation (FI) and reduced irrigation (RI). (**a**) Distribution of CLS lines and cv Regalona under FI and RI along principal components 1 and 2 (PC1 and PC2); (**b**) Contributions of traits to PC1 and PC2. Phenological traits included in the analysis were: days to milky grain stage (MG), days to doughy grain stage (DG), and days to physiological maturity (PM). Traits measured at the visible inflorescence stage included in the analysis were plant height (I-PH), stem diameter (I-SD), stem biomass (I-SB), leaf biomass (I-LB), leaf relative water content (I-RWC), stem water potential (I-SWP), thermal index 1 (I-TI1), and quantum efficiency of photosystem II in low-light (I-PhiPSII-low) and high-light (I-PhiPSII-high) conditions. Traits measured at harvest included in the analysis were plant height (H-PH), yield (H-Y), seed weight per plant (H-SWplant), 1000 seed weight (H-1kSW), 1.7 mm caliber seed weight (H-SW 1.7 mm), 1.4 mm caliber seed weight (H-SW 1.4 mm), 1.18 mm caliber seed weight (H-SW 1.18 mm), panicle length (H-PL) and panicle width (H-PW).

3. Discussion

Tolerance of quinoa to abiotic stresses such drought, salinity, low soil fertility and frost has been well documented, making it a target crop for addressing future food security in the context of a climate crisis [13,19,20,31,65]. Studies have recorded significant yield deficits, especially under low soil water availability and high vapor pressure deficit, high temperatures and nitrogen deficiency [15,66–70]. Inability to attain the full yield potential has been attributed to sink limitations, while higher yields could be obtained in quinoa if reproductive partitioning is increased [71]. Chilean quinoa varieties from the coastal lowland ecotype have become the basis of most breeding programs aimed at production in temperate environments such as northern Europe, due to their photoperiod insensitivity [32,34]. A high-quality reference genome assembly [72] based on the sequencing of the Chilean coastal quinoa accession PI 614886 (also known as NSL 106399 and QQ74) will contribute significantly to deeper understanding of the genetic attributes of quinoa and to further improvement of future breeding programs. The CLS lines grown in this study all belonged to the coastal lowland ecotype and were selected based on their yield potential across different geographical locations and seasons, architectural traits, and broad adaptability to adverse environmental conditions.

In quinoa, the flowering and grain-filling stages are considered the most critical for yield determination and the most sensitive to stress, including drought and high temperatures [18,68,70,73,74]. Indeterminate grain development in a complex panicle structure coupled with uneven grain filling lies at the basis of this sensitivity [67]. Stress induced by reduced irrigation was well established at the flowering and seed filling stages in our field trial and resulted in significant reductions in yield for the majority of CLS lines, but not in cv Regalona. Lower individual seed weight and a shift in seed size distribution from larger to smaller seeds were also observed. Individual seed weight and seed size were consistently among the smallest in cv Regalona, but total seed number may have compensated for this, as yield in cv Regalona was, on average, similar to the CLS lines. Grain number was the major component in grain yield determination, while grain weight showed a weak to strongly negative association with grain number across a multi-environmental evaluation for grain yield and its physiological determinants [75]. Nevertheless, seed size is an important commercial trait in quinoa [76] and breeding potential is considered to exist for both seed number and seed size [29,30,77,78]. The CLS lines showed potential in field performance for both yield and seed size, even though coastal lowland lines are characterized by a seed caliber under 2 mm. Indeed, fully irrigated (FI) plants yielded 3.24 t ha^{-1} on average, which decreased to 2.45 t ha^{-1} in reduced irrigation (RI) conditions. Nevertheless, in southern Mediterranean conditions the total seed yield ranged from 0.70 t ha^{-1} to 3.25 t ha^{-1}, even across seasons [79]. Considering recent reports from arid regions, quinoa Q26 produced the highest seed yield in Bastam, Iran (1.29 t ha^{-1} on average), which was not significantly different from Q29 (1.24 t ha^{-1}), while in Damghan, the highest seed yield was achieved in Q26 (1.19 t ha^{-1}) [80]. Another report from an arid growing site in China recorded the 1000 seed weight differences for two seasons (2.12–2.03 g) when soil matric potential decreased (−55 kPa SMP), which was significantly lower than under −15 kPa SMP (2.28–2.21 g), well below the averages of 1000 seed weight determined in this study [81].

Yield was maintained under RI in CLS-1 and CLS-2 as well as in cv Regalona, and was relatively sustained in CLS-9 despite the reduced water supply. Higher yields were recorded for both CLS-2 and CLS-9 compared to cv Regalona, which was similar in yield to CLS-1. High yields were achieved using different strategies for leaf relative water content, stem water potential, biomass and seed number and size. This response was similar to observations in the cvs Illpa and Rainbow, which used different strategies in the face of water deficit stress to prevent decreases in grain yield and quality under drought conditions [25]. At the visible inflorescence stage, CLS-2 had one of the largest recorded leaf biomasses, but also suffered the largest reduction in shoot biomass in response to the reduced water supply. High yields in CLS-2 were achieved with medium to large seeds on a short, compact panicle. In cv Regalona, a larger proportion of smaller caliber seeds was produced on a relatively short panicle. CLS-1 produced a higher proportion of large seeds on a long panicle, but overall seed size was lessened by reduced irrigation. Notably, CLS-1 seeds took a longer time to reach the doughy stage of seed maturation, whereas the overall shortest seed maturation time was recorded for the relatively small seeds of cv Regalona. Variation in grain weight was found to be strongly correlated with the rate of grain filling, and weakly or not associated with grain-filling duration [67,73], but this was not observed in our trial. CLS-9 was high yielding, mainly in terms of seed number, as it had a larger proportion of small seeds on a short panicle, like the phenotypic observations of cv Regalona. Notably, the TI1 suggested higher rates of transpiration under FI conditions for CLS-2, CLS-9 and cv Regalona.

CLS-7 had one of the highest proportions of large seeds recorded for both FI and RI treatments, although these were less numerous overall and were produced on a small panicle. Yield was, on average, severely reduced in CLS-5. Plants were short compared to other lines at the visible inflorescence stage, which was partially compensated for by a large panicle at harvest, although a lower proportion of small seeds compared to other lines was observed. CLS-8 already showed reduced stem and leaf biomasses under conditions of RI at the visible inflorescence stage. These effects were more pronounced compared to corresponding observation in cv Regalona. At harvest, CLS-8 plant height was low compared to other CLS lines, despite having a large panicle in both FI and RI treatments. On average, the lowest 1000 seed weight was recorded for CLS-8 and CLS-6, but overall seed yield was not consistently reduced in plots compared to FI. CLS-6 produced a fair yield in terms of seed numbers, but with a higher proportion of small (less than 1.7 mm caliber) seeds. CLS-8 was considered drought tolerant per se, which, according to the TI1, may have been related to very sensitive stomatal closure. Finally, CLS-3 exhibited tall plants both at the visible inflorescence stage and at harvest. While vegetative biomass was not reduced, inflorescences were smaller under RI and yield was reduced mainly due to a reduction in seed number, as CLS-3 produced the overall highest proportion of large seeds and had the heaviest 1000 seed weight. For the CLS lines, plant height at harvest was not correlated with yield, whereas lines with the tallest plants under RI, CLS-3, CLS-7, and CLS-1 and CLS-5 produced the highest proportion of large seeds under both FI and RI conditions (R^2 of 39%). Lines that showed the highest average yields, CLS-6 and CLS-9, generally had a higher proportion of smaller seeds, with the exception of CLS-2, which produced medium to large seeds.

Correlations between traits measured at the visible inflorescence stage and those measured at harvest were weak overall, most likely due to the small number of measurements per line taken at the visible inflorescence stage and the environmental conditions between flowering and harvest that affect plant physiology and, consequently, seed maturation. Nevertheless, these correlations may give indications about the influence of reduced irrigation from the branching to the visible inflorescence stage on the final yield and seed weight, and about the potential to predict agronomical traits from morphological and physiological traits measured earlier in the growing season. Between 15 and 35% of the variation in plant height at harvest and 5 to 16% of the variation in yield could be explained by shoot biomass measured at the visible inflorescence stage. The PCA analysis also confirmed a relationship between shoot morphology traits at the onset of flowering and yield. Among the physiological traits, leaf relative water content, stem water potential and the TI1 showed significant positive correlations with yield, indicating that these traits contribute to effects of reduced irrigation on yield. Moreover, a significant correlation was found between the TI1 and stem and leaf water potential (R^2 of 0.49 and 0.39 respectively), supporting the value of thermal infrared imaging for detecting differences in water use behavior among genotypes.

Plant drought response mechanisms have been reported in quinoa and include reduced growth [23,69], stomatal closure associated with abscisic acid and hydraulic signaling [22,25,82–84], peroxisome abundance as a cellular sensor [68], the accumulation of osmoprotectants, antioxidant defense and membrane stabilization [19,20,26,27], and elevated recovery capacities of PSII and PSI photochemical activities after re-watering [85]. Understanding how the physiological mechanisms employed by quinoa in response to drought as well as specific strategies implemented by different genotypes influence the final yield is crucial for both crop management and breeding. Better comprehension of genotype by environment (G × E) interactions and the optimization of irrigation for a specific crop's needs or to address an irrigation deficit [19,25,29,67] will lead to improved crop management (M) and higher-yielding harvests, improving G × E × M interactions. Nevertheless, we were unable to make clear distinctions between quinoa lines employing profligate or conservative water-use strategies [86] using the data collected in this trial via more traditional methods, as the frequency and number of measurements were relatively low. Canopy temperature and vegetation indices collected using handheld, ground-based or remote sensors could distinguish irrigated from non-irrigated treatments in a study by Sankaran et al. [38]. A significant relationship between water availability and canopy temperature was also detected for a selection of quinoa genotypes grown under different water regimes in the Brazilian Cerrado region [70]. We therefore included thermal infrared and hyperspectral imaging to determine whether these systems could deliver proxies for water use, photosynthetic activity or yield, or otherwise reveal differences between lines that may or may not have been observed using standard methods [36,87–92].

Both imaging techniques were fast and delivered data with a high temporal resolution and an increased frequency for a potentially better comparison of physiological responses between genotypes and treatments. The Huasco experimental center located in the southern Atacama Desert provided the ideal environmental conditions for this field trial. This was especially apparent for the measuring period at the visible inflorescence stage, where stable conditions coupled with high light intensity and clear skies throughout the day were ideal for applying imaging techniques. In addition, all measuring days had similar diurnal temperature and vapor pressure deficit profiles. The main factor that influenced plants differently was included in the trial itself as the treatment, namely soil water content. The observed responses to acute water deficit and irrigation (re-watering) in the measurement of stem and leaf water potential were also seen in the thermal and vegetation indices as the measurement day had a significant effect.

Further valuable information on the sensitivity of physiological processes to soil water content in the different CLS lines could have been obtained by means of diurnal imaging. With regard to the methodology to obtain the thermal indices, we were fortunate to observe wet leaves (dew) until quite late in the morning, when other environmental conditions (air temperature, radiation, VPD) were already at or close to the maximum values for the day. Covering leaves with petroleum jelly to obtain dry reference temperatures is not ideal, albeit a commonly used method. A more pragmatic approach would be to image a black reference target that adapts quickly to the prevailing environmental conditions, such as a thin sheet of aluminum [93]. An even, black reference surface can be reliably selected with automated image processing procedures. Data obtained from diurnal measurements before and after irrigation, as well as visible loss of turgor, may also help in setting thresholds on indices similar to the upper and lower baselines used in thermal imaging for irrigation scheduling [94].

In both the vegetation indices and the analysis of selected wavelengths, differences between lines were larger than treatment effects. The latter may have been influenced by the measurement days, as these included times of acute water deficit stress and recovery from stress after irrigation. Overall, the mean reflection in the green wavelengths was substantially decreased in RI conditions for most lines, except CLS-2, CLS-4 and CLS-7. Lower green reflection has been previously described in maize under conditions of drought stress [55]. This means that leaves appear darker green, which may be related to a higher concentration of chlorophyll in reduced mesophyll cell volumes and an accumulation of protective pigments [58]. Higher mean NIR reflection has been attributed to a decreased leaf water content and leaf thickness [55,95], and was observed in CLS-5, CLS-6 and CLS-7 in the RI treatment. The blue and green regions of the spectrum contributed the most in distinguishing genotypes, whereas the effect of the measurement day was most clearly observed in the red, red-edge and NIR region of the spectrum (File S1). The time interval since the last irrigation event may therefore have influenced these wavelength regions. Vegetation indices that consist of these wavelengths were the most effective in showing the measurement day effect or distinguishing lines. They included MCARI2, RGI, RGRI and G, consisting of green and red wavelengths, followed by WCI (green, red and NIR), CRI2 (blue, green and red-edge) and BRI (blue and green). CLS-7 and CLS-3 differed from many of the other lines in the FI treatment, whereas they were preceded by CLS-2 and CLS-4 in the RI treatment. The reasons that CLS-3 and CLS-7 strongly differ from each other, the other CLS lines and cv Regalona, based on spectral reflectance, are currently unknown. Nevertheless, both produced the highest proportion of large seeds and the highest 1000 seed weight.

The two water-related VIs could consistently detect differences between FI and RI plots. For the WBI, this was the case in CLS-1, CLS-6 and CLS-7, and for WCI in CLS-3, CLS-7, CLS-8 and CLS-9. No treatment effect was found for the WBI in CLS-2, CLS-5 and CLS-8. The WBI is based on the extent of the water-sensitive depression between 900 nm and 970 nm in the NIR [59], whereas the WCI includes wavelengths in the green, red and NIR regions [55]. In a study by Hinojosa et al. [37] using a handheld multispectral radiometer, the NDVI was suggested as a proxy for yield in quinoa. We could not confirm this result because no consistent relationship was found between the highest yielding CLS lines and their respective NDVI values. We did observe the highest gNDVI under FI conditions for CLS-4, which is known for its dark green leaves and red pigmentation. A similar observation was noted in Hinojosa et al. [37] for a genotype with red shoot coloration. Here, the gNDVI for CLS-4 was high because of a particularly low reflectance in green compared to the other CLS lines and cv Regalona.

No correspondence was found between the classification of lines according to the DTI and YTI and the clustering of lines in the PCA of VIs. In addition, no significant correlation with yield was found for any of the VIs. The majority of significant correlations were observed for morphological and physiological traits, with VIs including blue and red wavelengths correlating with morphological traits and stem and leaf water potential, and those calculated using green wavelengths correlating with leaf relative water content. The PCs mainly consisted of VIs that were originally developed to detect differences in chlorophyll fluorescence and photosynthesis. Potentially, lines CLS-3 and CLS-7 differ from the other lines in these traits. On the other hand, these VIs are also correlated with stem and leaf water potential as the latter has been associated with wavelengths in the red region [55]. We can conclude from this study that hyperspectral imaging has a great potential in estimating traits contributing to yield and in distinguishing genotypes along these traits, rather than providing proxies for yield itself or distinguishing genotypes based on yield.

Summary and Future Directions

- Significant correlations were detected between morphological and physiological traits measured at the onset of flowering and at harvest.
- Lines CLS-1, CLS-2 and CLS-9 performed best when faced with a 50% reduction in irrigation and performed well in terms of seed traits and plasticity for hyper-arid regions.
- Imaging techniques show good potential for high-throughput phenotyping of quinoa in future studies. Additional data from larger field trials will be needed to improve the quantitative evaluation of quinoa genotypic responses and their relationship to specific traits of interest, including productivity and physiological traits.

4. Materials and Methods

4.1. Field Trial Setup

Nine novel quinoa genotypes (CLS-1 to CLS-9) and one commercial cultivar (Regalona Baer, referred to in this text as cv Regalona), were grown in the field to investigate responses to full and reduced irrigation. The CLS genotypes were initially selected from the INIA SeedBank Collection based on morphological (i.e., branching type and panicle shape) and yield traits (i.e., total seed weight and seed diameter) observed in the field, and further developed in a breeding program by INIA Chile through a combination of mass selection, self-pollination of individual lines (over the course of at least two seasons) and panicle-furrow selection. The field trial took place at the INIA-Huasco experimental center located in the southern Atacama Desert (Vallenar, Chile, 28°34′ S, 70°47′ W and 469 m.a.s.l.) during the 2019/2020 growing season (sowing on 26 September 2019, harvest on 21 February 2020). This location was particularly well-suited for this trial as the Atacama Desert is known for being one of the driest regions in the world, ensuring that rainfall would not hinder the planned irrigation protocol. The soil corresponded to La Compañía series, a sandy loam textural class of soil composed of sand (75.5%), silt (10.9%) and clay (13.7%), with a low organic matter content (2.0%), an alkaline pH (8.1), slight salinity (2.4 dS m^{-1}), and with a basic content of N-P-K (45-21-311 mg kg^{-1} respectively).

The experimental design was split-plot with irrigation treatment as the main plot (between factor, two levels), genotype as the split-plot (within factor, 10 levels) and a blocking factor with six levels (Figure S6). Each plot of 2 by 4 m contained four rows of quinoa with an inter-row spacing of 0.5 m [76]. Seeds were sown at a density of 15 kg ha^{-1} with >85% germination. Two irrigation treatments were tested, with full irrigation (FI) considering the reposition of ET_0 and the lack of a crop coefficient (K_c) for quinoa growing in the region. For the reduced irrigation (RI) treatment, we defined a severe reduction of irrigation time (50%) and kept the same schedule as for the full irrigation treatment (every second or third day) by means of pressurized drip lines. The RI treatment started at growth stage 20 (branching, extended BBCH [39]). Fertilizer was provided at sowing (75-120-60 N-P-K) and an additional 75 N was provided at growth stage 12 (second pair of leaves visible, extended BBCH [39]). Plots were regularly inspected for weeds and pests, which were controlled manually, and disease was managed with the application of chlorothalonil (2 L ha^{-1} BRAVO 720, Syngenta) from growth stage 16 (six pairs of leaves visible, extended BBCH [39]). A meteorological station located at the INIA-Huasco experimental station, approximately 300 m from the experimental setup, collected weather data, and soil sensors that monitored volumetric water content and temperature were set up in both treatments (Figure 6). Total monthly precipitation recorded for the field site over the course of the trial was as follows: 0.0 mm from 26 to 30 September, 1.3 mm for the month of October, 0.3 mm for the month of November and 0.2 mm for the month of December. No precipitation was recorded for the months of January and February.

Over the course of the entire growing season, plots were regularly inspected and phenology was determined visually according to the extended BBCH scale by Sosa-Zúñiga et al. [39]. When plants were in the visible inflorescence stage of development (extended BBCH 59 [39]), morphological and physiological traits were measured (described in detail in File S4). The timing corresponded to 46–50 days after sowing (DAS). Morphological traits and leaf relative water content were measured on one day only for all plots of both treatments (47 DAS). Because of practical constraints related to time and the labor-intensive nature of some of the methods, the measurements of the other physiological traits were grouped per block and treatment at 48, 49 and 50 DAS.

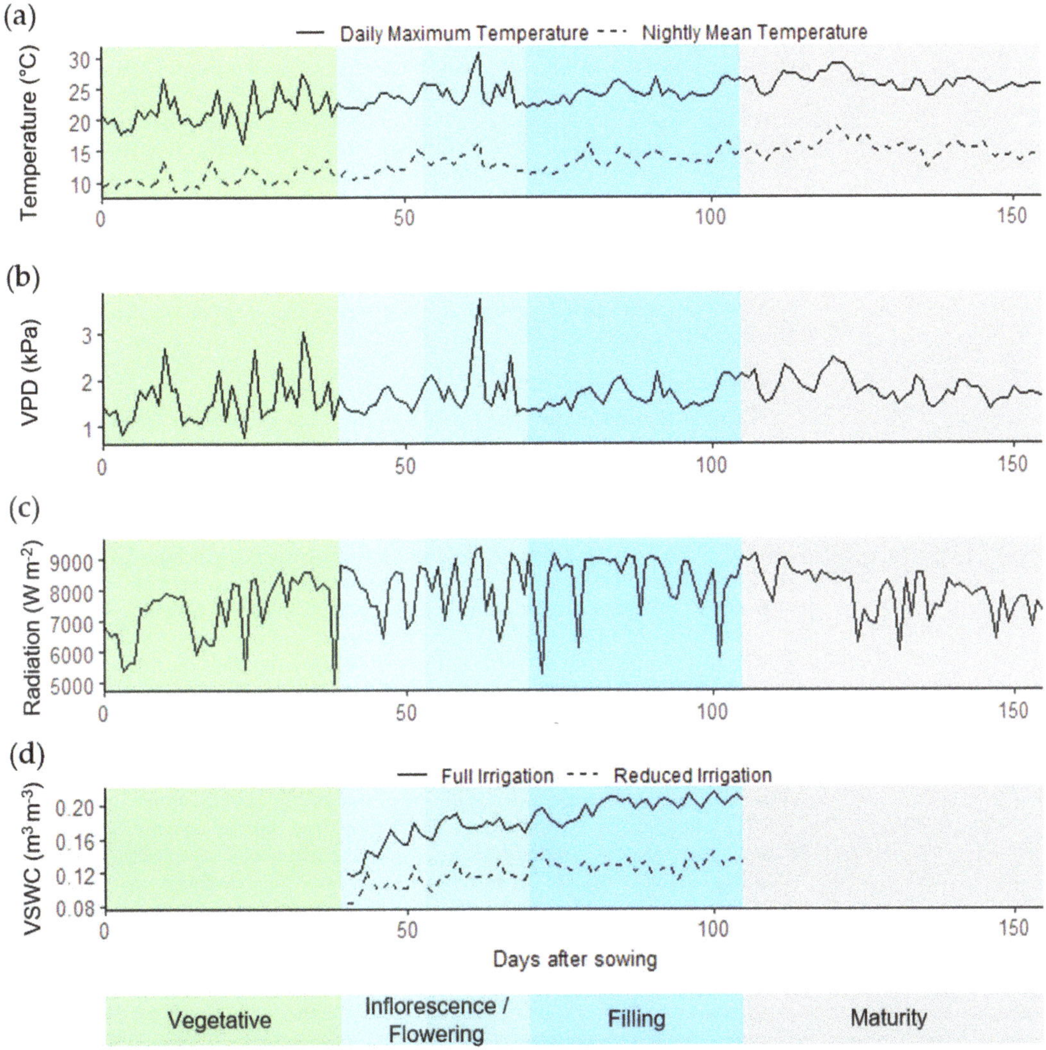

Figure 6. Weather data over the course of the field experiment from 26 September 2019 to 28 February 2020 collected by a meteorological station at the INIA-Huasco experimental center in the southern Atacama Desert (Vallenar, Chile, 28°34′ S, 70°47′ W and 469 m.a.s.l.) 300 m from the field trial and by soil sensors in the field trial; (**a**) daily maximum and night mean temperature; (**b**) daily maximum vapor pressure deficit (VPD); (**c**) daily total radiation; (**d**) daily mean volumetric soil water content (VSWC). The full irrigation and reduced irrigation treatments started at branching (35 days after sowing), but the sensor data were only available from 40 days after sowing. The colors in the figure represent the respective quinoa developmental stages.

4.2. Data Collected at Harvest

Plots were harvested during the senescence period once genotypes reached physiological maturity, i.e., when seeds from the main panicle became hard and resistant to pressure, which corresponds to a seed moisture content of about 20%. A 1 m^2 area consisting of two central rows was manually harvested in each plot. The plant number and height for the harvested area was recorded. Plant yield was determined as the total seed weight in one linear meter of two central rows per plot. Yield data are shown as the productivity in t ha^{-1} at a standard seed moisture content of 20%. Additionally, the total seed weight was divided by the number of plants sampled to obtain the seed weight per plant. The final height was measured for the same plants. Seed metrics included the 1000 seed weight, determined using a seed counting machine (S-JR, DATA Technologies, Kibbutz Tzora, Israel), and the proportional weight of a 10 g seed sample retained in sieves of different mesh openings (1.7, 1.4 and 1.18 mm) after 3 min of agitation at 65 rpm. Finally, panicle metrics were determined for 50 panicles of the central plot rows by imaging complete, intact panicles with a digital RGB camera (Nikon D3100, 16–55 mm lens), and measuring the panicle length and maximum width in the images using ImageJ [96]. A ruler was included in each image for calibration purposes. The drought tolerance index (DTI) and yield tolerance index (YTI) were calculated for each CLS genotype and cv Regalona as described in Ober et al. and del Pozo et al. [40,41].

4.3. Thermal Infrared Imaging

Thermal infrared imaging data were acquired using a CAT S60 smartphone equipped with a FLIR Lepton longwave infrared micro thermal camera module (https://www.catphones.com/ (accessed on 5 November 2019)). Images had a resolution of 320 × 480 pixels. Two images per plot were taken to cover the complete plot surface. Images overlapped at the center of the plot, where a fastened piece of crumpled aluminum foil (40 cm^2) and a leaf covered with petroleum jelly used as a dry reference could be included in both images. Images were acquired at the visible inflorescence stage of development for all plots in both the FI and RI treatments in the afternoon of 46 and 47DAS, whereas they were acquired in the afternoons of 48, 49 and 50 DAS for all plots of blocks 2 and 5, 1 and 4, and 3 and 6, respectively. Moreover, at 47 DAS, all plots were also imaged in the late morning when the leaves were still wet with morning dew. The derived temperature data were used as a wet reference. Images were processed in R using the 'Thermimage' package [97]. Settings for air temperature and relative humidity at the time of imaging were obtained from the collected environment data. Emissivity was set to 1 in the conversion of image data for measuring the mean temperature of the crumpled aluminum foil, which represented the reflected temperature. This was then applied together with an emissivity value of 0.96 to obtain the leaf temperature data. An ImageJ macro was used to semi-automatically determine regions of interest (ROIs) in the images and corresponding temperature data for the piece of aluminum foil, the dry reference leaf and patches of sunlit, exposed soil. The temperature data of these ROIs were then excluded from the image to obtain the mean leaf temperature data. Thermal index 1 (TI1) was calculated as follows:

$$TI1 = dTwet.m - dTm \quad (1)$$

The dTwet.m is the mean of the difference between the temperature of the wet leaves per plot at 47 DAS and the ambient air temperature at the time of imaging. The dTm is the difference between the mean leaf temperature per plot and the ambient air temperature at the time of imaging [88,89].

4.4. Hyperspectral Imaging

Hyperspectral image data were acquired using a Specim IQ (Specim Ltd., Oulu, Finland), a handheld push broom camera system with integrated operating system and controls [98]. The Specim IQ measures reflectance in the visible and near-infrared, i.e., from 400 to 1000 nm, with a spectral resolution (FWHM) of 7 nm, 204 spectral bands, and a spatial resolution of 512 × 512 $pixels^2$. The camera was mounted on a tripod at a height that allowed a complete individual plot to be captured in the image. The plots of blocks 1 and 4, and 2 and 5 were imaged at 49 and 48 DAS, respectively, between 15:00 and 16:00. The plots of blocks 3 and 6 were imaged at 49 DAS between 16:00 and 17:00. As blocks represented differences in the date and time of imaging, they were referred to as repetitions with blocks 1 and 4 assigned to repetition 1, blocks 2 and 5 to repetition 2, and blocks 3 and 6 to repetition 3. Plots were always captured in treatment pairs (RI after FI or vice versa). Each dataset contained a white reference tile, which was imaged simultaneously for data calibration. A dark reference, representing sensor noise without incoming light, was recorded automatically before each capture.

Upon image data acquisition, the Specim IQ integrated software allows for the selection of the white reference tile in the image based on its high reflectance values, in addition to automated calibration to obtain relative reflectance data. However, we noted that the white reference tile itself was not selected alone in some images, as other elements with high reflectance were present, such as pieces of crumpled aluminum foil (used for the measurement of stem water potential and thermal imaging). The calibration procedure was therefore redone in R after threshold-based selection of the white reference tile pixels using an ImageJ macro. All other hyperspectral data processing and analysis steps were also executed in R.

Plant pixels were segmented from the background using the Normalized Difference Vegetation Index (NDVI, Table S1) and a threshold level, which also excluded inflorescences and specular reflection. Shaded background and shaded plant parts with low reflectance were removed using a threshold in near-infrared (838 nm) and green (554 nm) wavelengths, respectively. Spectra were smoothed on the pixel level using the Savitzky–Golay smoothing filter [99] with a third order polynomial and a window size of 11 using the R package 'prospectr' [100].

A total of 41 published vegetation indices (VIs, Table S1) were calculated. By means of a cluster analysis, genotypes were grouped based on the similarity of VI data within the FI and RI treatment. Agglomerative hierarchical clustering was applied on the scaled VI mean observations for repetitions 1 to 3 using the 'agnes' function of the R package 'cluster' [101]. The trees were cut at five clusters. Pearson correlation coefficients were calculated to describe the linear relationships between traits measured at the visible inflorescence stage, plant morphology and performance traits measured at harvest, and VIs. In addition, differences in relative reflectance between genotypes and treatments, independent of VIs, were analyzed for a selection of wavelengths and wavelength bands. A selection was used because of the high degree of correlation or collinearity in the relative reflectance of mostly adjacent wavelengths. A Pearson correlation coefficient was calculated between the relative reflectance of all wavelengths. A threshold of 0.8 was then applied to split up the wavelength range in groups of high correlation. One wavelength was selected for further analysis per group. This yielded five wavelengths, 476 nm, 554 nm, 616 nm, 679 nm and 724 nm, in the blue, green, orange, red and red-edge regions of the spectrum, respectively. Furthermore, reflectance in the near-infrared region (NIR) was averaged and included in the selection.

4.5. Statistics

Statistics were performed in R (R version 4.0.3 [102]). Outliers were identified by applying the interquartile range method (R package 'rstatix' [103]). Data were checked for normality via visual inspection of the QQplot of the residuals of the model and by the Shapiro-Wilk test. Homogeneity of variances was determined by visually inspecting the residuals plot and by Levene's test.

The model 'trait ~ treatment × genotype + (1 | block)' was run for harvest data using the 'lmer' function in the R package 'lmerTest' [104]. Post-hoc pairwise comparisons were performed using estimated marginal means with a 95% confidence interval (R package 'emmeans' [105]). For thermal infrared imaging, a three-way ANOVA (TI1 ~ genotype × treatment × measurement day) was run. Significant main effects of treatment and measurement day, and their interaction, were followed by a one-way ANOVA to test the main effect of treatment at all levels of the measurement day, and vice versa.

For hyperspectral imaging, the VIs PSSRa, PSSRb, PSSRc, SR, SRChl, SRChlb and SRChltot, and relative reflectance data at 554 nm were log-transformed, a square root transformation was applied on data at 616 nm, 724 nm and the VI MCARI, and the reciprocal of the data of the VI WCI was used to improve normality before statistical testing could be performed. For VIs and each selected wavelength or wavelength band, a three-way ANOVA (vegetation index or wavelength ~ genotype × treatment × repetition) was run. Assumptions were checked based on the normal QQ plot and the residual plot. Statistical tests for normality or homogeneous variance could not be used because of the large sample size. The latter also affected the outcome of the ANOVA as all main and interaction effects were highly significant, and only very few non-significant contrasts (genotype and treatment) were detected. Effect sizes are therefore reported here. The three-way ANOVA was followed by a two-way ANOVA (vegetation index or wavelength ~ genotype × treatment or repetition). Effect sizes for the independent variable in the ANOVA models were the generalized eta squared, whereas they were calculated as pairwise differences of estimated marginal means, divided by the standard deviation of the population, for pairwise comparisons between genotypes and treatments.

Pearson or Spearman correlation coefficients, depending on the distribution of the trait data, were calculated to describe linear relationships between traits and indices measured at the visible inflorescence stage and agronomical traits measured at harvest (R packages 'Hmisc' [106] and 'corrplot' [107]). The mean values of traits per plot were used. A principal component analysis (PCA) was performed on the mean values of traits and indices per genotype and treatment using the 'prcomp' function in the 'stats' package [102]. The 'factoextra' package was used for the visualization of the PCA [108].

Supplementary Materials: The following are available online at https://www.mdpi.com/article/10.3390/plants11030323/s1, Figure S1: Seed metrics.pdf, Figure S2: Panicle metrics.pdf, Figure S3: Plant height at harvest.pdf, Figure S4: Vegetation indices-effect size.pdf, Figure S5: PCA of vegetation indices.pdf, Figure S6: Field trial plot layout and experimental design.pdf, Table S1: Vegetation indices.pdf, Table S2: Correlations between traits and vegetation indices.pdf, File S1: Analysis of selected wavelengths.pdf, File S2: Selected wavelengths-effect size-pairwise comparisons.xlsx, File S3: Phenology.xlsx, File S4: Morphological and physiological traits at the inflorescence visible stage.pdf.

Author Contributions: Conceptualization, K.D., N.W., F.F. and A.Z.-S.; methodology, K.D., N.W., F.F. and A.Z.-S.; formal analysis, K.D., N.W., F.F. and A.Z.-S.; investigation, K.D., N.W., F.F. and A.Z.-S.; writing—original draft preparation, K.D., N.W., F.F. and A.Z.-S.; writing—review and editing, K.D., N.W., C.A., D.C., F.F. and A.Z.-S.; project administration, C.A., D.C., F.F. and A.Z.-S.; funding acquisition, C.A., D.C., F.F. and A.Z.-S. All authors have read and agreed to the published version of the manuscript.

Funding: This research was funded by the German Ministry of Education and Research under the funding scheme Bioeconomy International 2015, project 'Quinoa Diversity' (FKZ 031B0293A to KD, FKZ 031B0293B to NW and FF) as well as Programa de Mejoramiento Genético Quínoa (PMG-Q 502181-70 INIA, Chile).

Institutional Review Board Statement: Not applicable.

Informed Consent Statement: Not applicable.

Data Availability Statement: The data presented in this study are available on request from the corresponding author. The data are not publicly available yet due to variety registration.

Acknowledgments: The authors would like to thank the PMG Quinoa Patricia Herrera for performing the planting and Sergio Rojas for his hard work maintaining the field trial. The authors would also like to thank Álvaro Castillo for providing the drone images of the field trial. The authors greatly appreciate the reviewers' comments on the previous version of this manuscript.

Conflicts of Interest: The authors declare no conflict of interest.

References

1. National Research Council. *Lost Crops of the Incas: Little-Known Plants of the Andes with Promise for Worldwide Cultivation*; National Academy Press: Washington, DC, USA, 1989; ISBN 978-0-309-04264-2.
2. Bazile, D.; Jacobsen, S.-E.; Verniau, A. The global expansion of quinoa: Trends and limits. *Front. Plant Sci.* **2016**, *7*, 850. [CrossRef] [PubMed]

3. Abugoch James, L.E. Quinoa (*Chenopodium quinoa* Willd.): Composition, chemistry, nutritional, and functional properties. *Adv. Food Nutr. Res.* **2009**, *58*, 1–31. [CrossRef] [PubMed]
4. Gordillo-Bastidas, E.; Díaz-Rizzolo, D.A.; Roura, E.; Massanés, T.; Gomis, R. Quinoa (*Chenopodium quinoa* Willd), from nutritional value to potential health benefits: An integrative review. *J. Nutr. Food Sci.* **2016**, *6*, 497. [CrossRef]
5. Lutz, M.; Martínez, A.; Martínez, E.A. Daidzein and genistein contents in seeds of quinoa (*Chenopodium quinoa* Willd.) from local ecotypes grown in arid chile. *Ind. Crop. Prod.* **2013**, *49*, 117–121. [CrossRef]
6. Paśko, P.; Zagrodzki, P.; Bartoń, H.; Chłopicka, J.; Gorinstein, S. Effect of quinoa seeds (*Chenopodium quinoa*) in diet on some biochemical parameters and essential elements in blood of high fructose-fed rats. *Plant Food. Hum. Nutr.* **2010**, *65*, 333–338. [CrossRef] [PubMed]
7. Vega-Gálvez, A.; Miranda, M.; Vergara, J.; Uribe, E.; Puente, L.; Martínez, E.A. Nutrition facts and functional potential of quinoa (*Chenopodium quinoa* Willd.), an ancient andean grain: A review. *J. Sci. Food Agric.* **2010**, *90*, 2541–2547. [CrossRef] [PubMed]
8. Antognoni, F.; Potente, G.; Biondi, S.; Mandrioli, R.; Marincich, L.; Ruiz, K.B. Free and conjugated phenolic profiles and antioxidant activity in quinoa seeds and their relationship with genotype and environment. *Plants* **2021**, *10*, 1046. [CrossRef]
9. Miranda, M.; Vega-Gálvez, A.; Uribe, E.; López, J.; Martínez, E.; Rodríguez, M.J.; Quispe, I.; Di Scala, K. Physico-chemical analysis, antioxidant capacity and vitamins of six ecotypes of chilean quinoa (*Chenopodium quinoa* Willd). *Proc. Food Sci.* **2011**, *1*, 1439–1446. [CrossRef]
10. Navruz-Varli, S.; Sanlier, N. Nutritional and health benefits of quinoa (*Chenopodium quinoa* Willd.). *J. Cereal Sci.* **2016**, *69*, 371–376. [CrossRef]
11. Santis, G.D.; Maddaluno, C.; D'Ambrosio, T.; Rascio, A.; Rinaldi, M.; Troisi, J. Characterisation of quinoa (*Chenopodium quinoa* Willd.) accessions for the saponin content in mediterranean environment. *Ital. J. Agron.* **2016**, *11*, 277–281. [CrossRef]
12. FAO; CIRAD. *State of the Art Report of Quinoa in the World in 2013*; Bazile, D., Bertero, D., Nieto, C., Eds.; FAO: Rome, Italy, 2015; ISBN 978-92-5-108558-5.
13. Zurita-Silva, A.; Fuentes, F.; Zamora, P.; Jacobsen, S.-E.; Schwember, A.R. Breeding quinoa (*Chenopodium quinoa* Willd.): Potential and perspectives. *Mol. Breed.* **2014**, *34*, 13–30. [CrossRef]
14. Hariadi, Y.; Marandon, K.; Tian, Y.; Jacobsen, S.-E.; Shabala, S. Ionic and osmotic relations in quinoa (*Chenopodium quinoa* Willd.) plants grown at various salinity levels. *J. Exp. Bot.* **2011**, *62*, 185–193. [CrossRef]
15. Razzaghi, F.; Ahmadi, S.H.; Jacobsen, S.-E.; Jensen, C.R.; Andersen, M.N. Effects of salinity and soil–drying on radiation use efficiency, water productivity and yield of quinoa (*Chenopodium quinoa* Willd.). *J. Agron. Crop Sci.* **2012**, *198*, 173–184. [CrossRef]
16. Ruiz-Carrasco, K.; Antognoni, F.; Coulibaly, A.K.; Lizardi, S.; Covarrubias, A.; Martínez, E.A.; Molina-Montenegro, M.A.; Biondi, S.; Zurita-Silva, A. Variation in salinity tolerance of four lowland genotypes of quinoa (*Chenopodium quinoa* Willd.) as assessed by growth, physiological traits, and sodium transporter gene expression. *Plant Physiol. Bioch.* **2011**, *49*, 1333–1341. [CrossRef]
17. Aly, A.A.; Al-Barakah, F.N.; El-Mahrouky, M.A. Salinity stress promote drought tolerance of *Chenopodium quinoa* Willd. *Commun. Soil Sci. Plan.* **2018**, *49*, 1331–1343. [CrossRef]
18. Eustis, A.; Murphy, K.M.; Barrios-Masias, F.H. Leaf gas exchange performance of ten quinoa genotypes under a simulated Heat wave. *Plants* **2020**, *9*, 81. [CrossRef]
19. Hinojosa, L.; González, J.A.; Barrios-Masias, F.H.; Fuentes, F.; Murphy, K.M. Quinoa abiotic stress responses: A review. *Plants* **2018**, *7*, 106. [CrossRef]
20. Morales, A.; Zurita-Silva, A.; Maldonado, J.; Silva, H. Transcriptional responses of chilean quinoa (*Chenopodium quinoa* Willd.) under water deficit conditions uncovers aba-independent expression patterns. *Front. Plant Sci.* **2017**, *8*, 216. [CrossRef]
21. Cocozza, C.; Pulvento, C.; Lavini, A.; Riccardi, M.; d'Andria, R.; Tognetti, R. Effects of increasing salinity stress and decreasing water availability on ecophysiological traits of quinoa (*Chenopodium quinoa* Willd.) grown in a mediterranean-type agroecosystem. *J. Agron. Crop Sci.* **2013**, *199*, 229–240. [CrossRef]
22. Jacobsen, S.-E.; Liu, F.; Jensen, C.R. Does root-sourced ABA play a role for regulation of stomata under drought in quinoa (*Chenopodium quinoa* Willd.). *Sci. Hortic* **2009**, *122*, 281–287. [CrossRef]
23. Sun, Y.; Liu, F.; Bendevis, M.; Shabala, S.; Jacobsen, S.-E. Sensitivity of two quinoa (*Chenopodium quinoa* Willd.) varieties to progressive drought stress. *J. Agron. Crop Sci.* **2014**, *200*, 12–23. [CrossRef]
24. Yang, A.; Akhtar, S.S.; Amjad, M.; Iqbal, S.; Jacobsen, S.-E. Growth and physiological responses of quinoa to drought and temperature stress. *J. Agron. Crop Sci.* **2016**, *202*, 445–453. [CrossRef]
25. Gámez, A.L.; Soba, D.; Zamarreño, Á.M.; García-Mina, J.M.; Aranjuelo, I.; Morales, F. Effect of water stress during grain filling on yield, quality and physiological traits of illpa and rainbow quinoa (*Chenopodium quinoa* Willd.) cultivars. *Plants* **2019**, *8*, 173. [CrossRef]
26. Szabados, L.; Savouré, A. Proline: A multifunctional amino acid. *Trends Plant Sci.* **2010**, *15*, 89–97. [CrossRef]
27. Sadak, M.S.; El-Bassiouny, H.M.S.; Dawood, M.G. Role of trehalose on antioxidant defense system and some osmolytes of quinoa plants under water deficit. *Bull. Natl. Res. Cent.* **2019**, *43*, 5. [CrossRef]
28. Garcia-Parra, M.; Stechauner-Rohringer, R.; Roa-Acosta, D.; Ortiz-González, D.; Ramirez-Correa, J.; Plazas-Leguizamón, N.; Colmenares-Cruz, A. Chlorophyll fluorescence and its relationship with physiological stress in *Chenopodium quinoa* Willd. *Not. Bot. Horti Agrobo.* **2020**, *48*, 1742–1755. [CrossRef]
29. Jacobsen, S.-E. The scope for adaptation of quinoa in northern latitudes of Europe. *J. Agron. Crop Sci.* **2017**, *203*, 603–613. [CrossRef]

30. Murphy, K.M.; Matanguihan, J.B.; Fuentes, F.F.; Gómez-Pando, L.R.; Jellen, E.N.; Maughan, P.J.; Jarvis, D.E. Quinoa breeding and genomics. In *Plant Breeding Reviews*; Goldman, I., Ed.; John Wiley & Sons, Ltd.: Hoboken, NJ, USA, 2018; Volume 42, pp. 257–320. ISBN 978-1-119-52135-8.
31. Ruiz, K.B.; Biondi, S.; Oses, R.; Acuña-Rodríguez, I.S.; Antognoni, F.; Martinez-Mosqueira, E.A.; Coulibaly, A.; Canahua-Murillo, A.; Pinto, M.; Zurita-Silva, A.; et al. Quinoa biodiversity and sustainability for food security under climate change. A review. *Agron. Sustain. Dev.* **2014**, *34*, 349–359. [CrossRef]
32. Bertero, H.D.; King, R.W.; Hall, A.J. Photoperiod-sensitive development phases in quinoa (*Chenopodium quinoa* Willd.). *Field Crop. Res.* **1999**, *60*, 231–243. [CrossRef]
33. Galwey, N.W. The potential of quinoa as a multi-purpose crop for agricultural diversification: A review. *Ind. Crop. Prod.* **1992**, *1*, 101–106. [CrossRef]
34. Jacobsen, S.-E. Adaptation of quinoa (*Chenopodium quinoa*) to northern european agriculture: Studies on developmental pattern. *Euphytica* **1997**, *96*, 41–48. [CrossRef]
35. Araus, J.L.; Cairns, J.E. Field high-throughput phenotyping: The new crop breeding frontier. *Trends Plant Sci.* **2014**, *19*, 52–61. [CrossRef] [PubMed]
36. Watt, M.; Fiorani, F.; Usadel, B.; Rascher, U.; Muller, O.; Schurr, U. Phenotyping: New windows into the plant for breeders. *Annu. Rev. Plant Biol.* **2020**, *71*, 689–712. [CrossRef] [PubMed]
37. Hinojosa, L.; Kumar, N.; Gill, K.S.; Murphy, K.M. Spectral reflectance indices and physiological parameters in quinoa under contrasting irrigation regimes. *Crop Sci.* **2019**, *59*, 1927–1944. [CrossRef]
38. Sankaran, S.; Espinoza, C.Z.; Hinojosa, L.; Ma, X.; Murphy, K. High-throughput field phenotyping to assess irrigation treatment effects in quinoa. *Agrosyst. Geosci. Environ.* **2019**, *2*, 180063. [CrossRef]
39. Sosa-Zuniga, V.; Brito, V.; Fuentes, F.; Steinfort, U. Phenological growth stages of quinoa (*Chenopodium quinoa*) based on the BBCH scale. *Ann. Appl. Biol.* **2017**, *171*, 117–124. [CrossRef]
40. Ober, E.S.; Clark, C.J.A.; Bloa, M.L.; Royal, A.; Jaggard, K.W.; Pidgeon, J.D. Assessing the genetic resources to improve drought tolerance in sugar beet: Agronomic traits of diverse genotypes under droughted and irrigated conditions. *Field Crop. Res.* **2004**, *90*, 213–234. [CrossRef]
41. del Pozo, A.; Yáñez, A.; Matus, I.A.; Tapia, G.; Castillo, D.; Sanchez-Jardón, L.; Araus, J.L. Physiological traits associated with wheat yield potential and performance under water-stress in a mediterranean environment. *Front. Plant Sci.* **2016**, *7*, 987. [CrossRef]
42. Aparicio, N.; Villegas, D.; Araus, J.L.; Casadesús, J.; Royo, C. Relationship between growth traits and spectral vegetation indices in durum wheat. *Crop Sci.* **2002**, *42*, 1547–1555. [CrossRef]
43. Blackburn, G.A. Spectral indices for estimating photosynthetic pigment concentrations: A test using senescent tree leaves. *Int. J. Remote Sens.* **1998**, *19*, 657–675. [CrossRef]
44. Broge, N.H.; Leblanc, E. Comparing prediction power and stability of broadband and hyperspectral vegetation indices for estimation of green leaf area index and canopy chlorophyll density. *Remote Sens. Environ.* **2001**, *76*, 156–172. [CrossRef]
45. Datt, B. A new reflectance index for remote sensing of chlorophyll content in higher plants: Tests using eucalyptus leaves. *J. Plant Physiol.* **1999**, *154*, 30–36. [CrossRef]
46. Datt, B. Remote Sensing of chlorophyll *a*, chlorophyll *b*, chlorophyll *a+b*, and total carotenoid content in eucalyptus leaves. *Remote Sens. Environ.* **1998**, *66*, 111–121. [CrossRef]
47. Daughtry, C.S.T.; Walthall, C.L.; Kim, M.S.; de Colstoun, E.B.; McMurtrey, J.E. Estimating corn leaf chlorophyll concentration from leaf and canopy reflectance. *Remote Sens. Environ.* **2000**, *74*, 229–239. [CrossRef]
48. Gamon, J.A.; Peñuelas, J.; Field, C.B. A narrow-waveband spectral index that tracks diurnal changes in photosynthetic efficiency. *Remote Sens. Environ.* **1992**, *41*, 35–44. [CrossRef]
49. Gamon, J.A.; Surfus, J.S. Assessing leaf pigment content and activity with a reflectometer. *New Phytol.* **1999**, *143*, 105–117. [CrossRef]
50. Gitelson, A.A.; Zur, Y.; Chivkunova, O.B.; Merzlyak, M.N. Assessing carotenoid content in plant leaves with reflectance spectroscopy. *Photochem. Photobiol.* **2002**, *75*, 272–281. [CrossRef]
51. Gitelson, A.A.; Merzlyak, M.N. Remote estimation of chlorophyll content in higher plant leaves. *Int. J. Remote Sens.* **1997**, *18*, 2691–2697. [CrossRef]
52. Haboudane, D.; Miller, J.R.; Pattey, E.; Zarco-Tejada, P.J.; Strachan, I.B. Hyperspectral vegetation indices and novel algorithms for predicting green LAI of crop canopies: Modeling and validation in the context of precision agriculture. *Remote Sens. Environ.* **2004**, *90*, 337–352. [CrossRef]
53. Jordan, C.F. Derivation of leaf-area index from quality of light on the forest floor. *Ecology* **1969**, *50*, 663–666. [CrossRef]
54. Lichtenthaler, H.K.; Lang, M.; Sowinska, M.; Heisel, F.; Miehé, J.A. Detection of vegetation stress via a new high resolution fluorescence imaging system. *J. Plant Physiol.* **1996**, *148*, 599–612. [CrossRef]
55. Mertens, S.; Verbraeken, L.; Sprenger, H.; Demuynck, K.; Maleux, K.; Cannoot, B.; Deblock, J.; Maere, S.; Nelissen, H.; Bonaventure, G.; et al. Proximal hyperspectral imaging detects diurnal and drought-induced changes in maize physiology. *Front. Plant Sci.* **2021**, *12*, 240. [CrossRef]
56. Oppelt, N. Monitoring of Plant Chlorophyll and Nitrogen Status Using the Airborne Imaging Spectrometer AVIS. Ph.D. Thesis, Ludwig-Maximilians-Universität München, München, Germany, 2002.

57. Peñuelas, J.; Filella, I.; Gamon, J.A. Assessment of photosynthetic radiation-use efficiency with spectral reflectance. *New Phytol.* **1995**, *131*, 291–296. [CrossRef]
58. Peñuelas, J.; Gamon, J.A.; Fredeen, A.L.; Merino, J.; Field, C.B. Reflectance indices associated with physiological changes in nitrogen- and water-limited sunflower leaves. *Remote Sens. Environ.* **1994**, *48*, 135–146. [CrossRef]
59. Peñuelas, J.; Filella, I.; Biel, C.; Serrano, L.; Save, R. The reflectance at the 950–970 nm region as an indicator of plant water status. *Int. J. Remote Sens.* **1993**, *14*, 1887–1905. [CrossRef]
60. Roujean, J.-L.; Breon, F.-M. Estimating PAR absorbed by vegetation from bidirectional reflectance measurements. *Remote Sens. Environ.* **1995**, *51*, 375–384. [CrossRef]
61. Rouse, J.W.; Haas, R.H.; Schell, J.A.; Deering, D.W.; Harlan, J.C. *Monitoring the Vernal Advancement and Retrogradation of Natural Vegetation*; NASA/GSFC: Greenbelt, MD, USA, 1974; pp. 1–371.
62. Zarco-Tejada, P.J.; Berjón, A.; López-Lozano, R.; Miller, J.R.; Martín, P.; Cachorro, V.; González, M.R.; de Frutos, A. Assessing vineyard condition with hyperspectral indices: Leaf and canopy reflectance simulation in a row-structured discontinuous canopy. *Remote Sens. Environ.* **2005**, *99*, 271–287. [CrossRef]
63. Zarco-Tejada, P.J.; Miller, J.R.; Noland, T.L.; Mohammed, G.H.; Sampson, P.H. Scaling-up and model inversion methods with narrowband optical indices for chlorophyll content estimation in closed forest canopies with hyperspectral data. *IEEE T. Geosci. Remote* **2001**, *39*, 1491–1507. [CrossRef]
64. Zarco-Tejada, P.J.; Miller, J.R. Land cover mapping at BOREAS using red edge spectral parameters from CASI imagery. *J. Geophys. Res.-Atmos.* **1999**, *104*, 27921–27933. [CrossRef]
65. Razzaghi, F.; Ahmadi, S.H.; Adolf, V.I.; Jensen, C.R.; Jacobsen, S.-E.; Andersen, M.N. Water relations and transpiration of quinoa (*Chenopodium quinoa* Willd.) under salinity and soil drying. *J. Agron. Crop Sci.* **2011**, *197*, 348–360. [CrossRef]
66. Bascuñán-Godoy, L.; Sanhueza, C.; Pinto, K.; Cifuentes, L.; Reguera, M.; Briones, V.; Zurita-Silva, A.; Álvarez, R.; Morales, A.; Silva, H. Nitrogen physiology of contrasting genotypes of *Chenopodium quinoa* Willd. (Amaranthaceae). *Sci. Rep.* **2018**, *8*, 17524. [CrossRef] [PubMed]
67. Bertero, H.D. Chapter 7—Quinoa. In *Crop Physiology Case Histories for Major Crops*; Sadras, V.O., Calderini, D.F., Eds.; Academic Press: Cambridge, MA, USA, 2021; pp. 250–281. ISBN 978-0-12-819194-1.
68. Hinojosa, L.; Sanad, M.N.M.E.; Jarvis, D.E.; Steel, P.; Murphy, K.; Smertenko, A. Impact of heat and drought stress on peroxisome proliferation in quinoa. *Plant J.* **2019**, *99*, 1144–1158. [CrossRef] [PubMed]
69. Miranda-Apodaca, J.; Agirresarobe, A.; Martínez-Goñi, X.S.; Yoldi-Achalandabaso, A.; Pérez-López, U. N metabolism performance in *Chenopodium quinoa* subjected to drought or salt stress conditions. *Plant Physiol. Bioch.* **2020**, *155*, 725–734. [CrossRef] [PubMed]
70. da Silva, P.C.; Ribeiro Junior, W.Q.; Ramos, M.L.G.; Celestino, S.M.C.; Silva, A.D.N.; Casari, R.A.D.C.N.; Santana, C.C.; de Lima, C.A.; Williams, T.C.R.; Vinson, C.C. Quinoa for the brazilian cerrado: Agronomic characteristics of elite genotypes under different water regimes. *Plants* **2021**, *10*, 1591. [CrossRef]
71. Gómez, M.B.; Castro, P.A.; Mignone, C.; Bertero, H.D.; Gómez, M.B.; Castro, P.A.; Mignone, C.; Bertero, H.D. Can yield potential be increased by manipulation of reproductive partitioning in quinoa (*Chenopodium quinoa*)? Evidence from gibberellic acid synthesis inhibition using paclobutrazol. *Funct. Plant Biol.* **2011**, *38*, 420–430. [CrossRef]
72. Jarvis, D.E.; Ho, Y.S.; Lightfoot, D.J.; Schmöckel, S.M.; Li, B.; Borm, T.J.A.; Ohyanagi, H.; Mineta, K.; Michell, C.T.; Saber, N.; et al. The genome of *Chenopodium quinoa*. *Nature* **2017**, *542*, 307–312. [CrossRef]
73. Bertero, H.D.; Ruiz, R.A. Determination of seed number in sea level quinoa (*Chenopodium quinoa* Willd.) cultivars. *Eur. J. Agron.* **2008**, *28*, 186–194. [CrossRef]
74. Geerts, S.; Raes, D.; Garcia, M.; Taboada, C.; Miranda, R.; Cusicanqui, J.; Mhizha, T.; Vacher, J. Modeling the potential for closing quinoa yield gaps under varying water availability in the bolivian altiplano. *Agr. Water Manag.* **2009**, *96*, 1652–1658. [CrossRef]
75. Curti, R.N.; de la Vega, A.J.; Andrade, A.J.; Bramardi, S.J.; Bertero, H.D. Multi-environmental evaluation for grain yield and its physiological determinants of quinoa genotypes across northwest argentina. *Field Crop. Res.* **2014**, *166*, 46–57. [CrossRef]
76. Bertero, H.D.; de la Vega, A.J.; Correa, G.; Jacobsen, S.E.; Mujica, A. Genotype and genotype-by-environment interaction effects for grain yield and grain size of quinoa (*Chenopodium quinoa* Willd.) as revealed by pattern analysis of international multi-environment trials. *Field Crop. Res.* **2004**, *89*, 299–318. [CrossRef]
77. Gomez-Pando, L. Quinoa Breeding. In *Quinoa: Improvement and Sustainable Production*; Murphy, K., Matanguihan, J., Eds.; John Wiley & Sons, Ltd.: Hoboken, NJ, USA, 2015; pp. 87–108. ISBN 978-1-118-62804-1.
78. Zurita-Silva, A.; Jacobsen, S.-E.; Razzaghi, F.; Alvarez-Flores, R.; Ruiz, K.B.; Morales, A.; Silva, H. Quinoa drought responses and adaptation. Chapter 2.4. In *FAO & CIRAD. State of the Art Report on Quinoa around the World in 2013*; Bazile, D., Bertero, D., Nieto, C., Eds.; FAO: Rome, Italy, 2015; pp. 157–171.
79. Granado-Rodríguez, S.; Aparicio, N.; Matías, J.; Pérez-Romero, L.F.; Maestro, I.; Gracés, I.; Pedroche, J.J.; Haros, C.M.; Fernandez-Garcia, N.; Navarro del Hierro, J.; et al. Studying the impact of different field environmental conditions on seed quality of quinoa: The case of three different years changing seed nutritional traits in Southern Europe. *Front. Plant Sci.* **2021**, *12*, 854. [CrossRef]
80. Nadali, F.; Asghari, H.R.; Abbasdokht, H.; Dorostkar, V.; Bagheri, M. Improved quinoa growth, physiological response, and yield by hydropriming under drought stress conditions. *Gesunde Pflanz.* **2021**, *73*, 53–66. [CrossRef]
81. Wang, X.; Vignjevic, M.; Jiang, D.; Jacobsen, S.; Wollenweber, B. Improved tolerance to drought stress after anthesis due to priming before anthesis in wheat (*Triticum aestivum* L.) var. Vinjett. *J. Exp. Bot.* **2014**, *65*, 6441–6456. [CrossRef]

82. González, J.A.; Bruno, M.; Valoy, M.; Prado, F.E. Genotypic variation of gas exchange parameters and leaf stable carbon and nitrogen isotopes in ten quinoa cultivars grown under drought. *J. Agron. Crop Sci.* **2011**, *197*, 81–93. [CrossRef]
83. Raney, J.A.; Reynolds, D.J.; Elzinga, D.B.; Page, J.; Udall, J.A.; Jellen, E.N.; Bonfacio, A.; Fairbanks, D.J.; Maughan, P.J. Transcriptome analysis of drought induced stress in *Chenopodium quinoa*. *AJPS* **2014**, *5*, 338–357. [CrossRef]
84. Razzaghi, F.; Jacobsen, S.-E.; Jensen, C.R.; Andersen, M.N. Ionic and photosynthetic homeostasis in quinoa challenged by salinity and drought—Mechanisms of tolerance. *Funct. Plant Biol.* **2015**, *42*, 136–148. [CrossRef]
85. Manaa, A.; Goussi, R.; Derbali, W.; Cantamessa, S.; Essemine, J.; Barbato, R. Photosynthetic performance of quinoa (*Chenopodium quinoa* Willd.) after exposure to a gradual drought stress followed by a recovery period. *Biochim. Biophys. Acta (BBA)-Bioenerg.* **2021**, *1862*, 148383. [CrossRef]
86. Blessing, C.H.; Mariette, A.; Kaloki, P.; Bramley, H. Profligate and conservative: Water use strategies in grain legumes. *J. Exp. Bot.* **2018**, *69*, 349–369. [CrossRef]
87. Kim, D.M.; Zhang, H.; Zhou, H.; Du, T.; Wu, Q.; Mockler, T.C.; Berezin, M.Y. Highly sensitive image-derived indices of water-stressed plants using hyperspectral imaging in SWIR and histogram analysis. *Sci. Rep.* **2015**, *5*, 15919. [CrossRef]
88. Maes, W.H.; Steppe, K. Estimating evapotranspiration and drought stress with ground-based thermal remote sensing in agriculture: A review. *J. Exp. Bot.* **2012**, *63*, 4671–4712. [CrossRef]
89. Perich, G.; Hund, A.; Anderegg, J.; Roth, L.; Boer, M.P.; Walter, A.; Liebisch, F.; Aasen, H. Assessment of multi-image unmanned aerial vehicle based high-throughput field phenotyping of canopy temperature. *Front. Plant Sci.* **2020**, *11*, 150. [CrossRef]
90. Yendrek, C.R.; Tomaz, T.; Montes, C.M.; Cao, Y.; Morse, A.M.; Brown, P.J.; McIntyre, L.M.; Leakey, A.D.B.; Ainsworth, E.A. High-throughput phenotyping of maize leaf physiological and biochemical traits using hyperspectral reflectance. *Plant Physiol.* **2017**, *173*, 614–626. [CrossRef]
91. Zarco-Tejada, P.J.; González-Dugo, V.; Williams, L.E.; Suárez, L.; Berni, J.A.J.; Goldhamer, D.; Fereres, E. A PRI-based water stress index combining structural and chlorophyll effects: Assessment using diurnal narrow-band airborne imagery and the CWSI thermal index. *Remote Sens. Environ.* **2013**, *138*, 38–50. [CrossRef]
92. Zhang, X.; He, Y. Rapid estimation of seed yield using hyperspectral images of oilseed rape leaves. *Ind. Crop. Prod.* **2013**, *42*, 416–420. [CrossRef]
93. Vialet-Chabrand, S.; Lawson, T. Thermography methods to assess stomatal behaviour in a dynamic environment. *J. Exp. Bot.* **2020**, *71*, 2329–2338. [CrossRef]
94. Payero, J.O.; Irmak, S. Variable upper and lower crop water stress index baselines for corn and soybean. *Irrig. Sci.* **2006**, *25*, 21–32. [CrossRef]
95. Neilson, E.H.; Edwards, A.M.; Blomstedt, C.K.; Berger, B.; Møller, B.L.; Gleadow, R.M. Utilization of a high-throughput shoot imaging system to examine the dynamic phenotypic responses of a C4 cereal crop plant to nitrogen and water deficiency over time. *J. Exp. Bot.* **2015**, *66*, 1817–1832. [CrossRef]
96. Schneider, C.A.; Rasband, W.S.; Eliceiri, K.W. NIH image to imagej: 25 years of image analysis. *Nat. Meth.* **2012**, *9*, 671–675. [CrossRef]
97. Tattersall, G.J. Thermimage: Thermal Image Analysis. Available online: https://CRAN.R-project.org/package=Thermimage (accessed on 27 February 2021).
98. Behmann, J.; Acebron, K.; Emin, D.; Bennertz, S.; Matsubara, S.; Thomas, S.; Bohnenkamp, D.; Kuska, M.T.; Jussila, J.; Salo, H.; et al. Specim IQ: Evaluation of a new, miniaturized handheld hyperspectral camera and its application for plant phenotyping and disease detection. *Sensors* **2018**, *18*, 441. [CrossRef]
99. Savitzky, A.; Golay, M.J.E. Smoothing and differentiation of data by simplified least squares procedures. *Anal. Chem.* **1964**, *36*, 1627–1639. [CrossRef]
100. Stevens, A.; Ramirez-Lopez, L. Miscellaneous Functions for Processing and Sample Selection of Vis-NIR Diffuse Reflectance Data. Available online: https://github.com/l-ramirez-lopez/prospectr (accessed on 30 November 2020).
101. Maechler, M.; Rousseeuw, P.; Struyf, A.; Hubert, M.; Hornik, K.; Studer, M.; Roudier, P.; Gonzalez, J.; Kozlowski, K.; Schubert, E.; et al. "Finding Groups in Data": Cluster Analysis Extended Rousseeuw et al. Available online: https://cran.r-project.org/web/packages/cluster/cluster.pdf (accessed on 26 February 2021).
102. R Development CORE TEAM. A Language and Environment for Statistical Computing. R Foundation for Statistical Computing, Vienna, Austria. 2020. Available online: https://www.R-project.org/ (accessed on 10 October 2020).
103. Kassambara, A. Pipe-Friendly Framework for Basic Statistical Tests. Available online: https://cran.r-project.org/web/packages/rstatix/rstatix.pdf (accessed on 25 February 2021).
104. Kuznetsova, A.; Brockhoff, P.B.; Christensen, R.H.B. LmerTest package: Tests in linear mixed effects models. *J. Stat. Softw.* **2017**, *82*, 1–26. [CrossRef]
105. Lenth, R.V.; Buerkner, P.; Herve, M.; Love, J.; Riebl, H.; Singmann, H. Estimated Marginal Means, aka Least-Squares Means. Available online: https://cran.r-project.org/web/packages/emmeans/emmeans.pdf (accessed on 25 February 2021).
106. Harrell Jr., F.E. Hmisc: Harrell Miscellaneous. Available online: https://CRAN.R-project.org/package=Hmisc (accessed on 12 August 2021).

107. Wei, T.; Simko, V.; Levy, M.; Xie, Y.; Jin, Y.; Zemla, J.; Freidank, M.; Cai, J.; Protivinsky, T. Corrplot: Visualization of a Correlation Matrix. Available online: https://CRAN.R-project.org/package=corrplot (accessed on 12 August 2021).
108. Kassambara, A.; Mundt, F. Factoextra: Extract and Visualize the Results of Multivariate Data Analyses. Available online: https://CRAN.R-project.org/package=factoextra (accessed on 12 August 2021).

plants

Article

Physicochemical Characterization of Thirteen Quinoa (*Chenopodium quinoa* Willd.) Varieties Grown in North-West Europe—Part II

Phara De Bock [1,*], Gerda Cnops [2], Hilde Muylle [2], Paul Quataert [2], Mia Eeckhout [1] and Filip Van Bockstaele [3]

[1] Research Unit of Cereal and Feed Technology, Department of Food Technology, Safety and Health, Faculty of Bioscience Engineering, Ghent University, 9000 Ghent, Belgium; mia.eeckhout@ugent.be
[2] Plant Sciences Unit, Flanders Research Institute for Agriculture, Fisheries and Food (ILVO), 9090 Melle, Belgium; gerda.cnops@ilvo.vlaanderen.be (G.C.); hilde.muylle@ilvo.vlaanderen.be (H.M.); paul.quataert@ilvo.vlaanderen.be (P.Q.)
[3] Food Structure and Function Research Group (FSF), Department of Food Technology, Safety and Health, Faculty of Bioscience Engineering, Ghent University, 9000 Ghent, Belgium; filip.vanbockstaele@ugent.be
* Correspondence: phara.debock@ugent.be

Abstract: Quinoa cultivation has gained increasing interest in Europe but more research on the characteristics of European varieties is required to help determine their end use applications. A comparative study was performed on 13 quinoa varieties cultivated under North-West European field conditions during three consecutive growing seasons (2017–2019). The seeds were milled to wholemeal flour (WMF) to evaluate the physicochemical properties. The WMFs of 2019 were characterized by the highest water absorption capacity (1.46–2.06 g/g), while the water absorption index (WAI) between 55 °C (2.04–3.80 g/g) and 85 °C (4.04–7.82 g/g) increased over the years. The WMFs of 2018 had the highest WAI at 95 °C (6.48–9.48 g/g). The pasting profiles were characterized by a high viscosity peak (1696–2560 mPa.s) and strong breakdown (−78–643 mPa.s) in 2017. The peak viscosity decreased in 2018 and 2019 (823–2492 mPa.s), while breakdown (−364–555 mPa.s) and setback (19–1037 mPa.s) increased. Jessie, Summer Red, Rouge Marie, Vikinga, and Zwarte WMFs were characterized by low WAIs and high shear resistance. Bastille WMF developed high viscosities and, along with Faro WMF, showed a high breakdown. The wide variation in physicochemical properties suggests that the potential food applications of WMFs depend on the variety and growing conditions.

Keywords: *Chenopodium quinoa* Willd.; North-West Europe; wholemeal flour; water holding capacity; water absorption index; pasting behavior

Citation: De Bock, P.; Cnops, G.; Muylle, H.; Quataert, P.; Eeckhout, M.; Van Bockstaele, F. Physicochemical Characterization of Thirteen Quinoa (*Chenopodium quinoa* Willd.) Varieties Grown in North-West Europe—Part II. *Plants* **2022**, *11*, 265. https://doi.org/10.3390/plants11030265

Academic Editors: Cataldo Pulvento and Didier Bazile

Received: 31 December 2021
Accepted: 17 January 2022
Published: 19 January 2022

1. Introduction

The pseudocereal quinoa (*Chenopodium quinoa* Willd.) belongs to the Amaranthaceae family. Quinoa was traditionally cultivated by pre-Hispanic civilizations in the Andean region [1]. The plant produces edible seeds that are small, flat, and round [2]. The seed color of most commercial varieties are white to grey or black but varieties with yellow or red–violet seeds exist as well [3,4]. After the recognition of its excellent nutritional value, quinoa gained increasing interest worldwide [1]. Quinoa is a promising crop for further expansion in many parts of the world [2]. Quinoa is mainly grown in South America (e.g., Peru, Bolivia, Ecuador), although it has been introduced worldwide in the last decades [5].

Quinoa is known for its excellent balance between carbohydrates, proteins, and lipids [6,7]. Quinoa's carbohydrate fraction mainly consists of starch (32–69 g/100 g), with a significant amount of resistant starch and dietary fiber (7.0–11.7 g/100 g) [5,6]. Albumins and globulins make up the main fraction of the total protein content (7–23 g/100 g). Quinoa is gluten-free, making it a suitable food source for people with celiac disease or other digestive disorders, such as gluten intolerance [6]. Quinoa proteins have a balanced composition of essential amino acids [7]. The protein fraction shows a high content of lysine, methionine, and threonine, the limiting amino acids in common cereals, such as wheat and maize [5]. All essential amino acids are present, meeting the requirements for adults suggested by FAO and WHO [5]. The protein digestibility is usually high, but may be affected by hydrolase inhibitors or enzyme inhibitory effects of endogenous phenolic compounds [6]. Besides its nutritional interesting protein content, quinoa has an interesting lipid content of approximately 5.5 to 7.4 g/100 g, which is higher than that of wheat, maize, and rice [5]. Quinoa seeds have a higher content of potassium, calcium, magnesium, iron, copper, and zinc than those found in common cereals [5]. Quinoa is a good source of vitamins with higher levels of α-tocopherol (vitamin E), riboflavin (B2), pyridoxine (B6), and folic acid than those of wheat [5]. Quinoa contains several antioxidants and phytochemicals with positive effects on human health and nutrition [5,8].

As starch is the major component in quinoa seed, it is expected that the properties of quinoa wholemeal flour (WMF) depend, to a large extent, on the composition and properties of starch [9]. Quinoa starch is present in the form of small polygonal granules of 1 to 3 μm in diameter [7,9,10]. The starch granules are present as single units or as spherical aggregates packed in the quinoa perisperm [7,11]. Quinoa starch has a low amylose content, which provides different physicochemical properties compared to cereal starches [12]. There is a considerable variability in the amylose content of quinoa [13]. The amylose content reported for quinoa starch ranges between 9 and 11%, although wider ranges of 3 to 20% have also been found [12,13]. It is unclear whether the reported variation in amylose content is truly a reflection of genetic variability, due to variations in agricultural practices or environment, the result of both, or attributable to differences in the quantification methods [13]. Quinoa amylopectin contains a large number of short chains from 8 to 12 units and a small number of longer chains from 13 to 20 units, as compared to the endosperm starches of cereals [13,14]. The low amylose content and unique amylopectin structure contribute to the unique physicochemical characteristics of quinoa starch [2,11]. It gelatinizes at relatively low temperatures, similar to the gelatinization temperature of wheat and potato starch. However, its pasting behavior is considerably different from that of wheat and potato starches [7,13]. Quinoa starch shows a higher water-binding capacity and swelling power compared to wheat, barley, or corn starch. Furthermore, it has excellent freeze–thaw stability [7,11–13]. Other main seed components, such as proteins and lipids, likely affect the physicochemical properties of the quinoa WMF as well [9]. Proteins can affect the water-binding capacity, pasting properties, and digestibility of the WMF [9]. Polar lipids could form V-type inclusion complexes with amylose, which influences starch gelatinization, retrogradation, and susceptibility to enzymatic breakdown [9]. Non-starch polysaccharides, such as dietary fiber, could increase gel viscosity or form a matrix with phenolic compounds, which may retard starch retrogradation and digestion [9].

The wide applicability of quinoa derives from its versatility as a food ingredient [1]. Quinoa is mostly consumed as a whole grain or milled as WMF [1]. As a whole grain, quinoa seeds are toasted, puffed, or boiled, and used in salads, cooked meals, breakfast, or soups [1,15]. The seeds can also be processed into dairy milk, fermented to make beer, or used as an ingredient in the traditional chicha drink [15,16]. Quinoa WMF, whether or not combined with wheat flour or corn meal, is mainly used for the production of baked (biscuits, cookies, bread, and gluten-free bakery products) or cooked (pasta, noodles) products [2,17]. However, end-product quality will greatly depend on the characteristics of the flour used [1]. According to previous studies, different varieties of quinoa have shown different results in similar food products [2]. Therefore, studying the physicochemical properties is important in determining the end uses of quinoa varieties, as these properties determine the changes that occur during processing [2].

The significant increase in demand for quinoa has stimulated efforts in Europe to develop new varieties that are suited to European growing conditions [2,18]. The characteristics of Andean quinoa varieties are relatively well studied, but far little information is available about European quinoa [1]. A better understanding of the individual variety characteristics will facilitate the selection of a variety for specific markets or products and will stimulate the utilization of European quinoa in the food industry [2]. To fill the gap in the current literature on European varieties and to assess their potential usefulness for farmers and end users, a comparative variety testing with ten European bred varieties and three Farm Original varieties (USA) was performed under field conditions in North-West Europe during three consecutive growing seasons (2017–2019). In a previous work [19], the agronomic performances of these varieties were evaluated, and the quinoa seeds were qualitatively characterized based on physical and nutritional properties. The present work focuses on the characterization of the physicochemical properties of WMFs obtained from these seeds. These properties were evaluated over the three growing seasons to estimate the impact of varying seed composition. In addition, principal components analysis (PCA) and hierarchical cluster analysis (HCA) were applied to group the WMFs into clusters with similar physicochemical properties.

2. Results and Discussion

2.1. Chemical Composition

The chemical composition of the quinoa varieties is summarized in Table 1 and visualized in Supplementary Materials (Figure S1). These data have been discussed in detail in Part I of this study [19]. In brief, the protein content of the quinoa seeds varied between 12.1 and 18.8 g/100 g dry matter (dm). All varieties obtained the highest protein concentration in 2019 due to favorable weather conditions. Among varieties, Atlas and Pasto were characterized by the highest protein content in combination with lower yields. Quinoa seeds contained a considerable amount of fat (5.42–8.54 g/100 g dm), with the highest levels measured in 2018. The seeds of Pasto, Puno, and Zwarte usually had lower fat levels than other varieties, while the highest amounts of fat were found in Rouge Marie, Summer Red, and Vikinga seeds. The main component of the quinoa seeds was starch (50.5 to 72.5 g/100 g dm). Jessie, Pasto, Titicaca, Vikinga and Zwarte obtained the lowest starch content in 2017, other varieties in 2019. Among varieties, Puno and Titicaca contained the highest amounts of starch, Faro the lowest. The quinoa seeds contained between 2.37 and 3.60 g minerals per 100 g dm. The seeds of 2019 usually had a lower ash content compared to corresponding seeds from 2017 or 2018. Puno and Titicaca contained low amounts of minerals, while Pasto seeds were characterized by the highest ash content.

2.2. Physicochemical Properties

2.2.1. Water Absorption Capacity

The water absorption capacity (WAC) is the ability of the WMF to physically hold water while exposed to a centrifugal force, a process in which macromolecules (e.g., carbohydrates and proteins) increase the water absorption by offering hydrophilic side chains [17,20]. Overall, the WAC ranged between 1.08 and 2.06 g/g (Figure 1), which corresponded with the findings of Ghumman et al. (1.22–2.17 g/g, [17]) and De Bock et al. (1.52–2.05 g/g, [20]). Vázquez-Luna et al. [21] reported a higher WAC for quinoa flour (2.11–2.55 g/g), while the WAC was lower according to Aluwi et al. (0.89–1.22 g/g, [2]). The growing season of the quinoa seeds had a significant impact on the WAC of the WMFs. The WAC of Faro, Oro de Valle, and Zwarte WMF improved every growing season, while other WMFs, except for Atlas, Jessie, Pasto, and Vikinga, obtained the lowest WAC in 2018. Jessie, Pasto, and Vikinga WMF showed a stable WAC during the first two seasons (i.e., 2017 and 2018). The WMFs of 2019 showed the highest ability to absorb and retain water, as the WAC varied between 1.46 and 2.06 g/g. The high protein levels in 2019 positively affected the WAC of the WMFs ($r = 0.563$, $p = 0.001$), as high protein contents lead to an improved potential in binding capacity [22,23]. However, the maximum WACs in 2019 were not linked to the WMFs with the highest protein content. Despite the high protein content, Atlas and Dutchess WMF were characterized by the lowest WAC in 2019. Vikinga, Jessie, Titicaca, and Puno, all WMFs rich in starch and low in minerals, showed the highest WAC for WMFs in 2019. The WAC might also be affected by the protein composition or the presence of fiber [23]. Moreover, differences in particle size and damaged starch content could also have an impact on the WAC [1,24]. Differences in WAC between varieties varied among growing seasons (variety × year: $p < 0.001$, Table S4). For example, Atlas was considered a WMF with low WAC in 2017 and 2019 whereas it had the highest WAC among WMFs in 2018. However, certain trends were observed over the different years. Jessie and Titicaca WMFs were generally characterized by a high WAC, while Bastille WMF showed a low capacity to absorb and retain water. WMFs from Pasto or dark colored seeds (i.e., Rouge Marie, Summer Red, and Zwarte) could be considered WMFs with medium–high WAC.

Table 1. Chemical composition (g/100 g dm) of 13 quinoa varieties grown under North-West European field conditions in 2017, 2018, and 2019.

Year	Variety	Protein (g/100 g dm) [1]	Fat (g/100 g dm) [1]	Starch (g/100 g dm) [1]	Ash (g/100 g dm) [1]
2017	Atlas	16.6 ± 0.2 fA	6.41 ± 0.04 cA	58.1 ± 0.6 dB	3.46 ± 0.02 gC
	Bastille	-	-	-	-
	Dutchess	13.8 ± 0.0 cA	7.03 ± 0.07 eA	59.6 ± 0.5 eB	3.35 ± 0.01 fB
	Faro	14.9 ± 0.3 dB	7.33 ± 0.09 fA	53.4 ± 0.7 aB	3.32 ± 0.02 efB
	Jessie	16.0 ± 0.2 eB	6.70 ± 0.04 dA	54.9 ± 0.7 bA	3.12 ± 0.02 dC
	Oro de Valle	12.1 ± 0.2 aA	6.18 ± 0.03 cA	67.2 ± 1.2 gC	2.86 ± 0.03 bA
	Pasto	15.5 ± 0.3 eA	5.42 ± 0.05 aA	56.5 ± 0.5 cA	3.47 ± 0.02 gA
	Puno	13.9 ± 0.2 cB	5.65 ± 0.02 abA	61.7 ± 0.4 fB	2.37 ± 0.02 aA
	Rouge Marie	-	-	-	-
	Summer Red	13.9 ± 0.2 cB	8.54 ± 0.08 iC	58.0 ± 0.2 dC	3.29 ± 0.01 eB
	Titicaca	14.8 ± 0.1 dB	7.61 ± 0.08 gB	60.7 ± 0.8 fA	2.84 ± 0.01 bAB
	Vikinga	14.9 ± 0.4 dB	8.04 ± 0.08 hB	57.8 ± 0.7 dA	3.02 ± 0.03 cB
	Zwarte	13.0 ± 0.3 bA	5.70 ± 0.03 bA	55.2 ± 0.7 bA	3.16 ± 0.03 dA
2018	Atlas	16.4 ± 0.1 fA	7.78 ± 0.03 eC	57.5 ± 0.5 aB	3.30 ± 0.02 cB
	Bastille	13.1 ± 0.2 bA	7.34 ± 0.01 dA	64.0 ± 0.3 eB	3.09 ± 0.03 bB
	Dutchess	14.4 ± 0.2 eB	7.75 ± 0.04 eB	61.4 ± 0.4 cC	3.32 ± 0.02 cB
	Faro	13.2 ± 0.2 bA	8.09 ± 0.08 fC	58.2 ± 0.4 abC	3.07 ± 0.01 bA
	Jessie	13.7 ± 0.1 cdA	7.93 ± 0.05 efB	59.1 ± 0.4 bC	3.08 ± 0.01 bB
	Oro de Valle	12.2 ± 0.1 aA	7.73 ± 0.06 eC	62.5 ± 0.6 dB	3.08 ± 0.01 bC
	Pasto	16.0 ± 0.2 fB	6.37 ± 0.04 aB	59.0 ± 0.6 bB	3.60 ± 0.02 dB
	Puno	12.5 ± 0.2 aA	6.67 ± 0.03 bC	63.7 ± 0.3 eC	2.90 ± 0.02 aC
	Rouge Marie	-	-	-	-
	Summer Red	13.3 ± 0.1 bcA	8.18 ± 0.01 fB	57.3 ± 0.3 aB	3.27 ± 0.03 cB
	Titicaca	13.2 ± 0.1 bA	7.08 ± 0.03 cA	63.9 ± 0.7 eB	2.88 ± 0.01 aB
	Vikinga	13.8 ± 0.4 dA	8.50 ± 0.09 gC	63.6 ± 0.2 eC	3.08 ± 0.03 bC
	Zwarte	-	-	-	-
2019	Atlas	18.5 ± 0.0 fgB	7.20 ± 0.14 dB	54.5 ± 0.4 cA	3.03 ± 0.02 cdA
	Bastille	16.6 ± 0.0 bcB	7.36 ± 0.19 deA	52.3 ± 0.5 bA	3.01 ± 0.03 cdA
	Dutchess	18.4 ± 0.1 fgC	7.17 ± 0.04 dA	54.8 ± 0.2 cdA	3.21 ± 0.01 fA
	Faro	17.6 ± 0.1 dC	7.61 ± 0.12 efB	50.5 ± 0.4 aA	3.09 ± 0.01 deA
	Jessie	18.8 ± 0.0 gC	6.70 ± 0.24 cA	55.6 ± 1.3 deB	2.84 ± 0.03 aC
	Oro de Valle	16.2 ± 0.0 abB	7.22 ± 0.07 dB	54.7 ± 0.8 cdA	2.91 ± 0.02 bB
	Pasto	-	-	-	-
	Puno	17.0 ± 0.1 cC	5.88 ± 0.24 aB	59.4 ± 1.1 fA	2.83 ± 0.01 aB
	Rouge Marie	17.8 ± 0.1 deA	7.65 ± 0.08 fA	53.9 ± 0.2 cA	3.10 ± 0.02 eA
	Summer Red	18.1 ± 0.2 efC	7.75 ± 0.09 fA	54.5 ± 0.4 cA	3.20 ± 0.01 fA
	Titicaca	16.0 ± 0.1 aC	7.25 ± 0.03 dA	72.5 ± 0.9 hC	2.83 ± 0.02 aA
	Vikinga	17.0 ± 0.1 cC	7.71 ± 0.04 fA	60.8 ± 0.5 gB	2.82 ± 0.02 aA
	Zwarte	16.0 ± 0.1 aB	6.36 ± 0.08 bB	56.2 ± 0.2 eB	3.29 ± 0.04 gB

[1] Within years, average values followed by the same lowercase letter are not significantly different ($p > 0.05$). Capital letters compare the three years for the same variety; average values followed by the same letter are not significantly different ($p > 0.05$).

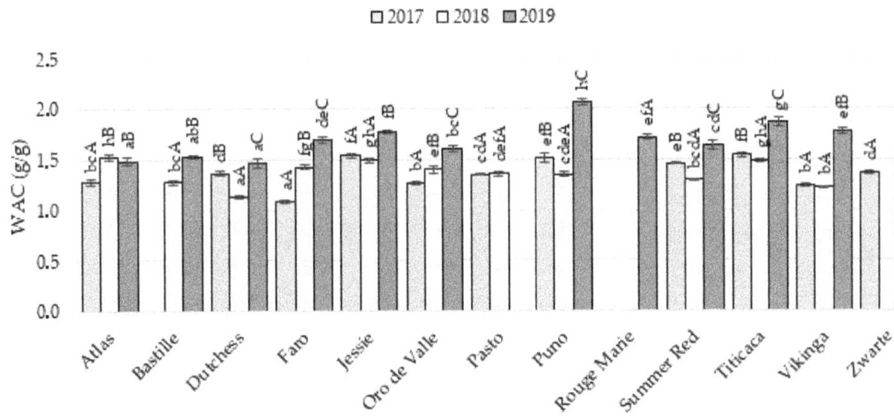

Figure 1. Water absorption capacity (WAC, g/g) of 13 quinoa wholemeal flours obtained from seeds cultivated under North-West European field conditions in 2017, 2018 and 2019 ($n = 3$). Within years, average values followed by the same lowercase letter are not significantly different ($p > 0.05$). Capital letters compare the three years for the same variety, average values followed by the same letter are not significantly different ($p > 0.05$).

2.2.2. Water Absorption Index

The water absorption index (WAI) measures a WMF's ability to absorb water and swell, and is a function of the gel-forming capacity of the WMF. It is in fact a property related to the characteristics of a thickening agent, as a high WAI characterizes a good binder [2,23]. The WAI was evaluated over a temperature range between 55 and 95 °C (Figure 2) and was significantly affected by a variety × year interaction ($p < 0.001$, Table S4). The increase in temperature had a positive impact on the WAI of the WMFs, as it promoted starch gelatinization and protein denaturation [1,23,25]. A steady increase in swelling behavior with rising temperature has been observed before [9,20]. Furthermore, more differentiation between the WMFs was observed at higher temperatures.

The WMFs of 2017 were arranged into three groups based on their swelling behavior (Figure 2). The first group consisted of WMFs with a high ability to swell, namely Faro, Oro de Valle, Puno, and Titicaca. At 65 °C, these WMFs already showed the strongest swelling behavior among all WMFs. The WAI only slightly increased at 75 °C but was still higher for Faro, Puno, and Oro de Valle compared to that of other WMFs. The largest increase in WAI was observed between 75 and 95 °C, where Titicaca WMF usually had the lowest WAI among these four WMFs. The WAI at 95 °C ranged between 7.25 and 7.76 g/g for Faro, Oro de Valle, and Puno (Table S1) and was higher compared to the WAI of other WMFs, except for Pasto WMF. Jessie and Vikinga WMF formed a second group among the WMFs of 2017. These WMFs had a higher WAI at 55 °C than most other WMFs. The WAI of Jessie and Vikinga WMF showed a continuous increase with temperature, eventually reaching the lowest WAI at 95 °C among all samples. The remaining WMFs, i.e., Atlas, Dutchess, Pasto, Summer Red, and Zwarte, were characterized by the lowest WAI at 85 °C, as the WAI increased only slightly between 65 and 85 °C. The strong improvement at 95 °C eventually led to a medium-high WAI for Atlas, Dutchess, and Zwarte WMF (6.68–6.94 g/g) and a high WAI for Pasto WMF (7.25 g/g, Table S1). Summer Red WMF showed a low WAI over the whole temperature range, but its WAI also showed a strong increase at 95 °C (6.00 g/g, Table S1).

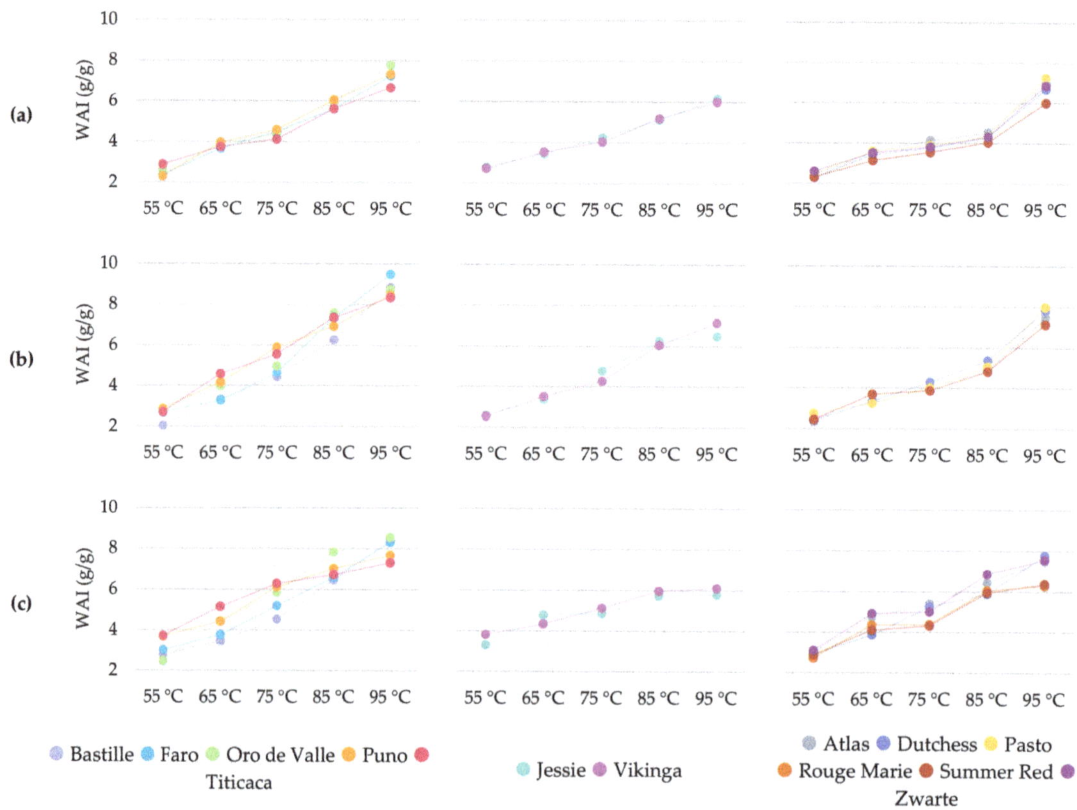

Figure 2. Water absorption index at 55, 65, 75, 85, and 95 °C (WAI, g/g) of 13 quinoa wholemeal flours obtained from seeds grown under North-West European field conditions in 2017 (**a**), 2018 (**b**), and 2019 (**c**) (n = 3).

Similar observations were found for the 2018 samples, although the first group was separated into two subgroups (Figure 2). The WAI of Puno and Titicaca WMF showed a continuous increase over the whole temperature range. At temperatures above 55 °C, the WAI of these WMFs was higher compared to that of the corresponding WMFs of 2017. Puno and Titicaca WMF were characterized by the highest WAI at 65 and 75 °C and also showed a high swelling behavior at higher temperatures. The WAI of Bastille, Faro, and Oro de Valle increased at a higher rate between 75 and 95 °C, when compared to the increase between 55 and 75 °C. The swelling behavior of Faro and Oro de Valle WMF was generally higher compared to that of the corresponding WMFs of 2017. However, Faro and Oro de Valle were considered WMFs with a high swelling behavior between 65 and 95 °C in 2017. In 2018, the WAI of Faro and Oro de Valle was only high compared to that of other WMFs at 85 and 95 °C. Between 55 and 65 °C, Jessie and Vikinga WMF showed a similar swelling behavior, but the WAI of Jessie was higher at 75 and 85 °C. However, the WAI of Jessie WMF deflected at 95 °C, resulting in a lower WAI than Vikinga. Nevertheless, both WMFs showed a low swelling behavior at 95 °C (6.48–7.14 g/g, Table S1) compared to most other WMFs (7.47–9.48 g/g, Table S1). Moreover, other studies reported a higher WAI at 95 °C for quinoa flour (8.11–9.75 g/g, [19]; 7.89–9.46 g/g, [9]). The swelling behavior of Jessie and Vikinga followed the same trend as observed in 2017, although the WAIs at 75 to 95 °C were higher in 2018. The WAI of Atlas, Dutchess, Pasto, and Summer Red WMF showed a continuous increase between 55 and 85 °C and was generally low compared to the WAI of other WMFs. A strong increase of the WAI, similar to what was observed in 2017, occurred at 95 °C. This resulted in medium-high WAIs for Atlas, Dutchess, and Pasto WMF (7.47–7.98 g/g, Table S1) but Summer Red WMF still had a low WAI at 95 °C (7.14 g/g, Table S1). Compared to the previous growing season (i.e., 2017), Atlas, Dutchess, Pasto, and Summer Red WMF of 2018 generally had higher WAIs between 75 and 95 °C.

The WMFs of 2019 showed a higher swelling behavior between 55 and 85 °C compared to the corresponding WMFs of previous years. Only at 95 °C were the WMFs of 2018 characterized by the highest WAIs (Table S1). In 2019, the group of Bastille, Faro, Oro de Valle, Titicaca, and Puno WMF was separated into three subgroups (Figure 2). Puno and Titicaca WMF no longer showed a continuous increase of WAI but the rate of increase decreased at 75 °C, resulting in medium-high WAI at 95 °C (7.31–7.66 g/g, Table S1). Puno and Titicaca WMF started off with a similar WAI at 55 °C but Titicaca had a higher WAI at 65 and 75 °C, and Puno at 85 and 95 °C. In contrast to previous years, the WAI of Oro de Valle WMF followed a different trend than the WAI of Bastille and Faro WMF. The WAI of Oro de Valle continuously increased between 55 and 85 °C but showed a deflection at 95 °C. This WMF was characterized by the highest WAI measured at 85 and 95 °C. Bastille and Faro WMF showed a low to medium-high swelling behavior between 55 and 85 °C, and up to 75 °C, the WAI of Bastille WMF was always lower compared to that of Faro. At 95 °C, both WMFs showed a similar and high WAI of 8.32 to 8.38 g/g (Table S1). While Jessie and Vikinga WMF showed a medium-high to high swelling behavior between 55 and 75 °C, its WAI was among the lowest at 85 and 95 °C. A deflection of the WAI at 95 °C was already observed for Jessie WMF in 2018, but was present for both WMFs in 2019. Furthermore, the WAI was improved less by increasing the temperature from 55 to 95 °C in 2019. In 2019, the strong increase in WAI between 85 and 95 °C was no longer observed for Atlas, Dutchess, Summer Red, and Zwarte WMF. In fact, the WAI of these WMFs followed a different trend compared to previous years, which separated the group into two subgroups. Atlas and Dutchess WMF showed a continuous increase in the WAI, with remarkably higher differences in WAI between both WMFs at 65 and 85 °C. The WAI of Rouge Marie, Summer Red, and Zwarte WMF showed a step-by-step increase, with the strongest improvements observed between 55 and 65 °C and between 75 and 85 °C. Rouge Marie and Summer Red WMF had a very similar swelling behavior between 75 and 95 °C and were considered WMF with a low WAI at these temperatures. Zwarte WMF had a higher WAI than Rouge Marie and Summer Red over the whole temperature.

As discussed above, quinoa WMFs showed a great diversity in their swelling behavior. Previous studies confirm that the swelling behavior of quinoa flour or starch is diverse and related this diversity to the variations in amylose content [9,13,26]. Lindeboom et al. compared quinoa lines with varying amylose content (3–20%) and noted a negative impact of the amylose content on the swelling power. Amylose is known for acting as a restraint to swelling, as it reinforces the internal structure of the starch granules [26]. The WMFs of Bastille, Faro, Oro de Valle, Puno, and Titicaca generally showed a strong improvement in the WAI as the temperature increased from 55 to 95 °C. Thus, this suggests a lower amylose content compared to WMFs, such as Jessie or Vikinga, which had a low WAI at 95 °C. The impact of the growing season suggests that the amylose content was also affected by the growing conditions. Nevertheless, Li et al. [9] noted that the swelling power of quinoa flour was considerably lower compared to that of its isolated starch. This indicates that other components, such as proteins and fat, could restrict the granule swelling as well.

2.2.3. Pasting Properties

The pasting profile of a WMF enables the relation of the functionality of starch to its structural characteristics. It determines potential industrial applications in products, as these depend on the viscosity and thickening behavior of the WMF [1]. The pasting profiles (Figure 3) showed that the WMFs differed in pasting and viscosity characteristics, and were affected by a significant interaction between variety and year ($p < 0.001$, Table S4). The pasting profile was defined by the pasting temperature, peak viscosity and temperature, breakdown, second peak viscosity, final viscosity, and total setback. These parameters are summarized in Table S2 in Supplementary Materials.

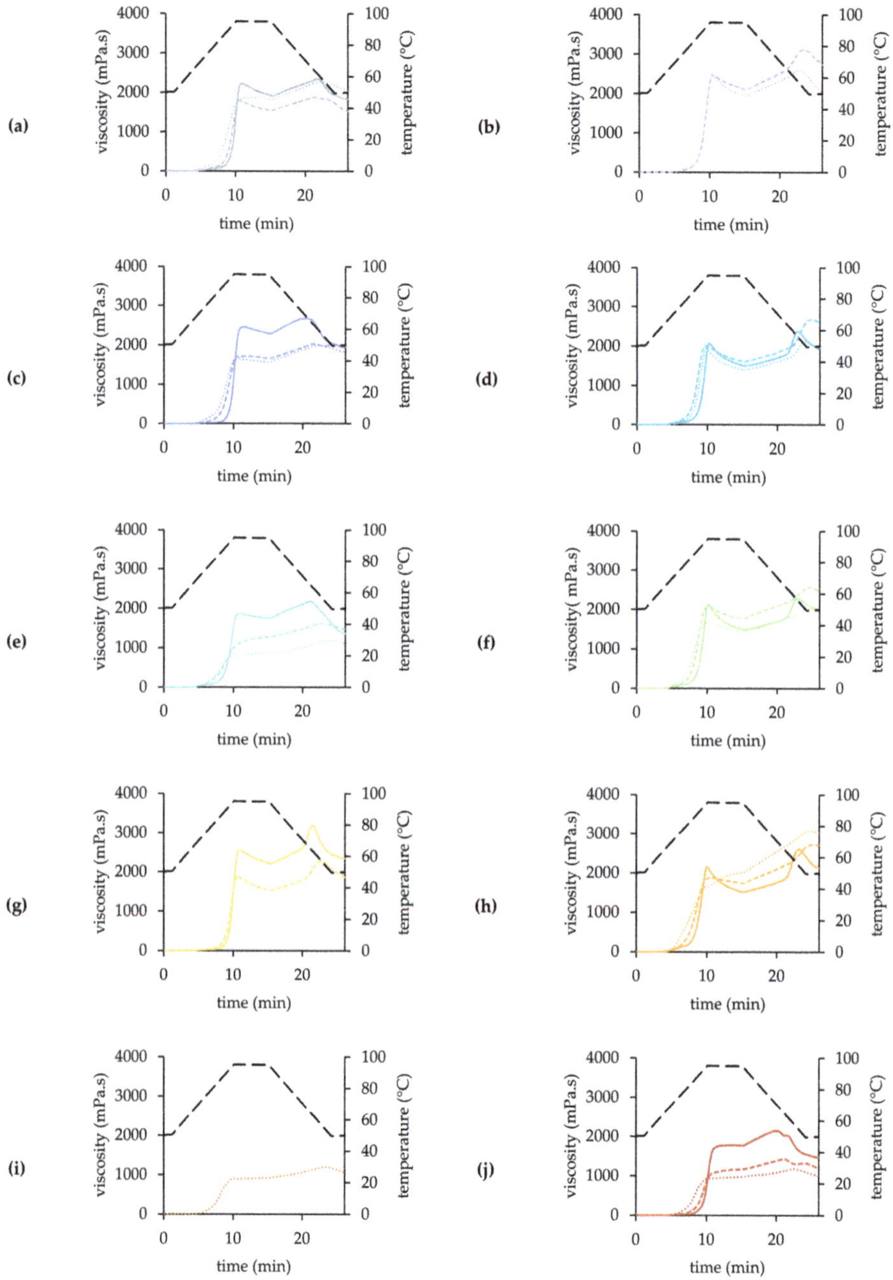

Figure 3. *Cont.*

(k)

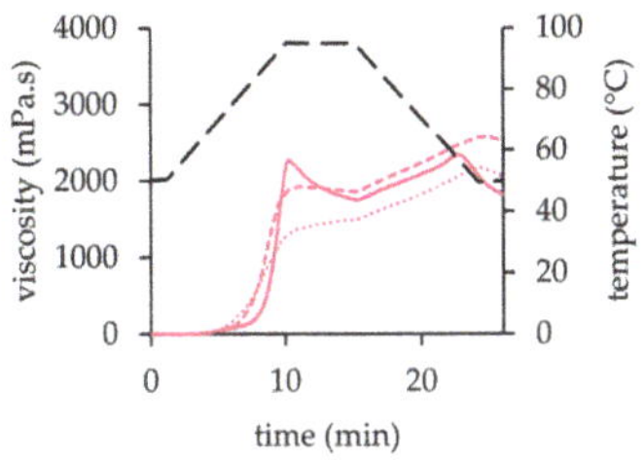

(l)

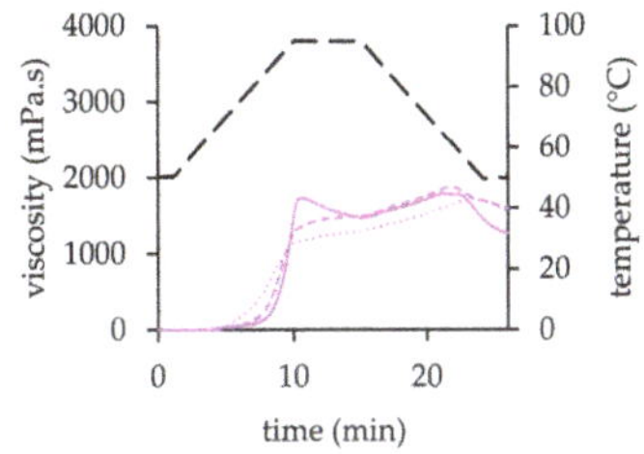

(m)

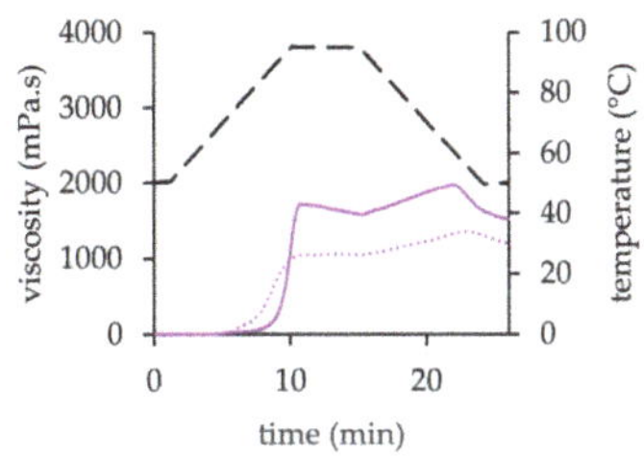

Figure 3. Pasting profile of 13 quinoa wholemeal flours ((**a**): Atlas, (**b**): Bastille, (**c**): Dutchess, (**d**): Faro, (**e**): Jessie, (**f**): Oro de Valle, (**g**): Pasto, (**h**): Puno, (**i**): Rouge Marie, (**j**): Summer Red, (**k**): Titicaca, (**l**): Vikinga, (**m**): Zwarte) obtained from seeds grown under North-West European field conditions in 2017 (—), 2018 (- - -) and 2019 (· · ·) (temperature: — — —, $n = 3$).

A certain temperature is required to launch the swelling process of the starch granules. This is the so-called pasting temperature, and is in fact the minimum temperature for starch gelatinization [1,27]. Over the years, the pasting temperature of the WMFs ranged between 57.70 and 67.11 °C (Table S2). Higher pasting temperatures have been reported for quinoa WMF (69.2 °C, [27]; 61.6–74.8 °C, [2]). The pasting temperature of Atlas, Bastille, Puno, and Vikinga WMF varied with the growing season. These WMFs generally had the highest pasting temperature in 2018, except for Puno. In 2018, a negative correlation between the pasting temperature and starch content ($r = -0.640$, $p = 0.034$) was observed. Tafadzwa et al. [23] confirmed that flours with a higher starch content had a lower pasting temperature. Oro de Valle and Titicaca were the only WMFs with a constant pasting temperature over the different years of cultivation. However, Dutchess, Jessie, Pasto, and Summer Red WMF showed a stable pasting temperature in 2017 and 2018, Faro and Zwarte WMF in 2017 and 2019. Pasting temperature correlated negatively with the WAI at 55 °C ($r = -0.546$, $p = 0.001$) and 65 °C ($r = -0.650$, $p < 0.001$). Thus, a high WAI indicates that the WMF is able to swell more freely, which results in a lower pasting temperature [20].

Peak viscosity occurs at an equilibrium point between granule swelling and polymer leaching. It is related to the water-binding capacity and the degree of swelling of the starch granules [9,27]. The peak viscosity can be indicative of end-product quality as well as the viscosity which is likely to be encountered by processes, such as mixing [2]. The peak viscosity showed a decreasing trend over the years of cultivation, and was affected by a significant variety × year interaction ($p < 0.001$, Table S4). The WMFs of 2017 had a peak viscosity between 1696 and 2560 mPa.s (Table S2) but higher values have been reported by Solaesa et al. (2690 mPa.s, [1]) and De Bock et al. (1418–2606 mPa.s, [20]). The high peak viscosities suggest a high degree of swelling of the starch granules during heating in 2017 [28]. However, WMFs of 2017 were characterized by low WAIs at 75 to 95 °C (Table S1), suggesting a low swelling behavior of these WMFs. This discrepancy might be related to the differences in measuring conditions (shear vs. no shear). Jessie, Summer Red, Vikinga, and Zwarte WMF developed the lowest peak viscosity that year, Pasto WMF the highest. Oro de Valle WMF showed a similar peak viscosity in 2017 and 2018, while the peak viscosity of other WMFs decreased (1030–2085 mPa.s, Table S2). Viscosity is directly related with the starch content, but the high starch levels did not result in the highest peak viscosity in 2018 [29]. The higher fat levels probably restricted the starch granule swelling due to complex formation with amylose upon heating, which reduced the viscosity of the WMFs of 2018 [29,30]. Jessie, Summer Red and Vikinga remained the WMFs with the lowest peak viscosity, while Bastille WMF reached the highest peak viscosity (2492 mPa.s) in 2018. Most WMFs had their lowest peak viscosity in 2019 (823–1959 mPa.s, Table S2), although the peak viscosity of Atlas (1871 mPa.s) and Bastille (2453 mPa.s) did not continue to decrease. Barak et al. [31] reported that flours with higher protein content reached lower peak viscosity. This might explain why the peak viscosities were generally lower compared to previous growing seasons in 2019 ($r = -0.350$, $p = 0.042$). Bastille remained the WMF with the highest peak viscosity, while the lowest viscosities were measured for Jessie, Rouge Marie, Summer Red, Titicaca, Vikinga, and Zwarte WMF.

The WMFs were characterized by a high peak temperature, which ranged between 93.69 and 95.22 °C (Table S2). The impact of growing season depended on the variety, indicating a significant variety × year interaction ($p < 0.001$, Table S4). The peak temperature of Faro and Puno WMF varied over the different growing seasons. Faro WMF obtained its lowest peak temperature in 2018, and differed significantly from all other WMFs of 2018. The growing season of 2019 resulted in the lowest peak temperature for Jessie, Oro de Valle, and Puno WMF. Jessie and Oro de Valle WMF were both characterized by a stable peak temperature in 2017 and 2018. The peak temperature of all other WMFs was not affected by the growing season. Peak temperature usually did not differ among varieties, except when the peak temperature was below 95 °C. Therefore, more variation in peak temperature was observed in 2019. The WMFs showed a high peak time, as peak viscosity was reached shortly before or during the 95 °C holding phase. Tiga et al. [27] attributed the high peak time of quinoa WMF to its high protein content, delaying the water absorption and swelling of the starch granules.

The breakdown measures the difference between the viscosity peak and the holding strength and is an indication of the stability against heat and shear [27]. The breakdown characterized the pasting profile of the WMF strongly, and was affected by a significant interaction between variety and year ($p < 0.001$, Table S4). All WMFs of 2017 were characterized by a positive breakdown (124–643 mPa.s, Table S2), except for Summer Red WMF. The high peak viscosities of these WMFs already suggested a high degree of starch granule swelling. The susceptibility to shear disruption increases as starch granules swell, which explains the high breakdown values in 2017 [13,28,32]. Puno and Oro de Valle WMF had the highest breakdown and showed the lowest stability against heat and shear. The lowest viscosity decreases were observed for Jessie, Dutchess, and Zwarte WMF. Summer Red WMF had a breakdown of −78 mPa.s, indicating that the viscosity continued to increase during the holding phase. This high stability against heat and shear is probably related to the low swelling behavior of Summer Red WMF (Table S1), as starch granules become increasingly susceptible to shear disruption as they swell [13]. The formation of amylose-lipid complexes may also play a role in the high shear stability, as Summer Red WMF contained high levels of fat [9]. The stability against heat and shear improved in 2018, except for Summer Red and Pasto WMF. The latter had a stable breakdown during the first two growing seasons. Jessie (−251 mPa.s) and Vikinga WMF (−187 mPa.s) also showed a negative breakdown in 2018 but the viscosity increase was much higher than the one observed for Summer Red WMF. The breakdown of other WMFs ranged between 65 and 410 mPa.s (Table S2), with the highest breakdown values measured for Pasto, Bastille and Faro WMF. The breakdown of Bastille, Faro, Jessie, Summer Red, and Vikinga WMF increased in 2019, but did not reach the same levels as in 2017. Bastille and Faro WMF were still characterized by the lowest stability against heat and shear. The breakdown of Atlas, Oro de Valle, Puno and Titicaca WMF continued to decrease in 2019, which resulted in negative breakdown values for Puno and Titicaca WMF. The latter showed the strongest viscosity increase among all WMFs with a negative breakdown. Puno and Titicaca WMF showed a medium-high swelling behavior in 2019, while their WAIs were relatively high compared to those of other WMFs in 2017 and 2018 (Table S1). The reduced swelling behavior in 2019 possibly promoted the shear and heat resistance during the holding phase [13]. The strong viscosity increase during the holding phase resulted in a unique pasting profile for Puno WMF of 2019 (Figure 3h), which significantly differed from that of previous growing seasons or other varieties. It is an indication that the water took longer to reach the starch granules and start the swelling process, and could be related to the increased protein content in 2019 [1]. However, Puno seeds were considerably smaller than the seeds of other varieties [19] and might show a different milling behavior due to the small seed size. Dziki et al. [33] concluded that small seeds were more difficult to reduce in size, resulting in a flour with a higher fraction of large particles. Solaesa et al. [1] concluded that the coarse fraction of quinoa flour showed no peak or valley in the pasting curve but a continuous increasing viscosity. Therefore, an increased share of large particles could possibly explain the unique pasting profile of Puno WMF in 2019. The breakdown value of Zwarte WMF was close to zero, indicating that the viscosity remained stable during the 95 °C holding phase. Aluwi et al. [2] also reported that the viscosity of Black WMF (possible origin for Zwarte [19]) remained relatively constant during the holding phase and increased shortly after.

The viscosity of the WMFs increased shortly after the holding phase. Due to the temperature decrease, starch chains associated and formed a more ordered structure, which increased the viscosity of the paste [34]. During this cooling phase, a second viscosity maximum, higher than the first peak, was observed in the pasting profile of all WMFs (Figure 3). The occurrence of a second peak has previously been described in other studies [1,2,20]. Over the years, this second peak viscosity ranged between 1183 and 3198 mPa.s (Table S2) and was closely related to the peak ($r = 0.856$, $p < 0.001$, Table S3) and final viscosity ($r = 0.902$, $p < 0.001$, Table S3). After the second peak, viscosity decreased and eventually reached final viscosity. This decrease might be attributed to the high protein levels in quinoa WMF, as the presence of proteins is known to reduce the resistance of gels to breakage [1]. The final viscosity indicates the ability of a WMF to form a viscous paste after cooking and cooling [27]. The final viscosity showed a wide variation over the different growing seasons and varieties, and varied between 1027 and 3047 mPa.s (Table S2). De Bock et al. [20] reported higher overall viscosities for WMFs from commercial quinoa seeds (1756–3274 mPa.s). The impact of the growing season strongly depended on the quinoa variety (variety × year: $p < 0.001$, Table S4). Most WMFs obtained the highest final viscosity in 2018. However, Pasto, Summer Red, and Zwarte WMF reached a higher final viscosity in 2017, Atlas and Puno WMF in 2019. The final viscosity of Dutchess, Faro, Oro de Valle, and Vikinga WMF did not vary every year, but showed a two-year stability over the experimental period from 2017 to 2019. Over the years, Jessie, Rouge Marie, Summer Red, and Zwarte were characterized as the WMFs with the lowest final viscosity. Puno and Bastille WMFs usually reached the highest final viscosities among all WMFs.

The total setback indicates the recovery of viscosity during cooling of the heated WMF suspension. It represents the tendency of the paste to retrograde and amylose to re-associate [11,27]. Atlas, Dutchess, Jessie, Summer Red, Vikinga, and Zwarte WMF were characterized by a negative setback in 2017, which ranged between −43 and −350 mPa.s (Table S2). Negative setback values for quinoa WMF have been reported before [35]. The negative values were related to a strong viscosity decrease after the second peak viscosity, which eventually resulted in a final viscosity lower than the holding strength. The setback of the other WMFs ranged between 108 and 654 mPa.s (Table S2), with the highest setback measured for Puno WMF. In 2018, the total setback ranged between 19 and 1037 mPa.s (Table S2), and had significantly increased compared to the previous growing season. This increase could be related to the reduced protein levels in the WMFs of 2018 [31]. Faro and Puno WMF showed a high setback, while the setback was low for Atlas and Summer Red WMF. The total setback of Atlas, Vikinga, and Zwarte WMF continued to increase in 2019, while it remained stable for Dutchess, Jessie, Puno, and Summer Red WMF. Summer Red WMF had the lowest setback (46 mPa.s), Puno WMF the highest (1030 mPa.s).

2.3. Principal Components Analysis and Hierarchical Cluster Analysis

For a more comprehensive understanding of the variation among WMFs, a PCA was performed on the macronutrient composition (starch, protein, fat, ash), WAIs (55, 75 and 95 °C) and pasting properties (first peak viscosity, breakdown, second peak viscosity, final viscosity and total setback). The variables plot of the first two PCs is visualized in Figure 4. The first and second PC accounted for 39.5 and 22.4% of the variance, respectively. Together, the first two PCs represented 61.9% of the total variance in the data. The first PC was mainly defined by the first and second peak viscosity, holding strength and final viscosity. These were all properties related to viscosity and closely inter-correlated (Table S3), as also concluded by Li et al. [9]. The second PC was defined by the WAI at 55 and 75 °C. A positive correlation between these parameters was observed ($r = 0.627$, $p < 0.001$, Table S3). A higher WAI at 55 °C indicated a higher water uptake by the WMF. A better hydration of the WMF possibly improved the starch gelatinization and, therefore, resulted in a higher WAI at 75 °C.

(a)

(b)

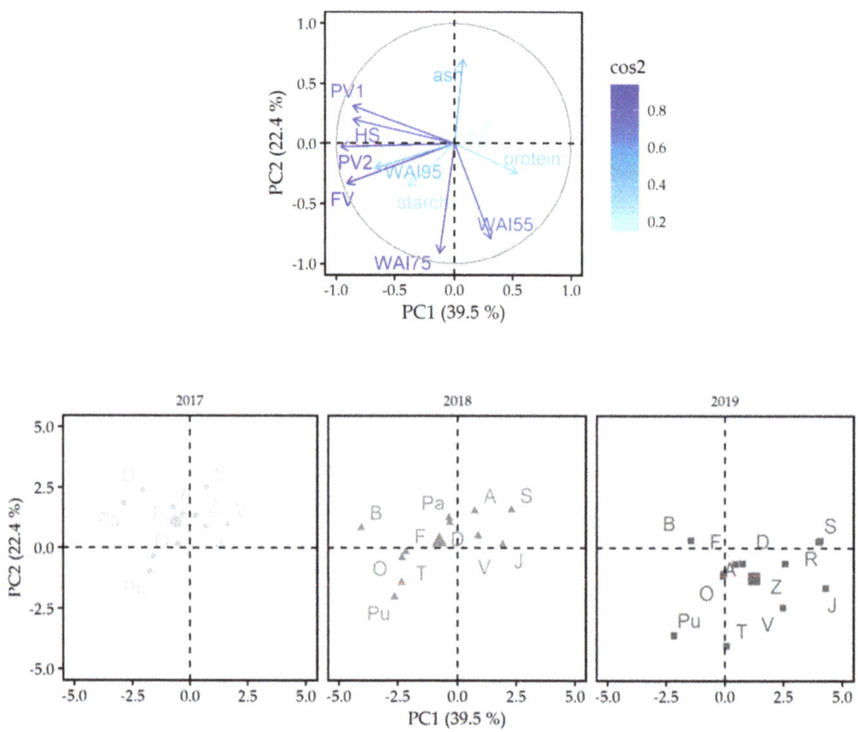

Figure 4. Principal components analysis (PCA; (**a**): variables plot, (**b**): score plot) of physicochemical properties of 13 quinoa wholemeal flours (A: Atlas, B: Bastille, D: Dutchess, F: Faro, J: Jessie, O: Oro de Valle, Pa: Pasto, Pu: Puno, R: Rouge Marie, S: Summer Red, T: Titicaca, V: Vikinga, Z: Zwarte) obtained from seeds grown under North-West European field conditions in 2017 (●), 2018 (▲) and 2019 (■). WAC: water absorption capacity, WAI: water absorption index, PV1: first peak viscosity, HS: holding strength, BD: breakdown, PV2: second peak viscosity, FV: final viscosity, TSB: total setback.

The score plot (Figure 4) showed that the grouping of the WMFs was generally consistent over different growing seasons. Within a given growing season, the WMFs were closely grouped with exception of some varieties. In 2017, Oro de Valle and Puno were separated from the other WMFs due to the low ash content and high WAI at 75 °C. The same WMFs, together with Bastille, Faro, and Titicaca WMF, showed a higher score on

PC1 compared to other WMFs of 2018. These high scores on PC1 were attributed to the high WAI at 95 °C, high second peak and final viscosity of these WMFs. The high swelling behavior at 75 °C separated Puno and Titicaca WMF from the other WMFs of 2019. The growing season caused a distinct shift in the scores of the WMFs. The WMFs of 2017 and 2018 were located on the upper side from the central axis of PC2, while most WMFs of 2019 were grouped in the lower right quadrant of the score plot. The WMFs of 2019 showed higher swelling behavior at 55 and 75 °C but developed lower pasting viscosities, which explains the positive scores on PC2 and the negative scores on PC1.

The macronutrient composition (starch, protein, fat, ash), WAIs (55, 75 and 95 °C), and pasting properties (first peak viscosity, breakdown, second peak viscosity, final viscosity and total setback) were further analyzed by HCA and the corresponding results were in good agreement with the PCA results. A total of four clusters were identified based on these parameters and the dendrogram of the hierarchical clustering is illustrated in Figure 5. The clusters (C1-C4) consisted of $n = 9$ WMFs, $n = 7$ WMFs, $n = 8$ WMFs and $n = 10$ WMFs, respectively. The first cluster grouped the WMFs which were mainly located in the lower right quadrant of the score plot of the PCA (Figure 4). This means that the cluster included all WMFs of 2019, except for Bastille, Puno, and Titicaca WMF. These WMFs were characterized by the highest WAI at 55 and 75 °C and generally showed lower peak viscosity and holding strength than the corresponding WMFs of 2017 and 2018. The second cluster grouped the WMFs with the lowest viscosity parameters in 2017 and 2018. This cluster included Summer Red and Vikinga WMF of 2017 and Atlas, Dutchess, Jessie, Summer Red, and Vikinga WMF of 2018. These WMFs were all characterized by a high score on PC1 (Figure 4). The third cluster mainly consisted of WMFs of 2017, Atlas, Dutchess, Faro, Jessie, Pasto, and Zwarte WMF, respectively. Overall, WMFs of 2017 were characterized by the highest first peak viscosity, the least stability against heat and shear and a low tendency to retrograde. Furthermore, Pasto WMF of 2018 and Bastille WMF of 2019 were included in cluster 3. The WMFs of the third cluster were generally located in the upper left quadrant of the score plot (Figure 4). The score plot showed that certain WMFs differentiated more from others despite the fact that the seeds were cultivated under the same growing conditions (Figure 4). These WMFs were grouped in the fourth cluster. This cluster included Oro de Valle, Puno, and Titicaca WMF of 2017, Bastille, Faro, Oro de Valle, Puno, and Titicaca WMF of 2018 and Puno and Titicaca WMF of 2019. The fact that all Puno and Titicaca WMFs were grouped within the same cluster indicates that the variety had a larger impact on the physicochemical properties of these WMFs than the growing season.

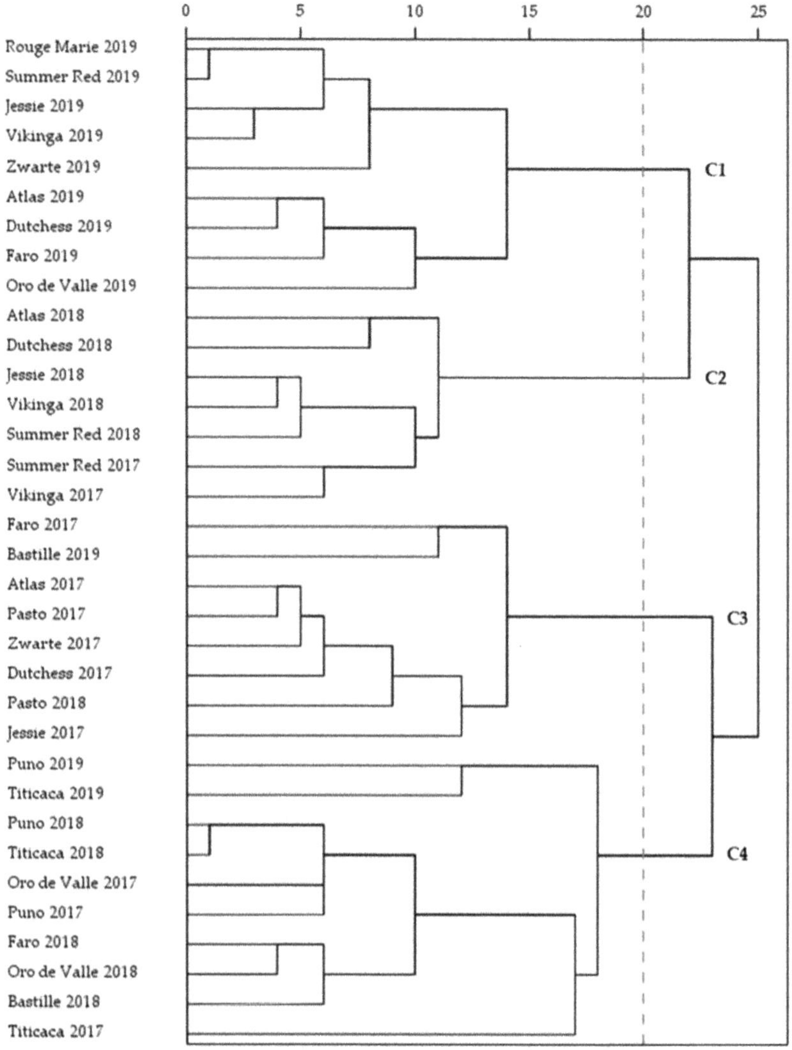

Figure 5. Dendrogram of the four clusters (C1–C4) based on hierarchical clustering.

3. Material and Methods

3.1. *Quinoa Wholemeal Flours*

Ten quinoa (*Chenopodium quinoa* Willd.) varieties were purchased at commercial companies (Gilbel sprl, Saint-Georges-Sur-Meuse, Belgium (previously a sublicensee of Radicle Crops, Wageningen, The Netherlands) and Quinoa Quality ApS, (Regstrup, Denmark) and three Farm Original varieties were obtained from De Nieuwe Tuin (De Klinge, Belgium). These 13 quinoa varieties were used for a comparative variety testing under Belgium field conditions during three consecutive growing seasons (2017–2019). Information on the variety origins and crop cultivation is described in detail in a previous work [19]. For chemical and physicochemical analyses, seeds were milled to WMF by a Hammertec mill (mesh size: 0.8 mm) (Foss, Hilleroed, Denmark).

3.2. Chemical Composition

Moisture (g/100 g) and ash (g/100 g dm) content were determined according to ICC methods no. 110 and 104/1, respectively. Total starch content (g/100 g dm) was analyzed as described by Englyst et al. [36]. The nitrogen content was determined by the use of a VarioMax C/N (Elementar Analysesystemen, Langenselbold, Germany) and converted to protein content (g/100 g dm) using a conversion factor of 6.25 [37]. Fat content (g/100 g dm) was determined by Soxhlet extraction with prior acid hydrolysis (ISO 6491). Lastly, crude fiber was calculated by the difference.

3.3. Physicochemical Properties

WAC and WAI of the WMFs were determined as described by De Bock et al. [20]. In brief, an aqueous suspension of 0.075 g WMF in 1.5 mL distilled water was shaken for 30 min at 1000 rpm in a ThermoMixer C (Eppendorf, Hamburg, Germany). For the determination of the WAC, the shaking temperature was 21 °C. The WAI was determined at a shaking temperature of 55, 65, 75, 85, and 95 °C. Subsequently, the suspension was centrifuged for 20 min at 8000× *g* and 21 °C. The supernatant was decanted and the sediment was weighed. WAC (g/g) was expressed as the amount of water absorbed by the WMF. WAI (g/g) was calculated as the weight of sediment per gram of WMF used.

Pasting properties of the WMFs were determined using a Rheometer MCR 102 (Anton Paar GmbH, Graz, Austria). Measurement was performed using 2.8 g WMF (based on 14% moisture) dispersed in 20 mL of distilled water. During the pre-shear phase, the suspension was heated to 50 °C while stirred at 960 rpm. The rotation speed was 160 rpm for the remainder of the test. The temperature was initially maintained at 50 °C for 1 min and then raised to 95 °C at a constant rate of 5 °C per min, held at 95 °C for 5 min, cooled to 50 °C at the same rate and finally held at 50 °C for 2 min. The parameters recorded were the pasting temperature (°C, temperature of initial viscosity increase), first peak viscosity (mPa.s, viscosity maximum during heating and/or holding phase), peak temperature (°C, temperature to reach first peak viscosity), holding strength (mPa.s, minimum viscosity at 95 °C), breakdown (mPa.s, difference between first peak viscosity and holding strength), second peak viscosity (mPa.s, viscosity maximum during cooling), final viscosity (mPa.s, viscosity at test finish) and total setback (mPa.s, difference between final viscosity and first peak viscosity) [38].

3.4. Statistical Analysis

The lme4 package in R (version 4.0.2, R Core Team, Vienna, Austria) was used to fit a linear model [39]. The fixed effects were variety and year, and the following model (1) was considered:

$$Y = \text{Variety} + \text{Year} + \text{Variety} \times \text{Year} \quad (1)$$

where 'Y' is the response variable, 'Variety' is the fixed effect of the variety, 'Year' is the fixed effect of the year and 'Variety × Year' is the interaction effect between variety and year. Then, an analysis of variance (ANOVA) was performed (Table S4) and a pairwise comparison between varieties and years was made with the Tukey test.

For a more comprehensive understanding of the variation among WMFs, a PCA was performed on the macronutrient composition (starch, protein, fat, ash), WAIs (55, 75 and 95 °C) and pasting properties (first peak viscosity, breakdown, second peak viscosity, final viscosity, and total setback). PCA was performed using the built-in R function 'prcomp' (R Foundation for Statistical Computing, Vienna, Austria). The graphs were made by the library 'factoextra' in combination with the ggplot2-package of R. Additionally, a Pearson correlation test was conducted to study the correlations between the physicochemical parameters more deeply (Table S3). The macronutrient composition (starch, protein, fat, ash), WAIs (55, 75 and 95 °C) and pasting properties (first peak viscosity, breakdown, second peak viscosity, final viscosity and total setback) were used to perform an HCA to group the WMFs with similar properties. HCA was performed with SPSS Statistics 27 (SPSS

Inc., Chicago, IL, USA) and the unweighted paired-group method with arithmetic mean was used. Distances among clusters were computed using Pearson correlation coefficients.

4. Conclusions

A comparative study was performed on ten European and three Farm Original (USA) quinoa varieties cultivated under North-West European field conditions during three consecutive growing seasons (2017–2019). Within the present study the quinoa seeds were milled to WMF, and their physicochemical properties were evaluated, more precisely the WAC, the WAI upon heating and the pasting properties. The results revealed large differences in physicochemical properties among quinoa WMFs and depended on the variety and the growing season. The WMFs of 2017 had the lowest WAIs at temperatures above 55 °C. The high peak viscosity of these WMFs was followed by a strong viscosity breakdown and the paste showed a low tendency to retrograde. The WAIs of the WMFs significantly increased in 2018. Moreover, the WMFs of 2018 were characterized by the highest WAI at 95 °C. This high swelling behavior was probably related to the high starch content of these WMFs. The pasting profiles were characterized by lower peak viscosities, improved heat and shear resistance, and higher setback values. The WMFs of 2019 contained the highest levels of protein, which improved the ability of these WMFs to absorb and retain water. This was reflected in the high WAC and WAIs at 55 and 65 °C, but resulted in more physical competition for water between proteins and starch. This probably restricted the starch granule swelling, which explains the low peak viscosity and holding strength of the WMFs of 2019. The impact of variety was less consistent than the effect of the growing season. However, certain trends were observed over the years. Jessie and Titicaca WMF had a high WAC while Oro de Valle WMF showed a high swelling behavior upon heating. Jessie, Summer Red, Rouge Marie, Vikinga, and Zwarte were WMFs characterized by a low swelling behavior. These WMFs developed low viscosities and had a high resistance to shear and heat. The pasting profile of Bastille WMF was generally characterized by a high peak and final viscosity and, together with Faro WMF, showed a high breakdown. Puno WMF developed a high final viscosity upon cooling, indicating a high tendency to retrograde.

The wide variation in physicochemical properties suggest that the potential food applications of WMFs depend on the quinoa variety. However, the growing conditions may significantly alter the macronutrient composition of the seeds, thereby affecting the physicochemical properties of the WMFs and eventually the end-product quality. Nevertheless, variations in the macronutrient composition only partially explained the differences in physicochemical properties. The impact of other components such as fiber, amylose/amylopectin, and damaged starch on the physicochemical properties of these quinoa WMFs might be the subject of further research.

Supplementary Materials: The following are available online at https://www.mdpi.com/article/10.3390/plants11030265/s1, Figure S1: Chemical composition of thirteen quinoa varieties ((**a**): Atlas, (**b**): Bastille, (**c**): Dutchess, (**d**): Faro, (**e**): Jessie, (**f**): Oro de Valle, (**g**): Pasto, (**h**): Puno, (**i**): Rouge Marie, (**j**): Summer Red, (**k**): Titicaca, (**l**): Vikinga, (**m**): Zwarte) grown under North-West European field conditions in 2017 (●), 2018 (●) and 2019 (●). Values are presented as the mean relative to the three-year mean., Table S1: Water absorption index at 55, 65, 75, 85, and 95 °C (WAI, g/g) of thirteen quinoa wholemeal flours obtained from seeds grown under North-West European field conditions in 2017, 2018, and 2019 (n = 3), Table S2: Pasting parameters of thirteen quinoa whole-meal flours obtained from seeds grown under North-West European field conditions in 2017, 2018 and 2019 (n = 3), Table S3: Pearson correlation analysis between physicochemical properties, Table S4: Results of ANOVA (factors, p value).

Author Contributions: Conceptualization, G.C., F.V.B., and P.D.B.; methodology, G.C., F.V.B., P.D.B.; software, P.Q. and H.M.; validation, P.D.B.; formal analysis, P.Q., H.M., G.C., and P.D.B.; investigation, P.D.B.; resources, G.C., F.V.B., and M.E.; data curation, H.M., P.Q., G.C., and P.D.B.; writing—original draft preparation, G.C. and P.D.B.; writing—review and editing, H.M., G.C., P.D.B., P.Q., M.E., and F.V.B.; visualization, P.D.B.; supervision, G.C., M.E., and F.V.B.; project administration, H.M., G.C., and P.D.B.; funding acquisition, G.C., F.V.B., and M.E. All authors have read and agreed to the published version of the manuscript.

Funding: This research was funded by a doctoral scholarship of the Ghent University Special Research Fund (DOC079-17 BOF) and VLAIO LA traject QUILO (Quinoa Lokaal, HBC.2018.2220).

Data Availability Statement: The data presented in this study are available in the article and Supplementary Materials.

Acknowledgments: The authors wish to thank Tania Roeges (LCA, Faculty of Bio-science Engineering, Ghent University) for the technical support while working with the VarioMax C/N. Furthermore, the authors thank master thesis students (Ghent University) Tine Vanhollebeke and Indy De Graeve, and bachelor thesis students (Ghent University) Jolien De Schaepmeester, Joëlle Temmerman and Lies Herman for their help with the nutritional and physicochemical characterization of the quinoa varieties.

Conflicts of Interest: The authors declare no conflict of interest. The funders had no role in the design of the study; in the collection, analyses, or interpretation of data; in the writing of the manuscript, or in the decision to publish the results.

References

1. Solaesa, Á.G.; Villanueva, M.; Vela, A.J.; Ronda, F. Protein and lipid enrichment of quinoa (cv.Titicaca) by dry fractionation. Techno-functional, thermal and rheological properties of milling fractions. *Food Hydrocoll.* **2020**, *105*, 105770. [CrossRef]
2. Aluwi, N.A.; Murphy, K.M.; Ganjyal, G.M. Physicochemical characterization of different varieties of quinoa. *Cereal Chem.* **2017**, *94*, 847–856. [CrossRef]
3. Nowak, V.; Du, J.; Charrondière, U.R. Assessment of the nutritional composition of quinoa (*Chenopodium quinoa* Willd.). *Food Chem.* **2016**, *193*, 47–54. [CrossRef]
4. Escribano, J.; Cabanes, J.; Jiménez-Atiénzar, M.; Ibañez-Tremolada, M.; Gómez-Pando, L.R.; García-Carmona, F.; Gandía-Herrero, F. Characterization of betalains, saponins and antioxidant power in differently colored quinoa (*Chenopodium quinoa*) varieties. *Food Chem.* **2017**, *234*, 285–294. [CrossRef] [PubMed]
5. Vilcacundo, R.; Hernández-Ledesma, B. Nutritional and biological value of quinoa (*Chenopodium quinoa* Willd.). *Curr. Opin. Food Sci.* **2017**, *14*, 1–6. [CrossRef]
6. Sobota, A.; Świeca, M.; Gęsiński, K.; Wirkijowska, A.; Bochnak, J. Yellow-coated quinoa (*Chenopodium quinoa* Willd)—physicochemical, nutritional, and antioxidant properties. *J. Sci. Food Agric.* **2020**, *100*, 2035–2042. [CrossRef] [PubMed]
7. Bhargava, A.; Srivastava, S. *Quinoa: Botany, production and uses*; CABI: Wallingford, UK, 2013; ISBN 1780642261.
8. Präger, A.; Munz, S.; Nkebiwe, P.M.; Mast, B.; Graeff-Hönninger, S. Yield and quality characteristics of different quinoa (*Chenopodium quinoa* willd.) cultivars grown under field conditions in southwestern Germany. *Agronomy* **2018**, *8*, 197. [CrossRef]
9. Li, G.; Zhu, F. Physicochemical properties of quinoa flour as affected by starch interactions. *Food Chem.* **2017**, *221*, 1560–1568. [CrossRef] [PubMed]
10. Bhargava, A.; Ohri, D. Origin of genetic variability and improvement of quinoa (*Chenopodium quinoa* Willd.). In *Gene Pool Diversity and Crop Improvement*; Springer: Berlin/Heidelberg, Germany, 2016; pp. 241–270.
11. Li, G.; Wang, S.; Zhu, F. Physicochemical properties of quinoa starch. *Carbohydr. Polym.* **2016**, *137*, 328–338. [CrossRef]
12. Contreras-Jiménez, B.; Torres-Vargas, O.L.; Rodríguez-García, M.E. Physicochemical characterization of quinoa (*Chenopodium quinoa*) flour and isolated starch. *Food Chem.* **2019**, *298*, 124982. [CrossRef]
13. Lindeboom, N. Studies on the Characterization, Biosynthesis and Isolation of Starch and Protein from Quinoa (*Chenopodium quinoa* Willd.). Ph.D. Thesis, University of Saskatchewan, Saskatoon, SK, Canada, 2005.
14. Filho, A.M.M.; Pirozi, M.R.; Borges, J.T.D.S.; Pinheiro Sant'Ana, H.M.; Chaves, J.B.P.; Coimbra, J.S.D.R. Quinoa: Nutritional, functional, and antinutritional aspects. *Crit. Rev. Food Sci. Nutr.* **2017**, *57*, 1618–1630. [CrossRef]
15. Bhargava, A.; Shukla, S.; Ohri, D. *Chenopodium quinoa*—An Indian perspective. *Ind. Crops Prod.* **2006**, *23*, 73–87. [CrossRef]
16. Pineli, L.L.O.; Botelho, R.B.A.; Zandonadi, R.P.; Solorzano, J.L.; de Oliveira, G.T.; Reis, C.E.G.; Teixeira, D. da S. Low glycemic index and increased protein content in a novel quinoa milk. *LWT—Food Sci. Technol.* **2015**, *63*, 1261–1267. [CrossRef]
17. Ghumman, A.; Mudgal, S.; Singh, N.; Ranjan, B.; Kaur, A.; Rana, J.C. Physicochemical, functional and structural characteristics of grains, flour and protein isolates of Indian quinoa lines. *Food Res. Int.* **2021**, *140*, 109982. [CrossRef] [PubMed]
18. Jacobsen, S.E. The scope for adaptation of quinoa in Northern Latitudes of Europe. *J. Agron. Crop Sci.* **2017**, *203*, 603–613. [CrossRef]
19. De Bock, P.; Van Bockstaele, F.; Muylle, H.; Quataert, P.; Vermeir, P.; Eeckhout, M.; Cnops, G. Yield and Nutritional Characterization of Thirteen Quinoa (*Chenopodium quinoa* Willd.) Varieties Grown in North-West Europe—Part I. *Plants* **2021**, *10*, 2689. [CrossRef]
20. De Bock, P.; Daelemans, L.; Selis, L.; Raes, K.; Vermeir, P.; Eeckhout, M.; Van Bockstaele, F. Comparison of the chemical and technological characteristics of wholemeal flours obtained from amaranth (*Amaranthus* sp.), quinoa (*chenopodium quinoa*) and buckwheat (*fagopyrum* sp.) seeds. *Foods* **2021**, *10*, 651. [CrossRef]

21. Vázquez-Luna, A.; Fuentes, F.; Rivadeneyra, E.; Hernández, C.; Díaz-Sobac, R. Nutrimental content and functional properties of Quinoa flour from Chile and Mexico. *Cienc. Investig. Agrar.* **2019**, *46*, 144–153. [CrossRef]
22. Coțovanu, I.; Batariuc, A.; Mironeasa, S. Characterization of quinoa seeds milling fractions and their effect on the rheological properties of wheat flour dough. *Appl. Sci.* **2020**, *10*, 7225. [CrossRef]
23. Tafadzwa, M.J.; Zvamaziva, J.T.; Charles, M.; Amiel, M.; Pepukai, M.; Shepherd, M. Proximate, physico-chemical, functional and sensory properties of quinoa and amaranth flour AS potential binders in beef sausages. *Food Chem.* **2021**, *365*, 130619. [CrossRef]
24. Srichuwong, S.; Curti, D.; Austin, S.; King, R.; Lamothe, L.; Gloria-Hernandez, H. Physicochemical properties and starch digestibility of whole grain sorghums, millet, quinoa and amaranth flours, as affected by starch and non-starch constituents. *Food Chem.* **2017**, *233*, 1–10. [CrossRef]
25. Ahmed, J.; Thomas, L.; Arfat, Y.A. Functional, rheological, microstructural and antioxidant properties of quinoa flour in dispersions as influenced by particle size. *Food Res. Int.* **2019**, *116*, 302–311. [CrossRef]
26. Rulahnia, K.; Khatkar, B.S. Isolation and Characterization of Starch from Quinoa (*Chenopodium Quinoa* Willd.). *Int. J. Innov. Res. Adv. Stud.* **2020**, *7*, 1–6.
27. Tiga, B.H.; Kumcuoglu, S.; Vatansever, M.; Tavman, S. Thermal and pasting properties of Quinoa—Wheat flour blends and their effects on production of extruded instant noodles. *J. Cereal Sci.* **2021**, *97*, 103120. [CrossRef]
28. Steffolani, M.E.; León, A.E.; Pérez, G.T. Study of the physicochemical and functional characterization of quinoa and kañiwa starches. *Starch/Staerke* **2013**, *65*, 976–983. [CrossRef]
29. Opazo-Navarrete, M.; Tagle Freire, D.; Boom, R.M.; Janssen, A.E.M.; Schutyser, M.A.I. Dry fractionation of quinoa sweet varieties Atlas and Riobamba for sustainable production of protein and starch fractions. *J. Food Compos. Anal.* **2018**, *74*, 95–101. [CrossRef]
30. Horstmann, S.W.; Belz, M.C.E.; Heitmann, M.; Zannini, E.; Arendt, E.K. Fundamental Study on the Impact of Gluten-Free Starches on the Quality of Gluten-Free Model Breads. *Foods* **2016**, *5*, 30. [CrossRef]
31. Barak, S.; Mudgil, D.; Khatkar, B.S. Relationship of gliadin and glutenin proteins with dough rheology, flour pasting and bread making performance of wheat varieties. *LWT-Food Sci. Technol.* **2013**, *51*, 211–217. [CrossRef]
32. Kaur, A.; Singh, N.; Ezekiel, R.; Guraya, H.S. Physicochemical, thermal and pasting properties of starches separated from different potato cultivars grown at different locations. *Food Chem.* **2007**, *101*, 643–651. [CrossRef]
33. Dziki, D.; Laskowski, J. Influence of kernel size on grinding process of wheat at respective grinding stages. *Polish J. food Nutr. Sci.* **2004**, *54*, 29–33.
34. Singh, J.; Singh, N. Studies on the morphological, thermal and rheological properties of starch separated from some Indian potato cultivars. *Food Chem.* **2001**, *75*, 67–77. [CrossRef]
35. Wu, G.; Morris, C.F.; Murphy, K.M. Evaluation of texture differences among varieties of cooked Quinoa. *J. Food Sci.* **2014**, *79*, S2337–S2345. [CrossRef]
36. Englyst, H.N.; Kingman, S.M.; Cummings, J.H. Classification and measurement of nutritionally important starch fractions. *Eur. J. Clin. Nutr.* **1992**, *46*, S33–S50. [CrossRef]
37. FAO Energy and Protein Requirements; Rome, 1973. Available online: https://www.fao.org/3/m2995e/m2995e00.htm (accessed on 28 August 2021).
38. Hellemans, T.; Abera, G.; De Leyn, I.; Van der Meeren, P.; Dewettinck, K.; Eeckhout, M.; De Meulenaer, B.; Van Bockstaele, F. Composition, Granular Structure, and Pasting Properties of Native Starch Extracted from *Plectranthus edulis* (*Oromo dinich*) Tubers. *J. Food Sci.* **2017**, *82*, 2794–2804. [CrossRef] [PubMed]
39. Bates, D.; Mächler, M.; Bolker, B.; Walker, S. Fitting Linear Mixed-Effects Models Using lme4. *J. Stat. Softw.* **2015**, *67*, 1–48. [CrossRef]

plants

MDPI

Article

Biogeographical Patterns of Herbivore Arthropods Associated with *Chenopodium quinoa* Grown along the Latitudinal Gradient of Chile

Rodrigo A. Chorbadjian [1,*], María I. Ahumada [1], Francisco Urra [2], Mario Elgueta [2] and Todd M. Gilligan [3]

1 Departamento de Ciencias Vegetales, Facultad de Agronomía e Ingeniería Forestal, Pontificia Universidad Católica de Chile, Santiago 7820436, Chile; miahumad@uc.cl
2 Sección Entomología, Museo Nacional de Historia Natural, Casilla 787, Santiago 7820436, Chile; francisco.urra@mnhn.gob.cl (F.U.); Mario.Elgueta@mnhn.cl (M.E.)
3 USDA-APHIS-PPQ-Science and Technology, Identification Technology Program, Fort Collins, CO 80526, USA; todd.m.gilligan@usda.gov
* Correspondence: rchorba@uc.cl

Abstract: Identifying the particular guilds of herbivore arthropods that affect the production of crops is key to developing sustainable pest-management strategies; however, there is incomplete information about the identity of herbivore arthropods that could potentially damage the production of both highland and lowland quinoa landraces grown in Chile. By both reviewing the literature and conducting field collections across a large latitudinal gradient, we generated an updated list of 43 herbivore arthropods associated with quinoa production in Chile. In general, most species are polyphagous feeders, and only seven are specialists. The number and identity of species varied in relation with the latitude, such that four distinctive assemblages of herbivores were identified, each containing 32, 27, 34, and 22 species between latitudes 18–26, 26–32, 32–40, and 40–44° S, respectively. The most northern production area (18–26° S) is affected by nine unique species, including the major quinoa pest *Eurysacca quinoae* Povolný (Lepidoptera: Gelechiidae). Similarly, the central area (32–40° S) contains four unique species, including *Eurysacca media* Povolný (Lepidoptera: Gelechiidae) and *Orthotylus flavosparsus* (Sahlberg) (Hemiptera: Miridae). The particular species assemblages described here will help further development of local pest-management practices.

Keywords: insects; pests; quinoa; distribution; Chile

Citation: Chorbadjian, R.A.; Ahumada, M.I.; Urra, F.; Elgueta, M.; Gilligan, T.M. Biogeographical Patterns of Herbivore Arthropods Associated with *Chenopodium quinoa* Grown along the Latitudinal Gradient of Chile. *Plants* **2021**, *10*, 2811. https://doi.org/10.3390/plants10122811

Academic Editors: Cataldo Pulvento and Didier Bazile

Received: 20 November 2021
Accepted: 16 December 2021
Published: 19 December 2021

1. Introduction

Quinoa (*Chenopodium quinoa* Willd.) is an annual plant mainly grown to obtain grains for human consumption. The interest in cultivating this crop has increased in recent years due to its high nutritional value and its tolerance to soil salinity and drought stress [1–3]. Quinoa is cultivated over a wide variety of environments in South America, extending from the high altitudes (>3500 m above sea level (m.a.s.l.)) of the Andean Altiplano areas of Bolivia, Chile, and Perú, to the lowland/coastal areas of Chile and Perú [4,5]. Two distinctive quinoa ecotypes are cultivated in Chile, the salares and the coastal ecotypes [4]. Plants belonging to the salares ecotype grow in the northern region of Chile (18–29° S), which is separated from the central and southern regions (33–43° S) by the Atacama Desert and is agroecologically more similar to the Altiplanos of Perú and Bolivia. In the central and southern production regions, the coastal ecotype of quinoa is produced along with many other agricultural vegetables, crops, and fruit trees. Regarding the distribution of quinoa cultivation in Chile, the northern zone (17–26° S) accounted for 31%, the central zone (29–36° S) accounted for 64%, and the southernmost zone (37–44° S) accounted for 5% of total production during the years 2015–2016 [6].

The variety of agroecological environments over which quinoa is cultivated can influence the diversity of arthropods that negatively affect its production. For instance, while species of *Eurysacca* (quinoa moth, Lepidoptera: Gelechiidae) are more frequently indicated as a major pest in the highlands [5,7,8], other species such as thrips (*Frankliniella occidentalis* (Pergande) (Thysanoptera: Thripidae)), aphids (*Macrosiphum euphorbiae* (Thomas) (Hemiptera: Aphididae)), and leafminers (*Liriomyza huidobrensis* Blanchard (Diptera: Agromyzidae)) add to the array of pest species than can reach high populations in lowland areas of Perú [9].

Understanding the herbivore communities is essential to developing sustainable insect pest management strategies in quinoa production; however, specific studies reporting insect species that feed on quinoa in Chile are scarce and incomplete. For instance, the general compendium of insect pests of economic importance in Chile published by Artigas [10] reported only five insect species that use quinoa as a host plant. A few years later, Lamborot et al. [11] reported five Lepidoptera species found in quinoa grown in central Chile. Later, another compendium of arthropod species found in Chilean agricultural plants reported only eight taxa [12]. Finally, Logarzo et al. [13] reported the finding of one leafhopper species on quinoa in the Chilean Altiplano area. Clearly, this is a low number of species when compared with the potential 78 arthropod taxa reported by Cruces et al. [5] in a recent compilation of quinoa pests that included: 29 species of Lepidoptera, 22 Hemiptera, 16 Coleoptera, 4 Orthoptera, 3 Thysanoptera, 3 Diptera, and 1 Acari.

Because many insect species previously reported to feed on quinoa in other countries are also present in Chile, we decided to review and update this information. The underlying hypothesis is that it will be possible to identify distinct clusters of herbivore arthropods along the latitudinal gradient where quinoas are cultivated in Chile. In addition, we expected to identify latitudinal patterns in host range use, feeding habit, and geographical origin of these arthropods. Hence, the objective of this work was to generate an updated list of arthropod species associated with quinoa production in Chile by both reviewing the literature and conducting field collections on quinoa fields across a latitudinal gradient in a variety of agroecosystems. To further characterize the community of herbivores, species were classified according to their geographical origin and host-range use (i.e., generalist or specialist).

2. Results

Bibliographical evidence, together with field collections conducted in this study, allowed us to construct an updated list of arthropod herbivores that feed on quinoa plants in Chile, resulting in a total of 43 arthropod taxa. Across all latitudes, Lepidoptera is represented with 20 taxa in 4 families (Coleophoridae, Crambidae, Gelechiidae, and Noctuidae), Hemiptera with 15 taxa in 8 families (Aphididae, Cicadellidae, Coreidae, Lygaeidae, Miridae, Pentatomidae, Rhopalidae, and Triozidae), Coleoptera with 3 taxa in 3 families (Chrysomelidae, Curculionidae, and Meloidae), Thysanoptera with 2 Thripidae species, 1 Diptera (Agromyzidae), 1 Orthoptera (Acrididae), and 1 mite species (Acari: Tetranychidae) (Table 1).

During field collections, we found and identified 19 taxa. Species identified included, *Achyra similalis* (Guenée) (Lepidoptera: Crambidae), *Agrotis ipsilon* (Hufnagel) (Lepidoptera: Noctuidae), *Aphis craccivora* Koch (Hemiptera: Aphididae), *Eurysacca quinoae* Povolný (Lepidoptera: Gelechiidae), *Frankliniella occidentalis* (Thysanoptera: Thripidae), *Feltia subterranea* (Fabricius) (Lepidoptera: Noctuidae), *Helicoverpa atacamae* Hardwick (Lepidoptera: Noctuidae), *Helicoverpa gelotopoeon* (Dyar) (Lepidoptera: Noctuidae), *Liorhyssus lineatoventris* (Spinola) (Hemiptera: Rhopalidae), *Liriomyza huidobrensis* (Diptera: Agromyzidae), *Macrosiphum euphorbiae* (Hemiptera: Aphididae), *Myzus persicae* (Sulzer) (Hemiptera: Aphididae), *Orthotylus flavosparsus* (Sahlberg) (Hemiptera: Miridae), *Tetranychus urticae* Koch (Acari: Tetranychidae), *Trichocyphus rubricollis* (Blanchard) (Coleoptera: Curculionidae), and *Trichoplusia ni* (Hübner) (Lepidoptera: Noctuidae). Three taxa were included only at the level of genera, as there was uncertainty regarding their species names. Specifically, further research is needed to clarify species identity for *Epitrix* sp. (Coleoptera: Chrysomelidae), *Pseudomeloe* sp. (Coleoptera: Meloidae), and *Copitarsia* spp. (Lepidoptera: Noctuidae).

Specimens of these three genera were collected in this study and saved for further taxonomic analyses.

Table 1. Species list and characteristics of herbivore arthropods that use quinoa as a host plant and are present in Chile.

Species	Distribution Range in Chile (° S)	Feeding Habit	Host Range	Geographical Origin	References
Orthoptera: Acrididae					
Dichroplus maculipennis (Blanchard, 1851)	29–56	Chewing	Generalist	Neotropic	[5,14,15]
Hemiptera: Cicadellidae					
Anacuerna centrolinea (Melichar, 1925)	19–22	Piercing–sucking	Generalist	Neotropic	[5,13]
Paratanus exitiosus Beamer, 1943	32–41	Piercing–sucking	Generalist	Neotropic	[5,7,10,16]
Hemiptera: Aphididae					
Aphis craccivora Koch, 1854	18–44 [a,b,c]	Piercing–sucking	Generalist	Palearctic	[5,7,10,15,16]
Aphis gossypii Glover, 1877	18–44	Piercing–sucking	Generalist	Unknown	[5,7,12,16,17]
Macrosiphum euphorbiae (Thomas, 1878)	18–56 [a,b,c,d]	Piercing–sucking	Generalist	Neotropic—Nearctic	[5,7,10,16–18]
Myzus persicae (Sulzer, 1776)	18–56 [b]	Piercing–sucking	Generalist	Indo-Malayan	[5,7,10,16,17]
Smynthurodes betae Westwood, 1849	18–22; 32–56	Piercing–sucking	Generalist	Unknown	[10,17]
Hemiptera: Triozidae					
Heterotrioza chenopodii (Reuter, 1876) (=*Trioza chenopodii* Reuter)	29–34	Piercing–sucking	Specialist	Palearctic	[5,15,19]
Hemiptera: Pentatomidae					
Nezara viridula (Linnaeus, 1758)	18–56	Piercing–sucking	Generalist	Unknown	[5,10,15]
Hemiptera: Lygaeidae					
Nysius simulans (Stål, 1859)	29–38	Piercing–sucking	Generalist	Neotropic	[10]
Oncopeltus miles (Blanchard, 1852)	19–40	Piercing–sucking	Specialist	Chile	[10,20]
Hemiptera: Coreidae					
Leptoglossus chilensis (Spinola, 1852)	26–44	Piercing–sucking	Generalist	Neotropic	[5,15,21]
Hemiptera: Rhopalidae					
Liorhyssus hyalinus (Fabricius, 1794)	18–38	Piercing–sucking	Generalist	Unknown	[10]
Liorhyssus lineatoventris (Spinola, 1852)	30–40 [b]	Piercing–sucking	Generalist	Neotropic	[22]
Hemiptera: Miridae					
Orthotylus flavosparsus (Sahlberg, 1841)	33–34 [b]	Piercing–sucking	Generalist	Unknown	[5,15,22]
Thysanoptera: Thripidae					
Frankliniella occidentalis (Pergande, 1895)	18–44 [b,c]	Cell puncturing	Generalist	Nearctic	[12,18]
Thrips tabaci Lindeman, 1889	18–40	Cell puncturing	Generalist	Unknown	[7,10,16,18]
Diptera: Agromyzidae					
Liriomyza huidobrensis Blanchard, 1926	18–49 [a,b,c]	Leaf mining	Generalist	Neotropic	[5,7,10,16]

Table 1. *Cont.*

Species	Distribution Range in Chile (° S)	Feeding Habit	Host Range	Geographical Origin	References
Coleoptera: Chrysomelidae					
Epitrix sp.	32–34 [b]	Chewing	Unknown	Neotropic—Nearctic	[5,7,10,12,16]
Coleoptera: Curculionidae					
Trichocyphus rubricollis (Blanchard, 1847)	18–26 [a]	Chewing	Unknown	Neotropic	[23]
Coleoptera: Meloidae					
Pseudomeloe sp.	19–22 [a]	Chewing	Unknown	Neotropic	[7]
Lepidoptera: Coleophoridae					
Coleophora versurella Zeller, 1849	32–40	Chewing	Specialist	Neotropic—Nearctic	[11,24,25]
Lepidoptera: Crambidae					
Achyra similalis (Guenée, 1854) (=*Loxostege similalis* (Guenée))	18–40 [b]	Chewing	Specialist	Neotropic—Nearctic	[10,11,25]
Spoladea recurvalis (Fabricius, 1794)	18–19	Chewing	Specialist	Neotropic	[5,7,10]
Lepidoptera: Gelechiidae					
Eurysacca media Povolný, 1986	32–34	Chewing	Specialist	Neotropic	[11,26,27]
Eurysacca quinoae Povolný, 1997	19 [a]	Chewing	Specialist	Neotropic	[28]
Lepidoptera: Noctuidae					
Agrotis experta (Walker, 1869) (=*Feltia experta* (Walker))	18–26	Chewing	Generalist	Neotropic	[5,7,10,16]
Agrotis ipsilon (Hufnagel, 1766)	18–44 [c]	Chewing	Generalist	Unknown	[5,7,10,16]
Agrotis malefida (Guenée, 1852)	32–56	Chewing	Generalist	Neotropic	[7,10]
Copitarsia spp. (species complex)	18–56 [a,b,c,d]	Chewing	Generalist	Neotropic	[5,10–12,16,29]
Chrysodeixis includens (Walker, 1858) (=*Pseudoplusia includens* (Walker)) (=*Phytometra oo* (Cramer))	18–26	Chewing	Generalist	Neotropic—Nearctic	[5,10,12]
Feltia subterranea (Fabricius, 1794) (=*Agrotis subterranea*)	18–40 [b,c]	Chewing	Generalist	Neotropic—Nearctic	[5,10]
Helicoverpa atacamae Hardwick, 1965	18–41 [a]	Chewing	Generalist	Neotropic	[5,16,30]
Helicoverpa gelotopoeon (Dyar, 1921)	18–40 [a]	Chewing	Generalist	Neotropic	[5,12,15,30]
Helicoverpa zea (Boddie, 1850)	18–44	Chewing	Generalist	Neotropic—Nearctic	[5,7,10,12,16,18]
Peridroma saucia (Hübner, 1808)	18–56	Chewing	Generalist	Unknown	[5,7,10,16]
Rachiplusia nu (Guenée, 1852)	18–44	Chewing	Generalist	Neotropic—Nearctic	[5,10,11,15]
Spodoptera eridania (Stoll, 1782)	18–33	Chewing	Generalist	Neotropic—Nearctic	[5,7,10,16]
Spodoptera frugiperda (J.E. Smith, 1797)	18–22	Chewing	Generalist	Neotropic—Nearctic	[5,7,10,15,16]
Spodoptera ochrea (Hampson, 1909)	18–22	Chewing	Generalist	Neotropic	[5,10]
Trichoplusia ni (Hübner, 1803)	18–44 [b]	Chewing	Generalist	Unknown	[12,18]
Acari: Tetranychidae					
Tetranychus urticae Koch, 1836	18–44 [b]	Cell puncturing	Generalist	Unknown	[5,10]

[a] Collected in this study in Tarapacá (19°24′ S, 68°35′ W); [b] Metropolitana (33°40′ S, 70°35′ W, or 33°29′ S, 70°36′ W); [c] O'Higgins (34°29′, 72°01′ W, or 34°15′ S, 71°47′ W); [d] Los Lagos (41°50′ S, 74°00′ W, or 42°00′ S, 73°53′ W).

The number of individual taxonomic units collected on each locality was 9 in the region of Tarapacá (Ancovinto site, 19° S), 13 in the Metropolitan region (Santiago and Pirque sites, 33° S), 7 in the O'Higgins region (Cahuil and Pailimo sites, 34° S), and 2 in the Los Lagos Region (Ancud sites, 42° S) (Table 1). Lepidoptera species were particularly abundant at the Ancovinto site. The identity of the quinoa moth *Eurysacca quinoae* was confirmed based on male genitalia structures that corresponded with its original description;

particularly, the parabasal processes with broadly rounded clavate tips, symmetrical lanceolate saccular processes with an acute tip, and a long and slender valve [28]. Two species of *Helicoverpa*, *H. atacamae* and *H. gelotopoeon*, were confirmed based on morphological structures in comparison with descriptions and illustrations in Hardwick [30]. In particular, the two can be separated by the length of the male valvae, the shape of the everted vesica, and differences in setae on the foretibia [30]. We also identified a new host record for *Trichocyphus rubricollis*. *Trichocyphus rubricollis* was originally described by Kuschel [31] as a variety of *T. formosus*, to later recognize them as different specific entities [32]; Although Lanteri [33] establishes the synonymy between both names, Elgueta and Marvaldi [23] consider both as valid species. A few adults of *T. rubricollis* were detected feeding on quinoa leaves, chewing from the external margin of the leaves towards the central vein. Other species detected in this area included *Copitarsia* sp., *Aphis craccivora*, *Macrosiphum euphorbiae*, *Liriomyza huidobrensis*, and *Pseudomeloe* sp.

In the central area (Metropolitana and O'Higgins areas), specimens collected included: *Achyra similalis*, *Agrotis ipsilon*, *Aphis craccivora*, *Copitarsia* spp., *Epitrix* sp., *Feltia subterranea*, *Frankliniella occidentalis*, *Liorhyssus lineatoventris*, *Liriomyza huidobrensis*, *Macrosiphum euphorbiae*, *Myzus persicae*, *Orthotylus flavosparsus*, *Tetranychus urticae*, and *Trichoplusia ni*. Among these species, a complex of chinch bugs (*L. lineatoventris* and *O. flavosparsus*), *Copitarsia* spp., and *L. huidobrensis* were more frequently collected. Our results also provide new information for the distribution range of *O. flavosparsus*, as it was known to be present in Chile [22], but no information on its distribution range and host use had been reported. In turn, the southernmost area of Ancud is characterized by the presence of fewer herbivore species. Here, we only collected *Copitarsia* spp. and *M. euphorbiae*. While only a few specimens of *M. euphorbiae* were detected, *Copitarsia* caterpillars were found more often but still in low numbers.

Regarding the total number of species expected to affect quinoa in Chile, we found that species assemblages varied in relation to the geographical region. Cluster analysis identified four groups with a percentage of similarity higher than 80% within each cluster (Figure 1). One cluster grouped the regions of Arica y Parinacota, Tarapacá, and Antofagasta (18–26° S) with 32 taxa; a second cluster included 27 taxa in the regions of Atacama and Coquimbo (26–32° S). In the central area of Chile, a third cluster included 34 taxa in the Regions of Valparaíso, Metropolitana, O'Higgins, Maule, Ñuble, Bio-Bío and Araucanía (32–40° S), and the fourth cluster assembled 22 taxa in the regions of Los Rios and Los Lagos (40–44° S) (Table 2). Interestingly, both Cluster 1 and 3 included unique species. Cluster 1 is defined by nine species that are uniquely found in the northernmost region, which is closest to the borders with Perú and Bolivia. This is the case for *Agrotis experta* (Walker) (Lepidoptera: Noctuidae), *Anacuerna centrolinea* (Melichar) (Hemiptera: Cicadellidae), *Chrysodeixis includens* (Walker) (Lepidoptera: Noctuidae), *Eurysacca quinoae*, *Pseudomeloe* sp., *Spodoptera frugiperda* (J.E. Smith) (Lepidoptera: Noctuidae), *Spodoptera ochrea* (Hampson) (Lepidoptera: Noctuidae), *Spoladea recurvalis* (Fabricius) (Lepidoptera: Crambidae), and *Trichocyphus rubricollis*. Similarly, Cluster 3 in the central region is uniquely defined by the presence of *Coleophora versurella* Zeller (Lepidoptera: Coleophoridae), *Epitrix* sp., *Eurysacca media* Povolný (Lepidoptera: Gelechiidae), and *Orthotylus flavosparsus*.

In general, most species that feed on quinoa in Chile have a wide range of host use, given that 33 of them are polyphagous (77%) and only 7 are specialists (16%) (Table 1). When their geographical distribution is analyzed as a variable, the proportion of specialist herbivores ranges between 5 and 15% between parallels 18 and 39° S, with the lowest proportion in the southernmost area delimited by parallels 40–44° S (Figure 2). The distribution of the specialist herbivores *Eurysacca quinoae*, *E. media*, *Coleophora versurella*, and *Heterotrioza chenopodii* (Reuter) (Hemiptera: Triozidae) highly influenced this pattern due to their absence in the southern latitudes of Chile.

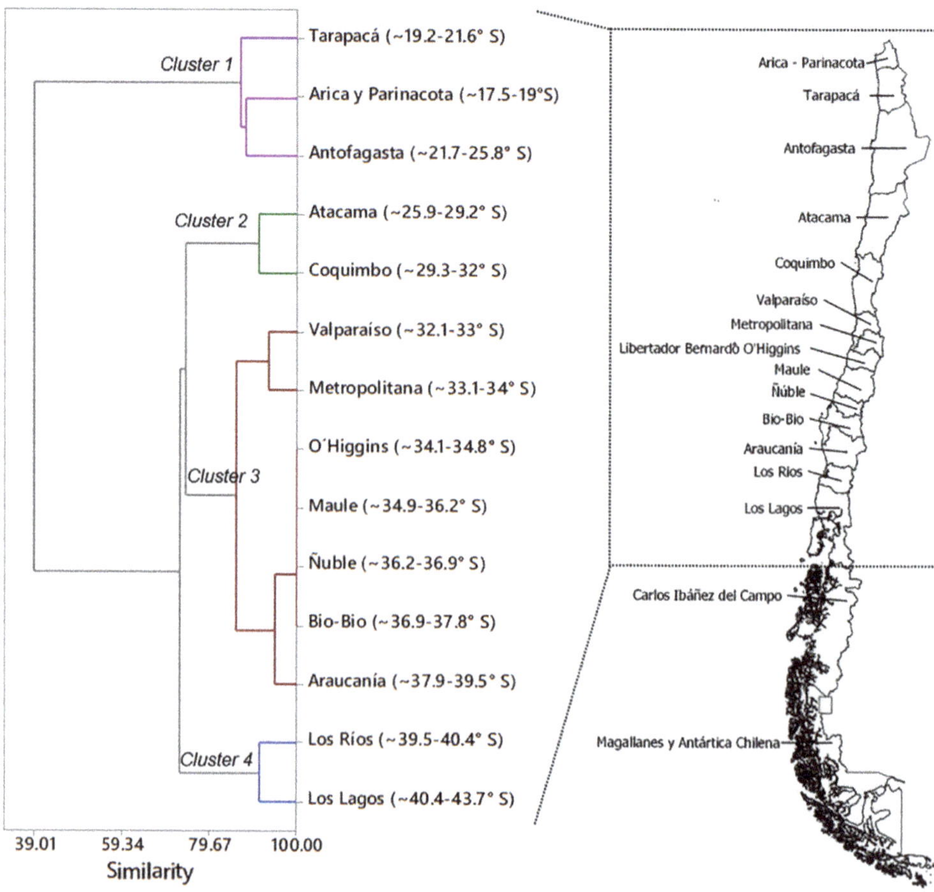

Figure 1. Dendrogram resulting from the multivariate cluster analysis of the geographical distribution variables of quinoa feeding arthropods present in Chile. Cluster description is presented in Table 2.

Table 2. Description of the clustering patterns of arthropod species associated with quinoa in Chile.

Cluster Group	Geographical Limits of Each Cluster (° S)	Species Present within the Delimited Area *
1	17.5–25.8	**32 species:** *Achyra similalis, Agrotis experta, Agrotis ipsilon, Anacuerna centrolinea, Aphis craccivora, Aphis gossypii, Chrysodeixis includens, Copitarsia* spp., *Eurysacca quinoae, Feltia subterranea, Frankliniella occidentalis, Helicoverpa atacamae, Helicoverpa gelotopoeon, Helicoverpa zea, Liorhyssus hialinus, Liriomyza huidobrensis, Macrosiphum euphorbiae, Myzus persicae, Nezara viridula, Oncopeltus miles, Peridroma saucia, Pseudomeloe* sp., *Rachiplusia nu, Smynthurodes betae, Spodoptera eridania, Spodoptera frugiperda, Spodoptera ochrea, Spoladea recurvalis, Tetranychus urticae, Thrips tabaci, Trichocyphus rubricollis, Trichoplusia ni*
2	25.9—32.0	**27 species:** *Achyra similalis, Agrotis ipsilon, Aphis craccivora, Aphis gossypii, Copitarsia* spp., *Dichroplus maculipennis, Feltia subterranea, Frankliniella occidentalis, Helicoverpa atacamae, Helicoverpa gelotopoeon, Helicoverpa zea, Heterotrioza chenopodii, Leptoglossus chilensis, Liorhyssus hialinus, Liorhyssus lineatoventris, Liriomyza huidobrensis, Macrosiphum euphorbiae, Myzus persicae, Nezara viridula, Nysius simulans, Oncopeltus miles, Peridroma saucia, Rachiplusia nu, Spodoptera eridania, Tetranychus urticae, Thrips tabaci, Trichoplusia ni*

Table 2. *Cont.*

Cluster Group	Geographical Limits of Each Cluster (° S)	Species Present within the Delimited Area *
3	32.1—39.5	**34 species:** *Achyra similalis, Agrotis ipsilon, Agrotis malefida, Aphis craccivora, Aphis gossypii, Coleophora versurella, Copitarsia* spp., *Dichroplus maculipennis, Epitrix* sp., *Eurysacca media, Feltia subterranea, Frankliniella occidentalis, Helicoverpa atacamae, Helicoverpa gelotopoeon, Helicoverpa zea, Heterotrioza chenopodii, Leptoglossus chilensis, Liorhyssus hialinus, Liorhyssus lineatoventris, Liriomyza huidobrensis, Macrosiphum euphorbiae, Myzus persicae, Nezara viridula, Nysius simulans, Oncopeltus miles, Orthotylus flavosparsus, Paratanus exitiosus, Peridroma saucia, Rachiplusia nu, Smynthurodes betae, Spodoptera eridania, Tetranychus urticae, Thrips tabaci, Trichoplusia ni*
4	39.5—43.7	**22 species:** *Achyra similalis, Agrotis ipsilon, Agrotis malefida, Aphis craccivora, Aphis gossypii, Copitarsia* spp., *Dichroplus maculipennis, Frankliniella occidentalis, Helicoverpa atacamae, Helicoverpa zea, Leptoglossus chilensis, Liorhyssus lineatoventris, Liriomyza huidobrensis, Macrosiphum euphorbiae, Myzus persicae, Nezara viridula, Paratanus exitiosus, Peridroma saucia, Rachiplusia nu, Smynthurodes betae, Tetranychus urticae, Trichoplusia ni*

* For information regarding species-specific distribution within each cluster, see Table 1.

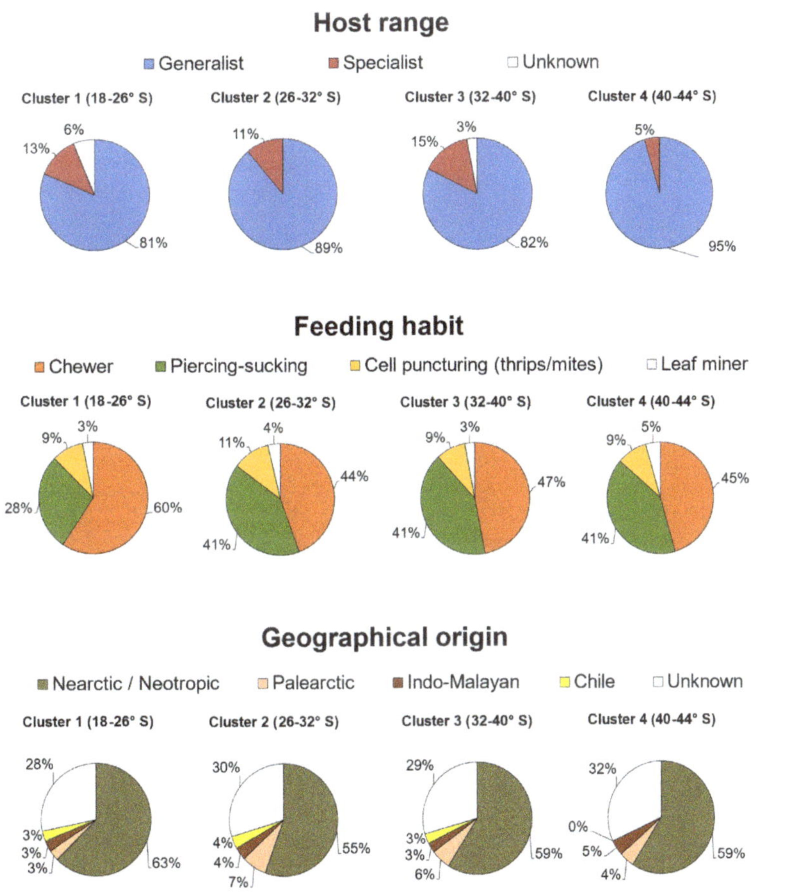

Figure 2. Latitudinal patterns of host range, feeding habit, and geographical origin of arthropod species that feed on quinoa in each latitudinal cluster of species.

The feeding habit of the arthropod species that attack quinoa in Chile is mostly dominated by chewing and piercing–sucking insects. The relative proportion of chewing insects ranges between 44 and 60%, and that of piercing–sucking hemipterans between 28 and 41%, depending on the geographical area (Figure 2). Only three species puncture plant cells, including two thrips (*F. occidentalis* and *Thrips tabaci* Lindeman) and the two-spotted spider mite (*T. urticae*). The leafminer *L. huidobrensis* was the only species identified with this strict feeding habit, which is distributed across the entire Chilean territory (Table 1). *Eurysacca* species were classified as chewer insects, although it has been reported that they have a leaf-mining habit during the early stages of larval development [7].

Regarding the geographical origin of the 43 arthropod species reported to feed on quinoa in Chile (Table 1), 30 species are native to the New World, while only 2 are Palearctic, and 1 comes from the Indo-Malayan realm. Interestingly, *Oncopeltus miles* (Blanchard) (Hemiptera: Lygaeidae) is the only native species reported to feed on quinoa in Chile [10]. In turn, 10 species have an uncertain geographical origin, as these are cosmopolitan agricultural pests (Table 1).

3. Discussion

This is the first study to compile a list of arthropods that feed on quinoa in Chile, which will help quinoa growers and future insect–plant interaction research. Until this study, only 10 insect species had been reported as quinoa feeders within the Chilean territory. Specifically, the general compendium of insect pests of economic importance in Chile reported one aphid (*Smynthurodes betae* Westwood (Hemiptera: Aphididae)), one chinchbug (*O. miles*), and three Lepidoptera species (*Helicoverpa zea* (Boddie), *Rachiplusia nu* Guenée, and *S. recurvalis*) [10]. Another study that focused on caterpillars reported five Lepidoptera species found in quinoa in central Chile, including *E. media*, *Copitarsia turbata* (Herrich-Schäffer), *R. nu*, *C. versurella*, and *A. similalis* [11]. The most recent compendium of arthropod species found in Chilean agricultural plants reported a total of 8 taxa, including *Copitarsia* sp., *H. zea*, *O. miles*, *R. nu*, *Sigelgaita chilensis* Heinrich, *S. betae*, *S. recurvalis*, and *Tapajosa* sp. [12]. However, we found a few incorrect species names and host use attributions in Klein-Koch and Waterhouse [12] and made appropriate corrections to construct Table 1. Specifically, *Sigelgaita chilensis* Heinrich does not feed on quinoa [10]; thus, it was not included here. Additionally, Klein-Koch and Waterhouse [12] lists *Tapajosa* sp. as a quinoa feeder, but as stated by Logarzo et al. [13], *Tapajosa* sp. was identified later as *A. centrolinea*. Finally, there is uncertainty about the identity of the species reported as *C. turbata* [11] because there is considerable confusion among *Copitarsia* species in South American literature [34]. Therefore, all individuals identified as *Copitarsia* during the conduction of this study are reported here as *Copitarsia* spp. and were saved for further taxonomic analysis.

It is possible that other insect species also feed on quinoa in Chile, but for which further studies are needed to clarify its potential presence on quinoas in Chile. For instance, this is the case for the genera *Rhinacloa* (Miridae), *Xenogenus* (Rhopalidae), *Empoasca* (Cicadellidae), *Bergallia* (Cicadellidae), *Conoderus* (Elateridae), *Cylydrorhinus* (formerly = *Adioristus*) (Curculionidae), *Tetraonyx* (Meloidae), and *Symmetrischema* (Gelechiidae). Specifically, regarding *Rhinacloa* sp. reported from the Altiplano area [7,16], there are five species in Chile, including *R. aricana* Carvalho, *R. azapa* Schuh and Schwartz, *R. incaicus* (Carvalho and Gomes), *R. penai* Schuh and Schwartz, and *R. peruana* Schuh and Schwartz [22]. The species *Xenogenus picturatum* Berg. was reported on quinoa [5,15], but only *Xenogenus gracilis* (Reed) is present in Chile [10]. Likewise, the genera *Empoasca* and *Bergallia* (Cicadellidae) have been reported attacking quinoa in Perú [35], for which *Empoasca curveola* Oman [10] and *Bergallia valdiviana* Berg [36] are found in Chile, but we did not find evidence of their association with the cultivation of quinoa in Chile. Similarly, Dughetti [15] and Cruces et al. [5] also report *Conoderus* sp. on quinoa; however, we did not find either *Conoderus chilensis* (Schwartz) or *C. rufangulus* (Gyllenhal), which are the two species present in Chile [10]. Another taxon previously reported only at the level of genus is

Adioristus sp. [5,7,16], but this name is a synonym of *Cylydrorhinus*, as stated by Wibmer and O'Brien [37]. Valoy et al. [38] reported *Tetraonyx* sp. with no details about species identity. For this genus in Chile, Elgueta and Arriagada [39] reported *Tetraonyx limbata*, *T. parviceps* and *T. septemguttata*, however there is no information that suggests quinoa as part of their host range. Regarding another potential quinoa pest, Dughetti [15] and Cruces et al. [5] also reported *Symmetrischema* sp., for which *S. nanum* Povolný, *S. striatella* (Murtfeldt), and *S. tangolias* (Gyen) are in Chile [26,40,41]. Undoubtedly, future samplings and taxonomic studies could expand the list of species reported in this study.

Most species that feed on quinoa are chewing stages belonging to Lepidoptera and a few species of Coleoptera and Orthoptera, followed by piercing–sucking Hemiptera. Insects with a chewing feeding habit feed on leaves, inflorescences, and developing grains. Flower- and seed-feeding insects, such as Gelechiidae and Noctuidae, often cause serious damage to quinoa production [5,7,16]. Piercing–sucking hemipterans, such as Aphididae and Cicadellidae, feed on phloem/xylem sap extracted from leaves, stems, and inflorescences, while chinch bugs in Heteroptera may also feed on immature grains [15]. In contrast, Thysanoptera species can puncture and extract cell contents of leaves, buds, inflorescences, and pollen. Agromyzidae species use a different feeding strategy, as the larvae of *L. huidobrensis* construct feeding galleries in the leaves of quinoa, and adult females puncture the leaves with their ovipositor to feed on cell content [5]. The larvae of a few species feed underground, such as *Epitrix* and those belonging to Anthomyiidae [5,18].

Although a variety of insects feed on quinoa, some are rarely seen in the field [9]. Additionally, only a few species are commonly observed in high population numbers, thereby causing concerns to growers about potential yield losses [5,7,9]. Typically, *Eurysacca melanocampta* (Meyrick) and *E. quinoae* are frequently cited as the most significant quinoa pests in Perú and Bolivia [5,7]. Nonetheless, the geographical location of quinoa production has been shown to influence species richness and its abundance, even within the same country. In a recent study conducted in two lowland sites (La Molina and Majes) and one highland site (San Lorenzo) of Perú, Cruces et al. [9] detected higher populations of *M. euphorbiae*, *E. melanocampta*, and *L. huidobrensis* in La Molina, as well as of *F. occidentalis* in the locality of Majes; only *E. melanocampta* was a major pest in San Lorenzo. Concordantly, we identified four distinct groups of species associated with quinoa along a latitudinal gradient of Chile (Table 2). Nine unique insect species feed on quinoa in the northern territory (Cluster 1), including *A. experta*, *A. centrolinea*, *C. includens*, *E. quinoae*, *Pseudomeloe* sp., *S. frugiperda*, *S. ochrea*, *S. recurvalis*, and *T. rubricollis*. This geographic area is close to the borders with Perú and Bolivia, and it is geographically isolated from the central and southern quinoa production areas by the Atacama Desert. Indeed, many of these species are also reported from the Altiplano areas of Bolivia and Perú [5]. During our observations in the quinoas grown in the altiplano area of Ancovinto, we found relatively high numbers of *E. quinoae*, *H. atacamae*, *H. gelotopoeon*, and *Copitarsia* sp. larvae feeding on leaves, flowers, and developing grains, as well as occasional clusters of the aphid *A. craccivora* during grain development, but otherwise, other species were uncommon.

Coastal ecotypes grown at higher latitudes are potentially affected by a distinct assemblage of insect species. Particularly, the central region (Cluster 3) concentrates the highest number of species, with unique species including *E. media*, *Epitrix* sp., *O. flavosparsus*, and *C. versurella*. Nonetheless, not all the species present in this area have been signified as major pests in other lowland areas of quinoa production [9]. Indeed, during our field studies, we only observed population outbreaks of *Copitarsia* sp., *L. lineatoventris*, and *O. flavosparsus*, as well as occasional infestations with *A. craccivora* and *L. huidobrensis*. In the southernmost production area (Cluster 4), the number of herbivorous arthropod species is the lowest, therefore representing potential advantages for the sustainable production of quinoa at higher latitudes.

Most species potentially found on quinoa in Chile show a wide range of host use. Generalist herbivores are often major pests in other agricultural crops, and therefore, they could potentially colonize quinoas grown near vegetables and other crop species as quinoa production areas diversify outside the highlands of the Andes. This has been reported in Perú, where lowland quinoas are negatively affected by polyphagous feeders such as *F. occidentalis*, *M. euphorbiae*, and *L. huidobrensis* [9]. Similarly, several polyphagous insect species negatively affect quinoas in other countries, such as Argentina [15], the United States [18], and Italy [42–44]. Interestingly, many of these generalist feeders are also cosmopolitan invasive pests that represent potential pest problems in other areas of the world where quinoa production is expanding.

4. Materials and Methods

4.1. Field Sampling and Species Identification

Commercial and experimental quinoa plantations were sampled periodically between 2015 and 2018 in 7 sampling sites within the 4 political regions named (abbr.) Tarapacá, Metropolitana, O'Higgins, and Los Lagos. For each region, sampling details are provided below.

- Tarapacá

  Ancovinto site with commercial plantations of salares ecotype (20 ha) (19°24′ S, 68°35′ W, 3720 m.a.s.l.). Inspected: 10 December 2016, 27 January 2017, 7 April 2017, and 29 January 2018.

- Metropolitana

  Pirque site. Research facility with experimental plantations (1 ha) of coastal ecotype (33°40′ S, 70°35′ W, 653 m.a.s.l.). Inspected: 22 December 2015, 14 January 2016, and 10 October 2018.
  Santiago site, research facility with demonstrative plantation of coastal ecotype (33°29′ S, 70°36′ W, 576 m.a.s.l.). Inspected on a monthly basis from November through April of 2016, 2017, and 2018.

- O'Higgins

  Cahuil site with commercial plantation of coastal ecotype (10 ha) (34°29′, 72°01′ W, 40 m.a.s.l.). Inspected: 12 October 2016, 12 December 2016, 12 January 2016, and 22 January 2017.
  Pailimo site with commercial plantation of coastal ecotype (5 ha) (34°15′ S, 71°47′ W, 242 m.a.s.l.). Inspected: 12 October 2016, 12 January 2016, and 22 January 2017.

- Los Lagos

  Ancud sites 1 and 2 with commercial plantations of coastal ecotypes (0.1 and 0.2 ha) (41°50′ S, 74°00′ W, 7 m.a.s.l., and 42°00′ S, 73°53′ W, 38 m.a.s.l.). Inspected: 16 December 2016, 13 January 2017, and 3 February 2017.

In these sites, quinoa plants were scouted by whole-plant visual inspections, plant beating, and using a sweeping net. Similar sampling efforts were devoted to each sampling date, which corresponded to 1 h of scouting. Special attention was given to collecting insects that were actively feeding. For inspections, each site was monitored in a random pattern, selecting at least 5 sectors where 5–10 plants were sampled per sector. Plant tissues affected by each species were annotated as leaves, stems, and/or panicles (flowers and/or seeds). Immature stages were brought to the laboratory and reared individually until adult emergence. Identifications of field-collected specimens were conducted on mounted adult specimens and comparing their morphological traits with available taxonomic publications [10,23,27,28,30–33,45–47], or directly with identified specimens in the Colección Nacional de Insectos, Museo Nacional de Historia Natural, Santiago, Chile (MNNC). Voucher specimens are conserved in the Entomological Collection of the Museo Nacional de Historia Natural, Santiago, Chile and in the Entomological Collection of Facultad de Agronomía e Ingeniería Forestal, Pontificia Universidad Católica de Chile, Santiago.

4.2. Total Number of Expected Species

First, we compiled records of insect and mite species to create a comprehensive list of arthropods reported to feed on quinoa around the world. We used the following scientific articles, books, and technical reports: [5,7,10,11,15,16,18,38,41,48]. Thereafter, this list was checked for the correct use of scientific names and updated. Arthropod species names were checked against currently available records for species present in Chile and updated with the species found in the surveys conducted in this study.

This checklist of arthropods associated with quinoa production in Chile was supplemented with bibliographical information about their geographical distribution in Chile, as well as their feeding habits and host range. Feeding habits were classified according to their mouthparts into chewing, piercing–sucking, cell-puncturing, and leaf-mining habits. Host range use was classified either as generalist or specialist if the species had been reported feeding on several non-related plant genera or only on a few related plant species [49]. Additionally, the species were classified according to their geographical origin into Nearctic (North America), Neotropic (Central and South America), Nearctic—Neotropic (American continent in general), Palearctic (Eurasia), Indo-Malayan (India—Asia), Chile (Native), or as unknown [10,13,20,27,28,50].

4.3. Data Analysis

Multivariate cluster analysis was conducted to find associations among species according to their presence/absence within each political region of Chile (latitudinal variables). Data matrix was constructed using the presence or absence (values 1 or 0, respectively) of the 43 arthropod taxa (reported in Table 1) in each of the 14 political regions that represent the variety of environments where quinoa is produced in Chile, specifically: Región de Arica y Parinacota (~17.5–19.1° S), Región de Tarapacá (~19.2–21.6° S), Región de Antofagasta (~21.7–25.8° S), Región de Atacama (~25.9–29.2° S), Región de Coquimbo (~29.3–32° S), Región de Valparaíso (~32.1–33° S), Región Metropolitana de Santiago (~33.1–34° S), Región del Libertador General Bernardo O'Higgins (~34.1–34.8° S), Región del Maule (~34.9–36.2° S), Región de Ñuble (~36.2–36.9° S), Región del Bio-Bio (~36.9–37.8° S), Región de La Araucanía (~37.9–39.5° S), Región de Los Ríos (~39.5–40.4° S), and Región de Los Lagos (~40.4–43.7° S). A dendrogram was constructed by distance correlation coefficient and complete linkage amalgamation steps functions using Minitab 17 software (Minitab Inc., State College, PA, USA). Specifically, for the linkage method, the distance between two clusters was calculated with the furthest-neighbor method, which is the maximum distance between variables of one cluster relative to another cluster. Distance between variables was calculated using the correlation method to consider positively correlated data to be closer than negatively correlated data, as it calculates distances between 0 and 1 for positive correlations and values between 1 and 2 for negative correlations. A total of 13 amalgamation steps were considered to create the dendrogram. Similarity levels ranged between 100 (13 clusters) and 91.3% (7 clusters) for the first 7 steps, and between 88.2 to 85.8% for steps 8 (6 clusters) to 10 (4 clusters), after which the similarity dropped below 74%.

Author Contributions: Conceptualization, methodology, formal analysis, data curation, writing—original draft preparation, R.A.C. and M.I.A.; resources, R.A.C.; investigation, writing—review and editing, R.A.C., M.I.A., F.U., M.E., and T.M.G. All authors have read and agreed to the published version of the manuscript.

Funding: This research was partially supported by funds provided by Fundación para la Innovación Agraria and Fondo de Innovación para la Competitividad (FIA EST-2016-0079, FIA EST-2016-0080, FIC PYT-2016-0450).

Data Availability Statement: Data is contained within the article.

Acknowledgments: We appreciate the help provided by Francisco Fuentes, Pablo Olguin, and Claudia Rojas to collect samples, as well as Audrey Grez for comments on an earlier version of this manuscript. We also thank Eduardo Faúndez for his assistance with Heteroptera. We also appreciate the following persons for allowing us to collect samples in their quinoa fields: Cecilia Guineo, Pilar Cárcamo, Lucia Araneda, Lautaro Díaz, and Abimael Gómez.

Conflicts of Interest: The authors declare no conflict of interest. The funders had no role in the design of the study; in the collection, analyses, or interpretation of data; in the writing of the manuscript, or in the decision to publish the results.

References

1. Fuentes, F.; Bhargava, A. Morphological analysis of quinoa germplasm grown under lowland desert conditions. *J. Agron. Crop Sci.* **2011**, *197*, 124–134. [CrossRef]
2. Miranda, M.; Vega-Gálvez, A.; Martinez, E.A.; López, J.; Marín, R.; Aranda, M.; Fuentes, F. Influence of contrasting environment on seed composition of two quinoa genotypes: Nutritional and functional properties. *Chil. J. Agric. Res.* **2013**, *73*, 108–116. [CrossRef]
3. Fuentes, F.; Paredes-González, X. Nutraceutical Perspectives of Quinoa: Biological Properties and Functional Applications. In *State of the Art Report on Quinoa around the World in 2013*; Bazile, D., Bertero, H.D., Nieto, C., Eds.; FAO & CIRAD: Rome, Italy, 2015; pp. 286–299.
4. Fuentes, F.F.; Martinez, E.A.; Hinrichsen, P.V.; Jellen, E.N.; Maughan, P.J. Assessment of genetic diversity patterns in Chilean quinoa (*Chenopodium quinoa* Willd.) germplasm using multiplex fluorescent microsatellite markers. *Conserv. Genet.* **2009**, *10*, 369–377. [CrossRef]
5. Cruces, L.M.; Callohuari, Y.; Carrera, C. *Quinua Manejo Integrado de Plagas. Estrategias en el Cultivo de la Quinua Para Fortalecer el Sistema Agroalimentario en las Zonas Andinas*; Organización de las Naciones Unidas para la Alimentación y la Agricultura: Santiago, Chile, 2016; pp. 1–198.
6. Fuentes, F.F.; Olguín, P.; Duarte, L.; Ojeda, M.; Figueroa, C.; Paredes-González, X.; Martínez, E.A. *Potencial Competitivo de la Quinua Chilena*; Publicaciones Fundación para la Innovación Agraria: Santiago, Chile, 2017; pp. 1–147.
7. Rasmussen, C.; Lagnaoui, A.; Esbjerg, P. Advances in the knowledge of quinoa pests. *Food Rev. Int.* **2003**, *19*, 61–75. [CrossRef]
8. Saravia, R.; Plata, G.; Gandarillas, A. *Plagas y Enfermedades del Cultivo de Quinua*; Fundación PROINPA: Cochabamba, Bolivia, 2014; pp. 1–148.
9. Cruces, L.; de la Peña, E.; De Clercq, P. Seasonal phenology of the major insect pests of quinoa (*Chenopodium quinoa* Willd.) and their natural enemies in a traditional zone and two new production zones of Peru. *Agriculture* **2020**, *10*, 644. [CrossRef]
10. Artigas, J.N. *Entomología Económica*; Ediciones Universidad de Concepción: Concepción, Chile, 1994; Volumes I and II, pp. 1–1126 and 1–943.
11. Lamborot, L.; Guerrero, M.A.; Araya, J.E. Lepidópteros asociados al cultivo de la quinoa (*Chenopodium quinoa* Willdenow) en la zona central de Chile. *Bol. San. Veg. Plagas* **1999**, *25*, 203–207.
12. Klein-Koch, C.; Waterhouse, D.F. Distribución e importancia de los artrópodos asociados a la agricultura y silvicultura en Chile. *ACIAR Monogr.* **2000**, *68*, 1–234.
13. Logarzo, G.A.; De León, J.H.; Triapitsyn, S.V.; González, R.H.; Virla, E.G. First report of a Proconiine Sharpshooter, *Anacuerna centrolinea* (Hemiptera: Cicadellidae), in Chile, with notes on its biology, host plants, and egg parasitoids. *Ann. Entomol. Soc. Am.* **2006**, *99*, 879–883. [CrossRef]
14. Elgueta, M.; Camousseight, A.; Carbonell, C.S. *Catálogo de Orthoptera (Insecta) de Chile*; Publicación Ocasional N° 54; Museo Nacional de Historia Natural: Santiago, Chile, 1999; pp. 1–60.
15. Dughetti, A.C. *Plagas de la Quinua y sus Enemigos Naturales en el Valle Inferior del Río Colorado, Buenos Aires, Argentina*, 1st ed.; Ediciones INTA: Buenos Aires, Argentina, 2015; pp. 1–63.
16. Gandarillas, A.; Saravia, R.; Plata, G.; Quispe, R.; Ortiz-Romero, R. Principales plagas y enfermedades de la quinua. In *State of the Art Report on Quinoa around the World in 2013*; Bazile, D., Bertero, H.D., Nieto, C., Eds.; FAO & CIRAD: Rome, Italy, 2015; pp. 192–215.
17. Nieto-Nafría, J.M.; Fuentes-Contreras, E.; Castro Colmenero, M.; Aldea Piera, M.; Ortego, J.; Mier Durante, M.P. Catálogo de los áfidos (Hemiptera, Aphididae) de Chile, con plantas hospedadoras y distribuciones regional y provincial. *Graellsia* **2016**, *72*, e050. [CrossRef]
18. Cranshaw, W.S.; Kondratieff, B.C.; Qian, T. Insect associated with Quinoa, *Chenopodium quinoa*, in Colorado. *J. Kans. Entomol. Soc.* **1990**, *63*, 195–199.
19. Burckhardt, D. Generic key to Chilean jumping plant-lice (Homoptera: Psylloidea) with inclusion of potential exotic pests. *Rev. Chil. Entomol.* **1994**, *21*, 57–67.
20. Faúndez, E.I.; Rocca, J. Descripción de un caso teratológico en *Oncopeltus* (*Erythrischius*) miles (Blanchard, 1852) (Heteroptera: Lygaeidae) con notas acerca de su distribución y biología. *Arq. Entomol.* **2016**, *15*, 39–43.
21. Faúndez, E.I.; Carvajal, M.A. Sobre la relación entre *Leptoglossus chilensis* (Spinola, 1852) y *Leptoglossus concaviusculus* Berg, 1892 stat. rest. (Heteroptera: Coreidae), con nuevos datos acerca de su morfología. *An. Inst. Patagon.* **2016**, *44*, 65–70. [CrossRef]
22. Prado, E. Conocimiento actual de Hemiptera—Heteroptera de Chile con lista de especies. *Bol. Mus. Nac. Hist. Nat. Chile* **2008**, *57*, 31–75.
23. Elgueta, M.; Marvaldi, A.E. Lista sistemática de las especies de *Curculionoidea* (Insecta: Coleoptera) presentes en Chile, con su sinonímia. *Bol. Mus. Nac. Hist. Nat. Chile* **2006**, *55*, 113–153.

24. Frías, D.; Henry, A.; Alviña, A.; Landry, J.F. Aspectos de la biología, taxonomía y control de las especies del género *Coleophora* (Lepidoptera: Coleophoridae) de distribución chilena. *Acta Entomol. Chil.* **1996**, *20*, 115–122.
25. Guerrero, M.A.; Lamborot, L.; Araya, J.E. Observaciones biológicas de *Achryra similalis* (Guenèe) (Pyralidae) y otros lepidópteros en amaranto, *Amaranthus cruentus* L. (Amaranthaceae), en la Región Metropolitana de Chile. *Bol. San. Veg. Plagas* **2000**, *26*, 591–598.
26. Cepeda, D.E. Contribution to the knowledge of Chilean Gelechiidae (Lepidoptera: Gelechioidea). *Insecta Mundi* **2017**, *584*, 1–8.
27. Povolný, D. Gnorimoschemini of southern South America II: The genus Eurysacca (Lepidoptera, Gelechiidae). *Steenstrupia* **1986**, *12*, 1–47.
28. Povolný, D. *Eurysacca quinoae* sp.n.—A new quinoa-feeding species of the tribe Gnorimoschemini (Lepidoptera, Gelechiidae) from Bolivia. *Steenstrupia* **1997**, *22*, 41–43.
29. Angulo, A.O.; Weigert, G.T. *Estados Inmaduros de Lepidópteros Nóctuidos de Importancia Económica en Chile y Claves Para su Determinación (Lepidoptera: Noctuidae)*; Publicación especial N°2; Sociedad de Biología de Concepción: Concepción, Chile, 1975; 153p.
30. Hardwick, D.F. The corn earworm complex. *Mem. Entomol. Soc. Can.* **1965**, *40*, 1–247. [CrossRef]
31. Kuschel, G. Los Curculionidae del extremo norte de Chile (Coleoptera. Curcul. Ap. 6°). *Acta Zool. Lilloana* **1949**, *8*, 5–54.
32. Kuschel, G. New rank for varieties of Trichocyphus formosus. In *Annotated Checklist of the Weevils (Curculionidae sensu lato) of South America (Coleoptera: Curculionoidea)*; Wibmer, G.J., O'Brien, N.C.W., Eds.; American Entomological Institute: Ann Arbor, MI, USA, 1986; Volume 39, p. 52.
33. Lanteri, A.A. Estudio sistemático de los géneros *Trichocyphus* Heller y *Mendozella* Hustache (Coleoptera: Curculionidae). *Bol. Soc. Biol. Concepción Chile* **1989**, *60*, 139–147.
34. Pogue, M.G. A review of the *Copitarsia decolora* (Guenée) (Lepidoptera: Noctuidae) species complex with the description of a new species from Chile and Argentina. *Neotrop. Entomol.* **2014**, *43*, 143–153. [CrossRef]
35. Yábar, E.; Gianoli, E.; Echegaray, E.R. Insect pest and natural enemies in two varieties of quinoa (*Chenopodium quinoa*) at Cusco, Peru. *J. Appl. Entomol.* **2002**, *126*, 275–280. [CrossRef]
36. Quiroga, N.; Longone, V.; González, X.; Zamorano, A.; Pino, A.M.; Picciau, L.; Alma, A.; Paltrinieri, S.; Contaldo, N.; Bertaccini, A.; et al. Transmission of 16SrIII-J phytoplasmas by the leafhoppers *Paratanus exitiousus* and *Bergallia valdiviana*. *Phytopathol. Mediter.* **2019**, *58*, 231–237.
37. Wibmer, G.J.; O'Brien, N.C.W. Annotated checklist of the weevils (*Curculionidae sensu* lato) of South America (Coleoptera: Curculionoidea). *Mem. Am. Entomol. Inst.* **1986**, *39*, 1–563.
38. Valoy, M.E.; Bruno, M.A.; Prado, F.E.; González, J.A. Insects associated to a quinoa crop in Amaicha del Valle, Tucumán, Argentina./Insectos asociados a un cultivo de quinoa en Amaicha del Valle, Tucumán, Argentina. *Acta Zool. Lilloana* **2011**, *55*, 16–22.
39. Elgueta, M.; Arriagada, G. Estado actual del conocimiento de coleópteros de Chile (Insecta: Coleoptera). *Rev. Chil. Entomol.* **1989**, *17*, 5–60.
40. Povolný, D. Gnorimoschemini of southern South America IV: The genus Symmetrischema (Lepidoptera, Gelechiidae). *Steenstrupia* **1989**, *15*, 57–104.
41. Povolný, D. Gnorimoschemini of southern South America VI: Identification keys, checklist of Neotropical taxa and general considerations (Insecta, Lepidoptera, Gelechiidae). *Steenstrupia* **1994**, *20*, 1–42.
42. Dioli, P.; Colamartino, A.D.; Negri, M.; Limonta, L. Hemiptera and coleoptera on *Chenopodium quinoa*. *Redia* **2016**, *99*, 139–141.
43. Reguzzi, M.C.; Nicoli Aldini, R.; Vercesi, A.; Ganimede, C.; Tabaglio, V.; Mazzoni, E.; Dioli, P. Quinoa, quali insetti infestanti sono presenti al Nord Italia. *L'Informatore Agrario* **2019**, *75*, 59–61.
44. Scaccini, D.; Furlan, L. *Nysius cymoides* (Hemiptera: Lygaeidae), a potential emerging pest: Overview of the information available to implement integrated pest management. *Int. J. Pest Manag.* **2021**, *67*, 73–88. [CrossRef]
45. Göllner-Scheiding, U. Revision der Gattungen *Liorhyssus* STÅL, 1870 (Heteroptera, Rhopalidae). *Dtsch. Entomol. Z.* **1975**, *23*, 181–206. [CrossRef]
46. Blackman, R.L.; Eastop, V.F. *Aphids on the World's Herbaceous Plants and Shrubs*; John Wiley & Sons: Chichester, UK, 2006; pp. 1–1456.
47. San Blas, G. *Agrotis* Ochsenheimer (Lepidoptera, Noctuidae): A systematic analysis of South American species. *Zootaxa* **2014**, *3771*, 1–64. [CrossRef]
48. Wrzesinska, D.; Wawrzyniak, M.; Gesinski, K. Population of true bugs (Heteroptera) on the inflorescences of quinoa (*Chenopodium quinoa* Willd.). *J. Plant Prot. Res.* **2001**, *41*, 333–336.
49. Schoonhoven, L.M.; van Loon, J.J.A.; Dicke, M. Herbivorous insects: Something for everyone. In *Insect-Plant Biology*, 2nd ed.; Oxford University Press Inc.: New York, NY, USA, 2005; pp. 5–28.
50. San Blas, G. A morphological phylogeny of *Agrotis* Ochsenheimer (Lepidoptera, Noctuidae), with emphasis on the South American species. *Zool. Scr.* **2014**, *44*, 153–164. [CrossRef]

plants

Article

Yield and Nutritional Characterization of Thirteen Quinoa (*Chenopodium quinoa* Willd.) Varieties Grown in North-West Europe—Part I

Phara De Bock [1,*], Filip Van Bockstaele [2], Hilde Muylle [3], Paul Quataert [3], Pieter Vermeir [4], Mia Eeckhout [1] and Gerda Cnops [3]

[1] Research Unit of Cereal and Feed Technology, Department of Food Technology, Safety and Health, Faculty of Bioscience Engineering, Ghent University, 9000 Ghent, Belgium; mia.eeckhout@ugent.be
[2] Food Structure and Function Research Group (FSF), Department of Food Technology, Safety and Health, Faculty of Bioscience Engineering, Ghent University, 9000 Ghent, Belgium; filip.vanbockstaele@ugent.be
[3] Plant Sciences Unit, Flanders Research Institute for Agriculture, Fisheries and Food (ILVO), 9090 Melle, Belgium; hilde.muylle@ilvo.vlaanderen.be (H.M.); paul.quataert@ilvo.vlaanderen.be (P.Q.); gerda.cnops@ilvo.vlaanderen.be (G.C.)
[4] Laboratory for Chemical Analysis (LCA), Department of Green Chemistry and Technology, Faculty of Bioscience Engineering, Ghent University, 9000 Ghent, Belgium; pieter.vermeir@ugent.be
* Correspondence: phara.debock@ugent.be; Tel.: +32-9-243-24-94

Citation: De Bock, P.; Van Bockstaele, F.; Muylle, H.; Quataert, P.; Vermeir, P.; Eeckhout, M.; Cnops, G. Yield and Nutritional Characterization of Thirteen Quinoa (*Chenopodium quinoa* Willd.) Varieties Grown in North-West Europe—Part I. *Plants* **2021**, *10*, 2689. https://doi.org/10.3390/plants10122689

Academic Editors: Cataldo Pulvento and Didier Bazile

Received: 19 October 2021
Accepted: 1 December 2021
Published: 7 December 2021

Abstract: The cultivation of quinoa has gained increasing interest in Europe. Different European varieties exist, but more research is required to understand the individual variety characteristics for end-use applications. The objective of this study is to evaluate the agronomic performance of 13 quinoa varieties under North-West European field conditions during three growing seasons (2017–2019). Furthermore, seeds were qualitatively characterized based on characteristics and composition. Yield differed among varieties and growing seasons (0.47–3.42 ton/ha), with lower yields obtained for late-maturing varieties. The saponin content varied from sweet to very bitter. The seeds contained high protein levels (12.1–18.8 g/100 g dry matter), whereas varieties had a similar essential amino acid profile. The main fatty acids were linoleic (53.0–59.8%), α-linolenic (4.7–8.2%), and oleic acid (15.5–22.7%), indicating a high degree of unsaturation. The clustering of varieties/years revealed subtle differences between growing seasons but also reflected the significant interaction effects of variety and year. Most varieties perform well under North-West European conditions, and their nutritional content is well within the values previously described for other cultivation areas. However, optimal yield and quality traits were not combined in one variety, illustrating the importance of breeding for adapted quinoa varieties.

Keywords: *Chenopodium quinoa* Willd.; North-West Europe; yield; saponins; amino acids; fatty acids; triacylglycerols; minerals

1. Introduction

Quinoa (*Chenopodium quinoa* Willd.) is a highly nutritional and resilient crop, native to the Andean region. It is a pseudocereal belonging to the *Amaranthaceae* family that was domesticated and cultivated more than 5000 years ago [1,2]. Unlike the other staple crops of the Andes, quinoa remained relatively unknown outside the Andean region until it was exported as a food product to the USA in the late 1970s. From that period onwards, quinoa consumption increased worldwide [2,3]. As the high nutritional value of quinoa seeds started to be recognized, this pseudocereal gained increasing global interest in the last decade [1,4]. The quinoa seeds can be used for the same purpose as cereals, such as wheat or rice, or as whole grains in salads, cooked meals, breakfast, or soups. They can be milled into flour to produce pasta, bread, biscuits, and pancakes. The seeds can be used for the production of dairy-free drinks or, in fermented form, for the production of beer or the traditional *chicha* drink [5,6].

The start of quinoa exports and consumer interest outside the Andean region rapidly increased the demand and price of quinoa in the main producing countries, Bolivia and Peru [2,4]. The subsequent expansion of the agricultural frontier and the unsustainable practices in South America emphasized the need to grow quinoa in other parts of the world [2,7]. In the 1980s, breeding programs were initiated in several European countries, i.e., the UK, Denmark, and the Netherlands. The research focused on the adaptation to European climatic conditions and the improvement of agronomic performance (e.g., early maturity, high yield, low saponin concentration, no head sprouting). Chilean varieties belonging to the coastal ecotype are less sensitive to photoperiods, an important requisite for northern latitudes, and were the ideal starting material for European varieties [3,4,8]. Nowadays, several Dutch and Danish varieties have been registered in Europe. The newest varieties show early maturity and are almost completely day-length neutral [3,9]. Moreover, Patiranage et al. [10] recently identified haplotypes that induce photoperiod insensitivity associated with early flowering under long day lengths. A haplotype-based breeding strategy of pyramiding favorable alleles is suggested by the authors to breed quinoa varieties that are early maturing under long day lengths and are, thus, suitable for cultivation in northern latitudes [10].

As a pseudocereal, quinoa has distinctive features from conventional cereals. Its protein content is substantially higher (7–23%) than that of the staple crops rice, barley, corn, and rye and slightly higher than that of wheat, with a better balanced essential amino acid profile [11,12]. Moreover, quinoa has a higher lipid content with a good representation of essential fatty acids, a higher ash content, and higher concentrations of vitamins E, C, B2, B6, and folic acid. Besides antioxidants, quinoa contains several other phytochemicals with positive effects on human health and nutrition. Furthermore, quinoa is generally safe to eat for patients with celiac disease [2,4,11–13].

Quinoa is characterized by a broad genetic diversity, which allows adaptability to diverse climates and cultivation in a wide range of environments such as highlands, coastal regions, subtropical environments, or arid regions. Furthermore, its ability to grow under various abiotic stress conditions (i.e., frost, salinity, drought) makes quinoa a promising crop for further expansion in many parts of the world to help feed the world's population in the context of a changing climate and food security [8,14–16].

Seed quality is a complex trait that results from the interaction between genetic and environmental factors, and, except for saponin content, it has gotten little attention in breeding programs [4,15]. There is also little information available on the nutritional quality of the different European varieties when produced in Europe, compared to quinoa produced in South America [4,15,17]. Insufficient knowledge of the characteristics and their consequences for potential end-uses complicate the use of European quinoa in the food industry. Therefore, more research is required to gain a better understanding of the individual variety characteristics for end-use applications and to bridge the gap between farmers and end-users [16]. The objective of this study is to perform comparative variety testing with 10 European bred varieties and 3 Farm Original varieties over 3 consecutive growing seasons (2017–2019) [18] to (i) evaluate their agronomic performance under North-West European field conditions and to (ii) qualitatively characterize the seed based on characteristics and composition. In addition, principal components analysis (PCA) and hierarchical cluster analysis (HCA) were employed to group the varieties/years into clusters with similar yield, characteristics, and macronutrient composition.

2. Results and Discussion

2.1. Yield

The overall mean and range of the yield (Figure 1) of the different varieties was strongly year-dependent, with the widest spread in 2017 (0.57–3.48 ton/ha). The ranking of the varieties was, however, mostly consistent over the tested growing seasons. Titicaca,

Bastille, Vikinga, and Dutchess were the best performing varieties under the North-West European weather conditions in 2017–2019. The three dark-colored (i.e., Rouge Marie, Summer Red, and Zwarte) varieties were less yielding compared to most white varieties and similar to Atlas and Pasto, with consistently lower yields. Moreover, the varieties clearly differed in the length of their growing season, which ranged between 111 and 187 growing days (GDs). Jessie, Vikinga, and Titicaca belonged to the early varieties, and Atlas and Pasto were consistently the latest to be harvested. The seed yields were comparable to the yields reported in other field trials [4,19]. In two previous studies, a small subset of the analyzed varieties was grown under European conditions, i.e., Puno, Titicaca, and Jessie in Germany [4] and Vikinga, Titicaca, and Puno in Spain [19]. The present seed yield data were in the range of the Spanish study (0.70–3.25 ton/ha, [19]) and higher than the German study (1.73–2.43 ton/ha, [4]). However, the same varieties needed similar GDs as in Germany [4] or Spain [19].

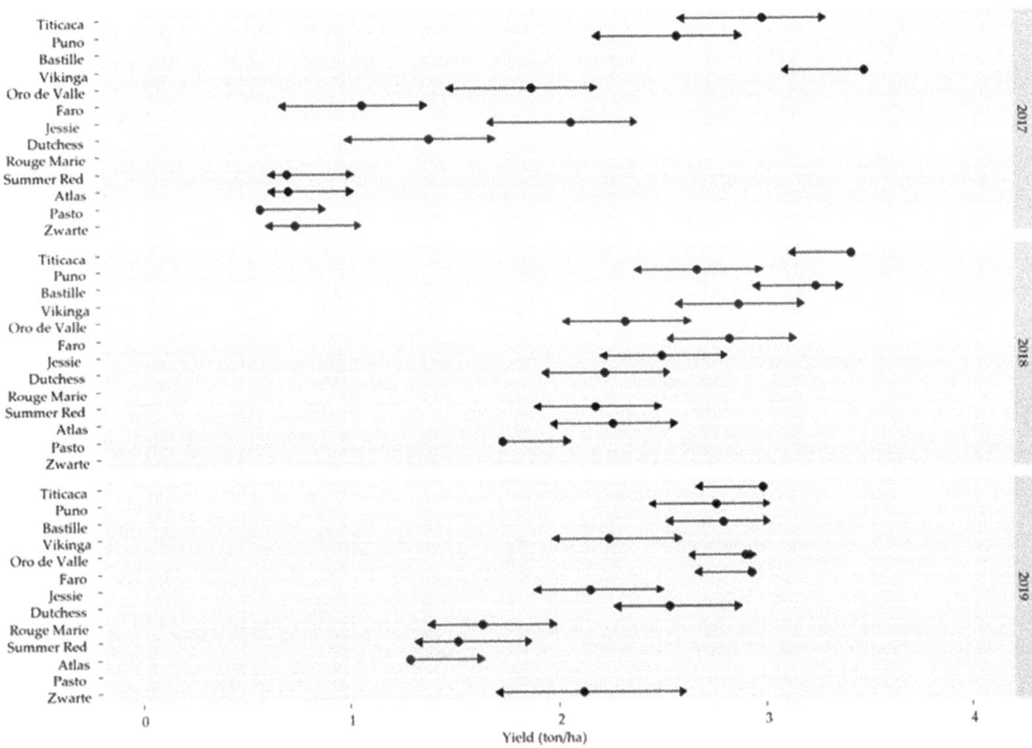

Figure 1. Yield (ton/ha) of 13 quinoa varieties grown under North-West European field conditions in 2017, 2018, and 2019 ($n = 3$).

A significant negative correlation ($r = -0.705$, $p < 0.001$) was found between yield and GDs (Figure S1) but was probably biased by the varieties Atlas and Pasto. These late varieties were not able to reach full maturity under the North-West European growth conditions before 15 September. Results from the EU project Protein2Food suggested that yield potential in quinoa can be influenced by the sowing date [20]. In the 2017–2019 trials, the sowing date was the same for early, mid, and late varieties and depended on the local field and weather conditions. An earlier sowing date for mid-late and, especially, late varieties might be beneficial and, in some years, even required for completing their life cycle in favorable weather conditions [20]. Earlier sowing dates (before 15 April), as recommended by Radicle Crops for their varieties (pers. comm.), might have resulted in higher yields, mainly for the late varieties (i.e., Pasto and Atlas) in 2018 and 2019. The European-bred varieties, except for the late varieties (i.e., Atlas and Pasto), did perform well under the North-West European growth conditions. The Farm Original varieties (i.e., Faro, Oro de Valle, and Zwarte) were taller, phenotypically less homogenous (data not shown), and mid-late maturing and, except for Zwarte, performed similarly to most European-bred varieties.

2.2. Seed Size and Weight

Seed size, thousand seed weight (TSW), and test weight differed significantly among varieties and years, with significant variety × year interactions. The quinoa seeds had a length of 1.65 to 2.25 mm and a width of 1.50 to 1.99 mm (Table 1), which corresponded to a length–width ratio (LWR) of 1.09 to 1.14 (Table S1). This is within the range of seed sizes reported by Bhargava et al. (1.34–2.21 mm, [17]) and Bertero et al. (1.8–2.2 mm, [21]). The ranking of the varieties according to seed size was inconsistent over the years. For example, the seeds of Puno and Vikinga were the smallest in 2019, while the growing season of 2018 resulted in the smallest seeds for Oro de Valle, Bastille, and Dutchess. The seed size of Faro, Jessie, and Summer Red remained stable in 2018 and 2019, while Pasto had a stable seed size in 2017 and 2018. In general, the growing season of 2017 resulted in the largest seeds. Among varieties, Puno and Pasto were characterized by the smallest seed size. Zwarte and Summer Red generally had the largest seeds. Over the years, Atlas and Jessie had comparable seed sizes.

The TSW of the quinoa varieties (Table 1) ranged between 1.90 and 3.68 g. These values corresponded with the ranges reported for studies in Germany (1.2–3.3 g, [4]), Italy (1.77–3.63 g, [22]), and Spain (1.7–3.4 g, [19]), in which four varieties were common to this study. Most varieties obtained the lowest TSW in 2018. This growing season was characterized by a prolonged dry period with 79% less overall rain compared to 2017 (Figure 5). Insufficient rain can reduce TSW during the seed filling phase [4]. However, the growth conditions in 2018 resulted in the largest seeds with the highest TSW for Titicaca. The TSW of Atlas remained stable at around 2.33 g in 2018 and 2019, while other varieties had a higher TSW in 2019. Overall, the highest TSW was obtained in 2017, except for Faro, Oro de Valle, and Titicaca. The two Farm Original varieties (i.e., Faro and Oro de Valle) reached the highest TSW under the growth conditions of 2019. Across growing seasons, the dark-colored varieties (i.e., Summer Red, Rouge Marie, and Zwarte) produced the largest seeds with the highest TSW, while the Puno seeds were the smallest and the lightest. The ranking of Puno, Jessie, and Titicaca (from small to high) was identical to the ranking by Präger et al. [4] in Germany. The present data support the influence of the genetic factor determining TSW. There is still room for increasing seed size and TSW through breeding using varieties from the coastal ecotype or incorporating this trait from other quinoa ecotypes, e.g., the *salares* ecotype (the real quinoa varieties from Bolivia), of which the seeds have the highest TSW [23,24].

The test weight (Table 1) varied between 68.9 and 82.5 kg/hL, which is higher compared to the test weight of quinoa grown in Bolivia (66.3–73.0 kg/hL, [25]) or the USA (66.1–75.7 kg/hL, [25]). In 2019, the test weight of the quinoa seeds was significantly lower compared to other growing seasons. However, Pasto and Faro had the lowest test weight in 2017. The highest test weights were measured in 2018 for all varieties, except for Atlas and Titicaca. The latter reached the highest test weight under the growth conditions of 2017. The test weight of Puno and Titicaca was usually high, especially in 2017 and 2018. In 2019, the highest test weights were measured for Zwarte and Faro. Wu et al. [25] concluded that Oro de Valle seeds had a higher test weight than Black seeds (possible origin for variety Zwarte). Within the present study, Zwarte seeds invariably had a higher test weight compared to Oro de Valle seeds. Despite the comparable seed size, Atlas and Jessie had different test weights in 2017 and 2018. These varieties also significantly differed in TSW, which suggests differences in seed composition. Dutchess invariably had the lowest test weight in comparison to the different quinoa varieties.

Table 1. Seed characteristics and saponin content (mg/g) of 13 quinoa varieties grown under North-West European field conditions in 2017, 2018, and 2019 ($n = 3$).

Year	Variety	Length (mm) [1]	Width (mm) [1]	TSW (g) [1,2]	Test Weight (kg/hL) [1]	Saponins (mg/g) [1,2]
2017	Atlas	$1.97 \pm 0.00^{d,B}$	$1.80 \pm 0.00^{f,B}$	$2.85 \pm 0.01^{e,B}$	$75.9 \pm 0.1^{e,C}$	<LoQ
	Bastille	-	-	-	-	-
	Dutchess	$1.96 \pm 0.00^{d,B}$	$1.80 \pm 0.01^{f,B}$	$2.94 \pm 0.01^{f,g,C}$	$71.9 \pm 0.2^{a,B}$	$0.1 \pm 0.0^{a,A}$
	Faro	$1.86 \pm 0.01^{c,B}$	$1.69 \pm 0.00^{c,B}$	$2.56 \pm 0.01^{c,B}$	$74.9 \pm 0.2^{c,d,A}$	$1.7 \pm 0.6^{c,d,A}$
	Jessie	$1.95 \pm 0.01^{d,B}$	$1.79 \pm 0.01^{e,f,C}$	$2.76 \pm 0.02^{d,C}$	$75.3 \pm 0.2^{d,B}$	<LoQ
	Oro de Valle	$1.97 \pm 0.01^{d,C}$	$1.76 \pm 0.01^{e,C}$	$2.98 \pm 0.01^{g,B}$	$74.6 \pm 0.1^{b,c,B}$	$1.1 \pm 0.2^{b,c,A}$
	Pasto	$1.77 \pm 0.00^{b,A}$	$1.61 \pm 0.01^{b,A}$	$2.37 \pm 0.01^{b,B}$	$74.7 \pm 0.2^{b,c,A}$	< LoQ
	Puno	$1.72 \pm 0.01^{a,C}$	$1.56 \pm 0.01^{a,B}$	$2.27 \pm 0.01^{a,C}$	$81.7 \pm 0.2^{g,B}$	$2.2 \pm 0.2^{d,A}$
	Rouge Marie	-	-	-	-	-
	Summer Red	$2.24 \pm 0.01^{e,B}$	$1.99 \pm 0.00^{g,B}$	$3.47 \pm 0.02^{h,C}$	$74.5 \pm 0.1^{b,B}$	<LoQ
	Titicaca	$1.88 \pm 0.01^{c,A}$	$1.69 \pm 0.01^{c,A}$	$2.87 \pm 0.03^{e,f,B}$	$82.5 \pm 0.1^{h,C}$	$0.4 \pm 0.1^{a,b,A}$
	Vikinga	$1.89 \pm 0.00^{c,C}$	$1.72 \pm 0.00^{d,C}$	$2.70 \pm 0.01^{d,C}$	$77.2 \pm 0.1^{f,B}$	< LoQ
	Zwarte	$2.25 \pm 0.02^{e,B}$	$1.98 \pm 0.02^{g,A}$	$3.68 \pm 0.02^{I,B}$	$77.6 \pm 0.2^{f,B}$	$0.8 \pm 0.2^{a,b,c,A}$
2018	Atlas	$1.89 \pm 0.01^{d,A}$	$1.74 \pm 0.01^{d,A}$	$2.32 \pm 0.02^{c,d,A}$	$75.6 \pm 0.1^{a,B}$	<LoQ
	Bastille	$1.87 \pm 0.00^{c,d,A}$	$1.68 \pm 0.01^{c,A}$	$2.49 \pm 0.02^{f,A}$	$78.1 \pm 0.3^{d,B}$	<LoQ
	Dutchess	$1.86 \pm 0.01^{c,A}$	$1.69 \pm 0.01^{c,A}$	$2.38 \pm 0.04^{d,e,A}$	$75.3 \pm 0.2^{a,C}$	<LoQ
	Faro	$1.80 \pm 0.01^{b,A}$	$1.64 \pm 0.00^{b,A}$	$2.28 \pm 0.03^{c,A}$	$79.6 \pm 0.1^{e,C}$	$5.2 \pm 0.6^{c,B}$
	Jessie	$1.90 \pm 0.02^{d,A}$	$1.75 \pm 0.02^{d,B}$	$2.24 \pm 0.02^{c,A}$	$76.8 \pm 0.1^{b,C}$	<LoQ
	Oro de Valle	$1.85 \pm 0.00^{c,A}$	$1.67 \pm 0.01^{c,A}$	$2.46 \pm 0.01^{f,A}$	$77.1 \pm 0.1^{b,C}$	$2.8 \pm 0.9^{b,B}$
	Pasto	$1.78 \pm 0.01^{b,A}$	$1.62 \pm 0.01^{b,A}$	$2.00 \pm 0.03^{b,A}$	$76.7 \pm 0.1^{b,B}$	<LoQ
	Puno	$1.68 \pm 0.01^{a,B}$	$1.54 \pm 0.01^{a,B}$	$1.90 \pm 0.03^{a,A}$	$81.6 \pm 0.1^{f,B}$	$5.1 \pm 1.2^{c,B}$
	Rouge Marie	-	-	-	-	-
	Summer Red	$2.05 \pm 0.01^{e,A}$	$1.88 \pm 0.01^{e,A}$	$2.95 \pm 0.04^{g,A}$	$78.3 \pm 0.2^{d,C}$	<LoQ
	Titicaca	$2.03 \pm 0.02^{e,B}$	$1.85 \pm 0.02^{e,C}$	$2.93 \pm 0.04^{g,C}$	$79.9 \pm 0.1^{e,B}$	$1.3 \pm 0.4^{a,B}$
	Vikinga	$1.85 \pm 0.02^{c,B}$	$1.68 \pm 0.02^{c,B}$	$2.43 \pm 0.01^{e,f,A}$	$77.6 \pm 0.2^{c,C}$	<LoQ
	Zwarte	-	-	-	-	-
2019	Atlas	$1.89 \pm 0.01^{c,A}$	$1.73 \pm 0.00^{c,d,A}$	$2.33 \pm 0.02^{b,A}$	$73.6 \pm 0.2^{c,d,A}$	<LoQ
	Bastille	$1.97 \pm 0.01^{d,B}$	$1.76 \pm 0.01^{d,B}$	$2.90 \pm 0.06^{e,B}$	$71.9 \pm 0.2^{b,A}$	<LoQ
	Dutchess	$2.00 \pm 0.01^{d,C}$	$1.83 \pm 0.01^{e,C}$	$2.77 \pm 0.04^{d,B}$	$68.9 \pm 0.1^{a,A}$	<LoQ
	Faro	$1.79 \pm 0.01^{b,A}$	$1.64 \pm 0.01^{b,A}$	$2.75 \pm 0.02^{d,C}$	$75.5 \pm 0.3^{g,B}$	$7.9 \pm 0.1^{e,C}$
	Jessie	$1.89 \pm 0.01^{c,A}$	$1.72 \pm 0.00^{c,A}$	$2.48 \pm 0.02^{c,B}$	$73.9 \pm 0.2^{d,e,A}$	< LoQ
	Oro de Valle	$1.89 \pm 0.01^{c,B}$	$1.73 \pm 0.00^{c,B}$	$3.04 \pm 0.02^{f,C}$	$73.4 \pm 0.2^{c,A}$	$5.4 \pm 0.4^{c,d,C}$
	Pasto	-	-	-	-	-
	Puno	$1.65 \pm 0.01^{a,A}$	$1.50 \pm 0.01^{a,A}$	$2.15 \pm 0.03^{a,B}$	$74.6 \pm 0.3^{f,A}$	$6.3 \pm 0.5^{d,C}$
	Rouge Marie	$1.98 \pm 0.01^{d,A}$	$1.83 \pm 0.01^{e,A}$	$3.27 \pm 0.05^{g,h,A}$	$73.5 \pm 0.3^{c,d,A}$	<LoQ
	Summer Red	$2.05 \pm 0.02^{e,A}$	$1.88 \pm 0.01^{f,A}$	$3.23 \pm 0.02^{g,B}$	$73.5 \pm 0.2^{c,d,A}$	<LoQ
	Titicaca	$1.88 \pm 0.01^{c,A}$	$1.71 \pm 0.01^{c,B}$	$2.79 \pm 0.01^{d,A}$	$74.5 \pm 0.2^{f,A}$	$3.2 \pm 0.4^{b,C}$
	Vikinga	$1.78 \pm 0.01^{b,C}$	$1.61 \pm 0.01^{b,A}$	$2.55 \pm 0.03^{c,B}$	$74.2 \pm 0.2^{e,f,A}$	$0.9 \pm 0.4^{a,A}$
	Zwarte	$2.19 \pm 0.01^{f,A}$	$1.99 \pm 0.01^{g,A}$	$3.35 \pm 0.05^{h,A}$	$75.9 \pm 0.1^{g,A}$	$4.8 \pm 0.4^{c,B}$

[1] Within years, average values followed by the same lowercase letter are not significantly different ($p > 0.05$). Capital letters compare the three years for the same variety; average values followed by the same letter are not significantly different ($p > 0.05$). [2] TSW: thousand seed weight, LoQ: limit of quantification (= 0.1 mg/g, [26]).

2.3. Seed Color

The seed color (L*a*b*) of the quinoa varieties is presented in Table S1. The variety × year interaction had a significant impact on the L*, a*, and b* values of the seed color. The luminosity (L*) of the white varieties ranged from 60.15 to 67.69 and was considerably higher than that of the dark-colored varieties (40.81–47.12). In 2017 and 2018, Pasto was characterized by the lowest luminosity among the white varieties, and Faro and Bastille by the highest. Moreover, Bastille had the clearest seeds in 2019. Lighter-colored seeds are generally preferred within white-seeded varieties [27]. The growing conditions in 2019 resulted in a higher luminosity for the dark-colored varieties. Among these varieties, Zwarte was characterized by the darkest seeds.

The red component of seed color was generally higher in 2019, although Faro seeds obtained the highest a* value in 2017. Rouge Marie had the reddest seeds (a* = 6.56) among all varieties, while Zwarte seeds were characterized by the lowest a* value (1.91). The a* value of the other varieties ranged between 3.72 and 5.42. The red component of Atlas and Puno seeds was generally lower compared to that of other white seeds. All varieties, except for Pasto and Vikinga, had comparable a* values in 2018.

The b* value was considerably lower for dark-colored seeds and varied between 0.69 and 5.48. Besides a lower a* value, Zwarte seeds were also characterized by a lower b* value than Rouge Marie or Summer Red. Zwarte is a very heterogeneous variety with a segregation for seed color that is genetically determined by two genes [28]. This results in a mixture of red, brown, and black seeds with low a* and b* values. The b* value ranged from 15.23 to 20.72 for the white varieties, with the highest values measured in 2019. Only Vikinga seeds had a higher b* value in 2017. On average, Oro de Valle, Faro, Jessie, and Bastille were characterized by the highest yellow pigment.

Granado-Rodriguez et al. [19] studied the color of Puno, Titicaca, and Vikinga seeds over three years of cultivation in Spain. These authors reported lower L* values (50.3–63.1) and higher b* values (19.6–24.2). Sobota et al. [29] described the color of Faro, Puno, and Titicaca seeds cultivated in Poland. The results were within the ranges reported within the present study.

2.4. Saponins

The saponin content of the quinoa varieties is listed in Table 1. Depending on the saponin content, quinoa is classified as sweet or bitter. Quinoa can be considered sweet when its saponin content is 1.1 mg/g or less [26]. Atlas, Bastille, Dutchess, Jessie, Pasto, Rouge Marie, and Summer Red produced sweet seeds in all growing seasons. The saponin content of these varieties, except for Dutchess in 2017 (0.1 mg/g), was too low in all years to quantify. Saponins have an undesired bitter taste, and the bitter components decrease the nutritional value of the seed [12]. In contrast to other seed characteristics, saponin concentration has a high heritability, which makes it an important breeding trait [15,30]. The WUR and AbbottAgra varieties were bred for sweetness, and the saponin concentrations are extremely low due to a genetic mutation in the T-SARL1 transcription factor [31]. The saponin concentration of the other varieties ranged from 0.1 to 7.9 mg/g, with Faro as the bitterest variety. Vikinga is another variety bred for sweetness. The present data confirmed this, but a higher saponin content of 0.9 mg/g was observed for Vikinga seeds from 2019. Granado-Rodríguez et al. [19] found an overall higher saponin content for Vikinga, with the highest content in 2019 (2.04 g/100 g) as well. This, together with the present data, suggests genetic segregation or cross-contamination rather than environmental conditions, especially because the Vikinga plots were less homogenous for other traits, such as plant height, stem color, and maturing in 2019 (data not shown). The lowest saponin concentrations were observed in 2017, whereas varieties classified as bitter had remarkably lower saponin content in 2018 and 2019. Oro de Valle, Titicaca, and Zwarte (0.4–1.1 mg/g) were even considered sweet in 2017. The highest saponin concentrations were measured in 2019, with ranges for the bitter varieties between 3.2 and 7.9 mg/g. Compared to varieties also analyzed by Präger et al. (0.0–3.4 mg/g, [4]), the saponin concentrations in the present study were in the same line for Jessie and Titicaca but higher for Puno (2018 and 2019). Medina-Meza et al. [32] qualified Oro de Valle and Black (possible origin for variety Zwarte) as low saponin content varieties and Puno and Titicaca as bitter varieties. This is in accordance with the present study, except for the saponin content of Titicaca seeds from 2017.

2.5. Chemical Composition

2.5.1. Macronutrients

The macronutrient composition (Table S2) of the quinoa seeds varied with variety and growing season, with a significant interaction effect between both factors. Macronutrient composition is presented in Figure 2 and discussed in the following paragraphs.

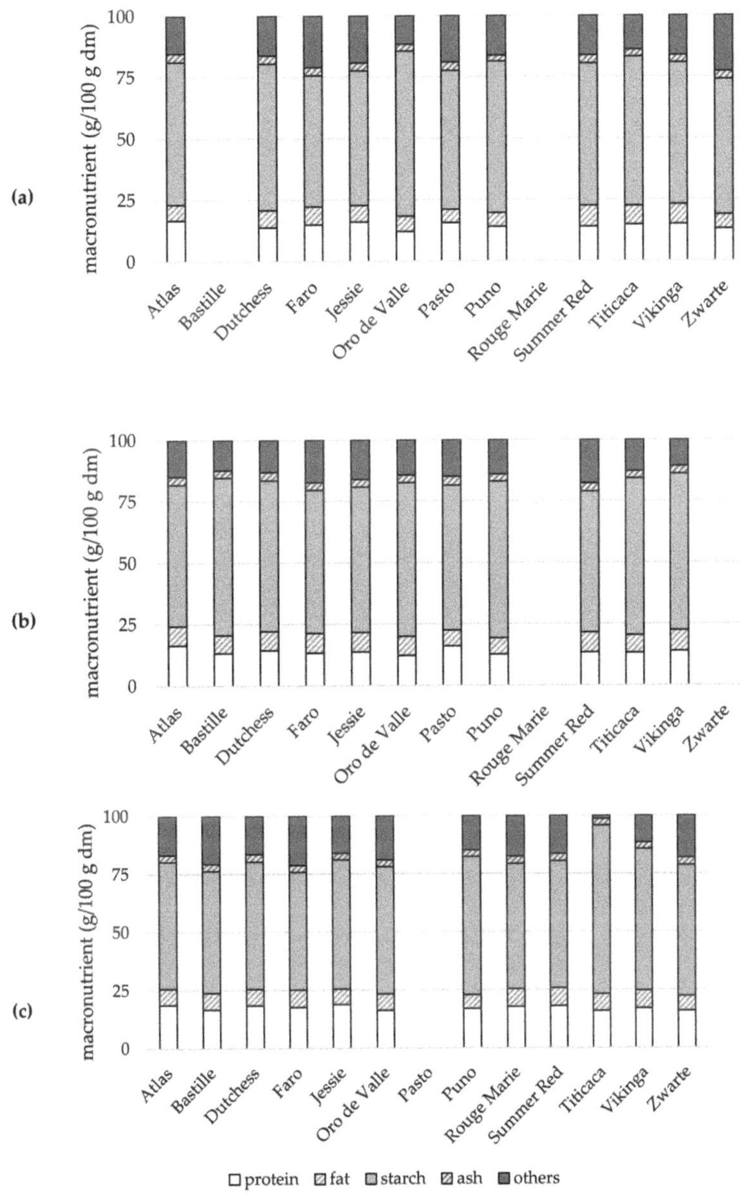

Figure 2. Macronutrient composition (g/100 g dm) of 13 quinoa varieties grown under North-West European field conditions in 2017 (**a**), 2018 (**b**), and 2019 (**c**) ($n = 3$).

2.5.2. Protein and Amino Acids

The protein content ranged from 12.1 to 18.8 g/100 g dry matter (dm) (Table S2) and was strongly year-dependent, as previously reported by Granado-Rodríguez et al. [19]. The protein contents determined in this study were similar to those reported by Bhargava et al. (12.6–21.0 g/100 g, [17]) and Granado-Rodríguez et al. (13.8–19.1 g/100 g, [19]). Lower values were achieved by Präger et al. (11.9–16.1 g/100 g, [4]) and Miranda et al. (11.3–16.1 g/100 g dm, [23]). The higher protein content of quinoa, compared to other grain staple crops, is an important nutritional fact [12].

All varieties achieved the highest protein concentration in 2019 (16.0–18.8 g/100 g dm). In spite of the 35 units higher nitrogen fertilization in 2017, the protein concentrations did not reach those of 2019. The extra nitrogen fertilization was probably insufficiently available for the plants during growth due to the extremely dry spring in 2017 (Figure 5). A negative trend was observed between yield and protein content. However, this trend was only significant for 2019 ($r = -0.690$, $p = 0.020$). This negative correlation between yield and protein content was also reported by Präger et al. [4] and Granado-Rodríguez et al. [19]. In 2017 and 2018, Oro de Valle and Atlas showed stable protein contents around 12.1 and 16.5 g/100 g dm, respectively. Compared to 2017, the protein content of Pasto and Dutchess increased by 3.1% and 4.3% in 2018, whereas the other varieties showed a reduction in protein content. Averaged over 2017–2019, Atlas and Pasto showed a significantly higher protein content compared to other varieties. In 2018, these varieties had a similar protein content, but in other years, the protein portion was higher in Atlas seeds. In 2017 and 2019, Jessie was also characterized by high protein levels. According to the findings of Rodríguez Gómez et al. [33], Atlas, Pasto, and Jessie were seeds with similar protein content. In 2017 and 2019, Jessie had a protein content comparable to that of Pasto and Atlas, respectively. Both Präger et al. [4] and Granado-Rodríguez et al. [19] concluded that Puno and Titicaca seeds had similar protein levels, even over different growing seasons. However, Aluwi et al. [16] reported a higher protein content for Puno, which corresponded with the results from 2019 within the present study. Nevertheless, the protein content of Titicaca was higher compared to that of Puno in 2017 and 2018. The Farm Original varieties, except for Faro, were characterized by lower average protein content. Aluwi et al. [16] also classified Oro de Valle and Black (possible origin for variety Zwarte) as varieties with a lower protein content than Titicaca or Puno. In 2017 and 2018, the protein content of Oro de Valle and Zwarte ranged between 12.1 and 13.0 g/100 g dm. However, the protein content increased to 16.0–16.2 g/100 g dm in 2019, which reduced the gap with the values of the European-bred varieties (16.0–18.8 g/100 g dm). The other Farm Original variety Faro had medium-high protein content over the different growing seasons (15.2 g/100 g dm). The protein content showed distinct genetic variation. However, yearly differences in protein content can reduce these differences greatly, proving that agro-ecological conditions are an important factor too, as was previously reported by other studies [4,15,19,34].

In addition to protein quantity, amino acid composition is an important property for the evaluation of protein quality [34]. Quinoa seeds from the different growing seasons contained large amounts (76–183 mg/g protein) of glutamic acid, aspartic acid, and arginine, along with lesser amounts (42–98 mg/g protein) of leucine, valine, glycine, and lysine, constituting about 60% of the total amino acids (Tables S3 and S4). This is in accordance with the study of Gonzalez et al. [34]. Specifically, the essential amino acids were available in considerable amounts. In most plant proteins, these amino acids are present in insufficient quantities for a balanced diet [34]. Over the years, the total essential amino acid content has ranged from 338 to 443 mg/g protein (Table S4). Across varieties and growing seasons, lower ranges for total essential amino acids were reported during field trials in Germany (204–278 mg/g protein, [4]). Moreover, the content of total essential amino acids differed in comparison with results from Argentina (180–374 mg/g protein, [34]) or Chile (297–339 mg/g protein, [23]). The main constituents of the essential amino acid profile (Table 2 and Table S4) were leucine (62–98 mg/g protein), lysine (42–73 mg/g protein), and valine (46–64 mg/g protein), as confirmed by previous studies [4,23]. The distribution patterns of the essential amino acids were different among growing seasons yet similar among varieties. However, Präger et al. [4] reported remarkable variations in the composition of the essential amino acid profile across varieties. Reguera et al. [15] noted that varieties grown in Chile presented no differences in amino acid content, while cultivation in Spain resulted in different alanine, asparagine, isoleucine, lysine, and valine contents among varieties. Concentrations of aromatic essential amino acids were considerably higher in 2018 and ranged from 43 to 51 mg/g protein for phenylalanine and from 16 to 24 mg/g protein for tryptophan. In 2018, the levels of methionine and lysine were also higher, while the valine concentrations were reduced during this growing season. The varieties generally had the highest leucine and isoleucine content in 2019. The growth conditions in 2017 resulted in the highest levels of threonine and histidine.

Table 2. Amino acid composition (mg/g protein) of 13 quinoa varieties grown under North-West European field conditions in 2017, 2018, and 2019.

Variety [1]/Year [2]	*n*	Amino Acid (mg/g Protein) [3]								
		Phe	*Leu*	*Met*	*Ile*	*Val*	*Thr*	*His*	*Lys*	*Trp*
Atlas	3	45 ± 3 [a]	76 ± 9 [a]	20 ± 3 [a]	45 ± 8 [a]	58 ± 15 [a]	36 ± 3 [a]	30 ± 3 [a]	54 ± 7 [a]	17 ± 2 [a]
Bastille	2	42 ± 1	71 ± 12	22 ± 2	42 ± 5	53 ± 9	36 ± 1	25 ± 7	51 ± 2	19 ± 5
Dutchess	3	42 ± 4 [a]	74 ± 6 [a]	20 ± 3 [a]	42 ± 5 [a]	56 ± 13 [a]	38 ± 6 [a]	30 ± 3 [a]	52 ± 4 [a]	17 ± 2 [a]
Faro	3	42 ± 6 [a]	75 ± 4 [a]	23 ± 3 [a]	44 ± 2 [a]	60 ± 7 [a]	39 ± 4 [a]	29 ± 5 [a]	55 ± 4 [a]	17 ± 2 [a]
Jessie	3	43 ± 6 [a]	76 ± 5 [a]	24 ± 2 [a]	44 ± 2 [a]	61 ± 7 [a]	41 ± 3 [a]	30 ± 6 [a]	56 ± 1 [a]	17 ± 2 [a]
Oro de Valle	3	43 ± 6 [a]	74 ± 5 [a]	24 ± 2 [a]	43 ± 3 [a]	60 ± 6 [a]	41 ± 3 [a]	30 ± 6 [a]	57 ± 2 [a]	15 ± 4 [a]
Pasto [4]	2	42 ± 4	70 ± 12	24 ± 1	41 ± 3	58 ± 17	43 ± 11	29 ± 10	59 ± 7	16
Puno	3	44 ± 6 [a]	74 ± 5 [a]	26 ± 5 [a]	42 ± 4 [a]	59 ± 7 [a]	41 ± 3 [a]	30 ± 6 [a]	58 ± 3 [a]	15 ± 4 [a]
Rouge Marie	1	41	82	21	48	64	36	30	46	16
Summer Red	3	44 ± 6 [a]	74 ± 5 [a]	25 ± 6 [a]	42 ± 3 [a]	59 ± 7 [a]	40 ± 4 [a]	30 ± 6 [a]	57 ± 4 [a]	15 ± 4 [a]
Titicaca	3	42 ± 8 [a]	71 ± 7 [a]	24 ± 7 [a]	40 ± 5 [a]	56 ± 6 [a]	38 ± 2 [a]	29 ± 5 [a]	54 ± 5 [a]	15 ± 3 [a]
Vikinga	3	42 ± 7 [a]	73 ± 7 [a]	24 ± 6 [a]	42 ± 6 [a]	59 ± 4 [a]	40 ± 4 [a]	28 ± 5 [a]	56 ± 7 [a]	16 ± 4 [a]
Zwarte	2	36 ± 1	73 ± 1	23 ± 1	39 ± 0	64 ± 2	44 ± 1	32 ± 2	57 ± 3	15 ± 2
2017	9	38 ± 3 [A]	74 ± 6 [A,B]	23 ± 1 [A]	41 ± 5 [A]	63 ± 5 [B]	43 ± 3 [B]	34 ± 2 [C]	57 ± 4 [A,B]	15 ± 2 [A]
2018	9	48 ± 2 [C]	69 ± 5 [A]	28 ± 5 [B]	41 ± 3 [A]	51 ± 6 [A]	38 ± 3 [A]	23 ± 2 [A]	58 ± 4 [B]	21 ± 2 [B]
2019	9	43 ± 3 [B]	81 ± 7 [B]	22 ± 3 [A]	47 ± 3 [B]	65 ± 4 [B]	36 ± 2 [A]	32 ± 2 [B]	53 ± 3 [A]	15 ± 1 [A]

[1] Amino acid content per variety averaged over the different years. [2] Amino acid content per year averaged over the nine varieties grown in all three years. [3] Lowercase letters compare the varieties grown in all three years; average values followed by the same letter are not significantly different ($p > 0.05$). Capital letters compare the three years; average values followed by the same letter are not significantly different ($p > 0.05$). [4] No analysis of tryptophan for Pasto in 2017.

Albumin and globulins make up the main fraction of proteins in quinoa, and they are characterized by a balanced composition of the essential amino acids necessary for a healthy human diet, making quinoa one of the few plants with a complete protein source [12]. This and other studies showed that all essential amino acids were available in sufficient amounts according to the daily requirements, in conformance with FAO/WHO requirements for school kids and adults [12,19,35]. In the category of 6-month-old babies to 3-year-old children, lysine is the limiting amino acid for all varieties and years. Leucine mainly did not meet the standard for this age group in 2018. Small shortages for threonine were also observed for a period of one or two years in a few varieties. However, tryptophan reached the standard for each group in all varieties, in contrast to Granado-Rodríguez et al. [19].

2.5.3. Fat, Fatty Acids, and Triacylglycerols

Depending on the growing season, the fat content varied between 5.42 and 8.54 g/100 g dm (Table S2). The results of Pulvento et al. (7.7–7.9 g/100 g dm, [22]) were within the range of the present study, while considerably lower fat contents were reported for a study in Spain (3.90–5.10 g/100 g, [33]). Moreover, the fat content in seeds from Argentina (4.7–7.1 g/100 g dm, [36]) or Chile (2.97–5.65 g/100 g dm, [37]) was also within a lower range. The data showed a significant variety × year effect for fat content in quinoa. The seeds harvested in 2017 generally showed lower fat content compared to the other years of cultivation. However, the growth conditions in 2017 resulted in the highest fat content for Titicaca and Summer Red. In 2017 and 2019, Dutchess and Jessie showed a stable fat content of around 7.10 and 6.70 g/100 g dm, respectively. The quinoa varieties, with the exception of Titicaca and Summer Red, obtained the highest fat content in 2018 (6.37–8.50 g/100 g dm). In 2019, the fat content decreased from 5.2% to 15.6% compared to the growing season of 2018. However, the fat content of Bastille and Titicaca remained stable at around 7.35 and 7.16 g/100 g dm, respectively. Contrary to other varieties, the decrease in 2019 resulted in the lowest fat content among growing seasons for Summer Red. Averaged over the three years, Pasto, Puno, and Zwarte showed significantly lower fat content compared to the other varieties. Summer Red and Vikinga had the highest fat content among varieties, followed by Faro and Rouge Marie.

Quinoa has a healthy balance between proteins and fat, with an interesting nutritional lipid composition of which essential fatty acids, i.e., polyunsaturated fatty acids (PUFAs), are necessary for a healthy human diet [12]. Contrary to the amino acid profile, the composition of fatty acids varied significantly between varieties. The study by Tang et al. [38] confirmed that the fatty acid profile may vary among varieties grown under the same growth conditions. According to the study of Präger et al. [4], the fatty acid composition did not vary among varieties or years. The fatty acid profile of the quinoa varieties (Table 3 and S5) consisted of 60.4% to 70.6% PUFAs, which corresponded with the results of Rodríguez Gómez et al. (65–70%, [33]). The main PUFA in quinoa seeds was linoleic acid (C18:2 n-6), ranging from 53.0% to 59.8%. Lower values were reported by Tang et al. (44.9–56.6%, [38]) and Vidueiros et al. (43.0–57.5%, [36]). Rodríguez Gómez et al. [33] obtained 60.66% to 61.40% linoleic acid for Jessie, Puno, and Titicaca. The content of linoleic acid was stable over the years yet varied among the varieties. The lowest linoleic acid contents were measured for Bastille, Dutchess, Pasto, and Vikinga and the highest for Puno. Furthermore, the quinoa seeds contained 4.7% to 8.2% α-linolenic acid (C18:3 n-3). This corresponded with the studies of Rodríguez Gómez et al. (5.15–7.17%, [33]), Tang et al. (4.8–9.6%, [38]), and Vidueiros et al. (3.2–9.4%, [36]). Compared to 2019, the content of α-linolenic acid was lower in 2018. The red varieties Rouge Marie and Summer Red were characterized by the lowest α-linolenic acid content over the years. The high content of PUFAs makes quinoa lipids a high-quality edible oil, with a composition similar to that of maize and soybean oil [12]. The monounsaturated fatty acid (MUFA) oleic acid (C18:1c) was the second prominent fatty acid in quinoa seeds, ranging from 15.5% to 22.7%. This fatty acid had a large share in the total content of MUFAs (18.1–25.5%), which explains why the total MUFA content was also the highest in 2018. According to Tang et al. [38], the oleic acid content varied between 15.7% and 28.9%, while Rodríguez Gómez et al. [33] reported a lower range of 17.93% to 20.54%. Even higher percentages were obtained by Vidueiros et al. (19.8–33.1%, [36]). Comparing the varieties, Puno had both the lowest oleic acid and total MUFA content. Furthermore, the fat fraction of quinoa contained important amounts of docosadienoic acid (C22:2, 1.3–7.5%), eicosenoic acid (C20:1, 0.9–1.4%), and erucic acid (C22:1, 0.9–1.4%) (Table S5). The balance between $\Omega 6$ and $\Omega 3$ fatty acids is highly important in health risk reduction, and an optimal ratio should range between 1:1 and 4:1 [39]. Excessive amounts of $\Omega 6$ PUFAs and high $\Omega 6/\Omega 3$ ratios promote the pathogenesis of many diseases, including cardiovascular disease, cancer, and inflammatory and autoimmune diseases, whereas increased levels of $\Omega 3$ PUFAs (low $\Omega 6/\Omega 3$ ratio) exert suppressive effects [40]. The $\Omega 6/\Omega 3$ ratio of the fat fraction varied between 6.7 and 12.0 (Table S5), which is higher compared to the results of Tang et al. (5.3–10.6, [38]) but within the range reported by Vidueiros et al. (4.9–17.4, [36]). Moreover, these ratios are not within the optimal range to reduce health risk [39]. The growing season of 2018 resulted in the highest $\Omega 6/\Omega 3$ ratios. Among varieties, the highest $\Omega 6/\Omega 3$ ratios were measured for the red varieties Rouge Marie and Summer Red.

Table 3. Fatty acid composition (% fatty acid ethyl esters) of 13 quinoa varieties grown under North-West European field conditions in 2017, 2018, and 2019.

Variety [1]/Year [2]	*n*	Fatty Acid (% FAME) [3,4] *C16:0*	*C18:0*	*C18:1c*	*C18:2 n-6*	*C18:3 n-3*
Atlas	3	9.3 ± 0.5 ^{a,b}	0.7 ± 0.1 ^{e}	20.9 ± 0.3 ^{c}	55.3 ± 0.4 ^{a,b}	5.4 ± 0.3 ^{a,b}
Bastille	2	9.4 ± 0.1	0.6 ± 0.0	19.9 ± 0.6	54.6 ± 0.2	7.9 ± 0.3
Dutchess	3	9.4 ± 0.6 ^{a,b}	0.6 ± 0.1 ^{d,e}	20.6 ± 0.7 ^{c}	54.6 ± 1.0 ^{a}	5.6 ± 0.4 ^{a,b}
Faro	3	8.9 ± 0.1 ^{a}	0.6 ± 0.1 ^{c,d,e}	19.9 ± 0.8 ^{b,c}	57.4 ± 0.6 ^{c}	6.1 ± 0.7 ^{b,c}
Jessie	3	9.1 ± 0.1 ^{a,b}	0.5 ± 0.0 ^{a,b}	18.7 ± 1.3 ^{b}	57.3 ± 0.5 ^{c}	6.3 ± 0.7 ^{b,c}
Oro de Valle	3	9.1 ± 0.3 ^{a,b}	0.5 ± 0.0 ^{a,b,c}	20.6 ± 0.3 ^{c}	56.3 ± 0.9 ^{b,c}	6.7 ± 0.2 ^{c,d}
Pasto	2	9.6 ± 0.2	0.8 ± 0.2	19.4 ± 0.2	54.8 ± 0.7	6.8 ± 0.3
Puno	3	9.2 ± 0.3 ^{a,b}	0.5 ± 0.0 ^{a}	16.8 ± 1.2 ^{a}	59.0 ± 0.6 ^{d}	7.2 ± 0.4 ^{d}
Rouge Marie	1	10.0	0.6	19.7	56.3	4.8
Summer Red	3	9.5 ± 0.7 ^{a,b}	0.6 ± 0.1 ^{a,b,c,d,e}	20.2 ± 1.1 ^{b,c}	55.8 ± 0.7 ^{a,b}	4.8 ± 0.2 ^{a}
Titicaca	3	10.3 ± 1.4 ^{b}	0.6 ± 0.1 ^{b,c,d,e}	19.7 ± 0.8 ^{b,c}	56.7 ± 0.9 ^{b,c}	5.5 ± 0.6 ^{a,b}
Vikinga	3	10.4 ± 1.2 ^{b}	0.6 ± 0.1 ^{a,b,c,d}	21.2 ± 1.0 ^{c}	54.7 ± 0.9 ^{a}	5.8 ± 0.7 ^{b,c}
Zwarte	2	9.5 ± 0.2	0.7 ± 0.1	18.9 ± 0.5	56.6 ± 0.6	7.1 ± 0.0
2017	9	9.2 ± 0.4 ^{A}	0.6 ± 0.1 ^{B}	19.6 ± 1.3 ^{A}	55.9 ± 1.6 ^{A}	6.0 ± 0.7 ^{A,B}
2018	9	9.8 ± 1.2 ^{A}	0.6 ± 0.1 ^{A,B}	20.8 ± 1.2 ^{B}	56.4 ± 1.5 ^{A}	5.5 ± 0.8 ^{A}
2019	9	9.5 ± 0.4 ^{A}	0.5 ± 0.1 ^{A}	19.2 ± 1.6 ^{A}	56.8 ± 1.4 ^{A}	6.3 ± 0.8 ^{B}

[1] Fatty acid content per variety averaged over the different years. [2] Fatty acid content per year averaged over the nine varieties grown in all three years. [3] Lowercase letters compare the varieties grown in all three years; average values followed by the same letter are not significantly different ($p > 0.05$). Capital letters compare the three years; average values followed by the same letter are not significantly different ($p > 0.05$). [4] FAME: fatty acid methyl esters.

Only 10.2% to 15.0% of the total fatty acid content (Table S5) were saturated fatty acids (SFAs). Compared to 2017, the content of SFAs was higher in 2018. The highest SFA contents were found in Titicaca or Vikinga seeds, while Jessie and Faro contained the least SFAs. The most abundant SFAs in quinoa seeds were palmitic acid (C16:0) and stearic acid (C18:0) (Table 3). The levels of palmitic acid ranged between 8.6% and 13.1% and did not vary among growing seasons. According to Vidueiros et al. [36], the level of palmitic acid varied between 8.4% to 22.9%, while Rodríguez Gómez et al. [33] reported 9.02% to 10.37% of palmitic acid. Vikinga and Titicaca contained the highest amount of palmitic acid and Faro the lowest. The levels of stearic acid were considerably lower compared to those of palmitic acid and varied between 0.4% and 1.0%. The highest amounts of stearic acid were found in the lipid fractions of Pasto, Atlas, and Zwarte, while Puno contained the lowest amount of stearic acid.

The nutritional quality of the fat fraction is defined by its fatty acid composition. However, the triacylglycerol (TAG) composition will determine the physical and functional properties of the fat [41]. Table 4 and Table S6 give the distribution of the different TAGs in the fat fraction of the quinoa varieties. The predominant TAGs were LLO (19.14–32.24 area %), LLL (16.62–26.20 area %), LLP (9.45–13.93 area %), OOL (5.35–10.22 area %), OLP (5.78–9.85 area %), and LLLn (5.04–9.61 area %), where L is linoleic acid, O is oleic acid, P is palmitic acid, and Ln is linolenic acid. According to Jahaniaval et al. [42], the predominant TAGs in the quinoa fat fraction were LLO + OOLn (21.09 area %), LLL (19.22 area %), OOL + PoOO (12.78 area %), LLP + PLnO (11.34 area %), OLP (9.47 area %), and LLLn (6.04 area %). These values were in agreement with the ranges reported in the present study. Fanali et al. [41] identified LLO (18.10 area %), OOL (15.30 area %), OLP (12.93 area %), LLP (8.71 area %), LLL (6.22 area %), and OLLn (5.91 area %) as predominant TAGs. The reported levels of OOL, OLP, and OLLn were higher compared to those described in the present study, while the LLL and LLLn levels were lower [41]. The levels of LLL, LLnP, LLO, LLP, OLP, and PPL showed no differences among the varieties but varied among growing seasons. The levels of LLP, OLP, and PPL were significantly lower in 2017, while the LLL and LLO levels were higher. The highest levels of LLnP were measured in

2019. The content of LLLn, OLLn, and OOL showed more variation among the varieties. Puno was characterized by the highest levels of LLLn but had the lowest share of OOL. Summer Red contained the lowest levels of LLLn and OLLn.

Table 4. Triacylglycerol composition (area %) of 13 quinoa varieties grown under North-West European field conditions in 2017, 2018, and 2019.

Variety [1]/Year [2]	*n*	Triacylglycerol (Area %) [3,4]					
		LLLn	*LLL*	*LLO*	*LLP*	*OOL*	*OLP*
Atlas	3	6.08 ± 0.66 [a,b,c]	20.29 ± 3.46 [a]	24.00 ± 3.05 [a]	11.30 ± 1.04 [a]	8.71 ± 0.71 [b]	7.74 ± 1.55 [a]
Bastille	2	7.49 ± 0.54	17.96 ± 1.55	22.04 ± 1.46	11.98 ± 0.26	8.13 ± 0.33	8.34 ± 0.60
Dutchess	3	5.67 ± 0.88 [a,b]	19.70 ± 3.76 [a]	25.01 ± 5.65 [a]	11.53 ± 1.61 [a]	8.73 ± 0.60 [b]	8.16 ± 1.89 [a]
Faro	3	7.11 ± 1.05 [b,c,d]	20.04 ± 2.21 [a]	22.23 ± 1.40 [a]	11.57 ± 0.47 [a]	8.50 ± 0.59 [b]	7.78 ± 0.43 [a]
Jessie	3	7.24 ± 1.08 [c,d]	20.23 ± 1.62 [a]	21.88 ± 0.59 [a]	12.41 ± 0.20 [a]	7.68 ± 1.10 [a,b]	8.15 ± 0.81 [a]
Oro de Valle	3	8.26 ± 1.06 [d,e]	22.05 ± 1.38 [a]	22.41 ± 3.12 [a]	12.02 ± 1.55 [a]	7.51 ± 1.50 [a,b]	7.06 ± 0.57 [a]
Pasto	2	7.47 ± 0.60	19.79 ± 3.28	23.29 ± 3.15	11.47 ± 1.39	7.19 ± 0.52	7.20 ± 1.51
Puno	3	8.92 ± 0.67 [e]	23.28 ± 0.94 [a]	21.19 ± 1.33 [a]	13.18 ± 0.61 [a]	6.12 ± 0.67 [a]	6.93 ± 0.53 [a]
Rouge Marie	1	5.20	18.77	22.37	13.29	8.61	9.35
Summer Red	3	5.30 ± 0.26 [a]	20.98 ± 4.08 [a]	24.83 ± 3.31 [a]	12.36 ± 1.15 [a]	8.23 ± 1.16 [b]	8.27 ± 1.94 [a]
Titicaca	3	6.39 ± 0.58 [a,b,c]	20.62 ± 0.92 [a]	22.76 ± 0.62 [a]	13.16 ± 0.29 [a]	8.38 ± 0.45 [b]	8.59 ± 0.47 [a]
Vikinga	3	6.26 ± 0.90 [a,b,c]	19.04 ± 1.47 [a]	23.33 ± 1.68 [a]	12.08 ± 1.08 [a]	8.81 ± 1.26 [b]	8.41 ± 1.48 [a]
Zwarte	2	8.29 ± 0.28	20.36 ± 1.05	21.93 ± 1.67	12.02 ± 1.13	7.39 ± 0.07	7.63 ± 0.63
2017	9	6.98 ± 1.21 [A,B]	22.36 ± 2.92 [B]	25.38 ± 3.76 [B]	11.24 ± 1.28 [A]	7.65 ± 0.72 [A]	6.85 ± 1.04 [A]
2018	9	6.01 ± 1.08 [A]	19.40 ± 1.73 [A]	22.52 ± 0.52 [A]	12.42 ± 0.44 [B]	8.92 ± 0.91 [B]	8.87 ± 0.82 [C]
2019	9	7.41 ± 1.47 [B]	20.32 ± 2.17 [A]	21.32 ± 1.28 [A]	12.87 ± 0.84 [B]	7.65 ± 1.43 [A]	7.97 ± 1.01 [B]

[1] Triacylglycerol content per variety averaged over the different years. [2] Triacylglycerol content per year averaged over the nine varieties grown in all three years. [3] Lowercase letters compare the varieties grown in all three years; average values followed by the same letter are not significantly different ($p > 0.05$). Capital letters compare the three years; average values followed by the same letter are not significantly different ($p > 0.05$). [4] L: linoleic acid, O: oleic acid, P: palmitic acid, Ln: linolenic acid.

2.5.4. Starch

The quinoa seeds contained 50.5 to 72.5 g starch per 100 g dm (Table S2). Aluwi et al. [16] reported a starch content of 54.4 to 64.3 g/100 g dm. Most varieties obtained the highest starch content in 2018, with the exception of Summer Red, Oro de Valle, and Titicaca. The growth conditions in 2017 resulted in the highest starch content for Summer Red and Oro de Valle, while the amount of starch was the highest in 2019 for Titicaca. The starch content of Atlas seeds remained stable at around 57.8 g/100 g dm in 2017 and 2018 and reduced by 5.7% in 2019. The seeds of 2019 generally had the lowest starch content, although Jessie, Titicaca, Vikinga, and Zwarte seeds contained the lowest amount of starch in 2017. Averaged over the different years, Titicaca seeds were characterized by the highest starch content. The Farm Original variety Faro generally had the lowest starch content among varieties. Other Farm Original varieties had medium-high (Zwarte) to high (Oro de Valle) starch content.

2.5.5. Minerals

The ash content of the quinoa varieties ranged between 2.37 and 3.60 g/100 g dm (Table S2). A higher total mineral content was reported by Pulvento et al. (3.96–4.28 g/100 g dm, [22]), while Rodríguez Gómez et al. [33] obtained an ash content of 2.99 to 3.80 g/100 g. The test weight correlated negatively with the ash content of the seeds from 2017 ($r = -0.734$, $p = 0.010$) and 2018 ($r = -0.695$, $p = 0.018$). The growth conditions of 2017 resulted in the lowest ash content for Puno, Oro de Valle, Pasto, and Zwarte, while other varieties obtained the lowest ash content in 2019. The ash content of Faro seeds remained stable in 2018 and 2019, while Dutchess and Summer Red had a stable ash content in 2017 and 2018. The highest amounts of minerals were found in the seeds of 2018. Only Atlas, Faro, and Jessie contained higher levels of minerals in 2017. In 2018, ash content correlated positively with the seed protein content ($r = 0.768$, $p = 0.006$). Among varieties, Puno invariably had the lowest ash content. Pasto was characterized by the highest amount of minerals, followed by Dutchess, Atlas, and Summer Red. According to Rodríguez Gómez et al. [33], the ash content of Atlas, Jessie, and Pasto seeds was similar among varieties. Within the present study, Jessie invariably had a lower ash content compared to Atlas or Pasto. Varieties Black (possible origin for Zwarte), Oro de Valle, Puno, and Titicaca were included in the study by Aluwi et al. [16]. The authors concluded that Oro de Valle had the lowest ash content among these four varieties [16]. Within the present study, Oro de Valle generally had a higher ash content than Puno or Titicaca, but Zwarte invariably had the highest ash content among these varieties. Titicaca seeds had a lower ash content than Zwarte seeds, although Aluwi et al. [16] reported that these varieties had comparable ash content.

The quinoa seeds (Table 5 and Table S7) showed a high content of potassium (K, 8790–14053 mg/kg dm), phosphorus (P, 3720–6156 mg/kg dm), magnesium (Mg, 1866–2742 mg/kg dm), and calcium (Ca, 399–806 mg/kg dm). Rodríguez Gómez et al. [33] reported similar levels of K (9088–13236 mg/kg dm), P (5162–5829 mg/kg dm), and Mg (1823–2319 mg/kg dm) but higher levels of Ca (868–1035 mg/kg dm) for quinoa seeds cultivated in Spain. Furthermore, the authors concluded that varieties Atlas, Jessie, and Pasto contained similar levels of P, Ca, and Mg. However, Jessie was characterized by higher levels of K [33]. Within the present study, Pasto contained higher Ca and Mg levels compared to Atlas and Jessie, but its P content was lower. The K concentrations were higher in Pasto than in Jessie. Granado-Rodríguez et al. [19] analyzed Puno, Titicaca, and Vikinga seeds cultivated over three growing seasons in Spain. Over the years, these varieties did not differ based on their P, K, Ca, and Mg levels. However, Vikinga had a lower P content in 2017 compared to Titicaca and Puno [19]. The present data confirmed that Puno, Titicaca, and Vikinga seeds had similar P, K, Ca, and Mg concentrations over different years. According to Miranda et al. [37], Chilean quinoa stored considerable higher amounts of K (10638–27103 mg/kg dm) and Ca (1108–3020 mg/kg dm) but lower levels of P (2888–4602 mg/kg dm) and Mg (1648–1768 mg/kg dm). All these data suggest that the mineral content in quinoa is both variety-dependent and the result of environmental conditions, such as soil type and its mineral composition, the used fertilization, and growing season [33]. The levels of P and Ca in this study did not differ over the years but varied between the varieties. The dark-colored varieties (i.e., Rouge Marie, Summer Red, and Zwarte) and Pasto accumulated the highest levels of P; Puno and Titicaca accumulated the lowest. Dutchess, Faro, and Pasto had a higher Ca concentration compared to Summer Red and Titicaca. The P and Mg concentrations were higher in the seeds of 2018, but no differences between varieties were observed. However, Pasto seeds tended to contain higher levels of Mg compared to other varieties cultivated in 2017 or 2018. The iron (Fe) content of the quinoa seeds showed no significant effect of variety or year, while the sodium (Na) content was only affected by the year (Table 5 and S7). Granado-Rodríguez et al. [19] also observed no significant differences between varieties Puno, Titicaca, and Vikinga in the Fe or Na content. In 2017 and 2018, several varieties showed a remarkably higher Fe or Na content compared to other years or varieties, which suggests a significant variety × year interaction. The Fe content generally ranged between 50 and 91 mg/kg dm, but higher levels, up to 723 mg/kg dm, were occasionally measured in 2017 and 2018. Rodríguez Gómez et al. [33] obtained an Fe content of 72 to 94 mg/kg dm and concluded that Atlas, Jessie, and Pasto seeds contained similar levels of Fe. However, the present data showed that Pasto had remarkably higher Fe content than Atlas and Jessie in 2017. A wider range for Fe content was reported by Miranda et al. (57–342 mg/kg dm, [37]). The Na concentration varied from 42 to 165 mg/kg dm, which corresponded with the study by Rodríguez Gómez et al. (74–159 mg/kg dm, [33]). However, the Na content was considerably higher for Dutchess (297 mg/kg dm) and Zwarte (241 mg/kg dm) in 2017 and for Puno (226 mg/kg dm) in 2018. Rodríguez Gómez et al. [33] observed that Atlas seeds contained higher levels of Na than Pasto seeds. The opposite applied for the growing season of 2017 within the present study. The zinc (Zn) concentration ranged between 38 and 109 mg/kg dm (Table 5 and S7), with the highest levels measured for Atlas, Pasto, and Bastille. The results of Miranda et al. (76.7–95.6 mg/kg dm, [37]) were within this range. Furthermore, the quinoa seeds (Table S7) contained low levels of copper (Cu, 2–19 mg/kg dm), boron (B, 9–18 mg/kg dm), and manganese (Mn, 19–46 mg/kg dm). The Cu content did not vary among varieties or years, only Atlas showed a remarkably higher Cu concentration in 2018 (179 mg/kg dm). The varieties contained a similar level of B but the concentrations were slightly higher in 2019. The levels of Mn were higher in seeds from 2018 or in Pasto seeds.

Table 5. Mineral composition (mg/kg dm) of 13 quinoa varieties grown under North-West European field conditions in 2017, 2018, and 2019.

Variety [1]/Year [2]	*n*	Mineral (mg/kg dm) [3]						
		Ca	*Fe*	*K*	*Mg*	*Na*	*P*	*Zn*
Atlas	3	$613 \pm 54^{a,b}$	77 ± 3^{a}	$11055 \pm 1232^{a,b}$	2299 ± 186^{a}	97 ± 28^{a}	5716 ± 370^{a}	60 ± 4^{b}
Bastille	2	636 ± 56	392 ± 468	11397 ± 460	2166 ± 138	74 ± 30	4492 ± 410	76 ± 47
Dutchess	3	686 ± 84^{b}	110 ± 67^{a}	$12055 \pm 1233^{a,b}$	2308 ± 123^{a}	160 ± 129^{a}	5344 ± 584^{a}	45 ± 3^{a}
Faro	3	686 ± 50^{b}	203 ± 236^{a}	$10634 \pm 846^{a,b}$	2486 ± 191^{a}	73 ± 26^{a}	5324 ± 560^{a}	48 ± 5^{a}
Jessie	3	$507 \pm 25^{a,b}$	73 ± 8^{a}	$10222 \pm 415^{a,b}$	2314 ± 149^{a}	73 ± 21^{a}	5533 ± 653^{a}	47 ± 3^{a}
Oro de Valle	3	$633 \pm 59^{a,b}$	85 ± 36^{a}	$11029 \pm 537^{a,b}$	2090 ± 241^{a}	74 ± 18^{a}	4597 ± 705^{a}	48 ± 4^{a}
Pasto	2	777 ± 41	161 ± 99	12539 ± 396	2706 ± 50	122 ± 61	5281 ± 130	67 ± 0
Puno	3	$576 \pm 70^{a,b}$	66 ± 5^{a}	9844 ± 1303^{a}	2164 ± 177^{a}	124 ± 90^{a}	4322 ± 633^{a}	44 ± 1^{a}
Rouge Marie	1	515	64	12025	2000	57	4766	44
Summer Red	3	474 ± 66^{a}	58 ± 4^{a}	12550 ± 409^{b}	2157 ± 100^{a}	88 ± 39^{a}	5103 ± 378^{a}	44 ± 3^{a}
Titicaca	3	450 ± 45^{a}	176 ± 142^{a}	9869 ± 954^{a}	2152 ± 152^{a}	80 ± 24^{a}	4671 ± 542^{a}	45 ± 6^{a}
Vikinga	3	$604 \pm 97^{a,b}$	73 ± 12^{a}	$10168 \pm 399^{a,b}$	2203 ± 106^{a}	71 ± 17^{a}	4630 ± 370^{a}	47 ± 4^{a}
Zwarte	2	658 ± 107	55 ± 7	13717 ± 476	2049 ± 35	143 ± 138	4515 ± 1	41 ± 2
2017	9	564 ± 81^{A}	101 ± 90^{A}	11076 ± 1596^{A}	2197 ± 187^{A}	137 ± 75^{B}	$4959 \pm 608^{A,B}$	48 ± 5^{A}
2018	9	545 ± 98^{A}	138 ± 133^{A}	10259 ± 922^{A}	2396 ± 139^{B}	$91 \pm 22^{A,B}$	5556 ± 444^{B}	46 ± 5^{A}
2019	9	634 ± 105^{A}	69 ± 7^{A}	11140 ± 764^{A}	2132 ± 92^{A}	53 ± 7^{A}	4564 ± 503^{A}	49 ± 7^{A}

[1] Mineral content per variety averaged over the different years. [2] Mineral content per year averaged over the nine varieties grown in all three years. [3] Lowercase letters compare the varieties grown in all three years; average values followed by the same letter are not significantly different ($p > 0.05$). Capital letters compare the three years; average values followed by the same letter are not significantly different ($p > 0.05$).

As described, the effects of variety or year were not always present for the different minerals. Except for Na, Granado-Rodríguez et al. [19] did report significant effects of both variety and year on the mineral content of quinoa. Except for Zn, Reguera et al. [15] also described a significant effect of the growth location. Besides proteins and fats, minerals are an essential quality of quinoa seeds. The mineral content for Ca, Mg, Fe, Cu, and Zn are higher than in common cereals and are, in this study, except for Ca, similar or higher than those reviewed by Bhargava et al. [12]. The minerals Ca, Mg, Fe, and K are present in sufficient amounts and available for a healthy human diet in bioavailable forms [19].

2.6. Principal Components Analysis and Hierarchical Cluster Analysis

For a more comprehensive understanding of the variation among quinoa seeds, a PCA was performed on yield, seed characteristics (seed size, TSW, test weight), and macronutrient composition (protein, fat, starch, ash). The PCA revealed three principal components (PCs) with eigenvalues greater than one, which explained 74.5% of the total variance. The loading plot of the first two PCs and the score plot with the varieties/years are illustrated in Figure 3. The first and second PCs accounted for 39.7% and 21.0% of the variance, respectively. The first PC was defined by seed length and width and TSW (Table S8). These were all properties related to the physical properties of the seed, of which the TSW correlated positively with seed length ($r = 0.873$, $p < 0.001$) and width ($r = 0.854$, $p < 0.001$). A strong positive association between TSW and seed size was also reported by Bhargava et al. [17]. The second PC was defined by test weight and protein and starch content (Table S8). The protein content correlated negatively with the starch content ($r = -0.543$, $p = 0.001$) and test weight ($r = -0.607$, $p < 0.001$). Protein-rich seeds showed a lower starch content, which suggests that the share of perisperm is smaller in these seeds. The smaller ratio of perisperm to bran and embryo could affect the test weight of the seeds, explaining the negative relation between the protein content and test weight [43,44]. The third PC explained 13.8% of the variance and was representative for the yield, ash, and fat content (Table S8).

Based on the yield, seed characteristics, and macronutrient composition, the varieties/years were classified into five groups (I–V) in the score plot (Figure 3b). Group I had the highest positive score on PC1 and included Summer Red of 2017 and 2018 and Zwarte of 2017 and 2019. These varieties were, within the given growing season, characterized by the largest seeds with the highest TSW. Group II only included varieties of 2019, namely, Atlas, Bastille, Dutchess, Faro, Jessie, Rouge Marie, and Summer Red. This growing season was characterized by protein-rich seeds with low test weight. However, the seed size and TSW of these varieties were diverse, as seen in the spread in scores on PC1. Puno 2019 had a high negative score on PC1 due to the small size of the seeds and was, therefore, included in Group III. The varieties in Group IV had a high positive score on PC2, indicating high test weight and starch content. This group included Bastille 2018, Oro de Valle 2017 and 2018, Puno 2017 and 2018, and Titicaca 2017 and 2018. The remaining varieties were classified in Group V and were, in fact, all close to the origin of the score plot.

(a)

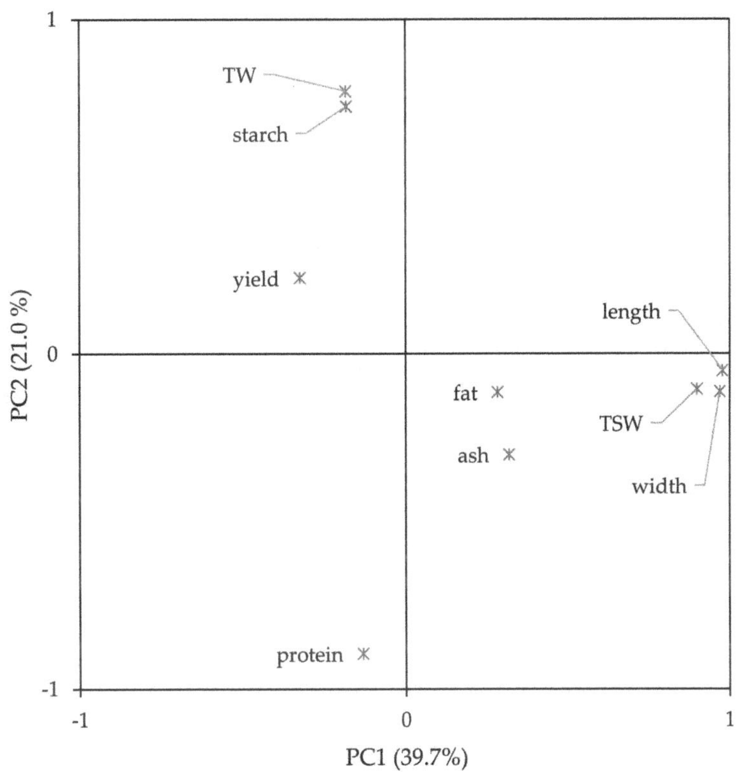

Figure 3. *Cont.*

(b)

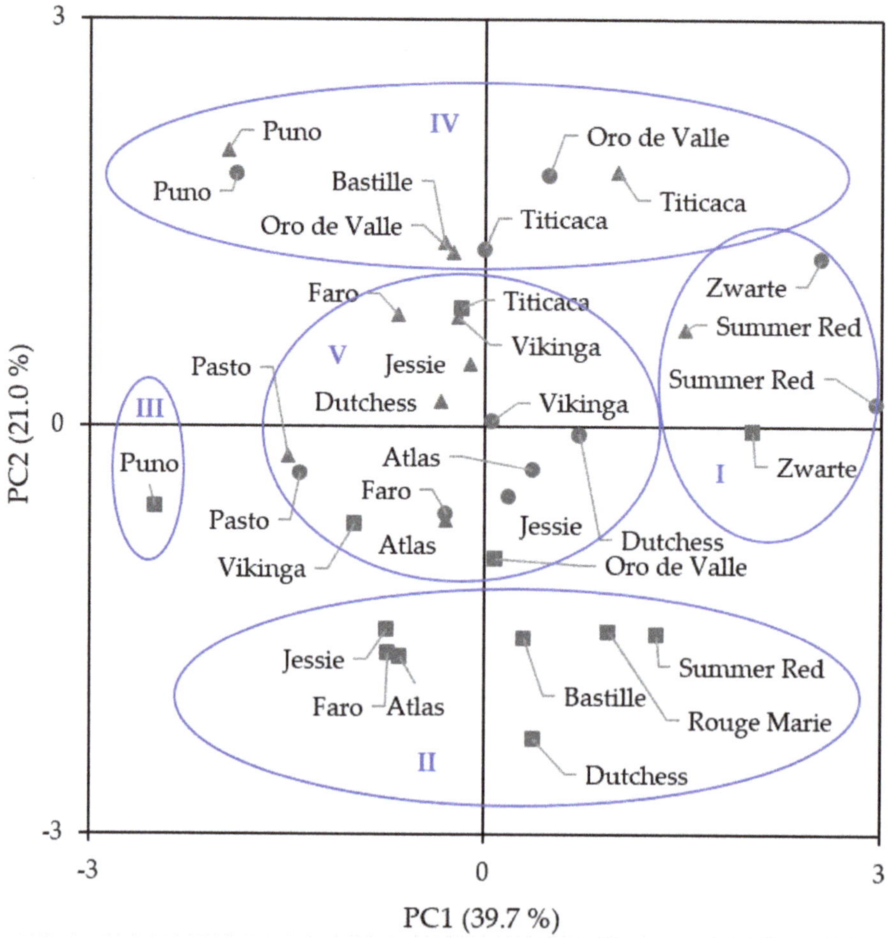

Figure 3. Principal components analysis (PCA—(**a**): loading plot, (**b**): score plot) of yield, seed characteristics, and composition of 13 quinoa varieties grown under North-West European field conditions in 2017 (•), 2018 (▲), and 2019 (■). TSW: thousand seed weight, TW: test weight.

A total of three clusters was identified based on the yield, seed characteristics, and macronutrient composition of the quinoa seeds by using HCA. The dendrogram of the hierarchical clustering is presented in Figure 4. The clustering revealed subtle differences between the growing seasons as most varieties of 2019 were grouped in the first cluster. However, the factor variety had a large impact on the clustering of Puno, Rouge Marie, Summer Red, Titicaca, Vikinga, and Zwarte. Hence, the significant interaction effects of year and variety were also reflected in the clustering of the varieties. Atlas, Bastille, Faro, and Oro de Valle seeds of 2019 showed more similarities to varieties of the same growing season than to seeds of the same variety from a different growing season and were grouped in the first cluster. Cluster 1 also included all dark-colored seeds (i.e., Rouge Marie, Summer Red, and Zwarte) and the seeds of Dutchess and Jessie grown in 2017 and 2019. Rouge Marie, Summer Red, and Zwarte were low-yielding over the years but provided large seeds with a high TSW. The second cluster was a small cluster ($n = 6$), only including seeds of 2017 and 2018. Atlas and Pasto seeds of the first two growing seasons were included in this cluster. Furthermore, Dutchess seeds of 2018 and Faro seeds of 2017 were grouped in the second cluster. All Puno, Titicaca, and Vikinga seeds were part of the third cluster. Titicaca and Vikinga were high-yielding varieties, while variety Puno produced small seeds with low levels of fat and ash. The Oro de Valle seeds of 2017 and 2018 were also grouped in the third cluster, as well as the Bastille, Faro, and Jessie seeds of 2018. The growing season helped to spread the seeds of Faro and Bastille over the three different groups, indicating a strong impact of environmental conditions.

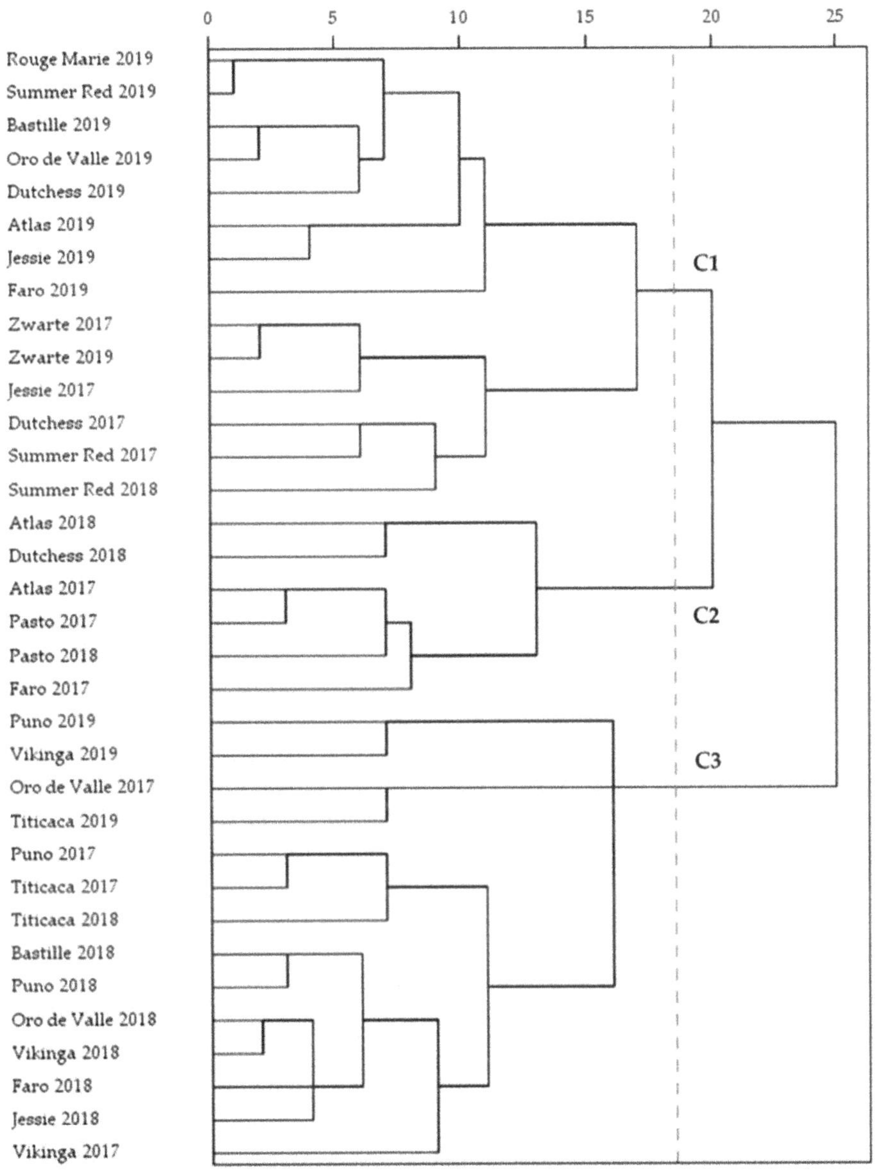

Figure 4. Dendrogram of the three clusters (C1–C3) based on hierarchical clustering.

3. Materials and Methods

3.1. Plant Varieties

Ten different quinoa (*Chenopodium quinoa* Willd.) varieties were obtained from commercial companies (Gilbel sprl, Saint-Georges-Sur-Meuse, Belgium (previously a sublicensee of Radicle Crops, Wageningen, The Netherlands) and Quinoa Quality ApS, Regstrup, Denmark) and three Farm Original varieties were purchased at De Nieuwe Tuin (De Klinge, Belgium). Besides white varieties, the set included two red varieties (i.e., Rouge Marie and Summer Red) and one black variety (i.e., Zwarte). An overview of the main characteristics of these varieties is listed in Table 6. The varieties of Atlas, Dutchess, and Pasto were bred at Plant Breeding, Wageningen Research (Wageningen, The Netherlands), and Bastille, Jessie, and Rouge Marie were bred by AbbottAgra (Longué-Jumelles, France). These varieties were area-adapted to the West European climate and photoperiod. Summer Red is the predecessor of Rouge Marie, bred by AbottAgra. Puno, Titicaca, and Vikinga were derived from crosses between Chilean and Peruvian material, followed by selection for long day length in Denmark [32]. The Farm Original varieties (i.e., Faro, Oro de Valle, and Zwarte) were adapted to/selected for the Belgian climate by De Nieuwe Tuin.

Table 6. Overview of quinoa varieties grown in North-West Europe.

Variety	Seed Source [1]	Origin	Year Data
Atlas	Gilbel	Wageningen University & Research, Wageningen, NL	3
Dutchess	Gilbel	Wageningen University & Research, Wageningen, NL	3
Pasto	Gilbel	Wageningen University & Research, Wageningen, NL	2
Bastille	Gilbel	AbbottAgra, Longué-Jumelles, FR	2
Jessie	Gilbel	AbbottAgra, Longué-Jumelles, FR	3
Rouge Marie	Gilbel	AbbottAgra, Longué-Jumelles, FR	1
Summer Red	Gilbel	AbbottAgra, Longué-Jumelles, FR	3
Puno	Quinoa Quality	University of Copenhagen, Copenhagen, DK [16]	3
Titicaca	Quinoa Quality	University of Copenhagen, Copenhagen, DK [16]	3
Vikinga	Quinoa Quality	University of Copenhagen, Copenhagen, DK [45]	3
Faro	De Nieuwe Tuin	Redwood Seeds, Manton, CA, USA	3
Oro de Valle	De Nieuwe Tuin	Wild Garden Seeds, Philomath, OR, USA [16]	3
Zwarte [2]	De Nieuwe Tuin	White Mountain Farm, Mosca, CO, USA [16]	2

[1] Gilbel sprl, Rue du Château d'Eau 54, 4470 Saint-Georges-Sur-Meuse, Belgium; Quinoa Quality ApS, Teglværksvej 10, 4420 Regstrup, Denmark; De Nieuwe Tuin, Trompwegel 27, 9170 De Klinge, Belgium; [2] The origin of variety Zwarte is unclear, but since *zwart* is the Dutch translation for black, it is likely that Zwarte originated from variety Black (White Mountain Farm, CO, USA).

3.2. Field Trials

Agronomy tests were conducted in a sandy loam soil on ILVO trial fields in Melle/Merelbeke, Belgium (50°58′52.39″ N, 3°46′34.605″ E). The varieties were sown with a Jacobi seeder in three (2017) or four repetitions (2018 and 2019) per variety on 12 April (2017), 24 April (2018), and 26 April (2019). The repetitions were arranged in blocks, with full randomization within the block. The field plots were 13 m^2 in size. The average temperature and precipitation during the growth period were 15.2 °C and 342 mm in 2017, 16.2 °C and 320 mm in 2018, and 15.2 °C and 379 mm in 2019 (Figure 5), respectively. The field plots were not irrigated. Maximum temperatures were generally higher in 2018, except for June and August. The precipitation in June and July was considerably lower in 2018 than in 2017 and 2019. Meteorological data were collected by the Royal Meteorological Institute of Belgium (RMI).

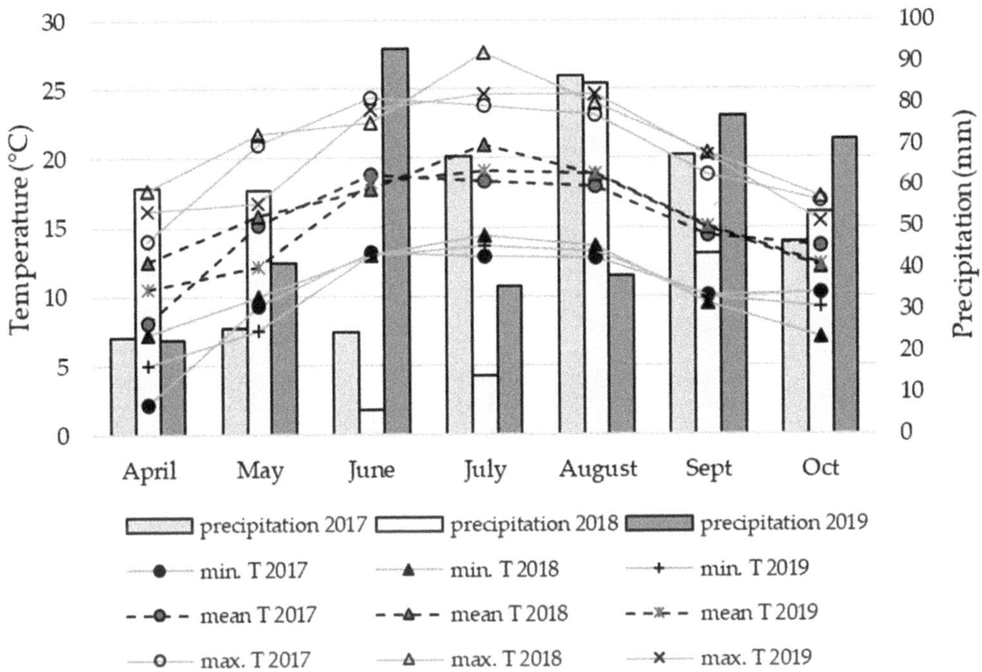

Figure 5. Total monthly precipitation (mm) and minimum, mean, and maximum monthly temperature (T, °C) during the experimental period in 2017, 2018, and 2019.

Sowing density was adjusted per seed lot to obtain a (theoretical) sowing density of 400 (2017 and 2018) or 300 (2019) germinating seeds per m^2. The inter-row distance was 26 cm, and the sowing depth was 1 to 2 cm. Calcium ammonium nitrate was applied before sowing at a rate of 135 units nitrogen in 2017 and 100 units nitrogen in 2018 and 2019. Weeding was done manually when necessary. Harvesting of the separate plots was done with a Wintersteiger harvester (Ried im Innkreis, Austria) according to the senescence stage of each variety (BBCH stage 95). After harvesting, seeds from each plot were collected in separate bags and dried using ventilation at 30 °C, cleaned with a Westrup LA-LS (Slagelse, Denmark), and color-sorted with a SEA Chromex (Cimbria, Thisted, Denmark) in 2017 and 2018 and an Os F color sorter (Petkus Selecta, Wutha-Farnroda, Germany) in 2019. Yield was determined after color sorting. Cleaned seeds were kept in a short storage room at 16 °C.

3.3. Seed Characteristics

Test weight (kg/hL) was determined by a grain analysis computer (GAC 2100, Dickey-john, Auburn, AL, USA). TSW (g) was calculated based on the weight of 300 seeds selected by a Contador seed counter (Pfeuffer GmbH, Kitzingen, Germany). Seed color, expressed as CIELAB coordinates L* (lightness), a* (redness to greenness), and b* (yellowness to blueness), was measured using a Konica Minolta CM-700d spectrophotometer (Konica Minolta, Tokyo, Japan). Seeds were scanned (dpi = 600) against a blue background (HP Scanjet 2400, Hewlett-Packard Company, Palo Alto, CA, USA). The resulting images were analyzed using SmartGRAIN software (version 1.1, National Institute of Agrobiological Sciences, Tsukuba, Japan) to estimate the length (mm), width (mm), and LWR of the seeds.

The saponin content of the seeds was estimated with Koziol's standard afrosimetric foam test [26] and calculated according to the following formula (1):

$$\text{saponin (mg/g)} = \frac{0.646 \times \text{H} - 0.104}{\text{W}} \tag{1}$$

where 'H' is the foam height in cm, and 'W' is the sample weight in g [26]. All seed characteristics were analyzed in triplicate.

3.4. Chemical Composition

For chemical analyses, seeds were milled to wholemeal flour (WMF) by a Hammertec mill (mesh size: 0.8 mm) (Foss, Hilleroed, Denmark).

3.4.1. Macronutrients

Moisture (g/100 g) and ash (g/100 g dm) content were determined according to ICC methods no. 110 and 104/1, respectively. Total starch content (g/100 g dm) was analyzed as described by Englyst et al. [46]. The nitrogen content was determined using aVarioMax C/N (Elementar Analysesystemen, Langenselbold, Germany) and converted to protein content (g/100 g dm) using a conversion factor of 6.25 [47]. Fat content (g/100 g dm) was determined by Soxhlet extraction with prior acid hydrolysis (ISO 6491). All macronutrient analyses were performed in triplicate.

3.4.2. Amino Acids

To determine the total standard amino acid composition (excluding tryptophan), acid hydrolysis was performed on the lyophilized samples using a hydrogen chloride and dithioglycolic acid solution for 4 h at 150 °C. Sample purification was done on an OASIS HLB cartridge (200 mg, 6 cc), followed by separation and detection by an LC–MS2 instrument (Nexera 8040, Shimadzu, Kyoto, Japan). Separation was done in normal phase mode on an Intrada amino acid column (3.0 µm, 100 × 3.0 mm) with acetonitrile (0.3% acetic acid) as solvent A and 20% acetonitrile-80% 100 mM ammonium formate as solvent B for the mobile phase gradient. The solvent gradient started at 80% solvent A and 20% solvent B. After 4 min, solvent B was gradually increased to 100% until the 14th min, at which it was held for 2 min before being returned to the initial gradient for 9 min. Amino acids were ionized by electrospray ionization (4.5 kV) followed by multiple reaction monitoring (MRM) on a triple quadrupole. All amino acids were measured in positive mode, except for aspartic acid, which was measured in negative mode. Methylvaline, homoarginine, and methyl-D3-methionine were used as internal standards.

To determine the tryptophan, alkaline hydrolysis was performed on lyophilized samples using a lithium hydroxide solution for 4 h at 150 °C. Sample purification was done on an OASIS HLB cartridge (200 mg, 6 cc), followed by separation and detection by an LC-MS2 instrument (Nexera 8040, Shimadzu Corporation, Kyoto, Japan). Separation was done in normal phase mode on an Intrada amino acid column (3.0 µm, 100 × 3.0 mm) with acetonitrile (0.3% acetic acid) as solvent A and 20% acetonitrile-80% 100 mM ammonium formate as solvent B for the mobile phase gradient. The solvent gradient started at 80% solvent A and 20% solvent B. After 4 min, solvent B was gradually increased to 100% until the 14th min, at which it was held for 2 min before being returned to the initial gradient for 9 min. Amino acids were ionized by electrospray ionization (4.5 kV), followed by MRM on a triple quadrupole. Tryptophan was measured in positive mode with methyl-tryptophan as the internal standard.

3.4.3. Fatty Acids

The fatty acid profile of the quinoa samples was determined as described by Foubert et al. [48]. Quinoa WMF was mixed with diethyl ether to extract the fat. The mixture was subsequently filtered over sodium sulfate, and diethyl ether was evaporated to retain the extracted fat. Four droplets of extracted quinoa fat were dissolved in 9 mL hexane and 1 mL

2 N potassium hydroxide/methanol reagent to produce fatty acid methyl esters (FAMEs). The blend was shaken for 30 s and allowed to settle. The FAMEs in the hexane layer were separated by a Varian 3380 gas chromatograph (Varian Inc., Palo Alto, CA, USA) equipped with a WCOT CP-sil 88 column and a flame ionization detector (FID). The conditions for the gas-chromatographic analysis were: temperature of injector: 250 °C, temperature of detector: 250 °C, flow rate of mobile phase (helium): 1 mL min^{-1}, flow rate of hydrogen: 40 mL min^{-1}, flow rate of air: 120 mL min^{-1}, injection volume: 1 mL, and column oven temperature: 120 °C for 2 min, followed by heating at 5 °C min^{-1} to 200 °C and holding at that temperature for 20 min.

3.4.4. Triacylglycerols

Analysis of the TAG species was performed as described by Rombaut et al. [49]. Quinoa WMF was mixed with diethyl ether to extract the fat. The mixture was subsequently filtered over sodium sulfate, and diethyl ether was evaporated to retain the extracted fat. The extracted fat was dissolved in a concentration of 5 mg/mL in dichloromethane/acetonitrile (30:70 *v*/*v*). Separation of the TAG species was performed on a Thermo Finnigan Surveyor RP-HPLC system (Thermo Electron Corporation, Brussels, Belgium) using a 150 × 3.0 mm Alltima HP C18 HL column with a 3 μm particle diameter (Grace Alltech, Lokeren, Belgium) and connected to an Alltech ELSD 2000 evaporative laser light scattering detector (Grace Alltech, Lokeren, Belgium). The injection volume was 25 μL.

3.4.5. Minerals

The mineral composition was determined via inductive coupled plasma-optical emission spectrometry (ICP-OES, IRIS Intrepid II XSP, Thermo Scientific, Waltham, MA, USA). Prior to the analysis, 1.0 g of WMF was ashed at 500 °C for 4 h in a muffle furnace. The ash was dissolved in hydrochloric acid during a 2 h reflux. The remaining residue was filtered (Whatman filter no. 5) before analysis.

3.5. Statistical Analysis

To determine the differences between varieties, years, and their interactions, the *lme4* package in R (version 4.0.2, R Core Team, Vienna, Austria) was used to fit a (mixed) linear model [50]. The fixed effects were variety and year, while block was considered a random effect. The following base model (2) was considered:

$$Y = \text{Variety} \times \text{Year} + \text{Random} \quad (2)$$

where 'Y' is the response variable, 'Variety' is the fixed effect of the variety, 'Year' is the fixed effect of the growing season, and 'Random' is the random effect of the block or measurement replicate.

Four different versions (3–6) were derived from the base model, as follows:

$$Y \sim \text{Random} \quad (3)$$

$$Y \sim \text{Variety} + \text{Random} \quad (4)$$

$$Y \sim \text{Year} + \text{Random} \quad (5)$$

$$Y \sim \text{Variety} + \text{Year} + \text{Random} \quad (6)$$

The five versions of the model (2–6) were tested, and the output was evaluated using Akaike information criteria (AIC) [51]. The best fit for each trait was then chosen based on the lowest AIC value (summarized in Table S9). Then, an analysis of variance (ANOVA) was performed, and pairwise comparisons between varieties and growing seasons were performed with the Tukey test. For the analyses of the seed characteristics and composition, one block was analyzed per variety. Therefore, the random effect of the measurement replicates was included in the base model instead of the block effect. In the case where 'Random' did not significantly attribute to the model, this effect was excluded from the model.

The analyses of amino acid, fatty acid, TAG, and mineral composition were performed without replicates. Therefore, the data did not allow the study of the effect of variety, year, or variety × year interaction. The data of all varieties common to all years were grouped to be able to make a comparison between the years. In a similar way, the data were grouped over the years to be able to estimate the variety effect.

A PCA and an HCA were performed with SPSS Statistics 27 (SPSS Inc., Chicago, IL, USA). For a more comprehensive understanding of the variation among quinoa varieties, a PCA was performed on the analyses of yield, seed characteristics, and macronutrient composition. Prior to analysis, the suitability of a PCA was assessed by testing the linearity between the variables and sample adequacy. Inspection of the correlation matrix showed that all variables had at least one correlation coefficient greater than 0.3. The overall Kaiser-Meyer-Olkin (KMO) measure was 0.641, with individual KMO measures all greater than 0.5. Bartlett's test of sphericity was statistically significant ($p < 0.001$), indicating that the data was likely factorizable. PCA revealed three components with eigenvalues greater than one, and visual inspection of the scree plot indicated that all three components should be retained. A Varimax orthogonal rotation with Kaiser normalization was employed to aid interpretability. The data on yield, seed characteristics, and macronutrient composition were used to perform an HCA to group the quinoa varieties with similar properties. The HCA was performed by using the unweighted paired-group method with an arithmetic mean. Distances among clusters were computed using Pearson correlation coefficients.

4. Conclusions

Thirteen varieties were evaluated for their agronomic performance under North-West European field conditions during three consecutive growing seasons (2017–2019). Additionally, the seeds were quantitatively characterized based on characteristics and composition. The study showed that all varieties, except for the very late ones, performed well under the environmental conditions in North-West Europe, thereby combining competitive yields with high seed quality over the years. Clustering of the varieties/years revealed subtle differences between growing seasons, but the factor of variety had a large impact on the clustering of Puno, Rouge Marie, Summer Red, Titicaca, Vikinga, and Zwarte. Hence, the significant interaction effects of variety and year were reflected in the clustering of the varieties.

Under North-West European environmental conditions, the studied varieties obtained nutritional values within the range of composition traits previously reported for European quinoa. The seeds contained high protein levels, and the varieties had a similar amino acid profile, containing all essential amino acids. The main fatty acids were linoleic, α-linolenic, and oleic acid, indicating a high degree of unsaturation. As a good source of minerals, the seeds were characterized by high K, P, Ca, and Mg levels and important amounts of Fe and Na. Varieties such as Bastille, Dutchess, Titicaca, and Vikinga were high-yielding and better adapted to North-West European environmental conditions. Atlas and Pasto did not reach full maturity under these environmental conditions but performed best regarding their protein and mineral content. The Farm Original varieties (i.e., Faro, Oro de Valle, and Zwarte) reached yields similar to those of most European-bred varieties. However, Oro de Valle and Zwarte were characterized by the lowest protein levels. Faro showed a more interesting nutritional profile as it was characterized by medium-high protein and high-fat content but was the bitterest variety.

Altogether, this study supports the huge potential of quinoa as a European crop. Thereby, the selection of the varieties should be well-informed and aligned with the production aim. However, these varieties are insufficiently studied to define their potential end-uses and facilitate their introduction in the European food industry. To gain more insight into the potential end-uses of North-West European quinoa, the seeds were milled to wholemeal flour and the physicochemical properties of these flours were determined. The physicochemical properties are described in Part II of this issue. Furthermore, the variability in yield, physical, and nutritional properties showed possibilities to stabilize or improve the yield and quality of European quinoa varieties using suitable agronomic practices. However, more research is needed to optimize these agronomic practices.

Supplementary Materials: The following are available online at https://www.mdpi.com/article/10.3390/plants10122689/s1, Figure S1. Correlation between yield and growing days ($r = -0.705$, $p < 0.001$) of 13 quinoa varieties grown under North-West European field conditions in 2017, 2018, and 2019, Table S1. Length-width ratio (LWR) and color parameters (L*a*b*) of 13 quinoa varieties grown under North-West European field conditions in 2017, 2018, and 2019 ($n = 3$), Table S2. Macronutrient composition (g/100 g dm) of 13 quinoa varieties grown under North-West European field conditions in 2017, 2018, and 2019 ($n = 3$), Table S3. Non- and semi-essential amino acid composition (mg/g protein) of 13 quinoa varieties grown under North-West European field conditions in 2017, 2018, and 2019 ($n = 1$), Table S4. Essential amino acid composition (mg/g protein) of 13 quinoa varieties grown under North-West European field conditions in 2017, 2018, and 2019 ($n = 1$), Table S5. Fatty acid composition (% fatty acid methyl ester) of 13 quinoa varieties grown under North-West European field conditions in 2017, 2018, and 2019 ($n = 1$), Table S6. Triacylglycerol composition (area %) of 13 quinoa varieties grown under North-West European field conditions in 2017, 2018, and 2019 ($n = 1$), Table S7. Mineral composition (mg/kg dm) of 13 quinoa varieties grown under North-West European field conditions in 2017, 2018, and 2019 ($n = 1$), Table S8. Rotated structure matrix for principal components analysis with Varimax rotation, Table S9. Results of analysis of variance (model, Akaike information criteria, p-value).

Author Contributions: Conceptualization, G.C., P.D.B., and F.V.B.; methodology, G.C., P.D.B. and F.V.B.; software, P.Q. and H.M.; validation, P.D.B. and G.C.; formal analysis, P.Q., H.M., G.C. and P.D.B.; investigation, G.C. and P.D.B.; resources, G.C., P.D.B., F.V.B., P.V., and M.E.; data curation, H.M., P.Q., G.C. and P.D.B.; writing—original draft preparation, H.M., G.C. and P.D.B.; writing—review and editing, H.M., G.C., P.D.B., P.Q., F.V.B., P.V., and M.E.; visualization, H.M., G.C. and P.D.B.; supervision, G.C., P.V., F.V.B. and M.E.; project administration, H.M., G.C. and P.D.B.; funding acquisition, G.C., F.V.B., and M.E. All authors have read and agreed to the published version of the manuscript.

Funding: This research was funded by a doctoral scholarship from the Ghent University Special Research Fund (DOC079-17 BOF) and VLAIO LA traject QUILO (Quinoa Lokaal, HBC.2018.2220).

Institutional Review Board Statement: Not applicable.

Informed Consent Statement: Not applicable.

Data Availability Statement: The data presented in this study are available in the article and Supplementary Materials.

Acknowledgments: The authors wish to thank Tania Roeges (LCA, Faculty of Bioscience Engineering, Ghent University) for the technical support while working with the VarioMax C/N. The authors are thankful for the help from Paul Provijn and Corine Loijson (FSF Research Group, Faculty of Bioscience Engineering, Ghent University) with the analysis of the fatty acids and TAG. The authors greatly thank the ILVO Plant technical team for the maintenance and processing of the field trials and Daphné Deloof and Bavo Dewitte (ILVO, Animal Sciences) for the analysis of amino acid composition. Furthermore, the authors thank master thesis students (Ghent University) Tine Vanhollebeke and Indy De Graeve and bachelor thesis students (Ghent University) Jolien De Schaepmeester, Joëlle Temmerman, and Lies Herman for their help with the nutritional and physicochemical characterization of the quinoa varieties.

Conflicts of Interest: The authors declare no conflict of interest. The funders had no role in the design of the study; in the collection, analyses, or interpretation of data; in the writing of the manuscript, or in the decision to publish the results.

References

1. Galwey, N.W. The potential of quinoa as a multi-purpose crop for agricultural diversification: A review. *Ind. Crops Prod.* **1992**, *1*, 101–106. [CrossRef]
2. Alandia, G.; Rodriguez, J.P.; Jacobsen, S.E.; Bazile, D.; Condori, B. Global expansion of quinoa and challenges for the Andean region. *Glob. Food Sec.* **2020**, *26*, 100429. [CrossRef]
3. Jacobsen, S.E. The scope for adaptation of quinoa in Northern Latitudes of Europe. *J. Agron. Crop Sci.* **2017**, *203*, 603–613. [CrossRef]
4. Präger, A.; Munz, S.; Nkebiwe, P.M.; Mast, B.; Graeff-Hönninger, S. Yield and quality characteristics of different quinoa (Chenopodium quinoa willd.) cultivars grown under field conditions in southwestern Germany. *Agronomy* **2018**, *8*, 197. [CrossRef]
5. Pineli, L.L.O.; Botelho, R.B.A.; Zandonadi, R.P.; Solorzano, J.L.; de Oliveira, G.T.; Reis, C.E.G.; Teixeira, D.d.S. Low glycemic index and increased protein content in a novel quinoa milk. *LWT-Food Sci. Technol.* **2015**, *63*, 1261–1267. [CrossRef]
6. Bhargava, A.; Shukla, S.; Ohri, D. Chenopodium quinoa—An Indian perspective. *Ind. Crops Prod.* **2006**, *23*, 73–87. [CrossRef]
7. CBI The European Market Potential for Quinoa. Available online: https://www.cbi.eu/market-information/grains-pulses-oilseeds/quinoa-grains/market-potential (accessed on 31 August 2021).
8. Craine, E.B.; Murphy, K.M. Seed Composition and Amino Acid Profiles for Quinoa Grown in Washington State. *Front. Nutr.* **2020**, *7*, 126. [CrossRef] [PubMed]
9. Mastebroek, H.D.; Van Loo, E.N.; Dolstra, O. Combining ability for seed yield traits of Chenopodium quinoa breeding lines. *Euphytica* **2002**, *125*, 427–432. [CrossRef]
10. Patiranage, D.S.R.; Asare, E.; Maldonado-Taipe, N.; Rey, E.; Emrani, N.; Tester, M.; Jung, C. Haplotype variations of major flowering time genes in quinoa unveil their role in the adaptation to different environmental conditions. *Plant. Cell Environ.* **2021**, *44*, 2565–2579. [CrossRef]
11. Vilcacundo, R.; Hernández-Ledesma, B. Nutritional and biological value of quinoa (Chenopodium quinoa Willd.). *Curr. Opin. Food Sci.* **2017**, *14*, 1–6. [CrossRef]
12. Bhargava, A.; Srivastava, S. *Quinoa: Botany, Production and Uses*; CABI: Wallingford, UK, 2013; ISBN 1780642261.
13. Zevallos, V.F.; Ellis, H.J.; Šuligoj, T.; Herencia, L.I.; Ciclitira, P.J. Variable activation of immune response by quinoa (Chenopodium quinoa Willd.) prolamins in celiac disease. *Am. J. Clin. Nutr.* **2012**, *96*, 337–344. [CrossRef] [PubMed]
14. Bazile, D.; Bertero, H.D.; Nieto, C. *State of the Art Report on Quinoa around the World in 2013*; FAO: Rome, Italy; CIRAD: Rome, Italy, 2015.
15. Reguera, M.; Conesa, C.M.; Gil-Gómez, A.; Haros, C.M.; Pérez-Casas, M.Á.; Briones-Labarca, V.; Bolaños, L.; Bonilla, I.; Álvarez, R.; Pinto, K.; et al. The impact of different agroecological conditions on the nutritional composition of quinoa seeds. *PeerJ* **2018**, *6*, e4442. [CrossRef]
16. Aluwi, N.A.; Murphy, K.M.; Ganjyal, G.M. Physicochemical characterization of different varieties of quinoa. *Cereal Chem.* **2017**, *94*, 847–856. [CrossRef]
17. Bhargava, A.; Shukla, S.; Ohri, D. Genetic variability and interrelationship among various morphological and quality traits in quinoa (Chenopodium quinoa Willd.). *Field Crops Res.* **2007**, *101*, 104–116. [CrossRef]
18. Murphy, K.M.; Bazile, D.; Kellogg, J.; Rahmanian, M. Development of a Worldwide Consortium on Evolutionary Participatory Breeding in Quinoa. *Front. Plant Sci.* **2016**, *7*, 608–614. [CrossRef]
19. Granado-Rodríguez, S.; Aparicio, N.; Matías, J.; Pérez-Romero, L.F.; Maestro, I.; Gracés, I.; Pedroche, J.J.; Haros, C.M.; Fernandez-Garcia, N.; Navarro del Hierro, J.; et al. Studying the Impact of Different Field Environmental Conditions on Seed Quality of Quinoa: The Case of Three Different Years Changing Seed Nutritional Traits in Southern Europe. *Front. Plant Sci.* **2021**, *12*, 649132. [CrossRef] [PubMed]
20. Crop Production—Protein2Food. Available online: https://www.protein2food.eu/research-activities/crop-production/ (accessed on 29 September 2021).
21. Bertero, H.D.; De La Vega, A.J.; Correa, G.; Jacobsen, S.E.; Mujica, A. Genotype and genotype-by-environment interaction effects for grain yield and grain size of quinoa (Chenopodium quinoa Willd.) as revealed by pattern analysis of international multi-environment trials. *Field Crops Res.* **2004**, *89*, 299–318. [CrossRef]
22. Pulvento, C.; Riccardi, M.; Lavini, A.; D'Andria, R.; Iafelice, G.; Marconi, E. Field Trial Evaluation of Two Chenopodium quinoa Genotypes Grown Under Rain-Fed Conditions in a Typical Mediterranean Environment in South Italy. *J. Agron. Crop Sci.* **2010**, *196*, 407–411. [CrossRef]
23. Miranda, M.; Vega-Gálvez, A.; Martinez, E.; López, J.; Rodríguez, M.J.; Henríquez, K.; Fuentes, F. Genetic diversity and comparison of physicochemical and nutritional characteristics of six quinoa (Chenopodium quinoa willd.) genotypes cultivated in Chile. *Food Sci. Technol.* **2012**, *32*, 835–843. [CrossRef]
24. Gomez-Pando, L. Quinoa breeding. *Quinoa Improv. Sustain. Prod.* **2015**, *42*, 87–108.
25. Wu, G.; Morris, C.F.; Murphy, K.M. Evaluation of texture differences among varieties of cooked Quinoa. *J. Food Sci.* **2014**, *79*, S2337–S2345. [CrossRef]
26. Koziol, M.J. Afrosimetric estimation of threshold saponin concentration for bitterness in quinoa (Chenopodium quinoa Willd). *J. Sci. Food Agric.* **1991**, *54*, 211–219. [CrossRef]

27. Wieme, R.A.; Carpenter-Boggs, L.A.; Crowder, D.W.; Murphy, K.M.; Reganold, J.P. Agronomic and economic performance of organic forage, quinoa, and grain crop rotations in the Palouse region of the Pacific Northwest, USA. *Agric. Syst.* **2020**, *177*, 102709. [CrossRef]
28. Peterson, A.; Jacobsen, S.E.; Bonifacio, A.; Murphy, K. A crossing method for Quinoa. *Sustainability* **2015**, *7*, 3230–3243. [CrossRef]
29. Sobota, A.; Świeca, M.; Gęsiński, K.; Wirkijowska, A.; Bochnak, J. Yellow-coated quinoa (Chenopodium quinoa Willd)—Physicochemical, nutritional, and antioxidant properties. *J. Sci. Food Agric.* **2020**, *100*, 2035–2042. [CrossRef]
30. Hinojosa, L.; González, J.A.; Barrios-Masias, F.H.; Fuentes, F.; Murphy, K.M. Quinoa abiotic stress responses: A review. *Plants* **2018**, *7*, 106. [CrossRef]
31. Jarvis, D.E.; Ho, Y.S.; Lightfoot, D.J.; Schmöckel, S.M.; Li, B.; Borm, T.J.A.; Ohyanagi, H.; Mineta, K.; Michell, C.T.; Saber, N.; et al. The genome of Chenopodium quinoa. *Nature* **2017**, *542*, 307–312. [CrossRef]
32. Medina-Meza, I.G.; Aluwi, N.A.; Saunders, S.R.; Ganjyal, G.M. GC–MS Profiling of Triterpenoid Saponins from 28 Quinoa Varieties (Chenopodium quinoa Willd.) Grown in Washington State. *J. Agric. Food Chem.* **2016**, *64*, 8583–8591. [CrossRef] [PubMed]
33. Rodríguez Gómez, M.J.; Matías Prieto, J.; Cruz Sobrado, V.; Calvo Magro, P. Nutritional characterization of six quinoa (Chenopodium quinoa Willd) varieties cultivated in Southern Europe. *J. Food Compos. Anal.* **2021**, *99*, 103876. [CrossRef]
34. Gonzalez, J.A.; Konishi, Y.; Bruno, M.; Valoy, M.; Prado, F.E. Interrelationships among seed yield, total protein and amino acid composition of ten quinoa (Chenopodium quinoa) cultivars from two different agroecological regions. *J. Sci. Food Agric.* **2012**, *92*, 1222–1229. [CrossRef]
35. FAO. *Dietary Protein Quality Evaluation in Human Nutrition*; FAO: Rome, Italy, 2013; Volume 92, ISBN 978-92-5-107417-6.
36. Vidueiros, S.M.; Curti, R.N.; Dyner, L.M.; Binaghi, M.J.; Peterson, G.; Bertero, H.D.; Pallaro, A.N. Diversity and interrelationships in nutritional traits in cultivated quinoa (Chenopodium quinoa Willd.) from Northwest Argentina. *J. Cereal Sci.* **2015**, *62*, 87–93. [CrossRef]
37. Miranda, M.; Vega-Gálvez, A.; Martínez, E.A.; López, J.; Marín, R.; Aranda, M.; Fuentes, F. Influence of contrasting environments on seed composition of two quinoa genotypes: Nutritional and functional properties. *Chil. J. Agric. Res.* **2013**, *73*, 108–116. [CrossRef]
38. Tang, Y.; Li, X.; Chen, P.X.; Zhang, B.; Liu, R.; Hernandez, M.; Draves, J.; Marcone, M.F.; Tsao, R. Assessing the fatty acid, carotenoid, and tocopherol compositions of amaranth and quinoa seeds grown in Ontario and their overall contribution to nutritional quality. *J. Agric. Food Chem.* **2016**, *64*, 1103–1110. [CrossRef] [PubMed]
39. Pellegrini, M.; Lucas-Gonzales, R.; Ricci, A.; Fontecha, J.; Fernández-López, J.; Pérez-Álvarez, J.A.; Viuda-Martos, M. Chemical, fatty acid, polyphenolic profile, techno-functional and antioxidant properties of flours obtained from quinoa (Chenopodium quinoa Willd) seeds. *Ind. Crops Prod.* **2018**, *111*, 38–46. [CrossRef]
40. Simopoulos, A.P. The importance of the ratio of omega-6/omega-3 essential fatty acids. *Biomed. Pharmacother.* **2002**, *56*, 365–379. [CrossRef]
41. Fanali, C.; Beccaria, M.; Salivo, S.; Tranchida, P.; Tripodo, G.; Farnetti, S.; Dugo, L.; Dugo, P.; Mondello, L. Non-polar lipids characterization of Quinoa (Chenopodium quinoa) seed by comprehensive two-dimensional gas chromatography with flame ionization/mass spectrometry detection and non-aqueous reversed-phase liquid chromatography with atmospheric pressure chemical ionization mass spectrometry detection. *J. Sep. Sci.* **2015**, *38*, 3151–3160. [CrossRef] [PubMed]
42. Jahaniaval, F.; Kakuda, Y.; Marcone, M.F. Fatty acid and triacylglycerol compositions of seed oils of five Amaranthus accessions and their comparison to other oils. *JAOCS J. Am. Oil Chem. Soc.* **2000**, *77*, 847–852. [CrossRef]
43. Dziki, D.; Laskowski, J. Influence of kernel size on grinding process of wheat at respective grinding stages. *Pol. J. Food Nutr. Sci.* **2004**, *13*, 29–33.
44. De Bock, P.; Van Bockstaele, F.; Raes, K.; Vermeir, P.; Van der Meeren, P.; Eeckhout, M. Impact of tempering process on yield and composition of quinoa flour. *LWT-Food Sci. Technol.* **2021**, *140*, 110808. [CrossRef]
45. Abbas, G.; Amjad, M.; Saqib, M.; Murtaza, B.; Asif Naeem, M.; Shabbir, A.; Murtaza, G. Soil sodicity is more detrimental than salinity for quinoa (Chenopodium quinoa Willd.): A multivariate comparison of physiological, biochemical and nutritional quality attributes. *J. Agron. Crop Sci.* **2021**, *207*, 59–73. [CrossRef]
46. Englyst, H.N.; Kingman, S.M.; Cummings, J.H. Classification and measurement of nutritionally important starch fractions. *Eur. J. Clin. Nutr.* **1992**, *46*, S33–S50. [CrossRef]
47. FAO; WHO. *Energy and Protein Requirements*; FAO: Rome, Italy; WHO: Geneva, Switzerland, 1973.
48. Foubert, I.; Dewettinck, K.; Huyghebaert, A. Modelling Isothermal Cocoa Butter Crystallization: Influence of Temperature and Chemical Composition. Ph.D. Thesis, Ghent University, Ghent, Belgium, 2003.
49. Rombaut, R.; De Clercq, N.; Foubert, I.; Dewettinck, K. Triacylglycerol analysis of fats and oils by evaporative light scattering detection. *JAOCS J. Am. Oil Chem. Soc.* **2009**, *86*, 19–25. [CrossRef]
50. Bates, D.; Mächler, M.; Bolker, B.; Walker, S. Fitting Linear Mixed-Effects Models Using lme4. *J. Stat. Softw.* **2015**, *67*, 1–48. [CrossRef]
51. Akaike, H. A new look at the statistical model identification. *IEEE Trans. Automat. Contr.* **1974**, *19*, 716–723. [CrossRef]

plants

MDPI

Review

Quinoa (*Chenopodium quinoa* Wild.) Seed Yield and Efficiency in Soils Deficient of Nitrogen in the Bolivian Altiplano: An Analytical Review

Jesús E. Cárdenas-Castillo [1,2], José Delatorre-Herrera [2,*], Luisa Bascuñán-Godoy [3] and Juan Pablo Rodriguez [4]

[1] Engineering Agriculture Department, Natural Sciences and Agriculture Faculty, Universidad Técnica de Oruro, Oruro 49, Bolivia; jecar65@hotmail.com
[2] Doctoral Program in Agriculture for Arid-Desert Environments, Faculty of Renewable Natural Resources, Desert Agriculture Area, Universidad Arturo Prat, Iquique 1100000, Chile
[3] Botany Departament, Faculty of Natural and Oceanographic Sciences, Universidad de Concepción, Concepción 4030000, Chile; lubascun@udec.cl
[4] Julius Kühn Institute (JKI)—Federal Research Centre for Cultivated Plants, Institute for Plant Protection in Horticulture and Forests, Messeweg 11/12, 38104 Braunschweig, Germany; juan-pablo.rodriguez-calle@julius-kuehn.de
* Correspondence: jodelato@unap.cl

Citation: Cárdenas-Castillo, J.E.; Delatorre-Herrera, J.; Bascuñán-Godoy, L.; Rodriguez, J.P. Quinoa (*Chenopodium quinoa* Wild.) Seed Yield and Efficiency in Soils Deficient of Nitrogen in the Bolivian Altiplano: An Analytical Review. *Plants* **2021**, *10*, 2479. https://doi.org/10.3390/plants10112479

Academic Editors: Cataldo Pulvento and Didier Bazile

Received: 25 August 2021
Accepted: 10 November 2021
Published: 16 November 2021

Abstract: Quinoa is a strategic crop due to its high N content and its adaptability to adverse conditions, where most of the soils are deficient of nitrogen (N). The central question in this review was the following: How can quinoa yield low levels of nitrogen in the soils of Altiplano? This question was unraveled based on different factors: (1) fertilization effect on productivity, (2) fertilization limits, (3) uptake and assimilation of nitrogen parameters, (4) monoculture practice effect, and (5) possible sources and strategies. One hundred eleven articles of different scientific platforms were revised and data were collected. Information from articles was used to calculate the partial factor productivity for nitrogen (PFPN), the apparent use efficiency of N ($APUE_N$), available nitrogen (AN), and nitrogen content harvested in grains (HarvN). Quinoa responds positively to fertilization, but differences in yield were found among irrigated and rainfed conditions. Quinoa can produce 1850 kg grains ha^{-1} with 50 kg N ha^{-1} under irrigated conditions, and 670 kg grains ha^{-1} with 15 kg N ha^{-1} in rainfed conditions. Quinoa increases seed yield and HarvN increases N fertilization, but decreases nitrogen efficiency. In Altiplano, without nitrogen fertilizer, the quinoa yield relies on between 500 and 1000 kg ha^{-1}, which shows that in the soil, there are other nitrogen sources.

Keywords: quinoa; nitrogen harvested by yield; apparent use efficiency of N; arid environments; Altiplano

1. Introduction

Integrated nutrient management for food production is an approach and paradigm that supports the food security, conservation, and sustainability of renewable natural resources [1]. Understanding nutrient cycles is essential for improving crop nutritional management. Particularly, in highland and arid agroecosystems such as the southern Bolivian Altiplano, nitrogen (N) supply limits plant growth and development [2]. No other element for life, such as nitrogen, takes so many chemical forms in the atmosphere, soil, and plants [3]. In the atmosphere, the most reactive are N and gas (N_2), while in soil, nitrogen oxide, NO, and nitrogen dioxide (NO_2) prevail; when fertilizer is used, forms such as ammonia (NH_3) can be found; while in water, nitrogen can be present in inorganic forms such as ammonia, ammonium, nitrate, and nitrite, and the organic form is present in proteins, amino acids, urea, and living or dead organisms [4].

In semi-arid and arid land regions, water resources are limited and have significant consequences on the soil nitrogen content [4]. The seasonal distribution of rainfall can affect the accumulation and emission of N in soils during the dry season [5,6]. Nitrogen is accumulated in the soil as wet and dry, and part of it is released to the atmosphere when pore spaces in the soil are filled with water, but this process depends on the soil type and climate [7].

Nitrogen use efficiency (NUE) determination in fragile soils such as the southern Bolivian Altiplano is significant for understanding soil NO_3–N converted into grain for quinoa (*Chenopodium quinoa* Wild), a rainfed crop. NUE can be expressed in several ways: grain production by unit of available N, or index of utilization, which is the absolute quantity of produced biomass per unit of available N [8]. The factors that influence this efficiency are edaphic structure, climatic conditions, interactions between soil and bacterial processes, nature of organic and inorganic nitrogen sources, and availability of N in the soil [9,10]. NUE denotes the relationships between total input compared to the nitrogen output. This is complex and involves absorption, metabolism, and redistribution in the plant. However, adopting a complete crop nutrition strategy allows efficiency, profitability, and sustainability to improve. NUE is a determined metric used to measure N management in the soil [11]. Moreover, NUE is the maximum economic yield produced per unit of N applied, absorbed, or utilized by the plant to produce grain and straw [12]. NUE is partitioned in two processes: (a) absorption efficiency, when the plant is able to remove the available N from the soil usually present as nitrate or ammonium ions, and (b) utilization efficiency, when the plant is able to transfer the available N to the grain as protein [13]. The absorption efficiency is of the utmost importance for predicting plant performance and yield. Most plants capture inorganic N as dissolved nitrate (NO_3^-) or ammonium (NH_4^+) from the soil through their roots [14]. Root architecture, morphology, rate of respiration, and transporter activity for available forms of N in the rhizosphere determine N uptake rate. The utilization efficiency requires the process of carbon fixation for nitrogen taken up, photosynthesis, canopy formation, and nutrient remobilization from all tissues to grain during seed filling [10]. The process is initiated once N is introduced into the plant cell and is reduced into organic molecules.

Quinoa is a strategic partner crop for food security as a plant-based protein source [15], and its adaptability to unfavorable growing conditions [16,17]. Quinoa is an Amarantaceae with an intermediate protein content, less than that of legumes and more than that of cereals [17]. The protein content in grain depends on the varieties and soil conditions, and it can reach up to 23%. This protein level requires a significant supply of nitrogen, which is not only essential for the grain, but also for plant growth and development. Its exceptional adaptations to limiting factors in the environment are tolerance to drought [18] and frost [19], as well as to saline and/or low-fertility soils, maintaining adequate yields [20,21]. The Intersalar region, in the southern part of the Bolivian Altiplano has an extreme climate, with rainfall from 150 to 300 mm per year, 200 days of night frost, strong winds, and intense solar radiation [22–24]. Furthermore, it has serious degradation problems and low nitrogen levels in the soil. Soil degradation is attributed to monoculture, the use of virgin soils to expand the agricultural frontier, the use of inadequate agricultural machinery (disc plough) in highly susceptible soils to wind erosion, traditional and manual harvesting, the use of left-harvested plants after grain threshing in camelid cattle, llama (*Lama glama*) or sheep (*Ovis aries*) feeding, neglect of traditional sowing in the traditional system of sectoral fallowing known as "mantos" [25], the practice of soil fallowing (two to three years without agriculture), and the lack of organic matter due to little or no incorporation of manure (reduction of llama and sheep livestock) and stubble leftover.

In the Intersalar, it has been observed that there are plots with more than 80 agricultural years of production under quinoa monoculture. There are no contributions of manure or other nitrogen mineral fertilizer sources, and the only form of cultural management is the practice of soil fallow (one to two years) [26]. However, acceptable and economically sustainable productivity is still utilized by farmers, and the yields are between 450 and 750 kg ha^{-1}, despite no application of nitrogen [27–30]. Due to their origin, and as a consequence of the abovementioned factors, the soils of the Intersalar zone are poor in N. Of these soils, 98% are classified as very low in N, while the remaining soils (2%) are classified as low [26,31]. With this position, we ask ourselves, how can quinoa be produced in the Bolivian Altiplano under low levels of nitrogen in the soil? This question was unraveled based on different factors: (1) the effect of fertilization on productivity under rainfed and irrigated agricultural conditions, (2) the top and bottom limits of fertilization, (3) the parameters related to the uptake and assimilation of N, and (4) the effect of monoculture on yield under rainfed agricultural conditions. This article aimed to explore theoretically the efficiency of nitrogen under rainfed and irrigated cultivation conditions to allow us to explain the performance of quinoa in Bolivian Altiplano soils without the application of fertilizer.

2. Results

2.1. The Effect of Fertilization on Productivity in Rainfed and Irrigated Quinoa Cultivation

Nitrogen fertilization in quinoa is an unsolved issue; the literature data show great variability in results, ranging from very low application (30 kg of N and production of 5.5 tons of grains per ha^{-1} [32]) to high application (175 kg ha^{-1} with 4.2 tons of grains per ha^{-1} [33]) (Table 1). The variation in the results regarding nitrogen fertilization over seed yield expressed in Table 1 are the consequences on cropping conditions, varieties, soil textures, irrigation systems, and management.

Table 1. Nitrogen dose and yield of grain in various quinoa cultivars under different growing conditions and soil textures.

Dose (kg ha^{-1})	Seed Yield (kg ha^{-1})	Cultivar	Soil Texture	Irrigation Type	Reference
0	1166	Blanca Junin	Sandy clay loam–sandy loam	Rainfed	Borda, 2018 [34]
0	1100	Regalona Baer	Silty clay	Rainfed	Campillo and Contreras, 2019 [35]
40	2093	KVL 8401	Clay loam	Rainfed	Jacobsen et al., 1994 [36]
80	2428	KVL 8401	Clay loam	Rainfed	Jacobsen et al., 1994 [36]
80	2140	Regalona Baer	Silty clay	Rainfed	Campillo and Contreras, 2019 [35]
120	3500	Cochabamba y Faro	Clay loam	Rainfed	Schulte et al., 2005 [37]
120	2685	KVL 8401	Clay loam	Rainfed	Jacobsen et al., 1994 [36]
160	2760	KVL 8401	Clay loam	Rainfed	Jacobsen et al., 1994 [36]
160	3000	Regalona Baer	Silty clay	Rainfed	Campillo and Contreras, 2019 [35]
240	3360	Regalona Baer	Silty clay	Rainfed	Campillo and Contreras, 2019 [35]
320	3540	Regalona Baer	Silty clay	Rainfed	Campillo and Contreras, 2019 [35]
400	3430	Regalona Baer	Silty clay	Rainfed	Campillo and Contreras, 2019 [35]
0	1068	Faro and UdeC10	Loam–silty loam	Supplementary	Berti el al., 2000 [38]
0	1700	Altiplano INIA, Salcedo INIA	Sandy loam	Supplementary	Mendoza Nieto et al., 2016 [39]
0	1868	Blanca Real	Sandy loam	Dripping	Llaca, 2014 [40]
0	981	Genotipo O3	Loamy	Surface	Franco, 2018 [41]
50	1848	Genotipo O3	Loamy	Surface	Franco, 2018 [41]
75	2112	Faro and UdeC10	Loam–silty loam	Supplementary	Berti et al., 2000 [38]
80	2240	Blanca Real	Sandy loam	Dripping	Llaca, 2014 [40]
100	2700	Altiplano INIA, Salcedo INIA	Sandy loam	Supplementary	Mendoza Nieto et al., 2016 [39]

Table 1. *Cont.*

Dose (kg ha^{-1})	Seed Yield (kg ha^{-1})	Cultivar	Soil Texture	Irrigation Type	Reference
100	2267	Genotipo O3	Loamy	Surface	Franco, 2018 [41]
120	3300	Titicaca	Sandy loam	Deficit irrigation	Razzaghi et al., 2012 [20]
120	3000	Titicaca	Sandy clay loam	Deficit irrigation	Razzaghi et al., 2012 [20]
120	2300	Titicaca	Sandy	Deficit irrigation	Razzaghi et al., 2012 [20]
150	2456	Faro and UdeC10	Loam–silty loam	Supplementary	Berti et al., 2000 [38]
150	2541	Genotipo O3	Loamy	Surface	Franco, 2018 [41]
180	3413	Salcedo INIA	Sandy loam	Dripping	Herreros, 2018 [42]
200	2800	Altiplano INIA, Salcedo INIA	Sandy loam	Supplementary	Mendoza Nieto et al., 2016 [39]
200	1659	Genotipo O3	Loamy	Surface	Franco, 2018 [41]
225	2912	Faro and UdeC10	Loam–silty loam	Supplementary	Berti et al., 2000 [38]
240	3240	Blanca Real	Sandy loam	Dripping	Llaca, 2014 [40]
270	4249	SalcedoINIA	Sandy loam	Dripping	Herreros, 2018 [42]
300	2600	Altiplano INIA, Salcedo INIA	Sandy loam	Supplementary	Mendoza Nieto et al., 2016 [39]
360	3783	Salcedo INIA	Sandy loam	Dripping	Herreros, 2018 [42]
400	2100	Altiplano INIA, Salcedo INIA	Sandy loam	Supplementary	Mendoza Nieto et al., 2016 [39]

Source: Prepared with data from several authors.

Table 1 shows the performance of several quinoa varieties cultivated even under the same nitrogen fertilizer rate, which influenced seed yield under rainfed and irrigated conditions. We can see in the table that not only did the nitrogen fertilization improve the seed yield, but irrigation, water quality, organic matter content, pests, weeds, planting density, and varieties also strongly influenced the seed yield. The average data for each dose of fertilizers presented in Table 1 were utilized to calculate and depict seed yield variation under irrigated and rainfed conditions in Figure 1.

From the data obtained, we can infer that these conform to a normal curve, with an increase in yield up to 240 kg ha^{-1} for applications of nitrogen fertilizer, after which it begins to become asymptotic and the yields begin to decrease after 300 kg ha^{-1} nitrogen fertilizers. The relationship between nitrogen uptake and nitrogen fertilizer rate (R^2 = 0.64, $p = 6.7 \times 10^{-8}$) was remarkably consistent (Figure 1).

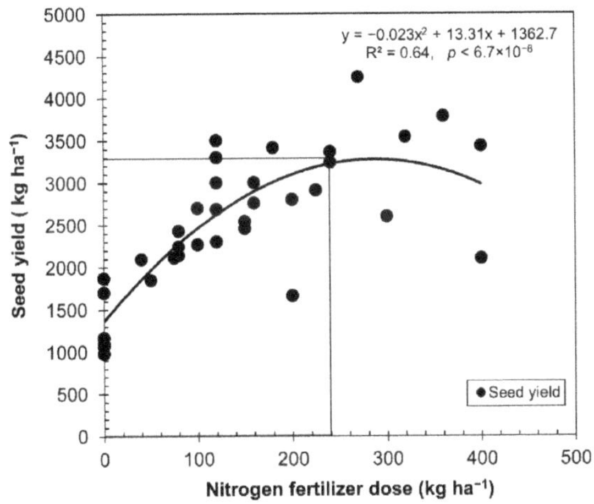

Figure 1. Relationship between nitrogen rates applied and seed yield. The data from Table 1 were utilized to calculate variations in seed yield in quinoa cultivated under rainfed and irrigated conditions. The symbols represent the average values of equal doses.

2.2. What Happens in the Altiplano Agroecosystem with a Quinoa Monoculture and without Nitrogen Fertilizer Applications under Rainfed Conditions?

The previous data and figure presented in the above section support the possibility that the performance of quinoa can increase under rainfed conditions, e.g., in the southern Bolivian Altiplano, which is an agroecosystem featuring sandy soils and a very low nitrogen content [26], with minimal use of manure and no chemical fertilizers. Figure 2 depicts the intensity of land use through several growing seasons in a plot exclusively for quinoa cultivation. The N available or contained in the soil decreased for the 2017 growing season. The curve shows the relationship of a monoculture practice in the same plot for several years, which decreased the nitrogen in the soil ($R^2 = 0.63$, $p = 1.2 \times 10^{-4}$) in the Intersalar area of the southern Bolivian Altiplano [26]. These data explain that in 90.9% of the farms, and according to the FAO (2013) [43], the soil nitrogen content is classified from very poor to poor (less than 0.15% N), with only 9.1% of the soils having a medium content (0.15–0.25% N).

Figure 2 shows the dynamics of total soil nitrogen and seed yield for several growing seasons. The data and curves presented in Figure 2 are based on 113 plots, with different years of quinoa monoculture use ($R^2 = 0.41$, $p = 0.009$), and only soil fallow of the plot without incorporation of organic matter or fertilizers and rainfed irrigation have been practiced [26]. Quinoa cultivation in the plots is practiced through the traditional Aymara system known as "mateado" or "speckled" and planted in holes ("kollas"), with planting densities of 1 × 1 m and three to four plants per "kolla" (i.e., planting rate). Figure 2 shows higher yields than 500 kg ha^{-1}; however, 60.9% of the farms evaluated do not exceed 1000 kg ha^{-1}, which is surprising due to the low nitrogen content of the soil (Figure 2), not exceeding 0.10–79.6% in the farms. Quinoa cultivated in the southern Altiplano of Bolivia, as through the case in Figure 2, denotes that in monoculture practice, the total soil nitrogen decreases over time, while seed production does not disappear, even though soil has a very low nitrogen content.

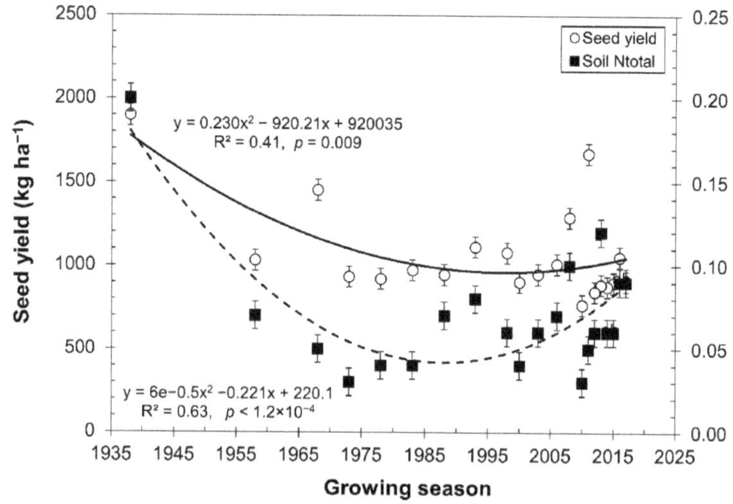

Figure 2. Seed yield and total soil nitrogen content from plots where quinoa monoculture has been practiced for several growing seasons (n = 80 years). Data adapted from Cardenas et al. [26].

When associating N values to the yield (Figure 3), it was found that in the trials with nitrogen fertilizer application, the curve trend showed an increase in yield at a higher concentration of nitrogen in the soil, which is asymptotic starting from 0.14% of total soil nitrogen.

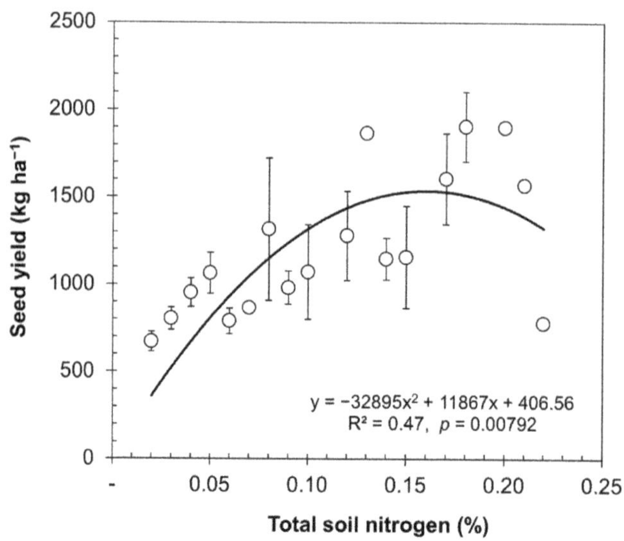

Figure 3. Relationship between seed yield and total soil N. The data were obtained by averaging the values of equal doses according to Table 1 (mean ± SE).

Figure 3 shows similar trends to Figure 2 for quinoa with fertilization, although Figure 3 depicts higher total soil nitrogen (R^2 = 0.47, p = 0.00792); however, the seed yield increased until it stabilized, and then it decreased. When analyzing the nitrogen use efficiency indicators (Table 2), deterioration can be observed.

Table 2. Determination of the efficiencies and indices according to the total soil nitrogen content in rainfed cultivation and non-fertilized soils.

Total Soil N (%)	Available Nitrogen (AN) (kg ha^{-1})	Yield Average (kg Grains ha^{-1})	Nitrogen Harvested by Yield (kg ha^{-1})	PFP_N (kg Grains kg^{-1}) (AN)	$APUE_N$ (%)
0.02	14.9	670.3	18.2	45.1	122.8
0.03	22.3	802.9	21.8	36.0	98.0
0.04	29.7	953.9	25.9	32.1	87.4
0.05	37.1	1063.6	28.9	28.6	77.9
0.06	44.6	789.2	21.5	17.7	48.2
0.07	51.9	865.5	23.5	16.7	45.3
0.08	59.4	1316.0	35.8	22.2	60.3
0.09	66.8	948.6	25.8	14.2	38.6
0.1	74.2	1070.4	29.1	14.4	39.2
0.12	89.1	1277.9	34.8	14.3	39.0
0.13	96.5	1865.9	50.8	19.3	52.6
0.14	103.9	1145.6	31.2	11.0	30.0
0.15	111.4	1156.1	31.4	10.4	28.2
0.17	126.2	604.9	43.7	12.7	34.6
0.18	133.7	1904.0	51.8	14.2	38.7
0.2	148.5	1899.2	51.7	12.8	34.8
0.21	155.9	1567.3	42.6	10.1	27.3
0.22	163.4	777.3	21.1	4.8	12.9

Source: Elaborated from Cardenas et al. [26].

According to Table 2, quinoa is more efficient at using nitrogen when there is less N in the soil. For example, in soils with very low nitrogen values such as 14.9 kg ha^{-1}, 45.1 kg of grain can be obtained for each kilogram of nitrogen, meaning an $APUE_N$ of 122.8%, which is difficult to explain without considering an extra contribution of nitrogen to the crop. However, soils with 0.22% of total soil nitrogen or 163 kg of available nitrogen produce only 4.8 kg of grain for every kilogram of available nitrogen.

3. Discussion

3.1. The Effect of Fertilization on Productivity in Irrigated and Rainfed Cultivation

The average data for each dose of fertilizer presented in Table 1 were used to determine the nitrogen uptake, expressed as kilograms of N to produce one ton of grain, as described in Figure 4. The relationship between nitrogen use efficiency and nitrogen fertilizer rate was remarkably consistent ($R^2 = 0.88$, $p = 2.2 \times 10^{-4}$).

From the data obtained, we can infer a normal curve, with an increase in nitrogen uptake per ton of produced grain, up to applications of 260 kg ha^{-1}, after which it started to become asymptotic, and over 400 kg ha^{-1}, the yields began to decrease as further nitrogen fertilizers were incorporated. In Figure 4, we demonstrate that nitrogen uptake increased when the nitrogen fertilizer rate increased from 35 kg of N per ton of with no nitrogen fertilizer, until reaching the optimum 90 kg of N per ton of produced grain with 260 kg N ha^{-1}. Alvar-Beltran et al. used three doses of nitrogen fertilization (25, 50, and 100 kg N ha^{-1}) and the extraction was 25 kg N per ton of grain produced (1:40 ratio) [44].

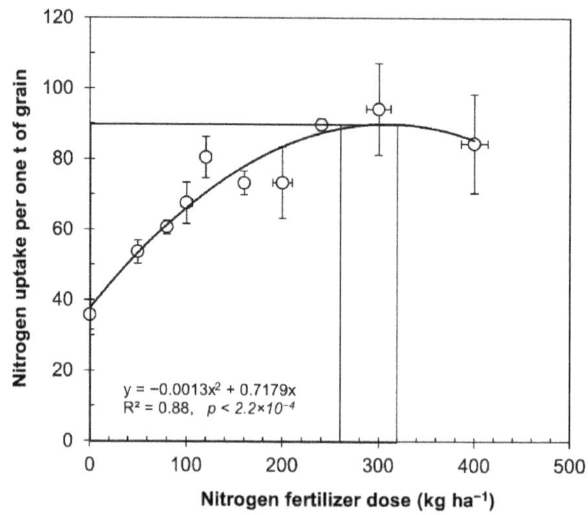

Figure 4. Relationship between nitrogen rates applied and uptake of nitrogen from experiments under irrigated and rainfed conditions. The data from Table 1 were utilized to calculate variations in seed yield in cultivated quinoa under rainfed and irrigated conditions. The symbols represent the average values of equal doses (mean ± SE).

3.2. *The Limits of Fertilization in Quinoa*

Table 3 shows the efficiency indicators based on the average yields from each nitrogen fertilization rate, according to the data and average values in Table 1.

Table 3. Efficiency indicators according to various fertilization tests under irrigated and rainfed conditions.

Nitrogen Fertilizer Dose (kg ha^{-1})	Average Yield (kg Grains ha^{-1})	Nitrogen Harvested by Yield (kg ha^{-1})	Partial Factor Productivity of Nitrogen (PFP_N) (kg Grains kg $^{-1}$) (AN)	Apparent Use Efficiency of N ($APUE_N$) (%)
50	1848	50.3	37.0	100.5
60	1771	48.2	29.5	80.3
75	2112	57.4	28.2	76.6
80	2314	62.9	28.9	78.7
100	2483	67.5	24.8	67.5
120	2749	74.8	22.9	62.3
150	2453	66.7	16.4	44.5
160	2882	78.4	18.0	49.0
180	3413	92.8	19.0	51.6
200	2193	59.6	11.0	29.8
225	2912	79.2	12.9	35.2
240	3300	89.8	13.8	37.4
300	2600	70.7	8.7	23.6
320	3540	96.3	11.1	30.1
360	3783	102.9	10.5	28.6
400	2765	75.2	6.9	18.8
Average	2695	73.3	19.7	50.9
R^2	0.83 *	0.88 **	0.83 **	0.77 *

Significant at the * 0.05 and ** 0.01 probability levels. Source: Calculated from Table 1.

By increasing the nitrogen fertilizer rates, the seed yield increased, reaching an optimum production at 130 kg of N ha^{-1}. However, after this point, the seed yield decreased, as depicted in Figure 5. Similar results were found in trials with a quinoa genotype O3 and two other cultivars [39,45]. The results were adjusted, with a good correlation to the law of diminishing returns ($R^2 = 0.83$, $p = 0.00513$) [45] and agreement with Pandey et al. [46], who indicated that high rates of nitrogen in crops cause a depressive effect.

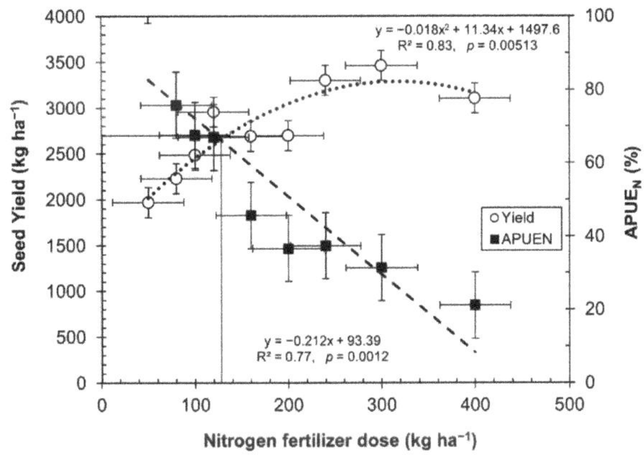

Figure 5. Break-even point between nitrogen fertilizer dose, seed yield, and $APUE_N$ (mean ± SE). These data were obtained from Table 2.

Figure 5 depicts the break-even point at which quinoa is efficient enough under certain levels of nitrogen fertilization. Based on the data in Table 3, Figure 5 shows a higher efficiency in the use of nitrogen in quinoa grown in soils even under very low nitrogen levels. A balance point appeared, indicating that up to 130 kg ha^{-1} of nitrogen is enough to produce 2700 kg ha^{-1}. This point is a balance between $APUE_N$ and seed yield, even when quinoa is grown with high rates of nitrogen fertilizer. There is a remarkable relationship between decreasing $APUE_N$ and an increasing amount of nitrogen fertilizer ($R^2 = 0.77$, $p = 0.0012$).

3.3. Parameters Related to the Uptake and Assimilation of Nitrogen

The PFP_N will depend on the physiological efficiency of the cultivar, that is, the proportion of available N absorbed by the crop, and on the losses during the cycle [43]. Nitrogen use efficiency averages 33% in cereals, indicating significant potential for improvement [47]. For quinoa grown on dry land and without an extra contribution of nitrogen fertilization, the PFP_N decreases when the nitrogen content in the soil increases (Table 3). With values for PFP_N ranging from 59.6 for very nitrogen-poor soil (0.02%) to 6.3 for very nitrogen-rich soil (0.22%), this means a loss of 89.4% in N efficiency.

In a trial with quinoa and five levels of N (0, 40, 80, 120, and 160 kg N ha^{-1}), the highest PFP_N was recorded with 40 kg N ha^{-1} and 30.52 kg of grains produced per kg of N applied [13]. Another study showed that the efficiency of the use of nitrogen in the yield of quinoa with N doses of 0, 50, 100, 150, and 200 kg N ha^{-1} was affected by higher availability of N in the soil [42]. The data in Figure 2 show a deterioration in PFP_N and $APUE_N$ when higher doses of fertilizers are included, although the yields increased. For example, applications of 40 kg N ha^{-1} produced 52 kg of grain for each kilogram of fertilizer applied. This is in contrast to doses of 160 kg N ha^{-1}, where only 17 kg of grain were produced for each kilogram of fertilizer applied, and with 400 kg N ha^{-1}, which produced only 5 kg of grains for each kilogram of nitrogen, that is, a 90.4% efficiency deterioration. The apparent use efficiency of nitrogen ($APUE_N$) shows that the applied or available nitrogen was multiplied 1.4 times in harvested grains. Differently, when 400 kg N ha^{-1} was applied, only 0.145 times the applied dose was harvested. The 1.4-fold increase in nitrogen content is striking, which could be explained by the presence of microorganisms or the contribution of rain, or by the deepening of the roots to increase the volume of soil to explore.

In addition, Figure 5 shows the balance point between yield and $APUE_N$, which means that the optimal ratio of yield versus nitrogen dose was around 130 kg N ha^{-1}.

The data obtained are consistent with those of Franco Alvarado (2018) [41], who applied up to 200 kg N ha^{-1}, finding that the absorption efficiency use of nitrogen ($APUE_N$) decreased as the applied dose of N increased. Without the application of nitrogen fertilizer, it reached the highest $APUE_N$. In contrast, upon application of 200 kg N ha^{-1}, the seed yield decreases. Franco Alvarado [41] found that the optimal dose of available N (62 kg N ha^{-1}) in the soil achieved the highest productivity in quinoa crop. This deterioration in the efficiency indicators indicates that increasing application of nitrogen fertilizers in quinoa is not used to produce grains, it could be derived from the production of biomass [48], or else there is a significant loss of this element by leaching. It has been estimated that between 50% and 70% of the applied nitrogen is lost from the soil–plant system, by surface runoff or leachate or by microbial denitrification, a process by which nitrate is converted to nitrogen oxides (N_2O and NO) and elemental nitrogen (N_2) is also lost by volatilization [44]. The loss of N by drainage (19.7 g N m^{-2}) represents the main output and the volatilization of urea (8.65 g N m^{-2}) [17].

The efficiency in nitrogen uptake and transfer to grains ($APUE_N$) explains the total nitrogen harvested in the grain compared to the total nitrogen uptake per ton of grain. Table 3 shows that plants with nitrogen deficiency stress have a higher $APUE_N$. The quinoa plants used the little available nitrogen better to produce grains, with a lower yield. The nitrogen-deficient plants showed a decrease in aerial and root biomass and a lower seed yield, but a greater efficiency in the use of nitrogen. Similarly, Calvache and Valle [48] found that as nitrogen increases, the aboveground biomass also increases (Table 4).

Table 4. Effect of nitrogen fertilizer dose application on the production of aboveground dry matter (kg ha^{-1}) in three quinoa varieties grown under irrigated conditions in Ecuador.

DAS	20	40	60	80	100	120
Nitrogen Fertilizer Dose (kg ha^{-1})						
0	166.6	712.7	1407.3	1835.0	3967.5	4524.8
75	183.4	948.9	2226.8	3650.8	7065.4	7832.9
150	221.7	1055.7	2659.7	5002.4	9943.7	11,366.1

Source: Adapted from Calvache and Valle [48]. DAS, days after sowing.

Our data resemble those of Alvar-Bertran et al. [44], who compared height and canopy in plants with seed yield. The highest seed yield was concentrated in plants of 40–60 cm with a 3–5% canopy. Calvache and Valle [48] compared the biomass produced by quinoa and seed yield as a function of the nitrogen dose under irrigated or rainfed condition (Figure 6). Unfortunately, the data only reached doses of 150 kg ha^{-1}, which did not allow one to establish, in higher doses, what the real behavior would be. Figure 6 shows that as the dose of nitrogen increased, the production of biomass also increased, while under rainfed and irrigation conditions, the rate of biomass production decreased.

Higher doses of nitrogen were derived by the quinoa plants to increase the above vegetative growth rather than to grain production (Figure 6), while decreasing the efficiency of nitrogen for grain production.

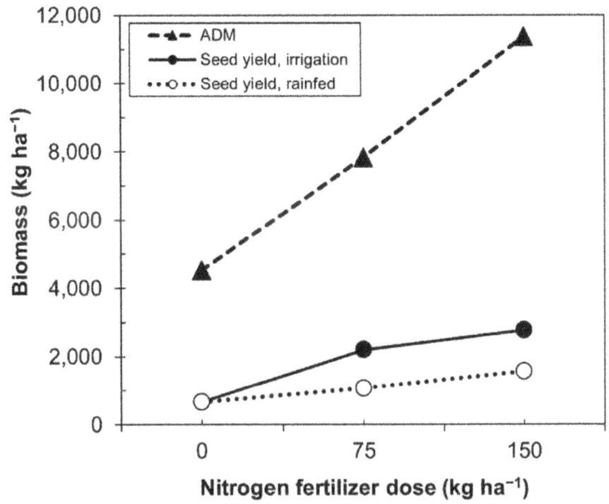

Figure 6. Relationship between the fertilizer dose and quinoa seed yield under rainfed and irrigation conditions. ADM, aboveground dry matter.

The accumulation and redistribution of N are important for the yield and quality of grain [49]. The supply of quinoa grain, like all seed-producing plants, depends on the N accumulated before anthesis. In wheat, approximately 50–95% of the N in grain at harvest comes from the remobilization of the N stored in the shoots and roots before anthesis [49,50]. In quinoa, these values have not been determined; however, data from Calvache and Valle [48] are depicted in Figure 7, where the nitrogen content in the panicle begins to increase from 80 days after sowing, while it remains the same in the stems and decreases in the leaves.

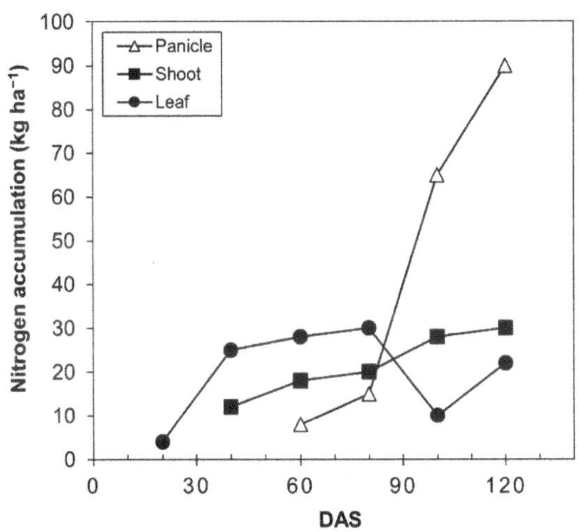

Figure 7. Nitrogen accumulation curve in quinoa plants (*Chenopodium quinoa* Willd) of the Imbaya variety grown under irrigated conditions. Source: Calvache and Valle [48]. DAS, days after sowing.

In all vegetable species, plant stress accelerates senescence and alters the source–sink relationship, resulting in a significant reduction in crop yield [51]. The southern Bolivian Altiplano has in average of 237 mm of annual rainfall (SENAMHI, Bolivia [52], (www.senamhi.gob.bo, accessed on 15 January 2021), and this is not enough for the cultivation of quinoa. Plants grow under conditions of water deficit, which, added to other environmental factors, creates a condition that accelerates plant senescence because of stress. Naturally, senescence induces the cessation of vegetative growth, accelerates the flowering and fruiting process, changes the plant metabolism, and alters the redistribution and partition of nutrients [53]. Stress senescence affects agronomic characteristics, including the efficiency and yield of carbohydrate/nitrogen use (C/N) and the C/N balance in the source–sink relationship [54,55]. The nitrogen remobilization efficiency (NRE; corresponds to the proportion of N in the crop) depends on the amount of N remobilized to the grain in the period after anthesis and the amount of N stored in the vegetative parts during anthesis. It is important to guarantee that stress senescence has not started prematurely, as nitrogen transportation into the grain will be affected [56]. After the plant takes up nitrogen and metabolizes it into plant proteins, this nitrogen is remobilized to the developing grain [57–59].

The growth of fruits and seeds indicates a new sink that competes with the rest of the plant for nutrients. At this point, the nitrogen partition process is important. Bascuñán-Godoy et al. [60] found that the total protein content in quinoa decreases with stress and increases when irrigated again. This decrease correlates with increases in NO_3^- and NH_4^+. The increase in NO_3^- could be associated with a marked stress-induced decrease in nitrate reductase (NR) activity, and the increase in NH_4^+ is probably associated more with the improvement in the protein degradation and re-assimilation processes of N [61]. Although, it can also be associated with the availability of water, which allows mobilization in the soil to the rhizosphere, improving nitrogen absorption and the presence of microorganisms that provide nitrogen to the plant.

3.4. *The Effect of Monoculture on Yielding in Non-Fertilizer Rainfed Cultivation in Bolivian Altiplano*

Our study demonstrated how a low soil nitrogen content, as in the southern Bolivian Altiplano, is associated with similar studies [44]. Of the Intersalar soils in the southern Bolivian Altiplano, 91% are sandy loam and sand [26]. The soil texture affects the availability of N by inducing the mineralization and the depth and distribution of the rooting system. Therefore, the application of 120 kg N ha^{-1} in plots with different soil textures results in differences in nitrogen absorption in quinoa: 134 kg ha^{-1} for sandy clay loam, 102 kg ha^{-1} for sandy loam, and 77 kg ha^{-1} for sand under full irrigation [20]. This situation is important, since the N from deep soil can be absorbed by diffusion and is an important part of the total absorption [62]. Based on applications of 25, 50, and 100 kg N ha^{-1} in a quinoa crop in Burkina Faso, Alvar-Beltran et al. [44] determined that the nitrogen concentration decreases from 0.051% to 0.037%, in depths from 20 to 60 cm, respectively, for applications of 25 kg ha^{-1}, while for 100 kg ha^{-1}, it decreases from 0.035% to 0.029%.

Under adequate water conditions, the quinoa seed yield increased with higher doses of N, as well as the harvest of N per hectare. In Figure 3, we show that the amount of N absorbed in the grain increased to 90 kg of N per ton of grain produced with 240 kg of nitrogen fertilizer, but the $APUE_N$ decreased from 100.5% with 50 kg of nitrogen fertilizer to 37% with 240 kg of nitrogen fertilizer (Table 3). When analyzing how the nitrogen deficit affects the conditions of the southern Bolivian highlands, from Table 2, it can be seen that very low values of soil total N (0.02%) equivalent to 14.4 kg N available ha^{-1} produce average yields of 670 kg of grain ha^{-1} with an $APUE_N$ of 122%, which is surprising, since it contained 22.8% more nitrogen than that provided by the soil. The maximum yield point was obtained with 0.13% of total N in soil, equivalent to 96.5 kg of nitrogen ha^{-1}; with this amount, 1866 kg grain ha^{-1} was produced, but the $APUE_N$ decreased to 52.6. However, these values are close to the equilibrium point shown in Figure 5 for fertilized and irrigated crops (130 kg N ha^{-1}). These values agree with those of Cassman et al. [57], who showed that a low content of N in the soil contributes to an increase in the efficiency of N.

3.5. Sources and Strategies to Improve N Supply and Efficiency in Quinoa in Non-Fertilized Soil in the Altiplano

3.5.1. Sources

Another source of soil N content comes from the atmosphere, which contains 79% by volume of nitrogen, making it a source of great reserve for the system, since it feeds the nitrogen cycle. In the Bolivian Altiplano, there is no history or records about the contribution of nitrogen by rainwater and the atmosphere [63]. The rainfall in the 2016 and 2017 seasons was between 194 and 280 mm year^{-1} (SENAMHI, Bolivia, www.senamhi.gob.bo, accessed on 15 January 2021). This low level of rainfall makes it difficult to evaluate the contribution of N. There are many controversies about the amount of N deposited through this way in soil. In temperate climates, it can fluctuate between 0.74 and 21 kg N ha^{-1} year^{-1} [64], and 15 kg N ha^{-1} year^{-1} [65] could be considered the average. These amounts would be higher in tropical climates, i.e., between 6.5 and 72 kg N ha^{-1} year^{-1} [64].

In the Bolivian Altiplano, electric shocks can be intense [66], but the amount of rain is much less. It is unknown whether the factor of electrical discharge and/or static electricity results in a higher nitrogen contribution, or why a scarcity of nitrogen is observed in Altiplano soils, which is significant in the soil nitrogen balance [63]. The electrical discharge that occurs during storms synthesizes nitrogen oxides from nitrogen (N_2) and oxygen (O_2) in the air, being driven into the ground by rain [67,68]. The quantity of nitrate produced across the world is estimated to 7.5 million ton per year.

3.5.2. Strategies

Interestingly, other parameters such as root biomass only correlate with the seed yield under low nitrate conditions, but not at sufficient levels of nitrate [69]. It has been published that root biomass is not important for the uptake of N [70], or even that plants cannot uptake N during grain filling [71]. However, Mi et al. [72] reported that root biomass is an important attribute for N uptake in corn at low nitrate levels (but not at sufficient nitrate levels). The roots of maize can take up N even during the reproductive phase [10]. Coke and Gallais [73] estimated that 62% of the N in the kernel originates from N remobilization, and 38% is derived from post-silking root N uptake. Recently, it has been reported that the increase in the number of secondary roots is related to the upregulation of nitrate transporter gene (Cq*NRT2*) under low nitrate conditions in quinoa seedlings of both Socaire (an Andean landrace) and Faro [74]. This indicates that a low amount of N induces in quinoa a series of mechanisms to cope with low N.

There are antecedents that relate nitrogen deficiency with other active compounds such as Strigolactones (SL). These hormones act by activating the signaling pathways that allow lipid catabolism to be the main carbon source in fungi. Under nutrient deprivation conditions, the production of large amounts of SL leads to the suppression of shoot branching and stimulates symbiosis [75,76]. Strigolactones promote the modification of the architecture of roots and shoots and stimulate a symbiosis of rhizobia bacteria and AMF fungi, and SLs play a crucial role in nitrogen and phosphorus deficiency.

Another of the strategies used by halophytes to capture nutrients is the association with soil microorganisms, especially arbuscular mycorrhizal fungi (AMF), which promotes growth and development under stressful conditions [77–79], and plant growth-promoting rhizobacteria (PGPR), with the ability to colonize the roots of many plant species, contributing to their development and survival [44].

The participation of arbuscular mycorrhizal fungi (AMF) in quinoa, a facultative halophyte, is debatable, since the presence of root symbiont fungi in Bolivian Andean quinoa plants is insignificant [80], and plant growth responses could be considered a mutualism–parasitism continuum [81]. However, some research, e.g., in the desert zone of Chile, has determined that there is a high presence of mycotrophic plant species with a high variation in the degree of mycorrhization in the root (mycorrhizal colonization and the mycorrhizal medium), through the production of resistance spores and extraradical mycelium [82]. Despite the low level of AMF colonization, it has been proposed that quinoa could be an interesting component for crops rotation to improve and increase N cycling in soils compared to other crops [83].

In quinoa, in particular, there are very few investigations on the presence of fungi and their contribution to growth or to withstand stressful conditions. The dominant fungal genera (*Penicillium*, *Phoma*, and *Fusarium*) have been detected in the roots of quinoa [84]; for example, Macia-Vicente et al. [85] and Khan et al. [86] previously found them as root inhabitants in several plant species. These fungal genera play a positive role in plant growth and tolerance to abiotic stress. The endophyte fungus community has been recognized as one of the Chilean quinoa ecotypes [84]. Despite a relatively high diversity of endophytic root fungi associated with quinoa plants, the dominant fungal community consists of only *Ascomycotaphyla*. The most abundant fungal genera in quinoa are *Penicillium*, *Phoma*, and *Fusarium*, which are common endophytes in plant roots, highlighting endophytic root fungi as a new additional performer [85].

Furthermore, there is a history of the participation of bacterial endophytes associated with quinoa [85,86]; 100% of quinoa seeds are inhabited by several bacteria from the genus *Bacillus* [85], which probably induces a state of natural readiness in quinoa plants, allowing them to overcome extreme environmental situations. Among the best-known microorganisms with PGPR activity are species of the genera *Rhizobium* sp., *Azospirillum* sp., and *Pseudomonas* sp. [87,88]. There are several mechanisms by which bacteria contribute to the germination, growth, and survival of plants, including biological nitrogen fixation, solubilization of phosphates, production of siderophores, biosynthesis of phytohormones (auxins, cytokines, and gibberellins), synthesis of antibiotics, and induction of systemic resistance [89,90]. Under low nitrogen concentrations, auxin biosynthesis and transcriptional accumulation are induced, thus regulating lateral root formation. Conversely, lateral root growth can be inhibited with a higher than optimal supply of N [91,92]. Nitrate transporter (NRT) genes have also been reported to be responsible for the high affinity of the NO_3^- transport system, which is related to the growth of lateral roots [74,93,94]. Based on the information provided, it is possible to assume that part of the nitrogen supply of quinoa in conditions of deficit of this element, is supplied by the interaction that occurs with these microorganisms.

4. Material and Methods

A systematic review (SR) was used to identify, select, and critically appraise the relevant primary information. The data included several studies and information extracted and analyzed according to the proposed methodology [95]. Throughout this review, the primary question was "How does quinoa obtain the necessary nitrogen for production under rainfed conditions with nitrogen deficiency?"

The databases Google Scholar, EBSCO, ISI Web of Science, Scopus, Elsevier Science Direct, Oxford Journals, Wiley Online Library, Springer, and Nature were used. Furthermore, project reports, university theses from national and international organizations such as FAO, and World Bank were used to find information. The keywords used were "quinoa," "yield," "Intersalar," "southern Altiplano," "efficient use of N," and "fertilization with N." The analysis focused on the Intersalar zone in the southern Bolivian Altiplano. A total of 1060 abstract articles were downloaded and read, and then 125 were reviewed, before finally selecting 54 articles. Finally, 10 articles were considered to determine the fertilization rates, as presented in Table 1. The average, standard deviation, and standard error values were calculated with INFOSTAT v2020 [96], and regression linear and scatter plots with OriginPro v2019 (OriginLab Corporation, Northampton, MA, USA) [97].

The collected data were evaluated with the indicators nitrogen use efficiency (NUE), partial factor productivity for nitrogen (PFP_N, kg grain kg^{-1} N_{fert}), and apparent use efficiency of N ($APUE_N$,%) according to the methodology described by Zhang et al. [98].

PFP_N is the ratio of crop yield per unit of nitrogen fertilizer applied by Zhang et al. and Kuosmanen [99,100], or that available in the soil, as per Equation (1):

$$PFP_N = Y_g / N_{fert}, \quad (1)$$

where Y_g and N_{fert} are the grain yield (kg ha^{-1}) and N_{fert} is the application rate of nitrogen or the available nitrogen (NA) under non-application of fertilizer conditions.

$APUE_N$ was calculated according to the recommendation by Al Naggar et al. [101], as per Equation (2):

$$APUE_N = (N_{kg}/N_{fert\ or\ NA}) \times 100 \quad (2)$$

where N_{kg} is the amount of N in quinoa grains (kg N grains ha^{-1}).

In soils without fertilizer, the data of total nitrogen (N_t, %) were used to calculate the available nitrogen (AN), and the results were transformed to kilograms per hectare (kg ha^{-1}) [102]. A weight of 1 ha of soil was determined, with a depth of 30 cm and an apparent density of 1.65 t m^{-3}. The density of the soils in the Intersalar zone is related to its texture. Therefore, 91% of the soils had sandy loam and sand texture with a bulk density of 1.4–1.9 g cm^{-3}, with a mean of 1.65 g cm^{-3}. For the estimation of AN, Equation (3) was used:

$$AN = Nt \times 0.015, \quad (3)$$

Soil nitrogen has two components, organic N and inorganic N. Soil microorganisms convert organic forms to inorganic forms, which plants absorb. It has been estimated that between 1.5% and 3% of the total N in the soil corresponds to inorganic N. We worked with 1.5% (or 0.015); with this, the available nitrogen (AN) was determined following the methodology of Aguilar et al. and Castellanos et al. [103,104].

The extracted or harvested nitrogen content ($Harv_N$) was estimated using an average value in quinoa of 16.6% protein, according to data taken from Abugoch et al. [105] and Covarrubias et al. [106], with a protein content of 16% nitrogen [107] and a conversion factor of 0.0272 used for the calculation of $Harv_N$. The nitrogen content harvested in grains ($Harv_N$) was determined with Equation (4):

$$Harv_N = \text{Yield grain} \times 0.0272 \quad (4)$$

5. Conclusions

Globally, quinoa cultivation is of interest due to its versatility, resilience, and provider of a nutritious grain. Nevertheless, quinoa cultivated in Andean countries, particularly in Bolivia are facing several constraints as low fertility in soils. Data obtained from this review provided better knowledge about increasing yield of quinoa (i) the optimal amount under these conditions was determined to be 240 kg of N with 3600 kg grain ha^{-1} (ii) the fertilization limits were obtained from the intersection of $APUE_N$ and seed yield, finding that the equilibrium point appeared to be up to 130 kg ha^{-1} of nitrogen, which is enough to produce 2700 kg ha^{-1}, (iii) the nitrogen uptake and assimilation parameters presented an inverse relationship, with higher doses of fertilizers showing a lower efficiency of nitrogen utilization in grain production. The information obtained shows that in the event of an increase in nitrogen fertilization, an important part of the N is destined for vegetative growth. (iv) Under monoculture practice without nitrogen fertilizers, as occurs in the Bolivian Altiplano, the response found in terms of yields is similar to the practice of irrigation and fertilization, although in much smaller proportions, in soils with very low nitrogen (0.02% total N), equivalent to 14.9 kg ha^{-1} N available. Meanwhile, 670 kg grains ha^{-1} and 45.1 kg were obtained for grains for each kilogram of available nitrogen, meaning an $APUE_N$ of 122%. The effect of monoculture on yield may be associated with a greater exploration of soil volumes when the roots grow very deeply, the expression of high-affinity N transporters and a series of orchestrated changes in the metabolism for the recycling of N. (v) These results, found in the Bolivian Altiplano, lead to the search for possible strategies, such as the distribution of N in plants and possible sources of N contribution. In addition, the contribution of nitrogen from rain and the presence of microorganisms such as AMF or PGPR and endophytes favor the uptake and fixation of N in the growth of quinoa roots.

Theoretical data from this review support the hypothesis that quinoa can produce grains in soil with a lower N content, as occurs in the Bolivian highlands. For optimal performance, we also recommend that organic amendments, endophytes (bacteria and fungi), and rotation with lupines should be considered to improve N in poor soils such as those of the southern Altiplano of Bolivia.

Author Contributions: Conceptualization, J.E.C.-C. and J.D.-H.; methodology, J.E.C.-C. and J.D.-H.; formal analysis, J.D.-H., J.P.R. and L.B.-G.; investigation, J.E.C.-C., J.P.R. and J.D.-H.; resources, J.E.C.-C.; data curation, J.E.C.-C., J.P.R., L.B.-G. and J.D.-H.; writing—original draft preparation, J.E.C.-C. and J.D.-H.; writing—review and editing, J.P.R. and L.B.-G.; visualization, J.D.-H. and J.P.R.; supervision, J.D.-H. All authors have read and agreed to the published version of the manuscript.

Funding: This research was funded by Doctoral Program in Desert and Arid Agriculture, Universidad Arturo Prat and project FONDEQUIP EQM1900088. LBG thanks to ANID Fondecyt Regular N° 1211473.

Institutional Review Board Statement: Not applicable.

Informed Consent Statement: Not applicable.

Data Availability Statement: The data presented in this study are available in this article.

Acknowledgments: We also appreciate the valuable contributions of the reviewers of this manuscript, they have been very useful to improve its quality.

Conflicts of Interest: The authors declare no conflict of interest.

References

1. Gutiérrez Castorena, E.V.; Gutiérrez Castorena, M.d.C.; Ortiz Solorio, C.A. Manejo integrado de nutrientes en sistemas agrícolas intensivos: Revisión. *Rev. Mex. Cienc. Agrícol.* **2015**, *6*, 201–215.
2. Kraiser, T.; Gras, D.E.; Gutierrez, A.G.; Gonzalez, B.; Gutierrez, R.A. A holistic view of nitrogen acquisition in plants. *J. Exp. Bot.* **2011**, *62*, 1455–1466. [CrossRef]
3. Robertson, G.P.; Groffman, P.M. Chapter 14: Nitrogen Transformations. In *Soil Microbiology, Ecology and Biochemistry*, 4th ed.; Paul, E.A., Ed.; Academic Press: Burlington, MA, USA, 2015; pp. 421–446. [CrossRef]
4. Delon, C.; Galy-Lacaux, C.; Boone, A.; Liousse, C.; Serça, D.; Adon, M.; Diop, B.; Akpo, A.; Lavenu, F.; Mougin, E.; et al. Atmospheric nitrogen budget in Sahelian dry savannas. *Atmos. Chem. Phys.* **2010**, *10*, 2691–2708. [CrossRef]
5. Austin, A.T.; Yahdjian, L.; Stark, J.M.; Belnap, J.; Porporato, A.; Norton, U.; Ravetta, D.A.; Schaeffer, S.M. Water pulses and biogeochemical cycles in arid and semiarid ecosystems. *Oecologia* **2004**, *141*, 221–235. [CrossRef] [PubMed]
6. Jaeglé, L.; Martin, R.V.; Chance, K.; Steinberger, L.; Kurosu, T.P.; Jacob, D.J.; Modi, A.I.; Yoboué, V.; Sigha-Nkamdjou, L.; Galy-Lacaux, C. Satellite mapping of rain-induced nitric oxide emissions from soils. *J. Geophys. Res. Atmos.* **2004**, *109*, 1–10. [CrossRef]
7. Galy-Lacaux, C.; Delon, C.; Solmon, F.; Adon, M.; Yoboué, V.; Mphepya, J.; Pienaar, J.; Diop, B.; Sigha, L.; Dungall, L.; et al. Dry and Wet Atmospheric Nitrogen Deposition in West Central Africa. In Proceedings of the 6th International Nitrogen Conference, Kampala, Uganda, 18 November 2013.
8. Karunarathne, S.D.; Han, Y.; Zhang, X.-Q.; Li, C. Advances in Understanding the Molecular Mechanisms and Potential Genetic Improvement for Nitrogen Use Efficiency in Barley. *Agronomy* **2020**, *10*, 662. [CrossRef]
9. Cabrera, M.L. Mineralización y nitrificación: Procesos claves en el ciclo del nitrógeno. In Proceedings of the Simposio Fertilidad 2007. IPNI Cono Sur-Fertilizar AC. Seminario Internacional de Nutricion Vegetal 2007. Rosario, Acassuso, Argentina, 10–11 May 2007; pp. 1–9.
10. Masclaux-Daubresse, C.; Daniel-Vedele, F.; Dechorgnat, J.; Chardon, F.; Gaufichon, L.; Suzuki, A. Nitrogen uptake, assimilation and remobilization in plants: Challenges for sustainable and productive agriculture. *Ann. Bot.* **2010**, *105*, 1141–1157. [CrossRef] [PubMed]
11. Congreves, K.A.; Otchere, O.; Ferland, D.; Farzadfar, S.; Williams, S.; Arcand, M.M. Nitrogen Use Efficiency Definitions of Today and Tomorrow. *Front. Plant Sci.* **2021**, *12*, 637108. [CrossRef]
12. Fageria, N.K.; Baligar, V.C. Enhancing Nitrogen Use Efficiency in Crop Plants. *Adv. Agron.* **2005**, 97–185. [CrossRef]
13. Grahmann, K.; Verhulst, N.; Buerkert, A.; Ortiz-Monasterios, I.; Govaerts, B. Nitrogen use efficiency and optimization of nitrogen fertilization in conservation agriculture. *CAB Rev.* **2013**, *53*, 6. [CrossRef]
14. Eshel, A.; Beeckman, T. *Plant Roots: The Hidden Half*, 4th ed.; CRC Press, Taylor and Francis Group: Abingdon, UK, 2013; p. 848. [CrossRef]
15. Alandia, G.; Rodriguez, J.P.; Jacobsen, S.E.; Bazile, D.; Condori, B. Global expansion of quinoa and challenges for the Andean region. *Glob. Food Secur.* **2020**, *26*, 100429. [CrossRef]

16. FAO/ORALC. *La Quinua: Cultivo Milenario para Contribuir a la Seguridad Alimentaria Mundial*; Oficina Regional para América Latina y el Caribe: Santiago, Chile, 2011.
17. Rodriguez, J.P.; Rahman, H.; Thushar, S.; Singh, R.K. Healthy and Resilient Cereals and Pseudo-Cereals for Marginal Agriculture: Molecular Advances for Improving Nutrient Bioavailability. *Front. Genet.* **2020**, *11*, 49. [CrossRef]
18. Geerts, S.; Raes, D.; Garcia, M.; Mendoza, J.; Huanca, R. Crop water use indicators to quantify the flexible phenology of quinoa (*Chenopodium quinoa* Willd.) in response to drought stress. *Field Crop. Res.* **2008**, *108*, 150–156. [CrossRef]
19. Jacobsen, S.E. Adaptación y Posibilidades para la Quinua en las Latitudes Septentrionales de Europa. In *Estado del Arte de la Quinua en el Mundo en 2013*; Bazile, D., Ed.; Food and Agriculture Organization of the United Nations: Rome, Italy, 2014; pp. 520–533.
20. Razzaghi, F.; Plauborg, F.; Jacobsen, S.-E.; Jensen, C.R.; Andersen, M.N. Effect of nitrogen and water availability of three soil types on yield, radiation use efficiency and evapotranspiration in field-grown quinoa. *Agric. Water Manag.* **2012**, *109*, 20–29. [CrossRef]
21. Rodriguez, J.P.; Ono, E.; Abdullah, A.M.S.; Choukr-Allah, R.; Abdelaziz, H. Cultivation of Quinoa (*Chenopodium quinoa*) in Desert Ecoregion. In *Emerging Research in Alternative Crops*; Hirich, A., Choukr-Allah, R., Ragab, R., Eds.; Environment & Policy; Springer International Publishing: Cham, Switzerland, 2020; Volume 58, pp. 145–161.
22. Geerts, S.; Raes, D.; Garcia, M.; Del Castillo, C.; Buytaert, W. Agro-climatic suitability mapping for crop production in the Bolivian Altiplano: A case study for quinoa. *Agric. For. Meteorol.* **2006**, *139*, 399–412. [CrossRef]
23. Andressen, R.; Monasterio, M.; Terceros, L.F. Regímenes climáticos del altiplano sur de Bolivia: Una región afectada por la desertificación. *Rev. Geogr. Venez.* **2007**, *48*, 11–32.
24. Winkel, T.; Alvarez-Flores, R.; Bommel, P.; Bourliaud, J.; Chevarria Lazo, M.; Cortes, G.; Cruz, P.; Del Castillo, C.; Gasselin, P.; Joffre, R.; et al. Altiplano Sur de Bolivia. In *Estado del Arte de la Quinua en el Mundo en 2013'*; Bazile, D., Bertero, D., Nieto, C., Eds.; FAO: Montpellier, France; CIRAD: Santiago, Chile, 2014; pp. 432–449.
25. Kerssen, T.M. Food sovereignty and the quinoa boom: Challenges to sustainable re-peasantisation in the southern Altiplano of Bolivia. *Third World Q.* **2015**, *36*, 489–507. [CrossRef]
26. Cárdenas, J.; Choque, W.; Guzmán, R. *Fertilidad, Uso y Manejo de Suelos en la Zona del Intersalar, Departamentos de: Oruro y Potosí*; Fundación FAUTAPO: La Paz, Bolivia, 2008; p. 113.
27. Ministerio de Desarrollo Rural y Tierras (MDRyT); Consejo Nacional de Comercializadores y Productores de Quinua (CONACO-PROQ). *Política Nacional de la Quinua*; Ministerio de Desarrollo Rural y Tierras (MDRyT): La Paz, Bolivia; Consejo Nacional de Comercializadores y Productores de Quinua (CONACOPROQ): La Paz, Bolivia, 2009; p. 134.
28. Echalar, M.; Torrico, J.; Martínez, F. Analisis de la sostenibilidad de la produccion de quinua (*Chenopodium quinoa*) en el intersalar Boliviano. *Cienciagro* **2011**, *2*, 303–312.
29. Miranda, R.; Carlesso, R.; Huanca, M.; Mamani, P.; Borda, A. Yield and nitrogen accumulation in quinoa (*Chenopodium quinoa* Willd.) produced with manure and supplementary irrigation. *Venesuelos* **2012**, *20*, 21–29.
30. Del Barco-Gamarra, M.T.; Foladori, G.; Soto-Esquivel, R. Insustentabilidad de la producción de quinua en Bolivia. *Estud. Sociales. Rev. Aliment. Contemp. Desarro. Reg.* **2019**, *29*, 2–26. [CrossRef]
31. Alejo Inda, R. Evaluación del Comportamiento del Nitrógeno, en Parcelas con Cultivo de Quinua bajo Diferente Manejo de Suelos (Municipios Salinas de Garci Mendoza), Oruro. Engineering Thesis, Universidad Mayor de San Andrés, Facultad de Agronomía, La Paz, Bolivia, 2010.
32. Delatorre, J.; Lanino, M.; Bar, Y. Ensayo de fertilización con nitrógeno y fósforo en quinua (*Chenopodium quinoa* Willd). *Rev. Agric. Desierto* **2001**, *2*, 31–38.
33. Rimac Mendez, L. Determinacion optima de niveles de fertilizacion para incrementar el rendimiento de quinua (*Chenopodium quinoa* W.) Cultivar Hualhuas en el Callejon de Huaylas. Engineering Thesis, Universidad Nacional Santiago Antúnez de Mayolo, Huaraz, Peru, 2015.
34. Borda López, F.M. Niveles de Urea y Guano de Isla con y sin Zeolita en el Rendimiento de Quinua (*Chenopodium quinoa* Willd) Canaria 3200 msnm-Ayacucho. Engineering Thesis, Universidad Nacional de San Cristobal de Huamanga, Ayacucho, Peru, 2018.
35. Campillo, R.; Contreras, G. Capitulo 4: Gestión nitrogenada y potásica del cultivo de quinoa en La Araucanía. In *Quinoa del Sur de Chile: Alternative Productiva y Agroindustrial de Alto Valor*; Diaz, S.J., Ed.; Colección Libros INIA-Instituto de Investigaciones Agropecuarias. Centro Regional de Investigacion Carillanca: Temuco, Chile, 2019; pp. 45–68.
36. Jacobsen, S.E.; Jørgensen, I.; Stølen, O. Cultivation of quinoa (*Chenopodium quinoa*) under temperate climatic conditions in Denmark. *J. Agric. Sci.* **2009**, *122*, 47–52. [CrossRef]
37. Auf'm Erley, G.S.; Kaul, H.-P.; Kruse, M.; Aufhammer, W. Yield and nitrogen utilization efficiency of the pseudocereals amaranth, quinoa, and buckwheat under differing nitrogen fertilization. *Eur. J. Agron.* **2005**, *22*, 95–100. [CrossRef]
38. Berti, M.; Wilckens, R.; Hevia, F.; Serri, H.; Vidal, I.; Méndez, C. Nitrogen fertilization in quinoa (*Chenopodium quinoa* Willd). *Ciencia e Investigación Agraria* **2000**, *27*, 81–90. [CrossRef]
39. Mendoza Nieto, E.; Luis Olivas, D.; Mejía Domínguez, C.M.; García Cochagne, J. Fertilización nitrogenada en el rendimiento de dos variedades de quinua. *Infinitum* **2016**, *6*, 11–15. [CrossRef]
40. Llaca Ninaja, G.C. Influencia de la Fertilización Nitrogenada y Fosfórica en el Rendimiento de Quinua (*Chenopodium quinoa* Willd) en el Proter Sama, Región Tacna. Engineering Thesis, Universidad Nacional Jorge Basadre Grohmann-Tacna, Tacna, Peru, 2014.

41. Franco Alvarado, L.A. Eficiencia de Utilizacion del Nitrogeno en el Rendimiento de Quinua (*Chenopodium quinoa* Willd) Adaptada a la Zona Norte de la Provincia de Los Rios. Engineering Thesis, Universidad Tecnica Estatal de Quevedo, Quevedo, Los Rios, Ecuador, 2018.
42. Herreros Quispe, A.L. Fertilización Nitrogenada y Fosfórica en Quinua (*Chenopodium quinoa* willd.) CV. Salcedo INIA" bajo Riego a Goteo en Zona Árida. Engineering Thesis, Universidad Nacional de San Agustin de Arequipa, Arequipa, Peru, 2018.
43. Food and Agriculture Organization of the United Nations (FAO). *El Manejo del Suelo en la Producción de Hortalizas con Buenas Prácticas Agrícolas*; Food and Agriculture Organization of the United Nations: Rome, Italy, 2013.
44. Alvar-Beltrán, J.; Saturnin, C.; Dao, A.; Dalla Marta, A.; Sanou, J.; Orlandini, S. Effect of drought and nitrogen fertilisation on quinoa (*Chenopodium quinoa* Willd.) under field conditions in Burkina Faso. *Ital. J. Agrometeorol.* **2019**, *1*, 33–43. [CrossRef]
45. Oyarzun Arrechea, M. Respuesta Productiva de un Cultivo de Maíz ("*Zea mays*" L. Var. Dracma) a Distintas Dosis de Nitrógeno con dos Tipos de Riego (Aspersión e Inundación) y Efecto sobre la Lixiviación de Nitratos. Engineering Thesis, Universidad Publica de Navarra, Navarra, Spain, 2010.
46. Pandey, R.K.; Maranville, J.W.; Bako, Y. Nitrogen fertilizer response and use efficiency for three cereal crops in Niger. *Commun. Soil Sci. Plant Anal.* **2001**, *32*, 1465–1482. [CrossRef]
47. Quintero, C.; Boschetti, G. Eficiencia de uso del Nitrógeno en Trigo y Maíz en la Región Pampeana Argentina. Facultad de Ciencias Agropecuarias UNER. 2009. Available online: https://www.researchgate.net/profile/Cesar-Quintero-4/publication/266357692_Eficiencia_de_uso_del_Nitrogeno_en_Trigo_y_Maiz_en_la_Region_Pampeana_Argentina/links/58e3c7c0458515b725b038bf/Eficiencia-de-uso-del-Nitrogeno-en-Trigo-y-Maiz-en-la-Region-Pampeana-Argentina.pdf (accessed on 7 January 2021).
48. Calvache Ulloa, M.; Valle, L. Índice de cosecha con macro-nutrientes en grano de quinua (*Chenopodium quinoa* Willd). *Rev. Alfa* **2021**, *5*, 15–28. [CrossRef]
49. Critchley, C.S. A Physiological Explanation for the Canopy Nitrogen Requirement of Winter Wheat. Ph.D. Thesis, University of Nothingham, Nottingham, UK, 2001. Available online: http://eprints.nottingham.ac.uk/12834/1/368341.pdf (accessed on 21 January 2021).
50. Kichey, T.; Hirel, B.; Heumez, E.; Dubois, F.; Le Gouis, J. In winter wheat (*Triticum aestivum* L.), post-anthesis nitrogen uptake and remobilisation to the grain correlates with agronomic traits and nitrogen physiological markers. *Field Crop. Res.* **2007**, *102*, 22–32. [CrossRef]
51. Sade, N.; Del Mar, M.R.-W.; Umnajkitikorn, K.; Blumwald, E. Stress-induced senescence and plant tolerance to abiotic stress. *J. Exp. Bot.* **2018**, *69*, 845–853. [CrossRef]
52. SENAMHI. *Yearly Data on Rainfall, Oruro*. 2017. Available online: http://senamhi.gob.bo/index.php/boletines (accessed on 15 January 2021).
53. Guiboileau, A.; Sormani, R.; Meyer, C.; Masclaux-Daubresse, C. Senescence and death of plant organs: Nutrient recycling and developmental regulation. *Comptes Rendus Biol.* **2010**, *333*, 382–391. [CrossRef]
54. Thomas, H.; Howarth, C.J. Five ways to stay green. *J. Exp. Bot.* **2000**, *51*, 329–337. [CrossRef] [PubMed]
55. Gregersen, P.L.; Culetic, A.; Boschian, L.; Krupinska, K. Plant senescence and crop productivity. *Plant Mol. Biol.* **2013**, *82*, 603–622. [CrossRef] [PubMed]
56. Gaju, O.; Allard, V.; Martre, P.; Le Gouis, J.; Moreau, D.; Bogard, M.; Hubbart, S.; Foulkes, M.J. Nitrogen partitioning and remobilization in relation to leaf senescence, grain yield and grain nitrogen concentration in wheat cultivars. *Field Crop. Res.* **2014**, *155*, 213–223. [CrossRef]
57. Cassman, K.G.; Dobermann, A.; Walters, D.T. Agroecosystems, nitrogen-use efficiency, and nitrogen management. *Ambio* **2002**, *31*, 132–140. [CrossRef]
58. Stewart, W. Consideraciones en el uso eficiente de nutrientes. *Inf. Agronóm.* **2007**, *67*, 1–10.
59. Puentes-Páramo, Y.; Menjivar-Flores, J.; Aranzazu-Hernández, F. Eficiencias en el uso de nitrógeno, fósforo y potasio en clones de cacao (*Theobroma cacao* L.). *Bioagro* **2014**, *26*, 99–106.
60. Bascunan-Godoy, L.; Reguera, M.; Abdel-Tawab, Y.M.; Blumwald, E. Water deficit stress-induced changes in carbon and nitrogen partitioning in *Chenopodium quinoa* Willd. *Planta* **2016**, *243*, 591–603. [CrossRef]
61. Reguera, M.; Peleg, Z.; Abdel-Tawab, Y.M.; Tumimbang, E.B.; Delatorre, C.A.; Blumwald, E. Stress-induced cytokinin synthesis increases drought tolerance through the coordinated regulation of carbon and nitrogen assimilation in rice. *Plant Physiol.* **2013**, *163*, 1609–1622. [CrossRef]
62. Kristensen, H.L.; Thorup-Kristensen, K. Root Growth and Nitrate Uptake of Three Different Catch Crops in Deep Soil Layers. *Soil Sci. Soc. Am. J.* **2004**, *68*, 529–537. [CrossRef]
63. Hervé, D.; Mita, V.; Coûteaux, M.-M. Construcción de un balance de nitrógeno en cultivos de papa bajo rotación con largo descanso. *Ecol. Boliv.* **2006**, *41*, 133–153.
64. Fassbender, H.W. *Química de Suelos, con Énfasis en Suelos de América Latina*; IICA: San Jose, Costa Rica, 1986; p. 398.
65. Wolf, J.; Van Keulen, H. Modeling long-term crop response to fertilizer and soil nitrogen. *Plant Soil* **1989**, *120*, 23–38. [CrossRef]
66. VAISALA. *Lightning like Never Before. Annual Lightning Report 2020*; VAISALA: Vaanta, Finland, 2021; p. 25.
67. Hill, R.D.; Rinker, R.G.; Wilson, H.D. Atmospheric Nitrogen Fixation by Lightning. *J. Atmos. Sci.* **1980**, *37*, 179–192. [CrossRef]
68. Drapcho, D.L.; Sisterson, D.; Kumar, R. Nitrogen fixation by lightning activity in a thunderstorm. *Atmos. Environ.* **1983**, *17*, 729–734. [CrossRef]

69. Bascunan-Godoy, L.; Sanhueza, C.; Pinto, K.; Cifuentes, L.; Reguera, M.; Briones, V.; Zurita-Silva, A.; Alvarez, R.; Morales, A.; Silva, H. Nitrogen physiology of contrasting genotypes of *Chenopodium quinoa* Willd. (Amaranthaceae). *Sci. Rep.* **2018**, *8*, 17524. [CrossRef]
70. Borrell, A.; Hammer, G.; Oosterom, E. Stay-green: A consequence of the balance between supply and demand for nitrogen during grain filling? *Ann. Appl. Biol.* **2001**, *138*, 91–95. [CrossRef]
71. Triboi, E.; Martre, P.; Girousse, C.; Ravel, C.; Triboi-Blondel, A.-M. Unravelling environmental and genetic relationships between grain yield and nitrogen concentration for wheat. *Eur. J. Agron.* **2006**, *25*, 108–118. [CrossRef]
72. Mi, G.H.; Chen, F.J.; Liu, J.A.; Tong, Y.P. Biological potential of nitrogen utilization in crops and its genetic improvement. In *Explore Biological Potential for Soil Nutrient Utilization and Maintain Nutrient Recycling in Soil Environment*; Li, Z., Ed.; China Agricultural Press: Beijing, China, 2004; pp. 202–216.
73. Coque, M.; Gallais, A. Genetic Variation for Nitrogen Remobilization and Postsilking Nitrogen Uptake in Maize Recombinant Inbred Lines: Heritabilities and Correlations among Traits. *Crop Sci.* **2007**, *47*, 1787–1796. [CrossRef]
74. Pinto-Irish, K.; Coba de la Pena, T.; Ostria-Gallardo, E.; Ibanez, C.; Briones, V.; Vergara, A.; Alvarez, R.; Castro, C.; Sanhueza, C.; Castro, P.A.; et al. Seed characterization and early nitrogen metabolism performance of seedlings from Altiplano and coastal ecotypes of Quinoa. *BMC Plant Biol.* **2020**, *20*, 343. [CrossRef]
75. Umehara, M.; Hanada, A.; Yoshida, S.; Akiyama, K.; Arite, T.; Takeda-Kamiya, N.; Magome, H.; Kamiya, Y.; Shirasu, K.; Yoneyama, K.; et al. Inhibition of shoot branching by new terpenoid plant hormones. *Nature* **2008**, *455*, 195–200. [CrossRef]
76. Gomez-Roldan, V.; Fermas, S.; Brewer, P.B.; Puech-Pages, V.; Dun, E.A.; Pillot, J.P.; Letisse, F.; Matusova, R.; Danoun, S.; Portais, J.C.; et al. Strigolactone inhibition of shoot branching. *Nature* **2008**, *455*, 189–194. [CrossRef] [PubMed]
77. Hildebrandt, U.; Janetta, K.; Ouziad, F.; Renne, B.; Nawrath, K.; Bothe, H. Arbuscular mycorrhizal colonization of halophytes in Central European salt marshes. *Mycorrhiza* **2001**, *10*, 175–183. [CrossRef]
78. Plenchette, C.; Duponnois, R. Growth response of the saltbush *Atriplex nummularia* L. to inoculation with the arbuscular mycorrhizal fungus *Glomus intraradices*. *J. Arid. Environ.* **2005**, *61*, 535–540. [CrossRef]
79. Gill, S.; Alshankiti, A.; Shahid, S.A.; Rodriguez, J.P. Amending Soil Health to Improve Productivity of Alternate Crops in Marginal Sandy Soils of the UAE. In *Emerging Research in Alternative Crops*; Hirich, A., Choukr-Allah, R., Ragab, R., Eds.; Environment & Policy; Springer International Publishing: Cham, Switzerland, 2020; pp. 93–123.
80. Urcelay, C.; Acho, J.; Joffre, R. Fungal root symbionts and their relationship with fine root proportion in native plants from the Bolivian Andean highlands above 3,700 m elevation. *Mycorrhiza* **2011**, *21*, 323–330. [CrossRef]
81. Kellogg, J.A.; Reganold, J.P.; Murphy, K.M.; Carpenter-Boggs, L.A. A Plant-Fungus Bioassay Supports the Classification of Quinoa (*Chenopodium quinoa* Willd.) as Inconsistently Mycorrhizal. *Microb. Ecol.* **2021**, *82*, 135–144. [CrossRef]
82. Santander, C.; Olave, J.; García, S.; Vidal, C.; Aguilera, P.; Borie, F.; Cornejo, P. Micorrizas arbusculares y su efecto nodriza en condiciones hídricas limitantes. *Exp. Rev. Transf. Cient.* **2014**, *4*, 59–61.
83. Wieme, R.A.; Reganold, J.P.; Crowder, D.W.; Murphy, K.M.; Carpenter-Boggs, L.A. Productivity and soil quality of organic forage, quinoa, and grain cropping systems in the dryland Pacific Northwest, USA. *Agric. Ecosyst. Environ.* **2020**, *293*, 106838. [CrossRef]
84. Gonzalez-Teuber, M.; Vilo, C.; Bascunan-Godoy, L. Molecular characterization of endophytic fungi associated with the roots of *Chenopodium quinoa* inhabiting the Atacama Desert, Chile. *Genom. Data* **2017**, *11*, 109–112. [CrossRef]
85. Pitzschke, A. Developmental Peculiarities and Seed-Borne Endophytes in Quinoa: Omnipresent, Robust Bacilli Contribute to Plant Fitness. *Front. Microbiol.* **2016**, *7*, 2. [CrossRef] [PubMed]
86. Ortuño, N.; Claros, M.; Gutiérrez, C.; Angulo, M.; Castillo, J. Bacteria associated with the cultivation of quinoa in the Bolivian Altiplano and their biotechnological potential. *Rev. Agric.* **2014**, *53*, 53–61.
87. Bashan, Y.; De-Bashan, L. Plant growth-promoting. *Encycl. Soils Environ.* **2005**, *1*, 103–115.
88. Glick, B.R. The enhancement of plant growth by free-living bacteria. *Can. J. Microbiol.* **1995**, *41*, 109–117. [CrossRef]
89. GarciA, G.N.; Navarro García, S. *Química Agrícola: Química del Suelo y de los Nutrientes Esenciales para las Plantas*; Mundi-Prensa Libros: Madrid, Spain, 2013.
90. Ribaudo, C.M.; Riva, D.S.; Curá, J.A.; Ponds, C.; Granell-Richard, A.; Cantora, M. Etileno como mediador de los mecanismos directos e indirectos de la promoción del crecimiento vegetal ejercido por rizobacterias. In *Rizósfera, Biodiversidad y Agricultura Sustentable*; García de Salamone, I.E., Vázquez, S., Penna, C., Cassán, F., Eds.; Asociación Argentina de Microbiología Pág: Buenos Aires, Argentina, 2013; pp. 215–240.
91. Guo, Y.; Chen, F.; Zhang, F.; Mi, G. Auxin transport from shoot to root is involved in the response of lateral root growth to localized supply of nitrate in maize. *Plant Sci.* **2005**, *169*, 894–900. [CrossRef]
92. Gruber, B.D.; Giehl, R.F.; Friedel, S.; von Wiren, N. Plasticity of the Arabidopsis root system under nutrient deficiencies. *Plant Physiol.* **2013**, *163*, 161–179. [CrossRef]
93. Vidal, E.A.; Araus, V.; Lu, C.; Parry, G.; Green, P.J.; Coruzzi, G.M.; Gutierrez, R.A. Nitrate-responsive miR393/AFB3 regulatory module controls root system architecture in *Arabidopsis thaliana*. *Proc. Natl. Acad. Sci. USA* **2010**, *107*, 4477–4482. [CrossRef]
94. Hu, S.; Zhang, M.; Yang, Y.; Xuan, W.; Zou, Z.; Arkorful, E.; Chen, Y.; Ma, Q.; Jeyaraj, A.; Chen, X.; et al. A novel insight into nitrogen and auxin signaling in lateral root formation in tea plant [*Camellia sinensis* (L.) O. Kuntze]. *BMC Plant Biol.* **2020**, *20*, 232. [CrossRef] [PubMed]
95. Sellami, M.H.; Pulvento, C.; Lavini, A. Agronomic Practices and Performances of Quinoa under Field Conditions: A Systematic Review. *Plants* **2021**, *10*, 72. [CrossRef]

96. Di Rienzo, J.A.; Casanoves, F.; Balzarini, M.G.; Gonzalez, L.; Tablada, M.; Robledo, C.W. *Infostat Version 2020*; Centro de Transferencia InfoStat: Cordoba, Argentina; Universidad Nacional de Cordoba: Cordoba, Argentina, 2020.
97. OriginPro 2019. *OriginLab Corporation*; OriginPro: Northampton, MA, USA, 2019.
98. Zhang, X.; Bol, R.; Rahn, C.; Xiao, G.; Meng, F.; Wu, W. Agricultural sustainable intensification improved nitrogen use efficiency and maintained high crop yield during 1980–2014 in Northern China. *Sci. Total Environ.* **2017**, *596–597*, 61–68. [CrossRef]
99. Zhang, F.; Cui, Z.; Fan, M.; Zhang, W.; Chen, X.; Jiang, R. Integrated soil-crop system management: Reducing environmental risk while increasing crop productivity and improving nutrient use efficiency in China. *J. Environ Qual.* **2011**, *40*, 1051–1057. [CrossRef]
100. Kuosmanen, N. Estimating stocks and flows of nitrogen: Application of dynamic nutrient balance to European agriculture. *Ecol. Econ.* **2014**, *108*, 68–78. [CrossRef]
101. Al-Naggar, A.M.M.; Shabana, R.; El-Aleem, M.M.A.; El-Rashidy, Z. Selection Criteria for High Nitrogen Use Efficiency in Wheat (*Triticum aestivum* L.) Parents and their F1 and F2 Progenies. *J. Agric. Ecol. Res. Int.* **2015**, *5*, 1–15. [CrossRef]
102. Cardenas, J.; Alvarez, F. *Investigación en Riego y Fertilización Orgánica Sostenible para la Quinua en la Sede Challapata, FCA y V-UTO. Informe de Avance Técnico y Económico de Actividades del Proyecto con Recursos IDH—Semestre I-2018*; Facultad de Ciencias Agricolas y Veterinarias, Universidad Tecnica de Oruro: Oruro, Bolivia, 2018; p. 28.
103. Aguilar Santelises, A. *Análisis Químico para Evaluar la Fertilidad del Suelo*; CONACYT: Mexico City, Mexico, 1987.
104. Castellanos, J.; Uvalle-Bueno, J.; Aguilar-Santelises, A. *Manual de Interpretación de Análisis de Suelos y Aguas Agrícolas, Plantas y ECP*; Instituto de Capacitación para la Productividad Agrícola: Mexico City, Mexico, 2000.
105. Abugoch, L.E.; Romero, N.; Tapia, C.A.; Silva, J.; Rivera, M. Study of some physicochemical and functional properties of quinoa (*Chenopodium quinoa* willd) protein isolates. *J Agric Food Chem* **2008**, *56*, 4745–4750. [CrossRef] [PubMed]
106. Covarrubias, N.; Sandoval, S.; Vera, J.; Núñez, C.; Alfaro, C.; Lutz, M. Contenido de humedad, proteínas y minerales en diez variedades de quínoa chilena cultivadas en distintas zonas geográficas. *Revista Chilena de Nutrición* **2020**, *47*, 730–737. [CrossRef]
107. Gavidia-Valencia, J.G.; Venegas-Casanova, E.A.; Ríos, M.; Uribe-Villarreal, J.C.; Gutierrez-Mendoza, D.D.; Rengifo-Penadillos, R.A.; Jara-Aguilar, D.R.; Martínez, J.L. Determinación del factor de conversión de nitrógeno a proteína en huevos de *Coturnix coturnix* L. (codorniz japonesa). *Arch. Venez. Farmacol. Ter.* **2020**, *39*, 706–711. [CrossRef]

plants

Review

Botany, Nutritional Value, Phytochemical Composition and Biological Activities of Quinoa

M. Iftikhar Hussain [1,2,*], Muhammad Farooq [3], Qamar Abbas Syed [4], Anum Ishaq [5], Abdullah Ahmed Al-Ghamdi [6] and Ashraf A. Hatamleh [6]

1 Department of Plant Biology & Soil Science, Universidad de Vigo, As Lagoas, Marcosende, s/n, 36310 Vigo, Spain
2 CITACA, Agri-Food Research and Transfer Cluster, Campus da Auga, Universidad de Vigo, 32004 Ourense, Spain
3 Department of Plant Sciences, College of Agricultural and Marine Sciences, Sultan Qaboos University, Al-Khoudh 123, Oman; farooqcp@gmail.com
4 National Institute of Food Science and Technology, Faculty of Food, Nutrition & Home Sciences University of Agriculture, Faisalabad 38000, Pakistan; Qamar.Abbas@uaf.edu.pk
5 School of Food and Agricultural Sciences, University of Management & Technology, Johar Town, Lahore 54770, Pakistan; anum1797@yahoo.com
6 Department of Botany and Microbiology, College of Science, King Saud University, Riyadh 11451, Saudi Arabia; abdaalghamdi@ksu.edu.sa (A.A.A.-G.); ahatamleh@ksu.edu.sa (A.A.H.)
* Correspondence: mih786@gmail.com

Abstract: Quinoa is a climate-resilient food grain crop that has gained significant importance in the last few years due to its nutritional composition, phytochemical properties and associated health benefits. Quinoa grain is enriched in amino acids, fiber, minerals, phenolics, saponins, phytosterols and vitamins. Quinoa possesses different human-health promoting biological substances and nutraceutical molecules. This review synthesizes and summarizes recent findings regarding the nutrition and phytochemical properties of quinoa grains and discusses the associated biological mechanisms. Quinoa grains and grain-based supplements are useful in treating different biological disorders of the human body. Quinoa is being promoted as an exceptionally healthy food and a gluten-free super grain. Quinoa could be used as a biomedicine due to the presence of functional compounds that may help to prevent various chronic diseases. Future research needs to explore the nutraceutical and pharmaceutical aspects of quinoa that might help to control different chronic diseases and to promote human health.

Keywords: quinoa grains; phytochemicals; anti-oxidants; nutrients; phytosterols

Citation: Hussain, M.I.; Farooq, M.; Syed, Q.A.; Ishaq, A.; Al-Ghamdi, A.A.; Hatamleh, A.A. Botany, Nutritional Value, Phytochemical Composition and Biological Activities of Quinoa. *Plants* **2021**, *10*, 2258. https://doi.org/10.3390/plants10112258

Academic Editors: Cataldo Pulvento and Didier Bazile

Received: 28 August 2021
Accepted: 11 October 2021
Published: 22 October 2021

1. Introduction

The current need of the food industry is the development of innovative and alternative products. However, the sector is saturated due to new products and invasive, foreign low-quality products. Therefore, the use of alternative ingredients compared to those commonly used is the only way to gain new slices of the market. The evolution of the concept of "food" has been profound; it is traditionally understood as the set of food and nutrient elements taken by the body for the normal performance of vital functions. Nonetheless, in addition to satisfying the nutritional and sensory needs, the food should have the capacity to influence the psycho-physical well-being of the organism, preventing the risk of the onset of certain pathologies. In this regard, some pseudo-cereals, such as quinoa, are a potential option. Several biomolecules from quinoa including carotenoids and essential amino acids can be exploited in the manufacturing of food items with a high dietetic–nutritional value.

Quinoa (*Chenopodium quinoa* Willd.) is an annual herbaceous plant. It is considered a pseudo-cereal because the plant belongs to the family of spinach and sugar beets. Quinoa is a native grain-like crop grown originally in the Andean region of South America including Peru, Bolivia, Ecuador, Colombia and Chile. It is a gluten-free food grain crop that is safe for celiac patients [1]. Almost 3000 to 4000 years ago, this wonderful crop was domesticated for human consumption and livestock feed [2]. In recent years, it has been introduced in other regions of the world such as Europe, North America, Australia, China and Japan. Quinoa can adapt well to different environments where the survival of other field crops is extremely difficult. For these reasons, the FAO declared the year 2013 as the International Year of Quinoa.

Quinoa has been the main food of the Inca civilization. In the Quechua language of the Incas, quinoa is called the chisaya mama—"mother of all the seeds", because, at the time of the Incas and the Aztecs, this food represented, together with the maize and the potato, an essential element of their diets. It was considered a source of life for its beneficial and healing properties and was even venerated as a sacred plant, and was, therefore, known as "Inca gold". However, following the Spanish conquest, the cultivation of quinoa could not maintain its momentum [3] and is currently limited to circumscribed areas of South America. In recent years, there has been a re-evaluation of quinoa for its agronomic and nutritional characteristics. The possibility of increasing the production of quinoa in other parts of the world, such as in Canada, is being considered and also the possibility of guaranteeing production in terms of quantity and quality to meet the demands of the food industries.

The quinoa plant has a broad genetic diversity that allows it to be highly resistant to cold, salt and drought conditions, with ecotypes growing well at high altitudes and in poor soils, where other cereal crops, such as wheat, rice and maize, do not growth well [4]. Quinoa is a hardy, drought-tolerant plant with a combined precipitation and irrigation requirement of 25–38 cm per years, which is far less than the water requirements of other cereals such as wheat and rice [4]. As long as the soil is naturally moist, plants should not be irrigated until the seedlings show two or three leaves. On the other hand, over-watering during the seedling stages can cause damping off and severe stunting of quinoa, whereas excessive irrigation, after the plant establishment, may result in lodging. In the Andean region, quinoa is usually cultivated in rotation with potato or cereals, without the use of fertilizers or manures. In other countries, quinoa responds well to nitrogen fertilizer; however, over-fertilization may cause excessive vegetative growth leading to lodging [5].

Despite the relatively lower quinoa grain yields ranging from 0.23 t/ha in Mauritania to 7.5 t/ha in Lebanon compared to that of common cereals, quinoa production has intensified quickly in recent years due to the increasing prices in the international market [4]. The price of the quinoa sold by farmers has almost tripled from 2004 to 2012, and is three times the price of soybean and five times the price of wheat [6]. The higher economic profits compared with those from common cereals drive farmers to expand the existing plant scale [7]. High quinoa prices will be sustained (including in China) because the production of quinoa cannot meet the demand in a short-term period from an increasing number of people who intend to include quinoa in their diets [6].

Quinoa leaves and seeds are edible, with the latter being the principal form for human consumption (Figure 1). Three main storage compartments can be distinguished within the mature quinoa seed (from center to edge): a large central perisperm, a peripheral embryo and a one to two cell layered endosperm only in the micropyle region surrounding the hypocotyl–radicle axis of the embryo [8]. The quinoa endosperm has 1–2 layers; however, the starch is principally stored in the non-living perisperm that occupies around 40% of the volume of the quinoa seed, although small amounts also exist in the embryo but not in the endosperm [8]. The quinoa embryo and endosperm are rich sources of minerals, proteins and lipids [8]. The seeds are round and flattened, about 1.5–4.0 mm in diameter and 0.5 mm in thickness; around 350 seeds weigh one gram, and the seed color ranges from black, white and gray-purple to yellow, red and violet. The varietal identification in quinoa is based on the plant morphology, color of the plant and seeds [2,9]. Quinoa seeds are being used for salad dressings or are cooked similar to rice and are used in soups and/or breakfast cereals. Quinoa use is common in various toasted and baked goods (bread, noodles, cakes, cookies, biscuits, flakes, pancakes and tortillas) [10].

Figure 1. Quinoa plants with two different colors of grains (yellow and purple red).

The interest in quinoa stems from its high capacity to adapt to different environments to make it competitive with common cereals. Quinoa can withstand extreme climates such as from sea level to 4000 m above sea level or from 40° South to 2° North. Moreover, it can be cultivated in arid, semi-arid and tropical environments. It is also resistant to frosts and can grow even in drought conditions as well as on cold mountains [11]. Currently there are over 250 varieties of quinoa, the most cultivated varieties include Bear, Cherry Vanilla, Cochabamba, Dave 407, Gossi, Isluga, Kaslala, Kcoito, Linares, Puno, Titicaca, Rainbow, Red lighthouse, Red head and Temuco [12,13].

Quinoa is a rich source of protein, fibers, minerals, vitamins and lipids. Quinoa grains contain essential amino acids and polyphenols [14]. The composition of essential amino acids in quinoa is similar to the amino acid requirement pattern and is higher than that in whole grain and refined wheat. Additionally, quinoa contains a significant amount of minerals. A significant quantity of polyphenols is also present in quinoa seeds [14]. The polyphenols show different pharmacological properties such as anti-allergic, antiviral, anti-inflammatory, cardiovascular protective and anticarcinogenic activity [15].

Origin and Dissemination

Quinoa originated from the Andean regions such as Peru, Bolivia and Ecuador, where it is still the staple food for the population. Slowly, other foreign crops colonized the most fertile lands, erasing the small terraces occupied by the quinoa and other traditional crops and distorting the landscape; the quinoa plant then ended up being despised by the local population because it was considered food for the poorest, as were the remaining indigenous crops. Ironically, the revaluation of quinoa began abroad, far from its typical areas of cultivation. Nowadays, it is above all thanks to fair trade that quinoa is experiencing a new renaissance, becoming not only a simple product of subsistence for local populations but also an export product. The quinoa of fair trade generally comes from organic farming, and this market is the result of collaboration between the Third World Cooperative Altromercato (CTM), also known as Altromercato, and the ANAPQUI (Asociaciòn Naciònal Productores de Quinoa) of Bolivia. Altromercato is the first association that manages the import of fair-trade products in Italy, and the second in the world, in terms of size and turnover (www.wikipedia.org/altromercato; accessed on 13 April 2019). The association, Altromercato, aims to base its activity exclusively on processes of solidarity, economy and responsible consumption, trying to ensure the suppliers of raw materials a fair price and transparent long-term relationships and also supporting organic farming.

Altromercato opera ANAPQUI (Asociaciòn Naciònal Productores de Quinoa) is an association founded in 1983 in the area of Salar, on the Bolivian plateau. It is a federation of seven regional associations that brings together a total of 5000 small producers of quinoa. The association has the task of buying quinoa from farmers and exporting it to various organizations, including those of fair trade. ANAPQUI's main aim is to offer producers the opportunity to bring their products to market on more favorable terms, given their inability to provide all the necessary marketing and exportation procedures autonomously. For this purpose, ANAPQUI promotes autonomous forms of organization of producers with a permanent training and information activity. The enhancement of quinoa in the food sector has also been encouraged by the FAO, which has massively promoted the cultivation and diffusion of this pseudo-cereal.

2. Quinoa Botany and Morphological Characteristics

Quinoa is a dicotyledonous annual herbaceous plant belonging to the subfamily of the Chenopodiaceae that has some starchy seeds (achenes). Its height ranges from 3 to 7 feet. Depending upon the variety, its woody central stem may be branched or unbranched and varying in color as well (green, red or purple). The flowering panicles appear from the leaf axils along the stem, or they may arise from the top of the plant. Generally, the flowers are self-fertilized though crosspollination may also occur [3].

2.1. Leaves

The leaves are made of the petiole and the lamina. The petioles are long, thin and grooved. The length of the petioles varies according to the variety but may vary even within the same plant. The leaf blade is also polymorphic in the same plant and can be diamond-shaped, triangular or lanceolate, flat or wavy [16]. Inside the leaves are calcium oxalate crystals that reduce excessive perspiration allowing the maintenance of adequate humidity inside the plant. The color of the leaves is very variable, from the green in the younger plants to the red or violet with different shades in the more mature plants; they have very pronounced and easily visible veins, deriving from the stem and are generally three in number. There are genotypes with more leaves and others with less; in general, the "Valley" quinoa has abundant foliage, allowing the use of the plant as fodder. In many areas of the Andean region, the young leaves before flowering are considered suitable for human consumption, due to their high nutritional value attributed to their content of vitamins, minerals and proteins.

2.2. Root

The plant is deep-rooted, with a highly branched system, which gives it good stability and allows it to withstand drought resistance [17]. The color varies according to the type of soil in which it grows. During the germination, the first part that develops is the radicle, which continues to grow both laterally and vertically reaching up to 180 cm. The root length and the distribution of the lateral roots vary with the genotype and allow a strong anchoring to the ground and a good resistance of the plant to water scarcity. The size of the root is closely related to the height of the plant and can exceed 30 cm, in general; however, it tends to grow deep to form a highly branched system to increase the resistance of the plant to drought.

2.3. Stem

The plant has an erect woody stem, which can be ramified or non-branched, with heights ranging from 30 cm to 3 m, depending on the variety of cultivated quinoa [17] and climatic conditions. It is generally cylindrical, with a resistant bark when the plants are young and a porous one when they mature. The diameter of the stem varies according to the genotype, the sowing distances, the fertilization and the cultivation conditions and ranges from 1 to 8 cm. The ramifications of the configuration of the plant can be modified by insect attack, mechanical damage or by some cultural practices. As genotypes, seeding density, nutrient availability and growth areas vary, the plants may be widely branched ("Sea-level" quinoa), without branching (plateau) or intermediately branched.

The color of the stem, similar to that in amaranth, may vary from pale yellow to red, depending upon the genotypes and the phenological phases [18]. The color often shows striations, with the leaf axils and branches being red or purple. The stem has a cutinized epidermis, with membranes of compact cellulose. Within it, the stem contains a marrow that disappears upon ripening, leaving the stem dry and empty. This stem, owing to high pectin and cellulose contents, can be used in the manufacturing of paper and cardboard. Wide leaves, attached to the stem alternating on four levels, are characterized by polymorphism. The uppermost leaves are small and lanceolate, while those at the bottom are large, rhomboidal or triangular [19].

2.4. *Inflorescences and Flowers*

The quinoa inflorescence is a typical panicle, consisting of a central axis and secondary and tertiary branches with the pedicels holding the glomeruli. The panicle may have sparse (amarantiforme) or compact (glomerular) inflorescences. There are also forms that present transitional characteristics between the two groups [17]. An inflorescence in compact groups with close pedicels is called glomerular, where the inflorescence with elongated glomeruli and several secondary and tertiary branches attached to central axis with flowers grouped in fairly loose tufts is termed as amarantiforme. This inflorescence type has been named so due to its resemblance to the inflorescence of the genus *Amaranthus*.

The length of the panicle is variable, depending on the genotype, the type of quinoa and the fertility conditions of the soil, and ranges from 30 to 80 cm in length and from 5 to 30 cm in diameter. The number of glomeruli per panicle varies from 80 to 120 and the number of seeds per panicle can range from 100 to 3000. The large panicles may yield up to 500 g of seed per inflorescence.

The plants produce panicle inflorescences having hermaphrodite and pistillate flowers. The hermaphrodite flower possesses five imbricate sepals and five antisepalous stamens; its filaments extend laterally forming a ring of nectariferous tissue surrounding the ovary having modified stomata. The gynoecium is bicarpellate with unilocular ovary having a single campylotropous ovule [17]. Quinoa plants are also hermaphrodites, which mean they are characterized by self-pollination [20]; however, cross-pollination (10–15%) can also take place [21]. The hermaphroditic flowers are located at the distal end, while the female flowers are in the proximal end [22]. They are self-fertile, and pollination is generally anemophilous (through the wind). Upon ripening, the plant produces a panicle containing seeds, called an achene, which resemble those of millet in appearance.

2.5. *Fruit*

The fruit is an indehiscent achene, with a spherical, conical or ellipsoidal shape, with a diameter varying between 1.0 and 2.6 mm. The coloring of the pericarp varies extremely from yellowish to gray with magenta. It is held with five lobes of the perianth, which are easily removable by abrasion. The fruits are small in size, and the weight of 1000 achenes ranges from 1.9 to 4.3 g [23]. The seeds have three major components viz. episperm, embryo and perisperm. A thin layer called the pericarp surrounds the achene. The pericarp has a rough and brittle outer surface that can be easily removed upon rubbing. This part contains the saponins, which give a bitter taste to the seed. The episperm is located below the pericarp.

The embryo is made up of two cotyledons and constitutes 30% of the total volume of the seed, surrounding the perisperm like a ring, with a curvature of 320°. Compared to other seeds, the embryo contains 35–40% of the total seed proteins, while the perisperm presents only 6.3–8.3% of the total proteins [8]. The perisperm is whitish in color and acts as the main storage fabric and consists mainly of starch granules. This represents almost

60% of the seed surface. The quinoa endosperm is composed of several layers around the embryo. After seed hydration, endosperm cells come into contact with the embryo and are rapidly consumed during its growth.

3. Nutritional Profile

Quinoa is pseudo-cereal rich in proteins, lipid, fiber, vitamins and minerals [24]. Quinoa grains are a good source of natural compounds with anti-oxidant activities and biological properties [25]. However, the composition of different genotypes varies significantly [26]. Chen et al. [27] classified quinoa into two groups. Group A containing higher phytochemicals and polyunsaturated fatty acids and group B containing higher linolenic and long-chain fatty acids. Both groups can be utilized in food products. Quinoa grain has an excellent nutritional profile, starch (32–60%), protein (10–18%) and fat (4.4% to 8.8%), while the ashes, formed mainly from potassium and phosphorus, constitute 2.4% to 3.7% and the fiber ranges from 1.1% to 13.4% [28,29]. The quinoa grains also contain a fair amount of vitamin B and vitamin E, a fat-soluble anti-oxidant vitamin.

The use of quinoa grains in food requires a series of operations aimed at the removal of fractions rich in antinutritional compounds, in particular saponins, located in the perianth and the pericarp, which are responsible for the sensation of bitterness and have negative effects on the intestinal mucosa. To remove the saponins, the quinoa grains are washed with tap water followed by hand rubbing and drying. However, this process may require several cycles [29–32]. The drying step is carried out under controlled conditions to avoid modifications of the quinoa seed constituents (such as starch gelatinization and protein denaturation), which may lead to significant changes in the nutritional and rheological properties of the quinoa flour. The summary of the impact of quinoa products on human health is given in Table 1.

Table 1. The effect of quinoa products in human health.

Study Participants	Treatment	Duration	Outcomes	Conclusions	References
50–65-month-old boys	100 g quinoa-added to baby food/day	15 days	Increase in plasma level of insulin-like growth factor-1	Potential role in reducing childhood malnutrition	[33]
18–45-year-old students	Quinoa cereal bars	30 days	Decrease in total cholesterol, triglycerides and low-density lipoprotein levels	Potential role in preventing cardiovascular disease	[34]
Post-menopausal, overweight women	25 g quinoa flakes and cornflakes/day	4 weeks	Reduction of total cholesterol, triglycerides and low-density lipoprotein levels and thiobarbituric acid reactive substances Increase of urinary secretion of enterolignans	Beneficial effect on metabolic parameter modulation	[35]
Celiac patients	50 g quinoa/day	6 weeks	Improved histological and serological parameters Mild hypocholesterolemic effect	Quinoa consumption is safe for celiac individuals	[36]
Overweight and obese participants	50 g quinoa/day	12 weeks	Reduction of serum triglyceride level Reduction of metabolic syndrome	Potential role in preventing obesity	[37]
35–70-year-old healthy, overweight males	Quinoa-enriched bread (with 20 g quinoa flour)/day	4 weeks	Reduction in blood glucose and low-density lipoprotein levels	Potential role in preventing cardiovascular disease	[38]

3.1. Protein

The quinoa grain is an important source of dietary proteins [31]. According to the reports of Abugoch [32], the proteins present in quinoa grains include albumins (35%) and globulins (37%) and a lower percentage of prolamins. The quality of proteins present in quinoa is comparable to milk protein (casein). Quinoa proteins constitute all essential amino acids such as tryptophan, histidine, isoleucine, leucine, lysine, methionine, phenylalanine, threonine, tryptophan, tyrosine and valine) [2,32]; this is why, this is considered a complete food [1].

The intake of proteins is an issue for the populations who rarely consume animal proteins, and therefore, their diets should include proteins of plant origin. Quinoa-based products are an apt option in this regard, because unlike other conventional cereals, quinoa does not suffer significant losses in protein content during industrial processing. Moreover, the amino acid composition of the quinoa proteins makes the organic value of quinoa superior to that of other traditional cereals. Electrophoretic studies showed that quinoa proteins consist of two large fractions—11S-globulin and 2S-protein. 11S-globulin, also called chenopodium [39,40], and represents about 37% of the total protein. This fraction contains polypeptides with a molecular weight of 22–23 and 32–39 kDa and has a relatively low content of amino acids (methionine and cysteine) [40]. The 2S-protein fraction has a molecular weight of 9 kDa and has a high content in cysteine, arginine and histidine, but it is relatively poor in methionine [39].

Quinoa has more than twice the lysine content of wheat, maize and rice. The lysine content in quinoa grains is equal to 96% of the quantity sufficient to reach the FAO standards and, therefore, remains a limiting amino acid. However, this is the highest value ever among all cereals. Leucine is the second limiting amino acid, but even in this case, it is a moderate limitation. According to the FAO standards, the content of this amino acid is 91% of the optimal quantity. Several traditional and modern methods are employed for protein isolation and characterization. This may include wet fractionation and dry fractionation methods. The first one involves huge consumption of water, chemicals and energy [41]. As compared to wet fractionation, dry fractionation was more efficient because it keeps the nutritional properties intact [42]. Motta et al. [43] reported the glutamic acid contents in steamed (2.0 g/100 g) and boiled (2.1 mg/100 g) quinoa seeds. Histidine and aromatic amino acids have also been reported in significant quantities.

3.2. Lipids

Quinoa was considered an alternative to oil seeds due to its lipid composition. The fat content of quinoa is quite high (5–10%) compared to that in common cereals and is mainly localized in the embryo [8]. The polar lipids represent about 25% of the total lipids consisting mainly of phospholipids (lysophosphatidyl ethanolamine and choline). Quinoa oil possesses high anti-oxidant activity, high contents of polyunsaturated fatty acids (63% of total) and a significant amount of tocopherols (2.5 mg/g of oil) [44]. Of the neutral lipids instead (glycerides and sterols), triglycerides account for 74% and diglycerides for 20%, while monoglycerides and waxes represent 3%, respectively. Several fatty acids such as linoleic acid (18:2n-6), linolenic acid (20:3n-6) (55–60%) and oleic acid (18:1 *cis*-9) (30%) constitute the main composition of fatty acids. Quinoa seed is also the main source of different essential fatty acids including omega-6 and omega-3 fatty acids. Linoleic acid is metabolized to arachidonic acid and linolenic acid to eicosapentaenoic acid (EPA) and docosahexaenoic acid (DHA) [45]. EPA and DHA play important roles in the metabolism of prostaglandins, thrombosis and atherosclerosis.

Quinoa seed exhibits a significant amount of oil that contains monounsaturated (oleic) and polyunsaturated (linoleic and linolenic) fatty acids. The oil is particularly stable due to the presence of a high content of natural anti-oxidants, such as α-tocopherols (69–75 mg/100 g of oil) and γ-tocopherols (76–93 mg/100 g of oil). The content of these compounds decreases up to 45 and 23 mg, respectively, after oil refining [23]. The quinoa grain also has high quality and quantity of oil due to which it has been defined by some authors as an oily pseudo-seed [23]. The high level of unsaturation may significantly increase the susceptibility of the lipid fraction to oxidative rancidity, but the presence of natural anti-oxidants especially tocopherols limit the oxidation initiation by acting as a natural defense against it [10]. Quinoa contains different phytosterols especially squalene that has shown anti-oxidant activity and helps to maintain cardiovascular health and to cure tumors in human beings. Quinoa seed showed a significant portion of phytosterols including b-sitosterol that possess functions to decrease the low-density lipoproteins [45]. Meanwhile, several other biological properties of phytosterols including anti-inflammatory, anti-oxidant and antitumor activity and cholesterol reduction have also been reported [46].

3.3. *Carbohydrates*

Based on the degree of polarization, carbohydrates can be grouped into simple sugars (monosaccharides, disaccharides), oligosaccharides and polysaccharides (starches). Carbohydrates play a vital role in nutrition and have different effects on metabolic processes such as diabetes, blood glucose, protein glycosylation. Quinoa seed has the characteristic of lowering hypoglycemic effects and decreasing free fatty acids and, thus, serves to exert a food nutraceutical impact through its carbohydrates. As compared to pasta and gluten-free bread, the glycemic index is lower and, hence, significantly reduces the free fatty acids.

According to reports of Jancurová et al. [47], quinoa has a significant amount of carbohydrates that are in the range of 67–74% of total dry matter, and the amylose contents are near 11%. Starch granules are lower than those reported for maize (range 1–2 μm) or wheat (2–40 μm). Meanwhile, some other carbohydrates are also documented, which include monosaccharides (2%) and disaccharides (2.3%), crude fiber (2.5–3.9%) and pentosans (2.9–3.6%) [48]. The small granule diameter of quinoa starch is useful for enhancing binding and reducing breakability [49]. Several researchers reported that quinoa is a good thickener for soups, sauces and flours. The gelling point of quinoa is sufficiently less and is more durable at low storage temperatures. A smooth texture and creamy nature can be obtained from quinoa seeds [2]. The carbohydrate contents of some quinoa genotypes are reported in Table 2. Sucrose is present in significant quantities compared to other sugars [50]. Despite this apparent difference, it can be said that the sugar content in quinoa is very similar to that of amaranth [51].

Table 2. Proximate composition of quinoa grains (g/100 g) [47–50].

Quinoa Genotypes	Moisture	Proteins	Fat	Crude Fiber	Ash	Carbohydrates
Ccoito	8.47 ± 0.08	14.72 ± 0.11	5.33 ± 0.06	1.81 ± 0.02	2.83 ± 0.00	68.1
INIA-415 Pasankalla	9.76 ± 0.07	12.69 ± 0.06	6.85 ± 0.10	2.2 ± 0.02	2.49 ± 0.03	67
Roja de Copotaque	8.3 ± 0.07	11.51 ± 0.10	5.22 ± 0.08	2.26 ± 0.02	2.93 ± 0.05	70.8
Witulla	8.81 ± 0.08	12.28 ± 0.00	5.32 ± 0.01	2.62 ± 0.02	2.57 ± 0.04	69.5
03-21-0093	8.47 ± 0.07	11.79 ± 0.11	nd	Nd	2.76 ± 0.02	nd
Salcedo INIA	8.26 ± 0.05	13.23 ± 0.01	5.3 ± 0.09	5.3 ± 0.2	2.37 ± 0.05	70
Commercial 1	10.13 ± 0.05	13.18 ± 0.01	6.51 ± 0.04	6.51 ± 0.03	2.34 ± 0.10	63.6
Commercial 2	11.51 ± 0.04	13.48 ± 0.06	6.34 ± 0.07	6.34 ± 0.03	2.27 ± 0.10	59.4
Huaripongo	10.34 ± 0.02	11.32 ± 0.01	6.14 ± 0.01	6.14 ± 0.01	2.92 ± 0.04	67.8
03-21-1181	9.37 ± 0.06	11.89 ± 0.02	3.95 ± 0.03	3.95 ± 0.01	3.12 ± 0.02	69.8
Mean ± SD	9.34 ± 1.1 ^{a}	12.61 ± 1.1 ^{a}	5.66 ± 0.09 ^{a}	5.66 ± 1.7 ^{a}	2.66 ± 0.3 ^{a}	67.3 ± 3.7 ^{a}

Means within a column with the same superscript letter are not significantly different ($p < 0.05$).

3.4. *Starch*

Starch is the key constituent in the quinoa seed and constitutes up to 70% of seed biomass. In starch, several polysaccharides joint together along with several glucose subunits. These molecules are mostly joined together through glycoside bonds. It is composed of two polymers: amylose, which is a linear polymer in which the glucose units are linked together with α $(1 \rightarrow 4)$, and amylopectin bonds, which is a branched polymer that has basic structure chains similar to amylose that arrange themselves to form a branched structure through the insertion of side chains through α bonds ($1 \rightarrow 6$). Several studies have been undertaken to characterize quinoa starch [52,53]. The granules are located in the perisperm, singly or aggregated to form more complex structures, formed by hundreds of individual granules [54]. The starch in quinoa is highly branched, with a minimum polymerization of 4600 units of glucan and a maximum of 161,000 units. The length of the chain may depend on the variety of quinoa but is in the order of 500–6000 glucose units. The average degree of polymerization of quinoa amylose (900) is lower than that of barley (1700). Amylose has an average of 11.6 chains per molecule; the length of the quinoa amylopectin chain is about 6700 glucan units. Quinoa amylopectin consists of short chains (8 to 12 units) and longer chains (13–20 units), compared to the other starches of the endosperm cereals. Quinoa starch has low digestibility and extractability due to protein binding [55].

Several studies documented the presence of amylose in quinoa, ranging from 7% to 27%, which is much higher compared to that in barley and rice [56]. The starch gelatinization properties indirectly depend upon different factors that include the size, the proportion and the type of crystalline structure and the ultrastructure of the starch granules. Quinoa starch gelatinizes at relatively low temperatures, similar to those for the gelatinization of wheat and potato starch, but lower than the gelatinization temperature of other large granules.

3.5. Fiber

Quinoa is generally considered an important source of fiber. The quinoa washing and abrasion processes to eliminate saponins do not significantly influence the fiber content [57]. According to some reports [58], dietary fiber contents in quinoa are equal to those of cereal and legume grains, but their amount is less than buckwheat. According to research conducted on animal models, quinoa grains exhibit arabinans and rhamnogalacturonan-I and have shown stomach metabolic disorders in rats [59]. Although, the fiber composition of quinoa is different from that of other cereal grains, the biochemical and therapeutic potential of quinoa should be studied to understand its specific physiological impact.

3.6. Minerals

The outer layers of the pericarp possess several minerals such as potassium, calcium, magnesium and phosphorus [29,52]. Based on the recommended daily dose (RDA), quinoa is a good source of magnesium, phosphorus and iron. In this context, the iron content (8–9 mg/100 g) is higher than that in common cereals; however, the presence of antinutritional compounds such as saponins and phytic acid can significantly reduce the bioavailability of some minerals [60]. Different treatments such as decortication and ash washing reduce the mineral content of 12–15% for iron, zinc and potassium, while it is 27% and 3% for copper and magnesium, respectively.

3.7. Vitamins

Quinoa is a good source of vitamins E and B and folic acid, which are concentrated in the embryo [61,62]. Quinoa possesses a significant amount of vitamin C, riboflavin, folic acid and thiamine compared to other cereals. The dietary requirements of an adult can be met through the consumption of 100 g of quinoa, which may provide the required pyridoxine and folic acid levels. Consumption of 100 g of quinoa grains may fulfill 80% of the DR of children and 40% of the protease, cellulase and hemicellulase requirements [63]. Mechanical abrasion increases the α-amylase and protease activity [63]. In contrast, the levels of cellulase and hemicellulase decrease with abrasion, thus suggesting the presence of the same in the pericarp.

4. Secondary Metabolites

4.1. Phytoecdysteroids

Among the secondary metabolites, phytoecdysteroids defend plants from insect pests and nematodes [64]. Phytoecdysteroids are polyhydroxylated compounds having a cyclopentanoperhydrophenanthrene ring. However, phytoecdysteroids present in the plants have a quite diverse structural configuration [65]. Phytoecdysteroids are mostly present in the bran portion of the grain as free and polar/non-polar conjugated compounds and are classified based on the number of carbon atoms in the structure as C27- and C28-phytoecdysteroids [66]. Quinoa is the only pseudo-cereal that possesses a significant quantity of phytoecdysteroids ranging from 138 to 570 μg/g [67]. The quinoa plant contains about 36 different types of phytoecdysteroids [68], out of which, C-27 phytoecdysteroids, which have several health benefits [68], are found in the highest concentration [45]. For instance, phytoecdysteroids possess anti-aging properties and are beneficial to prevent skin aging because of their anti-oxidant potential due to their metal ion chelating ability and free radical scavenging activity [69]. Moreover, phytoecdysteroids have also been used as a safe and effective replacement for anabolic steroids [70]. Additionally, these bioactive compounds are quite helpful in boosting the development of skeletal muscles and, thus, play an important role in increasing physical performance [71]. Likewise, several in vivo investigations have confirmed the role of quinoa phytoecdysteroids in preventing the problem of obesity. In a study, the supplementation of quinoa extract to mice fed with a high-fat diet was reported to be beneficial in controlling obesity [72]. The reduction in fat mass due to the dietary administration of quinoa is mainly due to elevated oxidation of carbohydrates and fecal defecation of lipids. Furthermore, the potential of quinoa phytoecdysteroids in preventing diabetes through reduced oxidative degeneration and improved blood glucose transportation has also been reported [73].

4.2. Saponins

Saponins are bioactive compounds present in the pericarp of quinoa grains that impart a bitter taste. Saponins are chemically composed of a triterpenoid or steroidal aglycone with one or more sugar moieties such as glucose, galactose, arabinose, xylose and glucuronic acid [74]. Saponins present in the quinoa grains are highly diverse in their structural characteristics, and studies reveal that about 68 different types of saponin compounds have been identified in crude quinoa seed extract by using nano-HPLC electrospray ionization multistage tandem mass spectrometry (nLC-ESI-MS/MS) [75]. The amount of saponins present in quinoa grains depends on its cultivar and can be classified into 'sweet' (<0.11% of saponins) or 'bitter' (>0.11% saponins) [76]. Meanwhile, bitter-flavored varieties of quinoa contain higher saponin concentrations as compared to sweet varieties [48]. Saponins are beneficial for crop protection against insect/bird herbivore and microbial infection, which accelerates the organic production of this crop. Saponins also possess several health-promoting impacts due to their vast pharmacological functionalities. For instance, dietary incorporation of saponins helps to lower the blood cholesterol levels due to their hemolytic action after directly interacting with the blood cells [77]. Additionally, saponins do not impart any toxic or harmful effect on human health after ingestion. However, saponin concentrations should be lowered during the processing of quinoa seeds as they impose an inhibitory effect on the digestibility and bioavailability of quinoa proteins. Saponin concentrations can be reduced by treating the quinoa seeds with cold, alkaline water followed by mechanical abrasion [78]. Moreover, several breeding strategies have also been adopted to develop new varieties of quinoa with lower saponin contents (<0.11% free saponins). The key functional aspects of quinoa saponins in improving human health include antiviral activity, antifungal ability, anticancer properties, antithrombotic effects, hypocholesterolemic potentiality, diuretic potential, hypoglycemic action and anti-inflammatory characteristics [2,75]. The antifungal effect of quinoa's saponins has been reported against *Candida albicans* [79]. The anti-inflammatory action of saponins is mainly attributed to the 3-O-β-D-glucopyranosyl oleanolic acid. Quinoa saponins decrease the levels of the inflammatory mediators and inhibit the release of inflammatory cytokines including interleukin-6 (IL-6) and tumor necrosis factor- (TNF-) and in lipopolysaccharide-induced RAW264.7 cells [46]. Due to their surfactant property, saponins can boost drug absorption through mucosal membranes. The property of insoluble complex formation of some saponins with minerals such as zinc and iron decrease the absorption and bioavailability of these minerals in the gut [47].

4.3. Phenolic Compounds

Phenolics consist of a large and diverse class of compounds comprising the hydroxyl group(s) attached to at least one aromatic hydrocarbon ring. Phenolics have high structural stability, which determines the strong anti-oxidant potential of these compounds [80]. Quinoa grains contain free phenolic compounds in the range of 167.2–308.3 mg gallic acid equivalents per 100 g of dry weight [81,82]. The range of a free fraction of total phenolic contents in seven varieties of quinoa was determined to be from 53.5% to 78.0%, among which ferulic acid and gallic acid were prominent compounds. Flavonoids in quinoa seeds include quercetin, rutin and kaempferol derivatives. The quinoa grains also possess bound phenolics, which are so named because they are bound with structures of the cell wall such as pectin, lignin cellulose, hemicellulose (arabinoxylans) and also with rod-shaped structural proteins [83–98]. Quinoa grains contain a lower amount of bound phenolics as compared to the free phenolics [14,81,85]. Bound phenolic compounds are mostly present in the leaves of quinoa but not in the seeds [84].

The phenolic compounds present in different varieties of quinoa play a vital role in preventing the problem of diabetes and obesity [81]. Over the last two decades, phenolic compounds have gained much interest due to their chronic-disease-prevention ability and health benefits [82]. Dietary phenolics help in maintaining gut health by regulating the microbial balance of the gut. Individual phenolic acids contribute to improving the metabolism and cell signaling and, thus, exert significant anticancer, anti-inflammatory, anti-obesity, antidiabetic and cardioprotective effects [80,83].

4.3.1. Phenolic Acids

There are free and chemically bound phenolic acids in the cell wall of quinoa leaves [84,85]. Quinoa leaves and seeds contain different types of phenolic acids and their derivatives hydroxycinnamic acid and hydroxybenzoic acid, which possess considerable health-promoting activities such as antihypertensive, anti-oxidative, antidiabetic, anti-inflammatory and anticarcinogenic effects. The concentration of phenolic acids in quinoa seeds varies from one to other variety [98–104] (Table 3 and Figure 2).

Table 3. Secondary metabolites identified from different quinoa varieties.

Variety	Observed Compounds/Reports	References
Minttumatilda	Vanillic acid, gallic acid, p-benzoic acid, syringaldehyde, ferulic acid	[99]
Four varieties of white (Q-w1), red (Q-r1 and Q-r2) and black (Q-b1) and three varieties of white (Q-w2), red (Q-r3) and black (Q-b2)	Gallic acid, caffeic acid, ferulic acid, p-coumaric acid, p-OH benzoic acid, vanillic acid, protocatechuic acid	[99]

Table 3. *Cont.*

Variety	Observed Compounds/Reports	References
Q25, Q50, Q100, Q25S, Q50S, Q100S	1-O-galloyl-β-D-glucose, acacetin/questin/apigenin-7-methyl ether, protocatechuic acid 4-O-glucoside, vanillic glucoside, penstebiosided, canthoside A/2-hydroxybenzoate 2-O-β-D-apiofuranosyl-(1→6)-O-β-Dglucopyranoside, ferulic acid 4-O-glucoside, ethyl-m-digallate, (epi) gallocatechin, quercetin 3-O-(2,6-di-α-L-rhamnopyranosyl)-β-D-galactopyranoside, kaempferol 3-O-β-D-apiofuranosyl(1‴→2″-O-[α-L-rhamnopyranosyl(1‴→6″]-β-D-galactopyranoside, kaempferol 3-O-β-D-apiofuranosyl(1‴→2″-O-[α-L-rhamnopyranosyl(1‴→ 6″]-β-D-galactopyranoside isomer, kaempferol 3-O-(2,6-di-α-L-rhamnopyranosyl)-β-D-galactopyranoside (mauritianin), quercetin 3-O-[β-D-apiofuranosyl(1‴→2″)]-β-D-galactopyranoside, rutin, quercetin glucuronide, quercetin 3-O-glucoside	[85]
Red quinoa (RQ), white quinoa (WQ), Mengli 1 (gray quinoa, GQ), Altiplano, djulis cultivar	Protocatechuic acid, p-hydroxybenzoic acid, vanillic acid, caffeic acid, syringic acid, p-coumaric acid, ferulic acid, isoferulic acid	[100]
White and red quinoa	Gentisic acid, L-alpha-hydroxy isovaleric acid, 3-hydroxybenzoic acid, p-coumaric acid, methyl-b-D-galactopyranoside, 3-(3,4-dihydroxyphenyl)propionic acid hesperidin	[100]
Puno, Titicaca	Ferulic acid, 5-O-caffeoylquinic acid, gentisic acid, p-coumaric acid, ellagic acid, pterostilbene, coniferyl aldehyde	[13]
White quinoa, red quinoa, black quinoa	Protocatechuic acid, p-hydroxybenzoic acid, vanillic acid, syringic acid, p-coumaric acid, ferulic acid, sinapic acid, isoferulic acid	[6]

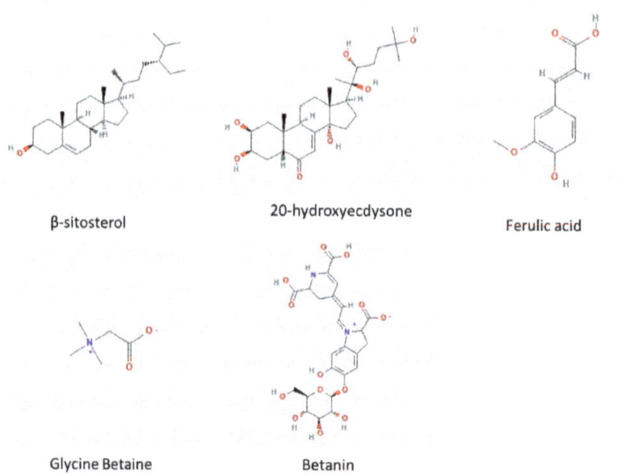

Figure 2. Secondary metabolites and bioactive molecules present in the quinoa grains.

4.3.2. Flavonoids

The quinoa plant has a vast variety of flavonoids including benzoic acid, vanillic acid, syringic acid, p-coumaric acid and ferulic acid [105]. Flavonoids are considered the second most abundantly present components in the quinoa seeds [81] (Table 4).

Table 4. Secondary metabolites (flavonoid) identified from different quinoa varieties.

References		Flavonoids
[101]		Quercetin 3-O-(2,6-di-α-Lrhamnopyranosyl)-β-Dgalactopyranoside
[85]		Quercetin 3-O-(2,6-di-α-Lrhamnopyranosyl)-β-Dgalactopyranoside
[85]		Quercetin-3-O-(2″-apiosyl)-rutinoside
[85]		Quercetin glucuronide
[101,102]		Kaempferol 3-O-(2,6-di-α-Lrhamnopyranosyl)-β-Dgalactopyranoside
[101,102]		Kaempferol 3-O-(β-D-apiofuranosyl (1″→2″)-α-L-rhamnopyranosyl) β- Dgalactopyranoside
[101]		Kaempferol 3-O-[β-D-apiofuranosyl (1″→2″)] β-D-galactopyranoside
[101]		Kaempferol-3-O-(α-Lrhamnopyranosyl) -β- Dgalactopyranoside, kaempferol glucuronide
[14]		Kaempferol 3-O-glucoside, kaempferol pentosyl rhamnoside, kaempferol
[13]	Puno, Titicaca	Quercetin, isorhamnetin, quercetin−3-Ogalactoside, isorhamnetin−3-Orutinoside, rutin, naringin, aesculin, phlorizin, eriodictyol
[103]	BRS Piabiru	Quercetin, kaempferol 3-O-rutinoside, quercetin 3-O-(2″,6″-di-O-α-L-rhamnoside)-β-D-galactoside, quercetin 3-O-(2″-O-β-apioside-6″-O-α-rhamnoside)-β-galactoside, kaempferol 3-O-(2″,6″-di-O-α-rhamnoside)-β-galactoside, kaempferol 3-O-(2″,6″-di-O-α-rhamnoside)-β-glucoside

4.3.2.1 Flavonol Glycosides

The flavonol glycosides are the most abundant flavonoids in the quinoa leaves and seeds [85]. Quinoa contains 12 different types of flavonol glycosides comprised of kaempferol and quercetin derivatives, with an average individual concentration of 839 μg/g on a dry weight basis [86].

4.3.2.2. Isoflavones

Plant-based isoflavones showed significant beneficial health impact to humans. Isoflavones were firstly identified in quinoa plants by Lutz et al. [87], whom confirmed the presence of daidzein in the range of 0.70 to 2.05 mg/100 g and genistein from 0.05–0.41 mg/100 g among seeds of 10 distinct varieties of quinoa.

4.4. Betalains

Quinoa contains betalain, which is a water-soluble phytochemical that acts as a natural anti-oxidant and, thus, helps in the prevention of cancer [88]. The red, black and yellow colors of quinoa seeds and their vegetative parts are due to betalain [10]. Betalain pigment contains nitrogen aromatic indole compounds derived from tyrosine (betacyanins that are red-violet and betaxanthins that are orange-red in color) [14]. Betalain is present in different varieties of quinoa in a diversified range. The betalain content (sum of betaxanthins and betacyanins) in some quinoa seeds ranges between 0.15 and 6.10 mg/100 g [89], while other varieties do not even contain detectable levels of betalain [90]. Betanin and isobetanin are most prominently present in the quinoa seeds and have similar health promoting activities such as anti-oxidant and antimicrobial and anti-inflammatory activity [14]. However, as compared to polyphenols, betalain shows higher anti-oxidant activity [91]. Betalain is used as an ingredient in functional foods due to its anticancer, antimicrobial, antilipidemic and anti-oxidant activities [92]. Microencapsulation has been studied recently to stabilize these compounds [93]. Maltodextrin microencapsulations having low saponins and high betacyanins show unique health-promoting activities. Betalain can be used as a natural dye due to its stability at pH 3–7. The United States Food and Drug Authority (U.S. FDA) and the European Union have approved betalain as a natural colorant with an E-number (E-162) in soups, sauce, dairy products pharmaceuticals and cosmetics [94]. Quinoa seed hulls are rich in betalain, which in combination with saponins produces a highly beneficial ingredient for food and pharmaceutical industries [95].

5. Conclusions

Quinoa is an important grain crop with excellent nutritive characteristics and phytochemical composition. Gluten-free quinoa grains are a valuable source of energy for several actors of society including children, the elderly and high-performance athletes. Quinoa has been a unique source of nutrition and food supplement to thousands of malnourished humans globally. Current knowledge regarding the use of quinoa and its human health benefits should be enhanced among all players and actors of society. Quinoa provides important fat-soluble anti-oxidant vitamins (vitamin group B and vitamin E). The availability and concentration of many minerals in quinoa are higher compared to those in other cereals. However, antinutritional compounds (saponins) should be removed to avoid the sensation of bitterness. Future research should include more and comprehensive clinical trials to gain more knowledge and understanding regarding possible action mechanisms of phenolics and flavonoids in human metabolic disorders and diseases.

Author Contributions: M.I.H., M.F., Q.A.S. and A.I. conceived and designed the study; M.I.H., M.F., Q.A.S., A.I., A.A.A.-G. and A.A.H. collected the literature, screened and draft the manuscript; M.I.H., M.F., Q.A.S. and A.I. wrote and revised the manuscript. A.A.A.-G. and A.A.H. help in correction and revision of the manuscript. All authors have read and agreed to the published version of the manuscript.

Funding: The authors extend their appreciation to the Deanship of Scientific Research at King Saud University for funding the work through research group project No. RG-1440-054.Funding source had no involvement in any phase of the study.

Institutional Review Board Statement: Not applicable.

Informed Consent Statement: Not applicable.

Data Availability Statement: This is a review article. All the data is presented in the manuscript.

Acknowledgments: The authors extend their appreciation to the Deanship of Scientific Research at King Saud University for funding the work through research group project No. RG-1440-054.

Conflicts of Interest: The authors declare no conflict of interest. The funders had no role in the design of the study; in the collection, analysis, or interpretation of data; in the writing of the manuscript, or in the decision to publish the results.

References

1. Maradini-Filho, A.M. Quinoa: Nutritional aspects. *J. Nutrac. Food Sci.* **2017**, *2*, 1–5.
2. Vega-Galvez, A.; Miranda, M.; Vergara, J.; Uribe, E.; Puente, L.; Martinez, E.A. Nutrition facts and functional potential of quinoa (*Chenopodium quinoa* Willd.). an ancient Andean grain: A review. *J. Sci. Food Agric.* **2010**, *90*, 2541–2547. [CrossRef]
3. National Research Council. *Lost Crops of the Incas: Little-Known Plants of the Andes with Promise for Worldwide Cultivation*; The National Academies Press: Washington, DC, USA, 1989. [CrossRef]
4. Jacobsen, S.-E.; Mujica, A.; Jensen, C.R. The resistance of quinoa (*Chenopodium quinoa* Willd.) to adverse abiotic factors. *Food Rev. Inter.* **2003**, *19*, 99–109. [CrossRef]
5. Saeidi, S.; Siadat, S.A.; Moshatati, A.; Moradi-Telavat, S.N. Effect of sowing time and nitrogen fertilizer rates on growth, seed yield and nitrogen use efficiency of quinoa (*Chenopodium quinoa* Willd) in Ahvaz. *Iran. J. Crop Sci.* **2020**, *21*, 354–367.
6. Gamboa, C.; Schuster, M.; Schrevens, E.; Maertens, M. Price volatility and quinoa consumption among smallholder producers in the Andes. *Sci. Agropecu.* **2020**, *11*, 113–125. [CrossRef]
7. Liu, M.; Zhu, K.; Yao, Y.; Chen, Y.; Guo, H.; Ren, G.; Yang, X.; Li, J. Antioxidant, anti-inflammatory, and antitumor activities of phenolic compounds from white, red, and black *Chenopodium quinoa* seed. *Cereal Chem.* **2020**, *97*, 703–713. [CrossRef]
8. Prego, I.; Maldonado, S.; Otegui, M. Seed structure and localization of reserves in *Chenopodium quinoa*. *Ann. Bot.* **1998**, *82*, 481–488. [CrossRef]
9. Ballester-Sánchez, J.; Gil, J.V.; Haros, C.M.; Fernández-Espinar, M.T. Effect of incorporating white, red or black quinoa flours on free and bound polyphenol content, antioxidant activity and colour of bread. *Plant Foods Human Nutr.* **2019**, *74*, 185–191. [CrossRef] [PubMed]
10. Bhargava, A.; Shukla, S.; Ohri, D. *Chenopodium quinoa*—An Indian perspective. *Ind. Crops Prod.* **2006**, *23*, 73–87. [CrossRef]

11. Katwal, T.B.; Bazile, D. First adaptation of quinoa in the Bhutanese mountain agriculture systems. *PLoS ONE* **2020**, *15*, e0219804. [CrossRef]
12. Sobota, A.; Świeca, M.; Gęsiński, K.; Wirkijowska, A.; Bochnak, J. Yellow-coated quinoa (*Chenopodium quinoa* Willd)—Physicochemical, nutritional, and antioxidant properties. *J. Sci. Food Agric.* **2020**, *100*, 2035–2042. [CrossRef]
13. Stikić, R.I.; Milinčić, D.D.; Kostić, A.Ž.; Jovanović, Z.B.; Gašić, U.M.; Tešić, Ž.L.; Djordjević, N.Z.; Savić, S.K.; Czekus, B.G.; Pešić, M.B. Polyphenolic profiles, antioxidant, and in vitro anticancer activities of the seeds of Puno and Titicaca quinoa cultivars. *Cereal Chem.* **2020**, *97*, 626–633. [CrossRef]
14. Tang, Y.; Li, X.; Chen, P.X.; Zhang, B.; Hernandez, M.; Zhang, H.; Marcone, M.F.; Liu, R.; Tsao, R. Characterisation of fatty acid, carotenoid, tocopherol/tocotrienol compositions and antioxidant activities in seeds of three *Chenopodium quinoa* Willd. genotypes. *Food Chem.* **2015**, *174*, 502–508. [CrossRef]
15. Scalbert, A.; Manach, C.; Morand, C.; Remesy, C.; Jimenez, L. Dietary polyphenols and the prevention of diseases. *Crit. Rev. Food Sci. Nutr.* **2005**, *45*, 287–306. [CrossRef]
16. Christensen, S.A.; Pratt, D.B.; Pratt, C.; Nelson, P.T.; Stevens, M.R.; Jellen, E.N.; Coleman, C.E.; Fairbanks, D.J.; Bonifacio, A.; Maughan, P.J. Assessment of genetic diversity in the USDA and CIP-FAO international nursery collections of quinoa (*Chenopodium quinoa* Willd.) using microsatellite markers. *Plant Genetic Resour.* **2007**, *5*, 82–95. [CrossRef]
17. Mujica, A.; Jacobsen, S.E.; Ezquierdo, J.; Marathee, J.P. *Resultados de la Prueba Americana y Europes de la Quinoa*; FAO: Rome, Italy, 2001; p. 51.
18. Mabry, T.J.; Taylor, A.; Turner, B.L. The betacyanins and their distribution. *Phytochemistry* **1963**, *2*, 61–64. [CrossRef]
19. Hunziker, A. Las especies alimenticias de Amaranthus y Chenopodium cultivadas por los indios de América. *Rev. Argent Agron.* **1943**, *10*, 297–354.
20. Risi, J.C.; Galwey, N.W. The pattern of genetic diversity in the Andean grain crop quinoa (*Chenopodium quinoa* Willd). I. Associations between characteristics. *Euphytica* **1989**, *41*, 147–162. [CrossRef]
21. Ward, S.M.; Johnson, D.L. Cytoplasmic Male Sterile Quinoa. U.S. Patent US5304718A, 19 April 1994.
22. Gandarillas, H. Botanica. In *Quinoa Kaniwa*; Tapia, M.E., Ed.; Cultivos Andinos Instituto Interamericano de Ciencias Agricolas: Bogota, Colombia, 1979; pp. 20–44.
23. Kozioł, M.J. Chemical composition and nutritional evaluation of quinoa (*Chenopodium quinoa* Willd.). *J. Food Compos. Analysis* **1992**, *5*, 35–68. [CrossRef]
24. Navruz-Varli, S.; Sanlier, N. Nutritional and health benefits of quinoa (*Chenopodium quinoa* Willd.). *J. Cereal Sci.* **2016**, *69*, 371–376. [CrossRef]
25. Pellegrini, M.; Lucas-Gonzales, R.; Ricci, A.; Fontecha, J.; Fernández-López, J.; Pérez-Álvarez, J.A.; Viuda-Martos, M. Chemical, fatty acid, polyphenolic profile, techno-functional and antioxidant properties of flours obtained from quinoa (*Chenopodium quinoa* Willd) seeds. *Ind. Crops Prod.* **2018**, *111*, 38–46. [CrossRef]
26. Nowak, V.; Du, J.; Charrondière, U.R. Assessment of the nutritional composition of quinoa (*Chenopodium quinoa* Willd.). *Food Chem.* **2016**, *193*, 47–54. [CrossRef]
27. Chen, Y.-S.; Aluwi, N.A.; Saunders, S.R.; Ganjyal, G.M.; Medina-Meza, I.G. Metabolic fingerprinting unveils quinoa oil as a source of bioactive phytochemicals. *Food Chem.* **2019**, *286*, 592–599. [CrossRef] [PubMed]
28. Romano, N.; Ureta, M.M.; Guerrero-Sánchez, M.; Gómez-Zavaglia, A. Nutritional and technological properties of a quinoa (*Chenopodium quinoa* Willd.) spray-dried powdered extract. *Food Res. Inter.* **2020**, *129*, 108884. [CrossRef] [PubMed]
29. Mhada, M.; Metougui, M.L.; El Hazzam, K.; El Kacimi, K.; Yasri, A. Variations of saponins, minerals and total phenolic compounds due to processing and cooking of quinoa (*Chenopodium quinoa* Willd.) seeds. *Foods* **2020**, *9*, 660. [CrossRef] [PubMed]
30. Lalaleo, L.; Hidalgo, D.; Valle, M.; Calero-Cáceres, W.; Lamuela-Raventós, R.M.; Becerra-Martínez, E. Differentiating, evaluating, and classifying three quinoa ecotypes by washing, cooking and germination treatments, using 1H NMR-based metabolomic approach. *Food Chem.* **2020**, *331*, 127351. [CrossRef]
31. Fotschki, B.; Juśkiewicz, J.; Jurgoński, A.; Amarowicz, R.; Opyd, P.; Bez, J.; Muranyi, I.; Petersen, I.L.; Llopis, M.L. Protein-rich flours from quinoa and buckwheat favourably affect the growth parameters, intestinal microbial activity and plasma lipid profile of rats. *Nutrients* **2020**, *12*, 2781. [CrossRef]
32. Miranda, M.; Vega-Galvez, A.; Martinez, E.; Lopez, J.; Rodriguez, M.J.; Henriquez, K.; Fuentes, F. Genetic diversity and comparison of physiochemical and nutritional characteristics of six quinoa (*Chenopodium quinoa* Willd.) genotypes cultivated in Chile. *Cienc. Technol. Aliment. Campinas* **2012**, *32*, 835–843. [CrossRef]
33. Ruales, J.; de Grijalva, Y.; Lopez-Jaramillo, P.; Nair, B.M. The nutritional quality of an infant food from quinoa and its effect on the plasma level of insulin-like growth factor-1 (IGF-1) in undernourished children. *Int. J. Food Sci. Nutr.* **2002**, *53*, 143–154. [CrossRef]
34. Farinazzi-Machado, F.M.V.; Barbalho, S.M.; Oshiiwa, M.; Goulart, R.; Pessan Junior, O. Use of cereal bars with quinoa (*Chenopodium quinoa* W.) to reduce risk factors related to cardiovascular diseases. *Cienc. Technol. Aliment. Campinas* **2012**, *32*, 239–244. [CrossRef]
35. De Carvalho, F.G.; Ovidio, P.P.; Padovan, G.J.; Jordao Junior, A.A.; Marchini, J.S.; Navarro, A.M. Metabolic parameters of postmenopausal women after quinoa or corn flakes intake-a prospective and double-blind study. *Int. J. Food Sci. Nutr.* **2014**, *65*, 380–385. [CrossRef]
36. Zevallos, V.F.; Herencia, I.L.; Chang, F.; Donnelly, S.; Ellis, J.H.; Ciclitira, P.J. Gastrointestinal effects of eating quinoa (*Chenopodium quinoa* Willd.) in celiac patients. *Am. J. Gastroenterol.* **2014**, *109*, 270–278. [CrossRef] [PubMed]

37. Navarro-Perez, D.; Radcliffe, J.; Tierney, A.; Jois, M. Quinoa seed lowers serum triglycerides in overweight and obese subjects: A dose-response randomized controlled clinical trial. *Curr. Dev. Nutr.* **2017**, *1*, e001321. [CrossRef] [PubMed]
38. Li, L.; Lietz, G.; Bal, W.; Watson, A.; Morfey, B.; Seal, C. Effects of quinoa (*Chenopodium quinoa* Willd.) consumption on markers of CVD risk. *Nutrients* **2018**, *10*, 777. [CrossRef]
39. Brinegar, C.; Sine, B.; Nwokocha, L. High-cysteine 2S seed storage proteins from quinoa (*Chenopodium quinoa*). *J. Agric. Food Chem.* **1996**, *44*, 1621–1623. [CrossRef]
40. Brinegar, C.; Goundan, S. Isolation and characterization of chenopodin, the 11S seed storage protein from quinoa (*Chenopodium quinoa* Willd). *J. Agric. Food Chem.* **1993**, *41*, 182–185. [CrossRef]
41. Wang, J.; de Wit, M.; Schutyser, M.A.I.; Boom, R.M. Analysis of electrostatic powder charging for fractionation of foods. *Innov. Food Sci. Emerg. Technol.* **2014**, *26*, 360–365. [CrossRef]
42. Pelgrom, P.J.M.; Vissers, A.M.; Boom, R.M.; Schutyser, M.A.I. Dry fractionation for production of functional pea protein concentrates. *Food Res. Int.* **2013**, *3*, 232–239. [CrossRef]
43. Motta, C.; Castanheira, I.; Gonzales, G.B.; Delgado, I.; Torres, D.; Santos, M.; Matos, A.S. Impact of cooking methods and malting on amino acids content in amaranth, buckwheat and quinoa. *J. Food Comp. Anal.* **2019**, *76*, 58–65. [CrossRef]
44. Benito-Román, O.; Rodríguez-Perrino, M.; Sanz, M.T.; Melgosa, R.; Beltrán, S. Supercritical carbon dioxide extraction of quinoa oil: Study of the influence of process parameters on the extraction yield and oil quality. *J. Supercritical Fluids* **2018**, *139*, 62–71. [CrossRef]
45. Graf, B.L.; Rojo, L.E.; Delatorre-Herrera, J.; Poulev, A.; Calfio, C.; Raskin, I. Innovations in health value and functional food development of quinoa (*Chenopodium quinoa* Willd.). *Compr. Rev. Food Sci Safety.* **2015**, *14*, 431–445. [CrossRef]
46. Yao, Y.; Yang, X.; Shi, Z.; Ren, G. Anti-inflammatory activity of saponins from quinoa (*Chenopodium quinoa* Willd.) seeds in lipopolysaccharide-stimulated RAW 264.7 macrophages cells. *J. Food Sci.* **2014**, *79*, H1018–H1023. [CrossRef]
47. Jancurová, M.; Minarovičová, L.; Dandár, A. Quinoa—A review. *Czech J. Food Sci.* **2009**, *27*, 71–79. [CrossRef]
48. Valencia-Chamorro, S.A. Quinoa. In *Encyclopedia of Food Science and Nutrition*; Caballero, B., Ed.; Academic Press: Amsterdam, The Netherlands, 2003; pp. 4895–4902.
49. Jan, K.N.; Panesar, P.S.; Singh, S. Effect of moisture content on the physical and mechanical properties of quinoa seeds. *Int. Agrophys.* **2019**, *33*, 41–48. [CrossRef]
50. González, J.A.; Roldán, A.; Gallardo, M.; Escudero, T.; Prado, F.E. Quantitative determinations of chemical compounds with nutritional value from Inca crops: *Chenopodium quinoa* ("quinoa"). *Plant Foods Hum. Nutr.* **1989**, *39*, 331–337. [CrossRef] [PubMed]
51. Gross, R.; Koch, F.; Malaga, I.; De Miranda, A.F.; Schoeneberger, H.; Trugo, L.C. Chemical composition and protein quality of some local Andean food sources. *Food Chem.* **1989**, *34*, 25–34. [CrossRef]
52. Contreras-Jiménez, B.; Torres-Vargas, O.L.; Rodríguez-García, M.E. Physicochemical characterization of quinoa (*Chenopodium quinoa*) flour and isolated starch. *Food Chem.* **2019**, *298*, 124982. [CrossRef]
53. Kierulf, A.; Whaley, J.; Liu, W.; Enayati, M.; Tan, C.; Perez-Herrera, M.; You, Z.; Abbaspourrad, A. Protein content of amaranth and quinoa starch plays a key role in their ability as pickering emulsifiers. *Food Chem.* **2020**, *315*, 126246. [CrossRef]
54. Lorenz, K. Quinoa (*Chenopodium quinoa* Willd) starch: Physico-chemical properties and functional characteristics. *Starch/Staerke* **1990**, *42*, 81–86. [CrossRef]
55. Ruales, J.; Nair, B. Properties of starch and dietary fibre in quinoa (*Chenopodium quinoa* Willd) seeds. *Plant Foods Hum. Nutr.* **1994**, *45*, 223–246. [CrossRef]
56. Goyat, J.; Rudra, S.G.; Suri, S.; Passi, S.J.; Dutta, H. Nutritional, functional and sensory properties of ready-to-eat chia and quinoa mix enriched low amylose rice based porridge mixes. *Curr. Res. Nutr. Food Sci. J.* **2019**, *7*, 399–414. [CrossRef]
57. Ranhotra, G.; Gelroth, J.; Glaser, B.; Lorenz, K.; Johnson, D. Composition and protein nutritional quality of quinoa. *Cereal Chem.* **1993**, *3*, 303–305.
58. Alvarez-Jubete, L.; Arendt, E.K.; Gallagher, E. Nutritive value and chemical composition of pseudocereals as gluten-free ingredients. *Inter. J. Food Sci. Nutr.* **2009**, *60*, 240–257. [CrossRef] [PubMed]
59. Lamothe, L.M.; Srichuwong, S.; Reuhs, B.L.; Hamaker, B.R. Quinoa (*Chenopodium quinoa* W.) and amaranth (*Amaranthus caudatus* L.) provide dietary fibres high in pectic substances and xyloglucans. *Food Chem.* **2015**, *15*, 490–496. [CrossRef]
60. Ruales, J.; Nair, B. Content of fat, vitamins and minerals in quinoa (*Chenopodium quinoa*, Willd) seeds. *Food Chem.* **1993**, *48*, 131–136. [CrossRef]
61. Basantes-Morales, E.R.; Alconada, M.M.; Pantoja, J.L. Quinoa (*Chenopodium quinoa* Willd) Production in the Andean Region: Challenges and potentials. *J. Exper. Agric. Int.* **2019**, *36*, 1–18. [CrossRef]
62. Abugoch, L.E. Quinoa (*Chenopodium quinoa* Willd.): Composition, chemistry, nutritional and functional properties. *Adv. Food Nutr. Res.* **2009**, *58*, 1–31.
63. Mohyuddin, S.G.; Riaz, A.; Qamar, A.; Ali, S.H.; Hu, C.; Wu, L.; Yu, T.; Ju, X.H. Quinoa is beneficial to the comprehensive nutritional value of potential health. *Pakistan J. Sci.* **2019**, *70*, 69–74.
64. Lorenz, K.; Nyanzi, F. Enzyme activities in quinoa (*Chenopodium quinoa* Willd). *Int. J. Food Sci. Technol.* **1989**, *24*, 543–551. [CrossRef]
65. Dinan, L.; Lafont, R. Innovative and future applications for ecdysteroids. In *Ecdysone: Structures and Functions*; Smagghe, G., Ed.; Springer: Dordrecht, The Netherlands, 2009; pp. 551–578.

66. Bajguz, A.; Bąkała, I.; Talarek, M. Ecdysteroids in plants and their pharmacological effects in vertebrates and humans. In *Studies in Natural Products Chemistry*; Rahman, A., Ed.; Elsevier: New York, NY, USA, 2015; Volume 45, pp. 121–145.
67. Kumpun, S.; Maria, A.; Crouzet, S.; Evrard-Todeschi, N.; Girault, J.P.; Lafont, R. Ecdysteroids from *Chenopodium quinoa* Willd., an ancient Andean crop of high nutritional value. *Food Chem.* **2011**, *125*, 1226–1234. [CrossRef]
68. Graf, B.L.; Rojo, L.E.; Delatorre-Herrera, J.; Poulev, A.; Calfio, C.; Raskin, I. Phytoecdysteroids and flavonoid glycosides among Chilean and commercial sources of *Chenopodium quinoa*: Variation and correlation to physico-chemical characteristics. *J. Sci. Food Agric.* **2016**, *96*, 633–643. [CrossRef]
69. Lin, M.; Han, P.; Li, Y.; Wang, W.; Lai, D.; Zhou, L. Quinoa secondary metabolites and their biological activities or functions. *Molecules* **2019**, *24*, 2512. [CrossRef]
70. Nsimba, Y.R.; Kikuzaki, H.; Konishi, Y. Ecdysteroids act as inhibitors of calf skin collagenase and oxidative stress. *J. Biochem. Mol. Toxicol.* **2008**, *22*, 240–250. [CrossRef]
71. Bathori, M.; Tóth, N.; Hunyadi, A.; Márki, A.; Zádor, E. Phtoecdysteroids and anabolic-androgenic steroids—Structure and effects in humans. *Curr. Med. Chem.* **2008**, *15*, 75–91. [CrossRef]
72. Gorelick-Feldman, M.D.; Maclean, D.; Ilic, N.; Poulev, A.; Lila, M.A.; Cheng, D.; Raskin, I. Phytoecdysteroids increase protein synthesis in skeletal muscle cells. *J. Agric. Food Chem.* **2008**, *56*, 3532–3537. [CrossRef]
73. Foucault, A.S.; Even, P.; Lafont, R.; Dioh, W.; Veillet, S.; Tomé, D.; Huneau, J.F.; Hermier, D.; Quignard-Boulangé, A. Quinoa extract enriched in 20-hydroxyecdysone affects energy homeostasis and intestinal fat absorption in mice fed a high-fat diet. *Physiol. Behav. J.* **2014**, *128*, 226–231. [CrossRef] [PubMed]
74. Xia, X.; Zhang, Q.; Liu, R.; Wang, Z.; Tang, N.; Liu, F.; Huang, G.; Jiang, X.; Gui, G.; Wang, L.; et al. Effects of 20-hydroxyecdysone on improving memory deficits in streptozotocin-induced type 1 diabetes mellitus in rat. *Eur. J. Pharmacol.* **2014**, *740*, 45–52. [CrossRef] [PubMed]
75. Yendo, A.C.A.; De Costa, F.; Gosmann, G.; Fett-Neto, A.G. Production of plant bioactive triterpenoid saponins: Elicitation strategies and target genes to improve yields. *Mol. Biotechnol.* **2010**, *46*, 94–104. [CrossRef]
76. Dini, I.; Tenore, G.C.; Dini, A. Antioxidant compound contents and antioxidant activity before and after cooking in sweet and bitter *Chenopodium quinoa* seeds. *LWT-Food Sci. Tech.* **2010**, *43*, 447–451. [CrossRef]
77. Gomez-Caravaca, A.M.; Iafelice, G.; Lavini, A.; Pulvent, C.; Caboni, M.F.; Marconi, E. Phenolic compounds and saponins in quinoa samples (*Chenopodium quinoa* Willd.) grown under different saline and nonsaline irrigation regimens. *J. Agric. Food Chem.* **2012**, *60*, 4620–4627. [CrossRef] [PubMed]
78. Macarena, S.; Ricardo, S.N. Antifungal properties of quinoa (*Chenopodium quinoa* Willd) alkali treated saponins against *Botrytis cinerea*. *Ind. Crops Prod.* **2008**, *27*, 296–302. [CrossRef]
79. Vilche, C.; Gely, M.; Santalla, E. Physical properties of quinoa seeds. *Biosyst. Engin.* **2003**, *86*, 59–65. [CrossRef]
80. Woldemichael, G.M.; Wink, M. Identification and biological activities of triterpenoid saponins from *Chenopodium quinoa*. *J. Agric. Food Chem.* **2001**, *49*, 2327–2332. [CrossRef]
81. Harborne, J.B.; Williams, C.A. Advances in flavonoid research since 1992. *Phytochemistry* **2000**, *55*, 481–504. [CrossRef]
82. Han, Y.; Chi, J.; Zhang, M.; Zhang, R.; Fan, S.; Dong, L.; Huang, F.; Liu, L. Changes in saponins, phenolics and antioxidant activity of quinoa (*Chenopodium quinoa* willd) during milling process. *LWT Food Sci. Technol.* **2019**, *114*, 108381. [CrossRef]
83. Okarter, N.; Liu, R.H. Health benefits of whole grain phytochemicals. *Crit. Rev. Food Sci. Nutr.* **2010**, *50*, 193–208. [CrossRef]
84. Da-Silva, W.S.; Harney, J.W.; Kim, B.W.; Li, J.; Bianco, S.D.; Crescenzi, A.; Christoffolete, M.A.; Huang, S.A.; Bianco, A.C. The small polyphenolic molecule kaempferol increases cellular energy expenditure and thyroid hormone activation. *Diabetes* **2007**, *56*, 767–776. [CrossRef]
85. Renard, C.M.; Wende, G.; Booth, E.J. Cell wall phenolics and polysaccharides in different tissues of quinoa (*Chenopodium quinoa* Willd). *J. Sci. Food Agric.* **1999**, *79*, 2029–2034. [CrossRef]
86. Gómez-Caravaca, A.M.; Segura-Carretero, A.; Fernández-Gutiérrez, A.; Caboni, M.F. Simultaneous determination of phenolic compounds and saponins in quinoa (*Chenopodium quinoa* Willd) by a liquid chromatography–diode array detection–electrospray ionization–time-of-flight mass spectrometry methodology. *J. Agric. Food Chem.* **2011**, *59*, 10815–10825. [CrossRef]
87. Lutz, M.; Martínez, A.; Martínez, E.A. Daidzein and genistein contents in seeds of quinoa (*Chenopodium quinoa* Willd.) from local ecotypes grown in arid Chile. *Ind. Crops Prod.* **2013**, *49*, 117–121. [CrossRef]
88. Samtiya, M.; Aluko, R.E.; Dhewa, T.; Moreno-Rojas, J.M. Potential health benefits of plant food-derived bioactive components: An overview. *Foods* **2021**, *10*, 839. [CrossRef] [PubMed]
89. Escribano, J.; Cabanes, J.; Jiménez-Atiénzar, M.; Ibañez-Tremolada, M.; Gómez-Pando, L.R.; García-Carmona, F.; Gandía-Herrero, F. Characterization of betalains, saponins and antioxidant power in differently colored quinoa (*Chenopodium quinoa*) varieties. *Food Chem.* **2017**, *234*, 285–294. [CrossRef] [PubMed]
90. Abderrahim, F.; Huanatico, E.; Segura, R.; Arribas, S.; Gonzalez, M.C.; Condezo-Hoyos, L. Physical features, phenolic compounds, betalains and total antioxidant capacity of coloured quinoa seeds (*Chenopodium quinoa* Willd.) from Peruvian Altiplano. *Food Chem.* **2015**, *183*, 83–90. [CrossRef] [PubMed]
91. Ann-Mari Repo-Carrasco-Valencia, R.; Lesli Astuhuaman, S. Quinoa (*Chenopodium quinoa*, Willd.) as a source of dietary fiber and other functional components. *Food Sci. Technol.* **2011**, *31*, 225–230. [CrossRef]
92. Neagu, C.; Barbu, V. Principal component analysis of the factors involved in the extraction of beetroot betalains. *J. Agroaliment. Proc. Technol.* **2014**, *20*, 311–318.

93. Gengatharan, A.; Dykes, G.A.; Choo, W.S. Betalains: Natural plant pigments with potential application in functional foods. *LWT Food Sci. Technol.* **2015**, *64*, 645–649. [CrossRef]
94. Aguilar-Tuesta, S.; Mamani-Navarro, W.; Espinoza-Silva, C.; Basilio-Atencio, J.; Condezo-Hoyos, L. Microencapsulated betacyanin from colored organic quinoa (*Chenopodium quinoa* Willd.): Optimization, physicochemical characterization and accelerated storage stability. *J. Sci. Food Agric.* **2018**, *98*, 5873–5883. [CrossRef] [PubMed]
95. Esatbeyoglu, T.; Wagner, A.E.; Schiniatbeyo, V.B.; Rimbach, G. Betanin—A food colorant with biological activity. *Mol. Nutr. Food Res.* **2015**, *59*, 36–47. [CrossRef] [PubMed]
96. Laqui-Vilca, C.; Aguilar-Tuesta, S.; Mamani-Navarro, W.; Montaño-Bustamante, J.; Condezo-Hoyos, L. Ultrasound-assisted optimal extraction and thermal stability of betalains from colored quinoa (*Chenopodium quinoa* Willd) hulls. *Ind. Crops Prod.* **2018**, *111*, 606–614. [CrossRef]
97. Tang, Y.; Zhang, B.; Li, X.; Chen, P.X.; Zhang, H.; Liu, R.; Tsao, R. Bound phenolics of quinoa seeds released by acid, alkaline, and enzymatic treatments and their antioxidant and α-glucosidase and pancreatic lipase inhibitory effects. *J. Agric. Food Chem.* **2016**, *64*, 1712–1719. [CrossRef]
98. Chen, P.X.; Tang, Y.; Marcone, M.F.; Pauls, P.K.; Zhang, B.; Liu, R.; Tsao, R. Characterization of free, conjugated and bound phenolics and lipophilic antioxidants in regular-and non-darkening cranberry beans (*Phaseolus vulgaris* L.). *Food Chem.* **2015**, *185*, 298–308. [CrossRef]
99. Acosta-Estrada, B.A.; Gutiérrez-Uribe, J.A.; Serna-Saldívar, S.O. Bound phenolics in foods, a review. *Food Chem.* **2014**, *152*, 46–55. [CrossRef] [PubMed]
100. Multari, S.; Marsol-Vall, A.; Keskitalo, M.; Yang, B.; Suomela, J.P. Effects of different drying temperatures on the content of phenolic compounds and carotenoids in quinoa seeds (*Chenopodium quinoa*) from Finland. *J. Food Compos. Anal.* **2018**, *72*, 75–82. [CrossRef]
101. Zhang, Q.; Xing, B.; Sun, M.; Zhou, B.; Ren, G.; Qin, P. Changes in bio-accessibility, polyphenol profile and antioxidants of quinoa and djulis sprouts during in vitro simulated gastrointestinal digestion. *Food Sci. Nutr.* **2020**, *8*, 4232–4241. [CrossRef] [PubMed]
102. De Simone, F.; Dini, A.; Pizza, C.; Saturnino, P.; Schettino, O. Two flavonol glycosides from *Chenopodium quinoa*. *Phytochemistry* **1990**, *29*, 3690–3692. [CrossRef]
103. Sampaio, S.L.; Fernandes, A.; Pereira, C.; Calhelha, R.C.; Sokovic, M.; Santos-Buelga, C.; Ferreira, I.C.F.R. Nutritional value, physicochemical characterization and bioactive properties of Brazilian quinoa BRS Piabiru. *Food Funct.* **2020**, *11*, 2969–2977. [CrossRef] [PubMed]
104. Gorinstein, S.; Lojek, A.; Číž, M.; Pawelzik, E.; Delgado-Licon, E.; Medina, O.J.; Moreno, M.; Salas, I.A.; Goshev, I. Comparison of composition and antioxidant capacity of some cereals and pseudocereals. *Inter. J. Food Sci. Technol.* **2008**, *43*, 629–637. [CrossRef]
105. Tsao, R. Chemistry and biochemistry of dietary polyphenols. *Nutrients* **2010**, *2*, 1231–1246. [CrossRef]

plants MDPI

Article

Structural Characterization and Antioxidant Capacity of Quinoa Cultivars Using Techniques of FT-MIR and UHPLC/ESI-Orbitrap MS Spectroscopy

Miguel García-Parra [1,*], Diego Roa-Acosta [2], Víctor García-Londoño [3], Brigitte Moreno-Medina [4] and Jesús Bravo-Gomez [2]

1 Doctoral Program in Agriculture and Agroindustrial Science, Agriculture Department, Universidad del Cauca, Popayán 190002, Colombia
2 Agroindustry Department, Faculty of Agricultural Sciences, Universidad del Cauca, Popayán 190002, Colombia; droa@unicauca.edu.co (D.R.-A.); jebravo@unicauca.edu.co (J.B.-G.)
3 Institute of Polymer and Nanotechnology, Facultad de Arquitectura Diseño y Urbanismo, University of Buenos Aires-CONICET, Intendente Güiraldes 2160, Buenos Aires C1428EGA, Argentina; vgarcia@qo.fcen.uba.ar
4 Facultad de Ciencias Agropecuarias, Universidad Pedagógica y Tecnológica de Colombia (UPTC), Tunja 150002, Colombia; brigitte.moreno@uptc.edu.co
* Correspondence: miguelgarciap@unicauca.edu.co

Abstract: The existence of more of 16,000 varieties of quinoa accessions around the world has caused a disregard on their structural and phytochemical characteristics. Most of such accessions belong to cultivars settled in Colombia. The goal of this research was to evaluate the structural attributes and antioxidant capacities from six quinoa cultivars with high productive potential from central regions in Colombia. This study used middle-range infrared spectroscopy (IR-MIR) to determine the proteins, starch and lipids distinctive to quinoa grains. Ultra-high-performance liquid chromatography electrospray ionization Orbitrap, along with high-resolution mass spectrometry (UHPLC/ESI-Orbitrap MS), were also used to identify the existence of polyphenols in cultivars. The antioxidant capacity was determined through DPPH, ABTS and FRAP. The spectrums exhibited significant variances on the transmittance bands associated with 2922 cm^{-1}, 1016 cm^{-1} and 1633 cm^{-1}. Moreover, the intensity variations on the peaks from the secondary protein structure were identified, mainly on the bands associated with β-Sheet-1 and -2, random coil α elice and β-turns-2 and -3. Changes found in the ratios 996 cm^{-1}/1014 cm^{-1} and 1041 cm^{-1}/1014 cm^{-1} were associated with the crystalline/amorphous affinity. Regarding the antioxidant capacity, great differences were identified ($p < 0.001$) mainly through FRAP methods, while the phenolic acids and flavonoids were determined by UHPLC/ESI-Orbitrap MS techniques. The presence of apigenin and pinocembrin on grains was reported for the first time. Titicaca and Nariño were the most phytochemically diverse quinoa seeds.

Keywords: Amaranthaceae; chromatography; polyphenol; quinoa starch; secondary protein

Citation: García-Parra, M.; Roa-Acosta, D.; García-Londoño, V.; Moreno-Medina, B.; Bravo-Gomez, J. Structural Characterization and Antioxidant Capacity of Quinoa Cultivars Using Techniques of FT-MIR and UHPLC/ESI-Orbitrap MS Spectroscopy. *Plants* **2021**, *10*, 2159. https://doi.org/10.3390/plants10102159

Academic Editors: Cataldo Pulvento and Didier Bazile

Received: 3 August 2021
Accepted: 28 August 2021
Published: 12 October 2021

Publisher's Note: MDPI stays neutral with regard to jurisdictional claims in published maps and institutional affiliations.

1. Introduction

According to different scientific reports, quinoa had its origins in South America between Colombia and Chile. However, its greater abundance in terms of morphologic, ecophysiology and nutritional diversity happens in countries like Ecuador, Bolivia, Perú and Chile. The distribution of cultivars happens according to their source of origin as quinoa from: the altiplano Inter-Andean Valley, from the coast and from Salares and Yungas [1]. Quinoa species show a great adaptability to harsh edaphoclimatic conditions, mainly related to dramatic changes in the temperature, salinity stress and lower availability of water and nutrients [2].

Along with quinoa, amaranth, buckwheat and chia were underutilized for the effect of introduction in crops such as rice, wheat, soybean and barley. However, underutilized species are important components of the local agriculture and comprise a broad variety of crops that are traditionally used and that may have potential for adaptation to climate change, medicinal properties and functional food development [3]. However, the quinoa cultivars incorporated into genetic improvement can reduce the richness of this species [4,5].

Under tropical conditions, grains were the main nutritional source for many pre-Hispanic cultures. Nowadays, quinoa is recognized in the food industry for some of its nutritional and techno-functional properties linked to its lipidic, protein and starchy nature [6,7].

Since 2013, the FAO boost production, transformation and commercialization of quinoa crops has caused an exponential rise in grain consumption, as well as its byproducts. Such a fact has benefited the increase of crops areas to select those with better agroclimatic adaptability, therefore improving hectare production [3,8], while the high genetic diversity could allow identifying the most suitable cultivar for each agroecological condition, considering its productive and nutritional performance [1].

Furthermore, it is relevant to consider the influence of different external conditions of biotic, abiotic and genetic attributes of such cultivars in relation to the nutritional characteristics of grains to which some researchers reported their findings. Authors like Reguera et al. (2018) [9] explained the results under torrid and outer equatorial conditions. García-Parra et al. (2019) [10] addressed the outcome of using different fertilization levels and sources. Lesjak & Calderini (2017) [11] reported modifying nighttime temperatures. Miranda et al. (2013) [12] evaluated environmental conditions over two consecutive years. These findings about analyses related to the techno-functional characteristics of quinoa grains should be applied to all cultivars according to the agroclimatic conditions.

Agro-industrial and bioactive characteristics in Colombian quinoa cultivars have been understudied; they lack information or show discrepancies between cultivars [13,14]. According to previously described information, all properties must be taken into account at the time to evaluate the nature of quinoa for its use in the food industry.

The increase in demand means that the prices of quinoa seeds and flours has increased rapidly worldwide, also increasing the risk of possible adulterations, requiring rapid and efficient analytical techniques for this detection. Additionally, the use of quinoa to produce nutraceuticals and functional foods requires a rapid and efficient analysis of their characterizations and even the agricultural production in areas of appropriate agroecological conditions [15].

In parallel, some research had made progress by studying the agro-industrial quinoa's precursor germplasm, as described by Rodríguez et al. (2019) [16]. They handled short and mid-range infrared spectroscopy to single out carbohydrates, lipids and proteins present from quinoa belonging to five different geographic areas. There is also an interest that has become relevant over the recent years around examining bioactive compounds and the antioxidant capacity of seeds [17,18].

Techniques such as mid-range spectroscopy (400–4000 cm^{-1}) allow to pinpoint protein and starch secondary structures present in different quinoa cultivars. Along with other procedures to determine the phenolic compounds and antioxidant capacity, this practice grants a better understanding to determine the grain potential towards functional food. Therefore, this research hypothesis brings up a significant variation between the spectroscopy characteristics and bioactive compounds from different quinoa cultivars.

2. Results

2.1. Quinoa Grain Characteristics

Color has recently become a classification factor for quinoa grains, as well as a quality attribute (Figure 1). Colombian cultivars show a cream-white color associated with genomic quinoas from the inter-Andean Valley region, with typical short days with a high sensitivity to luminosity [8,19]. However, the genetic diversity and increment of its consumption has brought an ease to its spreading to new cultivars, showing attributes linked to pericarpic coloring, dimensions and constitutive molecules.

The evaluated cultivar grains showed significant differences between the L*a*b* coordinates linked to external integument coloration. Such a condition happened to all the cultivars except Titicaca and Puno (on L*) and Salcedo and Puno on a*. The proteins, carbohydrates and fats were studied on all the cultivars. Titicaca and Pasankalla had greater protein values; Puno and Nariño were rich in carbohydrates, while the fats were greater in Puno region cultivars (Table 1).

Figure 1. Macroscopic grains image from different cultivars in Colombia.

Table 1. Physical–chemical characteristics in Colombian quinoa grains.

Cultivars	L*	a*	b*	Protein	Carbohydrates	Fat
Titicaca	53.50 ± 0.58 d	4.05 ± 0.08 b	23.84 ± 0.04 b	14.63 ± 0.30 a	49.36 ± 0.70 c	5.7 ± 0.30 c
Salcedo	73.59 ± 0.52 a	1.51 ± 0.03 e	18.14 ± 0.05 e	13.36 ± 0.47 b	59.63 ± 0.47 ab	6.76 ± 0.25 ab
Soracá	69.40 ± 0.90 b	1.93 ± 0.16 d	21.12 ± 0.21 c	14.10 ± 0.26 ab	57.30 ± 0.70 b	6.1 ± 0.1 c
Pasankalla	30.78 ± 0.46 e	10.4 ± 0.24 a	11.92 ± 0.26 f	14. 46 ± 0.14 a	57.30 ± 1.05 b	6.76 ± 0.25 ab
Puno	54.03 ± 0.22 d	1.32 ± 0.19 e	18.99 ± 0.16 d	11.73 ± 0.30 c	61.06 ± 0.51 a	7 ± 0.1 a
Nariño	62.29 ± 0.34 c	3.56 ± 0.04 c	25.59 ± 0.04 a	11.36 ± 0.30 c	62.53 ± 2.05 a	6.26 ± 0.15 bc

Results are expressed as the mean ± standard deviation (n = 3). Different letters had significant differences ($p < 0.05$) using Tukey's test.

2.2. *Infrared Spectroscopy*

The results allowed to identify a 1633-cm^{-1} transmittance linked to C=O stretching, which is also related to its protein composition [20]. Table 2 shows meaningful differences of the transmittance values from the cultivars. Since there is a direct relationship between the transmittance and protein, it is possible to say that Pasankalla is the one with a lesser protein value in contrast to Nariño, Salcedo and Puno, with greater amounts.

This way, band differences associated with C-H stretching and C-O-C and C-O folding were observed. About starch, a major proportion was observed on the Titicaca and Puno cultivars, while, in Soracá and Pasankalla, the rates were lower. The stretching of the C-H path group linked to lipids made it possible to identify the major differences among the cultivars.

Table 2. Band intensities distinctive to lipids, proteins and starch.

Cultivars	C-H Stretching (2922 cm^{-1})	C-H Stretching; C-O-C; C-O Bending (1016 cm^{-1})	C=O Stretching (1633 cm^{-1})
Titicaca	0.65 ± 0.02 b	0.0011 ± 0.00006 c	0.47 ± 0.04 b
Salcedo	0.68 ± 0.01 ab	0.0021 ± 0.0001 b	0.34 ± 0.06 c
Soracá	0.69 ± 0.008 ab	0.003 ± 0.0004 a	0.55 ± 0.04 ab
Puno	0.68 ± 0.04 ab	0.0009 ± 0.0004 c	0.30 ± 0.02 c
Pasankalla	0.76 ± 0.04 a	0.0031 ± 0.0001 a	0.61 ± 0.03 a
Nariño	0.68 ± 0.01 ab	0.0021 ± 0.0001 b	0.32 ± 0.005 c

Results are expressed as the mean ± standard deviation (n = 3). Different letters had significant differences ($p < 0.05$) using Tukey's test.

2.2.1. Secondary Protein Structure by FTIR Spectroscopy

Table 3 shows the bands after performing a deconvolution on amide I in the region between 1600 and 1700 cm^{-1}. This is the most spectral-sensitive area to secondary protein structures but also responsible for the vibration and strain of the peptic C=O bond [21]. The amide I band spectrum allowed to determine the secondary structure protein for all six cultivar quinoa grains, as in the previously described parameter analysis protocols.

Table 3. Secondary protein structures in six Colombian quinoa cultivars.

Cultivars	β-Sheet-1 (1624 cm^{-1})	β-Sheet-2 (1627 cm^{-1})	β-Sheet-3 (1635 cm^{-1})	Random Coil (1648 cm^{-1})	α Elice (1656 cm^{-1})	β-Turns-1 (1667 cm^{-1})	β-Turns-2 (1675 cm^{-1})	β-Turns-3 (1680 cm^{-1})
Titicaca	2.99 ± 0.13 a	2.87 ± 0.07 a	1.09 ± 0.08 c	3.08 ± 0.07 b	1.33 ± 0.07 e	3.53 ± 0.06 a	1.36 ± 0.05 b	0.26 ± 0.05 e
Salcedo	2.14 ± 0.05 b	1.89 ± 0.06 c	3.08 ± 0.07 b	3.35 ± 0.14 ab	3.07 ± 0.07 b	2.14 ± 0.14 b	0.92 ± 0.08 d	0.77 ± 0.05 c
Soracá	0.37 ± 0.04 e	1.49 ± 0.04 d	3.44 ± 0.12 a	3.59 ± 0.14 a	4.97 ± 0.09 a	1.46 ± 0.07 c	0.59 ± 0.01 e	0.33 ± 0.01 de
Puno	0.75 ± 0.04 d	0.17 ± 0.02 e	1.11 ± 0.19 c	1.24 ± 0.1 d	2.56 ± 0.14 c	1.66 ± 0.09 c	1.07 ± 0.04 c	1.31 ± 0.01 b
Pasankalla	0.58 ± 0.02 de	0.08 ± 0.009 e	0.95 ± 0.04 c	2.53 ± 0.1 c	0.14 ± 0.009 f	1.47 ± 0.05 c	0.51 ± 0.03 e	0.5 ± 0.08 d
Nariño	1.1 ± 0.12 c	2.36 ± 0.1 b	3.38 ± 0.13 ab	0.72 ± 0.09 e	1.73 ± 0.07 e	0.57 ± 0.1 d	1.9 ± 0.01 a	3.37 ± 0.09 a

Results are expressed as the mean ± standard deviation (n = 3). Different letters had significant differences ($p < 0.05$) using Tukey's test.

2.2.2. Secondary Starch Structure by FTIR Spectroscopy

There were noticeable peaks observed on 800 cm^{-1} and 1300 cm^{-1} attributed to starch and corresponding to a flexing COH vibration. According to the data, the evaluated cultivars showed distinguished results ($p < 0.05$) on each region, the most intense at 996 cm^{-1} and 1076 cm^{-1}. As for the 1041-cm^{-1} crystalline structure band, it had a greater intensity in Pasankalla and Nariño, while the 1014-cm^{-1} region had more shapeless structural intensity in Nariño (Table 4).

Table 4. Band intensities of starch from different Colombian quinoa cultivars.

Cultivars	996 cm^{-1}	1014 cm^{-1}	1041 cm^{-1}	1076 cm^{-1}	1099 cm^{-1}	1145 cm^{-1}	996/1014	1041/1014
Titicaca	0.9 ± 0.01 b	0.11 ± 0.01 c	0.14 ± 0.004 c	0.99 ± 0.007 a	0.39 ± 0.01 ab	0.53 ± 0.01 a	8.23 ± 0.79 b	1.3 ± 0.08 d
Salcedo	0.98 ± 0.01 a	0.06 ± 0.004 d	0.13 ± 0.006 c	0.96 ± 0.02 b	0.38 ± 0.01 ab	0.46 ± 0.01 b	15.28 ± 1.1 a	2.04 ± 0.12 b
Soracá	0.82 ± 0.01 c	0.06 ± 0.01 d	0.26 ± 0.01 b	0.98 ± 0.005 a	0.33 ± 0.03 c	0.53 ± 0.02 a	14 ± 2.1 a	4.02 ± 0.27 a
Puno	0.97 ± 0.01 a	0.16 ± 0.01 b	0.23 ± 0.01 b	0.86 ± 0.005 c	0.34 ± 0.01 bc	0.4 ± 0.01 c	6.11 ± 0.32 bc	1.48 ± 0.16 cd
Pasankalla	0.81 ± 0.01 c	0.17 ± 0.01 b	0.32 ± 0.004 a	0.99 ± 0.002 a	0.31 ± 0.05 c	0.53 ± 0.01 a	4.81 ± 0.34 c	1.91 ± 0.1 bc
Nariño	0.98 ± 0.01 a	0.22 ± 0.02 a	0.34 ± 0.01 a	0.97 ± 0.005 ab	0.41 ± 0.01 a	0.52 ± 0.01 a	4.51 ± 0.53 c	1.58 ± 0.16 bcd

Results are expressed as the mean ± standard deviation ($n = 3$). Different letters had significant differences ($p < 0.05$) using Tukey's test.

2.3. *Polyphenols and Antioxidant Capacity by ABTS, DPPH and FRAP*

Quinoa crops showed polyphenol values of 0.6681–1.7737-mg AG/g/sample, which were quite different among them ($p < 0.001$). A richer concentration was found in the Nariño cultivar. About the antioxidant capacity, according to the DPPH technique, Nariño had the highest value. Based on the FRAP technique, Titicaca and, on the ABTS method, both cultivars had the greatest values (Table 5).

Table 5. Phenolic composition and antioxidant activity in quinoa cultivars in Colombia.

Cultivars	Polyphenols (mg AG/g)	ABTS (µmol de T/g)	DPPH (µmol de T /g)	FRAP (µmol de AA/g)
Titicaca	1.0409 ± 0.0202 b	10.8536 ± 0.1317 a	2.3488 ± 0.0177 d	3.2626 ± 0.0585 a
Salcedo	0.9303 ± 0.0103 c	8.2387 ± 0.2718 c	2.7893 ± 0.0627 c	1.9886 ± 0.0643 c
Soracá	0.735 ± 0.0212 d	5.6546 ± 0.1075 d	2.6638 ± 0.0648 c	0.8063 ± 0.0215 f
Puno	0.6782 ± 0.0198 e	7.9468 ± 0.1222 c	2.8127 ± 0.0830 c	1.6469 ± 0.0159 d
Pasankalla	0.6681 ± 0.001 e	9.4534 ± 0.3435 b	3.2587 ± 0.0697 b	1.2685 ± 0.0373 e
Nariño	1.7737 ± 0.009 a	11.0023 ± 0.0857 a	3.8935 ± 0.0454 a	2.6356 ± 0.0212 b

Results are expressed as the mean ± standard deviation ($n = 3$). Different letters had significant differences ($p < 0.001$) using Tukey's test. AG: gallic acid; T: Trolox; AA: ascorbic acid.

2.4. *Phenolic Compounds under UHPLC-ESI-Orbitrap MS*

The results showed phenolic acid, xanthine, flavonoids and anthocyanin varieties among farms. Contrasting values were observed ($p < 0.001$), highlighting the Nariño cultivar for its apigenin (0.08 ± 0.01) presence and the highest value (42%) of identified elements. Salcedo had an elevated value of caffeic and vanillic acid and Soracá of *p*-coumaric acid and pinocembrin (Table 6).

Table 6. Phenolic compound contents (mg/Kg) in quinoa seed samples using UHPLC/ESI+-Orbitrap/MS.

Cultivars	Caffeine	Caffeic Acid	Vanillic Acid	*p*-Coumaric Acid	Ferulic Acid	Pinocembrin	Quercetin-3-Glucoside
Salcedo	0.12 ± 0.002 d	2.41 ± 0.02 a	4.02 ± 0.01 a	1.92 ± 0.02 d	5.58 ± 0.1 f	0.04 ± 0.01 bc	0.53 ± 0.09 c
Puno	0.13 ± 0.003 c	1.88 ± 0.01 b	3.66 ± 0.01 b	2.04 ± 0.02 d	6.71 ± 0.09 e	0.01 ± 0.01 c	1.23 ± 0.11 b
Pasankalla	0.15 ± 0.003 a	1.43 ± 0.02 d	3.02 ± 0.01 e	2.6 ± 0.02 c	9.26 ± 0.1 c	0.01 ± 0.01 c	0.37 ± 0.14 cd
Soracá	0.14 ± 0.003 b	1.26 ± 0.01 e	3.49 ± 0.01 c	14 ± 0.2 a	11.83 ± 0.09 b	0.05 ± 0.02 ab	0.1 ± 0.1 d
Titicaca	0.15 ± 0.003 a	1.49 ± 0.01 c	1.94 ± 0.02 f	5.53 ± 0.2 b	7.42 ± 0.1 d	0.01 ± 0.009 c	1.79 ± 0.12 a
Nariño	0.13 ± 0.002 c	1.84 ± 0.01 b	3.07 ± 0.01 d	14.41 ± 0.02 a	14.67 ± 0.1 a	0.07 ± 0.01 a	1.49 ± 0.11 b

Results are expressed as the mean ± standard deviation ($n = 3$). Different letters had significant differences ($p < 0.001$) using Tukey's test.

2.5. *Phytochemical Multivariate Structure Analysis*

The results showed organization in relation to the phenolic components. As noticed in the heatmap, the groupings showed a high diversity in bioactive compounds (Figure 2A). The bootstrap exhibited two main groups: the first includes one xanthine and one anthocyanin and an antioxidant capacity method, while the second combined phenolic acid, a flavonoid, total polyphenols and an antioxidant technique (Figure 2B).

An analysis of the quinoa cultivars' main components allowed to evaluate the differences in their bioactive composition (Figure 3). The cumulative variance of the two first components was 67.2% (CP); however, when including a third component, the CP variance was 90.8%. It is possible to notice a discrimination of the first two CP from the variance of all six cultivars. The first group's Nariño cultivar showed the highest amounts of pinocembrin, *p*-coumaric, total polyphenol and high antioxidant capacity with DPPH. Secondly, Titicaca quinoa had greater values of quercetin-3-glucoside, caffeine and antioxidant capacity with ABTS and FRAP. Pasankalla and Soracá presented an observable similarity when adding PC3 as an average cultivar. Finally, Salcedo and Puno were similar because of the presence of some phenolic acids.

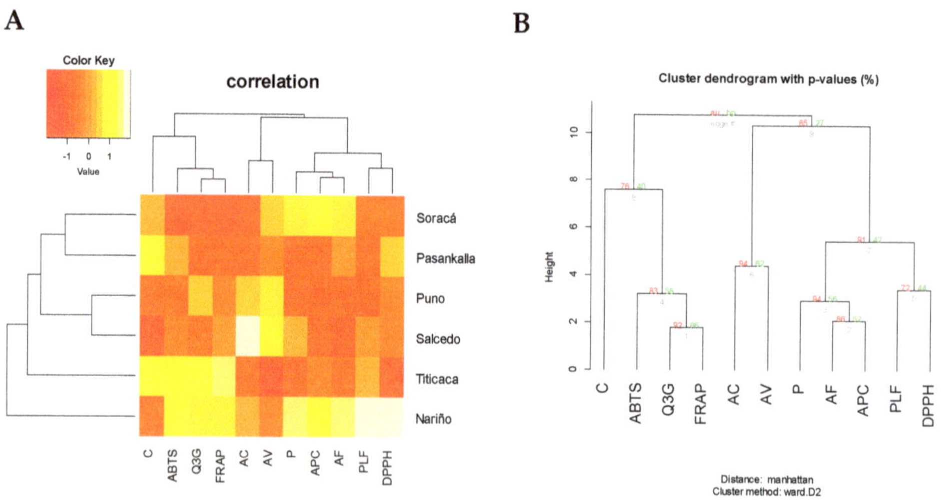

Figure 2. Sample groups according to bioactive compositions from different quinoa seeds. (**A**) Heatmap analysis according to Manhattan distance. (**B**) Boostrap grouping. AU: ap-proximately unbiased. BP: bootstrap probability. C: caffeine, Q3G: quercetin-3-glucoside. AC: caffeic acid; AV: vanillic acid. P: pinocembrin; AF: ferulic acid. APC: *p*-coumaric acid and, PLF: total polyphenols.

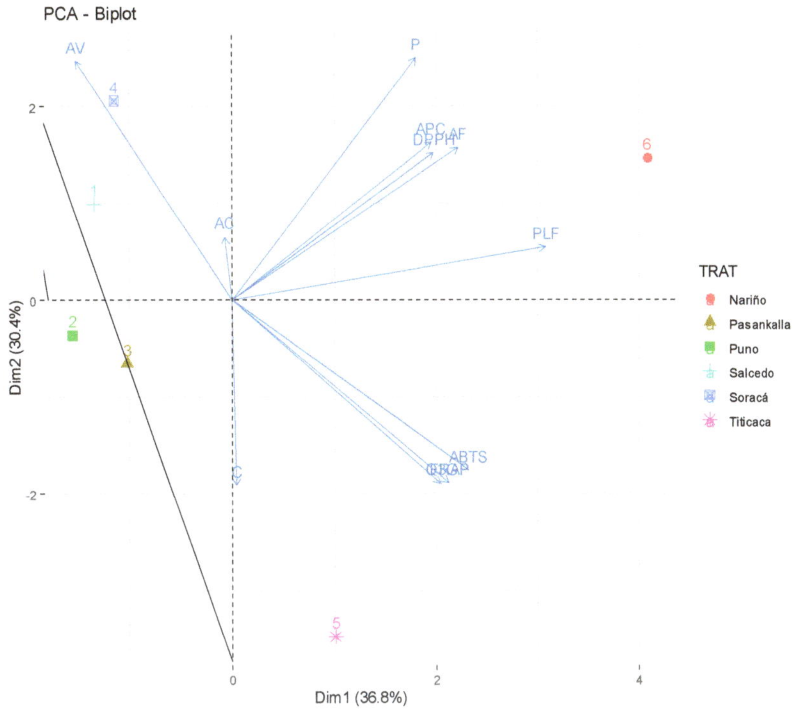

Figure 3. Plot analysis of principal components (PCA) of the bioactive characteristics and antioxidant capacity from six quinoa cultivars in Colombia.

3. Discussion

The visual aspects of quinoa seeds are determinant on its variety and even compositional differentiation [22]. Over recent years, research evaluating colorimetric coordinates allowed compound identification to boost quinoa cultivars with black, pink, purple and red seeds. Furthermore, it brings about the relationship between tannins, saponins and phytic acid compounds [23,24].

In this sense, protein, lipidic and starchy attributes on grains are fundamental to consider agro-industrial processing or even more, when aiming towards hyper-protein, starch and fat enriched for highly functional quality food [14,25].

To elaborate a structural analysis with the goal to deeply learn about the secondary protein structures of each cultivar, spectrum deconvolution techniques were performed on 1633 cm^{-1} [26]. Deconvolution of the 1016 cm^{-1} band can relate to the crystallinity of a shapeless starch granule [27]. This turned out to be an important fact, because the functional flour does not only depend on the protein and starch concentrations but, also, on its organization and molecular conformation.

The intensity from the amide I band from quinoa flour resulted in 1632 cm^{-1} and 1645 cm^{-1}. This condition is relevant to crops' characterization and differentiation. Nariño and Pasankalla offer a band intensity at 1645 cm^{-1}, Puno and Soracá at 1643 cm^{-1}, Salcedo at 1632 cm^{-1} and Titicaca at 1633 cm^{-1}.

According to Yang et al. (2015) [26], the absorbance peak at the amide I band withstood a deconvolution process to determine the structures like β-sheet (1624–1642 cm^{-1}), Random coil (1645–1648 cm^{-1}), α-elice (1653–1656 cm^{-1}) and β-turns (1667–1694 cm^{-1}). All the cultivars displayed contrasting intensity peaks on the secondary structures (Figure 4). Titicaca had a greater β-sheet-1 and -2 proportions, Soracá β-sheet-3, Random coil and α-elice. Titicaca had β-turns-1, while there was β-turns-2 and -3 in Nariño.

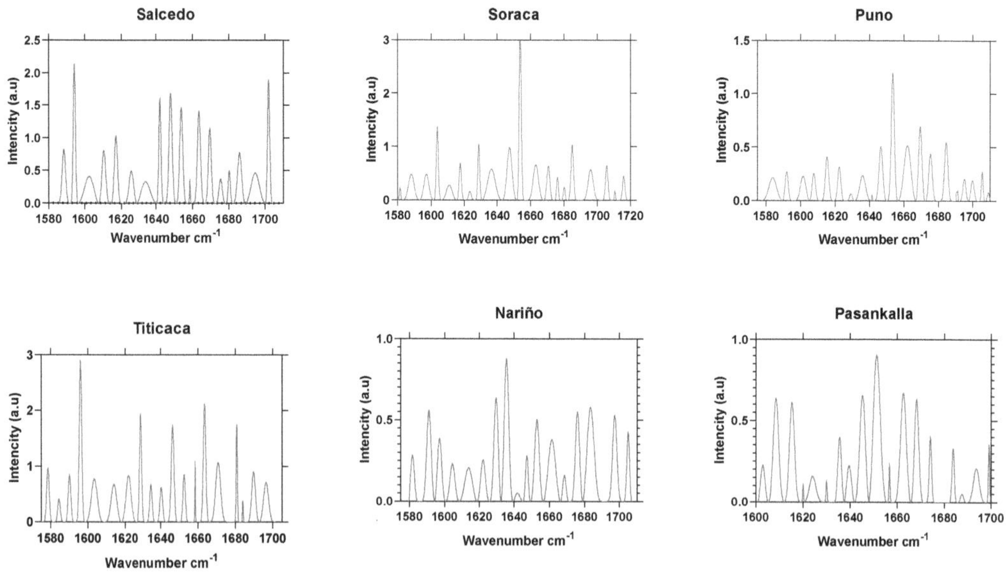

Figure 4. Protein secondary structures of quinoa grains from Colombian cultivars through the deconvolution FTIR technique.

In this sense, the discovered results were dramatically different according to other research because of its major manifestation, the peak intensity was found using random coil and α-elice. The results obtained agree with what was stated by Wang, Zhao, and Yuan (2020) [28], who found secondary structures of proteins in the same peak position in the original spectra in this research. Wolkers et al. (1998) [29] explained that such variability could be associated with effects on its production regions because of the edaphoclimatic conditions.

It is also important to outline that secondary structure spectrums could be also used like a fingerprint distinct to each cultivar. Improving the spectroscopy methods could contribute to identify not only the genetic variability but to offer quality control towards adulterated flour [30].

Then, when looking at each cultivar protein fingerprint, it could be possible to assert that the structural differences directly affect the technical and functional capacity to produce gel, foams or emulsions [31].

Ratios 1041/1014 and 996/1014 cm^{-1} were used with greater frequency to measure the crystallinity range in starch. Such a fact made it possible to establish that the range 1041/1014 cm^{-1} varied among all six evaluated cultivars (Figure 5). The organized structure of granules indicates that starch changes mostly because of the two conditions determining photoassimilate transports from organs to grains. The first was related to K^+ movement on apoplast, determining the flow pressure. The second, the invertase acid activity influencing the transforming saccharose in hexoses and restructuring dispose organs for the nutritional quality of seeds [32,33].

Such results are quite important, because to obtain gel or translucent films, is important to use starch with a high crystallinity index, to which the Soracá cultivar is the most suitable (1041/1014 = 4.02). Retrograde phenomenon must be avoided when using flour as the complementing nutritional value. Then, starch with a lower crystallinity index should be used, in which 1.3 and 1.48 indexes from Pasankalla and Nariño are the most appropriate.

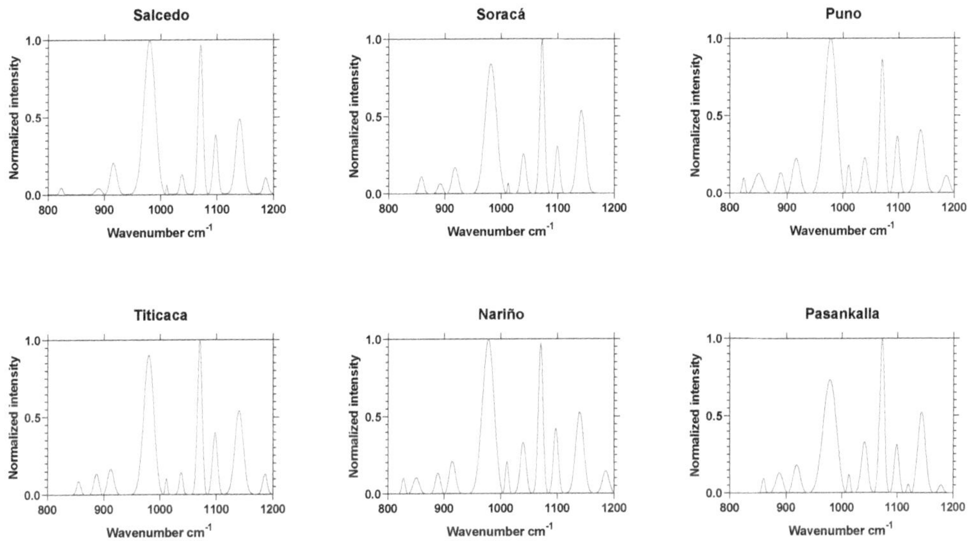

Figure 5. Deconvolution of the starch FTIR band in six Colombian quinoa grain cultivars.

In this sense, the high diversity of quinoa makes it possible to find variations in the foliar area of the plants, which is related to the CO_2 net assimilation and to obtain triose phosphate in the CO_2 fixation phase, which allows the obtaining of starches for their translocation to sink organs [34,35]. Additionally, Reguera and coworkers found changes in the synthesis of secondary metabolites of three quinoa cultivars and highlighted that these compounds seem to be genotype-specific and could vary significantly under stressful conditions [9].

The polyphenol values vary in all quinoa cultivars and are linked to climatic conditions, agronomical practices and genetic interaction [17]. Li, Lietz & Seal (2021) [36] reported 1.44-mg AG/g of total phenols and antioxidant activity between 8.61 and 522 µm of Trolox/g under ABTS and DPPH. Such results are different from the findings in this paper. The Pasankalla and Salcedo farms reported an antioxidant activity of 5.6- and 3.18-µm Trolox/g under DPPH [37] and a 27.9-µm Trolox/g value with the ABTS technique on a Bolivian farm [38]. Such results are meaningfully high compared to the ones observed here.

This parameter turns to be important to perform a study on how to enrich nutritional models with a low antioxidant capacity, as well as to identify cultivars that might bring about this quality [39]. Likewise, the cultivars with the highest antioxidant capacity were Nariño and Titicaca, because their measurements were highly recognized over other samples during at least two different techniques.

Polyphenols have been recently studied in quinoa, because their diversity is equal or greater than other cereals. Findings of 0.14–0.33 mg/kg of caffeic acid and 0.4- and 0.94-mg/kg vanillic acid were in ten different accessions in Chilean quinoa [18].

Reference Abdelaleem & Elbassiony (2021) [17] reported Egyptian cultivars lacking caffeine but with values of caffeic acid, vanillic acid and apigenin up to 0.49 mg/kg, 30.5 mg/kg and 0.44 mg/kg, accordingly. The quercetin-3-glucosid concentrations observed were considerably low in this research, and the *p*-coumaric acid values reported were similar to those by Tang & Tsao (2017) [40].

The results observed by the heatmap and bootstrap determined the bioactive differences from each evaluated cultivar. It is relevant to mention that many polyphenols are not synthetized on a regular basis in plants, but this could be a result from external conditions such as biotic and abiotic factors [41].

The similarities between quercetin-3-glucoside, FRAP and ABTS happen because of the high capacity from anthocyanins to capture free radicals. In the same way, the concentration resemblance in some phenolic acids occurs because of methylation from a few compounds like caffeic acid, the forefather of vanillic acid [42]. The correlation between ferulic acid and *p*-coumaric acid is a consequence, because they both belong to the hydroxycinnamic group.

It is accurate to mention that this paper reports, for the first time, the presence of pinocembrin and apigenin in quinoa compounds. For other species, its synthesis is attributed to pathogen agents affecting the plant during primary production. Additionally, the influence of edaphoclimatic conditions like salinity, water deficit or very cold weather in amaranth plants was found [43,44].

High contributions on the PCA of ferulic acid and vanillic acid, among other acids, coincide with the ones reported by Antognoni et al. (2021) [45]. They identified that Titicaca has the greatest polyphenol values expressed by these two acids and that its concentration could be greater than 100 mg/kg in quinoa flour.

Carciochi, Manrique & Dimitrov (2014) [46] found a relationship between ferulic acid and vanillic acid concentrations in seeds. Outlining that the two elements are the ones supporting the seed antioxidant activity, in parallel, its content value rises along with the plant growing towards germination.

4. Materials and Methods

4.1. Materials

We used Salcedo, Puno, Pasankalla, Soracá, Titicaca and Nariño cultivars collected from crops established in the municipalities of Oicatá, Ventaquemada and Tuta in the Department of Boyacá, and others were taken from the seed collection of the "Laboratorio de Biotecnología Vegetal de la Gobernación de Boyacá". The cultivars were grown in a greenhouse located at Victoria Granja Agroecologica Company, Ventaqumada, Colombia (5°22′47″ N, 73°30′10″ W). The seeds were germinated in 10-kg soil pots. The soil corresponded to an Andosol with the following physicochemical composition: pH 6.1; electrical conductivity 6.3 ds m^{-1}; organic matter 8.8%; cations (cmol Kg^{-1}) Al^{3+} 1, Ca 36.4, Mg 7.2, K 10.3 and Na 1.1 and micronutrients (mg Kg^{-1}) Fe 93.8, Mn 11.8, Cu 9.4, Zn 9.4 and B 1.5. During the study, the irrigation and preventive sanitary management were carried out in order to maintain plant health. The seeds were manually cleaned to remove foreign matter and broken and immature seeds.

The reagents to determine the antioxidant capacity were: free radical DPPH (1,1-diphenyl-2-picryl-hydrazyl), free radical ABTS (2,2′-azinobis(3-ethylobenzothiazoline-6-sulphonate)), iron III chloride (FeCl3); triphenyl tetrazolium chloride (TPTZ), sodium acetate, ascorbic acid, potassium persulfate, Trolox (97%) and ethanol (analytic grade).

The total polyphenol reagents were Folin Ciocalteau, sodium acid carbonate (99%) and gallic acid.

The standard phenolic reference: caffeine (Part N° C8960250G), theobromine (Part N° T4500-25G); theophylline (Part N° T163325G), (±)-catechin (C) (Part N° C1788-500MG), (-)-epigallocatechin gallate (EGCG) (Part N° E4143-50MG), (-)-epicatechin (EC) (Part N° E1753-1G), (-)-epicatechin gallate (ECG) (Part N° E3893-10MG), (-)-epigallocatechin (EGC) (Part N° E3768-5MG), caffeic acid (Part N° C0625), p-coumaric acid (Part N° C9008) and vanillic acid (Part N° 06380590-50MG).

Ferulic acid (Part N° 52229-50MG), rosmarinic acid (Part N° 536954-5G), quercetin (Part N° Q4951-10G) and naringenin (Part N° N5893-1G). Luteolin (Part N° L9283-10MG), kaempferol (Part N° K0133-50MG), ursolic acid (Part N° U6753-100MG), pinocembrin (Part N° P5239) and carnosic acid (Part N° C060910MG).

Apigenin (Part N° A3145-25MG), cyanidin 3-rutinoside (Part N° G36428) and pelargonidin 3-glucoside (Part N° 53489). All of them were bought from Sigma-Aldrich (St. Louis, MI, USA). Quercetin-3-glucoside (Part N° 89230) and Kaempferol-3-glucósido (Part N° 89237) were acquired from Phytolab.

4.2. Proximal Characteristics

To determinate the protein content, the Kjeldahl method was used. AOAC 960.52 was multiplied by a conversion factor of 6.25 to measure the total raw protein from the seeds. Fats were defined by the AOAC 922.06 method using Soxhlet (Soxtec 2050). Finally, the total carbohydrates were calculated by the difference (i.e., protein + ash + fat + moisture − 100).

*4.3. CIEL*a*b* Coordinates Determination*

The spectrophotometer used was Konica Minolta (Model CM-5; Tokyo, Japan) to determine the seed color hue from the quinoa cultivars. The results were recorded as L* a* b* values.

4.4. Mid-Range Infrared Spectroscopy

IR spectrum was obtained using IRAFFINITY-1S equipment (Shimadzu Corp., Kyoto, Japan). The spectrum was performed by the reflection mode between 400 cm^{-1} and 4000 cm^{-1} and proportional over 32 scans with 4 cm^{-1} and a temperature of 25 °C. In the previous measurements, each cultivar had a blank background spectrum.

A spectrum analysis was made by using OriginPro, 7th version. Measurements were fixed and standardized over a baseline between 0 and 1 (represented in the figures) to show the highest transmittance peaks. Deconvolution was performed on a spectrum using Fourier transform to determine the protein secondary structure changes (band 1600–1700 cm^{-1}). Later, a modeling of the Gaussian function spectrum was applied. The starch concentration was studied from its formation on a short-range band between 875 cm^{-1} and 1175 cm^{-1}. Lipids were present on the 2800–2900 cm^{-1} range. All samples were analyzed in triplicate.

4.5. Polyphenols

From each cultivar, 0.5 g of quinoa flour were mixed. Then, to each sample was added 30 mL of sodium hydroxide 2M and stirring for 10 s with vortex (Fisher Scientific, G560, Bohemia, Waltham, MA, USA). It was left for extraction over 24 h on amber glass at room temperature (25 °C) and mixed in darkness. After that, 5.7 mL of hydrochloride acid were added to the samples upon reaching a pH value of 2. Finally, the samples were centrifugated for 15 min with a rotational speed of 10,000 rpm (HERMLE model Z306, La-Bortechnik GmbH, Wasserburg, Germany), then vacuum-filtrated with Whatman paper (No 1).

Phenolic compounds were determined through the Folin Ciocalteau (FC) method. Triple samples were disposed with 40 µL on a separate 1800 µL with FC reagent. For 15 min, a vortex mix was performed to later rest for 5 min. Afterwards, 1200 µL of sodium acid carbonate ($NaHCO_3$) at 7.5% (m/v) was added and let to settle for 60 min. Absorbance at 765 nm was measured on a V-630 UV-VIS model spectrophotometer (JASCO Inc., Easton, MD, USA). The phenol amounts were measured by sample grams expressed in gallic acid milligrams (mg AG/g).

4.6. Antioxidants

4.6.1. FRAP

In a 50-mL falcon tube, 0.4 g of the sample were placed. Then, 10 mL of ethanol were added to begin the extraction process. This was mixed by vortex for 15 s and left to later extract the liquid–solid in the mixer over 30 min at 37 °C. After, the samples were centrifugated (15 min. at 5000 rpm and 4 °C) and later vacuum-filtered with Whatman paper (No 1).

Test tubes with a lid were prepared for 3 different samples: blank, experiment and calibration. For neutral (blank) reagents, the sample had 60 µL of methanol. To begin the reaction, 180 µL of distill water were put in each tube. Then, there was 15 s of vortexing to later add 1800 µL of the FRAP reagent. Next, it was incubated at 37 °C for 30 min.

Finally, 595-nm absorbance was measured on a V-630 UV-VIS model spectrophotometer (JASCO Inc., Easton, MD, USA). The antioxidant activity was expressed with μmol of ascorbic acid/g of the sample (μmol de AA/g).

4.6.2. ABTS

In a 50-mL falcon test tube, 0.4 g from the sample was disposed. To begin the extraction, 10 mL of ethanol were added, vortexed for 15 s and left for solid–liquid extraction over 16 h at 4 °C. Subsequently, the samples were centrifugated over 15 min at 10,000 rpm, then vacuum-filtered with Whatman paper (No 1).

In the test tube, 4 mL of ABTS solution was placed and totally covered with aluminum foil. To begin the process, 135 μL of standard solution were added and then mixed by vortexing for 5 s. Blank reagent consisted of 4 mL of buffer acetate and 135 μL of ethanol. The zero point was mixed with 4.5-mL ABTS solution and 135 μL of ethanol. The test tube was closed and left to react for30 min to finally measure the absorbance at a 729.7-nm wavelength.

4.6.3. DPPH

In a 50-mL falcon test tube, 0.4 g were measured before adding 10 mL of ethanol. Vortexing for 15 s was done and then, the solid–liquid extract was left to settle for 16 h at 4 °C. After such a time, the samples were centrifugated for 15 min at 10,000 rpm, then vacuum-filtered with Whatman paper (No 1).

In a test tube, 3.9 mL of DPPH solution and 100 μL of standard solution were applied to launch a reaction by vortex stirring for 5 s. Blank reagents (control) were carried out with ethanol. The zero point was adjusted by adding 3.9 mL of DPPH solution and 100-μL ethanol. The sample was covered for 30 min to initiate the reaction to later measure the absorbance at 517 nm.

4.7. Phenolic Compounds Determination by UHPLC/ESI-Orbitrap MS

The seeds were macerated for four minutes with a blade grinder (Krups; Solingen, Germany). Later, 100 mL of 80% ethanol at 60 °C was added to 5.0 g of quinoa flour. The resultant extract was centrifugated at 10,000 rpm over 10 min. The leftover sample was recovered and stored at 4 °C for later use.

The extract was analyzed on a liquid high-efficiency chromatograph (UHPLC) Dionex Ultimate 3000 (Thermo Scientific, Sunnyvale, CA, USA, EE.UU.). This equipment consists of a binary gradient pump (HP G3400RS) and automatic sample injector (WPS 300TRS) and a thermostatic unit for a TCC 3000 column. The interface of LC-MS was electrospray ionization (ESI). The spectrometer Orbitrap had high mass resolution with ion current detecting system. Chromatography separation was performed by Hypersil GOLD Aq 100 × 2.1 mm, 1.9 μm (Thermo Scientific, Sunnyvale, CA, USA, EE.UU.) column at 30 °C.

Mobile phase A consisted of a 0.2% aqueous solution of ammonium formate. Mobile phase B: 0.2% on acetonitrile ammonia formate. The initial gradient condition was 100%. A moved linearly towards 100% of B in 8 min. Such a phenomena was held for 4 min, then returned to the initial conditions in one minute. The total running time lasted 13 min, plus 3 post-running additional minutes.

Mass spectrometer Orbitrap (Exactive Plus, Thermo Scientific, Sunnyvale, CA, USA, EE.UU.) connected through the electrospray (HESI) interface was handled under a positive capillary voltage of 4.5 kV. Nitrogen gas was used as the dry element. The mass spectrum range was m/z 60–900. The Orbitrap mass detector was calibrated with certified reference solutions: Ultramark^{TM} 1621 Mass Spec. (AB172435, ABCR GmbH & Co. KG, Karlsruhe, Germany), sodium dodecyl sulfate (SDS) (L4509, Sigma-Aldrich) and sodium taurocholate hydrate (T4009, Sigma-Aldrich).

All compound identification was achieved by using the full scan mode and ion extraction (EIC) corresponding to the $[M+H]^+$ of the target compounds. A mass assessment was performed with the accuracy precision of $\Delta_{ppm} < 0.001$ and using a standard mix solution of phenolic compounds. To quantify the samples, the external standard calibration was applied using the results factor (Rf) stablished by the standard solutions analysis under different concentrations.

4.8. Statistical Analysis

The values presented are the means ± standard deviation of three replicates. The data were analyzed using the Shapiro–Wilk normality test and Bartlett's homogeneity test, and the variance analysis (ANOVA) by Tukey's test was done to observe significant differences ($p < 0.05$). A principal component analysis (PCA) was collected to stablish the relationship of the antioxidant and polyphenol changes among the cultivars using the *Factoextra* and *FactoMinerR* libraries. A bootstrap analysis was elaborated using the *pvclust* function [47]. A double-grouping analysis was performed by Manhattan distance with the *gplots* and *RColorBrewer* libraries [48]. All the analyses were conducted using R version 3.6.1 software.

5. Conclusions

A spectroscopy analysis of different quinoa cultivars showed significant variations attributed to their genetic character. This showed that not only external factors but, also, morphological and physiological diversity are influential on the structural configuration of proteins. The same conditions happen to the crystalline/amorphous properties of starch granules.

It is relevant to mention that some alloys like caffeic acid, quercetin 3-glucosid, caffeine and pinocembrin are not very well-identified in quinoa grains; therefore, the influence over its antioxidant activity is still uncertain.

The teamwork between the spectrum deconvolution technique, chromatographic analysis and multivariate analysis are strong tools used for the seed characterization of cultivars from the Colombian Inter-Andean Valley.

Author Contributions: Conceptualization: M.G.-P., D.R.-A. and B.M.-M.; methodology: D.R.-A., M.G.-P., V.G.-L. and B.M.-M.; software, M.G.-P. and D.R.-A.; validation: J.B.-G., V.G.-L. and D.R.-A.; formal analysis: M.G.-P. and J.B.-G.; investigation, M.G.-P. and D.R.-A. and resources: M.G.-P., B.M.-M. and D.R.-A. All authors have read and agreed to the published version of the manuscript.

Funding: The authors express gratitude to the minciencias (Ministerio de Ciencia and Tecnología e Innovación) No. 779/2017. We are also grateful to the Boyacá Department of the government and the Universidad del Cauca, project ID 4854.

Data Availability Statement: The data generated within this work are open access and available to be shared with interested persons.

Conflicts of Interest: The authors declare no conflict of interest.

References

1. Bazile, D.; Pulvento, C.; Verniau, A.; Al-Nusairi, M.S.; Ba, D.; Breidy, J.; Hassan, L.; Mohammed, M.I.; Mambetov, O.; Otambekova, M.; et al. Worldwide Evaluations of Quinoa: Preliminary Results from Post International Year of Quinoa FAO Projects in Nine Countries. *Front Plant Sci.* **2016**, *7*, 850. [CrossRef]
2. García-Parra, M.; Zurita-Silva, A.; Stechauner-Rohringer, R.; Roa-Acosta, D.; Jacobsen, S.-E. Quinoa (*Chenopodium quinoa* Willd.) and its relationship with agroclimatic characteristics: A Colombian perspective. *Chil. J. Agric. Res.* **2020**, *80*, 290–302. [CrossRef]
3. Ruiz, K.B.; Biondi, S.; Oses, R.; Acuña-Rodríguez, I.S.; Antognoni, F.; Martinez-Mosqueira, E.A.; Coulibaly, A.; Canahua-Murillo, A.; Pinto, M.; Zurita-Silva, A.; et al. Quinoa biodiversity and sustainability for food security under climate change. A review. *Agron. Sustain. Dev.* **2014**, *34*, 349–359. [CrossRef]
4. Pinedo-Taco, R.; Gómez-Pando, L.; Julca-Otiniano, A. Sostenibilidad ambiental de la producción de quinua (*Chenopodium quinoa* Willd.) en los valles interandinos del Perú. *Cienc. Tecnol. Agropecu.* **2020**, *21*, 1–17.
5. Bazile, D.; Martínez, E.A.; Fuentes, F. Diversity of quinoa in a biogeographical Island: A review of constraints and potential from arid to temperate regions of Chile. *Not. Bot. Horti Agrobot.* **2014**, *42*, 289–298. [CrossRef]
6. Curti, R.; Carmen-Sanahuja, M.; Vidueiros, S.; Curti, C.; Pallaro, A.; Bertero, H. Oil quality in sea level quinoa as determined by cultivar-specific responses to temperature and radiation conditions. *J. Sci. Food Agric.* **2019**, *100*, 1358–1361. [CrossRef]
7. Li, G.; Zhu, F. Quinoa starch: Structure, properties, and applications. *Carbohydr. Polym.* **2018**, *181*, 851–861. [CrossRef]
8. Bazile, D.; Bertero, H.D.; Nieto, C. *Estado del Arte de la Quinua en el Mundo 2013*; FAO: Santiago, Chile; CIRAD: Montpellier, Francia, 2014; Volume 1, 724p.

9. Reguera, M.; Conesa, C.M.; Gil-Gómez, A.; Haros, C.M.; Pérez-Casas, M.Á.; Briones-Labarca, V.; Bolaños, L.; Bonilla, I.; Álvarez, R.; Pinto, K.; et al. The impact of different agroecological conditions on the nutritional composition of quinoa seeds. *PeerJ* **2018**, *14*, 1–20. [CrossRef]
10. García-Parra, M.; García-Molano, J.; Deaquiz-Oyola, Y. Physiological performance of quinoa (*Chenopodium quinoa* Willd.) under agricultural climatic conditions in Boyaca, Colombia. *Agron. Colomb.* **2019**, *37*, 160–168. [CrossRef]
11. Lesjak, J.; Calderini, D.F. Increased Night Temperature Negatively Affects Grain Yield, Biomass and Grain Number in Chilean Quinoa. *Front. Plant Sci.* **2017**, *8*, 1–11. [CrossRef]
12. Miranda, M.; Vega-Gálvez, A.; Martínez, E.; López, J.; Marín, R.; Aranda, M.; Fuentes, F. Influence of contrasting environments on seed composition of two quinoa genotypes: Nutritional and functional properties. *Chil. J. Agric. Res.* **2013**, *73*, 108–116. [CrossRef]
13. García-Salcedo, A.J.; Torres-Vargas, O.L.; Ariza-Calderón, H. Physical-chemical characterization of quinoa (*Chenopodium quinoa* Willd.), amaranth (*Amaranthus caudatus*, L.), and chia (*Salvia hispanica*, L.) flours and seeds. *Acta Agron.* **2018**, *67*, 215–222.
14. Roa-Acosta, D.F.; Bravo-Gómez, J.E.; García-Parra, M.A.; Rodríguez-Herrera, R.; Solanilla-Duque, J.F. Hyper-protein quinoa flour (*Chenopodium Quinoa* Wild): Monitoring and study of structural and rheological properties. *LWT-Food Sci. Technol.* **2020**, *121*, 1–7. [CrossRef]
15. Kwil, I.; Piwowar-Sulej, K.; Krzywonos, M. Local entrepreneurship in the context of food production: A review. *Sustainability* **2020**, *12*, 424. [CrossRef]
16. Rodríguez, S.D.; López-Fernández, M.P.; Maldonado, S.; Buera, M.P. Evidence on the discrimination of quinoa grains with a combination of FT-MIR and FT-NIR spectroscopy. *J. Food Sci. Technol.* **2019**, *56*, 4457–4464. [CrossRef]
17. Abdelaleem, M.A.; Elbassiony, K.R.A. Evaluation of phytochemicals and antioxidant activity of gamma irradiated quinoa (*Chenopodium quinoa*). *Braz. J. Biol.* **2021**, *81*, 806–813. [CrossRef]
18. Carrasco-Sandoval, J.; Rebolledo, P.; Peterssen-Fonseca, D.; Fischer, S.; Wilckens, R.; Aranda, M.; Henríquez-Aedo, K. A fast and selective method to determine phenolic compounds in quinoa (*Chenopodium quinoa* Will) seeds applying ultrasound-assisted extraction and high-performance liquid chromatography. *Chem. Pap.* **2020**, *75*, 431–438. [CrossRef]
19. Bazile, D.; Jacobsen, S.-E.; Verniau, A. The Global Expansion of Quinoa: Trends and Limits. *Front. Plant Sci.* **2016**, *7*, 1–6. [CrossRef]
20. Barth, A. Infrared spectroscopy of proteins. *Biochim. Biophys. Acta Bioenerg.* **2007**, *1767*, 1073–1101. [CrossRef]
21. Sadat, A.; Joye, I.J. Peak fitting applied to fourier transform infrared and raman spectroscopic analysis of proteins. *Appl. Sci.* **2020**, *10*, 5918. [CrossRef]
22. Escribano, J.; Cabanes, J.; Jiménez-Atiénzar, M.; Ibañez-Tremolada, M.; Gómez-Pando, L.R.; García-Carmona, F.; Gandía-Herrero, F. Characterization of betalains, saponins and antioxidant power in differently colored quinoa (*Chenopodium quinoa*) varieties. *Food Chem.* **2017**, *234*, 285–294. [CrossRef]
23. Medina, W.; Skurtys, O.; Aguilera, J.M. Study on image analysis application for identification Quinoa seeds (*Chenopodium quinoa* Willd) geographical provenance. *LWT-Food Sci. Technol.* **2010**, *43*, 238–246. [CrossRef]
24. Abderrahim, F.; Huanatico, E.; Segura, R.; Arribas, S.; Gonzalez, M.C.; Condezo-Hoyos, L. Physical features, phenolic compounds, betalains and total antioxidant capacity of coloured quinoa seeds (*Chenopodium quinoa* Willd.) from Peruvian Altiplano. *Food Chem.* **2015**, *183*, 83–90. [CrossRef]
25. Li, G.; Zhu, F. Rheological properties in relation to molecular structure of quinoa starch. *Int. J. Biol. Macromol.* **2018**, *114*, 767–775. [CrossRef]
26. Yang, H.; Yang, S.; Kong, J.; Dong, A.; Yu, S. Obtaining information about protein secondary structures in aqueous solution using Fourier transform IR spectroscopy. *Nat. Protoc.* **2015**, *10*, 382–396. [CrossRef]
27. Roa-Acosta, D.F.; Solanilla-Duque, J.F.; Agudelo-Laverde, L.; Villada-Castillo, H.S.; Tolaba, M.P. Structural and thermal properties of the amaranth starch granule obtained by high-impact wet milling. *Int. J. Food Eng.* **2020**, *16*, 1–12. [CrossRef]
28. Wang, X.; Zhao, R.; Yuan, W. Composition and secondary structure of proteins isolated from six different quinoa varieties from China. *J. Cereal Sci.* **2020**, *95*, 103036. [CrossRef]
29. Wolkers, W.F.; Bochicchio, A.; Selvaggi, G.; Hoekstra, F.A. Fourier Transform Infrared Microspectroscopy Detects Changes in Protein Secondary Structure Associated with Desiccation Tolerance in Developing Maize Embryos. *Plant Physiol.* **1998**, *116*, 1169–1177. [CrossRef] [PubMed]
30. Zeng, H.Y.; Cai, L.H.; Cai, X.L.; Wang, Y.J.; Li, Y.Q. Structure characterization of protein fractions from lotus (*Nelumbo nucifera*) seed. *J. Mol. Struct.* **2011**, *1001*, 139–144. [CrossRef]
31. Sánchez-Mendoza, N.A.; Ruiz-Ruiz, J.C.; Dávila, G.; Jiménez-Martínez, C. Propiedades tecnofuncionales y biológicas de harina, aislado y fracciones proteicas mayoritarias de semillas de *Inga paterno*. *CyTAJ Food* **2017**, *15*, 400–408. [CrossRef]
32. Taiz, L.; Zeiger, E. *Fisiología Vegetal*; Universitat Jaume: Castellón de la Plana, Spain, 2006; Volume 1, 782p.
33. Smith, M.R.; Rao, I.M.; Merchant, A. Source-Sink Relationships in Crop Plants and Their Influence on Yield Development and Nutritional Quality. *Front. Plant Sci.* **2018**, *9*, 1–10. [CrossRef] [PubMed]
34. Hussain, M.I.; Al-Dakheel, A.J.; Reigosa, M.J. Genotypic differences in agro-physiological, biochemical and isotopic responses to salinity stress in quinoa (*Chenopodium quinoa* Willd.) plants: Prospects for salinity tolerance and yield stability. *Plant Physiol. Biochem.* **2018**, *129*, 411–420. [CrossRef] [PubMed]
35. Cai, Z.-Q.; Gao, Q. Comparative physiological and biochemical mechanisms of salt tolerance in five contrasting highland quinoa cultivars. *BMC Plant Biol.* **2020**, *20*, 70. [CrossRef] [PubMed]

36. Li, L.; Lietz, G.; Seal, C.J. Phenolic, apparent antioxidant and nutritional composition of quinoa (*Chenopodium quinoa* Willd.) seeds. *Int. J. Food Sci. Technol.* **2021**, *56*, 3245–3254. [CrossRef]
37. Choque-Quispe, D.; Ligarda-Samanez, C.A.; Ramos-Pacheco, B.S.; Leguía-Damiano, S.; Calla-Florez, M.; Zamalloa-Puma, L.M.; Colque-Condeña, L. Phenolic compounds, antioxidant capacity, and protein content of three varieties of germinated quinoa (*Chenopodium quinoa* willd). *Ing. Investig.* **2021**, *41*, 1–7.
38. Paśko, P.; Bartoń, H.; Zagrodzki, P.; Gorinstein, S.; Fołta, M.; Zachwieja, Z. Anthocyanins, total polyphenols and antioxidant activity in amaranth and quinoa seeds and sprouts during their growth. *Food Chem.* **2009**, *115*, 994–998. [CrossRef]
39. Ballester-Sánchez, J.; Gil, J.V.; Haros, C.M.; Fernández-Espinar, M.T. Effect of Incorporating White, Red or Black Quinoa Flours on Free and Bound Polyphenol Content, Antioxidant Activity and Colour of Bread. *Plant Foods Hum. Nutr.* **2019**, *74*, 185–191. [CrossRef]
40. Tang, Y.; Tsao, R. Phytochemicals in quinoa and amaranth grains and their antioxidant, anti-inflammatory, and potential health beneficial effects: A review. *Mol. Nutr. Food Res.* **2017**, *61*, 1–16. [CrossRef]
41. Hinojosa, L.; González, J.; Barrios-Masias, F.; Fuentes, F.; Murphy, K. Quinoa Abiotic Stress Responses: A Review. *Plants* **2018**, *7*, 106. [CrossRef]
42. Šamec, D.; Karalija, E.; Šola, I.; Vujcic, V.; Salopek-Sondi, B. The Role of Polyphenols in Abiotic Stress Response: The Influence of Molecular Structure. *Plants* **2021**, *10*, 118. [CrossRef]
43. Rasul, A.; Millimouno, F.M.; Eltayb, W.A.; Ali, M.; Li, J.; Li, X. Pinocembrin: A Novel Natural Compound with Versatile Pharmacological and Biological Activities. *BioMed Res. Int.* **2013**, *2013*, 379850. [CrossRef]
44. Sarker, U.; Oba, S. Salinity stress enhances color parameters, bioactive leaf pigments, vitamins, polyphenols, flavonoids and antioxidant activity in selected Amaranthus leafy vegetables. *J. Sci. Food Agric.* **2018**, *99*, 2275–2284. [CrossRef]
45. Antognoni, F.; Potente, G.; Biondi, S.; Mandrioli, R.; Marincich, L.; Ruiz, K.B. Free and conjugated phenolic profiles and antioxidant activity in quinoa seeds and their relationship with genotype and environment. *Plants* **2021**, *10*, 1046. [CrossRef] [PubMed]
46. Carciochi, R.A.; Manrique, G.D.; Dimitrov, K. Changes in phenolic composition and antioxidant activity during germination of quinoa seeds (*Chenopodium quinoa* Willd.). *Int. Food Res. J.* **2014**, *21*, 767–773.
47. Manjarres-Hernández, E.; Morillo-Coronado, A.; Ojeda-perez, Z.; Cárdenas-Chaparro, A.; Arias-Moreno, D. Characterizacion of the yield components and selection of materials for breeding programs of quinoa (*Chenopodium quinoa* Willd.). *Euphytica* **2021**, *217*, 101. [CrossRef]
48. Landinez-Torres, A.; Panelli, S.; Picco, A.M.; Comandatore, F.; Tosi, S.; Capelli, E. A meta-barcoding analysis of soil mycobiota of the upper Andean Colombian agro-environment. *Sci. Rep.* **2019**, *9*, 10085. [CrossRef] [PubMed]

plants

MDPI

Article

Genotype-Dependent Variation of Nutritional Quality-Related Traits in Quinoa Seeds

Sara Granado-Rodríguez [1], Susana Vilariño-Rodríguez [2], Isaac Maestro-Gaitán [1], Javier Matías [3], María José Rodríguez [4], Patricia Calvo [4], Verónica Cruz [3], Luis Bolaños [1] and María Reguera [1,*]

[1] Departamento de Biología, Campus de Cantoblanco, c/Darwin 2, Universidad Autónoma de Madrid, 28049 Madrid, Spain; sara.granado@uam.es (S.G.-R.); isaac.maestro@uam.es (I.M.-G.); luis.bolarios@uam.es (L.B.)

[2] Vitrosur Lab SLU, Algodonera del Sur, Carretera Trebujena C-441 (km 5.5), Lebrija, 41740 Sevilla, Spain; svilarino@algosur.com

[3] Agrarian Research Institute "La Orden-Valdesequera" of Extremadura (CICYTEX), 06187 Badajoz, Spain; javier.matias@juntaex.es (J.M.); veronica.cruz@juntaex.es (V.C.)

[4] Technological Institute of Food and Agriculture of Extremadura (CICYTEX), 06007 Badajoz, Spain; mariajose.rodriguezg@juntaex.es (M.J.R.); patricia.calvo@juntaex.es (P.C.)

* Correspondence: maria.reguera@uam.es; Tel.: +34-914978189

Abstract: Exploiting the relationship between the nutritional properties of seeds and the genetic background constitutes an essential analysis, which contributes to broadening our knowledge regarding the control of the nutritional quality of seeds or any other edible plant structure. This is an important aspect when aiming at improving the nutritional characteristics of crops, including those of *Chenopodium quinoa* Willd. (quinoa), which has the potential to contribute to food security worldwide. Previous works have already described changes in the nutritional properties of quinoa seeds due to the influence of the environment, the genotype, or their interaction. However, there is an important limitation in the analyses carried out, including the outcomes that can be translated into agronomical practices and their effect on seed quality. In the present study, several seed nutritional-related parameters were analyzed in 15 quinoa cultivars grown in a particular environmental context. Important agronomical and nutritional differences were found among cultivars, such as variations in mineral or protein contents and seed viability. More importantly, our analyses revealed key correlations between seed quality-related traits in some cultivars, including those that relate yield and antioxidants or yield and the germination rate. These results highlight the importance of considering the genotypic variation in quinoa when selecting improved quinoa varieties with the best nutritional characteristics for new cultivation environments.

Keywords: quinoa; genotype; nutritional traits; seed quality

Citation: Granado-Rodríguez, S.; Vilariño-Rodríguez, S.; Maestro-Gaitán, I.; Matías, J.; Rodríguez, M.J.; Calvo, P.; Cruz, V.; Bolaños, L.; Reguera, M. Genotype-Dependent Variation of Nutritional Quality-Related Traits in Quinoa Seeds. *Plants* **2021**, *10*, 2128. https://doi.org/10.3390/plants10102128

Academic Editors: Cataldo Pulvento, Didier Bazile and Alberto Gianinetti

Received: 7 September 2021
Accepted: 5 October 2021
Published: 7 October 2021

1. Introduction

Quinoa (*Chenopodium quinoa* Willd.) is a halophytic crop that belongs to the *Amaranthaceae* family. It can be adapted to a wide variety of agroecosystems and is resistant to stressful environmental conditions [1,2]. Moreover, quinoa possess excellent nutritional properties [1,3]. All these characteristics have resulted in a global expansion of quinoa cultivation during the past three decades [4], and consequently, this explains why quinoa can be found from the harsh climatic conditions associated with high altitudes of the Andean Altiplano (reaching over 3500 m above sea level) to coastal areas.

Quinoa was first domesticated by pre-Columbian cultures more than 7000 years ago, when it became one of their main sources of nutrients, given the lack of animal protein. After the Spanish conquest, quinoa was highly rejected but maintained by indigenous farmers, despite the introduction of Old-World species. These farmers domesticated the cultivars, preserving the genetic diversity found currently in quinoa [5]. This genetic diversity can be divided into different ecotypes that include thousands of accessions (16,422) [6] that reflect the diffusion from the center of origin of the crop around Lake Titicaca [5].

Currently, quinoa is still the principal protein source in many areas of the Altiplano. The nutritional value of quinoa seeds was rediscovered during the last decades of the 20th century, leading to a renewal of its production [7]. This led to a rapid spread in its cultivation, from very few countries around the Andean Altiplano in the 1980s to 123 countries in 2018 [4]. The success in its international acceptance has been possible in part due to the nutritional characteristics of the seeds. Quinoa seeds are considered pseudocereals because they resemble cereal grains in their high starch content and overall morphology. However, quinoa seeds are gluten-free and have a low glycemic index, being low in sugar and calories. They present a high protein content with an excellent balance of essential amino acids, as well as high contents of fiber, lipids, carbohydrates, minerals, and bioactive compounds, such as vitamins (B2 and E), carotene, tocopherols, and other molecules with antioxidant properties, like flavonoids and other phenolic compounds [8–12].

Antioxidants are of economic interest since they can minimize rancidity and increase the shelf-life of food products [13]. Moreover, they are also of nutritional interest due to their health-related benefits. Antioxidants have been found to reduce the risks of cancer and cardiovascular disease and to present anti-inflammatory and anti-microbial activity [14,15]. Quinoa also shows unique fiber, lipid (with a high ratio of omega-6:omega-3), and micro- and macronutrient profiles (often higher than cereal-based products) that give quinoa seeds beneficial characteristics, such as decreasing the risk of cancer and cardiovascular and inflammatory diseases; decreasing blood pressure, diabetes, and development of hemorrhoids; and weight control [16], thus improving intestinal health [9]. Overall, quinoa seeds provide nutritional and health benefits, which is why quinoa is considered a "superfood of the future" [17,18].

Furthermore, quinoa is offered as a nutritious food for low-income countries and constitutes a crop able to grow on marginal lands (including those with limited rainfalls or poor soil quality) not suitable for other major crops [8]. This brings interesting opportunities for the agriculture of low-income countries and, generally, for those countries where agricultural water supply is (or will be soon) limited. These include Mediterranean countries where there is an urgent need to develop sustainable practices to mitigate the impacts of climate change and human pressure on soil resources [8]. This is especially relevant within the current climate change and food security context [19]. Besides, it should be noted that quinoa is not only consumed by humans, as its different plant parts can be used as a nutritionally valuable forage crop, apt for feeding sheep, pigs, cattle, poultry, and horses [20].

Importantly, it should be noted that quinoa exhibits a strong variability in cultivar-specific responses to environmental variation. There are reports of different environmental conditions impacting some seed quality-related parameters in quinoa, including seed size and protein or mineral contents, depending on the specific genotype [21–25]. Thus, different cultivars of quinoa have shown substantial differences in the nutritional characteristics, which also vary with the environment. However, it is still unclear if the parameters that were evaluated are stable among cultivars at different locations or if steady correlations can be found between nutritional-related parameters. A recent work by Granado-Rodriguez et al. [26] showed that some quinoa cultivars, Titicaca and Vikinga, present better quality-related traits (including higher protein contents), despite not being the most productive cultivar when growing in the Northwestern part of Spain. In line with this, selecting the best adapted genotypes for a particular cultivation environment is key, in terms of yield potential and biotic and abiotic stress tolerance but also considering the different nutritional traits [27]. Thus, we need to better understand the genetic and environmental factors determining the nutritional characteristics of quinoa. This will be achieved through the use of conventional and molecular tools that will help to unlock the rich biodiversity and cultivation potential of this crop [28].

Therefore, in the present study, we aimed at analyzing differences in seed quality among the 15 quinoa genotypes examined. We analyzed a variety of nutritional-related parameters with cultivation potential in the Southern region of Spain. Differences in the parameters analyzed were found among varieties for most of the parameters analyzed, including plant height, panicle length, mildew incidence, lodging, seed weight and area, protein content, germination rates, seed viability, mineral contents (except for Mg), antioxidants, and saponin contents. Furthermore, important correlations between different seed quality-related parameters dependent on the genotype (including yield and antioxidants or germination rates) support the presence of genetic determinants of nutritional quality in quinoa.

2. Results

2.1. Plant Performance and Physiological and Agronomical Traits

In this study, we aimed at analyzing different nutritional traits under filed conditions. Seeds of 15 different quinoa cultivars were sown on 27th January 2018, and plants were harvested on either July 18th or August 1st (Supplementary Figure S1). Cultivars 'A-SE-03', 'A-SE-06', 'A-SE-07', 'A-SE-09', 'A-SE-12', 'A-SE-13', and 'A-SE-15' were harvested in July (*Harvesting 1*, Supplementary Figure S1), while cultivars 'A-SE-01', 'A-SE-02', 'A-SE-04', 'A-SE-05', 'A-SE-08', 'A-SE-10', 'A-SE-11', and 'A-SE-14' presented longer life cycles, and they were harvested later (*Harvesting 2*, Supplementary Figure S1). Total seed yield varied among cultivars, with A-SE-08 cv. Being the cultivar that presented the highest seed yield (4.7 t/ha), followed by A-SE-11 cv. and A-SE-05 cv. (with 3.5 and 3.4 t/ha, respectively). A-SE-01 was the cultivar with the lowest seed yield (0.96 t/ha). Precipitation across the growing season was concentrated in the first months of cultivation (February to April), coinciding with plant nascence and emergence (Supplementary Figure S1), while scarce precipitation was registered from flowering to harvesting time (May–July). Daily mean temperatures varied from 8.7 °C (at sowing) to 25.7 °C and 30.7°C (at harvesting of short- and long-life-cycle cultivars, respectively), and they increased progressively along the growing season (Supplementary Figure S1). Inflorescences started appearing in May. At the seed maturation stage, temperatures higher than 25°C were registered for all cultivars.

Plant height also showed significant variations among cultivars (Figure 1). At early stages (82 days after sowing (d.a.s)), A-SE-05 cv., A-SE-03 cv., and A-SE-06 cv. plants were the tallest, with plant lengths of 54 $\pm$ 5.76 cm, 51 $\pm$ 4.79 cm, and 52.4 $\pm$ 5.6 cm, respectively, and, at middle stages (100 d.a.s.), A-SE-03 cv. and A-SE-06 cv. plants were still the tallest, with heights of 130.4 + 8.94 cm and 130.4 $\pm$ 6.33 cm, respectively. At the latest stages (128 d.a.s), A-SE-07 cv. and A-SE-15 cv. were the tallest plants, presenting as 170.90 $\pm$ 22.87 cm and 163.40 $\pm$ 31.99 cm in height, respectively. On the other hand, A-SE-04 cv. was the shortest cultivar throughout the season, going from 14.9 $\pm$ 5.27 cm tall at the early stage, to 69.10 $\pm$ 11.74 cm tall at the middle stage, and to 117.30 $\pm$ 25.12 cm at the end of the life cycle. In line with this parameter, lodging was also evaluated in this study. Thus, it was observed that the cultivars A-SE-07 and A-SE-03 presented greater lodging resistance (3% of lodging plants), while the cultivar A-SE-12 showed great sensitivity to lodging, with 36% of affected plants (Supplementary Figure S2).

Panicle length was determined in the different cultivars, evaluated at 128 d.a.s. Most of the cultivars showed panicle lengths between 30 cm and 40 cm, with A-SE-01 cv. having the lowest values (21.70 $\pm$ 2.53 cm) and A-SE-13 cv. showing the highest (40.40 $\pm$ 3.03 cm) (Figure 2). In addition to the panicle length, the weight of those panicles was also measured (Supplementary Figure S2). The results indicated that having larger panicles usually correlated with heavier weights, although some exceptions were observed (i.e., A-SE-02 cv. was 33.8 $\pm$ 3.67 cm long but presented the biggest weight of 3.20 kg/25 panicles). Interestingly, lodging resistance was not related to panicle length, as the cultivars with contrasting lodging resistance (A-SE-07, A-SE-03, and A-SE-12) did not show significant differences in panicle lengths.

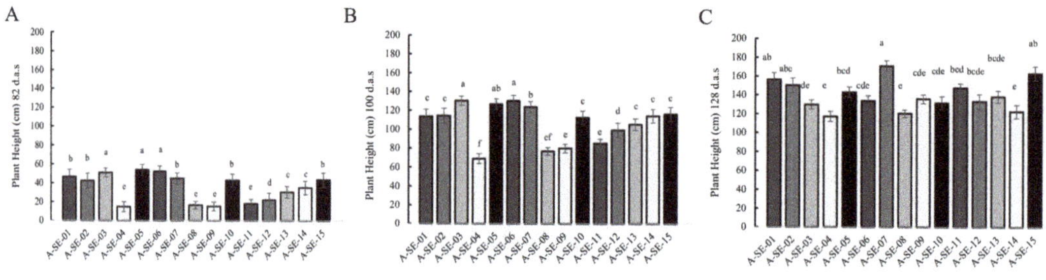

Figure 1. Plant height. Plant height (cm) was determined in the 15 cultivars analyzed at (**A**) 82, (**B**) 100, and (**C**) 128 d.a.s. Error bars represent the standard deviation. Bars that do not share the same letters show statistically significant differences following the Kruskal–Wallis test at a p-value < 0.05 for 82 d.a.s. and 100 d.a.s. and the ANOVA test and Tukey post-hoc test at a p-value < 0.05 for 128 d.a.s. (days after sowing).

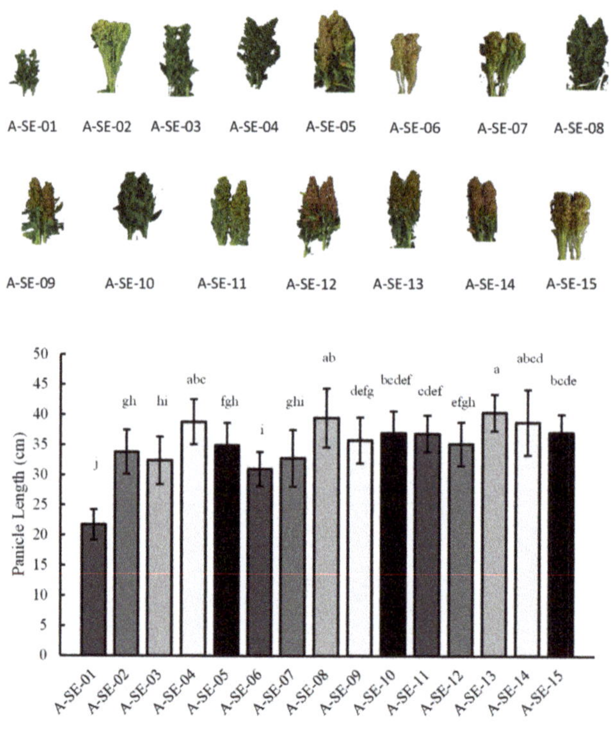

Figure 2. Panicle lenght. Panicle length (cm) was determined in the 15 cultivars analyzed. Error bars represent the standard deviation. Bars that do not share the same letters show statistically significant differences following the Kruskal–Wallis test at a p-value < 0.05.

Mildew incidence and severity were analyzed throughout the experiment (at 82, 100, and 128 d.a.s.) (Figure 3). A-SE-03 cv. and A-SE-08 cv. were the less-affected cultivars at the early stages (82 and 100 d. a. s.), and A-SE-06 cv. and A-SE-12 cv. were less affected at the later stages (128 d. a. s.). Meanwhile, A-SE-09 cv., A-SE-10 cv., and A-SE-11 cv. were the most afflicted ones in terms of severity, especially at 128 d.a.s.

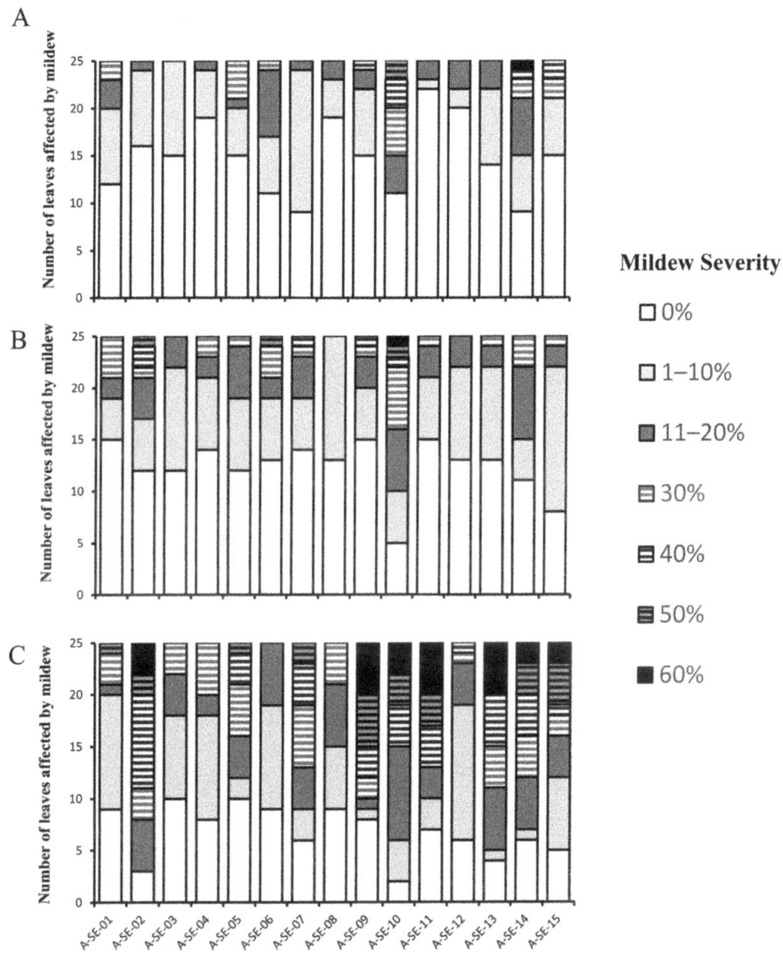

Figure 3. Mildew incidence and severity. Mildew incidence and severity were determined as described in the Methods section. Mildew severity degreewas expressed as the percentage (%) of leaf affected by the pathogen (leaf area converage of 0%, 1–10%, 11–20%, 30, 40%, 50%, or more than 60%). Mildew incidence and severity were evaluated at different developmental stages: at (**A**) 82 days after sowing (d.a.s.) (upper panel), (**B**) 100 d.a.s. (middle panel), and (**C**) 128 d.a.s (bottom panel).

Seed weight exhibited an effect that was related to the cultivar (Figure 4A). The cultivars A-SE-05 cv., A-SE-13 cv., and A-SE-15 cv. presented the heaviest seeds, while A-SE-04 cv., A-SE-10 cv., and A-SE-09 cv. showed the lightest seed weights.

Seed area showed a high correlation with seed weight (Supplementary Figure S3). A-SE-05 cv., A-SE-03 cv., and A-SE-15 cv. presented the largest seeds, while A-SE-04 cv., A-SE-08 cv., A-SE-09 cv., and A-SE-10 cv. had the narrowest seed areas (Figure 4B).

2.2. Germination Rates and Seed Viability

Germination rates were determined from the seeds harvested to evaluate the germination capacity (Figure 5A). Noticeably, differences were found in the germination rates of the various cultivars analyzed. A-SE-04 cv. and A-SE-15 cv. showed germination rates above 50% at 3 d.a.s, and A-SE-04 cv. reached 80% germination rate at 7 d.a.s (Figure 5B). Meanwhile, A-SE-03 seeds were unable to germinate, and A-SE-06 cv. and A-SE-01 cv. did not overtake 20% germination rates at 3 d.a.s. On the other hand, even though A-SE-12 cv. seeds showed a germination delay, they were able to reach a germination rate of almost 50% at 7 d.a.s, being close to the A-SE-05 cv. seeds at 7 d.a.s.

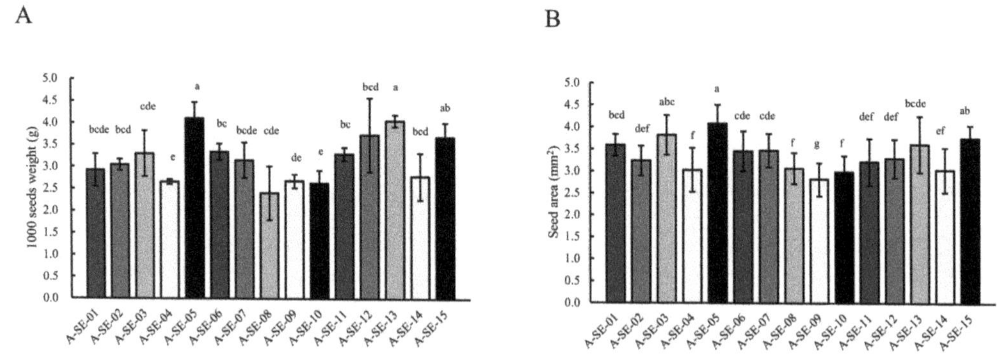

Figure 4. Seed weight and seed area. (**A**) Seed weight (g) and (**B**) area (mm^2) were determined among the different cultivars studied. Error bars represent the standard deviation. Bars that do not Scheme 0.

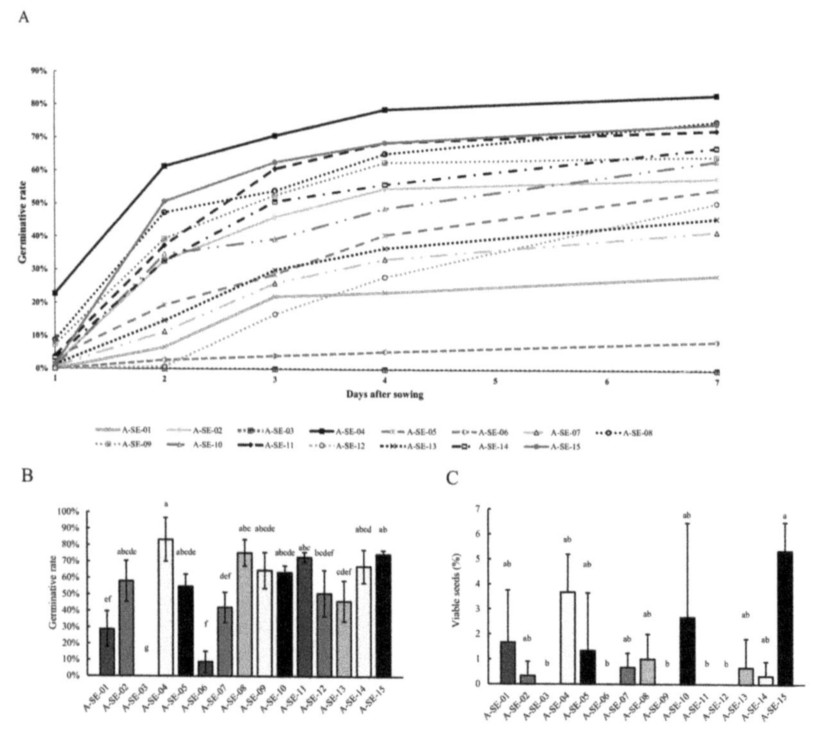

Figure 5. Germination rates (%) and seed viability. (**A**) Time course of the germination percentage (%) of quinoa seeds 1, 2, 3, 4, 5, 6, and 7 seven days after sowing (d.a.s.); (**B**) germination rate percentage (%) 7 d.a.s.; and (**C**) percentage (%) of viable seeds. Error bars represent the standard deviation. Bars that do not share the same letters show statistically significant differences following a Kruskal–Wallis test by ranks for multiple comparisons at a p-value < 0.05.

Seed viability was determined to complete the physiological analysis of the seeds (Figure 5C). For most of the cultivars, except for A-SE-15 cv., A-SE-04 cv., A-SE-03 cv., and

A-SE-06 cv., seed viability showed no correlation with seed germination, as it was generally severely reduced in most of the seeds tested.

2.3. Protein Content

Total protein contents in seeds revealed variations among cultivars (Figure 6). The cultivars A-SE-15 cv. and A-SE-02 cv. showed the highest contents, while, in contrast, A-SE-06 cv. and A-SE-03 cv. showed the lowest. Interestingly, as shown in Supplementary Figure S3, protein content positively correlated with TPC, FRAP, saponin content, Zn, Mg, and P and negatively with C/N ratio and Na content. Regarding the agronomical parameters, protein content correlated negatively with panicle height and panicle biomass and positively with the germination rate (Supplementary Figure S3).

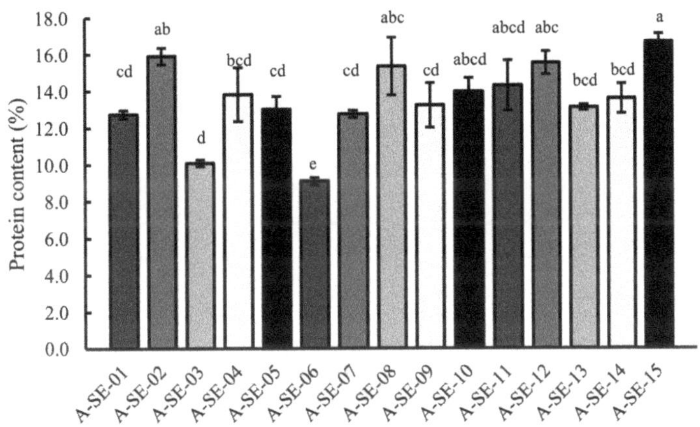

Figure 6. Seed protein content (%). Protein content was determined in seeds of the 15 cultivars evaluated. Error bars represent the standard deviation. Bars that do not share the same letters show statistically significant differences following the Kruskal–Wallis test by ranks for multiple comparisons at a p-value < 0.05.

2.4. Mineral Content

The total contents (as %) of phosphorous (P), potassium (K), calcium (Ca), magnesium (Mg), (and as mg/Kg) sodium (Na), copper (Cu), iron (Fe), manganese (Mn), and zinc (Zn) in quinoa seeds were determined to analyze the effect of the genotype on this nutritional-related parameter (Table 1). Some mineral nutrients, such as Mg, did not show significant variation among genotypes or, as in the case of K, showed a small fluctuation. On the contrary, minerals such as Zn showed a steeper variation. A-SE-12 cv., A-SE-15 cv., and A-SE-13 cv. showed the highest Zn contents, while A-SE-03 cv. and A-SE-09 cv. presented the lowest. Among cultivars, it should be noted that A-SE-12 cv. presented higher contents of P, Ca, Fe, and Zn and intermediate levels of the rest of minerals, while A-SE-15 cv. presented the highest contents of P, Cu, and Zn and the lowest of Ca, Na, and Fe. At the same time, A-SE-03 cv., A-SE-04 cv., and A-SE-06 cv. had higher contents of Ca and Na and lower contents of P and Cu.

Table 1. Mineral seed contents. Mean ± SD mineral contents are presented as a percentage of seed weight (P, K, Ca, and Mg) or as mg/Kg (Na, Fe, Cu, Mn, and Zn). Statistical analysis following a Kruskal–Wallis test by ranks (P, K, Mg, Na, Mn, Cu, and Zn content) or a Welch's ANOVA with a Games–Howell post-hoc test (Ca and Fe contents) were performed. Different letters under each mineral content show statistically significant differences between samples.

	Ash (%)	Nitrogen (%)	C/N ratio	P (%)	K (%)	Ca (%)	Mg (%)	Na (ppm)	Fe (ppm)	Mn (ppm)	Cu (ppm)	Zn (ppm)
A-SE-01	3.48 ± 0.02 b	2.04 ± 0.04 cd	18.45 ± 0.35 bc	0.29 ± 0.00 bc	1.14 ± 0.01 ab	0.30 ± 0.02 abc	0.21 ± 0.00 -	146.98 ± 5.99 bcdefg	40.41 ± 0.27 b	19.57 ± 0.30 bcdef	9.30 ± 1.42 bcde	29.18 ± 0.34 efg
A-SE-02	3.22 ± 0.10 abc	2.55 ± 0.07 ab	14.81 ± 0.56 de	0.27 ± 0.00 bc	1.03 ± 0.01 b	0.28 ± 0.01 bc	0.19 ± 0.01 -	116.63 ± 7.66 gh	38.98 ± 1.72 b	11.69 ± 0.05 f	11.70 ± 1.73 bc	29.56 ± 0.44 ef
A-SE-03	3.47 ± 0.07 abc	1.62 ± 0.03 d	22.59 ± 0.28 b	0.17 ± 0.00 d	1.26 ± 0.01 ab	0.38 ± 0.02 a	0.19 ± 0.00 -	233.42 ± 55.08 ab	36.05 ± 1.35 b	16.65 ± 0.99 cdef	7.54 ± 0.56 cdef	23.93 ± 0.38 i
A-SE-04	3.03 ± 0.33 abc	2.21 ± 0.23 bcd	17.47 ± 1.56 bcd	0.22 ± 0.02 cd	0.98 ± 0.12 ab	0.35 ± 0.02 ab	0.20 ± 0.02 -	279.7 ± 37.63 a	32.85 ± 3.45 bc	13.83 ± 0.95 ef	7.86 ± 1.21 bcdef	29.95 ± 0.52 def
A-SE-05	3.66 ± 0.27 abc	2.08 ± 0.11 cd	18.21 ± 0.92 bcd	0.27 ± 0.02 bcd	1.24 ± 0.10 ab	0.35 ± 0.04 abcd	0.18 ± 0.01 -	173.49 ± 21.92 abcdef	29.73 ± 2.74 bc	13.42 ± 0.72 def	9.34 ± 0.80 bcde	31.60 ± 0.74 bcd
A-SE-06	3.11 ± 0.17 abc	1.46 ± 0.03 e	25.34 ± 0.44 a	0.15 ± 0.00 e	1.13 ± 0.03 ab	0.36 ± 0.12 abcdef	0.17 ± 0.02 -	194.52 ± 59.36 abcde	35.08 ± 15.60 bc	14.32 ± 6.60 cdef	9.40 ± 3.17 bcdef	25.53 ± 1.23 ghi
A-SE-07	3.18 ± 0.09 bc	2.04 ± 0.03 cd	18.40 ± 0.45 bc	0.26 ± 0.01 bcd	1.10 ± 0.02 ab	0.18 ± 0.02 def	0.19 ± 0.00 -	125.43 ± 14.68 efgh	42.71 ± 3.77 abc	19.92 ± 0.18 bcde	10.28 ± 0.31 bc	30.85 ± 0.22 cde
A-SE-08	3.43 ± 0.36 abc	2.46 ± 0.25 abc	15.65 ± 1.65 bcde	0.28 ± 0.03 bc	1.15 ± 0.12 ab	0.21 ± 0.02 cde	0.22 ± 0.02 -	141.79 ± 12.98 cdefgh	43.18 ± 3.60 abc	22.90 ± 0.38 ab	10.32 ± 0.44 bcd	31.28 ± 0.80 bcd
A-SE-09	3.50 ± 0.47 abc	2.12 ± 0.19 cd	17.90 ± 1.50 bcd	0.26 ± 0.03 bcd	1.16 ± 0.16 ab	0.27 ± 0.01 bcde	0.18 ± 0.02 -	153.07 ± 32.93 bcdefgh	39.54 ± 4.68 abc	22.19 ± 1.50 ab	7.08 ± 0.53 def	23.80 ± 0.25 i
A-SE-10	3.68 ± 0.34 abc	2.24 ± 0.12 abcd	16.95 ± 0.99 bcde	0.24 ± 0.04 bcd	1.22 ± 0.17 ab	0.25 ± 0.04 abcdef	0.19 ± 0.03 -	191.24 ± 21.87 abcd	47.12 ± 8.56 abc	20.77 ± 2.11 abcd	6.41 ± 1.01 ef	25.32 ± 1.36 hi
A-SE-11	3.62 ± 0.63 abc	2.29 ± 0.22 abcd	16.74 ± 1.57 bcde	0.28 ± 0.03 bcd	1.23±0.21 ab	0.24 ± 0.03 abcdef	0.21 ± 0.03 -	194.76 ± 22.11 abc	39.56 ± 6.24 abc	20.05 ± 1.63 abcd	6.55 ± 0.68 f	28.76 ± 0.30 fgh
A-SE-12	3.26 ± 0.03 c	2.48 ± 0.10 abc	15.13 ± 0.59 cde	0.31 ± 0.01 b	1.12 ± 0.01 ab	0.36 ± 0.01 a	0.19 ± 0.00 -	139.22 ± 16.03 defgh	56.39 ± 1.56 a	17.62 ± 0.51 cdef	8.98 ± 0.76 bcdef	35.23 ± 0.21 a
A-SE-13	3.73 ± 0.03 a	2.10 ± 0.02 bcd	17.93 ± 0.28 bcd	0.27 ± 0.01 bcd	1.39 ± 0.03 a	0.10 ± 0.03 ef	0.18 ± 0.01 -	121.64 ± 15.62 fgh	35.17 ± 3.55 bc	22.17 ± 1.15 abc	19.52 ± 0.60 a	32.45 ± 0.58 bc
A-SE-14	3.34 ± 0.21 abc	2.17 ± 0.13 bcd	17.52 ± 1.00 bcd	0.25 ± 0.01 bcd	1.16 ± 0.06 ab	0.10 ± 0.02 f	0.20 ± 0.01 -	125.35 ± 15.11 fgh	33.19 ± 2.41 bc	23.16 ± 1.16 a	18.46 ± 0.88 ab	29.37 ± 0.97 efg
A-SE-15	3.19 ± 0.08 bc	2.67 ± 0.07 a	14.39 ± 0.48 e	0.35 ± 0.01 a	1.12 ± 0.02 ab	0.09 ± 0.01 f	0.21 ± 0.01 -	100.87 ± 8.96 h	25.10 ± 1.29 c	20.34 ± 0.50 abcd	19.75 ± 0.96 a	34.09 ± 0.81 b

2.5. Antioxidant Capacity

We evaluated the antioxidant capacity of the seeds by performing an FRAP assay and a quantification of total polyphenols (TPC) and flavonoids (TFC) contents (Figure 7). Among cultivars, A-SE-04 cv. showed the highest antioxidant capacity and phenolic and flavonoid contents, followed by A-SE-15 cv. On the contrary, A-SE-01 cv., together with A-SE-06 cv., showed the lowest values. The rest of cultivars showed distinct patterns, presenting changes among the antioxidant-related parameters here evaluated. For instance, A-SE-10 cv. showed intermediate and high FRAP and TFC values, respectively, and low TPC levels.

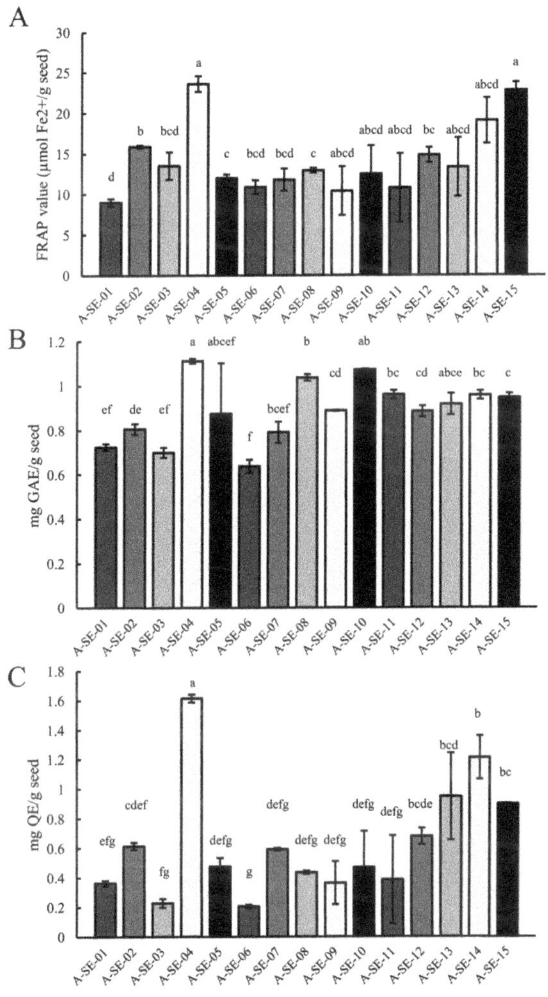

Figure 7. Antioxidant capacity of quinoa seeds. (**A**) Antioxidant power, determined by an FRAP assay, is expressed as μmol of Fe^{2+} per gram of seed. Statistical differences were analyzed through a Welch's ANOVA test, followed by a Games–Howell post-hoc test. (**B**) TPC is expressed as milligrams of gallic acid equivalents (GAE) per gram of seeds. The statistical analysis performed was a Welch's ANOVA test, followed by a Games–Howell post-hoc test. (**C**) TFC is expressed as milligrams of quercetin equivalents (QE) per gram of seeds. A Kruskal–Wallis test by ranks was performed for multiple comparisons. Bars that do not share the same letters show statistically significant differences at a p-value < 0.05. Error bars represent the standard deviation.

2.6. *Saponin Content*

Saponin content was quantified in the cultivars studied (Figure 8). A-SE-06 was the cultivar showing the lowest saponin content, while A-SE-10 was the cultivar with the highest saponin level. All of the cultivars exceed the limit of 0.11%, established to classify quinoa varieties as sweet [29]; however, none of them presented a content higher than 1%, which is usually overtaken by bitter quinoa seeds [30].

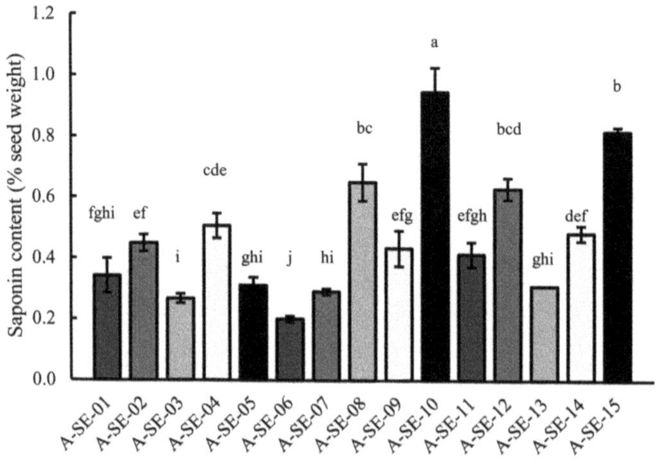

Figure 8. Saponin content. Saponin content was determined in seeds of the 15 cultivars evaluated. Error bars represent the standard deviation. Bars that do not share the same letters show statistically significant differences following the Kruskal–Wallis test by ranks at a p-value < 0.05.

2.7. *Principal Components Analysis (PCA)*

A Pearson's correlation coefficient test was performed to analyze the correlation between variables (Supplementary Figure S3) and a principal component analysis (PCA) to reduce the number of variables. This analysis identified five principal components that were able to explain 74.76% of the variance. Component 1, which contributed to 21.31% of the variance, was mainly explained by the protein and saponin contents and most minerals' contents (P, Ca, Mg, Mn, Cu, and Zn contributed positively, and Ca and Na contributed negatively) and by the germination rate, lodging, plant height, and mildew severity at 128 d.a.s. For this new variable, A-SE-12 cv. and A-SE-15 cv. showed high values, while A-SE-03 cv. and A-SE-06 cv. presented the lowest (Figure 9). There were correlations between most of these variables, but those between protein content and germination rate (r = 0.801), protein and P contents (r = 0.846), P and Zn contents (r = 0.728), and protein and saponin contents (r = 0.695) were the strongest (Supplementary Figure S3). Component 2 contributed to the variance with 18.30%, and it comprised panicle length and biomass, plant dry weight, yield, germinative rate of seeds, and total phenolic content, and inversely, it comprised plant height (at three time-points) and seed area. Plant height at early stages (82 and 100 d.a.s.) correlated negatively with the final plant biomass, yield, germinative rate of seeds, and protein and phenolic contents, while the panicle height and plant weight at 128 d.a.s. correlated positively with these parameters. Strong correlations were found between the phenolic content and the germinative rate and panicle height (r = 0.884 and r = 0.780, respectively) (Supplementary Figure S3). For this component, there were high values in A-SE-08 cv. and A-SE-04 cv. and low values in A-SE-01 cv (Figure 9). Component 3 explained 12.19% of the variance and comprised the viability rate, flavonoid contents, and antioxidant capacity, and inversely, it comprised ash, K, and Fe contents. Both viability and germinative rate correlated with each other and with the antioxidant capacity and phenolic and flavonoid contents. There was also a strong correlation between ash and K content (r = 0.851) since K is the main mineral present in quinoa (Table 1). A-SE-04 cv. and A-SE-15 cv. showed high component 3 values, and A-SE-09 cv., A-SE-11 cv., and A-SE-13 showed low values. Area and seed weight and K and Cu contents contributed positively to component 4 (explaining 11.96% of variance), and Fe and Ca contents contributed negatively. Area and seed weight showed a strong correlation (r = 0.748). A-SE-13 cv. presented the highest, and A-SE-01 cv., A-SE-09 cv., and A-SE-10 cv. presented the lowest values for component 4. Component 5 (11.00% of variance) comprised saponin content in seeds, panicle length, and mildew severity at three stages. There was a correlation between mildew severity at 82 and 100 d.a.s. but not with severity at 128 d.a.s. Saponin content and panicle height also showed a strong correlation (r = 0.655). A-SE-10 cv. showed high component 5 values, while A-SE-12 cv. showed the lowest. Life cycle duration correlated with yield, germination rate of seeds, and their phenolic contents.

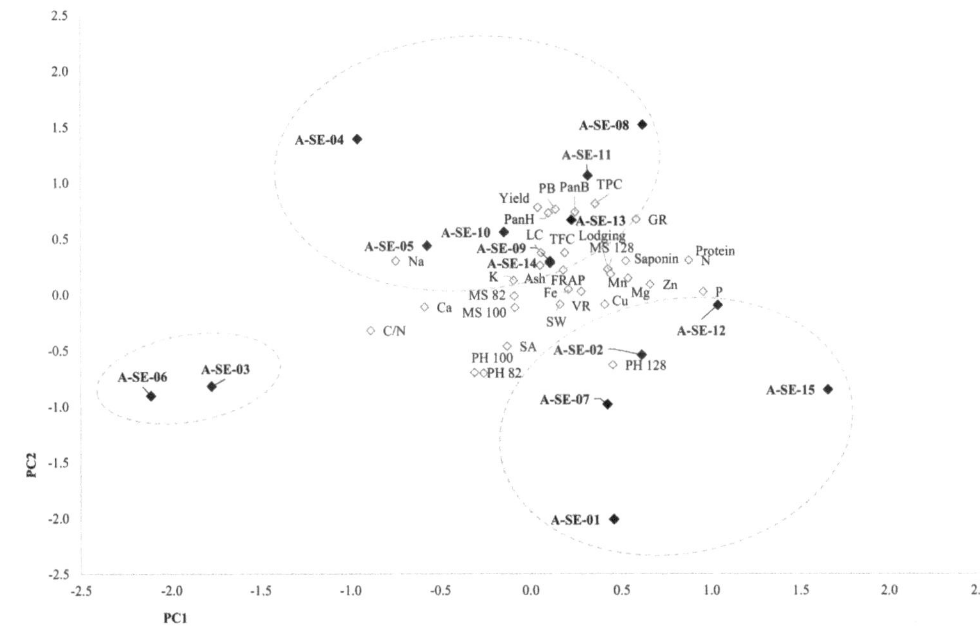

Figure 9. Principal components analysis (PCA). Biplot of main components 1 and 2 for the cultivars sown and for the variables tested. Component 1 (X-axis) was contributed mainly by protein and saponin contents, germination rate, minerals (P, Mg, Ca (-), Na (-), Mn, Cu, Zn), lodging, plant height at 128 d.a.s., and mildew at 128 d.a.s. Component 2 (Y-axis) included germination rate, total phenolics content, plant height (-), plant biomass, panicle height, and panicle weight (-).

Plotting component 1 against component 2 revealed three clusters of cultivars (Figure 9). The first cluster was made up of A-SE-03 and A-SE-06, and it was low for both components 1 and 2, which means they had taller plants at early stages but low yields; germinative capacity; and protein, P, saponin, and phenolic contents. These cultivars were also low for component 3, and they showed the lowest viability rate and low antioxidant capacity but a high ash content. The second cluster, comprising A-SE-01 cv., A-SE-02 cv., A-SE-07 cv., A-SE-12 cv., and A-SE-15 cv., had low component 2 values, especially for A-SE-01 cv., so they were tall plants with smaller panicles and lower yields but higher component 1 values than the first cluster, so generally, they had higher protein, P, and Cu contents and lower Ca and Na contents than A-SE-03 cv. and A-SE-06 cv. A-SE-05 cv., A-SE-08 cv., A-SE-09 cv., A-SE-10 cv., A-SE-11 cv., A-SE-13 cv., and A-SE-14 cv. comprised the third cluster, which had high component 2 values, having shorter plants and larger panicles, heavier plants, higher yields and germination rates, and higher protein and phenolic contents. A-SE-04 cv. and A-SE-15 cv. showed the highest component 3 values, with the largest viability rates and FRAP values, but A-SE-04 cv. had lower component 1 values, with high Ca and Na contents and lower Cu and Zn contents. A-SE-15 cv. had high component 1 values, and it had high germination rates and protein, P, Cu, and Zn contents and low Ca and Na contents.

A path analysis was also performed to define the direct and indirect contributions of each trait to seed yield (Supplementary Tables S1 and S2). First, a predictive multiple linear regression model was performed following the stepwise method in order to find traits with a direct effect on germination rates (Supplementary Table S1) and yield (Supplementary Table S2). As shown in Table S2, germination rates would be affected positively by the phenols (TPC) and P contents and indirectly by physiological or agronomical parameters, such as seed area or panicle biomass, or by biochemical properties of seeds, such as protein or saponin content. On the other hand, yield would be explained in a negative way by the seed weight and panicle height and positively by the panicle biomass and total biomass, meaning that these parameters may directly impact the seed yield of the quinoa varieties here analyzed.

3. Discussion

Quinoa is often compared to cereals and even considered a 'pseudocereal' due to the similarities in the composition and uses of their seeds [31]. It presents unique nutritional properties that make its cultivation very interesting [32]. However, its full yield potential is not yet reached for new cultivation areas, with levels similar to those of cereals such as wheat or rice before the Green Revolution [2]. The center of origin of quinoa is the Andean Altiplano [33], but in the last decades, quinoa has been introduced as an alternative emerging crop in more than 75 countries [5]. Along with its expansion, it has been observed that the establishment and adaptation of quinoa cultivars to these new agroclimatic contexts can result in changes in the nutritional properties of quinoa seeds, which are associated with variations in the genotype (G), the environment (E), and their interaction (GXE) [14,21]. Thus, there is still much left for researchers and breeders to do in order to develop quinoa cultivars that are better adapted to specific locations that present high yields while also maintaining or even improving the nutritional value of the seeds. In this study, we evaluated physiological and agronomical characteristics together with different nutritional-related traits of seeds harvested from 15 different quinoa cultivars grown in southern Spain, aiming to expand our knowledge of the relationship between yield and the nutritional quality of quinoa seeds and therefore, contribute to the selection of quinoa cultivars more appropriate for cultivation in a particular area of interest.

As previously described, through the PCA, we could classify the cultivars into three clusters, depending on their distinct characteristics (Figure 9). At earlier stages of development (82 and 100 d.a.s.), plants from clusters 1 (A-SE-03 cv. and A-SE-06 cv.) and 2 (A-SE-01, A-SE-02 cv., A-SE-07 cv., and A-SE-15 cv.) were the tallest, but only those from cluster 2 remained taller at a later developmental stage (128 d.a.s., beginning of grain maturation) (Figure 1). Cluster 1 and cluster 2 plants also presented lower panicle lengths and biomass (Figure 2) and smaller seed yields. Seeds from these cultivars did not show higher nor lower seed weights or areas, and there was no correlation between these variables and yield (Supplementary Figure S3). From this, it can be assumed that plants from clusters 1 and 2 invested more resources towards growing at earlier stages of development but they invested fewer resources in the development of the panicle and seed biomass. It should be noted that, for these plants, lower yields were not correlated with reduced seed weights, but with smaller panicles producing fewer seeds. This negative relation between plant height and seed yield and positive relation between panicle size (biomass and height) and yield (Table S2, Supplementary Figure S3) have been previously described [34–36]. Furthermore, Gómez et al. [37] reported a correlation between plant height and yield in quinoa. They postulated that quinoa's low yield can be explained by its low sink capacity and that an increase in reproductive partitioning, reducing plant height, could positively impact yield in this crop, as had happened previously to wheat and rice during the Green Revolution [38]. Furthermore, some reports have found a positive correlation between yield and plant height, pointing as well to the influence of the environment in controlling this trait. Therefore, the implication of the environment should be investigated for the varieties here analyzed [39]. Nonetheless, plant height can be an important trait for breeding. Thus, further research analyzing endogenous factors that may control quinoa height and its relationship with yield and lodging (i.e., phytohormones) should be considered in quinoa.

Intriguingly, previous works have observed correlations between yield and seed nutritional-related traits, like the positive correlation found between yield and the antioxidant capacity or the K content or the negative correlation between yield and protein content or the amount of different amino acids [22,26,40]. However, these studies compared the nutritional profiles of quinoa seeds harvested from different cultivars but grown in different environmental conditions (with variations in the sowing date, the cultivation location, and/or the year of cultivation). Thus, the variations on the nutritional traits of the seeds in these cases were mainly determined by differences in the environmental conditions at the seed-filling stage, which affect both yield and seed-quality traits [26]. In the present study, only the genetic factor was evaluated, so the lack of correlations between yield and seed nutritional-related traits suggests that there might be no link between them. Therefore, those relations are only relevant when introducing the cultivars to new environments where they are not yet adapted.

Downey mildew, caused by the fungus *Peronospora variabilis* Gäum, is one of the main diseases affecting quinoa on a global scale [41]. Optimal conditions for mildew development are found at high humidity (>80% RH) and moderate temperatures (between 18 °C and 22 °C), but its expansion can be interrupted by long periods of sunny and dry conditions [41]. In this study, high RH was found in March, when most of the precipitation occurred and plants were still emerging or developing their first true leaves (Supplementary Figure S1). However, temperatures at that time were lower than 18 °C, which is suboptimal for mildew development. Mildew produces chlorotic patches on leaves, which may result in premature defoliation by the plant as a defense mechanism. This reduction of the photosynthetic area can lead to an atrophied development and smaller panicles, which in turn, lowers seed yield [42]. When the infection occurs at early stages of development of quinoa, 20–40% yield penalties have been estimated for mildew-resistant cultivars [43], and losses of up to 99% have been estimated in susceptible cultivars [42]. However, the impact of mildew on well-established mature plants is less important than abiotic stresses [44], since the defoliation caused by the disease and by the natural senescence of the plant overlap [41]. In the present study, mildew incidence and severity were limited at early stages (82 and 100 d.a.s.), with severities lower than 10% in most cultivars (Figure 3), despite not using plant protectants. However, severity increased at a later stage (128 d.a.s.), with A-SE-02 cv. (cluster 2), A-SE-09 cv., A-SE-10 cv., A-SE-11 cv., and A-SE-13 cv. (cluster 3) being the most affected ones and A-SE-01 cv., A-SE-12 cv. (cluster 2), A-SE-03 cv., A-SE-06 (cluster 1), A-SE-04 cv., and A-SE-08 cv. (cluster 3) being the least affected (Figure 3). Mildew severity did not correlate to yield nor to any seed nutritional traits (Supplementary Figure S3) [44], which suggests that the cultivars tested were resistant to mildew and did not suffer significant yield losses related to this disease.

Saponins are secondary metabolites found in the pericarp of quinoa seeds, which cause a bitter taste when they are present in substantial amounts. They also have a negative effect on the bioavailability of minerals like Fe and Zn [45]. For these two reasons, saponins are considered as "anti-nutrients". Different breeding programs have been focused on the development of cultivars with very low seed saponin contents (sweet quinoa cultivars) [28]. Koziol [46] established the limit between sweet seeds and bitter seeds at 0.11% of seed weight, while Mastebroek et al. [47] considered those with saponin contents between 0.02% and 0.04% as sweet seeds and those with contents between 0.47% and 1.13% as bitter seeds. All of our samples fitted the definition of a bitter seed by Koziol [46], but only A-SE-04, A-SE-08, A-SE-10 cv. (cluster 3), A-SE-12 cv., and A-SE-15 cv. (cluster 2) would be considered bitter following the Mastebroek et al. [47] criterion, and all samples would be 'low-saponin' seeds according to Medina-Meza et al. [30]. In this regard, it should be noted that sweet varieties are normally preferred, since the elimination process of saponins is avoided. However, some farmers prefer bitter cultivars, because saponins may confer resistance to biotic stresses [28]. Although saponins have been hypothesized to also give mildew resistance to quinoa [48], since they possess antifungal activities [49], no correlation has been found between seed saponin contents and mildew resistance in quinoa [50] (Supplementary Figure S3).

Saponin is a highly genotype-dependent seed trait in quinoa [51], and no correlation to other seed traits had been found previously. In the present study, we found, for the first time to our knowledge, a correlation between saponin content a other seed quality-related traits, like germination, protein content, and flavonoid content (Supplementary Figure S3) [52–54].

Germination capacity is an important seed characteristic for breeding programs. This is because any genetic potential achieved through breeding efforts cannot be exploited if seed establishment in the field is not successful. In this study, most cultivars surpassed the 50% germination rate, but A-SE-13, A-SE-07, A-SE-01, and especially A-SE-03 and A-SE-06 (cluster 1), showed very low germination rates (Figures 5 and 9). Interestingly, a correlation between seed germination rates and the panicle's characteristics and seed yield of the mother plants was found [26] (Supplementary Figure S3), but germination rates were also influenced by nutritional traits of seeds. Both the correlation and pathway analyses showed a strong effect of the phenolic compounds and the P contents on the germination capacity of seeds (Supplementary Figure S3, Table S1). These results are in part supported by previous works, as a positive correlation between phenolic compounds and the germination capacity of quinoa seeds has previously been found [26]. Furthermore, a stimulating effect of these compounds on the germination capacity has been reported as well in the close quinoa relative species *Chenopodium album* L. [55]. On the other hand, P is present in quinoa seeds, mainly as phytate [56], a form of P storage not bioavailable for many monogastric animals, including humans [57,58]. During germination, however, the phytase activity catalyzes the hydrolysis of the phytate [59], providing inorganic phosphate essential for the metabolism of the seed at the beginning of germination [60,61]. According to Nadeem et al., [62], a higher phytate content also means more hydrolysis and thus, higher phosphate available during germination. This may explain the correlation between P content and germination (Supplementary Figure S3, Table S1), since higher P contents in seeds are related to faster germination and better establishment of seeds in the field [63]. A strong positive correlation was also found between the germination capacity and protein content (Supplementary Figure S3), probably associated with the role that storage proteins play in germination [64]. Noteworthily, the environmental conditions may differently impact the cultivars included in this study and, consequently, result in variations in seed germination [25,65]. To further explore this aspect, the cultivation in consecutive years should be considered in future works.

In the present study, the protein content of seeds varied depending on the cultivar, with values ranging between 12.7% and 16.7% with the exception of two cultivars, the low-performing A-SE-03 and A-SE-06 (cluster 1). These two cultivars presented seed protein contents of 9–10%, closer to the values found in cereals like maize and barley and lower than the values found in wheat [66]. The contents of the rest of the cultivars fell within the range expected for quinoa seeds, with A-SE-02, A-SE-12, A-SE-15 (cluster 2), and A-SE-08 (cluster 3) exceeding 15% [66]. However, it should be noted that the importance of the quinoa protein does not only rely on the quantity, but also on the quality. Quinoa seed proteins contain all amino acids, and they are present in a proper balance, similar to the complete amino acid profile found in cow's milk and close to the ideal equilibrium recommended by the FAO for human consumption [7,67].

The ash content ranged from 3.03% to 3.73%, depending on the cultivar, although few significant differences were found among cultivars (Table 1). These were normal values for quinoa but generally higher than those of cereals like wheat or rice [66,67]. The minerals that were present in higher amounts were K, P, Ca, and Mg, while Na, Fe, Zn, Mn, and Cu contents were the lowest (Table 1) [67]. All these minerals fell within the ranges previously reported for quinoa seeds [26,66], and some of them, like K, Ca, Mg, and Na, were higher than those found in cereals like maize, barley, rice, and wheat [7]. The high contents of Fe, Ca, and Mg are especially important, since they are minerals less present in gluten-free products, and thus, quinoa seeds can be an important source of these minerals for people with coeliac disease [67].

Noteworthily, the contents were significantly different among cultivars for all minerals except for Mg (Table 1). The variations in mineral contents in quinoa seeds had previously been reported to be cultivar-dependent, but they also respond to environmental differences during plant growth [22,26]. For instance, the cultivar A-SE-03 showed high Ca and Na and low P and Zn contents, while A-SE15 cv. showed high P, Cu, and Zn contents and low Ca, Na, and Fe contents (Table 1). P content was high, but according to Konishi et al. [56], P is mostly found in quinoa seeds as phytic acid, which can form complexes with Fe, Zn, Mg, and Ca, reducing their bioavailability for human digestion [68]. Interestingly, Ruales and Nair [45] noted that, in feeding experiments with rats, there were no differences in Fe availability in quinoa-supplemented diets compared to those supplemented with $FeSO_4$. Thus, further evaluation of the actual effect of quinoa seeds' phytic acid on Fe, Zn, Mg, and Ca availability should be conducted in order to elucidate which percentage of these minerals' contents is actually taken up during human digestion and if these contents reach the human nutritional requirements [69].

A correlation between phenolic compounds and flavonoids contents and the antioxidant capacity was expected (Supplementary Figure S3) [26,70]. The antioxidant capacity and phenolic compounds content are genotype-dependent in quinoa seeds [71], although they can also change depending on the environmental context [21,26,72]. In the present study, the antioxidant capacity, TPC, and TFC were comparable to those found in previous studies (Figure 7) [15,21,26,71] and changed depending on the cultivar. For instance, the cultivars A-SE-04 and A-SE-15 and A-SE-04, A-SE-08, and A-SE-10 showed the highest levels of antioxidant capacity and TPC, respectively, while A-SE-01 presented the lowest antioxidant capacity, and A-SE-03 and A-SE-06 exhibited the lowest TPC and TFC (Figure 7). These results correlated well with other seed-related traits, like protein content and germination rates, and with seed yield (Supplementary Figure S3). This, together with the overall health benefits of antioxidants, make TPC and TFC interesting traits for quinoa breeding programs.

Targeting phenotypic traits such as physiological, agronomical, or seed nutritional-related parameters might be very useful when aiming at performing phenotyping screenings or for breeding programs. The selection of the best quinoa cultivars for production can be based on the results obtained for some of the parameters here discussed (including the results of their correlation). For instance, the present work highlights that the cultivar A-SE-08 is the most promising one amongst the 15 cultivars studied, based on the higher protein contents, yield, germination rate, and P and phenolic contents. Targeting these traits can be very useful for selecting the best adapted varieties for a particular area of cultivation.

4. Materials and Methods

4.1. Plant Material, Experimental Design, and Location

Field trials were conducted in a field experimental station located in Lebrija (Seville, Spain, 36.88° N, 6.13° W) in clay-loam soil. Sowing to harvesting dates took place from January to August of 2018. Fifteen different quinoa cultivars, given by Algosur S. A. (Lebrija, Spain), were used in this study, encoded as follows: 'A-SE-01', 'A-SE-02', 'A-SE-03', 'A-SE-04', 'A-SE-05', 'A-SE-06', 'A-SE-07', 'A-SE-08', 'A-SE-09', 'A-SE-10', 'A-SE-11', 'A-SE-12', 'A-SE-13', and 'A-SE-14'. (Supplementary Figure S4).

Each cultivar was sown on January 27th in two replicates of non-randomized plots, with dimensions of 4.5 m × 266 m, spacing between rows of 0.75 m, and 0.02 m within rows. The plot dimension was large enough to ensure uniformity according to our field experiments previously performed. A drilling machine was used to sow with a density of seeds of 2 $kg \cdot ha^{-1}$.

During the experiment, different measurements of agronomical traits were taken. Plant height and downy mildew incidence and severity were measured at 82, 100, and 128 d.a.s, which corresponded to different developmental stages: fully emerged plants, panicle emergence, and beginning of seed ripening, respectively. Furthermore, at 128 d.a.s., the panicle length and weight of 25 plants per cultivar were measured. Plant harvesting took place when plants had naturally dried out at different time points: July 18th (172 days after sowing) for the cultivars 'A-SE-03' cv. 'A-SE-06' cv., 'A-SE-07' cv., 'A-SE-09' cv., 'A-SE-12' cv., 'A-SE-13' cv., and 'A-SE-15' cv. and on August 1st (186 d.a.s.) for 'A-SE-01' cv., 'A-SE-02' cv., 'A-SE-04' cv., 'A-SE-05' cv., 'A-SE-08' cv., 'A-SE-10' cv., 'A-SE-11' cv., and 'A-SE-14' cv. Total seed yield was quantified from an 11.25 m^2 plot for each cultivar, and the dry weight of 20 plants was measured.

Climatological data, including total precipitation, relative humidity (RH), and temperature, were obtained daily from a local climatological station (Supplementary Figure S1). Sprinkler irrigation was supplemented at different developmental stages: at seed sowing (30 L/m^2), 5 days after sowing (30 L/m^2), at the beginning of branching (30 L/m^2), at flowering (50 L/m^2), and during grain filling (50 L/m^2).

4.2. Seed Weight and Area

Seeds were manually counted and weighed in an analytical balance. The seed area was analyzed using the open-source software ImageJ (http://rsbweb.nih.gov/ij/ accessed on 12 November 2020). Images were taken using an Olympus SZ61 stereomicroscope (Olympus Corporation, Shinjuku, Tokyo, Japan) and processed with the AnalySIS GetIT image software (analysis getIT 5.1, Olympus Corporation, Shinjuku, Tokyo, Japan). To determine seed weight, 1000 seeds were used per replication, and 3 replications were utilized. For seed area measurement, 50 seeds were used per replication, and 3 replications were utilized

4.3. Seed Germination Rate

Quinoa seeds (using 50 seeds per replication and 3 replications per cultivar) were sterilized by soaking first in ethanol 70% for two minutes, next in bleach 50% with a droplet of Tween-20 for two minutes, and then rinsing several times in distilled water (H_2O). Sterilized seeds were sown on a double layer of filter paper, wet with distilled water, on Petri dishes and then transferred to a growth chamber under darkness and a controlled temperature of 23 °C. The germination rate was counted daily for the first week after sowing. Seeds were considered as germinated when the radicle protrusion was longer than 2 mm.

4.4. Seed Viability

Seed viability tests were performed using the tetrazolium method (2,3,5-triphenyl-2H-tetrazolium chloride). First, seeds (using 100 seeds per replication and 3 replications per cultivar) were imbibed in distilled water at 30 °C for an hour to facilitate longitudinal and superficial cuts of the embryo and to ensure a homogeneous dying of the seed tissues. After cutting, seeds were submerged in 1% tetrazolium chloride at 30 °C for two hours. Seeds with more than 50% staining in the embryonic tissue were considered viable.

4.5. Saponin Content

To determine saponin content, 20 mL of 50 % ethanol was added to 1 g of powdered sample and left to macerate for 72 h at room temperature. Then, the extracts were filtered into 20 mL volumetric flasks. The samples were then filtered using a 0.45 μm nylon Filter-Lab syringe and analyzed by High-Performance Liquid Chromatography Fluorescence and Diodo Array Detection (HPLCDAD, Serie 1100, Agilent Technologies, Waldbronn,

Germany) at 225 nm [73]. Saponin (Merck, Germany) was used as a standard. The results were expressed in g saponin 100 g^{-1} of fresh weight.

4.6. Protein Content

The protein content was determined according to AOAC Official Methods [74], using an elemental analyzer Leco TruSpec (LECO TruSpec (LECO, MI, USA)) and considering a conversion factor of 6.25 [75].

4.7. Mineral Content

The mineral content was analyzed following the official methods of analysis of the Spanish Ministry of Agriculture [76]. Phosphorus content was determined using a spectrophotometer UV-VIS (Hitachi U-2810, Tokyo, Japan) (yellow coloration, 430 nm). Potassium was determined through flame atomic emission spectroscopy. Calcium, magnesium, sodium, iron, copper, manganese, and zinc contents were assessed using flame atomic absorption spectroscopy (AAS) (SpectrAA 110, Agilent Technologies Inc., Palo Alto, CA, USA) after mineralizing the samples with H_2O and HCl (35%).

4.8. Ferric Reducing Antioxidant Power (FRAP) Assay

To obtain total extracts, seeds were ground to a fine powder, and 100 mg of the flour was homogenized in 1 mL of an extraction buffer, consisting of methanol (50%), acetic acid (1%), and distilled water (49%). These samples were vortexed for 2 min and kept in the dark at 4 °C for 48 h before centrifugation for 15 min at 13500 rpm. The supernatants were stored at −20 °C until their use in the FRAP and flavonoid content assays.

The antioxidant capacity of seeds was determined following the procedure described by Benzie and Strain [77]. The FRAP reagent consisted of a mix of 300 mM acetate buffer (pH 3.6), with 10 mM TPTZ in 40 mM HCl and 20 mM $FeCl_3 \cdot 6H_2O$ at a ratio of 10:1:1 (*v*/*v*/*v*). A total of 20 µL of the sample extract and 180 µL of the FRAP reagent were added into a 96-well microplate and incubated for 4 min. Absorbance was read at 593 nm using a microplate reader Lector Multi-ModalSynergy HTX (BioTek Instruments, Inc., Winooski, VT, USA). The antioxidant capacity was calculated from a calibration curve obtained with iron (II) sulfate ($FeSO_4$). The FRAP value was expressed as µmol of Fe^{2+} g^{-1} of seed.

4.9. Total Phenol Content (TPC)

Extracts were obtained after homogenizing 100 mg of seed flour in 1 mL of ice-cold methanol (95%). The mix was vortexed, and centrifuged at 13,500 rpm for 5 min after 48 h kept in the dark at 4 °C.

The content of polyphenols was measured following the protocol described by Ainsworth and Gillespie [78]. Briefly, 100 µL of the sample extract or standard were added to 200 µL of the Folin–Ciocalteu reagent (10%) and the mixture was vortexed for 1 min. Next, 800 µL of sodium carbonate (7.5%) were added. The mix was then incubated for 2 h in the dark. The samples were centrifuged in order to eliminate precipitates. Absorbance was read at 765 nm using a microplate reader Lector Multi-ModalSynergy HTX (BioTek Instruments, Inc., Winooski, VT, USA). Concentrations of gallic acid between 20 $\mu g \cdot mL^{-1}$ and 200 $\mu g \cdot mL^{-1}$ in methanol (95%) were used as a standard, and thus, the TPC was expressed as mg of gallic acid equivalents per grams of quinoa seed (mg $GAE \cdot g^{-1}$).

4.10. Total Flavonoid Content (TFC)

Flavonoid content was determined following the procedure described by Valenzuela [79]. The same extracts as in the FRAP assay were used. Briefly, 30 µL of the sample extract or standard, 10 µL of aluminum chloride ($AlCl_3$) 10%, 10 µL of sodium acetate ($NaC_2H_3O_2$) 1 M, and 250 µL of dH2O were mixed and incubated for 30 minutes. The absorbance was read at 415 nm using a microplate reader Lector Multi-ModalSynergy HTX (BioTek Instruments, Inc., Winooski, VT, USA). Quercetin dissolved in ethanol (80%) was used as a standard, with concentrations ranging from 10 μg·mL^{-1} to 140 μg·mL^{-1}. The results were expressed in mg of quercetin equivalents per gram of quinoa seed (mg QE·g^{-1}).

4.11. Statistical Analysis

To analyze the differences between cultivars, different one-way ANOVA tests were performed. For variables where normality and equal variances could be assumed, a one-way ANOVA test was performed, followed by a Tukey post-hoc test to perform multiple comparisons at a probability level of 5% ($p < 0.05$). A one-way ANOVA on ranks (Kruskal–Wallis test by ranks) was performed when data did not present a normal distribution, and a Welch's ANOVA test followed by a Games–Howell post-hoc test was performed when variances were not equal, both at a probability level of 5% ($p < 0.05$). Normality and equality of variances of the data were tested through a Kolmogorov–Smirnov's test and a Levene's test, respectively. A principal component analysis (PCA) was performed for plant parameters, like plant height at three stages, panicle length and biomass, plant biomass, mildew severity at different stages, resistance to lodging, life-cycle length, and yield, and for seed parameters, like viability and germination rates, 1000 seeds' weight, seed area, saponin content, N and protein content and C-N ratio, FRAP value, phenols and flavonoids contents, and mineral contents. Correlations amongst variables were evaluated with a Pearson's correlation coefficient test (Supplementary Figure S3). A sequential path analysis was performed to evaluate the specific contribution of different traits to yield or germination rate (Supplementary Tables S1 and S2). This analysis allowed ordering different variables as predictors of yield of the first, second, or third-order [80]. For this purpose, a stepwise multiple linear regression procedure was used, where variables that showed a weak contribution ($p > 0.05$) to the dependent variable (yield or germination rate) or high multicollinearity were automatically dropped from the model. The variables entered into the model were considered as first-order predictors, and the procedure was repeated using these variables as the response variable to identify traits that function as second-order predictors of yield. The tolerance and variance inflation factor (VIF) were used to measure the level of multicollinearity for each predictor trait. Tolerance lower than 0.1 or VIF values higher than 10 were considered as high levels of collinearity. Tolerance (1- R2i, where R2i is the coefficient of determination for the prediction of variable i by the predictor variables) is the amount of variance of the selected independent variable not explained by other independent variables. VIF (1/Tolerance) indicates the extent of the effects of other independent variables on the variability of the selected independent variable. The SPSS Statistics for Windows (Version 24.0., IBM Corp., Armonk, NY, USA) package was used for the statistical analyses.

5. Conclusions

This study revealed differences among cultivars for each physiological, agronomical, and seed nutritional-related trait analyzed, although there were similarities among some cultivars. For instance, A-SE-03 and A-SE-06, which clustered together in the PCA, showed taller plants at early stages of development but shorter plants with smaller panicles and lower yields at maturity (Figures 1 and 9, Supplementary Figure S3). Regarding seed traits, these cultivars presented lower germination rates and lower protein, P, phenols, flavonoids, and saponins contents (Figures 5 and 7–9, Table 1). On the contrary, the most promising cultivars for this agroclimatic context were those included in cluster 3, due to the higher yields, germination rates, and TPC (Figure 9). Since quinoa seeds are well known to be an excellent source of high-quality protein of non-animal origin, this is one of the main traits that makes quinoa a crop with a high nutritional quality, important for achieving food security locally and globally [81]. In this study, the higher protein contents were shown by A-SE-15 cv., A-SE-02 cv., A-SE-12 cv., and A-SE-08 cv. Therefore, overall, the cultivar A-SE-08 (cluster 3) can be considered the most promising cultivar for this particular area, since it not only presented higher protein contents, but also larger germination rates and P and phenolic contents. Having all these traits positively correlated can greatly facilitate the development of a better adapted cultivar. However, it should be noted that saponin content was also higher in this cultivar (Figure 9). Considering that reducing the saponin contents can improve nutritional quality and flavor [28,82], it would be interesting to explore the possibilities offered by agronomical management practices that allow the reduction of saponins [54,83].

Therefore, the results here presented highlight the importance of considering the genotypic variation in quinoa when selecting improved quinoa varieties with better nutritional characteristics for new cultivation environments. Further studies are required to determine which exact parameters are genotype-variable and which ones show genotypic stability.

Supplementary Materials: The following are available online at https://www.mdpi.com/article/10.3390/plants10102128/s1, Figure S1:]Climatological conditions and general crop phenology. Figure S2: Fresh and Dry weight of 25 panicles. Figure S3: Correlogram of variables measured. Figure S4: Pictures of seeds of the different cultivars harvested and analyzed in this study. Table S1: Direct effects of predictor variables of the first-, second-, third-, and fourth-order on germination rate, tolerance, and variance inflation factor of the path analysis. Table S2: Direct effects of predictor variables of the first-, second-, third-, and fourth-order on germination rate, tolerance, and variance inflation factor of the path analysis.

Author Contributions: M.R., J.M. and S.V.-R. conceived and planned the experiments. S.G.-R., J.M., S.V.-R., I.M.-G., M.J.R. and P.C. carried out the experiments. M.R., S.V.-R., S.G.-R., I.M.-G., J.M., M.R., V.C. and L.B. contributed to the interpretation of the results. M.R., I.M.-G. and S.G.-R. took the lead in writing the manuscript. All authors provided critical feedback and helped shape the research, analysis, and manuscript. All authors have read and agreed to the published version of the manuscript.

Funding: This work was supported by the Ministerio de Ciencia e Innovación (MICINN, Spain) (PID2019-105748RA-I00), the Madrid Government (Comunidad de Madrid-Spain) under the Multi-annual Agreement with Universidad Autónoma de Madrid in the line of action encouraging youth research doctors, in the context of the V PRICIT (Regional Program of Research and Technological Innovation) (SI1/PJI/2019-00124), the CYTED (ValSe-Food 119RT0567), the FPI UAM Fellowship Programme 2019 (to SG-R), the CM research assistant fellowship (to I. M. G.), and the Ramón y Cajal Programme 2019 (to MR).

Data Availability Statement: The data that support the findings of this study are available upon request.

Acknowledgments: The authors greatly thank ALGOSUR S.A. for providing the quinoa seeds used in this study.

Conflicts of Interest: The authors declare no conflict of interest.

References

1. Zou, C.; Chen, A.; Xiao, L.; Muller, H.M.; Ache, P.; Haberer, G.; Zhang, M.; Jia, W.; Deng, P.; Huang, R.; et al. A high-quality genome assembly of quinoa provides insights into the molecular basis of salt bladder-based salinity tolerance and the exceptional nutritional value. *Cell Res.* **2017**, *27*, 1327–1340. [CrossRef]
2. Ruiz, K.B.; Biondi, S.; Oses, R.; Acuña-Rodríguez, I.S.; Antognoni, F.; Martinez-Mosqueira, E.A.; Coulibaly, A.; Canahua-Murillo, A.; Pinto, M.; Zurita-Silva, A.; et al. Quinoa biodiversity and sustainability for food security under climate change. A review. *Agron. Sustain. Dev.* **2014**, *34*, 349–359. [CrossRef]
3. Wu, Q.; Bai, X.; Wu, X.; Xiang, D.; Wan, Y.; Luo, Y.; Shi, X.; Li, Q.; Zhao, J.; Qin, P.; et al. Transcriptome profiling identifies transcription factors and key homologs involved in seed dormancy and germination regulation of *Chenopodium quinoa*. *Plant Physiol. Biochem.* **2020**, *151*, 443–456. [CrossRef]
4. Alandia, G.; Rodriguez, J.P.; Jacobsen, S.-E.; Bazile, D.; Condori, B. Global expansion of quinoa and challenges for the Andean region. *Glob. Food Sec.* **2020**, *26*, 100429. [CrossRef]
5. Bazile, D.; Jacobsen, S.E.; Verniau, A. The Global Expansion of Quinoa: Trends and Limits. *Front. Plant Sci.* **2016**, *7*, 622. [CrossRef] [PubMed]
6. Rojas, W.; Pinto, M.; Alanoca, C.; Gómez-Pando, L.R.; León-Lobos, P.; Alercia, A.; Diulgheroff, S.; Padulosi, S.; Bazile, D. Quinoa genetic resources and ex situ conservation. In *State of the Art Report of Quinoa in the World in 2013*; Bazile, D., Bertero, H.D., Nieto, C., Eds.; FAO & CIRAD: Rome, Italy, 2015; pp. 56–82.

7. Repo-Carrasco, R.; Espinoza, C.; Jacobsen, S.E. Nutritional value and use of the andean crops quinoa (*Chenopodium quinoa*) and kañiwa (*Chenopodium pallidicaule*). *Food Rev. Int.* **2003**, *19*, 179–189. [CrossRef]
8. Jarvis, D.E.; Ho, Y.S.; Lightfoot, D.J.; Schmöckel, S.M.; Li, B.; Borm, T.J.A.; Ohyanagi, H.; Mineta, K.; Michell, C.T.; Saber, N.; et al. The genome of *Chenopodium quinoa*. *Nature* **2017**, *542*, 307–312. [CrossRef]
9. Liu, Y.; Kong, Z.; Liu, J.; Zhang, P.; Wang, Q.; Huan, X.; Li, L.; Qin, P. Non-targeted metabolomics of quinoa seed filling period based on liquid chromatography-mass spectrometry. *Food Res. Int.* **2020**, *137*, 109743. [CrossRef]
10. Asher, A.; Galili, S.; Whitney, T.; Rubinovich, L. The potential of quinoa (*Chenopodium quinoa*) cultivation in Israel as a dual-purpose crop for grain production and livestock feed. *Sci. Hortic.* **2020**, *272*, 109534. [CrossRef]
11. Rodríguez Gómez, M.J.; Matías Prieto, J.; Cruz Sobrado, V.; Calvo Magro, P. Nutritional characterization of six quinoa (*Chenopodium quinoa* Willd) varieties cultivated in Southern Europe. *J. Food Compos. Anal.* **2021**, *99*, 1–7. [CrossRef]
12. Tang, Y.; Tsao, R. Phytochemicals in quinoa and amaranth grains and their antioxidant, anti-inflammatory, and potential health beneficial effects: A review. *Mol. Nutr. Food Res.* **2017**, *61*, 1600767. [CrossRef] [PubMed]
13. Vega-Gálvez, A.; Miranda, M.; Vergara, J.; Uribe, E.; Puente, L.; Martínez, E.A. Nutrition facts and functional potential of quinoa (*Chenopodium quinoa* willd.), an ancient Andean grain: A review. *J. Sci. Food Agric.* **2010**, *90*, 2541–2547. [CrossRef]
14. Benavente-García, O.; Castillo, J. Update on uses and properties of citrus flavonoids: New findings in anticancer, cardiovascular, and anti-inflammatory activity. *J. Agric. Food Chem.* **2008**, *56*, 6185–6205. [CrossRef] [PubMed]
15. Park, J.H.; Lee, Y.J.; Kim, Y.H.; Yoon, K.S. Antioxidant and antimicrobial activities of Quinoa (*Chenopodium quinoa* Willd.) Seeds Cultivated in Korea. *Prev. Nutr. Food Sci.* **2017**, *22*, 195–202. [CrossRef]
16. Muziri, T.; Chaibva, P.; Chofamba, A.; Madanzi, T.; Mangeru, P.; Mudada, N.; Manhokwe, S.; Mugari, A.; Matsvange, D.; Murewi, C.T.F.; et al. Using principal component analysis to explore consumers' perception toward quinoa health and nutritional claims in Gweru, Zimbabwe. *Food Sci. Nutr.* **2021**, *9*, 1025–1033. [CrossRef]
17. Groeniger, J.O.; van Lenthe, F.J.; Beenackers, M.A.; Kamphuis, C.B.M. Does social distinction contribute to socioeconomic inequalities in diet: The case of 'superfoods' consumption. *Int. J. Behav. Nutr. Phys. Act.* **2017**, *14*. [CrossRef]
18. Bedoya-Perales, N.S.; Pumi, G.; Talamini, E.; Padula, A.D. The quinoa boom in Peru: Will land competition threaten sustainability in one of the cradles of agriculture? *Land Use Policy* **2018**, *79*, 475–480. [CrossRef]
19. Cramer, W.; Guiot, J.; Fader, M.; Garrabou, J.; Gattuso, J.-P.; Iglesias, A.; Lange, M.A.; Lionello, P.; Llasat, M.C.; Paz, S.; et al. Climate change and interconnected risks to sustainable development in the Mediterranean. *Nat. Clim. Chang.* **2018**, *8*, 972–980. [CrossRef]
20. El Estado Mundial de la Agricultura y la Alimentación, 1994. Available online: http://www.fao.org/3/T4450S/t4450s05.htm826 (accessed on 26 July 2021).
21. Aloisi, I.; Parrotta, L.; Ruiz, K.B.; Landi, C.; Bini, L.; Cai, G.; Biondi, S.; del Duca, S. New Insight into Quinoa Seed Quality under Salinity: Changes in Proteomic and Amino Acid Profiles, Phenolic Content, and Antioxidant Activity of Protein Extracts. *Front. Plant Sci.* **2016**, *7*, 656. [CrossRef]
22. Reguera, M.; Conesa, C.M.; Gil-Gómez, A.; Haros, C.M.; Pérez-Casas, M.Á.; Briones-Labarca, V.; Bolaños, L.; Bonilla, I.; Álvarez, R.; Pinto, K.; et al. The impact of different agroecological conditions on the nutritional composition of quinoa seeds. *PeerJ* **2018**, *6*, e4442. [CrossRef]
23. Gonzalez, J.A.; Konishi, Y.; Bruno, M.; Valoy, M.; Prado, F.E. Interrelationships among seed yield, total protein and amino acid composition of ten quinoa (*Chenopodium quinoa*) cultivars from two different agroecological regions. *J. Sci. Food Agric.* **2011**, *92*, 1222–1229. [CrossRef]
24. Bertero, H.D.; De La Vega, A.J.; Correa, G.; Jacobsen, S.E.; Mujica, A. Genotype and genotype-by-environment interaction effects for grain yield and grain size of quinoa (*Chenopodium quinoa* Willd.) as revealed by pattern analysis of international multi-environment trials. *F. Crop. Res.* **2004**, *89*, 299–318. [CrossRef]
25. Matías, J.; Rodríguez, M.J.; Cruz, V.; Calvo, P.; Reguera, M. Heat stress lowers yields, alters nutrient uptake and changes seed quality in quinoa grown under Mediterranean field conditions. *J. Agron. Crop Sci.* **2021**, *207*, 481–491. [CrossRef]
26. Granado-Rodríguez, S.; Aparicio, N.; Matías, J.; Pérez-Romero, L.F.; Maestro, I.; Garcés, I.; Pedroche, J.J.; Haros, C.M.; Fernández-García, N.; Navarro del Hierro, J.; et al. Studying the Impact of Different Field Environmental Conditions on Seed Quality of Quinoa: The Case of Three Different Years Changing Seed Nutritional Traits in Southern Europe. *Front. Plant Sci.* **2021**, *12*, 649132. [CrossRef]
27. Hussain, M.I.; Muscolo, A.; Ahmed, M.; Asghar, M.A.; Al-Dakheel, A.J. Agro-Morphological, Yield and Quality Traits and Interrelationship with Yield Stability in Quinoa (*Chenopodium quinoa* Willd.) Genotypes under Saline Marginal Environment. *Plants* **2020**, *9*, 1763. [CrossRef] [PubMed]
28. Zurita-Silva, A.; Fuentes, F.; Zamora, P.; Jacobsen, S.E.; Schwember, A.R. Breeding quinoa (*Chenopodium quinoa* Willd.): Potential and perspectives. *Mol. Breed.* **2014**, *34*, 13–30. [CrossRef]
29. Koziol, M.J. Afrosimetric estimation of threshold saponin concentration for bitterness in quinoa (*Chenopodium quinoa* Willd). *J. Sci. Food Agric.* **1991**, *54*, 211–219. [CrossRef]
30. Medina-Meza, I.G.; Aluwi, N.A.; Saunders, S.R.; Ganjyal, G.M. GC-MS Profiling of Triterpenoid Saponins from 28 Quinoa Varieties (*Chenopodium quinoa* Willd.) Grown in Washington State. *J. Agric. Food Chem.* **2016**, *64*, 8583–8591. [CrossRef]

31. Machado Alencar, N.M.; de Carvalho Oliveira, L. Advances in Pseudocereals: Crop Cultivation, Food Application, and Consumer Perception. In *Bioactive Molecules in Food. Reference Series in Phytochemistry*; Mérillon, J.M., Ramawat, K., Eds.; Springer: Cham, Switzerland, 2019; pp. 1695–1713.
32. Abugoch James, L.E. Quinoa (*Chenopodium quinoa* Willd.): Composition, Chemistry, Nutritional, and Functional Properties. In *Advances in Food and Nutrition Research*; Elsevier Inc.: Santiago, Chile, 2009; Volume 58, pp. 1–31.
33. Fuentes, F.; Bhargava, A. Morphological Analysis of Quinoa Germplasm Grown Under Lowland Desert Conditions. *J. Agron. Crop Sci.* **2011**, *197*, 124–134. [CrossRef]
34. Miranda, M.; Vega-Gálvez, A.; Martinez, E.; López, J.; Rodríguez, M.J.; Henríquez, K.; Fuentes, F. Genetic diversity and comparison of physicochemical and nutritional characteristics of six quinoa (*Chenopodium quinoa* willd.) genotypes cultivated in Chile. *Food Sci. Technol.* **2012**, *32*, 835–843. [CrossRef]
35. Benlhabib, O.; Boujartani, N.; Maughan, P.J.; Jacobsen, S.E.; Jellen, E.N. Elevated genetic diversity in an F2:6 population of quinoa (*Chenopodium quinoa*) developed through an inter-ecotype cross. *Front. Plant Sci.* **2016**, *7*, 1–9. [CrossRef]
36. Pulvento, C.; Riccardi, M.; Lavini, A.; D'Andria, R.; Iafelice, G.; Marconi, E. Field Trial Evaluation of Two *Chenopodium quinoa* Genotypes Grown Under Rain-Fed Conditions in a Typical Mediterranean Environment in South Italy. *J. Agron. Crop Sci.* **2010**, *196*, 407–411. [CrossRef]
37. Gómez, M.B.; Castro, P.A.; Mignone, C.; Bertero, H.D. Can yield potential be increased by manipulation of reproductive partitioning in quinoa (*Chenopodium quinoa*)? Evidence from gibberellic acid synthesis inhibition using Paclobutrazol. *Funct. Plant Biol.* **2011**, *38*, 420–430. [CrossRef] [PubMed]
38. Sakamoto, T.; Matsuoka, M. Generating high-yielding varieties by genetic manipulation of plant architecture. *Curr. Opin. Biotechnol.* **2004**, *15*, 144–147. [CrossRef] [PubMed]
39. Maliro, M.F.A.; Guwela, V.F.; Nyaika, J.; Murphy, K.M. Preliminary studies of the performance of quinoa (*Chenopodium quinoa* Willd.) genotypes under irrigated and rainfed conditions of central Malawi. *Front. Plant Sci.* **2017**, *8*, 227. [CrossRef]
40. Curti, R.N.; Sanahuja, M.d.C.; Vidueiros, S.M.; Pallaro, A.N.; Bertero, H.D. Trade-off between seed yield components and seed composition traits in sea level quinoa in response to sowing dates. *Cereal Chem.* **2018**, *95*, 734–741. [CrossRef]
41. Gandarillas, A.; Rojas, W.; Ojeda, N. Quinoa in Bolivia: The Proinpa foundation's perspective. In *State of the Art Report on Quinoa around the World in 2013*; Bazile, D., Daniel Bertero, H., Nieto, C., Eds.; FAO & CIRAD: Rome, Italy, 2015; pp. 344–361.
42. Danielsen, S.; Munk, L. Evaluation of disease assessment methods in quinoa for their ability to predict yield loss caused by downy mildew. *Crop Prot.* **2004**, *23*, 219–228. [CrossRef]
43. Danielsen, S.; Bonifacio, A.; Ames, T. Diseases of quinoa (*Chenopodhim quinoa*). *Food Rev. Int.* **2003**, *19*, 43–59. [CrossRef]
44. Jacobsen, S.E. The scope for adaptation of quinoa in Northern Latitudes of Europe. *J. Agron. Crop Sci.* **2017**, *203*, 603–613. [CrossRef]
45. Ruales, J.; Nair, B.M. Content of fat, vitamins and minerals in quinoa (*Chenopodium quinoa*, Willd) seeds. *Food Chem.* **1993**, *48*, 131–136. [CrossRef]
46. Wahli, C. *Quinua: Hacia su Cultivo Comercial*; Wahli, C., Ed.; Latinreco S. A.: Quito, Ecuador, 1990; ISBN 9978990135.
47. Mastebroek, H.D.; Limburg, H.; Gilles, T.; Marvin, H.J.P. Occurrence of sapogenins in leaves and seeds of quinoa (*Chenopodium quinoa* Willd). *J. Sci. Food Agric.* **2000**, *80*, 152–156. [CrossRef]
48. Khalifa, W.; Thabet, M. Variation in downy mildew (Peronospora variabilis Gäum) resistance of some quinoa (*Chenopodium quinoa* Willd) cultivars under Egyptian conditions. *Middle East J. Agric. Res.* **2018**, *7*, 671–682.
49. Woldemichael, G.M.; Wink, M. Identification and biological activities of triterpenoid saponins from *Chenopodium quinoa*. *J. Agric. Food Chem.* **2001**, *49*, 2327–2332. [CrossRef]
50. Colque-Little, C.; Abondano, M.C.; Lund, O.S.; Amby, D.B.; Piepho, H.P.; Andreasen, C.; Schmöckel, S.; Schmid, K. Genetic variation for tolerance to the downy mildew pathogen Peronospora variabilis in genetic resources of quinoa (*Chenopodium quinoa*). *BMC Plant Biol.* **2021**, *21*, 41. [CrossRef] [PubMed]
51. De Santis, G.; Maddaluno, C.; D'Ambrosio, T.; Rascio, A.; Rinaldi, M.; Troisi, J. Characterisation of quinoa (*Chenopodium quinoa* Willd.) accessions for the saponin content in Mediterranean environment. *Ital. J. Agron.* **2016**, *11*, 277–281. [CrossRef]
52. Escribano, J.; Cabanes, J.; Jiménez-Atiénzar, M.; Ibañez-Tremolada, M.; Gómez-Pando, L.R.; García-Carmona, F.; Gandía-Herrero, F. Characterization of betalains, saponins and antioxidant power in differently colored quinoa (*Chenopodium quinoa*) varieties. *Food Chem.* **2017**, *234*, 285–294. [CrossRef]
53. Präger, A.; Munz, S.; Nkebiwe, P.M.; Mast, B.; Graeff-Hönninger, S. Yield and quality characteristics of different quinoa (*Chenopodium quinoa* willd.) cultivars grown under field conditions in southwestern Germany. *Agronomy* **2018**, *8*, 197. [CrossRef]
54. Gómez-Caravaca, A.M.; Iafelice, G.; Lavini, A.; Pulvento, C.; Caboni, M.F.; Marconi, E. Phenolic compounds and saponins in quinoa samples (*Chenopodium quinoa* Willd.) grown under different saline and nonsaline irrigation regimens. *J. Agric. Food Chem.* **2012**, *60*, 4620–4627. [CrossRef]
55. Reigosa, M.J.; Souto, X.C.; Gonz´lez, L. Effect of phenolic compounds on the germination of six weeds species. *Plant Growth Regul.* **1999**, *28*, 83–88. [CrossRef]
56. Konishi, Y.; Hirano, S.; Tsuboi, H.; Wada, M. Distribution of minerals in quinoa (*Chenopodium quinoa* Willd.) seeds. *Biosci. Biotechnol. Biochem.* **2004**, *68*, 231–234. [CrossRef]
57. Kumar, V.; Sinha, A.K.; Makkar, H.P.S.; Becker, K. Dietary roles of phytate and phytase in human nutrition: A review. *Food Chem.* **2010**, *120*, 945–959. [CrossRef]

58. Holm, P.B.; Kristiansen, K.N.; Pedersen, H.B. Transgenic approaches in commonly consumed cereals to improve iron and zinc content and bioavailability. *J. Nutr.* **2002**, *132*, 514–516. [CrossRef]
59. Honke, J.; Kozłowska, H.; Vidal-Valverde, C.; Frias, J.; Górecki, R. Changes in quantities of inositol phosphates during maturation and germination of legume seeds. *Eur. Food Res. Technol.* **1998**, *206*, 279–283. [CrossRef]
60. Hall, J.R.; Hodges, T.K. Phosphorus Metabolism of Germinating Oat Seeds. *Plant Physiol.* **1966**, *41*, 1459–1464. [CrossRef]
61. White, P.J.; Veneklaas, E.J. Nature and nurture: The importance of seed phosphorus content. *Plant Soil* **2012**, *357*, 1–8. [CrossRef]
62. Nadeem, M.; Mollier, A.; Morel, C.; Vives, A.; Prud'homme, L.; Pellerin, S. Relative contribution of seed phosphorus reserves and exogenous phosphorus uptake to maize (*Zea mays* L.) nutrition during early growth stages. *Plant Soil* **2011**, *346*, 231–244. [CrossRef]
63. Grant, C.A.; Flaten, D.N.; Tomasiewicz, D.J.; Sheppard, S.C. The importance of early season phosphorus nutrition. *Can. J. Plant Sci.* **2001**, *81*, 211–224. [CrossRef]
64. Koyro, H.W.; Eisa, S.S. Effect of salinity on composition, viability and germination of seeds of *Chenopodium quinoa* Willd. *Plant Soil* **2008**, *302*, 79–90. [CrossRef]
65. Pilco-Quesada, S.; Tian, Y.; Yang, B.; Repo-Carrasco-Valencia, R.; Suomela, J.P. Effects of germination and kilning on the phenolic compounds and nutritional properties of quinoa (*Chenopodium quinoa*) and kiwicha (*Amaranthus caudatus*). *J. Cereal Sci.* **2020**, *94*, 102996. [CrossRef]
66. Koziol, M.J. Chemical Composition and Nutritional Evaluation of Quinoa (*Chenopodium quinoa* Willd.). *J. Food Compos. Anal.* **1992**, *5*, 35–68. [CrossRef]
67. Maradini Filho, A.M.; Ribeiro Pirozi, M.; da Silva Borges, J.T.; Pinheiro Sant'Ana, H.M.; Paes Chaves, J.B.; Dos Reis Coimbra, J.S. Quinoa: Nutritional, functional, and antinutritional aspects. *Crit. Rev. Food Sci. Nutr.* **2017**, *57*, 1618–1630. [CrossRef]
68. Schlemmer, U.; Frølich, W.; Prieto, R.M.; Grases, F. Phytate in foods and significance for humans: Food sources, intake, processing, bioavailability, protective role and analysis. *Mol. Nutr. Food Res.* **2009**, *53*, S330–S375. [CrossRef] [PubMed]
69. World Health Organization and United Nations University; FAO. *Vitamin and Mineral Requirements in Human Nutrition*, 2nd ed.; WHO: Geneva, Switzerland, 2004; ISBN 92-4-154612-3.
70. Ismail, H.; Dragišic Maksimovic, J.; Maksimovic, V.; Shabala, L.; Branka, D.Ž.; Tian, Y.; Jacobsen, S.; Shabala, S. Rutin, a fl avonoid with antioxidant activity, improves plant salinity tolerance by regulating K+ retention and Na+ exclusion from leaf mesophyll in quinoa and broad beans. *Funct. Plant Biol.* **2016**, *43*, 75–86. [CrossRef]
71. Abderrahim, F.; Huanatico, E.; Segura, R.; Arribas, S.; Gonzalez, M.C.; Condezo-Hoyos, L. Physical features, phenolic compounds, betalains and total antioxidant capacity of coloured quinoa seeds (*Chenopodium quinoa* Willd.) from Peruvian Altiplano. *Food Chem.* **2015**, *183*, 83–90. [CrossRef] [PubMed]
72. Bascuñán-Godoy, L.; Reguera, M.; Abdel-tawab, Y.M.; Blumwald, E. Water deficit stress-induced changes in carbon and nitrogen partitioning in *Chenopodium quinoa* Willd. *Planta* **2015**, *243*, 591–603. [CrossRef]
73. Lozano, M.; Tícona, E.; Carrasco, C.; Flores, Y.; Almanza, G.R. Cuantificación de saponinas en residuos de quinua real chenopodium quinoa willd. *Rev. Boliv. Química* **2012**, *29*, 131–138.
74. Proceedings of the AOAC International Annual Meeting & Exposition—AOAC International, Denver, CO, USA, 6–12 September 2019.
75. Nascimento, A.C.; Mota, C.; Coelho, I.; Gueifão, S.; Santos, M.; Matos, A.S.; Gimenez, A.; Lobo, M.; Samman, N.; Castanheira, I. Characterisation of nutrient profile of quinoa (*Chenopodium quinoa*), amaranth (*Amaranthus caudatus*), and purple corn (*Zea mays* L.) consumed in the North of Argentina: Proximates, minerals and trace elements. *Food Chem.* **2014**, *148*, 420–426. [CrossRef] [PubMed]
76. MAPA Ministerio de Agricultura, Pesca y Alimentación. *Métodos Oficiales de Análisis. Tomo III*; Ministerio de Agricultura, Pesca y Alimentación: Madrid, Spain, 1995; ISBN 84-491-0003-0008.
77. Benzie, I.F.F.; Strain, J.J. The ferric reducing ability of plasma (FRAP) as a measure of "antioxidant power": The FRAP assay. *Anal. Biochem.* **1996**, *239*, 70–76. [CrossRef]
78. Ainsworth, E.A.; Gillespie, K.M. Estimation of total phenolic content and other oxidation substrates in plant tissues using Folin-Ciocalteu reagent. *Nat. Protoc.* **2007**, *2*, 875–877. [CrossRef]
79. Valenzuela Bustamante, P.D. Evaluación de la Actividad Antioxidante y Determinación del Contenido de Fenoles Totales y Flavonoides de Hojas de Diferentes Genotipos de Ugni Molinae Turcz. Bachelor's Thesis, University of Chile, Santiago, Chile, 2015; p. 65.
80. Mohammadi, S.A.; Prasanna, B.M. Analysis of Genetic Diversity in Crop Plants—Salient Statistical Tools and Considerations. *Crop Sci.* **2003**, *43*, 1235–1248. [CrossRef]
81. Martínez, E.A. Quinoa: Nutritional aspects of the rice of the Incas. In *State of the Art Report of Quinoa in the World in 2013*; Bazile, D., Bertero, H.D., Nieto, C., Eds.; FAO & CIRAD: Rome, Italy, 2015; pp. 278–285.
82. Ruales, J.; Nair, B.M. Saponins, phytic acid, tannins and protease inhibitors in quinoa (*Chenopodium quinoa* Willd) seeds. *Food Chem.* **1993**, *48*, 137–143. [CrossRef]
83. Pulvento, C.; Riccardi, M.; Lavini, A.; Iafelice, G.; Marconi, E.; D'Andria, R. Yield and Quality Characteristics of Quinoa Grown in Open Field Under Different Saline and Non-Saline Irrigation Regimes. *J. Agron. Crop Sci.* **2012**, *198*, 254–263. [CrossRef]

plants

MDPI

Article

Phytopathological Threats Associated with Quinoa (*Chenopodium quinoa* Willd.) Cultivation and Seed Production in an Area of Central Italy

Giovanni Beccari, Mara Quaglia *, Francesco Tini, Euro Pannacci and Lorenzo Covarelli

Department of Agricultural, Food and Environmental Sciences, University of Perugia, Borgo XX Giugno, 74, 06121 Perugia, Italy; giovanni.beccari@unipg.it (G.B.); francesco.tini@collaboratori.unipg.it (F.T.); euro.pannacci@unipg.it (E.P.); lorenzo.covarelli@unipg.it (L.C.)

* Correspondence: mara.quaglia@unipg.it

Abstract: In 2017, in a new *Chenopodium quinoa* cultivation area (Central Italy), emergence failures of the Titicaca, Rio Bamba, and Real varieties, whose seeds were obtained the previous year (2016) in the same location, were observed. Moreover, leaf disease symptoms on the Regalona variety, whose seeds came from Chile, were detected. Visual and microscopic analyses showed the presence of browning/necrotic symptoms on the seeds of the three varieties whose emergence in the field had failed. In addition, their in vitro germination rates were strongly compromised. *Fusarium* spp. was isolated with high incidence from Titicaca, Rio Bamba, and Real seeds. Among the detected *Fusarium* species, in the phylogenetic analysis, the dominant one clustered in the sub-clade *Equiseti* of the *Fusarium incarnatum-equiseti* (FIESC) species complex. Instead, the pathogen associated with Regalona leaf symptoms was identified, by morphological and molecular features, as *Peronospora variabilis*, the causal agents of downy mildew. This is the first report of both *P. variabilis* and *F. equiseti* on *C. quinoa* in Italy. Species-specific primers also detected *P. variabilis* in Regalona seeds. These results underline the importance of pathogen monitoring in new quinoa distribution areas, as well as of healthy seed production and import for successful cultivation.

Keywords: downy mildew; *Fusarium*; FIESC; *Peronospora variabilis*; quinoa; seed

Citation: Beccari, G.; Quaglia, M.; Tini, F.; Pannacci, E.; Covarelli, L. Phytopathological Threats Associated with Quinoa (*Chenopodium quinoa* Willd.) Cultivation and Seed Production in an Area of Central Italy. *Plants* **2021**, *10*, 1933. https://doi.org/10.3390/plants10091933

Academic Editors: Cataldo Pulvento and Didier Bazile

Received: 27 July 2021
Accepted: 14 September 2021
Published: 16 September 2021

1. Introduction

Quinoa (*Chenopodium quinoa* Willd.) is an annual dicotyledonous seed-producing plant belonging to the *Amaranthaceae* family [1]. This species originates from South America, where it was first domesticated, presumably more than 7000 years ago, in the Andean region, near Lake Titicaca [2]. The same area acts also as a natural quinoa germplasm bank, since several quinoa varieties have here been differentiated and preserved over time by the indigenous populations [3]. Due to the high level of genetic diversity, the crop is highly resilient to agro-ecological extremes (soil, rainfall, temperature, and altitude) and it is tolerant to frost, drought, and soil salinity [2,4,5]. For this reason, quinoa can be grown on marginal lands unsuitable for other major crops, providing in these disadvantaged areas food of high nutritional value [6]. Indeed, quinoa seeds are gluten-free, with a low glycemic index, and contain an excellent balance of all nine essential amino acids, with high concentrations of histidine, lysine, and methionine; moreover, they are rich in fiber, lipids, carbohydrates, vitamins, minerals (including calcium, magnesium, and iron), and health-promoting compounds (flavonoids) [1,7,8]. Due to its agricultural and nutritional aspects, since the end of the 20th century, quinoa has gained international attention [9] to such an extent that the United Nations declared 2013 as the "International Year of Quinoa" [10] and in 2017 its genome was completely sequenced [6]. Therefore, to address the major market requests, a significant increase in the cropped area and production has been recorded during the last ten years [11]. Until 2018, quinoa was considered a major crop only in Bolivia (70,763 t, 111,605 ha), Peru (86,011 t, 64,660 ha), and Ecuador (2146 t, 2048 ha) [11,12]. However, quinoa is currently cultivated in all continents [13] because there have been numerous attempts at quinoa introduction from the area of origin to other countries, including the European ones. The number of countries growing the crop has increased from 8 in 1980, to 40 in 2010, to 75 in 2014 [2]. In 2018, quinoa was present for research and production in 123 countries, even if 74% of global exports are still supplied by Peru and Bolivia [14]. In Italy, quinoa was introduced in the early 2000s thanks to several research projects dealing with the adaptation of this species to the peninsula agricultural system [15–17]. Results indicated that, given its high resiliency and tolerance to abiotic stresses, in particular salinity and drought, quinoa could represent a good alternative to the traditional Mediterranean crops in light of the current climate change [18].

However, quinoa is susceptible to several biotic stresses that could strongly impair yield, both in the area of origin and in those where the crop has spread. Phytopathogenic fungi and oomycetes are among the main biotic stress factors affecting quinoa crops. In the Andean region, quinoa yield losses caused by several phytopathogens have been well documented [19,20]. In this geographical area, downy mildew is undoubtedly the most diffused and well-known disease and the epidemics, caused by the heterothallic oomycete *Peronospora variabilis* (Gaüm), greatly reduce seed yield [21]. In addition to downy mildew, other quinoa diseases were also detected in the Andean area, such as leaf spots caused by *Ascochyta hyalospora* (Cooke and Ell.) and *Ascochyta chenopodii* Rostr., black stem caused by *Ascochyta caulina* [P. Karst. (v.d. Aa and v. Kest.)], leaf spot caused by *Cercospora* spp., brown stalk rot by *Phoma exigua* var. *foveata* [(Foister) Boerema], and damping-off of root/seedlings by *Rhizoctonia* spp., *Fusarium* spp., and *Pythium* spp. [12,19,20,22]. The incidence and the severity of the various diseases caused by fungi and oomycetes depend on different factors, such as environmental conditions, the cultivated variety, and the phenological stage of the crop during which the infection occurs [20]. With the rapid expansion of quinoa growing areas, the problems of the negative impact of some of the above-reported phytopathogens have arisen also in the new countries [23]. Therefore, with the extension of quinoa cultivation, traditional and/or new pathogens may threaten the spread of this species. Consequently, understanding the phytosanitary problems related to this crop in new cultivation areas is a key step to efficiently counteract and manage them.

For this reason, in a new quinoa cultivation area (Umbria, Central Italy), after observing in 2017 emergence failures of seedlings belonging to the Titicaca, Rio Bamba, and Real varieties, as well as leaf symptoms on the Regalona variety, a series of phytopathological surveys, both on seed lots and on leaves, were carried out. The results of this survey are presented in this study and can be useful to better manage quinoa diseases in Italy, both for healthy seed production, as well as for its successful cultivation.

2. Results

2.1. Examination of Sampled Material

2.1.1. Seed Material

A combination of visual and stereomicroscopic observations of seed material allowed the detection of the presence of browning/necrotic symptoms on the seeds of the Titicaca, Rio Bamba, Real, and Regalona varieties, with the seeds of the first three varieties showing symptom incidences of 93%, 77%, and 70%, respectively, which were significantly higher ($p \leq 0.05$) than those recorded in the seeds of the Regalona variety (17%) (Figure 1). In the first three varieties, seed symptoms were uniformly distributed almost on the whole seed surface and were more severe compared to those recorded on the Regalona ones, showing only small browned/necrotic areas (Figure 1).

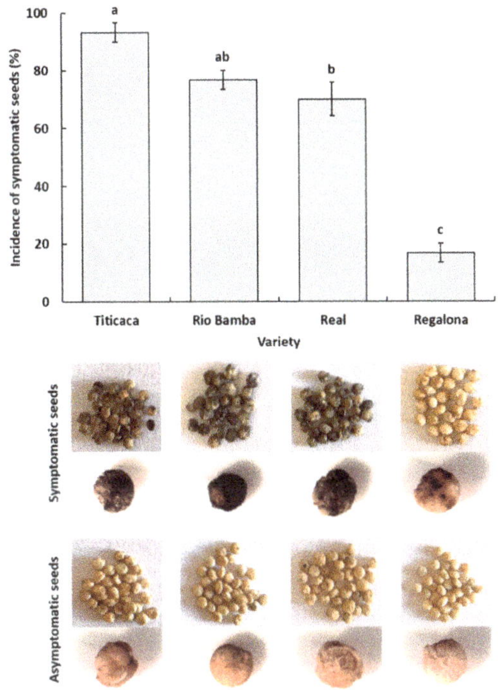

Figure 1. Incidence (%) of symptomatic seeds for each quinoa variety, as assessed by visual and stereomicroscopic observations. Columns represent the average (±standard error) of the three replicates, of 10 seeds each, for a total of 30 observed seeds for each quinoa variety. Columns with the same letter were not different at $p \leq 0.05$ based on the Tukey HSD for multiple comparisons. Under each column, pictures of symptomatic (browned/necrotic) and asymptomatic seeds of the corresponding variety are shown.

2.1.2. Plant Material

Visual examination of Regalona plant material showed the presence of foliar chlorosis/yellowing symptoms (Figure 2a,b). In detail, a combination of visual and stereomicroscopic observation showed that the upper side of symptomatic leaves was characterized by yellow/chlorotic spots with corresponding black/grey efflorescence on the adaxial surface (Figure 2b–d). Microscopic observation of this efflorescence showed the presence of colorless zoosporangiophores subdichotomously branched, slightly curved, with obtuse tips, each carrying a single pale brown ellipsoidal zoosporangium with a short pedicel (Figure 2e). These features led us to hypothesize that the pathogen belonged to the *Peronospora* genus inside the *Oomycetes* class of the *Pseudofungi* subphylum of the *Chromista* kingdom [24], thus molecular investigations were carried out to complete the identification (Section 2.5).

2.2. Seed Germination Rate

Within each seed category (symptomatic and asymptomatic), significant differences ($p \leq 0.05$) in germination rate among the four varieties were detected (Figure 3). In detail, considering asymptomatic seeds, the Regalona variety showed the highest ($p \leq 0.05$) germination rate (100%), followed by Real (70%), Rio Bamba (40%), and Titicaca (36.6%) (Figure 3). Only the germination rate of asymptomatic seeds of the Real variety was not significantly different ($p \geq 0.05$) from that of Regalona (Figure 3). Considering symptomatic seeds, no germination was recorded in the Titicaca, Rio Bamba, and Real varieties, while symptomatic seeds of the Regalona variety showed a germination rate of 73.3% (Figure 3).

Figure 2. Field view of quinoa plants with foliar chlorosis/yellowing detected in June 2017 on Regalona plots in a trial located at Papiano (Perugia, Umbria, Central Italy) (**a**); yellow chlorotic spots on the upper side of Regalona leaves (**b**) to which corresponded a black-grey efflorescence on the adaxial surface (**c**,**d**) consisting of zoosporangiophores and zoosporangia of the pathogen (**e**, scale bar 50 μm).

Within each variety, significant differences ($p \leq 0.05$) in germination rate were also recorded among the two seed categories (Figure 3). In all varieties, asymptomatic seeds showed significantly higher ($p \leq 0.05$) germination rate levels than those of symptomatic ones.

2.3. Seed Mycological Analysis

The combination of visual, stereo-, and light-microscope observations directly on isolation plates allowed us to identify the mycoflora composition of each variety and seed category. In general, *Alternaria* and *Fusarium* were the most representative fungal genera, being detected in each seed category of the Titicaca, Rio Bamba, and Real varieties (Figure 4). With a lower incidence, the genera *Aspergillus* and *Penicillium* were also detected in the seeds of these three varieties, with *Aspergillus* detected only on asymptomatic seeds and never on the symptomatic seeds of these three lots, in which, however, healthy seeds were absent. On the contrary, apart from the negligible presence of *Alternaria* and *Penicillium* genera, a high incidence of healthy seeds was detected in each seed category of the Regalona variety, where *Fusarium* and *Aspergillus* were not detected (Figure 4).

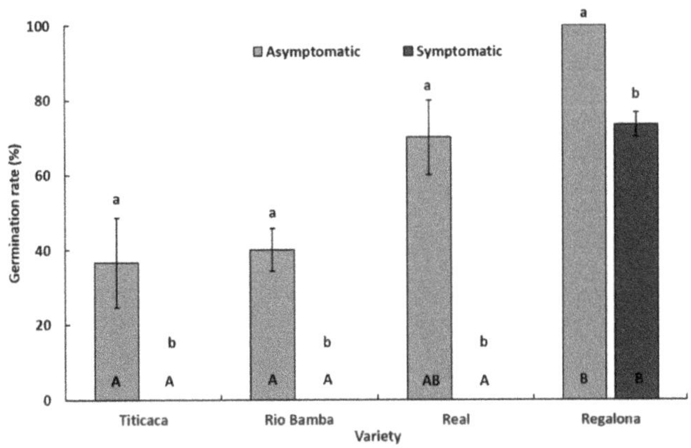

Figure 3. Germination rate (%) of asymptomatic and symptomatic seeds of Titicaca, Rio Bamba, Real, and Regalona varieties. Columns represent the average (±standard error) of the three replicates, each consisting of 10 observed seeds, for a total of 30 observed seeds of each seed category (asymptomatic, symptomatic) per quinoa variety. Within each variety (a and b) or seed category (A and B), averages with the same letter were not different at $p \leq 0.05$ based on the Tukey HSD for multiple comparisons.

Significant differences ($p \leq 0.05$) between the incidences of different fungal genera within each "variety–seed category" combination were recorded (Figure 4). Indeed, in the seeds of Titicaca, Rio Bamba, and Real varieties, colonies belonging to the genera *Alternaria* and *Fusarium* always showed an incidence higher than those of the other genera (Figure 4). These differences were "mitigated" in the asymptomatic seed category but, conversely, they were stronger in the symptomatic seed category, where, in addition, *Fusarium* incidence (87%, 70%, and 97% for Titicaca, Rio Bamba, and Real varieties, respectively) was always significantly higher ($p \leq 0.05$) than those of *Alternaria* (40%, 27%, and 40% for Titicaca, Rio Bamba, and Real varieties, respectively) (Figure 4). *Fusarium* was absent in all the two categories of Regalona seeds where, differently from the other varieties, the incidences of healthy seeds were always significantly higher in each seed category (90% and 73% for asymptomatic and symptomatic seeds, respectively) (Figure 4).

Significant differences ($p \leq 0.05$) among the incidences of different fungal genera within each "variety–fungal genera" combination were also detected. In particular, *Fusarium* incidence in symptomatic seeds of Titicaca, Rio Bamba, and Real varieties (87%, 70%, and 97%, respectively, as above reported) was significantly higher ($p \leq 0.05$) than that recorded in asymptomatic seeds (33%, 37%, and 27%, respectively). No significant differences ($p \geq 0.05$) of *Aspergillus* and *Penicillium* incidences were detected among the different seed categories of Titicaca, Rio Bamba, and Real. Similarly, the incidence of healthy seeds was not significantly different ($p \leq 0.05$) among the two different seed categories in the Regalona variety (Figure 4).

Finally, significant differences ($p \leq 0.05$) between the incidences of the different fungal genera within each "fungal genera–seed category" combination were also recorded. The incidences of the *Alternaria* genus in both seed categories of Titicaca, Rio Bamba, and Real varieties were significantly ($p \leq 0.05$) higher than those detected in the same two seed categories of Regalona varieties. In addition, *Fusarium* incidence in the symptomatic seed of the Rio Bamba variety was significantly lower ($p \leq 0.05$) than those recorded in the same category of Titicaca and Real varieties.

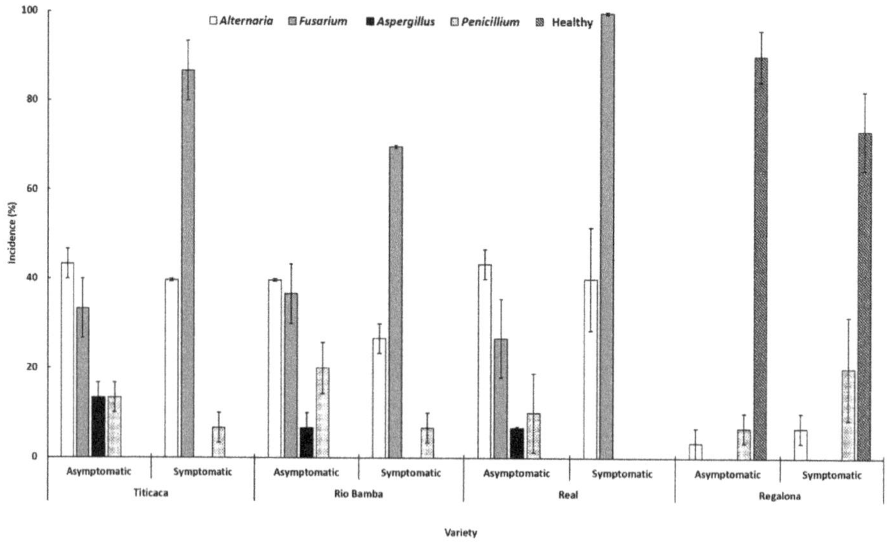

Figure 4. Incidence (%) of different fungal genera in asymptomatic and symptomatic seeds of the four quinoa varieties. Columns represent the average (±standard error) of the three replicates, 10 observed seeds per replicate, for a total of 30 observed seeds of each seed category (asymptomatic, symptomatic) per variety. Data were subject to one-way analysis of variance within the "variety–seed category", "variety–fungal genera" and "fungal genera–seed category" combinations. To assess pairwise contrasts, Tukey HSD ($p \leq 0.05$) was used. The main significant differences are discussed in the text.

2.4. Molecular Identification of Fusarium *spp. Associated to Quinoa Seeds*

The BLAST analysis indicated that the *Fusarium* community isolated from quinoa seed lots of the Titicaca, Rio Bamba, and Real varieties, multiplied in 2016 at Papiano (42°57′ N, 12°22′ E, 165 m a.s.l., Perugia, Umbria, Central Italy), was composed of a total of four species: *Fusarium incarnatum-equiseti* species complex (FIESC), *Fusarium avenaceum* [(Fr.) Sacc.], *Fusarium culmorum* [(Wm.G.Sm.) Sacc.], and *Fusarium sporotrichioides* (Sherb.) (Figure 5). Interestingly, FIESC was the most frequent species of the *Fusarium* complex in each variety–seed combination category ($p \leq 0.05$). Indeed, the other three species were detected with an incidence that ranged from 1% to 4% (Figure 5).

Therefore, FIESC incidence was investigated more in-depth. In detail, within each seed category (symptomatic, asymptomatic), no significant differences ($p \geq 0.05$) in FIESC incidence between the three varieties (Titicaca, Rio Bamba, and Real) affected by FIESC were detected (Figure 6), while, within each variety, significant differences ($p \leq 0.05$) in FIESC incidence between the two seed categories were observed. In particular, in Titicaca, Rio Bamba, and Real varieties, symptomatic seeds showed FIESC incidences (80, 66, and 86%, respectively) significantly higher ($p \leq 0.05$) than those recorded in asymptomatic seeds (30, 33, and 26%, respectively) (Figure 6).

To assess the impact of FIESC incidence on quinoa seeds' germination rate, the correlation between these two parameters was evaluated (Figure 7). Taking the Titicaca, Rio Bamba, and Real varieties (the only ones in which FIESC were detected) and each seed category together, seed germination was negatively and significantly related to FIESC incidence ($R^2 = 0.62$; $p = 0.004$). Considering individual varieties, in Rio Bamba and Titicaca, the negative association between germination rate and FIESC incidence was statistically significant ($p = 0.02$ and 0.03, respectively). Finally, considering each seed category individually, the strongest negative association was observed for symptomatic seeds ($p = 6 \times 10^{-11}$), since the absence of germination and the highest FIESC incidence was detected in the seeds of this category for the Titicaca, Rio Bamba and Real varieties.

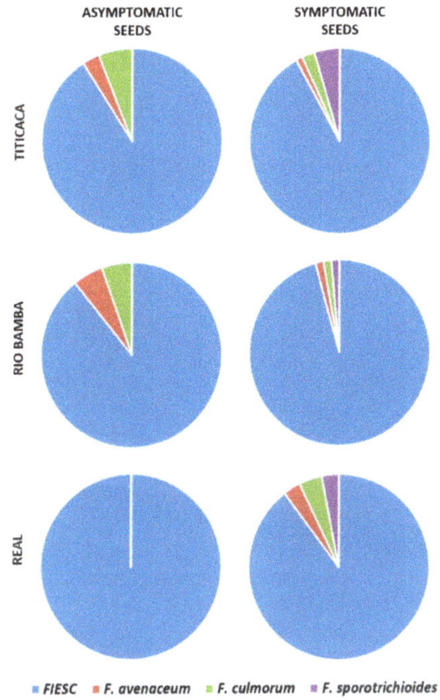

Figure 5. Incidence (%) of the different *Fusarium* species obtained from the asymptomatic and symptomatic seeds of the Titicaca, Rio Bamba, and Real varieties. No *Fusarium* isolates were obtained from the Regalona variety. Pie charts show the average of the *Fusarium* species incidence on the total of *Fusarium* isolates developed from three replicates, with 10 seeds for each replicate, for a total of 30 seeds per seed category (asymptomatic, symptomatic) for each quinoa variety. FIESC = *Fusarium incarnatum-equiseti* species complex.

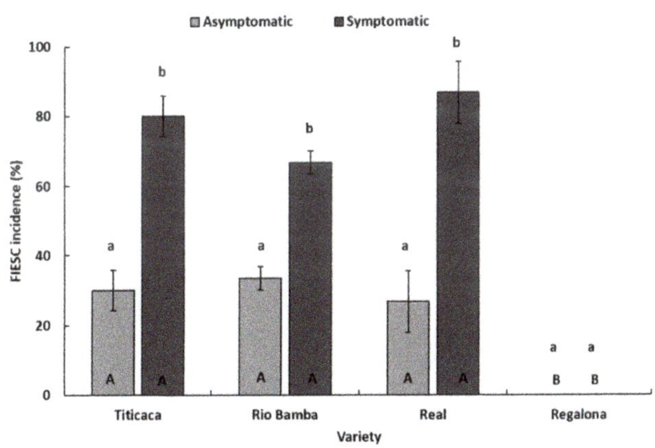

Figure 6. Incidence (%) of *Fusarium incarnatum-equiseti* species complex in asymptomatic and symptomatic seeds of Titicaca, Rio Bamba, Real, and Regalona varieties. Columns represent the average (±standard error) of the three replicates, 10 observed seeds for each replicate, for a total of 30 observed seeds per seed category (asymptomatic, symptomatic) for each quinoa variety. Within each variety (a and b) or seed category (A and B), averages with the same letter were not different at $p \leq 0.05$ based on the Tukey HSD for multiple comparisons.

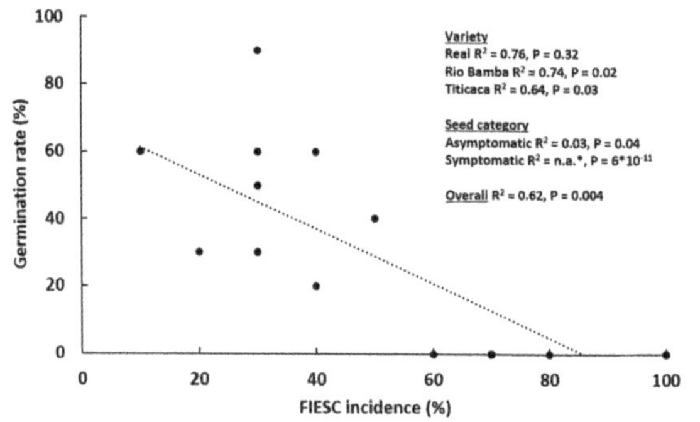

Figure 7. Correlation between seed germination rate (%) of the Titicaca, Rio Bamba, and Real quinoa varieties and *Fusarium incarnatum-equiseti* species complex (FIESC) incidence (%). The three replicates for each "seed category–variety" combination are shown separately. * No germination was detected in symptomatic seeds and no R^2 was available.

Due to the negative impact of FIESC on the germination rate of the quinoa seeds belonging to the varieties Titicaca, Rio Bamba, and Real, the FIESC representative isolate, denominated FIESC-PG-Q1, was further molecularly characterized (Figure 8). Amplification of the *translation elongation factor 1α* (*tef1α*) of FIESC-PG-Q1 isolate produced a fragment of 700 bp. According to Reference [25], in the phylogram constructed on the concatenated sequences of *tef1α* region of the validated phylogenetic species of FIESC, two major clades emerged: *Equiseti* clade, including the phylogenetically validated species from FIESC 1-a to FIESC 14-a and 30-a (Supplementary Table S1) and *Incarnatum* clade, including the phylogenetic validated species from FIESC 15-b to FIESC 29-a. Isolate FIESC-PG-Q1 (MZ191105 GenBank accession, Supplementary Table S1) clustered in a sub-clade of the *Equiseti* clade together with the phylogenetically validated species FIESC 5 (a, b, c, d, and f) (Figure 8).

2.5. Molecular Identification of Pathogen Infecting Quinoa Leaves and Its Detection Seeds

Amplification of the *Internal Transcribed Spacer* (*ITS*) region of PV-PG-Q1 isolate produced a fragment of about 280 bp. Through preliminary BLAST analysis, this isolate was attributed both to *Peronospora variabilis* and to *Peronospora farinosa*, thus, a phylogenetic analysis was carried out to uniquely attribute it to a certain species. In the phylogram built on the concatenated sequences of the *ITS* region of validated *Peronospora* species from *C. quinoa* and other plant species belonging to the *Amaranthaceae* family (Supplementary Table S2), three major clades (A, B and C) emerged (Figure 9). Clade A contained isolates of *P. variabilis* from *Chenopodium album* L. and *C. quinoa*; clade B contained isolates of *Peronospora chenopodii* Schltdl. from *Chenopodium hybridum* L., *Peronospora boni-henrici* Gäum. from *Chenopodium bonus-henricus* L., *Peronospora effusa* (Greville) Cesati from *Spinacia oleracea* L., and *Peronospora farinosa* from *Atriplex* spp.; clade C contained isolates of *Peronospora chenopodii-polyspermi* Gäum. from *Chenopodium polyspermum* L. (Figure 9). Isolate PV-PG-Q1 (MZ191106 GenBank accession, Supplementary Table S2) clustered together with *P. variabilis* isolates from *C. album* and *C. quinoa* (Figure 9).

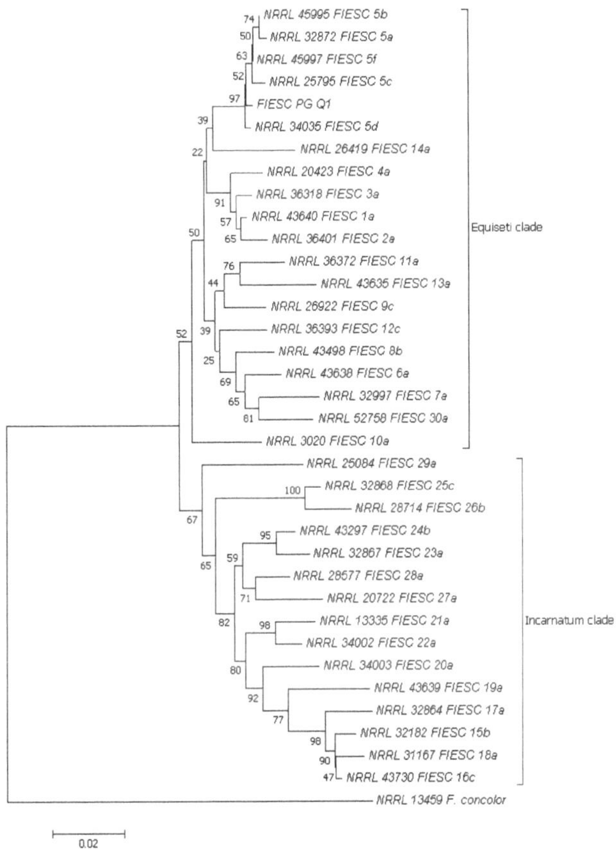

Figure 8. Neighbor-joining tree based on partial *translation elongation factor 1 α* (*tef1α*) gene sequences, showing the phylogenetic relationships between the FIESC-PG-Q1 isolate obtained in the present research and validated phylogenetic species of the *Fusarium incarnatum-equiseti* species complex (FIESC). The *Fusarium concolor* isolate NRRL 13459 was included as an outgroup. The optimal tree with the sum of branch length = 0.64901758 is shown. The percentage of replicate trees in which the associated taxa clustered together in the bootstrap test (1000 replicates) is shown next to the branches [26]. The tree is drawn to scale, with branch lengths in the same units as those of the evolutionary distances used to infer the phylogenetic tree. The evolutionary distances were computed using the maximum composite likelihood method [27] and are in the units of the number of base substitutions per site. The analysis involved 36 nucleotide sequences. All positions containing gaps and missing data were eliminated. There was a total of 545 positions in the final dataset. Evolutionary analyses were conducted in MEGA7 [28].

Finally, the PCR protocol used with PV6F and PV6R primers also showed the presence of the same foliar pathogen in the seeds of the Regalona variety (presence of specific amplification product at 278 bp) belonging to the same batch used to set up the experimental field (Supplementary Figure S1).

Figure 9. Neighbor-joining tree based on partial *internal transcribed spacer* (*ITS*) region sequences, showing the phylogenetic relationships between the PV-PG-Q1 isolate obtained in the present research from *Chenopodium quinoa* plants of the Regalona variety and validated phylogenetic *Peronospora* species from *C. quinoa* and other plants species belonging to the genus *Chenopodium*. The *Peronospora manshurica* isolate KUS-F17669 from *Glycine soja* was included as an outgroup. The optimal tree with the sum of branch length = 0.19783452 is shown. The percentage of replicate trees in which the associated taxa clustered together in the bootstrap test (1000 replicates) is shown next to the branches [26]. The tree is drawn to scale, with branch lengths in the same units as those of the evolutionary distances used to infer the phylogenetic tree. The evolutionary distances were computed using the maximum composite likelihood method [27] and are in the units of the number of base substitutions per site. The analysis involved 37 nucleotide sequences. All positions containing gaps and missing data were eliminated. There was a total of 229 positions in the final dataset. Evolutionary analyses were conducted in MEGA7 [28].

3. Discussion

The rapid expansion of quinoa from its area of origin into new cultivation areas could be accompanied by the crop exposition to the damaging effects of known and/or new pathogens [23]. In this context, monitoring quinoa in new growing areas is a determining factor to identify the phytosanitary threats of this crop and to plan efficient management strategies. For this reason, this work shows the results of a series of phytopathological surveys carried out after observing, in a new cultivation site (Umbria, Central Italy): (a) severe crop emergence failures in a number of quinoa varieties and (b) severe leaf disease symptoms on one of the surveyed varieties.

Our investigations, carried out on the same seed batches used to set up the field trial in which emergence failures were observed, allowed us to detect the presence of browning/necrotic symptoms on seeds, with the highest incidence and severity detected on the seeds of the three varieties (Titicaca, Rio Bamba, and Real) showing emergence failures. The germination rate of Titicaca, Rio Bamba, and Real seeds was strongly compromised if compared with that of Regalona variety, not only in symptomatic seeds but also in the asymptomatic ones. In addition, the mycobiota composition associated with the seeds showed substantial differences between Titicaca/Rio Bamba/Real varieties and the Regalona variety, with the latter almost free from fungal pathogens except for low levels of *Alternaria* spp. and *Penicillium* spp. Conversely, the fungal community isolated from Titicaca, Rio Bamba, and Real seeds showed a high incidence of microorganisms belonging to the genera *Fusarium* and *Alternaria*, with the first ones particularly abundant in symptomatic seeds. This difference in composition/incidence of the mycobiota associated with quinoa seeds could be attributable to the different areas in which the seeds were obtained. Indeed, as mentioned above, the seeds of Titicaca, Rio Bamba, and Real varieties were all obtained in the new Italian cultivation site (Umbria, Central Italy) in the year (2016) preceding the one (2017) in which emergence failures were observed, while the *Fusarium*-free Regalona seeds originated from Chile. Thus, the seed production area affected the composition of the fungal community associated with quinoa seeds.

The *Fusarium* community associated with seeds of Titicaca, Rio Bamba, and Real varieties was dominated by FIESC, which was particularly present in the symptomatic lots. In vitro seed germination rates, especially those of asymptomatic and symptomatic seeds, were negatively correlated to FIESC incidence, suggesting that this *Fusarium* complex could have potentially been implicated in the emergence failure observed in the field. In fact, the negative effect of *Fusarium* species on seed germination has been well documented [29,30].

However, the presence of other fungal genera and other *Fusarium* species detected during the present work makes it difficult to define with certainty the amount of seed germination and seedling emergence losses caused by FIESC and observed in the field. In addition, fungal pathogens could be only one of the factors that influence the germination and emergence of quinoa seedlings. Indeed, in addition to biotic factors (pathogens, but also pests), a complex of abiotic factors and agronomic conditions, such as soil condition, soil preparation before sowing, sowing depth, environmental conditions, low soil moisture content, and crusty topsoil [12,31,32], can also strongly reduce quinoa seed germination and plant emergence [33]. Further investigations could be useful to assess the impact of FIESC on seed germination and on quinoa emergence.

FIESC members are generally associated with diseases of several crops, particularly cereals [25,34,35]. They are considered "sporadic" and "weak" causal agents of *Fusarium* head blight (FHB) of wheat and barley. For example, in the same Italian cultivation area of the quinoa experiment described in this paper, FIESC has been often detected as a "minor" component of the FHB complex of wheat and barley [36–38]. In addition, leaf spots caused by FIESC have been observed in different Italian areas on leafy vegetable hosts (i.e., lettuce, rocket, spinach, etc.) [35]. FIESC members have also the ability to biosynthesize mycotoxins, including type A and type B trichothecenes, zearalenone, beauvericin, fusaric acid, and moniliformin [39,40], representing a risk for human and animal health.

FIESC is a phylogenetically species-rich complex that includes over 30 cryptic phylogenetic species, making identification based on phenotypic characteristics problematic [41]. The molecular analysis conducted in this study was able to ascribe the representative isolate FIESC-PG-Q1 in a sub-clade within the *Equiseti* clade together with the phylogenetic validated species FIESC 5. At least two distinct morphotypes have been reported in the literature for *Equiseti* clade species: morphotype I and II, with short or long apical cells, respectively [42]. Morphotype II is predominant in southern Europe [43] and has been associated to FIESC 5 [25]. Although FIESC 5 has also been previously detected by Reference [44] in Italian soil, to the best of our knowledge, this is the first report of FIESC 5 on *C. quinoa* seeds in Italy. *F. equiseti* has been reported in *C. quinoa* also in Brazil [45].

FIESC members are usually seed-borne pathogens [46] and *C. quinoa* seed production in Italy could be threatened by them. In fact, similarly to the main FHB agents, FIESC accumulation in cereal grains occurred following field infection at the head level (during the anthesis stage) and it could have been then exacerbated by inappropriate storage conditions. The presence of FIESC in *C. quinoa* seeds may follow a similar pathway; however, specific studies should be performed to assess the infection process of FIESC in quinoa inflorescence and/or seeds. In addition, the ability of this pathogen to produce mycotoxins reveals a possible threat to the safety of quinoa seeds destined for human consumption.

Several reports on *Fusarium* spp. isolation from diseased quinoa roots and young seedlings are available. *Fusarium* spp. were reported as one of the quinoa root rot and damping-off casual agents, often in association with *Rhizoctonia solani* (Cooke) Wint., *Pythium* spp., *Alternaria* spp., and *Acremonium* spp. [19,33,47,48]. Focusing on the genus *Fusarium*, damping-off and root rot diseases of quinoa are currently reported to be associated with *Fusarium solani* [(Mart.) Sacc.] and *Fusarium oxysporum* (von Schlechtendal) in Egypt [47,48] and *Fusarium avenaceum* in the Czech Republic [33]. The results of the present study contribute to expanding the knowledge of the *Fusarium* species that could be associated with *C. quinoa* in new cultivation areas. Microorganisms belonging to this genus could be common soil-inhabiting fungi that colonize the roots [35]. However, our results show that, in addition to soil-borne diseases, *Fusarium* spp. could also cause quinoa seed-borne diseases, as they were isolated from quinoa seeds before sowing.

The seeds of the Regalona variety analyzed during this study were completely free of *Fusarium* spp. However, the developed Regalona plants, which did not have any emergence problems, showed severe leaf symptoms (chlorosis and necrosis) and signs (sporangiophores) attributable to a downy mildew infection. The morphological features of the representative isolate PV-PG-Q1 obtained from symptomatic Regalona leaves matched with those described by References [49,50] for the species *Peronospora variabilis*. The identification of isolate PV-PG-Q1 as *P. variabilis* was also confirmed by molecular analysis.

As reported by several authors [20,22,51,52], *P. variabilis* has been specifically detected in *C. quinoa* in Bolivia, Chile, Colombia, Ecuador, Peru, USA, India, Canada, Turkey, Denmark, Korea, and Egypt, but not in Italy. The same pathogen was also reported on *C. album* in several countries (Peru, Argentina, Romania, Germany, Latvia, the Netherlands, Ireland, Korea, China, Turkey, Denmark) [21,22,50,53,54], including Italy [49]. Thus, to the best of our knowledge, this is the first report of *P. variabilis* on *C. quinoa* in Italy. Downy mildew is reported as one of the main causes of quinoa yield reduction all over the world; the losses depend on various parameters, such as the plant's phenological stages at the infection time, presence of favorable weather conditions for the pathogen, and the level of cultivar susceptibility [55]. On a susceptible cultivar, in favorable weather conditions, a *P. varaibilis* attack during the first phenological stages can lead to total yield loss. As, unlike *Fusarium* spp., *P. variabilis* does not produce mycotoxins and does not affect the healthiness of the food product, the loss it causes is mainly quantitative.

Since the field experiment analyzed in this study had been sown with Regalona variety seeds, the presence of the pathogen in the seeds was hypothesized and confirmed by PCR assays carried out with species-specific primers (PV6F and PV6R), confirming *P. variabilis* as a seed-borne oomycete pathogen [20,22]. Indeed, while during *C. quinoa* growth in the field,

P. variabilis are easily transmitted by low-distance zoospore dissemination by wind and rain, at long distances and between successive crop cycles the pathogen spreads by oospores [20]. However, given that *P. variabilis* is heterothallic, the presence of the two mating types P1 and P2 is required for sexual reproduction and, thus, oospores formation [20]. Previous research showed that oospores of *P. variabilis* were present in the seed pericarp [55]. Another study revealed that they were mainly localized in the perianth and seed coat (>85%), while only a very small percentage (<5%) were detected in the embryo and perisperm [56]. From the seeds, *P. variabilis* can move inside *C. quinoa* tissues, causing a systemic host colonization. A few days after seed germination, oospores were detected in the cortex of hypocotyls and the mesophyll of cotyledons [22,57]. The efficacy of infection transmission from the seed to seedling is favored by the high relative humidity at the sowing time, as well as by large oospore density [58]. This suggests that, in addition to the use of tolerant varieties, also avoiding the excess of water in the field and adjusting the space between rows, making the area less dense, are cultural practices that could contribute to reducing the risk of *P. variabilis* proliferation [20].

Thus, our investigation showed FIESC and *P. variabilis* are also important threats to quinoa cultivation in Italy and underlines the importance of healthy seed production, import, and use for successful quinoa cultivation. As already practiced for other crops (i.e., sunflower, cereals, etc.), the import and commercialization on a national and global scale of seed materials dressed with fungicides or, for organic farming, with bio-fungicides or heat-treated, would be desirable to avoid the spread of FIESC and *P. variabilis*, as well as of other seed-borne pathogens [20].

However, given the current lack of registered fungicides on quinoa in Italy, the management of these two pathogens (as well as of other pathogens) can negatively affect quinoa cultivation in both organic and integrated disease management approaches. Moreover, even if, to the best of our knowledge, no pathogenicity tests have ever been carried out to verify the infectivity of *P. variabilis* isolates from *C. album* on quinoa, the elimination of *C. album* weeds on the quinoa field would be advisable.

In fact, *C. album* is frequently infected by downy mildew throughout Europe because it is conspecific with the *P. variabilis* from *C. quinoa*; therefore, it is likely to be a reservoir for the pathogen and an alternative host [22]. Other *Chenopodium* species were reported to host the pathogen but these studies need further investigation. Cross-infection of *P. variabilis* on *C. album* and *C. quinoa* is the only instance that has been reported to date [22]. Mechanical and manual weed control is the most suitable method to control *C. album*, both in organic farming and in integrated pest management, because there are no selective herbicides registered for quinoa and other minor crops in Europe [17,59,60].

With the rapid expansion of quinoa growing areas, the negative impact of some other pathogens can arise in quinoa cultivation in Italy as well. For example, in addition to downy mildew, in the new cultivation regions, other fungal diseases such as gray mold caused by *Botrytis cinerea* (Pers.) [23], black stem caused by *Ascochyta caulina* [61], leaf spot caused by *Heterosporicola beijingense* sp. nov. [62], and anthracnose caused by *Colletotrichum nigrum* (Ellis and Halst) and *Colletotrichum truncatum* (Schwein) [22] were observed in the UK (gray mold), China (black stem and leaf spot), and USA (anthracnose), showing that diseases may easily occur when quinoa is introduced into nontraditional cultivation areas.

In conclusion, the introduction of a new plant species like *C. quinoa* in a new cultivation area requires the constant monitoring of its pathogens, with particular attention to those coming from other areas through infected seeds and those that arise as new threats in the new geographical distribution zone. In detail, these results highlight a particular need for the development of rapid, sensitive, and reliable methods to screen for quinoa seeds and plant pathogens. As suggested by Reference [22], this could be useful for the early detection of casual agents before diseases become too developed, but also to ensure certified pathogen-free quinoa seeds, which is an important requirement to be achieved at the global level.

4. Materials and Methods

4.1. Field Observations, Samples Collection, and Examination of Sampled Materials

In April 2017, in a quinoa field trial located at Papiano (42°57′ N, 12°22′ E, 165 m a.s.l.), near Perugia (Umbria, Central Italy; Figure 10), emergence failures of the Titicaca, Rio Bamba, and Real varieties were observed. Seed multiplication of these varieties had been performed during the previous year (2016) in the same location. Conversely, an optimal emergence was recorded for the seeds belonging to the Regalona variety, imported from Chile. However, on Regalona plants, foliar chlorotic symptoms were detected in the same field trial in June 2017, stimulating a series of phytopathological investigations on seed lots of the four varieties and on the leaves of the Regalona variety.

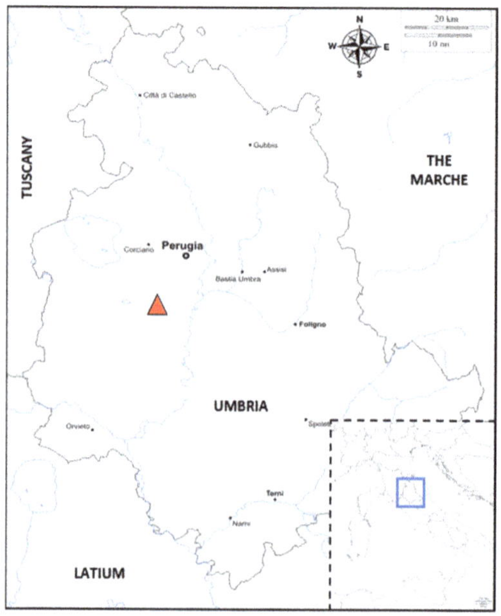

Figure 10. Map of the Umbria region (Central Italy, grey line) showing the experimental field trial location (red triangle) and neighboring regions. In the bottom right corner (outlined box), a map of Italy indicates the region's location (blue box) in the national geographical context.

Thus, representative samples (30 g each corresponding to about 15,000 seeds on average) of the same four seed batches used to set up the field trial and a total of 30 plants of the Regalona variety, randomly chosen during the field trial, were collected. All samples (seeds and plants) were subject to a combination of visual and stereomicroscopic (SZX9, Olympus, Tokyo, Japan) observations to detect symptoms and, if present, signs of possible pathogens.

Concerning the four seed batches, during visual and stereomicroscopic observations, the incidence (%) of symptomatic seeds was assessed on a subsample of 30 randomly chosen seeds divided into three replicates (10 seeds per replicate).

For the Regalona samples, the formation of pathogen signs was promoted by placing a total of 10 symptomatic leaves, randomly collected from sampled plants, in humid chambers obtained by using 40 mL of sterile 1% water–agar (Biolife Italiana, Milan, Italy) in Petri dishes (150 mm diameter; Nuova Aptaca, Canelli, Italy). After 24 h of incubation under natural lighting, chlorotic leaves in humid chambers were observed as previously described. Moreover, pathogen signs were observed in more detail and photographed by a light microscope (Axiophot, Zeiss, Oberkochen, Germany).

4.2. Seed Germination Test

A total of 10 g (about 5000 seeds) of the above-reported seed batches of the Titicaca, Rio Bamba, Real, and Regalona varieties were randomly collected and used to assess the germination rate (%). In detail, for each batch, a total of 30 symptomatic and 30 asymptomatic seeds were randomly selected and divided into 3 replicates of 10 seeds each that were placed into 3 Petri dishes (90 mm diameter; Nuova Aptaca) onto two layers of sterile filter paper (Whatman N. 1, Maidstone, UK), previously added to 10 mL of sterile deionized water. Before assessing the incidence of germinated seed, Petri dishes were sealed with parafilm (Bemis Amcor, Oshkosh, WI, USA), to avoid water evaporation, and incubated for 6 days at 22 °C, in the dark.

4.3. Seed Mycological Analysis

For each of the four varieties (Titicaca, Rio Bamba, Real, and Regalona) a total of 30 symptomatic and 30 asymptomatic seeds were taken from a randomly collected seed sample of 10 g and used to assess the mycobiota associated with each seed batch. In detail, seed mycological analysis was performed as previously indicated by Reference [36], adapting this method to quinoa seeds. Briefly, seeds were externally disinfected for 2 min using a water–ethanol (95%, Sigma Aldrich, Saint Louis, MO, USA)-sodium hypochlorite (7%, Carlo Erba Reagents, Milan, Italy) solution (82:10:8% vol.) and rinsed with deionized sterile water for 1 min. Thirty seeds were placed onto potato dextrose agar (PDA; Biolife Italiana) pH 5.7, supplemented with streptomycin sulfate (0.16 g/L, Sigma Aldrich) into 3 Petri dishes (90 mm diameter, Nuova Aptaca) (replicates) containing 10 seeds each, for a total of three replicated plates per variety. The dishes were incubated at 22 °C in the dark.

After 6 days of incubation, a combination of visual and stereomicroscopic (SZX9, Olympus) observations were carried out on each seed to assess the presence of the different developed fungal genera. Light-microscopic (Axiophot, Zeiss) observations of fungal structures characterizing the developed colonies were also performed.

4.4. Molecular Identification of *Fusarium* spp. Associated to Quinoa Seeds

Fungal isolates obtained from seed samples were transferred in pure cultures into new plates containing PDA and placed at 22 °C, in the dark. After 10 days of incubation, *Fusarium* cultures developed from each single seed category (symptomatic and asymptomatic) of Titicaca, Rio Bamba, and Real varieties were assigned to a particular "morphotype" according to colony color and shape on PDA, as well as to the morphology of reproductive structures as observed by microscopic analysis (Axiophot, Zeiss). This selection allowed us to obtain a subset of isolates composed of one representative isolate per morphotype. These representative isolates (four in total), after obtaining monosporic culture, were placed into new PDA plates at 22 °C in the dark for two weeks. Their mycelium was then scraped from the PDA and placed into 2 mL sterile plastic tubes (Eppendorf, Hamburg, Germany) at −80 °C, freeze-dried with a lyophilizer (Heto Powder Dry LL3000; Thermo Fisher Scientific, Waltham, MA, USA), and the mycelium was finely ground with a grinding machine (MM60, Retsch, Dusseldorf, Germany) for 5 min with a frequency of 25 Hz. DNA extraction was performed as described in Reference [63], with modifications reported in Reference [37]. Extracted genomic DNA was visualized on a 1% agarose, trizma base-glacial acid acetic-ethylenediamine-tetraacetic acid disodium salt dihydrate (TAE; all from Sigma Aldrich, Merck KGaA, St. Louis, MO, USA) gel in TAE buffer (1X) containing 500 μL/L of RedSafe (iNtRON Biotechnology, Burlington, MA, USA). DNA fragments were separated in 10 cm-long agarose gels, with an electrophoresis apparatus (Eppendorf), applying a tension of 110 V for ~30 min. Electrophoretic runs were visualized using an ultraviolet transilluminator (Euroclone, Milan, Italy). DNA concentration was estimated by comparison with a 1 kb gene ruler (Thermo Fisher Scientific, Milan, Italy). DNA was diluted in DNase-free sterile water for molecular biology use (5prime, Hilden, Germany) to obtain a concentration of ~30 ng/μL and stored at −20 °C until use.

The DNA extracted from *Fusarium* representative isolates was subject to partial *tef1α* gene amplification, purification, and sequencing. A PCR protocol was adopted using a total reaction volume of 50 μL. Each reaction contained 29 μL of sterile water for molecular biology use, 5 μL of 10X Dream Taq Buffer + magnesium chloride (Thermo Fisher Scientific), 3.75 μL of cresol red (Sigma Aldrich), 5 μL of dNTP mix 10 mM (Microtech, Naples, Italy), 2.5 μL of 10 μM EF1 and EF2 primers [64,65], 0.25 μL of 5 U/μL Dream Taq Polymerase (Thermo Fisher Scientific), and 2 μL of template DNA. The PCR cycle consisted of an initial denaturation step (94 °C for 5 min), followed by 30 cycles of denaturation (94 °C for 1 min), annealing (53 °C for 1 min) and extension (72 °C for 1 min), and a final extension (72 °C for 10 min). PCR assays were performed on a T-100 thermal cycler (Bio-Rad, Hercules, CA, USA). PCR fragments were visualized on TAE 1X agarose gel (2%) containing 500 μL/L of RedSafe. DNA fragments were separated at 110 V for ~40 min. Electrophoretic runs were observed with an ultraviolet transilluminator. The size of the amplified fragments was obtained by comparison with HyperLadder 100–1000 bp (Bioline Meridian Bioscience, Cincinnati, OH, USA). PCR fragments were purified and sequenced by an external sequencing service (Genewiz Genomics Europe, Takeley, UK). The sequences obtained were verified by Chromatogram Explorer Lite v4.0.0 (Heracle Biosoft srl 2011) and analyzed by the BLAST database (National Center for Biotechnology Information, http://blast.ncbi.nih.gov (accessed on 18 March 2021)). The sequence of the most representative *Fusarium* isolate (denominated FIESC-PG-Q1) was used to build a phylogenetic tree. Phylogenetic analyses were performed by MEGA software version 7.0 [28] according to [25] by using partial *tef1α* region sequence of the FIESC-PG-Q1 obtained in the present study and of validated phylogenetic species of the FIESC reported in GenBank (Supplementary Table S1); *Fusarium concolor* Reinking NRRL 13459 was chosen as the outgroup [25]. Sequences were aligned, nucleotide gaps and missing data were deleted, and a phylogenetic tree was built using the neighbor-joining method [66] with the bootstrap test for 1000 replicates [26]. The evolutionary distances were computed using the maximum composite likelihood method [27].

4.5. Molecular Identification of the Pathogen Associated with Quinoa Leaves

A combination of visual, stereomicroscopic, and microscopic observations allowed us to hypothesize that the symptoms associated with Regalona variety leaves were caused by a pathogen belonging to the oomycetes class. For this reason, a sample of 10 leaves was randomly collected from sampled Regalona plants, bulked together and placed into a 50 mL sterile plastic tube (Falcon, Corning, Glendale, AR, USA) at −80 °C, freeze-dried with a lyophilizer (Heto Powder Dry LL3000), and finely ground with a grinding machine (MM60, Retsch) for 5 min with a frequency of 25 Hz. A sub-sample of 1 g was placed into a new 50 mL tube (Falcon, Corning) and subject to DNA extraction using the method previously used by Reference [67] for wheat grains and successfully adapted to quinoa leaves. Total extracted genomic DNA was visualized and quantified as previously described (Section 4.4). The whole DNA extracted was subject to partial ribosomal DNA (rDNA) *ITS* gene amplification of oomycetes. A PCR protocol was adopted as previously described (see Section 4.4) using ITS6 and ITS7 primers [68,69] targeting the *ITS* region of rDNA in oomycetes. The PCR cycle consisted of an initial denaturation step (94 °C for 5 min), followed by 35 cycles of denaturation (94 °C for 30 s), annealing (57 °C for 30 s) and extension (72 °C for 1 min), and a final extension (72 °C for 10 min). PCR assays were performed as previously described (Section 2.4). PCR fragments were purified and sequenced by an external sequencing service (Genewiz Genomics Europe, Takeley, UK). The sequences obtained were verified by Chromatogram Explorer Lite v4.0.0 and analyzed by the BLAST database (National Center for Biotechnology Information, http://blast.ncbi.nih.gov (accessed on 19 March 2021).

According to References [49,50], phylogenetic analyses were performed as previously described (Section 4.4) by using partial *ITS* region sequences of the isolate obtained in the present research and denominated PV-PG-Q1 and those of validated *Peronospora* species from *C. quinoa* and other plant species belonging to the genera *Chenopodium*, *Spinacia*, and *Atriplex* in the *Amaranthaceae* family reported in GenBank (Supplementary Table S2). The *Peronospora manshurica* (Naumov) Syd. isolate KUS-F17669 from *Glycine soja* Hort. was included as an outgroup [49,50].

Finally, to detect if the pathogen was seed-transmitted, a sample of 10 g of Regalona seeds was randomly collected and finely ground with a grinding machine (MM60, Retsch) for 5 min with a frequency of 25 Hz. A sub-sample of 4 g was placed into a 50 mL tube (Falcon, Corning) and subject to DNA extraction using the CTAB method previously used by [66] for wheat grains and successfully adapted to quinoa seeds. Total extracted genomic DNA was visualized and quantified as previously described (Section 4.4). The whole DNA extracted was used to carry out a PCR protocol as previously described (Section 4.4) but using PV6F and PV6R species-specific primers [22] for *P. variabilis* detection. The PCR cycle consisted of an initial denaturation step (94 °C for 2 min), followed by 10 cycles of denaturation (95 °C for 30 s), annealing (66 °C for 45 s, −1 °C per cycle) and extension (72 °C for 1.5 min), 22 cycles of denaturation (95 °C for 30 s), annealing (63 °C for 45 s) and extension (72 °C for 90 s), and final extension (72 °C for 5 min). PCR assays were performed and amplified fragments were visualized as described before (Section 4.4). DNA from Titicaca seeds and asymptomatic and symptomatic (infected by *P. variabilis*) leaves of Regalona were also added as a control to this PCR assay.

4.6. Statistical Analysis

- Incidence (%) of asymptomatic and symptomatic seeds is indicated for each variety as the average (±standard error, SE) of three biological replicates. Data were subject to one-way analysis of variance by considering "variety" as a factor and "incidence" as a variable.
- The germination rate (%) was indicated for each variety as the average (± SE) of three biological replicates (Petri dishes), both for the asymptomatic and symptomatic selected seeds. Data were subject to one-way analysis of variance by considering, within a variety or between varieties, "seed category (asymptomatic, symptomatic)" as a factor and "germination rate" as a variable.
- The incidences (%) of each fungal genus recovered during the entire survey are expressed as the average (± SE) of three biological replicates (Petri dishes), both for the asymptomatic and symptomatic selected seeds. Data were subject to one-way analysis of variance by considering: within the variety–seed category combination, "fungal genera" as a factor and "incidence" as a variable; within the variety–fungal genera combination, "seed category (asymptomatic, symptomatic)" as a factor and "incidence" as a variable; and within the fungal genera–seed category combination, "variety" as a factor and "incidence" as a variable.
- The incidences (%) of each *Fusarium* species recovered during the entire survey were calculated as the incidence of isolates belonging to the morphotype from which the identified isolate was sub-sampled and are expressed as the average of three biological replicates (Petri dishes), both for the asymptomatic and symptomatic selected seeds. Data were subject to one-way analysis of variance by considering, within the variety–seed combination category, "*Fusarium* species" as a factor and "incidence" as a variable;
- The incidence (%) of FIESC recovered during the entire survey is expressed as the average (± SE) of three biological replicates (Petri dishes) both for the asymptomatic and symptomatic selected seeds. Data were subject to one-way analysis of variance by considering: within the variety–seed combination category, "fungal genera" as a factor and "incidence" as a variable; within the variety–fungal genera combination, "seed category (asymptomatic, symptomatic)" as a factor and "incidence" as a variable; within the fungal genera–seed combination category, "variety" as a factor and "incidence" as a variable.

In all cases, to assess pairwise contrasts, Tukey's honestly significant difference (HSD) ($p \leq 0.05$) was used. All statistical analyses were performed with the Microsoft Excel Macro "DSAASTAT" ver. 1.0192 [70]. Finally, the correlations between seed germination rate (%) and FIESC incidence (%) in each quinoa variety and seed category were studied using the coefficient of determination (R^2), followed by a Student *t*-test.

Supplementary Materials: The following are available online at https://www.mdpi.com/article/10.3390/plants10091933/s1, Table S1: Isolates of *Fusarium* species belonging to the *Fusarium incarnatum-equiseti* species complex (FIESC) used in the phylogenetic analysis and related GenBank accession numbers of *tef1α* region and clade. Table S2: Isolates of *Peronospora* spp. used in the phylogenetic analysis and related host, origin, and GenBank accession numbers of the *ITS* region. Figure S1: Electrophoresis gel showing the result of the PCR assay for the detection of *Peronospora variabilis* in Regalona seeds with the use of species-specific primers (PV6F and PV6R). Lane 1, Titicaca seeds; lane 2, Regalona seeds; lane 3, Regalona asymptomatic leaves; lane 4, Regalona leaves showing downy mildew symptoms from *P. variabilis* infection (positive control). According to [21], product size was approximately 278 bp.

Author Contributions: Conceptualization, G.B., M.Q., E.P., and L.C.; formal analysis, G.B., M.Q., and F.T.; investigation, G.B. and M.Q.; resources, L.C.; writing—original draft preparation, G.B and M.Q.; writing—review and editing, F.T., E.P., and L.C.; visualization, G.B., M.Q., and F.T.; supervision, L.C. All authors have read and agreed to the published version of the manuscript.

Funding: This research received no external funding.

Acknowledgments: This research was realized within the framework of the "Rural Development Programs 2014–2020 for the Umbria region, Project QUINUmbriA (Mis 16.2.2)". Graphical abstract was created with BioRender.com.

Conflicts of Interest: The authors declare no conflict of interest.

References

1. Vega-Galvez, A.; Miranda, M.; Vergara, J.; Uribe, E.; Puente, L.; Martinez, E.A. Nutrition facts and functional potential of quinoa (*Chenopodium quinoa* Willd.) an ancient Andean grain: A review. *J. Sci. Food Agric.* **2010**, *90*, 2541–2547. [CrossRef] [PubMed]
2. Bazile, D.; Jacobsen, S.E.; Verniau, A. The global expansion of quinoa: Trends and limits. *Front. Plant Sci.* **2016**, *7*, 622. [CrossRef] [PubMed]
3. FAO. Catalogue of Commercial Varieties of Quinoa in Peru. Available online: http://www.fao.org/documents/card/en/c/f351e07b-bd68-4724-a849-778dd44e6358/ (accessed on 10 March 2021).
4. Ruiz, K.B.; Biondi, S.; Oses, R.; Acuña-Rodriguez, I.S.; Antognoni, F.; Martinez-Mosquiera, E.A.; Coulibaly, A.; Canahua-Murillo, A.; Pinto, M.; Zurita-Silva, A.; et al. Quinoa biodiversity and sustainability for food security under climate change. A review. *Agric. Sustain. Dev.* **2014**, *34*, 349–359. [CrossRef]
5. Ruiz, K.B.; Biondi, S.; Martínez, E.A.; Orsini, F.; Antognoni, F.; Jacobsen, S.E. Quinoa—A model crop for understanding salt tolerance mechanisms in halophytes. *Plant Biosyst.* **2015**, *150*, 357–371. [CrossRef]
6. Jarvis, D.E.; Ho, Y.S.; Lightfoot, D.J.; Schmöckel, S.M.; Li, B.; Borm, T.J.A.; Ohyanagi, H.; Mineta, K.; Michell, C.T.; Saber, N.; et al. The genome of *Chenopodium quinoa*. *Nature* **2017**, *542*, 307–312. [CrossRef]
7. Gordillo-Bastidas, E.; Díaz-Rizzolo, D.A.; Roura, E.; Massanés, T.; Gomis, R. Quinoa (*Chenopodium quinoa* Willd.), from nutritional value to potential health benefits: An integrative review. *J. Nutr. Food Sci.* **2016**, *6*, 497.
8. Wright, K.H.; Pike, O.A.; Fairbanks, D.J.; Huber, S.C. Composition of *Atriplex hortensis*, sweet and bitter *Chenopodium quinoa* seeds. *J. Food Sci.* **2002**, *67*, 1383–1385. [CrossRef]
9. Repo-Carrasco, R.; Espinoza, C.; Jacobsen, S.E. Nutritional value and use of the Andean crops quinoa (*Chenopodium quinoa*) and kañiwa (*Chenopodium pallidicaule*). *Food Rev. Int.* **2003**, *19*, 179–189. [CrossRef]
10. FAO. The International Year of Quinoa. Available online: http://www.fao.org/quinoa-2013/en/ (accessed on 10 March 2021).
11. FAO. FAOSTAT. Statistic Division. Food and Agriculture Organization of the UN. Database 2018. Available online: http://faostat.fao.org (accessed on 29 October 2020).
12. Basantes-Morales, E.R.; Alconada, M.M.; Pantoja, J.L. Quinoa (*Chenopodium quinoa* Willd.) production in the Andean region: Challenges and potentials. *J. Exp. Agric. Int.* **2019**, *36*, 1–18. [CrossRef]
13. García-Parra, M.; Zurita-Silva, A.; Stechauner-Rohringer, R.; Roa-Acosta, D.; Jacobsen, S. Quinoa (*Chenopodium quinoa* Willd.) and its relationship with agroclimatic characteristics: A Colombian perspective. *Chil. J. Agric. Res.* **2020**, *80*, 290–302. [CrossRef]
14. Alandia, G.; Rodriguez, J.P.; Jacobsen, S.E.; Bazile, D.; Condori, B. Global expansion of quinoa and challenges for Andean region. *Glob. Food Secur.* **2020**, *26*, 100429. [CrossRef]
15. Lavini, A.; Pulvento, C.; d'Andria, R.; Riccardi, M.; Choukr-Allah, R.; Balhabib, O.; Yazar, A.; Incekaya, C.; MetinSezen, S.; Qadir, M.; et al. Quinoa's potential in the Mediterranean region. *J. Agron. Crop Sci.* **2014**, *200*, 344–360. [CrossRef]

16. Pulvento, C.; Riccardi, M.; Biondi, S.; Orsini, F.; Jacobsen, S.E.; Ragab, R.; D'Andria, R.; Lavini, A. Quinoa in Italy: Research and perspectives. In *FAO & CIRAD*; State of the Art Report on Quinoa in the World in 2013; Bazile, D., Bertero, D., Nieto, C., Eds.; Food and Agriculture Organization of the United Nations (FAO): Rome, Italy, 2015; Chapter 6.13; pp. 192–217.
17. Damiani, F.; Brunetti, M.; Pannacci, E. Quinoa cultivation in Italy. In *Food Science and Technology—Quinoa Cultivation, Nutritional Properties and Effect on Health*; Peiretti, P.G., Gai, F., Eds.; Nova Science Publisher Inc.: New York, NY, USA, 2019; pp. 1–31.
18. Bilalis, D.J.; Roussis, I.; Kakabouki, I.; Folina, A. Quinoa (*Chenopodium quinoa* Willd.) crop under Mediterranean conditions: A review. *Int. J. Agric. Nat. Resour.* **2019**, *46*, 51–68. [CrossRef]
19. Danielsen, S.; Bonifacio, A.; Ames, T. Diseases of Quinoa (*Chenopodium quinoa*). *Food Rev. Int.* **2003**, *19*, 43–59. [CrossRef]
20. Gandarillas, A.; Saravia, R.; Plata, G.; Quispe, R.; Ortiz-Romero, R. Principle quinoa pests and diseases. In *FAO & CIRAD*; State of the Art Report on Quinoa in the World in 2013; Bazile, D., Bertero, D., Nieto, C., Eds.; Food and Agriculture Organization of the United Nations (FAO): Rome, Italy, 2015; Chapter 2.6; pp. 192–217.
21. Testen, A.L.; Jiménez-Gasco, M.M.; Ochoa, J.B.; Backman, P.A. Molecular detection of *Peronospora variabilis* in quinoa seed and phylogeny of the quinoa downy mildew pathogen in South America and the United States. *Phytopathology* **2014**, *104*, 379–386. [CrossRef] [PubMed]
22. Colque-Little, C.; Amby, D.B.; Andreasen, C. A review of *Chenopodium quinoa* (Willd.) diseases—An updated perspective. *Plants* **2021**, *10*, 1228. [CrossRef]
23. Li, J.; Zhou, X.; Huang, H.; Li, G. Diseases characteristic and control measurements for *Chenopodium quinoa* Willd. *Adv. Eng. Res.* **2017**, *143*, 305–308.
24. Cavalier-Smith, T. Kingdom Chromista and its eight phyla: A new synthesis emphasizing periplastid protein targeting, cytoskeletal and periplastid evolution, and ancient divergences. *Protoplasma* **2018**, *255*, 297–357. [CrossRef] [PubMed]
25. Castellá, G.; Cabañes, F.J. Phylogenetic diversity of *Fusaium incarnatum-equiseti* species complex isolated from Spanish wheat. *Antoine van Leeuwenhoek* **2014**, *106*, 309–317. [CrossRef] [PubMed]
26. Felsenstein, J. Confidence limits on phylogenies: An approach using the bootstrap. *Evolution* **1985**, *39*, 783–791. [CrossRef]
27. Tamura, K.; Nei, M.; Kumar, S. Prospects for inferring very large phylogenies by using the neighbor-joining method. *Proc. Natl. Acad. Sci. USA* **2004**, *101*, 11030–11035. [CrossRef] [PubMed]
28. Kumar, S.; Stecher, G.; Tamura, K. MEGA7: Molecular evolutionary genetics analysis version 7.0 for bigger dataset. *Mol. Biol. Evol.* **2016**, *33*, 1870–1874. [CrossRef]
29. Gilbert, J.; Tekauz, A. Effect of fusarium head blight and seed treatment on germination, emergence and seedling vigor of spring wheat. *Can. J. Plant Pathol.* **1995**, *3*, 252–259. [CrossRef]
30. Suthar, R.; Bhatt, D.P.; Bhatt, P.N. Effect of culture filtrate of *Fusarium equiseti* on seed germination and seedling growth of cumin (*Cuminum cyminum*). *Indian Phytopathol.* **2014**, *67*, 193–194.
31. Causin, H.F.; Bordón, D.A.E.; Burrieza, H. Salinity tolerance mechanisms during germination and early seedling growth in *Chenopodium quinoa* Wild. Genotypes with different sensitivity to saline stress. *Environ. Exp. Bot.* **2020**, *172*, 103995. [CrossRef]
32. Jacobsen, S.E.; Bach, A.P. The influence of temperature on seed germination rate in quinoa (*Chenopodium quinoa* Willd.). *SST* **1998**, *26*, 515–523.
33. Drimalkova, M.; Veverka, K. Seedlings damping-off of *Chenopodium quinoa* Willd. *Plant Prot. Sci.* **2004**, *40*, 5–10. [CrossRef]
34. Villani, A.; Moretti, A.; De Saeger, S.; Han, Z.; Di Mavungu, J.D.; Soares, C.M.G.; Proctor, R.H.; Venâncio, A.; Lima, N.; Stea, G.; et al. A polyphasic approach for characterization of a collection of cereal isolates of the *Fusarium incarnatum-equiseti* species complex. *Int. J. Food Microbiol.* **2016**, *234*, 24–35. [CrossRef] [PubMed]
35. Matić, S.; Tabone, G.; Guarnaccia, V.; Gullino, M.L.; Garibaldi, A. Emerging leafy vegetable crop diseases caused by the *Fusarium incarnatum-equiseti* species complex. *Phytopathol. Mediterr.* **2020**, *59*, 303–317.
36. Beccari, G.; Caproni, L.; Tini, F.; Uhlig, S.; Covarelli, L. Presence of *Fusarium* species and other toxigenic fungi in malting barley and multi-mycotoxin analysis by liquid chromatography-high-resolution mass spectrometry. *J. Agric. Food Chem.* **2016**, *64*, 4390–4399. [CrossRef]
37. Beccari, G.; Senatore, M.T.; Tini, F.; Sulyok, M.; Covarelli, L. Fungal community, Fusarium head blight complex and secondary metabolites associated with malting barley grains harvested in Umbria, Central Italy. *Int. J. Food Microbiol.* **2018**, *271*, 33–42. [CrossRef]
38. Beccari, G.; Prodi, A.; Senatore, M.T.; Balmas, V.; Tini, F.; Onofri, A.; Pedini, L.; Sulyok, M.; Brocca, L.; Covarelli, L. Cultivation area affects the presence of fungal communities and secondary metabolites in Italian durum what grains. *Toxins* **2020**, *12*, 97. [CrossRef]
39. Villani, A.; Proctor, R.H.; Kim, H.; Brown, D.W.; Logrieco, A.F.; Amatulli, M.T.; Moretti, A.; Susca, A. Variation in secondary metabolite production potential in the *Fusarium incarnatum-equiseti* species complex revealed by comparative analysis of 13 genomes. *BMC Genom.* **2019**, *20*, 314. [CrossRef] [PubMed]
40. Avila, C.F.; Moreira, G.M.; Nicolli, C.P.; Gomes, L.B.; Abreu, L.M.; Pfenning, L.H.; Haidukowski, M.; Moretti, A.; Logrieco, A.; Del Ponte, E.M. *Fusarium incarnatum-equiseti* species complex associated with brazilian rice: Phylogeny, morphology and toxigenic potential. *Int. J. Food Microbiol.* **2019**, *306*, 108267. [CrossRef] [PubMed]
41. Xia, J.W.; Sandoval-Denis, M.; Crous, P.W.; Zhang, X.G.; Lombard, L. Numbers to names—Restyling the *Fusarium incarnatum-equiseti* species complex. *Persoonia* **2019**, *43*, 186–221. [CrossRef] [PubMed]

42. Samson, R.A.; Hoekstra, E.S.; Frisvad, J.C. *Introduction to Food- and Airborne Fungi*, 7th ed.; Centraalbureau voor Schimmelcultures: Utrech, The Netherlands, 2004.
43. Marín, P.; Moretti, A.; Ritieni, A.; Jurado, M.; Vázquez, C.; González-Jaén, M.T. Phylogenetic analysis and toxigenic profiles of *Fusarium equiseti* and *Fusarium incarnatum* isolates from Southern Europe. *Food Microbiol.* **2012**, *31*, 229–237. [CrossRef]
44. Balmas, V.; Migheli, Q.; Schern, B.; Garau, P.; O'Donnel, K.; Ceccherelli, G.; Kang, S.; Geiser, D. Multilocus phylogenetics shows high levels of endemic fusaria inhabiting Sardinian soils (Tyrrhenian Islands). *Mycologia* **2010**, *102*, 803–812. [CrossRef] [PubMed]
45. Mendes, M.A.S.; da Silva, V.L.; Dianese, J.C.; Ferreira, M.A.S.V.; dos Santos, C.E.N. *Fungos em Plants no Brasil*; Embrapa-SPI/Embrapa-Cenargen: Brasilia, Brazil, 1998; 555p.
46. Gilardi, G.; Pintore, I.; Gullino, M.L.; Garibaldi, A. Occurrence of *Fusarium equiseti* as a contaminant of *Diplotaxis tenuifolia* seeds. *J. Plant Pathol.* **2017**, *99*, 245–248.
47. Abd El Moity, T.H.; Badrawy, H.B.M.; Ali, A.M. Survey on diseases and pests attack quinoa in Egypt. In Proceedings of the Sixth International Scientific Agricultural Symposium, Jahorina, Bosnia and Herzegovina, 15–18 October 2015; pp. 868–876.
48. El-Hadidy, A.E. Effect of biocontrol agents on damping-off and root-rot diseases of quinoa (*Chenopodium quinoa* Willd.) seedlings. *Egypt. J. Desert Res.* **2019**, *69*, 21–38. [CrossRef]
49. Choi, Y.J.; Denchev, C.M.; Shin, H.D. Morphological and molecular analyses support the existence of host-specific *Peronospora species* infecting *Chenopodium*. *Mycopathologia* **2008**, *165*, 155–164. [CrossRef]
50. Choi, Y.; Danielsen, S.; Lübeck, M.; Hong, S.; Delhey, R.; Shin, H. Morphological and molecular characterization of the causal agent of downy mildew on Quinoa (*Chenopodium quinoa*). *Mycopathologia* **2010**, *169*, 403–412. [CrossRef] [PubMed]
51. Farr, D.F.; Rossman, A.Y.; Fungal Databases, U.S. National Fungus Collections, ARS, USDA. Retrieved 20 July 2021. Available online: https://nt.ars-grin.gov/fungaldatabases/ (accessed on 20 July 2021).
52. Jacobsen, S.E. The scope for adaptation of quinoa in Northern latitudes of Europe. *J. Agric. Crop Sci.* **2017**, *203*, 603–613. [CrossRef]
53. Danielsen, S.; Jacobsen, S.E.; Echegaray, J.; Ames, T. *Impact of Downy Mildew on the Yield of Quinoa*; Program Report 1999–2000; International Potato Center (Centro Internacional de la Papa): Lima, Peru; pp. 397–401.
54. Gul, E. First report of *Peronospora variabilis* on *Chenopodium quinoa* in Turkey. *J. Plant Pathol.* **2021**, *103*, 389–390. [CrossRef]
55. Danielsen, S.; Ames, T. *Mildew (Peronospora Farinosa) of Quinoa (Chenopodium Quinoa) in the Andean Region: Practical Manual for the Study of the Disease and Pathogen*; International Potato Center (Centro Internacional de la Papa): Lima, Peru, 2004.
56. El-Assiuty, E.M.; Taha, E.M.; Fahmy, Z.M.; Fahmy, G.M. Histological and molecular detections of *Peronospora variabilis* Gäum oospores in seeds of Quinoa (*Chenopodium Quino* L.). *Egypt. J. Exp. Biol.* **2019**, *15*, 197–203.
57. El-Assiuty, E.M.; Fahmy, G.M.; Taha, E.M.; Fahmy, Z.M.; Ismael, A.S.M.; Abd-Elghany, W.R.; Kafsheer, D.A. Microscopic visualization of *Peronospora variabilis* Gäum., the cause of Quinoa downy mildew in plant tissues at different stages of plant growth. *Int. J. Sci. Eng. Res.* **2019**, *10*, 1022–1033.
58. Danielsen, S.; Mercado, V.H.; Ames, T.; Munk, L. Seed transmission of downy mildew (*Peronospora farinosa* f. sp. *chenopodii*) in quinoa and effect of relative humidity on seedling infection. *Seed Sci. Technol.* **2004**, *32*, 91–98. [CrossRef]
59. Pannacci, E.; Lattanzi, B.; Tei, F. Non-chemical weed management strategies in minor crops: A review. *Crop Prot.* **2017**, *96*, 44–58. [CrossRef]
60. Pannacci, E.; Farneselli, M.; Ottavini, D.; Tei, F. Mechanical and chemical weed control in quinoa. In Proceedings of the XLVIII Conference of Italian Society for Agronomy, Perugia, Italy, 18–20 September 2019; Seddaiu, G., Benincasa, P., Eds.; 2019; pp. 100–101.
61. Yin, H.; Zhou, J.; Lv, H.; Qin, N.; Chang, F.J.; Zhao, X. Identification, pathogenicity, and fungicide sensitivity to *Ascochyta caulina* (Teleomorph: *Neocamarosporium calvescens*) associated with black stem on quinoa in China. *Plant Dis.* **2020**, *104*, 2585–2597. [CrossRef]
62. Brahmanage, R.S.; Liu, M.; Wanasinghe, D.N.; Dayarathne, M.C.; Mei, L.; Jeewon, R.; Li, X.; Hyde, K.D. *Heterosporicola beijingense* sp. nov. (*Leptosphariaceae, Pleosporales*) associated with *Chenopodium quinoa* leaf spots. *Phytopathol. Mediterr.* **2020**, *59*, 219–227.
63. Covarelli, L.; Beccari, G.; Prodi, A.; Generotti, S.; Etruschi, F.; Juan, C.; Ferrer, E.; Mañes, J. *Fusarium* species, chemotype characterization and trichothecene contamination of durum and soft wheat in an area of Central Italy. *J. Sci. Food Agric.* **2015**, *95*, 540–551. [CrossRef] [PubMed]
64. O'Donnell, K.; Kistler, H.C.; Cigelnik, E.; Ploetz, R.C. Multiple evolutionary origins of the fungus causing Panama disease of banana: Concordant evidence from nuclear and mitochondrial gene genealogies. *Proc. Natl. Acad. Sci. USA* **1998**, *95*, 2044–2049. [CrossRef]
65. Geiser, D.M.; Jimenez-Gasco, M.D.; Kang, S.C.; Makalowska, I.; Veeraghavan, N.; Ward, T.J.; Zhang, N.; Kuldau, G.A.; O'Donnel, K. FUSARIUM-ID v. 1.0: A DNA sequence for identifying *Fusarium*. *Eur. J. Plant Pathol.* **2004**, *110*, 473–479. [CrossRef]
66. Saitou, N.; Nei, M. The neighbor-joining method: A new method for reconstructing phylogenetic trees. *Mol. Biol. Evol.* **1987**, *4*, 406–425. [PubMed]
67. Beccari, G.; Arellano, C.; Covarelli, L.; Tini, F.; Sulyok, M.; Cowger, C. Effect of wheat infection timing on *Fusarium* head blight causal agents and secondary metabolites in grain. *Int. J. Food Microbiol.* **2019**, *290*, 214–225. [CrossRef] [PubMed]
68. Cooke, D.E.L.; Drenth, A.; Duncan, J.M.; Wagels, G.; Brasier, C.M. A molecular phylogeny of *Phytophtora* and related oomycetes. *Fungal Gen. Biol.* **2000**, *30*, 17–32. [CrossRef]
69. Sapkota, R.; Nicolaisen, M. An improved high throughput sequencing method for studying oomycete communities. *J. Microbiol. Methods* **2015**, *110*, 33–39. [CrossRef]
70. Onofri, A.; Pannacci, E. Spreadsheet tools for biometry classes in crop science programmes. *Commun. Biometry Crop. Sci.* **2014**, *9*, 43–45.

plants

MDPI

Article

Field Evaluation of Cypermethrin, Imidacloprid, Teflubenzuron and Emamectin Benzoate against Pests of Quinoa (*Chenopodium quinoa* Willd.) and Their Side Effects on Non-Target Species

Luis Cruces [1,2,*], Eduardo de la Peña [2,3] and Patrick De Clercq [2]

1 Department of Entomology, Faculty of Agronomy, Universidad Nacional Agraria La Molina, Lima 12-056, Peru
2 Department of Plants & Crops, Faculty of Bioscience Engineering, Ghent University, B-9000 Ghent, Belgium; eduardo.deLaPena@ugent.be (E.d.l.P.); patrick.declercq@ugent.be (P.D.C.)
3 Instituto de Hortofruticultura Subtropical y Mediterránea "La Mayora (IHSM-UMA-CSIC), Estación Experimental "La Mayora", 29750 Malaga, Spain
* Correspondence: luiscruces@lamolina.edu.pe; Tel.: +051-999-448427

Abstract: During the last few years, quinoa, a traditional Andean crop, has been cultivated at low elevations where pest pressure is high and farmers resort to intensive use of insecticides. This field study investigated the impact of four insecticides (cypermethrin, imidacloprid, teflubenzuron and emamectin benzoate) on insect pests of quinoa and their side effects on the arthropod community at the coastal level of Peru, by analysing the species composition, species diversity and population density. The arthropod community was examined with pitfall traps (for ground dwelling species), plant samplings (for pests and their natural enemies that inhabit the crop), and yellow pan traps (to catch flying insects). The results demonstrated that *Macrosiphum euphorbiae*, *Frankliniella occidentalis* and *Spoladea recurvalis* were efficiently controlled by cypermethrin and imidacloprid; the latter compound also showed long-term effects on *Nysius simulans*. Teflubenzuron and emamectin benzoate proved to be efficient to control *S. recurvalis*. Imidacloprid had the strongest adverse effects on the arthropod community in terms of species diversity, species composition and natural enemy density as compared to the other insecticides. Findings of this study may assist farmers intending to grow quinoa at the coastal level in selecting the most appropriate insecticides under an integrated pest management approach.

Keywords: insecticides; quinoa pests; side effects; natural enemy; IPM

Citation: Cruces, L.; de la Peña, E.; De Clercq, P. Field Evaluation of Cypermethrin, Imidacloprid, Teflubenzuron and Emamectin Benzoate against Pests of Quinoa (*Chenopodium quinoa* Willd.) and Their Side Effects on Non-Target Species. *Plants* **2021**, *10*, 1788. https://doi.org/10.3390/plants10091788

Academic Editors: Cataldo Pulvento and Didier Bazile

Received: 5 August 2021
Accepted: 23 August 2021
Published: 27 August 2021

1. Introduction

Quinoa is an Andean grain that has increasingly gained international interest due to its high nutritional value [1]. During the last decade, the cultivated area of quinoa has substantially increased in South American countries such as Ecuador, Chile and particularly Bolivia and Peru [2–4]. In the case of Peru, the production areas have extended to lower elevations, reaching even the coastal level [5]. Moreover, there have been attempts of cultivating quinoa outside of its Andean place of origin, in countries such as the United States of America, the United Kingdom and a number of other European countries [6,7]. Therefore, recent efforts have been made to adapt the cultivation of quinoa under non-Andean conditions, including studies regarding to the entomofauna associated with the crop [5,8–12].

In the Andes, at an altitude between 2300 and 3800 m a.s.l., quinoa is traditionally cultivated since ancient times [7]. In this region *Eurysacca melanocampta* Meyrick and *Eurysacca quinoae* Povolvý are the key pests of quinoa, causing damage by feeding on the developing grains; a range of other phytophagous insects are considered of minor importance [2,6].

At the coastal level, the number of relevant phytophagous insects infesting quinoa is substantially larger. These include species of wide distribution such as the cosmopolitan aphids (*Macrosiphum euphorbiae* (Thomas)), thrips (*Frankliniella occidentalis* (Pergande)) and leafminer flies (*Liriomyza huidobrensis* Blanchard); and also others of neotropical distribution such as certain true bugs (*Nysius simulans* Stål and *Liorhyssus hyalinus* (Fabricius)) and lepidopteran larvae (i.e., *E. melanocampta*, *Chloridea virescens* (Fabricius), *Spoladea recurvalis* Fabricius)) that feed on the developing grains [2,5,11]. Under this scenario, farmers may be prompted to apply more pesticides than quinoa growers from the highlands. Hence the need for exploring a range of chemical compounds that may be suitable for use in an integrated pest management (IPM) programme. As in other field crops [13], selective insecticides with a more favourable toxicological profile to the natural enemy community may be a valuable tool for IPM in quinoa.

Cypermethrin is commonly used by the quinoa growers [6]. This insecticide of the pyrethroid group is a nonpersistent sodium channel modulator, characterized by a broad spectrum activity. The compound acts by direct contact, causing neuronal hyperexcitation alongside the axon [10,14]. Due to its relatively short residual effects and lower price, this pesticide is often overused, causing environmental issues and promoting resistance in pest insects [5,10]. The adverse effects of cypermethrin on non-target organisms are widely documented [15,16].

Imidacloprid is one of the most widely used insecticides worldwide [17–19] and it is also commonly used by farmers in coastal areas of quinoa production [20]. This compound may effectively control a range of phytophagous insects noted to be pests of quinoa (including aphids, thrips, true bugs and some lepidopteran species). This neurotoxic insecticide of the neonicotinoid group, is an acetylcholine receptor agonist with broad spectrum and highly systemic activity, acting by ingestion and direct contact, causing neuronal hyperexcitation at the level of the synapses [5,21]. Toxicity of imidacloprid to non-target organisms, including beneficial species such as pollinators (i.e. bees) and natural enemies has been documented in different crops, but there are presently no reports for quinoa [15,16,22–27]. Due to these adverse effects, particularly to the bees, this active ingredient has been banned in Europe [28].

Teflubenzuron has a more favourable environmental profile, with lesser toxicity to a range of non-target organisms as compared to the broad-spectrum compounds, and thus may be considered as a tool for an IPM program in quinoa [29]. This insect growth regulator (IGR) of the benzoylphenylureas group is a highly active inhibitor of chitin synthesis, aimed mainly at lepidopteran larvae. This compound is considered to be safer to the beneficial fauna (especially in the adult stage) than pyrethroids and neonicotinoids, although there are reports of its toxicity towards a number of arthropod predators [16,30–32].

Emamectin benzoate, a neurotoxic insecticide of the avermectin group, is another insecticide reported to be more selective against lepidopteran larvae. Although toxicity to some natural enemies and non-target arthropods have been reported, this insecticide is considered less harmful to beneficial arthropods as compared with broad spectrum compounds [16,33]. The insecticide acts mainly by ingestion causing paralysis in the insect by activating allosterically the glutamate-gated chloride channels in the synapses [16,30,34–36].

The aim of the present field study was to examine the effects of two broad spectrum insecticides (cypermethrin and imidacloprid) and two selective insecticides (teflubenzuron and emamectin benzoate), with different modes of action and from different chemical groups, against quinoa pests in Peru and record their side effects on non-target arthropods by analysing the species composition, species diversity and population density in quinoa fields at the coastal level (a region with potential areas for quinoa production). To assess the effects of these insecticides on the arthropod community, including phytophagous and beneficial species, we combined three sampling techniques (i.e., pitfall traps, plant sampling and yellow pan traps) targeting groups from different ecological habitats. The findings of this field study should be of special interest to quinoa farmers and agricultural extensionists from Andean and non-Andean countries who are exploring the cultivation of quinoa, to use insecticides of higher compatibility with natural biological control as part of an IPM approach.

2. Results

2.1. Effects on the Composition of the Arthropod Fauna

The NMDS-plots show the distances between treatments concerning the composition of the arthropods collected from 24 September 2017 to 11 December 2017, with the different sampling methodologies (in Figure 1), based on the Bray–Curtis dissimilarity index. Ellipses are formed by the replications of each treatment (based on the presence and abundance of species) and the closeness or distances between them reflect their similarities and dissimilarities, respectively. Some morphospecies are shown in the NMDS-plots to depict the differences in the species composition collected in each insecticide treatment: the distance of a morphospecies to the centre of an ellipse (treatment) reflects its scarcity or even absence in the treatment.

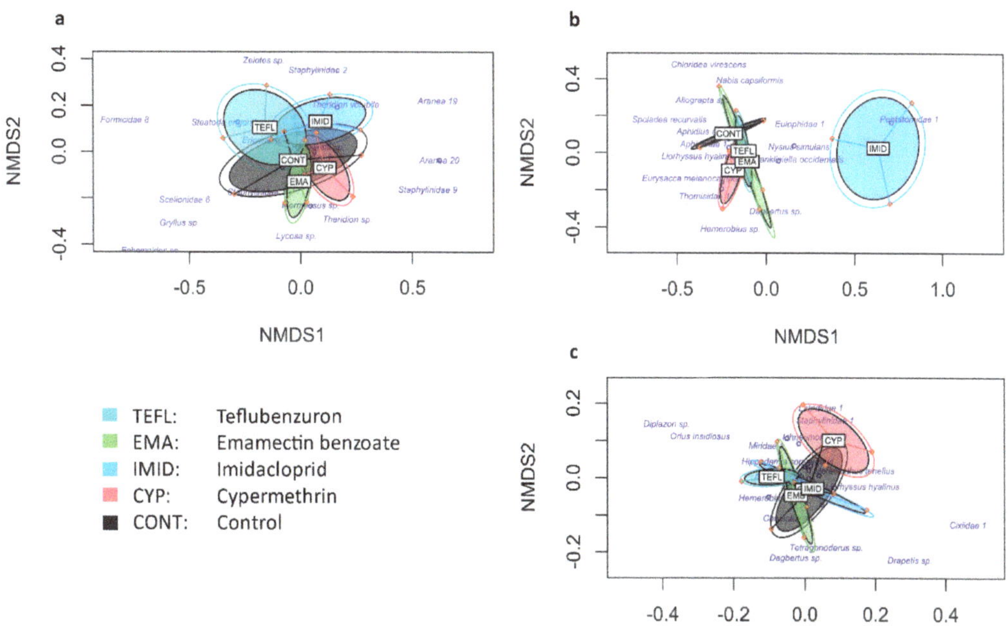

Figure 1. The NMDS plot showing the compositional distance between treatments for different sampling methods: (**a**). pitfall trapping; (**b**). plant sampling; (**c**). pan trapping. Plots are displayed by orange dots; plots that belong to the same treatment are fitted in a single ellipse. Each treatment is represented by a different colour.

As to the pitfall trap data, the proximity of the ellipses in the NMDS-plot (Figure 1a) suggests high similarity among the treatments in terms of the composition of ground dwelling species, which was confirmed by the PerMANOVA test that indicated no significant differences between treatments (F–model$_{4,10}$ = 0.7537, p = 0.819). Likewise, for the flying insects collected at the top of the canopy with the pan traps (Figure 1c), no significant differences between treatments were found (F–model$_{4,10}$ = 1.066, p = 0.441). For the plant sampling data (i.e., the specimens collected from the quinoa plants), however, the ellipse corresponding to the imidacloprid treatment in the NMDS-plot (Figure 1b) is separated from the others, suggesting high dissimilarity between imidacloprid and the other treatments; in this case, the test was significant (F–model$_{4,10}$ = 2.835, p = 0.001).

2.2. *Effects on Diversity of Arthropods*

2.2.1. Structure

Rank abundance curves of the morphospecies collected in the period between 24 September 2017 and 11 December 2017, were calculated for each insecticide treatment (Figure 2). For the pitfall trap data, the corresponding curves for the treatments and the untreated control have similar patterns, except for imidacloprid in which a slightly more pronounced slope can be observed (Figure 2a), indicating that the imidacloprid treatment affected the evenness of the ground dwelling arthropod community to a higher degree than the other insecticides. When applying the ANOVA, significant differences between treatments were found for the Shannon ($F_{4,8} = 4.109$, $p = 0.042$) and Simpson's dominance ($F_{4,8} = 4.038$, $p = 0.038$) indices. The Duncan test confirmed that imidacloprid had a significantly greater impact on the species equitability (with lowest value of Shannon index), preventing dominance of certain taxa (with the lowest value of Simpson's dominance index) (Table 1).

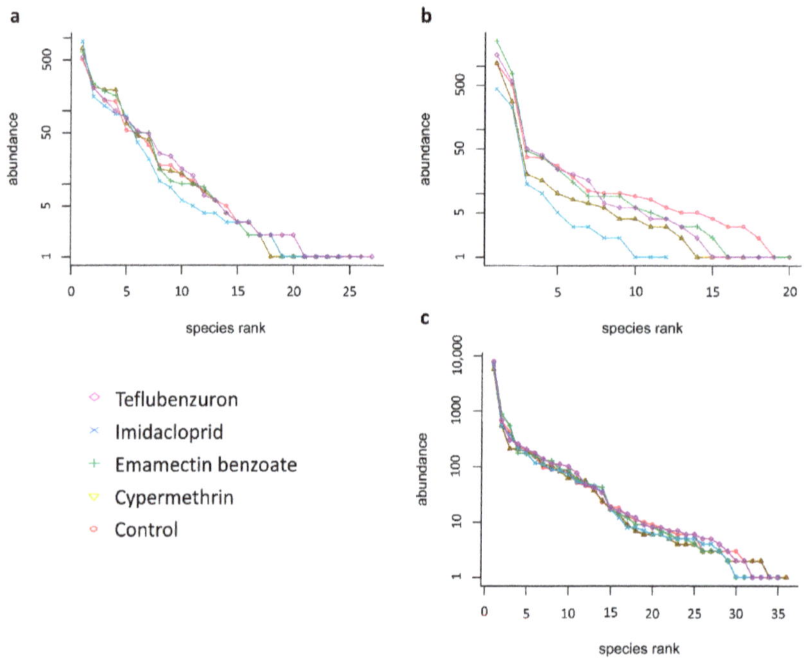

Figure 2. Rank abundance curves for the morphospecies found with the different sampling methodologies, per treatment (log series distribution): (**a**). pitfall trapping; (**b**). plant sampling; (**c**). pan trapping.

As to the specimens collected from the quinoa plants, the corresponding curve for imidacloprid markedly differs from the other treatments and the untreated control, due to the lower number of species collected (Figure 2b). However, no significant differences between treatments for the Shannon ($F_{4,8} = 2.57$, $p = 0.119$) and Simpson's dominance ($F_{4,8} = 1.81$, $p = 0.220$) indices were found (Table 1). For the pan trap data, the curves for all treatments and the control have similar patterns (Figure 2c), suggesting a similar distribution of flying species in the community over all plots. The ANOVA confirmed no significant differences between treatments in terms of the Shannon ($F_{4,8} = 0.34$, $p = 0.841$) and Simpson's dominance ($F_{4,8} = 0.45$, $p = 0.771$) indices.

Table 1. Diversity index (mean ± standard deviation) of morphospecies, according to the sampling methodology applied, collected after the first insecticide treatment (sampling period between 24 September 2017 and 11 December 2017).

Diversity Index	Treatments					F Value	p-Value
	Teflubenzuron	Emamectin Benzoate	Imidacloprid	Cypermethrin	Control		
Pitfall trapping							
Shannon	1.92 ± 0.11 [a]	1.80 ± 0.01 [a]	1.44 ± 0.11 [b]	1.73 ± 0.17 [a]	1.87 ± 0.25 [a]	4.11	0.042
Simpson's dominance	0.77 ± 0.03 [a]	0.75 ± 0.05 [a]	0.60 ± 0.05 [b]	0.73 ± 0.09 [a]	0.76 ± 0.09 [a]	4.04	0.038
Margalef	3.06 ± 0.16 [a]	2.85 ± 0.09 [ab]	2.48 ± 0.12 [c]	2.68 ± 0.21 [bc]	2.78 ± 0.16 [ab]	5.55	0.019
Plant sampling							
Shannon	0.99 ± 0.08 [a]	0.84 ± 0.15 [a]	0.94 ± 0.18 [a]	0.82 ± 0.13 [a]	1.13 ± 0.06 [a]	2.57	0.119
Simpson's dominance	0.49 ± 0.06 [a]	0.43 ± 0.09 [a]	0.52 ± 0.10 [a]	0.40 ± 0.08 [a]	0.55 ± 0.02 [a]	1.81	0.220
Margalef *	1.71 ± 0.13 [a]	1.71 ± 0.26 [a]	1.09 ± 0.27 [b]	1.78 ± 0.06 [a]	1.97 ± 0.52 [a]	4.76	0.029
Pan trapping							
Shannon	1.11 ± 0.14	1.13 ± 0.19	1.09 ± 0.01	1.18 ± 0.10	1.13 ± 0.05	0.34	0.841
Simpson's dominance	0.40 ± 0.06	0.43 ± 0.08	0.04 ± 0.01	0.44 ± 0.04	0.42 ± 0.04	0.45	0.771
Margalef *	3.52 ± 0.29	3.40 ± 0.08	3.50 ± 0.20	3.61 ± 0.15	3.61 ± 0.23	0.49	0.747

Different letters within a row indicate significant differences at $\alpha = 0.05$ (Duncan test), when the ANOVA was significant. * ANOVA run after using Box–Cox transformation method ($\gamma = 0.2$).

2.2.2. Species Richness

Significant differences between treatments were found in species richness of the ground dwelling arthropods, measured by the Margalef index ($F_{4,8} = 5.55$, $p = 0.019$). The lowest species richness was obtained with imidacloprid, being significantly inferior to that of the teflubenzuron and emamectin benzoate treatments and the untreated control, but without significant differences with cypermethrin. Significant differences were also found for the insects collected from the plants ($F_{4,8} = 4.76$, $p = 0.029$), with the imidacloprid treatment having the lowest value. For the pan trapping data, no significant differences in species richness of flying insects between treatments and the control were found ($F_{4,8} = 0.49$, $p = 0.747$) (Table 1).

2.3. *Effects on Functional Species Pools*

2.3.1. Phytophagous Group

Four herbivorous species infested the plots in relatively high abundance: *Spoladea recurvalis,* which appeared at the early stages of the crop phenology and *Macrosiphum euphorbiae*, *Frankliniella occidentalis* and *Nysius simulans*, the incidence of which in all plots was recurrent throughout the cropping season. The mean numbers of these species per plant were compared between treatments (Table 2).

The statistical analysis indicated that all insecticides were efficient to reduce *S. recurvalis* incidence after the first application as compared to the untreated control ($F_{4,8} = 7.73$, $p = 0.007$). Since this pest had disappeared in the treated plots, the effects after the second application could not be evaluated, neither at day 6 nor at day 69 after the application.

Significant differences in the numbers of *M. euphorbiae* were observed, 6 days after the first ($F_{4,8} = 28.73$, $p < 0.001$) and 6 days after the second applications ($F_{4,8} = 7.32$, $p = 0.008$), and also 69 days after the second application ($F_{4,8} = 7.80$, $p = 0.007$). Six days after the first application, the lowest mean number per plant was obtained with imidacloprid, differing significantly from the numbers observed in the teflubenzuron and emamectin benzoate treatments and in the untreated control, whereas the imidacloprid and cypermethrin treatments were similar. On day 6 after the second application, significantly lower aphid numbers were registered for the imidacloprid and cypermethrin treatments, whereas the teflubenzuron and emamectin benzoate treatments were similar to the untreated control. At the last sampling, 69 days after the second application, the aphid numbers were similar in the teflubenzuron, cypermethrin and the untreated plots; the highest aphid abundance was recorded with emamectin benzoate and the lowest with imidacloprid.

Table 2. Numbers of individuals of the major insect pests (mean no. per plant ± standard deviation) under different treatments.

Taxa	Treatments					F Value	p-Value
	Teflubenzuron	Emamectin Benzoate	Imidacloprid	Cypermethrin	Control		
Spoladea recurvalis							
1DBA	2.4 ± 1.39	2.8 ± 0.51	2.7 ± 0.33	3.2 ± 1.64	2.4 ± 1.02	0.15	0.956
1st application							
* 6DAA	0.4 ± 0.77 [b]	0.1 ± 0.19 [b]	0 ± 0.0 [b]	0.1 ± 0.19 [b]	2.0 ± 0.88 [a]	7.73	0.007
2nd application							
6DAA	0	0	0	0	0.67 ± 0.67	N.A.	N.A.
69DAA	0	0	0	0	0	N.A.	N.A.
Macrosiphum euphorbiae							
1DBA	11.4 ± 4.33	15.7 ± 9.17	13.1 ± 4.19	20.4 ± 10.49	12.3 ± 4.26	1.33	0.338
1st application							
** 6DAA	11.6 ± 4.74 [b]	7.8 ± 0.77 [b]	1.8 ± 2.04 [c]	2.9 ± 2.99 [c]	27.4 ± 10.83 [a]	28.73	<0.001
2nd application							
6DAA	6.9 ± 4.44 [a]	11.2 ± 3.56 [a]	0.2 ± 0.19 [b]	0.3 ± 0.33 [b]	6.6 ± 3.89 [a]	7.32	0.008
*** 69DAA	145 ± 40.19 ab	250 ± 104.46 a	36.2 ± 1.67 [c]	113.8 ± 25.06 b	86.1 ± 27.48 bc	7.80	0.007
Frankliniella occidentalis							
1DBA	2.2 ± 1.26	2.4 ± 1.26	2.9 ± 0.84	4.3 ± 0.58	1.8 ± 0.68	2.62	0.115
1st application							
6DAA	5.3 ± 1.20 [a]	4.7 ± 1.15 [a]	2.3 ± 0.33 [b]	1.4 ± 0.38 [b]	5.4 ± 1.17 [a]	9.47	0.004
2nd application							
*** 6DAA	5.1 ± 0.84 [b]	4.3 ± 1.15 [b]	2.1 ± 0.51 [c]	0.6 ± 0.19 [c]	9.89 ± 2.46 [a]	171.17	<0.001
69DAA	26.3 ± 11.1[b]	62.1 ± 6.50 [a]	7.2 ± 1.89 [d]	15.1 ± 5.42 [bc]	20.3 ± 11.98 bc	46.76	<0.001
Nysius simulans							
* 1DBA	0.1 ± 0.19	0.2 ± 0.19	0.1 ± 0.19	0.3 ± 0.33	0.8 ± 0.84	0.82	0.549
1st application							
6DAA	0 ± 0	0 ± 0	0.2 ± 0.19	0.1 ± 0.19	0.1 ± 0.19	1.75	0.232
2nd application							
* 6DAA	0.6 ± 0.38	0.6 ± 0.69	0.3 ± 0.33	0.1 ± 0.19	0 ± 0	1.62	0.261
69DAA	5.0 ± 0.67 [a]	3.6 ± 0.19 [b]	1.0 ± 0.67 [d]	2.0 ± 0.33 [c]	2.9 ± 0.19 [bc]	25.87	<0.001

DBA: days before application; DAA: days after application; N.A.: not applicable. Different letters within a row indicate significant differences at $\alpha = 0.05$ (Duncan test), when the ANOVA was significant. * ANOVA run after using Box–Cox transformation method $\gamma = -2.0$, ** $\gamma = 0.45$, *** $\gamma = 0.30$.

Significant differences in *F. occidentalis* numbers were observed 6 days after the first ($F_{4,8} = 9.47$, $p = 0.004$) and 6 days after second applications ($F_{4,8} = 171.17$, $p < 0.001$), and 69 days after the second application ($F_{4,8} = 46.76$, $p < 0.001$). On day 6 after the first application, the lowest mean values were observed with imidacloprid and cypermethrin, and no significant differences were found between teflubenzuron, emamectin benzoate and the untreated control. Six days after the second application, the lowest mean values were also obtained with imidacloprid and cypermethrin, and lower thrips numbers were found in the teflubenzuron and emamectin benzoate plots as compared to the untreated control. At the last sampling date, 69 days after the second application, the lowest thrips numbers were obtained with imidacloprid, whereas the cypermethrin and teflubenzuron treatments and the untreated control had similar numbers; the thrips numbers in the emamectin benzoate plots were significantly higher than in the other plots.

Significant differences in *N. simulans* numbers were not seen until 69 days after the second application ($F_{4,8}$ = 25.87, p < 0.001). The lowest mean values were obtained with imidacloprid and cypermethrin (in this order); the mean value obtained in the untreated control was similar to that in the cypermethrin and emamectin benzoate treatments, whereas the highest value was recorded in the teflubenzuron treatment.

The phytophagous insects were also examined in the pitfall and pan traps. The mean cumulative number of *N. simulans* and *F. occidentalis*, recorded from 24 September 2017 to 11 December 2017, were compared between treatments (Table 3). No significant differences between treatments were found as to *N. simulans* numbers trapped with pitfall traps ($F_{4,8}$ = 1.62, p = 0.261) nor as to those collected with pan traps ($F_{4,8}$ = 0.42, p = 0.792).

Table 3. Cumulative numbers (mean no. per trap ± standard deviation) of the most abundant phytophagous insects and natural enemies, collected with two sampling methodologies, after the second insecticide application (sampling period from 24 September 2017 to 11 December 2017).

Taxa	Treatments					F Value	*p*-Value
	Teflubenzuron	Emamectin Benzoate	Imidacloprid	Cypermethrin	Control		
Pitfall trapping							
Nysius simulans	181.3 ± 82.6	214.0 ± 83.5	301.3 ± 49.9	241.0 ± 113.9	169.0 ± 89.9	1.62	0.261
Laminacauda sp.	80.0 ± 30.3	115.3 ± 9.9	67.3 ± 11.8	135.7 ± 55.4	90.7 ± 26.7	2.21	0.159
Blennidus peruvianus	68.7 ± 28.0	79.7 ± 38.4	52.0 ± 3.46	65.3 ± 30.6	68.3 ± 4.6	0.61	0.669
Trimorus sp.	33.7 ± 16.3	33.3 ± 7.5	19.7 ± 4.7	35.7 ± 17.9	28.0 ± 10.6	0.78	0.568
Pan traps							
Nysius simulans	36.7 ± 4.73	29.0 ± 10.5	38.7 ± 12.3	36.3 ± 10.7	32.3 ± 11.6	0.42	0.792
Frankliniella occidentalis	2634.0 ± 188.9 [a]	2559.3 ± 84.1 [ab]	2276.3 ± 205.0 [b]	1892.0 ± 54.7 [c]	2311.0 ± 59.6 [b]	11.18	0.002
Dolichopodidae	100.0 ± 38.2	179.7 ± 121.3	124.3 ± 34.7	53.3 ± 42.3	138.7 ± 125.7	1.14	0.403
Syrphidae	25.3 ± 2.1	19.3 ± 4.7	11.7 ± 3.8	18.3 ± 8.4	15.0 ± 2.64	3.33	0.069
Aphidiinae *	33.3 ± 13.0 [ab]	51.0 ± 25.2 [a]	18.0 ± 8.7 [c]	20.7 ± 5.5 [bc]	27.0 ± 7.0 [abc]	5.33	0.022

Different letters within a row indicate significant differences at α = 0.05 (Duncan test), when the ANOVA was significant. * ANOVA run after using Box–Cox transformation method γ = −0.4.

There were significant differences in *F. occidentalis* numbers collected with pan traps ($F_{4,8}$ = 11.18, p = 0.002). The lowest mean values were observed in the cypermethrin treatments, whereas the highest thrips numbers were noted in the teflubenzuron plots; the emamectin benzoate, imidacloprid and untreated plots had similar values. Adults of *L. huidobrensis* were also collected in relatively high abundance in the pan traps, but as they appeared at the later stages of the crop phenology, leaf miner larvae were not observed in the examined leaves and therefore they were not considered in the analysis.

2.3.2. Natural Enemies

The most recurrent natural enemy groups found on the collected plants were Aphidiinae wasps (adult and parasitized aphids), predatory true bugs (*Metacanthus tenellus* Stål, *Rhinacloa* sp., and *Nabis capsiformis* Germar), syrphid larvae (*Allograpta* sp.) and chrysopid larvae. The mean cumulative numbers of individuals per plot of these groups were compared between treatments (Table 4). Other predators such as coccinellids and hemerobiids were also found in some plots, but their incidence was irregular throughout the monitoring and in small numbers; therefore, they were excluded from the analysis.

Table 4. Numbers of individuals of the most abundant insect natural enemies collected (mean no. per plant ± standard deviation) under different treatments.

Taxa	Treatments					F Value	*p*-Value
	Teflubenzuron	Emamectin Benzoate	Imidacloprid	Cypermethrin	Control		
Aphidiinae wasps							
1DBA	1.66 ± 1.53	1.67 ± 0.88	2.22 ± 1.35	1.44 ± 0.84	1.00 ± 0.67	0.96	0.481
1st application							
* 6DAA	0.22 ± 0.19	0.56 ± 0.19	0.11 ± 0.19	0.22 ± 0.39	3.44 ± 4.28	2.66	0.111
2nd application							
* 6DAA	0.33 ± 0.0	0.55 ± 0.69	0.0 ± 0.0	0.11 ± 0.19	0.11 ± 0.19	2.12	0.169
** 69DAA	0.22 ± 0.19 ab	0.11 ± 0.19 b	0.0 ± 0.0 b	0.22 ± 0.19 ab	0.56 ± 0.20 a	4.07	0.043
Predatory true bugs							
1DBA	0.0 ± 0.0	0.2 ± 0.17	0.23 ± 0.40	0.33 ± 0.35	0.0 ± 0.0	N.A.	N.A.
1st application							
* 6DAA	0.10 ± 0.10	0.03 ± 0.06	0.03 ± 0.06	0.0 ± 0.0	0.0 ± 0.0	N.A.	N.A.
2nd application							
6DAA	0.22 ± 0.39	0.0 ± 0.0	0.0 ± 0.0	0.0 ± 0.0	0.22 ± 0.19	N.A.	N.A.
69DAA	1.67 ± 0.67 a	1.78 ± 1.01 a	0.11 ± 0.19 b	0.22 ± 0.39 b	1.89 ± 0.69 a	5.48	0.020
Syrphid larvae							
1DBA	0.0	0.0	0.0	0.0	0.0	N.A.	N.A.
1st application							
* 6DAA	0.23 ± 0.40	0.33 ± 0.58	0.0 ± 0.0	0.10 ± 0.17	1.57 ± 1.25	3.09	0.082
2nd application							
* 6DAA	0.0 ± 0.0 b	0.67 ± 1.15 ab	0.0 ± 0.0 b	0.0 ± 0.0 b	0.56 ± 0.19 a	5.39	0.021
1DBA	0.0	0.0	0.0	0.0	0.0	N.A.	N.A.
Chrysopid larvae							
1DBA	0.0	0.0	0.0	0.0	0.0	N.A.	N.A.
2nd application							
6DAA	0.0	0.0	0.0	0.0	0.0	N.A.	N.A.
69DAA	1.0 ± 0.0 ab	0.53 ± 0.69 bc	0.0 ± 0.0 c	0.56 ± 0.51 bc	1.56 ± 0.51 a	4.35	0.037

DBA: days before application; DAA: days after application; NA: not applicable. Different letters within a row indicate significant differences at $\alpha = 0.05$ (Duncan test), when the ANOVA was significant. * ANOVA run after using Box–Cox transformation method $\gamma = -2.5$; ** $\gamma = 0.3$.

Numbers of Aphidiinae wasps collected in the plants did not differ between treatments, neither 6 days after the first application ($F_{4,8} = 2.66$, $p = 0.111$), nor 6 days after the second application ($F_{4,8} = 2.12$, $p = 0.169$). However, 69 days after the second application, significant differences were found ($F_{4,8} = 4.07$, $p = 0.043$), with zero specimens of Aphidiinae wasps (neither larvae in the mummified aphids nor adults) collected in the plots treated with imidacloprid.

The predatory heteropterans were scarce 6 days after the first and the second applications (with only seven specimens recorded over the different plots) and data recorded in these samplings were not subjected to ANOVA. On day 69 after the second application, the predatory true bugs became relatively more abundant, and significant differences between treatments and the control ($F_{4,8} = 5.48$, $p = 0.020$) were found, with the lowest numbers of heteropterans predators recorded in the imidacloprid and cypermethrin treatments, whereas values for the untreated control and the teflubenzuron and emamectin benzoate plots were similar.

No syrphid larvae were observed in the plots before the first application. Whereas six days after the first application, numbers of syrphid larvae did not differ among treatments ($F_{4,8} = 3.09$, $p = 0.082$), significant differences were found 6 days after second application ($F_{4,8} = 5.39$, $p = 0.021$), with zero specimens collected in the imidacloprid, cypermethrin and teflubenzuron treatments.

Since chrysopid larvae appeared at the later stages of the crop phenology, short-term effects of the insecticides on their numbers could not be observed. However, 69 days after the second application, significant differences in numbers of chrysopid larvae between the insecticide treatments and the untreated control were observed ($F_{4,8}$ = 4.35, p = 0.037), with zero specimens collected in the imidacloprid plots, and very low numbers as well in the cypermethrin and emamectin benzoate plots. Incidence of chrysopid larvae in the teflubenzuron treatment was similar to that in the control.

The natural enemies were also examined in the pitfall traps. The most abundant morphospecies collected from 24 September 2017 to 11 December 2017 were the arachnid *Laminacauda* sp., the ground beetle *Blennidus peruvianus* Dejean and the wasp *Trimorus* sp.; the mean cumulative number of these morphotypes was compared between treatments and control (Table 3). There were no significant differences between treatments in numbers of *Laminacauda* sp. ($F_{4,8}$ = 2.21, p = 0.159), *B. peruvianus* ($F_{4,8}$ = 0.61, p = 0.669) and *Trimorus* sp. ($F_{4,8}$ = 0.78, p = 0.568).

When examining the natural enemies collected in the pan traps (recorded from 24 September 2017 to 11 December 2017), three groups were the most abundant: Dolichopodidae, Syrphidae and Aphidiinae. There were no significant differences between treatments in the numbers of adults of Dolichopodidae ($F_{4,8}$ = 1.14, p = 0.403) and Syrphidae ($F_{4,8}$ = 3.33, p = 0.069). The number of Aphidiinae wasps was similar when comparing each treatment with the untreated control, but in the imidacloprid plot significantly lower numbers were found than in the teflubenzuron and emamectin benzoate treatments ($F_{4,8}$ = 5.33, p = 0.022).

3. Discussion

The side effects of insecticides applied in staple crops (i.e., vegetables, legumes, rice, maize, citrus) and industrial crops (i.e., cotton, sugarcane, sugar beet) have been widely studied [16]. Thus, relevant knowledge has been gained to improve integrated pest management schemes, taking the biological control services offered by a biodiverse agroecosystem into consideration [30]. Hitherto, however, little is known about the unintentional effects of insecticides used in quinoa on non-target organisms [8]. The present field study provides information about the effects of four insecticides from different chemical groups (teflubenzuron, emamectin benzoate, imidacloprid and cypermethrin) on target and non-target arthropods in quinoa, assessed with three sampling methodologies: pitfall trapping for the ground dwelling arthropods, plant sampling for those that dwell on the quinoa plants and pan trapping for the insects that fly just above the crop canopy.

When an insecticide is incorporated into the cropping system, changes in the structure, richness and composition of the plant dwelling arthropod community may occur, which may eventually lead to a disruption of the ecosystem services provided by the beneficial fauna [37–40]. In the present study, foliar application with imidacloprid appeared to have a higher impact on arthropods residing in the quinoa crop than the other insecticides. For example, the richness in plant dwelling species was significantly lower in the imidacloprid treatment and also the species composition differed significantly from that in the other treatments and the untreated control.

Given that the soil surface of an agricultural system is in permanent interaction with the higher strata, changes in the plant dwelling arthropod community tend to precede changes in the structure of the ground dwelling species community [23]. In this context, the reduction in species richness and the change in composition of species residing on the quinoa plants in the plots treated with imidacloprid may be related to the lower values of Shannon and Margalef indices found at the ground level as compared to the other treatments and the untreated control. The changes in the plant-associated community as a consequence of the insecticide application may have broken food webs that affected the incidence of a variety of species [41–43]. This may eventually be reflected in an altered species evenness and a reduction of species richness at the soil surface level, as observed in the present study [44]. On the contrary, no differences between treatments and control were found in terms of species composition and diversity of the insects collected in the yellow pan traps, probably due to the greater interaction at the top of the crop canopy with the areas surrounding the plots.

All of the tested insecticides substantially reduced densities of *S. recurvalis* larvae after the first application, which is in line with several previous studies that have demonstrated their efficacy against lepidopteran larvae [29,45–50]. As re-infestation by *S. recurvalis* larvae did not occur in any of the plots, including the control, long-term effects of the insecticide treatments to control this pest could not be judged.

Whereas the efficacy of imidacloprid to control lepidopteran larvae by direct contact action has been demonstrated, this insecticide is also known to have an excellent systemic activity and therefore, the target organisms are mainly sucking insects such as thrips, aphids, whiteflies, leafhoppers and true bugs [29]. Accordingly, the population densities of *M. euphorbiae, F, occidentalis* and *N. simulans,* which recurrently infested our plots, were significantly affected by the imidacloprid treatment.

The short-term effect of imidacloprid on *M. euphorbiae* was similar to that of the cypermethrin treatment. Differences between both treatments could be noted 69 days after the second application, with the imidacloprid plot having the lowest number of aphids per plant, likely due to its widely documented residual effects [23,29,51–54]. On the other hand, teflubenzuron and emamectin benzoate had lesser effects on the aphids as compared to imidacloprid and cypermethrin; their impact on the aphid population as compared to the untreated control was noted 6 days after the first application. Teflubenzuron is reported to have low contact activity, but due to its systemic action in the plant it may cause toxicity to aphids by ingestion [29,32,55,56]. Contrarily, emamectin benzoate has no systemic activity but can kill the exposed aphids by direct contact [29,57].

Imidacloprid and cypermethrin had similar effects on *F. occidentalis* numbers, 6 days after both the first and second application. However, the residual effect of imidacloprid appeared to have prevented the infestation to a higher degree 69 days after the second application, resulting in the lowest number of thrips per plant. The effects of teflubenzuron and emamectin benzoate treatments on *F. occidentalis* were observed 6 days after the second application. On day 69 after the second treatment the residual activity of teflubenzuron may explain the significantly lower mean number of thrips per plant (26.3 specimens) as compared to the emamectin benzoate treatment.

No visible short-term effects were observed on the population density of *N. simulans,* in any of the treatments, probably because the infestation was very low at the early stages of the crop when the insecticide treatments were done, and also because this pest was more abundant on the soil at this time. However, by its residual effect imidacloprid may have prevented a higher level of infestation by *N. simulans*, resulting in the lowest number of individuals per plant 69 days after the second application.

Given its broad-spectrum activity, imidacloprid may also affect non-target arthropods [29]. These non-target organisms may be exposed to imidacloprid by direct contact, but the compound being highly systemic, non-target omnivorous insects (including natural enemies) that feed on plant fluids or pollen may also be exposed, even a relatively long time after an application [58]. Furthermore, reduction of prey densities (i.e., the target organisms) may eventually affect the beneficial fauna that will not be able to find sufficient food, facing a greater intra- and interspecific competition [59]. In this context, with *M. euphorbiae* being highly affected by the imidacloprid treatment, there was a tendency towards lower numbers of individuals (even zero) of the aphidophagous guild on the plants as compared to the untreated control, both for the specialized natural enemies (such as the Aphidiinae wasps and predatory Syrphidae larvae) as for generalist predators (such as predaceous true bugs and chrysopid larvae). These observations are in line with previous studies indicating that imidacloprid affects key natural enemies of aphids such as coccinellids, anthocorids, geocorids, chrysopids, and Aphidiinae wasps [22,23,53,60,61].

Due to its broad-spectrum activity, cypermethrin is considered to be harmful to a range of natural enemies by direct contact, but its relatively short residual activity suggests that a recolonisation of the crop by these organisms may occur sometime after an application [29]. However, the observations in the present study indicate similar long-term effects of cypermethrin to those of imidacloprid on the different natural enemies collected.

Teflubenzuron and emamectin benzoate sprays, however, tended to have less harmful effects on the beneficial fauna than the broad-spectrum insecticides used, which is in line with the literature [29,32]; although environmental risks of emamectin benzoate such as toxicity to certain non-target arthropods and aquatic organisms have been reported [33].

One limitation of this study is that the field experiment was done over a single growing season only. To demonstrate the efficacy of an insecticide, it is recommended that trials be carried out in different locations or growing seasons [62]. On the other hand, data provided about the side effects of the insecticides on the non-target species, particularly natural enemies, are in line with standard methods [63], so this information is relevant for quinoa growers in order to take actions to improve their current use of the pesticides. Moreover, the data are in line with those of previous studies [5,11,64] and the crop management practices are representative for quinoa cultivation in Peru and neighbouring countries [2,8,65].

4. Materials and Methods

4.1. Location

The study was carried out in experimental fields belonging to the National Agrarian University La Molina in Lima, Peru (coordinates: 12°04′57.0″S, 76°56′49′W; altitude: 244 m).

4.2. Experimental Units

The field trial was conducted under a stratified–randomized design with three replications. Each experimental unit consisted of one plot of 21 m^2 (7 ridges of 0.75 cm width, 4 m length), with a plant density of 36 quinoa plants per linear meter (variety "Negra Collana") after seedling removal. Each plot (as shown in Figure 3) was surrounded by polypropylene films (0.5 mm thickness, 1.5 m height, black colour) three days before the treatment (15 September 2017) and maintained until the time of harvest. At the beginning of the grain filling stage (on 5 November 2017), the whole experiment was covered with anti-bird netting to protect the crop from bird damage. Growing specifications of the field site are described in Table 5.

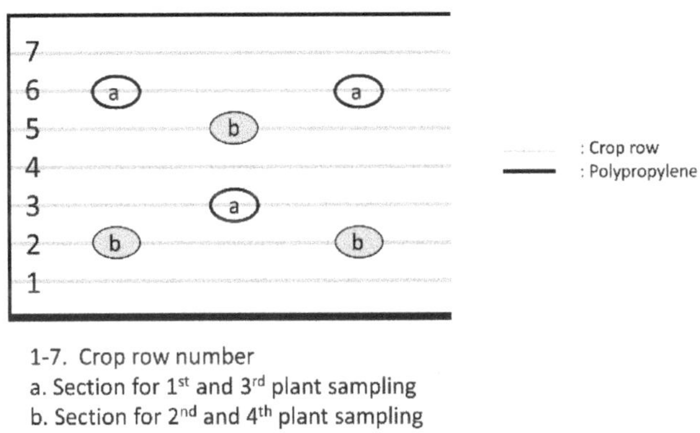

Figure 3. Schematic diagram of plant sampling setup for an experimental plot.

Table 5. Growing specifications of the experimental field.

	Specifications	Dates
Sowing Harvest	drilling sowing method threshing	19 July 2017 12 December 2017
Irrigation	surface irrigation	(20 July 2017; 17 August 2017; 7 September 2017; 28 September 2017; 25 October 2017; 9 November 2017)
Fertilisation doses (NKP)	160–80–160	19 July 2017
Soil type	clay loam	
Neighbouring crops	Quinoa (*Chenopodium quinoa*), Wheat (*Triticum* spp.) Corn (*Zea mays*) Kiwicha (*Amaranthus caudatus*)	
Fungicides	1° benzomyl 2° metalaxyl + mancozeb 3° dimetomorph	(24 July 2017) (7 August 2017) (25 August 2017)
Weed management	Manual control	(25 July 2017, 11 August 2017; 19 November 2017)
Previous crop	Fallow period of 4 months	

4.3. Insecticide Treatments

Treatments were done with four insecticides (as formulated materials): teflubenzuron (150 g/L), emamectin benzoate (50 g/kg), imidacloprid (350 g/L) and cypermethrin (250 g/L). Water was used as a negative control. Specifications of the insecticides are described in Table 6.

Table 6. Insecticide specifications used in the treatments.

Insecticide	Label Field Rate (g a.i. ha^{-1}) *	Chemical Group	Commercial Name	Company
Cypermethrin	75	Pyrethroid	Cypmor 25 EC	Jebsen and Jessen Peru S.A.C.
Teflubenzuron	33.75	Benzoylphenylurea	Mercury 150 SG	Point Andina S.A.
Emamectin benzoate	10	Avermectin	Olimpo 5% SG	Sharda Peru S.A.C.
Imidacloprid	131.25	Neonicotinoid	Phantom	Jebsen and Jessen Peru S.A.C.

a.i. = active ingredient. * spray liquid applied at a rate of 300 L/ha.

The insecticides were applied using a manual sprayer (SOLO461: pressure 3 bars, capacity 5 L) with a full cone nozzle (TeeJet®). The sprayer was calibrated to apply 0.65 L per plot (corresponding to 300 L per ha). Before insecticide dissolution, water was acidified to a pH range of 5.0 to 6.0 (as recommended on the labels), with an acidifying product (SUPER ACID, 43% of organic and inorganic acids) at 0.05%. Similarly, water was acidified for the negative control.

Two applications at the maximum recommended field rate (Table 6) were made at flowering stage on 18 September 2017 (61 days after sowing) and 3 October 2017 (76 days after sowing). At this crop phase, the plants had reached their maximum height (ca., 1.2 m).

4.4. Sampling Methodology

Three sampling techniques were used for studying the arthropod fauna (insects and arachnids): pitfall trapping, for ground dwelling species; plant sampling, for foliage dwelling species (phytophagous insects and natural enemies); and, pan traps placed at the level of the top canopy (1.2 m), for flying insects.

4.4.1. Pitfall Trapping

One pitfall trap (as an experimental unit) was installed in the middle of each experimental plot six days after the first insecticide application and maintained until one day before harvest (from 24 September 2017 to 11 December 2017). Traps consisted of a polypropylene container (transparent, Ø 10 cm at opening and at bottom, 12 cm deep) with a mix of water and 40% v/v formaldehyde (9:1), and a few drops of detergent. The pitfall trap content was periodically collected (a total of 5 times) in airtight recipients (of the same dimensions as the traps) and carefully labelled to be transported to the laboratory for further processing. Thereafter, the collection fluid was replaced.

4.4.2. Plant Sampling

At each experimental plot, four samplings were performed, i.e., one day before the first application (17 September 2017), 6 days after the first application (24 September 2017), 6 days after the second application (9 October 2017) and 69 days after the second application (11 December 2017). Sampling consisted of taking three plants from crop rows 3 and 6 (Figure 3) in the 1st and 3rd samplings, and from rows 2 and 5 in the 2nd and 4th samplings. Plants near the borders of the plots were always avoided.

In each plot, every plant was randomly selected and carefully collected: the plant (after cutting it at its base with scissors) was shaken over a container (width 26 cm × large 36 cm × height 20 cm) with a mix of water and 96% *v*/*v* ethanol (3:1), and some drops of liquid detergent. Thereafter, the sampled plants were carefully chopped into small pieces and the whole sample (including the liquid content) was transferred to an airtight container (volume 3l).

4.4.3. Pan Trapping

Each pan trap consisted of a yellow polypropylene container (Ø 20 cm at opening and 18 cm at bottom, 7 cm deep) with a mix of water and 40% *v*/*v* formaldehyde (9:1), and some drops of detergent. One pan trap was installed in the middle of each experimental plot, at a height of 1.2 m (the opening at the level of the top of the crop canopy), six days after the first insecticide application and maintained until one day before harvest (from 24 September 2017 to 11 December 2017). As these traps were exposed to desiccation, they were regularly inspected and when needed, they were re-filled with the same collection fluid.

The pan trap content was periodically collected (5 times) in airtight recipients (500 ml of capacity) and carefully labelled to be transported to the laboratory for further processing. Thereafter, the collection fluid was replaced.

4.5. Sample Processing

All samples were processed at the laboratories of the Museum of Entomology "Klaus Raven Büller" of the National Agrarian University La Molina, in Lima, Peru, where the collected specimens were deposited.

4.5.1. Sample Washing

The recipients containing pitfall trap and pan trap samples were poured onto a 1 mm mesh sieve and carefully washed with water, removing larger material such as stones, straw or leaves. The collected specimens were transferred to a labelled plastic container (Ø 5 cm, 6 cm length) containing 75% *v*/*v* ethanol for conservation and further processing (i.e., morphotyping).

The recipient with the plant samples was decanted through a 1 mm mesh sieve and carefully washed. Then, the plant parts (leaves, stem and panicle) were examined under a binocular stereoscope (Carl Zeiss: Stemi 508) to check for the presence of mines and to collect the insects that remained stuck to the plant. These specimens and those which easily detached from the plant materials were transferred to a labelled plastic container (Ø 5 cm, 6 cm length) containing 75% *v*/*v* ethanol for conservation and morphotyping.

4.5.2. Morphological Identification

The specimens were examined using a binocular stereoscope (Carl Zeiss: Stemi 508) and sorted on the basis of morphological characteristics as morphospecies [66,67]. Each new morphospecies was photographed and codified, facilitating comparison when a new similar morphospecies was found, and then placed in a glass vial (Ø 1.5 cm, 4 cm length) with 75% *v*/*v* ethanol for preservation. When necessary, the morphotypes were re-examined. Each morphospecies was counted and classified at family level with the help of taxonomic keys from the literature [68].

The most abundant morphospecies were identified to genus level and, when possible, to level species with a help of specific taxonomic keys as follows: *B. peruvianus* [69,70]; *L. hyalinus* [71]; *N. simulans* [72], *L. huidobrensis* [73,74], *Rhinacloa* sp. [75], *M. tenellus* [76,77]; *N. capsiformis* [78–80]; *S. recurvalis* [81]. The identification of the Araneae families and genera (i.e., *Laminacauda* sp.) was assisted by arachnologist Manuel Andía associated to the Museum of Entomology "Klaus Raven Büller".

4.5.3. Molecular Identification

DNA extraction and PCR procedures were performed at the Department of Plants and Crops of Ghent University in Belgium to identify and/or confirm the species *L. huidobrensis*, *M. euphorbiae*, *L. hyalinus* and *F. occidentalis*, following the protocols provided in the literature [82–86].

DNA was extracted from a single specimen (*M. euphorbiae*, *F. occidentalis*) or a leg (*L. hyalinus*, *L. huidobrensis*) that was removed from an adult specimen with a fine cutter. The sample was transferred to a 1.5 mL Eppendorf and then crushed with a plastic rod with 20 µL of STE-buffer (100 mM NaCl, 10 mM Tris-HCL, 1 mM EDTA, pH 8.0) and 2 µL of proteinase K (10 mg/mL). This mix was incubated at 60 ° C for 30 min. Then, the activity of the proteinase K was stopped at 95 °C for 5 min.

DNA samples were subjected to PCR analysis with the primers LCO1490 FW and HCO2198RV (for *L. hyalinus* and *L. huidobrensis*); MTD 7.2 F and COI-MTD 9.2 R (for *F. occidentalis*); and, C1-J-1718 and C1-N-2191 (for *M. euphorbiae*). Amplification was performed in 50 µL total mix reaction, containing 2 µL of DNA sample, 0.25 µL GoTaq® DNA Polymerase (5 u/µL), 3 µL MgCl2 solution (25 mM), 1 µL dNTPs (10 µM each), 2.5 µL forward primer, 2.5 µL reverse primer, 10 µL: 5× Colorless GoTaq® Flexi Buffer, 28.75 µL water. This solution was placed in a thermal cycler with the following parameters: 2 min at 95 °C, 35 cycles of 30 s at 95 °C, 30 s at 66 °C, 45 s at 72 °C, and a final extension of 5 min at 72 °C. After amplification, 10 µL of the PCR products were subjected to electrophoresis on a 2% agarose gel, and PRC products were purified using the EZNA® Cycle Pure Kit (Omega Bio-Tek) following the manufacturer's protocols. Bidirectional Sanger sequencing, using the PCR primers, was outsourced to LGC genomics (Germany).

4.6. Statistical Analysis

For each sampling methodology applied, differences in the collected species composition between treatments were evaluated with the Bray–Curtis dissimilarity index and NMDS (non-metric multidimensional scaling) using the presence and abundance of the morphospecies to detect distances between plots. Significant differences between treatments were assessed using the PerMANOVA test (999 permutations).

The effects of the insecticide treatments on the diversity of the arthropod community, per each sampling methodology, were analysed through (a) the rank abundance curves and the indices of Shannon and Simpson's dominance to evaluate the structure of the community (evenness and dominance of species) and (b) the Margalef index to assess the species richness. They were calculated for each experimental plot.

The diversity indices, mean numbers of the major pests and mean numbers of natural enemies were compared, according to each sampling methodology, between treatments by one-way ANOVA and Duncan tests, after having tested the normality and homoscedasticity of the data through Shapiro–Wilk and Bartlett tests, respectively. When the data did not meet the assumption of homogeneity of variances, the Box–Cox transformation method was applied to stabilize the variance; however, untransformed data are presented in the tables.

All statistical analyses were performed using R software, version 3.4.2 [87]. The tests were analysed at a significance level of $\alpha = 0.05$.

5. Conclusions

The results of this study indicate that due to the detrimental effects of imidacloprid on arthropod diversity, on the composition of species and specifically on the natural enemy population, foliar application of this active ingredient is less suitable for an IPM programme in quinoa as compared to the other insecticides, in spite of its good performance in the control of the target pests. Teflubenzuron and emamectin benzoate substantially suppressed *S. recurvalis* larvae, with less negative effects than imidacloprid and cypermethrin to the beneficial fauna; however, further research is needed to evaluate the efficacy of both selective insecticides against quinoa pests in order to be considered as an element of an IPM package in quinoa. Due to the negative impact of cypermethrin on the natural enemy complex, restricted use is recommended for the management of quinoa pests.

Author Contributions: Conceptualization, L.C., E.d.l.P. and P.D.C.; methodology, L.C. and P.D.C.; formal analysis, L.C., E.d.l.P. and P.D.C.; investigation, L.C., E.d.l.P. and P.D.C.; data curation, L.C.; writing—original draft preparation, L.C.; writing—review and editing, L.C., E.d.l.P. and P.D.C.; supervision, E.d.l.P. and P.D.C. All authors have read and agreed to the published version of the manuscript.

Funding: This research was funded by THE PROJECT 2: "Development of Value Chains for Biodiversity Conservation and Improvement of Rural Livelihoods"—Sub Project: Native Grains, VLIR-UOS IUC/UNALM.

Institutional Review Board Statement: Not applicable.

Informed Consent Statement: Not applicable.

Data Availability Statement: The data presented in this study are available upon request from the corresponding author.

Acknowledgments: We thank the professors from the National Agrarian University La Molina in Peru; Luz Gómez and Clorinda Vergara, for the facilities and permits in the Cereals and Native Grain programme and in the Museum of Entomology "Klaus Raven Büller", respectively. We also acknowledge VLIR-UOS/UNALM, for funding this study and Eddie Schrevens for his unwavering support. Finally, we thank the arachnologist Manuel Andía for the identification of the spiders.

Conflicts of Interest: The authors declare no conflict of interest.

References

1. Alandia, G.; Rodriguez, J.P.; Jacobsen, S.E.; Bazile, D.; Condori, B. Global expansion of quinoa and challenges for the Andean region. *Glob. Food Secur.* **2020**, *26*, 100429. [CrossRef]
2. Cruces, L.; Callohuari, Y.; Carrera, C. *Quinua Manejo Integrado de Plagas. Estrategias en El Cultivo de La Quinua Para Fortalecer El Sistema Agroalimentario en La Zona Andina*; Organización de las Naciones Unidas para la Alimentación y la Agricultura: Santiago, Chile, 2016; pp. 1–189.
3. Bedoya-Perales, N.S.; Pumi, G.; Mujica, A.; Talamini, E.; Domingos Padula, A. Quinoa expansion in Peru and its implications for land use management. *Sustainability* **2018**, *10*, 532. [CrossRef]
4. Hinojosa, L.; Leguizamo, A.; Carpio, C.; Muñoz, D.; Mestanza, C.; Ochoa, J.; Castillo, C.; Murillo, A.; Villacréz, E.; Monar, C.; et al. Quinoa in Ecuador: Recent advances under global expansion. *Plants* **2021**, *10*, 298. [CrossRef] [PubMed]
5. Cruces, L.; de la Peña, E.; De Clercq, P. Seasonal phenology of the major insect pests of quinoa (*Chenopodium quinoa* Willd.) and their natural enemies in a traditional zone and two new production zones of Peru. *Agriculture* **2020**, *10*, 644. [CrossRef]
6. Saravia, R.; Plata, G.; Gandarillas, A. *Plagas y Enfermedades del Cultivo de Quinua*; Fundación PROINPA: Cochabamba, Bolivia, 2014; pp. 1–82.
7. Gamboa, C.; Van den Broeck, G.; Maertens, M. Smallholders' preferences for improved quinoa varieties in the Peruvian Andes. *Sustainability* **2018**, *10*, 3735. [CrossRef]
8. Bazile, D.; Bertero, H.; Nieto, C. *Estado del Arte de La Quinua en El Mundo en 2013*; FAO: Santiago, Chile; CIRAD: Montpellier, France, 2014; pp. 1–724.

9. Bocchi, S.; Cinquanta, D.; Negri, M.; Dioli, P.; Limonta, L. *Nysius cymoides* (Spinola) on *Chenopodium quinoa* Willd. cultivated in Italy. *J. Entomol. Acarol. Res.* **2016**, *48*, 332–334. [CrossRef]
10. Latorre, J. Is Quinoa Cultivation on the Coastal Desert of Peru Sustainable? A Case Study from Majes, Arequipa. Master's Thesis, Aarhus University, Aarhus, Denmark, 2017.
11. Cruces, L.; de la Peña, E.; De Clercq, P. Insect diversity associated with quinoa (*Chenopodium quinoa* Willd.) in three altitudinal production zones of Peru. *Int. J. Trop. Insect Sci.* **2020**, *40*, 955–968. [CrossRef]
12. Jacobsen, S. The scope for adaptation of quinoa in northern latitudes of Europe. *J. Agron. Crop Sci.* **2017**, *203*, 603–613. [CrossRef]
13. Torres, J.B.; Bueno, A.D. Conservation biological control using selective insecticides–a valuable tool for IPM. *Biol. Control* **2018**, *126*, 53–64. [CrossRef]
14. Naumann, K. *Synthetic Pyrethroid Insecticides: Chemistry and Patents*; Springer: Heidelberg, Germany, 2012; pp. 1–390.
15. Desneux, N.; Decourtye, A.; Delpuech, J.M. The sublethal effects of pesticides on beneficial arthropods. *Annu. Rev. Entomol.* **2007**, *52*, 81–106. [CrossRef]
16. Stanley, J.; Preetha, G. *Pesticide Toxicity to Non-Target Organisms: Exposure, Toxicity and Risk Assessment Methodologies*; Springer: Berlin, Germany, 2016; pp. 1–502.
17. Zhu, Y.C.; Yao, J.; Adamczyk, J.; Luttrell, R. Feeding toxicity and impact of imidacloprid formulation and mixtures with six representative pesticides at residue concentrations on honey bee physiology (*Apis mellifera*). *PLoS ONE* **2017**, *12*, e0178421. [CrossRef]
18. Zhu, Y.C.; Yao, J.; Adamczyk, J.; Luttrell, R. Synergistic toxicity and physiological impact of imidacloprid alone and binary mixtures with seven representative pesticides on honey bee (*Apis mellifera*). *PLoS ONE* **2017**, *12*, e0176837. [CrossRef]
19. Dadther-Huaman, H.; Machaca-Paccara, A.; Quispe-Castro, R. Eficacia de nueve métodos de control de *Oregmopyga peruviana* (Granara de Willink & Diaz) (Hemiptera: Coccoidea: Eriococcidae) en *Vitis vinifera* L.'Negra Criolla' y 'Quebranta'. *Sci. Agropecu.* **2020**, *11*, 95–103.
20. Cáceres-del Carpio, C.; Alexandra, F.; Iannacone, J. Evaluación del riesgo ambiental por los insecticidas fipronil e imidacloprid en el camarón de río (*Cryphiops caementarius*). La Granja. *Rev. Cienc. Vida* **2021**, *33*, 104–114.
21. Jeschke, P.; Moriya, K. Five-membered Neonicotinoids: Imidacloprid and Thiacloprid. In *Modern Crop Protection Compounds*; Jeschke, P., Witschel, M., Krämer, W., Schirmer, U., Eds.; Wiley-VCH: Weinheim, Germany, 2019; Volume 3, pp. 1293–1308.
22. Prabhaker, N.; Castle, S.J.; Naranjo, S.E.; Toscano, N.C.; Morse, J.G. Compatibility of two systemic neonicotinoids, imidacloprid and thiamethoxam, with various natural enemies of agricultural pests. *J. Econ. Entomol.* **2011**, *104*, 773–781. [CrossRef]
23. El-Naggar, J.B.; Zidan, N.E. Field evaluation of imidacloprid and thiamethoxam against sucking insects and their side effects on soil fauna. *J. Plant Prot. Res.* **2013**, *53*, 375–387. [CrossRef]
24. Douglas, M.R.; Tooker, J.F. Meta-Analysis reveals that seed-applied neonicotinoids and pyrethroids have similar negative effects on abundance of arthropod natural enemies. *PeerJ* **2016**, *4*, e2776. [CrossRef] [PubMed]
25. Calvo-Agudo, M.; González-Cabrera, J.; Picó, Y.; Calatayud-Vernich, P.; Urbaneja, A.; Dicke, M.; Tena, A. Neonicotinoids in excretion product of phloem-feeding insects kill beneficial insects. *Proc. Natl. Acad. Sci. USA* **2019**, *116*, 16817–16822. [CrossRef] [PubMed]
26. Ricupero, M.; Desneux, N.; Zappalà, L.; Biondi, A. Target and non-target impact of systemic insecticides on a polyphagous aphid pest and its parasitoid. *Chemosphere* **2020**, *247*, 125728. [CrossRef]
27. Motaung, T.E. Chloronicotinyl insecticide imidacloprid: Agricultural relevance, pitfalls and emerging opportunities. *Crop Prot.* **2020**, *131*, 105097. [CrossRef]
28. Gasparic, H.V.; Grubelic, M.; Uzelac, V.D.; Bazok, R.; Cacija, M.; Drmic, Z.; Lemic, D. Neonicotinoid residues in sugar beet plants and soil under different agro-climatic conditions. *Agriculture* **2020**, *10*, 484. [CrossRef]
29. Jeschke, P.; Witschel, M.; Krämer, W.; Schirmer, U. *Modern Crop Protection Compounds*; Wiley-VCH: Weinheim, Germany, 2019; Volume 3, pp. 989–1642.
30. El-Wakeil, N.; Gaafar, N.; Sallam, A.; Volkmar, C. Side effects of insecticides on natural enemies and possibility of their integration in plant protection strategies. In *Insecticides-Development of Safer and More Effective Technologies*; Trdan, S., Ed.; InTech: Rijeka, Croatia, 2013; pp. 4–56.
31. Spomer, N.; Sheets, J. Chitin biosynthesis and inhibitors. In *Modern Crop Protection Compounds*; Jeschke, P., Witschel, M., Krämer, W., Schirmer, U., Eds.; Wiley-VCH: Weinheim, Germany, 2019; Volume 3, pp. 1067–1084.
32. Ishaaya, I.; Degheele, D. *Insecticides with Novel Modes of Action: Mechanisms and Application*; Springer: Heidelberg, Germany, 1998; pp. 1–289.
33. European Food Safety Authority. Conclusion on the Peer Review of the Pesticide Risk Assessment of the Active Substance Emamectin. *EFSA J.* **2012**, *10*, 1–89.
34. Jansson, R.K.; Dybas, R.A. Avermectins: Biochemical mode of action, biological activity and agricultural importance. In *Insecticides with Novel Modes of Action: Mechanisms and Application*; Ishaaya, I., Degheele, D., Eds.; Springer: Berlin, Germany, 1998; pp. 152–170.
35. Ishaaya, I.; Barazani, A.; Kontsedalov, S.; Horowitz, A.R. Insecticides with novel modes of action: Mechanism, selectivity and cros-resistance. *Entomol. Res.* **2007**, *37*, 148–152. [CrossRef]
36. Pitterna, T. Glutamate-gated chloride channel allosteric modulators: Avermectins and milbemycins. In *Modern Crop Protection Compounds*; Jeschke, P., Witschel, M., Krämer, W., Schirmer, U., Eds.; Wiley-VCH: Weinheim, Germany, 2019; Volume 3, pp. 1478–1501.

37. Brown, M.W.; Adler, C.R. Community Structure of phytophagous arthropods on apple. *Environ. Entomol.* **1989**, *18*, 600–607. [CrossRef]
38. Altieri, M.A. The ecological role of biodiversity in agroecosystems. *Agric. Ecosyst. Environ.* **1999**, *74*, 19–31. [CrossRef]
39. Letourneau, D.K.; Goldstein, B. Pest damage and arthropod community structure in organic vs. conventional tomato production in California. *J. Appl. Ecol.* **2001**, *38*, 557–570. [CrossRef]
40. Suttman, C.E.; Barrett, G.W. Effects of sevin on arthropods an agricultural and old-field plant community. *Ecology* **1979**, *60*, 628–641. [CrossRef]
41. Crowder, D.W.; Jabbour, R. Relationships between biodiversity and biological control in agroecosystems: Current status and future challenges. *Biol. Control* **2014**, *75*, 8–17. [CrossRef]
42. Montoya, J.M.; Pimm, S.L.; Solé, R. Ecological networks and their fragility. *Nature* **2006**, *442*, 259–264. [CrossRef]
43. Tylianakis, J.M.; Tscharntke, T.; Lewis, O.T. Habitat modification alters the structure of tropical host–parasitoid food webs. *Nature* **2007**, *455*, 202–205. [CrossRef]
44. Youming, H.; Xiongfei, P.; Guangwen, L.; Minsheng, Y. Effect of chemical insecticides on the diversity of arthropods in vegetable fields. *Acta Ecol. Sin.* **2001**, *21*, 1262–1268.
45. Clarke-Harris, D.; Fleischer, S.J.; Fuller, C.; Bolton, J. Evaluation of the efficacy of new chemistries for controlling major Lepidoptera pests on vegetable amaranth in Jamaica. *CARDI Rev.* **2004**, *4*, 12–19.
46. Shivankar, S.B.; Magar, S.B.; Shinde, V.D.; Yadav, R.G.; Patil, A.S. Field bio-efficacy of chemical, botanical and bio-pesticides against *Spodoptera litura* Fab. in sugar beet. *Ann. Plant Prot. Sci.* **2008**, *16*, 312–315.
47. Manjula, K.N.; Kotikal, Y.K. Evaluation of insecticides against *Agrotis segetum* (Denis and Schiffermuller) and *Spoladea recurvalis* (Fabricius) on fenugreek, *Trigoniella foenumgraecum* L. *J. Entomol. Zool. Stud.* **2018**, *6*, 1177–1182.
48. Mead, H.M.; Khedr, M.M. Role of teflubenzuron as a chitin synthesis inhibitor against *Spodoptera littoralis* larvae. *Egypt Acad. J. Biol. Sci.* **2018**, *10*, 49–58. [CrossRef]
49. Muralikrishna, P.; Mathew, T.B.; Paul, A.; Nithya, P.R. Evaluation of bio-efficacy of new generation insecticides, botanicals and microbial insecticides on leaf webber of amaranth. *J. Entomol. Zool. Stud.* **2019**, *7*, 516–520.
50. Ishaaya, I.; Navon, A.; Gurevitz, E. Comparative toxicity of chlorfluazuron (IKI-7899) and cypermethrin to *Spodoptera littoralis, Lobesia botrana* and *Drosophila melanogaster*. *Crop. Prot.* **1986**, *5*, 385–388. [CrossRef]
51. Neuen, R. Behaviour modifying effects of low systemic concentrations of imidacloprid on *Myzus persicae* with special reference to an antifeeding response. *Pestic. Sci.* **1995**, *44*, 145–153. [CrossRef]
52. Devine, G.J.; Harling, Z.K.; Scarr, A.W.; Devonshire, A.L. Lethal and sublethal effects of imidacloprid on nicotine-tolerant *Myzus nicotianae* and *Myzus persicae*. *Pestic. Sci.* **1996**, *48*, 57–62. [CrossRef]
53. Mohammed, A.A.; Desneux, N.; Fan, Y.; Han, P.; Ali, A.; Song, D.; Gao, X. Impact of imidacloprid and natural enemies on cereal aphids: Integration or ecosystem service disruption? *Entomol. Gen.* **2018**, *37*, 47–61. [CrossRef]
54. Elbert, A.; Nauen, R.; Leicht, W. Imidacloprid, a novel chloronicotinyl insecticide: Biological activity and agricultural importance. In *Insecticides with Novel Modes of Action: Mechanisms and Application*; Ishaaya, I., Degheele, D., Eds.; Springer: Heidelberg, Germany, 1998; pp. 50–73.
55. Sánchez-Bayo, F.; Tennekes, H.A.; Goka, K. Impact of systemic insecticides on organisms and ecosystems. In *Insecticides-Development of Safer and More Effective Technologies*; Trdan, S., Ed.; InTech: Rijeka, Croatia, 2013; pp. 365–414.
56. Chakraborty, S.; Chatterjee, M.L. Effect of four benzophenylureas on population of safflower aphid, *Dactynotus carthami* HRL. and lady bird predators, *Coccinella Septumpunctata* L. and *Coccinella* sp. *Indian J. Exp. Biol.* **1999**, *37*, 374–378.
57. Jain, P.; Singh, S.B.; Borban, K.; Badaya, A.K. Bio-efficacy of novel insecticides against chilli aphid, *Aphis gossypii* Glover and thrips, *Scirtothrips dorsalis* Hood in Malwa Region of Madhya Pradesh. *Ann. Plant Soil. Res.* **2018**, *20*, 172–177.
58. Pons, X.; Albajes, R. Density of epigeal predators on maize plants untreated and treated with imidacloprid. *IOBC WPRS Bull.* **2001**, *24*, 73–78.
59. Mills, N. Interspecific competition among natural enemies and single versus multiple introductions in biological control. In *Trophic and Guild in Biological Interactions Control*; Brodeur, J., Boivin, G., Eds.; Springer: Dordrecht, The Netherlands, 2006; pp. 191–220.
60. Varenhorst, A.J.; O'Neal, M.E. The response of natural enemies to selective insecticides applied to soybean. *Environ. Entomol.* **2012**, *41*, 1565–1574. [CrossRef] [PubMed]
61. Roubos, C.R.; Rodriguez-Saona, C.; Holdcraft, R.; Mason, K.S.; Isaacs, R. Relative toxicity and residual activity of insecticides used in blueberry pest management: Mortality of natural enemies. *J. Econ. Entomol.* **2014**, *107*, 277–285. [CrossRef] [PubMed]
62. FAO. *Guidelines on Efficacy Evaluation for the Registration of Plant Protection Products*; FAO Publications: Rome, Italy, 2006; pp. 1–61.
63. Hassan, S.A. Standard Methods to Test the Side-effects of Pesticides on Natural Enemies of Insects and Mites Developed by the IOBC/WPRS Working Group 'Pesticides and Beneficial Organisms'. *Eppo Bull.* **1985**, *15*, 214–255. [CrossRef]
64. Soca, N. Fluctuación Poblacional de Insectos Fitófagos Asociados al Cultivo de Quinua (*Chenopodium quinoa* Willd.) En La Molina. Master's Thesis, Universidad Nacional Agraria La Molina, Lima, Peru, 2021.
65. Gómez, L.; Aguilar, E. *Guía de Cultivo de La Quinua*; Organización de las Naciones Unidas para la Alimentación y la Agricultura, Universidad Nacional Agraria La Molina: Lima, Peru, 2016.
66. Oliver, I.; Beattie, A.J. A possible method for the rapid assessment of biodiversity. *Conserv. Biol.* **1993**, *7*, 562–568. [CrossRef]

67. Oliver, I.; Beattie, A.J. Invertebrate morphospecies as surrogates for species: A case study. *Conserv. Biol.* **1996**, *10*, 99–109. [CrossRef]
68. Triplehorn, C.A.; Johnson, N.F. *Borror and DeLong's Introduction to the Study of Insects*; Thompson Brooks/Cole: Belmont, CA, USA, 2005; pp. 1–864.
69. Moret, P. Contribution à la connaissance du genre néotropical *Blennidus* Motschulsky, 1865. *Bull. Soc. Entomol. Fr.* **1995**, *100*, 489–500.
70. Moret, P. Clave de identificación para los géneros de Carabidae (Coleoptera) presentes en Los Páramos del Ecuador y del sur de Colombia. *Rev. Colomb. Entomol.* **2003**, *29*, 185–190.
71. Göllner-Scheiding, U. Revision der gattung *Liorhyssus* Stål, 1870 (Heteroptera, Rhopalidae). *Dtsch. Entomol. Z.* **1976**, *23*, 181–206. [CrossRef]
72. Pall, J.P.; Kihn, R.G.; Diez, F.; Coscaron, M. A review of genus *Nysius* Dallas in Argentina (Hemiptera: Heteroptera: Orsillidae). *Zootaxa* **2016**, *4132*, 221–234. [CrossRef]
73. Korytkowski, C. Contribución al conocimiento de los Agromyzidae (Diptera: Muscomorpha) en el Perú. *Rev. Entomol.* **2014**, *49*, 1–106.
74. Spencer, K.A. *Agromyzidae (Diptera) of Economic Importance.*; Springer Science & Business Media: Dordrecht, The Netherlands, 1973; pp. 1–418.
75. Hernández, L.M.; Henry, T.J. *The Plant Bugs, or Miridae (Hemiptera: Heteroptera), of Cuba*; Pensoft: Sofia, Bulgaria, 2010; pp. 1–212.
76. Gross, G. The Stilt-Bugs (Heteroptera-Neididae) of the Australian and New Zealand Regions. *Rec. South Aust. Mus.* **1950**, *9*, 13–1326.
77. Henry, T.; Dellapé, P.; de Paula, A. The Big-Eyed Bugs, Chinch Bugs, and Seed Bugs (Lygaeoidea). In *True Bugs (Heteroptera) of the Neotropics*; Panizzi, A.R., Grazia, J., Eds.; Springer: Dordrecht, The Netherlands, 2015; pp. 459–514.
78. Kerzhner, I.M.; Henry, T.J. Three new species, notes and new records of poorly known species, and an updated checklist for the North American Nabidae (Hemiptera: Heteroptera). *Proc. Entomol. Soc. Wash.* **2008**, *110*, 988–1011. [CrossRef]
79. Cornelis, M.; Coscarón, M.C. The Nabidae (Insecta, Hemiptera, Heteroptera) of Argentina. *ZooKeys* **2013**, *333*, 1–30. [CrossRef] [PubMed]
80. Cornelis, M. Biodiversidad de Nabidae (Insecta: Heteroptera): Revisión Taxonómica y Análisis Cladístico del Género Nabis Latreille, 1802. Ph.D.Thesis, Universidad Nacional de La Plata, La Plata, Argentina, 2015.
81. Solis, M.A. Key to Selected Pyraloidea (Lepidoptera) Larvae Intercepted at US Ports of Entry: Revision of Pyraloidea in "Keys to Some Frequently Intercepted Lepidopterous Larvae" by Weisman 1986. USDA Systematic Entomology Laboratory 2006. Available online: https://digitalcommons.unl.edu/systentomologyusda/1/ (accessed on 1 January 2020).
82. Shufran, K.A.; Puterka, G.J. DNA Barcoding to identify all life stages of holocyclic cereal aphids (Hemiptera: Aphididae) on wheat and other poaceae. *Ann. Entomol. Soc. Am.* **2011**, *104*, 39–42. [CrossRef]
83. Nakamura, S.; Masuda, T.; Mochizuki, A.; Konishi, K.; Tokumaru, S.; Ueno, K.; Yamaguchi, T. Primer design for identifying economically important *Liriomyza* species (Diptera: Agromyzidae) by multiplex PCR. *Mol. Ecol. Resour.* **2013**, *13*, 96–102. [CrossRef]
84. Harbhajan, K.; Kaur, S. DNA barcoding of six species of family Rhopalidae (Insecta: Hemiptera: Heteroptera) from India. *Int. J. Life Sci.* **2017**, *5*, 517–526.
85. Ding, T.; Chi, H.; Gökçe, A.; Gao, Y.; Zhang, B. Demographic analysis of arrhenotokous parthenogenesis and bisexual reproduction of *Frankliniella occidentalis* (Pergande) (Thysanoptera: Thripidae). *Sci. Rep.* **2018**, *8*, 3346. [CrossRef]
86. Nguyen, V.H.; Jonckheere, W.; Nguyen, D.T.; de Moraes, G.J.; Van Leeuwen, T.; De Clercq, P. Phytoseiid mites prey effectively on thrips eggs: Evidence from Predation trials and molecular analyses. *Biol. Control* **2019**, *137*, 104012. [CrossRef]
87. R Core Team. *R: A Language and Environment for Statistical Computing. R Foundation for Statistical Computing*; R Core Team: Vienna, Austria, 2017; Available online: http://www.R-project.org/ (accessed on 30 June 2020).

plants

MDPI

Review

Quinoa Phenotyping Methodologies: An International Consensus

Clara S. Stanschewski [1], Elodie Rey [1], Gabriele Fiene [1], Evan B. Craine [2], Gordon Wellman [1], Vanessa J. Melino [1], Dilan S. R. Patiranage [1,3], Kasper Johansen [4], Sandra M. Schmöckel [5], Daniel Bertero [6], Helena Oakey [7], Carla Colque-Little [8], Irfan Afzal [9], Sebastian Raubach [10], Nathan Miller [11], Jared Streich [12], Daniel Buchvaldt Amby [8], Nazgol Emrani [3], Mark Warmington [13], Magdi A. A. Mousa [14,15], David Wu [16], Daniel Jacobson [12], Christian Andreasen [8], Christian Jung [3], Kevin Murphy [2], Didier Bazile [17,18], Mark Tester [1,*] and on behalf of the Quinoa Phenotyping Consortium †

1 Center for Desert Agriculture, Biological and Environmental Sciences and Engineering Division, King Abdullah University of Science and Technology (KAUST), Thuwal 23955, Saudi Arabia; clara.stanschewski@kaust.edu.sa (C.S.S.); elodie.rey@kaust.edu.sa (E.R.); gabriele.fiene@kaust.edu.sa (G.F.); gordon.wellman@kaust.edu.sa (G.W.); vanessa.melino@kaust.edu.sa (V.J.M.); d.sarange@plantbreeding.uni-kiel.de (D.S.R.P.)

2 Department of Crop and Soil Sciences, Washington State University, Pullman, WA 99164, USA; evan.craine@wsu.edu (E.B.C.); kmurphy2@wsu.edu (K.M.)

3 Plant Breeding Institute, Christian-Albrechts-University of Kiel, 24118 Kiel, Germany; n.emrani@plantbreeding.uni-kiel.de (N.E.); c.jung@plantbreeding.uni-kiel.de (C.J.)

4 Water Desalination and Reuse Center, King Abdullah University of Science and Technology (KAUST), Thuwal 23955, Saudi Arabia; kasper.johansen@kaust.edu.sa

5 Department Physiology of Yield Stability, Institute of Crop Science, University of Hohenheim, 70599 Stuttgart, Germany; sandra.schmoeckel@uni-hohenheim.de

6 Department of Plant Production, School of Agriculture, University of Buenos Aires, Buenos Aires C1417DSE, Argentina; bertero@agro.uba.ar

7 Robinson Research Institute, Adelaide Medical School, University of Adelaide, Adelaide, SA 5005, Australia; helena.oakey@adelaide.edu.au

8 Department of Plant and Environmental Sciences, University of Copenhagen, DK-2630 Taastrup, Denmark; cxl@plen.ku.dk (C.C.-L.); amby@plen.ku.dk (D.B.A.); can@plen.ku.dk (C.A.)

9 Department of Agronomy, University of Agriculture, Faisalabad 38000, Pakistan; irfanuaf@gmail.com

10 Department of Information and Computational Sciences, The James Hutton Institute, Invergowrie, Dundee AB15 8QH, UK; Sebastian.Raubach@hutton.ac.uk

11 Department of Botany, University of Wisconsin, 430 Lincoln Dr, Madison, WI 53706, USA; ndmill@gmail.com

12 Biosciences, Oak Ridge National Laboratory, Oak Ridge, TN 37831, USA; streichjc@ornl.gov (J.S.); jacobsonda@ornl.gov (D.J.)

13 Department of Primary Industries and Regional Development, Agriculture and Food, Kununurra, WA 6743, Australia; Mark.Warmington@dpird.wa.gov.au

14 Department of Arid Land Agriculture, Faculty of Meteorology, Environment and Arid Land Agriculture, King Abdulaziz University, Jeddah 21589, Saudi Arabia; mamousa@kau.edu.sa

15 Department of Vegetables, Faculty of Agriculture, Assiut University, Assiut 71526, Egypt

16 Shanxi Jiaqi Agri-Tech Co., Ltd., Taiyuan 030006, China; quinoaking@vip.163.com

17 CIRAD, UMR SENS, 34398 Montpellier, France; didier.bazile@cirad.fr

18 SENS, CIRAD, IRD, University Paul Valery Montpellier 3, 34090 Montpellier, France

* Correspondence: mark.tester@kaust.edu.sa

† Quinoa Phenotyping Consortium: Abdalla Dao [a], Abdelaziz Hirich [b], Achim Präger [c], Ada Luz Flores [d], Alipio Canahua Murillo [e], Amr S. Shams [f], Andrés Zurita-Silva [g], Arafet Manaa [h], Cataldo Pulvento [i], Christos Noulas [j], David E. Jarvis [k], Eric N. Jellen [k], Ernesto J. Chura [e], Hatem Ben Jouira [h], J.A. González [l], Karina B. Ruiz [m], Karl Schmid [c], Lavini Antonella [n], Luis Erazzú [o], Luz Gomez-Pando [p], Malia A. Gehan [q], Miguel García-Parra [r], Mohamed Houssemeddine Sellami [o], Olivia Eisner [s], Peter J. Maughan [k], Rakesh Kumar Singh [t], Ritva Repo-Carrasco-Valencia [p], Sergey Shabala [u], Simone Graeff-Hönninger [c], Sven-Erik Jacobsen [v], Ursula Steinfort [w], Walter Quadros Ribeiro Junior [x], Yuanyuan Li [y] [a] Institut de l'Environnement et Recherches Agricoles (INERA), Burkina Faso; [b] African Sustainable Agriculture Research Institute (ASARI), Mohammed VI Polytechnic University (UM6P), Laayoune, Morocco; [c] University of Hohenheim, Faculty of Agriculture, Stuttgart, Germany; [d] Universidad Andina Néstor Cáceres Velásquez, Juliaca, Peru; [e] Universidad Nacional del Altiplano, Puno, Peru; [f] Agricultural Research Center (ARC), Giza, Egypt; [g] Instituto de Investigaciones Agropecuarias, INIA Intihuasi, Colina San Joaquín s/n, Casilla 36-B, La Serena, Chile; [h] Laboratory of Extremophile Plants, Centre of Biotechnology of Borj Cedria, Hammam Lif, Tunisia; [I] University of Bari, Department of Agricultural and Environmental Science, Italy; [j] Institute of Industrial and

Citation: Stanschewski, C.S.; Rey, E.; Fiene, G.; Craine, E.B.; Wellman, G.; Melino, V.J.; Patiranage, D.S.R.; Johansen, K.; Schmöckel, S.M.; Bertero, D.; et al. Quinoa Phenotyping Methodologies: An International Consensus. *Plants* **2021**, *10*, 1759. https://doi.org/10.3390/plants10091759

Academic Editor: Ioannis Ganopoulos

Received: 5 July 2021
Accepted: 12 August 2021
Published: 24 August 2021

Forage Crops, Hellenic Agricultural Organization "Demeter", Larissa, Greece; [k] Plant and Wildlife Sciences Department, Brigham Young University, Provo, UT, USA; [l]Fundación Miguel Lillo, Instituto de Ecologí[a], Comportamiento y Conservación, Tucumán, Argentina; [m] Química y Farmacia, Facultad de Ciencias de la Salud, Universidad Arturo Prat, Iquique, Chile; [n] National Research Council of Italy (CNR), Institute for Agricultural and Forestry Systems in the Mediterranean (ISAFOM), Portici, NA, Italy; [o] National Institute of Agricultural Technology, INTA EEA Famaillá, Tucumán, Argentina; [p] Universidad Nacional Agraria La Molina, Lima, Peru; [q] Donald Danforth Plant Science Center, Saint Louis, Missouri, USA; [r] Department Agricultural Science, Cauca University, Research Group Agriculture, Organizations and Fruits AOF-JDC, Colombia; [s] Private Quinoa Founder, Argentina; [t] International Center for Biosaline Agriculture (ICBA), Dubai, UAE; [u] Tasmanian Institute of Agriculture, University of Tasmania, Hobart, Australia; [v] Quinoa Quality ApS, Regstrup, Denmark; [w] Department of Plant Sciences, Faculty of Agronomy and Forestry Engineering, Pontificia Universidad Católica de Chile, Santiago, Chile; [x] Empresa Brasileira de Pesquisa Agropecuária (EMBRAPA) Cerrados, Brazil, [y] CAS Center for Excellence in Molecular Plant Sciences, Institute of Plant Physiology and Ecology, Chinese Academy of Sciences, Shanghai, China.

Abstract: Quinoa is a crop originating in the Andes but grown more widely and with the genetic potential for significant further expansion. Due to the phenotypic plasticity of quinoa, varieties need to be assessed across years and multiple locations. To improve comparability among field trials across the globe and to facilitate collaborations, components of the trials need to be kept consistent, including the type and methods of data collected. Here, an internationally open-access framework for phenotyping a wide range of quinoa features is proposed to facilitate the systematic agronomic, physiological and genetic characterization of quinoa for crop adaptation and improvement. Mature plant phenotyping is a central aspect of this paper, including detailed descriptions and the provision of phenotyping cards to facilitate consistency in data collection. High-throughput methods for multi-temporal phenotyping based on remote sensing technologies are described. Tools for higher-throughput post-harvest phenotyping of seeds are presented. A guideline for approaching quinoa field trials including the collection of environmental data and designing layouts with statistical robustness is suggested. To move towards developing resources for quinoa in line with major cereal crops, a database was created. The Quinoa Germinate Platform will serve as a central repository of data for quinoa researchers globally.

Keywords: *Chenopodium quinoa*; descriptors; genetic diversity; scoring card; architecture; panicle; disease; high throughput seed phenotyping; remote sensing; database

1. Introduction

Food systems are experiencing intense pressure owing to, among other factors, increasing population and environmental change (such as increases in the frequency and severity of extreme weather events). Changes in environmental conditions are causing changes in where particular crops can be cultivated as well as the types of crops that can be planted in affected areas [1,2]. Uncertainty in climate and weather predictions highlight the need for crops and varieties that are stable across time and regions. Diversification of the crops grown is also important as increasing yield losses are projected to pose a serious threat to food security [3]. These challenges will become ever more paramount with a growing global population. Apart from the concerns of undernourishment in our population, many are not receiving adequate amounts or diversity of micronutrients in their diet; malnutrition is referred to as "hidden hunger" [4]. Dietary intake has been identified as a key factor in the treatment and prevention of numerous non-communicable diseases [5]. The threat posed by climate change to food security, in addition to the concerns of human nutrition, highlights the urgent need for diversification of the food system [6]. Numerous strategies may be used to respond to both hunger and hidden hunger. One such strategy is the reintroduction of genetic diversity to fully domesticated crops to increase variability in the genes responsible for the environmental adaptability, plasticity, and resilience that their wild ancestors still possess. An alternative approach lies in the domestication of plants not yet used in agriculture at the global level and focus breeding efforts in Neglected and Underutilized Species and plants that are only partially domesticated (orphan crops) and those that still harbor considerable genetic variability, which may contribute to improved yield and adaptability. An example of the latter approach is the use of quinoa (*Chenopodium quinoa* Willd.), a crop providing high nutritional content that has become increasingly popular in recent years.

Quinoa has not been domesticated to the same extent as the other major grain crops for global food and agriculture such as wheat, maize, or rice. Because of its cultivation in various agro-ecological conditions over the last few millennia, including mesothermal zones, highlands, salt flats, and subtropical zones of the Andes, quinoa experiences a wide range of challenging environmental conditions, resulting in high genetic diversity and tolerance to a range of biotic and abiotic stresses of this crop [7,8]. Furthermore, the seed is gaining popularity for its high nutritional value, an important feature of the UN's Sustainable Development Goals [9]. Despite the ability of quinoa to maintain yield under a wide range of environments and the growing worldwide interest in the crop, the primary locations for cultivation are still the countries that have a tradition of growing quinoa: Bolivia, Peru, Colombia, Ecuador, Argentina, and Chile [10]. In recent years, quinoa has been more widely cultivated at sites such as Europe, North America, and China [11]. A primary strategy developed to meet the increasing food demand is the expansion of cultivation through the identification of adaptable and high yielding quinoa cultivars for different agro-ecological zones [12]. Seed nutritional contents and interannual stability of cultivars are traits of much importance in quinoa [13]. Local production will reduce food miles, thereby reducing transportation costs and potentially environmental impacts of quinoa consumption. To achieve this aim, insights into the broad performance spectrum of different varieties grown in a range of environments as well as an improved understanding of the genes underlying traits of agronomic importance are needed.

Recently, genomic tools have emerged to support the development of quinoa germplasm for novel environments. Several novel genomic resources have been developed in the past decade, including bacterial artificial chromosome (BAC), expressed sequence tag (EST) libraries, and DNA-based molecular markers (see [14]), as well as, more recently, a chromosome-scale reference genome sequence of a coastal (or "lowland") quinoa accession (BioSample Accession Code QQ74) [15] and the resequencing of several wild and cultivated quinoas. Together, these are helping develop genomics-informed breeding programs and genetic studies to accelerate crop improvement [14]. Phenomics (i.e., high-throughput phenotyping) are currently a limiting factor in genetic analyses and genomic prediction, after recent advances in throughput and reduced genotyping costs realized over the past two decades [16]. Quinoa phenotyping strategies, however, have not been standardized. This is one of the limiting factors in the common characterization of quinoa and assessment of its adaptation to different environments, and which limits advancements in quinoa genetic research.

Yield is an important but complex trait in plant breeding because it embodies the link between the cumulative effects of all plant traits and economic value of the crop [17]. Owing to the additive and genetically complex nature of yield, the genetic architecture of yield cannot be easily analyzed, and is usually best analyzed through secondary traits that contribute to yield, such as harvest index and other developmental traits, and photosynthetic parameters and other physiological traits (see [18]). Crop improvement efforts through indirect selection that focuses on secondary traits which are correlated with yield but are potentially monogenic have long been proposed for crops [19,20] and have been successfully used for sunflower [21,22], and many other examples [23–29]. The assessment of differences in the response of genotypes to different environmental conditions facilitates the development of genotypes that lead to improved phenotypes in a specific environment [30] and also allow insights into the genetic architecture of a trait. To gain insights into complex traits, a study of large numbers of genotyped accessions across multiple environments is required to identify genotype-by-environment (G $\times$ E) interactions [31]. This is particularly important in quinoa, due to the large G $\times$ E interactions that have been reported [32,33]. For the generation of varieties for different situations, G × E interactions also need to be considered in association with agricultural practices that are shaped by growers and societies. Building collaborations between breeders and growers is important for the adoption of new varieties [34,35].

It is often difficult to determine the pattern of quinoa genotypic responses across environments. The biplot analysis may also provide a powerful solution to alleviate this problem. For effective cultivar evaluation, both the effect of genotype (G) and the interaction G × E must be considered simultaneously [36,37]. A G plus G × E (GGE) biplot was shown to effectively identify the G × E interaction pattern of the data and to show clearly which genotype won in which environments. In addition, the GGE biplot technique is useful in selecting superior genotypes and test environments for a given mega-environment; that is, a group of locations that consistently share the same best genotype or genotypes.

Given the need for data from multiple environments and the high costs incurred from large field trials, the benefits of robust collaborations are clear. In addition to the need for international collaboration for the exchange of genetic material [13], the establishment of an internationally accessible framework for quinoa phenotyping is crucial because it will facilitate the comparison and sharing of data from trials among researchers globally. The previously published "Descriptors for Quinoa and wild relatives" [38] has been a useful guide for the establishment of several phenotypes. We also note the EU Community Plant Variety Office publication protocol for tests on distinctness, uniformity, and stability: *Chenopodium quinoa* [39]. However, there is a need for more detailed explanations of many of the traits, as the current guide leaves scope for interpretation reducing comparability among trials. A lack of good guidelines for the recording of traits or a margin for interpretation, results in uncertainty around data sets that could be reduced through standardization and clear definitions.

Based on our experiences over the past six years since the publication of these descriptors with the International Year of Quinoa, we have identified several useful phenotypes covering the variability expressed by over 1000 quinoa genotypes across multiple environments (including some in which quinoa has not been cultivated extensively) and have adapted the description of some traits to clarify the definitions of the traits. We present this information in this article, resulting from discussion and general agreements among the Quinoa Phenotyping Consortium. Furthermore, this article provides guidelines for the entire process of a field trial, starting with the experimental design, providing advice on crop management decisions, and detailing the minimum environmental data which must be collected. Without standardized metadata information about a trial, experiments cannot be replicated, even when phenotyping standards are followed [40].

A consensus on phenotyping methods is presented in this article, starting with phenotyping methods that are performed throughout the growing seasons, including the international standard on the recording of phenological growth stages [41] and options for high-throughput phenotyping using remote sensing. This section is followed by detailed descriptions of the traits that can be assessed in mature plants. These traits are also summarized in phenotyping cards (Supplementary file S1) to aid phenotyping in the field. Next, we focus on the process of harvesting and describe post-harvest phenotyping options and methods. The next step in facilitating the creation of a collaborative network of trials is a platform that allows the viewing of trials undertaken by different researchers globally, through which data can then be easily shared and analyzed. This structure is provided by the quinoa Germinate platform. Thus, this article presents the collective efforts of a large number of quinoa researchers, reflecting the experience gained over many years of working with this crop, and aimed to establish a baseline to approach field trials and move toward an era of accelerated discoveries in the global quinoa research community. The traits presented herein represent the phenotyping scales that are currently in use for exploring the natural diversity across different environments. As quinoa research is progressing, these scales will need to be adapted to match new situations and applications. For this, we are aiming for the Quinoa Phenotyping Consortium to hold an annual meeting to refine the guidelines and procedures with the aim of both increasing the quality and standardization of phenotyping.

2. Quinoa Database

The establishment of an international consensus for our methods in phenotyping quinoa field trials is the first important step in driving international quinoa research forward through allowing comparisons among field trials and facilitating collaborations. An important component of the phenotyping operation is the recording of all information needed for further exploitation of the data. These records need to be complemented with a platform that allows sharing of data generated with the herein described methodologies and facilitates access to the available datasets.

Contribution of data to this database is of great value to (a) breeders and other people who want to start working with quinoa, (b) other researchers, and (c) to the team that shares their data. Considering the monetary and time cost of field trials, in addition to the phenotypic plasticity of quinoa, access to information on how quinoa varieties perform in different locations is highly valuable in providing decision-making support to growers and individuals who want to start working with quinoa. Access to data from trials in a similar environment would help them to make informed decisions on what varieties might be performing best in their region. For other researchers, as well as the team that contributes the data, the importance of data shared in a database lies in the opportunities for collaborations that it creates. Many analyses require datasets from multiple locations and years. It is through collaborations that the speed of discoveries is accelerated. If a team has not published their data yet and wants to keep it private, it can still be uploaded to the database under a privacy setting through which the data remains accessible only to the person who uploaded it and optionally a defined list of other individuals (i.e., the research team). In addition to these privacy settings, license terms can optionally be assigned to datasets before access to the data is granted. In this case, data would only be accessed after the license has been accepted.

Here, we present a central repository Quinoa Germinate Database. Available online: http://germinate.quinoadb.org (accessed on 15 August 2021), which was created using the Germinate platform [42,43]. The Germinate platform is used by many international organizations, including CIMMYT and data uploaded to Germinate will also become available through the BreedingAPI.

This database allows storing phenotypic and genotypic data as well as germplasm passport data and environmental information. To upload a dataset, users need to first register to the database and data can only be uploaded through the use of standard data templates (https://github.com/germinateplatform/germinate/tree/master/datatemplates accessed on 15 August 2021) to ensure consistency between datasets. For the quinoa community, some of these data templates have been customized to include the information that was highlighted in this paper, including the phenotypes that have been described. These quinoa datasets can be downloaded from the home page of germinate.quinoadb.org. Traits included in the quinoa-traits-data-template.xlsx can be amended to fit your data, but we encourage that all traits uploaded to the Quinoa Germinate Database be trait variables also described in the Quinoa Ontology database of the Crop Ontology (CO_367) of the Generation Challenge Program (GCP) (http://cropontology.org/ accessed on 15 August 2021). Data uploaded using the templates are checked for consistency and correctness by the program before it is imported into the database. A detailed report highlights issues within the data that require fixing before the dataset can be imported. Once added to Germinate, trial data along with climate information can be queried and visualized to discover and highlight patterns and correlations. Data across different trial sites, years, and treatments can be compared to gain a better insight into the effects on performance of germplasm. Customizable lists of germplasm facilitate the comparison of the performance of different accessions, as well as data export of subsets of interest.

All traits described in this manuscript are being uploaded to the Phenotype and Trait Ontologies of the Crop Ontology Curation Tool (https://www.cropontology.org/

accessed on 15 August 2021) to create a Quinoa Ontology database. Any traits measured in Quinoa trials should be included and described in this ontology database. The use of a standard nomenclature for phenotypic traits is another aim in the context of data reproducibility and reusability [44]. The use of standard ontology and recommended methods allows others to quickly interpret and compare data generated from other teams. Hence, we recommend using the described trait variables for quinoa (CO_367), and if a trait is not included in the already described, to add it to the ontology by filing a request (https://www.cropontology.org/add-ontology accessed on 15 August 2021). The ontology spreadsheet can be downloaded from the website and from Quinoa Germinate Database. Available online: http://germinate.quinoadb.org (accessed on 15 August 2021).

The step of using standardized templates and standard quinoa ontologies ensures that datasets from different groups have the same structure and include the information required by another team to use these data in their analyses. Uploaded datasets can be visualized and analyzed using different tools that are available through the platform. The platform's integrated tools for data exploration and analysis facilitate the process from data collection to the next exciting steps for new discoveries about quinoa.

3. Germplasm Selection

The quality of an experiment and its usability in an international framework requires careful consideration and planning, with the first step being the selection of genotypes. Based on the proposal of the Plurinational State of Bolivia, quinoa has been recognized by the United Nations (during its General Assembly in New York, 22/12/2011 [45]) as a potentially valuable crop for future generations. In recognition of the considerable genetic diversity created and maintained by the Andean civilization, the International Year of the Quinoa that was declared, with the FAO in charge as the office of secretary [46]. The exceptional genetic diversity resulted from the cultivation and domestication of quinoa for several thousand years in the harsh and diverse environments of the Andes, combined with the tradition of seed exchange by Andean growers. In 2013, Rojas et al. [7,47] estimated that "16,422 accessions of quinoa and its wild relatives, both closer and more distant (*C. quinoa*, *C. album*, *C. berlandieri*, *C. hircinum*, *C. petiolare*, *C. murale*, and *Chenopodium* sp.) were conserved in 59 genebanks distributed in 30 countries". However, most (88%) of these resources reside in genebanks from the Andean region, for which access is limited to people from those countries [48,49]. Increased international research partnerships with Andean countries is crucial to facilitate access to genetic resources in order to continue quinoa plant breeding and the generation of new varieties which are adapted to new regions. Ex situ collections of 987 and 229 *C. quinoa* accessions are publicly available from the Genebank Information System of the IPK Gatersleben (Available online: https://gbis.ipk-gatersleben.de/ (accessed on 15 August 2021)) and the USDA U.S. National Plant Germplasm System (Available online: https://npgsweb.ars-grin.gov/gringlobal/search.aspx accessed on 15 August 2021)) genebanks, respectively. They include germplasm from 15 countries, mostly the South American countries of Peru, Bolivia, Chile, Ecuador, and Argentina. The extent of duplication within these accessions is uncertain, although likely significant.

To select a panel of accessions representing quinoas genetic diversity, one can, for instance, choose accessions based on their origin, i.e., country/region of origin. Such selection is possible because country/region information is usually documented for accessions available in public genebanks—although it should be noted that the original country of origin can be ambiguous. We found that a substantial proportion of the ex situ collections have more precise information missing about the collection site (region, province, or closest city), preventing them from being selected on the basis of a strict geographical distribution.

Another option for selecting a subset of quinoa genetic diversity is to consider stable morphological characteristics, such as color and shape of leaf, stem, and seed, as a phenotypic passport for each accession. Such data are not always provided by the germplasm provider, and stability is rarely known, so screening of a large, diverse population is usually required before further selection is possible.

A third option is to use the phylogenetic relatedness and population structure information provided by recent genetic diversity analyses performed using either DNA molecular marker screening of the different quinoa populations or, more recently, whole-genome resequencing [50–59]. Whole genome resequencing will be stored in a separate database administered by David Jarvis, BYU (quinoadb.org). Although no comprehensive screening has been performed on the entire collection of quinoa resources available in public gene banks to date, several studies have provided estimations of the genetic diversity of each category of germplasm (in particular, highland vs. coastal (originally designated "sea level": [60]), the relatedness among accessions, and an estimation of the heterozygosity level at given sites within each population. Cumulatively, these parameters are important for establishing a suitable panel of quinoa genotypes for genetic analysis.

There is large genetic variation available in quinoa germplasm with a large reservoir of genes for conferring resistance to biotic and abiotic stresses necessary for quinoa adaptation to challenging and changing environments. Developing new plant breeding methods that can maintain a high level of genetic diversity from the population varieties (usually traditional varieties which are often highly heterogeneous material [61]) may confer more stability of seed yields and a higher resilience to the cropping system [62]. The genetic variability in quinoa germplasm is, however, also driving major challenges in fundamental genetic studies. The traditional practice of admixing, together with the possibility of quinoa to outcross, has resulted in extensive allelic richness translating to a relatively high level of genetic heterozygosity in some populations [50–59]. Owing to this genetic heterozygosity, quinoa genotypes collected in Andean countries and maintained in public gene banks are likely to be highly heterogeneous, necessitating several generations of selfing and single-seed descent to reduce the heterozygosity at potential loci of interest before these materials can be used in genetic studies. Considering the high level of heterozygosity, we recommend a minimum of two generations, and preferably five generations, of self-pollination before initiating any genetic studies to minimize phenotypic heterogeneity and genetic heterozygosity that would interfere with association analyses in genetic studies [63,64]. The use of doubled haploids to obtain homozygosity might also be possible, although this technique has not yet been developed in quinoa. Any accession studied must subsequently be maintained through single-seed descent and given a unique identifier, for it will likely become a selected genotype, genetically distinct from the original material. After several generations and confirmation of homozygosity, seed could then be bulked to facilitate larger trials, in larger plots and/or at multiple locations.

Another consequence of heterogeneity of seeds in public gene banks is the strong segregation and therefore divergence between plants that are likely to be selected for self-pollination for each quinoa accession in different laboratories worldwide. Ideally, this diversity could be counteracted by growing a subset of seeds from each accession and selecting plants for propagation that represent the majority of the accession. In reality, however, the phenotypic plasticity of quinoa resulting from strong genotype (G) × environment (E) interactions causes quinoa accessions with the same genetic background to produce drastically different phenotypes in different environments [32], making the creation of a phenotypic passport for each genotype difficult. Therefore, caution must be maintained when comparing results generated for a single accession from seed stocks maintained by different genebanks worldwide. These accessions should be considered as genotypes and must be clearly identified with an independent identifier because they might evolve over time into a new commercial variety considering the Distinction-Uniformity-Stability descriptors of the UPOV system (as defined by being material that is new, distinct, and uniform).

To maintain the relationship among quinoa accessions in different seed stocks worldwide, we propose that the identification of quinoa germplasm is standardized through the use of Digital Object Identifiers (DOIs), allowing each quinoa accession worldwide to be uniquely identified by an alphanumeric string that is assigned by a registration agency such as the Global Information System. Available online: https://ssl.fao.org/glis/ (accessed on 15 August 2021, which provides a persistent link to the location of information about the object on the Internet. This unique identifier co-exists with other identifiers such as those given by the gene banks and allows for unambiguous and permanent identifications of plant genetic resources, which can be exchanged across organizations, and therefore, facilitate the comparison of results obtained by different teams.

4. Experimental Design and Crop Management

The statistical design of a plant trial experiment will depend on the aim of the study and will dictate the way in which the data from the experiment are analyzed. Plant breeding and genetics rely thoroughly on large-scale variety trials to generate reliable phenotype data and the scale of those grows continuously [65]. For field trials, spatial designs (i.e., planting layouts in the field where genotypes are replicated and their positions along rows and columns are randomized) are recommended because they allow for better estimation of environmental variation in the field and increase the reliability of the experiments. In general, researchers/breeders should aim for balanced experiments to evaluate the germplasm/varieties in multiple locations and years. Statistical procedures based on mixed models are used to compute fixed effects (best linear unbiased estimate) and random effects (best linear unbiased predictions) from the phenotype data generated by field experiments. If the environment-effect on genotype (G × E) is not significant, no or little difference between BLUE and BLUP was observed. However, when G × E is significant, BLUP was superior to BLUE [66,67]. When a treatment is tested, the number of subjects required is double because treatments always need to be tested alongside control conditions, and both should be fully replicated and separately randomized [68]. There are different options to account for technical constraints that might need to be resolved. For example, a split-plot design might need to be used if a treatment factor, e.g., irrigation treatment, can only be applied to a larger area. A good resource for advice on several different designs is the book *Statistical Methods in Biology: Design and Analysis of Experiments and Regression* [68]. The inclusion of genetic relatedness in the design of early generation trials can also be considered for improvements of selection decisions at early stages in breeding programs [69]. Breeders can also optimize breeding programs to improve efficiency and effectiveness with the help of simulation software such as QU-GENE [70], AlphaSim [71], and DeltaGen [72]. Moreover, analysis can be improved into a new stage by incorporation of genotypic (SNP) and phenotypic data to perform genomic selection (GBLUP) and GWAS [65].

In early generation variety trials with large numbers of genotypes and small amounts of seed, it is often not technically feasible to implement a fully replicated design. Partially replicated designs [73], where at least 25% of the genotypes have two replicates, are recommended under such circumstances. If resources permit, fully replicated designs are preferred because the statistical power and confidence in the results increase as the number of replicates of each genotype increase. If breeding values (the estimation of the value that the genes of a variety would have if used as a parent in crosses) [74] are required, a pedigree analysis or a genomic selection approach is needed. Both require generation of a relationship matrix based on either the pedigree or genomic information. Replication in these approaches ensures that the total genetic variation can be partitioned into additive or non-additive variation, which is important for determining breeding values versus commercial values of genotypes [75]; thus, fully replicated trials are preferred. For genome-wide association studies (GWAS) aiming at identifying the genetic variation associated with a particular phenotype, it is important to ensure adequate genetic diversity, allowing enough statistical power to perform an association analysis. Because there is a clear relationship between the effective sample size and the statistical power of an association study, it has been recommended to use a minimum of 100 genotypes to perform GWAS analyses [76], although the minimum sample size also depends on the genetic diversity within the population, the number of markers used in the study, the trait considered, and where the minor QTLs are being sought. Once these data are obtained, replication should be prioritized over the inclusion of a larger number of accessions. In limited space, the inclusion of more genotypes but fewer replicates comes at the expense of obtaining an accurate estimation of their phenotypic performance, which in turn reduces the power of SNP-loci association calculations, and might lead to false positives. A good balance can be achieved with partially replicated designs which combine the advantages of high replication levels for a subset of the genotypes and a large enough genetic diversity in the panel of genotypes.

DiGGer [77] is a useful package for R software that can be used to create partially replicated and fully replicated spatial designs. It can be downloaded from NSW DPI Biometrics software download page. Available online: http://nswdpibiom.org/austatgen/software/ (accessed on 15 August 2021). Once in the R version 4.0 software environment, the file can be installed from the downloaded zip file (e.g., "digger.zip") using the following commands:

install.packages("C:/path/DiGGer.zip", repos = NULL, type = "source")

where "C:/path/" indicates the path to the folder in which the zip file has been saved. The downloaded zip file contains a manual with examples of the code required to create these designs. An example of a fully replicated and a partially replicated trial created in DiGGer is shown in Figure 1A,B respectively. Other packages with many useful functions are agricolae: Statistical Procedures for Agricultural Research (Available online: https://cran.r-project.org/web/packages/agricolae/index.html (accessed on 15 August 2021)) and ASReml-R. Available online: https://www.vsni.co.uk/software/asreml-r (accessed on 15 August 2021).

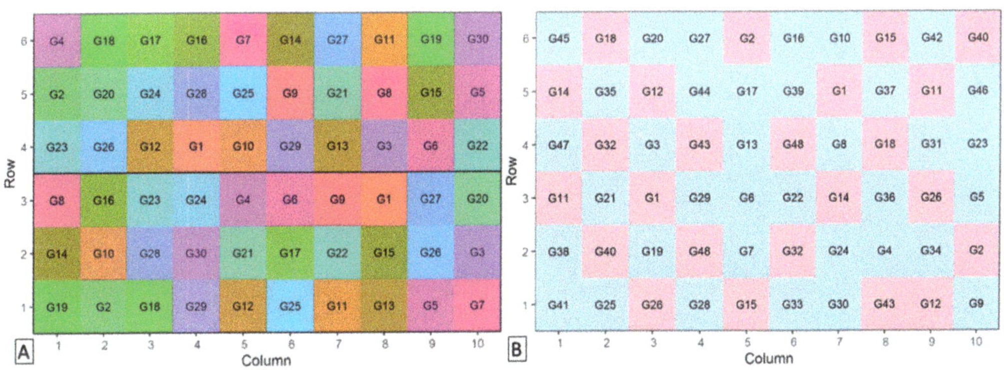

Figure 1. Example trial design generated using DiGGer of (**A**) a fully replicated trial with two replicates, including 60 plots arranged in six rows by 10 columns, with each replicate block of 30 genotypes (GXX) comprising three rows by 10 columns (color representing genotype), and (**B**) a partially replicated trial with 25% replication, including 48 genotypes (GXX) in total, with 12 genotypes having two replicates (pink) and 36 genotypes having one replicate (blue).

Although not included in the trial design examples shown, it is very important to include the rows and columns of border plants that are needed around the field to provide a "buffer area" for the experiment, thereby reducing the environmental border effects. This buffer area should be at least 1 m wide and should be planted with one or two well-performing genotypes. Border effects are also observed at the plot level, making the high phenotypic plasticity of quinoa immediately apparent. Plants on the edges of the plot commonly show higher levels of branching, which is probably a response to the increased space and resource availability. These plants can also differ greatly in height from plants in the middle of the plot, a phenomenon which may depend on their exposure to nitrogen and other nutrients, water, and varying ratios of red to far-red light. If there are maturity or height variations among genotypes, neighboring plots may also cause shading, which is an issue to be considered when planning the plot size.

Therefore, it is recommended to have plot sizes of approximately 4 m^2 to allow a "plot in plot" design. When recording observations, researchers can disregard the plants on the outer edges of the plot, which occupy approximately 0.5 m on each side, and focus only on the plants in the middle of each plot. The phenotypes observed in the middle of a plot are the phenotypes of interest because the proportion of plants along the edge becomes negligible when plants are grown in large fields at a commercial scale. Border rows may be used for bagging and seed multiplication. However, if bagging is not adequately performed, heterogeneous seeds may be obtained, which should not be used in future trials.

Regarding yield predictions, larger plots are more beneficial, and bed sizes such as 5 m long × 3 m wide (12 rows, 25 cm between rows) = 15 m^2 are recommended to ensure a more accurate representation of the behavior and expected yield from an accession if it were cultivated in a large field. A 10 cm spacing between plants within a row equates to 600 plants per plot. If 500 plots are used, the field area required for the plots is 7500 m^2. The total area required will be larger, after adding in the area needed for spacing between plots, and the inclusion of border plots around the field. Smaller plots are often required owing to individual constraints. We found plots sized 2 m × 2 m to be a practical size for multi-environment trials quantifying phenotypic traits to assess a large set of several hundred accessions, but plots this small should not be used for accurate measures of yield.

In any trial, growing conditions should be as close as possible to the actual condition under which the plants will be grown commercially, particularly if investigating yield or traits closely related to yield [78]. These conditions involve minimization of both abiotic stresses (such as water and nutrients) and biotic stresses (such as pests, weeds, and diseases). If stress effects are important, these conditions must be implemented in plants grown under otherwise optimal conditions to allow quantification of the specific responses to stress. Optimal conditions and crop management approaches are largely dependent on location, soil types, and genotypes. The following section provides some suggestions for planting and growing conditions. More detailed recommendations for specific situations can be found in the book "Quinoa: Improvement and Sustainable Production" [79]. Although there are no clear guidelines for some of the following crop management aspects yet, a research gap that needs to be addressed, it is advisable to follow strategies that are successful for others growing quinoa under similar environmental conditions and soil types [13,80].

4.1. Planting

Planting density has a significant effect on the phenotypes of quinoa (see, e.g., [81]), specifically on their branching habit, and should therefore be optimized to suit the cropping system. Soil types and irrigation systems are important factors to consider; for example, factors such as furrow irrigation can restrict row width. When assessing the effect of plant density, it is important to consider two distances: the distance between rows of a plot, plus the distance between plants in a row. Studies that have assessed the effects of planting density on quinoa phenotypes were conducted in a specific environment. Due to the use of different methodologies and genotypes, comparisons between environments are not possible to draw. It appears that the optimal planting density depends on multiple environmental and management factors, such as weed management, as well as genotype [82]. Considering the significant phenotypic plasticity of quinoa in response to planting density, it is clear that trials with similar planting densities are better in multi-environment analyses and that planting density always needs to be recorded. Moreover, plots with poor emergence should be recorded, and might need to be excluded from analyses. When assessing hundreds of genotypes in one trial, only one uniform planting density can be used, unless planting density is a criterion being tested.

Current and future cropping systems aim for the production of high yielding varieties with highly nutritive seeds, requiring low fertilizer and phytosanitary inputs. Optimum sowing density can vary depending on quinoa variety, which in turn can differ according to plant and panicle architectures, seed size, sowing technique (broadcast, rows, or grooves), and agroecosystem. Quinoa sown at high densities grow into less robust and smaller plants, with lower yield per plant, than those planted at low densities. However, the planting of too few plants per unit area may result in branched plants that may not mature before the first frost, and it may also provide more space for the growth of weeds. Therefore, Aguilar and Jacobsen, 2003 [83], have recommended a density of 40 plants per m^2, with 10 cm between plants in a row and 25 cm between rows, depending on practical details of the system. Quinoa seed viability is sometimes low; therefore, it is prudent to plant more seed (~3 kg/Ha without taking seed size into consideration), followed by a thinning of plants to achieve the desired distances between plants. Thinning (removing excess seedlings) should be performed when plants are 5 to 10 cm high, after the plants are established, but before signs of intense competition for light, such as elongated seedlings, are detected. Sowing a lot of seed followed by thinning to achieve the desired plant density is a method often popular in a field situation that does not allow the recording of seed viability in the field. Hence, germination rate of the seeds from each seed batched used in the field trial should be tested in petri dishes to calculate seed germination rate (%) and other germination parameters. This also allows more precise calculation of the amount of seed that needs to be sown out (see, e.g., [84]).

The above recommendations relate to ordinary flat fields. In the Andes highlands, farmers have through generations developed a completely different strategy where they sow numerous seeds in holes that are distant from each other. The outermost plants protect the plants in the middle from flying sands and strong winds while the holes also collect water.

Planting may be performed by hand, although this approach is labor-intensive and is not feasible for large trials. Alternatively, a hand push seeder can be used. The use of a mechanical seed planter (such as a cone seeder) allows even higher throughput, although this requires a skilled operator to ensure that seeds from one genotype are not carried over into adjacent plots. Seeds are planted at shallow depths. A recommendation is to plant at a depth that is three times the diameter of the seed. With seeds between 1.5 and 2 mm in diameter, the planting depth should be approximately 4.5–6 mm. Recommended planting depths can vary with location and soil types. Soil compaction is important because quinoa seeds germinate better in looser soils, so germination rates are better in sandy soils than in heavy clay soils [85]. Hydromorphic soil types are problematic for growing quinoa due to a high sensitivity to waterlogging [86]. Waterlogging was also found to negatively impact the percentage of emergence when seeds were planted too deeply, whereas shallow-planted seeds may be subject to drying [87]. A uniform planting depth of the seeds is important to reduce the risk of uneven field emergence.

Water availability and other environmental constraints can be considered in the decisions around the season for cultivation. Depending on environmental factors or for consideration of crop rotations, quinoa can be used as a winter or spring crop. The sowing date requires careful consideration as this decision impacts growth and productivity of the crop. Due to a range of factors from soil temperature for seed germination through to high temperatures inhibiting grain fill [88]. Consideration could also be given to staggering planting to account for differing times to maturity depending on the traits being measured in the experiment.

4.2. *Irrigation*

Irrigation is known to affect several aspects of quinoa phenotypes, from plant height [89] to seed saponin content [90] and yield [91,92]. Irrigation needs for optimal growth depend on soil type and environmental conditions and should therefore be calculated for each location. The irrigation requirements for quinoa can be estimated using the crop coefficients (K_c) for quinoa, as described below.

To calculate the evapotranspiration rates, a reference evapotranspiration rate (ET_0) is first calculated from a range of meteorological parameters using the only standard accepted method by the FAO, the Penman–Monteith equation [93]. To facilitate the calculation of ET_0, a calculator has been developed by the FAO for Windows OS [94]. Next, water requirements can be calculated by multiplying the ET_0 with quinoa crop coefficients ($K_c \times ET_0$), and finally, irrigation amounts are planned by subtracting any rainfall from the water requirements that were calculated. The requirements differ with growth stages, considering that the principal growth stages 6 (flowering) and 8 (ripening) are most sensitive to drought stress [95]. Rainfed crops may therefore need additional irrigation at several points of the growing cycle: before sowing, at the beginning of flowering, and at the start of seed filling. Proposed K_c values also reflect the differing requirements by providing three different values to choose from: one of the following three K_c values may be used to make a crude estimate of water requirements: 0.52 for initial stages, 1.00 for mid-season stages, and 0.70 during the principal growth stage 9 senescence [91], although compare [96]). More accurate estimates of water use can be performed by using a growth model relating the dynamics of leaf area and radiation interception. Integrated crop management tools such as the SALTMED model can be used for informing irrigation strategies, and has been calibrated and used for quinoa [6,97,98]. The K_c values used in these three model applications differed among them, suggesting that different K_c values yield better model predictions in different environments and for different genotypes. For example, planting densities were found to greatly affect K_c values, varying to such an extent that a single K_c value was difficult to assign [82].

Owing to strong phenotypic responses of quinoa to water availability, it is also crucial that the land used for the experiment is laser leveled before sowing to reduce the heterogeneity in the amount of water that plants receive across the field. Types of irrigation vary depending on local environmental conditions and technical possibilities. In any case, it is important to record details of irrigation schedules and the amount of water used. Even though quinoa is considered a facultative halophyte, salinity levels of the irrigation water are important to measure with precision as well, not only when salinity is a treatment.

4.3. Fertilization

There has been little research on the fertilization of quinoa. Most publications provide information on local recommendations but lack thorough physiological and biochemical characterization across genotypes. Several studies focus only on varying inorganic nitrogen supply following the conventional method for determination of optimal nitrogen availability by measuring responses according to yield [99] without examining growth responses to other important macro- and micro-nutrients or organic sources of these nutrients. Quinoa yield has been shown to positively respond to nitrogen supply [96,100,101]; however, it is important, when breeding quinoa for smallholder growers, to consider the selection of genotypes that can maintain their yield under nitrogen limiting conditions. Interestingly, quinoa cv. Titicaca has been shown to maintain size, weight, and nitrogen content of their seeds irrespective of the nitrogen supply [102,103].

4.4. Weeding, Pest and Disease Controls

There have not yet been any effective herbicides developed for quinoa and the crop is mainly cultivated under organic practices; therefore, weeding often needs to be performed by hand. Weeding is important because weeds compete with quinoa plants for nutrients, light, and water, confounding the results of systematic investigations on the culture of quinoa and potentially reducing yield. If machine weeding is available, spacing between rows/plots needs to be sufficient to enable movement of machinery.

For scientific studies in-field, preventative management of pests and diseases need to be conducted. For this, it is better to apply treatment on a regular schedule rather than waiting for problems to appear. Of course, for trials testing resistance to pests and diseases, such a regimen should not be conducted, nor for commercial fields using integrated pest management. In several countries, no approved products for quinoa are available; therefore, management must often be based on recommendations for similar crops such as beet, chard, or spinach. For preventing loss in quinoa yields, it is essential to be aware of the diseases and pest that may occur, but in the case of quinoa, there is high uncertainty about local ecological interactions when adapting the crop to new agro-ecological zones and environments. The only manual on quinoa pest and disease [104] was developed for Andean conditions and it was not adapted for different conditions of cultivation, nor translated into English. This point implies that we need entomologists and plant pathologists within the research teams for controlling and preventing damages in trials. Digital tools such as Platix [105] to detect diseases and foliar disorders are becoming increasingly powerful and can supplement expert input.

5. Environmental Variables

Although crop management techniques can be used to influence some environmental variables, such as when irrigation regimes are planned according to rainfall, virtually no environmental variables can be controlled in a field experiment. These variables include soil drainage, soil pH, and microtopographic effects, causing pooling in some locations [106,107]. However, environmental variables greatly affect phenotypes, and the degree of plasticity of a genotype to adapt to environmental conditions varies greatly. Thus, consideration of G × E interactions are of great importance in matching genotypes with the appropriate environments for achieving maximum yield. The investigation of these interactions also facilitates the uncovering of genetic correlations by plant geneticists [108]. To gain clearer insight into the input of genetic factors to a trait, potential variability introduced by environment (E) and management (M) needs to be reduced as much as possible. This can be achieved if as many variables as is reasonably possible are monitored. To limit the effects of management on the variables measured in a trial, the management practices should be harmonized by adopting the recommendations from the section above. Environmental variables can vary substantially among and even within the plots in a field. Therefore, it is important that researchers try to collect at least a minimal set of variables for a field site from a nearby weather station. The most important variables that should be collected for each trial are summarized in Table 1.

Effective nutrient and soil management relies on data from the testing of soil cores by internationally accredited laboratories. A soil core sample is collected using a hollow steel tube called a "core drill", which may be up to 40 cm in depth. It is best if cores from 0–20 cm deep are separated from those taken at 20–40 cm depth. There is considerable spatial variability in the physical and chemical properties of soil both horizontally and vertically, and therefore, large sampling regimes are recommended. Multiple core samples (25–30) should be collected at random sites from multiple, well-defined locations such as fence lines, tree lines, hills, or GPS coordinates.

A more strategic approach involves defining "zones" within the field where variations in management practices are predicted to be necessary owing to differences in slope or soil color or areas in which growth has been shown to vary in previous years. A good visual example is provided in [109]. This strategy can reduce sampling time by using more "cluster" sampling, i.e., using five cores per zone. Soil sampling depth depends on the rooting depth of quinoa accessions used in a specific soil and can be assessed using a soil core drill in the plot and looking for roots within the soil core. The soil type affects numerous factors, including the soil's water holding capacity, nutrient storage, and aeration. These factors affect crop productivity and phenotype. Soil texture is an important factor to include in the environmental information [110]. The USDA provides soil texture calculators to define a single point texture class based on the percentage of sand, silt, and clay. This calculator is available at the following URL: https://www.nrcs.usda.gov/wps/portal/nrcs/detail/soils/survey/?cid=nrcs142p2_054167 (accessed 15 August 2021).

Proximity to the study perimeter can also cause G × E effects on plant phenotypes. Outer plants incur the brunt of wind, causing high rates of evapotranspiration. In a randomized trial assessing diversity of phenotypic variation across the population, neighboring plots can impact the conditions of each plot. It is possible that larger and more heavily branched plants require more water than smaller plants, and therefore, irrigation systems could over- and under-water plants of different sizes. This issue again highlights the importance of replication in experimental design and the importance of measurement of environmental parameters. The outputs from measurement devices and plot locations in association models can be used as covariates as a broad control for known or unknown factors [106].

Table 1. Minimum set of environmental variables to record for quinoa trials.

Soil, To Be Measured Before and After the Field Season
Watering regime
Water holding capacity
Composition in terms of % sand, silt, organic matter, etc.
Nutrient and mineral composition—total nitrogen, organic carbon, phosphorus, potassium, sulfur, etc. Note: when measuring nitrates, the soil sample must be kept cold because nitrates are unstable
Soil physical properties affecting plant growth
pH
Apparent density
Electrical conductivity (EC), especially for salinity trials
Weather
Precipitation, and irrigation schedule
Temperature, at least daily T_{max} and T_{min}, but preferably recorded continuously throughout the day to enable calculation of degree-days to flowering and to maturity
Humidity—relative humidity/dewpoint temperature
Daily irradiance (mol m^{-2} d^{-1}), recorded continuously throughout the day
Wind speed (average daily speed)
Day length (including twilight time)

Depending on the aim of a trial, more detailed environmental measurements might be needed. Humidity and temperature, which can be combined into a heat index, can account for within-site environmental variation in studies of disease traits because fungal species generally prefer humid conditions. The cost of temperature–humidity meters ranges from less than USD 100 to several hundred dollars. Using plot position, i.e., two columns of vectors of site grid positions, as covariates in spatial models can account for some environmental effects. However, better corrections are achieved with measurements of environmental information as covariates. Small hand-held temperature–humidity devices such as Extech EA20 (Extech Instruments, Nashua, NH, USA, available online: www.extech.com (accessed on 15 August 2021)) or a UNI-T UT333 (Uni-Trend Technology, Dongguan City, China, available online: www.uni-trend.com (accessed on 15 August 2021)) and others may be used for measurements at a plot level. Trials examining crop responses to water also require measurements for atmospheric vapor pressure deficit (VPD), which is related to evaporative demand and thereby driving plant water transport. The higher the pressure deficit, the higher the water stress experienced by the plant [111,112].

The amount of soil moisture per plot can be used to correct environmental variations when water use efficiency and photosynthetic traits are measured. Soil moisture measurements for individual plots can be used as covariates to correct environmental variation in water use efficiency and photosynthetic studies, as can thermal measurements in heat stress conditions. Soil moisture sensors also have a wide range of costs and can provide much needed assessment of local soil factors such as temperature, pH, and water content that could drive some of the phenotypic variation in a study. Several inexpensive devices are available; however, advanced systems can cost approximately USD 1000, such as the Vegetronix VG-Meter-200 (Vegetronix, Riverton, Utah, USA, available online at: www.vegetronix.com (accessed on 15 August 2021)) is a mid-range soil data logger or APERA PH8500-SL (Apera Instruments, Wuppertal, Germany, available online at: www.aperainst.com (accessed on 15 August 2021)). Gravimetric measurements of soil water content can also give some indications of water availability for plants and can in some conditions be the only method practically feasible. Sensors measuring soil moisture frequently overestimate soil humidity in soils with high levels of iron [113] and in this case, water content can be measured gravimetrically.

Crop growth simulation models such as those of the Decision Support System for Agrotechnology Transfer (DSSAT) can be used to investigate G × E interactions. Crop models can be used to facilitate crop management decisions by evaluating multiple scenarios. The integration of quinoa into the DSSAT system was initiated with the calibration of the CROPGRO template for quinoa [114]. However, more work on getting DSSAT models calibrated and usable for quinoa is required. Other crop growth models or crop water models such as SALTMED and AquaCrop are good resources for quinoa and already applied (e.g., [97,98,115–119]). The minimal requirements necessary to use these models must be considered (for AquaCrop see [120]; [121]; for SALTMED see [122]).

To facilitate the reuse of data and establishment of multiple local networks of phenotypic data, Minimum Information About a Plant Phenotyping Experiment (MIAPPE) metadata standard for plant phenotyping must be followed [40]. A MIAPPE spreadsheet template is published online for guidance, MIAPPE available online: https://github.com/MIAPPE/MIAPPE (accessed on 15 August 2021). Alternatively, we recommend downloading the data templates created for the Germinate platform, which are created in accordance with metadata standards, data templates available online: https://github.com/germinateplatform/germinate/tree/master/datatemplates (accessed on 15 August 2021). Templates specifically for quinoa are available at the Quinoa Germinate Database. Available online: http://germinate.quinoadb.org (accessed on 15 August 2021) In this manuscript (see Section 9 for more details), we also offer a platform for sharing quinoa datasets using the Germinate database structure [42,43]. To use this structure, the templates provided should be filled in with the relevant information. Therefore, it would be best to start using such templates from the initial planning of a trial and then continue with phenotypic observations.

The above list of environmental parameters is comprehensive and will be difficult to achieve for all experimental sites. Pragmatic decisions will often need to be made and can be guided by both the particular properties of the site and the scientific questions being addressed.

6. Observations during Growth

There are several measurements that can be conducted throughout the growing period.

6.1. Phenology over Time

Clearly defined phenological stages are of great importance for reproducible phenotyping. Multiple studies have investigated and described phenological stages in quinoa [41,123–127]. These studies have provided valuable information about the characterization of the crop; however, only one study has followed the complete international scale system proposed by the Biologische Bundesanstalt Bundessortenamt und Chemische Industrie (BBCH). Sosa-Zuniga et al., (2017) [41] provided the most recent and complete description according to the BBCH guidelines. Here, the main principal growth stages and their relevance to phenotyping are briefly summarized. An overview of the main growth stages is also shown in the phenotyping cards (Supplementary File S1: Phenotyping cards).

Sosa-Zuniga et al., (2017) [41] describe eight major growth stages. All developmental stages from seed germination until cotyledon emergence belong to principal growth stage 0 (BBCH 00–09). The sowing date is the most critical part of this stage because it should be optimized for local conditions. The next principal growth stage 1 covers all stages of leaf development (BBCH 10–19). Fully emerged cotyledons can be observed in BBCH 10. This stage is considered the field emergence stage, and it is necessary to record both the time and the emergence percentage. The emergence percentage can be measured by assessing a randomly selected 1 m^2 area from the middle part of a plot. At BBCH 11, true leaves develop, and the time from stage 10 to 11 is essential for estimating the early seedling vigor of quinoa accessions.

At principal growth stage 2, branches start to grow from the basal leaves. This stage can occur before the principal growth stage 5, depending on the accession. The inflorescence emergence stage (principal growth stage 5) is one of the most critical stages in quinoa development. At the initial inflorescence emergence stage, floral buds are enclosed in leaves (BBCH 50) and are thus difficult to phenotype. At BBCH 51, buds are visible from the top but still surrounded by the leaf primordia. This stage should be recorded and can be represented as days to inflorescence emergence.

Principal growth stage 6 in quinoa is its flowering stage. Because yield and many other agronomically important traits are highly correlated with flowering, this stage is vital. According to Sosa-Zuniga et al., (2017) [41], stage 6 is divided into three BBCH stages. BBCH 60 is the beginning of anthesis, which is marked by flowers opening and extruding their anthers. BBCH 67 is the beginning of senescence of anthers. BBCH 69 is the completion of senescence of anthers. The length of time between BBCH 60 and BBCH 69 depends on the accession and can vary widely. Phenotyping of BBCH 67 and 69 can be imprecise; therefore, reproducible phenotyping attempts in the flowering stage should be focused on recording BBCH 60. This stage is used to score the flowering time of an accession. A panicle can be classified as BBCH 60 as soon as at least one flower with extruded anthers is visible. When scoring flowering time in the field, a plot may be scored for the onset of anthesis, once >50% of plants in the plot have reached this stage. The number of days to flowering among quinoa accessions is highly diverse.

The next principal growth stage is the fruit development stage BBCH 70. At this stage, ovary thickening occurs and can be identified by naked eye, as shown in the phenotyping cards (Supplementary File S1: Phenotyping cards). Depending on the accession, BBCH 70 and late flowering stages could overlap. Therefore, it is challenging to record this stage, and these difficulties might result in imprecise phenotyping. Nonetheless, this is a crucial stage because it is the beginning of the seed set, and therefore, careful and frequent observation should be performed to record the BBCH 70 stage.

Principal growth stage 8 comprises three ripening stages: BBCH 81, 85, and 89. At BBCH 81, onset of seed filling can be observed, and seeds at this stage appear as milky grains owing to the nature of soft fruits. Thick and fully ripened grains are present in the stages BBCH 85 and 89, respectively. When scoring plants at this principal growth stage, it is important to carefully examine the seeds by sampling some from the upper middle half of the panicle and crushing the seed by hand, typically with a fingernail. Quinoa accessions show varying degrees of ripening panicles, and a single panicle may show all stages at a single time point, indicating a highly unsynchronized panicle ripening. When the complete panicle is at BBCH 89, it is ready to harvest, but the plant is still not fully senesced. Mechanical harvesting may be difficult at this stage. BBCH 89 is considered the stage of physiological maturity. Sometimes, BBCH 89 is not possible to score, particularly with day-length sensitive plants that can exhibit regrowth in the panicle.

Senescence is the next growth stage (9)—although sometimes this can occur before BBCH 89. Five BBCH scores are assigned to this stage. Senescence starts from the basal leaves (BBCH 91) and spreads upward in a plant. At BBCH 95, all leaves are dead, and the stems turn yellow to beige in color. Plants are ready for harvest when the whole plant is dead at BBCH 99. However, depending on the accession, it may be advisable to harvest before plants reach this final stage to prevent seed loss from shattering or feeding birds.

For the precise recording of principal growth stages, observations must be performed two to three times a week. All plots must be assessed on the same day, and observations should be evenly spaced throughout the growth cycle. At least 50% of the plants in the inner part of a plot should have reached the corresponding stage for the recording of the growth stage of the plot, as judged overall by eye. In practice, recording all stages may not be feasible, especially when the trials are large. In this case, it is necessary to record at least the sowing date, field emergence (BBCH 09), emergence of first true leaves (BBCH 11), floral bud emergence (BBCH 51), anthesis (BBCH 60), physiological maturity (BBCH 89), and harvest date.

Note that this approach for the scoring of phenology is distinct to the measuring of more specific (often physiological) traits, where it is usually best for at least 3–5 plant per plot be measured and marked for resampling of the same plants (not randomly selected plants each time). It is best if these plants are within the final harvest sector or at least comply with requisites for sampling area (e.g., not border plants).

Apart from recording the timing at which each genotype reaches the selected phenological growth stages, another phenotype that must be assessed during growth are leaf area-related measurements, such as radiation interception efficiency. Because the last main stem leaves of quinoa plants appear around the beginning of anthesis (BBCH 60), whereas those on branches continue expanding until the end of flowering [127], the best growth stage to count leaf numbers and areas is after the anthesis growth stage (BBCH 70). Of course, an ideal approach would be to count leaf numbers as soon as leaves emerge throughout the growth period because by anthesis, many would have fallen from the plant. However, this is highly labor-intensive and often not feasible.

6.2. Radiation Capture and Efficiency of Use

The yield potential of a crop in field trials is correlated with the amount of photosynthetically active radiation (PAR) available and the plants' efficiency in capturing it [128]. Variability in radiation capture owing to differences in the architectural traits of genotypes and efficiency of use of radiation in response to environmental conditions have been proposed as heritable traits which could be used to reduce the complexity of phenotypic responses to the environment. The results of genetic analyses of these traits, in combination with environmental and genotype information, can be used in yield simulations. This approach requires data from experiments performed in different years and locations for the validation of the simulation results, although most other traits have to be assessed in the context of much larger networks of experiments across multiple years and locations [129,130].

Using a resource capture and efficiency approach, biomass accumulation by a crop can be modeled as follows:

$$biomass = resource\ availability \times resource\ capture\ (or\ uptake)\ efficiency \times resource\ use\ efficiency$$

where the resource can be radiation, water, nitrogen, or another nutrient. From a management perspective, radiation availability can be partially achieved by the selection of sowing dates and location and selection of genotypes with suitable cycle durations, whereas water and nutrients can be managed by altering irrigation or fertilization practices.

One of the main determinants of the uptake of any resource is radiation interception efficiency, i.e., the proportion of incident radiation intercepted by the crop. This property affects both radiation capture and the partitioning of water use between potential transpiration and evaporation [131]. Interception efficiency (IE) is determined by the leaf area index (LAI), the leaf area per unit soil area, and k, the light extinction coefficient. The association between IE and k is described by Beer's law [132] as follows:

$$IE = 1^{(-k \times LAI)}$$

The association between LAI and IE is affected by variation in the light extinction coefficient k. In quinoa, variation was detected according to genotype and plant density, with a positive correlation between k and plant density, suggesting rearrangement of leaves when the density was modified. However, a common k of 0.59, resulting in a critical (95% IE) LAI value of 5.09 can be used across genotypes and densities [127]. If this k value is assumed, IE can be calculated with LAI estimations from emergence to critical LAI. Estimations of LAI can be obtained through destructive leaf sampling or a portable, non-destructive, Plant Canopy Analyzer. As these two methods are low in throughput or not calibrated for quinoa, the use of remote sensing, as described later in this section, is a better alternative.

Leaf area indices can be calculated from measurements of IE, which can be estimated using a ceptometer, an instrument for measuring PAR (see details about its estimation in [127]), and by taking measurements above and below the canopy, the amount of light that is not used by the plant for photosynthesis can be determined. Results can be used as an estimation of radiation capture, integrating LAI development, senescence, and *k* variation aspects. Measurements using a ceptometer are time-consuming; however, because it can only be used for a couple of hours around midday, and for each replicate, six measurements need to be taken, two above the canopy to determine incident irradiation and four below the canopy along the ground to calculate the amount of intercepted radiation [133].

The last component of the biomass accumulation equation, resource use efficiency, which equals radiation use efficiency (RUE, g m^{-2} MJ^{-1} or [g mol m^{-2} d^{-1} $PFFD^{-1}$]; PPFD: photosynthetic photon flux density) from a carbon balance perspective, can also be calculated from IE measurements. Similar to LAI, RUE is associated with accumulated intercepted PAR ($\sum$ incident PAR $\times$ IE over the period among samplings). RUE might not be stable during the crop cycle of quinoa. Ruiz and Bertero (2008) [127] detected two different RUE estimates from emergence to end of flowering (BBCH 70), an initial low value of 1.25 g m^{-2} MJ intercepted $PPFD^{-1}$ and a second, higher RUE value of 2.78 g m^{-2} MJ intercepted $PPFD^{-1}$ with a breakpoint at 107 MJ m^{-2} [127]. This experiment was run using a mid-winter sowing time, and no similar differences in RUE across the cycle were detected for spring sowings [134]. Low RUEs under low temperature conditions are expected to occur in several quinoa growing environments, such as the high Andes, during early spring in Northern Europe, or autumn and winter in Mediterranean environments, which partially explains the low initial growth observed for quinoa in many environments.

However, the approaches for LAI and RUE measurements described here are not feasible for application in large field trials owing to several limitations, which make them low in throughput. This process is time-consuming and labor-intensive and also affects the quality of the data collected because measurements must be taken over a longer time span. Instead, unmanned aerial vehicles (UAVs) deployed with multi-spectral sensors may be used to collect image data for the entire trial area to estimate LAI of all quinoa plots. However, low-throughput field-based methods for measuring LAI are still required for modelling LAI from UAV-based imagery and for independent validation of the image-derived LAI measurements Estimates for RUE or radiation capture can also be obtained from the Photochemical Reflectance Index (PRI), which can be extracted from spectroradiometers (PRI) or normalized difference vegetation index (NDVI) estimates [135]. Several spectral reflectance indices can be obtained at a high throughput through the use of UAVs. Of these indices, NDVI and green normalized difference vegetation index (GNDVI) have been identified as the most informative for quinoa when assessing irrigation treatment effects [136,137]. Furthermore, spectral reflectance measurements have been identified as the most effective tool in the assessment of disease impacts in quinoa [138].

6.3. *Unmanned Aerial Vehicle-Based Phenotyping*

Phenotyping methods have been considered a bottleneck in plant genetic studies [16]; therefore, it is important to quickly progress to non-destructive high-throughput and high-precision options for phenotypic data collection. The use of UAV technology is emerging to be one of the most promising solutions. Recent advances in UAV technology and miniaturization of mountable sensors have facilitated accurate, consistent, and expansive high-throughput phenotyping of crops [139] and standardized data collection approaches and image processing workflows [140]. The use of UAV-derived data for crop phenotyping may increase the amount of data collected, frequency of data acquisition, consistency of information extracted, and the ability to undertake retrospective studies, while significantly reducing human labor. Red-Green-Blue (RGB) and multispectral cameras are used most frequently for crop phenotyping studies because of their relatively low costs and ease of use; however, Light Detection and Ranging (LiDAR), thermal infrared, and hyperspectral sensors have also been used to collect information on phenotypic traits [141].

UAV-based RGB and multispectral cameras have been found to be suitable for plant height estimation of maize [142], yield prediction of tomatoes [143], plant vigor assessment of barley [144], detection of pathogens, e.g., cotton root rot disease [145], yellow rust disease in winter wheat [146], and maize streak virus [147], and mapping of growth patterns of winter wheat [148]. UAV-captured data can also be used to calculate a crop water stress index [149] where leaf temperature is normalized using environmental conditions measured around the experiment along with leaf temperature. The use of digital photogrammetric processing approaches such as structure-from-motion and multi-view stereo reconstruction of overlapping optical images allows detailed surface elevation characterization and three-dimensional models to be generated [150]. Although LiDAR sensors have greater plant canopy penetration capabilities and provide height information with higher accuracy than photogrammetrically processed optical data, their weight, costs, and range currently limit their operational use in the agricultural industry [151].

UAV-derived thermal data have become increasingly popular for use in phenotyping applications to monitor canopy temperature, detect plant disease, and estimate yield [152]. However, camera effects such as vignetting, camera warming, and temperature drift as well as meteorological conditions such as ambient temperature, wind, and wind direction often affect the accuracy of UAV-derived at-surface temperature measurements [153,154]. Malbeteau et al., (accepted) [155] provide practical information and examples of how to overcome issues related to dynamic temperatures and wind effects during thermal UAV data collection to improve data consistency and accuracy of UAV-based orthomosaics of a field trial designed for phenotyping of tomato plants. UAV-based hyperspectral imagery contains hundreds of spectral bands that allow the collection of detailed spectral information on phenotypic traits [156]. Examples of agricultural hyperspectral data applications include estimation of chlorophyll, mapping of plant disease, detection of nutrient status, and assessment of plant nitrogen content [157]. Although UAV-based hyperspectral data provide information well-suited for phenotyping, data processing and analysis are often complex, requiring careful image calibration and pre-processing [158,159] to achieve spectrally accurate reflectance data suitable for scientific research and consistent monitoring [160]. Ivushkin et al., (2019) [161] used data from UAV-based multispectral, hyperspectral, thermal, and LiDAR sensors to discriminate between quinoa plants in a salt-treated plot and a control plot and found hyperspectral vegetation indices to be better suited than multispectral data for the spectral discrimination of quinoa plants between the two treatments. The LiDAR data were used to detect a lower plant height of salt-treated plants compared with the control plants. Temperature measurements could clearly discriminate the quinoa plants in the two treatments, when the thermal data were clustered based on plants with similar vegetation index values. These findings emphasize the potential of UAV-based phenotyping of quinoa plants.

The successful application of UAV-based sensing technologies for phenotyping in quinoa field trials depends on the seeding pattern and dimensions because single rows of quinoa plants with limited spacing between rows (Figure 2A) precludes separation of plots owing to lodging across neighboring plots. Shading from tall quinoa plants of neighboring shorter plants affects their spectral reflectance characteristics as observed from RGB, multispectral, and hyperspectral imagery and plant temperature measured by thermal infrared cameras. To reduce shadow effects, it is generally recommended to collect RGB, multispectral, and hyperspectral UAV imagery close to solar noon. Although rectangular and square plots of greater width (Figure 2B,C) improve the separation of quinoa plants from neighboring plots for image analysis, except along their perimeter, the larger the plots and the greater the separation (Figure 2D), the more feasible it becomes to derive image-based representative samples of individual plots for analysis of phenotypic traits.

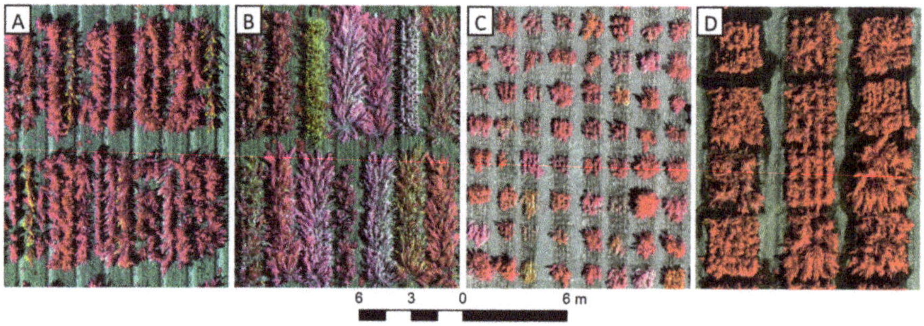

Figure 2. Quinoa plant trials showing (**A**) linear plots with a width of one plant; (**B**) rectangular plots with multiple plants next to each other; (**C**) square plots of 1 × 1 m^2; and (**D**) square plots of 2.5 × 2.5 m^2. All UAV images are displayed as false color composites.

The collection of UAV imagery throughout the growing season allows multi-temporal assessment to study plant growth, behavior, and phenology. Multi-temporal assessment generally requires accurate georeferencing of the image datasets acquired during the growing season, which can be achieved using GPS-surveyed ground control points [140]. Variations in solar elevation, irradiance, and atmospheric conditions alter the illumination conditions over time. To enable spectral characteristics to be compared among multiple datasets, normalization of the image digital numbers to a set standard, normally at-surface reflectance, is required [162]. For consistent multi-temporal results and to ensure similar spatial resolution of the imagery, it is also recommended to employ the same flight pattern, altitude, and type of camera because different cameras are sensitive to different spectral wavelengths [140]. Finally, it is important to collect field calibration and validation data of phenotypic traits to be mapped. While in situ calibration data allow relationships and models to be developed, e.g., for UAV-based image classification, independent validation data enable the accuracy of maps to be assessed [163].

The use of automated and active phenotyping systems using light-induced fluorescence transient (LIFT) canopy scans can be useful for linking of photosynthetic performance and canopy structure and offers great potential for plant breeding and crop growth modeling [164,165]. UAVs, automated greenhouse phenotyping facilities, and rapid handheld phenotyping devices can create time series of imagery of large populations of plants over the course of a season or developmental stage(s) [166]. The data collected from these approaches may include a range of imaging modalities, e.g., hyperspectral, thermal, RGB, LiDAR, etc. This results in large volumes of high dimensional data for each plant that also contain spatial and temporal components. These data inherently contain cryptic information about biochemical, physiological, and morphological information of plants and their variance over time, and their conditionality on environmental conditions and genotype. Data volumes from such approaches can reach tens of terabytes per day and thus require automated approaches for phenotype extraction. Machine learning approaches, including deep learning, can be trained in a supervised manner to recognize phenotypes of interest such as height, chlorophyl content, flowering, plant architecture, abiotic stress, pathogen detection, disease quantification, etc. [167]. This is typically achieved by researchers manually labelling datasets and training neural networks to be able to predict those labels from the image data. After adequate training, the resulting neural network models can be used to analyze large volumes of data to extract phenotypes in an automated manner. However, the training phase is labor-intensive and supervised learning approaches will not, by themselves, extract all meaningful information from these complex, high-dimension datasets. As an alternative, unsupervised approaches such as autoencoders can be used to learn the latent spaces in a dataset and the resulting patterns extracted for use as phenotypes [168]. Traditionally, the downside of such an unsupervised approach has been the "black box" nature of neural networks which makes the extracted latent spaces difficult to interpret.

However, recent advances in neural network architectures such as transformers with attention mechanisms show promise in relating the latent-space topology to the original features and thus producing interpretable phenotypes that are automatically extracted out from large datasets [169,170].

7. Phenotyping of Mature Plants

Toward the end of plant growth, before physiological maturity and harvest, it is useful to phenotype several traits at once in one large phenotyping event. Phenotyping at this stage can be seen as the cumulative effect of different developmental phases on the trait studied. Therefore, phenotyping at maturity can be a way to summarize the morphological strategy of the plant.

For most traits, the most useful stage for undertaking this intensive phenotyping is around mid-seed filling, during principal growth stage 8, i.e., ripening. It is important not to schedule this event too early because colors may not have developed or too late because during senescence, colors and leaves are lost. Depending on the number of traits decided upon for investigation and the size of the trial, this process is usually completed in 2–3 weeks. Preparation for the phenotyping event includes the set-up of a spreadsheet containing the field plan, plot numbers and associated accession information, and the traits that are to be described. An example of such a table lies in the templates linked on the home page of Quinoa Germinate Database. Available online: http://germinate.quinoadb.org (accessed on 15 August 2021). Not all phenotypes will be recorded in each trial; only phenotypes that show variation across the field should be selected. In addition to the descriptions, phenotyping cards are created (Supplementary File S1). These may be printed, laminated, and given to each person of the phenotyping team to carry while assessing plots within the field.

Multiple teams of individuals may divide the workload to shorten the time required for completing the phenotyping. However, this comes at a risk of introducing biases and errors. If the work is shared, it is crucial to train each person adequately and to ensure that all agree with the method using which traits are scored or measured. Hence, it is important that extra time for training each person is planned in for the first day of phenotyping. The first plots are scored by all people together. This gives an opportunity for ensuring that the scoring method for each phenotype is clear to each person. Next, around five plots should be scored independently by each person, and scores should be compared. Where differences in the given scores are identified, these should be discussed. The next plots can only be split up between the people scoring, once all are in agreement. At multiple times throughout each day in the field, each team should phenotype overlapping plots and compare the measurements to ensure that the measurements agree and traits persist. The following section describes in detail the phenotypes that we recommend to measure during phenotyping of mature plants.

7.1. Assessing the Quality of Phenotypic Data

To address questions that may arise during the analysis after termination of the trial, for instance, checking the data for outliers or apparent typos, photographs are taken of each plot. Four photographs are taken, as shown in the example images in Figure 3, starting with one picture of the entire plot. It is useful to have a pole used for height measurements included in the middle of the plot as a scale reference. Next, one representative and easily accessible plant (avoiding the outer edge plants) is chosen for a second picture, which should show the entire plant. For this picture and the following close-up picture of the panicle, a black background is needed to capture all details of the individual plant. This backdrop should be composed of cloth with a matte finish to prevent light reflectance and should be attached to a wooden frame that can be easily transported. A color rendition chart (e.g., ColorChecker Passport Photo 2, X-Rite Inc., Grand Rapids, MI, USA) is also important, to include color calibration because light conditions may vary highly throughout the day depending on the weather conditions. Therefore, a camera with high resolution and manual adjustments of the settings to permit adaptations to changing light conditions is needed. For example, the images in Figure 3B,C were taken with a Canon EOS 70D device (Canon, Tokyo, Japan), with manual settings for F-number (f/7.1) and exposure time (1/250). However, for the plot image, this exposure time resulted in an overexposed image (too bright). A second photograph was taken with exposure time 1/500 (Figure 3A). Different settings were used again for Figure 3D, the image of seeds on the blue card. This photograph was taken using the automatic setting "aperture priority" instead of a manual setting, and used a different F-number (f/9). For image analysis, images need to be stored either as raw data files or using lossless compression techniques to enable complete reconstruction of the data from compressed data. This is also important because the quality of the images taken determines whether they can later be used for image feature extraction. Algorithms for the extraction of plant morphological features from these types of images are currently being developed, and seed phenotypic characteristics can already be extracted from seed images by placing the seeds on a blue card background (Figure 3D). This technique is described in detail later (Section: Seed Phenotyping, Section 9.2.1.).

Figure 3. Example photos of images taken during mature plant phenotyping event. These include (**A**) a picture of the entire plot with reference for plant height, (**B**) a photograph of one representative plant for the plot in front of a black background with ColorChecker and ranging pole for height reference, (**C**) a close-up picture of the primary panicle, and (**D**) a picture of a seed sample (~20 seeds) on a 10 × 10 cm square blue card background (used for later image extractions of seed size and color.

7.2. Plot-Level Phenotypes

In addition to the photographs of plants, there are a number of phenotypes that should be scored to allow others to do quantitative analyses without the need to be involved in data collection. A number of quality control phenotypes are assessed as the percentage of the plot that is affected (Table 2). These phenotypes may also need to be considered in analyses of the trial data where the phenotypes might affect any of the later described "plant-level" phenotyping data in which one representative plant for the plot is chosen for further measurement.

Table 2. Overview and description of phenotypes that may be recorded at the plot level and record the majority of the plot. Scoring metrics followed by an asterisk(*) represent assessments of the percentage of the plot covered or affected by levels 1 (up to 20%), 3 (20–40%), 5 (40–60%), 7 (70%–80%), and 9 (over 80%). Border plants of each plot should not be considered when phenotyping. See Supplementary File S1: Phenotyping cards for visual examples for the traits.

Plot-Level Phenotype	Scoring Metric	Description
Plot coverage	1,3,5,7,9 *	Percentage of the plot covered, 1 = poor to 9 = good establishment
Plot population homogeneity	1,3,5,7	Judgment of homogeneity of the accession, 1 = homogeneous to 7 = mixed

Table 2. *Cont.*

Plot-Level Phenotype	Scoring Metric	Description
Branchiness	1,3,5,7	Score for the overall amount of side branches along the entire length of the stem, ignoring very small and spindly branches, ranging from 1 = no branches to 7 = bushy plant with many (i.e., greater than 7) major lateral branches
Growth habit	1,3,5,7	Four categories of growth habit described in images on the phenotyping card. Here the focus lies on whether branching is present in the bottom third of the stem from the base of the plant and if a main inflorescence can be identified
Stem breakage incidence	1,3,5,7,9 *	Stems are broken or detached, assessing the percentage of the plot affected
Stem lodging incidence	1,3,5,7,9 *	Plants are prostrate, on or near the ground, with intact stems; assessing the percentage of the plot affected
Stem lying incidence	1,3,5,7,9 *	Stem of the plant is not emerging straight up from the soil but has a kink at the base, growing along the ground before rising; assessing the percentage of the plot affected
Stem angle	1,3,5,7	The angle at which the majority of plants are leaning, measured between the vertical axis and the horizontal axis
Panicle axis angle	1,3,5,7	The angle at which the majority of panicle axes are leaning, measured between an upright panicle on the vertical and a panicle pointing towards the ground

7.2.1. Plot Population Homogeneity

The most important trait for quality control of data in the analyses of datasets is the score of heterogeneity of the phenotypes of the individual plants within a plot. The genetic diversity of quinoa is wide owing to less-intensive breeding events (and thus a relative paucity of population bottlenecks), and several quinoa accessions are landraces that produce a heterozygous phenotype. Quinoa is predominantly self-pollinating and has varying rates of natural hybridization of 10–17%, which are likely to be greater at lower plant spacings, and depending on the coincidence of flowering with the windiness of the site or the presence of other pollen vectors [7,171]. There is also a possibility for outcrossing if panicles are not isolated with a bag. The heterogeneity of a population can be beneficial in small-scale cultivation, where it might confer greater yield stability in unpredictable weather conditions. However, for genetic studies, heterogeneity poses considerable challenges because the phenotype must be correlated with the genotypic information. Hence, highly heterogeneous genotypes are not suitable for genetic studies, and plots need to be excluded from the analysis if >50% of the plants within a plot are segregating, i.e., they are observably different, and a plot must be excluded when the main phenotype of the accession is not identifiable within the plot (for example images of the different categories of heterogeneity scores, see Supplementary File S1: Phenotyping cards). If possible, producing inbred genotypes by isolating the panicles within 100 μm mesh pollination bags ("bagging") before the start of genetic studies is recommended. When bagging panicles in the field, bag size is important. Bags of size 10 × 15 cm were found to be best; larger bags can be caught more easily by the wind. After flowering (at BBCH 70), the bags should be removed to allow the panicles to expand and grow. It is important that the panicles remain tagged after the removal of the bags to keep information on which panicle is to be harvested for pure seeds. For example, bags can be tied to the plant underneath the panicle upon removal.

Heterogeneity is scored by assessing the percentage of plants in the plot that have a visibly distinct phenotype from the majority of plants in the plot, where four categories are described as follows (see also Supplementary File S1: Phenotyping cards):

- 1: Most plants are the same (up to 10% different).
- 3: Over half of plants are the same (10–30% different).
- 5: Less than half of plants are the same (30–50% different).
- 7: Over 50% of the plants are different, completely mixed plot; will need to be excluded from analysis.

7.2.2. Plot Coverage

Another quality control measure is plot coverage, which can give an indication of the spatial heterogeneity at the plot level. The importance of recording plot coverage lies in the aforementioned phenotypic plasticity of most quinoa accessions in response to the space available around them, which is why this information may be included as a covariate in analyses of responsive traits.

The plot is assessed using a scoring metric based on percentages, where:

- 1: Up to 20% of the plot is covered, plant establishment is very poor.
- 3: Less than half of the plot is covered, ~30% (20–40%).
- 5: Around half of the plot is covered, ~50% (40–60%).
- 7: Over half of the plot is covered, ~70% (60–80%).
- 9: Over 80% of the plot is covered, plant establishment is very good.

To make the assessment of plot coverage in percentage easier, and if the plot size allows it, plants can be counted to see how the number compared to the total amount of plants a plot should have. Alternatively, plot coverage information can be deduced from plant emergence or UAV data, if available.

7.2.3. Stem Breakage Incidence

Other factors such as plant damage should also be considered in subsequent analyses of traits. For instance, damage to the stems may be relevant for yield measurements when the damage causes a loss of yield from affected plants. Stem breaks are often caused by strong winds, but can also happen after insect damage or some fungal diseases (which should be checked when stem breakages are observed). Because panicles get detached from the plant, further progression of their life cycle is prevented. Depending on the timing of the damage event, the affected panicles may still be harvestable. Stem breakage is an undesirable trait in a cultivar and is assessed on the basis of the percentage of the plot affected, where:

- 1: Up to 20% of the plot is affected.
- 3: Up to half of the plot is affected, ~30% (20–40%).
- 5: Around half of the plot is affected, ~50% (40–60%).
- 7: Over half of the plot is affected, ~70% (60–80%).
- 9: Over 80% of the plot is affected.

7.2.4. Stem Lodging and Stem Angle

In contrast to the snapping of stems, stem lodging refers to bent plants lying on or near the ground, with intact stems. Although the panicles are still harvestable in this case, depending on the degree to which a plant lodges and how close it is to the ground the panicle is, they might not be picked up by a combine harvester. In addition, where panicles are lying on the ground, pre-harvest sprouting and fungal infections may arise. Often the proportion of the plot that is affected by entirely lodged plants varies. Hence, for scoring stem lodging, the percentage of the plot affected is recorded using the same percentage-based categories as described above.

Alternatively, stem vertical angle, i.e., the angle at which the majority of plants in the plot are leaning towards, measured from the vertical axis, may be scored (see Supplementary File S1: Phenotyping cards). Here, a scoring system is used, where:

- $1 < 22.5°$ inclination or deviation of the stem from the vertical (i.e., most plants are upright).
- $3 < 45°$.
- $5 < 67.5°$.
- $7 < 90°$ (i.e., most plants are on or very close to the ground).

7.2.5. Panicle Axis Angle

Depending on the environment, or if a treatment (for example salinity) is applied, genotypes can be observed to have panicles that droop towards the ground. A deviation from the vertical may also represent a heat avoidance strategy as was observed in sunflower [172]. Similar to stem vertical angle, panicle drooping is assessed for the majority of the plot and based on the degree at which a panicle deviates from the vertical. Because a panicle can droop towards the ground while its stem remains vertical, this scoring system goes up to 180° in this case:

- 1 < 45° inclination or deviation of the panicle from the vertical (i.e., most panicles are upright).
- 3 < 90°.
- 5 < 135°.
- 7 < 180° (i.e., most panicles are pointing towards the ground).

7.2.6. Stem Lying Incidence

Stem lying sometimes co-occurs with stem lodging. The cause of stem lying remains unclear, but it results in the lying of stems on the ground, which may occur at the seedling stage of emergence (BBCH 00–09). The length of the section of a stem that is growing along the ground varies, indicating that the plants with stem lying could be classified into distinct groups. However, during in-field experiments, with multiple plants in a plot, it is apparent that there are some complications with this phenotype. Some plants exhibit severe stem lying, whereas others with the same genotype in the same plot have none. Hence, when stem lying is noted in the plot, the percentage of the plot that is affected should be scored, not the severity of the lying itself. The scoring categories are therefore the same as those for stem breakage listed above. Severely affected plots may need to be excluded from subsequent analyses. Depending on the severity of this phenotype observed in the field, stem lying may not be necessary to record.

7.2.7. Growth Habit

The architecture of quinoa plants varies greatly, and because this phenotype is particularly responsive to environmental conditions, the growth habit category for an accession is also classified at the plot level. Depending on the experiment, growth habit may also be assessed for individual plants.

The categories shown in Figure 4 were drawn based on their presentation in the Quinoa Descriptors [38], but the description of the categories was adapted to include the extent of growth habits that we observed in our diversity panel of approximately 1000 accessions. The feature that differentiates the groups is branching in the lower-third of the plant as well as the size of the panicles on those branches.

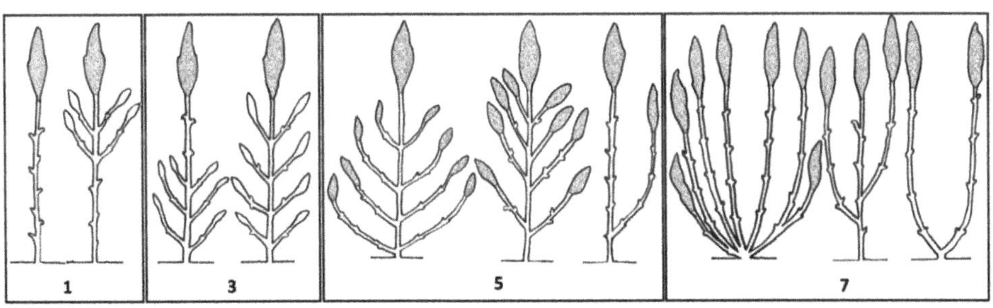

Figure 4. Growth habit of quinoa plants is grouped into four categories based on the branching habits at the base of the plant, and the size of the panicles of these branches: **1**—not branched at base; **3**—some branching from the base with no significant panicles on branches; **5**—branching from the base with more significant panicles; **7**—main panicle is difficult to identify.

- 1: Not branched at base, usually with a clearly defined terminal panicle.
- 3: Some branching from the base; no significant panicles on branches in the basal area (thus, this is not worth harvesting).
- 5: Branching from the base with more significant panicles.

- 7: Main panicle is difficult to identify.

7.2.8. Branchiness

Growth habit is focused on the branching habit at the base of the plant; however, in this category, we assess the degree of branching across the entire plant. The number of branches coming from the primary axis is easiest to assess when the plant is in principal growth stage 9, senescence. Because the phenotyping event is planned for a time where most plants are at principal growth stage 8, attention should be paid to avoid bias from the leafiness of the plant when scoring for branching degree. Plants from the middle of the plot are categorized into:

- 1: Low number or no secondary branches.
- 3: Some branches (30–50% of the primary branch length has secondary branching).
- 5: Branched (50–70% of the primary branch length has secondary branching).
- 7: Highly branched (above 70% of the primary branch length has secondary branching).

7.3. Plant-Level Phenotypes

After assessment of plot-level phenotypes, it is recommended that detailed observations of representative plants from each plot are made. As representative plants, again we select individuals from the middle of the plot (to avoid edge effects) which share features with plants across the entire plot, including the traits outlined in Tables 3 and 4. Depending on the time available for plant phenotyping, we select 1 to 3 individuals per plot. If there is a clear segregation of phenotypes in a plot, these plots are to be marked heterogeneous and are excluded from genetic studies. Representative individuals from each "type" can be separately phenotyped to maintain a record of how the accession segregated phenotypically. The phenotypes measured for each plant are divided into quantitative and categorical traits, as outlined in the following sections.

Table 3. Overview of the quantitative plant-level phenotypes that are measured for representative plants of a plot.

Plant-Level Phenotype	Unit	Description
Plant height	cm	Height of the most representative plants of the plot, usually from the middle of the plot, measured with a long measuring stick from soil to the tip of the panicle. If more than one distinct phenotype is present, more than one plant may be recorded in a new row of the spreadsheet, with all phenotypes that are differing recorded separately
Panicle length	cm	Length of the primary panicle measured with the same stick. Measured from the base of the panicle to the tip
Stem diameter near plant base	mm	Thickness of the stem measured with calipers at the middle of the bottom third of the plant stem
Stem diameter under panicle	mm	Thickness of the stem measured just underneath the panicle
Number of significant panicles	count	Count of the number of significant panicles, i.e., larger panicles, near the top of the plant, harvestable, that provide a major contribution to the seed harvested from the plant

Table 4. Overview of categorical plant-level phenotypes and description of the according scoring metrics.

Plant-Level Phenotype	Scoring Metric	Description
Growth stage	BBCH scale	Phenological growth stage; very important to record at mature phenotyping
Seed shattering	1,3,5,7	Grain persistence in the plant at physiological maturity. Assessing how easy seeds fall off the panicle upon light touch: 1 = no seeds falling to 7 = majority of seeds falling
Panicle shape	1,3,5	Classified into one of the three categories: glomerulate, intermediate, or amarantiform
Panicle density	1,3,5,7	Scored from 1 = lax (loose) with panicle axes easily visible to 7 = tight and compact panicles
Panicle leafiness	1,3,5,7	Scored from 1 = no leaves to 7 = many leaves
Panicle color	13,4,15,16,5,7	Categorized according to the color phenotyping card
Stem color	13,4,15	Categorized into green (13), red (4), or no pigmentation (15)
Stem striae	0,1	Presence (1) absence (0) scoring of stem streaks or stripes
Axil pigmentation	0,1	Presence (1) absence (0) scoring of pigmented axils
Stem leaf shape	1,2	Leaves of the stem are categorized into two groups: rhomboidal (1) and triangular (2)

7.3.1. Quantitative Plant-Level Phenotypes

An overview of the quantitative traits is shown in Table 3. They are presented based on the order used during phenotyping in the field.

Plant Height

Plant height is measured from the base of the plant at soil level to the tip of the primary panicle. We found that the use of metal ranging poles with alternating red and white colors to be the best option for a height reference because they are sturdy, yet not too heavy, metal poles with alternating red and white colors which work as height references. Extra markings and height labels need to be added to the poles using black tape and permanent markers. As ranging poles are made up of multiple parts that are screwed together, the height of the poles can be adjusted to the maximum height of the plant in the trial. If representative plants are not entirely upright, they can be held upright to record their height.

Panicle Length

Panicle length is measured for the primary panicle from the base of the panicle to its tip. In most commercial varieties, panicles are often extremely distinct and easily measured. However, in some quinoa accessions with wild phenotypes resembling *Chenopodium hircinum* accessions, panicles can be sparse, with the inflorescences spaced along the length of each branch. In these cases, we grasp the panicle below the lowest lateral branch with a large inflorescence and bring them together to get an estimate of the primary panicle.

Stem Diameter

Stem thickness is measured twice, once at the base of the plant and once just below the panicle. In our protocol, digital calipers are used to measure both. In dense plots or with large field trials, these measurements can be very time-consuming and are best applied for studies assessing the likelihood of stem lying, stem-breaking, and lodging.

Number of Significant Panicles

An attractive trait for a commercial variety that is harvested by machine is the presence of a single primary panicle. However, in many accessions, several additional panicles can be observed emerging from lateral branches. A significant panicle is a panicle which is large enough to make an important contribution to the yield obtained from a plant. These panicles are often on the upper half of the plant, near the primary panicle. The number of significant panicles should be counted, which is generally equivalent to the number of primary branches. The "significant panicles" should, together, contribute to an estimated 90% or more of total plant yield.

Categorical Plant-Level Phenotypes

Aside from the phenotypes that are measured, a number of traits are visually assessed and assigned to defined categories. An overview of the qualitative traits is shown in Table 4.

Seed Shattering

Dehiscence (or seed shattering) is a dispersal strategy of importance to wild plants, but this trait is a major cause for crop yield losses [173]. Increased persistence of the grain within the panicle is therefore a priority trait during the domestication process of a crop. This trait is assessed by lightly tapping 3–5 panicles while holding the other hand or a piece of paper underneath the panicle to catch the seed that falls off. The number of seeds that fell may be:

- 1: No seeds falling.
- 3: Some seeds falling.
- 5: Many seeds falling.

- 7: Majority of seeds is falling, "raining" seeds, and a large number of seeds present on the ground at measurement.

Panicle Shape

Three categories have been described for the overall shape of quinoa panicles [38], as illustrated in Figure 5. However, this trait has caused problems, as some panicles are not easily categorized into the described groups. The three groups were changed to scores of 1,3,5 instead of the scores previously named 1,2,3 in the Quinoa Descriptors [38] because quinoa panicles are so diverse that panicles cover a wide spectrum of shapes, rather than falling into the three distinct groups that are described below, or the two categories that were previously suggested, glomerulate and amarantiform [123]. The difficulty in categorizing is causing inconsistencies with scoring. Glomerulate panicles usually have clusters of glomerules at the end of a cluster of branches emerging from the secondary axis, as shown in Figure 6. To distinguish between the intermediate and amarantiform groups, the length of the secondary axis, which is usually packed tightly with glomerules up to the junction (resembling "fingers"), should be considered. Additionally, intermediate panicles can have short tertiary branches emerging from the secondary axes, usually from the bottom half of a "finger". The glomerules inserted into the short tertiary axes create "bulbous clusters" and lead to the presence of both, glomerulate and amarantiform features in an intermediate trait. Generally, it was also observed that panicles fitting into the intermediate category have their elongated glomerules only starting from a lower half of the panicle, while the top resembles a triangular shape of glomerulate glomerules. To improve this classification system in the future, image analysis algorithms are being developed for the identification of new groups from images of panicle inflorescences.

- 1: Glomerulate—glomerules with globose shape, resembling "bulbous clusters".
- 3: Intermediate—panicles have both amarantiform and glomerulate traits, resembling fingers with glomerules.
- 5: Amarantiform—glomerules with elongated shape, resembling "fingers".

Figure 5. Three distinct groups of observable panicle shapes: **1**. Glomerulate, **3**. Intermediate, and **5**. Amarantiform.

Figure 6. An example of a glomerulate panicle with a visible network of branches.

Panicle Density

Panicle density, a trait that can vary greatly with environment (such as temperature), contributes to the complications with defining panicle shape and is scored separately as follows:

- 1: Lax (loose)—glomerules sparsely spaced, panicle axes easily visible.
- 3: Intermediate—glomerules tighter but with panicle axes still visible.
- 5: Primary axis rarely visible.
- 7: Compact—glomerules tightly packed, no panicle axes visible.

Panicle Leafiness

Within the inflorescences, there are often leaves growing among the flowers (see Supplementary File S1: Phenotyping cards). Variability between genotypes and environments is observed for this trait and can be scored as follows:

- 1: Leaves are present in less than one-third of the panicles.
- 3: Leaves are present in more than one-third but less than three-fourths of the primary, sporadic, and not dense panicles.
- 5: Leaves present in three-fourths to of the entire primary axis, frequent but not dense leafiness.
- 7: Many leaves present throughout the primary axis.

Panicle Color

Color code descriptors with 15 colors for quinoa panicle scoring were previously provided by [38]; however, scoring for color is highly subjective. We find disagreement among individuals scoring panicle colors, particularly for differentiating red, pink, and purple. This problem highlights the need for providing color cards (e.g., Royal Horticultural Society Colour Chart, Methuen Handbook of Colour, or Munsell Color Chart for Plant Tissues) as a direct reference when accurate classification of colors is of interest, and in this case, it is advisable to extract color information from images. In most field trials, detailed color recording may not be a priority, but recording the most prevalent color is a useful indicator of phenotypic segregation and quality control as color is a dominant morphological marker [174]. Therefore, we reduced the number of categories to the following six:

- Green (13);
- Green with Purple (16);
- Pink/Purple/Red (4);
- Orange/Yellow (5);

- Dark colored (7);
- Beige/White (i.e., no pigmentation, mostly for mature plants) (15).

Stem Color

For stems, 11 colors were proposed [38]. This wide color variation has not been observed in the field among our diversity panel and is difficult to correctly identify under field conditions. Stem color is useful to serve as a quality control for validating protocols in genetic analyses because the loci associated with stem color have already been identified [15]. Therefore, plants only need to be categorized into the following:

- Green (13);
- Red (4);
- No pigmentation (beige, white, yellow) (15).

Stem color can also provide information about the homozygosity of an accession, and accessions have been observed with red coloration at the leaf–stem intersection, which may be used as a control for whether crosses have been successful [174]. A lack of pigmentation is based on panicles and stems that have lost their color owing to senescence. It is not always possible to time the scoring campaign for color across all accessions because there can be great variability in the number of days required to achieve maturity.

Stem Striae and Axil Pigmentation

The presence of stronger pigmentation forming stripes on the stems and pigmented axils are traits that persist and may be used for example for the identification of successful crosses. If desired, stem striae color can also be recorded. Separate genetic mechanisms to color may likely regulate the location of the pigment accumulation or synthesis. Hence, it should be sufficient to score for:

- Presence (1);
- Absence (0).

Stem Leaf Shape Characteristics

Leaf shape characteristics are heritable traits that may be useful to breeders for variety identification. These traits may also be correlated with irradiation capture, water use stability or yield stability. They need to be recorded at flowering. As described by [38] stem leaf shape can be categorized into two groups:

- Rhomboidal (1);
- Triangular (2).

The leaf margin refers to the shape of the edges of the leaf. This trait also summarizes teeth number, another leaf shape characteristic which can be scored separately if more detail is required. Leaf margin is categorized into:

- Entire (1);
- Dentate (3);
- Serrate (5).

8. Phenotyping of Disease

Depending on the environment in which a trial is conducted, quinoa may also be affected by diseases that influence plant health and yield negatively. Therefore, recording of disease occurrence and scoring for disease severity might be necessary in field trials where diseases are observed despite taking pest and disease control measures. As infections can progress rapidly, it is important that the plots of a trial are regularly assessed for disease throughout the growing season, and controlled for [175]. Trials where disease is not controlled but representing the studied treatment are independent trials designed for the phenotyping of disease, typically by the selection of genetically diverse accessions with variable disease resistance. Disease assessment includes measuring the incidence (the number of affected plants out of the total assessed) and severity (proportion of plant area or fruit volume destroyed by a pathogen) [176]. Observations must be conducted at least three times: once during phenological growth stage 1, when plants have around nine true leaves, and before branches develop; a second time during either principal development 4 (development of harvestable vegetative parts), principal growth stage 5 (inflorescence is visible), or principal growth stage 6 (flowering); and a third time during principal growth stage 8. To avoid difficulties arising with senescing leaves, observations should be conducted before principal growth stage 9.

Accurate diagnostics of quinoa diseases, however, are complicated because multiple pathogens often appear in communities [177]. A field situation is a complex interaction between the plant and its microbiome. Therefore, incorrect identification of the pathogen involved can occur if only a single organism is considered. Some plant disease agents can be identified through their symptoms and classified from infected tissue by skilled plant pathologists. However, numerous pathogens cannot be distinguished from each other based on the visual assessment of disease symptoms. In fact, molecular tools and clear distinctions among quinoa plant diseases were lacking in the past, with only a few examples properly described. The pathogens identified to affect quinoa include the fungal pathogens *Ascochyta caulina, Cercospora* cf. *chenopodii, Colletotrichum nigrum, C. truncatum, Fusarium* spp., and *Pseudomonas syringae* [175]. The predominant and most well-described pathogen is the oomycete *Peronospora variabilis:* it causes downy mildew, and its impact is considered to be one of the most economically important [51,178–186]. The complexity of diseases can lead to inaccurate diagnosis in field trials. Therefore, we propose for the in-field phenotyping of disease symptoms to assess all three parts of the plant: foliage, stem, and panicles. Leaf symptoms should be well-described with respect to detectable changes in lesion shape (irregular, blotch, spots) (Figure 7(Aa,b); 7(Ca,b), color (pink, bronze, chlorotic, mix) (Figure 7(Aa–d), other symptoms (surrounding halos, concentric rings) (Figure 7(Aa,d)), and distinctive signs such as sporulation usually present on the abaxial side of the leaf (Figure 7(Ba,b)), and chlorotic leaf veins (Figure 7(Bc)). Next, the type of lesions on a leaf (dots, diffuse or extensive) are recorded (Figure 7(Ca–d)), followed by the amount of spread which is measured by the percentage of leaf area affected (Figure 7(D1–5)) in relation to its total area. The scoring could be based on a scale from 1 to 5 (1 = 0–10%; 2 = 11–25%; 3 = 26–50%; 4 = 51–70%; 5 = 71–100%), which is frequently used as a measurement of disease caused by various pathogens including severity of downy mildew of quinoa [187].

For the estimation of severity in the field, we suggest a selection of 3–10 representative leaves. Disease symptoms are often influenced by the age of the plant and position of the tissue [176,188]. Leaves at the base of the plant could be displaying symptoms of senescence. In contrast, leaves toward the apex of the plant may display induced resistance which often occurs not only at the site of the initial infection but also in distal uninfected parts [189,190]. Therefore, we propose that samples selected from the middle part of the plants best represent the infection. In the case of stems and panicles, similar principles should be applied. Examples for visible changes in panicles and stems brought upon by pathogens are shown in Figure 8. Panicles and seeds can be affected by various pathogens causing different amounts of rotting, which can result in total harvest loss and contaminated seed. The example for stem disease in Figure 8B shows a mature stem with necrosis of diamond shape (which can cause plant lodging), concentric rings, and visible pycnidia. Another young stem shows pink coloration and white mycelia.

Figure 7. (**A**) Lesions on leaf surface: (**a**) pale or yellow chlorotic lesions with or without a halo and occasional pink-orange discoloration caused by leaf pathogens, (**b**) bronze irregular lesions caused by *Alternaria* sp., (**c**) diffuse chlorotic spots caused by *P. variabilis* and (**d**) concentric and chlorotic halo under artificial inoculation with *Alternaria* sp. (**B**) Sporulation on lower side leaf surface with (**a**) black dots showing downy mildew sporangia, (**b**) dark gray-violaceous sporulation caused by *P. variabilis* and (**c**) vein discoloration, general chlorosis and pink-orange spots caused by *Fusarium* sp. (**C**) Lesion type on upper surfaces and amount of disease ranging from (**a**) dots, (**b**) dots expanding, (**c**) diffuse, and (**d**) extensive. (**D**) Severity phenotyping scale used for assessing the percentage of the leaf area affected where 1 = 0–10%, 2 = 11–25%, 3 = 26–50%, 4 = 51–70% and 5 = 71–100%. Leaf examples given represent different degrees of severity during the infection of *Alternaria* spp. (Colque-Little and some images previously published in [51]).

For accurate disease diagnostic, further procedures are needed: samples should be saved for isolation, microscopic analysis, and molecular identification of the causal agent. Koch's postulates, guidelines for determining a causative relationship between a microorganism and a disease, should be validated. The steps required for the standardized procedure include (a) description of disease symptoms, (b) isolation of the disease agent, (c) artificial inoculation of quinoa tissue with the isolated agent, (d) recording of symptoms on the infected quinoa tissue followed by (e) a re-isolation of the microorganism from the infected tissue [133,176,191]. More ongoing research is required to identify possible pathogens, especially newly emerging ones, behind quinoa diseases. Advancement of molecular diagnostics such as development of rapid DNA extractions and newly designed species-specific primers along with advanced remote sensor-based image techniques are expected to be helpful for fast and more accurate disease detection in the near future.

Figure 8. (**A**) Examples for panicle diseases, with (**a**) a panicle infected with *Alternaria spp*., and (**b**), a panicle predominantly infected with *Cladosporium* spp. at the end of the season. (**B**) Examples for stem diseases, showing (**a**) quinoa black stem caused by *Ascochyta caulina* with presence of dark structures (pycnidia), (**b**) pink stem and light pink mycelia corresponding to *Fusarium* spp. (pictures by Colque-Little, (**B**)(a) courtesy of [182].

9. Harvest and Post-Harvest

When all other phenotypic traits have been collected and plants have reached maturity (BBCH 89), they may be harvested and prepared for post-harvest data collection. Depending on the trial size, type, and resources available, harvest may be approached using different methodologies. Irrespective of the method selected, it is important that an indication of the number of plants that contributed to the yield is available, by counting the plants when hand-harvesting, harvesting only a set number of plants per plot, or harvesting the entire plot and using previous field emergence information. Recording the plant number allows the yields obtained in one trial to be compared with results from another trial. This is crucial information to collect for international collaborative quinoa research. The recommended harvest and post-harvest traits are listed in Table 5.

Table 5. Overview of harvest and post-harvest traits.

Harvest and Post-Harvest	Unit	Description
Number of plants harvested	count	
Above-ground dry biomass	grams	Cutting plants at the very base with secateurs and drying the entire plant in an oven until mass is constant. Recording total dry weight
Below-ground biomass	grams	If possible, root biomass could also be measured (especially when plants are growing in sandy soil)
Seed yield for representative plants	grams	Seed mass of approximately four representative plants that were harvested from the center of the plot (seed should be dried to constant weight)
Total seed yield per plot	grams $\times$ m^{-2}	Harvesting the panicles remaining per plot while excluding borders, and adding the weight to that from the four representative plants, seed dried in oven to constant weight
Seed yield per plant	grams	Total harvested seed mass per plant may be calculated from the seed weight of all plants in the plot divided by the number of plants harvested
Harvest index		Yield/ above-ground biomass
Seed weight (TGW)	grams/1000 seeds	Thousand Grain Weight (TGW), the weight of 1000 seeds
Seed hectoliter weight	grams/100mL	Estimation of density, determined by weighing all seeds fitting into a 100 mL volume
Seed size (average area; average perimeter)	millimeter	Seed size outputs from image analysis separated by semicolon (method options described in Section 9.2)
Seed color (average red; average green; average blue)	Numeric RGB equivalent	Seed color output values for red, green, and blue components, semi colon separated (obtained from image analysis methods, see Section 9.2)

9.1. Harvest Protocols

When developing harvest protocols, it is important to consider edge effects. Therefore, when choosing a plant for phenotyping, it is best to select only plants from the inside of the plot, while disregarding unusually small plants or the typically larger and more branched individuals at the borders, which might cause a bias in the yield predictions. Similarly, in plots with uneven emergence, unusually large individuals should not be selected. Depending on the plot size, it is recommended to harvest a larger number of representative individuals (20–30 plants) because smaller sample sizes may lead to less accurate predictions for yield. The number of harvested plants as well as the plot coverage should always be noted. Main panicles may be harvested separately from secondary panicles of a plant to obtain an indication about the distribution of seed on a plant.

For calculation of the harvest index, i.e., the ratio of harvested seeds to total dry above-ground biomass, a subset of plants should be selected for harvest as entire plants and cut at the base of the plant at soil level using secateurs. Depending on the plot size, 4–6 plants may be used for this evaluation. The plants should be placed in bags and dried in in an oven at no less than 60 °C for several days (until weight is constant). Once plants are dry, they can be weighed to obtain shoot biomass before proceeding to threshing and weighing the seed. This information is extremely important for identifying the best-performing genotypes, which are those that invest more resources in their seeds rather than to their above-ground biomass.

Threshing and winnowing are, together, the process of separating the seed from the chaff or straw and is easiest when the panicle is dry. Before mechanically threshing, the seed should be loosened from panicles and chaff by hand to facilitate the threshing and winnowing processes and reduce seed loss during machine threshing. After threshing, the seeds are weighed to obtain yield. A sample can be taken and weighed to obtain the Thousand Grain Weight (g/1000 seeds). However, for higher throughput and more detailed information, the seed scanning method described in the following section is recommended.

9.2. Seed Phenotyping

Because the seed is the final product, seed properties, especially their nutritional properties, are important to be considered when selecting varieties of interest. Seed color was not found to be correlated with significant differences in most nutritional properties, except perhaps protein and carbohydrate contents. Pereira et al., (2019) [192] have reported that white seeds had the lowest protein and highest carbohydrate contents compared with red and black seeds. However, color is of interest because large white seeds are in demand in the market for quinoa. In some countries, such as Bolivia and Peru, large red or large black seeds are also desirable. Irrespective of the color, large grain size is the highest priority for international market [7].

The phenotypes of seed color and grain shape were previously divided into different categories [38]. However, visual assessment of the seed and assigning it one of four not easily distinguishable categories for shape is not very precise. With color identified as a highly subjective trait, in the following section, we present an alternative that saves considerable time spent on phenotyping these traits without the need for expensive equipment.

9.2.1. In-Field Seed Morphology Descriptors

An image feature extraction algorithm was developed to produce the following seed morphological descriptors from an image: seed area, perimeter, color, and counts. Although centralized image collection with controlled lighting and fixed resolution provides the highest quality images, the collection of images in the field can save time and resources. This algorithm also provided accurate results on images that were obtained using the camera of a mobile phone. However, adjustments to the feature extraction methods are required because images from these cameras are distorted.

Apart from a camera, a backdrop for quinoa seeds is needed. A blue card of fixed size is best. The blue color provides a color contrast with the quinoa seeds, and the fixed size allows pixel to mm conversion. A template for this 10×10 cm^2 square blue card is provided (Supplementary File S2: Blue card template) and should be cut from weighted paper of the reference color. Overall, 10–50 seeds are placed on the card, and a photograph is taken. In the case of in-field image collection, the image may be taken with a camera held in one hand, while the card with the seeds is held in the other, as shown in Figure 3D.

The measurement of seeds on the blue card comprises three steps: isolation of the blue card from the scene, rectification of the card to a square of known pixel count, and segmentation and measurement of the seeds from the card. The image analysis software CyVerse was made available to the community on CyVerse [193]. In the CyVerse Discovery Environment, images can be analyzed using the Phytomorph Image Phenomics Toolkit. After clicking on the app, an analysis name can be assigned and single images or folders that are uploaded to the CyVerse Data Environment can be selected for analysis. For this, the required image analysis algorithm must be selected. The algorithm for the method described here is called "Quinoa Seed Card". Once the analysis is completed, the user is notified via email and can return to the Discovery Environment where the outputs are stored in the user's database. The outputs can be browsed there or downloaded. The results include values for average area, average perimeter, and average red, average green, and average blue components. In addition, images are returned with seeds that were detected by the algorithm as single seeds, used for analysis, and highlighted in red. This allows confirmation that the algorithm showed expected performance.

9.2.2. Seed Scanning

For more accurate measurements of seed morphological characteristics, including size, shape, and color, a high-throughput seed scanning system has been established in the Sustainable Seed Systems Laboratory (SSSL) at Washington State University (WSU) to capture images of quinoa seeds. Similar systems may be established elsewhere; however, users must consider their needs. The SSSL system uses eight flatbed scanners, which capture images at a resolution of 1200 dots per inch (dpi). System design is focused on the ability to queue and initiate four samples on one set of scanners, with the process being repeated on other set of scanners while the first set of images are captured. This system allows the analysis of approximately 50 samples per hour and supports data collection for thousands of samples in a year.

Users should consider their desired throughput when designing a system to best meet their needs because this will determine the number of scanners required and their orientation in the workspace. Certain pieces of equipment are required, regardless of throughput (Supplementary File S3: Equipment List File). The SSSL system uses Epson Perfection V39 flatbed scanners (Epson America, Inc., Long Beach, CA 90806), which balance affordability (USD 50–100) with capability (up to 4800 dpi optical resolution; 600 dpi color scan in 30 s). A small subsample (approximately 1–2 g) of each sample is taken for seed scanning. A sample splitter or other comparable equipment should be used to collect a representative subsample. Seed samples are weighed before scanning; however, scanners with integrated balances are available. The SSSL uses an Ohaus Scout SPX123 Analytical Balance (Parsippany, NJ, USA), with a capacity of 120 g and a precision of 0.001 g. Seeds are carefully scattered on the scanner, within the field of view, to limit the number of seeds that are touching. Then, the scanner lid is closed. The underside of the scanner lid, acting as the image background, has been painted using the flat, matte paint color "Blueberry Festival", which can be characterized by red, green, and blue values of 73, 139, and 184, respectively, and HEX #498BB8 (Valspar Paints, Cleveland, OH, USA).

Scanners are connected to a Dell Optiplex 9010 SFF Computer (Intel Core i5-3470 3.2 GHz, 16GB RAM, 2 TB HDD) with USB cables. This machine runs a Linux operating system (Ubuntu 18.04) to support simultaneous scanner operation and image filing using Python shell scripts. The system is command-line driven, where QR codes are scanned using a wireless 1D/2D barcode scanner. The QR codes have embedded information related to the scan command, scanner identity, and sample identity to facilitate high-throughput operation and reduce the risk of human errors associated with manual information entry. Images captured on each scanner are saved, backed up on an external hard drive, and uploaded to the CyVerse Discovery Environment for image analysis.

Subsequent image analysis provides a robust data set for use in phenotyping and quantitative analyses. Image analysis is performed in the CyVerse Discovery Environment using the Phytomorph Image Phenomics Toolkit, the same application that was used for the analysis of the blue card images described previously. First, a single high-throughput file path is created with a set of images. Next, the high-throughput file path is selected for analysis with either the Arabidopsis Seed Method (black background) or All Grains (blue, lilac, green or white background) single file tool in the Phytomorph Image Phenomics Toolkit. A high-throughput multifile path is created from the output using .json files. This high-throughput multipath is then selected for analysis using the JSON compiler in the multifile tools in the Phytomorph Image Phenomics Toolkit. Finally, a .csv file of results for each image is available for download. The data include measurements of each seed in the image, with mean and standard deviations. A tutorial demonstrating this process is available on the SSSL YouTube channel, SSSL Training - Image Analysis. Available online: https://www.youtube.com/watch?v=9SK4vkfeJHI (accessed on 15 August 2021).

The CSV file from the seed image analysis contains a large amount of information. Seed size, shape, and color are quantified. The number of seeds in each subsample is automatically counted, and this information is combined with the weight of the subsample to calculate the weight of 1000 seeds (Figure 9). Seed size includes seed area and the length of the major and minor axis, which are used to calculate seed shape (i.e., eccentricity). Seed color includes red, green, and blue values that can be used to produce any color. Seed count is reported as the number of objects in the image. If seeds are touching, the size of that shape is divided by the average size seed to determine the number of seeds in the aggregate. Means and standard deviations are reported for each measurement. The mean values represent an average calculated using the values for each seed, with the standard deviation representing the variability within the sample. For some measurements, such as color, a standard deviation of standard deviations is reported; a standard deviation for the sample is calculated using the standard deviations of all the seeds in the sample (Supplementary File S4: QuinoaCollab_NIR_Scan File).

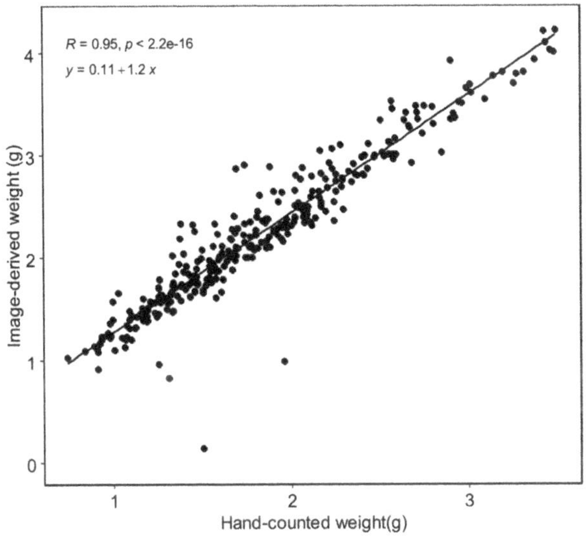

Figure 9. Correlation between thousand seed weight as estimated using image analysis, where the number of seeds is counted and the weight of those seeds is known, and hand-counted thousand seed weight.

9.2.3. Seed Nutritional Phenotyping

Near-Infrared Spectroscopy

Quinoa seed composition, especially protein content and composition, can be highly variable [194,195]. Seed composition can vary depending on processing, when certain parts of the seed are removed or modified through washing, abrasion, or milling [196]. Therefore, seed composition analyses and their results must be considered in the context of processing and physical state of the quinoa seed before analysis. One method to achieve HT analysis of quinoa seed composition is the use of near-infrared (NIR) spectroscopy, in which, in theory, any sample measurement can be predicted as long as the sample's spectral data are correlated with the desired measurement [197]. This technology has diverse capabilities, such as phenomic selection and prediction of maize yield from kernels [198], and has been successfully applied to predict amino acid content in quinoa [199]. In addition to the NIR methodology, mid-infrared should be considered as a useful technology with practical applications in quinoa. For example, mid-infrared has been used to classify groups of quinoa [200] and characterize rheological properties [201].

An NIR calibration requires careful selection of samples to serve as the reference data, from which multivariate regression equations will be created. Spectral data are used to predict the target measurement(s). Sample selection should represent the target population of spectra [197]. One approach is to collect spectral data for as many samples as possible. These samples should represent the genotypes and environments that will be routinely analyzed. Next, principal component analysis of the spectral data, combined with the Kennard–Stone method of sample selection, can identify the best candidates to include in the calibration [202]. The R package prospectr provides multiple options for selecting samples for calibration and validation sets using a multivariate spectral data set. Another approach is to select samples representative of a normal distribution for the desired measurement, such as protein content.

Reference data, such as protein content and amino acid composition, should be measured according to Official Methods of Analysis [203]. Blind duplicates, i.e., multiple samples from the same seed sample submitted for analysis and acting as quality control, should be included when possible to estimate and account for errors during calibration development. Various methods can be used to develop calibration equations, such as partial-least squares regressions. The calibration metrics, such as standard error of prediction, should be quantified through either internal or external validation processes [198,204]. The development of calibration processes is ongoing. The initial calibration should be updated over time by incorporating outliers into the reference data set or additional products (market classes, flours, etc.) into the crop analysis profile to improve the calibration(s).

NIR—An Example of Calibration Development for Quinoa

When beginning to develop an NIR calibration, one should consider how the quinoa samples will be processed before analysis and determine the desired outputs from the analysis. For example, the WSU SSSL has initiated NIR calibration development with raw whole quinoa seed, which are unprocessed but cleaned of non-seed material to predict crude protein, crude fat, ash, and moisture content, in addition to a complete amino acid profile using a DA7250 (PerkinElmer, Waltham, MA, USA) with an NIR range of 950–1650 nm. The advantage of this approach is that samples require minimal processing before analysis. The disadvantage is that sample homogenization through milling could provide a more accurate representation of seed composition because the entire seed is made available for spectral reflectance. Scatter corrections, such as standard normal variate (SNV) or multiplicative scatter correction (MSV) can be applied during pre-processing of spectral data to correct for differences in the sample matrix (i.e., seed surface) [205]. A non-destructive approach ensures that seed viability is maintained to support subsequent breeding activities, such as greenhouse seed increases, field trials, and further research and analyses, such as food science studies.

The SSSL has improved the stock DA7250 NIR calibration twice. The stock calibration included 27 samples and reference data for moisture, protein, and ash. However, it was not robust and poorly predicted novel quinoa samples; NIR analysis is best at interpolation rather than extrapolation. The second version (V2) added samples from the WSU breeding program materials and field research trials [206]. These samples were randomly selected across a normal distribution of crude protein content predicted using the stock calibration. The current NIR calibration (V3) incorporated 37 samples selected from a collection of diverse genotypes grown in Australia in 2018. Samples were selected using the Kennard–Stone method, with principal component analysis of the spectral data. The reference data collection included 10 blind duplicates to measure and account for laboratory standard error in the calibration.

Eight-fold cross validation was performed in triplicate to measure calibration prediction accuracy metrics, which are reported as an average measure (Table 6). These metrics provide an indication of how well the calibration may perform. For example, a large range in reference data values is beneficial for encompassing the possible values of experimental samples that may be analyzed, and usually contributes to higher prediction accuracy as measured by the correlation coefficient between measured and predicted values. The metrics related to the cross validation provide various measures of the calibration prediction accuracy. Although the current calibration includes various seed colors, calibrations specific to the major quinoa seed colors—white, black, and red—may be more appropriate. NIR calibrations exist for developed market classes in other crops, such as red and white wheat. The SSSL will continue to analyze diverse genotypes grown in varying environments for identifying candidate samples with the potential to improve the accuracy of the NIR calibration for predicting quinoa seed composition. This will be achieved by either including samples with novel spectral signatures, or by including data that increase the range of reference data for particular seed components.

Table 6. Washington State University Sustainable Seed Systems lab NIR calibration (V3) metrics. The range, minimum (min), and maximum (max) are calculated using reference data for quinoa samples included in the calibration (n = 175). Calibration prediction accuracy metrics are reported as an average measure of 8-fold cross validation in triplicate.

	Stats from WSU Calibration V3 Data (g 100g^{-1} Protein)							
	Range	**Min**	**Max**	**RMSECV**	**SECV**	**Robust SECV**	**RPDCV**	**R2CV**
Alanine	1.99	2.89	4.88	0.022	0.022	0.018	3.036	0.892
Arginine	4.68	4.58	9.25	0.053	0.053	0.044	4.308	0.946
Aspartic acid	3.22	5.51	8.73	0.039	0.040	0.036	3.768	0.930
Cysteine	0.76	1.31	2.07	0.010	0.010	0.010	3.188	0.902
Glutamic acid	7.04	8.22	15.26	0.093	0.093	0.086	3.802	0.931
Glycine	1.33	4.78	6.11	0.041	0.041	0.036	2.447	0.834
Histidine	1.05	1.96	3.01	0.015	0.015	0.014	4.564	0.952
Isoleucine	1.51	2.89	4.41	0.021	0.022	0.019	3.392	0.913
Leucine	2.55	4.3	6.85	0.031	0.032	0.029	3.473	0.917
Lysine	3.14	3.45	6.59	0.029	0.029	0.033	3.290	0.908
Methionine	1.15	1.31	2.46	0.012	0.012	0.009	2.955	0.886
Phenylalanine	1.57	2.71	4.28	0.019	0.019	0.018	3.889	0.934
Proline	1.68	2.80	4.48	0.023	0.023	0.018	2.556	0.847
Serine	1.39	2.89	4.28	0.019	0.019	0.016	3.176	0.901
Taurine	1.96	0.82	2.79	0.012	0.012	0.009	1.669	0.645
Threonine	1.60	2.43	4.02	0.017	0.017	0.016	3.015	0.890
Tryptophan	0.93	0.55	1.48	0.012	0.012	0.009	1.681	0.647
Tyrosine	0.93	2.12	3.05	0.014	0.014	0.013	3.393	0.913
Valine	1.84	3.36	5.20	0.024	0.024	0.023	3.260	0.906
Hydroxylysine	0.18	0.05	0.23	0.004	0.004	0.003	1.591	0.605
Hydroxyproline	0.93	0.29	1.21	0.010	0.010	0.011	1.821	0.699
	Stats from WSU calibration V3 data (g 100g^{-1} sample)							
Crude protein	11.95	6.82	18.77	0.394	0.395	0.406	5.521	0.967
Ash	3.32	2.21	5.53	0.154	0.154	0.129	3.084	0.895
Crude fat	6.95	0.00	6.95	0.310	0.311	0.316	3.883	0.934
Crude fiber	13.67	1.44	15.11	0.442	0.443	0.377	4.904	0.958
Moisture	3.76	6.41	10.17	0.183	0.183	0.159	6.579	0.977
TotalAA	10.06	5.84	15.90	0.413	0.413	0.328	4.018	0.938
	Range	Min	Max	RMSECV	SECV	Robust SECV	RPDCV	R2CV

Hydroxylysine and hydroxyproline are poorly predicted. Lanthionine and ornithine were eliminated from the calibration owing to limited lab analysis. RMSECV = root mean square error of cross validation; SECV = standard error of cross validation; RPDCV = ratio of reference data standard deviation to standard error of prediction; R2CV = coefficient of determination of cross validation.

The Nutritional Phenotyping Pipeline at Washington State University

HT analysis of seed composition and characteristics is performed by SSSL at WSU using a Nutritional Phenotyping Pipeline. The system has been applied to barley, perennial grains, quinoa, and camelina and has the potential to be applied to diverse crops. The system is flexible and can accommodate varying sample amounts. Seed characteristics can be measured using image analysis with as little as 1–2 g of quinoa. With approximately 10 g of quinoa, seed composition can be estimated using a NIR analyzer; however, larger seed samples of 100–300 g ensure a more robust and representative analysis. This flexibility supports the analysis of single plant samples up to plot-level samples. Moreover, analysis of mineral content and composition in whole quinoa seeds is under development using energy-dispersive X-ray spectroscopy. Sample organization and tracking is maintained throughout the Nutritional Phenotyping Pipeline using a system that relies on 2D digital barcodes (QR codes), barcode scanners, and USB drivers. The system is designed for quality assurance and quality control by automating sample data entry and processing. Additional information on the pipeline workflow and development as well as videos detailing each step in the process can be found on the SSSL YouTube channel. Available online: https://youtube.com/playlist?list=PLdKoK4IZoGTAFYZWCev4vteOErik3CKcS (accessed on 15 August 2021)

9.2.4. Detection of Saponins in Quinoa

Most quinoa seeds also contain a large variety of compounds called saponins, some of which have shown to harbor antinutritive properties, thus making them undesirable for human consumption. The predominant form of saponins in quinoa are triterpenoid glycosides [207–209]. Saponins have foaming characteristics and are bitter in taste. They have been found to largely localize on the outside of the seed, making it possible to wash them off or remove them by abrasion before consumption (reviewed in [210]). The production of quinoa without those bitter saponins has been a breeding target, and some naturally non-bitter quinoa accessions have been identified [15]. For breeding and phenotyping purposes, testing for saponins is desirable. Over 90 different saponins have been found in quinoa [208]. Saponins are composed of an aglycone backbone with sugar moieties. The combination of those two make up the large variety. Four of the common aglycones in quinoa are oleanolic acid, hederagenin, phytolaccagenic acid, and serjanic acid [208].

For phenotyping purposes, a simple detection test may be used that utilizes the foaming characteristics of saponins. When shaken in water, saponins foam, a property that is used for the afrosimetric method [211]. Colorimetric methods such as the use of spectrophotometry can also be used for saponin detection (e.g., [212]). However, these methods also detect the phytosterols in plants. Colorimetric methods are, therefore, not worth the extra effort, while it is easier and cheaper to do the afrosimetric test. Jarvis et al., (2017) [15] validated the results of an afrosimetric test using a more specific detection method, i.e., gas chromatography–mass spectrometry (GC–MS), on a mapping population, which segregated for saponins. The afrosimetric test and GC–MS method corroborated the absence or presence of saponins.

The afrosimetric test can be easily performed as part of a field trial on a large number of samples according to the following method: five quinoa seeds, free from loose hull, are placed in a 1.5 mL microcentrifuge tube containing 500 µL of double-distilled water. The tubes are shaken by hand vigorously for 30 s until a foam appears that is stable in height, as shown in Figure 10. If foam occurs, it may be semi-quantified by using a caliper and measuring the foam height. A vortex swirls the sample and does not, on its own, lead to stable foam heights. The afrosimetric method is not suitable for the quantification of saponins or investigation of the types of saponins.

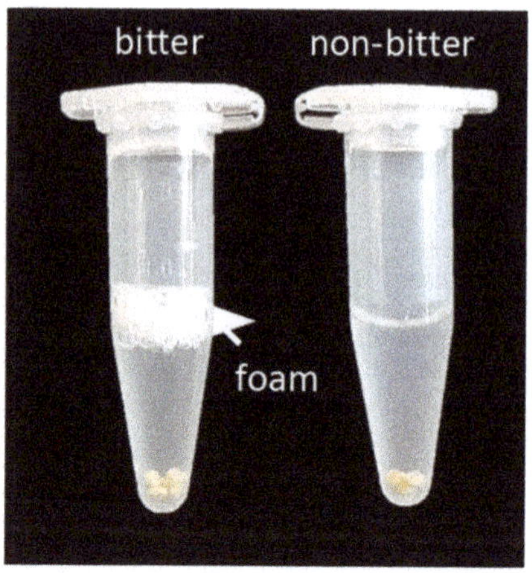

Figure 10. Foaming test (afrosimetric method) for presence or absence of saponins on quinoa seeds.

To quantify saponins, GC–MS or liquid chromatography–mass spectrometry (LC–MS) can be used with appropriate standards. GC–MS includes a derivatization step, which removes the sugar moieties from the aglycone backbone; hence, GC–MS allows quantification based on the aglycone backbone [15]. LC–MS allows detection of the individual saponins; however, this quantification is limited by the availability of standards. Both GC–MS and LC–MS are more laborious and expensive than the afrosimetric test.

9.2.5. Quinoa Seed Longevity

Every stage of seed production, from field selection to harvesting and processing to seed storage, is crucial for the quality management of seed [213]. Seed longevity is related to the prediction of seed viability in a storage environment, and depends greatly on its composition, the environment during seed maturation, and harvesting [214,215]. Moisture content, temperature, and oxygen are fundamental factors that control seed longevity [216]. Among these factors, elevated seed moisture is the main culprit for loss of seed quality during storage [217,218].

Both natural and economic resources are wasted owing to inadequate seed storage if seeds of poor quality are sown [219]. Thus, high quality seeds are ensured at the time of planting if the seeds maintain their quality during seed production and at the time of harvesting, processing, and storage. Seeds having higher initial quality have greater longevity than seeds from the same genotype of lower initial quality [219]. Environmental conditions, especially higher temperatures and long photoperiods, during seed development promote dormancy after harvest [220,221]. As quinoa belongs to the Amaranthaceae family, its varieties proceed through different types of seed dormancy or sometimes have no dormancy [222]. The influence of maternal environment on the seed-coat's characteristics is associated with the level of dormancy in Chenopodium seeds. The Chilean accession showing higher level of dormancy had a significantly thicker episperm for all sowing dates [221]. Some quinoa varieties have no dormancy and in wet environments, seeds may germinate inside the panicle before harvest [222]. This condition can lead to large yield losses and the desiccation intolerance of unorthodox seeds leads to different storage requirements. It is, therefore, important that quinoa varieties are also evaluated for their dormancy type and that preharvest sprouting is recorded when observed. After harvest, the viability of seed should be monitored with Tetrazolium tests on regular basis (see [221]).

Cultivated quinoa has small, flat seeds that are highly hygroscopic in nature and absorb water very quickly, within a day, owing to the porosity of its integument. This moisture gain can be used as an indicator to predict seed longevity [220]. Despite the potential to grow quinoa under adverse environmental conditions, its seed quality deteriorates with inadequate storage conditions, particularly at high temperatures and relative humidity [223,224]. Quinoa seed loses viability extremely quickly compared with conventional cereals such as maize, wheat, and rice [225]. There is a need to explore the physiological and biochemical changes associated with seed longevity under ambient storage conditions. Castellión et al., (2010) [226] found a strong association between quinoa seed aging and the accumulation of Maillard reaction products formed by a reaction between amino acids and reducing sugars, which is responsible for protein aggregation and insolubility. Thus, protein insolubility and water mobility through the multilayers of the seeds are key indicators for the prediction of seed longevity in quinoa germplasm [220]. Both pericarp and seed coat are comprised of two layers of cells. In the pericarp, the inner layer is discontinuous and its cells are tangentially stretched while the large cells of the outer layer are papillose in shape. The seed coat also consists of two cell layers, the exotesta and the endotegmen [227]. In contrast, lipid peroxidation is not a good indicator for seed longevity because of the high oxidative stability of the lipids associated with the high vitamin E content of quinoa seeds [228].

The pattern of loss of viability among quinoa accessions depends on post-harvest management, seed provenance, germplasm, and conditions prevailing during seed development, where seed maturation is the most sensitive phase for seed viability [222]. Low seed moisture content and temperature are basic principles for the storage of orthodox seeds such as many quinoa varieties [219]. The best approach is Dry Chain Technology, which is aimed at proper drying (natural or artificial drying to safe moisture limits) of seeds after harvest followed by hermetic packaging to keep it dry until used in the value chain [218]. Hermetic bags are composed of a plastic that resists the exchange of moisture and gases, thus a modified atmosphere can be created by depleting oxygen and enriching carbon dioxide inside the bags [217]. The popular hermetic bags are Super Bags (GrainPro, Washington, D.C., USA) and Purdue Improved Crop Storage (PICS) bags, and are being used in >80 countries to protect grains, legumes, and industrial commodities [229]. A PICS bag is comprised of a double layer of high-density polyethylene (HDPE) liners inside of a woven bag, while a Super Bag consists of a single HDPE layer to control post-harvest losses in cereals [230]. In a recent study conducted at agro-climatic conditions of Pakistan, quality of quinoa seeds stored in hermetic bags at 8% initial seed moisture content is preserved in terms of higher germination and vigor and negligible seed deterioration compared with traditional storage under diverse ambient conditions. Rapid loss of seed viability in traditional porous bags was owing to moisture absorption from the ambient high relative humidity, which resulted in seed deterioration [231].

Quinoa seed of initial seed quality of 80% germination and 8% moisture content can be stored hermetically for six months without loss of viability under ambient conditions (25–40 °C and 50–60% RH) while after one-year storage germination declines to 15% [231]. Quinoa seed maintains physiological quality for longer periods (up to 300 days) in impermeable packaging and under low temperature (4 ± 2 °C) [232]. It is also reported that quinoa seed with 5% moisture content at 5 °C can be stored for 8 months with maximum viability [233], and this moisture content is also ideal for long term storage. Seed of 5% moisture content can be stored for one year at 25 °C with only one percent loss of germination (according to seed viability equation given by [234]). The moisture content below which quinoa seed longevity is not further improved lies at 4.1% when stored at 65 °C [234].

10. Conclusions

Advances in quinoa crop improvement can be accelerated through the development of an international network of quinoa researchers and trial datasets. Collaborations across members of the Global Collaborative Network on Quinoa (GCN-Quinoa). Available online: gcn-quinoa.org (accessed on 15.08.2021) would be highly facilitated if all of members will be able to use the same language, tools and methods for establishing trials, collecting data, and then sharing datasets. This network initiated in 2015 has connected all the participants of different FAO–TCP programs. Now, the GCN-Quinoa links 296 members from all around the world in more than 75 countries. The sharing of data among global and regional research groups allows deeper exploration of each dataset in the context of its environment. Owing to the significant effects of environment and management practices on quinoa phenotypes, the reusability of each dataset greatly depends on the quality of the metadata recorded, including a detailed profile of environmental parameters, and on maintaining as many factors constant as is possible. The most important variables that need consistency among trials are the methods of data collection, which can be achieved by an international agreement on phenotyping methods describing common protocols for establishing a dataset with comparable standards, as have been proposed in this article. The Germinate database is available at Quinoa Germinate Database. Available online: http://germinate.quinoadb.org (accessed on 15 August 2021) to further facilitate standardizing global quinoa dataset structures and sharing and analyzing of data.

Quinoa is a remarkable crop with many valuable properties, but it is also a crop that still needs significant amounts of research and breeding to facilitate its move to become a major or widely cultivated food crop. It is hoped that this paper will facilitate these efforts by providing a framework for globally consistent phenotyping, benchmarking the phenotyping of quinoa plants and easing the comparison of results obtained around the world.

Supplementary Materials: The following are available online at https://www.mdpi.com/article/10.3390/plants10091759/s1, S1: Phenotyping cards, S2: Blue card template, S3: Scanner Equipment, S4: QuinoaCollab_NIR_Scan File.

Author Contributions: C.S.S.: conceptualization, writing—original draft preparation, writing—review and editing. E.R.: supervision, writing of Section 3, writing—review and editing. G.F.:

conceptualization and supervision. E.B.C.: writing Sections 9.2.2 and 9.2.3. G.W.: writing Section 7.3.1 and conceptualization. V.J.M.: supervision and writing—review and editing. D.S.R.P.: writing Section 6.1. K.J.: writing Section 6.3. S.M.S.: conceptualization and writing Section 9.2.4. D.B. (Daniel Bertero): writing Section 6.2. H.O.: writing Section 4. C.C.-L.: writing Section 8. I.A.: writing Section 9.2.5. S.R.: software—Section 3. N.M.: software—Section 9. J.S.: writing Section 5. D.B.A.: supervision Section 8. N.E.: supervision Section 6.1. M.W.: resources. M.A.A.M.: resources. D.W.: resources. D.J.: writing of last paragraph in Section 6.3 and supervision Section 5. C.A.: supervision Section 8. C.J.: supervision Section 6.1. K.M.: supervision Sections 9.2.2 and 9.2.3 D.B. (Didier Bazile): project administration, consortium formation. QPC review and approval of the manuscript. M.T.: conceptualization, supervision, funding acquisition, writing—review and editing. All authors have read and agreed to the published version of the manuscript.

Funding: The research reported in this publication from the lead author's laboratory was supported by King Abdullah University of Science and Technology (KAUST), including with a grant from the Office of Sponsored Research (OSR) under Award No. CRG5-2966. A section of this publication is based upon work supported by the National Science Foundation Graduate Research Fellowship under Grant No. 1842493. As a disclaimer, any opinion, findings, and conclusions or recommendations expressed in this material are those of the authors(s) and do not necessarily reflect the views of the National Science Foundation. Two authors are supported by the Center for Bioenergy Innovation, a US Department of Energy Bioenergy Research Center supported by the Office of Biological and Environmental Research in the DOE Office of Science. The manuscript was coauthored by UT-Battelle, LLC under Contract No. DE-AC05-00OR22725 with the US Department of Energy. The publisher, by accepting the article for publication, acknowledges that the US Government retains a nonexclusive, paid-up, irrevocable, worldwide license to publish or reproduce the published form of this manuscript, or allow others to do so, for US Government purposes. The Department of Energy will provide public access to these results of federally sponsored research in accordance with the DOE Public Access Plan. Available online: http://energy.gov/downloads/doe-public-access-plan (accessed on 15 August 2021).

Data Availability Statement: Data available upon request.

Acknowledgments: We thank Briony Hughes for the illustrations for plot population homogeneity, growth habit, stem angle, panicle axis angle, and panicle shape (Supplementary File S1: Phenotyping cards).

Conflicts of Interest: The authors declare no conflict of interest.

References

1. Cho, S.J.; McCarl, B.A. Climate change influences on crop mix shifts in the United States. *Sci. Rep.* **2017**, *7*, 40845. [CrossRef] [PubMed]
2. King, M.; Altdorff, D.; Li, P.; Galagedara, L.; Holden, J.; Unc, A. Northward shift of the agricultural climate zone under 21st-century global climate change. *Sci. Rep.* **2018**, *8*, 1–10. [CrossRef]
3. Challinor, A.; Watson, J.; Lobell, D.; Howden, S.; Smith, D.R.; Chhetri, N. A meta-analysis of crop yield under climate change and adaptation. *Nat. Clim. Chang.* **2014**, *4*, 287–291. [CrossRef]
4. Burchi, F.; Fanzo, J.; Frison, E. The Role of Food and Nutrition System Approaches in Tackling Hidden Hunger. *Int. J. Environ. Res. Public Heal.* **2011**, *8*, 358–373. [CrossRef]
5. Boushey, C.J.; Coulston, A.M.; Delahanty, L.; Ferruzzi, M. *Nutrition in the Prevention and Treatment of Disease*, 4th ed.; Elsevier: Amsterdam, The Netherlands, 2017.
6. IPCC. *2019: Climate Change and Land: An IPCC Special Report on Climate Change, Desertification, Land Degradation, Sustainable Land Management, Food Security, and Greenhouse Gas Fluxes in Terrestrial Ecosystems, CH5*; Shukla, P.R., Skea, J., Calvo Buendia, E., Masson-Delmotte, V., Pörtner, H.-O., Roberts, D.C., Zhai, P., Slade, R., Connors, S., van Diemen, R., et al., Eds.; IPCC: Geneva, Switzerland, 2019.
7. Bazile, D.; Bertero, D.; Nieto, C. *State of the Art Report of Quinoa in the World in 2013*; FAO & CIRAD: Rome, Italy, 2015.
8. Ruiz, K.B.; Biondi, S.; Oses, R.; Acuña-Rodríguez, I.S.; Antognoni, F.; Martinez-Mosqueira, E.A.; Coulibaly, A.; Canahua-Murillo, A.; Pinto, M.; Zurita-Silva, A.; et al. Quinoa biodiversity and sustainability for food security under climate change. A review. *Agron. Sustain. Dev.* **2014**, *34*, 349–359. [CrossRef]
9. United Nations. SD Goal 2 Department of Economic and Social Affairs. Retrieved 22 December 2020. Available online: https://sdgs.un.org/goals/goal2 (accessed on 15 August 2021).
10. FAO Secretariat, 2013 International Year of Quinoa. Distribution and Production. Retrieved 22 December 2020. Available online: http://www.fao.org/quinoa-2013/what-is-quinoa/distribution-and-production/en/ (accessed on 15 August 2021).
11. Alandia, G.; Rodriguez, J.; Jacobsen, S.-E.; Bazile, D.; Condori, B. Global expansion of quinoa and challenges for the Andean region. *Glob. Food Secur.* **2020**, *26*, 100429. [CrossRef]

12. Angeli, V.; Silva, P.M.; Massuela, D.C.; Khan, M.W.; Hamar, A.; Khajehei, F.; Graeff-Hönninger, S.; Piatti, C. Quinoa (Chenopodium quinoa Willd.): An Overview of the Potentials of the "Golden Grain" and Socio-Economic and Environmental Aspects of Its Cultivation and Marketization. *Foods* **2020**, *9*, 216. [CrossRef]
13. Bazile, D.; Pulvento, C.; Verniau, A.; Al-Nusairi, M.S.; Ba, D.; Breidy, J.; Hassan, L.; Mohammed, M.; Mambetov, O.; Otambekova, M.; et al. Worldwide Evaluations of Quinoa: Preliminary Results from Post International Year of Quinoa FAO Projects in Nine Countries. *Front. Plant. Sci.* **2016**, *7*, 850. [CrossRef] [PubMed]
14. Rey, E.; Jarvis, D.E. Structural and Functional Genomics of Chenopodium quinoa. In *The Quinoa Genome*; Schmöckel, S.M., Ed.; Springer: Cham, Switzerland, 2021. [CrossRef]
15. Jarvis, D.E.; Ho, Y.S.; Lightfoot, D.; Schmöckel, S.M.; Li, B.; Borm, T.J.A.; Ohyanagi, H.; Mineta, K.; Michell, C.; Saber, N.; et al. The genome of Chenopodium quinoa. *Nat. Cell Biol.* **2017**, *542*, 307–312. [CrossRef]
16. Cobb, J.N.; Declerck, G.; Greenberg, A.; Clark, R.; McCouch, S. Next-generation phenotyping: Requirements and strategies for enhancing our understanding of genotype–phenotype relationships and its relevance to crop improvement. *Theor. Appl. Genet.* **2013**, *126*, 867–887. [CrossRef]
17. Reynolds, M.; Chapman, S.; Herrera, L.A.C.; Molero, G.; Mondal, S.; Pequeno, D.N.; Pinto, F.; Pinera-Chavez, F.J.; Poland, J.; Rivera-Amado, C.; et al. Breeder friendly phenotyping. *Plant. Sci.* **2020**, *295*, 110396. [CrossRef] [PubMed]
18. Morton, M.J.L.; Awlia, M.; Al Tamimi, N.; Saade, S.; Pailles, Y.; Negrão, S.; Tester, M. Salt stress under the scalpel–Dissecting the genetics of salt tolerance. *Plant. J.* **2019**, *97*, 148–163. [CrossRef]
19. Khush, G.S. Strategies for increasing the yield potential of cereals: Case of rice as an example. *Plant. Breed.* **2013**, *132*, 433–436. [CrossRef]
20. Reynolds, M.; Foulkes, J.; Furbank, R.; Griffiths, S.; King, J.; Murchie, E.; Parry, M.; Slafer, G. Achieving yield gains in wheat. *Plant. Cell Environ.* **2012**, *35*, 1799–1823. [CrossRef]
21. González, F.G.; Capella, M.; Ribichich, K.F.; Curin, F.; Giacomelli, J.I.; Ayala, F.; Watson, G.; Otegui, M.E.; Chan, R.L. Field-grown transgenic wheat expressing the sunflower gene HaHB4 significantly outyields the wild type. *J. Exp. Bot.* **2019**, *70*, 1669–1681. [CrossRef]
22. González, F.G.; Rigalli, N.; Miranda, P.V.; Romagnoli, M.; Ribichich, K.F.; Trucco, F.; Portapila, M.; Otegui, M.E.; Chan, R.L. An Interdisciplinary Approach to Study the Performance of Second-generation Genetically Modified Crops in Field Trials: A Case Study With Soybean and Wheat Carrying the Sunflower HaHB4 Transcription Factor. *Front. Plant. Sci.* **2020**, *11*. [CrossRef] [PubMed]
23. Ertiro, B.T.; Olsen, M.; Das, B.; Gowda, M.; Labuschagne, M. Efficiency of indirect selection for grain yield in maize (Zea mays L.) under low nitrogen conditions through secondary traits under low nitrogen and grain yield under optimum conditions. *Euphytica* **2020**, *216*, 1–12. [CrossRef]
24. Fernandes, S.B.; Dias, K.O.D.G.; Ferreira, D.F.; Brown, P.J. Efficiency of multi-trait, indirect, and trait-assisted genomic selection for improvement of biomass sorghum. *Theor. Appl. Genet.* **2018**, *131*, 747–755. [CrossRef]
25. Lozada, D.N.; Godoy, J.V.; Ward, B.P.; Carter, A.H. Genomic Prediction and Indirect Selection for Grain Yield in US Pacific Northwest Winter Wheat Using Spectral Reflectance Indices from High-Throughput Phenotyping. *Int. J. Mol. Sci.* **2019**, *21*, 165. [CrossRef]
26. Musvosvi, C.; Setimela, P.S.; Wali, M.C.; Gasura, E.; Channappagoudar, B.B.; Patil, S.S. Contribution of Secondary Traits for High Grain Yield and Stability of Tropical Maize Germplasm across Drought Stress and Non-Stress Conditions. *Agron. J.* **2018**, *110*, 819–832. [CrossRef]
27. Sra, S.K.; Sharma, M.; Kaur, G.; Sharma, S.; Akhatar, J.; Sharma, A.; Banga, S.S. Evolutionary aspects of direct or indirect selection for seed size and seed metabolites in Brassica juncea and diploid progenitor species. *Mol. Biol. Rep.* **2019**, *46*, 1227–1238. [CrossRef]
28. Zaman, S.U.; Malik, A.I.; Kaur, P.; Ribalta, F.M.; Erskine, W. Waterlogging Tolerance at Germination in Field Pea: Variability, Genetic Control, and Indirect Selection. *Front. Plant. Sci.* **2019**, *10*, 953. [CrossRef]
29. Ziyomo, C.; Bernardo, R. Drought Tolerance in Maize: Indirect Selection through Secondary Traits versus Genomewide Selection. *Crop. Sci.* **2013**, *53*, 1269–1275. [CrossRef]
30. Malosetti, M.; Ribaut, J.-M.; Van Eeuwijk, F.A. The statistical analysis of multi-environment data: Modeling genotype-by-environment interaction and its genetic basis. *Front. Physiol.* **2013**, *4*, 44. [CrossRef]
31. White, J.W.; Andrade-Sanchez, P.; Gore, M.A.; Bronson, K.F.; Coffelt, T.A.; Conley, M.M.; Feldmann, K.A.; French, A.; Heun, J.T.; Hunsaker, D.J.; et al. Field-based phenomics for plant genetics research. *Field Crop. Res.* **2012**, *133*, 101–112. [CrossRef]
32. Bertero, H.; de la Vega, A.; Correa, G.; Jacobsen, S.; Mujica, A. Genotype and genotype-by-environment interaction effects for grain yield and grain size of quinoa (Chenopodium quinoa Willd.) as revealed by pattern analysis of international multi-environment trials. *Field Crop. Res.* **2004**, *89*, 299–318. [CrossRef]
33. Curti, R.; De La Vega, A.; Andrade, A.; Bramardi, S.; Bertero, H. Adaptive responses of quinoa to diverse agro-ecological environments along an altitudinal gradient in North West Argentina. *Field Crop. Res.* **2016**, *189*, 10–18. [CrossRef]
34. Desclaux, D.; Nolot, J.M.; Chiffoleau, Y.; Gozé, E.; Leclerc, C. Changes in the concept of genotype × environment interactions to fit agriculture diversification and decentralized participatory plant breeding: Pluridisciplinary point of view. *Euphytica* **2008**, *163*, 533–546. [CrossRef]

35. Leclerc, C.; D'Eeckenbrugge, G.C. Social Organization of Crop Genetic Diversity. The G × E × S Interaction Model. *Divers* **2011**, *4*, 1–32. [CrossRef]
36. Yan, W.; Hunt, L.; Sheng, Q.; Szlavnics, Z. Cultivar Evaluation and Mega-Environment Investigation Based on the GGE Biplot. *Crop. Sci.* **2000**, *40*, 597–605. [CrossRef]
37. Yan, W.; Kang, M.S. GGE Biplot Analysis. 2002. Available online: https://doi.org/10.1201/9781420040371 (accessed on 15 August 2021).
38. Bioversity International; FAO. Descriptors for quinoa (Chenopodium quinoa Willd.) and wild relatives. In *Bioversity International, FAO, PROINPA, INIAF and IFAD. 2013. Descriptors for Quinoa (Chenopodium quinoa Willd.) and Wild Relatives*; Bioversity International: Rome, Italy; Fundación PROINPA: La Paz, Bolivia; Instituto Nacional de Innovación Agropecuaria y: La Paz, Bolivia, 2013; Volume 1.
39. CPVO. Protocol for Tests on Distinctness, Uniformity and Stability Chenopodium Quinoa Willd. 2021. Available online: https://cpvo.europa.eu/sites/default/files/documents/chenopodium.pdf (accessed on 15 August 2021).
40. Papoutsoglou, E.A.; Faria, D.; Arend, D.; Arnaud, E.; Athanasiadis, I.N.; Chaves, I.; Coppens, F.; Cornut, G.; Costa, B.V.; Ćwiek-Kupczyńska, H.; et al. Enabling reusability of plant phenomic datasets with MIAPPE 1.1. *New Phytol.* **2020**, *227*, 260–273. [CrossRef]
41. Sosa-Zuniga, V.; Brito, V.; Fuentes, F.; Steinfort, U. Phenological growth stages of quinoa (Chenopodium quinoa) based on the BBCH scale. *Ann. Appl. Biol.* **2017**, *171*, 117–124. [CrossRef]
42. Raubach, S.; Kilian, B.; Dreher, K.; Amri, A.; Bassi, F.M.; Boukar, O.; Cook, D.; Cruickshank, A.; Fatokun, C.; El Haddad, N.; et al. From bits to bites: Advancement of the Germinate platform to support prebreeding informatics for crop wild relatives. *Crop. Sci.* **2020**, *61*, 1538–1566. [CrossRef]
43. Shaw, P.D.; Raubach, S.; Hearne, S.; Dreher, K.; Bryan, G.; McKenzie, G.; Milne, I.; Stephen, G.; Marshall, D. Germinate 3: Development of a Common Platform to Support the Distribution of Experimental Data on Crop Wild Relatives. *Crop. Sci.* **2017**, *57*, 1259–1273. [CrossRef]
44. Shrestha, R.; Matteis, L.; Skofic, M.; Portugal, A.; McLaren, G.; Hyman, G.; Arnaud, E. Bridging the phenotypic and genetic data useful for integrated breeding through a data annotation using the Crop Ontology developed by the crop communities of practice. *Front. Physiol.* **2012**, *3*, 326. [CrossRef]
45. UN. Resolution Adopted by the General Assembly on 22 December 2011. 2012. Available online: https://www.un.org/ga/search/view_doc.asp?symbol=A/RES/66/221&referer=/english/&Lang=E (accessed on 15 August 2021).
46. Assessment of the International Year of Quinoa 2013 Executive Summary. 2014. Available online: http://www.fao.org/quinoa-2013/iyq/en/ (accessed on 15 August 2021).
47. Rojas, W.; Pinto, M.; Alanoca, C.; Gómez-Pando, L.; León-Lobos, P.; Alercia, A.; Diulgheroff, S.; Padulosi, S.; Bazile, D. Quinoa genetic resources and ex situ conservation. In *State of the Art Report on Quinoa around the World in 2013*; Didier, B., Daniel, B.H., Carlos, N., Eds.; FAO: Rome, Italy, 2015; pp. 56–82.
48. Bazile, D. Fair and sustainable expansion of traditional crops-lessons from quinoa. *Farming Matters* **2016**, *32.2*, 36–39.
49. Bazile, D.; Jacobsen, S.-E.; Verniau, A. The Global Expansion of Quinoa: Trends and Limits. *Front. Plant. Sci.* **2016**, *7*, 622. [CrossRef]
50. Christensen, S.A.; Pratt, D.B.; Pratt, C.; Nelson, P.T.; Stevens, M.R.; Jellen, E.; Coleman, C.E.; Fairbanks, D.J.; Bonifacio, A.; Maughan, P.J. Assessment of genetic diversity in the USDA and CIP-FAO international nursery collections of quinoa (Chenopodium quinoa Willd.) using microsatellite markers. *Plant. Genet. Resour.* **2007**, *5*, 82–95. [CrossRef]
51. Colque-Little, C.; Abondano, M.C.; Lund, O.S.; Amby, D.B.; Piepho, H.-P.; Andreasen, C.; Schmöckel, S.; Schmid, K. Genetic variation for tolerance to the downy mildew pathogen Peronospora variabilis in genetic resources of quinoa (Chenopodium quinoa). *BMC Plant. Biol.* **2021**, *21*, 1–19. [CrossRef]
52. Tártara, S.M.C.; Manifesto, M.M.; Bramardi, S.J.; Bertero, H.D. Genetic structure in cultivated quinoa (Chenopodium quinoa Willd.), a reflection of landscape structure in Northwest Argentina. *Conserv. Genet.* **2012**, *13*, 1027–1038. [CrossRef]
53. Fuentes, F.; Martinez, E.A.; Hinrichsen, P.; Jellen, E.; Maughan, P.J. Assessment of genetic diversity patterns in Chilean quinoa (Chenopodium quinoa Willd.) germplasm using multiplex fluorescent microsatellite markers. *Conserv. Genet.* **2009**, *10*, 369–377. [CrossRef]
54. Mason, S.L.; Stevens, M.R.; Jellen, E.N.; Bonifacio, A.; Fairbanks, D.J.; Coleman, C.E.; Mccarty, R.R.; Rasmussen, A.G.; Maughan, P.J. Development and Use of Microsatellite Markers for Germplasm Characterization in Quinoa (Chenopodium quinoa Willd.). *Crop. Sci.* **2005**, *45*, 1618–1630. [CrossRef]
55. Mizuno, N.; Toyoshima, M.; Fujita, M.; Fukuda, S.; Kobayashi, Y.; Ueno, M.; Tanaka, K.; Tanaka, T.; Nishihara, E.; Mizukoshi, H.; et al. The genotype-dependent phenotypic landscape of quinoa in salt tolerance and key growth traits. *DNA Res.* **2020**, *27*. [CrossRef]
56. Patiranage, D.S.R.; Rey, E.; Emrani, N.; Wellman, G.; Schmid, K.; Schmöckel, S.M.; Tester, M.; Jung, C. Genome-wide association study in the pseudocereal quinoa reveals selection pattern typical for crops with a short breeding history. *bioRxiv* **2020**. [CrossRef]
57. Rana, T.S.; Narzary, D.; Ohri, D. Genetic diversity and relationships among some wild and cultivated species of Chenopodium L. (Amaranthaceae) using RAPD and DAMD methods. *Curr. Sci.* **2010**, *98*, 840–846.
58. Salazar, J.; Torres, M.D.L.; Gutierrez, B.; Torres, A.F. Molecular characterization of Ecuadorian quinoa (Chenopodium quinoa Willd.) diversity: Implications for conservation and breeding. *Euphytica* **2019**, *215*, 60. [CrossRef]

59. Zhang, T.; Gu, M.; Liu, Y.; Lv, Y.; Zhou, L.; Lu, H.; Liang, S.; Bao, H.; Zhao, H. Development of novel InDel markers and genetic diversity in Chenopodium quinoa through whole-genome re-sequencing. *BMC Genom.* **2017**, *18*, 1–15. [CrossRef] [PubMed]
60. Tapia, M.E.; Mujica, A.; Canahua, A. *Origin, Geographic Distribution and Production System of Quinoa (Chenopodium Quinoa)*; Publicacion-Universidad Nacional Tecnica del Altiplano: Puno, Peru, 1980.
61. Chable, V.; Thommen, A.; Goldringer, I.; Infante, V.; Levillain, T.; Lammerts Van Bueren, E. Report on the Definitions of Varieties in Europe, of Local Adaptation, and of Varieties Threatened by Genetic Erosion. 2008. Available online: https://hal.inrae.fr/hal-02820022 (accessed on 15 August 2021).
62. Murphy, K.M.; Bazile, D.; Kellogg, J.; Rahmanian, M. Development of a Worldwide Consortium on Evolutionary Participatory Breeding in Quinoa. *Front. Plant. Sci.* **2016**, *7*. [CrossRef] [PubMed]
63. Bonifacio, A.; Proinpa, F. Improvement of Quinoa (Chenopodium quinoa Willd.) and Qañawa (Chenopodium pallidicaule Aellen) in the context of climate change in the high Andes. *Cien. Inv. Agr.* **2019**, *46*, 113–124. [CrossRef]
64. Jacobsen, S.-E. Developmental stability of quinoa under European conditions. *Ind. Crop. Prod.* **1998**, *7*, 169–174. [CrossRef]
65. Mackay, I.; Piepho, H.; Garcia, A.A.F. Statistical Methods for Plant Breeding. In *Handbook of Statistical Genomics*; Wily: Hoboken, NJ, USA, 2019; pp. 501–520. [CrossRef]
66. Molenaar, H.; Boehm, R.; Piepho, H.-P. Phenotypic Selection in Ornamental Breeding: It's Better to Have the BLUPs Than to Have the BLUEs. *Front. Plant. Sci.* **2018**, *9*. [CrossRef] [PubMed]
67. Piepho, H.P.; Möhring, J.; Melchinger, A.E.; Büchse, A. BLUP for phenotypic selection in plant breeding and variety testing. *Euphytica* **2008**, *161*, 209–228. [CrossRef]
68. Welham, S.J.; Gezan, S.A.; Clark, S.J.; Mead, A. *Statistical Methods in Biology: Design and Analysis of Experiments and Regression*; CRC Press: Boca Raton, FL, USA, 2014.
69. Cullis, B.R.; Smith, A.B.; Cocks, N.A.; Butler, D.G. The Design of Early-Stage Plant Breeding Trials Using Genetic Relatedness. *J. Agric. Biol. Environ. Stat.* **2020**, *25*, 553–578. [CrossRef]
70. Podlich, D.W.; Cooper, M. QU-GENE: A simulation platform for quantitative analysis of genetic models. *Bioinformatics* **1998**, *14*, 632–653. [CrossRef] [PubMed]
71. Faux, A.; Gorjanc, G.; Gaynor, R.; Battagin, M.; Høj-Edwards, S.; Wilson, D.L.; Hearne, S.; Gonen, S.; Hickey, J.M. AlphaSim: Software for Breeding Program Simulation. *Plant. Genome* **2016**, *9*. [CrossRef] [PubMed]
72. Jahufer, M.Z.Z.; Luo, D. DeltaGen: A Comprehensive Decision Support Tool for Plant Breeders. *Crop. Sci.* **2018**, *58*, 1118–1131. [CrossRef]
73. Cullis, B.R.; Smith, A.B.; Coombes, N.E. On the design of early generation variety trials with correlated data. *J. Agric. Biol. Environ. Stat.* **2006**, *11*, 381–393. [CrossRef]
74. Falconer, D.S.; Mackay, T.F.C. Introduction to Quantitative Genetics (Fourth Edition). In *Trends in Genetics*; Elsevier: Amsterdam, The Netherlands, 1996.
75. Oakey, H.; Verbyla, A.; Pitchford, W.; Cullis, B.; Kuchel, H. Joint modeling of additive and non-additive genetic line effects in single field trials. *Theor. Appl. Genet.* **2006**, *113*, 809–819. [CrossRef]
76. Hong, E.P.; Park, J.W. Sample Size and Statistical Power Calculation in Genetic Association Studies. *Genom. Inform.* **2012**, *10*, 117–122. [CrossRef]
77. Coombes, N.E. DiGGer, a spatial design program. In *Biometric Bulletin*; NSW Department of Primary Industries: Orange, Australia, 2009.
78. van Ittersum, M.; Rabbinge, R. Concepts in production ecology for analysis and quantification of agricultural input-output combinations. *Field Crop. Res.* **1997**, *52*, 197–208. [CrossRef]
79. Quinoa: Improvement and Sustainable Production. 2015; Available online: https://doi.org/10.1002/9781118628041 (accessed on 15 August 2021).
80. Sellami, M.H.; Pulvento, C.; Lavini, A. Agronomic Practices and Performances of Quinoa under Field Conditions: A Systematic Review. *Plants* **2020**, *10*, 72. [CrossRef]
81. Eisa, S.S.; Abd El-Samad, E.H.; Hussin, S.A.; Ali, E.A.; Ebrahim, M.; González, J.A.; Ordano, M.; Erazzú, L.E.; El-Bordeny, N.E.; Abdel-Ati, A.A. Quinoa in Egypt-Plant Density Effects on Seed Yield and Nutritional Quality in Marginal Regions. *Middle East J. Appl. Sci.* **2018**, *8*, 515–522.
82. Ahmadi, S.H.; Solgi, S.; Sepaskhah, A.R. Quinoa: A super or pseudo-super crop? Evidences from evapotranspiration, root growth, crop coefficients, and water productivity in a hot and semi-arid area under three planting densities. *Agric. Water Manag.* **2019**, *225*, 105784. [CrossRef]
83. Aguilar, P.C.; Jacobsen, S.-E. Cultivation of Quinoa on the Peruvian Altiplano. *Food Rev. Int.* **2003**, *19*, 31–41. [CrossRef]
84. Abdelaziz, H.; Redouane, C.-A. *Phenotyping the Combined Effect of Heat and Water Stress on Quinoa*; Springer: Cham, Switzerland, 2020; pp. 163–183. [CrossRef]
85. Aufhammer, W.; Kaul, H.-P.; Kruse, M.; Lee, J.; Schwesig, D. Effects of sowing depth and soil conditions on seedling emergence of amaranth and quinoa. *Eur. J. Agron.* **1994**, *3*, 205–210. [CrossRef]
86. Hinojosa, L.; González, J.A.; Barrios-Masias, F.H.; Fuentes, F.; Murphy, K.M. Quinoa Abiotic Stress Responses: A Review. *Plants* **2018**, *7*, 106. [CrossRef] [PubMed]
87. Oelke, E.A.; Putnam, D.H.; Teynor, T.M.; Oplinger, E.S. Alternative Field Crops Manual: Quinoa; University of Wisconsin-Extension 1992, Cooperative Extension University of Minnesota: Center for Alternative Plant & Animal Products and the

Minnesota Extension Service. 1992. Available online: https://hort.purdue.edu/newcrop/afcm/quinoa.html (accessed on 15 August 2021).
88. Hirich, A.; Jacobsen, S.-E.; Choukr-Allah, R. Quinoa in Morocco-Effect of Sowing Dates on Development and Yield. *J. Agron. Crop. Sci.* **2014**, *200*, 371–377. [CrossRef]
89. Yang, A.; Akhtar, S.S.; Amjad, M.; Iqbal, S.; Jacobsen, S.-E. Growth and Physiological Responses of Quinoa to Drought and Temperature Stress. *J. Agron. Crop. Sci.* **2016**, *202*, 445–453. [CrossRef]
90. Präger, A.; Munz, S.; Nkebiwe, P.M.; Mast, B.; Graeff-Hönninger, S. Yield and Quality Characteristics of Different Quinoa (Chenopodium quinoa Willd.) Cultivars Grown under Field Conditions in Southwestern Germany. *Agronomy* **2018**, *8*, 197. [CrossRef]
91. Garcia, M.; Raes, D.; Jacobsen, S.-E. Evapotranspiration analysis and irrigation requirements of quinoa (Chenopodium quinoa) in the Bolivian highlands. *Agric. Water Manag.* **2003**, *60*, 119–134. [CrossRef]
92. Jayme-Oliveira, A.; Júnior, W.Q.R.; Ramos, M.L.G.; Ziviani, A.C.; Jakelaitis, A. Amaranth, quinoa, and millet growth and development under different water regimes in the Brazilian Cerrado. *Pesquisa Agropecuária Brasileira* **2017**, *52*, 561–571. [CrossRef]
93. Allen, R.G.; Pereira, L.S.; Raes, D.; Smith, M. Crop evapotranspiration: Guidelines for computing crop requirements. In *Irrigation and Drainage Paper No. 56*; FAO: Rome, Italy, 1998. [CrossRef]
94. FAO. ETo Calculator. In *Land and Water Digital Media Series*; FAO: Rome, Italy, 2012.
95. Geerts, S.; Raes, D.; Garcia, M.; Vacher, J.; Mamani, R.; Mendoza, J.; Huanca, R.; Morales, B.; Miranda, R.; Cusicanqui, J.; et al. Introducing deficit irrigation to stabilize yields of quinoa (Chenopodium quinoa Willd.). *Eur. J. Agron.* **2008**, *28*, 427–436. [CrossRef]
96. Razzaghi, F.; Plauborg, F.; Jacobsen, S.-E.; Jensen, C.R.; Andersen, M.N. Effect of nitrogen and water availability of three soil types on yield, radiation use efficiency and evapotranspiration in field-grown quinoa. *Agric. Water Manag.* **2012**, *109*, 20–29. [CrossRef]
97. Fghire, R.; Wahbi, S.; Anaya, F.; Ali, O.I.; Benlhabib, O.; Ragab, R. Response of Quinoa to Different Water Management Strategies: Field Experiments and Saltmed Model Application Results. *Irrig. Drain.* **2015**, *64*, 29–40. [CrossRef]
98. Pulvento, C.; Riccardi, M.; Lavini, A.; D'Andria, R.; Ragab, R. Saltmed Model to Simulate Yield And Dry Matter for Quinoa Crop And Soil Moisture Content Under Different Irrigation Strategies In South Italy. *Irrig. Drain.* **2013**, *62*, 229–238. [CrossRef]
99. Bertero, H.D. Quinoa. In *Crop Physiology Case Histories for Major Crops*; Academic Press: Cambridge, MA, USA, 2020. [CrossRef]
100. Geren, H.; Kavut, Y.; Altınbaş, M. Effect of different row spacings on the grain yield and some yield characteristics of quinoa (Chenopodium quinoa Wild.) under Bornova ecological conditions. *Ege Üniversitesi Ziraat Fakültesi Dergisi* **2015**, *52*, 69–78.
101. Erley, G.S.A.; Kaul, H.-P.; Kruse, M.; Aufhammer, W. Yield and nitrogen utilization efficiency of the pseudocereals amaranth, quinoa, and buckwheat under differing nitrogen fertilization. *Eur. J. Agron.* **2005**, *22*, 95–100. [CrossRef]
102. Alandia, G.; Jacobsen, S.-E.; Kyvsgaard, N.C.; Condori, B.; Liu, F. Nitrogen Sustains Seed Yield of Quinoa Under Intermediate Drought. *J. Agron. Crop. Sci.* **2016**, *202*, 281–291. [CrossRef]
103. Bascuñán-Godoy, L.; Sanhueza, C.; Pinto, K.; Cifuentes, L.; Reguera, M.; Briones-Labarca, V.; Zurita-Silva, A.; Álvarez, R.; Morales, A.; Silva, H. Nitrogen physiology of contrasting genotypes of Chenopodium quinoa Willd. (Amaranthaceae). *Sci. Rep.* **2018**, *8*, 17524. [CrossRef]
104. Cruces, L.; Delgado, P.; Santivañez, T.; Jara, B.; Vernal, P. Guía de Identificación y Control de las Principales Plagas que Afectan a la Quinua en la Zona Andina. 2016. Available online: https://bivica.org/files/quinua-plagas.pdf (accessed on 15 August 2021).
105. Rupavatharam, S.; Kennepohl, A.; Kummer, B.; Parimi, V. Automated plant disease diagnosis using innovative android App (Plantix) for farmers in Indian state of Andhra Pradesh. *Phytopathology TSI* **2018**, *108*, 10.
106. Brachi, B.; Aime, C.; Glorieux, C.; Cuguen, J.; Roux, F. Adaptive Value of Phenological Traits in Stressful Environments: Predictions Based on Seed Production and Laboratory Natural Selection. *PLoS ONE* **2012**, *7*, e32069. [CrossRef]
107. Rausher, M.D. The Measurement of Selection on Quantitative Traits: Biases Due to Environmental Covariances between Traits and Fitness. *Evolution* **1992**, *46*, 616–626. [CrossRef] [PubMed]
108. De Leon, N.; Jannink, J.-L.; Edwards, J.W.; Kaeppler, S.M. Introduction to a Special Issue on Genotype by Environment Interaction. *Crop. Sci.* **2016**, *56*, 2081–2089. [CrossRef]
109. Condon, J. Effective Soil Sampling–High and Low Cost Options to Gain Soil Fertility Information for Management. GRDC. 2019. Available online: https://grdc.com.au/resources-and-publications/grdc-update-papers/tab-content/grdc-update-papers/2019/02/effective-soil-sampling-high-and-low-cost-options-to-gain-soil-fertility-information-for-management. (accessed on 15 August 2021).
110. Xu, Y. Envirotyping for deciphering environmental impacts on crop plants. *Theor. Appl. Genet.* **2016**, *129*, 653–673. [CrossRef] [PubMed]
111. Fricke, W. Water transport and energy. *Plant. Cell Environ.* **2017**, *40*, 977–994. [CrossRef]
112. Zhang, D.; Du, Q.; Zhang, Z.; Jiao, X.; Song, X.; Li, J. Vapour pressure deficit control in relation to water transport and water productivity in greenhouse tomato production during summer. *Sci. Rep.* **2017**, *7*, 43461. [CrossRef]
113. Kargas, G.; Londra, P.; Anastasatou, M.; Moustakas, N. The Effect of Soil Iron on the Estimation of Soil Water Content Using Dielectric Sensors. *Water* **2020**, *12*, 598. [CrossRef]
114. Präger, A.; Boote, K.J.; Munz, S.; Graeff-Hönninger, S. Simulating Growth and Development Processes of Quinoa (Chenopodium quinoa Willd.): Adaptation and Evaluation of the CSM-CROPGRO Model. *Agronomy* **2019**, *9*, 832. [CrossRef]

115. Alvar-Beltrán, J.; Gobin, A.; Orlandini, S.; Marta, A.D. AquaCrop parametrisation for quinoa in arid environments. *Ital. J. Agron.* **2020**, *16*. [CrossRef]
116. Geerts, S.; Raes, D.; Garcia, M. Using AquaCrop to derive deficit irrigation schedules. *Agric. Water Manag.* **2010**, *98*, 213–216. [CrossRef]
117. Geerts, S.; Raes, D.; Garcia, M.; Miranda, R.; Cusicanqui, J.A.; Taboada, C.; Mendoza, J.; Huanca, R.; Mamani, A.; Condori, O.; et al. Simulating Yield Response of Quinoa to Water Availability with AquaCrop. *Agron. J.* **2009**, *101*, 499–508. [CrossRef]
118. Geerts, S.; Raes, D.; Garcia, M.; Taboada, C.; Miranda, R.; Cusicanqui, J.; Mhizha, T.; Vacher, J. Modeling the potential for closing quinoa yield gaps under varying water availability in the Bolivian Altiplano. *Agric. Water Manag.* **2009**, *96*, 1652–1658. [CrossRef]
119. Kaoutar, F.; Abdelaziz, H.; Ouafae, B.; Redouane, C.-A.; Ragab, R. Yield and Dry Matter Simulation Using the Saltmed Model for Five Quinoa (Chenopodium Quinoa) Accessions Under Deficit Irrigation in South Morocco. *Irrig. Drain.* **2017**, *66*, 340–350. [CrossRef]
120. FAO. Required Input for Simulations with AquaCrop. 2016. Available online: http://www.fao.org/3/i6050e/i6050e.pdf (accessed on 15 August 2021).
121. Vanuytrecht, E.; Raes, D.; Steduto, P.; Hsiao, T.C.; Fereres, E.; Heng, L.K.; Garcia-Vila, M.; Moreno, P.M. AquaCrop: FAO's crop water productivity and yield response model. *Environ. Model. Softw.* **2014**, *62*, 351–360. [CrossRef]
122. Ragab, R. A holistic generic integrated approach for irrigation, crop and field management: The SALTMED model. *Environ. Model. Softw.* **2002**, *17*, 345–361. [CrossRef]
123. Bertero, D.; Medan, D.; Hall, A.J. Changes in Apical Morphology during Floral Initiation and Reproductive Development in Quinoa (Chenopodium quinoaWilld.). *Ann. Bot.* **1996**, *78*, 317–324. [CrossRef]
124. Curti, R.; de la Vega, A.; Andrade, A.; Bramardi, S.; Bertero, H. Multi-environmental evaluation for grain yield and its physiological determinants of quinoa genotypes across Northwest Argentina. *Field Crop. Res.* **2014**, *166*, 46–57. [CrossRef]
125. Jacobsen, S.-E.; Stølen, O. Quinoa-Morphology, phenology and prospects for its production as a new crop in Europe. *Eur. J. Agron.* **1993**, *2*, 19–29. [CrossRef]
126. Mujica, A.; Canahua, A. Fases fenológicas del cultivo de la quínua (Chenopodium quinoa Willd.). In Proceedings of the Curso Taller 1989, Fenología de Cultivos Andinos y Uso de La Información Agrometeorológica, Salcedo, Puno, Peru 7–10 August 1989; pp. 23–27.
127. Ruiz, R.; Bertero, H. Light interception and radiation use efficiency in temperate quinoa (Chenopodium quinoa Willd.) cultivars. *Eur. J. Agron.* **2008**, *29*, 144–152. [CrossRef]
128. Tardieu, F. Plant response to environmental conditions: Assessing potential production, water demand, and negative effects of water deficit. *Front. Physiol.* **2013**, *4*, 17. [CrossRef]
129. Tardieu, F.; Bosquet, L.C.; Welcker, C. Model assisted dissection of the Genotype x Environment interaction. In Proceedings of the ASA 2012, CSSA and SSSA International Annual Meetings, Cincinnati, OH, USA, 21–24 October 2012.
130. Tardieu, F.; Tuberosa, R. Dissection and modelling of abiotic stress tolerance in plants. *Curr. Opin. Plant. Biol.* **2010**, *13*, 206–212. [CrossRef]
131. Passioura, J.; Angus, J. Improving Productivity of Crops in Water-Limited Environments. *Adv. Agron.* **2010**, *106*, 37–75. [CrossRef]
132. Trapani, N.; Hall, A.; Sadras, V.; Vilella, F. Ontogenetic changes in radiation use efficiency of sunflower (*Helianthus annuus* L.) crops. *Field Crop. Res.* **1992**, *29*, 301–316. [CrossRef]
133. Grimes, D.J. Koch's Postulates—Then and Now. *Microbe Mag.* **2006**, *1*, 223–228. [CrossRef]
134. Gómez, M.B.; Castro, P.A.; Mignone, C.; Bertero, H.D. Can yield potential be increased by manipulation of reproductive partitioning in quinoa (Chenopodium quinoa)? Evidence from gibberellic acid synthesis inhibition using Paclobutrazol. *Funct. Plant. Biol.* **2011**, *38*, 420–430. [CrossRef] [PubMed]
135. Garbulsky, M.; Peñuelas, J.; Gamon, J.; Inoue, Y.; Filella, I. The photochemical reflectance index (PRI) and the Remote Sensing of leaf, canopy and ecosystem radiation use efficiencies: A review and meta-analysis. *Remote Sens. Environ.* **2011**, *115*, 281–297. [CrossRef]
136. Hinojosa, L.; Kumar, N.; Gill, K.S.; Murphy, K.M. Spectral Reflectance Indices and Physiological Parameters in Quinoa under Contrasting Irrigation Regimes. *Crop. Sci.* **2019**, *59*, 1927–1944. [CrossRef]
137. Sankaran, S.; Espinoza, C.Z.; Hinojosa, L.; Ma, X.; Murphy, K. High-Throughput Field Phenotyping to Assess Irrigation Treatment Effects in Quinoa. *Age* **2019**, *2*, 1–7. [CrossRef]
138. Danielsen, S.; Munk, L. Evaluation of disease assessment methods in quinoa for their ability to predict yield loss caused by downy mildew. *Crop. Prot.* **2004**, *23*, 219–228. [CrossRef]
139. Xie, C.; Yang, C. A review on plant high-throughput phenotyping traits using UAV-based sensors. *Comput. Electron. Agric.* **2020**, *178*, 105731. [CrossRef]
140. Tmušić, G.; Manfreda, S.; Aasen, H.; James, M.R.; Gonçalves, G.; Ben-Dor, E.; Brook, A.; Polinova, M.; Arranz, J.J.; Mészáros, J.; et al. Current Practices in UAS-based Environmental Monitoring. *Remote Sens.* **2020**, *12*, 1001. [CrossRef]
141. Yang, G.; Liu, J.; Zhao, C.; Li, Z.; Huang, Y.; Yu, H.; Xu, B.; Yang, X.; Zhu, D.; Zhang, X.; et al. Unmanned Aerial Vehicle Remote Sens.ing for Field-Based Crop Phenotyping: Current Status and Perspectives. *Front. Plant. Sci.* **2017**, *8*, 1111. [CrossRef] [PubMed]
142. Ziliani, M.G.; Parkes, S.D.; Hoteit, I.; McCabe, M.F. Intra-Season Crop Height Variability at Commercial Farm Scales Using a Fixed-Wing UAV. *Remote Sens.* **2018**, *10*, 2007. [CrossRef]

143. Johansen, K.; Morton, M.J.L.; Malbeteau, Y.; Aragon, B.; Al-Mashharawi, S.; Ziliani, M.G.; Angel, Y.; Fiene, G.; Negrão, S.; Mousa, M.A.A.; et al. Predicting Biomass and Yield in a Tomato Phenotyping Experiment Using UAV Imagery and Random Forest. *Front. Artif. Intell.* **2020**, *3*. [CrossRef]
144. Di Gennaro, S.F.; Rizza, F.; Badeck, F.-W.; Berton, A.; Delbono, S.; Gioli, B.; Toscano, P.; Zaldei, A.; Matese, A. UAV-based high-throughput phenotyping to discriminate barley vigour with visible and near-infrared vegetation indices. *Int. J. Remote Sens.* **2018**, *39*, 5330–5344. [CrossRef]
145. Wang, T.; Thomasson, J.A.; Yang, C.; Isakeit, T.; Nichols, R.L. Automatic Classification of Cotton Root Rot Disease Based on UAV Remote Sensing. *Remote Sens.* **2020**, *12*, 1310. [CrossRef]
146. Su, J.; Liu, C.; Coombes, M.; Hu, X.; Wang, C.; Xu, X.; Li, Q.; Guo, L.; Chen, W.-H. Wheat yellow rust monitoring by learning from multispectral UAV aerial imagery. *Comput. Electron. Agric.* **2018**, *155*, 157–166. [CrossRef]
147. Chivasa, W.; Mutanga, O.; Biradar, C. UAV-Based Multispectral Phenotyping for Disease Resistance to Accelerate Crop Improvement under Changing Climate Conditions. *Remote Sens.* **2020**, *12*, 2445. [CrossRef]
148. Holman, F.H.; Riche, A.B.; Castle, M.; Wooster, M.J.; Hawkesford, M.J. Radiometric Calibration of 'Commercial off the Shelf' Cameras for UAV-Based High-Resolution Temporal Crop Phenotyping of Reflectance and NDVI. *Remote Sens.* **2019**, *11*, 1657. [CrossRef]
149. Jackson, R.D.; Idso, S.B.; Reginato, R.J.; Pinter, P.J. Canopy temperature as a crop water stress indicator. *Water Resour. Res.* **1981**, *17*, 1133–1138. [CrossRef]
150. Turner, D.; Lucieer, A.; Watson, C. An Automated Technique for Generating Georectified Mosaics from Ultra-High Resolution Unmanned Aerial Vehicle (UAV) Imagery, Based on Structure from Motion (SfM) Point Clouds. *Remote Sens.* **2012**, *4*, 1392–1410. [CrossRef]
151. Shendryk, Y.; Sofonia, J.; Garrard, R.; Rist, Y.; Skocaj, D.; Thorburn, P. Fine-scale prediction of biomass and leaf nitrogen content in sugarcane using UAV LiDAR and multispectral imaging. *Int. J. Appl. Earth Obs. Geoinf.* **2020**, *92*, 102177. [CrossRef]
152. Messina, G.; Modica, G. Applications of UAV Thermal Imagery in Precision Agriculture: State of the Art and Future Research Outlook. *Remote Sens.* **2020**, *12*, 1491. [CrossRef]
153. Aragon, B.; Johansen, K.; Parkes, S.; Malbeteau, Y.; Al-Mashharawi, S.; Al-Amoudi, T.; Andrade, C.F.; Turner, D.; Lucieer, A.; McCabe, M.F. A Calibration Procedure for Field and UAV-Based Uncooled Thermal Infrared Instruments. *Sensors* **2020**, *20*, 3316. [CrossRef] [PubMed]
154. Kelly, J.; Kljun, N.; Olsson, P.-O.; Mihai, L.; Liljeblad, B.; Weslien, P.; Klemedtsson, L.; Eklundh, L. Challenges and Best Practices for Deriving Temperature Data from an Uncalibrated UAV Thermal Infrared Camera. *Remote Sens.* **2019**, *11*, 567. [CrossRef]
155. Malbeteau, Y.; Johansen, K.; Aragon, B.; Al-Mashhawari, S.K.; McCabe, M.F. Overcoming the challenges of thermal infrared orthomosaics using a swath-based approach to correct for dynamic temperature and wind effects. *Remote Sens.* **2021**, *13*, 3255. [CrossRef]
156. Aasen, H.; Bolten, A. Multi-temporal high-resolution imaging spectroscopy with hyperspectral 2D imagers-From theory to application. *Remote Sens. Environ.* **2018**, *205*, 374–389. [CrossRef]
157. Adão, T.; Hruška, J.; Pádua, L.; Bessa, J.; Peres, E.; Morais, R.; Sousa, J.J. Hyperspectral Imaging: A Review on UAV-Based Sensors, Data Processing and Applications for Agriculture and Forestry. *Remote Sens.* **2017**, *9*, 1110. [CrossRef]
158. Angel, Y.; Turner, D.; Parkes, S.; Malbeteau, Y.; Lucieer, A.; McCabe, M.F. Automated Georectification and Mosaicking of UAV-Based Hyperspectral Imagery from Push-Broom Sensors. *Remote Sens.* **2020**, *12*, 34. [CrossRef]
159. Barreto, M.A.P.; Johansen, K.; Angel, Y.; McCabe, M.F. Radiometric Assessment of a UAV-Based Push-Broom Hyperspectral Camera. *Sensors* **2019**, *19*, 4699. [CrossRef]
160. Hassler, S.C.; Baysal-Gurel, F. Unmanned Aircraft System (UAS) Technology and Applications in Agriculture. *Agronomy* **2019**, *9*, 618. [CrossRef]
161. Ivushkin, K.; Bartholomeus, H.; Bregt, A.K.; Pulatov, A.; Franceschini, M.H.; Kramer, H.; van Loo, E.N.; Roman, V.J.; Finkers, R. UAV based soil salinity assessment of cropland. *Geoderma* **2019**, *338*, 502–512. [CrossRef]
162. Holman, F.H.; Riche, A.B.; Michalski, A.; Castle, M.; Wooster, M.J.; Hawkesford, M.J. High Throughput Field Phenotyping of Wheat Plant Height and Growth Rate in Field Plot Trials Using UAV Based Remote Sensing. *Remote Sens.* **2016**, *8*, 1031. [CrossRef]
163. Galli, G.; Sabadin, F.; Costa-Neto, G.M.F.; Fritsche-Neto, R. A novel way to validate UAS-based high-throughput phenotyping protocols using in silico experiments for plant breeding purposes. *Theor. Appl. Genet.* **2020**, *134*, 715–730. [CrossRef]
164. Keller, B.; Matsubara, S.; Rascher, U.; Pieruschka, R.; Steier, A.; Kraska, T.; Muller, O. Genotype Specific Photosynthesis x Environment Interactions Captured by Automated Fluorescence Canopy Scans Over Two Fluctuating Growing Seasons. *Front. Plant. Sci.* **2019**, *10*, 1482. [CrossRef]
165. Raesch, A.R.; Muller, O.; Pieruschka, R.; Rascher, U. Field Observations with Laser-Induced Fluorescence Transient (LIFT) Method in Barley and Sugar Beet. *Agriculture* **2014**, *4*, 159–169. [CrossRef]
166. Pieruschka, R.; Schurr, U. Plant Phenotyping: Past, Present, and Future. *Plant. Phenomics* **2019**, *2019*, 1–6. [CrossRef]
167. Mochida, K.; Koda, S.; Inoue, K.; Hirayama, T.; Tanaka, S.; Nishii, R.; Melgani, F. Computer vision-based phenotyping for improvement of plant productivity: A machine learning perspective. *GigaScience* **2018**, *8*. [CrossRef]
168. Ubbens, J.; Cieslak, M.; Prusinkiewicz, P.; Parkin, I.; Ebersbach, J.; Stavness, I. Latent Space Phenotyping: Automatic Image-Based Phenotyping for Treatment Studies. *Plant. Phenomics* **2020**, *2020*, 1–13. [CrossRef]

169. Brunner, G.; Liu, Y.; Pascual, D.; Richter, O.; Ciaramita, M.; Wattenhofer, R. On Identifiability in Transformers. *arXiv* **2019**, arXiv:1908.04211.
170. Chefer, H.; Gur, S.; Wolf, L. Transformer Interpretability Beyond Attention Visualization. 2021. Available online: https://openaccess.thecvf.com/content/CVPR2021/html/Chefer_Transformer_Interpretability_Beyond_Attention_Visualization_CVPR_2021_paper.html (accessed on 15 August 2021).
171. Mastebroek, H.; van Loo, E.; Dolstra, O. Combining ability for seed yield traits of Chenopodium quinoa breeding lines. *Euphytica* **2002**, *125*, 427–432. [CrossRef]
172. Ploschuk, E.L.; Hall, A. Capitulum position in sunflower affects grain temperature and duration of grain filling. *Field Crop. Res.* **1995**, *44*, 111–117. [CrossRef]
173. Dong, Y.; Ewang, Y.-Z. Seed shattering: From models to crops. *Front. Plant. Sci.* **2015**, *6*, 476. [CrossRef] [PubMed]
174. Peterson, A.; Jacobsen, S.-E.; Bonifacio, A.; Murphy, K. A Crossing Method for Quinoa. *Sustainability* **2015**, *7*, 3230–3243. [CrossRef]
175. Colque-Little, C.; Amby, D.; Andreasen, C. A Review of *Chenopodium quinoa* (Willd.) Diseases—An Updated Perspective. *Plants* **2021**, *10*, 1228. [CrossRef] [PubMed]
176. Agrios, G. *Plant Pathology*; Elsevier Academic Press: Cambridge, MA, USA, 2005.
177. Lamichhane, J.R.; Venturi, V. Synergisms between microbial pathogens in plant disease complexes: A growing trend. *Front. Plant. Sci.* **2015**, *6*, 385. [CrossRef]
178. Danielsen, S.; Jacobsen, S.-E.; Hockenhull, J. First Report of Downy Mildew of Quinoa Caused by Peronospora farinosa f. sp. chenopodii in Denmark. *Plant. Dis.* **2002**, *86*, 1175. [CrossRef] [PubMed]
179. Testen, A.; McKemy, J.M.; Backman, P.A. First Report of Passalora Leaf Spot of Quinoa Caused by Passalora dubia in the United States. *Plant. Dis.* **2013**, *97*, 139. [CrossRef]
180. Testen, A.; McKemy, J.M.; Backman, P.A. First Report of Ascochyta Leaf Spot of Quinoa Caused by Ascochyta sp. in the United States. *Plant. Dis.* **2013**, *97*, 844. [CrossRef]
181. Testen, A.; Jiménez-Gasco, M.D.M.; Ochoa, J.B.; Backman, P.A. Molecular Detection of Peronospora variabilis in Quinoa Seed and Phylogeny of the Quinoa Downy Mildew Pathogen in South America and the United States. *Phytopathology* **2014**, *104*, 379–386. [CrossRef]
182. Yin, H.; Zhou, J.; Lv, H.; Qin, N.; Chang, F.J.; Zhao, X.J. Identification, Pathogenicity, and Fungicide Sensitivity of Ascochyta caulina (Teleomorph: Neocamarosporium calvescens) Associated with Black Stem on Quinoa in China. *Plant. Dis.* **2020**, *104*. [CrossRef] [PubMed]
183. Dřímalková, M.; Veverka, K. Seedlings damping-off of Chenopodium quinoa Willd. *Plant. Prot. Sci.* **2004**, *40*, 5–10. [CrossRef]
184. Fonseca-Guerra, I.; Chiquillo, C.; Padilla, M.J.; Benavides-Rozo, M. First report of bacterial leaf spot on Chenopodium quinoa caused by Pseudomonas syringae in Colombia. *J. Plant. Dis. Prot.* **2021**, *128*, 871–874. [CrossRef]
185. Isobe, K.; Sugiyama, T.; Katagiri, M.; Ishizuka, C.; Tamura, Y.; Higo, M.; Fujita, Y. Study on the Cause Damping-off in Quinoa (Chenopodium quinoa Willd.) and a Method for Suppressing its Occurrence. *Jpn. J. Crop. Sci.* **2019**, *88*, 117–124. [CrossRef]
186. Pal, N.; Testen, A.L. First Report of Quinoa Anthracnose Caused by Colletotrichum nigrum and C. truncatum in the United States. *Plant. Dis.* **2021**, *105*, 705. [CrossRef]
187. Danielsen, S.; Ames, T. *Mildew (Peronospora farinosa) of quinua (Chenopodium quinoa) in the Andean Region: Practical Manual for the Study of the Disease and Pathogen*; International Potato Center: Lima, Peru, 2004.
188. Staub, J.; Bacher, J.; Poetter, K. Sources of Potential Errors in the Application of Random Amplified Polymorphic DNAs in Cucumber. *HortScience* **1996**, *31*, 262–266. [CrossRef]
189. Conrath, U. Molecular aspects of defence priming. *Trends Plant. Sci.* **2011**, *16*, 524–531. [CrossRef]
190. Hammerschmidt, R.; M'Etraux, J.-P.; Van Loon, L. Inducing Resistance: A Summary of Papers Presented at the First International Symposium on Induced Resistance to Plant Diseases, Corfu, May 2000. *Eur. J. Plant. Pathol.* **2001**, *107*, 1–6. [CrossRef]
191. Grogan, R.G. The Science and Art of Plant-Disease Diagnosis. *Annu. Rev. Phytopathol.* **1981**, *19*, 333–351. [CrossRef]
192. Pereira, E.; Encina-Zelada, C.; Barros, L.; Gonzales-Barron, U.; Cadavez, V.; Ferreira, I.C. Chemical and nutritional characterization of Chenopodium quinoa Willd (quinoa) grains: A good alternative to nutritious food. *Food Chem.* **2019**, *280*, 110–114. [CrossRef]
193. Merchant, N.; Lyons, E.; Goff, S.; Vaughn, M.; Ware, D.; Micklos, D.; Antin, P. The iPlant Collaborative: Cyberinfrastructure for Enabling Data to Discovery for the Life Sciences. *PLoS Biol.* **2016**, *14*, e1002342. [CrossRef]
194. Nowak, V.; Du, J.; Charrondière, U.R. Assessment of the nutritional composition of quinoa (Chenopodium quinoa Willd.). *Food Chem.* **2016**, *193*, 47–54. [CrossRef] [PubMed]
195. Valencia-Chamorro, S.A. QUINOA. In *Encyclopedia of Food Sciences and Nutrition*; Elsevier: Amsterdam, The Netherlands, 2003; pp. 4895–4902.
196. Aluwi, N.A.; Gu, B.-J.; Dhumal, G.S.; Medina-Meza, I.G.; Murphy, K.M.; Ganjyal, G.M. Impacts of Scarification and Degermination on the Expansion Characteristics of Select Quinoa Varieties during Extrusion Processing. *J. Food Sci.* **2016**, *81*, E2939–E2949. [CrossRef] [PubMed]
197. Foley, W.J.; McIlwee, A.; Lawler, I.; Aragones, L.; Woolnough, A.P.; Berding, N. Ecological applications of near infrared reflectance spectroscopy-a tool for rapid, cost-effective prediction of the composition of plant and animal tissues and aspects of animal performance. *Oecologia* **1998**, *116*, 293–305. [CrossRef] [PubMed]

198. Lane, H.M.; Murray, S.C.; Montesinos-López, O.A.; Montesinos-López, A.; Crossa, J.; Rooney, D.K.; Barrero-Farfan, I.D.; De La Fuente, G.N.; Morgan, C.L.S. Phenomic selection and prediction of maize grain yield from near-infrared reflectance spectroscopy of kernels. *TPPJ* **2020**, *3*, e20002. [CrossRef]
199. Escuredo, O.; Martín, M.I.G.; Moncada, G.W.; Fischer, S.; Hierro, J.M.H. Amino acid profile of the quinoa (Chenopodium quinoa Willd.) using near infrared spectroscopy and chemometric techniques. *J. Cereal Sci.* **2014**, *60*, 67–74. [CrossRef]
200. Rodríguez, S.D.; Rolandelli, G.; Buera, M.P. Detection of quinoa flour adulteration by means of FT-MIR spectroscopy combined with chemometric methods. *Food Chem.* **2019**, *274*, 392–401. [CrossRef]
201. Roa-Acosta, D.F.; Bravo-Gómez, J.E.; García-Parra, M.A.; Rodríguez-Herrera, R.; Solanilla-Duque, J.F. Hyper-protein quinoa flour (Chenopodium Quinoa Wild): Monitoring and study of structural and rheological properties. *LWT* **2020**, *121*, 108952. [CrossRef]
202. Morais, C.L.M.; Santos, M.C.D.; Lima, K.M.G.; Martin, F.L. Improving data splitting for classification applications in spectrochemical analyses employing a random-mutation Kennard-Stone algorithm approach. *Bioinformatics* **2019**, *35*, 5257–5263. [CrossRef]
203. Horwitz, W. *Official Methods of Analysis of AOAC International*; Association of Official Analytical Chemists International: Gaithersburg, MD, USA, 2019.
204. Agelet, L.E.; Hurburgh, C.R. A Tutorial on Near Infrared Spectroscopy and Its Calibration. *Crit. Rev. Anal. Chem.* **2010**, *40*, 246–260. [CrossRef]
205. Rinnan, Å.; Berg, F.V.D.; Engelsen, S.B. Review of the most common pre-processing techniques for near-infrared spectra. *TrAC Trends Anal. Chem.* **2009**, *28*, 1201–1222. [CrossRef]
206. Craine, E.B.; Murphy, K.M. Seed Composition and Amino Acid Profiles for Quinoa Grown in Washington State. *Front. Nutr.* **2020**, *7*, 126. [CrossRef]
207. Kuljanabhagavad, T.; Thongphasuk, P.; Chamulitrat, W.; Wink, M. Triterpene saponins from Chenopodium quinoa Willd. *Phytochemistry* **2008**, *69*, 1919–1926. [CrossRef]
208. Madl, T.; Sterk, H.; Mittelbach, M.; Rechberger, G.N. Tandem mass spectrometric analysis of a complex triterpene saponin mixture of Chenopodium quinoa. *J. Am. Soc. Mass Spectrom.* **2006**, *17*, 795–806. [CrossRef] [PubMed]
209. Woldemichael, G.M.; Wink, M. Identification and biological activities of triterpenoid saponins from Chenopodium quinoa. *J. Agric. Food Chem.* **2001**, *49*, 2327–2332. [CrossRef]
210. Otterbach, S.; Wellman, G.; Schmöckel, S.M. Saponins of Quinoa: Structure, Function and Opportunities. In *The Quinoa Genome*; Schmöckel, S.M., Ed.; Springer: Cham, Swizterland, 2021; pp. 119–138. [CrossRef]
211. Koziol, M.J. Afrosimetric estimation of threshold saponin concentration for bitterness in quinoa (Chenopodium quinoa Willd). *J. Sci. Food Agric.* **1991**, *54*, 211–219. [CrossRef]
212. Hirich, A.; Rafik, S.; Rahmani, M.; Fetouab, A.; Azaykou, F.; Filali, K.; Ahmadzai, H.; Jnaoui, Y.; Soulaimani, A.; Moussafir, M.; et al. Development of Quinoa Value Chain to Improve Food and Nutritional Security in Rural Communities in Rehamna, Morocco: Lessons Learned and Perspectives. *Plants* **2021**, *10*, 301. [CrossRef] [PubMed]
213. Afzal, I.; Kamran, M.; Basra, S.M.A.; Khan, S.H.U.; Mahmood, A.; Farooq, M.; Tan, D.K. Harvesting and post-harvest management approaches for preserving cottonseed quality. *Ind. Crop. Prod.* **2020**, *155*, 112842. [CrossRef]
214. Hong, T.D.; Linington, S.; Ellis, R.H. Seed Storage Behaviour: A Compendium Handbooks for Genebanks No. 4. In *Ecology and Classification of North American Freshwater Invertebrates*; International Plant Genetic Resources Institute: Rome, Italy, 1996; Available online: https://cgspace.cgiar.org/handle/10568/105158 (accessed on 15 August 2021).
215. Hong, T.D.; Ellis, R.H. *A Protocol to Determine Seed Storage Behaviour*; IPGRI Technical Bulletin No. 1; International Plant Genetic Resources Institute (IPGRI): Rome, Italy, 1996.
216. Roberts, E.H.; Ellis, R. Water and Seed Survival. *Ann. Bot.* **1989**, *63*, 39. [CrossRef]
217. Afzal, I.; Bakhtavar, M.A.; Ishfaq, M.; Sagheer, M.; Baributsa, D. Maintaining dryness during storage contributes to higher maize seed quality. *J. Stored Prod. Res.* **2017**, *72*, 49–53. [CrossRef]
218. Bradford, K.J.; Dahal, P.; Van Asbrouck, J.; Kunusoth, K.; Bello, P.; Thompson, J.; Wu, F. The dry chain: Reducing postharvest losses and improving food safety in humid climates. *Trends Food Sci. Technol.* **2018**, *71*, 84–93. [CrossRef]
219. De Vitis, M.; Hay, F.; Dickie, J.B.; Trivedi, C.; Choi, J.; Fiegener, R. Seed storage: Maintaining seed viability and vigor for restoration use. *Restor. Ecol.* **2020**, *28*. [CrossRef]
220. Ceccato, D.; Delatorre-Herrera, J.; Burrieza, H.; Bertero, D.; Martínez, E.; Delfino, I.; Moncada, S.; Bazile, D.; Castellión, M. Seed physiology and response to germination conditions. In *State of the Art Report on Quinoa around the World in 2013*; FAO: Santiago, Chile, 2015; pp. 131–142.
221. Ceccato, D.V.; Bertero, H.D.; Batlla, D. Environmental control of dormancy in quinoa (Chenopodium quinoa) seeds: Two potential genetic resources for pre-harvest sprouting tolerance. *Seed Sci. Res.* **2011**, *21*, 133–141. [CrossRef]
222. McGinty, E.; Murphy, K.; Hauvermale, A. Seed Dormancy and Preharvest Sprouting in Quinoa (*Chenopodium quinoa* Willd). *Plants* **2021**, *10*, 458. [CrossRef] [PubMed]
223. Ellis, R.H.; Hong, T.D.; Jackson, M.T. Seed Production Environment, Time of Harvest, and the Potential Longevity of Seeds of Three Cultivars of Rice (Oryza sativa L.). *Ann. Bot.* **1993**, *72*, 583–590. [CrossRef]
224. Romero, G.; Heredia, A.; Chaparro-Zambrano, H.N. Germinative potential in quinoa (Chenopodium quinoa Willd.) seeds stored under cool conditions. *Rev. UDCA Actual. Divulg. Científ.* **2018**, *21*, 341–350. [CrossRef]

225. Spehar, C.R.; Santos, R.L.D.B. Quinoa BRS Piabiru: Alternative for diversification of cropping systems. *Pesqui. Agropecuária Brasileira.* **2002**, *37*, 809–893. [CrossRef]
226. Castellión, M.; Matiacevich, S.; Buera, P.; Maldonado, S. Protein deterioration and longevity of quinoa seeds during long-term storage. *Food Chem.* **2010**, *121*, 952–958. [CrossRef]
227. Prego, I.; Maldonado, S.; Otegui, M. Seed Structure and Localization of Reserves inChenopodium quinoa. *Ann. Bot.* **1998**, *82*, 481–488. [CrossRef]
228. Ng, S.-C.; Anderson, A.; Coker, J.; Ondrus, M. Characterization of lipid oxidation products in quinoa (Chenopodium quinoa). *Food Chem.* **2007**, *101*, 185–192. [CrossRef]
229. Baributsa, D.; Njoroge, A. The use and profitability of hermetic technologies for grain storage among smallholder farmers in eastern Kenya. *J. Stored Prod. Res.* **2020**, *87*, 101618. [CrossRef]
230. Kiobia, D.; Silayo, V.; Mutabazi, K.; Graef, F.; Mourice, S. Performance of hermetic storage bags for maize grains under farmer-managed conditions: Good practice versus local reality. *J. Stored Prod. Res.* **2020**, *87*, 101586. [CrossRef]
231. Bakhtavar, M.A.; Afzal, I. Climate smart Dry Chain Technology for safe storage of quinoa seeds. *Sci. Rep.* **2020**, *10*, 1–12. [CrossRef]
232. de Jesus Souza, F.I.; Devilla, I.A.; de Souza, R.T.; Teixeira, I.R.; Spehar, C.R. Physiological quality of quinoa seeds submitted to different storage conditions. *Afr. J. Agric. Res.* **2016**, *11*, 1299–1308. [CrossRef]
233. Mammadi, A.; Afshari, R.T. Modeling of quinoa (Chenopodium quinoa) seed viability with probit analysis. *Iran. J. Field Crop. Sci.* **2018**, *49*, 49–57. [CrossRef]
234. Ellis, R.; Hong, T.D.; Roberts, E.H. A Low-Moisture-Content Limit to Logarithmic Relations Between Seed Moisture Content and Longevity. *Ann. Bot.* **1988**, *61*, 405–408. [CrossRef]

plants

MDPI

Article

Quinoa for the Brazilian Cerrado: Agronomic Characteristics of Elite Genotypes under Different Water Regimes

Patrícia Carvalho da Silva [1], Walter Quadros Ribeiro Junior [2,*], Maria Lucrecia Gerosa Ramos [1], Sonia Maria Costa Celestino [2], Alberto do Nascimento Silva [1], Raphael Augusto das Chagas Noqueli Casari [3], Charles Cardoso Santana [1], Cristiane Andrea de Lima [1], Thomas Christopher Rhys Williams [4] and Christina Cleo Vinson [1]

[1] Faculdade de Agronomia e Medicina Veterinária, Universidade de Brasília, Brasília 70910970, DF, Brazil; patriciacarvalhoagro@gmail.com (P.C.d.S.); lucreciaunb@gmail.com (M.L.G.R.); albertosilvaagro@gmail.com (A.d.N.S.); ccsantana.agro@gmail.com (C.C.S.); agro.cristiane@gmail.com (C.A.d.L.); ccvinson@unb.br (C.C.V.)
[2] Embrapa Cerrados, Empresa Brasileira de Pesquisa Agropecuária, Planaltina 73310970, DF, Brazil; sonia.celestino@embrapa.br
[3] Instituto de Geociências, Universidade de Brasília, Brasília 70910970, DF, Brazil; casari.raphael@gmail.com
[4] Departamento de Botânica, Instituto de Ciências Biológicas, Universidade de Brasília, Brasília 70910970, DF, Brazil; tcrwilliams@unb.br
* Correspondence: walter.quadros@embrapa.br; Tel.: +55-(61)-3388-9886

Citation: da Silva, P.C.; Ribeiro Junior, W.Q.; Ramos, M.L.G.; Celestino, S.M.C.; Silva, A.d.N.; Casari, R.A.d.C.N.; Santana, C.C.; de Lima, C.A.; Williams, T.C.R.; Vinson, C.C. Quinoa for the Brazilian Cerrado: Agronomic Characteristics of Elite Genotypes under Different Water Regimes. *Plants* **2021**, *10*, 1591. https://doi.org/10.3390/plants10081591

Academic Editors: Cataldo Pulvento and Didier Bazile

Received: 7 June 2021
Accepted: 19 July 2021
Published: 2 August 2021

Abstract: Quinoa stands out as an excellent crop in the Cerrado region for cultivation in the off-season or irrigated winter season. Here, we tested the effects of different water regimes on the agronomic characteristics, physiology, and grain quality of different elite quinoa genotypes under field conditions. The experiment was conducted under field conditions at Embrapa Cerrados (Planaltina, DF, Brazil). The experimental design was in randomized blocks, in a split-plot scheme, with four replications. The plots were composed of 18 quinoa genotypes and modified BRS Piabiru (the currently used genotype), and the split-plots were divided into 4 different water regimes. The following variables were evaluated: productivity and productivity per unit of applied water (PUAA), plant height, flavonoids, anthocyanins, gas exchange, chlorophyll, leaf proline, and relative water content. Our results showed that water regimes between 309 and 389 mm can be recommended for quinoa in the Cerrado region. CPAC6 and CPAC13 presented the highest yield and PUAA under high and intermediate WRs, and hence were the most suitable for winter growth under irrigation. CPAC17 is most suitable for off-season growth under rainfed conditions, as it presented the highest PUAA under the low WRs (247 and 150). CPAC9 stood out in terms of accumulation of flavonoids and anthocyanins in all WRs. Physiological analyses revealed different responses of the genotypes to water restriction, together with symptoms of stress under lower water regimes. Our study reinforces the importance of detailed analyses of the relationship between productivity, physiology, and water use when choosing genotypes for planting and harvest in different seasons.

Keywords: *Chenopodium quinoa*; water use efficiency; phenolic compounds; gas exchange

1. Introduction

Crop development and yield are affected by different environmental factors, and water restriction is the most important constraint on agricultural yield [1,2]. This is a particular problem in the Brazilian Cerrado, which has a tropical climate with an average of 1500 mm of rain, but where approximately 90% of precipitation occurs during the rainy season (from October to April). The rainy season is followed by a dry season (from May to September), during which the relative humidity is low, the evaporation very high, and precipitation is rare. There are three harvest periods in the Brazilian Cerrado: (1) The main crop season, which occurs during the wet season from October to January; (2) The off-season crop, which is grown at the end of wet season without irrigation, is planted between the months of January to March [3,4], and is harvested in May during the dry season; and (3) the winter season crop, which is cultivated under irrigated conditions, with the crop being both planted (April to May) and harvested (August to September) during the dry season. Both the off-season and winter season require careful selection of genotypes for grain production; drought-tolerant genotypes (DT) should be selected for the off-season crop, and high productivity per unit of applied water (PUAA) genotypes are needed for winter crops as they are grown under irrigation. Obtaining genotypes that are better adapted to stressful edaphoclimatic conditions in order to resist periods with water deficiency whilst maintaining the highest possible productivity for each condition is therefore of great importance in plant breeding programs [5].

Quinoa (*Chenopodium quinoa* Willd.) is a pseudocereal rich in natural antioxidants, flavonoids, and anthocyanins [6,7], and these compounds may protect plants against biotic and abiotic stress [8]. Water stress increases leaf temperature, and reduces crop height, stomatal conductance, plant biomass, and yield [9,10].

Quinoa has been cultivated for millennia under conditions of low rainfall, as it has physiological and morphological strategies to overcome water deficit [11]. Moreover, this crop has been cultivated in different agroclimatic zones as it is well adapted to a variety of different environments due to its high genetic diversity [11]. Quinoa has mainly been cultivated in Argentina, Bolivia, Chile, Colombia, Ecuador, and Peru, though high productivity has also been observed when planted in Kenya as well as the Himalayas and northern plains of India [12].

In Brazil, research carried out in Embrapa Cerrados led to the selection of the genotype BRS Piabiru [3], which is the first cultivar in use for quinoa grain cultivation that is adapted to Cerrado conditions. Although planting quinoa during the main crop is not recommended due to the high water availability during the harvesting period (which can potentially result in panicle seed germination [3,5]), quinoa is recommended for growth during the off-season or irrigated winter season due to its high water use efficiency, drought tolerance, and adaptation to different environmental conditions.

Here, we therefore analysed growth, yield, and physiological and functional grain parameters, with the aim of identifying genotypes with potential for growth during the off-season and winter season for the Cerrado region.

2. Material and Methods

2.1. Experimental Area Characterization

The experiment was conducted between May and September 2017 at the experimental station of Embrapa Cerrados-Planaltina DF (Brazil), located between the geographic coordinates: 15°35′ S and 47°42′ W with an approximate altitude of 1200 m.

The climate of the region is tropical wet–dry (Aw), according to Köppen's classification [13], with 2 well-defined seasons (dry and rainy). The average annual temperature is 21.3 °C and the mean annual rainfall 1400 mm. The experiment was conducted in winter due to the absence of rainfall during this period, which allowed for effective control of the amount of water applied. The long-term temperature and rainfall data from the last 20 years (1997 to 2017) are shown in Figure 1 and illustrate the annual dry season during the months of May to September. Temperature and rainfall data from the study period are in good agreement with long-term patterns (Figure 1).

The soil of the experimental area is classified as clayey Oxisol (Typic Haplustox) (Soil Survey Staff 2014), with a soft undulating relief and a clayey texture. The soil has the following characteristics in the 0–20 cm layer: pH (in water) = 5.71; Ca (mg kg^{-1}) = 543.08; Mg (mg kg^{-1}) = 103.32; K (mg kg^{-1}) = 56; H + Al (mg kg^{-1}) = 56.82; p (mg kg^{-1}) = 22.87; organic matter (g kg^{-1}) = 24.7.

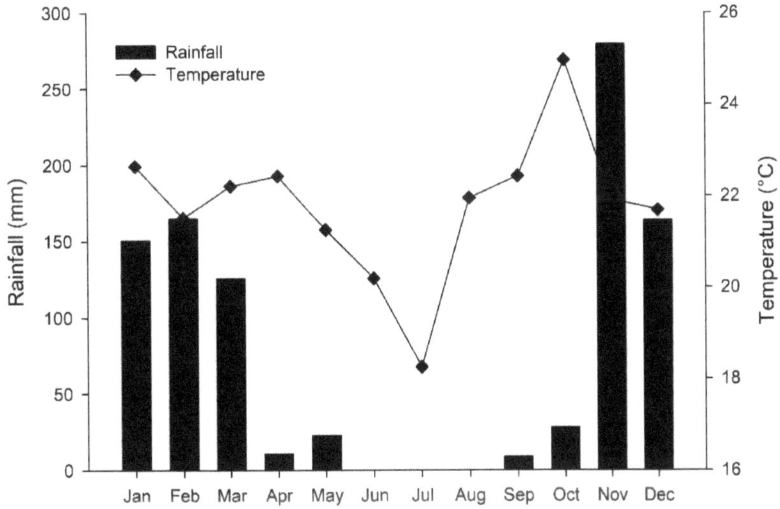

Figure 1. Mean rainfall and temperature in long-term data from the last 20 years (1997–2017).

2.2. Experimental Design

Tests with 980 selected genotypes were previously conducted in Cristalina-GO, and genotypes with an early cycle were selected; early cycles were defined as those of equal to or less than 110 days as this is the period for growth in the off-season or irrigated winter season. The experiment was repeated over 4 years; genotypes were sown in the off-season and winter of 2011, 2012, 2013, and 2014, and morphological characteristics were evaluated to obtain morphological descriptors of the genotypes. As a result, 18 genotypes with life cycle equal to or less than 110 days and with seeds larger than 2 mm in diameter were selected for use in this study: CPAC1, CPAC2, CPAC3, CPAC4, CPAC5, CPAC6, CPAC8, CPAC10, CPAC11, CPAC12, CPAC13, CPAC14, CPAC16, CPAC17, CPAC18, CPAC19, and CPAC20, together with the BRS Piabiru cultivar, which was modified by excluding high-cycle individuals.

The experimental design was in randomized blocks with split-plots with 4 replications. The plots were composed of the 18 quinoa genotypes and modified BRS Piabiru each separated by a distance of 1 m, and the split-plots were composed of 4 water regimes (WRs 150, 247, 389, and 480 mm). Seeds of the different genotypes of quinoa were sown manually under a no-tillage system. After soil analysis, fertilization was carried out in the furrows at a level of 400 kg ha^{-1} of the formula 04-30-16 (NPK). Sowing was done with a density of 0.7 g of seeds per linear meter and line spacing of 0.5 m. Nitrogen topdressing was carried out 30 days after seedling emergence at a dose of 100 kg ha^{-1} N in the form of urea. In order to avoid the competition of invasive plants, manual weeding was carried out.

The different water regimes were obtained using a sprinkler irrigator bar 40 m wide, connected to a spool with adjustable speed. Irrigation was performed homogeneously for the first 30 days after emergence in order to establish growth of all plants, with a total applied depth of 126 mm. After this period, the modified line source methodology was applied [14], using sprinklers with decreasing water flows from the central area to the end of the bar. The sprinklers were overlapped in order to promote a decreasing gradient of water from the central area to the edge of the bar.

Each water regime consisted of an experimental unit of 2.5 m^2 in area within the plot (subplot), formed by 5 lines, with 0.5 m spacing. Over the stress treatment 10 irrigations were carried out, and the accumulated uniform irrigation plus the variable irrigations gave a total of 150, 247, 389, and 480 mm for the 4 water regimes. The highest level of irrigation was determined as described in the irrigation monitoring program in the Cerrado [15], through replacement of evapotranspiration using agrometeorological indicators of the region, the soil type, the date of full emergence of plants, and wheat as a reference. Although our experiment was carried out with quinoa, we used wheat as a reference crop for irrigation, understanding that it has similar water consumption. Irrigation was carried out approximately every 5 days, according to the climatic conditions and crop phenological phase. To measure the water depth applied in each irrigation 2 rows of collectors parallel to the irrigation line were set up.

Thermal images were acquired using a thermal infrared camera (FLIR® T420, FLIR Systems, Wilsonville, OR, USA) with the following characteristics: thermal spatial resolution of 320 × 240 pixels, spectral response of 7.5 to 13 μm, thermal sensitivity of pixels from 0.045 °C to 30 °C, and temperature accuracy of ±2. The camera was mounted on an unmanned aerial vehicle (XFly, O X800, Bauru, SP, Brazil) at a height of 60 m from the canopy and the images were performed at the same time and day as the physiological variables. For the processing of images and to obtain temperature data in the canopies of quinoa plants, the QGIS software was used [16]. The RGB (red, green and blue) and thermal images were geo-referenced for the generation of an orthomosaic and later classified to select only the areas of plant canopies in the useful parcel, avoiding picking up exposed soil.

2.3. *Variables Analyzed*

2.3.1. Grain Analysis

Quinoa grain productivity and productivity per unit of applied water (PUAA) were evaluated for all studied genotypes under 4 water regimes (WRs 150, 247, 389 and 480 mm). The panicles were harvested by hand, and later placed in a Wintersteiger plot harvester. In each subplot an area of 2.5 m^2 was harvested. Grain moisture was corrected to 13% and the results were analysed as ton ha^{-1}. The productivity per unit of applied water (PUAA) was determined by the ratio between the grain yield and the amount of water applied during the whole crop cycle, analysed in kg ha^{-1} mm^{-1}. Productivity per unit of applied water (PUAA) was also evaluated, and calculated using the following equation:

$$\text{PUAA} = \text{Prod}/\text{LTD}$$

where

PUAA: Productivity per unit of water applied (kg ha mm^{-1});

Prod: Grain productivity, in kg ha^{-1}; and

LTD: Total irrigation depth available referring to the amount of water applied during the crop cycle, in mm.

The weight of 1000 grains was evaluated for 4 genotypes (CPAC4, CPAC11, CPAC19, and the early BRS Piabiru). The weight of 1000 grains was determined according to the method described by [17], in which 100 grains from each subplot are weighed using a semi-analytical balance and the result obtained is multiplied by 10.

The functional quality of the grains was evaluated for all studied genotypes. Functional quality was evaluated by quantification of flavonoids and total anthocyanins and flavonoids [18]. The grains were ground; subsequently, 5 g was weighed and the anthocyanin and flavonoids from the grains were extracted with 30 mL of a solution of ethyl alcohol and 1.5 N HCl. After homogenization for 1 min, the volume was made up to 50 mL in volumetric flasks. This mixture remained at 8 °C for 16 h. After filtration, absorbances (Abs) were taken at 535 nm for anthocyanin and 374 nm for flavonoids. The results were expressed in mg/100 g of quinoa grains. The contents of flavonoids 9 factor 76.6) and of anthocyanins (factor 98.2) were calculated according to the formulas: Dilution Factor (DF) = (100 × volume (mL))/weight (g)); Content of flavonoids/anthocyanins = (Abs × DF)/factor. Results were expressed in mg/100 g of quinoa grains.

2.3.2. Physiological Analysis

During the flowering period of the crops, 60 days after emergence the following variables were evaluated: gas exchange (photosynthesis; stomatal conductance, internal CO_2 concentration in the leaves, and transpiration), leaf chlorophyll (*a*, *b*, total), chlorophyl fluorescence, proline content, relative water content, and plant height. These variables were analysed for 4 genotypes (CPAC4, CPAC11, CPAC19, and the early BRS Piabiru), with exception of plant height, which was analysed in all genotypes. Plant height was measured from the base of the stem to the peak of its inflorescence using a measuring tape. Five plants per subplot were evaluated.

Gas exchange was measured using an IRGA infrared gas analyzer model LI-6400XT (LI-COR, Inc., Lincoln, NE, USA) between 08:00 and 11:00 on the youngest fully expanded leaves. Ambient air temperature and relative humidity were used and the CO_2 concentration in the chamber was maintained at 400 μmol mol^{-1} together with a saturating photosynthetically active photon flux density (PPFD) of 2000 μmol m^{-2} s^{-1}. To determine light saturation, a PPFD curve of 0, 20, 60, 100, 250, 500, 1000, 2000, 2250, 2500, and 3000 μmol m^{-2} s^{-1} was obtained under ambient CO_2. Maximum fluorescence (F′m) and basal fluorescence (F′0) were evaluated for light-adapted leaves using the portable open flow gas exchange system (IRGA-LI-6400XT; LI-COR Inc., Lincoln, NE, USA). The effective quantum yield of Photosystem II is given by, $F'v/F'm = (Fm' - F0')/Fm'$, according to [19].

The chlorophyll index (*a*, *b*, *a* + *b* and *a/b*) was evaluated in the morning with the aid of a digital ChlorofiLOG, model CFL 1030, from Falker in the youngest fully expanded leaves, and 10 readings were taken for each treatment.

The relative water content (RWC) was determined according to the methodology of [20], using leaf discs of 1 cm in diameter obtained from leaves collected predawn between 03:00 and 05:00. The discs were immediately weighed after collection to determine the fresh mass (FM). The turgid mass (TM) of the leaf discs was obtained by hydration in distilled water for 24 h and the dry mass (DM) was obtained after drying in a forced air oven at 60 °C. The RWC was calculated according to the formula.

$$RWC = ((FM - DM)/(TM - DM)) \times 100\%,$$

The proline content was determined according to [21]. Completely expanded young leaves were collected between 13:30 and 15:00. Samples were extracted by grinding 0.5 g of fresh leaf material in 10 mL of 3% sulfosalicylic acid. Then, 500 μL of the extract of each sample was placed in a test tube containing 2.75 mL of distilled water, with the addition of 2 mL of acid ninhydrin, 2 mL of glacial acetic acid, and 100 microliters of glycine (126 mM). Acid ninhydrin was prepared by warming 1.25 g in 30 mL glacial acetic acid and 20 mL 6 M phosphoric acid. Then, the samples were kept for 1 h in a water bath at 100 °C [21]. After cooling the samples by immersion in an ice bath the optical density was evaluated at 515 nm using a Pharmacia Ultrospect III spectrophotometer. The absorbance obtained was compared with the standard proline curve and the results expressed as μmol of proline/g of fresh mass (FM).

2.4. Statistical Analysis

Data were subjected to analysis of variance at 5% probability by the F test and the comparison of means was carried out using the Scott–Knott test and Tukey's test. The statistical model was adjusted using the Proc Mixed from SAS [22] using the restricted maximum likelihood method. For non-significant interactions, factors were analysed separately (genotype and water regime). Figures were produced using the software Sigma Plot, version 10 [23].

3. Results

3.1. Productivity

In general, the genotypes responded differently to water regimes, reinforcing the idea that the productive potential, productivity per unit of applied water (PUAA), and responsiveness to irrigation depend on the genotype (Tables 1 and 2). We found a significant interaction between quinoa genotypes and WRs for productivity (Table 1). For most quinoa genotypes, the yield was reduced with WR 247, with exception of CPAC3, CPAC8, CPAC9, CPAC14, and CPAC19 which had reduced yield with WR 389 mm.

Table 1. Productivity (t ha^{-1}) of 18 quinoa genotypes and BRS Piabiru under 4 water regimes.

Genotypes	Water Regime (mm)			
	480	389	247	150
CPAC1	7.82 aA	8.34 bA	4.06 bB	1.56 bC
CPAC2	8.32 aA	7.89 bA	3.94 cB	1.83 bC
CPAC3	8.02 aA	7.81 bB	6.25 aC	1.94 bD
CPAC4	8.40 aA	8.23 bA	5.21 aB	1.95 bC
CPAC5	8.17 aA	7.90 bA	5.22 aB	1.94 bC
CPAC6	8.50 aA	8.84 aA	5.68 aB	2.46 aC
CPAC8	8.64 aA	7.07 cB	4.50 aC	1.58 bD
CPAC9	8.21 aA	7.01 cB	5.75 aC	2.11 bD
CPAC10	8.56 aA	8.36 bA	5.12 aB	2.61 aC
CPAC11	5.66 bB	6.80 cA	5.40 aB	2.38 aC
CPAC12	8.57 aA	7.83 bA	5.82 aB	2.58 aC
CPAC13	8.85 aA	9.73 aA	4.17 bB	2.60 aC
CPAC14	9.21 aA	6.51 cB	3.88 bB	1.61 bB
BRS Piabiru	7.58 aA	8.14 bA	5.40 aB	1.84 bC
CPAC16	8.51 aA	7.46 cA	4.74 aB	2.33 aC
CPAC17	8.92 aA	8.44 bA	4.80 aB	3.64 aC
CPAC18	8.96 Aa	9.16 aA	4.75 aB	2.21 bC
CPAC19	7.71 aA	6.42 cB	2.53 cC	1.79 bC
CPAC20	7.97 aA	7.08 cA	3.93 bB	2.09 bC

Means followed by the same lowercase letter (column) or uppercase letter (line), do not differ according to the Scott–Knott test at a 5% probability.

Table 2. Productivity per unit of applied water (kg $ha^{-1}mm^{-1}$) of 18 quinoa genotypes and BRS Piabiru under 4 water regimes.

Genotypes	Water Regime (mm)			
	480	389	247	150
CPAC1	17.18 aB	22.2 aA	16.43 cB	10.41 cC
CPAC2	15.89 aB	21.0 aA	13.02 dB	13.0 cB
CPAC3	18.59 aB	20.8 aB	25.29 aA	13.5 cD
CPAC4	16.89 aB	21.15 aA	21.1 bA	13.03 cC
CPAC5	16.9 aB	21.0 aA	23.9 bA	12.1 cC
CPAC6	17.7 aB	22.72 aA	23.0 aA	16.4 bB
CPAC8	18.01 aA	18.18 bA	18.23 cA	10.52 bB
CPAC9	17.73 aB	18.01 bB	23.3 aA	14.10 cC
CPAC10	17.83 aB	21.49 aA	20.6 bA	15.38 cB
CPAC11	11.1 bC	18.0 bB	21.89 bA	15.85 bA
CPAC12	18.0 aB	20.1 aB	23.56 aA	17.24 bD
CPAC13	19.83 aB	25.9 aA	19.1 bB	17.6 bB
CPAC14	12.69 aB0	16.3 bA	15.71 cA	10.1 cC
BRS Piabiru	15.79 aB	20.92 aA	21.88 bA	14.37 cB
CPAC16	17.74 aA	19.8 bA	19.18 bA	15.57 bA
CPAC17	18.58 aB	21.70 aA	19.44 bB	24.25 aA
CPAC18	18.66 aB	23.56 aA	16.89 cB	14.71 cC
CPAC19	16.08 aA	16.50 bA	11.6 dB	10.25 cB
CPAC20	17.82 aA	18.82 bA	15.94 cA	11.89 cB

Means followed by the same lowercase letter (column) or uppercase letter (line) do not differ according to the Scott–Knott test at a 5% probability.

Under the highest WR (480 mm) all genotypes obtained similarly high yields, meaning that the genotypes could not be distinguished. The only exception was CPAC11 (a dwarf genotype), which presented significantly lower productivity (Table 1) and the lowest PUAA (Table 2). However, in the intermediate WR (389 mm), there were major differences among genotypes, with the genotypes CPAC6, CPAC13, and CPAC18 standing out as having the highest productivity (Table 1). Under the water deficit regimes there was a clearer distinction between the genotypes, which were divided into three groups: high, medium, and low productivity. Under the 247 mm WR, with a reduction of 49% in applied water, the genotypes presented a high productivity of between 2.53 and 6.26 t ha^{-1}. On the other hand, under the extremely low water regime (150 mm), the genotypes CPAC6, CPAC10, CPAC11, CPAC12, CPAC13, CPAC16, and CPAC17 were more productive. The genotypes CPAC3, CPAC8, CPAC9, CPAC14, and CPAC19 were more sensitive to mild water deficit (WR 389), showing reduced productivity under this water regime, whilst CPA11 showed the opposite response, with increased productivity with 91 mm less water applied (Table 1).

3.2. Productivity Per Unit of Applied Water (PUAA)

In general, the lowest PUAA was found for WR 480, with only the genotypes CPAC8, CPAC16, CPAC19, and CPAC20 presenting similar PUAA at 480 mm and under deficit conditions (380 and 247 mm, Table 2). The highest PUAA occurred between 247 and 389 mm WR for most quinoa genotypes (Table 2). Comparing WR 480 mm with the intermediate 389 mm and 247 mm WRs, there was a reduction of 36 and 49% in volume of total water applied and an increase by 23 and 21% in PUAA, respectively (Table 2). The genotypes with the highest PUAA in 247 mm were CPAC3, CPAC6, CPAC9, and CPAC12, and with the exception of CPAC6, the other genotypes had low PUAA in the other WRs (Table 2).

Under the low water regime of 150 mm the PUAA for CPAC17 was significantly higher than the other genotypes, and importantly, this variable was similar under intermediate and low water regimes, meaning that regardless of the water regime its PUAA did not alter. Although CPAC17 had the highest productivity under WR 150 mm (3.64 ton ha^{-1}) compared to all studied genotypes (Table 1), it had greater productivity under intermediate (8.44 ton ha^{-1}) and high (8.92 ton ha^{-1}) WRs. Under the lowest WR (128 mm), however, overall efficiency was very low as the stress was very high.

3.3. Thousand Grain Weight

The weight of 1000 grains (TGW) of quinoa was evaluated, and there was no significant interaction between the factors. The effect of the water regimes on TGW is shown in Figure S1A), where there is an increase in TGW up to WR 389 mm and from this level, there was no increase in grain weight with the increase in water availability. The CPAC11 and early Piabiru genotypes showed better results, and CPAC19 had lower grain weights compared to other genotypes. In this study, a TGW of 2.35 to 2.84 was obtained, depending on the genotype (Figure S1B).

3.4. Functional Quality

Our study shows a significant interaction between quinoa genotypes and water regimes and the abundance of flavonoids (Table 3) and total anthocyanins (Table 4). For flavonoid contents, BRS Piabiru and CPAC2 did not respond to all WRs, but showed lower values than CPAC9, which was influenced by WRs. For anthocyanin contents, CPAC1, CPAC11, CPAC16, CPAC18, CPAC19, and CPAC20 presented similar values in all WRs, and these genotypes had lower values than CPAC9.

Table 3. Total flavonoid concentrations (mg/100 g) in grains of 18 quinoa genotypes and BRS Piabiru under 4 water regimes.

Genotypes	Water Regime (mm)			
	480	389	247	150
CPAC1	81.59 cB	101.62 dA	66.21 dC	88.34 dA
CPAC2	95.97 bA	97.23 dA	85.71 dA	92.21 dA
CPAC3	98.83 bA	51.54 eB	97.12 cA	113.25 cA
CPAC4	80.02 cB	98.31 dA	91.64 cA	96.23 dA
CPAC5	84.07 cB	103.0 dA	74.58 dB	86.80 dB
CPAC6	83.65 cC	115.23 cA	82.81 dC	100.05 dB
CPAC8	96.95 bB	110.72 dA	114.25 bA	93.58 dB
CPAC9	215.22 aB	171.32 aC	205.11 aB	226.02 aA
CPAC10	85.48 cC	110.0 dB	96.32 cC	210.81 bA
CPAC11	89.00 cB	118.59 cA	100.97 cB	113.25 cA
CPAC12	96.17 bB	120.98 cA	84.82 dB	112.51 cA
CPAC13	99.47 bA	111.33 dA	84.73 dB	102.33 dA
CPAC14	115.19 bA	119.74 cA	84.94 dB	104.71 cA
BRS Piabiru	104.15 bA	105.89 dA	101.98 cA	104.56 cA
CPAC16	105.09 bB	124.80 cA	99.48 cB	110.93 cB
CPAC17	115.73 bC	145.76 bA	97.05 cC	118.86 cB
CPAC18	88.27 cB	62.76 eC	94.25 cB	113.57 cA
CPAC19	109.63 bB	119.74 cA	102.59 cB	100.01 dB
CPAC20	119.93 bB	139.03 bA	114.40 bB	115.41 cB

Means followed by the same lowercase letter (column) or uppercase letter (line) do not differ according to the Scott–Knott test at a 5% probability.

Table 4. Total anthocyanin concentrations (mg/100 g) in grains of 18 quinoa genotypes and BRS Piabiru under 4 water regimes.

Genotypes	Water Regime (mm)			
	480	389	247	150
CPAC1	0.80 cA	0.76 dA	0.76 cA	0.80 dA
CPAC2	0.63 dA	0.68 dA	0.62 dA	0.40 cB
CPAC3	0.63 dD	1.0 bB	0.83 cC	1.16 cA
CPAC4	0.59 dB	0.65 eB	0.59 dB	0.76 dA
CPAC5	0.74 cA	0.59 eB	0.54 dB	0.70 dA
CPAC6	0.60 dB	0.73 dA	0.65 dB	0.84 dA
CPAC8	0.58 dB	1.07 bA	0.55 dB	0.57 eB
CPAC9	1.72 aC	1.89 aB	1.44 aD	2.05 aA
CPAC10	1.16 bA	0.73 dB	0.76 cB	0.91 dB
CPAC11	0.61 dA	0.48 fA	0.54 dA	0.61 eA
CPAC12	0.55 dB	0.68 dA	0.50 dB	0.65 eB
CPAC13	0.66 dB	0.63 eB	0.50 dC	0.73 dA
CPAC14	0.83 cB	0.82 cB	0.68 cC	0.94 cA
BRS Piabiru	0.89 cB	1.08 bA	0.61 dD	0.74 dC
CPAC16	0.72 cA	0.65 eA	0.81 cA	0.80 dA
CPAC17	1.03 bB	0.86 cC	0.79 cC	1.18 bA
CPAC18	0.68 dA	0.61 eA	0.72 cA	0.60 eA
CPAC19	0.64 dA	0.63 eA	0.71 cA	0.66 eA
CPAC20	1.13 bA	0.99 bA	1.06 bA	1.05 cA

Means followed by the same lowercase letter (column) or uppercase letter (line) do not differ according to the Scott–Knott test at 5% probability.

The accumulation of flavonoids in CPAC9 was greater than in BRS Piabiru and all other genotypes in all water regimes, reaching approximately double the concentration of flavonoids in most genotypes in both higher and lower water regimes (Table 3).

Similar to that observed for flavonoid accumulation, CPAC9 also accumulated higher anthocyanin content and had greater concentrations than BRS Piabiru and other genotypes under all water regimes, accumulating approximately double the concentration of anthocyanins (Table 4). In addition, CPAC9 was among the genotypes with the highest productivity under both high (8.21 ton ha^{-1}) and intermediate (7.01 ton ha^{-1}) WRs (Table 1).

3.5. Plant Height

Figure 2A shows quinoa plant height in response to water regimes. In the higher water regimes (389 and 480 mm), the plants presented greater heights (1.50 and 1.51 m) than the intermediate (247 mm) and low (150 mm) water regimes, which presented plants with lower heights of 1.40 and 1.17 m, respectively (Figure 2B). Severe water deficit (128 mm) reduced plant height by 22%. Comparing among genotypes, genotype CPAC11 was significantly shorter than all others, with an average height of 0.94 m (Figure 2B).

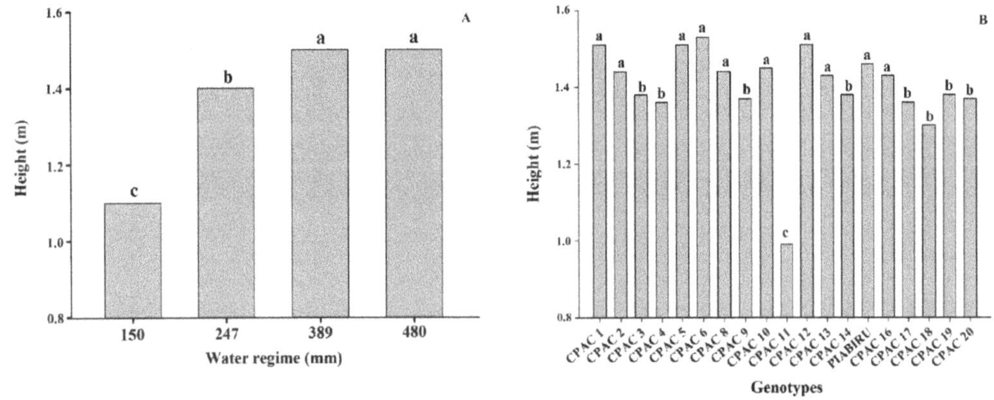

Figure 2. Height of different quinoa genotypes under 4 water regimes. (**A**) Effect of 4 water regimes in plant height in 20 quinoa genotypes. (**B**) Means followed by the same lowercase letter do not differ according to the Scott–Knott test at 5% probability.

3.6. Canopy Temperature

For the leaf temperature variable, no interaction between the factors was obtained and no significant difference was observed between the genotypes. Thus, the effect of water regimes on the temperature of the plants was subjected to regression analysis. This analysis showed a significant relationship between water availability and leaf temperature ($p < 0.05$), with the maximum temperature value (35.1 °C) for the lowest WR (150 mm) and the minimum (25.5 °C) in WR 480 mm (Figure S2), demonstrating that under severe drought the canopy of quinoa genotypes showed an increase in temperature of 9.6 °C.

3.7. Leaf Gas Exchange

The stomatal conductance and transpiration, as expected, were higher in the high-water regime and decreased as water restriction occurred, with a reduction of 94, 88, 92, and 91 % for the genotypes CPAC4, CPAC11, and CPAC19 and the BRS Piabiru precoce, respectively (Table 5). The different water conditions affected the four analysed genotypes differently, and CPAC9 had the highest stomatal conductance in the intermediate WR (247 mm), which represents off-season conditions. The CPAC11 genotype was more efficient in maintaining stomatal conductance and transpiration, as it had lower values in the high water regimes but little decrease in the intermediate WR (Table 5).

Table 5. Stomatal conductance (g_s–mol m^{-2} s^{-1}), transpiration rate (E–mmol m^{-2} s^{-1} Internal carbon (C_i-µmol m^{-2} s^{-1}), and photosynthetic rate (A-µmol m^{-2} s^{-1}) of 4 quinoa genotypes (CPAC4, CPAC9, CPAC11, and Piabiru), under 4 water regimes (150, 247, 389, 480 mm).

Genotypes		Water Regime (mm)			
		150	247	389	480
gs	CPAC4	0.034 cD	0.235 bC	0.470 bB	0.583 bA
	CPAC11	0.040 cC	0.232 bB	0.358 cA	0.342 cA
	BRS Piabiru	0.058 aD	0.218 bC	0.516 abB	0.629 aA
	CPAC19	0.050 bC	0.329 aB	0.552 aA	0.594 abA
E	CPAC4	1.45 bcD	6.15 aC	8.96 aB	9.88 aA
	CPAC11	1.20 cD	4.87 bC	6.43 cA	5.67 cB
	BRS Piabiru	3.11 aD	4.76 bC	8.20 aB	9.46 aA
	CPAC19	1.71 bD	5.83 aC	7.27 bB	8.38 bA
Ci	CPAC 4	144.0 aC	195.3 aB	239.9 bA	241.2 bA
	CPAC11	112.6 bC	198.6 aB	253.6 aA	259.2 aA
	BRS Piabiru	129.0 abD	191.3 aC	257.0 aB	270.3 aA
	CPAC19	82.5 cC	198.4 aB	255.9 aA	256.9 abA
A	CPAC4	4.9 cD	19.0 cC	22.0 bB	28.1 bA
	CPAC11	5.9 bC	20.5 bcB	22.9 bA	20.0 cB
	BRS Piabiru	9.0 aD	22.4 bC	32.9 aB	35.4 aA
	CPAC19	8.9 aC	27.0 aB	32.9 aA	34.3 aA

Means followed by the same lowercase letter (column) or uppercase letter (line) do not differ according to the Tukey test at a 5% probability.

We detected increased internal carbon (Ci) under the high water regimes and decreased Ci under severe stress, which agrees with the stomatal conductance and transpiration data as the absorption of CO_2 is directly related to the degree of stomatal opening. (Table 5). Similarly, photosynthesis of the four genotypes decreased gradually with reduced water availability. In the comparison between the highest (480 mm) and the lowest (150 mm) WR, reductions in photosynthetic rates of 83, 70, 74, and 75% were observed for the genotypes CPAC4, CPAC11, CPAC19, and BRS Piabiru precoce, respectively. An interesting pattern was observed for net photosynthesis, as under the high WR there were two groups of genotypes. CPAC19 and the BRS Piabiru obtained high values whilst CPAC4 and CPAC11 obtained lower values; however, their values were similar under intermediate WR and then suffered a drastic reduction in the severe WR. Ci proved to be the least sensitive parameter and could only distinguish the genotypes under the most severe water stress regime (WR 150).

3.8. *Effective Quantum Yield of Photosystem II and Chlorophyll Index*

For the F'v/F'm ratio, the factors varied independently, and no significant difference was observed between the genotypes. The effect of water regimes on the F'v/F'm ratio of quinoa plants was therefore analysed using regression. This analysis indicated a significant relationship between water availability and F'v/F'm ($p < 0.05$), with a maximum value (0.52) of F'v/F'm in the highest WR (480 mm) and the minimum (0.43) in WR 150 mm (Figure S3), indicating that the efficiency of PSII was significantly reduced when the plants were submitted to the severe stress of WR150 mm. In contrast, chlorophyll index was not affected by the water regimes, but there were differences between the genotypes, with CPAC19 having the greatest total chlorophyll index and CPAC11 the lowest principally due to differences in chlorophyll a. CPAC19 also stood our as having the lowest chlorophyll a to b ratio (Supplemental Table S1).

3.9. *Leaf Proline and Relative Water Content*

There was a significant interaction between the quinoa genotypes and the water regimes for leaf proline content (Table S2). However, in general, proline concentrations were relatively insensitive to water restriction, as concentrations only increased under severe stress (Table S2). We detected no effects of genotype on leaf relative water content (RWC). Regression analysis revealed a significant relationship between water regime and RWC and a maximal RWC of 81% at WR 480 mm and minimal RWC of 58% at WR 150 mm (Figure S4). The 19% reduction in water application of the WR 389 mm resulted in a minimal reduction in RWC (Figure S4).

4. Discussion

4.1. Productivity and Productivity Per Unit of Applied Water (PUAA)

Irrigation water use efficiency refers to the yield obtained per unit of applied water [24] and is a fundamental physiological parameter that indicates the ability of crops to conserve water in a region under water stress due to drought resistance and high potential productivity [25]. In our study, the low WR of 150 mm resulted in lower PUAA because under severe water restriction the quinoa genotypes cannot express their productive potential, whilst at high WRs (above 480 mm) plants also had a lower PUAA due to an inability to absorb all supplied water and potentially an intolerance to excess water (Table 2). Under the high WR, genotypes showed high productivity; however, the grain dry matter per unit of applied water was low, indicating that there was no consistent relationship between crop yield and PUAA for this WR (Tables 1 and 2). Thus, the 389 mm and 247 mm WRs showed the highest PUAA, but the highest productivity was observed under the WR 480 mm and 389 mm. Thus, WR 389 mm can be indicated for cultivating quinoa under an irrigated system in the Cerrado, as there is a trade-off in the relationship between productivity and water saving.

Quinoa plants can control the relationship between photosynthetic rate and transpiration, even with low leaf water potentials [26], and by limiting transpiration and inducing stomatal closure, they can increase PUAA and influence productivity under water stress [27]. In our study, with a 49% reduction in water applied throughout the crop cycle there was a 42% average yield loss over all genotypes (Table 1), indicating both drought tolerance and efficient water use, as this grain yield was obtained using half the water normally needed to meet the demands of the crop. Quinoa seeds can be obtained when little water is available in the vegetative stage, producing an average of 1.2 to 2.0 t ha^{-1} with half the required irrigation [28,29]. On the other hand, when a low irrigation strategy was used during all phenological stages this resulted in a 75% reduction in seed yield of the quinoa cultivar 'Belen 2000' [28,29].

Our work obtained higher yield values than others reported in the literature, and under all WRs the genotypes with high productivity were CPAC13, CPAC6, CPAC3, CPAC12, and CPAC17 and the genotypes with lower yield potential were CPAC19, CPAC11, and CPAC14 [30,31]. Under higher WR (480 mm) the genotypes did not differ, with the exception of CPAC11, which presented the lowest productivity and low PUAA; however, CPAC11 was also the only dwarf material used in this study (see below) (Figure 2, Tables 1 and 2). For the 150 mm WR the CPAC17 genotype was superior to the other genotypes, and whilst productivity was altered there were no changes in efficiency between WR 150 and WR 389 mm (Table 1), meaning that it is a suitable genotype for use in situations with limited water availability such as the off-season.

Under high and intermediate water regimes, the highest PUAA was observed for CPAC3, CPAC6, and CPAC12 between WR 480 and WR 247 mm (Table 2) and considering that they were amongst the genotypes with highest productivity, these genotypes are suggested for the winter season. Specifically, CPAC6 exhibited reduced productivity only under the 247 mm WR (Table 1) and presented the highest PUAA of 26.7 kg ha^{-1} mm^{-1}. This value is 32% higher than the PUAA of WR 480 mm (Table 2). CPAC13 presented higher productive potential and productivity than BRS Piabiru under a moderate water regime (389 mm), with 9.73 t ha^{-1} for CPAC13 and 8.14 t ha^{-1} for BRS Piabiru, respectively.

4.2. The Effects of Water Regime and Genotype on Grain Quality Indicators

In addition to productivity indicators of grain quality such as the concentrations of flavonoids and anthocyanins and 1000-grain weight should also be taken into account when selecting genotypes (Tables 3 and 4, Figure S1). Our results show that the accumulation of flavonoids and anthocyanins in quinoa plants was more influenced by genotype than by the WRs. In particular, CPAC9 accumulated these compounds under both higher and lower water regimes and accumulated nearly double the concentration of the other genotypes under all WRs. In addition, CPAC9 is among the genotypes with greater productivity and PUAA under high and intermediate WRs (Tables 1 and 2). The authors of [32], when studying the levels of flavonoids, phenolic acids, and betaines in the Andean grains of quinoa, kaniwa, and kiwicha, found flavonoid contents ranging from 36.2 to 144.3 mg/100 g, which are similar than those found here. In other crops such as peanuts, depending on the genotype and drought period, concentrations of phenolic compounds in seeds may be between 60 and 220 mg/100 g [33]. Further studies focusing on the biosynthesis of phenolic compounds and oxidation processes under water stress will provide more information on the genotypic variation of phenolic content in grains [34]. Grain weight is also affected by water restriction. With a water supply of 150 mm, there was a 14% decrease in TGW for the four genotypes analysed, similar to a previous study where TGW in irrigated plants was significantly higher (5.5 g) than in rainfed plants (4.2 g) [28,29]. Indeed, when water stress is applied during the grain filling period, it generally reduces the grain yield, the number of grains per plant, and the individual weight of the grains [35].

4.3. Low Water Availability Leads to Reduced Plant Height

The heights of the quinoa genotypes in this study ranged from 0.99 to 1.53 m, which are superior to those in literature [36]. Our results support previous studies demonstrating that a reduction in irrigation resulted in a significant decrease in plant height by 0.55 and 0.80 m for Bolivian quinoa [37]. Here, we detected a decrease in cultivated quinoa plant height under 70% irrigation deficit in relation to control plants, similar to that observed by [38]. The reduction of plant size and leaf area under stress conditions is directly related to a decline in dry mass when compared to plants maintained under adequate soil water potential conditions [39]. The low water potential of the soil limits water absorption capacity, and this quickly suppresses the rate of cell expansion and division of growing tissues [40]. Moreover, under severe water deficit stomata close and the consequent reduction of photosynthesis results in a decrease in dry mass production [39]. Plant height can also be used as a criterion to determine the susceptibility of quinoa genotypes to drought, as the longer the cycle, the larger the plant. Short-to-medium cycle cultivars are desirable, and would provide the possibility of more crops per year in irrigated systems [2] and enable off-season and winter planting. A shorter crop cycle also represents a method of escape from water stress under Cerrado conditions, besides contributing to the definition of the sowing season, such that the grain maturation occurs when the humidity is reduced, thus avoiding seed deterioration [2]. Amongst the studied genotypes, CPAC11 was significantly shorter. Whilst this is ideal for avoiding lodging, this genotype also showed the lowest productivity in most WRs under the planting density we employed. This may be related to reduced cell expansion, which results in a reduction in leaf area and, consequently, less photoassimilates for translocation to the grain [41]. In addition, water deficit affects carbohydrate utilization, altering the efficiency with which photoassimilates are used in the growth and development of new plant organs [41]. This genotype (CPAC11), however, may not have expressed its productive potential considering that it is the only dwarf material used in this study (Figure 2) and may need a higher planting density per square meter.

4.4. Physiological Parameters

Abiotic stresses generally result in reduced rates of photosynthesis and transpiration that can ultimately contribute to reduced growth and productivity [41], making assessment of physiological parameters in the field an important tool in screening for stress tolerance.

For this reason, we analysed a number of physiological parameters in a subset of the quinoa genotypes. Remote sensing of leaf temperature by thermal imaging can be a reliable way to detect changes in the physiological state of plants in response to different biotic and abiotic stresses [42]. Indeed, canopy temperature has been used successfully in breeding programs for drought-prone environments [43,44], as genotypes that maintain transpiration will tend to have lower canopy temperature than other genotypes under the same environmental conditions in the field [45]. Here, reduced water availability led to increased canopy temperature, reflecting the stomatal closure and reduced transpiration that we detected using gas exchange analysis. However, we were not able to detect differences between the genotypes using this parameter (Figure S2), despite the fact that differences in transpiration were detected between genotypes at each WR (Table 5). It may be the case that larger differences in transpiration between genotypes are required to result in alterations in leaf temperature that can be measured using this technique.

Stomatal closure is a common response to water restriction where water deficit affects guard cell turgidity and stomatal aperture, resulting in decreases in the rates of transpiration and assimilation of CO_2 and increased leaf temperature [46–48]. Gas exchange parameters such as net photosynthesis, transpiration and stomatal conductance are therefore sensitive indicators of water deficit in plants that are useful in the evaluation of genotypes adapted for growing in environments with limited water availability. Our data are consistent with previous studies, since the photosynthetic rate presented a behaviour similar to that observed for stomatal conductance and transpiration, reflecting the opening and closing of stomata under different water conditions [49], and quinoa is known to suffer both stomatal and mesophyll limitations under drought stress [50] These results also suggest that the reduction in photosynthetic rates in the most stressed water regimes are related to partial stomatal closure and a consequent reduction in CO_2 assimilation (Table 5). Water stress is considered to be severe when stomatal conductance values are below 0.1 $mol.m^{-2}\ s^{-1}$ [47], and therefore in this study all genotypes were under severe stress in the 150 mm WR (Table 5). Similar values for stomatal conductance and effects of water restriction in quinoa have been reported elsewhere; for example stomatal conductance decreased from 0.213 $mol.m^{-2}\ s^{-1}$ under irrigation to 0.091 $mol.m^{-2}\ s^{-1}$ under water deficit in one study [51] whilst an investigation of 10 genotypes in the field without irrigation that received around 160 mm of water reported values between 1.0 and 0.18 $mol.m^{-2}\ s^{-1}$ and a clear relationship between maintenance of stomatal conductance and higher photosynthetic rate [52].

Analysis of gas exchange also revealed differences in the responses of the genotypes to alterations in water availability and relationships with productivity. CPAC4 and CPAC11 appear most sensitive to water restriction as they presented the lowest net photosynthesis under the 150 mm and 247 mm regimes. Interestingly though, whilst CPAC19 showed greater photosynthesis under all water regimes, this was not reflected in greater productivity, indicating that other factors, such as plant morphology and the capacity to use photoassimilates for grain filling also has an important impact. This is also seen in the relationship between water regime, photosynthesis and productivity for individual genotypes; for example, the increased photosynthesis shown by CPAC4 between WR389 and WR480 did not translate into greater productivity, reinforcing the importance of parameters such as PUAA for selecting genotypes under irrigated conditions (Tables 1, 2 and 5).

Gas exchange measurements may not always be able to detect the deleterious effects of water restriction on chloroplast function parameters such as the effective quantum yield of photosystem II potentially useful tools [53]. Here we detected decreased F'v/F'm with low water availability (WR 150 mm, Figure S3); this decrease in plants stressed by drought, in comparison with well-hydrated plants, is mainly due to the lack of CO_2 inside the leaf and it is under this WR that we detected large decreases in Ci for all genotypes (Table 5). However, this parameter responded little to the median level of stress (389 mm), despite the fact that this WR provoked changes in several gas exchange parameters (Table 5), and furthermore we did not detect any differences between the genotypes (Figure S3). Indeed, measurements of chlorophyll a fluorescence tend to have low sensitivity to mild stresses, for example, 18 days of suspension of irrigation were required to reduce Fv/Fm in two greenhouse-grown quinoa genotypes [53]. In contrast to F′v/F′m, chlorophyll indices proved to be unaffected by WR but showed differences between genotypes (Table S1), and lack of an effect of drought and flooding on chlorophyll in quinoa has previously been reported [52]. Despite the lack of an effect of stress on chlorophyll, these indices may prove useful for selection of genotypes due to the fact that chlorophyll abundance is typically positively related to photosynthetic potential and productivity [54]. In this sense of the four genotypes analysed CPAC19 stood out due to greater total chlorophyll and a lower chlorophyll a:b ratio that may indicate greater light absorption capacity by photosystem II [55].

A lack of water in the soil increases the risk of the rate of transpiration exceeding the rate of water absorption and transport, resulting in a situation of water deficit. Partial stomatal closure can reduced transpiration, but under water stress, plants often also accumulate compatible solutes or osmoprotectors including proline, glycine, betaine, and sugars [56]. The accumulation of compatible solutes reduces cellular osmotic potential, thereby permitting water absorption and maintaining turgor pressure and physiological processes [57]. The accumulation of proline may therefore be an important characteristic for the selection of drought-tolerant plants [58] and indeed seed or leaf treatment of quinoa plants with free proline can increase growth under water stress [59] whilst a number of studies have connected compatible proline accumulation with drought and salt tolerance in this species [49]. However, whilst we detected increased proline concentrations in quinoa in response to water stress, this only occurred under the most severe water regimes, meaning that it could not be used to discriminate between the genotypes under water regimes that reflect Cerrado conditions (Table S2). Despite morphological alterations, stomatal control and osmotic adjustment water restriction may eventually affect leaf water status. Leaf relative water content (RWC) can therefore be used to indicate the balance between water supply and transpiration [60], and in the case of F′v/F′m, we did not detect differences between the genotypes for this parameter (Figure S4). However, RWC did serve to indicate the degree of stress to which the plants were subject, as the RWC values detected below 389 mm correspond to those associated with the beginning of wilting (Figure S4, [20] and are similar to those observed in greenhouse grown plants during suspension of irrigation [53].

5. Conclusions

Through experiments performed under different water regimes here we have shown that quinoa has excellent potential for planting as an off-season and winter crop in the Cerrado region. Several genotypes presented advantages in relation to the currently used BRS Piabiru genotype; the choice of genotype will depend on farming practices, nutritional content, and weather conditions. CPAC13 and CPAC6 are particularly suited to growth as a winter crop under irrigated conditions, and CPAC17 under off-season rain-fed conditions, whilst CPAC9 appears advantageous in terms of phenolic compounds in the grains. The accumulation of flavonoids and anthocyanins in quinoa genotypes was more influenced by quinoa genotype than by the WRs. Analysis of physiological parameters provided information regarding the mechanisms involved in stress tolerance in different quinoa genotypes, which is essential if we are to develop strategies to maintain or increase plant productivity in environments with water limitation. The results of this work show that the water regimes for quinoa can be reduced without a significant reduction in grain yield. This increase in dry matter accumulation efficiency per unit of water applied in quinoa means it is a crop that can be cultivated under Cerrado conditions, for both the off-season and winter season, under relatively low levels of irrigation whilst obtaining high yields. This fact, coupled with proper water management this can result in higher yield per area, which is desirable for areas under irrigated cultivation where irrigation is a costly practice.

Supplementary Materials: The following are available online at https://www.mdpi.com/article/10.3390/plants10081591/s1, Figure S1. Effects of water regimes in quinoa genotypes on 1000 grain weight- TWG (A). Weight of 1000 grains- TWG in four quinoa genotypes under four water regimes (B). Means followed by the same letter do not differ according to the Tukey test at 5% probability. Figure S2. Effects of water regimes (150, 247, 389 and 480 mm) on leaf temperature. Figure S3. Effects of water regimes (150, 247, 389 and 480 mm) on the effective quantum yield of photosystem II (F'v/F'm) in quinoa geno-types. Figure S4. Effects of water regimes (150, 247, 389 and 480 mm) on the relative water con-tent (RWC) in quinoa genotypes. Table S1: Chlorophyll *a*, *b*, *a*/*be a* + *b* measured in four quinoa genotypes independent of water regime (150, 247, 389 and 480 mm). Table S2: Proline content (μ mol g^{-1} FM) in leaves of four quinoa genotypes (CPAC4, CPAC 9, CPAC 11 e BRS Piabiru Pre-coce) under four water regimes (150, 247, 389, 480 mm).

Author Contributions: Conceptualization, P.C.d.S., W.Q.R.J. and M.L.G.R.; methodology P.C.d.S., W.Q.R.J., C.A.d.L., M.L.G.R., R.A.d.C.N.C. and S.M.C.C.; formal analysis P.C.d.S., A.d.N.S., T.C.R.W. and C.C.V.; software R.A.d.C.N.C., C.C.S.; writing—original draft preparation, P.C.d.S., M.L.G.R., W.Q.R.J., T.C.R.W. and C.C.V. All authors have read and agreed to the published version of the manuscript.

Funding: This research received no external funding.

Data Availability Statement: Data is contained within the article and Supplementary Material.

Acknowledgments: To Conselho Nacional de Desenvolvimento Científico e Tecnológico (CNPq), for the scientific productitvity fellowships granted to Maria L G Ramos. Furthermore, to Coordenação de Aperfeiçoamento de Pessoal de Nível Superior for the Master and Pos Doc fellowship granted to Patricia C. Silva and Christina C. Vinson, respectively.

Conflicts of Interest: The authors declare that they have no conflict of interest.

References

1. Scheelbeek, P.F.; Bird, F.A.; Tuomisto, H.L.; Green, R.; Harris, F.B.; Joy, E.J.M.; Chalabi, Z.; Allemn, E.; Haines, A.; Dangour, A.D. Effect of environmental changes on vegetable and legume yields and nutritional quality. *Proc. Natl. Acad. Sci. USA* **2018**, *115*, 6804–6809. [CrossRef]
2. Soares, G.F.; Ribeiro, W.Q., Jr.; Pereira, L.F.; Lima, C.A.; Soares, D.D.S.; Muller, O.; Rascher, U.; Ramos, M.L.G. Characterization of wheat genotypes for drought tolerance and water use efficiency. *Sci. Agric.* **2021**, *78*, e20190304. [CrossRef]
3. Spehar, C.R.; Santos, R.L.B. Quinoa BRS Piabiru: Alternativa para diversificar os sistemas de produção de grãos. *Pesq. Agropec. Bras.* **2001**, *37*, 889–893. [CrossRef]
4. Garcia, R.A.; Ceccon, G.; Sutier, G.A.D.; Santos, A.L.F.D. Soybean-corn succession according to seeding date. *Pesq. Agropec. Bras.* **2018**, *53*, 22–29. [CrossRef]
5. Präger, A.; Munz, S.; Nkebiwe, P.M.; Mast, B.; Graeff-Hönninger, S. Yield and quality characteristics of different quinoa (*Chenopodium quinoa* Willd.) cultivars grown under field conditions in Southwestern Germany. *Agronomy.* **2018**, *8*, 197. [CrossRef]
6. Saad-Allah, K.M.; Youssef, M.S. Phytochemical and genetic characterization of five quinoa (*Chenopodium quinoa* Willd.) genotypes introduced to Egypt. *Physiol. Mol. Biol. Plants* **2018**, *24*, 617–629. [CrossRef]
7. Wang, D.; Cao, D.; Yao, Y.; Wang, J.; Li, Z.; Liu, B. Understanding the checmial foundation and genetic mecanism of the black grain trait in quinoa by integrating metabolome and transcriptome analyses. *Biotechnol. Biotechnol. Equip.* **2020**, *34*, 1095–1103. [CrossRef]
8. Razzaghi, F.; Bahadori-Ghasroldashti, M.R.; Henriksen, S.; Sepaskhah, A.R.; Jacobsen, S.E. Physiological characteristics and irrigation water productivity of quinoa (*Chenopodium quinoa* Willd.) in response to deficit irrigation imposed at different growing stages—A field study from Southern Iran. *J. Agron. Crop Sci.* **2020**, *206*, 390–404. [CrossRef]
9. Landi, M.; Tattini, M.; Gould, K.S. Multiple functional roles of anthocyanins in plant-environment interactions. *Environ. Exp. Bot.* **2015**, *119*, 4–17. [CrossRef]
10. Jayme-Oliveira, A.; Ribeiro, W.Q.; Ramos, M.L.G.; Ziviani, A.C.; Jakelaitis, A. Amaranth, quinoa, and millet growth and development under different water regimes in the Brazilian Cerrado. *Pesq. Agropec. Bras.* **2017**, *52*, 561–571. [CrossRef]
11. Bazile, D.; Pulvento, C.; Verniau, A.; Al-Nusairi, M.S.; Ba, D.; Breidy, J.; Hassan, L.; Mohammed, M.I.; Mambetov, O.; Otambekova, M.; et al. Worldwide evaluations of quinoa: Preliminary results from post international year of quinoa FAO projects in nine countries. *Front. Plant Sci.* **2016**, *7*, 850. [CrossRef]
12. FAO. *The State of Food Insecurity in the World 2014*; Food and Agriculture Organization of the United Nations: Rome, Italy, 2021. Available online: http://www.fao.org/publications/sofi/en/ (accessed on 11 March 2021).
13. Alvares, C.A.; Stape, J.L.; Sentelhas, P.C.; Gonçalves, J.D.M. Köppen's climate classification map for Brazil. *Meteorol. Z.* **2013**, *22*, 711–728. [CrossRef]
14. Hanks, R.J.; Keller, J.; Rasmussen, V.P.; Wilson, G.D. Line source sprinkler for continuous variable irrigation crop production studies. *Soil Sci. Soc. Am. J.* **1976**, *40*, 426–429. [CrossRef]

15. Embrapa—Empresa Brasileira de Pesquisa Agropecuária. Monitoramento de Irrigação no Cerrado. 2016. Available online: http://hidro.cpac.embrapa.br (accessed on 25 May 2017).
16. QGIS Development Team; QGIS Geographic Information System. *Open Source Geospatial Foundation Project*; QGIS Development Team: Chur, Switzerland, 2021.
17. Secretaria de Defesa Agropecuária, Brasil Ministério da Agricultura, Pecuária e Abastecimento. *Regras para Análise de Sementes*; Brasil Ministério da Agricultura, Pecuária e Abastecimento: Brasília, Brazil, 2009.
18. Lees, D.H.; Francis, F.J. Standardization of pigment analyses in cranberries. *HortScience* **1972**, *7*, 83–84.
19. Genty, B.; Briantais, J.M.; Baker, N.R. The relationship between the quantum yield of photosynthetic electron transport and quenching of chlorophyll fluorescence. *Biochim. Biophys. Acta Gen. Subj.* **1989**, *990*, 87–92. [CrossRef]
20. Barrs, H.D.; Weatherley, P.E. A re-examination of the relative turgidity technique for estimating water deficits in leaves. *Aust. J. Biol. Sci.* **1962**, *15*, 413–428. [CrossRef]
21. Bates, L.S.; Waldren, R.P.; Teare, I.D. Rapid determination of free proline for water-stress studies. *Plant Soil* **1973**, *39*, 205–207. [CrossRef]
22. SAS-Statistical Analyses System. *Statistical Analysis System User's Guide*; Institute, Statistical Analyses System: Cary, NC, USA, 2008.
23. SYSTAT Software. *Scientific Graphing Software: Sigma Plot, Versão 10.0*; HearneScientific Software: San Rafael, CA, USA, 2006.
24. Howell, T.A. Irrigation Efficiency. In *Encyclopedia of Soil Science*, 2nd ed.; CRC: Boca Raton, FL, USA, 2003.
25. Zhang, M.; Duan, L.; Tian, X.; He, Z.; Li, J.; Wang, B.; Li, Z. Uniconazole-induced tolerance of soybean to water deficit stress in relation to changes in photosynthesis, hormones and antioxidant system. *J. Plant Physiol.* **2007**, *164*, 709–717. [CrossRef] [PubMed]
26. Vacher, J.J. Responses of two main Andean crops, quinoa (*Chenopodium quinoa* Willd) and papa amarga (*Solanum juzepczukii* Buk.) to drought on the Bolivian Altiplano: Significance of local adaptation. *Agric. Ecosyst. Environ.* **1998**, *68*, 99–108. [CrossRef]
27. Yousfi, S.; Serret, M.D.; Araus, J.L. Shoot δ(15)N gives a better indication than ion concentration or Δ(13)C of genotypic differences in the response of durum wheat to salinity. *Funct. Plant Biol.* **2009**, *36*, 144–155. [CrossRef] [PubMed]
28. Geerts, S.; Raes, D.; Garcia, M.; Condori, O.; Mamani, J.; Miranda, R.; Cusicanqui, J.; Taboada, C.; Yucra, E.; Vacher, J. Could deficit irrigation be a sustainable practice for quinoa (*Chenopodium quinoa* Willd.) in the Southern Bolivian Altiplano? *Agric. Water Manag.* **2008**, *95*, 909–917. [CrossRef]
29. Geerts, S.; Raes, D.; Garcia, M.; Vacher, J.; Mamani, R.; Mendoza, J.; Huanca, R.; Morales, B.; Miranda, R.; Cusicanqui, J.; et al. Introducing deficit irrigation to stabilize yields of quinoa (*Chenopodium quinoa* Willd.). *Eur. J. Agron.* **2008**, *28*, 427–436. [CrossRef]
30. Martínez, E.A.; Veas, E.; Jorquera, C.; San Martín, R.; Jara, P. Re-introduction of quinoa into Arid Chile: Cultivation of two lowland races under extremely low irrigation. *J. Agron. Crop Sci.* **2009**, *195*, 1–10. [CrossRef]
31. Delgado, A.I.; Palacios, J.H.; Betancourt, C. Evaluation of 16 genotypes of sweet quinoa (*Chenopodium quinoa* Willd.) in the municipality of Iles, Nariño (Colombia). *Agron. Colomb.* **2009**, *27*, 159–167.
32. Repo-Carrasco-Valencia, R.; Hellström, J.K.; Pihlava, J.M.; Mattila, P.H. Flavonoids and other phenolic compounds in Andean indigenous grains: Quinoa (*Chenopodium quinoa*), kañiwa (*Chenopodium pallidicaule*) and kiwicha (*Amaranthus caudatus*). *Food Chem.* **2010**, *120*, 128–133. [CrossRef]
33. Aninbon, C.; Jogloy, S.; Vorasoot, N.; Patanothai, A.; Nuchadomrong, S.; Senawong, T. Effect of end of season water deficit on phenolic compounds in peanut genotypes with different levels of resistance to drought. *Food Chem.* **2016**, *196*, 123–129. [CrossRef]
34. Martini, D.; Taddei, F.; Nicoletti, I.; Ciccoritti, R.; Corradini, D.; D'Egidio, M.G. Effects of genotype and environment on phenolic acids content and total antioxidant capacity in durum wheat. *Cereal Chem.* **2014**, *91*, 310–317. [CrossRef]
35. Gámez, A.L.; Soba, D.; Zamarreño, Á.M.; García-Mina, J.M.; Aranjuelo, I.; Morales, F. Effect of water stress during grain filling on yield, quality and physiological traits of Illpa and Rainbow quinoa (*Chenopodium quinoa* Willd.) cultivars. *Plants* **2019**, *8*, 173. [CrossRef]
36. Maliro, M.F.; Guwela, V.F.; Nyaika, J.; Murphy, K.M. Preliminary studies of the performance of quinoa (*Chenopodium quinoa* Willd.) genotypes under irrigated and rainfed conditions of central Malawi. *Front. Plant Sci.* **2017**, *8*, 227. [CrossRef]
37. Talebnejad, R.; Sepaskhah, A.R. Effect of deficit irrigation and different saline groundwater depths on yield and water productivity of quinoa. *Agric. Water Manag.* **2015**, *159*, 225–238. [CrossRef]
38. Yang, A.; Akhtar, S.S.; Amjad, M.; Iqbal, S.; Jacobsen, S.E. Growth and physiological responses of quinoa to drought and temperature stress. *J. Agron. Crop Sci.* **2016**, *202*, 445–453. [CrossRef]
39. Endres, L. Photosynthesis and water relations in Brazilian sugarcane. *Open Agric. J.* **2010**, *4*, 31–37. [CrossRef]
40. Shao, H.B.; Chu, L.Y.; Shao, M.A.; Jaleel, C.A.; Hong-mei, M. Higher plant antioxidants and redox signaling under environmental stresses. *Comptes Rendus Biol.* **2008**, *331*, 433–441. [CrossRef] [PubMed]
41. Jordan, W.R. Whole Plant Response to Water Deficits: An Overview. *Limit. Effic. Water Use Crop. Prod.* **1983**, 289–317. [CrossRef]
42. Li, C.; Liu, S.; Berninger, F. Picea seedlings show apparent acclimation to drought with increasing altitude in the eastern Himalaya. *Trees Struct. Funct.* **2004**, *18*, 277–283. [CrossRef]
43. Balota, M.; Payne, W.A.; Evett, S.R.; Peters, T.R. Morphological and physiological traits associated with canopy temperature depression in three closely related wheat lines. *Crop Sci.* **2008**, *48*, 1897–1910. [CrossRef]
44. Jones, H.G.; Serraj, R.; Loveys, B.R.; Xiong, L.; Wheaton, A.; Price, A.H. Thermal infrared imaging of crop canopies for the remote diagnosis and quantification of plant responses to water stress in the field. *Funct. Plant Biol.* **2009**, *36*, 978–989. [CrossRef]

45. Blum, A. Effective use of water (EUW) and not water-use efficiency (WUE) is the target of crop yield improvement under drought stress. *Field Crops Res.* **2009**, *112*, 119–123. [CrossRef]
46. Waseem, M.; Ali, A.; Tahir, M.; Nadeem, M.A.; Ayub, M.; Tanveer, A.; Ahmad, R.; Hussain, M. Mechanism of drought tolerance in plant and its management through different methods. *Cont. J. Agric. Sci.* **2011**, *5*, 10–25.
47. Flexas, J.; Diaz-Espejo, A.; Gago, J.; Gallé, A.; Galmés, J.; Gulías, J.; Medrano, H. Photosynthetic limitations in Mediterranean plants: A review. *Environ. Exp. Bot.* **2014**, *103*, 12–23. [CrossRef]
48. Tatagiba, S.D.; Pezzopane, J.E.M.; Reis, E.F. Fotossíntese em Eucalyptus sob diferentes condições edafoclimáticas. *Rev. Eng. Agric. RevEng* **2015**, *23*, 336–345. [CrossRef]
49. Hinojosa, L.; González, J.A.; Barrios-Masias, F.H.; Fuentes, F.; Murphy, K.M. Quinoa Abiotic Stress Responses: A Review. *Plants* **2018**, *7*, 106. [CrossRef]
50. Killi, D.; Haworth, M. Diffusive and metabolic constraints to photosynthesis in quinoa during drought and salt stress. *Plants* **2017**, *6*, 49. [CrossRef] [PubMed]
51. Hussin, S.; Khalifa, W.; Geissler, N.; Koyro, H.W. Influence of the root endophyte Piriformospora indica on the plant water relations, gas exchange and growth of *Chenopodium quinoa* at limited water availability. *J. Agron. Crop Sci.* **2017**, *203*, 373–384. [CrossRef]
52. González, J.A.; Gallardo, M.; Hilal, M.B.; Rosa, M.D.; Prado, F.E. Physiological responses of quinoa (*Chenopodium quinoa*) to drought and waterlogging stresses: Dry matter partitioning. *Bot. Stud.* **2009**, *50*, 35–42.
53. Morales, A.; Zurita-Silva, A.; Maldonado, J.; Silva, H. Transcriptional responses of Chilean quinoa (*Chenopodium quinoa* Willd.) under water deficit conditions uncovers ABA-independent expression patterns. *Front. Plant Sci.* **2017**, *8*, 216. [CrossRef]
54. O'Neill, P.M.; Shanahan, J.F.; Schepers, J.S. Use of chlorophyll fluorescence assessments to differentiate corn hybrid response to variable water conditions. *Crop Sci.* **2006**, *46*, 681–687. [CrossRef]
55. Belous, O.; Klemeshova, K.; Malyarovskaya, V. Photosynthetic Pigments of Subtropical Plants. In *Photosynthesis—From Its Evolution to Future Improvements in Photosynthetic Efficiency Using Nanomaterials*; IntechOpen Limited: London, UK, 2018; pp. 31–52.
56. Nounjan, N.; Chansongkrow, P.; Charoensawan, V.; Siangliw, J.L.; Toojinda, T.; Chadchawan, N.; Chansongkrow, P.; Charoensawan, V.; Siangliw, J.L.; Toojinda, T.; et al. High performance of photosynthesis and osmotic adjustment are associated with salt tolerance ability in rice carrying drought tolerance QTL: Physiological and co-expression network analysis. *Front. Plant Sci.* **2018**, *9*, 1135. [CrossRef] [PubMed]
57. Pintó-Marijuan, M.; Munné-Bosch, S. Ecophysiology of invasive plants: Osmotic adjustment and antioxidants. *Trends Plant Sci.* **2013**, *18*, 660–666. [CrossRef]
58. Ben Rejeb, K.; Abdelly, C.; Savouré, A. How reactive oxygen species and proline face stress together. *Plant Physiol. Biochem. PPB* **2014**, *80*, 278–284. [CrossRef] [PubMed]
59. Yaqoob, H.; Akram, N.A.; Iftikhar, S.; Ashraf, M.; Khalid, N.; Sadiq, M.; Alyemeni, M.N.; Wijaya, L.; Ahmad, P. Seed pretreatment and foliar application of proline regulate morphological, physio-biochemical processes and activity of antioxidant enzymes in plants of two cultivars of quinoa (*Chenopodium quinoa* Willd.). *Plants* **2019**, *8*, 588. [CrossRef] [PubMed]
60. Lugojan, C.; Ciulca, S. Evaluation of relative water content in winter wheat. *For. Biotechnol.* **2011**, *15*, 173–177.

plants

MDPI

Article

Phenotypic Characterization of Quinoa (*Chenopodium quinoa* Willd.) for the Selection of Promising Materials for Breeding Programs

Elsa Helena Manjarres-Hernández [1,*], Diana Marcela Arias-Moreno [2], Ana Cruz Morillo-Coronado [1], Zaida Zarely Ojeda-Pérez [2] and Agobardo Cárdenas-Chaparro [3]

[1] Grupo CIDE Competitividad Innovación y Desarrollo empresarial, Universidad Pedagógica y Tecnológica de Colombia, Tunja 150003, Colombia; ana.morillo@uptc.edu.co
[2] Grupo de Investigación BIOPLASMA, Universidad Pedagógica y Tecnológica de Colombia, Tunja 150003, Colombia; diana.arias04@uptc.edu.co (D.M.A.-M.); zaida.ojeda@uptc.edu.co (Z.Z.O.-P.)
[3] Grupo de Química—Física Molecular y Modelamiento Computacional QUIMOL, Universidad Pedagógica y Tecnológica de Colombia, Tunja 150003, Colombia; acardenasc@unal.edu.co
* Correspondence: elsa.manjarres@uptc.edu.co; Tel.: +57-315-309-7182 (ext. 150003)

Abstract: Quinoa is an ancestral crop in the Andean region, characterized by its adaptability to different agroclimatic conditions, great nutritional value, and broad genetic variability. A preliminary approach for understanding the genetics of quinoa materials entails a morphologic characterization, which can provide the basis for the selection of materials that satisfy the needs of farmers and consumers. Therefore, this study aimed to evaluate the phenotypic characteristics of thirty genetic *C. quinoa* accessions for the selection of outstanding accessions in terms of yield and grain quality. A randomized complete block design was used, with nine replications for each accession under greenhouse conditions. Nine quantitative and twelve qualitative descriptors were evaluated with descriptive analysis, Spearman correlation variance, and multivariate and cluster analysis. The results showed that the accessions with heights greater than the average (>176.72 cm) and long panicles (>57.94 cm) presented lower yields and smaller seed sizes, thus decreasing the grain quality. The multivariate and cluster analyses established groups of accessions with good yields (>62.02 g of seeds per plant) and stable morphological characteristics. The proposed selection index, based on yield components and morphological descriptors, indicated four accessions as potential parents for quinoa breeding programs in Colombia.

Keywords: grain quality; ancestral crop; *Chenopodium quinoa*; morphologic descriptors; selection index; yield; pseudocereal

Citation: Manjarres-Hernández, E.H.; Arias-Moreno, D.M.; Morillo-Coronado, A.C.; Ojeda-Pérez, Z.Z.; Cárdenas-Chaparro, A. Phenotypic Characterization of Quinoa (*Chenopodium quinoa* Willd.) for the Selection of Promising Materials for Breeding Programs. *Plants* **2021**, *10*, 1339. https://doi.org/10.3390/plants10071339

Academic Editors: Cataldo Pulvento and Didier Bazile

Received: 25 May 2021
Accepted: 24 June 2021
Published: 30 June 2021

1. Introduction

Quinoa (*Chenopodium quinoa* Willd.) is a pseudocereal that is considered one of the most complete foods for humans. It is grown in South America, from Colombia to southern Chile. However, the greatest diversity is found in Peru and Bolivia [1,2]. This species presents high phenotypic variability that can be easily recognized by the pigmentation of the plant, inflorescences and seeds, earliness diversity, shape and size of grain, compaction of the panicles, and resistance to adverse factors, such as drought, frost, excess humidity, salinity, and diseases among others. This variability explains the ability of this species to adapt to different agroclimatic conditions [3–5].

According to the Food and Agriculture Organization of the United Nations (FAO) [6], in recent years (2000–2019), there has been a significant global increase in the area cultivated with quinoa crops, mainly in Peru and Bolivia, with increases between 36% and 72%, respectively [4,7]. In Colombia, quinoa production has increased by more than 1100 tons since 2017 [8]. The departments where quinoa is grown include Cundinamarca, Cauca,

Nariño, and Boyacá. The latter is in the central, eastern part of Colombia and has a collection of quinoa germplasm [9]. However, information on the morphological characteristics and grain yield of these accessions is lacking.

Therefore, morphoagronomic characterizations using quantitative and qualitative descriptors of domestic and introduced quinoa materials in Colombia are essential to efficiently using genetic variability to increase the productivity of crops under different environmental conditions for the commercialization of grains or derived products in response to global demand for this pseudocereal.

In Colombia, studies on the morphoagronomic characterization of quinoa are scarce. Torres and collaborators carried out a morphological evaluation of 19 accessions on the Bogotá Savanna, finding high variability in terms of grain yield, biomass, and earliness [10]. Veloza and collaborators found differences in the materials Piartal, Nariño and Bolivia in terms of yield, protein content, stalk coloration, panicle shape and physiological maturity [11]. In the Department of Boyacá, Infante and collaborators [12] carried out a morphological characterization of six varieties of quinoa, finding that adult plants have constant morphological characteristics, such as the presence of striae, pigmented axillae, and number of teeth on leaves. Morillo et al., 2020 [9] reported the existence of high morphological variability in 19 quinoa materials from the Department of Boyacá, where the more variable characteristics were the color of axillae and striae, plant height, number of panicles, seed yield per plant, and weight of 1000 grains.

The morphoagronomic characterization of quinoa accessions will facilitate the selection of materials that will improve production in regions with specific environmental conditions, initiate certified seed registration processes, discriminate accessions, determine potential uses, form core collections, identify duplicates in collections, and promote use in conservation and genetic improvement programs [9]. This study aimed to evaluate phenotypic characteristics with qualitative and quantitative descriptors of thirty accessions of *C. quinoa* to select outstanding accessions from the seed collection in the Department of Boyacá, Colombia.

2. Results

The climatic conditions of the study region were typical of tropical zones, characterized by few fluctuations in the photoperiod and average temperatures. The minimum temperature during this study ranged between 7.43 and 10.00 °C, and the maximum was between 17.80 and 20.00 °C, with an average temperature between 13.43 and 14.75 °C, and the average relative humidity was 78%. The daily illumination that the accessions received during the experiment was approximately 12 h.

2.1. Morphologic Characterization Using Quantitative Descriptors

The nine quantitative descriptors evaluated in the quinoa accessions had broad variation. The plant height had a mean of 176.7 cm, where the Quinua beteitiva accession exhibited the highest height (PH = 248.2 cm), and the largest panicle diameter (PD = 35.4 cm) (Table 1). However, the variables yield, seed weight, and seed diameter were zero because grains were not formed. In contrast, the accession with the lowest height was Quinua Peruana (PH = 111.9 cm), with PL = 39.0 cm, PD = 20.0 cm, NP = 6.8, and NT = 9.8, the lowest values for these variables, with respect to the other evaluated accessions. However, this accession presented the largest seed diameter (GD = 2.63 mm) and a higher-than-average yield value (Y = 62.02 g). On the other hand, Amarilla de maranganí had the highest weight of 1000 seeds with a value of 0.40 g and a seed diameter higher than average (GD = 2.57 mm). Quinua Blanca de Jericó Tuta2 presented the longest panicle length (PL = 72.4 cm) and plant height, 220.4 cm, but the yield was below average (Y = 18.17 g) (Table 1).

Table 1. Averages, standard deviation, and Tukey multiple comparison (MC) for the quantitative variables of the thirty quinoa accessions.

Accessions	PH (cm) (S.D.)	SD (cm) (S.D.)	PL (cm) (S.D.)	PD (cm) (S.D.)	NP (#) (S.D.)	NT (#) (S.D.)	Y (g) (S.D.)	WS (g) (S.D.)	GD (mm) (S.D.)
Quinoa real	128.6 (14.35) g, h, j	3.2 (0.62) b, d	43.6 (10.73) a, c, e, g	26.6 (7.06) a, b, c, d, e	12.3 (2.35) a, b, c, d, e	13.8 (6.82) a, b, d	68.55 (37.62) b e	0.32 (0.02) f	2.16 (0.09) a e g
Quinoa aurora	182.9 (23.60) a, b, d, f, i	3.5 (0.49) a, b, c, d	57.7 (12.14) a, b, c, d, e, f, g	30.2 (2.91) a, b, c, d, e	12.7 (2.55) b, c, d, e	17.3 (4.12) a, b, c, d	27.08 (17.55) a d f g i	0.23 (0.03) a b d g h j	2.05 (0.08) b c f g
Quinoa ceniza	148.6 (28.80) a, c, g, h, j, k	3.2 (0.44) b, d	56.8 (20.42) a, b, c, d, e, f, g	24.8 (4.97) a, c, e	11.9 (3.79) a, b, c, e	21.6 (5.46) a, c	19.40 (9.48) d f g i	0.21 (0.03) a c g i j	1.87 (0.14) b c f h
Quinoa beteitiva	**248.2 (11.29) e**	4.0 (0.50) a, c, d	71.6 (13.64) b, d	**35.4 (4.88) d**	17.7 (5.10) b, d	17.1 (6.57) a, b, c, d	**0.00**	**0.00**	**0.00**
Quinoa sotaquirá	202.9 (25.17) b, f, i	3.9 (0.62) a, b, c, d	64.9 (13.33) b, d, f	31.7 (8.49) b, c, d, e	11.2 (4.63) a, c, e	20.0 (5.57) a, c, d	41.17 (20.70) a b c d f g h i	0.25 (0.03) a b d g h	2.23 (0.09) a e g
Quinoa negra	138.7 (18.99) c, g, h, j, k	**3.1 (0.35) b**	40.2 (8.80) e, g	27.4 (7.75) a, b, c, d, e	13.8 (3.27) b, d, e	21.8 (8.15) a, c	44.34 (20.63) a	0.16 (0.01) i	1.75 (0.05) h
Tunkahuan ICA	186.3 (13.25) a, b, d, f, i	3.6 (0.32) a, b, c, d	52.3 (6.04) a, b, c, d, e, f, g	27.1 (5.69) a, b, c, d, e	12.0 (2.12) a, b, c, e	20.3 (4.74) a, c, d	36.73 (18.45) a b c d f g h i	0.27 (0.06) b d f h	2.21 (0.07) a e g
Blanca de Jericó Tuta	192.2 (22.92) b, d, f, i	3.7 (0.48) a, b, c, d	64.2 (14.33) b, d, f	25.1 (7.67) a, b, c, e	13.3 (2.65) b, d, e	22.2 (8.09) a, c	34.00 (11.09) a c d f g i	0.22 (0.03) a c d g h j	2.15 (0.11) a e g
Amarilla de maranganí	120.3 (11.12) g, j	3.6 (0.49) a, b, c, d	41.1 (7.18) a, e, g	22.4 (6.56) a, c	7.2 (2.17) a, c	12.0 (9.27) b, d	46.46 (14.49) a b c f h i	**0.40 (0.02) e**	2.57 (0.10) d
Quinoa colorada	175.4 (18.60) a, c, d, f, i	3.7 (0.42) a, b, c, d	59.6 (6.65) a, b, c, d, f, g	33.8 (4.99) b, d, e	13.1 (2.57) b, d, e	19.3 (4.58) a, c, d	26.89 (19.12) a d f g i	0.23 (0.03) a b d g h j	2.17 (0.14) a e g
Blanca dulce de Soracá	169.2 (25.57) a, c, d, f, h, i, k	3.5 (0.62) a, b, c, d	51.4 (10.79) a, b, c, e, f, g	30.0 (7.81) a, b, c, d, e	12.9 (2.80) b, c, d, e	20.4 (5.08) a, c, d	47.53 (24.32) a b c f h i	0.24 (0.03) a b d g h	2.14 (0.16) a g
Piartal Chocotá	183.0 (20.11) a, b, d, f, i	3.5 (0.36) a, b, c, d	64.7 (7.58) b, d, f	29.1 (5.53) a, b, c, d, e	11.1 (3.86) a, c, e	16.9 (7.75) a, b, c, d	34.42 (16.09) a c d f g h i	0.25 (0.02) a b d g h	2.16 (0.07) a e g
Quinoa dulce de Tuta	183.8 (20.25) a, b, d, f, i	3.8 (0.62) a, b, c, d	65.2 (12.85) b, d, f	28.8 (5.09) a, b, c, d, e	13.3 (3.91) b, d, e	22.4 (7.54) a, c	55.49 (24.48) a b c e h	0.26 (0.03) b d g h	2.28 (0.20) a e
Quinoa semiamarga	189.2 (14.53) a, b, d, f, i	4.0 (0.49) a, c, d	61.1 (9.36) a, b, c, d, f	33.0 (5.41) b, d, e	14.0 (3.84) b, d, e	**24.2 (3.53) c**	66.63 (27.87) b e h	0.27 (0.02) b f h	2.24 (0.05) a e g
Quinoa peruana	**111.9 (16.17) j**	3.3 (0.38) a, b, d	**39.0 (6.86) e**	**20.0 (4.66) a**	**6.8 (2.39) a**	9.8 (7.90) b	62.02 (16.69) b c e h	0.39 (0.01) e	**2.63 (0.09) d**
Quinoa siachoque	207.9 (25.78) b, e, i	3.9 (0.54) a, b, c, d	70.8 (7.36) b, d	34.7 (7.94) b, d, e	14.6 (4.25) b, d, e	20.7 (6.40) a, c, d	29.08 (18.02) a d f g i	0.23 (0.02) a b d g h j	2.18 (0.09) a e g
Blanca de Jericó Tuta2	220.4 (32.14) b, e	3.8 (0.42) a, b, c, d	**72.4 (14.98) d**	33.6 (2.74) b, d, e	12.1 (3.82) a, b, c, d, e	20.7 (3.46) a, c, d	18.17 (9.63) d f g i	0.20 (0.03) a c i j	2.08 (0.08) a f g
Piartal Tibasosa	169.8 (32.82) a, c, d, f, i, k	3.5 (0.45) a, b, c, d	53.3 (10.58) a, b, c, d, e, f, g	29.0 (6.80) a, b, c, d, e	14.4 (3.28) b, d, e	18.2 (3.67) a, b, c, d	48.75 (17.54) a b c f h	0.27 (0.02) b d f h	2.21 (0.06) a e g
Blanca de Jericó Tunja	196.6 (36.14) b, d, f, i	3.6 (0.36) a, b, c, d	71.1 (15.38) b, d	26.3 (5.52) a, b, c, d, e	11.1 (2.93) a, c, e	15.3 (5.00) a, b, c, d	23.07 (14.00) d f g i	0.22 (0.03) a c d g j	2.12 (0.08) a g
Blanca de Jericó Toca	179.9 (24.10) a, b, d, f, i	3.6 (0.30) a, b, c, d	65.3 (7.16) b, d, f	30.4 (7.95) b, c, d, e	12.6 (2.46) a, b, c, d, e	16.9 (7.94) a, b, c, d	33.48 (12.47) a c d f g i	0.25 (0.02) a b d g h	2.13 (0.11) a g
Cremosa malvinas	169.8 (18.64) a, c, d, f, i, k	3.9 (0.50) a, b, c, d	60.3 (12.31) a, b, c, d, f, g	31.0 (4.44) b, c, d, e	15.2 (5.19) b, d, e	18.2 (5.87) a, b, c, d	44.64 (23.46) a b c f g h i	0.27 (0.02) b d f h	2.17 (0.06) a e g
Tunkahuan Tibasosa	187.0 (18.30) a, b, d, f, i	4.1 (0.37) a, c	57.8 (10.03) a, b, c, d, e, f, g	32.4 (4.88) b, c, d, e	14.0 (1.66) b, d, e	19.1 (8.43) a, c, d	41.59 (23.84) a b c d f g h i	0.28 (0.02) b f	2.17 (0.06) a e g
Tunkahuan siachoque	185.1 (46.69) a, b, d, f, i	3.6 (0.89) a, b, c, d	60.9 (20.03) a, b, c, d, f	28.3 (8.06) a, b, c, d, e	11.3 (5.70) a, c, e	14.0 (8.54) a, b, d	15.44 (14.15) d g i	0.22 (0.03) a c d g j	2.06 (0.11) b f g
Mezcla Siachoque 1	194.9 (21.17) b, d, f, i	3.8 (0.46) a, b, c, d	62.9 (8.88) b, c, d, f	31.4 (4.59) b, c, d, e	11.3 (3.28) a, c, e	22.9 (9.01) a, c	29.60 (22.49) a d f g i	0.22 (0.03) a c d g h j	2.06 (0.07) b c f g
Mezcla Siachoque 2	200.1 (32.27) b, d, f, i	**4.3 (0.43) c**	64.9 (13.12) b, d, f	27.6 (7.45) a, b, c, d, e	**17.9 (6.97) d**	19.1 (6.33) a, c, d	12.63 (13.85) d g	0.16 (0.05) c i	1.87 (0.20) b c h
Mezcla Siachoque 3	198.4 (19.14) b, d, f, i	4.1 (0.40) a, c	67.7 (9.75) b, d, f	33.2 (7.14) b, d, e	10.9 (4.20) a, c, e	17.6 (5.73) a, b, c, d	28.85 (17.68) a d f g i	0.22 (0.04) a c d g j	1.85 (0.12) c h
Quinoa primavera	133.3 (17.87) g, h, j, k	3.5 (0.67) a, b, c, d	42.3 (5.15) a, e, g	24.4 (6.39) a, c, e	12.3 (3.00) a, b, c, d, e	16.2 (6.36) a, b, c, d	**87.53 (40.89) e**	0.32 (0.04) f	2.24 (0.15) a e g
Quinoa Galindo	159.1 (21.00) a, c, d, g, h, k	3.7 (0.38) a, b, c, d	47.8 (6.55) a, c, e, f, g	35.2 (5.29) b, d	15.0 (2.00) b, d, e	11.8 (4.74) b, d	35.63 (16.16) a c d f g h i	0.28 (0.02) b f	2.36 (0.09) e
Col-quinua	165.8 (34.01) a, c, d, f, h, k	3.7 (0.52) a, b, c, d	47.6 (17.83) a, c, e, f, g	28.9 (8.99) a, b, c, d, e	14.7 (1.94) b, d, e	15.8 (5.33) a, b, c, d	21.46 (12.66) d f g i	0.17 (0.06) c i j	1.87 (0.23) b c h
Susunaga	172.2 (24.10) a, c, d, f, i, k	4.3 (0.66) c	59.8 (15.44) a, b, c, d, f, g	33.9 (7.77) b, d, e	11.4 (3.43) a, c, e	15.1 (4.26) a, b, c, d	12.28 (9.21) d	0.21 (0.04) a c g i j	2.08 (0.15) a g
Mean	176.7	3.7	57.9	29.5	12.7	18.0	38.20	0.25	2.14

PH Plant height, SD Stem diameter, PL Panicle length, PD Panicle diameter, NP N ° plant panicles, NT N ° teeth per leaf, Y Yield per plant, WS Weight of 1000 seeds, GD Grain diameter. Averages in each column with the same letters do not differ statistically (Tukey $p < 0.05$). The maximum and minimum values for each variable are in bold.

The Tukey test ($p < 0.05$) for all the quantitative variables, shows differences between the accessions evaluated. The best yields per quinoa plant were achieved by the accessions Quinoa primavera with 87.53 g, Quinoa real with 68.55 g, Quinoa semiamarga with 66.63 g, and Quinoa peruana with 62.02 g. The variables weight of the seeds (WS) and diameter of the seeds (GD) presented lower standard deviations (Table 1).

The analysis of variance detected statistically significant differences ($p < 0.05$) between the evaluated accessions for the characteristics plant height, stem diameter, length and diameter of the panicle, number of panicles, number of teeth on the leaf, yield, and seed weight and diameter.

The Spearman correlation analysis ($p \leq 0.05$) between the quantitative variables showed that there were high and significant correlations between the weight (WS) and diameter (GD) of the seeds (r = 0.89), the length (PL) and height (PH) of the plants (r = 0.88), seed weight (WS) and yield (Y) (r = 0.82), and seed diameter (GD) and yield (Y) (r = 0.71). There were also negative correlations between yield (Y) and panicle length (PL) (r = −0.51), and yield (Y) and plant height (PH) (r = −0.50) (Figure 1).

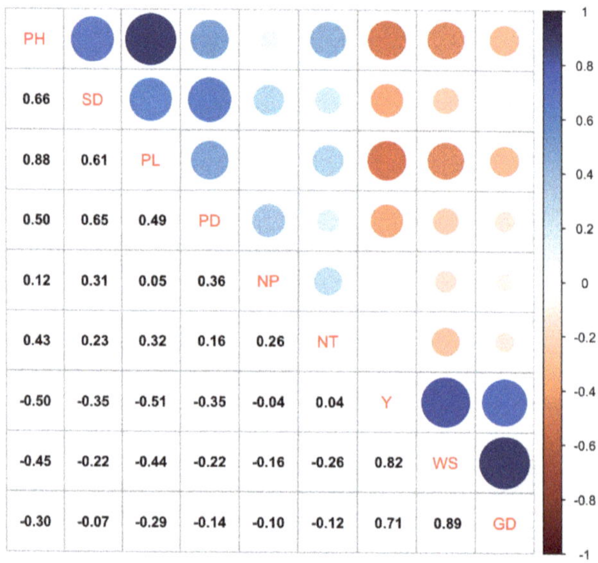

Figure 1. Spearman correlation analysis among the quantitative variables in the 30 quinoa accessions.

The principal components analysis showed that 68.6% of the total variance was explained by the first two components (CP1 = 54.6% and CP2 = 14.0%) (Figure 2a). The variables that made the greatest contribution to the variation of CP1 included plant height, length, diameter, and number of panicles. The weight of 1000 seeds, seed diameter and stem diameter stem contributed more to the variation in CP2. On the other hand, the variables associated with weight and seed diameter correlated more with yield than the variables plant height, stem diameter, number of teeth on the leaves, and the variables associated with the panicle (Figure 2a).

The cluster analysis of the quantitative variables grouped the accessions into five clusters (Figure 2b). However, the groups were not established according to the collection area or place of origin. The first group had the accessions that presented the highest values for seed weight (0.32–0.40 g), seed diameter (2.16–2.63 mm) and yield (46.47–87.53 g). The second group had the Quinua beteitiva accession, which had the greatest height (PH = 248.2 cm) and panicle diameter (PD = 35.4 cm) but did not develop seeds. The accessions in the third group presented yields between 12.28 and 48.75 g, seed diameters from 1.87 to 2.36 mm, weights of 1000 seeds from 0.17 to 0.28 g, and stem diameters from

3.2 to 4.3 cm. The accessions in the fourth group included Quinua dulce de Tuta and Quinua semiamarga, with heights between 183.8 and 189.2 cm, panicle diameters between 2.2 and 2.3 mm, and yields between 55.49 and 66.63 g. Finally, group five had the accessions with the lowest values for yield (12.28 to 41.17 g), seed diameter (1.85 to 2.23 mm), and seed weight (0.16 to 0.25 g). These analyses were consistent with the principal component analysis.

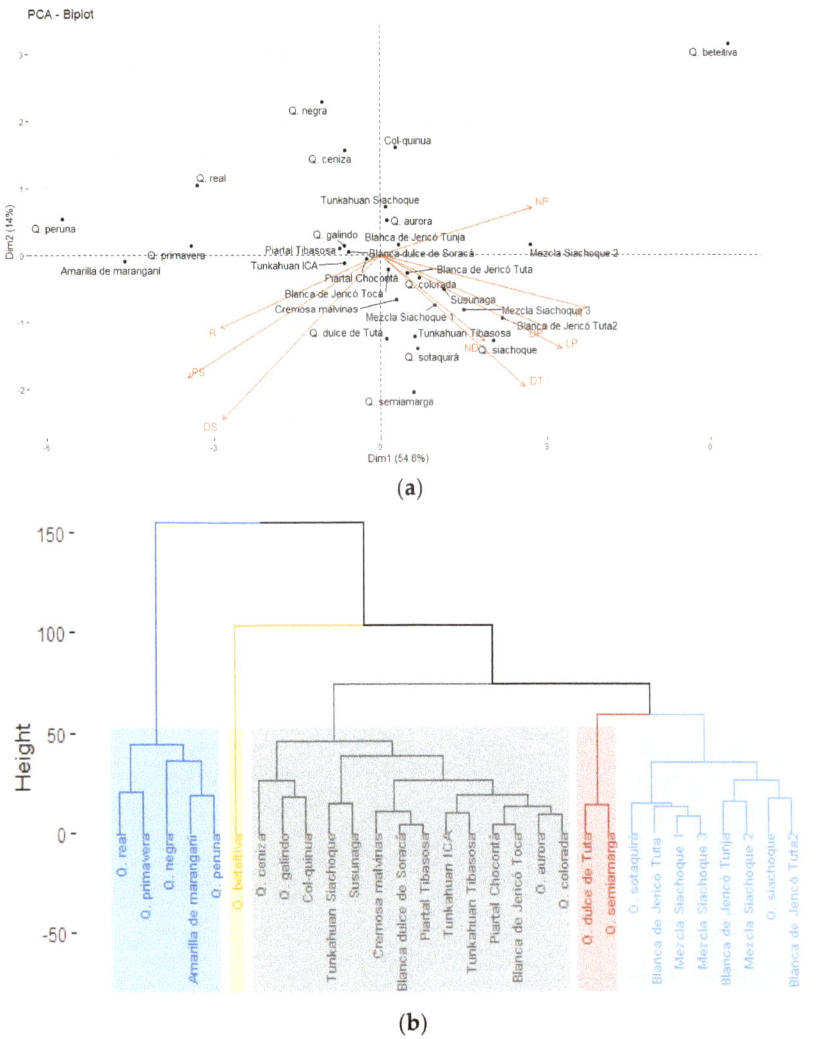

Figure 2. (**a**) Principal component analysis biplot. The variables that contributed to yield were weight and diameter of the seeds; (**b**) Hierarchical cluster analysis of quinoa accessions considering quantitative variables.

These results were used to infer that the quantitative variables of yield, plant height and variables associated with the seeds, such as weight and diameter, were the most discriminative parameters, differentiating between the accessions evaluated in this study. Therefore, with these variables, it is possible to select materials to start breeding programs for quinoa.

2.2. Morphologic Characterization Using Qualitative Descriptors

When evaluating the qualitative variables, it was observed that, after germination and during the development of seedlings, the stem color was generally green, although, in some accessions, the color changed during flowering to shades of purple (Q. ceniza, Q. Sotaquirá, Tunkahuan—ICA, Amarilla de maranganí, Q. Peruana, Blanca de Jericó de Toca, Tunkahuan Tibasosa). For stem shape, all accessions had angular stems with striae (Figure 3a,b), which were green in 99% of the plants. Only three plants of the Quinua ceniza presented a purple color. Further, 20% of the evaluated plants had pigmented axillae (Figure 3c) that were purple in all cases (Tunkahuan, Amarilla de maranganí, Blanca dulce Soracá, Peruana, Quinua Siachoque, Piartal de Tibasosa, Blanca de Jericó de Toca, Quinua cremosa Malvinas). The seedlings had calcium oxalates that varied between white, pink, purple and purple (Figure 3h,i). The purple accessions included Quinua ceniza, Quinua colorada, Tunkahuan, and Quinua Siachoque. These oxalates were observed only until the flowering stage began and subsequently disappeared.

The leaves were green until physiological maturity. Then, in all accessions, they became yellow, starting from the basal leaves towards the apical ones, until senescence. The shape of the leaves had four types: lanceolate, rhomboidal, triangular, or oval, which is a polymorphic characteristic in the same plant. The most common forms were lanceolate in the apical branches, and triangular, oval, or rhomboidal in the basal leaves. (Figure 3g). The edge of the leaves was serrated, entire and dentate. In the basal leaves, 93% had serrated leaves, while in the apical leaves, 83% of the plants were serrated. The most common growth habit was branched up to the first third (61%), and a simple habit was observed at 37%.

Figure 3. (**a**) Angled stem with green striae; (**b**) Purple striae; (**c**) Stem with green striae and purple axillae; (**d**) Stem with green striae and no axillae; (**e**) Apical leaves with lanceolate and entire shape; (**f**) Leaves with rhomboid shapes and serrated edge; (**g**) Triangular shaped leaves with serrated edge; (**h**) White calcium oxalates on the upper surface and underside of the leaf; (**i**) Pink calcium oxalates on the upper surface of the leaf.

The panicles demonstrated high variation for color that depended on the accessions and stage of development. Thus, the panicle colors observed at physiological maturity were: purple, pink, yellow, orange, red, green, and a mixture between these colors (Figure 4a–e).

For panicle shape, 93% of the plants were glomerulate, while the intermediate and amarantiform forms were less common, appearing in only 7% of the plants. Panicle density was 61% loose, 29% intermediate, and 10% compact. Further, 95% of the plants had yellow flowers and 5% had white flowers.

There were four grain shape types: lenticular, cylindrical, ellipsoidal, and conical. Eighty three percent of the plants had a cylindrical grain shape, 7% were ellipsoidal, 6% were lenticular, and 4% were conical. For grain edge, 48% of the plants had a wavy edge, 26% were smooth, and 21% were intermediate (Figure 4g–i). The color of the episperm was also variable: 69% of the plants had a transparent episperm, 13% were white, 7% were black, and 6% were beige. The color of the perigonium presented orange, black, brown, and beige variations (Figure 4f). For seed germination capacity, good vigor was observed in 86% of the accessions. However, in accessions such as Quinua negra, the seeds required approximately 15 days for germination, while the average for the other accessions was 24 h.

Figure 4. (**a**) Lax green panicles; (**b**) Compact, amarantiform, and purple; (**c**) Loose, and green and purple; (**d**) Lax with a mix of green and yellow; (**e**) Compact purple panicle; (**f**) Color of the seed episperm; (**g**) Seed with a beige intermediate border; (**h**) Seed with beige wavy edge; (**i**) Smooth edge seed with brown perigonium.

The multiple correspondence analysis showed that 36.1% of the total variance was explained by the first two components, CP1 (19.0%) and CP2 (17.1%). The first component grouped the accessions according to characteristics such as fuchsia calcium oxalates, amarantiform panicle shapes, orange perigonia pink granules at bloom, and pink panicles at bloom. The second component was grouped according to transparent episperm, intermediate density, compact panicle, and white granules at flowering. Figure 5 shows the distribution of variables according to their contribution to the total variance in the first two components.

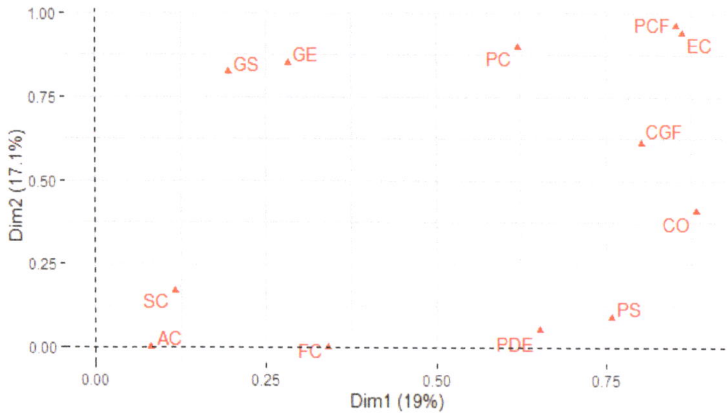

Figure 5. Multiple correspondence analysis showing the contribution of variables, ordering the accessions according to the qualitative variables.

The cluster analysis of the qualitative variables formed seven groups (Figure 6). The first one had the accession Quinua beteitiva, which did not develop grains but developed panicles, which were lax and fragile. The second group was defined by orange panicles at flowering, purple granules at flowering, purple calcium oxalates, purple striae, and compact panicles. The third group was characterized by green panicles at bloom, green granules at bloom, white calcium oxalates, and intermediate panicle density. The fourth group was represented by Quinua negra, which had a conical grain shape and black perigonium and episperm. The fifth group had Quinua primavera and Quinua real, grouped by pink panicle at physiological maturity and intermediate panicles. Group six was characterized by green panicles at green bloom and wavy grain borders. Group seven contained 15 of the 30 accessions and did not present qualitative variables that defined the grouping, but most of the plants had a beige perigonium color, glomerulate panicle shape, white oxalate color, and lax panicle density.

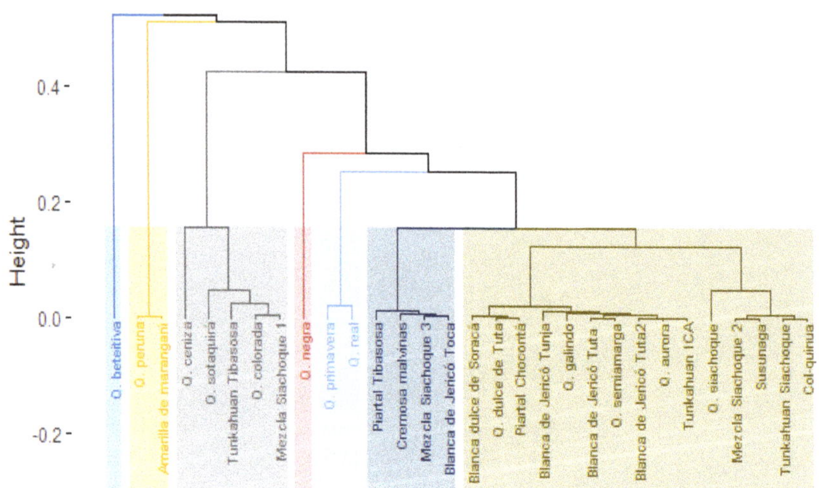

Figure 6. Cluster analysis showing seven groups of quinoa accessions formed according to the qualitative variables.

In general terms, the qualitative variables that contributed significantly to the selection or discrimination of accessions included grain color and panicle density because of the presence and coloration characteristics of structures, such as axillae, striae, and panicles, that were highly variable between and within the evaluated quinoa accessions.

2.3. Morphologic Characterization Taking into Account the Joint Analysis of Qualitative and Quantitative Descriptors

The factorial analysis of mixed data considered all quantitative and qualitative variables and discriminated the accessions with outstanding morphologic characteristics. This analysis showed that the contribution of the variables to the first two components was 38.36%. The variables that contributed positively to CP1 (25.81%) included yield, weight, and diameter of the seeds (quantitative variables), yellow striae, white calcium oxalates, and purple axillae (qualitative variables) (Figure 7a). For CP2 (13.05%), the quantitative variables were yield and number of teeth on the leaves, and the qualitative variables were fuchsia calcium oxalates, uncolored axillae, and beige perigonia.

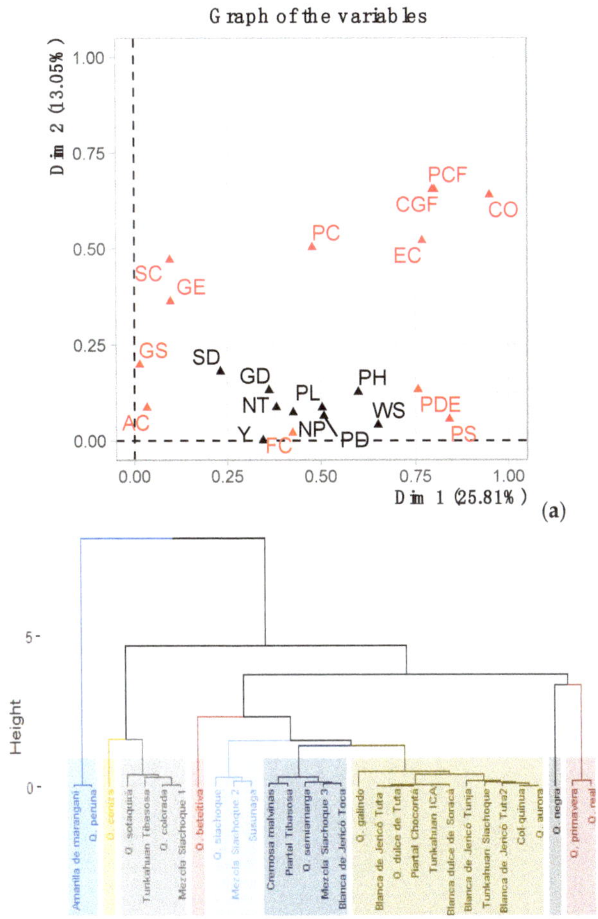

Figure 7. (**a**) Mixed factor analysis of the contribution of the variables, ordering the accessions according to the qualitative and quantitative variable; (**b**) Cluster analysis, showing nine groups of accessions formed according to the qualitative and quantitative variables.

The cluster analysis formed nine groups (Figure 7b). The first group had the accessions Amarilla de Maranganí and Quinua peruana, which were characterized by below average seed weight and diameter (0.39 g and 2.59 mm), and panicle height and length (PH = 176.7 cm and PL = 57.9 cm). In the second group, ash quinoa was characterized by purple striae, grey perigonia and black episperm. Group three was characterized by plant heights between 175.4 to 202.9 cm, seed diameters from 2.06 to 2.23 mm, purple oxalates, and lax panicles. The fourth group had the accession Quinua beteitiva, which had a plant height that was higher than average, 248.2 cm, and did not produce grain. Group five was made up of the accessions Quinua Siachoque, Mezcla Siachoque 2, and Susunaga, which presented yields between 12.28 and 29.08 g of seeds per plant, weights of 1000 seeds from 0.16 to 0.23 g, and glomerulate, lax panicles. The accessions in group six were characterized by yellow striae, an average plant height of 181.4 cm, 10–15 panicles per plant, and seed diameters of 0.26 mm. Group 7 was made up of eleven accessions and did not present qualitative and quantitative variables that defined this grouping. However, these accessions had lower than average yields, that is, less than 37.16 g of seeds per plant, plant heights between 159.1 and 220.4 cm, and seed diameters from 1.87 to 2.36 mm. Quinua negra represented group 8, characterized by lower-than-average stem and seed diameters (3.1 cm and 1.7 mm), with a conical grain shape. Finally, group 9 was made up of the accessions Quinua primavera and Quinua real, which presented, on average, Y = 78.04 g of seeds per plant, PH = 130.9 cm, and PL = 42.9 cm.

The joint analysis of the morphologic descriptors proved to be robust and grouped the accessions according to the principal qualitative and quantitative variables, identifying promising accessions with the potential to start a breeding for quinoa in Colombia.

2.4. Promising Accessions Selection Index in Breeding Programs

The selection index confirmed the results obtained with the factorial analysis of mixed data, which established that the high-yield accessions were Quinua primavera, Quinua Peruana, Quinua real, and Amarilla de maranganí. These accessions were characterized by small sizes, large grains and good yield, which are ideal characteristics for the commercialization of quinoa grain in Colombia. The accessions Q. siachoque, Mezcla Siachoque 2, Blanca de Jericó Tuta2, and Q. beteitiva did not meet the needs of farmers because they had higher plant heights with no or little grain production (Figure 8).

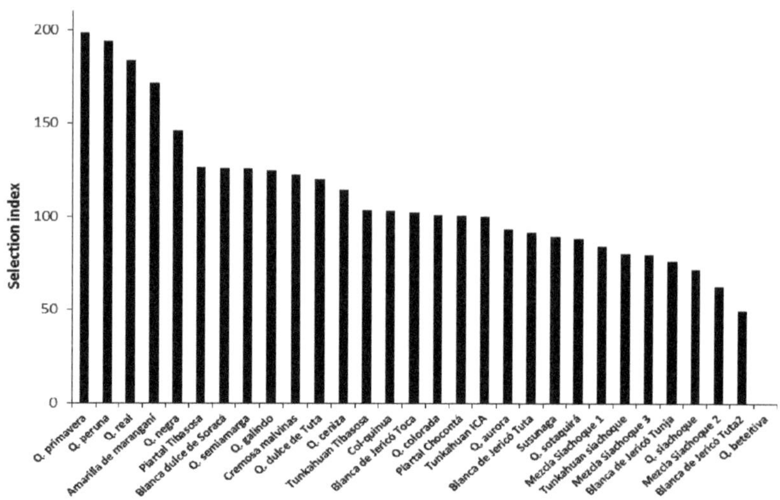

Figure 8. Values of the selection index for each of the evaluated quinoa accessions (*C. quinoa*), showing the discrimination of accessions according to agronomic characteristics.

3. Discussion

The quinoa from the seed collection of the Department of Boyacá, Colombia were phenotypically characterized to select elite accessions to improve quinoa production in the region. The life cycle of the evaluated quinoa accessions was approximately six months, with a photoperiod and average temperature that were constant in the study area. The yield had high variability between the accessions. In general, the accessions with plant heights higher than average (>176.7 cm) presented lower yields, as reported by [13]), who, when evaluating quinoa materials in southern Italy, found negative correlations between plant height and yield. [14] also observed a decrease in growth with the formation of grains in the Department of Cauca, Colombia in the Tunkahuan, Blanca de Jericó, and Blanca de Soracá varieties. This was probably due to the fact that the increase in plant growth coincides with the beginning of the reproductive stage, where the energy produced by plants is distributed in the growth and the beginning of panicles. Some genotypes use this energy only for growth, and grains do not develop properly because of the source–sink relationship [15,16].

It has been reported that longer panicles could provide higher grain yield than shorter ones [17,18]. However, this trend was not observed in the studied accessions. The Blanca de Jericó Tuta2 accession presented the longest panicle length, PL = 72.4 cm, and a below average yield (Y = 18.17 g). Quinua Peruana had the shortest panicle length, PL = 39.0 cm, and obtained an above average yield (Y = 62.02 g). This suggests that this descriptor could be useful in the selection of accessions with better yield. Additionally, the seed characteristics, such as weight and diameter, were correlated with yield. It should be noted that these two variables are important for the commercialization of quinoa and are commonly used as criteria for the selection of materials for the improvement of quinoa [19].

On the other hand, although qualitative variables constitute a fundamental tool to determine the adaptation strategies of plants and are used as varietal descriptors [20], in this study, these traits had broad genetic variability, as represented in the different colorations of the striae, axillae, panicles and seeds. In addition, these traits were highly variable within the same accessions, that is, there was high heterogeneity in the characteristics associated with these variables. This behavior was also observed in quinoa materials evaluated in the Rio Grande do Sul region of Brazil [21] and in cultivars of Quinua Blanca de Jericó, expressed in different pigmentations within individuals in structures such as panicles and stems. These variations allow plants to adapt more quickly to environmental conditions [22]. However, these variables are the basis for genetic improvement programs because, if there is no variability, no selection can be made, since all individuals respond in the same way to the evaluated conditions. Therefore, the existence of phenotypic variability associated with qualitative or quantitative morphological characteristics will allow the selection of materials that respond to the needs of farmers, producers, and consumers.

The joint analysis of the quantitative and qualitative variables differentiated the accessions with higher yields and identified promising genetic lines. The groupings of the accessions Quinua amarilla de maranganí, Quinua peruana, Quinua primavera, and Quinua real remained constant in all analyses, meaning they have high potential for the extensive production of quinoa grain because their yields are higher and they present stable morphological characteristics, such as grain color and panicle density. For phenotyping quinoa with important agronomic traits, the quantitative variables that should be considered are yield, plant height, stem diameter, panicle length, weight of 1000 seeds, and grain diameter, while qualitative variables are panicle density and grain color since they are useful for the selection of materials with potential for quinoa production.

The broad variability in the morphological traits observed in this study may have been due to the facts that farmers maintain mixtures of different materials in the same crop. In Colombia, selection parameters have not been defined for quinoa materials because of a lack of knowledge. Therefore, farmers have marketing difficulties because of a loss of grain quality and decreases in production caused by a lack of selection of planting material and pure materials. Thus, these findings regarding the high variability in both qualitative and quantitative descriptors suggest that they can be very useful in breeding programs. In addition to increasing the productivity of crops, it is possible for accessions to demonstrate great capacity to adapt to different environmental conditions, because the evaluated materials have the advantage of agroecological adaptation to the region since they have been cultivated in the region [23].

However, there are additional morphological characteristics for the good commercialization of the quinoa grain in Colombia, such as small size of plant, erectness, and uniformity, which facilitate the harvesting process, a single terminal panicle with compact glomeruli, and large, white grains [9,24]. Therefore, the proposed selection index confirmed the outstanding accessions (Quinua primavera, Quinua Peruana, Quinua real, and Amarilla de maranganí) using these descriptors since these characteristics are decisive when estimating the commercial quality of crops, with the advantage that these accessions have already been cultivated in the environmental conditions of the region.

Finally, our results suggest that, for Colombia, it is essential to continue with the characterizations of quinoa accessions using morphological descriptors and to include biochemical and molecular descriptors because of the large number of mixtures present in cultivars, allowing efficient selection if pollination is controlled and accessions that exhibit undesirable characteristics to be eliminated, thereby obtaining a variety that responds to the needs of farmers, producers, and consumers through the development and implementation of adequate breeding schemes.

4. Materials and Methods

A total of 30 accessions of quinoa (*C. quinoa*) were evaluated, which belong to the seed collection of the Department of Boyacá (Table 2). The morphoagronomic characterization was carried out under greenhouse conditions in the city of Tunja, located at an altitude of 2690 m.a.s.l., with an average temperature of 13 °C, relative humidity of 78%, and a 12:12 photoperiod. The germination of the seeds was carried out in the nursery with a mixture of humus and peat in a 2:1 ratio. By accessions, 16 alveoli were sown whereby three seeds were placed that were taken randomly, after 20 days of growth when the seedlings had six true leaves. They were transplanted to the greenhouse beds, and thinning was carried out when more than two plants grew per alveolus. The accessions were sown under a randomized complete block design (RCB) of three plants per block (three blocks) for a total of nine repetitions for each accession, with conventional agronomic management. The harvest was carried out manually when the plants reached physiological maturity.

Table 2. Sites of origin of the evaluated quinoa (*C. quinoa*) accessions.

	Accessions	Location	Coordinates
1	*Quinoa real*	Ventaquemada	5°22′00.4″ N 73°31′16.9″ W
2	*Quinoa aurora*	Soracá	5°30′06.9″ N 73°20′00.5″ W
3	*Quinoa ceniza*	La colorada Tunja	5°34′44.7″ N 73°20′36.0″ W
4	*Quinoa beteitiva*	Beteitiva	5°54′39.1″ N 72°48′31.2″ W
5	*Quinoa sotaquirá*	Sotaquirá. Vereda Bociga	5°45′57.6″ N 73°14′52.2″ W
6	*Quinoa negra*	La colorada Tunja	5°34′44.7″ N 73°20′36.0″ W
7	*Tunkahuan ICA*	ICA Surbatá	5°47′45.5″ N 73°04′20.2″ W
8	*Blanca de Jericó Tuta*	Tuta	5°41′26.6″ N 73°13′39.1″ W
9	*Amarilla de maranganí*	Pasca	4°18′32.8″ N 74°17′59.6″ W
10	*Quinoa colorada*	La colorada Tunja	5°34′44.7″ N 73°20′36.0″ W
11	*Blanca dulce de Soracá*	Soracá	5°30′06.9″ N 73°20′00.5″ W
12	*Piartal Chocotá*	Chocontá	5°08′44.3″ N 73°41′07.0″ W
13	*Quinoa dulce de Tuta*	Tuta	5°41′26.6″ N 73°13′39.1″ W
14	*Quinoa semiamarga*	Duitama	5°49′36.3″ N 73°02′03.9″ W
15	*Quinoa peruna*	Cómbita	5°38′01.9″ N 73°19′28.4″ W
16	*Quinoa siachoque*	Siachoque	5°30′45.3″ N 73°14′44.3″ W
17	*Blanca de Jericó Tuta2*	Tuta	5°41′26.6″ N 73°13′39.1″ W

Table 2. *Cont.*

	Accessions	Location	Coordinates
18	*Piartal Tibasosa*	Tibasosa	5°44′40″ N 73°14′16″ W
19	*Blanca de Jericó Tunja*	Tunja	5°31′4″ N 73°23′48″ W
20	*Blanca de Jericó Toca*	Toca-Vda Tuaneca	5°34′03.6″ N 73°11′24.2″ W
21	*Cremosa malvinas*	Siachoque	5°31′00.8″ N 73°14′59.7″ W
22	*Tunkahuan Tibasosa*	Tibasosa	5°44′40″ N 73°14′16″ W
23	*Tunkahuan siachoque*	Siachoque-Finca San Antonio	5°31′55.6″ N 73°16′10.6″ W
24	*Mezcla Siachoque 1*	Siachoque	5°31′00.8″ N 73°14′59.7″ W
25	*Mezcla Siachoque 2*	Siachoque	5°31′00.8″ N 73°14′59.7″ W
26	*Mezcla Siachoque 3*	Siachoque	5°31′00.8″ N 73°14′59.7″ W
27	*Quinoa primavera*	Siachoque-Sabana de Bogotá	4°24′56.3″ N 74°06′06.0″ W
28	*Quinoa galindo*	Cómbita	5°38′01.9″ N 73°19′28.4″ W
29	*Col-quinua*	Cómbita	5°38′01.9″ N 73°19′28.4″ W
30	*Susunaga*	Cómbita	5°38′01.9″ N 73°19′28.4″ W

Twenty-one morphologic descriptors were evaluated, of which nine were quantitative and twelve qualitative, defined by the FAO for quinoa [6] (Table 3). Measurements were taken on nine individuals of each accession.

Table 3. Morphologic descriptors used for the characterization of quinoa accessions from the Department of Boyacá.

Qualitative	Acronyms	Quantitative	Acronyms	Unit of Measurement
Calcium oxalates color	(CO)	Plant height	(PH)	cm
Strie color	(SC)	Stem diameter	(SD)	cm
Axil color	(AC)	N ° teeth per leaf	(NT)	#
Color of the granules at flowering	(CGF)	Panicle length	(PL)	cm
Panicle color at flowering	(PCF)	Panicle diameter	(PD)	cm
Flower color	(FC)	N ° plant panicles	(NP)	#
Panicle shape	(PS)	Yield per plant	(Y)	g
Panicle density	(PDE)	Weight of 1000 seeds	(WS)	g
Grain shape	(GS)	Grain diameter	(GD)	mm
Episperm color	(EC)			
Grain edge	(GE)			
Perigonium color	(PC)			

For the selection of accessions with important and highly productive agronomic characteristics, the program R was used to analyze selection indices (RIndSel: R software to analyze Selection Indices) to obtain Smith's linear phenotypic selection index [25]. The following variables were weighted: yield, plant height, seed diameter and grain color, as described below:

$$\text{I.S} = \textit{Yield (0.95)} - \textit{Plant height (0.94)} + \textit{Grain diameter (3.70)} + \textit{Grain color (1.31)}.$$

The variables yield, grain diameter, and grain color were expressed positively since plants with more grams of seeds per plant, a greater seed diameter and light grain colors are sought. The plant height was expressed negatively since low-bearing accessions are sought.

Statistical Analysis

For the quantitative variables, a descriptive analysis was carried out. Then, the assumptions for the parametric analyzes were verified, and the analysis of variance (ANOVA) was carried out. To determine the significant differences between treatments, a Tukey multiple comparison test was performed with $p < 0.05$. These analyses were performed using R Core Team [26] and the missMDA package [27]. The Spearman correlation was estimated and plotted using the R package "corrplot": Visualization of a Correlation Matrix (Version 0.84) [28]. For the multivariate analysis, a hierarchical grouping with principal components (HCPC) was carried out with the algorithms in the factoextra package of the R program [29], which were plotted on a two-dimensional plane using the FactoMineR package [30]. The dendrogram was done using the main components, Euclidean distance, and Ward's minimum variance hierarchical grouping method with the FactoMineR package [30] in the R program [26].

For the qualitative variables, frequency analyses were performed with Infostat [31], a multiple correspondence analysis with the algorithms in R program's factoextra package [29], and the dendrogram obtained with the components, Euclidean distance, and Ward's minimum variance hierarchical grouping method, using the FactoMineR package [30] in the R program [26]. For the joint analysis of the quantitative and qualitative variables, a factorial analysis of mixed data was carried out with the factoextra package in the R program. Additionally, a dendrogram was generated using the Euclidean distance and hierarchical grouping method of Ward's minimum variance with the FactoMineR package [30]. Finally, the selection index was calculated with an accessions percentage of 5% and the variance-covariance matrix for the variables yield, plant height, seed diameter, and grain color using the RindSel program. The index for each of the accession was plotted in Microsoft Excel 2013.

5. Conclusions

The characteristics, such as seed diameter, panicle density, plant height, and grain color, allowed for the selection of quinoa accessions with better yield and desirable agronomic characteristics. The broad phenotypic variability in the accessions, in terms of both grain quality and yield, constitutes a fundamental tool for recording varieties that improve the positioning of quinoa in Colombia. In addition, this study revealed relationships between yield and morphological characteristics, which could be useful for the selection of parental lines in future breeding programs that seek to generate quinoa hybrids in Colombia. Our results indicated that the accessions Quinua primavera, Quinua Peruana, Quinua real, and Amarilla de maranganí represent the most promising parents for the future development of breeding programs that aim to respond to the needs of farmers, producers, and consumers in Colombia.

Author Contributions: Conceptualization, E.H.M.-H. and A.C.M.-C.; methodology, E.H.M.-H., A.C.M.-C. and D.M.A.-M.; software, E.H.M.-H.; validation, E.H.M.-H. and D.M.A.-M.; formal analysis, E.H.M.-H. and D.M.A.-M.; investigation, E.H.M.-H. and A.C.M.-C.; resources, E.H.M.-H. and A.C.M.-C.; data curation, E.H.M.-H. and D.M.A.-M.; writing—original draft preparation, E.H.M.-H.; writing—review and editing, E.H.M.-H., A.C.M.-C., D.M.A.-M., Z.Z.O.-P. and A.C.-C.; visualization, E.H.M.-H. and D.M.A.-M.; supervision, A.C.M.-C.; project administration, A.C.M.-C.; funding acquisition, A.C.M.-C. All authors have read and agreed to the published version of the manuscript.

Funding: This research was funded by "Patrimonio Autónomo Fondo Nacional de Financiamiento para la Ciencia, la Tecnología y la Innovación Francisco José de Caldas-MinCiencias. Cód. 63924".

Institutional Review Board Statement: Not applicable.

Informed Consent Statement: Not applicable.

Data Availability Statement: The data generated within this work are open access and available to be shared with interested persons.

Acknowledgments: Al grupo CIDE Competitividad Innovación y Desarrollo Empresarial de la Facultad de Ciencias Agropecuarias de la UPTC y al grupo Bioplasma de la Facultad de Ciencias de la UPTC.

Conflicts of Interest: The authors declare no conflict of interest.

References

1. López, M.L.; Recalde, M.A. The first quinoa (*Chenopodium quinoa* Willd.) macrobotanical remains at Sierras del Norte (Central Argentina) and their implications in pre-Hispanic subsistence practices. *J. Archaeol. Sci. Rep.* **2016**, *8*, 426–433. [CrossRef]
2. Vargas, P.; Arteaga, R.; Cruz, L. Análisis bibliográfico sobre el potencial nutricional de la quinua (*Chenopodium quinoa*) como alimento funcional. *Rev. Cent. Azúcar* **2009**, *46*, 89–100.
3. García-Parra, M.; Zurita-Silva, A.; Stechauner-Rohringer, R.; Roa-Acosta, D.; Jacobsen, S.-E. Quinoa (*Chenopodium quinoa* Willd.) and its relationship with agroclimatic characteristics: A Colombian perspective. *Chil. J. Agric. Res.* **2020**, *80*, 290–302. [CrossRef]
4. Jaikishun, S.; Li, W.; Yang, Z.; Song, S. Quinoa: In Perspective of Global Challenges. *Agronomy* **2019**, *9*, 176. [CrossRef]
5. Romaní-Morón, M.; Valdez-Arana, J. Effect of habitat change on the nutritional and functional characteristics of 16 quinoa (*Chenopodium quinoa* Willd.) accessions cultivated in the Peruvian Coast. *Sci. Agropecu.* **2019**, *10*, 293–302. [CrossRef]
6. Bioversity International; Organización de las Naciones Unidas para la Agricultura y la Alimentación (FAO); Fundación PROINPA; Instituto Nacional de Innovación Agropecuaria y Forestal (INIAF); Fondo Internacional de Desarrollo Agrícola (FIDA). *Descriptores Para Quinua Y Sus Parientes Silvestres*; Bioversity International: Rome, Italy, 2013; p. 52.
7. García, M.A.; Roa, D.F.; Stechauner, R.; Molano, F.G.; Bazile, D.; Plazas, N. Effect of temperature on the growth and development of quinoa plants (*Chenopodium quinoa* Willd.): A review on a global scale. *Sylwan* **2020**, *165*, 23.
8. Santos, T. *Mercadeo Potencial Para la Quinua Colombiana*; Universitaria Agustiniana: Bogotá, Colombia, 2019.
9. Morillo, A.; Manjarres, E.; Morillo, Y. Evaluación morfoagronómica de 19 materiales de Chenopodium quinoa en el Departamento de Boyacá. *Biotecnol. Sect. Agropecu. Y Agroind.* **2020**, *18*, 84–96. [CrossRef]
10. Torres, J.; Vargas, H.; Corredor, G.; Marina, S. Caracterización morfoagronómica de diecinueve cultivares de quinua (*Chenopodium quinoa* Willd.) en la Sabana de Bogotá. *Agron. Colomb.* **2000**, *17*, 60–68.
11. Veloza, C.; Romero, G.; Gómez, J. Respuesta morfoagronómica y calidad en proteína de tres accesiones de quinua (*Chenopodium quinoa* Willd.) en la sabana norte de Bogotá. *Rev. UDCA Actual. Divulg. Cient.* **2016**, *19*, 325–332.
12. Infante, I.; Albesiano, S.; Arrienta, L.; Gomez, N. Morphological characterization of varieties of Chenopodium quinoa cultivated in the department of Boyacá, Colombia. *Rev. UDCA Actual. Divulg. Cient.* **2018**, *21*, 329–339. [CrossRef]
13. De Santis, G.; D'Ambrosio, T.; Rinaldi, M.; Rascio, A. Heritabilities of morphological and quality traits and interrelationships with yield in quinoa (*Chenopodium quinoa* Willd.) genotypes in the Mediterranean environment. *J. Cereal Sci.* **2016**, *70*, 177–185. [CrossRef]
14. García-Parra, M.Á.; Stechauner-Rohringer, R.; García-Molano, J.F.; Ortiz, D. Análise do crescimento e desempenho morfofisiológico de três cultivares de quinoa colombiana cultivadas em casa de vegetação. *Rev. Ciênc. Agrovet.* **2020**, *19*, 73–83. [CrossRef]
15. Do, T.H.T.; Martinoia, E.; Lee, Y. Functions of ABC transporters in plant growth and development. *Curr. Opin. Plant Biol.* **2018**, *41*, 32–38. [CrossRef] [PubMed]
16. Egli, D. *Seed Biology and the Yield of Grain Crops*; CABI: Wallingford, UK, 2017.
17. De Santis, G.; Ronga, D.; Caradonia, F.; Ambrosio, T.D.; Troisi, J.; Rascio, A.; Fragasso, M.; Pecchioni, N.; Rinaldi, M. Evaluation of two groups of quinoa (*Chenopodium quinoa* Willd.) accessions with different seed colours for adaptation to the Mediterranean environment. *Crop. Pasture Sci.* **2018**, *69*, 1264. [CrossRef]
18. Maliro, M.F.A.; Guwela, V.F.; Nyaika, J.; Murphy, K.M. Preliminary Studies of the Performance of Quinoa (*Chenopodium quinoa* Willd.) Genotypes under Irrigated and Rainfed Conditions of Central Malawi. *Front. Plant Sci.* **2017**, *8*, 1–9. [CrossRef] [PubMed]
19. Zurita-Silva, A.; Fuentes, F.; Zamora, P.; Jacobsen, S.-E.; Schwember, A.R. Breeding quinoa (*Chenopodium quinoa* Willd.): Potential and perspectives. *Mol. Breed.* **2014**, *34*, 13–30. [CrossRef]
20. Katwal, T.B.; Bazile, D. First adaptation of quinoa in the Bhutanese mountain agriculture systems. *PLoS ONE* **2020**, *15*, e0219804. [CrossRef]
21. Vergara, R.D.O.; Martins, A.B.N.; Soares, V.N.; Carvalho, I.R.; Barbosa, M.H.; Conte, G.G.; Gadotti, G.I.; Ludke, R.; Villela, F.A. Agronomic and morphological characteristics of quinoa grown in the southern region of Rio Grande do Sul State. *Rev. Bras. Eng. E Sustentabilidade* **2020**, *8*, 18–25. [CrossRef]
22. Bazile, D.; Bertero, D.; Nieto, C. *State of the Art Report of Quinoa in the World in 2013*; FAO & CIRAD: Rome, Italy, 2015.
23. Allende, M.J. Caracterización Morfológica y Molecular de Accesiones de Quinua (*Chenopodium quinoa* Willd.) Para Estimar Variabilidad Genética. 2017. Available online: http://repositorio.lamolina.edu.pe/handle/UNALM/2935 (accessed on 25 March 2021).
24. Rodríguez, J.P.; Rahman, H.; Thushar, S.; Singh, R.K. Healthy and Resilient Cereals and Pseudo-Cereals for Marginal Agriculture: Molecular Advances for Improving Nutrient Bioavailability. *Front. Genet.* **2020**, *11*, 49. [CrossRef]
25. Smith, H.F. A discriminant function for plant selection. *Annal. Eugen.* **1936**, *7*, 240–250. [CrossRef]
26. The R Development Core Team. *R: A Language and Environment for Statistical Computing*; R Foundation for Statistical Computing: Vienna, Austria, 2020.
27. Josse, J.; Husson, F. missMDA: A Package for Handling Missing Values in Multivariate Data Analysis. *J. Stat. Softw.* **2016**, *70*, 1–31. [CrossRef]
28. Taiyun, W.; Simko, V. R package "corrplot": Visualization of a Correlation Matrix (Version 0.84). *Statistician* **2017**, *56*, e24.
29. Kassambara, A.; Mundt, F. Factoextra: Extract and Visualize the Results of Multivariate Data Analyses. *R Package Version* **2020**, *1*, 337–354.

30. Le, S.; Josse, J.; Husson, F. FactoMineR: An R Package for Multivariate Analysis. *J. Stat. Softw.* **2008**, *25*, 1–18. [CrossRef]
31. Di Rienzo, J.A.; Casanoves, F.; Balzarini, M.G.; Gonzalez, L.; Tablada, M.; Robledo, C.W. *InfoStat, Versión 2020*; Universidad Nacional de Córdoba: Córdoba, Argentina, 2020.

plants

MDPI

Review

A Review of *Chenopodium quinoa* (Willd.) Diseases—An Updated Perspective

Carla Colque-Little, Daniel Buchvaldt Amby and Christian Andreasen *

Department of Plant and Environmental Sciences, University of Copenhagen, Højbakkegaard Allé 13, DK2630 Taastrup, Denmark; cxl@plen.ku.dk (C.C.-L.); amby@plen.ku.dk (D.B.A.)
* Correspondence: can@plen.ku.dk; Tel.: +45-51322551

Abstract: The journey of the Andean crop quinoa (*Chenopodium quinoa* Willd.) to unfamiliar environments and the combination of higher temperatures, sudden changes in weather, intense precipitation, and reduced water in the soil has increased the risk of observing new and emerging diseases associated with this crop. Several diseases of quinoa have been reported in the last decade. These include *Ascochyta caulina*, *Cercospora* cf. *chenopodii*, *Colletotrichum nigrum*, *C. truncatum*, and *Pseudomonas syringae*. The taxonomy of other diseases remains unclear or is characterized primarily at the genus level. Symptoms, microscopy, and pathogenicity, supported by molecular tools, constitute accurate plant disease diagnostics in the 21st century. Scientists and farmers will benefit from an update on the phytopathological research regarding a crop that has been neglected for many years. This review aims to compile the existing information and make accurate associations between specific symptoms and causal agents of disease. In addition, we place an emphasis on downy mildew and its phenotyping, as it continues to be the most economically important and studied disease affecting quinoa worldwide. The information herein will allow for the appropriate execution of breeding programs and control measures.

Keywords: causal agents; downy mildew; pathogenicity; *Peronospora*; resistance factors; severity; quinoa diseases; quinoa disease assessment

Citation: Colque-Little, C.; Amby, D.B.; Andreasen, C. A Review of *Chenopodium quinoa* (Willd.) Diseases—An Updated Perspective. *Plants* **2021**, *10*, 1228. https://doi.org/10.3390/plants10061228

Academic Editors: Cataldo Pulvento and Didier Bazile

Received: 9 April 2021
Accepted: 7 June 2021
Published: 16 June 2021

1. Introduction

Agriculture is affected by global climate change. Non-traditional crops with high nutritional value and the ability to cope with abiotic stress are of special interest in today's world. Quinoa (*Chenopodium quinoa* Willd.) is an ancient crop that exhibits remarkable tolerance to frost, salt, and drought. Moreover, it is highly nutritious and has a vast genetic diversity resulting from its fragmented and localized production over the Andean region. The recent introduction and cultivation of quinoa in novel environments has resulted in a wider spectrum and higher intensity of infectious diseases. Oomycetes and fungi are the two most important eukaryotic plant pathogens [1]; their predominance on the quinoa pathobiome is also evident.

Diseases of quinoa have been reviewed previously [2–6]. However, an update is necessary because new emerging diseases of the quinoa mycobiome are being discovered. Taxonomy based on the morphological characteristics and nomenclature of fungi is relatively conserved and informative when high-level classifications (genus level) are considered. However, there is uncertainty when lower-level phylogenies (species level) are considered due to the fast-evolving traits and phenotypic plasticity of fungi [7]. As a result, DNA and molecular sequence-database comparisons techniques have been employed, along with various DNA fingerprinting and more advanced and complex methods such as whole-genome sequencing, for the identification of plant pathogens [8,9].

The universal nuclear ribosomal primers developed by White et al. (1990) for PCR amplification of the internal transcribed spacer (ITS) region have become a key component the description and characterization of fungal diversity [10]. In addition to ITS, various other markers exist for multi-locus sequencing. It is commonly used by combining ITS with other relevant genomic regions (e.g., COX I, calmodulin, and TEF1 gene regions). It has proven helpful and necessary for the accurate identification of microbial plant pathogens [11–13]. Such molecular approaches should be paired with pathogenicity assessments, including the description of disease symptoms, isolation and artificial inoculation of quinoa tissue, recording of symptoms, and re-isolation. These tests are known as Koch's postulates [14–16]; their validation discriminates an opportunistic association from a pathogenic-type interaction.

This review aims to provide an updated overview of microbial plant pathogens causing disease in quinoa, focusing on the morphological characterization and molecular identification of the causal agents. Research carried out in the Andean countries some decades ago provides insightful and valuable reports, described herein. We compiled and analyzed existing information, with a marked emphasis on downy mildew.

2. Downy Mildew of Quinoa

2.1. Nomenclature and Distribution

The oomycete *Peronospora variabilis* Gäum. 1919 [17] is the causal agent for downy mildew on *C. quinoa* (www.indexfungorum.org, accessed on 10 June 2021) and *C. album* L. The genus *Peronospora* belongs to the *Peronosporaceae* family (Peronosporales order), which are highly physiologically specialized, biotrophic organisms. Phytopathogenic oomycetes are eukaryotic microbes with filamentous vegetative growth and spores for reproduction (fungus-like). Molecular analysis revealed they are among the *Stramenopiles* (or heterokont), closely related to golden-brown algae and diatoms [1,18–20]. Fundamental features are:

1. Oomycetes cell walls are mostly composed of glucans, in contrast to chitin from fungi [14].
2. Most oomycetes are insensitive to azole fungicides (e.g., ketoconazole) because they do not have the ergosterol pathway needed to activate the azole-fungicide mode of action [21–23].
3. During their vegetative state, oomycetes are diploid compared to haploid or dikaryotic fungi [1].

Due to taxonomic confusion, downy mildew was previously classified as *Peronospora farinosa* and considered as such by most studies for about 50 years [24–26]. Byford (1967a,b) [27,28] investigated cross-inoculation experiments and concluded the division of three *formae speciales* (f. spp.) Table 1.

Later, a phylogenetic study on *P. variabilis* of *C. quinoa* and *C. album* from different geographical regions showed that both are located in the same phylogenetic cluster with no evidence to separate them into different taxa [29–32]. Morphological, molecular, and biological host specialization analyses revealed that a narrow species concept is more appropriate for the downy mildews. The available evidence strongly suggests that the host range of *P. variabilis* is limited to *C. quinoa* and *C. album* [29], that of *P. effusa* is limited to spinach [33,34], and more recently that of *P. chenopodii* has been shown to be limited to *C. hybridum* L. (maple leave goosefoot), *P. chenopodii-ambrosioides* to *C. ambrosioides* L. (Jesuit's tea, Payqu), *P. chenopodii-ficifolii* to *C. ficifolium* Sm. (fig leave goosefoot) [13], *P. chenopodii-polyspermi* to *C. polyspermum* L. (many-seeded goosefoot), and *P. schachtii* to sugar beet [26]. In older literature, *P. farinosa* was used as the causal agent of downy mildew of quinoa. However, the species name "*farinosa*" had been ascribed to an unrelated genus (*Atriplex*) and is no longer valid as a species name for *Peronospora* [35].

Table 1. *Peronospora* species current identity and classification by Byford [27,28].

Host (Genus/Species)	Pathogen Current Identity	Byford Classification (f. spp.) *P. farinosa formae speciales*
Beta spp.	*P. schachtii* [26]	*P. farinosa* f. sp. *betae*
C. álbum + *C. quinoa*	*P. variabilis* [29,30]	*P. farinosa* f. sp. *chenopodii*
Spinacia oleracea	*P. effusa* [33,34]	*P. farinosa* f. sp. *spinaciae*

The earliest report of downy mildew infecting quinoa in South America came from Martin Cardenas (1941), who found it infecting quinoa in Cochabamba, Bolivia, and identified it as *P. farinosa* [36]. *P. variabilis* has been documented throughout the world (Figure 1) [26,29–32,37–52] wherever quinoa is cultivated. It is expected to become ubiquitous in all quinoa cropping areas as oospores found in seeds have also been seen in old dried leaves [32,53,54]. Moreover, *C. album* (known as goosefoot, fat hen, or lamb's quarters) [55] is frequently infected by downy mildew throughout Europe because it is conspecific [56] with the *P. variabilis* from *C. quinoa*. Therefore, it is likely to be a reservoir for the pathogen and an alternative host [29,52,56]. Other *Chenopodium* species, such as *C. murale* L. (nettle leaf goosefoot), *C. ambrosoides* L. (Indian goosefoot, Mexican tea), *C. berlandieri* Moq. (pit seed goosefoot), and *C. ficifolium* Sm. (fig leaf goosefoot), were reported to harbor the pathogen based on morphological identification [39,45,57] and molecular COX2 bar coding for *C. berlandieri* var macrocalycium (Table 2). These reports require further investigation to confirm the accurate identity of the pathogen. Cross-infection reported so far is solely that of *P. variabilis* isolated from *C. album* and *C. quinoa* [52].

Figure 1. Downy mildew disease of *Chenopodium* spp. distribution map (CAB international, last modified 21 November 2019 via www.cabi.org/isc/datasheet/39704 (accessed on 10 June 2021).

Table 2. Documented reports for downy mildew on *C. quinoa* and weedy *Chenopods*.

Country	*C. quinoa* Leaves (√), Seed (x)		*C.album* Leaves		*C. berlandieri* var. Macrocalycium		*C. murale* Leaves		*C. ambrosoides* Leaves		*C. ficifolium* Leaves		Researcher	Year	[Ref]
	Mor.	Mol.	Mor.	Mol.	Mor.	Mol.	Mor.	Mol.	Mor.	Mol.	Mor.	Mol.			
Bolivia	√												Martin Cardenas	1941	[36]
Peru	√												G. Garcia	1947	[37]
Canada	√												JF.Tewari	1990	[38]
Peru	√		√				√		√				L.Aragon	1992	[39]
Ecuador	√												Jose Ochoa	1999	[40]
Denmark	√												S. Danielsen	2002	[41]
Poland	√												Panka	2004	[42]
India	√												A. Kumar	2006	[43]
Bolivia	√	√											Erica Swenson	2006	[44]
Argentina China Ireland South Korea Netherlands Germany Latvia Romania Italy			√ √ √ √ √ √ √ √ √	√ √ √ √ √ √ √ √ √									Y.J. Choi	2008 2010	[26] [29]
Peru Ecuador Denmark		√ √ √											S. Danielsen	2010	[30]
India							√						P. Baisvar	2010	[45]
USA (Pennsylvania)	√	√											Ana Testen	2012	[46]
Bolivia Ecuador USA		x x x											Ana Testen	2014	[32]
Korea	√	√											Y.J. Choi	2014	[47]

Table 2. *Cont.*

Country	*C. quinoa* Leaves (√), Seed (x)		*C.album* Leaves		*C. berlandieri* var. Macrocalycium		*C. murale* Leaves		*C. ambrosoides* Leaves		*C. ficifolium* Leaves		Researcher	Year	[Ref]
	Mor.	**Mol.**	**Mor.**	**Mol.**	**Mor.**	**Mol.**	**Mor.**	**Mol.**	**Mor.**	**Mol.**	**Mor.**	**Mol.**			
Morocco	-	-											Manal Mhada	2014	[48]
Egypt	√	√											Walaa Khalifa	2018	[49]
USA (N. Hampshire)	√	√	√	√	x√*	x√*						√**	Helen Nolen	2019	[50]
Turkey			√	√									M.Kara	2020	[31]
Turkey	√	√											Esra Gül	2021	[51]
Denmark	√	√	√	√									C. Colque-Little	2021	[52]

Mor. = morphological characterization; Mol. = molecular identification. Source: elaborated from references on the column [Ref]. x√* Koch postulates failed; √** corresponds to a field population.

2.2. Infection Biology and Disease Symptoms

Based on various scientific studies, we assembled a hypothetical disease cycle for *P. variabilis* (Figure 2).

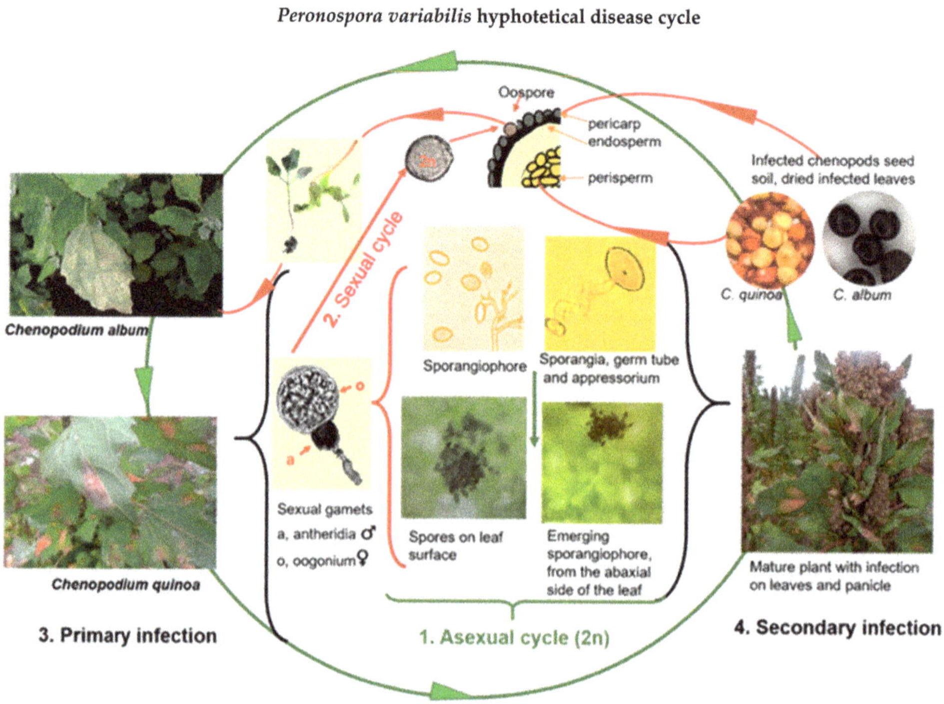

Figure 2. Proposed disease cycle of quinoa downy mildew caused by *Peronospora variabilis* (Photos: C. Colque-Little). Picture of haploid gamets adapted from Judelson [58].

When mature sporangia fall on compatible leaf tissue with free moisture and relative humidity (more than 85%), the infection begins. Spores from pathogenic oomycetes produce an adhesive vesicle on the spore side in contact with the host (ventral) at early infection stages (Figure 3F). Next, a germ tube that faces the host is produced and grows chemotropically toward a suitable penetration site. In most downy mildews, the hyphae enter the leaf via stomatal pores [58] (Figure 3C). The formation of an appressorium-like swelling (penetration structures that exert pressure) on histopathological samples was observed under a microscope [52,59]. It penetrated the stomata (Figure 3B,D) but did not directly penetrate the cuticle [49,59]. Spores are chemotaxically and mechanically dependent on the stomatal aperture [60,61].

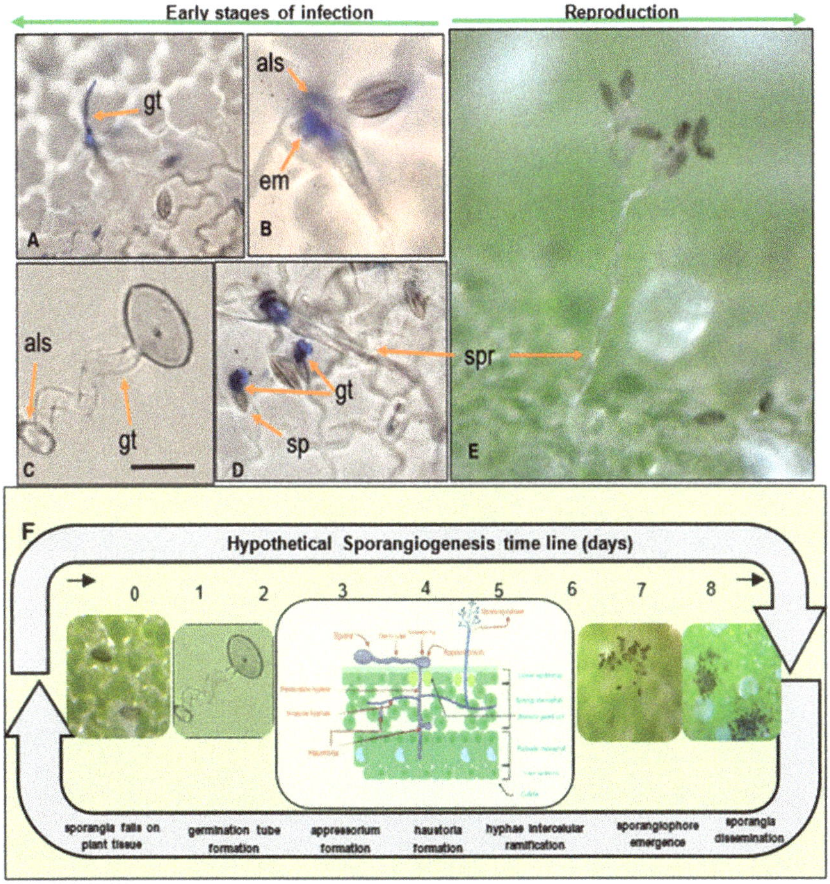

Figure 3. Quinoa leaf infections caused by *P. variabilis* sporangiogenesis during the early stages of asexual reproduction. (**A**) Sporangium forming germ tube (gt) and faint penetration hyphae towards the mesophyll. (**B**) Extracellular matrices (em) secreted from germinating sporangium (sp) and appressorium-like (als) structure penetrating stomata. (**C**) Sporangium, forming germ-tube (gt) and appressorium like structure in water. (**D**) Sporangiophore (spr) emerging from stomata. (**E**) Sporangiophore holding sporangia, emerging from lower epidermis. Scale bar: 20 μm. (**F**) Hypothetical *P. variabilis* sporangiogenesis timeline (Photos: C. Colque-Little). Illustration in timeline created with Biorender.com.

Stomata colonization happens relatively quickly. Once an appressorium is established, the secretion of extracellular matrices during the germination of the sporangia appears, as reported elsewhere (Figure 3B) [59]. The hyphae ramify intercellularly, forming haustoria (feeding structures) through the leaf tissue five to six days after penetration (Figure 3A–F). The sporangiophores emerge from the leaf's surface around the seventh day, carrying asexual lemon-shaped sporangia (Figure 3E,F). Seven to ten days after the primary infection, sporangia are disseminated to other leaves by wind and water [40,59] (Figure 3F). They are assumed to be of importance for spreading the disease during the growing season at this stage [62]. In general, *Peronospora* species require moderate temperatures (10 °C–20 °C) for optimal sporulation [63,64]. While the disease is developing, several asexual cycles (reproduction of sporangia) may occur. Secondary infection demonstrated that the disease could spread rapidly in the field if the optimal conditions are present [54].

Infected leaf tissue manifests lesions and signs on both sides of the leaf. Sporulation becomes apparent mostly on the leaf surface. Symptoms on infected plants vary depending on genotype, growth stage, and environmental conditions (Figure 4A–D). Classic symptoms include pale or yellow chlorotic lesions on the leaf surface (Figure 3E) and dark gray-violaceous sporulating areas, mostly on the lower surface (Figure 3F). The lesions can be several and small in some cultivars, whereas in others the lesions are extensive, diffuse, and irregular (Figure 4G). Lesions turn pink, red, purple, or light-brown, depending on the plant's pigments (red-violet and yellow betalains [65,66]). A hypersensitive response has also been observed (Figure 4E,G). The sporulation presence differs considerably, probably due to cultivar responses and the pathogenic capability of the specific isolate [52,54,67].

Figure 4. (**A**) Quinoa crop severely damaged by downy mildew. (**B–D**). Infected varieties in the fields of the main quinoa growing areas of Bolivia. (**E**) Adaxial leaf side belonging to different quinoa genotypes artificially infected with downy mildew. (**F**) Abaxial side of the leaves showing sporulation. (**G**) Differences in disease symptoms, ranging from hypersensitive reactions causing pale yellowish spots (**left**) to high susceptibility with chlorotic lesions covering the whole leaf (**right**) (Photos: C. Colque-Little).

Downy mildew primarily affects the foliage, but it is possible to find it colonizing different organs and tissues of quinoa plants. However, its symptoms are less obvious and sporulation is inexistent. Therefore, polymerase chain reaction (PCR) was used to amplify *P. variabilis* DNA. Taha (2019) gathered a composite of quinoa seedlings at different growth stages, subdivided them into different organs, and detected *P. variabilis* DNA on 0.8% of the root samples, 83% on the cotyledon and leaf, and 42% on steam samples. The PCR was also positive for 60–80-day-old plants' inflorescences [68]. In addition, scanning electron microscopy was capable of visualizing *P. variabilis* on petioles [59], and the mycelium was seen as in the intercellular spaces of the leaf midrib of 80-day old plants [69]. Since the pathogen was detected at early and late growth stages of the quinoa plant, it was thought to present a systemic mode of infection [68]. However, other researchers argue [69] that the germinated oospores-mycelium spreads through intercellular parenquimatic spaces (next to xylem but not wood vessels) of the hypocotyl acropetally, towards the plant's aerial parts, and is finally inserted into the developing seed. For clarification of the mode of infection of *P. variabilis*, more research is needed.

2.3. Morphology and Reproduction

Peronospora variabilis hyphae are coenocytic (hyphae without septae) and multinucleate, resulting from nuclear divisions within the cell without an accompanying division of the cytoplasm (cytokinesis). Sporangiophores are 240–580 μm long, slender, arborescent, dichotomously ramified five to six times in a sharp angle, ending in two to three straight to slightly curved branches (Figure 5A). Ultimate branchlets are in pairs or single, flexuous to curved 8–23 (av. 12.3) μm long, with obtuse tips (Figure 5C) [29]. Sporangia are pedicellate, deciduous, olivaceous with a grayish tint, broadly ellipsoidal to ellipsoidal (av. 27.7 μm long × 21.0 μm wide (Table 3), ending in an apical translucid papilla [47]. Taxonomic measurements such as spore lengths and widths can vary depending on the homogeneity of the conidium population, the origin of the isolates, the spore subpopulation, or different roles or times in the pathogen's life history [70]. Measuring that variability under the microscope allows researchers to estimate the mean length/width with a reasonable level of resolution when a minimum of 41–71 spores are measured for the *Peronospora* genus [71]. Even though *P. variabilis*, infecting *C. quinoa* and *C. album*, is conspecific [56], sporangia found in *C. album* were slightly bigger. Further research is needed to figure out why this difference exists. Table 3 illustrates this variability from measurements taken by various researchers [26,29,31,46,47,49,51,54,72] and the average of their measurements is provided as a reference (Figure 5B,D).

Table 3. *Peronospora variabilis* sporangium sizes when isolated from *C. quinoa* and *C. album*.

***P. variabilis* Sporangium Isolation Origin**		
C. quinoa **av. Length × Width (μm)**	***C. album*** **av. Length × Width (μm)**	**Reference**
25.5 × 17.5		Khalifa and Thabet 2018 [49]
22 × 23.13		Yin et al., 2018 [72]
27.5 × 20		Gül, 2021b [51]
28.8 × 21.8		Danielsen & Ames, 2004 [54]
30.7 × 23.8		Choi et al., 2010 [29]
31 × 23		Testen et al., 2012 [46]
28.5 × 23.5		Choi et al., 2014 [47]
	29.5 × 23	Choi et al., 2008 [26]
	30 × 25	Kara et al., 2020 [31]
27.7 × 21.0	30.1 × 24	av. size
1.32	1.25	av.ratio

Peronospora variabilis can reproduce asexually (sporangiogenesis) and sexually (oospore formation and germination). It has been reported to be heterothallic and requires two compatible partners for oospore formation (mating). When eight single-lesion isolates coming from different regions of Peru and Bolivia were crossed in all possible combinations using a detached leaf assay, the existence of two mating types, P1 and P2, was apparent [73]. Sexual cycles start with a male (antheridium) and a female (oogonia) gametangia. These structures can be observed in the leaf mesophyll of plants sown 45 days earlier [69] and have the appearance of swollen hyphal tips [74]. Once in contact, both swell, especially the oogonia. Next, synchronous meiosis occurs within each one, and a pore develops between them. A single haploid nucleus is then transmitted from a male to a female. After fecundation, the development of an oospore starts by establishing a thick multi-layered wall. During maturation, the ribosomes and cytochromes disappear. The combination of their lowered metabolism, thick wall, and lipid-rich cytoplasm make them effective resting structures. Walls are usually hyaline, yet contain a brownish pigment, and their thickness ranges between 3 and 6 μm in most *Peronosporales* [58]. *Peronospora variabilis* oogonia (isolated from *C. album*) are subglobose with an average diameter of 43.5 μm [26,31]. The oospore shape is globose to ovoid; their color varies from transparent to golden brown to brown [53,75].

Ooospore diameter has been reported to range from 18.2 to 44.5 μm on average [53,69,75] when isolated from *C. quinoa*, compared to 25 to 44.5 μm on average when isolated from *C. album* [26,54]. These differences may be due to interactions with the host, environmental conditions, the age of the spore, or the pathogen races [53] (Figure 5 Aa,D). Oospores can survive inhospitable environments, such as freezing, desiccation, starvation, and microbial degradation [19]. They permit the completion of the pathogen life cycle and enhance its fitness by providing a mechanism for genetic variation [58]. Resting structures are often the source of initial infection. El-Assiuty (b) et al. (2019) hypothesize that oospores bearing tissues (cotyledons, leaves, and the perianths of seeds) shed during the life cycle of quinoa plants may play a role in the persistence of oospores in soil.

Danielsen and Ames (2004) [54] detected oospores in the pericarp (external tegument of the episperm) using ultra-microtome cuts (Figure 5F). El-Assiuty (a) et al. (2019) confirmed their occurrence in examined seed samples, revealing a 90% presence in the perianth, 87% in the seed coat, 3% in the embryo, and 2% in the perisperm [53].

To follow the passage of *P. variabilis* inside tissues, El-Assiuty (b) et al. (2019) conducted histopathological/microscopic investigations at different plant growth stages. After planting the surface-sterilized seed of a downy mildew susceptible variety, the observations started. Oospores were present in the radicle-pith three days after germination, inside the cortex of hypocotyls, and in the mesophyll of cotyledons seven days after planting. Oospore germination started with two undulating germ-tubes located opposite to one another. They develop in the cortex tissue of juvenile seedlings 15 days post-planting [69]. This research is consistent with what has been found for other downy mildew diseases such as *Plasmopara viticola*, the mature oospores of which germinated for 3–7 days under a favorable regime of rainfall and temperature [76].

Moreover, oospores were detected in all tissues of quinoa plants that had been sown 45–120 days previously [53,69]. They were also seen on leaves of senescent infected plants, artificially inoculated with a single Danish isolate under greenhouse conditions, suggesting that the isolate had both mating types present (Figure 5E. Colque-Little, unpublished). In addition, they have been found in old infected leaves collected in Andean regions of quinoa production (Peru, Bolivia) [54] and fresh leaf tissue collected in Pennsylvania, USA [77].

Greenhouse experiments with oospore-infected seed samples sown in high and low relative humidity showed a significant difference in visible seedling infections among samples under high humidity and with a large oospore density in most cases. However, oospore density seems to be more critical for seedling infections when the relative humidity is low [75,78,79]. The number of oospores can be estimated using the seed washing method [54,80,81]. Briefly, the seed is soaked in water under agitation. Seeds are removed with cheesecloth, the solution centrifuged, the supernatant discarded, and the pellet is dissolved in sterile water. The number of oospores is counted using a hemocytometer under the microscope. Calixtro (2017) quantified the number of oospores present on susceptible seeds and found it was three times greater than the number on tolerant varieties demonstrating that host genotype is an important factor [82].

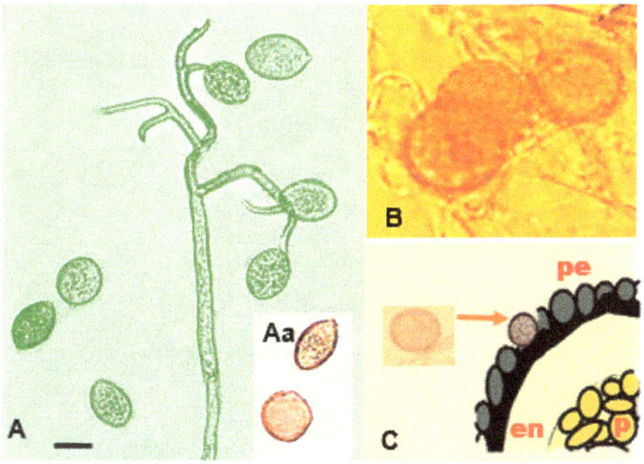

Figure 5. *Peronospora variabilis* spores isolated from *C. quinoa*. (**A**) Sporangiophore with lemon-shaped sporangia. (**Aa**) Oospores–sporangia size comparison. (**B**) Oospore on top of dried leaf tissue. Scale bars: 20 µm (Photos: C. Colque-Little). (**C**) Schematic representation of oospore localization in quinoa seed. o = oospore; pe = pericarp; en = endosperm; p = perisperm (illustration adapted from Danielsen and Ames (2004) [54] and Prego et al. (1998) [78]).

2.4. Peronospora variabilis Genotypic Diversity and Virulence Profiling

Peronospora variabilis is a genetically diverse group [30] with multiple population structures, in light of three facts:

1. Chenopod hosts have a vast degree of genetic diversity and plasticity [83,84].
2. *Peronospora variabilis* has great adaptability (climatically and geographically), hence its worldwide geographic presence [54].
3. The occurrence of sexual reproduction permits genotypic pathotype expansion [4].

Quinoa cultivation areas of the Andean region have resulted in severe infections under field conditions. Swenson (2006) collected 43 isolates from eight Bolivian regions [44]. Phylogenetic fingerprinting relationships revealed high genotypic diversity within a geographical region. The most recent fungal and oomycete identification initiatives were carried out using DNA sequencing [12]. A group of *P. variabilis* herbarium and isolates from different geographic locations (Argentina, Bolivia, Denmark, Ecuador, and Peru) were phylogenetically analyzed based on ITS rDNA sequences. The majority of the Danish and South American isolates were separated into two major clusters [29]. *P.variabilis* was detected in 31 out of 33 quinoa seed lots destined for human consumption and originated in six different countries. Subsequently, ITS and Cox2 phylogenetic relationships were examined to determine whether geographical differences occurred. ITS-derived phylogeny showed no genetic differences, but the Cox2 phylogeny indicated that geographical differences existed between US and South American samples [32]. In another study, researchers characterized 40 isolates from *P. variabilis* originating in the Andean highlands (Peru and Ecuador) and Denmark (Jutland, Sealand) using universally primed PCR (UP-PCR) fingerprinting analysis. A separation between the Danish and Andean isolates in two distinctive clusters was found, together with genotypic variations between isolates within each cluster [30].

In the future, the next step might be the virulence profiling of *P. variabilis*, achieved through the sequencing of its genome, followed by transcriptomic analysis. Progress in genome sequencing technologies can provide genome data to better understand how microbes live, evolve, and adapt. Indeed, the genome of three races of *P. effusa* (downy mildew of spinach) was recently sequenced, assembled, and annotated to gain insights into its gene repertoire and identify infection-related genes [9]. The genomes of microbial pathogens can vary greatly in size and composition; this also includes when closely related species are considered. In the case of *Peronospora*, species greatly vary between 45.6 to 159.9 Mb when estimates are made using image analysis of nuclear Feulgen staining [85]. Whether genome sizes have an impact on the lifestyle of *Peronospora* species is still unknown [86].

Another way to elucidate genotypic and phenotypic variation within pathogen populations is to use virulence-phenotypic assays with a standard set of differential hosts. Spinach downy mildew has such a set composed of 11 cultivars, maintained with the help of the international working group on *P. effusa* (IWGP) [86–88].

This organization invites researchers to use the set to identify new isolates that can later be nominated, tested for various criteria, and then given a race designation [86–88]. An international system for monitoring the virulence of *P. variabilis* has not yet been developed. However, Ochoa et al. (1999) made the first step towards this from a collection of twenty *P. variabilis* isolates that corresponded to different Ecuatorian ecoregions [40]:

1. An area where quinoa cultivation was not regularly practised. The least virulent strains were present here and were identified as virulence group 2 (V2).
2. A region where landraces and newly released cultivars were introduced. Only the most virulent strains belonging to group 4 (V4) were present here.
3. Fields located where landraces and newly released cultivars have been cultivated for many years. Here, all four virulence groups were present.

Ochoa et al. (1999) investigated seedlings under controlled environments from 60 selected genotypes and the above-mentioned *P. variabilis* collection; quinoa lines were selected for consistent compatible/incompatible reactions. Based on these results, four resistance factors (R1, R2, R3, and R4) were postulated [40]. It was most likely that two mating types are present. areasall difference exists. However, these genotypes are exclusive to the National Ecuadorian collection and thus not available for research. The measurements of severity and sporulation of downy mildew from reference cultivars (Puno, Titicaca, and Vikinga) and many other genotypes used by Colque-Little et al. (2021) are comparable to the 1−5 scale developed by Ochoa et al. (1999) (Table 4 and Figure 6). Therefore, we suggest that the presence of resistance factors could be preliminarily hypothesized on reference cultivars. Importantly, the seed of these cultivars is commercially available (Quinoaquality.com, Denmark) and could be established as an international reference set.

Table 4. Set of quinoa cultivars and *Chenopodium album* postulated for profiling the virulence of *Peronospora variabilis*.

Cultivar	Hypothesized Resistance Factors	Response to Downy Mildew		Origin
		% Severity	% Sporulation	
C. album	R1, R2, R3, R4	5	0.04	Denmark
Puno	R1, R2, R3, R4	11	0.2	Denmark
Rosa Blanca	R1, R2, R3	32	17	Bolivia
Blanca	R1, R2, R3	46	47	Bolivia
Titicaca	R1, R2	52	40	Denmark
Vikinga	R1	70	69	Denmark

Figure 6. Set of reference cultivars postulated for profiling the virulence of *Peronospora variabilis,* including *C. album* and two Bolivian cultivars with intermediate reactions. Leaves from three-week-old artificially inoculated plants. Numbers in red indicate the percentage of severity on the adaxial side, and those in purple indicate the percentage of sporulation on the abaxial side [52]. Numbers in green (incompatible response) and blue (compatible response) correspond to Ochoa's scale: 0 = no symptoms; 1 = 2–5-mm lesion with truncated mycelium in the mesophyll of the leaf; 2 = 4–8-mm chlorotic lesions with minor sporulation; 3 = medium-sized and confined chlorotic lesions with sporulation mainly on the abaxial side of the leaf; 4 = large, not clearly confined chlorotic lesions with sporulation mainly on the abaxial side of the leaf; 5 = mild chlorosis with abundant sporulation on both adaxial and abaxial sides of the leaf (Ochoa et al., 1999 [40], Colque-Little et al., 2021 [52]). Both assessments are comparable in terms of severity and sporulation; thus, the existence of resistance factors is hypothesized in this set of reference cultivars.

2.5. Disease Assessment under Controlled Conditions and in the Field

Reliable identification, followed by the assessment of disease, is the first step in efficient management. It is also an important component in the development of disease-tolerant quinoa varieties. It allows for crop-loss assessments and screening for host–pathogen interactions. Assessment methods must be in close agreement with the goals of the trial(s). Evaluations might differ according to the experimental setup. For seedlings, detached leaves and plantlets under controlled conditions and a disease assessment scale can be used (Figure 6). For the assessment of diseased plants in the field, it is necessary to take into account:

1. The phenological stage of the plants. Age-related resistance becomes relevant for biotrophic pathogens, which require healthy plant tissue to complete their cycle. [89,90]. The observation of symptoms should reflect the progression of the disease through periodical records, rather than observing its percentage of occurrence or incidence. For the quinoa/downy mildew interaction, it has been demonstrated that disease incidence has a low heritability H^2 = 0.4 and a low correlation with severity and

sporulation (0.67 and 0.65, respectively) [52]. Therefore, incidence or whole plant scores are unsuitable for this type of trial. To measure the area under the disease curve progression (AUDPC), a minimum record of three to four observations of disease severity is essential. A similar study has highlighted the importance of measuring the disease severity over time for other interactions, such as *Phytophthora infestans* infecting potatoes. The objective is to capture low, medium, and high infection levels in all the genotypes, including the susceptible ones [91].

Calixtro (2017) recorded high variability in the area under the disease progress curve (AUDPC) within the same quinoa accession during different phenological stages. The higher AUDPC values were seen at 104 days after sowing with favorable disease conditions [82].

Therefore, we suggest assessing downy mildew as soon as the first symptoms of the disease are visible. The first reading could be when nine pairs of leaves (BBCH 1–1.9) have emerged or beforehand in cases where disease symptoms are visually observed. Time intervals among subsequent readings depend on whether the disease advances slowly or quickly [91]. Other observation points could be during development (BBCH = 4) or visible inflorescence (BBCH = 5–5.9) and the last one at complete anthesis (BBCH 6–6.9). The phenological growth stages mentioned here correspond to the international quinoa-based coding system BBCH (Biologische Bundesanstalt Bundessortenamt und Chemische Industrie) [92].

2. The vegetative cycle of the plants. Late-maturing quinoa genotypes will display some degree of resistance [93] by increasing the latent period of the pathogen. Thus, readings for severity were taken ten days after infection instead of five in a recent study, in which a late cultivar Blanca was compared with the Danish cultivars Puno, Titicaca, and Vikinga [52] (Figure 7). Puno matures ten days later than Titicaca [94]. Cultivar Blanca is considered susceptible when additional time is given [52] (Figure 7). Vegetative cycle effects were also shown in another study that analyzed the mean-based cluster of inter-ecotype F2:6 population crosses and identified the following three clusters [48,95,96]:
   (a) Cluster one: consisting of late, mildew-resistant, high-yielding lines;
   (b) Cluster two: consisting of semi-late lines with intermediate yield and mildew susceptibility;
   (c) Cluster three: consisting of early to semi-late accessions with low yield and mildew susceptibility.

Therefore, for a proper comparison, quinoa lines with similar vegetative cycles should be screened in the same experiment or statistical adjustments should be carried out as part of the analysis. In addition, a positive control (susceptible variety) and a negative control (resistant variety) might be beneficial in the analysis.

3. Sampling method and sample size. Depending on the size of the experiment, there is no need to take severity readings in all the quinoa plants. Instead, consider the plot level and take readings on representative samples. Normally, 6−10 plants per plot are sufficient [54,91]. Next, an estimation of the percentage of affected foliage is required. Given the size of the plants and abundant foliage, it is not feasible to analyze the entire foliage; thus, it is recommended to perform scoring on individual leaves of the chosen plants [54]. Danielsen and Munk [97] evaluated various field assessment methods to predict yield losses due to downy mildew. The three-leaf method resulted in the highest negative correlation to yield (r = −0.736). Furthermore, disease progression relies on the successful infection of the host. It is often assumed that the susceptibility of host tissue is constant. However, in reality, it is a function of plant age and leaf position [14,98,99]. These responses might result from inducible plant defense responses, which occurs at the starting interaction site but also in distal, uninfected parts [99–101]. For these reasons, we suggest randomly choosing three leaves from the middle part (lower third, middle third, and upper third), as illustrated in Figure 8. Avoid the lower and upper extremities of the plant because they are prone to senescence/defoliation [97] and plant

defense responses, respectively. Next, estimate the percentage of affected leaf area using the attached scale [79] (Figure 9). The average value from the score of the three leaves becomes the percentage of severity for each plant.

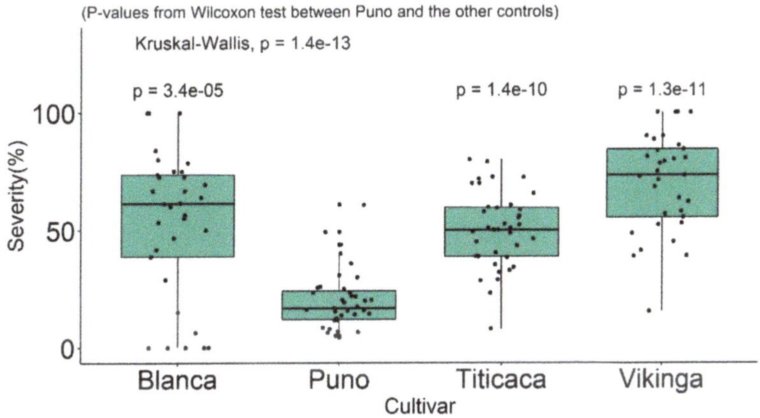

Figure 7. Reference cultivars' responses to infection with *Peronospora variabilis,* measured in mean severity under greenhouse conditions. Source: Colque-Little et al. [52].

2.6. *Yield Losses and Management*

The losses caused by downy mildew depend on the plant's phenological phase at the time of infection and the amount of resistance that the cultivar has [102]. Infection of susceptible cultivars may result in severe yields losses if the pathogen has favorable weather conditions, particularly high relative humidity [54]. If the infection occurs in the plant's initial growth stages, susceptible crops could completely fail; in less susceptible cultivars, the loss may fluctuate between 20% and 40% [4]. In a conventional intensive agriculture system of Cajamarca (Peru), between five to seven fungicide applications were needed to control the infection during the agricultural campaign [103].

Due to the high capability of *P. variabilis* for the proliferation and latent infection on *C. quinoa* and *C. album*, the scenario for low-input farming has only two options for disease control:

1. tolerant crop varieties; and
2. cultural practices (options on the list below).

Alternative cultural practices:

(a) Policymakers, smallholder farmers, and other stakeholders need resources for collective action for the establishment of a seed supply chain with quality standards (low levels of key seed-borne diseases). Experiences with complementary intervention such as capacity building and technical assistance have shown this influence in an appropriate conceptual model of sustainable production [104].

(b) The detection of *P. variabilis* on the seed is achieved using a simple method [32,54]. In the case of the presence of an oospore, treat the seed with a systemic fungicide [105]. For small samples, alternative treatments such as a hot water bath (50 °C–60 °C) could be considered for 10 to 30 min, as this method has been applied successfully to eradicate seed-borne pathogens of spinach [106]. After or without treatment, the addition of beneficial microbes by priming the seed with products such as commercially available *Trichoderma* can enhance the growth of the plants [107].

(c) Adjusting the space between rows and individuals, making the area less dense and increasing space between plants. In areas where the RH is as high as 80%, the minimum should be a 0.5-m space between rows and 0.15 m between plants [5].

(d) Avoiding excess water in the field;
(e) Implementing effective weed control, especially of alternate host *C. album*;
(f) Practicing crop rotation;
(g) Spraying the plants around 45 days after planting in areas with endemic infection as a preventive measure [69]. Use oomycete sensitive chemical control measures (e.g., Alietti) at principal growth stages, e.g., leaf development, inflorescence emergence, flowering, and fruit development [14,54,92]. Fungicides could be applied, alternating between systemic and contact products, starting with systemic products. Bio-pesticide or plant extracts could replace fungicides with a uniform and preventive application [5]. Inducers of resistance are an alternative [108].

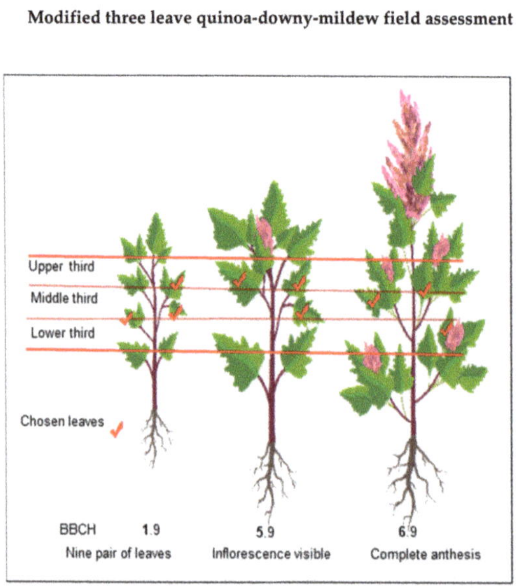

Figure 8. Modified from Danielsen and Munk (2004) [97]. Three-leaf field assessment method for quinoa-downy mildew at different growth stages.

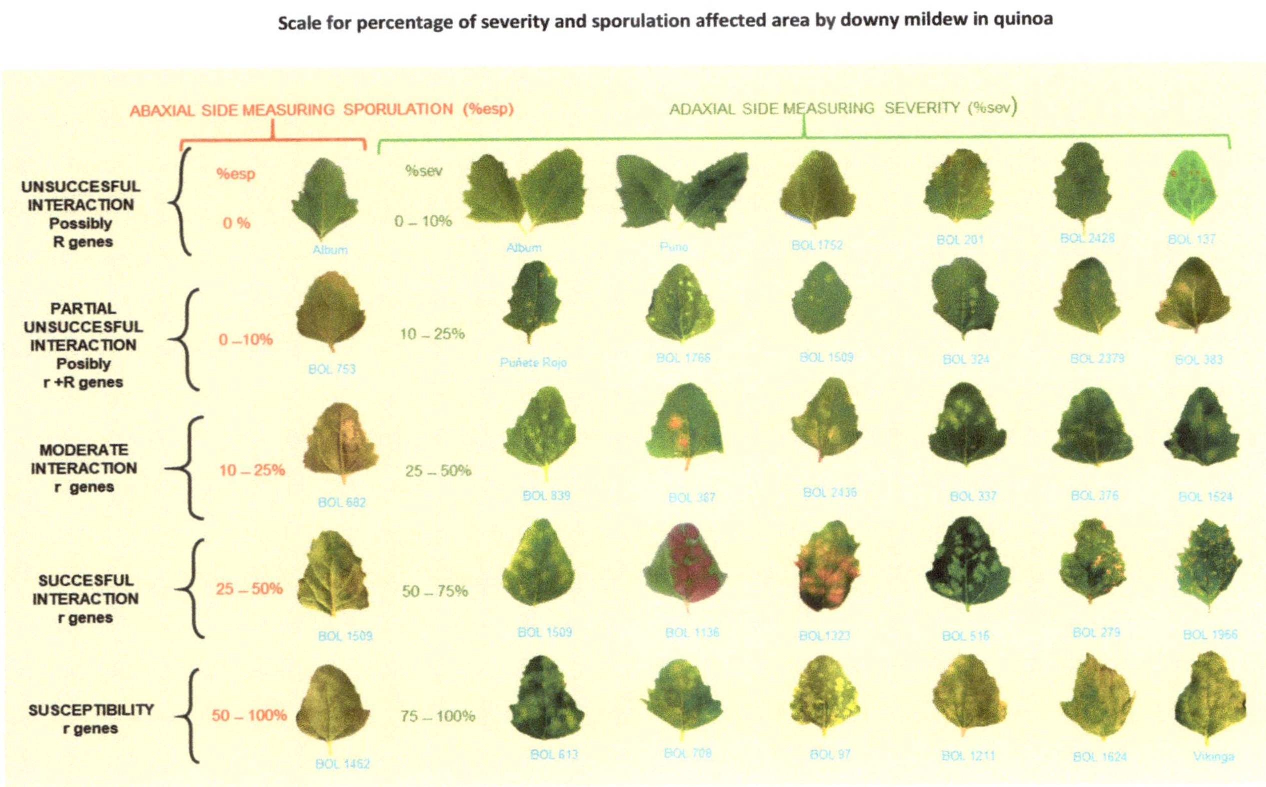

Figure 9. Scale for percentage of severity and sporulation area affected by downy mildew in quinoa. r = postulated minor genes; R = hypothesized major genes. BOL = accession numbers. **Note**: Percentage of sporulation is estimated on the abaxial leaf side area covered by visible lesion. It is not estimated on the total abaxial side leaf area (Colque-Little et al., 2021) [52]. Photos by Colque-Little.

2.7. Genetics of Resistance to Downy Mildew

For agriculture, field or host resistance is still the most important way of controlling diseases because it leads to the most cost-effective ratio for the grower [109–111].The response to downy mildew in a diversity panel of 132 quinoa genotypes resulted in strong phenotypic variation with high disease trait heritability (H^2 = 0.78 for severity, H^2 = 0.82 for sporulation). This variability was paired with the analysis of 603,871 SNPs in 61 genotypes with FarmCPU [52]. A single variant on chromosome 4, located above a threshold with a lack of siginificant marker trait wide associations. A single variant on chromosome 4, located above a threshold with a lack of significant marker-trait wide associations, suggested a polygenic architecture for the downy mildew interaction in agreement with other studies [43,48,49,95,104,112–115]. However the interactions of the host resistance pathway with a biotrophic pathogen (e.g., *P. variabilis*) are complex. The interaction oscillates between compatible (susceptible) and incompatible (resistant) states, because the genes involved can introduce quantitative variations, adding different levels of reactio to the extreme responses adding different levels of reaction to the extreme responses [116]. The same study phenotyped hypersensitive responses, most probably corresponding to R-genes, and very low sporulation on resistant genotypes, which could correspond to defeated R-genes [52] (Figure 3E,G). Indeed, Gabriel et al. (2012) characterized the quinoa/downy mildew pathosystem in field experiments and discussed the presence of R-genes, multiple r genes, defeated R-genes, and combinations, with the most common interaction being that corresponding to field resistance [93]. The deployment of different genes depends on many factors, such as pathogen isolate aggressivity [40] and environmental conditions. For quinoa downy mildew, it was demonstrated that the variance for genotype-by-experiment interaction σ2 G E was large, reflecting that even minute environmental changes can trigger a genotype to respond differently to the disease (Figure 10) [52]. The degree of resistance that the plant displays is determined by these changes interacting with the host genetic composition [116]. Furthermore, segregation in an F2 mapping population derived from a cross of saponin-free and bitter genotypes suggested that downy mildew resistance has a dominant inheritance [117].

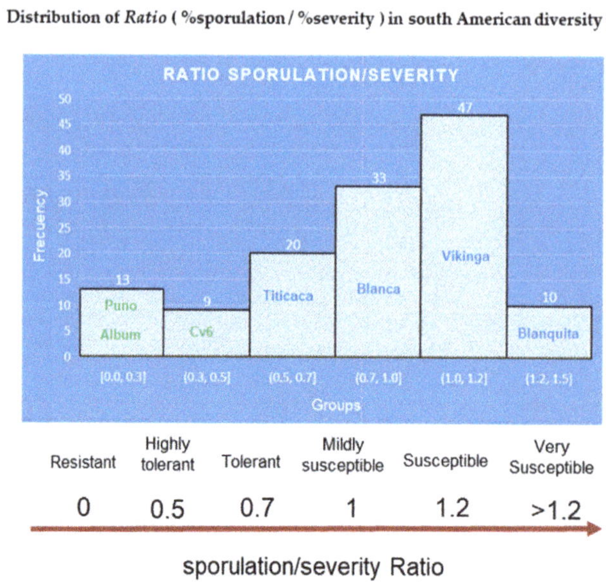

Figure 10. Ratio calculated from mean averages of sporulation/severity for the South American diversity panel. The names inside the histogram bars correspond to reference and representative cultivars for each group. Source: calculated with the data set from Colque-Little et al. (2021) [52].

Therefore, field phenotyping experiments of *P. variabilis* infections using diverse quinoa genotypes should include multiple environments and points in time. Using mixed modeling to detect quantitative trait loci (QTLs) by considering them as random samples from a population of target environments and time could be one alternative [118]. Under controlled conditions, it would make sense to use elite diversity panels with replicates, reference cultivars, and genetically diverse pathogen isolates in a series of experiments that are designed randomly.

The characterization of a south American panel demonstrated robust differences between the genotypes for all disease traits [52] (Figure 11). Moreover, at least five cultivars that were released by Bolivian breeding programs [112,119], which included downy mildew tolerance, showed moderate to low severity and reduced the reproduction of the pathogen. Interestingly, the incidence (Figure 11A,C) and severity of cultivars 6, 17, and 18 might have classified them as susceptible, but their ability to prevent the pathogen from multiplying conferred them some degree of resistance [110] (Table 5). Indeed, the Danish variety Titicaca was classified as susceptible through the solely detached leaf sporulation assay [48]. When the assessment was done as a function of both parameters, by calculating the ratio (R = %sporulation/%severity), Titicaca's $R = (40/52) = 0.77$ showed that it is not completely susceptible (Table 5). This finding suggests that the scoring of both parameters in plantlets can contribute to better disease assessments of cultivars.

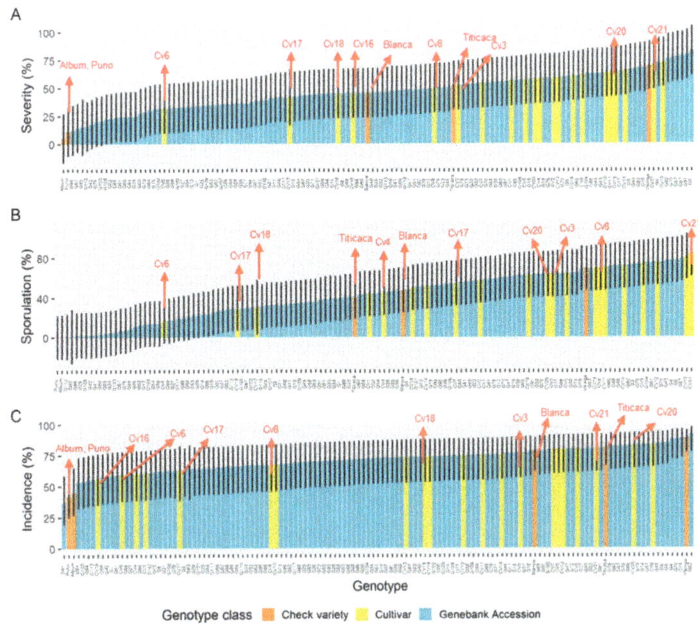

Figure 11. Disease traits estimated means fitted on a generalized linear mixed model (GLMM) for a diversity panel, comprising gene bank accessions (landraces), cultivars (Bolivian-bred cultivars), and check varieties (reference cultivars). (**A**) Severity of infection, (**B**) sporulation, and (**C**) incidence of infection. Error bars represent 95% confidence intervals. Adapted from Colque-Little et al. (2021) [52].

Table 5. Phenotypic infection traits and Ratio for representative cultivars and reference varieties.

Name	% Severity	% Sporulation	Spo/Sev Ratio	% Incidence	Ratio Based Classification
C. album	5	0.4	0.08	45	Resistant
Puno	11	0.2	0.02	42	Resistant
Cv6 (Rosa Blanca)	32	17	0.53	59	Highly tolerant
Cv17 (Canchis)	41	30	0.73	73	Mildly resistant
Cv18 (Pandela Roja)	45	29	0.64	74	Mildly resistant
Cv16 (Kurmi)	45	50	1.1	56	Susceptible
Blanca	46	47	1	79	Mildly susceptible
Cv8 (Blanquita)	50	69	1.4	67	Very susceptible
Titicaca	52	40	0.77	81	Mildly resistant
Cv3 (Ayrampu)	52	63	1.2	77	Susceptible
Cv20 (Aynoka)	58	63	1.1	83	Susceptible
Cv21 (Mariqueña)	71	84	1.2	82	Susceptible

Therefore, we propose using the (R = % sporulation/% severity) ratio to better rate elite genotypes in breeding programs. Using the data set from a previous study [52], the ratio was calculated. Histograms separated the diversity panel into six groups and derived a ratio-based scale (Figure 10). The bimodal distribution displayed by the histograms is consistent with previous findings for *P. variabilis* field interactions [44].

Quinoa cultivation in South America occurs in agro-climatological polar regions. These regions have been classified according to their soil type, rainfall, and temperature as Northern, Central, Southern highlands, and Andean slopes (Table 6) [4,96,112,119]. The Andes have heterogenic topography; their altitude ranges between 3200 and 6500 m above sea level; hence, there are variations in temperature and humidity [120]. Indeed, temperature decreases at a rate of 0.7 °C for every 100-m increase in altitude in Chile's Tarapaca region. Therefore the coastal Atacama littoral plains differ from mountain sites (e.g., Los Condores) which enjoy fog oases and lomas vegetation [121]. A similar situation is expected for the slopes in the Andes of Bolivia, Peru, and Ecuador.

Table 6. Eco-regions for quinoa production in South America.

Eco-Region	Soil	Altitude m.a.s.l	Rainfall (mm)	Temperature Max.	Temperature Min.	Temperature Av.
Northern Highland shores of Lake Titicaca	Rich in organic matter	3500–4000	500	14	4	7
Central Highland	Slightly acid	3300–4100	350	17.7	−2	8.7
Southern Highland	Arid and poor soils	3200–4000	50–200	18	−11	5.7
Andean Slopes	Variable	800–3200	3500–700	12	3	7.6
Coastal/Lowland Northern, Central, and Southern	Variable	Sea level to Mountain range	40 > 2000	23 21 17	−8 7 6	4.5 14 11

Within the sub-regions, temperatures vary depending on location (coast or foothills), not shown. Source: elaborated with information from Gandarillas et al. (2015) [112]; Seiler et al. (2013) [120]; Cereceda et al. (2008) [121]; http://germoplasma.iniaf.gob.bo (accessed on 15 April 2020).

Quinoa ecoregions were inferred from information provided by passport data (germplasma.iniaf.gob.bo-GRIN global, accessed 15 April, 2020) and the characterization of Bolivian and Coastal ecoregions [112,120,121]. The information is summarized in Table 5. Disease traits data (mean values of severity and sporulation) from the South American diversity panel [52] were analyzed with the Tuckey test for their relationship with the seed-ecoregion collection site. For this analysis, we used Inti-Yupana for R [122], and the results pointed at significant differences for the variables. The graph represents data from means of sporulation only because data from means of severity was very similar (Figure 12).

Even though the sample size from the central highlands was overrepresented and the Coastal sample size was underrepresented, significant differences ($p = 0.05$) for severity and sporulation were detected. The most resistant genotypes from the South American diversity panel came from the coastal/lowland and northern highlands ecoregions. The northern highlands are the most humid since they are close to Lake Titicaca. This ecoregion is suitable for pathogen infections and disease pressure. This outcome is in agreement with previous reports [4,48,54,93,95]. The Danish cultivar Puno was found to be resistant, as reported elsewhere [48]. Moreover, principal component analysis of genome-wide association studies (GWASs) demonstrated that Puno is genetically close to Chilean coastal lines and separated from highlander genotypes [52]. The central highlands showed the largest quantity of susceptible genotypes, likely because they were also the most numerous. However, a few genotypes with a large amount of sporulation came from the southern highlands of Bolivia (i.e., G16, G17, and G82) [52] (Figures 11B and 12B).

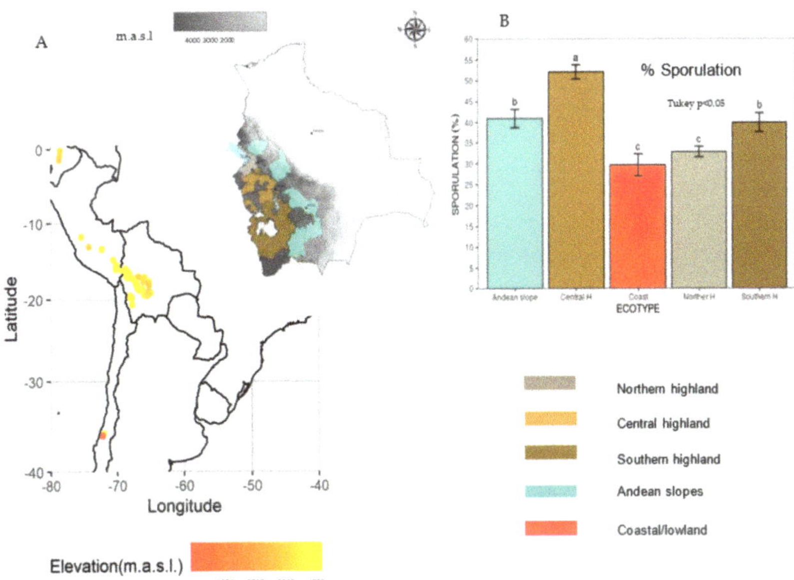

Figure 12. (**A**) Modified map of germplasm bank accession across South America by elevation Source: Colque-Little et al. (2021) [52] and modified map of Bolivian ecoregions for quinoa production. Source: Gandarillas et al. (2015) [6]. (**B**) Mean sporulation on diversity panel related to quinoa ecoregions calculated with Tukey test ($p = 0.05$). Different letters (a,b,c) represent significant differences between the sporulation produced by genotypes coming from different ecoregions when infected with *P. variabilis*.

In conclusion, the genetic improvement of quinoa for downy mildew tolerance is possible because resistance is present in multiple genotypes, but a virulent pathotype might overcome it. Other options to consider are discovering, transforming, and deploying resistant alleles existent in wild species such as *C. albums* [52,123,124]. Because tolerant varieties seem to delay and reduce the disease progression, inducers of resistance [125,126] could be a feasible option [108].

3. Ascomycete Fungi

3.1. Fungi Identified by Molecular and Morphological Approaches

3.1.1. Ascochyta Leaf Spot and Black Stem (*Ascochyta hyalospora* and *A. chenopodii*)

At least two *Ascochyta* species infect quinoa, causing quinoa leaf spot (described below) and black stem (described in next section). Quinoa leaf spot is either caused by *Ascochyta hyalospora* or *A. chenopodii*. *A. hyalospora* Coole and Ellis is an *Ascomycete*, class Dothideomycetes, order Pleosporales. It was first found as a seed-borne pathogen of *C. quinoa* from the Bolivian central highlands, for which a blotter test (seed incubation method on well-soaked filter papers [127]) revealed 8%–26% of infection. It was identified morphologically, followed by pathogenicity tests causing whitish leaf spots 5 dpi, followed by pycnidia at 10 dpi, and necrosis on leaves and stem of *C. quinoa* and *C. album* plants [128]. Testen et al. (2013) isolated a fungal pathogen from quinoa fields in Pennsylvania, USA, and through DNA sequencing of the ITS1-2 region matched it to *Ascochyta sp.*, and reported that it resembled the morphological characteristics of *A. chenopodii* and *A. caulina*, which at the time of identification had no DNA bar-codes available for comparison. However, the ITS1-2 sequences from Testen et al. (2013) were not released as GenBank sequence data [129]. Thus, it is still not possible to make the comparison.

Ascochyta hyalospora pycnidia are globose to subglobose, usually 17.5 to 25 μm in diameter [128], and contain sub-hyaline to light-brown-colored conidia. The conidia are cylindrical to ovoid, measuring 19 × 7.5 μm [129] and 25 × 10 μm [128] on average. They often have one to two septa and less commonly have three septa. Boerema (1977) noted that the conidia formed on leaf spots after artificial inoculation were longer (35 μm) and often had two or three septa (Figure 13E,F). Lesions on the leaves are of irregular shape, and are bronze to reddish-brown with darker edges. Spots eventually turn necrotic. Thereafter, numerous black pycnidia, distributed randomly in each lesion, can be seen [129].

Figure 13. Leaves showing symptoms of infection caused by *A. hyalospora* (**A**) on the adaxial side of the leaf and (**B**) on the abaxial side. (**C**) Stems showing pycnidia and brown stalk. (**D**) Stem showing pycnidia. (**E**) *A. hyalospora* conidia (Photos: Testen, 2020) [77]. (**F**) *A. hyalospora*: (**a**) pycnidium (×200); (**b**) conidiogenous cells of pycnidium (×1000); (**c**) conidia from pycnidium (×400); (**d**) bi and tri-septate conidia from pycnidium on an inoculated stem of *C. quinoa* (×400); (**e**) conidia from pycnidium on leafspot of inoculated leaf of *C.quinoa* (×400). Source: photos (**A**–**E**) provided by A.L. Testen. F. Adapted from [128].

The stems show necrosis, and the pycnidia are visible to the naked eye (Figure 13A–D). The seeds turn brown, and pycnidia are observed at the stereomicroscope [128,130]. *Ascochyta* leaf spots have been considered of minor importance in the Andean region [3,4,6,131]. In 2014, large-scale cultivation (12,000 ha) of quinoa started in China [132], where the production was affected. Infected foliage decays and falls, leaving the plant defoliated [5]. Effects on quinoa production in the USA have not yet been assessed [129]. Experiments in Bolivia showed that the germination rates of seeds from infected plants were reduced by 6% to

10%. Moreover, the disease was transmitted to seedlings [5]. One possibility for control would be the use of high-quality seeds, since the pathogen is seed-borne [130].

3.1.2. Quinoa Black Stem (*Ascochyta caulina*)

Molecular and phylogenetic analysis of representative isolates from quinoa black-stem revealed that its causal agent is *Ascochyta caulina* (van der Aa and van Kesteren 1979). Its sexual teleomorph stage is called *Neocamarosporium calvescens* (de Gruyter et al. 2009), previously known as *Pleospora calvescens*. The taxonomic status of *P. calvescens* has changed recently, based on multigene analyses. It has been established in the genus *Neocamarosporium* Crous and Wingfield in 2014, which comprises 15 species, including *N. betae*, *N. chenopodii*, and *N. calvescens*. These species share the same large phylogenetic branch with *N. calvescens* [133–137]. *Ascohyta caulina* in its asexual form belongs to the family Didymellaceae and has often been confused with *A. hyalospora* [137]. Previously, it has also been found to infect eight species of *Atriplex* and eight species of *Chenopodium*, including *C. album* [138].

Another report [139] on *A. caulina* was accomplished through a morphological description of the isolate found on quinoa seeds of cv. Cochabamba of Bolivian origin (stored at the Gene Bank of the Research Institute of Crop Production in Prague-Ruzyně). For pathogenicity tests, the isolate was inoculated in seedlings, and symptoms were reproduced. Interestingly, quinoa seeds from the University of Copenhagen, Denmark, analyzed simultaneously, were free of *A. caulina* [139]. This finding might indicate that the disease is not present in Denmark. *A. hyalospora* pycnidia are rigid structures, grayish-white or light brown, spherical or pear-shaped, and have a single chamber. They are 162 × 134 μm in size, on average. Conidia are elliptical or fusiform, light brown, oblong at the top and flat at the base, and measure 17 × 6 μm on average [137]. Conidia usually have one septum, which is erect or curved (Figure 14D). The optimal conditions for its germination are between 15–25 °C, RH = 60%. Compared to *A. hyalospora* leaf spots, black stem lesions were more likely to develop under cooler conditions [140]. Pathogenicity tests on detached stems of *C. quinoa* showed typical symptoms 10 dpi and were densely covered with pycnidia. At 15 dpi, typical symptoms appeared on the stems of plants inoculated in outdoor conditions. Detached inoculated leaves of *C. quinoa* developed visible symptoms 8 dpi and were grayish white. However, necrotic lesions are rarely seen on the leaves in the field [137] (Figure 14A).

Figure 14. Typical symptoms of quinoa black stem in the fields of China. (**A**) Symptoms induced by inoculation of *A. caulina* on *C. quinoa* (left of midrib) and on *C. album* (right of midrib). (**B**) 10 dpi diamond shaped lesion on quinoa stem with presence of pycnidia. (**C**) Necrotic quinoa stem prior to lodging; (**D**) morphological characteristics of conidia and pycnidia of *A. caulina*. Source: illustrations based on pictures from Yin et al. [114].

Quinoa black stem primarily infects the stem; lesions are recorded at the flowering stage up to maturity. Symptoms first appear at the lower and middle parts of the stalk, subsequently moving upwards. They are diamond-shaped, pale or tan, and present slight depressions, as the plants are prone to drying and consequent shrinkage. The diameter of the lesion averages 7.9 cm.

The stem lesions turn necrotic in later stages and are accompanied by abundant small round protrusions of black pycnidia (Figure 14B,C). In severe cases, lesions wrap around the stem, causing lodging, foliar chlorosis, leaf abscission, and the development of "empty" and sterile grains on the panicle [137].

Quinoa black stem is considered a newly emerging disease in Chinese regions (Jingle County, Shanxi province), where the disease was severe. The incidence was around 80% and the yield was reduced by 45% [137,140]. The fungicides mancozeb and azoxystrobin are shown to have a strong inhibitory effect on conidia germination, whereas tebuconazole and difenoconazole were most effective towards mycelial growth in tests performed in vitro [137].

Sixteen European countries concentrated integrative approaches for the biological control of the weed *C. album* from 1994 to 1999. The European Research Programme (COST-816) concerted the use of a combination of *A. caulina* with ascaulitoxin for this purpose [141,142]. Experiments using *A. caulina* as a microbial herbicide were up to 70% successful in reducing field conditions, as it was able to kill its host in one week [143–146].

3.1.3. Cockerel Eye/Quinoa Cercorporoid Leaf Spot

Quinoa leaf spot was first reported in Ecuador (2009) and given the Spanish common name "ojo de gallo", or cockerel eye, because of the symptoms exhibiting a dark center and round shape. It was then associated with *Cercospora* spp. [147]. The genus *Cercospora* was established by Fresenius (1863) and belongs to the family Mycosphaerellaceae, class Ascomycota. A comprehensive list of cercorporoids assembled in Poland included a species under the name *Cercospora chenopodii* Fresenus, 1863, found on *C. album* [148].

Testen et al. (2013) amplified the ITS1-2 region of strains isolated from quinoa field plots located in Pennsylvania, USA, and identified them as *Passalora dubia* (Riess) U. Braun (GenBank EF535655). Conidia were septate, hyaline, and measured 25−98 μm long × 5−10 μm wide—with an average of six cells per conidium (Figure 15D). Disease symptoms of leaves were round to oval with a diameter of less than 1 cm, and were brown to gray-black with darker brown or reddish borders (Figure 15A−C). In addition to quinoa, *P. dubia* has also been isolated from *C. album* [77,149].

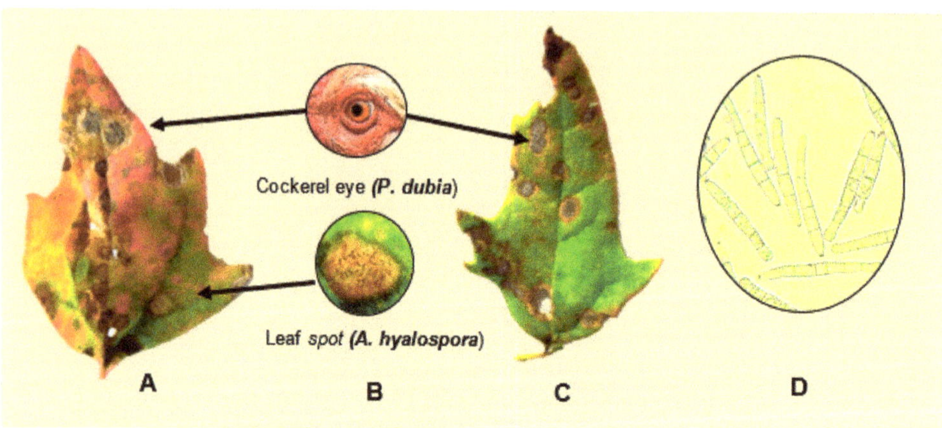

Figure 15. (**A**,**C**) depict symptoms of *P. dubia* on leaf tissue. (**B**) Comparison of "cockerel eye" and leaf spot symptoms. (**D**) Conidia of *P. dubia*. Source: pictures (**A–C**) provided by Testen and (**D**) Testen [77].

A pathogen identified as *P. dubia* has been tested as a microbial herbicide for the biocontrol of *C. album* in Europe. It was shown to reduce *C. album*'s dry weight by 20% [143].

3.1.4. Cercospora Leaf Spot Caused by *Cercospora* cf. *chenopodii*

Cercospora leaf spot, infecting quinoa in Shanxi, China, was classified as *Cercospora* cf. *chenopodii* based on multi-loci sequencing and phylogenetic analysis using LSU rpb2 and ITS as target genes. The qualifier "cf" indicates a provisional identification [150], even though most diagnostic characteristics correspond to *C. chenopodii*. At the early onset, the lesions were nearly round and pale yellow to light brown. Later, the lesion became grayish brown, with a slightly elevated surface, a yellow halo, and an average diameter of 5.4 mm. The pathogen's conidia were observed to be septate and hyaline to brown. They were 40.01 × 7.99 μm on average. They contain an average of four cells per conidium (Figure 16C). Spore suspensions made in glycerin causes disease symptoms 5 dpi, spreads quickly, and produces large yellow lesions, which causes defoliation 10 dpi. Optimum temperatures for infections are 22–26 °C, with a high relative humidity (75−80%) [151]. Based on multigene phylogeny (LSU, rpb2, ITS, cmdA, and other genes), various *Passalora* species have been proposed to be re-classified as *Cercospora* Fresen. *P. dubia* is included in this phylo-group and is considered synonymous with *Cercospora* cf. *chenopodii* [151–153].

Figure 16. (**A**,**B**) Foliar symptoms of *Cercospora* leaf spot. (**C**) Condidia of *Cercospora*. Source: illustrations adapted from Yin et al. (2019) [151].

3.1.5. Quinoa Anthracnose Caused by *Colletotrichum nigrum* and *C. truncatum*

Stem lesions have been observed on quinoa plants growing in Ames, Iowa (USA). Symptoms are recognized as oval to linear, slightly narrow at the ends, light in color, silvery-white to dark gray, and are slightly sunken in lesions. They contain setose acervuli. Two isolates (CQ1, CQ2) were cultured in V8 media for the subsequent examination for their morphological characteristics and DNA barcoding [154].

CQ1 mycelia were gray, sparse and flat. They produced abundant sclerotia and conidia. The conidia were cylindrical, hyaline, and aseptate. The size of 50 conidia averaged 21 × 4.3 μm. CQ2 mycelia were gray to dark and fluffy. They produced abundant sclerotia, acervuli, and conidia. The conidia were falcate, hyaline, and aseptate. The size, averaged from 50 conidia, was 26.8 × 2.4 μm [154]. Both isolates have been identified by multigene sequencing, and the multiple sequence alignment of vouchered CBS isolates generated a maximum likelihood phylogenetic tree. Based on this information, CQ1 was identified as *Colletotrichum nigrum* and CQ2 as *Colletotrichum truncatum*. The sequences' GenBank vouchers are: MN581860, MK675238, MF682518, and MK118057 [154].

For the completion of Koch postulates, 40-day-old quinoa plants (PI 634920) were inoculated on stems and leaves. Two weeks later, the stems showed bleached to tan sunken areas on wounded sites. After an extra week under humid conditions, the plants inoculated with *C. nigrum* produced acervuli (asexual stage) and sclerotia, whereas *C. truncatum* produced only acervuli. Infected stems were cultured in artificial media. The morphological characteristics of grown mycelia matched those of the initial inoculum used on the plants. Inoculated detached leaves developed brownish, circular lesions. This disease may cause lodging and emerge in new quinoa production areas, resulting in yield losses [154].

3.2. Fungi Identified by Morphological Approaches

3.2.1. Brown Stalk Rot

Brown stalk rot was observed in *C. quinoa* growing in rotation with potatoes in the highlands of Puno, Peru, in 1974 and 1975. The organism was isolated from diseased stems of *C. quinoa* bearing pycnidia. As a practical first step for identification, the alkaline substance 1 *M* NaOH was added dropwise. Its purpose was to demonstrate the presence of substance "E" (a colorless metabolite from exigua) in malt extract agar cultures of the fungus to distinguish it from *Phoma exigua* var. *exigua* [155]. The test gave a positive result for *P. exigua* var. *foveata*, and comparative morphological characteristics with the causal agent of potato gangrene were carried out [156]. As both were similar and pathogenicity tests on potatoes were positive, the quinoa brown stalk rot's causal agent was identified as *Phoma exigua* var. *foveata* (Foister) Boerema. Furthermore, isolates were sent to the Dutch Protection Service and the Commonwealth Mycological Institute (UK) for final confirmation [157].

Symptoms were described as follows: small lesions on the higher third of the stem progress until reaching the upper part. At this stage, pycnidia are visible, the foliage wilts, the panicle does not form grain, and the brown stalk is prone to break (Figure 17A). The pycnidia are globose and dark brown; their size ranges between 101−116 μm in diameter. The ostiole is 30 μm in diameter, and the pycnidiospores are hyaline, ellipsoidal, unicellular, and biguttulated (small drop-shaped).

Figure 17. (**A**) Brown stalk rot. (**B**) Diamond-shaped symptoms bearing pycnidia. Source: illustrations adapted from Alandia et al. [4].

Their average size ranged between 6 × 2.2 μm in artificial media and 6.8 × 2.3 μm when coming from infected stems. Cross-inoculations, aided (and not aided) with mechanical wounds, were performed on potato plants and tubers, tomato plants, beetroot, sugar beets, and quinoa. Quinoa plants showed symptoms 3 dpi, potatoes and tomato plants showed foliar blight, potato tubers got black rot, whereas beetroot and sugar beets showed no symptoms. Overall, mechanical wounds increased the rate of infection, but pycnidia were rarely observed. The disease developed better at 3–5 °C than at 15–20 °C [157].

Based on in vitro experiments, it was hypothesized [157] that the highlands of South America are the geographic origin for the potato gangrene fungus *Phoma exigua* var. *foveata* because it is as pathogenic to potatoes as the virulent European isolates. However, on *C. quinoa* and *C. album*, it was more pathogenic. After inoculation, it caused a brown discolored area of rotting tissue, 1−3 cm long on both hosts, four dpi. On older leaves of both *Chenopodium* spp., concentric leaf spots of 0.5−1.0 cm in diameter were visible. The European strain caused similar spots, but one week later [158].

3.2.2. Quinoa Diamond Black Stem/"Mancha Ojival del Allo"

Diamond black stem was observed in *C. quinoa* in the highlands of Puno, Peru, in 1974 and 1975. The disease is primarily present in the stem, with diamond-shaped lesions (2−3 cm), whitish to gray in the center, with brown edges and a vitreous halo. They bear pycnidia. At a later stage, the lesions join around the stem, causing it to collapse [133] (Figure 17B).

3.2.3. Sclerotium in Quinoa

Stem rot affecting quinoa plants was observed at the Experimental Station of Kayr'a (Cuzco, Peru) during 1997. The mycelium was cultured, and pathogenicity tests were carried out on three-month-old plants of quinoa, amaranth, potato, frejol, sunflower, and *Lupinus mutabilis*. All plants were infected, and quinoa was the most susceptible. Morphological comparison with *Sclerotinia* from potatoes allowed the morphological identification of *Sclerotinia* sp., currently known as *Whetzelinia sp*. Inoculation with ascospores was followed by mycelial growth after 17 days. Dark sclerotia measuring between 4−9 mm appeared five dpi in PDA cultured at 10 °C. Apothecia developed 53 dpi at 16 °C and 12 days later produced ascospores. The fungi caused dry rot in the stem's neck in quinoa, leaves wilted, and the disease moved towards the panicle [159].

3.2.4. Damping-Off

1. Sensitivity of *Pythium zingiberum* and *P. butleri* oospores: Soil inoculation of oospores of *P. zingiberum* and *P.butleri* on soil caused damping-off of susceptible *C. quinoa* seedlings after ten days of incubation at 30 °C [160].
2. Seedling damping-off caused by Fusarium avenacearum and Pytium aphanidermatum: The fungi were isolated from infected stems of quinoa seedlings grown in a greenhouse. Microbes were morphologically described and the cultured fungi were inoculated on *C. quinoa* cv. Cochabamba. Pathogenicity tests confirmed that *P. aphanidermatum* and *F. avenaceum* were the causal agents of the damping-off of quinoa seedlings under greenhouse conditions. The seedling infection was significantly higher up to the first pair of leaves, showing that quinoa is most susceptible to the pathogens before emergence. However, the sum of post-emergence damping-off was significantly lower than that observed in sugar beets and higher than that observed in cabbage plants, except for *F. avenacearum*, which also produced marked susceptibility at the first true leaves stage. In addition to the two pathogens, *Ascochyta caulina*, *Fusarium spp*., and *Alternaria spp* were also isolated from infected tissue but could not infect quinoa seedlings during pathogenicity tests [139].
3. Pathogenicity tests on seedlings infected by *Rhizoctonia solani* and *Fusarium* spp. *Rhizoctonia solani* was isolated from the field in Peru. Pathogenicity tests performed in a greenhouse showed that *R. solani* prevented seed germination. It also created

sunken lesions on the stems of old plants at ground level. *Fusarium spp.* reproduced wilting in old plants [4,161]. Quinoa seedling damping-off (Figure 18A) was observed during field experiments conducted at the experimental station of Nihon (Japan). It occurred from emergence until the four-leaf stage and increased under high soil moisture conditions. *Rhizoctonia* spp. (Figure 18B) and *Fusarium* spp. (Figure 18C) were identified morphologically from the symptomatic lesions [162].

4. Pathogenicity tests on seedlings caused by *Sclerotium rolfsii* Sacc *Sclerotium rolfsii* was isolated from diseased seedlings of *C. quinoa* in a field of Southern California. The susceptibility of *C. quinoa* to *S. rolfsii* was demonstrated in vitro and under greenhouse conditions [163].

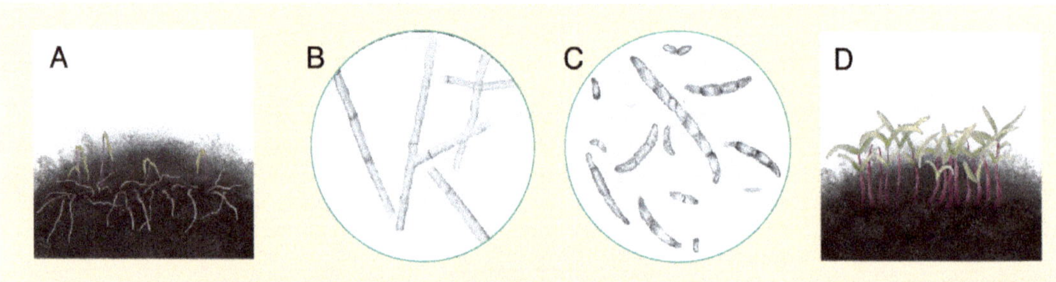

Figure 18. (**A**) Quinoa seedlings affected by damping-off. (**B**) *Rhizoctonia* spp hyphae. (**C**) *Fusarium* spp. spores. (**D**) Healthy quinoa seedlings growing under low soil moisture conditions. Source: illustration adapted from Isobe et al. (2019) [155].

4. Chemical Control for Oomycetes and Fungi

The control of oomycete and fungal diseases continues to rely mainly upon chemical measures for conventional agriculture. Fungicides can be effective only on a few closely related pathogens, in which case they are designated as narrow -spectrum fungicides, are often systemic, and usually have a single-site activity. In contrast, broad-spectrum fungicides can control a wide range of unrelated pathogens. In the case of oomycetes, 16 chemicals with different modes of action, translocations in the plant, types of activity, and risks of developing resistance, are available. A summary is presented in Table 7. However, sexual recombination, resulting in high pathogenic genetic diversity, as well as the migration rate, including low dispersal (within a few meters), increases local epidemics and the appearance of new pathogen genotypes in local populations. These facts will continuously affect sensitivity to fungicides, requiring the repeated adaptation of control strategies [105]. Several different target-site fungicides can achieve the chemical control of fungal pathogens in a mixture or in an alternating regime on the same crop. The most recent fungicides of this type are phenyl-pyrroles (P.P. fungicides) and dimethylation inhibitors (DMIs). They are considered the most effective chemicals registered to control diseases caused by *Ascomycetes* [164–166], depicted in blue on Table 7. The table aims to provide a general reference. The choice of fungicide is highly dependent on the availability and conditions of the particular fields to be treated.

Table 7. Major fungicide groups and key active ingredients, application site, and resistance risk. Adapted from Gisi and Zierotski (2015) [105]; Lebeda and Cohen (2021) [165]; Plimmer, (2003) [166]; and Masielo et al., (2019) [164]. Rows in blue correspond to fungicides that are effective against *Ascomycetes*.

Mode of Trans-location	Fungicide Group and Key Active Ingredients	Resistance Risk [a]	Foliar	Seed	Soil	Type of Activity	Translocation in Plants	Biochemical Mode of Action
Fully Systemic	Phenylamides: Metalaxyl, mefenoxam, oxadixyl, benalaxyl, kiralaxyl, ofurace	High	√	√		Preventive, curative, eradica-tive	Apoplastic, symplastic, translami-nar	Inhibition of rRNA synthesis
Partially Systemic	[b] Quinone outside inhibitors: Azoxystrobin, fenamidone, famox, adone, trifloxystrobin: kresoxin-methyl, Pyraclostrobin		√	√		Preventive	translaminar apoplastic	Inhibition of mitochondrial respiration at enzyme complex III
Non-Systemic	[b] Multisites: For example, mancozeb; chlorothalonil, copper, cu-oxychloride, cu-hydroxide; folpet; thiram, chlorothalonil	Low	√			Preventive		Multi-site inhibition
Non-Systemic	Carboxylic acid amides: Dimethomorph, flumorph; iprovalicarb, benthiavalicarb; mandipropamid	Moderate	√			Preventive	Translaminar	Cell wall synthesis, Ces3A cellulose synthase inhibition
Fully Systemic	Cyanoacetamide, oximes (cymoxanil)	Moderate	√	√		Preventive, curative	Apoplastic, symplastic, translami-nar	Inhibition of mitochondrial respiration at the enzyme complex III
Non-Systemic	Dinitroanilines (fluazinam)	Moderate				Preventive		Inhibition of ATP production
Fully Systemic	Phosphonates (fosetyl-Al)	Moderate	√			Preventive, curative	Apoplastic, symplastic,	Inhibition of spore germination, retardation of mycelia
Partially Systemic	Quinone inside respiration inhibitors: Cyazofamid, amisulbrom	Medium to hight	√	√		Preventive, curative, eradica-tive/	Translaminar	
Fully Systemic	Benzamides (fluopicolide)	Mod.	√		√	Preventive, curative	Apoplastic, symplastic, translami-nar	Delocalization of spectrin-like proteins
	Benzamides, carboxamides Ethaboxam, zoxamide	Low						
Systemic	Hymexaxol (heteroaromatics)			√	√			Fungal RNA and DNA syntheses
Contact	[b] Thiadiazoles (Etridiazole)				√	Preventive, curative		Lipid structure of Mitochondria
Resistance inducer	Acibenzolar-S-methyl.			√				

Table 7. *Cont.*

Mode of Trans-location	Fungicide Group and Key Active Ingredients	Resistance Risk [a]	Foliar	Seed	Soil	Type of Activity	Translocation in Plants	Biochemical Mode of Action
	[b] Demethylation inhibitor fungicides (DMIs): Imidazoles, triazolinthiones, triazoles prothioconazole, prochloraz, terbuconazole, difenoconazole		√	√		Preventive, curative		Sterole biosynthesis in membranes
	[b] PP fungicides (Phenylpyrroles) phenylpyrroles Fludioxonil		√	√		Preventive, curative		Signal transduction
Fully Systemic	Carbamates: Propamocarb, prothiocarb					Preventive, eradica-tive	Apoplastic	Multi-site inhibition Affecting the membrane

[a] Nomenclature according to Fungicide Resistance Action Committee mode of action code list, 2014, www.frac.info (accessed on 10 June 2021). [b] Quinone outside inhibitors and multi-sites are broad-spectrum fungicides, including activity against fungi.

5. Bacteria

5.1. Bacterial Leaf Spot Caused by *Pseudomonas* spp.

Bacterial leaf symptoms are small irregular spots both in leaves and stems. In leaves, they turn dark brown with concentric rings and a wet halo; in stems, they become necrotic, causing a deep lesion and wilting [133].

5.2. Bacterial Leaf Spot Caused by *Pseudomonas syringae*

Bacteria were isolated from symptomatic leaves and inoculates on surface sterilized leaves of quinoa cv. Piartal. Between three to fice dpi, leaf spots were visible (Figure 19). The bacteria colonies were identified at the species level via morphology and molecula tools using a Bruker Daltonik MALDI Biotyper system (Germany). The coucher for the identified bacteria was uploades to the NCBI database as txid317 [167].

Figure 19. Symptoms of bacterial leaf spot on quinoa. Source: illustration adapted from Fonseca-Guerra et al. [167].

6. Viruses

Pathogenicity assays for the identification of viruses under greenhouse conditions require indicator plants. These plants show distinctive and consistent reactions to virus infections. Many plant viruses can be transmitted to indicator plants via mechanical infection or insects. *Nicotiana* (tobacco) and *Chenopodium* are hosts for a great number of viruses [168]. Therefore, *C. quinoa* could be infected with the viruses that infect host plants that grow next to it.

1. Chenopodium mosaic virus: Seedlings of *C. quinoa* were found to contain a highly infectious, seed-borne virus that may remain latent. The virus was restricted to the *Chenopodiaceae* and was similar to the soybean mosaic virus in morphology and physio-chemical properties [169].
2. Amaranthus leaf mottle virus (ALMV): Successful infections were achieved on *C. quinoa*, which exhibited chlorotic local lesions and severe systemic mosaic, leaf deformation, wilting, stunning, and finally collapse of the plants. Transmission via *Aphis gossypii* was confirmed 2 to 3 weeks after the 1-day inoculation access period [170].
3. Arracacha virus A: AVA is common in arracacha (*Arracacia xanthorrhiza*) in the region of the Peruvian Andes. AVA was not transmitted by *Myzus persicae*, but was transmitted by the inoculation of sap and is best propagated in *C. quinoa* and *Nicotiana clevelandii* [171–173].
4. Ullucus virus C: UVC is a comovirus prevalent in *Ullucus tuberosus* grown at high altitudes in the Bolivian and Peruvian Andes. It was transmitted mechanically to *C. amaranticolor* and *C. quinoa*. It caused a systemic infection. UVC was not transmitted by either aphid species (*Aphis gossypii* or *Myzus persicae*) or through seeds of *C. quinoa*. However, it was transmitted through leaf contact between infected and healthy plants, causing chlorosis [173].
5. Potato virus S (PVS): *Chenopodium quinoa* plants displayed symptoms of PVS infection 14 days after artificial inoculation with PVS [174,175].
6. Potato Andean latent virus: APLV was found to infect both *C. quinoa* and *C. amaranticolor* [176].
7. Cucumber mosaic virus (CMV): Partially purified extracts from leaves of *Phytolacca americana* caused marked inhibition of CMV infection on *C. quinoa* [177].
8. Tobacco mosaic virus: TMV has successfully infected *C. quinoa* [178].
9. Passiflora latent virus (PLV): *Chenopodium quinoa* plants presenting systemic symptoms after inoculation with PLV showed high concentrations of virus particles in their cytoplasm, mitochondria, and chloroplasts [179]
10. Plantago asiatica mosaic virus (PlAMV): Mechanical inoculation with infected sap of *Lilium* leaves on *C. quinoa* yielded chlorotic or necrotic local lesions [180].
11. Carnation latent virus: *C. quinoa* is an indicator species for the carnation latent virus [181].
12. Chlorotic leaf spot virus: Sap inoculation on *C. quinoa* resulted in a satisfactory infection [182].

7. Conclusions and Future Directions

The growing interest in quinoa has prompted research on all aspects of this crop. From the perspective of phytopathology, it is essential to collaborate as quinoa cultivation has been introduced to many countries worldwide and continues to enter new regions. Therefore, it faces different challenges in each area. The impact on final seed yield has not been quantified for many diseases yet, as they have only been identified causing symptoms on plant tissue, but it is essential to turn our attention to this aspect.

Determining the mycobiota in quinoa grain food is of prime importance. The presence of seed pathogens associated with mycotoxins is concerning. These secondary metabolites are generally produced by fungi belonging to the genera (*Alternaria, Aspergillus, Fusarium* and *Penicillium)* [183,184]. The latter two pathogens from this list have been isolated from quinoa plants. Thus, mycotoxin production may occur in the field or during post-harvest, storage, or processing [185]. Indeed, a recent comparative study [183] of mycotoxin occurrence in quinoa grains cultivated in South America, and North Europe found a large array of mycotoxins on Northern European grain. Mycotoxins were predominantly associated with *Fusarium* spp. (e.g., butenolid, aurofusarin, equisetin, culmorin), *Alternaria* spp. (e.g., tenuazonic acid and altersetin), *Cladosporium* spp. (e.g., Cladosporim), and *Penicillium* spp. (e.g., ochratoxin A, flavogaucin, and mycophenolic acid). Unspecific metabolites were also found in modest amounts. Cleaning seeds provided a considerable reduction (ca. 50%) in the content of mycotoxins, but overall the North European grains had considerably more mycotoxins compared to South American grains even after cleaning. Weather conditions, cultivation method and post-harvest treatments could explain mycotoxins array presence differences on grain examined. The resilience of Andean grains to the growth of mycotoxin-producing fungi could be due to their adaptation to their natural centre of origin. Something that drastically changes when quinoa is cultivated in other latitudes [186,187]. It could also be argued that high saponin-containing quinoa grains may prevent the growth of fungi [188] or serve as a fungistatic. Therefore, monitoring seed quality during post-harvest should become a routine procedure. The implementation of this practice will highlight the fragility of organic quinoa production in new temperate environments.

It is essential to standardize the descriptions of diseases, taking into account the following suggestions:

- Morphological identification paired with molecular tools for accurate descriptions of causal agents, published in scientific journals, as well as the sharing of knowledge within quinoa networks and conferences.
- The performance of inclusive pathogenicity tests and Koch's postulates to clarify the type of interaction observed (e.g., pathogenic, endophytic/symbiotic, or saprophytic.).
- Standardized protocols for disease propagation and assessment methods for severity after infection.
- The development of strategies for seed sanitation.
- There exist several research centers located in areas where quinoa is traditionally grown, and recently a pilot global collaborative network on quinoa (GCN-Quinoa) (www.gcn-quinoa.org, accessed on 10 June 2021) has been established [189]. These networks primarily share knowledge on cultivation and plant breeding. Knowledge sharing in relation to quinoa diseases should also be considered.
- More research on methodologies for the rapid, high throughput screening of quinoa seeds and plants for the presence of economically important pathogens of quinoa is needed. This would be useful for detecting causal agents early in disease development and ensuring certified pathogen-free quinoa seeds. Moreover, phone apps with deep learning models for diagnosing various plant diseases and pest attacks are becoming interesting tools, which may be useful in the future.

Author Contributions: Writing—original draft preparation, C.C.-L.; writing—review and editing, C.C.-L., D.B.A. and C.A.; visualization, C.C.-L.; supervision, D.B.A. and C.A.; project administration, C.C.-L. and C.A.; funding acquisition, C.C.-L. All authors have read and agreed to the published version of the manuscript.

Funding: This research was funded by the Plurinational State of Bolivia with the Programme of Scholarships for Scientific Sovereignty and partly by the University of Copenhagen, Denmark.

Institutional Review Board Statement: Not applicable.

Informed Consent Statement: Not applicable.

Data Availability Statement: Not applicable.

Acknowledgments: We express our gratitude to Anna L. Testen and Yin et al. for sharing pictures, to Sven-Erick Jacobsen for advice and making the seed of Titicaca, Puno and Vikinga available, and for the illustrations, we thank Ana Sofia Patiño and Lucia Little.

Conflicts of Interest: The authors declare no conflict of interest. The funders had no role in the design of the study; in the collection, analyses, or interpretation of data; in the writing of the manuscript, or in the decision to publish the results.

References

1. Latijnhouwers, M.; de Wit, P.J.G.M.; Govers, F. Oomycetes and Fungi: Similar Weaponry to Attack Plants. *Trends Microbiol.* **2003**, *11*, 462–469. [CrossRef] [PubMed]
2. Salas, B.; Otazu, V. Enfermedades. In *Quinua y Kañiwa*; IICA: Bogota, Colombia, 1976; pp. 137–148.
3. Alandia, S.; Otazu, V.; Salas, B. Enfermedades. In *Quinua y Kañiwa, Cultivos Andinos*; Tapia, M.E., Ed.; Libros educativos; Instituto Interamericano de ciencias Agricolas IICA: Bogota, Colombia, 1979; Volume 49, pp. 137–148.
4. Danielsen, S.; Bonifacio, A.; Ames, T. Diseases of Quinoa (*Chenopodium Quinoa*). *Food Rev. Int.* **2003**, *19*, 43–59. [CrossRef]
5. Saravia, R.; Plata, G.; Gandarillas, A. *Plagas y Enfermedades Del Cultivo de Quinua*; Fundación PROINPA: Bolivia, Colombia, 2014.
6. Gandarillas, A.; Saravia, R.; Quispe, R.; Ortiz-Romero, R. Principle quinoa pests and diseases. In *State of the Art. Report on Quinoa around the World in 2013*; FAO Regional Office for Latin America and the Caribean: Rome, Italy, 2015; pp. 192–217.
7. Peintner, U.; Kuhnert-Finkernagel, R.; Wille, V.; Biasioli, F.; Shiryaev, A.; Perini, C. How to Resolve Cryptic Species of Polypores: An Example in Fomes. *IMA Fungus* **2019**, *10*, 1–21. [CrossRef] [PubMed]
8. Bruns, T.D.; White, T.J.; Taylor, J.W. Fungal Molecular Systematics. *Annu. Rev. Ecol. Syst.* **1991**, *22*, 525–564. [CrossRef]
9. Klein, J.; Neilen, M.; van Verk, M.; Dutilh, B.E.; Van den Ackerveken, G. Genome Reconstruction of the Non-Culturable Spinach Downy Mildew Peronospora Effusa by Metagenome Filtering. *PLoS ONE* **2020**, *15*, e0225808. [CrossRef]
10. Yahr, R.; Schoch, C.L.; Dentinger, B.T. Scaling up Discovery of Hidden Diversity in Fungi: Impacts of Barcoding Approaches. *Philos. Trans. R. Soc. B Biol. Sci.* **2016**, *371*, 20150336. [CrossRef]
11. Tekpinar, A.D.; Kalmer, A. Utility of Various Molecular Markers in Fungal Identification and Phylogeny. *Nova Hedwig.* **2019**, *109*, 187–224. [CrossRef]
12. Seifert, K.A. Progress towards DNA Barcoding of Fungi. *Mol. Ecol. Resour.* **2009**, *9*, 83–89. [CrossRef]
13. Lee, J.S.; Shin, H.-D.; Choi, Y.-J. Rediscovery of Seven Long-Forgotten Species of Peronospora and Plasmopara (Oomycota). *Mycobiology* **2020**, *48*, 331–340. [CrossRef]
14. Agrios, G.N. *Plant Pathology*, 5th ed.; Elservier: Amsterdam, The Netherlands, 2005.
15. Grogan, R.G. The Science and Art of Plant-Disease Diagnosis. *Annu. Rev. Phytopathol.* **1981**, *19*, 333–351. [CrossRef]
16. Grimes, D.J. Koch's Postulates-Then and Now. *Microbe Am. Soc. Microbiol.* **2006**, *1*, 223. [CrossRef]
17. Gäumann, E. Zur Kenntnis Der Chenopodiaceen Bewohnenden Peronospora-Arten. *JSTOR* **1920**, 155.
18. Baldauf, S.L. The Deep Roots of Eukaryotes. *Science* **2003**, *300*, 1703–1706. [CrossRef]
19. Kamoun, S. Plant Pathogens: Oomycetes (water mold). In *Encyclopedia of Microbiology*, 3rd ed.; Schaechter, M., Ed.; Academic Press: Oxford, UK, 2009; pp. 689–695. ISBN 978-0-12-373944-5.
20. Lee, S.C.; Ristaino, J.B.; Heitman, J. Parallels in Intercellular Communication in Oomycete and Fungal Pathogens of Plants and Humans. *PLoS Pathog.* **2012**, *8*, e1003028. [CrossRef]
21. Griffith, J.M.; Davis, A.J.; Grant, B.R. Target Sites of Fungicides to Control Oomycetes. *Target Sites Fungic. Action* **1992**, 69–100.
22. Madoui, M.-A.; Bertrand-Michel, J.; Gaulin, E.; Dumas, B. Sterol Metabolism in the Oomycete Aphanomyces Euteiches, a Legume Root Pathogen. *New Phytol.* **2009**, *183*, 291–300. [CrossRef]
23. Warrilow, A.G.; Hull, C.M.; Rolley, N.J.; Parker, J.E.; Nes, W.D.; Smith, S.N.; Kelly, D.E.; Kelly, S.L. Clotrimazole as a Potent Agent for Treating the Oomycete Fish Pathogen Saprolegnia Parasitica through Inhibition of Sterol 14α-Demethylase (CYP51). *Appl. Environ. Microbiol.* **2014**, *80*, 6154–6166. [CrossRef]
24. Desmazières, J.B.H.J. *Notice Sur Quelques Plantes Cryptogames Nouvellement Découvertes En France: Et Qui Vont Paraître, En Nature, Dans La Collection Publiée Par l'auteur*; Annales des Sciences Naturelles Botanique: Paris, France, 1837.
25. Yerkes, W.D.; Shaw, C.G. Taxonomy of the Peronospora Species on Cruciferaceae and Chenopodiaceae. *Phytopathology* **1959**, 49.
26. Choi, Y.-J.; Denchev, C.M.; Shin, H.-D. Morphological and Molecular Analyses Support the Existence of Host-Specific Peronospora Species Infecting Chenopodium. *Mycopathologia* **2008**, *165*, 155. [CrossRef]
27. Byford, W.J. Field Experiments on Sugar-Beet Downy Mildew (*Peronospora Farinosa*). *Ann. Appl. Biol.* **1967**, *60*, 97–107. [CrossRef]
28. Byford, W.J. Host Specialization of Peronospora Farinosa on Beta, Spinacia and Chenopodium. *Trans. Br. Mycol. Soc.* **1967**, *50*, 603–607. [CrossRef]
29. Choi, Y.-J.; Danielsen, S.; Lübeck, M.; Hong, S.-B.; Delhey, R.; Shin, H.-D. Morphological and Molecular Characterization of the Causal Agent of Downy Mildew on Quinoa (*Chenopodium Quinoa*). *Mycopathologia* **2010**, *169*, 403–412. [CrossRef]
30. Danielsen, S.; Lübeck, M. Universally Primed-PCR Indicates Geographical Variation of *Peronospora Farinosa* Ex. *Chenopodium Quinoa*. *J. Basic Microbiol.* **2010**, *50*, 104–109. [CrossRef]
31. Kara, M.; Soylu, E.M.; Uysal, A.; Kurt, Ş.; Choi, Y.-J.; Soylu, S. Morphological and Molecular Characterization of Downy Mildew Disease Caused by *Peronospora Variabilis* on *Chenopodium Album* in Turkey. *Australas. Plant Dis. Notes* **2020**, *15*, 1–3. [CrossRef]
32. Testen, A.L.; del Mar Jiménez-Gasco, M.; Ochoa, J.B.; Backman, P.A. Molecular Detection of *Peronospora Variabilis* in Quinoa Seed and Phylogeny of the Quinoa Downy Mildew Pathogen in South America and the United States. *Phytopathology* **2014**, *104*, 379–386. [CrossRef]

33. Kandel, S.L.; Mou, B.; Shishkoff, N.; Shi, A.; Subbarao, K.V.; Klosterman, S.J. Spinach Downy Mildew: Advances in Our Understanding of the Disease Cycle and Prospects for Disease Management. *Plant Dis.* **2019**, *103*, 791–803. [CrossRef]
34. Choi, Y.-J.; Hong, S.-B.; Shin, H.-D. Re-Consideration of Peronospora Farinosa Infecting Spinacia Oleracea as Distinct Species, Peronospora Effusa. *Mycol. Res.* **2007**, *111*, 381–391. [CrossRef]
35. Choi, Y.-J.; Thines, M. (2288) Proposal to Reject the Name Botrytis Farinosa (Peronospora Farinosa)(Peronosporaceae: Oomycetes). *Taxon* **2014**, *63*, 675–676. [CrossRef]
36. Stevenson, J.A.; Cardenas, M. Lista Preliminar de Los Hongos de Bolivia. *Lilloa* **1949**, *21*, 130.
37. García Rada, G. *Fitopatología Agrícola del Perú; Una Completa Recopilación de las Enfermedades Fungosas, Bacterianas y de Virus de las Plantas, Constatadas en el país, a Través de más de 20 años de Investigaciones*; Estación Experimental Agrícola de la Molina: Lima, Peru, 1947.
38. Tewari, J.P.; Boyetchko, S.M. Occurrence of Peronospora Farinosa f. Sp. *Chenopodii on Quinoa in Canada. Can. Plant Dis. Surv.* **1990**, *70*, 127–128.
39. Aragón, L.; Gutiérrez, W. Downy Mildew on Four Chenopodium Species. *Fitopatología* **1992**, *27*, 104–109.
40. Jacobs, T. Postulation of Virulence Groups and Resistance Factors in the Quinoa/Downy Mildew Pathosystem Using Material from Ecuador. *Plant Pathol.* **1999**, *48*, 425–430. [CrossRef]
41. Danielsen, S.; Jacobsen, S.-E.; Hockenhull, J. First Report of Downy Mildew of Quinoa Caused by *Peronospora Farinosa* f. Sp. *Chenopodii in Denmark. Plant Dis.* **2002**, *86*, 1175. [CrossRef]
42. Pańka, D.; Lenc, L.; Gęsiński, K. Preliminary Observations on Quinoa (Chenopodium Quinoa) Health Status in Poland. *Phytopathol. Pol.* **2004**, *31*, 61–66.
43. Kumar, A.; Bhargava, A.; Shukla, S.; Singh, H.B.; Ohri, D. Screening of Exotic *Chenopodium Quinoa* Accessions for Downy Mildew Resistance under Mid-Eastern Conditions of India. *Crop Prot.* **2006**, *25*, 879–889. [CrossRef]
44. Swenson, E.M. Genetic Diversity of Bolivian Peronospora Farinosa f.Sp. Chenopodii (Downy Mildew) and Quinoa's Resistance Response. Master's Thesis, Brigham Young University, Provo, UT, USA, 2006.
45. Baiswar, P.; Chandra, S.; Kumar, R.; Ngachan, S.V. Peronospora Variabilis on Chenopodium Murale in India. *Australas. Plant Dis. Notes* **2010**, *5*, 45–47. [CrossRef]
46. Testen, A.L.; McKemy, J.M.; Backman, P.A. First Report of Quinoa Downy Mildew Caused by Peronospora Variabilis in the United States. *Plant Dis.* **2012**, *96*, 146. [CrossRef]
47. Choi, Y.J.; Choi, I.Y.; Kim, J.S.; Shin, H.-D. First Report of Quinoa Downy Mildew Caused by *Peronospora Variabilis* in Republic of Korea. *Plant Dis.* **2014**, *98*, 1003. [CrossRef]
48. Mhada, M.; Ezzahiri, B.; Benlhabib, O. Assessment of Downy Mildew Resistance (Peronospora Farinosa) in a Quinoa (Chenopodium Quinoa Willd.) Germplasm. *Int. J. Biol. Med. Res.* **2015**, *6*, 4748–4752.
49. Khalifa, W.; Thabet, M. Variation in Downy Mildew (*Peronospora Variabilis* Gäum) Resistance of Some Quinoa (*Chenopodium Quinoa* Willd) Cultivars under Egyptian Conditions. *Middle East J. Agric.* **2018**, *70*, 671–682.
50. Nolen, H.B. Assessing Disease Concerns on Quinoa and Evaluating Sources of Disease Resistance in Chenopodium Species in New England. Master's Thesis, University of new Hampshire, Durham, NH, USA, 2019.
51. Gül, E. First Report of Peronospora Variabilis on Chenopodium Quinoa in Turkey. *J. Plant Pathol.* **2021**, *103*, 389–390. [CrossRef]
52. Colque-Little, C.; Abondano, M.C.; Lund, O.S.; Amby, D.B.; Piepho, H.-P.; Andreasen, C.; Schmöckel, S.; Schmid, K. Genetic Variation for Tolerance to the Downy Mildew Pathogen Peronospora Variabilis in Genetic Resources of Quinoa (Chenopodium Quinoa). *BMC Plant Biol.* **2021**, *21*, 41. [CrossRef]
53. El-Assiuty, E.M.; Taha, E.M.; Fahmy, Z.M.; Fahmy, G.M. Histological and Molecular Detections of *Peronospora Variabilis* Gäum Oospores in Seeds of Quinoa (*Chenopodium Quino* L.). *Egypt. J. Exp. Biol. (Bot.)* **2019**, *15*, 197–203.
54. Danielsen, S.; Ames, T. *Mildew (Peronospora Farinosa) of Quinua (Chenopodium Quinoa) in the Andean Region: Practical Manual for the Study of the Disease and Pathogen*; International Potato Center: Lima, Peru, 2004.
55. CABI Invasive Species Compendium 2020. Available online: https://www.cabi.org/publishing-products/invasive-species-compendium/ (accessed on 1 April 2021).
56. Thines, M.; Choi, Y.-J. Evolution, Diversity, and Taxonomy of the Peronosporaceae, with Focus on the Genus *Peronospora*. *Phytopathol.* **2016**, *106*, 6–18. [CrossRef]
57. Verma, S.C.; Chauhan, L.S.; Mathur, R.S. *Peronospora Farinosa* (Fr.) Fr. on *Chenopodium Murale* L.-A New Record for India. *Curr. Sci.* **1964**, *33*, 720–721.
58. Judelson, H.S. Sexual Reproduction in Oomycetes: Biology, Diversity and Contributions to Fitness. In *Oomycete Genetics and Genomics: Diversity, Interactions and Research Tools*; John Wiley & Sons Inc.: Hoboken, NJ, USA, 2009; pp. 121–138.
59. Kitz, L. *Evaluation of Downy Mildew (Peronospora Farinosa f.Sp. Chenopodii) Resistance among Quinoa Genotypes and Investigation of P. Farinosa Growth Using Scanning Electron Microscopy*; Brigham Young University: Provo, UT, USA, 2008.
60. Cahill, D.M.; Hardham, A.R. Exploitation of Zoospore Taxis in the Development of a Novel Dipstick Immunoassay for the Specific Detection of Phytophthora Cinnamomi. *Phytopathology* **1994**, *84*, 193–200. [CrossRef]
61. Kiefer, B.; Riemann, M.; Büche, C.; Kassemeyer, H.-H.; Nick, P. The Host Guides Morphogenesis and Stomatal Targeting in the Grapevine Pathogen Plasmopara Viticola. *Planta* **2002**, *215*, 387–393. [CrossRef]
62. Michelmore, R.W.; Ilott, T.; Hulbert, S.H.; Farrara, B. The downy mildews. In *Advances in Plant Pathology*; Elsevier: Amsterdam, The Netherlands, 1988; Volume 6, pp. 53–79.

63. Frinking, H.D.; Linders, E.G.A. A Comparison of Two Pathosystems: Downy Mildew on *Spinacia Oleracea* and on *Chenopodium Album*. *Neth. J. Plant Pathol.* **1986**, *92*, 97. [CrossRef]
64. Doshi, A.; Thakore, B.B.L. Effect of Certain Environmental Factors on Germination of Conidia of *Peronospora Arborescens* (Berk) de Bary. *J. Phytol. Res.* **1993**, *6*, 59–62.
65. Imamura, T.; Takagi, H.; Miyazato, A.; Ohki, S.; Mizukoshi, H.; Mori, M. Isolation and Characterization of the Betalain Biosynthesis Gene Involved in Hypocotyl Pigmentation of the Allotetraploid Chenopodium Quinoa. *Biochem. Biophys. Res. Commun.* **2018**, *496*, 280–286. [CrossRef]
66. Polturak, G.; Aharoni, A. "La Vie En Rose": Biosynthesis, Sources, and Applications of Betalain Pigments. *Mol. Plant* **2018**, *11*, 7–22. [CrossRef]
67. Nielsen, C.; Hajek, A.E.; Humber, R.A.; Bresciani, J.; Eilenberg, J. Soil as an Environment for Winter Survival of Aphid-Pathogenic Entomophthorales. *Biol. Control* **2003**, *28*, 92–100. [CrossRef]
68. Taha, E.M. Molecular Detectionand Phylogeny of *Peronospora Variabilis* Gäum., The Causal Agent of Downy Mildew Disease of Quinoa at Different Growth Stages. *Plant Cell Biotechnol. Mol. Biol.* **2019**, *20*, 1189–1200.
69. El-Assiuty, E.M.; Fahmy, G.M.; Taha, E.M.; Fahmy, Z.M.; Ismael, A.S.M.; Abd-Elghany, W.R.; Kafsheer, D.A. Microscopic Visualization of Peronospora Variabilis Gäum., the Cause of Quinoa Downy Mildew in Plant Tissues at Different Stages of Plant Growth. *Int. J. Sci. Eng. Res.* **2019**, *10*.
70. Chapela, I.H. Spore Size Revisited: Analysis of Spore Populations Using Automated Particle Size. *Sydowia* **1991**, *43*, 1–14.
71. Hamilton, A.J.; Cunnington, J.H. Calculating Minimum Sample Sizes for Taxonomic Measurements: Examples Using Gaumann's Peronospora Spore Data. *Mycotaxon* **2006**, *95*, 189–194.
72. Yin, H.; Zhou, J.B.; Chang, F.J.; Lu, H.; Gong, L.; Zhao, X. Identification of the pathogen causing downy mildew on *Chenopodium quinoa*. *Acta Phytopathol. Sin.* **2018**, *48*, 413–417. [CrossRef]
73. Danielsen, S. Heterothallism in Peronospora Farinosa f. Sp. Chenopodii, the Causal Agent of Downy Mildew of Quinoa (Chenopodium Quinoa). *J. Basic Microbiol. Int. J. Biochem. Physiol. Genet. Morphol. Ecol. Microorg.* **2001**, *41*, 305–309.
74. Fabritius, A.-L.; Cvitanich, C.; Judelson, H.S. Stage-Specific Gene Expression during Sexual Development in Phytophthora Infestans. *Mol. Microbiol.* **2002**, *45*, 1057–1066. [CrossRef]
75. Danielsen, S.; Mercado, V.H.; Ames, T.; Munk, L. Seed Transmission of Downy Mildew (*Peronospora Farinosa* f.Sp. *Chenopodii*) in Quinoa and Effect of Relative Humidity on Seedling Infection. *Seed Sci. Technol.* **2004**, *32*, 91–98. [CrossRef]
76. Vercesi, A.; Tornaghi, R.; Sant, S.; Burruano, S.; Faoro, F. A Cytological and Ultrastructural Study on the Maturation and Germination of Oospores of Plasmopara Viticola from Overwintering Vine Leaves. *Mycol. Res.* **1999**, *103*, 193–202. [CrossRef]
77. Testen, A. *Microbial Approaches to Support Andean Quinoa Production*; The Pennsylvania State University: State College, PA, USA, 2012.
78. Inaba, T.; Takahashi, K.; Morinaka, T. Seed Transmission of Spinach Downy Mildew. *Plant Dis.* **1983**, *67*, 1139–1141. [CrossRef]
79. Inaba, T.; Hino, T. Production of Oospore-Encrusted Seeds by Inoculation of Conidia of Downy Mildew Fungus, *Peronospora Manshurica*, to Soybean Flower and Pod. *Jpn. J. Phytopathol.* **1980**, *46*, 533–538. [CrossRef]
80. Risco Mendoza, A. Severidad de *Peronospora variabilis* Gäum en *Chenopodium quinoa* Willd. "Pasankalla" como Respuesta a Plicaciones de Fungicidas y Bioestimulantes. Master's Thesis, Universidad Nacional Agraria La Molina, Lima, Peru, 2014.
81. Risco, A.; Mattos, L. Severidad de Peronospora Variabilis GÄUM. En Chenopodium Quinoa WILLD.'Pasankalla'Como Respuesta a Aplicaciones de Fungicidas Sintéticos y Bioestimulantes. *An. Científicos* **2015**, *76*, 383–392. [CrossRef]
82. Calixtro, G. Respuesta de 100 Accesiones de Guinua a la Infección Natural de Mildiu (*Peronospora variabilis* Gäum) en el Valle del Mantaro. Bachelor's Thesis, Universidad Nacional Agraria La Molina, Lima, Peru, 2017.
83. Risi, J.C.; Galwey, N.W. The Pattern of Genetic Diversity in the Andean Grain Crop Quinoa (*Chenopodium Quinoa* Willd). I. Associations between Characteristics. *Euphytica* **1989**, *41*, 147–162. [CrossRef]
84. Mandák, B.; Krak, K.; Vít, P.; Lomonosova, M.N.; Belyayev, A.; Habibi, F.; Wang, L.; Douda, J.; Štorchová, H. Hybridization and Polyploidization within the Chenopodium Album Aggregate Analysed by Means of Cytological and Molecular Markers. *Mol. Phylogenetics Evol.* **2018**, *129*, 189–201. [CrossRef]
85. Voglmayr, H.; Greilhuber, J. Genome Size Determination in Peronosporales (Oomycota) by Feulgen Image Analysis. *Fungal Genet. Biol.* **1998**, *25*, 181–195. [CrossRef] [PubMed]
86. Feng, C.; Saito, K.; Liu, B.; Manley, A.; Kammeijer, K.; Mauzey, S.J.; Koike, S.; Correll, J.C. New Races and Novel Strains of the Spinach Downy Mildew Pathogen Peronospora Effusa. *Plant Dis.* **2018**, *102*, 613–618. [CrossRef]
87. Irish, B.M.; Correll, J.C.; Koike, S.T.; Morelock, T.E. Three New Races of the Spinach Downy Mildew Pathogen Identified by a Modified Set of Spinach Differentials. *Plant Dis.* **2007**, *91*, 1392–1396. [CrossRef]
88. Correll, J.; Koike, S.; Smilde, D.; de Visser, J. *Race 15: New Spinach Downy Mildew Race Confirmed*; Agriculture and Natural Resources, University of California: Los Angeles, CA, USA, 2014.
89. Farber, D.H.; Mundt, C.C. Effect of Plant Age and Leaf Position on Susceptibility to Wheat Stripe Rust. *Phytopathology* **2017**, *107*, 412–417. [CrossRef]
90. Panter, S.N.; Jones, D.A. Age-Related Resistance to Plant Pathogens. *Adv. Bot. Res.* **2002**, *38*, 251–280.
91. Forbes, G.; Perez, W.; Andrade Piedra, J. *Field Assessment of Resistance in Potato to Phytophthora Infestans: International Cooperators Guide*; International Potato Center: Lima, Peru, 2014.

92. Sosa-Zuniga, V.; Brito, V.; Fuentes, F.; Steinfort, U. Phenological Growth Stages of Quinoa (*Chenopodium Quinoa*) Based on the BBCH Scale. *Ann. Appl. Biol.* **2017**, *171*, 117–124. [CrossRef]
93. Gabriel, J.; Luna, N.; Vargas, A.; Magne, J.; Angulo, A.; La Torre, J.; Bonifacio, A. Quinua de Valle (*Chenopodium Quinoa* Willd.): Fuente Valiosa de Resistencia Genética al Mildiu (*Peronospora Farinosa* Willd.). *J. Selva Andin. Res. Soc.* **2012**, *3*, 27–44. [CrossRef]
94. Jacobsen, S.-E. Adaptation and scope for quinoa in northern latitudes of Europe. In *State of the Art Report on Quinoa Around the World in 2013*; FAO Regional Office for Latin America and the Caribean: Rome, Italy, 2015; pp. 431–446.
95. Benlhabib, O.; Boujartani, N.; Maughan, P.J.; Jacobsen, S.E.; Jellen, E.N. Elevated Genetic Diversity in an F2: 6 Population of Quinoa (*Chenopodium Quinoa*) Developed through an Inter-Ecotype Cross. *Front. Plant Sci.* **2016**, *7*, 1222. [CrossRef]
96. Bonifacio, A. *Chenopodium* Sp.: Genetic Resources, Ethnobotany, and Geographic Distribution. *Food Rev. Int.* **2003**, *19*, 1–7. [CrossRef]
97. Danielsen, S.; Munk, L. Evaluation of Disease Assessment Methods in Quinoa for Their Ability to Predict Yield Loss Caused by Downy Mildew. *Crop Prot.* **2004**, *23*, 219–228. [CrossRef]
98. Staub, J.; Bacher, J.; Poetter, K. Sources of Potential Errors in the Application of Random Amplified Polymorphic DNAs in Cucumber. *HortScience* **1996**, *31*, 262–266. [CrossRef]
99. Hammerschmidt, R.; Métraux, J.-P.; Van Loon, L.C. Inducing Resistance: A Summary of Papers Presented at the First International Symposium on Induced Resistance to Plant Diseases, Corfu, May 2000. *Eur. J. Plant Pathol.* **2000**, *107*, 1–6. [CrossRef]
100. Hammerschmidt, R. *Induced Resistance to Disease in Plants*; Springer: Berlin/Heidelberg, Germany, 1995.
101. Strobel, N.E.; Ji, C.; Gopalan, S.; Kuc, J.A.; He, S.Y. Induction of Systemic Acquired Resistance in Cucumber by Pseudomonas Syringae Pv. Syringae 61 HrpZPss Protein. *Plant J.* **1996**, *9*, 431–439. [CrossRef]
102. Colque-Little, C.X. Characterization of New Emerging Diseases Infecting Quinoa (Chenopodium Quinoa Willd.) and Search for Sources of Resistance to Downy Mildew (Peronospora Variavilis Gäum). Ph.D. Thesis, University of Copenhagen, Copenhagen, Denmark, 2021.
103. La Torre-Farfan, J.P. Is Quinoa Cultivation on the Coastal Desert of Peru Sustainable? A Case Study from Majes, Arequipa. Master's Thesis, Aarhus University, Aarhus, Denmark, 2017.
104. Colque-Hinojosa, A.O. Collective Action and Contract Farming for Enhancing Smallholder Access to Markets: Case Studies of Farmer Organizations Participating in Rural Alliances Project in Bolivia. Master's Thesis, Institute of Food and Resource Economics, University of Copenhagen, Copenhagen, Denmark, 2011.
105. Gisi, U.; Sierotzki, H. Oomycete fungicides: Phenylamides, quinone outside inhibitors, and carboxylic acid amides. In *Fungicide Resistance in Plant Pathogens*; Springer: Berlin/Heidelberg, Germany, 2015; pp. 145–174.
106. Du Toit, L.J.; Derie, M.L.; Brissey, L.M.; Holmes, B.J.; Gatch, E. Efficacy of Organic and Conventional Seed Treatments for Management of Verticillium in Spinach Seed. *Phytopathology* **2009**, *99*, S30.
107. Leon Ttacca, B.; Ortiz Calcina, N.; Condori Ticona, N.; Chura Yupanqui, E. Cepas de Trichoderma Con Capacidad Endofitica Sobre El Control Del Mildiu (*Peronospora Variabilis* Gäum.) y Mejora Del Rendimiento de Quinua. *Rev. Investig. Altoandinas* **2018**, *20*, 19–30. [CrossRef]
108. Aguilar, R.; More-Yarleque, M.M.; Rafael-Rutte, R.; Maldonado, E. Inductores de Defensa En El Control Del Mildiu (Peronospora Variabilis Gaum.) En El Cultivo de Quinua: Detección, Epidemiología, Síntomas, Características y Control. *Sci. Agropecu.* **2020**, *11*, 555–563. [CrossRef]
109. Kiraly, Z.; Ersek, T.; Barna, B.; Adam, A.; Gullner, G. Pathophysiological Aspects of Plant Disease Resistance. *Acta Phytopathol. Entomol. Hung.* **1991**, *26*, 233–250.
110. Van der Plank, J.E. *Disease Resistance in Plants*; Elsevier: Amsternad, The Netherlands, 1984.
111. Gamboa, C.; Van den Broeck, G.; Maertens, M. *Adoption of Improved Quinoa Varieties among Smallholder Farmers in the Peruvian Andes*; CIP: Lima, Peru, 2018.
112. Gandarillas, A.; Rojas, W.; Bonifacio, A.; Ojeda, N. Quinoa in Bolivia: The PROINPA foundation's perspective. Chapter 5.1.a. In *State of the Art Report on Quinoa around the World in 2013*; Bazile, D., Bertero, D., Nieto, C., Eds.; FAO: La Paz, Bolivia, 2015; pp. 344–361.
113. Gandarillas, H. Genetica y Origen. In *Quinua y Kañiwa, Cultivos Andinos*; Tapia, E.M., Ed.; Libros educativos; Instituto Interamericano de Ciencias Agricolas—IICA: Bogota, Colombia, 1979; Volume 49, pp. 45–64.
114. McElhinny, E.; Mazon, N.; Rivera, M.; Peralta, I. Lineas promisorias de quinua con resistencia cuantitativa al mildiu en Ecuador. In *Agro Biodiversidad y Produccion de Semilla con el Sector Informal a Traves del Mejoramiento participativo en la Zona Andina*; PREDUZA, INIAP: Lima, Peru, 2003; pp. 40–47.
115. Curti, R.N.; De la Vega, A.J.; Andrade, A.J.; Bramardi, S.J.; Bertero, H.D. Multi-Environmental Evaluation for Grain Yield and Its Physiological Determinants of Quinoa Genotypes across Northwest Argentina. *Field Crops Res.* **2014**, *166*, 46–57. [CrossRef]
116. Poland, J.A.; Balint-Kurti, P.J.; Wisser, R.J.; Pratt, R.C.; Nelson, R.J. Shades of Gray: The World of Quantitative Disease Resistance. *Trends Plant Sci.* **2009**, *14*, 21–29. [CrossRef]
117. Mastebroek, H.D.; Limburg, H.; Gilles, T.; Marvin, H.J.P. Occurrence of Sapogenins in Leaves and Seeds of Quinoa (Chenopodium Quinoa Willd). *J. Sci. Food Agric.* **2000**, *80*, 152–156. [CrossRef]
118. Piepho, H.-P. A Mixed-Model Approach to Mapping Quantitative Trait Loci in Barley on the Basis of Multiple Environment Data. *Genetics* **2000**, *156*, 2043–2050. [CrossRef]

119. Rojas, W.; Risi, J.; Bonifacio, A.; Gandarillas, A. El cultivo de la quinua en Bolivia. Capitulo 2. In *Produccion y Mercado de la Quinua en Bolivia*; Risi, J., Rojas, W., Pacheco, M., Eds.; IICA, Insitituo Interamericano de Investigacion Agricola: La Paz, Bolivia, 2015; pp. 33–67.
120. Seiler, C.; Hutjes, R.W.; Kabat, P. Climate Variability and Trends in Bolivia. *J. Appl. Meteorol. Climatol.* **2013**, *52*, 130–146. [CrossRef]
121. Cereceda, P.; Larrain, H.; Osses, P.; Farías, M.; Egaña, I. The Climate of the Coast and Fog Zone in the Tarapacá Region, Atacama Desert, Chile. *Atmos. Res.* **2008**, *87*, 301–311. [CrossRef]
122. Lozano-Isla, F. *Inti: Tools and Statistical Procedures in Plant Science. R Package Version 0.1.3*; R Foundation: Vienna, Austria, 2021.
123. Jellen, R.; Maughan, P.; Fuentes, F.; Kolano, B. Botany, phylogeny and evolution. Chapter 1.1. In *State of the Art Report on Quinoa Around the World in 2013*; Bazile, D., Bertero, D., Nieto, C., Eds.; FAO Regional Office for Latin America and the Caribean: Katowice, Poland, 2015; pp. 12–23.
124. López-Marqués, R.L.; Nørrevang, A.F.; Ache, P.; Moog, M.; Visintainer, D.; Wendt, T.; Østerberg, J.T.; Dockter, C.; Jørgensen, M.E.; Salvador, A.T. Prospects for the Accelerated Improvement of the Resilient Crop Quinoa. *J. Exp. Bot.* **2020**, *71*, 5333–5347. [CrossRef]
125. Kim, Y.C.; Blee, K.A.; Robins, J.; Anderson, A.J. Oxycom^{TM} under Field and Laboratory Conditions Increases Resistance Responses in Plants. *Eur. J. Plant Pathol.* **2001**, *107*, 129–136. [CrossRef]
126. Bokshi, A.I.; Jobling, J.; McConchie, R. A Single Application of Milsana® Followed by Bion® Assists in the Control of Powdery Mildew in Cucumber and Helps Overcome Yield Losses. *J. Hortic. Sci. Biotechnol.* **2008**, *83*, 701–706. [CrossRef]
127. Mathur, S.B.; Kongsdal, O. *Common Laboratory Seed Health Testing Methods for Detecting Fungi*; International Seed Testing Association: East Lansing, MI, USA, 2003.
128. Boerema, G.H.; Mathur, S.B.; Neergaard, P. *Ascochyta Hyalospora* (Cooke & Ell.) Comb. Nov. in Seeds of *Chenopodium Quinoa*. *Neth. J. Plant Pathol.* **1977**, *83*, 153–159.
129. Testen, A.L.; McKemy, J.M.; Backman, P.A. First Report of Ascochyta Leaf Spot of Quinoa Caused by *Ascochyta* Sp. in the United States. *Plant Dis.* **2013**, *97*, 844. [CrossRef]
130. Otazu, V.; Aguilar, P.C.; Canahua, A. Resistance of Chenopodium quinoa to mildew (Peronospora effusa). *Fitopatologia* **1976**, *11*, 47–49.
131. Fries, A.M.; Tapia, M.E. *Guía de Campo de Los Cultivos Andinos*; FAO, ANPE-PERÚ: Rome, Italy, 2007.
132. Xiu-shi, Y.; Pei-you, Q.; Hui-min, G.; Gui-xing, R. Quinoa Industry Development in China. *Int. J. Agric. Nat. Resour.* **2019**, *46*, 208–219. [CrossRef]
133. Boerema, G.H.; Loerakker, W.M.; Hamers, M.E. Check-List for Scientific Names of Common Parasitic Fungi. Supplement Series 2a (Additions and Corrections): Fungi on Field Crops: Beet and Potato; Caraway, Flax and Oil-Seed Poppy. *Neth. J. Plant Pathol.* **1987**, *93*, 1–20. [CrossRef]
134. De Gruyter, J.; Woudenberg, J.H.C.; Aveskamp, M.M.; Verkley, G.J.M.; Groenewald, J.Z.; Crous, P.W. Redisposition of Phoma-like Anamorphs in Pleosporales. *Stud. Mycol.* **2013**, *75*, 1–36. [CrossRef]
135. Crous, P.W.; Shivas, R.G.; Van der Quaedvlieg, W.; Van der Bank, M.; Zhang, Y.; Summerell, B.A.; Guarro, J.; Wingfield, M.J.; Wood, A.R.; Alfenas, A.C. Fungal Planet Description Sheets: 214–280. *Pers. Mol. Phylogeny Evol. Fungi* **2014**, *32*, 184. [CrossRef]
136. Papizadeh, M.; Wijayawardene, N.N.; Amoozegar, M.A.; Saba, F.; Fazeli, S.A.S.; Hyde, K.D. Neocamarosporium Jorjanensis, N. Persepolisi, and N. Solicola Spp. Nov.(Neocamarosporiaceae, Pleosporales) Isolated from Saline Lakes of Iran Indicate the Possible Halotolerant Nature for the Genus. *Mycol. Prog.* **2018**, *17*, 661–679. [CrossRef]
137. Yin, H.; Zhou, J.; Lv, H.; Qin, N.; Chang, F.J.; Zhao, X. Identification, Pathogenicity, and Fungicide Sensitivity of *Ascochyta Caulina* (Teleomorph: *Neocamarosporium Calvescens*) Associated with Black Stem on Quinoa in China. *Plant Dis.* **2020**, *104*, 2585–2597. [CrossRef]
138. Van der Aa, H.A.; Van Kesteren, H.A. Some Pycnidial Fungi Occurring on Atriplex and Chenopodium. *Pers. Mol. Phylogeny Evol. Fungi* **1979**, *10*, 267–276.
139. Dřímalková, M.; Veverka, K. Seedlings Damping-off of *Chenopodium Quinoa* Wild. *Plant Protect. Sci* **2004**, *40*, 5–10. [CrossRef]
140. Li, J.; Zhou, X.; Huang, H.; Li, G. Diseases Characteristic and Control Measurements for Chenopodium Quinoa Willd. In *Proceedings of the 2017 6th International Conference on Energy and Environmental Protection (ICEEP 2017)*; Zhubai, China, 29–30 June 2017, Atlantis Press: Paris, France, 2017.
141. Evidente, A.; Capasso, R.; Cutignano, A.; Taglialatela-Scafati, O.; Vurro, M.; Zonno, M.C.; Motta, A. Ascaulitoxin, a Phytotoxic Bis-Amino Acid N-Glucoside from Ascochyta Caulina. *Phytochemistry* **1998**, *48*, 1131–1137. [CrossRef]
142. Muller-Scharer, H.; Scheepens, P.C.; Greaves, M.P. Biological Control of Weeds in European Crops: Recent Achievements and Future Work. *Weed Res.* **2000**, *40*, 83–98. [CrossRef]
143. Scheepens, P.C.; Kempenaar, C.; Andreasen, C.; Eggers, T.H.; Netland, J.; Vurro, M. Biological Control of the Annual Weed *Chenopodium Album*, with Emphasis on the Application of *Ascochyta Caulina* as a Microbial Herbicide. *Integr. Pest Manag. Rev.* **1997**, *2*, 71–76. [CrossRef]
144. Kempenaar, C.; Wanningen, R.; Scheepens, P.C. Control of Chenopodium Album by Soil Application of Ascochyta Caulina under Greenhouse Conditions. *Ann. Appl. Biol.* **1996**, *129*, 343–354. [CrossRef]
145. Kempenaar, C.; Horsten, P.J.; Scheepens, P.C. Growth and Competitiveness of Common Lambsquarters (Chenopodium Album) after Foliar Application of Ascochyta Caulina as a Mycoherbicide. *Weed Sci.* **1996**, 609–614. [CrossRef]

146. Kempenaar, C.; Horsten, P.; Scheepens, P.C. Effects of Ascochyta Caulina on Photosynthesis of Leaves of Chenopodium Album. *New Phytol.* **1996**, *132*, 453–457. [CrossRef]
147. Peralta, I. *La Quinua En Ecuador." Estado Del Arte"*; INIAP: Quito, Ecuador, 2009.
148. Świderska-Burek, U. Preliminary List of Cercosporoid Fungi from Poland. *Mycotaxon* **2007**, *102*, 5–8.
149. Testen, A.L.; McKemy, J.M.; Backman, P.A. First Report of Passalora Leaf Spot of Quinoa Caused by *Passalora Dubia* in the United States. *Plant Dis.* **2013**, *97*, 139. [CrossRef]
150. Sigovini, M.; Keppel, E.; Tagliapietra, D. Open Nomenclature in the Biodiversity Era. *Methods Ecol. Evol.* **2016**, *7*, 1217–1225. [CrossRef]
151. Yin, H.; Zhou, J.; Lü, H.; Chang, F.; Qin, N.; Zhai, S.; Xing, K.; Zhao, F.; Zhao, X. Identification of the pathogen causing Cercospora leaf spot on quinoa. *Acta Phytopathol. Sin.* **2019**, *49*, 408–414.
152. Videira, S.I.R.; Groenewald, J.Z.; Nakashima, C.; Braun, U.; Barreto, R.W.; de Wit, P.J.; Crous, P.W. Mycosphaerellaceae–Chaos or Clarity? *Stud. Mycol.* **2017**, *87*, 257–421. [CrossRef]
153. Videira, S.I.R.; Groenewald, J.Z.; Braun, U.; Shin, H.-D.; Crous, P.W. All That Glitters Is Not Ramularia. *Stud. Mycol.* **2016**, *83*, 49–163. [CrossRef]
154. Pal, N.; Testen, A.L. First Report of Quinoa Anthracnose Caused by Colletotrichum Nigrum and C. Truncatum in the United States. *Plant Dis.* **2020**, *105*, 3. [CrossRef]
155. Boerema, G.H.; Höweler, L.H. Phoma Exigua Desm. and Its Varieties. *Pers. Mol. Phylogeny Evol. Fungi* **1967**, *5*, 15–28.
156. Logan, C.; Khan, A.A. Comparative Studies of Phoma Spp. Associated with Potato Gangrene in Northern Ireland. *Trans. Br. Mycol. Soc.* **1969**, *52*, 9–17. [CrossRef]
157. Otazu, V.; Salas, B. Brown Stalk Rot of *Chenopodium Quinoa* Caused by Phoma Exigua Var. Foveata. *Fitoatologia* **1977**, *12*, 54–58.
158. Otazú, V.; Boerema, G.H.; Mooi, J.C.; Salas, B. Possible Geographical Origin of Phoma Exigua Var. Foveata, the Principal Causal Organism of Potato Gangrene. *Potato Res.* **1979**, *22*, 333–338. [CrossRef]
159. Daza Vera, E. Enfermedad Micotica En Quinua. *SITUA* **1995**, 29–32.
160. Ikeda, A.; Ichitani, T. Density of *Pythium Zingiberum* Oospores in Relation to Infection of *Chenopodium Quinoa* Seedlings. *Bull. Univ. Osaka Prefect.* **1985**, *37*, 13–18.
161. Barbosa, J.; Ames, T.; Jacobsen, S. *Reduccion de La Germinacion y Muerte de Plantas de Chenopodium Quinoa Causadas Por Rhizoctonia y Fusarium*; Centro Internacional de la Papa, Universidad Agraria la Molina: Lima, Peru, 2000.
162. Isobe, K.; Sugiyama, T.; Katagiri, M.; Ishizuka, C.; Tamura, Y.; Higo, M.; Fujita, Y. Study on the Cause Damping-off in Quinoa (Chenopodium Quinoa Willd.) and a Method for Suppressing Its Occurrence. *Jpn. J. Crop Sci.* **2019**, *88*, 117–124. [CrossRef]
163. Beckman, P.M.; Finch, H.C. Seed Rot and Damping-off of *Chenopodium Quinoa* Caused by *Sclerotium Rolfsii*. *Plant Dis.* **1980**, *64*, 497–498. [CrossRef]
164. Masiello, M.; Somma, S.; Ghionna, V.; Logrieco, A.F.; Moretti, A. In Vitro and in Field Response of Different Fungicides against Aspergillus Flavus and Fusarium Species Causing Ear Rot Disease of Maize. *Toxins* **2019**, *11*, 11. [CrossRef]
165. Lebeda, A.; Cohen, Y. Fungicide resistance in Pseudoperonospora cubensis, the causal pathogen of cucurbit downy mildew. In *Fungicide Resistance in Crop Protection: Risk and Management*; CABI: Wallingford, UK, 2012; pp. 44–63.
166. Plimmer, J.R. *Encyclopedia of Agrochemicals*; Wiley-Interscience: Hoboken, NJ, USA, 2003.
167. Fonseca-Guerra, I.; Chiquillo, C.; Padilla, M.J.; Benavides-Rozo, M. First Report of Bacterial Leaf Spot on Chenopodium Quinoa Caused by Pseudomonas Syringae in Colombia. *J. Plant Dis. Prot.* **2021**, *128*, 871–874. [CrossRef]
168. Walkey, D.G. *Applied Plant Virology*; Springer Science & Business Media: Berlin, Germany, 2012.
169. Dias, H.F.; Waterworth, H.E. The Identity of a Seed-Borne Mosaic Virus of *Chenopodium Amaranticolor* and *C. Quinoa*. *Can. J. Bot.* **1967**, *45*, 1285–1295. [CrossRef]
170. Segundo, E.; Lesemann, D.E.; Martín, G.; Carmona, M.P.; Ruiz, L.; Cuadrado, I.M.; Velasco, L.; Janssen, D. Amaranthus Leaf Mottle Virus: 3′-End RNA Sequence Proves Classification as Distinct Virus and Reveals Affinities within the Genus Potyvirus. *Eur. J. Plant Pathol.* **2007**, *117*, 81–87. [CrossRef]
171. Kenten, R.H.; Jones, R.A.C. Arracacha Virus B, a Second Isometric Virus Infecting Arracacha (*Arracacia Xanthorrhiza; Umbelliferae*) in the Peruvian Andes. *Ann. Appl. Biol.* **1979**, *93*, 31–36. [CrossRef]
172. Jones, R.A.C.; Kenten, R.H. A Strain of Arracacha Virus B Infecting Oca (*Oxalis Tuberosa*: *Oxalidaceae*) in the Peruvian Andes. *Phytopathol. Z.* **1981**, *100*, 88–95. [CrossRef]
173. Brunt, A.A.; Barton, R.J.; Phillips, S.; Jones, R.A.C. Ullucus Virus C, a Newly Recognised Comovirus Infecting *Ullucus Tuberosus* (*Basellaceae*). *Ann. Appl. Biol.* **1982**, *101*, 73–78. [CrossRef]
174. Hiruki, C.; Shukla, P. Intracellular Location of Potato Virus S in Leaf Tissue of *Chenopodium Quinoa*. *Can. J. Bot.* **1973**, *51*, 1699–1702. [CrossRef]
175. Herrera, V.V.; Juárez, E.R.; Muñiz, R.B. Detección de Virus Por Serología y Plantas Indicadoras En El Tubérculo-Semilla y Plantas de Cultivo de Meristemos En Papa (*Solanum Tuberosum* L.) Var. Alfa. *Rev. Mex. De Fitopatol.* **2003**, *21*, 176–180.
176. Gibbs, A.J.; Hecht-Poinar, E.; Woods, R.D.; McKee, R.K. Some Properties of Three Related Viruses: Andean Potato Latent, Dulcamara Mottle, and Onions Yellow Mosaic. *Microbiology* **1966**, *44*, 177–193. [CrossRef]
177. Tomlinson, J.A.; Walker, V.M.; Flewett, T.H.; Barclay, G.R. The Inhibition of Infection by Cucumber Mosaic Virus and Influenza Virus by Extracts from *Phytolacca Americana*. *J. Gen. Virol.* **1974**, *22*, 225–232. [CrossRef]

178. Murray, R.R.; Emblow, M.S.; Hetherington, A.M.; Foster, G.D. Plant Virus Infections Control Stomatal Development. *Sci. Rep.* **2016**, *6*, 34507. [CrossRef]
179. Bos, L.; Rubio-Huertos, M. Intracellular Accumulation of Passiflora Latent Virus in *Chenopodium Quinoa*. *Neth. J. Plant Pathol.* **1971**, *77*, 145–153. [CrossRef]
180. Hammond, J.; Bampi, D.; Reinsel, M.D. First Report of Plantago Asiatica Mosaic Virus in Imported Asiatic and Oriental Lilies (Lilium Hybrids) in the United States. *Plant Dis.* **2014**, *99*, 292. [CrossRef]
181. Hollings, M.; Stone, O.M. *Chenopodium Quinoa* Willd. as an Indicator Plant for Carnation Latent Virus. *Plant Pathol.* **1965**, *14*, 66–68. [CrossRef]
182. Pfaeltzer, H.J. *Chenopodium Quinoa*, a Herbaceous Test Plant for Chlorotic Leaf Spot Virus in Apple. *Neth. J. Plant Pathol.* **1968**, *74*, 12–16. [CrossRef]
183. Jelinek, C.F.; Pohland, A.E.; Wood, G.E. Worldwide Occurrence of Mycotoxins in Foods and Feeds—an Update. *J. Assoc. Off. Anal. Chem.* **1989**, *72*, 223–230. [CrossRef]
184. Miller, J.D. Mycotoxins in Small Grains and Maize: Old Problems, New Challenges. *Food Addit. Contam.* **2008**, *25*, 219–230. [CrossRef]
185. Magan, N.; Aldred, D. Post-Harvest Control Strategies: Minimizing Mycotoxins in the Food Chain. *Int. J. Food Microbiol.* **2007**, *119*, 131–139. [CrossRef]
186. Ramos-Diaz, J.M.; Sulyok, M.; Jacobsen, S.-E.; Jouppila, K.; Nathanail, A.V. Comparative Study of Mycotoxin Occurrence in Andean and Cereal Grains Cultivated in South America and North Europe. *Food Control* **2021**, 108260. [CrossRef]
187. Kolakowski, B.; O'ROURKE, S.M.; Bietlot, H.P.; Kurz, K.; Aweryn, B. Ochratoxin A Concentrations in a Variety of Grain-Based and Non–Grain-Based Foods on the Canadian Retail Market from 2009 to 2014. *J. Food Prot.* **2016**, *79*, 2143–2159. [CrossRef]
188. Woldemichael, G.M.; Wink, M. Identification and Biological Activities of Triterpenoid Saponins from Chenopodium Quinoa. *J. Agric. Food Chem.* **2001**, *49*, 2327–2332. [CrossRef]
189. Murphy, K.M.; Bazile, D.; Kellogg, J.; Rahmanian, M. Development of a Worldwide Consortium on Evolutionary Participatory Breeding in Quinoa. *Front. Plant Sci.* **2016**, *7*, 608. [CrossRef]

plants

Article

How Does Mechanical Pearling Affect Quinoa Nutrients and Saponin Contents?

Sifeddine Rafik [1,2], Mohamed Rahmani [2], Juan Pablo Rodriguez [3], Said Andam [2], Amine Ezzariai [4], Mohamed El Gharous [1], Salwa Karboune [5], Redouane Choukr-Allah [6] and Abdelaziz Hirich [4,*]

[1] Agricultural Innovation and Technology Transfer Center (AITTC), Mohammed VI Polytechnic University (UM6P), Ben Guerir 43150, Morocco; sifeddine.rafik@um6p.ma (S.R.); mohamed.elgharous@um6p.ma (M.E.G.)
[2] Department of Food and Nutritional Sciences, Section of Agricultural and Food Industries, Agronomic and Veterinary Medicine Hassan II Institute, Rabat 10112, Morocco; rahmanimohammed@yahoo.fr (M.R.); andamsaiid@gmail.com (S.A.)
[3] International Center for Biosaline Agriculture, Directorate of Programs, Dubai 14660, United Arab Emirates; jprodriguezcalle@yahoo.fr
[4] African Sustainable Agriculture Research Institute (ASARI), Mohammed VI Polytechnic University (UM6P), Laayoune 70000, Morocco; Amine.Ezzariai@um6p.ma
[5] Department of Food Science and Agricultural Chemistry, McGill University, 21111 Lakeshore Road, Sainte-Anne-de-Bellevue, QC H9X 3V9, Canada; salwa.karboune@mcgill.ca
[6] Department of Horticulture, Agronomic and Veterinary Medicine Hassan II Institute, Rabat 10112, Morocco; redouane53@yahoo.fr
* Correspondence: abdelaziz.hirich@um6p.ma

Citation: Rafik, S.; Rahmani, M.; Rodriguez, J.P.; Andam, S.; Ezzariai, A.; El Gharous, M.; Karboune, S.; Choukr-Allah, R.; Hirich, A. How Does Mechanical Pearling Affect Quinoa Nutrients and Saponin Contents?. *Plants* **2021**, *10*, 1133. https://doi.org/10.3390/plants10061133

Academic Editors: Cataldo Pulvento and Didier Bazile

Received: 31 March 2021
Accepted: 12 May 2021
Published: 3 June 2021

Abstract: Agriculture is facing many challenges, such as climate change, drought, and salinity, which call for urgent interventions for fast adaptation and crop diversification. The introduction of high-value and stress tolerant crops such as quinoa would be a judicious solution to overcome constraints related to abiotic stress and to increase land productivity and farmers' incomes. The implementation of quinoa in Morocco has not been supported by a full valorization program to control the quality of quinoa seeds. The novelty of this work is to assess the pearling operation as an efficient method of saponins removal as well as the determination of total residual saponins. This study aimed to evaluate the effects of several pearling durations on nutrient and saponin content of quinoa seeds of three tested varieties (Puno, Titicaca, and ICBA-Q5). Five pearling durations were tested (0, 2, 4, 6, 7, and 8 min) using a locally manufactured pearling machine. The results indicated that a pearling duration of two minutes was enough to reduce total saponin content from 0.49% to 0.09% for Puno variety, from 0.37% to 0.07% for Titicaca variety, and from 0.57% to 0.1% for ICBA-Q5 variety. Our results showed that pearling slightly reduced protein, total fat, and moisture contents for all varieties except for Puno, where total fat content slightly increased with the pearling. Puno variety had the highest seed content in terms of protein and total fat; the ICBA-Q5 variety had the lowest. Titicaca had the highest bran content in terms of protein and total fat, ICBA-Q5 had the highest bran content in terms of ash and the lowest bran content in terms of protein and total fat, and Puno had the lowest bran content in terms of ash. Pearling had no significant effect on macronutrient contents in the processed seed, but it resulted in a very highly significant difference for most of them in the bran except for Mg and S. Regarding seed content in terms of micro-nutrients, statistical analysis showed significant differences between varieties in terms of Zn, Cu, and Mn contents, but no significant difference was recorded for Fe or B. Pearling had no significant effect on seed micronutrient contents. Therefore, to retain maximum nutritional content in the quinoa and maintain quinoa integrity, it is necessary to limit the pearling duration of quinoa to two minutes, which is enough to reduce saponin content below the Codex Standard threshold (0.12%).

Keywords: pearling; postharvest; saponin; minerals; processing; seed bran; quinoa seeds quality; nutrition

1. Introduction

Water scarcity and drought are becoming direct consequences of climate change on a global scale in addition to soil salinization, especially in the Middle East and North Africa region (MENA) [1,2]. The introduction of salt-tolerant crop species is being considered as a great alternative to overcoming these environmental challenges. In this regard, quinoa (*Chenopodium quinoa* Willd.) is one of the plants with significant potential to maintain food security in countries suffering from arid conditions. Several research works have been conducted to introduce, develop, and promote quinoa crops around the world. The number of countries cultivating this staple crop has increased rapidly, from eight in 1980 to more than 120 countries in 2018 [3].

Regarding Morocco, quinoa was introduced for the first time in the Khénifra region in 1999 within the framework of the BAFI/BYU-IAV Hassan II project where 14 accessions were tested for adaptation trials [4]. Then, several research trials were carried out, especially in Agadir [5–7] and Rehamna [8–11], where several parameters were taken into account, such as irrigation frequencies, water quality, and soil nature. The Food and Agriculture Organization of the United Nations (FAO) recognized, in 2013, quinoa as a leading crop for food security and sustainability in the context of global change [12]; this initiative aimed to improve access and awareness of the quinoa value chain and accelerate the development of this crop around the world.

Quinoa is not only receiving attention worldwide due to its adaptability to different agro-environmental growth conditions, but also its high nutritional quality [13]. Over the last 20 years, quinoa (*Chenopodium quinoa* Willd.) has become a popular food, particularly in Europe and North America [14], because of the growing interest in vegetarian diets, its high nutritional quality, and the increase in people suffering from celiac disorders. Several studies have shown that quinoa is a very interesting food, because of its complete nutritional characteristics [15]. It is a starchy dicotyledonous seed named a pseudo-cereal. The protein content of quinoa seeds is substantial (12% to 20%) compared to corn (10%), rice (8%), and wheat (13%). In addition, a valuable characteristic of quinoa is the quality of its amino acid composition as well—especially the presence of lysine (5.1–6.4%) and methionine (0.4–1.0%). These amino acids are not abundant in vegetable diets [16,17]. Furthermore, starch is the major component of quinoa seeds and represents about 52–69.2% of the dry weight basis [18,19]; it is genuinely located in the perisperm. Quinoa is also considered a good source of fat; fat varies from 2% to 10% of quinoa depending on the variety [20]. Compared to other cereals, quinoa has high levels of calcium, phosphorus, magnesium, iron, zinc, potassium, and copper [16,21]. Some studies showed that the antioxidant properties of quinoa seeds and flour can be used as taken advantage of when used as ingredients to enrich food preparations. For instance, they can be used in meat industries that require antioxidants from natural sources to substitute synthetic antioxidants because of their negative effects [22,23].

The quality of quinoa is considered as one of the most crucial aspects that should be taken into consideration in order to justify its consumption and its manufacturing on an industrial scale. Nevertheless, quinoa quality in Morocco is facing many technical constraints, such as the bitter taste, which is a major organoleptic default. This sensorial limitation is explained by the lack of control of operations that aim to remove saponin from seeds and the dosage of residual saponin. Quinoa's saponins are considered anti-nutritional factors that cause bitterness. Additionally, saponins may cause digestive irritation [24]. Besides, saponins' concentration and distribution in seeds are variables depending on varieties and climatic conditions [25]. According to the quinoa standard (step 8 before publication), the Environmental Working Group (EWG) agreed to a saponin content of 0.12% as a maximum. This threshold was adopted from the Bolivian standard "Andean Standard NB 0038 for processed quinoa seeds" [26], measured through the afrosimetric method (foam test) developed by Koziol et al. [27]. However, a standardized method for total saponin quinoa dosage has not yet been officially published. The presence of saponins requires processing in order to eliminate saponins partially or totally. They are mainly present in the pericarp (86%), which explains why external abrasion is one of the best options for saponin removal [28,29]. Many methods have been developed for saponin removal; the most commonly adopted one is the mechanical abrasion process. It is an operation based on physical frictions to remove the bran. It allows obtaining a by-product (seed bran) which is rich in saponins and other nutrients. This process must eliminate saponins while preserving seed nutrients and physical properties. The optimization of the process can be done by adjusting a set of parameters involved in the mechanical abrasion operation, such as the duration of the abrasion (pearling time) and milling performance. This research focused on pearling duration as a principal factor to optimize the pearling process. The optimal pearling duration will allow saponin elimination, and preservation of the overall nutrient profile (macronutrients and micronutrients) and morphological aspects. Moreover, this research investigated the behavior of three quinoa cultivars in response to the pearling process.

2. Results

2.1. *ANOVA Results*

Table 1 summarizes results of the ANOVA (analysis of variance) of investigated parameters for both processed seeds and seed bran. Pearling had no significant effect on macronutrient contents in the processed seeds of any variety. It caused a very highly significant difference ($p < 0.001$) for most of the micronutrients except for Mg, S, and Fe. Pearling duration caused significant differences ($p < 0.001$) in 1000 seed weight, moisture, and saponin content for all quinoa varieties. However, there was no significant effect on the nutritional contents of quinoa seed varieties.

2.2. *Physical Parameters*

2.2.1. Pearling Efficiency

Pearling efficiency (PE) is defined as the ratio between processed seeds and raw seeds. Figure 1 shows the variation in PE after 8 min of pearling. Presented data indicate that pearling efficiency varies from one variety to another, and Puno variety has the highest pearling rate and Titicaca has the lowest.

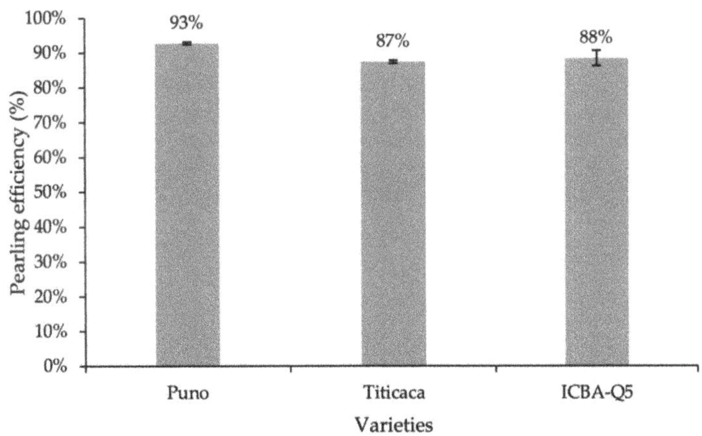

Figure 1. Variation of pearling efficiency per variety. Error bars indicate the standard deviation. Treatments without a common letter are significantly different at $p < 0.05$.

Table 1. Results of ANOVA (analysis of variance) for processed seeds and seed bran—the effects of pearling on nutritional and saponin contents of quinoa seeds.

Product	Factor	1000 SW	PR	A	TF	M	Macro-Elements							Micro-Elements					SP
							N	P	K	Mg	Na	Ca	S	Fe	Zn	Cu	Mn	B	
Processed seeds	V	***	***	***	***	***	***	***	***	NS	***	***	NS	NS	***	*	**	NS	NS
	PD	***	NS	NS	NS	***	NS	NS	NS	NS	NS	NS	NS	NS	NS	NS	NS	NS	***
	V × PD	NS	NS	NS	NS	*	NS	NS	NS	NS	NS	NS	NS	NS	NS	NS	NS	NS	*
	p-value	0.223	0.439	0.855960	0.133	0.022630	0.439	0.358	0.904432	0.9115	0.939080	0.968891	0.889	0.961	0.145	0.7194	0.73797	0.9343	0.0469
Seed bran	V	-	***	NS	***	***	***	***	**	*	***	***	***	***	***	**	***	**	-
	PD	-	***	***	***	***	***	***	***	NS	***	***	NS	NS	***	NS	NS	NS	-
	V × PD	-	***	NS	***	**	***	***	*	NS	**	NS	NS	*	NS	NS	NS	NS	-
	p-value		6.09×10^{-5}	0.164	1.01×10^{-7}	0.001727	6.09×10^{-5}	5.82×10^{-6}	0.01169	0.9239	0.00407	0.737929	0.095	0.0317	0.568709	0.25700	0.097694	0.1639	-

V: variety; PD: pearling duration; NS: non significant; -: no data available; *: significant difference; **: highly Significant; ***: very highly significant; SW: seed weight; PR: protein; A: ash; TF: total fat; M: moisture; N: nitrogen; P: phosphorus; K: potassium; Mg: magnesium; Na: sodium; Ca: calcium; S: sulfur; Fe: iron; Zn: zinc; Cu: copper; Mn: manganese; B: boron; SP: saponin.

2.2.2. 1000 SW

Figure 2 presents the variation of 1000 SW in response to pearling time. The obtained data clearly indicate that average SW decreases with the pearling as a result of seed polishing and bran removal. After 2 min of pearling, average seed weight decreased by 10, 3, and 19% for Puno, Titicaca, and ICBA-Q5 varieties, respectively. However, it decreased by 18, 21 and 23% after 8 min of pearling for Puno, Titicaca, and ICBA-Q5 varieties, respectively.

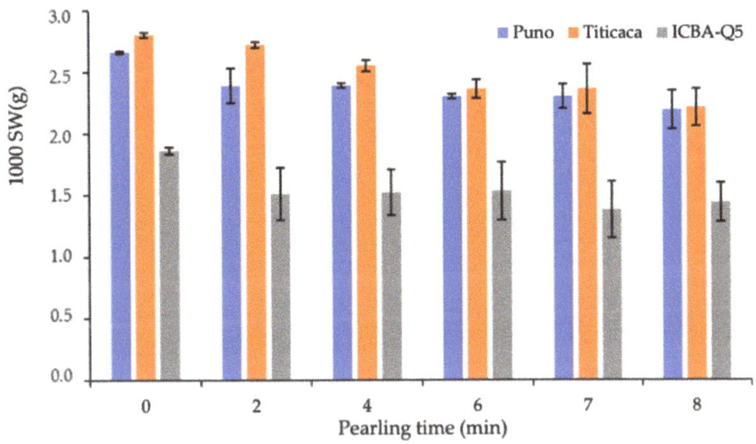

Figure 2. Variation of 1000 SW in response to pearling duration. Error bars indicate the standard deviation. Treatments without a common letter are significantly different at $p < 0.05$.

2.2.3. Morphological Aspects

Figure 3 shows morphological aspects of Puno seed in response to the pearling process. The obtained pictures show the effects of mechanical friction during the abrasion. After 6 min of pearling, quinoa seeds start to have some damage, and the abrasions mill the embryo and start damaging the perisperm. After 2 min of pearling, the pericarp is entirely removed, and the seed becomes transparent.

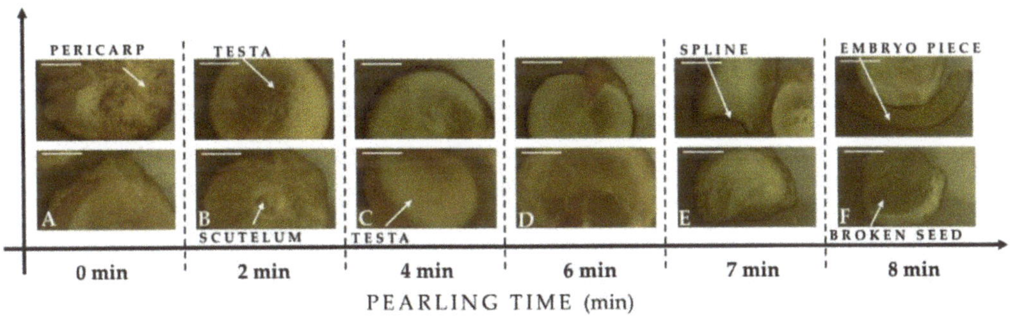

Figure 3. Optical microscopy images of raw (0 min) and treated seeds of Puno Variety after different pearling durations. Pearling duration: (**A**). Raw seed (0 min); (**B**) 2 min; (**C**) 4 min; (**D**) 6 min; (**E**) 7 min; (**F**) 8 min. (Bar length: 500 μm).

2.3. Chemical Parameters

Table 2 summarizes the results of Tukey's HSD post hoc test, indicating homogenous groups using small letters.

Table 2. Results of Tukey's HSD post hoc test for both processed seeds and seed bran—the effects of pearling on the nutritional content of quinoa seeds.

Variety/Product	PD	Proximate Analysis				Macro-Nutrients							Micro-Nutrients				
		PR	A	TF	M	N	P	K	Mg	Na	Ca	S	Fe	Zn	Cu	Mn	B
Puno																	
Processed seeds	0	a	a	b	a	a	a	a	a	a	a	a	a	b	a	a	a
	2	a	b	ab	ab	a	a	b	ab	ab	b	a	a	ab	a	bc	ab
	4	a	b	a	b	a	a	b	ab	b	b	a	a	ab	a	ab	ab
	6	a	b	a	bc	a	a	b	ab	b	b	a	a	ab	a	bc	ab
	7	a	b	a	c	a	a	b	ab	ab	b	a	a	ab	a	bc	ab
	8	a	b	ab	bc	a	a	b	b	ab	b	a	a	ab	a	c	b
Seed bran	2	d	a	c	a	c	a	a	a	a	a	c	a	a	a	b	a
	4	cd	b	c	a	c	a	b	ab	a	a	bc	a	a	a	ab	b
	6	bc	c	b	a	b	a	c	b	a	b	ab	a	a	a	a	b
	7	ab	c	ab	a	ab	b	c	b	a	b	a	a	a	a	ab	b
	8	a	c	a	a	a	b	c	b	a	b	a	a	a	a	ab	b
ICBA-Q5																	
Processed seeds	0	a	a	a	a	a	a	a	a	a	a	a	a	a	a	a	a
	2	a	a	a	ab	a	a	a	a	a	a	a	a	a	a	a	a
	4	a	a	a	ab	a	a	a	a	a	a	a	a	a	a	a	a
	6	a	a	a	ab	a	a	a	a	a	a	a	a	a	a	a	a
	7	a	a	a	b	a	a	a	a	a	a	a	a	a	a	a	a
	8	a	a	a	b	a	a	a	a	a	a	a	a	a	a	a	a
Seed bran	2	a	a	a	a	a	a	a	a	a	a	a	a	a	a	a	a
	4	a	a	a	a	a	a	a	a	ab	a	a	a	a	a	a	a
	6	a	a	a	a	a	a	a	a	b	a	a	a	a	a	a	a
	7	a	a	a	a	a	a	a	a	ab	a	a	a	a	a	a	a
	8	a	a	a	a	a	a	a	a	ab	a	a	a	a	a	a	a
Titicaca																	
Processed seeds	0	a	Ash	b	a	a	a	a	a	a	a	a	a	b	a	a	a
	2	a	b	ab	ab	a	a	b	ab	ab	b	a	a	ab	a	bc	ab
	4	a	b	ab	b	a	a	b	ab	b	b	a	a	ab	a	ab	ab
	6	a	b	ab	bc	a	a	b	ab	b	b	a	a	ab	a	bc	ab
	7	a	b	ab	c	a	a	b	ab	ab	b	a	a	ab	a	bc	ab
	8	a	b	ab	bc	a	a	b	b	ab	b	a	a	ab	a	c	b
Seed bran	2	d	a	c	a	d	b	a	a	a	a	a	a	b	c	ab	a
	4	cd	b	c	a	cd	b	b	b	b	b	a	a	b	bc	a	a
	6	bc	c	b	a	bc	a	c	b	bc	c	a	a	ab	ab	ab	a
	7	ab	c	ab	a	ab	a	c	b	c	d	a	a	ab	a	ab	a
	8	a	c	a	a	a	a	c	b	c	e	a	a	a	a	b	a

PD: pearling duration; PR: protein; A: ash; TF: total fat; M: moisture; N: nitrogen; P: phosphorus; K: potassium; Mg: magnesium; Na: sodium; Ca: calcium; S: sulfur; Fe: iron; Zn: zinc; Cu: copper; Mn: manganese; B: boron. Treatments without a common letter are significantly different at $p < 0.05$.

2.3.1. Proximate Analysis

Figure 4 shows the variations of protein, ash, total fat, and moisture content in processed seeds and seed bran (by-product) in response to pearling duration. Statistical analysis revealed very highly significant differences ($p < 0.001$) between tested varieties in terms of protein, ash, and total fat, but pearling duration affected only seed moisture. Pearling slightly reduced protein, total fat, and moisture contents for all varieties except for Puno, where total fat content slightly increased with the pearling. However, for quinoa bran, the findings are different from the processed seeds: both variety and pearling showed very highly significant differences for most of the parameters. Protein and total fat content increased when pearling duration increased for all tested varieties, whereas ash content decreased with pearling duration. The data indicate that the Puno variety had the highest protein and total fat contents in the seeds; the ICBA-Q5 variety recorded the lowest. Titicaca had the highest bran contents of protein and total fat; ICBA-Q5 had the highest bran content of ash, and the lowest bran content of protein and total fat; and Puno had the lowest bran content of ash.

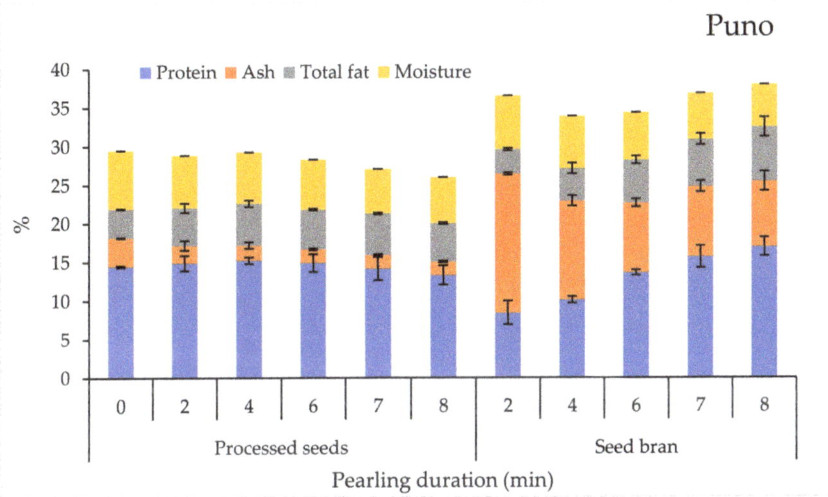

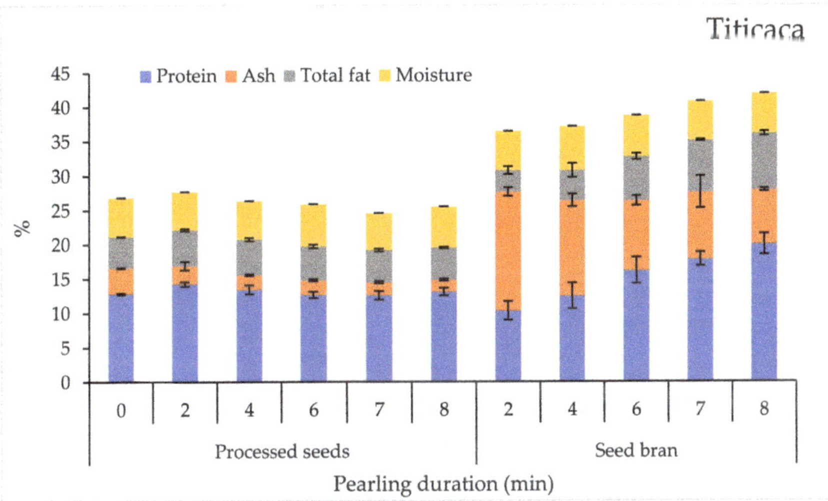

Figure 4. *Cont.*

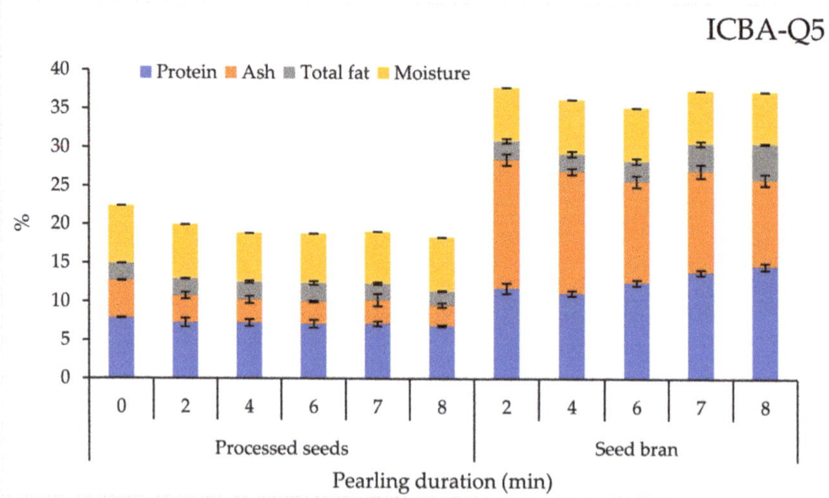

Figure 4. Variations of protein, ash, total fat, and moisture contents in processed seeds and seed bran in response to pearling duration. Error bars indicate the standard deviations.

2.3.2. Macronutrient Content

Macro-nutrient variations in response to pearling duration are presented in Figure 5. Pearling had no significant effect on macronutrient contents in the processed seeds, but it had resulted in very highly significant differences ($p < 0.001$) for most of the macronutrient contents in the bran except Mg and S. The variety effect was obvious and significant for all macronutrients and for both processed seeds and bran. Regarding processed seeds, the highest contents in terms of N, P, and S were recorded for Titicaca; the highest content in terms of K, Mg, and Na for ICBA-Q5; and the highest content in terms of Ca for Puno. Most of the macronutrient contents in seed bran decreased with pearling, except N and P whose contents increased with the pearling.

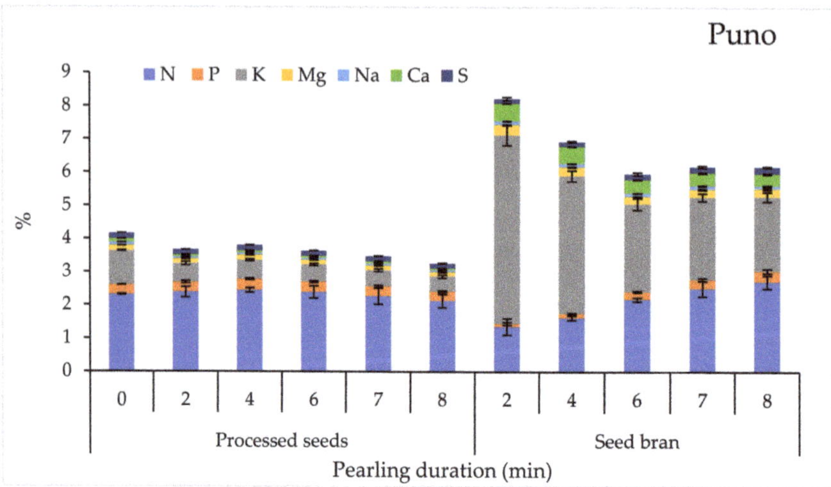

Figure 5. *Cont.*

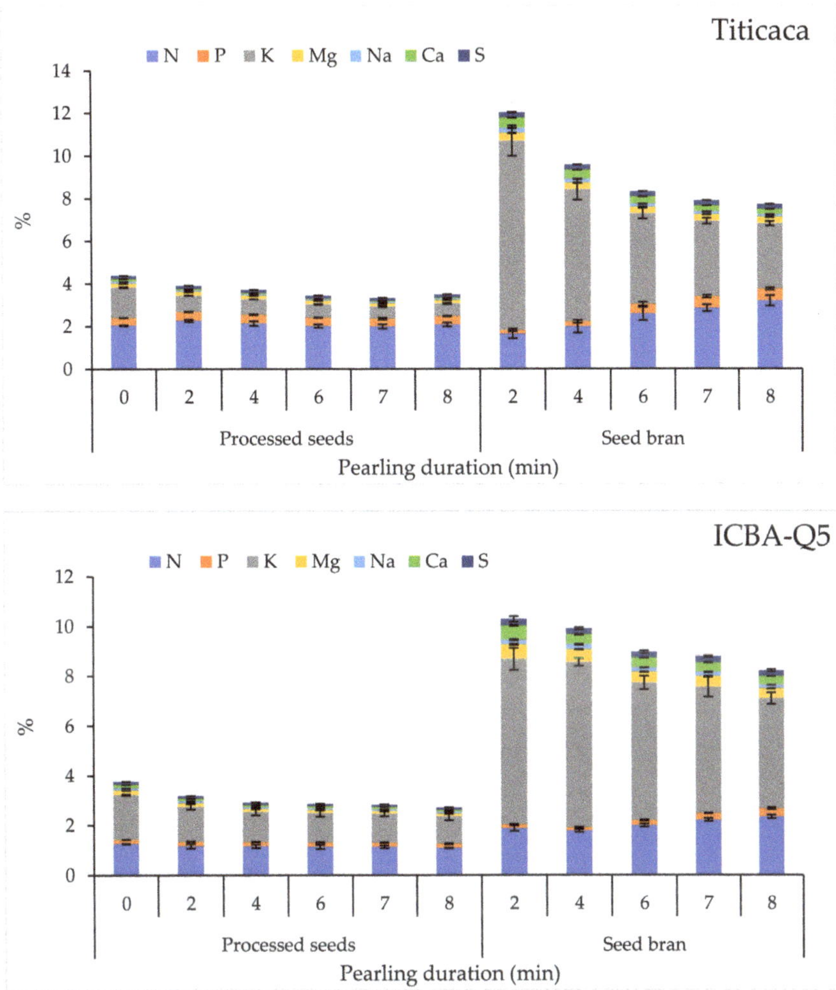

Figure 5. Variations of macronutrient contents in processed seeds and seed bran in response to pearling duration. Error bars indicate the standard deviations.

2.3.3. Micronutrient Contents

The effects of pearling on micronutrient contents for the tested quinoa varieties are presented in Figure 6. Statistical analysis showed significant differences ($p < 0.001$) between varieties in terms of Zn, Cu, and Mn contents in processed seeds. Pearling duration had no significant effect on micronutrient contents in processed seeds. Nevertheless, for seed bran, there were very highly significant differences between tested varieties in terms of micronutrients. Pearling had a significant effect only on Zn, which increased with the increased pearling duration.

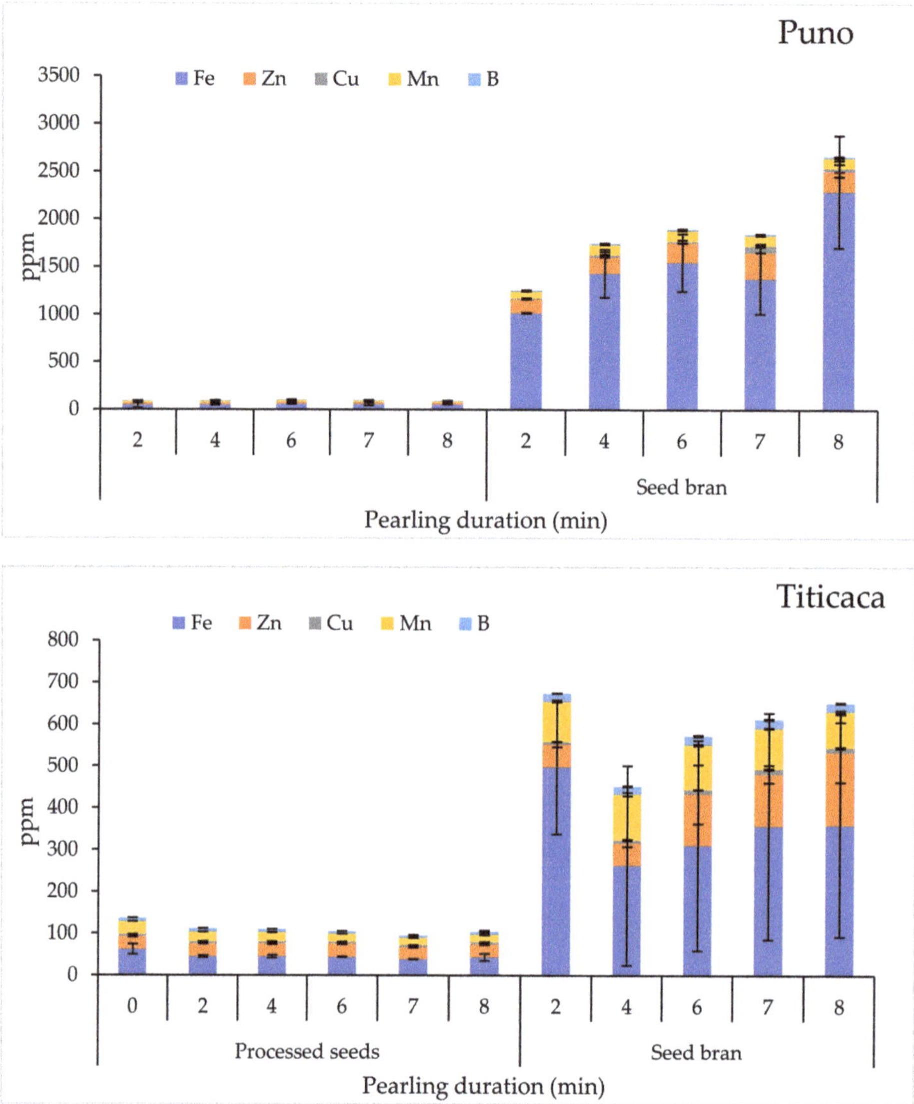

Figure 6. *Cont.*

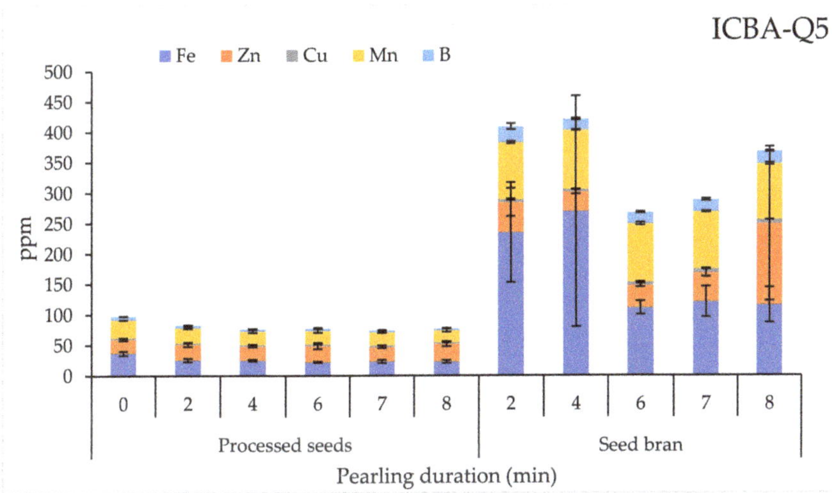

Figure 6. Variations of micronutrient contents in processed seeds and seed bran in response to pearling duration. Error bars indicate the standard deviations.

2.3.4. Saponin Content

Figure 7 shows the variations of saponin contents for processed seeds in response to several pearling durations. According to statistical analysis, saponin content in processed seeds was affected only by pearling duration; no significant difference was recorded between varieties. However, the interaction between variety and pearling duration factors was significant, which indicates that the tested varieties responded differently to the pearling process. For instance, the ICBA-Q5 variety responded very well to pearling, as it had the highest saponin content in raw seeds (0 min) and the lowest in processed seeds (6 min), and consequently the highest saponin removal.

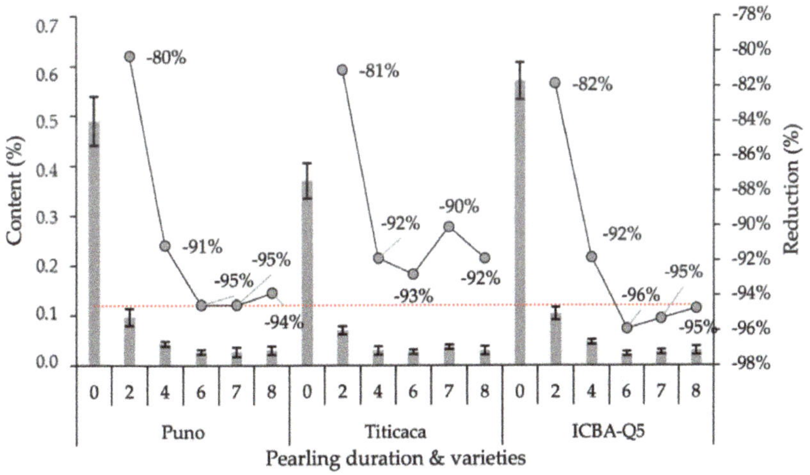

Figure 7. Variations of saponin contents in processed seeds in response to pearling duration. Error bars indicate the standard deviation. Treatments without a common letter are significantly different at $p < 0.05$.

2.4. Pearson Correlation

Pearson's correlation analysis was conducted for all investigated nutrients in processed seeds and seed bran separately, by correlating all the physicochemical parameters (productibility and growth) with the varieties used. The obtained results are shown in Figure 8.

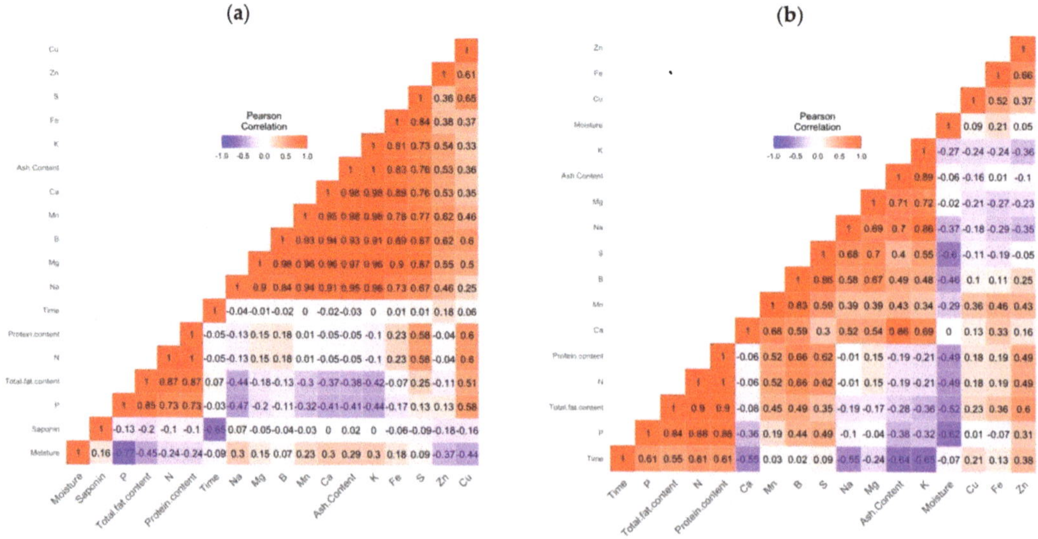

Figure 8. Pearson correlations among all chemical parameters for processed seeds (**a**) and seed brans (**b**).

The analysis revealed a strong positive correlation between groups of parameters G1 (N, protein content) and G2 (Na, Mg, Mn, Ca, ash content, and K). Strong negative correlations were found between moisture and P, saponin, and time. On the other hand, no correlation between time, Mn, K, saponin, Ca, and K was observed (Figure 8a).

For the obtained seed brans (Figure 8b), a strong correlation was found between N, protein content, and total fat. A strong negative correlation was observed between time, ash content, K, moisture, P, and S. No correlation between moisture and Ca was observed.

2.5. Principal Component Analysis (PCA)

Results of PCA (Figure 9) indicate that the first two components represent over 69% of the variability. PCA axis 1 was largely determined by B, Mn, S, Ca, Na, and K. Among them, B and Ca had the highest positive correlation (r = 0.927–0.975). Variation along PCA axis 2 was highly determined by P, fat content, N, and protein content. P and fat content had the highest positive correlation (r = 0.932–0.937).

The PCA graph of variables showed that the pearling duration was correlated negatively with saponin, moisture, Ca, Na, Mg, and Ca. On the other hand, this key parameter was correlated positively with the rest of the parameters.

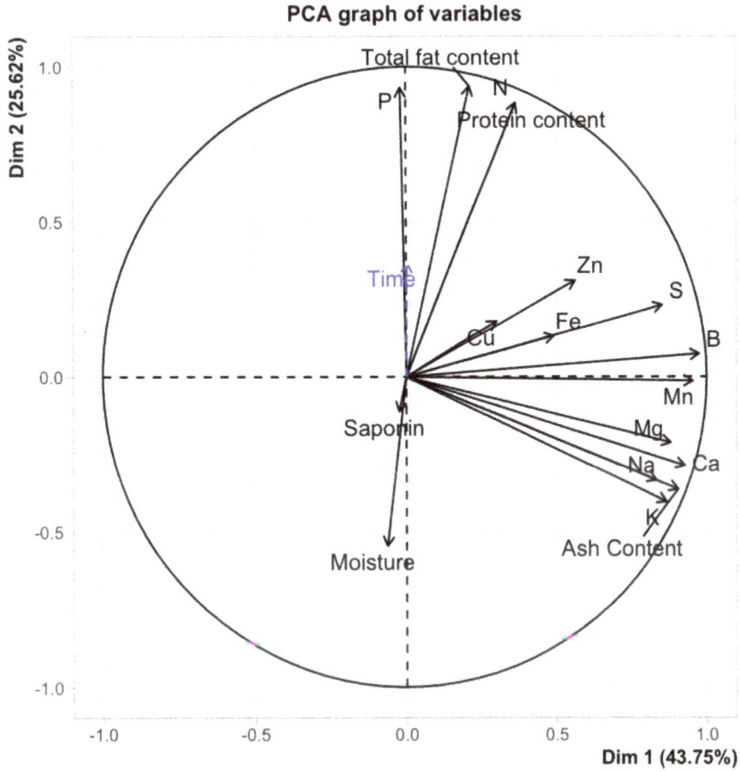

Figure 9. PCA analysis of investigated parameters for processed seeds.

3. Discussion

Since its introduction to Morocco in the 2000s, Quinoa has been seen as a rustic and stress-tolerant crop with several potentialities to replace cereals and other traditional crops in marginal environments, mainly those affected by drought and salinity. A deep understanding of the nutritional profiling of quinoa and the high amount of saponin reveal the importance of processing in order to remove saponins and produce a palatable seed. Regarding the morphological aspects of quinoa seeds, the quinoa seed is disc-shaped, approximately 2 mm in diameter, and 0.5 mm in thickness. The size of a quinoa seed depends on several factors, such as quinoa cultivar and abiotic conditions (climate, soil, etc.). In fact, the whole quinoa seed is divided into three main structural components: perisperm, embryo, and bran. The bran is defined as the out layer that surrounds the embryo. The embryo entirely covers the perisperm, the innermost layer, as a headband. The bran is eliminated during abrasion (manual or mechanical) because it contains 86% saponins compared to other grain fractions [29]. The embryo and perisperm contain 11% and 3% of saponins, respectively [29].

The special structure of quinoa seeds requires special scarification methods, and the dehulling rates change when milled for the same rate of duration using different equipment [30]. Hence, it is necessary to have a balance between the pearling duration and the nutritional contents of quinoa. For all used varieties, the thousand-seed weight of quinoa seeds decreased with increase pearling duration due to the removal of the external layer of quinoa [31]. Some research works have mentioned that the average diameter can reveal some quinoa characteristics, such as hardness coefficient and saponin content. Otherwise, the higher the hardness coefficient of quinoa, the more saponins are present [32].

The average thousand-seed weight was negatively correlated with the pearling duration. Our finding indicates that after 2 min of mechanical abrasion, a high percentage of saponins were removed.

There are two main processes of saponin removal. The first one is a wet method based on washing. It increases seed moisture, which then requires drying to prevent mold growth, and enhances the seed's shelf life (while raising the cost of seed processing) [10]. The second method, which is the commonest one, is a dry method based on abrasion that removes the outer bran layer by mechanical scarification. Often, processing operations involve a preliminary washing step following abrasion to further reduce the level of saponins [33]. Nevertheless, for some bitter quinoa varieties that contain higher amounts of saponins, if a wash operation is skipped, the embryo may be damaged as the next outer layer after the bran, which is totally eliminated [34]. Quinoa's embryo is the richest nutritional source; it contains 57% of the protein, 49% of the fat, 20% of the sugar, 45% of the dietary fiber, and 51% of the ash of the whole grain [29]. Hence, removal techniques that keep the embryo intact preserve the complete nutritional profile of quinoa. Abrasion alone after 30 s reduced saponin levels from 1.19% to 0.45%. However, after an additional 2 min, the saponin level was reduced slowly to 0.21% [33]. Our results showed the same tendency of reducing saponins for the three studied varieties. Two minutes of mechanical abrasion was enough to reduce saponins below 0.12% which is the accepted threshold for quinoa consumption [35]. On the other hand, washing alone augmented the saponin percentage of oleanolic acid, while diminishing phytolaccagenic acid. Although abrasion decreased the percentage of oleanolic acid, it increased the percentage of phytolaccagenic acid [33]. These variations are correlated with the chemical structure of the specific saponin and the location within the bran layer. Furthermore, hydrating the quinoa without manual abrasion increased the saponin levels from 3.3% to 3.6% [32]. The mechanical polishing permitted a reduction of saponins from 2% to 0.4% for Titicaca variety, and from 1.4% to 0.5% for Puno, decreases of 80% and 64% of the initial saponin respectively [36].

The protein contents of the studied quinoa cultivars were equal to 14.4%, 12.8%, and 7.9% (dry weight basis) for Puno, Titicaca, and ICBA-Q5, respectively. Puno and Titicaca's protein contents were consistent with the protein content (9.5–15.7%, dry weight) reported by Nowak Verena [37]. However, protein content for each variety did not change during the mechanical abrasion. The protein loss increased in terms of seed weight when the pearling duration was higher (after 4 min of mechanical abrasion). The decrease in weight can be explained by the fact that the protruding scutellum of the pericarp of quinoa was scoured away [38,39]. Furthermore, when pearling duration was greater than 6 min, protein loss became greater, since it is primarily present in an embryo, which accounts for 57% of total quinoa protein [37]. Our finding indicates that the embryo and endosperm were milled at this stage of the pearling process.

The fat contents of quinoa cultivars were equal to 3.7%, 4.5%, and 2.2% (dry weight basis) for Puno, Titicaca, and ICBA-Q5, respectively. All contents fit within the range of the total lipid content of quinoa [40]. The increase of total fat after 2 min of mechanical abrasion is explained by the structure and nutritional distribution of quinoa kernels, as 49% of the total fat content of quinoa is present in the embryo, whereas the pericarp contains just 5% of the total fat [29,30]. Moreover, we noticed a decrease in fat content after 6 min of mechanical abrasion, as the bran was removed and the embryo was ground. The obtained results are aligned with another study showing that a dehulling rate of 8.6% can decrease fat content by 7.6% [41].

The ash contents for the three studied varieties fit within the range reported in the literature [40]. The ash content of treated seeds decreased with the increase of the pearling duration. After 2 min of mechanical abrasion, the ash content decreased sharply for all studied varieties. The decrease of ash content is explained by the fact that the pericarp, where most of the minerals are mostly concentrated, was removed. On the other hand, the seed bran obtained from two minutes of mechanical abrasion is highly rich in minerals, and the ash contents were equal to 18.0%, 17.3%, and 16.7% (dry weight basis) for Puno,

Titicaca, and ICBA-Q5, respectively. In another study, Ando et al. [29] reported that the ash content was about 9.2% (dry weight basis) in the bran after pearling. That is relatively lower than the content found in this work, which can be explained by the use of different pearling processes and equipment.

A suitable pearling duration is the one that results in saponin content lower than the threshold required from the CODEX standard [35]. Thus, two minutes of mechanical abrasion was the best pearling duration. In this study, we found that K was the most abundant macroelement for Puno, Titicaca, and ICBA-Q5. The remaining macroelements, by abundance, were P, followed by Mg, Ca, S, and then Na. Regarding micronutrients, Fe was the most abundant element with concentrations in raw seeds of 55.2 mg/kg for Puno, 61.8 mg/kg for Titicaca, and 36.81 mg/kg for ICBA-Q5. Our findings in this regard are in agreement with results reported by many other studies where K and P were the dominant macroelements, followed by Mg and then Ca. Fe was the dominant micronutrient [29,36,42]. The quinoa cultivated in Morocco is characterized by higher amounts of K, Ca, and Mg compared to other studies focusing on mineral profiling of quinoa seeds [29,42,43]. Mhada et al. [36] explained that these high contents are mainly due to calcareous soil rich in K, Ca, and other elements.

Our results suggest that protein and mineral contents were not affected by pearling, which indicates that protein and most of minerals are mostly located in the endosperm and embryo, which confirm the finding reported by Prego et al. [44]. According to Konishi et al. [45], the P, K, and Mg are located in embryonic tissue. In particular, P's origin is attributed to phytic acid, and the origins of Mg and K to phytate. The same authors also reported that K and Ca were present in the pericarp [45]. Furthermore, they found that Ca is scarcely found in embryonic tissues of quinoa seeds. It occurred mostly in the pericarp and seed coat, as well as the boundary between perisperm and embryo. It is associated with carboxyl groups of pectin molecules in the cell wall to form Ca-pectin complexes [44], which confirms the finding of this study that showed decreases in Ca of 58%, 64%, and 14% for Puno, Titicaca, and ICBA-Q5, respectively. Moreover, the decreases in K (56%, 56%, and 16% for Puno, Titicaca, and ICBA-Q5, respectively) can be explained by the fraction located in the pericarp; the K fraction located in the embryo was not eliminated. According to Ando et al. [29], milled quinoa had lower K, Mg, and Ca, and higher P. Nevertheless, the authors also found that Fe, Zn, Cu, and Mn contents remained stable [29]. This disagreement may be explained by some differences in terms of genetic material, since they worked with another variety (quinoa Real) and a lower degree of polishing by the pearling machines used for seed processing [29,36]. According to a previous study conducted with the same pearling machine, the authors found that the pearling process had no effect on macronutrients and protein content, except for Ca, which declined by 28% for Puno variety [10,11]. However, micronutrient contents were significantly reduced. The pearling process resulted in significant losses for most micronutrients—by 60%, 57%, 2%, and 6% for Cu, B, Fe, and Zn, respectively; however, there was no significant reduction in terms of Mn [10,11].

During the pearling process, an important portion of the nutrients is wasted in the produced seed bran; therefore, its valorization is essential. All seed brans generated from the three studied varieties are rich in nutrients compared to the raw seeds. This is explained by the fact that the pericarp has high mineral content, as all produced seed brans contain high ash content. In fact, the research findings demonstrate the richness of quinoa seed brans. However, their valorization will mainly depend on several parameters (saponin contents, microelement contents, and microelement contents), and their usage scope is vast: they can be valorized in food, cosmetic, pharmaceutical, or agronomic industries. Recently, many studies were conducted to valorize saponins in different sectors. For instance, the effect of saponins on pests was studied. Some recent studies suggested the implementation of saponins as an ingredient for the storage of food and grains in order to preserve their sanitary quality. In fact, saponins rupture the internal mucous cells of pest's intestines by forming complexes with digestive enzymes such as proteases [46].

According to Zegarra [47], the acaricidal activity was evaluated from saponins present in bitter quinoa varieties. The author indicated that the Markjo variety shows bioactivity close to a commercial acaricide product. Furthermore, saponins have a tremendous inhibitory effect on fungi such as botrytis cinerea [48]. Quinoa saponins and polyphenols have high antioxidant effects and free radical scavenging activities, as well as an anti-inflammatory effect [49]. Drugs that contain saponins improve the body's immune response significantly through the enhancement of absorption and their hemolytic properties. Saponins grabbed worldwide attention due to their anti-carcinogenic characteristics and cholesterol-lowering effects [50]. In food industries, saponins have a positive effect on the physical stabilization of oil/water (O/W) emulsions. Moreover, they contribute to preserving liquids from oxidation, and they can be incorporated as foaming agents in beverage industries [50,51].

4. Materials and Methods

4.1. Seed Material

Quinoa seeds (*Chenopodium quinoa* Willd) grown in Berrechid, Morocco (2018 harvest) were used in this experiment. Seeds of three varieties of quinoa (Puno, Titicaca, and ICBA-Q5) were used after threshing and winnowing operations to study the effects of several pearling durations on seed nutritional and anti-nutritional properties. Quinoa seeds were not unisized before the pearling process, since it is not important to ensure this additional postharvest operation as well as simulating the real conditions that will be adopted by farmers.

4.2. Quinoa Pearling Machine Design

The pearling machine used in the experiment was locally manufactured by Benrim farm (Figure 10). It is equipped with two motors. The first one is designed to turn a drum with a rotation speed of 750 rpm. The second one is more powerful (3000 rpm) and designed to extract the fine dust produced during the pearling process. The rotating drum is made of an 80 cm long perforated inox steel and has six baffles distributed throughout the drum.

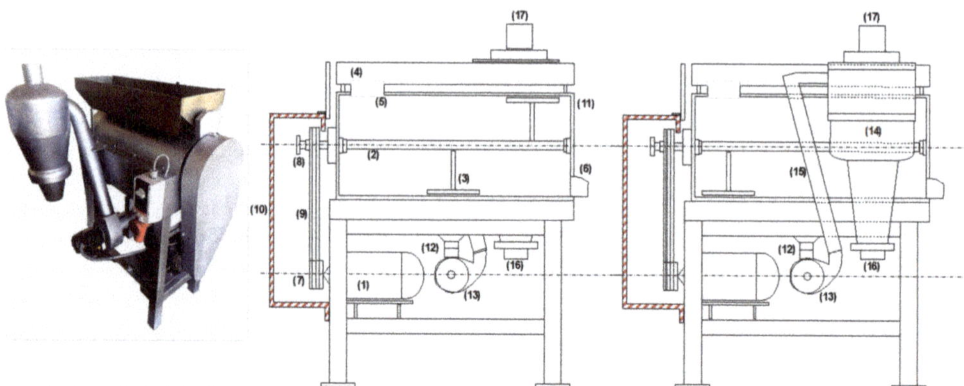

a) The mechanical abrasion machine developed within the Quinoa Rehamna project

b) Cross-sectional view (level 1) of the mechanical abrasion machine

c) Cross-sectional view (level 2) of the mechanical abrasion machine

(1) : Motor Rotating Drum (1500 rpm); (2) : Rotating drum axis; (3) : Baffle (3x2); (4) : Feed hopper; (5) : Input; (6) : Output; (7) : Driving pulley; (8) : Receiving pulley; (9) : Transmission belt; (10) : Safety device; (11) : Perforated rotating drum (80 cm length); (12) : Co-product output; (13) : Vacuum motor (3000 rpm) ; (14) : Cyclone separator; (15) : Outlet line; (16) : Quinoa husk output; (17) : fine dust output;

Figure 10. Design of the quinoa pearling machine used in the present experiment.

4.3. Pearling Treatments and Machine Operating Mode

The machine operates on a semi-industrial scale—a minimal quantity of 8 kg per sample is required. In this trial, five pearling durations were tested (0, 2, 4, 6, 7, and 8 min) for the three selected varieties; 0 min corresponds to raw seeds. Each combination pearling duration x variety was repeated 3 times. For each duration, a 500 g sample was taken, and a quantity of the seed bran recovered for each step was measured. During seed processing, the speed of rotation and friction (seed-seed, seed-drum, and seed-baffles) gradually increased the temperature of the seeds to 40 °C and decreased the moisture from 13% to 10%. After carrying out the pearling process, we collected 99 samples, including 45 processed quinoa samples, 45 seed bran samples, and 9 samples of the untreated raw seeds of the three tested varieties.

4.4. Physical Analysis

A precision electronic balance (reading to 0.001 g) was used to determine the 1000 SW by weighing 100 seeds in triplicate and then extrapolating this weight to get 1000 SW [52]. Pictures of seeds were taken using optical microscopy Nikon Eclipse Lv100nd.

4.5. Chemical Analysis

After processing, quinoa samples were ground to a fine powder using FOSS CT 293 Cyclotec grinder with a range of screens of 0.3 mm.

The moisture content was measured by drying 100 g of the sample at 105 °C for 48 h according to ICC Standard 110/1. The protein, fat, and ash contents were determined according to AOAC procedures [53].

For the evaluation of protein content, a macro-Kjeldahl method (N × 6.25) was used. The fat content was determined by Soxhlet method, using hexane solvent for the extraction for 4 h [54].

The ash content was determined according to the AOAC 923.03 standard. Micronutrients were determined after sample mineralization. Representative samples (0.25 g) were digested with 7.5 mL of nitric acid (HNO3) in a DigiPrep system for two hours at 100 °C. After digestion, the solutions were filtered through 45 µm filters, and the filtrates were diluted to 50 mL with deionized water and acidified (2% HNO3) in order to undergo the analysis by ICP-OES using Agilent technologies 5110 ICP-OES [10].

All measurements were performed in triplicate.

4.6. Saponin Determination by Liebermann–Buchard Colorimetric Assay

The protocol allows the determination of triterpene saponins from quinoa seeds. The principle of the assay is based on the reaction between the Liebermann–Burchard (LB) reagent and the triterpene molecules by developing a color [55].

4.6.1. Determining the Wavelength for the Dosage

A series of absorbance readings were taken in different wavelengths to get the wavelength corresponding to the maximum absorbance. We found that the maximal absorbance corresponds to 528 nm.

4.6.2. Dilutions for Range Preparation

According to Table 3, a series of standard solutions (0.2, 0.3, 0.5, and 0.8 mL) were taken, and the mixed solutions were prepared by the method described above for measuring absorbance values.

Table 3. Dilution series for the standard curve determination.

0.8 mg/mL	$Fd = \frac{Vf}{Vi} = \frac{1\left(\frac{mg}{ml}\right)}{0.8\left(\frac{mg}{ml}\right)} = 1{,}3$	1 mL of the stock solution and 0.3 mL of ethanol.	1 mL of the diluted solution is taken.
0.5 mg/mL	$Fd = \frac{Vf}{Vi} = \frac{1\left(\frac{mg}{ml}\right)}{0.5\left(\frac{mg}{ml}\right)} = 2$	0.5 mL of the stock solution and 0.5 mL of ethanol.	1 mL of the diluted solution is taken.
0.3 mg/mL	$Fd = \frac{Vf}{Vi} = \frac{1\left(\frac{mg}{ml}\right)}{0.3\left(\frac{mg}{ml}\right)} = 3.3$	0.3 mL of the stock solution and 0.7 mL of ethanol.	1 mL of the diluted solution is taken.
0.2 mg/mL	$Fd = \frac{Vf}{Vi} = \frac{1\left(\frac{mg}{ml}\right)}{0.2\left(\frac{mg}{ml}\right)} = 5$	0.5 of the stock solution and 4.5 mL of ethanol.	1 mL of the diluted solution is taken.

4.6.3. Saponin Extraction

Five grams of the defatted quinoa powder (Soxhlet extraction by using hexane solvent for 4 h and repeated three times) was taken in 50 mL of ethanol. The mixture was then shaken vigorously for 30 min to extract total saponins. Then, the mixture was filtered using a filter paper (*Fironi standard—Plisse*). The filtrate was collected and topped up to 50 mL with ethanol.

4.6.4. Development of the color

The saponin content determination was carried out according to the method applied by Irigoyen et al. [55], by adding 1 mL of the diluted solution to 3.5 mL of the Lieberman-Buchard reagent (16.7% acetic anhydride in concentrated sulfuric acid). The solution was vortexed and stored in the dark for 30 min at room temperature. The absorbance of the solution was measured at 528 nm in a spectrophotometer (Agilent 8453 Spectrophotometer, Los Angeles, CA, USA). Figure 11 shows the development of the color for each dilution.

Figure 11. Different concentrations of the colored solutions.

4.6.5. Calibration Curve Preparation

Regarding the preparation of the stock solution of the standard, oleanolic acid (97%) (Sigma-Aldrich, St. Louis, MO, USA) was used as the pure standard for triterpene saponin; 10 mg of OA was dissolved in 10 mL of pure ethanol to obtain a stock solution concentration of 1 mg/mL [56].

The range of absorbance obtained after dilutions was between 0.2 and 1 mg/mL which corresponds to an absorption of 0.246 to 1.055 (Figure 12). Beer–Lambert's law is applicable in the straight part of the trace, and the range of linearity is between 0.2 and 0.8 absorbance units.

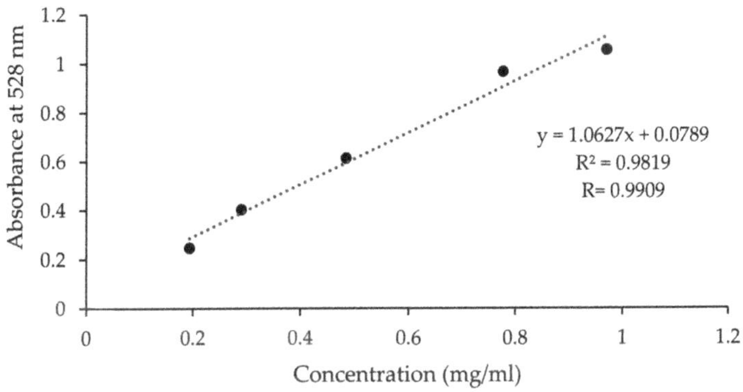

Figure 12. Calibration curve of oleanolic acid at 528 nm.

From the calibration curve, X is calculated as the following:

$$x\left(\frac{mg}{ml}\right) = \frac{y - 0.0789}{1.0627} \quad (1)$$

The concentration of saponins is calculated according to the following equation:

$$Concentration\left(\frac{g}{100g}DM\right) = \frac{x\left(\frac{mg}{ml}\right) \times V_D \times 10^{-3} \times 100 \times 100}{(1 + x(Fat)) \times Sample\ weight\ (g) \times (100 - Moisture)} \quad (2)$$

4.7. Statistical Analysis

Statistical analysis was carried out using R version 3.6.2 software. A two-way analysis of variance (ANOVA) was used to assess the effects of the pearling process on each variety (physical and chemical parameters). The level of significance was set to $p < 0.05$.

Pearson's correlation coefficient was evaluated according to the linear correlations between any two parameters. If the obtained coefficient was −1 or +1, there was a perfect negative or positive linear relationship, respectively. The value of 0 denotes no correlation between the two variables. Generally, a correlation could be statistically significant at a 95% confidence interval ($p < 0.05$).

Principal component analysis (PCA) was used to investigate the potential correlation matrix between all the measured parameters and the pearling time, as well as the data groupings after the pearling treatment.

5. Conclusions and Recommendations

One of the limiting factors facing quinoa valorization and consumption in Morocco and other countries is the seed content of saponin. Furthermore, traditional ways of removing saponins using manual abrasion and washing require intensive labor and thus increase production costs. Therefore, developing mechanical dry pearling tools for saponin removal will solve two interlinked challenges—reducing the processing costs and saving water, thereby increasing farmer's income.

In our study, we tested three quinoa varieties grown in Morocco and polished with a locally manufactured machine. As most cleaning and saponin removal is done under wet or humid methods, our method by dry abrasion had advantageous results, which can be beneficial for farmers and processors.

The findings of this study clearly indicate that two-minute pearling was enough to remove saponins and keep their concentration below the consumption threshold, while at the same time preserving both physical and nutritional quality of the seeds. This study highlights the potentialities of the locally manufactured machine to be used in other regions in Morocco or in other countries, especially as it is a low-cost tool and could be easily operated by smallholder farmers.

Author Contributions: S.R. investigated physical parameters, conducted chemical analysis, analyzed the data, and wrote the manuscript; M.R. supervised chemical analysis and saponin analysis assays; J.P.R. contributed to manuscript writing-review and editing; S.A. carried out macronutrient analysis; A.E. conducted data statistical analysis; M.E.G. supervised micronutrient analysis at AITTC laboratory; S.K. conducted saponin analysis and interpretation; R.C.-A. supervised morphological aspects' results and wrote the manuscript; A.H. supervised the research activities, analyzed the data, and wrote the manuscript. All authors have read and agreed to the published version of the manuscript.

Funding: This study was carried out within an IDRC (International Development Research Center-Canada) funded project entitled: Scaling up quinoa value chain to improve food and nutritional security in poor rural communities of Morocco under the agreement number 108637-001.

Institutional Review Board Statement: Not applicable.

Informed Consent Statement: Not applicable.

Data Availability Statement: The data generated within this work are open access and available to be shared with interested persons according to signed agreement with IDRC.

Acknowledgments: First, we thank our sponsor, the International Development Research Center (IDRC)—Canada, for providing fund to conduct the project activities. We also appreciate the efforts deployed by the technical team of UM6P—Agricultural Innovation and Technology Transfer Center (AITTC), and the technical support of the International Center of Biosaline Agriculture (ICBA). Special thanks to Benrim Farm (Haj Mohamed Ibrahimi) for his cooperation. Our gratitude goes to our local partners and facilitators (farmers, female cooperatives, startups, OCP Group, DRA Marrakech-Safi, DPA Rehamna, ONCA Rehamna, ONSSA Rehamna. etc.).

Conflicts of Interest: The authors declare that they have no conflict of interest.

References

1. Jacobsen, S.E.; Jensen, C.R.; Liu, F. Improving crop production in the arid Mediterranean climate. *F. Crop. Res.* **2012**, *128*, 34–47. [CrossRef]
2. Choukr-Allah, R.; Rao, N.K.; Hirich, A.; Shahid, M.; Alshankiti, A.; Toderich, K.; Gill, S.; Butt, K.U.R. Quinoa for marginal environments: Toward future food and nutritional security in MENA and central Asia regions. *Front. Plant Sci.* **2016**, *7*. [CrossRef] [PubMed]
3. Bazile, D.; Jacobsen, S.-E.; Verniau, A. The Global Expansion of Quinoa: Trends and Limits. *Front. Plant Sci.* **2016**, *7*, 1–6. [CrossRef] [PubMed]
4. Benlhabib, O. Les cultures alternatives: Quinoa, amarante et épeautre. *Transf. Technol. Agric.* **2005**, *133*, 1–4.
5. Hirich, A.; Choukr-Allah, R.; Jacobsen, S.E.; El-Youssfi, L.; El-Omari, H. Using deficit irrigation with treated wastewater in the production of quinoa (*Chenopodium quinoa* Willd.) in Morocco. *Rev. Cient. UDO Agric.* **2012**, *12*, 570–583.
6. Hirich, A.; Jelloul, A.; Choukr-Allah, R.; Jacobsen, S.E. Saline water irrigation of quinoa and chickpea: Seedling rate, stomatal conductance and yield responses. *J. Agron. Crop Sci.* **2014**, *200*, 378–389. [CrossRef]
7. Hirich, A.; Choukr-Allah, R.; Jacobsen, S.E. Deficit irrigation and organic compost improve growth and yield of Quinoa and Pea. *J. Agron. Crop Sci.* **2014**, *200*, 390–398. [CrossRef]
8. Filali, K. Caractérisation et évaluation du rendement de lignées de quinoa (Chenopodium quinoa) dans la région de Rhamna. Amélioration Génétique des Productions Agricoles. Master's Thesis, Institute Agronomique et Vétérinaire Hassan II, Rabat, Morocco, 2011; p. 109.
9. Rafik, S.; Chaoui, M.; Assabban, Y.; Jazi, S.; Choukr-Allah, R.; El Gharouss, M.; Hirich, A. Quinoa value chain, adoption, and market assessment in Morocco. *Environ. Sci. Pollut. Res.* **2020**, 1–12. [CrossRef] [PubMed]
10. Rafik, S.; Rahmani, M.; Choukr-Allah, R.; El Gharous, M.; Calle, J.P.R.; Filali, K.; Hirich, A. Techno-economic assessment of quinoa production and transformation in Morocco. *Environ. Sci. Pollut. Res.* **2021**, 1–16. [CrossRef]
11. Hirich, A.; Rafik, S.; Rahmani, M.; Fetouab, A.; Azaykou, F.; Filali, K.; Ahmadzai, H.; Jnaoui, Y.; Soulaimani, A.; Moussafir, M.; et al. Development of Quinoa Value Chain to Improve Food and Nutritional Security in Rural Communities in Rehamna, Morocco: Lessons Learned and Perspectives. *Plants* **2021**, *10*, 301. [CrossRef]

12. Ruiz, K.B.; Biondi, S.; Oses, R.; Acuña-Rodríguez, I.S.; Antognoni, F.; Martinez-Mosqueira, E.A.; Coulibaly, A.; Canahua-Murillo, A.; Pinto, M.; Zurita-Silva, A.; et al. Quinoa biodiversity and sustainability for food security under climate change. A review. *Agron. Sustain. Dev.* **2014**, *34*, 349–359. [CrossRef]
13. Ruiz, K.B.; Khakimov, B.; Engelsen, S.B.; Bak, S.; Biondi, S.; Jacobsen, S.E. Quinoa seed coats as an expanding and sustainable source of bioactive compounds: An investigation of genotypic diversity in saponin profiles. *Ind. Crops Prod.* **2017**, *104*, 156–163. [CrossRef]
14. Abugoch James, L.E. Quinoa (*Chenopodium quinoa* Willd.): Composition, chemistry, nutritional, and functional properties. *Adv. Food Nutr. Res.* **2009**, *58*. [CrossRef]
15. Lindeboom, N. Studies on the characterization, biosynthesis and isolation of starch and protein from quinoa (*Chenopodium quinua* Willd.). Ph.D. Thesis, University of Saskatchewan, Saskatoon, SK, Canada, 2005; p. 170.
16. Jancurová, M.; Minarovičová, L.; Dandár, A. Quinoa—A review. *Czech J. Food Sci.* **2009**, *27*, 71–79. [CrossRef]
17. Bhargava, A.; Shukla, S.; Ohri, D. *Chenopodium quinoa*—An Indian perspective. *Ind. Crops Prod.* **2016**, *23*, 73–87. [CrossRef]
18. Li, G.; Wang, S.; Zhu, F. Physicochemical properties of quinoa starch. *Carbohydr. Polym.* **2016**, *137*, 328–338. [CrossRef]
19. Comai, S.; Bertazzo, A.; Bailoni, L.; Zancato, M.; Costa, C.V.L.; Allegri, G. The content of proteic and nonproteic (free and protein-bound) tryptophan in quinoa and cereal flours. *Food Chem.* **2007**, *100*, 1350–1355. [CrossRef]
20. Repo-Carrasco-Valencia, R.; Hellström, J.K.; Pihlava, J.M.; Matilla, P.H. Flavonoids and other phenolic compounds in Andean indigenous grains: Quinoa (*Chenopodium quinoa*), kañiwa (*Chenopodium pallidicaule*) and kiwicha (*Amaranthus caudatus*). *Food Chem.* **2010**, *120*, 128–133. [CrossRef]
21. Palombini, S.V.; Claus, T.; Maruyama, S.A.; Gohara, A.K.; Souza, A.H.P.; de Souza, N.E.; Visentainer, J.V.; Gomes, S.T.M.; Matsushita, M. Evaluation of nutritional compounds in new amaranth and quinoa cultivars. *Food Sci. Technol.* **2013**, *33*, 339–344. [CrossRef]
22. Fernández-López, J.; Viuda-Martos, M.; Sayas-Barberá, M.E.; Navarro-Rodríguez de Vera, C.; Lucas-González, R.; Roldán-Verdú, A.; Botella-Martínez, C.; Pérez-Alvarez, J.A. Chia, Quinoa, and Their Coproducts as Potential Antioxidants for the Meat Industry. *Plants* **2020**, *9*, 1359. [CrossRef]
23. Pellegrini, M.; Lucas-Gonzales, R.; Ricci, A.; Fontecha, J.; Fernández-López, J.; Pérez-Álvarez, J.A.; Viuda-Martos, M. Chemical, fatty acid, polyphenolic profile, techno-functional and antioxidant properties of flours obtained from quinoa (*Chenopodium quinoa* Willd) seeds. *Ind. Crops Prod.* **2018**, *111*, 38–46. [CrossRef]
24. De Costa, F.; Yendo, A.C.A.; Fleck, J.D.; Gosmann, G.; Fett-Neto, A.G. Immunoadjuvant and Anti-Inflammatory Plant Saponins: Characteristics and Biotechnological Approaches Towards Sustainable Production. *Mini-Rev Med. Chem.* **2011**, *11*, 857–880. [CrossRef] [PubMed]
25. Cheok, C.Y.; Abdel, H.; Salman, K.; Sulaiman, R. Extraction and quanti fi cation of saponins: A review. *Food Res Int.* **2014**, *59*, 16–40. [CrossRef]
26. Norma Boliviana. NB NA 0038 Granos Andinos—Pseudo Cereales—Quinua en grano Clasificación y Requisitos. 2007. Available online: http://www.conal.gob.ar/CONASE/actas/ActaCONASE_2013_04Abr18_AnexoI.pdf (accessed on 18 April 2021).
27. Koziol, M.J. Afrosimetric estimation of threshold saponin concentration for bitterness in quinoa (*Chenopodium quinoa* Willd). *J. Sci. Food Agric.* **1991**, *54*, 211–219. [CrossRef]
28. Gargiulo, L.; Grimberg, Å.; Repo-Carrasco-Valencia, R.; Carlsson, A.S.; Mele, G. Morpho-densitometric traits for quinoa (*Chenopodium quinoa* Willd.) seed phenotyping by two X-ray micro-CT scanning approaches. *J. Cereal Sci.* **2019**, *90*. [CrossRef]
29. Ando, H.; Chen, Y.-C.; Tang, H.; Shimizu, M.; Watanabe, K.; Mitsunaga, T. Food Components in Fractions of Quinoa Seed. *Food Sci. Technol. Res.* **2002**, *8*, 80–84. [CrossRef]
30. D'Amico, S.; Jungkunz, S.; Balasz, G.; Foeste, M.; Jekle, M.; Tömösközi, S.; Schoenlechner, R. Abrasive milling of quinoa: Study on the distribution of selected nutrients and proteins within the quinoa seed kernel. *J. Cereal Sci.* **2019**, *86*, 132–138. [CrossRef]
31. Czekus, B.; Pećinar, I.; Petrović, I.; Paunović, N.; Savić, S.; Jovanović, Z.; Stikić, R. Raman and Fourier transform infrared spectroscopy application to the Puno and Titicaca cvs. of quinoa seed microstructure and perisperm characterization. *J. Cereal Sci.* **2019**, *87*, 25–30. [CrossRef]
32. Nickel, J.; Spanier, L.P.; Botelho, F.T.; Gularte, M.A.; Helbig, E. Effect of different types of processing on the total phenolic compound content, antioxidant capacity, and saponin content of *Chenopodium quinoa* Willd grains. *Food Chem.* **2016**, *209*, 139–143. [CrossRef]
33. Ridout, C.L.; Price, K.R.; Dupont, M.S.; Parkerc, M.L.; Fenwick, G.R. Quinoa Saponins -Analysis and Preliminary Investigations into the Effects of Reduction by Processing. *J. Sci. Food Agric.* **1991**, *54*, 165–176. [CrossRef]
34. Bazile, D.; Bertero, H.D.; Nieto, C. *State of the Art Report on Quinoa Around the World in 2013*; United Nations Organization of Food and Agriculture: Rome, Italy, 2015.
35. CODEX Standard for Quinoa-CXS 333–2019. 2019, CODEX Aliment.-Int. Food Stand. United Nations Organization of Food and Agriculture (FAO), Rome, Italy. Available online: http://www.fao.org/fao-who-codexalimentarius/sh-proxy/en/?lnk=1&url=https%253A%252F%252Fworkspace.fao.org%252Fsites%252Fcodex%252FStandards%252FCXS%2B333-2019%252FCXS_333e.pdf (accessed on 18 April 2021).
36. Mhada, M.; Metougui, M.L.; El Hazzam, K.; El Kacimi, K.; Yasri, A. Variations of Saponins, Minerals and Total Phenolic Compounds due to Processing and Cooking of Quinoa (*Chenopodium quinoa* Willd.) Seeds. *Foods* **2020**, *9*, 660. [CrossRef]

37. Nowak, V.; Du, J.; Charrondière, U.R. Assessment of the nutritional composition of quinoa (*Chenopodium quinoa* Willd.). *Food Chem.* **2016**, *193*, 47–54. [CrossRef]
38. Riccardi, M.; Mele, G.; Pulvento, C.; Lavini, A.; D'Andria, R.; Jacobsen, S.E. Non-destructive evaluation of chlorophyll content in quinoa and amaranth leaves by simple and multiple regression analysis of RGB image components. *Photosynth. Res.* **2014**, *120*, 263–272. [CrossRef]
39. Valcárcel-Yamani, B.; Lannes, S.C.D.S. Applications of Quinoa (*Chenopodium Quinoa* Willd.) and Amaranth (*Amaranthus* spp.) and Their Influence in the Nutritional Value of Cereal Based Foods. *Food Public Heal.* **2012**, *2*, 265–275. [CrossRef]
40. Koziol, M.J. Chemical Composition and Nutritional Evaluation of Quinoa (*Chenopodium quinoa* Willd.). *J. Food Compos. Anal.* **1992**, *5*, 35–68. [CrossRef]
41. Wu, L.G.; Wang, A.; Shen, R.; Qu, L. The effect of slight milling on nutritional composition and morphology of quinoa (*Chenopodium quinoa*) grain. *Int. J. Food Eng.* **2020**, *16*. [CrossRef]
42. Prado, F.E.; Fernández-Turiel, J.L.; Tsarouchi, M.; Psaras, G.K.; González, J.A. Variation of seed mineral concentrations in seven quinoa cultivars grown in two agroecological sites. *Cereal Chem.* **2014**, *91*, 453–459. [CrossRef]
43. Nascimento, A.C.; Mota, C.; Coelho, I.; Gueifão, S.; Santos, M.; Matos, A.S.; Gimenez, A.; Lobo, M.; Samman, N.; Castanheira, I. Characterisation of nutrient profile of quinoa (*Chenopodium quinoa*), amaranth (*Amaranthus caudatus*), and purple corn (*Zea mays* L.) consumed in the North of Argentina: Proximates, minerals and trace elements. *Food Chem.* **2014**, *148*, 420–426. [CrossRef]
44. Prego, I.; Maldonado, S.; Otegui, M. Seed structure and localization of reserves in *Chenopodium quinoa*. *Ann. Bot.* **1998**, *82*, 481–488. [CrossRef]
45. Konishi, Y.; Hirano, S.; Tsuboi, H.; Wada, M. Distribution of minerals in quinoa (*Chenopodium quinoa* Willd.) seeds. *Biosci. Biotechnol. Biochem.* **2004**, *68*, 231–234. [CrossRef]
46. Balwinder, S.; Amritpal, K. Control of insect pests in crop plants and stored food grains using plant saponins: A review. *LWT* **2018**, *87*, 93–101. [CrossRef]
47. Zegarra Vilchez, G.H. Actividad Deterrente y Acaricida de Principios Activos de Quinuas Amargas, Aceites Esenciales y Tarwi. Bachelor's Thesis, Facultad de Ciencias e Ingeniería, Pontificia Universidad Católica del Peru, Lima, Peru, 2011.
48. Guzmán, B.; Tenorio, R.; Cruz, D.; Espinal, C.; Alvarado, J.; y Mollinedo, P. Saponins from *Chenopodium quinoa* Willd. and *Chenopodium pallidicaule* Aellen as biocontrollers of Phytopathogen fungi and Hemolysis agents. *Revista Boliviana de química* **2015**, *32*, 8–14.
49. Karimian, J.; Abedi, S.; Shirinbakhshmasoleh, M.; Moodi, F.; Moodi, V.; Ghavami, A. The effects of quinoa seed supplementation on cardiovascular risk factors: A systematic review and meta-analysis of controlled clinical trials. *Phytother. Res.* **2020**, 1–9. [CrossRef]
50. Güçlü-Üstündag, O.; Mazzaa, G. Saponins: Properties, Applications and Processing. *Crit. Rev. Food Sci. Nutr.* **2007**, *47*, 231–258. [CrossRef]
51. Tippel, J.; Gies, K.; Harbaum-Piayda, B.; Steffen-Heins, A.; Drusch, S. Composition of Quillaja saponin extract affects lipid oxidation in oil-in-water emulsions. *Food Chem.* **2016**, *221*, 386–394. [CrossRef]
52. Vilche, C.; Gely, M.; Santalla, E. Physical properties of quinoa seeds. *Biosyst. Eng.* **2003**, *86*, 59–65. [CrossRef]
53. AOAC. *AOAC Official Methods of Analysis*, 20th ed.; AOAC International: Rockville, MD, USA, 2016; Volume 20.
54. Marmouzi, I.; El Madani, N.; Charrouf, Z.; Cherrah, Y.; El Abbes Faouzi, M.Y. Proximate analysis, fatty acids and mineral composition of processed Moroccan *Chenopodium quinoa* Willd. and antioxidant properties according to the polarity. *Phytotherapie* **2015**, *13*, 110–117. [CrossRef]
55. Irigoyen, R.M.T.; Giner, S.A. Extraction kinetics of saponins from quinoa seed (*Chenopodium quinoa* Willd.). *Int. J. Food Stud.* **2018**, *7*, 76–88. [CrossRef]
56. Chen, Q.; Zhang, Y.; Zhang, W.; Chen, Z. Identification and quantification of oleanolic acid and ursolic acid in Chinese herbs by liquid chromatography-ion trap mass spectrometry. *Biomed. Chromatogr.* **2011**, *25*, 1381–1388. [CrossRef] [PubMed]

plants

MDPI

Article

Free and Conjugated Phenolic Profiles and Antioxidant Activity in Quinoa Seeds and Their Relationship with Genotype and Environment

Fabiana Antognoni [1,*], Giulia Potente [1], Stefania Biondi [2], Roberto Mandrioli [1], Lorenzo Marincich [1] and Karina B. Ruiz [3]

[1] Department for Life Quality Studies, Rimini Campus, Alma Mater Studiorum—University of Bologna, Corso d' Augusto 237, 47921 Rimini, Italy; giulia.potente@unibo.it (G.P.); roberto.mandrioli@unibo.it (R.M.); lorenzo.marincich2@unibo.it (L.M.)

[2] Department of Biological, Geological and Environmental Sciences, Alma Mater Studiorum—University of Bologna, Via Irnerio 42, 40126 Bologna, Italy; stefania.biondi@unibo.it

[3] Química y Farmacia, Facultad de Ciencias de la Salud, Universidad Arturo Prat, Av. Arturo Prat Ch. 2120, Iquique 1100000, Chile; karuiz@unap.cl

* Correspondence: fabiana.antognoni@unibo.it; Tel.: +39-0541-434623

Citation: Antognoni, F.; Potente, G.; Biondi, S.; Mandrioli, R.; Marincich, L.; Ruiz, K.B. Free and Conjugated Phenolic Profiles and Antioxidant Activity in Quinoa Seeds and Their Relationship with Genotype and Environment. *Plants* **2021**, *10*, 1046. https://doi.org/10.3390/plants10061046

Academic Editors: Cataldo Pulvento and Didier Bazile

Received: 23 April 2021
Accepted: 19 May 2021
Published: 21 May 2021

Abstract: The nutraceutical interest in quinoa (*Chenopodium quinoa* Willd.) seeds is associated with the presence of macronutrients, micronutrients, minerals, vitamins, and polyphenols. In particular, polyphenols contribute to the health-promoting effects of this food crop, and their levels are influenced by environmental conditions. Production of quinoa is recently being explored in temperate climate areas, including Italy. The aim of this research was to assess the profile of bioactive compounds in seeds of two quinoa varieties, Regalona-Baer and Titicaca, grown in northern Italy, compared to that of seeds of those varieties grown in Chile and Denmark, respectively. High-performance liquid chromatography-diode array detector (HPLC-DAD) analysis of phenolic acid and flavonoid profiles, both in their free and soluble conjugated forms, showed that the main differences between Regalona grown in Chile and Italy were for the free vanillic acid and daidzein contents, while the two Titicaca samples mainly differed in quercetin derivative levels. The total phenolic index was comparable in Titicaca and Regalona, and only a slight decrease in this parameter was found in seeds of the two varieties grown in Italy. The in vitro antioxidant activity of seed extracts, evaluated by means of three different assays, indicated that it correlated with flavonol (quercetin derivative) levels. In conclusion, the results indicate that, although environmental conditions alter the polyphenolic profile and biological activities, it is possible to grow good-quality quinoa in northern Italy.

Keywords: antioxidant activity; *Chenopodium quinoa*; conjugated phenolics; flavonoids; phenolic acids; nutraceutical properties

1. Introduction

Quinoa (*Chenopodium quinoa* Willd.) seeds have an excellent nutritional profile [1,2], due to their remarkably high protein content, well-balanced content of essential amino acids comparable, in terms of quality, to that of casein [3], and the presence of dietary fiber, vitamins, and unsaturated fatty acids, e.g., linoleic and α-linolenic acids [2–6]. Quinoa seeds are considered an excellent example of a "functional food", able to exert health-promoting effects also due to the presence of secondary metabolites [6]. These include phenolic acids and flavonoids [7], as well as terpenoids, steroids, and large amounts of α- and γ-tocopherol, which are known antioxidants [8]. For some of these specialized metabolites, a clear role in preventing various human diseases, such as neurological and other chronic disorders, has been recognized, as a result of their effects on cell signalling and metabolism [6,9,10].

Quinoa cultivation remained, for a long time, exclusive to Andean populations, especially from Bolivia and Peru [11], and, to a lesser extent, from Chile and Ecuador. The wide genetic diversity of this species, resulting from its fragmented and localized cultivation over the centuries in the Andean region, together with its high tolerance to extreme environmental conditions and abiotic stresses [12,13], has led to the differentiation of five major ecotypes, based on their ability to adapt to specific agroecological conditions [11]. Nevertheless, in recent decades, with the boom in demand for "superfoods", the interest in this ancient Andean halophytic seed-producing crop has become global, and its cultivation has been growing accordingly. In less than twenty years, the global demand for quinoa grew to such an extent that it led to the triplication of the Andean areas dedicated to its cultivation [14]. Moreover, the adaptability and stress tolerance of this plant is being exploited to establish cultivations outside the Andean territories [11,15]. As a consequence, the number of countries growing quinoa has rapidly risen starting from the 1980s, and, nowadays, more than 95 countries are cultivating or testing quinoa as a crop [16]. In the early 1990s, quinoa began to be cultivated in Europe, and the crop has been successfully tested in several Mediterranean countries, such as Greece, Morocco, Spain, and Italy [17–20].

The main concern about the introduction of this crop was related to its high sensitivity to photoperiods during the reproductive phase [21]. The best results were achieved with the sea level/coastal ecotype from central and southern Chile, since it is the best adapted to temperate environments. Using these Chilean quinoa lines, a new variety, called Titicaca, was bred in Denmark [22]. Titicaca is one of four European quinoa cultivars (with Puno, Jessie, and Zeno) originating from different gene pools and already tested in various countries, including Germany [23] and Italy, where several field trials have been carried out at different latitudes to test the adaptability of the crop to varying environments [21,23,24]. Pulvento et al. [25] reported good plant performance and tolerance to high temperatures and water deficit, typical conditions of southern Italy, for cultivars Regalona-Baer and Titicaca.

Several studies point to the importance of environmental and/or agronomical factors in affecting the nutritional properties of quinoa. For example, Reguera et al. [26] showed that amino acid profiles, total protein content, mineral composition, and phytate content varied in seeds of three cultivars (Salcedo-INIA, Titicaca, Regalona) grown in three different countries (Spain, Peru, and Chile). However, the impact of agroecological conditions and crop management techniques on quinoa seed quality in terms of functional bioactive compounds (e.g., phenolics) has been little explored. Miranda et al. [27] reported that seeds of several quinoa genotypes grown in different geographical locations in Chile contained variable amounts of phenolics and flavonoids. Considering that these secondary metabolites are plastically produced in plant tissues as an acclimation response to environmental cues, their concentrations in plant tissues are the combined result of several factors, including genetic components, environmental conditions, and the complex interplay between the two [28–30]. In the present work, we checked this hypothesis by analysing the phenolic profiles of quinoa seeds of the same variety from different agroecological environments and verified if the antioxidant activity (AA) of seed extracts likewise changed. Such analyses are useful to identify the environmental conditions that modulate the health-promoting characteristics of quinoa seeds. To this purpose, we evaluated the composition in bioactive free and soluble-conjugated phenolic acids and flavonoids and the AA of seeds of two quinoa varieties, the Chilean Regalona-Baer and the Danish-bred Titicaca, grown in northern Italy and compared them with those of seeds of the same varieties grown in Chile and Denmark. The present results could help to establish the best conditions (genotype/geographical cultivation zone) leading to seed enrichment in functional compounds and to reinforce the notion that good-quality quinoa can be grown in Italy.

2. Results and Discussion

2.1. Free and Soluble-Conjugated Total Phenolic Content (TPC) and Total Flavonoid Content (TFC) in Quinoa Seed Extracts

Phenolic acids and flavonoids in seed extracts were quantified by analysing the free and soluble-conjugated forms (and not the insoluble ones) because they represent the fractions of dietary phenolic compounds that are more rapidly absorbed by the stomach and small intestine and for which important health benefits have been reported [31].

Total phenolic content (TPC) and total flavonoid content (TFC) were evaluated in extracts of the four seed samples enriched in the solvent-extractable (free; F) and in both the acid-hydrolysed (AH) and base-hydrolysed (BH) soluble-conjugated forms (Figure 1).

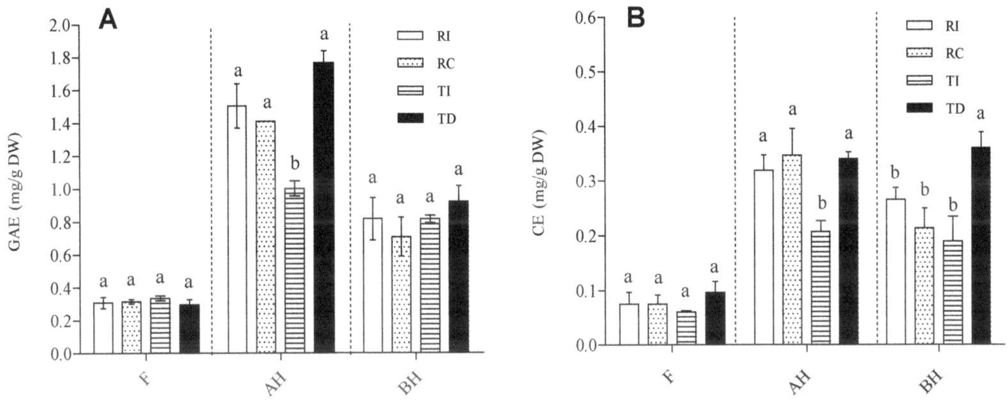

Figure 1. Total phenolic content (**A**) and total flavonoid content (**B**) in free (F), acid-hydrolysed (AH), and base-hydrolysed (BH) fractions of quinoa seed extracts. (**A**) TPC is expressed as mg gallic acid equivalents (GAE)/g DW seed powder. (**B**) TFC is expressed as catechin equivalents (CE)/g DW seed powder. RI: Regalona grown in Italy; RC: Regalona grown in Chile; TI: Titicaca grown in Italy; TD: Titicaca grown in Denmark. Data are the means ± standard error of two independent determinations with three biological replicates. Different letters within the same parameter indicate statistically significant differences at $p \leq 0.05$.

As concerns TPC (Figure 1A), AH was the richest fraction, with levels reaching 1.7–1.8 mg GAE/g DW of seed powder, followed by the BH fraction (0.6–0.8 mg GAE/g DW), while much lower levels were observed in the free form (0.2–0.3 mg/g DW). In both the F and BH fractions, no significant differences in TPC were observed between RI, RC, TI, and TD. By contrast, the Danish-grown Titicaca seeds (TD) had about twice the concentration of AH soluble-conjugated phenolics as compared with the same cultivar grown in Italy. As regards Regalona, the TPC in this fraction (AH) was very similar in both Chilean- and Italian-grown seeds, and not significantly different from that of TD (Figure 1A). This suggests that the conjugated form of soluble phenolics is mainly linked through ether bonds, such as the glycosidic one, which is acid hydrolysable. The distribution of TPC into the three fractions in the quinoa seeds analysed here is in line with that reported by Tang et al. [32] in a white seed quinoa, but not with that by Gómez-Caravaca et al. [33] for genotypes Kancholla and Witulla. These authors reported that most phenolic compounds present in their seed extracts were in the free form, but it should be underlined that these authors used a different analytical method, i.e., HPLC-DAD-ESI-TOF-MS, which could explain the discrepancies. Although Repo-Carrasco-Valencia et al. [7] observed a huge variation in the proportion of soluble (free and conjugated) phenolics (ranging from 7 to 61%), among the ten genotypes taken into consideration, we did not observe large differences in the total amounts of soluble phenolics between the two T and R varieties. It should be pointed out that the TPC in the AH fraction might be overestimated, since several authors [32,34] indicated that, under acid hydrolysis, the degradation of free sugars can give rise to furan derivatives, which react with the Folin–Ciocalteu's reagent. Thus, the distribution of phenolic compounds between the free and AH conjugated forms, resulting from the TPC determination, only provides a rough estimate, which must be confirmed through more specific analytical techniques.

As concerns TFC, both the relative distribution in the three forms and the pattern within each fraction were similar to those of TPC; the conjugated forms were more abundant than the free ones, even though, in this case, the levels of AH and BH soluble flavonoids were comparable (Figure 1B). In plants, it is known that flavonoids are mostly conjugated to various types of molecules. The abundance of flavonoids in both the AH and BH fractions indicates that these were linked via both ether and ester bonds; in the former category, glycosides represent the most widespread form, while the latter suggest their linkage to organic acids and proteins, as demonstrated by Koistinen et al. [35]. As concerns the comparison among samples, the pattern of TFC in the free and AH fractions was very similar to that of TPC, with AH forms again higher in TD than in TI. Danish-grown Titicaca showed the highest TFC in the BH fraction, while no differences were found between the other three samples (Figure 1B). Based on these results, both genotype and environmental conditions seem to exert their effect mainly on the conjugated form of phenolics, leaving the free form almost unchanged. Moreover, they indicate that TI had the lowest level of phenolics conjugated through ether linkages.

2.2. Chromatographic Analysis of Free and Soluble-Conjugated Phenolic Compounds in Quinoa Seed Extracts

The three fractions were subjected to high-performance liquid chromatography-diode array detector (HPLC-DAD) analysis to identify and quantify single phenolic compounds. These include phenolic acids (both hydroxycinnamic and dihydroxybenzoic acids), flavonols, flavan-3-ols, and isoflavones (Figure 2).

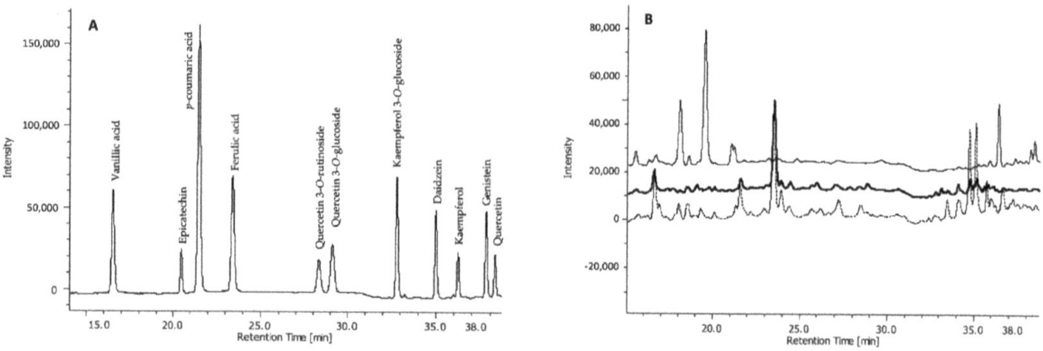

Figure 2. Chromatogram of HPLC-DAD separation of (**A**) standard phenolic acids and flavonoids, and (**B**) representative chromatographic profiles of free (upper line), base-hydrolysed (middle line), and acid-hydrolysed (bottom line) fractions of one seed extract (TI).

In the free fraction, vanillic and ferulic acids were the major hydroxycinnamic acids detected in all samples, the former being about 10-fold more concentrated than the latter (Figure 3). Seed extracts of Chilean Regalona had a vanillic acid content about twice that of the same genotype grown in Italy, and of both TD and TI, whose concentrations were very similar. As for ferulic acid levels, no differences were observed between the two cultivars, no matter where they were grown (Figure 3).

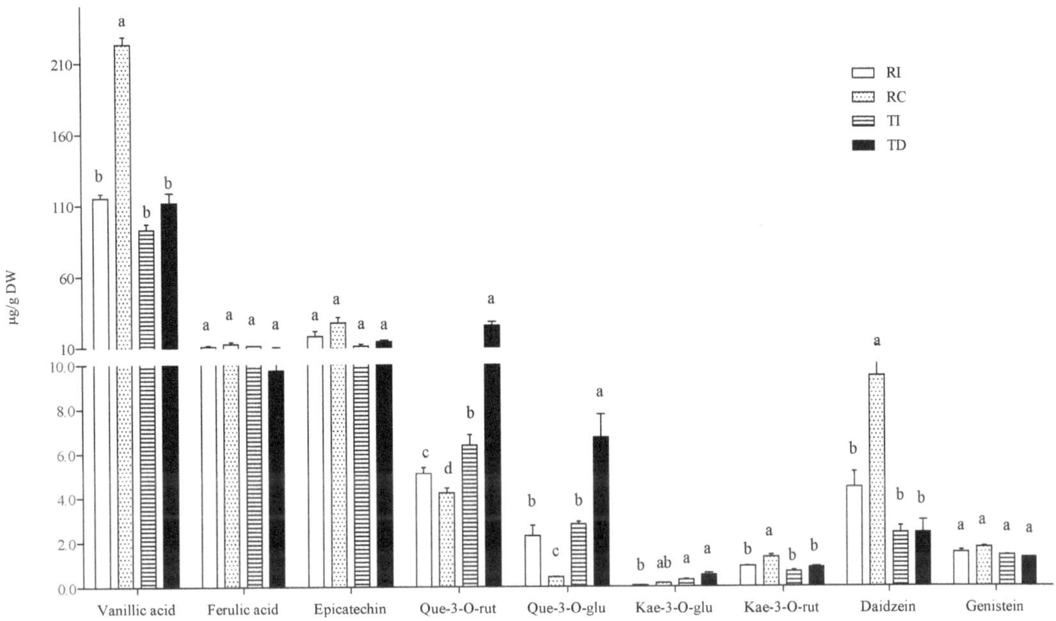

Figure 3. Free phenolic acid and flavonoid concentrations in quinoa seed extracts. RI: Regalona grown in Italy; RC: Regalona grown in Chile; TI: Titicaca grown in Italy; TD: Titicaca grown in Denmark. Data are the means ± standard error of two independent determinations with three biological replicates. Different letters within the same parameter indicate statistically significant differences at $p \leq 0.05$.

The prevalence of vanillic acid as the main representative of the phenolic acid class in the free form is in line with the results of Tang et al. [32] in a white seed quinoa, even though significant variations in the relative amounts of single phenolic acids in different quinoa genotypes and varieties were also reported by Repo-Carrasco-Valencia et al. [7]. Nevertheless, a comparison with the results from these authors is not easy, since they reported the soluble phenolic acid content without distinguishing between the free and the soluble-conjugated fraction.

As concerns flavonoids, epicatechin, quercetin, and kaempferol derivatives (quercetin-3-*O*-rutinoside, quercetin-3-*O*-glucoside, kaempferol-3-*O*-glucoside, and kaempferol-3-*O*-rutinoside) and the isoflavones daidzein and genistein were detected in the free fraction (Figure 3). Significant differences among samples were observed only in some cases. In particular, TD seeds showed higher levels of quercetin-3-*O*-glucoside and quercetin-3-*O*-rutinoside compared to seeds of the same cultivar grown in Italy (two- and four-fold, respectively), and compared to both Regalona extracts. In Regalona seeds, a higher level of these two quercetin derivatives was found in Italian samples as compared to Chilean ones. Conversely, the latter showed the highest daidzein levels among all samples, and no differences in this isoflavone were found among other seeds. All seed extracts contained very low amounts of kaempferol derivatives, whose differences were of no biological relevance, nor were any differences observed as concerns genistein and epicatechin (Figure 3). Thus, the agroclimatic conditions in which the Chilean Regalona seeds were produced seem to be particularly favourable for the accumulation of daidzein since, in this sample, the highest daidzein/genistein ratio (=6) was found. This value is very close to that observed by Lutz et al. [36] in some commercial quinoa seeds and suggests that, under certain environmental conditions, a seed with a good phytoestrogen-like activity can be obtained.

In the AH conjugated fraction, *p*-coumaric and ferulic acids did not vary considerably among samples (Figure 4). The ferulic acid concentration was very similar to that found in the free fraction, while *p*-coumaric acid was not detected in the free form. As concerns flavonoids, the aglycones quercetin and kaempferol were found as the main representatives, and TD was confirmed to be the richest in these flavonols (about three-fold higher than in TI), while no differences were detected between the two Regalona samples (Figure 4).

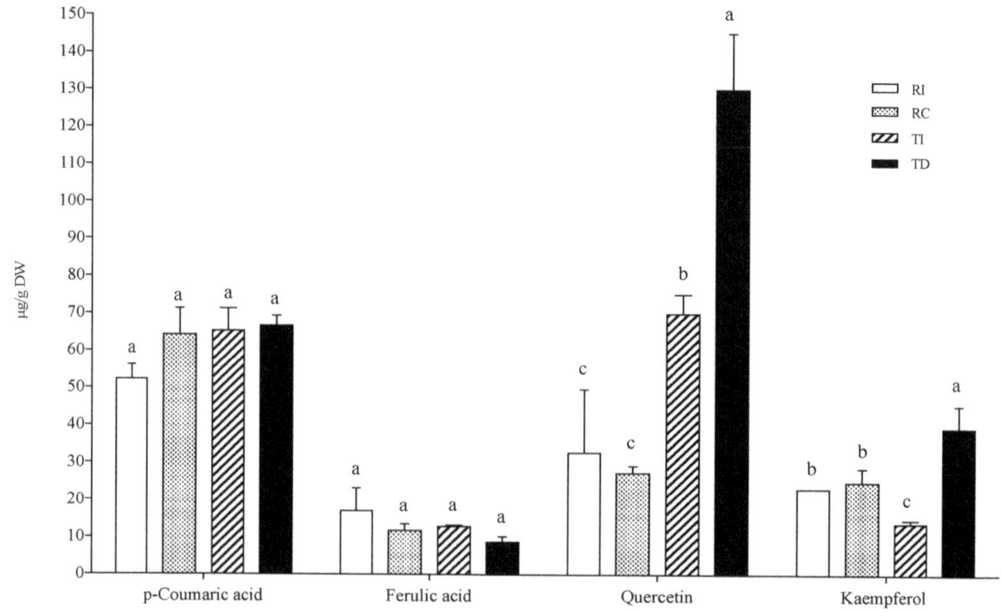

Figure 4. Phenolic acid and flavonoid concentrations in the acid-hydrolysable soluble-conjugated fraction of quinoa seed extracts. RI: Regalona grown in Italy; RC: Regalona grown in Chile; TI: Titicaca grown in Italy; TD: Titicaca grown in Denmark. Data are the means ± standard error of two independent determinations with three biological replicates. Different letters within the same parameter indicate statistically significant differences at $p \leq 0.05$.

Phenolic acids and flavonoids were also detected in the BH soluble conjugated fraction. Both *p*-coumaric and ferulic acids were present, and the latter was detected at higher levels than in the AH fraction (Figure 5).

Vanillic acid was present at similar levels to those of the free fraction, while conjugated ferulic acid was over ten-fold more concentrated than in the free form. Thus, the results indicate that a relevant portion of these polyphenols is conjugated through ester bonds with soluble cellular components, such as short peptides and/or low-molecular weight oligosaccharides. It is known that ferulic and *p*-coumaric acids are covalently bound to mixed-linkage (1→3, 1→4)-β-D-glucans and to hemicelluloses, forming the soluble dietary fiber [37]. Being a dicot species, quinoa cell walls contain mainly xyloglucans in their matrix. The extractability of hemicelluloses during the preparation of crude seed extracts using sonication has been demonstrated [38] as well as the binding of ferulic acid to short-chain hemicelluloses [39]. Thus, the crude extracts used in this study probably contain soluble feruloylated and/or coumaroylated xylo-oligosaccharides, which could explain the presence of high levels of aglycone components in our BH fraction. Several health-promoting activities have been reported for these short-chain feruloylated oligosaccharides, including an immunomodulatory effect [40]. As concerns the comparison among samples, the results indicate that concentrations of vanillic and *p*-coumaric acids did not differ significantly between Regalona and Titicaca, no matter where they were grown, while for ferulic acid, a 55% higher concentration was found in RC compared to seeds of the same genotype grown in Italy; no differences were found between it and TD or TI seeds.

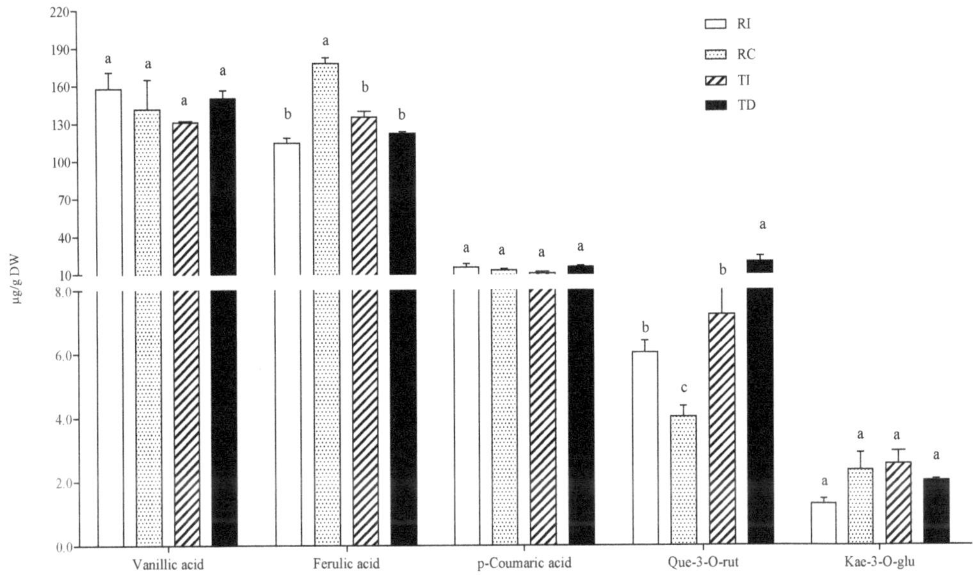

Figure 5. Phenolic acid and flavonoid concentrations in the base-hydrolysed solubleconjugated fraction of quinoa seed extracts. RI: Regalona grown in Italy; RC: Regalona grown in Chile; TI: Titicaca grown in Italy; TD: Titicaca grown in Denmark. Data are the means ± standard error of two independent determinations with three biological replicates. Different letters within the same parameter indicate statistically significant differences at $p \leq 0.05$.

Quercetin-3-*O*-rutinoside (or rutin) and kaempferol-3-*O*-glucoside were identified as the main flavonols in the BH fraction. As regards the former, the pattern was similar to those in the free and AH conjugated fractions, with DT being richer than the Italian-grown counterpart, while RI had a slightly, but significantly, higher rutin content than RC; no differences were observed for kaempferol-3-*O*-glucoside among the four samples (Figure 4).

Some authors have investigated the phenolic composition of different quinoa varieties, and phenolic acids and flavonoids were reported to exist both in their aglycone and glycosidic forms [32,41,42]. On average, the concentrations of these compounds found in our work were similar to those found by other authors [41,43], but lower compared to other reports [33]. These discrepancies are not surprising, since several parameters were demonstrated to have a great impact on the extraction efficiency of these metabolites from plant material, such as sampling, solvent-to-solid ratio used for extraction, duration of extraction, and sonication treatment applied, among others [44]. The use of more advanced separative techniques, such as the HPLC-DAD-ESI-TOF-MS utilized by Gómez-Caravaca et al. [33], allowed these authors to identify more compounds, which are not detectable with other techniques, and this can also explain the differences.

Hirose et al. [45] reported that quinoa and buckwheat contain several quercetin and kaempferol derivatives, which are not present in any of the widely consumed cereals. They also compared the flavonoid pattern and in vitro antioxidant capacity of seeds cultivated in Japan with those of seeds from Bolivia and Peru. Results indicated that the major differences between the two groups of seeds were found for quercetin and kaempferol derivatives, which were significantly higher in Japanese-grown seeds compared to South American ones, and this was accompanied by a significantly higher antioxidant capacity. Seasonal variations in the pattern of flavonol glycosides observed by these authors allowed them to hypothesize that sunlight was the factor that mostly influenced the accumulation of quercetin glycosides [45]. Thus, it is possible that variations in quercetin and its derivatives between Titicaca and Regalona grown at different latitudes may be due to differences in day length and light quality/intensity. Even though the effect of latitude on flavonoid accumulation has yet to be fully clarified, due to the complexity of latitude itself as a parameter to be investigated [29], many studies have concluded that northern climates may have a positive impact on the biosynthesis of flavonoids in plants, although there are variations in the response among species and within individual flavonoid groups. Flavonols, especially moieties of quercetin or kaempferol, have been reported to accumulate in response to increased UV-B radiation [46], which typically characterizes the higher latitudes in the Northern Hemisphere. The prominent role for this subclass of flavonoids in regulating plant–environment interactions has been demonstrated, particularly as concerns the acclimation of plants to different light exposures [47]. In general, flavonoids may play prominent roles as scavengers of reactive oxygen species (ROS) generated by adverse conditions [48,49], and quercetin derivatives, due to their chemical characteristics, are particularly efficient in buffering alterations in ROS homeostasis [28]. Moreover, several reports indicate that they strongly affect phytohormone (auxin and ABA) signalling, due to their ability to affect the activity of a wide range of proteins [50,51]. Thus, the higher accumulation of quercetin derivatives in Titicaca seeds grown in Denmark compared to seeds of the same cultivar grown in Italy and to both Regalona samples might be the result of an acclimation mechanism to long light exposures, typical of the higher latitudes. The bioavailability and therapeutic potentials of quercetin and its derivatives in plant-based foods have been extensively investigated [52], and several studies demonstrated that they play relevant roles both in the prevention and treatment of chronic diseases, including cardiovascular and neurodegenerative ones and some types of cancer [53–55]. Thus, the enrichment of quinoa seeds in flavonol derivatives could represent an added value able to reinforce the well-known nutraceutical properties of this plant.

2.3. Antioxidant Activity of Quinoa Seed Extracts

The AA of the three fractions of Regalona and Titicaca seed extracts was evaluated using three different in vitro assays, i.e., FRAP, TEAC, and ORAC (Figure 6).

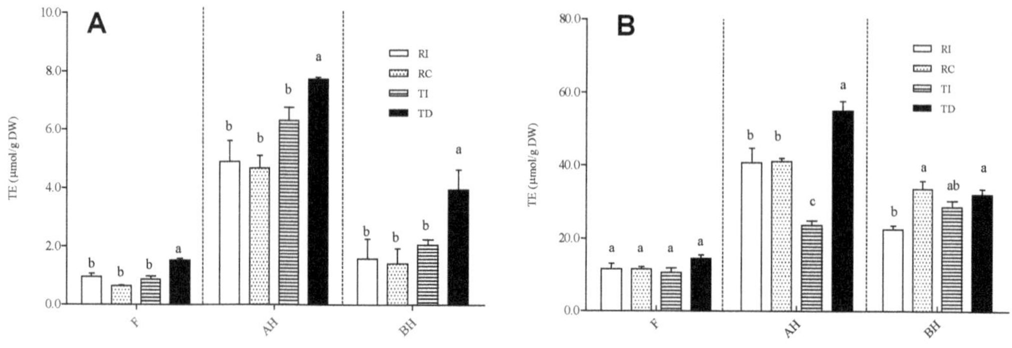

Figure 6. Antioxidant activity, assayed by TEAC (**A**) and ORAC (**B**), in free (F), acid-hydrolysed (AH), and base-hydrolysed (BH) fractions of seed extracts from different quinoa samples. Results are expressed as Trolox equivalents (TE)/g DW seed powder. RI: Regalona grown in Italy; RC: Regalona grown in Chile; TI: Titicaca grown in Italy; TD: Titicaca grown in Denmark. Data are the means ± standard error of two independent determinations with three biological replicates. Different letters within the same parameter indicate statistically significant differences at $p \leq 0.05$.

The results obtained by the TEAC and ORAC assays univocally show that the fractions enriched in soluble-conjugated phenolic acids and flavonoids possess a higher AA compared to the free fraction (Figure 6A,B). FRAP provided the same results (data not shown). Although this may derive, at least in part, from the above-cited overestimation due to the formation of 5-hydroxymethyl-2-furfural, which displays AA [34], it has been widely demonstrated that soluble-conjugated polyphenols and/or feruloylated/coumaroylated short oligosaccharides are able to greatly contribute to the antioxidant capacity of grains [56,57]. As concerns the comparison among samples, the results obtained with the TEAC/FRAP and ORAC assays are slightly different. For all fractions (free, AH, BH), the former assay (Figure 6A; FRAP data not shown) revealed that TD had a slightly, but significantly, higher AA compared to TI and to both Regalona extracts, which, conversely, did not differ from each other. According to ORAC, no differences among samples were observed in the free fraction, while for the AH one, TD was confirmed to have the highest AA; no differences were found between the two Regalona samples grown in Chile and in Italy (Figure 6B). In the BH fraction, a statistically significant difference emerged between the Italian-grown Regalona and both RC and TD, even though the biological relevance of this difference is questionable. Thus, it can be concluded that the higher amount of flavonol derivatives in Danish-grown Titicaca might contribute to the higher antioxidant activity, given the strong redox capacity associated to this subgroup of flavonoids.

For an overall comparison among samples, the total phenolic index (TPI) and the total antioxidant index (TAI) were calculated as the sum of the individual phenolic compounds and AA, respectively, detected in all fractions.

As shown in Figure 7A, despite a different distribution between the free and soluble conjugated forms, the results clearly indicate that the Chilean Regalona and the Danish Titicaca show a very similar TPI, while seeds of these cultivars grown in Italy respond to the different climatic/environmental conditions with a 20% lower accumulation of the main phenolic compounds compared to their counterparts. The results of the two-way ANOVA indicate that the area of cultivation significantly contributed to the variance of TPI (78%, $p = 0.005$), while neither the genotypes nor their interaction had a significant impact on the variation ($p = 0.13$ and 0.56, respectively). As for the TAI resulting from the ORAC assay (Figure 7B), it was only slightly lower (15%) in RI compared to RC, while a greater difference (40%) was observed between TD and TI; similar results were found for the TEAC assay (data not shown). These results confirm that the total antioxidant activity only weakly correlates with the TPI, as it is more tightly related to the specific composition in antioxidants; in fact, single compounds or classes of compounds can provide a greater contribution compared to others. The two-way ANOVA indicated that the total antioxidant index did not vary significantly depending on genotype ($p = 0.61$), while the cultivation area, as well as its interaction with genotype, provided a significant contribution ($p < 0.0001$ and $p = 0.0031$, respectively).

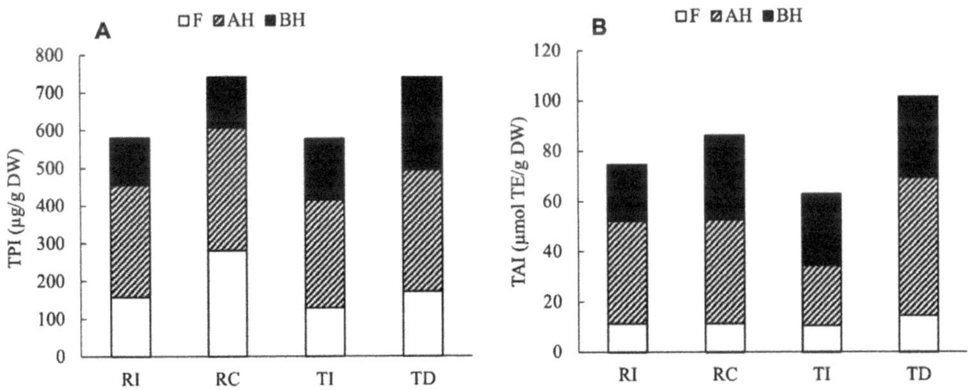

Figure 7. Total phenolic index (TPI, **A**) and total antioxidant index (TAI, **B**) in the four seed extracts of quinoa. TPI was calculated as the sum of individual phenolic compounds detected, and TAI as the sum of ORAC-assayed AA in each fraction.

In conclusion, the present results suggest that, both in terms of phenolic profiles and AA, genotype-dependent differences are not very relevant as far as Titicaca and Regalona are concerned, both of which are adapted to temperate climates, belong to the coastal ecotype, and were bred from gene pools originating from southern/central Chile. The results also indicate that agroecological conditions can, to some extent, alter these phytochemical profiles and the biological activities; in particular, light/UV-B intensity may have contributed to higher flavonol levels in seeds from Denmark [58,59]. Nonetheless, the changes are quite small and corroborate previous data [25], demonstrating that both Titicaca and Regalona can be successfully cultivated in Italy, where they maintain good growth, productivity, and nutritional properties.

3. Materials and Methods

3.1. Chemicals and Reagents

The following chemicals and reagents were purchased from Sigma-Aldrich Italia (Milan, Italy): Folin–Ciocalteu's phenol reagent, 6-hydroxyl-2,5,7,8-tetramethyl-chroman-2-carboxylic acid (Trolox), 2,2-azinobis (3-ethylbenzothiazoline-6-sulfonic acid; ABTS), $AlCl_3$, $NaNO_2$, $FeCl_2$, $FeCl_3$, FerroZine®, fluorescein, 2,2-azobis(2-amidinopropane) dihydrochloride (AAPH), phosphoric acid (85–87%, *w*/*w*), hydrochloric acid (37%, *w*/*w*), monobasic sodium phosphate (>98%), sodium hydroxide beads (>98%), and HPLC-grade solvents. Pure standards of phenolic acids (4-hydroxybenzoic, gallic, caffeic, chlorogenic, ferulic, *p*-coumaric, sinapic, syringic, trans-cinnamic, and vanillic acids) and flavonoids (quercetin, quercetin-3-*O*-glucoside, quercetin-3-*O*-rutinoside, quercetin-3-*O*-galactoside, kaempferol, kaempferol-3-*O*-rutinoside, catechin, epicatechin, daidzein, and genistein) were purchased from Extrasynthese (Genay Cedex, France). The IUPAC names for these compounds are indicated in Table S1. All standards (>99.5% purity in powder form) were prepared as stock solutions at 1 mg/mL in methanol and stored in the dark at −18 °C for less than three months.

3.2. Plant Material and Extraction Procedure

Seeds of quinoa cultivars Regalona-Baer (opaque white) and Titicaca (pale yellow) harvested in 2017 were used. Seeds of cv. Regalona collected from plants grown in Italy (RI) were purchased from Dall'Ara & Lolli farm (Campiano, Ravenna, Italy; 44°18′10″ N, 12°12′07″ E; mean annual temperature 14.4 °C, max 19.0 °C, min 8.3 °C; total annual precipitation 646.0 mm), while seeds of the same cultivar grown in Chile (RC) were purchased from the seed company Semillas Baer (Temuco, Chile; 38°44′23″ S, 72°36′00″ W; mean annual temperature 10.9 °C, max 17.3 °C, min 5.9 °C; total annual precipitation 1355.2 mm). Seeds of quinoa cv. Titicaca grown in Italy (TI) were kindly supplied by D. Vanuzzi (Tuttoquinoa) and collected from plants grown in Sale (Alessandria, Italy; 44°58′54″ N, 8°48′37″ E; mean annual temperature 13.5 °C, max 18.6 °C, min 8.5 °C; total annual precipitation 501.5 mm), while Titicaca seeds from plants grown in Denmark (TD) were kindly supplied by S. Jacobsen (Copenhagen, Denmark; 55°38′46″ N, 12°17′53″ E; mean annual temperature 8.9 °C, max 19.2 °C, min −3.4 °C; total annual precipitation 848.8 mm).

The extractions were performed in triplicate, following the procedure described by Tang et al. [32], with slight modifications. Seeds were ground in a knife mill for 4 × 30 s (A11 basic, IKA Werke GmbH & Co. KG, Staufen, Germany) and in a mortar to obtain a fine and homogeneous powder. The seed powder was then subjected to the "coning and quartering" sampling procedure, and three technical replicates were carried out. A 3-g aliquot of fine powdered sample was transferred in a 50-mL tube, mixed with 10 mL of 70% MeOH acidified with HCl (0.1%, *v*/*v*), and kept on an orbital shaker (Duomax 1030, Heidolph Instruments, Schwalbach, Germany) at 200 rpm for 3 h at RT; after a 15-min ultrasound extraction (Elma Schmidbauer GmbH, Singen, Germany), the mixture was centrifuged for 30 min at 4000 rpm, and the supernatant, containing all extractable compounds, was collected. The procedure was repeated twice, and supernatants were combined to form the crude extract (CE).

Preparation of Free (F), Base-Hydrolysed (BH), and Acid-Hydrolysed (AH) Soluble Fractions

To obtain the free (F) fraction, the CE was evaporated to dryness, re-suspended with 2 mL of acidified water (pH = 2), and subsequently extracted with 2 mL of a diethyl ether/ethyl acetate mixture (1:1, *v*/*v*) three times. The organic phases were merged, evaporated to dryness, re-suspended in 2 mL of 70% MeOH, filtered through 0.2-μm nylon syringe filters, and stored at −80 °C until analysis.

The aqueous phases were also pooled and subjected to base and acid hydrolyses to obtain the base-hydrolysed (BH) and acid-hydrolysed (AH) fractions, respectively. A 2-mL volume of 10 N NaOH was added to 8 mL of aqueous phase, and the mixture was stirred for 1 h under N_2 flow and brought to pH 2 with concentrated HCl. The resulting solution was then subjected to the extraction with diethyl ether/ethyl acetate mixture as previously described. The combined organic phases were evaporated to dryness, re-suspended with 2 mL of 70% MeOH, filtered through 0.2-μm nylon syringe filters, and stored at −80 °C until analysis.

The same volume of the aqueous phase was subjected to acid hydrolysis, by adding 1.6 mL of 12 N HCl, and the mixture was stirred for 1 h at 85 °C in a water bath. The resulting solution was extracted three times as previously described, and the combined organic phases were evaporated to dryness, re-suspended with 2 mL of 70% MeOH, filtered through 0.2-μm nylon syringe filters, and stored at −80 °C until analysis.

3.3. Determination of Total Phenolic Content and Total Flavonoid Content

The total phenolic content (TPC) assay was carried out using the Folin–Ciocalteu reagent according to the procedure described by Singleton and Rossi [60], with modifications. A mixture containing 100 μL of diluted extract fractions or standard and 440 μL of Folin–Ciocalteu's reagent (diluted 1:10 with water) was incubated for 10 min at RT. Then, 440 μL of 7.5% sodium carbonate was added and the mixture was incubated in the dark for 60 min at RT. Gallic acid was used as a standard, and a calibration curve was built, in the 1–25 ppm range. Absorbance was measured at 765 nm using a double-beam spectrophotometer (V-630 Jasco, Jasco Europe s.r.l., Cremella, Italy) and used to calculate TPC, expressed as mg gallic acid equivalents (GAE) g^{-1} DW of seed powder.

The total flavonoid content (TFC) assay was carried out according to the procedure described by Zou et al. [61], with modifications. A mixture containing 100 μL of diluted extract fractions or standard and 440 μL of 0.066 M $NaNO_2$ was left to react for 5 min at RT. Then, 60 μL of 0.75 M $AlCl_3$ was added, and the mixture was incubated for 5 min. Lastly, 400 mL of 0.5 M NaOH was added, and the mixture was incubated for 6 min at RT. The absorbance was measured at 510 nm, and the TPC of samples was calculated by interpolating with the calibration curve built with catechin as a standard, in the 1–25 ppm concentration range. Results are expressed as mg catechin equivalents (CE) g^{-1} DW of seed powder.

3.4. HPLC Determination of Phenolic Compounds

The chromatographic method was adapted from Tang et al. [32], with some modifications. The extracts were injected into a Jasco (Tokyo, Japan) HPLC-DAD system, which consisted of a PU-4180 pump, an MD-4015 PDA detector, and an AS-4050 autosampler. The stationary phase was an Agilent (Santa Clara, CA, USA) Zorbax Eclipse Plus C18 reverse-phase column (100 × 3 mm I.D., 3.5 μm). The mobile phase was a mixture of solvent A (water/formic acid 95/5, *v*/*v*) and solvent B (methanol/acetonitrile 95/5, *v*/*v*), with a composition gradient ranging from 95% to 5% of solvent A and flowing at 0.7 mL/min. Injection volume was 20 μL for all determinations, and analyte detection was carried out with a diode array detector (DAD) by monitoring at 280, 329, and 360 nm. Quantification was performed with pure standards using calibration curves ranging between 1.25 and 30 μg mL^{-1} ($r^2 \geq 0.9634$).

3.5. In Vitro Antioxidant Activity Assays

The ABTS assay was carried out essentially as described by Thaipong et al. [62]. After incubating 950 μL of 7.4 mM 2,2-azinobis(3-ethylbenzothiazoline-6-sulfonic acid) (ABTS) in methanol with either 50 μL of methanolic Trolox solution at different concentrations (0.05–1.00 mM), methanol (blank solution), or the diluted sample in the dark at RT for 2 h, the absorbance of the solution was read at 734 nm in the V630 spectrophotometer. Calibration curves were set up by plotting the discoloration ratio (i.e., $[Abs_{\text{without TX}}/Abs_{\text{with TX}}] - 1$) as a function of Trolox concentration. The antioxidant capacity of the sample, expressed as Trolox equivalents (TEs), was calculated by interpolating with the calibration curve. The oxygen radical absorbance capacity (ORAC) assay was carried out on sample fractions essentially as described by Moore et al. [63], using a Viktor X3 multilabel plate reader (Perkin Elmer, Turku, Finland). Trolox equivalents (TEs) were calculated from the relative area under the curve of the emission intensity vs. time plots.

The ferric reducing antioxidant power-ferrozine (FRAP-FZ) assay was performed as described by Mandrone et al. (2015) [64].

3.6. Statistical Analysis

Statistical analysis was performed using GraphPad (San Diego, CA, USA) Prism Software v. 5.0. Comparison among samples was conducted through one-way analysis of variance (ANOVA) with Tukey's multiple comparison test, and values of $p \leq 0.05$ were considered statistically significant. Two-way ANOVA was carried out to analyse the genotype x cultivation area interaction.

Supplementary Materials: The following are available online at https://www.mdpi.com/article/10.3390/plants10061046/s1, Table S1: Common and IUPAC names of the phenolic compounds analysed in the study.

Author Contributions: Conceptualization, F.A., R.M., and K.B.R.; data curation, F.A., G.P., and R.M.; funding acquisition, F.A., S.B., R.M., and K.B.R.; investigation, G.P. and L.M.; methodology, G.P. and L.M.; supervision, F.A. and R.M.; writing—original draft, F.A. and G.P.; writing—review and editing, S.B. and R.M. All authors have read and agreed to the published version of the manuscript.

Funding: This research was funded by Ricerca Fondamentale Orientata (RFO) grants from the University of Bologna to F.A., R.M., and S.B. FONDECYT INI 11190834 grant from ANID (Chile) to K.B.R. also funded the research.

Institutional Review Board Statement: Not applicable.

Informed Consent Statement: Not applicable.

Data Availability Statement: Data presented in this study are available upon request.

Acknowledgments: The authors would like to thank Dario Vanuzzi (Tuttoquinoa, Italy), S.E. Jacobsen (Copenhagen), Giorgio Dall'Ara (Quinoa Romagna, Italy), Ingrid von Baer and Ricardo Bonert (Semillas Baer, Chile) for kindly providing Titicaca and Regalona seeds. The authors also thank Giuseppe Gabellini and Elena Lombardini (undergraduate students) for their technical contribution.

Conflicts of Interest: The authors declare no conflict of interest.

References

1. Vega-Gálvez, A.; Miranda, M.; Vergara, J.; Uribe, E.; Puente, L.; Martínez, E.A. Nutrition Facts and Functional Potential of Quinoa (*Chenopodium quinoa* Willd.), an Ancient Andean Grain: A Review. *J. Sci. Food Agric.* **2010**, *90*, 2541–2547. [CrossRef] [PubMed]
2. Nowak, V.; Du, J.; Charrondière, U.R. Assessment of the Nutritional Composition of Quinoa (*Chenopodium quinoa* Willd.). *Food Chem.* **2016**, *193*, 47–54. [CrossRef] [PubMed]
3. Mahoney, A.W.; Lopez, J.G.; Hendricks, D.G. Evaluation of the Protein Quality of Quinoa. *J. Agric. Food Chem.* **1975**, *23*, 190–193. [CrossRef]

4. Saleh, A.S.M.; Zhang, Q.; Chen, J.; Shen, Q. Millet Grains: Nutritional Quality, Processing, and Potential Health Benefits. *Compr. Rev. Food Sci. Food Saf.* **2013**, *12*, 281–295. [CrossRef]
5. Filho, A.M.M.; Pirozi, M.R.; Borges, J.T.D.S.; Pinheiro Sant'Ana, H.M.; Chaves, J.B.P.; Coimbra, J.S.D.R. Quinoa: Nutritional, Functional, and Antinutritional Aspects. *Crit. Rev. Food Sci. Nutr.* **2017**, *57*, 1618–1630. [CrossRef]
6. Lin, M.; Han, P.; Li, Y.; Wang, W.; Lai, D.; Zhou, L. Quinoa Secondary Metabolites and Their Biological Activities or Functions. *Molecules* **2019**, *24*, 2512. [CrossRef] [PubMed]
7. Repo-Carrasco-Valencia, R.; Hellström, J.K.; Pihlava, J.-M.; Mattila, P.H. Flavonoids and Other Phenolic Compounds in Andean Indigenous Grains: Quinoa (Chenopodium quinoa), Kañiwa (Chenopodium pallidicaule) and Kiwicha (Amaranthus caudatus). *Food Chem.* **2010**, *120*, 128–133. [CrossRef]
8. Frankel, E.N. The Antioxidant and Nutritional Effects of Tocopherols, Ascorbic Acid and Beta-Carotene in Relation to Processing of Edible Oils. *Bibl. Nutr. Dieta* **1989**, *43*, 297–312. [CrossRef]
9. Bassani, B.; Rossi, T.; Stefano, D.D.; Pizzichini, D.; Corradino, P.; Macrì, N.; Noonan, D.M.; Albini, A.; Bruno, A. Potential chemopreventive activities of a polyphenol rich purified extract from olive mill wastewater on colon cancer cells. *J. Funct. Foods* **2016**, *27*, 236–248. [CrossRef]
10. Silva, R.F.M.; Pogačnik, L. Polyphenols from Food and Natural Products: Neuroprotection and Safety. *Antioxidants* **2020**, *9*, 61. [CrossRef]
11. Bazile, D.; Jacobsen, S.-E.; Verniau, A. The Global Expansion of Quinoa: Trends and Limits. *Front. Plant. Sci.* **2016**, *7*. [CrossRef]
12. Ruiz, K.B.; Biondi, S.; Oses, R.; Acuña-Rodríguez, I.S.; Antognoni, F.; Martinez-Mosqueira, E.A.; Coulibaly, A.; Canahua-Murillo, A.; Pinto, M.; Zurita-Silva, A.; et al. Quinoa Biodiversity and Sustainability for Food Security under Climate Change: A Review. *Agron. Sustain. Dev.* **2014**, *34*, 349–359. [CrossRef]
13. Ruiz, K.B.; Biondi, S.; Martínez, E.A.; Orsini, F.; Antognoni, F.; Jacobsen, S.-E. Quinoa—A Model Crop for Understanding Salt-Tolerance Mechanisms in Halophytes. *Plant. Biosyst. Int. J. Deal. All Asp. Plant Biol.* **2016**, *150*, 357–371. [CrossRef]
14. Bazile, D.; Bertero, H.D.; Nieto, C. *State of the Art Report on Quinoa around the World in 2013*; FAO & CIRAD: Rome, Italy, 2015.
15. Jacobsen, S.-E. The Situation for Quinoa and Its Production in Southern Bolivia: From Economic Success to Environmental Disaster: Quinoa Production in Southern Bolivia. *J. Agron. Crop Sci.* **2011**, *197*, 390–399. [CrossRef]
16. Bazile, D.; Baudron, F. The Dynamics of the Global Expansion of Quinoa Growing in View of Its High Biodiversity. In *State of the Art Report on Quinoa around the World in 2013*; FAO: Santiago du Chili, Chile, 2015.
17. Iliadis, C.; Karyotis, T.; Mitsibonas, T. Research on Quinoa (Chenopodium quinoa) and Amaranth (Amaranthus caudatus) in Greece. In Proceedings of the COST-Workshop, CPRO-DLO, Wageningen, The Netherlands, 24–25 October 1997; pp. 24–25.
18. Iliadis, C.; Karyotis, T.; Jacobsen, S.E. Adaptation of Quinoa under Xerothermic Conditions and Cultivation for Biomass and Fibre Production. In *Memorias, Primer Taller Internacional sobre Quinua–Recursos Geneticos y Sistemas de Producción, UNALM*; International Potato Center (CIP): Lima, Peru, 2001; pp. 371–378.
19. Lavini, A.; Pulvento, C.; d'Andria, R.; Riccardi, M.; Choukr-Allah, R.; Belhabib, O.; Yazar, A.; İncekaya, Ç.; Sezen, S.M.; Qadir, M.; et al. Quinoa's Potential in the Mediterranean Region. *J. Agron. Crop Sci.* **2014**, *200*, 344–360. [CrossRef]
20. Taviani, P.; Rubini, A.; Menconi, L.; Pieroni, G.; Damiani, F. Introduzione Di Nuove Colture: La Quinoa (Chenopodium quinoa Willd.) Progetto Co. *Al. Ta. Ii. Presentati Nell'ambito Del Progett. Di. Al. Ta. Ii. Divulg. Delle Colt. Altern. Al Tob. Retrieved Febr.* **2008**, *10*, 2019.
21. De Santis, G.; D'Ambrosio, T.; Rinaldi, M.; Rascio, A. Heritabilities of Morphological and Quality Traits and Interrelationships with Yield in Quinoa (*Chenopodium Quinoa* Willd.) Genotypes in the Mediterranean Environment. *J. Cereal Sci.* **2016**, *70*, 177–185. [CrossRef]
22. Jacobsen, S.-E. The Scope for Adaptation of Quinoa in Northern Latitudes of Europe. *J. Agron. Crop Sci.* **2017**, *203*, 603–613. [CrossRef]
23. Casini, P. Possibilita Di Introdurre La Quinoa Negli Ambienti Mediterranei. *Inf. Agrar.* **2002**, *58*, 29–34.
24. De Santis, G.; Maddaluno, C.; D'Ambrosio, T.; Rascio, A.; Rinaldi, M.; Troisi, J. Characterization of Quinoa (*Chenopodium quinoa* Willd.) Accessions for the Saponin Content in Mediterranean Environment. *Ital. J. Agron.* **2016**, *11*. [CrossRef]
25. Pulvento, C.; Riccardi, M.; Lavini, A.; D'Andria, R.; Iafelice, G.; Marconi, E. Field Trial Evaluation of Two Chenopodium Quinoa Genotypes Grown Under Rain-Fed Conditions in a Typical Mediterranean Environment in South Italy: Quinoa in the Mediterranean. *J. Agron. Crop Sci.* **2010**, *196*, 407–411. [CrossRef]
26. Reguera, M.; Conesa, C.M.; Gil-Gómez, A.; Haros, C.M.; Pérez-Casas, M.Á.; Briones-Labarca, V.; Bolaños, L.; Bonilla, I.; Álvarez, R.; Pinto, K.; et al. The Impact of Different Agroecological Conditions on the Nutritional Composition of Quinoa Seeds. *PeerJ* **2018**, *6*, e4442. [CrossRef] [PubMed]
27. Miranda, M.; Vega-Gálvez, A.; Martinez, E.; López, J.; Rodríguez, M.J.; Henríquez, K.; Fuentes, F. Genetic Diversity and Comparison of Physicochemical and Nutritional Characteristics of Six Quinoa (*Chenopodium quinoa* Willd.) Genotypes Cultivated in Chile. *Food Sci. Technol.* **2012**, *32*, 835–843. [CrossRef]
28. Brunetti, C.; Fini, A.; Sebastiani, F.; Gori, A.; Tattini, M. Modulation of Phytohormone Signaling: A Primary Function of Flavonoids in Plant–Environment Interactions. *Front. Plant Sci.* **2018**, *9*, 1–8. [CrossRef] [PubMed]
29. Jaakola, L.; Hohtola, A. Effect of Latitude on Flavonoid Biosynthesis in Plants: Effect of Latitude on Flavonoid Biosynthesis. *Plant Cell Environ.* **2010**, *33*, 1239–1247. [CrossRef]

30. Lätti, A.K.; Jaakola, L.; Riihinen, K.R.; Kainulainen, P.S. Anthocyanin and Flavonol Variation in Bog Bilberries (*Vaccinium uliginosum* L.) in Finland. *J. Agric. Food Chem.* **2010**, *58*, 427–433. [CrossRef] [PubMed]
31. Chandrasekara, A.; Shahidi, F. Bioactivities and Antiradical Properties of Millet Grains and Hulls. *J. Agric. Food Chem.* **2011**, *59*, 9563–9571. [CrossRef]
32. Tang, Y.; Li, X.; Zhang, B.; Chen, P.X.; Liu, R.; Tsao, R. Characterisation of Phenolics, Betanins and Antioxidant Activities in Seeds of Three Chenopodium quinoa Willd. Genotypes. *Food Chem.* **2015**, *166*, 380–388. [CrossRef] [PubMed]
33. Gómez-Caravaca, A.M.; Segura-Carretero, A.; Fernández-Gutiérrez, A.; Caboni, M.F. Simultaneous Determination of Phenolic Compounds and Saponins in Quinoa (*Chenopodium quinoa* Willd) by a Liquid Chromatography–Diode Array Detection–Electrospray Ionization–Time-of-Flight Mass Spectrometry Methodology. *J. Agric. Food Chem.* **2011**, *59*, 10815–10825. [CrossRef] [PubMed]
34. Chen, P.X.; Tang, Y.; Zhang, B.; Liu, R.; Marcone, M.F.; Li, X.; Tsao, R. 5-Hydroxymethyl-2-Furfural and Derivatives Formed during Acid Hydrolysis of Conjugated and Bound Phenolics in Plant Foods and the Effects on Phenolic Content and Antioxidant Capacity. *J. Agric. Food Chem.* **2014**, *62*, 4754–4761. [CrossRef] [PubMed]
35. Koistinen, K.M.; Soininen, P.; Venäläinen, T.A.; Häyrinen, J.; Laatikainen, R.; Peräkylä, M.; Tervahauta, A.I.; Kärenlampi, S.O. Birch PR-10c Interacts with Several Biologically Important Ligands. *Phytochemistry* **2005**, *66*, 2524–2533. [CrossRef] [PubMed]
36. Lutz, M.; Martínez, A.; Martínez, E.A. Daidzein and Genistein Contents in Seeds of Quinoa (*Chenopodium quinoa* Willd.) from Local Ecotypes Grown in Arid Chile. *Ind. Crop. Prod.* **2013**, *49*, 117–121. [CrossRef]
37. Ishii, T. Structure and Functions of Feruloylated Polysaccharides. *Plant Sci.* **1997**, *127*, 111–127. [CrossRef]
38. Hromádková, Z.; Ebringerová, A. Ultrasonic Extraction of Plant Materials—Investigation of Hemicellulose Release from Buckwheat Hulls. *Ultrason. Sonochem.* **2003**, *10*, 127–133. [CrossRef]
39. Saulnier, L.; Thibault, J.-F. Ferulic Acid and Diferulic Acids as Components of Sugar-Beet Pectins and Maize Bran Heteroxylans. *J. Sci. Food Agric.* **1999**, *79*, 396–402. [CrossRef]
40. Fang, H.-Y.; Chen, Y.-K.; Chen, H.-H.; Lin, S.-Y.; Fang, Y.-T. Immunomodulatory Effects of Feruloylated Oligosaccharides from Rice Bran. *Food Chem.* **2012**, *134*, 836–840. [CrossRef] [PubMed]
41. Dini, I.; Carlo Tenore, G.; Dini, A. Phenolic Constituents of Kancolla Seeds. *Food Chem.* **2004**, *84*, 163–168. [CrossRef]
42. Gómez-Caravaca, A.M.; Iafelice, G.; Verardo, V.; Marconi, E.; Caboni, M.F. Influence of Pearling Process on Phenolic and Saponin Content in Quinoa (*Chenopodium quinoa* Willd). *Food Chem.* **2014**, *157*, 174–178. [CrossRef] [PubMed]
43. Alvarez-Jubete, L.; Wijngaard, H.; Arendt, E.K.; Gallagher, E. Polyphenol Composition and in Vitro Antioxidant Activity of Amaranth, Quinoa Buckwheat and Wheat as Affected by Sprouting and Baking. *Food Chem.* **2010**, *119*, 770–778. [CrossRef]
44. Michiels, J.A.; Kevers, C.; Pincemail, J.; Defraigne, J.O.; Dommes, J. Extraction Conditions Can Greatly Influence Antioxidant Capacity Assays in Plant Food Matrices. *Food Chem.* **2012**, *130*, 986–993. [CrossRef]
45. Hirose, Y.; Fujita, T.; Ishii, T.; Ueno, N. Antioxidative Properties and Flavonoid Composition of Chenopodium quinoa Seeds Cultivated in Japan. *Food Chem.* **2010**, *119*, 1300–1306. [CrossRef]
46. Treutter, D. Significance of Flavonoids in Plant Resistance and Enhancement of Their Biosynthesis. *Plant Biol.* **2005**, *7*, 581–591. [CrossRef] [PubMed]
47. Hectors, K.; Van Oevelen, S.; Geuns, J.; Guisez, Y.; Jansen, M.A.K.; Prinsen, E. Dynamic Changes in Plant Secondary Metabolites during UV Acclimation in *Arabidopsis thaliana*. *Physiol. Plant.* **2014**, *152*, 219–230. [CrossRef]
48. Agati, G.; Stefano, G.; Biricolti, S.; Tattini, M. Mesophyll Distribution of 'Antioxidant' Flavonoid Glycosides in Ligustrum Vulgare Leaves under Contrasting Sunlight Irradiance. *Ann. Bot.* **2009**, *104*, 853–861. [CrossRef]
49. Agati, G.; Tattini, M. Multiple Functional Roles of Flavonoids in Photoprotection. *New Phytol.* **2010**, *186*, 786–793. [CrossRef] [PubMed]
50. Peer, W.A.; Murphy, A.S. Flavonoids as Signal Molecules: Targets of Flavonoid Action. In *The Science of Flavonoids*; Grotewold, E., Ed.; Springer: New York, NY, USA, 2006; pp. 239–268. ISBN 978-0-387-28822-2.
51. Lewis, D.R.; Ramirez, M.V.; Miller, N.D.; Vallabhaneni, P.; Ray, W.K.; Helm, R.F.; Winkel, B.S.J.; Muday, G.K. Auxin and Ethylene Induce Flavonol Accumulation through Distinct Transcriptional Networks. *Plant Physiol.* **2011**, *156*, 144–164. [CrossRef] [PubMed]
52. Lesjak, M.; Beara, I.; Simin, N.; Pintać, D.; Majkić, T.; Bekvalac, K.; Orčić, D.; Mimica-Dukić, N. Antioxidant and Anti-Inflammatory Activities of Quercetin and Its Derivatives. *J. Funct. Foods* **2018**, *40*, 68–75. [CrossRef]
53. Serban, M.-C.; Sahebkar, A.; Zanchetti, A.; Mikhailidis, D.P.; Howard, G.; Antal, D.; Andrica, F.; Ahmed, A.; Aronow, W.S.; Muntner, P.; et al. Effects of Quercetin on Blood Pressure: A Systematic Review and Meta-Analysis of Randomized Controlled Trials. *J. Am. Heart Assoc.* **2016**, *5*, 1–15. [CrossRef]
54. Russo, M.; Spagnuolo, C.; Tedesco, I.; Bilotto, S.; Russo, G.L. The Flavonoid Quercetin in Disease Prevention and Therapy: Facts and Fancies. *Biochem. Pharm.* **2012**, *83*, 6–15. [CrossRef] [PubMed]
55. Zhang, H.; Zhang, M.; Yu, L.; Zhao, Y.; He, N.; Yang, X. Antitumor Activities of Quercetin and Quercetin-5′,8-Disulfonate in Human Colon and Breast Cancer Cell Lines. *Food Chem. Toxicol.* **2012**, *50*, 1589–1599. [CrossRef] [PubMed]
56. Garcia-Conesa, M.T.; Wilson, P.D.; Plumb, G.W.; Ralph, J.; Williamson, G. Antioxidant Properties of 4, 4′-Dihydroxy-3, 3′-Dimethoxy-β, B′-Bicinnamic Acid (8-8-Diferulic Acid, Non-Cyclic Form). *J. Sci. Food Agric.* **1999**, *79*, 379–384. [CrossRef]
57. Adom, K.K.; Liu, R.H. Antioxidant Activity of Grains. *J. Agric. Food Chem.* **2002**, *50*, 6182–6187. [CrossRef] [PubMed]
58. Zhang, W.J.; Björn, L.O. The Effect of Ultraviolet Radiation on the Accumulation of Medicinal Compounds in Plants. *Fitoterapia* **2009**, *80*, 207–218. [CrossRef] [PubMed]

59. Zivcak, M.; Brückova, K.; Sytar, O.; Brestic, M.; Olsovska, K.; Allakhverdiev, S.I. Lettuce Flavonoids Screening and Phenotyping by Chlorophyll Fluorescence Excitation Ratio. *Planta* **2017**, *245*, 1215–1229. [CrossRef] [PubMed]
60. Singleton, V.L.; Rossi, J.A. Colorimetry of Total Phenolics with Phosphomolybdic-Phosphotungstic Acid Reagents. *Am. J. Enol. Vitic.* **1965**, *16*, 144–158.
61. Zou, Y.; Chang, S.K.C.; Gu, Y.; Qian, S.Y. Antioxidant Activity and Phenolic Compositions of Lentil (Lens culinaris Var. Morton) Extract and Its Fractions. *J. Agric. Food Chem.* **2011**, *59*, 2268–2276. [CrossRef]
62. Thaipong, K.; Boonprakob, U.; Crosby, K.; Cisneros-Zevallos, L.; Hawkins Byrne, D. Comparison of ABTS, DPPH, FRAP, and ORAC Assays for Estimating Antioxidant Activity from Guava Fruit Extracts. *J. Food Compos. Anal.* **2006**, *19*, 669–675. [CrossRef]
63. Moore, J.; Hao, Z.; Zhou, K.; Luther, M.; Costa, J.; Yu, L.L. Carotenoid, Tocopherol, Phenolic Acid, and Antioxidant Properties of Maryland-Grown Soft Wheat. *J. Agric. Food Chem.* **2005**, *53*, 6649–6657. [CrossRef] [PubMed]
64. Mandrone, M.; Lorenzi, B.; Venditti, A.; Guarcini, L.; Bianco, A.; Sanna, C.; Ballero, M.; Poli, F.; Antognoni, F. Antioxidant and Anti-Collagenase Activity of *Hypericum hircinum* L. *Ind. Crop. Prod.* **2015**, *76*, 402–408. [CrossRef]

plants MDPI

Article

Heat Stress Impact on Yield and Composition of Quinoa Straw under Mediterranean Field Conditions

Javier Matías [1,*], Verónica Cruz [1] and María Reguera [2]

[1] Agrarian Research Institute "La Orden-Valdesequera" of Extremadura (CICYTEX), 06187 Badajoz, Spain; veronica.cruz@juntaex.es

[2] Department of Biology, Universidad Autónoma de Madrid, c/Darwin 2, Campus de Cantoblanco, 28049 Madrid, Spain; maria.reguera@uam.es

* Correspondence: javier.matias@juntaex.es; Tel.: +34-924-014-046

Abstract: Quinoa (*Chenopodium quinoa* Willd.) is receiving increasing attention globally due to the high nutritional value of its seeds, and the ability of this crop to cope with stress. In the current climate change scenario, valorization of crop byproducts is required to support a climate-smart agriculture. Furthermore, research works characterizing and evaluating quinoa stems and their putative uses are scarce. In this work, straw yield and composition, and the relative feed value of five quinoa varieties, were analyzed in two consecutive years (2017–2018) under field conditions in Southwestern Europe. High temperatures were recorded during the 2017 growing season resulting in significantly decreased straw yield and improved feed value, associated with compositional changes under elevated temperatures. Crude protein, ash, phosphorus, and calcium contents were higher under high temperatures, whereas fiber contents decreased. The relative feed value was also higher in 2017 and differed among varieties. Differences among varieties were also found in straw yield, and contents of phosphorus, potassium, and calcium. Overall, the results presented here support a sustainable quinoa productive system by encouraging straw valorization and shedding light on the mechanisms underlying heat-stress responses in this crop.

Keywords: quinoa; stems; high temperatures; food security; climate smart agriculture; quinoa by-products

Citation: Matías, J.; Cruz, V.; Reguera, M. Heat Stress Impact on Yield and Composition of Quinoa Straw under Mediterranean Field Conditions. *Plants* **2021**, *10*, 955. https://doi.org/10.3390/plants10050955

Academic Editor: Cataldo Pulvento

Received: 9 April 2021
Accepted: 7 May 2021
Published: 11 May 2021

1. Introduction

In the twenty-first century, climate change has become one of the most critical global challenges, and is particularly crucial for the agricultural and livestock sectors [1–4]. This phenomenon is associated with increments of atmospheric greenhouse gas (GHG) emissions and the increase in the mean global temperature, in addition to alterations in the precipitation regime, affecting food and feed production, and its nutritional composition [1,3,5–8]. Because these changes represent a serious threat to global food security, it is necessary to understand how crops respond to elevated temperatures and how their tolerance to heat stress can be improved [9]. In particular, the Mediterranean climate comprises the western edges of continents between latitudes of 30° and 45° [10–12]. Regions with a Mediterranean climate are characterized by hot and dry summers, mild wet winters, and low and irregular rainfall. Because of climate change, severe droughts, increased temperatures, and salinity problems are predicted to be more frequent in the near future in these areas [13].

In addition, the current rapid growth in the global population is expected to lead to a sharp increase in demand for food in coming decades [14]. Furthermore, available land for agriculture and water resources is increasingly scarce following soil degradation and other land uses [4,15]. Hence, agriculture in the twenty-first century must meet increasing food demand with fewer resources, while GHG emissions should be reduced. Within this context, it is necessary to introduce crops with lower irrigation requirements (e.g., in South

Europe alone, more than 60% of freshwater is used by agriculture [13]) and better adapted to less favorable soil conditions, in addition to valorizing crop byproducts. The latter can be achieved if these materials are first well characterized, shedding light on the effects of climate change on the nutritional value of plants.

Quinoa (*Chenopodium quinoa* Willd.) is an ancient crop from South America that has expanded globally because of the increasing interest in the nutritional composition of its seeds. This has contributed to its selection as one of the crops destined to contribute to food security in the next century, according to the FAO (Food and Agricultural Organization) [16]. Quinoa has also been studied for its potential use as a forage crop because of the high nutritional value of the whole plant for livestock [17]. However, this use has been little explored. In contrast, quinoa seeds are well known for being gluten-free, because of their high-quality protein and because they are one of the few plant foods that contain all nine essential amino acids, with a wider amino acid profile than cereals or legumes [18]. Quinoa is an annual C3 plant belonging to the Amaranthaceae family [19], with remarkable adaptability to unfavorable growing conditions [20]. It is a facultative halophytic plant species that is also tolerant to the combined effect of high temperatures and salinity [21–23], and is considered an environmental stress-resilient crop plant [17].

Quinoa has been cultivated for thousands of years in the Andean region comprising Peru, Bolivia, Colombia, and Ecuador although currently its cultivation has expanded globally [6,24]. Indeed, there is a remarkable interest in growing quinoa at European latitudes [25]. In the Mediterranean region, quinoa has been cultivated in Spain, Italy, Greece, and Portugal. In Spain, quinoa cultivation has been expanded significantly in recent years, particularly in the southern part of the country. Interestingly, one of the consequences of the massive expansion of quinoa cultivation is the generation of residues composed of quinoa plant straw, for which possible uses have been little explored [15,26]. Only recently has the use of quinoa straw been evaluated for animal feed [15,17] or generation of renewable energy [15,26].

Thus, in this study, aiming to further explore possible uses of quinoa crop byproducts, and therefore contribute to the implementation of sustainable quinoa cultivation under Mediterranean conditions, the straw production and composition of five quinoa varieties were evaluated in a field located in Southwestern Spain, where episodes of elevated temperatures are frequent. Differences in straw biomass and composition were expected to occur based on the developmental differences of the genotypes tested. Furthermore, the impact of high temperatures under field conditions was analyzed. These results will contribute to a better understanding of how temperature stress might influence straw production.

2. Results

2.1. Straw Yield and Biomass Partitioning

As observed in Figure 1, the straw yield varied significantly according to the year and variety. Interaction between both factors was also significant. The mean straw yield was significantly higher in 2018 (2.2 t ha^{-1}) than that in 2017 (1.7 t ha^{-1}), except for Roja and Duquesa, which achieved lower yields in 2018. Marisma reached the highest mean stem yield among varieties (2.8 t ha^{-1}), and was remarkably high in 2018 (3.9 t ha^{-1}). Biomass partitioning was evaluated by determining the HI index, which was significantly influenced by the variety and the year (Figure 2). The average HI in 2017 (0.49) was lower than that in 2018 (0.53), mainly due to the 20% HI reduction achieved in 2017 by Roja (0.40) and Duquesa (0.45). The highest average HI (0.55) was achieved by Jessie (short cycle), whereas the lowest HI was obtained by the varieties with longer life cycles (Roja: 0.48 and Duquesa: 0.50). HI was not correlated with straw yield.

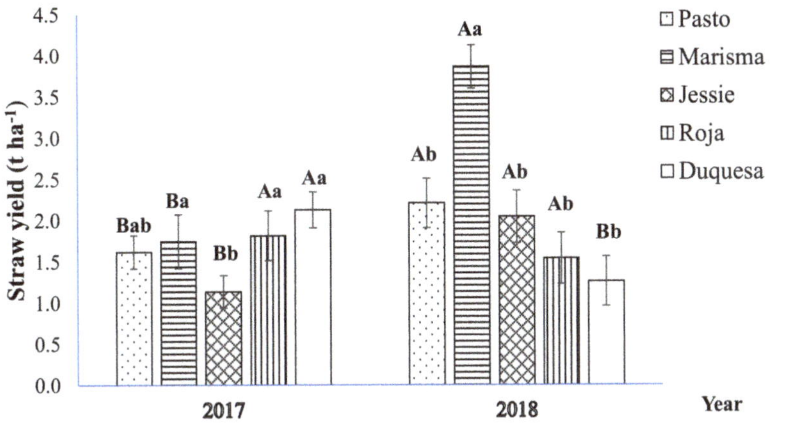

Figure 1. Straw yield (t. ha^{-1}) of five quinoa varieties during the two years (2017 and 2018) of field experiments. Error bars represents the standard deviation. Different uppercase letters in the same variety indicate a significant difference between years according to Student's *t*-test at $p < 0.05$. In each year, different lowercase letters indicate significant differences among varieties according to Tukey's test at $p < 0.05$.

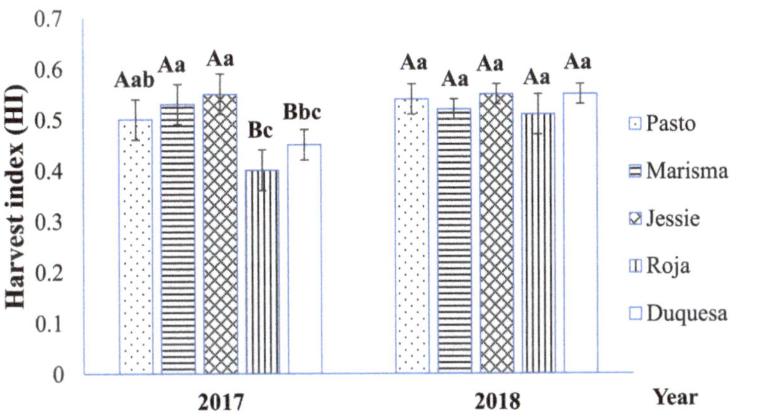

Figure 2. Harvest index (HI) of five quinoa varieties during the two years (2017 and 2018) of field experiments. Error bars represent the standard deviation (SD). Different uppercase letters in the same variety indicate a significant difference between years according to Student's *t*-test at $p < 0.05$. In each year, different lowercase letters indicate significant differences among varieties according to Tukey's test at $p < 0.05$.

2.2. Crude Protein (CP), Crude Fibre (CF), and Ash Content

The CP and CF contents were significantly influenced by the year, whereas no significant differences were found according to the variety (Table 1). The CP content in 2017 (12.8%) was 52.3% higher than that in 2018 (8.4%). On the contrary, the CF content was significantly higher in 2018 (33.8%) compared to 2017 (28.0%). The average ash content differed significantly according to the variety. The average ash content varied from 11.5% (Roja) to 16.6% (Jessie) with no different found between years.

Table 1. Crude protein (CP), crude fiber (CF), and ash content (%) of straw of five quinoa varieties during the two years (2017 and 2018) of field experiments.

Variety	CP[1] (%)			CF[2] (%)			Ash (%)		
	2017	2018	Mean	2017	2018	Mean	2017	2018	Mean
Pasto	11.7	7.6	9.7	29.8	32.5	31.1	15.1	15.7 ab	15.4 abc
Marisma	13.5	8.1	10.8	27.8	32.3	30.0	16.7	14.6 ab	15.6 ab
Jessie	12.7	10.5	11.6	24.7	30.1	27.4	17.4	15.9 a	16.6 a
Roja	13.2	6.9	10.0	30.3	39.2	34.7	13.0	12.6 b	12.8 c
Duquesa	13.1	8.4	10.8	27.6	35.0	31.3	14.4	13.0 ab	13.7 bc
Mean	12.8 A	8.4 B	10.6	28.0 B	33.8 A	30.9	15.3	14.3	14.8
HSD	4.0	5.3	4.4	6.5	16.2	7.8	4.6	3.2	2.6
Significance									
Year (Y)			**			*			n.s.
Variety (V)	n.s.	n.s.	n.s.	n.s.	n.s.	n.s.	n.s.	*	**
Y × V			n.s.			n.s.			n.s.

CP[1]: crude protein; CF[2]: crude fiber. Variety means denoted by different lowercase letters in the same column are significantly different at $p < 0.05$ according to Tukey's test. Year means followed by different uppercase letters are significantly different at $p < 0.05$ according to Tukey's test. HSD: critical value for comparison. n.s.: not significant; significant at * $p < 0.05$; ** $p < 0.01$, respectively.

2.3. Fibre Composition

Significant higher neutral detergent fiber (NDF), acid detergent fiber (ADF), and acid detergent lignin (ADL) contents were determined in 2018 (55.4%; 40.5%; 6.5%, respectively) compared to 2017 (44.1%; 30.8%; 5.2%, respectively), as shown in Table 2. The variety showed a significant influence on the ADF and the ADL content. The average ADF content ranged from 31.6% (Jessie) to 40.9% (Roja), whereas the average ADL content varied from 5.3% (Marisma) to 6.5% (Roja). Interactions between the year and variety were not significant. The hemicellulose (HEM) contents only showed significant differences in 2017 according to the variety, ranging from 11.2% (Marisma) to 15.2% (Jessie). Significant differences were also found in cellulose (CEL) contents in 2017 according to the variety, ranging from 20.6% (Jessie) to 29.2% (Pasto). The mean CEL content was significantly affected by the year and the variety. It was higher in 2018 (34.0%) than in 2017 (25.6%), and ranged from 25.3% in Jessie to 34.5% in Roja. Neither the HEM nor the CEL contents were significantly affected by the year × variety interaction.

Table 2. Fiber composition of straw of five quinoa varieties during the two years (2017 and 2018) of field experiments.

Variety	NDF[1] (%)			ADF[2] (%)			ADL[3] (%)			HEM[4] (%)			CEL[5] (%)		
	2017	2018	Mean	2017	2018	Mean	2017	2018	Mean	2017	2018	Mean	2017	2018	Mean
Pasto	47.2	52.2	49.7	34.2	38.0	36.1 ab	5.1 ab	6.0	5.5 ab	13.0 ab	14.2	13.6	29.2 a	32.0	30.6 ab
Marisma	41.4	53.7	46.7	30.3	37.9	34.1 ab	4.5 b	6.1	5.3 b	11.2 b	15.8	13.5	25.8 ab	31.8	28.8 ab
Jessie	41.8	51.7	47.6	26.6	36.6	31.6 b	6.0 a	6.6	6.3 a	15.2 a	15.0	15.1	20.6 b	30.0	25.3 b
Roja	48.3	63.6	46.7	34.0	47.8	40.9 a	5.9 a	7.0	6.5 a	14.4 a	15.8	15.1	28.1 ab	40.8	34.5 a
Duquesa	42.1	55.7	48.9	28.8	42.1	35.5 ab	4.6 ab	7.0	5.8 ab	13.3 ab	13.6	13.5	24.2 ab	35.1	29.7 ab
Mean	44.1 B	55.4 A	49.8	30.8 B	40.5 A	35.6	5.2 B	6.5 A	5.9	13.4	14.9	14.2	25.6 B	34.0 A	29.8
HSD	11.4	19.5	10.0	9.3	16.2	8.3	1.4	1.4	0.9	3.1	4.2	2.5	8.3	15.1	7.7
Significance															
Year (Y)			**			**			**			n.s.			**
Variety (V)	n.s.	n.s.	n.s.	n.s.	n.s.	*	*	n.s.	**	*	n.s.	n.s.	*	n.s.	*
Y × V			n.s.			n.s.			n.s.			n.s.			n.s.

NDF[1]: neutral detergent fiber, ADF[2]: acid detergent fiber, ADL[3]: acid detergent lignin, HEM[4]: hemicellulose, CEL[5]: cellulose. Variety means denoted by different lowercase letters in the same column are significantly different at $p < 0.05$ according to Tukey's test. Year means followed by different uppercase letters are significantly different at $p < 0.05$ according to Tukey's test. HSD: critical value for comparison. n.s.: not significant; significant at * $p < 0.05$; ** $p < 0.01$, respectively.

2.4. Mineral Composition

As can be observed in Table 3, the N, P, and Ca contents differed significantly according to year, and all of them were higher in 2017. The variety had a significant influence on the P, K, and Ca contents, but not on the N and Mg average contents, which were 1.7% and 0.64%, respectively. The average P content ranged from 0.17% (Roja and Duquesa) to 0.24% (Jessie). K showed average contents from 4.4% (Roja) to 6.0% (Jessie), and the average of Ca content varied between 0.8% (Roja) and 1.6% (Pasto).

Table 3. Mineral composition of straw of five quinoa varieties during the two years (2017 and 2018) of field experiments.

Variety	N (%)			P (%)			K (%)			Ca (%)			Mg (%)		
	2017	2018	Mean	2017	2018	Mean	2017	2018	Mean	2017	2018	Mean	2017	2018	Mean
Pasto	1.9	1.2	1.5	0.23	0.15 b	0.19 ab	4.3 b	5.1	4.7 ab	1.6 a	1.5 a	1.6 a	0.62	0.68	0.65
Marisma	2.2	1.3	1.7	0.28	0.15 b	0.22 ab	4.6 ab	5.0	4.8 ab	1.8 a	1.1 abc	1.5 ab	0.82	0.69	0.76
Jessie	2.0	1.7	1.9	0.23	0.25 a	0.24 a	6.1 a	5.8	6.0 a	1.1 b	1.2 ab	1.2 bc	0.63	0.77	0.70
Roja	2.1	1.2	1.6	0.25	0.09 b	0.17 b	4.6 ab	4.1	4.4 b	0.9 b	0.6 c	0.8 d	0.60	0.45	0.53
Duquesa	2.1	1.3	1.7	0.22	0.12 b	0.17 b	5.1 ab	4.9	5.0 ab	1.1 b	0.9 bc	1.0 cd	0.63	0.55	0.59
Mean	2.0 A	1.3 B	1.7	0.24 A	0.15 B	0.20	5.0	5.0	5.0	1.3 A	1.1 B	1.2	0.66	0.63	0.64
HSD	0.6	1.0	0.5	0.10	0.09	0.06	1.7	2.6	1.4	0.4	0.5	0.3	0.38	0.34	0.26
Significance															
Year (Y)			**			***			n.s.			*			n.s.
Variety (V)	n.s.	n.s.	n.s.	n.s.	**	*	*	n.s.	*	***	**	***	n.s.	n.s.	n.s.
Y × V			n.s.			**			n.s.			**			n.s.

Variety means denoted by different lowercase letters in the same column are significantly different at $p < 0.05$ according to Tukey's test. Year means followed by different uppercase letters are significantly different at $p < 0.05$ according to Tukey's test. HSD: critical value for comparison. n.s.: not significant; significant at * $p < 0.05$; ** $p < 0.01$ and *** $p < 0.001$, respectively.

Interactions between year and variety were significant for the P and Ca contents. In addition, the P content was relatively similar in both years in the short cycle variety (Jessie), whereas in the other varieties, the P content was considerably higher in 2018, especially in Roja (medium-long cycle), for which differences were found to be higher than 60%. Regarding the Ca content, differences between years were small in Jessie and Pasto, whereas in 2018 the Ca content decreased by more than 18% in the other varieties.

2.5. Relative Feed Value

The year showed a significant influence on the digestible dry matter (DDM), the dry matter intake (DMI) and on the relative feed value (RFV), which were higher in 2017 (Table 4). The variety significantly influenced the DMI and RFV values, achieving the highest values in the short cycle variety Jessie (64.3 and 131.0, respectively), and the lowest values in the medium-long cycle varieties (Roja; 57.1 and 99.4, respectively). Interaction between the year and the variety was not significant.

Table 4. Straw forage quality of five quinoa varieties during the two years (2017 and 2018) of field experiments.

Variety	DDM[1]			DMI[2]			RFV[3]		
	2017	2018	Mean	2017	2018	Mean	2017	2018	Mean
Pasto	2.6	2.3	2.4	62.3	59.3	60.8 ab	124.2	106.3	115.2 ab
Marisma	2.9	2.3	2.6	65.3	59.4	62.3 ab	147.5	104.1	125.8 ab
Jessie	2.9	2.3	2.6	68.2	60.3	64.3 a	152.7	109.3	131.0 a
Roja	2.5	1.9	2.2	62.4	51.7	57.1 b	120.9	77.9	99.4 b
Duquesa	2.9	2.2	2.5	66.5	56.1	61.3 ab	147.1	94.9	121.0 ab
Mean	2.7 A	2.2 B	2.5	64.9 A	57.4 B	61.1	138.5 A	98.5 B	118.5
HSD	0.7	0.7	0.5	7.2	12.7	6.5	49.6	51.4	31.5
Significance									
Year (Y)			**			**			**
Variety (V)	n.s.	n.s.	n.s.	n.s.	n.s.	*	n.s.	n.s.	*
Y × V			n.s.			n.s.			n.s.

DDM[1]: digestible dry matter; DMI[2]: dry matter intake; RFV[3]: relative feed value. Variety means denoted by different lowercase letters in the same column are significantly different at $p < 0.05$ according to Tukey's test. Year means followed by different uppercase letters are significantly different at $p < 0.05$ according to Tukey's test. HSD: critical value for comparison. n.s.: not significant; significant at * $p < 0.05$; ** $p < 0.01$, respectively.

3. Discussion

The use of quinoa straw has the potential to contribute to the development of a cleaner agriculture. This byproduct has been used and studied for agricultural purposes as feed and bedding material, in addition to bioenergy or biomaterial production [27,28]. Quinoa cultivation has experienced a rapid increase in the past decade, in parallel with its expansion to many different geographical areas around the world [29]. However, quinoa straw use has been little explored because the interest in this crop has focused mainly on seed yield. Thus, the current study aimed to evaluate variations in quinoa straw characteristics linked to genetic differences under Mediterranean field conditions, to implement alternative uses of this crop and therefore contribute to the development of sustainable agriculture. This study was performed in two consecutive years showing different climatic conditions, which included average differences in Tmax during the quinoa growth period of 5 °C. Generally, elevated temperatures inhibit quinoa plant growth by impacting quinoa flowering [19,30–32]. Heat stress can limit the source and sink capacity of plants, reducing growth and development [33]. Indeed, the higher temperatures registered during the first year resulted in a reduction of straw yield in 2017 (1.7 t ha^{-1}) compared to 2018 (2.1 t ha^{-1}) (Figure 1), especially in short-medium cycle varieties, such as Marisma, which doubled its yield in 2018. On the contrary, long-cycle varieties, such as Roja or Duquesa, showed lower straw yields in 2018, which may point to a differential response depending on the developmental stage affected by high temperatures. In cereals, such as wheat, it has been reported that heat stress is an important factor that reduces straw yield, because photosynthesis is altered [34,35]. Nonetheless, further analysis should be performed to determine the exact impact of heat stress on the vegetative growth and its influence on straw yield, particularly in quinoa. It should be noted that the high straw yield achieved by Marisma (2.8 t ha^{-1}, on average), especially in 2018 (3.9 t ha^{-1}), was well correlated with a higher seed yield, as previously reported [36], which is interesting from an agronomical perspective because seed yield penalties should be avoided. Intriguingly, the two-year average straw yield (1.9 t ha^{-1}) achieved in this work was lower than those previously reported for Mediterranean field conditions (3.0 t ha^{-1} for Titicaca; 7.4 t ha^{-1} for Regalona) [37], and similar to those obtained by Asher et al. (2020), although at the bottom of the range (0.5–9.1 t ha^{-1}). The results reported earlier can be partially explained by the lower HI obtained in these works, which resulted in higher straw yield per seed yield. This was probably a consequence of the different environmental conditions, and differences in the planting density and/or in the genotypes used, that resulted in detrimental effects on seed yield but positive effects on vegetative growth [38,39]. In the current study, the higher temperatures registered in 2017 also had a significant impact on the HI in medium-long cycle varieties (Roja and Duquesa). Due to the longer vegetative stage of Roja and Duquesa, flowering coincided with higher Tmax and lower RH, increasing the probability of flower damage causing a HI decrease (Figure 2). On the contrary, the shortest-cycle genotype Jessie showed the highest HI (0.55) in 2017, which was related to its ability to escape from high temperatures at flowering stage, which resulted in the maintenance of its sink capacity but larger straw yield penalties. In general, as shown in Figure 2, the average HI was 0.51, which is similar to the HI reported in modern wheat varieties (0.3–0.6) [39,40], and higher than those reported for grain legumes and canola [41]. Nonetheless, no correlation between straw yield and HI was found, in contrast to the findings of previous studies [42,43].

The straw composition differed between years as observed in Tables 1–4, probably influenced by the high temperatures of 2017. This effect has been previously studied in the main products of different crops [44–48], although the impact on byproducts, such as straw, has been little explored. In fact, the current study is pioneering in its evaluation of the effect of high temperatures on quinoa straw yield and composition. Furthermore, the CP content range in the straws obtained in this study (6.9–13.5%) was slightly higher than that obtained by Asher et al. (2020) under Mediterranean conditions (5.1–10.6%).

As can be observed in Table 1, the CP content was considerably higher in 2017 (52.7%). Because of the higher average straw yield of 2018, the lower straw CP content achieved in 2018 could have been a consequence of a dilution effect. However, in 2017 the CP content was also higher in the medium-long cycle varieties, which also achieved higher straw yields. Therefore, high temperatures could lead to higher straw CP content independently of the straw yield. Interestingly, no correlation was found between CP content and the other parameters analyzed in this study, except for the straw N content. It should be noted that the high CP content observed here (particularly in 2017) was also higher than the contents found in cereal (2.9%) or legume (7.4%) straws [49], which is important when the aim is to use this byproduct for animal feed. In contrast, the CF content was reduced about 20% in 2017 without a significant effect linked to the variety (Table 1). Plant fiber includes the cell wall used for providing mechanical support to the plant, and the vascular system in which fluids are transported [50,51]. The response observed in the CF content was likely related to a change in composition due to the heat stress impact on the biosynthesis-related pathways of cell wall components [52–54]. Actually, a steeper decrease was detected in the ADF content (with an average reduction of 24%), and in the cellulose content (which showed an average decrease of 25%) (Table 2). The two-year average of NDF was slightly lower than the values determined by Asher et al. 2020 (41.4–63.6% and 44.4–71.3%, respectively), whereas the lignin was higher than the values reported by these authors (5.3–7.0% and 5.03–7.83%, respectively). This effect raises an interesting aspect related to the straw composition as lignocellulosic biomass, which is now considered an important fiber resource for renewable energy and biomaterial production [55]. By comparison, the ash content did not change between years, achieving values close to 15%, restricting quinoa straw's use as a solid biofuel (for which ash should not exceed 1.5%) [56].

When analyzing the increased mineral and protein contents in 2017, the changes could point to a detrimental effect of higher temperatures in nutrient translocation into the seed as observed in wheat [33]. However, the protein and mineral contents were also higher in quinoa seeds in 2017 [36]. It is known that the environmental conditions can affect nutrient levels of quinoa seeds [57]. In the current study, the straw N and P contents were significantly lower in 2018, which could be related to a dilution effect linked to higher yields, similar to that described in cereal grains [58]. However, those differences were even higher in Roja and Duquesa, which achieved higher yields in 2017. Furthermore, the straw Ca content was significantly lower in 2017, especially in Roja and Duquesa. Therefore, differences in mineral nutrient contents were probably more related to a still unknown heat-induced adaptation mechanism (and/or to the effect caused by the interaction between nutrients) than to a direct effect caused by nutrient dilution. The effect of heat stress on cellular osmotic adjustments due to the increased transpiration rates when elevated temperatures occur should also not be discounted; this could result in increments in the mineral and protein content [59,60]. Interestingly, the mineral composition of quinoa straw was richer from a nutritional perspective compared with the mineral composition of cereal straws commonly used for animal feed [61]. For instance, Ca or Mg contents in wheat straws were reported to be, on average, 0.18% and 0.06%, respectively whereas quinoa contents were, on average, 1.2% and 0.64%, respectively (Table 3).

The relative feed value (RFV) developed by the Universities of Minnesota and Wisconsin and the American Forage and Grassland Council (AFGC), is an index widely used to determine the forage quality, which combines significant nutritional factors (including voluntary intake and digestibility) [62,63]. When evaluating the changes in composition of the quinoa straw, a significant impact on the RFV was detected, which resulted in the increase in this index in 2017, due to the lower NDF and ADF, improving the straw nutrient intake and digestibility (Table 4). Considering the AFGC classification method [62], the mean value for the RFV in 2017 (138.5) would indicate that this byproduct would be classified as a forage type I (125–151), which is considered acceptable. In 2018, the mean value (98.5) would correspond to a low-quality forage (being classified as a type IV forage (87–102). Therefore, based on these results, in addition to the higher CP and mineral contents (higher than those found in winter cereals, which is the straw most frequently used in animal feed), it can be considered that the quinoa straw obtained in 2017 possesses a higher nutritional value. Nonetheless, further research is required to determine the exact impact of heat stress on straw yield and composition. Based on the results presented here, elevated temperatures, which are expected to continue increasing due to the continuation of the warming period [64], may positively impact straw composition in a genotype-dependent manner.

Overall, the results here presented show that Marisma is the variety with the best cultivation potential in terms of straw yield for this particular area (Southwestern Spain), despite suffering important yield penalties linked to the elevated temperatures suffered in 2017 [36]. Furthermore, the straw composition analysis performed shows that quinoa straw is a valuable resource for animal feed, biofuel, or biomaterial production. To enhance quinoa straw valorization, it is crucial to evaluate the straw yield and compositional changes related to the genotype and environment, and to understand the relationships among biomass, seed, and straw yield. Our results, together with the previously reported changes in chemical composition of cereal straws, highlight that quinoa straw yield and composition may be affected by agronomic and genotypic factors, and environmental and climate conditions [50,55]. Within the current climate context in which more frequent episodes of elevated temperatures in the Mediterranean area are expected to occur, and considering that the effects of heat stress on plants trigger complex responses that result in the alteration of growth and development, thereby changing physiological functions and reducing seed/grain formation and plant yield [33,65], further studies should be undertaken to evaluate the impact of heat stress on straw yield and composition, with the aim of selecting the best adapted quinoa cultivars for a particular area of cultivation. More importantly, this work points to the many possibilities offered by the use of quinoa straw as an agricultural byproduct that can greatly contribute to sustainable agriculture.

4. Materials and Methods

4.1. Location, Climate and Soil Characteristics of the Experimental Site

A two-year field experiment was conducted during 2017–2018 at the experimental farm of the Center for Scientific and Technological Research of Extremadura (CICYTEX), located in Southwest Spain (lat. 38°51′10″ N; long. 6°39′10″ W). Data of monthly mean minimum and maximum temperature (Tmin and Tmax), and the rainfall during the crop cycle (Supplementary Figure S1) were obtained from the weather station located at the experimental farm. The soil was a sandy loam, neutral (pH 6.9), presenting 0.38% organic matter, 0.24% total N, and 93.4 ppm and 57.9 ppm of available P and K, respectively.

4.2. Layouts of Experiments, Plant Material and Crop Management

Five European varieties of quinoa (Pasto, Marisma, Jessie, Roja, Duquesa) were evaluated in a randomized complete block design with four replications: Jessie (short cycle; ~120 d); Pasto and Marisma (medium cycle; ~135 d); Roja and Duquesa (medium-long cycle; ~145 d). The plot size was formed by four rows 0.75 m apart and 10 m long. Sowing was conducted mechanically in early February, at a dose of 4 kg ha^{-1}. Weeding was carried out by hand when required. Irrigation was carried out by sprinkling to maintain the soil under non-limiting water conditions. The crop was fertilized at the rate of 150, 100, and 100 kg/ha of N, P_2O_5, K_2O, respectively. Plants were harvested by hand at physiological maturity (in the middle of June for Jessie, and early July for the other four varieties). The sampling area was 3 m^2. A stationary thresher was used to collect the seed. The straw samples were ground through a 1 mm screen for further analysis.

4.3. Analysis and Measurements

Data were expressed on a dried weight (dW) basis. Analysis of moisture, crude protein (CP), and ash contents were conducted following the AOAC Official Methods [66]. The mineral content was assessed following the official methods of the Spanish Ministry of Agriculture [67]. P was analyzed by a UV-VIS spectrophotometer (Hitachi U-2810). K was determined using flame atomic emission spectroscopy, and Ca and Mg by flame atomic absorption (SpectrAA 110, Agilent). The crude fiber (CF), neutral detergent fiber (NDF), acid detergent fiber (ADF), and acid detergent lignin (ADL) values were analyzed following the Ankom procedure (ANKOM Technology, Fairport, NY, USA), using a fiber analyzer (ANKOM 2000) and F57 Ankom filter bags (porosity: 25 μm). Hemicellulose (Hem) and cellulose (Cel) contents were calculated as follows:

$$\text{Hemicellulose (Hem)} = [\text{NDF\%} - \text{ADF\%}]$$

$$\text{Cellulose (Cel)} = [\text{ADF\%} - \text{ADL\%}]$$

Relative feed value (RFV) was calculated as calculated from the digestible dry matter (DDM) and the dry matter intake (DMI) (live weight: LW, %) according to the following equations:

$$\text{DDM}\ (\%) = 88.9 - [0.779 \times \text{ADF}\%]$$

$$\text{DMI}\ (\text{Live Weight}:\ \text{LW}\ \%) = 120/[\text{NDF}\%]$$

$$\text{RFV} = [\text{DMD} \times \text{DMI}]/1.29$$

Harvest index (HI) was calculated as the ratio between the grain yield (G) and the yields of straw plus grain (G + S), in order to determine the biomass partitioning index, as previously reported [36].

4.4. Statistical Analysis

All measured and derived data were processed using a two-way analysis of variance (ANOVA), including the year, the variety, and their interactions in the model. The year was treated as a fixed factor. Normality and equal variances could be assumed, according to the results of the Shapiro Wilk test and Levene's test, respectively. For better interpretation of the data, additional one-way ANOVA was carried out for each year separately. When the F ratio was significant ($p < 0.05$), Tukey's test was performed and used to compare means. Student's *t*-test was used to determine statistically significant differences between years for each variety. Analyses were performed using the Statistix 8 analytical software.

Supplementary Materials: The following are available online at https://www.mdpi.com/article/10.3390/plants10050955/s1, Figure S1: Monthly rainfall and maximum (Tmax) and minimum (Tmin) mean temperature registered at the experimental station during the quinoa growth period of 2017 and 2018.

Author Contributions: Conceptualization, J.M. and M.R.; methodology J.M., V.C. and M.R.; software, J.M., V.C. and M.R.; validation, J.M. and M.R.; formal analysis, J.M. and M.R.; investigation, J.M., V.C. and M.R.; resources, J.M., V.C. and M.R.; data curation, J.M., V.C. and M.R.; writing—original draft preparation, J.M., V.C. and M.R.; writing—review and editing, J.M., V.C. and M.R.; visualization, J.M., V.C. and M.R.; supervision, J.M. and M.R. All authors have read and agreed to the published version of the manuscript.

Funding: The authors gratefully acknowledge the financial support received from the Junta de Extremadura, the European FEDER Funds (MESOCEX), the Ministerio de Ciencia e Innovación (MICINN, Spain) (PID2019-105748RA-I00) and the Comunidad de Madrid-Universidad Autónoma de Madrid (SI1/PJI/2019-00124).

Data Availability Statement: The data that support the findings of this study are available upon request.

Acknowledgments: The authors greatly thank Susana Vilariño (Algosur, Spain) for her support helping to conceive and plan the experiments and for providing the quinoa seeds used in this study.

Conflicts of Interest: The authors declare no conflict of interest.

References

1. Godde, C.M.; Mason-D'Croz, D.; Mayberry, D.; Thornton, P.; Herrero, M. Impacts of Climate Change on the Livestock Food Supply Chain, a Review of the Evidence. *Glob. Food Sec.* **2021**, *28*, 100488. [CrossRef] [PubMed]
2. Naylor, A.; Ford, J.; Pearce, T.; Van Alstine, J. Conceptualizing Climate Vulnerability in Complex Adaptive Systems. *One Earth* **2020**, *2*, 444–454. [CrossRef]
3. Shayanmehr, S.; Rastegari Henneberry, S.; Sabouhi Sabouni, M.; Shahnoushi Foroushani, N. Climate Change and Sustainability of Crop Yield in Dry Regions Food Insecurity. *Sustainability* **2020**, *12*, 9890. [CrossRef]
4. Abumhadi, N.; Todorovska, E.; Assenov, B.; Tsonev, S.; Vulcheva, D.; Vulchev, D.; Atanasova, L.; Savova, S.; Atanassov, A.; Keith, W. Agricultural Research in 21 St Century: Challenges Facing the Food Security under the Impacts of Climate Change. *Bulg. J. Agric. Sci.* **2012**, *18*, 801–818.
5. Wu, J.-Z.; Zhang, J.; Ge, Z.-M.; Xing, L.-W.; Han, S.-Q.; Shen, C.; Kong, F.-T. Impact of climate change on maize yield in China from 1979 to 2016. *Integr. Agric.* **2021**, *20*, 289–299. [CrossRef]

6. Bazile, D.; Jacobsen, S.-E.; Verniau, A. The Global Expansion of Quinoa: Trends and Limits. *Front. Plant Sci.* **2016**, *7*, 622. [CrossRef]
7. Blakstad, M.M.; Smith, E.R. Climate Change Worsens Global Inequity in Maternal Nutrition. *Lancet Planet. Health. E* **2020**, *4*, e547–e548. [CrossRef]
8. Farooq, M.; Bramley, H.; Palta, J.A.; Siddique, K.H.M. Heat Stress in Wheat during Reproductive and Grain-Filling Phases. *Crit. Rev. Plant Sci.* **2011**, *30*, 491–507. [CrossRef]
9. Halford, N.G. New Insights on the Effects of Heat Stress on Crops. *J. Exp. Bot.* **2009**, *60*, 4215–4216. [CrossRef]
10. Scocco, P.; Piermarteri, K.; Malfatti, A.; Tardella, F.M.; Catorci, A. Increase of Drought Stress Negatively Affects the Sustainability of Extensive Sheep Farming in Sub-Mediterranean Climate. *J. Arid Environ.* **2016**, *128*, 50–58. [CrossRef]
11. Lionello, P.; Malanotte-Rizzoli, P.; Boscolo, R.; Alpert, P.; Artale, V.; Li, L.; Luterbacher, J.; May, W.; Trigo, R.; Tsimplis, M.; et al. The Mediterranean Climate: An Overview of the Main Characteristics and Issues. *Dev. Earth Environ. Sci.* **2006**, *4*, 1–26. [CrossRef]
12. Joffre, R.; Rambal, S. Mediterranean Ecosystems. In *Encyclopedia of Life Sciences*; Mcmillan Publishers, Ltd.: New York, NY, USA, 2001; pp. 1–7. [CrossRef]
13. Jacobsen, S.-E.; Jensen, C.R.; Liu, F. Improving Crop Production in the Arid Mediterranean Climate. *Field Crop. Res.* **2012**, *128*, 34–47. [CrossRef]
14. Lam, D. How the World Survived the Population Bomb: Lessons From 50 Years of Extraordinary Demographic History. *Demography* **2011**, *48*, 1231–1262. [CrossRef]
15. Filik, G. Biodegradability of Quinoa Stalks: The Potential of Quinoa Stalks as a Forage Source or as Biomass for Energy Production. *Fuel* **2020**, *266*, 117064. [CrossRef]
16. Jacobsen, S.E.; Mujica, A.; Jensen, C.R. The Resistance of Quinoa (*Chenopodium quinoa* Willd.) to Adverse Abiotic Factors. *Food Rev. Int.* **2003**, *19*, 99–109. [CrossRef]
17. Asher, A.; Galili, S.; Whitney, T.; Rubinovich, L. The Potential of Quinoa *Chenopodium Quinoa*. Cultivation in Israel as a Dual-Purpose Crop for Grain Production and Livestock Feed. *Sci. Hortic.* **2020**, *272*, 109534. [CrossRef]
18. Ruales, J.; Nair, B.M. Nutritional Quality of the Protein in Quinoa (*Chenopodium quinoa*, Willd.) Seeds. *Plant Foods Hum. Nutr.* **1992**, *42*, 1–11. [CrossRef]
19. Hinojosa, L.; González, J.A.; Barrios-Masias, F.H.; Fuentes, F.; Murphy, K.M. Plants Quinoa Abiotic Stress Responses: A Review. *Plants* **2018**, *7*, 106. [CrossRef]
20. Razzaghi, F.; Ahmadi, S.H.; Jacobsen, S.-E.; Jensen, C.R.; Andersen, M.N. Effects of Salinity and Soil-Drying on Radiation Use Efficiency, Water Productivity and Yield of Quinoa (*Chenopodium quinoa* Willd.). *J. Agron. Crop Sci.* **2012**, *198*, 173–184. [CrossRef]
21. Rezzouk, F.Z.; Shahid, M.A.; Elouafi, I.A.; Zhou, B.; Araus, J.L.; Serret, M.D. Agronomic Performance of Irrigated Quinoa in Desert Areas: Comparing Different Approaches for Early Assessment of Salinity Stress. *Agric. Water Manag.* **2020**, *240*, 106205. [CrossRef]
22. Iqbal, S.; Basra, S.M.A.; Afzal, I.; Wahid, A. Exploring Potential of Well Adapted Quinoa Lines for Salt Tolerance. *Int. J. Agric. Biol.* **2017**, *19*, 933–940. [CrossRef]
23. Adolf, V.I.; Jacobsen, S.E.; Shabala, S. Salt Tolerance Mechanisms in Quinoa (*Chenopodium quinoa* Willd.). *Environ. Exp. Bot.* **2013**, *92*, 43–54. [CrossRef]
24. Ahmadi, S.H.; Solgi, S.; Sepaskhah, A.R. Quinoa: A Super or Pseudo-Super Crop? Evidences from Evapotranspiration, Root Growth, Crop Coefficients, and Water Productivity in a Hot and Semi-Arid Area under Three Planting Densities. *Agric. Water Manag.* **2019**, *225*, 105784. [CrossRef]
25. Präger, A.; Munz, S.; Nkebiwe, P.; Mast, B.; Graeff-Hönninger, S. Yield and Quality Characteristics of Different Quinoa (*Chenopodium quinoa* Willd.) Cultivars Grown under Field Conditions in Southwestern Germany. *Agronomy* **2018**, *8*, 197. [CrossRef]
26. Xie, W.; Ren, Y.; Jiang, F.; Liang, J.; Du, S. kui. Pretreatment of Quinoa Straw with 1-Butyl-3-Methylimidazolium Chloride and Physiochemical Characterization of Biomass. *Renew. Energy* **2020**, *146*, 1364–1371. [CrossRef]
27. She, D.; Sun, R.C.; Jones, G.L. Chemical Modification of Straw as Novel Materials for Industries. In *Cereal Straw as a Resource for Sustainable Biomaterials and Biofuels*; Elsevier: Amsterdam, The Netherlands, 2010; pp. 209–217. [CrossRef]
28. Wang, K.; Sun, R.C. Biorefinery Straw for Bioethanol. In *Cereal Straw as a Resource for Sustainable Biomaterials and Biofuels*; Elsevier: Oxford, UK, 2011; pp. 267–287. [CrossRef]
29. Alandia, G.; Rodriguez, J.P.; Jacobsen, S.-E.; Bazile, D.; Condori, B. Global Expansion of Quinoa and Challenges for the Andean Region. *Glob. Food Sec.* **2020**, *26*, 100429. [CrossRef]
30. Tovar, J.C.; Quillatupa, C.; Callen, S.T.; Elizabeth Castillo, S.; Pearson, P.; Shamin, A.; Schuhl, H.; Fahlgren, N.; Gehan, M.A. Heating Quinoa Shoots Results in Yield Loss by Inhibiting Fruit Production and Delaying Maturity. *Plant J.* **2019**, *102*, 1058–1073. [CrossRef]
31. Bertero, H.D. Environmental Control of Development. In *State of the Art Report on Quinoa around the World in 2013*; Bazile, D., Bertero, D., Nieto, C., Eds.; FAO & CIRAD: Rome, Italy, 2015; pp. 120–130.
32. Hinojosa, L.; Matanguihan, J.B.; Murphy, K.M. Effect of High Temperature on Pollen Morphology, Plant Growth and Seed Yield in Quinoa (*Chenopodium quinoa* Willd.). *J. Agron. Crop Sci.* **2019**, *205*, 33–45. [CrossRef]
33. Akter, N.; Rafiqul Islam, M. Heat Stress Effects and Management in Wheat. A Review. *Agron. Sustain. Dev.* **2017**, *37*, 1–17. [CrossRef]

34. Feng, B.; Liu, P.; Li, G.; Dong, S.T.; Wang, F.H.; Kong, L.A.; Zhang, J.W. Effect of Heat Stress on the Photosynthetic Characteristics in Flag Leaves at the Grain-Filling Stage of Different Heat-Resistant Winter Wheat Varieties. *J. Agron. Crop Sci.* **2014**, *200*, 143–155. [CrossRef]
35. Mahrookashani, A.; Siebert, S.; Hüging, H.; Ewert, F. Independent and Combined Effects of High Temperature and Drought Stress around Anthesis on Wheat. *J. Agron. Crop Sci.* **2017**, *203*, 453–463. [CrossRef]
36. Matías, J.; Cruz, V.; Rodríguez, M.R.; Calvo, P. Heat stress lowers yields, alters nutrient uptake, and changes seed quality in quinoa grown under Mediterranean field conditions. *J. Agron. Crop Sci.* **2021**, accepted, in press. [CrossRef]
37. Lavini, A.; Pulvento, C.; d'Andria, R.; Riccardi, M.; Choukr-Allah, R.; Belhabib, O.; Yazar, A.; İncekaya, Ç.; Metin Sezen, S.; Qadir, M.; et al. Quinoa's Potential in the Mediterranean Region. *J. Agron. Crop Sci.* **2014**, *200*, 344–360. [CrossRef]
38. Hammer, G.L.; Broad, I.J. Genotype and Environment Effects on Dynamics of Harvest Index during Grain Filling in Sorghum. *Agron. J.* **2003**, *95*, 199–206. [CrossRef]
39. Dai, J.; Bean, B.; Brown, B.; Bruening, W.; Edwards, J.; Flowers, M.; Karow, R.; Lee, C.; Morgan, G.; Ottman, M.; et al. Harvest Index and Straw Yield of Five Classes of Wheat. *Biomass Bioenerg.* **2016**, *85*, 223–227. [CrossRef]
40. Hay, R.K.M. Harvest Index: A Review of Its Use in Plant Breeding and Crop Physiology. *Ann. Appl. Biol.* **1995**, *126*, 197–216. [CrossRef]
41. Unkovich, M.; Baldock, J.; Forbes, M. Variability in Harvest Index of Grain Crops and Potential Significance for Carbon Accounting: Examples from Australian Agriculture. *Adv. Agron.* **2010**, *105*, 173–219. [CrossRef]
42. Ali, S.; Chattha, M.U.; Hassan, M.U.; Khan, I.; Chattha, M.B.; Iqbal, B.; Rehman, M.; Nawaz, M.; Amin, M.Z. Growth, Biomass Production, and Yield Potential of Quinoa (*Chenopodium quinoa* Willd.) as Affected by Planting Techniques Under Irrigated Conditions. *Int. J. Plant Prod.* **2020**, *14*, 427–441. [CrossRef]
43. De Santis, G.; Ronga, D.; Caradonia, F.; Ambrosio, T.D.; Troisi, J.; Rascio, A.; Fragasso, M.; Pecchioni, N.; Rinaldi, M. Evaluation of Two Groups of Quinoa Chenopodium Quinoa Willd.. Accessions with Different Seed Colours for Adaptation to the Mediterranean Environment. *Crop Pasture Sci.* **2018**, *69*, 1264–1275. [CrossRef]
44. Assefa, Y.; Purcell, L.C.; Salmeron, M.; Naeve, S.; Casteel, S.N.; Kovács, P.; Archontoulis, S.; Licht, M.; Below, F.; Kandel, H.; et al. Assessing Variation in US Soybean Seed Composition Protein and Oil. *Front. Plant Sci.* **2019**, *10*, 298. [CrossRef]
45. Gibson, L.R.; Mullen, R.E. Soybean Seed Composition under High Day and Night Growth Temperatures. *J. Am. Oil Chem. Soc.* **1996**, *73*, 733–737. [CrossRef]
46. Thomas, J.M.G.; Boote, K.J.; Allen, L.H.; Gallo-Meagher, M.; Davis, J.M. Elevated Temperature and Carbon Dioxide Effects on Soybean Seed Composition and Transcript Abundance. *Crop Sci.* **2003**, *43*, 1548–1557. [CrossRef]
47. Wolf, R.B.; Cavins, J.F.; Kleiman, R.; Black, L.T. Effect of Temperature on Soybean Seed Constituents: Oil, Protein, Moisture, Fatty Acids, Amino Acids and Sugars. *J. Am. Oil Chem. Soc.* **1982**, *59*, 230–232. [CrossRef]
48. Habermann, E.; Dias de Oliveira, E.A.; Contin, D.R.; Delvecchio, G.; Viciedo, D.O.; de Moraes, M.A.; de Mello Prado, R.; de Pinho Costa, K.A.; Braga, M.R.; Martinez, C.A. Warming and Water Deficit Impact Leaf Photosynthesis and Decrease Forage Quality and Digestibility of a C4 Tropical Grass. *Physiol. Plant.* **2019**, *165*, 383–402. [CrossRef] [PubMed]
49. López, S.; Davies, D.R.; Giráldez, F.J.; Dhanoa, M.S.; Dijkstra, J.; France, J. Assessment of Nutritive Value of Cereal and Legume Straws Based on Chemical Composition and in Vitro Digestibility. *J. Sci. Food Agric.* **2005**, *85*, 1550–1557. [CrossRef]
50. Liu, L.; Ye, X.P.; Womac, A.R.; Sokhansanj, S. Variability of Biomass Chemical Composition and Rapid Analysis Using FT-NIR Techniques. *Carbohydr. Polym.* **2010**, *81*, 820–829. [CrossRef]
51. Sorieul, M.; Dickson, A.; Hill, S.J.; Pearson, H. Plant Fibre: Molecular Structure and Biomechanical Properties, of a Complex Living Material, Influencing Its Deconstruction towards a Biobased Composite. *Materials* **2016**, *9*, 618. [CrossRef]
52. Le Gall, H.; Philippe, F.; Domon, J.M.; Gillet, F.; Pelloux, J.; Rayon, C. Cell Wall Metabolism in Response to Abiotic Stress. *Plants* **2015**, *4*, 112–166. [CrossRef]
53. Houston, K.; Tucker, M.R.; Chowdhury, J.; Shirley, N.; Little, A. The Plant Cell Wall: A Complex and Dynamic Structure as Revealed by the Responses of Genes under Stress Conditions. *Front. Plant Sci.* **2016**, *10*, 984. [CrossRef]
54. Lima, R.B.; Dos Santos, T.B.; Vieira, L.G.E.; Ferrarese, M.D.L.L.; Ferrarese-Filho, O.; Donatti, L.; Boeger, M.R.T.; Petkowicz, C.L.D.O. Heat Stress Causes Alterations in the Cell-Wall Polymers and Anatomy of Coffee Leaves *Coffea arabica* L. *Carbohydr. Polym.* **2013**, *93*, 135–143. [CrossRef]
55. Collins, S.R.; Wellner, N.; Martinez Bordonado, I.; Harper, A.L.; Miller, C.N.; Bancroft, I.; Waldron, K.W. Variation in the Chemical Composition of Wheat Straw: The Role of Tissue Ratio and Composition. *Biotechnol. Biofuels* **2014**, *7*, 121. [CrossRef]
56. Platače, R.; Adamovičs, A. The Evaluation of Ash Content in Grass Biomass Used for Energy Production. *WIT Trans. Ecol. Environ.* **2014**, *190*, 1057–1065. [CrossRef]
57. Reguera, M.; Conesa, C.M.; Gil-Gómez, A.; Haros, C.M.; Pérez-Casas, M.Á.; Briones-Labarca, V.; Bolaños, L.; Bonilla, I.; Álvarez, R.; Pinto, K.; et al. The Impact of Different Agroecological Conditions on the Nutritional Composition of Quinoa Seeds. *Peer J.* **2018**, *14*, e4442. [CrossRef]
58. Murphy, K.M.; Reeves, P.G.; Jones, S.S. Relationship between Yield and Mineral Nutrient Concentrations in Historical and Modern Spring Wheat Cultivars. *Euphytica* **2008**, *163*, 381–390. [CrossRef]
59. Patakas, A.; Zioziou, E.; Nikolaou, N.; Radoglou, K.; Noitsakis, B. The role of organic solute and ion accumulation in osmotic adjustment in drought-stressed grapevines. *Plant Sci.* **2002**, *163*, 361–367. [CrossRef]

60. Argentel-Martínez, L.; Garatuza-Payan, J.; Yepez, E.A.; Arredondo, T.; De los Santos-Villalobos, S. Water regime and osmotic adjustment under warming conditions on wheat in the Yaqui Valley, Mexico. *Peer J.* **2019**, *7*, e7029. [CrossRef]
61. Pearce, G.R.; Beard, J.; Hilliard, E.P. Variability in the Chemical Composition of Cereal Straws and in Vitro Digestibility with and without Sodium Hydroxide Treatment. *Aust. J. Exp. Agric.* **1979**, *19*, 350–353. [CrossRef]
62. Linn, J.G.; Martin, N.P. Forage Quality Analyses and Interpretation. *Vet. Clin. North Am. Food Anim. Pract.* **1991**, *7*, 509–523. [CrossRef]
63. Moore, J.E.; Burns, J.C.; Fisher, D.S. Multiple Regression Equations for Predicting Relative Feed Value of Grass Hays. In Proceedings of the Proc. American Forage and Grassld, Vancouver, BC, Canada, 13–15 June 1996; Williams, M.J., Ed.; pp. 135–139.
64. Wheeler, T.; Von Braun, J. Climate Change Impacts on Global Food Security. *Science* **2013**, *341*, 508–513. [CrossRef]
65. Mondal, S.; Singh, R.P.; Crossa, J.; Huerta-Espino, J.; Sharma, I.; Chatrath, R.; Singh, G.P.; Sohu, V.S.; Mavi, G.S.; Sukaru, V.S.P.; et al. Earliness in Wheat: A Key to Adaptation under Terminal and Continual High Temperature Stress in South Asia. *F. Crop. Res.* **2013**, *151*, 19–26. [CrossRef]
66. Association of Official Agricultural Chemist (AOAC). *Official Methods of Analysis of AOAC International*, 17th ed.; Horwitz, W., Ed.; AOAC: Arlington, TX, USA, 2000.
67. MAPA, Ministerio de Agricultura, Pesca y Alimentación. *Métodos Oficiales de Análisis*; Tomo III; MAPA: Madrid, Spain, 1995; pp. 205–285.

plants

MDPI

Article

The Importance of Non-Diffusional Factors in Determining Photosynthesis of Two Contrasting Quinoa Ecotypes (*Chenopodium quinoa* Willd.) Subjected to Salinity Conditions

José Delatorre-Herrera [1,*], Karina B. Ruiz [2] and Manuel Pinto [3]

[1] Doctoral Program in Agriculture for Arid-Desert Environments, Faculty of Renewable Natural Resources, Desert Agriculture Area, Universidad Arturo Prat, Iquique 1100000, Chile
[2] Facultad de Ciencias de la Salud, Universidad Arturo Prat, Iquique 2120, Chile; karuiz@unap.cl
[3] Plant Physiology Laboratory, Institute of Agronomic and Veterinary Sciences, Universidad de O´Higgins, Rancagua 2820000, Chile; manuel.pinto@uoh.cl
* Correspondence: jodelato@unap.cl

Abstract: The broad distribution of quinoa in saline and non-saline environments is reflected in variations in the photosynthesis-associated mechanisms of different ecotypes. The aim of this study was to characterize the photosynthetic response to high salinity (0.4 M NaCl) of two contrasting Chilean genotypes, Amarilla (salt-tolerant, *salares* ecotype) and Hueque (salt-sensitive, coastal ecotype). Our results show that saline stress induced a significant decrease in the K^+/Na^+ ratio in roots and an increase in glycine betaine in leaves, particularly in the sensitive genotype (Hueque). Measurement of the photosynthesis-related parameters showed that maximum CO_2 assimilation (A_{max}) in control plants was comparable between genotypes (ca. 9–10 μmol CO_2 m^{-2} s^{-1}). However, salt treatment produced different responses, with A_{max} values decreasing by 65.1% in the sensitive ecotype and 37.7% in the tolerant one. Although both genotypes maintained mesophyll conductance when stomatal restrictions were removed, the biochemical components of Amarilla were impaired to a lesser extent under salt stress conditions: for example, the maximum rate of ribulose-1,5-bisphosphate carboxylase/oxygenase (RubisCO; V_{cmax}) was not as affected in Amarilla, revealing that this enzyme has a higher affinity for its substrate in this genotype and, thus, a better carboxylation efficiency. The present results show that the higher salinity tolerance of Amarilla was also due to its ability to control non-diffusional components, indicating its superior photosynthetic capacity compared to Hueque, particularly under salt stress conditions.

Keywords: Na^+, K^+, CO_2 assimilation; stomatal restrictions; non-diffusional; diffusional; RubisCO activity

Citation: Delatorre-Herrera, J.; Ruiz, K.B.; Pinto, M. The Importance of Non-Diffusional Factors in Determining Photosynthesis of Two Contrasting Quinoa Ecotypes (*Chenopodium quinoa* Willd.) Subjected to Salinity Conditions. *Plants* **2021**, *10*, 927. https://doi.org/10.3390/plants10050927

Academic Editors: Cataldo Pulvento and Didier Bazile

Received: 19 March 2021
Accepted: 27 April 2021
Published: 6 May 2021

1. Introduction

At present, about one-third of the world's irrigated land [1] is affected by salinity, which reduces plant growth and crop yield [2]. As the main food crops are rather sensitive to this stress, tolerance to salinity has become an important agronomical trait for breeders, physiologists, and agronomists [3]. The salinity of soils is predominantly caused by salt transported by irrigation water [4]. On the other hand, due to climate change, well water is becoming the main water resource for irrigation in many parts of the world, and several studies have indicated that this method is linked to potential risks of soil salinization [5–7]. An increase in the soil concentration of Na^+ and other ions, such as Ca^{2+}, Cl^-, and K^+, causes a decrease in the soil water potential, which limits the water absorbed by roots and induces water stress [8]. Specific ion toxicities are due to the accumulation of sodium, chloride, and/or boron in the tissue of transpiring leaves to damaging levels. The accumulation of injurious ions may inhibit photosynthesis and protein synthesis, inactivate enzymes, and damage chloroplasts and other organelles [9]. Under normal physiological conditions, plants maintain a high cytosolic K^+/Na^+ ratio. Given the difference in negative membrane potential at the plasma membrane (–140 mV), a rise in extracellular Na^+ concentration will establish a large electrochemical gradient that favors the passive transport of Na^+ into cells through the activation of K^+ transporters/channels and through non-selective channels that are sensitive to Ca^{2+}. The permeation of Na^+ via voltage-independent cation (VIC) channels is inhibited by an increase in extracellular Ca^{2+} concentration [10]. Overaccumulation of Na^+ in the cytosol inhibits protein synthesis, enzyme activity [1], and many photosynthetic processes [11,12]. Therefore, maintaining its water supply and excluding Na^+ from photosynthetic organs are crucial mechanisms used by tolerant plants to ensure an adequate rate of carbon fixation under salt stress [13]. It is well known that a reduction in stomatal conductance negatively affects the CO_2 assimilation rate as well as the water balance in leaves [3,14].

Excess salt affects plant growth due to an increase in soil osmotic pressure and interference with plant nutrition. A high salt concentration in the soil solution reduces the ability of plants to acquire water, which is referred to as the osmotic or water-deficit effect of salinity [9]. A decrease in CO_2 assimilation is a widely reported effect of salt [15]. In many cases, this decrease is associated with the negative effect of salinity on diffusional mechanisms. These mechanisms depend on the gradient between the external and internal CO_2 concentrations (mesophyll and chloroplasts), on the stomatal conductance in the gaseous phase (g_s), and on the mesophyll conductance in the liquid phase (g_m) [15–18].

However, salt also affects non-diffusional mechanisms, which are typically associated with electron transport and the activity of enzymes in carboxylation cycles; key enzymes in these processes include ribulose-1,5-bisphosphate carboxylase/oxygenase (RubisCO) [19,20] and those involved in the regeneration of ribulose-1,5-bisphosphate (RuBP) [21,22].

Most authors agree that diffusional factors are predominant in affecting the CO_2 assimilation rate [15,23]. For example, Flexas et al. [24] reported that the conductance of the mesophyll in grapevine was strongly correlated with the rate of photosynthesis. In olive trees (*Olea europaea*), this occurs in the leaves of stressed plants with a conductance greater than 50 $\mu mol\ m^{-2}\ s^{-1}$ [22]. According to the aforementioned study, the combined reduction in stomatal and mesophyll conductance in different salt-stressed olive cultivars increases the difference in CO_2 concentration between the environment and chloroplasts. These results indicate that the low concentration of CO_2 in chloroplasts caused by decreased stomatal and mesophyll conductance is the main limiting factor of photosynthesis.

Quinoa (*Chenopodium quinoa* Willd.) has evolved a number of adaptive responses to saline stress conditions, such as the increased elasticity of cell walls, low water potential, osmotic adjustment, reduction in the foliar area, the presence of *papillae* with calcium oxalate, the production of organic solutes [25–28], the adaptation of physiological functions such as photosynthesis [2,29], the regulation of the water status [29], ionic partitioning [30,31], and changes in stomatal conductance [3,16,27].

There are an increasing number of studies on CO_2 assimilation in quinoa [2,16,26,32,33], although few have characterized the importance of diffusional and non-diffusional mechanisms. This species is particularly diverse at the genetic level [34], a consequence of its broad distribution in both saline and non-saline environments. This diversity could allow some ecotypes to thrive under saline stress conditions due to improved rates of photosynthesis [30].

Thus, it is possible to postulate that saline-tolerant ecotypes have better control of CO_2 diffusion mechanisms than those that are less tolerant to this stress. Therefore, the aim of this study was to evaluate the differences in the photosynthetic processes between two contrasting quinoa ecotypes, as well as to analyze the relative importance of non-diffusional factors in the two ecotypes when subjected to salinity conditions.

2. Materials and Methods

2.1. Plant Material

The two quinoa ecotypes used in this study, Amarilla and Hueque, have been previously characterized as tolerant and sensitive to salinity, respectively [30]. The Amarilla ecotype comes from the salt flat area in the northern highlands, and Hueque originates from the wet area at sea level in southern Chile.

Disinfected seeds (2% sodium hypochlorite for 7 min) were germinated in 330 mL pots using perlite as the substrate. Pots containing 4 plants each were arranged over a tray containing water or solution nutritive and connected to it by means of a cotton wick to ensure water supply by capillarity. The plants were watered with sodium-free nutrient solution or 0 M (0.25 dS m^{-1}) until the third pair of leaves developed (45 days after sowing (DAS)). At this stage, half of the pots of each ecotype were transferred to a modified Hoagland 2 solution containing 0.4 M NaCl (38.1 dS m^{-1}). This concentration corresponds to the LD_{50max} for quinoa, as determined previously by Delatorre-Herrera and Pinto [30]. To avoid a saline shock, 0.4 M NaCl was applied in increments of 0.1 M per day. Control plants were irrigated with solution nutritive in a semi-hydroponic system (Schlick and Bubenheim [35], modified by Delatorre and Pinto [30]); for this purpose, filtered water with an electrical conductivity of 0.25 dS m^{-1} with a pH of 6 was used. When the plants developed their fourth or fifth pair of true leaves (70–75 DAS), they were placed in a shaded field, where the average temperature of the day was 20 ± 5 °C, and the maximum light intensity was 1500 µmol $m^{-2}s^{-1}$ of photosynthetically active radiation (PAR).

2.2. Na^+ and K^+

Sodium (Na^+) and potassium (K^+) contents were determined in samples of root, stems, and leaves taken from the bottom, middle, and top parts of the stem. Once collected, the material was dried at 70 °C for 48 h and then finely ground. Na^+ and K^+ were extracted by digesting 0.1 g of each sample in 15 mL of 0.5 M HCl for 2 days. The concentrations of Na^+ and K^+ were determined according to the procedure described by Hunt [36] using a flame photometer (Jenway Model PFP 7, Cole Parmer, Vernon Hills, IL, USA).

2.3. Proline and Glycine Betaine

Proline (Pro) content was determined spectrophotometrically (Spectrophotometer, Genesys, Thermo Scientific, Waltham, MA, USA) in tissue with a dry weight of 100 mg using the method described by Bates et al. [37]. A standard curve was constructed with proline (M.W. 115.13 g mol^{-1}, Sigma-Aldrich, Santiago, Chile).

The content of Glycine Betaine (GB) was determined according to Grieve and Grattan [38] but modified for quinoa. Briefly, tissue with a dry weight of 100 mg was stirred in 4 mL of water for 24 h at 25 °C, after which it was filtered and stored at 4 °C until analysis. For the determination of quaternary compounds, 50 µL samples were thawed, diluted in 50 µL of 2 N sulfuric acid, and cooled to 0 °C for 1 h. Then, 40 µL of $KI\text{-}I_2$ reagent (15.7 g iodine and 20 g KI in 100 mL of water) was added and gently stirred in a vortex. This solution was stored at −4 °C for 16 h, after which it was centrifuged at 10,000 rpm for 15 min. The supernatant was carefully removed, and the precipitate was dissolved in 1.6 mL of 1,2-dichloroethane, shaken vigorously by vortexing, and left to stand for 2.5 h at room temperature. GB content was detected at 365 nm. The standard curve was constructed with betaine (M.W. 117.15 g mol^{-1}, Sigma-Aldrich, Santiago, Chile).

2.4. Determination of CO_2 Assimilation

Gas Exchange

Gas exchange measurements were performed between 09:00 and 15:00 using a gas exchange chamber connected to a portable infrared analyzer (Licor 6200, LI-COR, Lincoln, NE, USA). The relative humidity of the chamber ranged from 40% to 50%, and the leaf temperature was 25 ± 2 °C. Measurements were made on fully expanded mature leaves (middle part of the stem). Each measurement was repeated three times on the same leaf (one leaf per plant and 6 plants per treatment). These measurements were taken when 50% of the plants had grown the fourth or fifth pair of true leaves (70–75 DAS).

(i) A/PFD curves

The photon flows (PFDs) used were 0, 20, 90, 120, 150, 300, 500, 700, 1200, 1500, and 2500 µmol photons $m^{-2}s^{-1}$. The light source was a halogen lamp, and the different PFDs were obtained by placing neutral filters between the lamp and the photosynthetic chamber.

The temperature in the photosynthetic chamber was 25 ± 2 °C. During measurements, the CO_2 and the O_2 concentrations in the chamber were maintained at 360 µL L^{-1} and 20%, respectively (CO_2 and O_2, certified gas, INDURA S.A, Alto Hospicio, Chile).

The gross photosynthesis rate (A), the apparent quantum efficiency (Φ), photosynthetically active radiation (PAR), the rate of assimilation at saturating light intensity (A_{max}), and mitochondrial respiration in darkness (Rd) were obtained from the A/PFD curves adjusted to a non-rectangular hyperbole-type equation, according to the procedure described by Lambers et al. [39]. The straight section of the curve was adjusted to a polynomial equation whose intersection on the *x*-axis corresponds to the light compensation point, and the slope is the apparent quantum yield.

(ii) A/C_i Curves

Measurements were made at light saturation (1500 µmol photons $m^{-2}s^{-1}$). The relative humidity of the chamber ranged from 40% to 50%, and the leaf temperature was 25 ± 2 °C. The CO_2 concentrations used were 50, 350, 500, 700, 900, 1200, and 1500 µL L^{-1}, and the assimilation response to these intensities was adjusted to the Farquhar model [40,41] with PHOTOSYN software (version 1.1.2, Dundee Scientific Ltd., Dundee, UK). Parameters such as the maximum rate of RubisCO activity ($V_{c,max}$), the transport of electrons at light saturation (J_{max}), and the triose phosphate transport rate (TPU) were obtained from this model, which expresses the ratio of the assimilation rate (A) to the internal concentration of CO_2 (C_i) for each of the three factors affecting assimilation.

When carboxylation is limited only by the activity of RubisCO, then W_c can be described by the model developed by Farquhar et al. [40], which is based on gaseous exchange measurements.

When the transport of electrons limits photosynthesis due to the effect of RuBP regeneration, W_j can be expressed according to Farquhar and Von Caemmerer [41]. The potential rate of electron transport (J_p) was calculated using the expression developed by Harley et al. [42].

(iii) Determination of the linear transport rate of electrons

To measure photosystem II (PSII) efficiency, the IRGA 6200 chamber was adapted by introducing a Hansantech PEA modulated pulse fluorometer sensor at one end. The parameters were calculated according to the methodology described by Maxwell and Johnson [43].

Quantum yield (ΦPSII) can be used to calculate the linear transport rate of electrons (*J*), which corresponds to photosynthetic capacity in vivo, according Genty et al. [44]. A factor of 0.84 is assumed for leaf absorbance in C3 plants [45], although this absorbance may change slightly with temperature (this experiment was conducted at foliar temperatures of 25 ± 3 °C).

The photochemical energy used for photosynthesis or photochemical quenching (qP) and non-photochemical quenching (NPQ) were calculated using equations described by Genty et al. [44].

2.5. Removal of Stomatal Effects

To remove the stomatal effect on CO_2 assimilation (A), the methodology described by Centritto et al. [15] was used. For this purpose, a set of A/C_i curves were generated for both ecotypes. The procedure employed can be divided into four phases: In phase (I), stomatal restriction (R) was induced by lowering the concentration of CO_2 from 350 to 50 µL L^{-1} and allowing g_s to decrease to values close to 15 mmol $m^{-2}s^{-1}$. Subsequently, in phase (II), the CO_2 concentration was raised to 1500 µL L^{-1}, and measurements of the A/Ci curve were made until 50 µL L^{-1}. In phase (III), a CO_2 concentration of 50 µL L^{-1} was maintained for a minimum of 40 min, which depended on the stomatal behavior of each ecotype and treatment; in this way, stomatal opening was activated. Phase (IV), the final phase, was initiated once stomatal conductance began to rise above 600 mmol m^{-2} s^{-1}. This raised the CO_2 concentration to 1500 µL L^{-1}, and A/C_i was measured with no stomatal restrictions (nR). A value of 600 mmol m^{-2} s^{-1} was used based on the data of Jacobsen et al. [26], who reported a g_s of 600 mmol H_2O m^{-2} s^{-1} in quinoa without water restrictions. The times used in each stage are detailed in Table 1.

Table 1. Times of application of different concentrations of environmental CO_2 to induce the closure and opening of stomata.

Stage I		Stage II		Stage III		Stage IV	
Initial Period Closed Stomata		With Stomatal Restriction (R)		Induction Period Open Stomata		No Stomatal Restriction (nR)	
Time min	$[CO_2]$	Time min	$[CO_2]$ µL L^{-1}	Time min	$[CO_2]$ µL L^{-1}	Time mins	$[CO_2]$ µL L^{-1}
0	350	21	1500	107	50	152	1500
8	350	28	1200	132	50	162	1200
13	50	32	900	147	50	167	900
		37	700			172	700
		47	500			177	500
		54	350			182	350
		60	200			187	200
		67	100			192	100
		72	50			204	50

2.6. Determination of Mesophyll Conductance (g_m)

Mesophyll conductance (g_m) was determined according to the procedure described by Harley et al. [42] using the following parameters: assimilation rate (A), light compensation point (Γ), electron transport rate (J), and dark respiration (Rd).

2.7. Rate Carboxylation and Electron Transport In Vivo

The RubisCO activity rate at RuBP saturation ($V_{c,max}$), the maximum electron transport rate (J_{max}), the maximum rate of assimilation at saturating PFD, and CO_2 (A_{max}) and mitochondrial respiration in light per unit of foliar area (R_d) were obtained using the model described by Farquhar et al. [40]. From this model, A was determined using Equation (1):

$$A = vc - 0.5vo - Rd \quad (1)$$

where vc and vo are the rates of carboxylation and oxygenation of RubisCO, respectively, and 0.5 is the stoichiometric relationship between the O_2 absorbed by RuBP oxygenase and the photorespiratory evolution of CO_2 [41]. Additionally, vc cannot be greater than the minimum rate of carboxylation of RubisCO (A_c), and the rate of electron transport limits the regeneration of RuBP (A_j).

2.8. Data Analysis

A 2 × 2 factorial design was used, arranged in completely randomized blocks (B_i). The first factorial level was NaCl concentration (C_j) (0 M and 0.4 M NaCl), and the second level was the ecotype (A_k) (Amarilla and Hueque). Each treatment had 6 replications, and the experimental unit consisted of 2 pots, each with 4 plants (8 plants). Duncan's multiple comparison one-way analysis of variance (ANOVA) was performed using Infostat V 2016 (Universidad de Cordoba, Cordoba, Argentina) $p < 0.01$ was considered significant.

3. Results

3.1. Stress Indicators

When comparing the control plants of the two ecotypes, Hueque had a higher concentration of GB in the roots. When exposed to 0.4 M NaCl, the GB content increased in both ecotypes, and it was higher in Hueque. Figure 1a shows that in the salt-tolerant ecotype (Amarilla), the GB content was higher in the leaves than in the roots. In the roots, salt stress had no effect on GB content, whereas in the leaves, GB content increased significantly. The leaf/root ratio of GB content in Amarilla control plants was 3.64, while in stressed plants, it rose slightly to 4.34, which could indicate that this ecotype is not severely stressed by high salinity. In the salt-sensitive ecotype (Hueque), the constitutive concentration of GB was also higher in the leaves than in the roots. In the roots, the salt stress caused a decrease in the GB content; in contrast, GB rose significantly in the leaves. The leaf/root ratio of GB was 2.12 in Hueque control plants and 7.95 in stressed plants, which means that the GB synthesis in leaves was more sensitive to salt signals coming from roots.

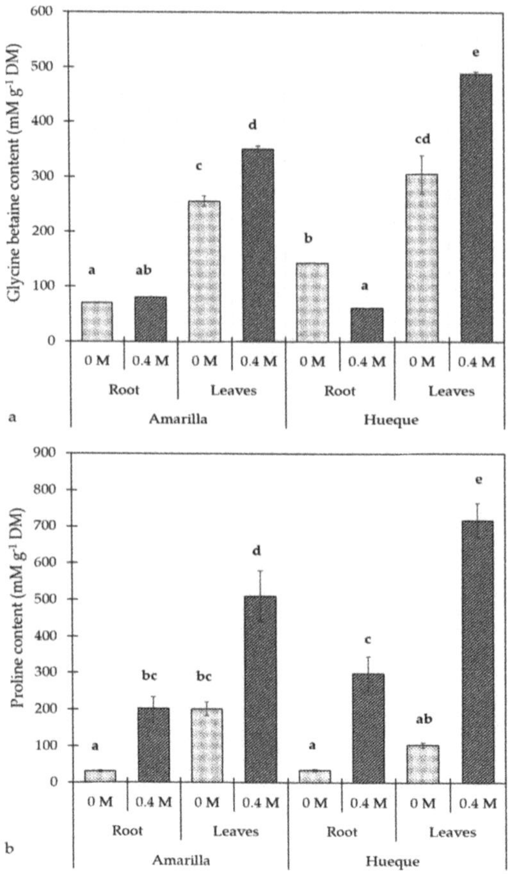

Figure 1. Effect of salinity on glycine betaine (**a**) and proline (**b**) content in the roots and leaves of quinoa plants. Different letters denote significant differences ($p \leq 0.05$). Average values were calculated based on 4 samples per treatment (mean ± SE).

The Pro content in the control plants did not differ much between the two ecotypes. Exposure to 0.4 M NaCl increased the Pro content in all tissues tested in both ecotypes, and it was highest in Hueque. Figure 1b shows that in the tolerant ecotype, the content of Pro was significantly higher in the leaves than in the roots under both control and saline conditions. The leaf/root ratio of Pro in control plants was 6.53, and in salt-exposed plants, it fell to 2.53; this decrease was mainly due to the Pro increase in the roots. In Hueque, the content of Pro was greater in the leaves than in the roots. Applying salt did not cause changes in Pro levels in the roots, whereas in leaves, the Pro content rose significantly. The leaf/root ratio of Pro in control plants was 3.02, and when exposed to 0.4 M NaCl, it was 2.40.

3.2. *Na^+ and K^+ Content*

The tolerant ecotype accumulated in root 64.6% more Na^+ than the sensitive one in control conditions. Under salt stress, this difference decreased to 33% in both ecotypes because the sensitive ecotype accumulated proportionally more sodium than the tolerant one (7.4 versus 3.4 times, respectively) (Figure 2a). However, the plants showed no changes in the leaf Na^+ level, regardless of salt exposure. The K^+ content in roots was similar in the control plants of both tolerant and sensitive ecotypes and decreased significantly under salt stress. Under high concentrations of soil salt, the K^+ content decreased in the roots and was maintained in the leaves in both ecotypes (Figure 2b).

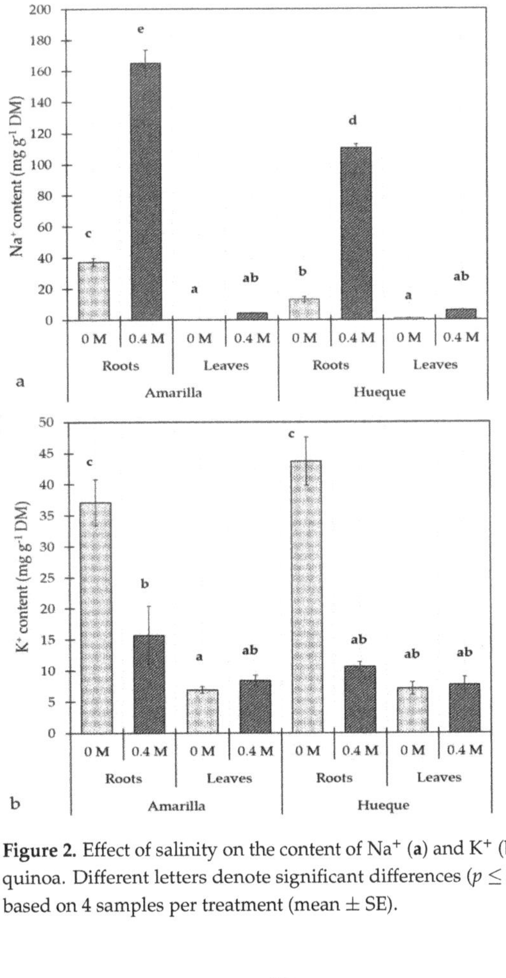

Figure 2. Effect of salinity on the content of Na^+ (**a**) and K^+ (**b**) in Amarilla and Hueque ecotypes of quinoa. Different letters denote significant differences ($p \leq 0.05$). Average values were calculated based on 4 samples per treatment (mean ± SE).

3.3. *Effect of Salinity on CO_2 Assimilation under Different Light Intensities*

The A/PFD curves are shown in Figure 3. In the absence of salt stress, both ecotypes had a similar quantum yield performance and assimilation rate at light saturation. The A_{max} values for Amarilla and Hueque were 9.98 and 9.05 µmol CO_2 $m^{-2}s^{-1}$, respectively (Table 2). However, the intensities at which light saturation reached were significantly different between the ecotypes: CO_2 assimilation saturated at 957 PAR µ$m^{-2}s^{-1}$ in Hueque and at 1262 µmol PAR $m^{-2}s^{-1}$ in Amarilla (Table 2).

Table 2. The effect of salinity on different photosynthetic values of Amax and g_s were obtained from curves of Figure 3 determined at the light saturation point and 350 µL CO_2 L^{-1}.

Ecotype	NaCl (M)	Saturation Intensity (µmoles PAR m^{-2} s^{-1})	Amax µmol CO_2 m^{-2} s^{-1}	% Fall	g_s mmol H_2O m^{-2} s^{-1}	% Fall
Amarilla	0	1262 c	9.98		310	
Amarilla	0.4	685 ab	6.22	37.7	256	17.4
Hueque	0	957 b	9.05		471	
Hueque	0.4	420 a	3.16	65.1	201	57.3

* Different letters indicate significant differences ($p \leq 0.01$).

Under saline conditions (0.4 M NaCl), both ecotypes presented a drastic decrease in CO_2 assimilation rates with respect to control plants (Figure 3). For example, the decrease in A_{max} was 37.7% in the tolerant ecotype but 65.1% in the sensitive one (Table 3). On the other hand, at light intensities exceeding the corresponding light saturation point, saline conditions induced photoinhibition in both ecotypes. This was very drastic in the sensitive ecotype, in which assimilation was almost null at 2500 µmol photons $m^{-2}s^{-1}$ (Figure 3).

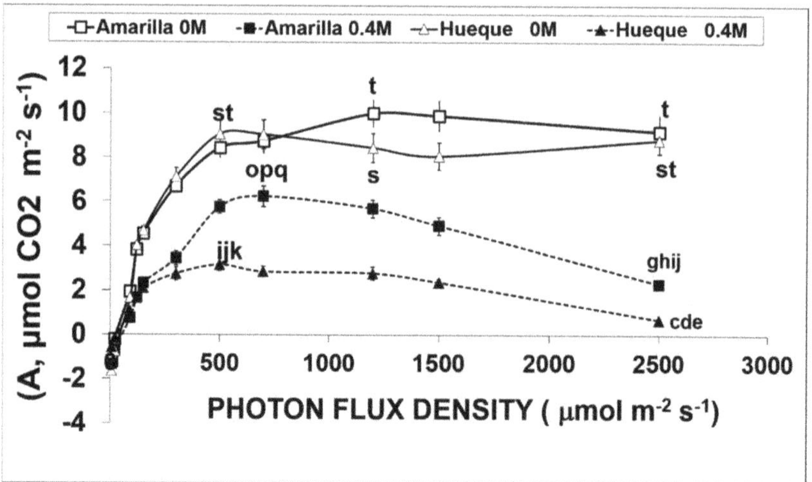

Figure 3. CO_2 assimilation curves of two quinoa ecotypes subjected to salinity conditions. Empty symbols and continuous lines correspond to treatments without salt, while filled symbols with dotted lines correspond to 0.4 M NaCl treatments. Different letters denote significant differences ($p \leq 0.05$). Average values were calculated based on 3 leaves taken from 6 plants per treatment (mean ± SE).

The analysis of the main photochemical parameters (Table 3) shows that the photochemical efficiency (ΦPSII) and rate of linear electron transport (J) were negatively affected by salt stress in both ecotypes. ΦPSII decreased by 18.1% for Amarilla and 29.2% for Hueque. J changed by 17.4% and 27.7% for Amarilla and Hueque, respectively.

Table 3. Harvested energy efficiency (ΦPSII) and rate of linear electron transport (J).

Ecotype	NaCl	ΦPSII	Significance	J	Significance
Amarilla	0 M	0.27	c	167.6	c
Amarilla	0.4 M	0.22	b	138.4	b
Hueque	0 M	0.24	bc	147.9	bc
Hueque	0.4 M	0.17	a	106.9	a

* Different letters indicate significant differences ($p < 0.01$).

3.4. Determination of Non-Restrictive (nR) and Restrictive (R) Stomatal Conductance

Figure 4 shows the evolution of Ci and g_s at different timepoints, CO_2 concentrations, and NaCl treatments for both ecotypes according to the procedure proposed by Centritto et al. [21]. This enabled us to determine the times and CO_2 concentrations required to induce the rates of stomatal conductance, both restrictive (R) and non-restrictive (nR).

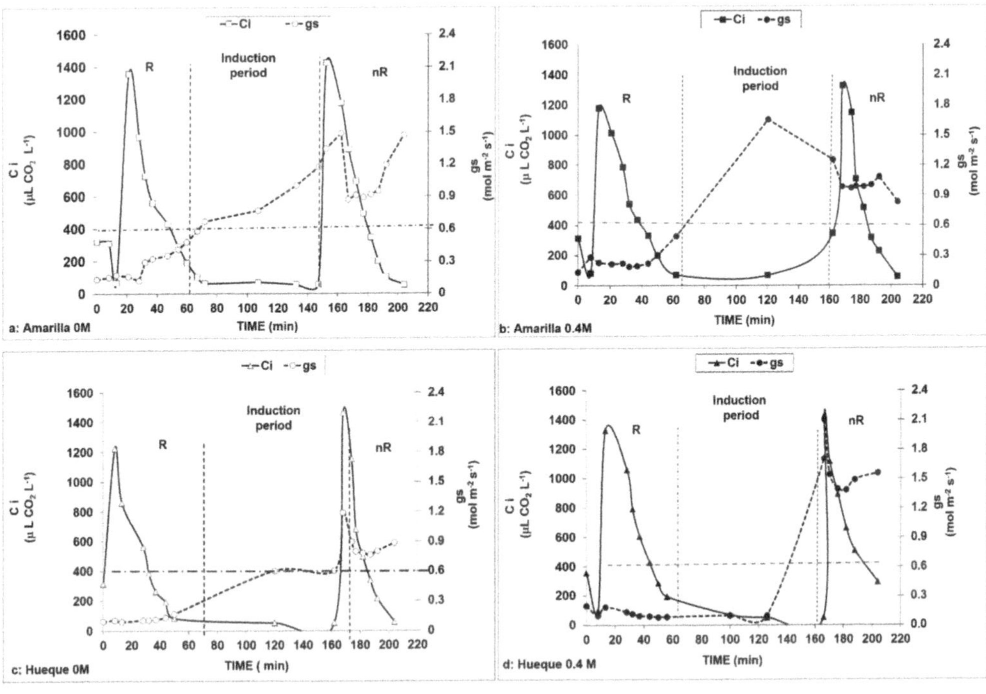

Figure 4. Curves of each ecotype used to determine the times necessary to induce stomatal opening. The continuous line represents the internal concentration of CO_2 (Ci), and the dotted line is stomatal conductance. The line parallel to the x-axis represents the minimum limit required for open stomata. Lines parallel to the y-axis define the measurement periods. R—measurements with stomatal constraints; nR—measurements with no stomatal constraints. The induction period corresponds to the time interval required to induce stomata opening by applying a concentration of 50 µL CO_2 L^{-1}. Internal concentrations correspond to IRGA measurements with respect to the application of a specific environmental CO_2 concentration, applied according to the time sequences described in Table 1.

Once the CO_2 concentration was lowered to 50 µL L^{-1} (phase III), the stomata opened after approximately 1.5 h (90 min). This allowed similar stomatal conductance behaviors to be maintained between non-stress and 0.4 M NaCl conditions in both ecotypes. The maintenance of Ci at values equal to or less than 50 µL L^{-1} caused a rapid response in the tolerant ecotype in control and saline conditions (Figure 4a,b). Thus, the method initially reduced the conductance to values close to 0.1 mol H_2O $m^{-2}s^{-1}$, which then increased to 1.5 and 1.6 mol H_2O $m^{-2}s^{-1}$ in the tolerant ecotype (at 0 and 0.4 M NaCl, respectively). In Hueque, the initial value was 0.09 mol H_2O $m^{-2}s^{-1}$, which then increased to 1.2 mol H_2O $m^{-2}s^{-1}$ at 0 M NaCl and 2.1 mol H_2O $m^{-2}s^{-1}$ at 0.4 M NaCl. It was also observed that Hueque presented a higher g_s as well as a greater duration of stomatal opening in saline conditions (Figure 4c,d).

These data are consistent with the assimilation rates observed in plants with stomatal restriction (R) and without stomatal restriction (nR) (Table 4). In the salt-tolerant ecotype under control and nR condition, A_{max} increased from 13.8 to 21.43 μmol CO_2 m^{-2} s^{-1}, which means that, in this case, the elimination of the stomatal restriction accounted for 55.7% of the assimilation rate. This effect was repeated in plants subjected to saline stress under the nR condition, but in this case, stomatal restriction accounted for only 25.4% of the increase in A_{max} (Table 4).

In the sensitive ecotype under non-salt conditions, the elimination of stomatal restriction did not induce a significant increase in A_{max} (6.7%), which suggests that this ecotype involves another resistance mechanism, such as mesophyll resistance. However, when salt-sensitive plants were treated with salt, elimination of stomatal restriction induced a 41.2% increase in A_{max}. In both ecotypes, elimination of stomatal restriction induced a decrease in gross respiration under both control and salt stress conditions (Table 4).

Table 4. The effect of salinity Amax, Gross Respiration and Carboxylation efficiency parameters of tolerant and sensitive ecotypes of quinoa, obtained from A/Ci curves (Figure 5).

Ecotype	NaCl	CONDITION	Amax		Sig.	Gross Resp.		Sig.	Carbox.		Sig.
	M		μmol CO_2 m^{-2} s^{-1}	± SD		μmol CO_2 $m^{-2}s^{-1}$	± SD		Efic.	± SD	
Amarilla	0	R	13.8	1.03	cd	−4.09	1.39	ab	0.135	0.036	d
Amarilla	0.4	R	12.8	0.94	c	−4.8	1.15	a	0.117	0.034	d
Amarilla	0	nR	21.49	2.93	f	−2,9	1.12	cd	0.127	0.026	d
Amarilla	0.4	nR	16.05	2.46	e	−3.42	1.3	bc	0.071	0.033	b
Hueque	0	R	13.75	2.82	cd	−3.25	1.11	cd	0.096	0.026	c
Hueque	0.4	R	7.3	0.82	a	−3.2	0.31	cd	0.023	0.005	a
Hueque	0	nR	14.67	3.2	de	−1.95	0.95	e	0.07	0.027	b
Hueque	0.4	nR	10.31	1.62	b	−2.51	0.98	de	0.036	0.008	a

R: with stomatal restriction; nR no stomatal restriction; Amax: Maximum assimilation rate; Gross resp.: Gross respiration (dark and photorespiration); Carbox. Efic.: Carboxylation efficiency; Different letters indicate Significance level ($p \leq 0.01$).

The analysis of carboxylation parameters (Table 4) showed that the carboxylation efficiency in Amarilla control plants under non-restrictive stomatal conditions (nR) was not significantly altered ($p > 0.05$). However, in salt-stressed plants with nR, carboxylation efficiency was reduced by 44.1% compared to that observed in control plants, which may be associated with the effect of mesophyll conductance (g_m). In the salt-sensitive ecotype, salt stress caused decreases in carboxylation efficiency of 76% and 48% for R and nR, respectively.

On the other hand, the unaltered maximum rates of RubisCO activity (Vcmax) and Triose Phosphate Transport (TPU) (Table 5) indicate that, in the tolerant ecotype, the reduction in CO_2 assimilation caused by salt was not due to RubisCO activity or the supply of inorganic phosphate to chloroplasts. Differently, Vcmax, TPU in Hueque fell by 36.6% and 17.9% in R and nR conditions, respectively, showing that CO_2 assimilation was affected by these factors (Table 5).

Table 5. The effect of salinity on the main photosynthetic parameters of tolerant and sensitive ecotypes of quinoa using the model of Farquhar et al. (1980).

Ecotype	NaCl	CONDITION	Jmax		Sig.	Vcmax		Sig.	TPU		Sig.
	M		μmol CO_2 m^{-2} s^{-1}	± SD	$p \leq 0.01$	μmol CO_2 m^{-2} s^{-1}	± SD	$p \leq 0.01$		± SD	$p \leq 0.01$
Amarilla	0	R	152.66	62.62	d	32.76	6.41	c	10.56	2.77	d
Amarilla	0.4	R	120.53	20.61	c	29.75	4.57	bc	9.27	1.49	cd
Amarilla	0	nR	122.52	23.61	c	33.4	9.77	c	9.93	1.83	cd
Amarilla	0.4	nR	122.59	24.82	c	29.55	5.06	bc	9.69	1.57	cd
Hueque	0	R	115.15	24.66	c	27.68	4.94	b	8.96	1.83	bc
Hueque	0.4	R	72.29	13.55	a	18.86	2.72	a	5.68	1.26	a
Hueque	0	nR	94.66	21.29	b	25.67	6.55	b	7.88	2.07	b
Hueque	0.4	nR	75.67	17.51	a	19.26	5.62	a	6.47	1.74	a

R: with stomatal restriction; nR without stomatal restriction; Vc,max is: Maximum rate of RubisCO activity; Jmax is the transport of electrons at light saturation and TPU is triose phosphate transport rate. Different letters indicate Significance level ($p \leq 0.01$).

The quantum efficiency of PSII (ΦPSII) determined by the ratio of variable fluorescence (FV) and maximum fluorescence (Fm) (FV/FM) did not reveal an effect of salinity on the photochemical components of photosynthesis in either ecotype subjected to salt stress (Table 6). This means that photosystems I and II were not damaged. Therefore, it is unlikely that the decreases in the assimilation of CO_2 and TPU are affected by the supply of ATP or NADPH.

Table 6. Quantum efficiency of PSII.

ECOTYPE	M NaCl	ΦPSII	Significance *
Amarilla	0	0.82	b
Amarilla	0.4	0.78	ab
Hueque	0	0.79	ab
Hueque	0.4	0.77	a

* Different letters indicate significant differences ($p \leq 0.01$).

4. Discussion

4.1. Effect of Salt on Stress Indicators

With respect to stress indicators, the present results confirm those of the previous studies in quinoa using the same or other ecotypes under similar salt conditions [16,30,46]. Pro accumulation has been reported in several southern Chilean quinoa accessions in response to 300 and 450 mM NaCl; genotypes considered tolerant to salt stress accumulated 3–5-fold more Pro than control plants, while the more sensitive ones exhibited moderate increases [19,46–48]. Several reports on glycophytes and halophytes have indicated that accumulation of GB depends on the genotype and salt stress intensity, and a positive correlation between GB levels and salt tolerance has been reported [49,50]. Our results concur with these previous studies, as increases in the GB and Pro contents were observed in the leaves of both ecotypes in response to salt stress; however, in this case, it was the sensitive ecotype that presented a greater accumulation of both osmolytes. While many studies have indicated a positive relationship between GB and/or Pro accumulation and plant stress tolerance, some have argued that increases in these compounds are a product of, non- adaptive response to, salt stress [16]. In drought-stressed corn, increased ABA content was followed by an elevation in betaine aldehyde dehydrogenase (BADH) activity, leading to GB accumulation [51]. Moreover, BADH and P5CS genes, related to GB and Pro biosynthesis, respectively, were upregulated in response to treatments with salt or chromium combined with salt in quinoa [50,52,53]. In addition, exogenous applications of GB and Pro in crop species such as rice and tomato help mitigate the effects of environmental stresses, including water and salt stress [47], meaning that they may have a role in adaptive responses to abiotic stresses. The salt-sensitive ecotype (Hueque) had a higher constitutive concentration of GB, and under salt stress conditions, the GB content in leaves was significantly enhanced and was higher than that in Amarilla. Under saline conditions, GB biosynthesis in most plant species occurs in the chloroplast [48]. Thus, GB production may be overstimulated in leaves, thereby increasing the shoot/root ratio of GB, as observed in Hueque. In this salt-sensitive ecotype, high foliar levels of GB may contribute to the maintenance of growth by acting as a nitrogen donor [54], contributing to osmotic homeostasis [55], and protecting the plant from oxidative damage, in combination with Pro [56]. In the sensitive ecotype (Hueque), the situation seems to be similar at the root level because, while GB decreased, Pro increased to levels that were 50% higher than those in the tolerant plant, which maintained its GB content under salt stress.

Our results for Na^+ and K^+ contents agree with those reported by Adolf et al. [3] in two contrasting quinoa varieties, salt-tolerant and -sensitive, namely, Titicaca and Utusaya, respectively. They found that the Na^+ content in the leaves increased drastically in plants treated with 0.4 M NaCl, and the increase was more pronounced in the salt-tolerant plant (Titicaca) compared to the more sensitive one (Utusaya). The significant increase found in leaf Na^+ content in these two genotypes could be due to the duration of salt treatment (42 days of salt treatment vs. 15 days in our experimental design).

The salt treatment also caused a significant increase in the K^+ concentration in the xylem of both ecotypes, but there were no differences in K^+ between treated and untreated plants. Our findings are similar in that there were no differences in leaf K^+ content between control and salt-treated plants. Our findings are similar to those reported by Orsini et al. [16] in that there were no differences in leaf K^+ content between control and up to 450 mM NaCl salt-treated plants of the coastal ecotype accession BO78. Potassium ions are essential for enzyme activity, protein synthesis, photosynthesis, osmoregulation, transport of phloem solutes, and the maintenance of the cation–anion balance in the cytosol and vacuoles. Recently, K^+ was also proposed to be a secondary messenger [57]. Thus, the ability of plants to retain K^+ under salt stress has emerged as an important trait for salt tolerance. By contrast, Na^+ is toxic in glycophytic plants, but not in halophytes [58]. Sodium ions cause multifactorial responses, such as osmotic stress [59,60], the inhibition of vital enzymes [61,62], and competition with K^+. Under salt stress, accumulated ions, such as Na^+, Cl^-, and K^+, are used for osmotic adjustment in the aerial portions of halophytic plants [63], thus facilitating water uptake and transport and, presumably, lowering the metabolic cost required to produce large amounts of organic osmolytes, as previously described by Hariadi et al. for other quinoa genotypes [31].

At the root level, both tolerant and sensitive plants had similar responses in salinity conditions, decreasing the uptake of K^+ and increasing the absorption of Na^+; the latter was especially pronounced in the tolerant ecotype (Amarilla). In some cases, Na^+ can replace K^+, particularly in its osmotic functions in vacuoles under K^+ starvation conditions. Halophytes require less K^+ for growth than glycophytes [1,10], demonstrating a link between the ability to replace K^+ with Na^+ and salt tolerance [64], which is consistent with the results found in this study, in which the salt-tolerant ecotype had higher levels of Na^+. In barley, a high concentration of Na^+ allows plants to osmotically adapt to and maintain turgor under high salinity, which is a metabolically inexpensive mechanism for osmotic adaptation. However, this new Na^+/K^+ homeostasis could create a greater demand for organic solutes for osmotic adjustment, thereby compromising the energy balance of the plant [65]. In salt-tolerant barley cultivars [66], K^+ was reported to be the main contributor to cytoplasmic osmolality, whereas in salt-sensitive genotypes, GB and Pro compensated for reduced cytosolic K^+ levels.

4.2. CO_2 Assimilation and Stomatal and Mesophyll Conductance

In both ecotypes, the decrease in CO_2 assimilation due to salinity was accompanied by a decrease in stomatal conductance. However, this decrease in g_s was greater in Hueque, (57%, Table 2), than in Amarilla, where was only 17.4%. These values are consistent with those reported by Centritto et al. in olive [15] and by Killi and Haworth in quinoa [2]. These authors concluded that the reduction in stomatal conductance in these species is the principal factor in reducing CO_2 diffusion, inducing a decrease in the internal CO_2 gas pressure and thus reducing the rate of photosynthesis. In Hueque, the magnitude of the decrease in g_s (57.3%) due to salt treatment was proportional to the decrease in net CO_2 assimilation (65.1%). However, in Amarilla, reduction in CO_2 assimilation (37.7%) was not accompanied by a proportional reduction in g_s (17.4%). Thus, assuming a direct relationship between CO_2 assimilation and g_s, in the tolerant ecotype, only 46% of the reduction in CO_2 assimilation caused by salt can be explained by the reduction in stomatal conductance. By contrast, in the case of Hueque, 90% of the reduction in CO_2 assimilation can be associated to the g_s reduction. This finding suggests that in the tolerant ecotype the CO_2 assimilation rate depends less on g_s than in the case of the sensitive one. This in turn suggests that, in this ecotype, another diffusional mechanism may be involved in controlling CO_2 assimilation, such as mesophyll conductance (g_m) for example. This possibility is also supported by Bongi and Loreto [19], who indicated that g_m was reduced under saline stress conditions in olive.

Despite the different behaviors of the ecotypes, g_s was low in both when treated with NaCl, suggesting that Na^+ did not interfere with K^+ stomatal signaling or guard cell osmoregulation [61], which is consistent with the exclusion of Na^+ from salt-stressed quinoa [62]. This concurs with the data shown in Figure 2, which shows that the K^+ content of leaves did not change under saline conditions.

Calculations of mesophyll conductance (Table 7) in control conditions indicate that at 350 μL CO_2 L^{-1} and light saturation, g_m was 57 mmol H_2O $m^{-2}s^{-1}$ in Amarilla, representing 18% of g_s, but was 38 mmol H_2O $m^{-2}s^{-1}$ (< 8% g_s) in Hueque. This indicates that in favorable conditions, the relative importance of g_m in CO_2 diffusion is greater in the tolerant ecotype than in the sensitive one. However, with the application of saline stress, the g_m values were significantly reduced in both ecotypes, falling by 49% in Amarilla (57 to 29 mmol m^{-2} s^{-1}) and 47% (38 to 20 mmol $m^{-2}s^{-1}$) in Hueque (Table 7). Nevertheless, these g_m values only account for 11% and 10% of g_s, respectively, which shows that under salt stress conditions, in both ecotypes, stomatal conductance exerts much greater control over CO_2 assimilation, and g_m is of marginal importance.

Table 7 also shows g_m values obtained under conditions without stomatal restriction. In both ecotypes and conditions (R and nR), the increase in salinity (0.4 M NaCl) caused a reduction in g_m, results similar to those found by Delfine et al. [23] in spinach (*Spinacia oleracea*). However, after eliminating stomatal constraints under non-salt conditions, g_m increased in both ecotypes, but the change was more significant in the tolerant plants (61% in Amarilla versus 21% in Hueque). This is consistent with Delfine et al. [23], who demonstrated that g_m was not irreversible in olive, as we show here for quinoa [2]. On the other hand, under saline conditions and with stomatal restrictions, both ecotypes had g_m values that were similar to those obtained without stomatal restriction. This finding confirms that, in quinoa, g_m is not a relevant factor in determining the diffusion of CO_2 in salt conditions.

Table 7. The effect of salinity on mesophyll conductance (g_m) determined at the light saturation point, with and without stomatal restriction.

Ecotype	NaCl (M)	With Stomatal Restriction (R)		No Stomatal Restriction (nR)	
		g_m mmol H_2O m^{-2} s^{-1}	% Fall	g_m mmol H_2O m^{-2} s^{-1}	% Fall
Amarilla	0	57		92	
Amarilla	0.4	29	49.0	32	65.2
Hueque	0	38		46	
Hueque	0.4	20	47.4	20	56.5

4.3. Effect of Salinity on Non-Diffusional Parameters

The effect of salinity on CO_2 assimilation, measured by the internal CO_2 concentration and the effect of diffusion factors, is shown in Figure 5a,b. In both ecotypes stomatal limitations (R) strongly affected assimilation rates, especially upon the addition of 0.4 M NaCl. However, the ecotypes had different responses: while A increased in Amarilla, with the removal of stomatal constraints (nR), Hueque did not show the same response. This is similar to the observations reported in Table 2, which shows that Hueque has little control over stomatal opening mechanisms, which is apparently an important factor in decreasing the CO_2 assimilation rate. However, above 900 μmol CO_2, Amarilla had a CO_2 assimilation rate in saline conditions similar to that of control plants in unrestricted conditions. Thus, the better response of Amarilla may be due to greater photochemical efficiency and greater RubisCO activity, while in Hueque, difficulties arise as a result of diffusion and lower energy efficiency (Tables 5 and 6).

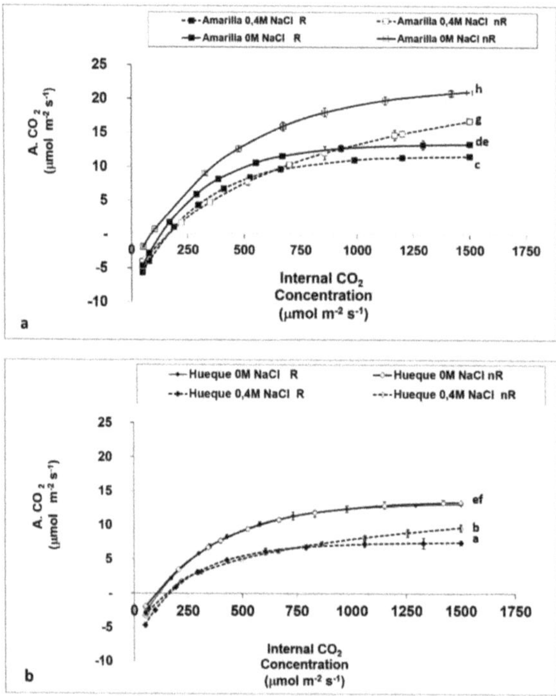

Figure 5. Evolution of CO_2 assimilation based on internal CO_2 concentration in the leaves of the quinoa ecotypes Amarilla (**a**) and Hueque (**b**) under limiting and non-limiting stomatal conditions. Different letters denote significant differences ($p \leq 0.01$). Average values were calculated based on 3 leaves taken from 6 plants per treatment (mean ± SE).

The results for CO_2 assimilation in both ecotypes under restrictive (R) and non-restrictive (nR) conditions (Figure 4) indicate that, under restrictive conditions, there is little variation between salt-free and saline treatments in the tolerant ecotype (with conductance less than 300 mmol H_2O $m^{-2}s^{-1}$). This result reaffirms earlier observations in this work: in the tolerant ecotype, the assimilation rate does not appear to be limited by stomatal conductance since A_{max} is only reduced by 7.2%. This is confirmed by the data of the A/Ci curve under non-restrictive conditions (Figure 3). The values show an increase of 55.7% in A_{max} without salt, whereas with salt, there is a smaller increase in the assimilation rate of 8.7% (Table 5). This result shows that the main limiting factor in the tolerant ecotype is non-diffusional. This finding, in addition to the resistance of the mesophyll, which contributes to the decrease in A, is consistent with the first conclusions of this study (Table 2).

Table 8 shows some photochemical parameters obtained from A/PFD curves: Quantum Requirement (QR), Light compensation points (LCP) and Rate of dark respiration (DR). The obtained values reveal that the quantum requirement rose with salinity, and the increase was greater in Hueque (168%) than in Amarilla (27%). This demonstrates that salt induced a decrease in the effectiveness of photosynthesis in the sensitive ecotype, which was caused by the increase in the QR and decrease in Jmax, Vcmax, and TPU. These results are similar to those found by Killi and Haworth [2], where salinity resulted in a lower Vcmax of RubisCO and a lower Jmax for the regeneration of RuBP.

Table 8. Effect of salinity on the QR, LCP and DR of two quinoa ecotypes.

Parameters	Amarilla (Tolerant)				Hueque (Sensitive)			
	0 M	0.4 M	Difference	%	0 M	0.4 M	Difference	%
Quantum Requirement (QR) μmol photons/μmol CO_2	35.2	44.6	9.43	27	28.7	76.9	48.27	168
Light compensation points(LCP) Photons	15.2	59.5	44.3	291	15.9	15.6	−0.30	−2
Rate of dark respiration (DR) μmol CO_2	−0.432	−1.33	−0.898	208	−0.553	−0.202	0.351	−63

Light compensation point (LCP) also differs between the two ecotypes. Hueque shows virtually no changes with the addition of salt, which together with decreased mitochondrial respiration, could reflect the absence of mechanisms to tolerate salinity. In contrast, the tolerant increased LCP and the mitochondrial respiration, which reflect that this ecotype increases its energy requirements to activate possible mechanisms that allow it to tolerate this environmental stress. The greater tendency towards photoinhibition and the lower LCP of Hueque, reveals an acclimatization to the lower light intensities typical from the southern part of Chile, compared to the high light intensities found in northern highlands where Amarilla is acclimated. To verify this assertion, several photochemical parameters were determined. The data show that control Amarilla and Hueque plants have statistically equal quantum efficiencies. While there is a tendency to decrease ΦPSII, salt application does not cause significant changes in quantum efficiency (FV/Fm) in either ecotype with respect to its respective control.

Table 5 shows the Jmax values found in Amarilla (120–152 μmol $m^{-2}s^{-1}$) are very similar to the average of 109 C_3 species (134 μmol $m^{-2}s^{-1}$; [67]) but much higher than those found by Centritto et al. [15] in olive trees (79.5 μmol $m^{-2}s^{-1}$). Killi and Haworth [2] observed values from 72–152 μmol $m^{-2}s^{-1}$ in quinoa, and significant falls in saline conditions. In our study, the salt sensitive ecotype has the lower values (72–115 μmol m^{-2} s^{-1}), demonstrating that in this ecotype exposure to saline stress causes reductions in Jmax, and that the decrease in A could also be associated with the transport of electrons. It is striking that comparing this parameter between 0 M and 0.4 M NaCl with and without stomatal restriction produces a significant reduction in Hueque, but not in Amarilla, indicating that the difficulties in the sensitive ecotype could be due to carboxylation, given by an effect on Jmax, which would indicate a problem on the supply of ATP and NADPH. According to Killi and Haworth [2], the reductions in Vcmax found in salt-stressed quinoa would be compatible with the altered carboxylation of RubisCO [68,69] and/or by the reduced content of RubisCO [70]. Salt stress also reduced the regenerative capacity of RuBP in quinoa indicative of reduced RuBP availability [69] and expression [71,72], particularly in salinity.

In the case of Hueque, the parameters evaluated highlight that in addition to the high incidence of g_s on the rate of CO_2 assimilation, there is also an effect caused by RubisCO, which is seen by comparing carboxylation efficiencies and Vcmax, between salt and control growth conditions. Similarly, Jmax and the TPU also play a role.

The Vcmax values in Amarilla quinoa under control conditions found here were lower (29–33 μmol $m^{-2}s^{-1}$) than those reported by Wullschleger [67] (average values of 64 μmol $m^{-2}s^{-1}$ for 109 species) and Killi and Haworth [2] (values from 60 to 160 μmol $m^{-2}s^{-1}$ in quinoa). However, the decreases in Vcmax observed in plants sensitive to saline stress in our study, as well as the results of Killi and Haworth [2] in quinoa treated with salt, indicate that differences in reported values may be associated with the particular characteristics of the different ecotypes used. The tolerant plants had higher Vcmax values than sensitive ecotypes. In this regard, Manter and Kerrigan [73] indicated that, in woody species, the values ranged between 31.2 and 42.2 μmol $m^{-2}s^{-1}$ and were associated with low mesophilic conductance, similar to the two Chilean quinoa ecotypes used in our study. In addition, salinity also induced reductions in Vcmax and Jmax, corroborating the loss of photosynthetic ability. The reductions in Vcmax found in quinoa with salt stress are compatible with poor carboxylation by RubisCO [2,69,74].

The ability to maintain intact membranes and photosystems may enable Amarilla to maintain photosynthesis even in saline conditions. In the cells of photosynthetic organisms, salt stress leads to a decrease in cell volume, induces osmotic stress, and inhibits the photosynthetic electron transfer process [75–77], and in *Synechococcus* cells, PSII and PSI are both inactivated due to changes in the K^+/Na^+ ratio [76]. In our experiments, salt treatment led to a 10.5-fold reduction in the level of K^+ in the roots of the salt-tolerant ecotype (Amarilla) and a 34.6-fold reduction in the more sensitive ecotype (Hueque). The lower supply of this element could mean that PSII is more inactivated in Hueque than in Amarilla, as shown by the CO_2 assimilation rates (Table 2), resulting in high Na^+ and Cl^- fluxes into cells, disrupting ion homeostasis, and leading to the accumulation of reactive oxygen species (ROS) [76,78,79], which is associated with membrane lipid peroxidation [80,81] and can adversely affect photosynthesis [81].

Control of oxidation is achieved through the synthesis of antioxidants such as polyphenols, which are divided into several subgroups, among which are the flavonoids (including flavonol glycosides and isoflavones) [82]. Tocopherol and carotenoids are known to be very important for the scavenging of lipid peroxides in *Synechocystis* 6803 [83]. Flavonol glycosides constitute the most abundant phenolics in quinoa seeds and leaves [84]. Several phenolic acids, including hydroxycinnamic acid and hydrobenzoic acid derivatives, have been identified in quinoa seeds and leaves [85]. Furthermore, the highest activity was observed in red-violet quinoa varieties containing both betacyanins and betaxanthins, with significant capacity/activity also exhibited by the yellow ecotype. These varieties or ecotypes are characterized by a high dopaxanthin content, whose dihydroxylated substructure is a powerful antioxidant [83]. Our results show that both ecotypes significantly increase the GB content as a mechanism to protect photosynthetic activity. Some plant varieties are able of biosynthesize GB, exhibiting a greater tolerance to abiotic stress, and often have enhanced growth and yield relative to varieties that do not accumulate GB [86]. The increased GB accumulation mainly occurs in the chloroplast and is responsible for initiating a network of interactions between the plant's photosynthetic apparatus, its "stress" and "growth" hormones, and reactive oxygen species. The increased abiotic stress tolerance of plants able to accumulate GB appears in large part to be due to the ability of chloroplast-produced GB to protect the photosynthetic apparatus [86]. In particular, the accumulation of GB in the chloroplast in response to a stress signal protects enzymes and lipids that are required to maintain both the flow of electrons through thylakoid membranes and the continued assimilation of CO_2 [87,88].

It is probable that the Amarilla ecotype also has other mechanisms for membrane protection, such as the presence of trehalose [89]. Trehalose can act as a structural component when incorporated into glycolipids, thereby stabilizing membranes [50,89,90].

5. Conclusions

The Calvin cycle requires energy inputs that come from the photochemical phase, and the data show that electron transport is not strongly affected in the tolerant ecotype, which had higher J and Jmax values than those found in the sensitive ecotype. This allows NADPH and ATP to maintain their contributions to the production of triose phosphates (TPU), which, according to our results, do not differ from the treatment without salt. This suggests that the Amarilla ecotype maintains its rate of RuBP under salinity conditions. On the other hand, the Quantum Requirement (QR), light compensation points, and the dark respiration rate are increased, which may be the result of an adaptation of the photochemical apparatus through membrane protection, as seen by the increase in GB.

In the sensitive ecotype (Hueque), the CO_2 assimilation rate was affected in both the biochemical and photochemical components; in this respect, Vcmax, TPU, Jmax, J, ΦPSII, and QR dramatically decreased. Light compensation points and the dark respiration rate were not affected.

Another mechanism that was activated in the short term in response to salt treatment in both ecotypes was the exclusion of Na^+ towards the leaves and growth centers. Both ecotypes retained Na^+ in the roots and restricted its entry to the leaves. This mechanism must be associated with the compartmentalization and blocking of excess Na^+ not only in the vacuoles of quinoa leaves (tissue tolerance) but also in those of root cells (ion exclusion), mechanisms that are important for protection in response to ionic toxicity induced by salt at the cellular level.

Author Contributions: Conceptualization, J.D.-H. and M.P.; methodology, M.P.; formal analysis, J.D.-H.; investigation, J.D.-H.; resources, J.D.-H.; data curation, J.D.-H.; writing—original draft preparation, J.D.-H. and K.B.R.; writing—review and editing, K.B.R.; visualization, K.B.R.; supervision, M.P.; project administration, J.D.-H. All authors have read and agreed to the published version of the manuscript.

Funding: This research was funded by Doctoral Program in Desert and Arid Agriculture, Universidad Arturo Prat and project FONDEQUIP EQM1900088- ANID. K.B.R. was supported by grant PAI 79170093.

Institutional Review Board Statement: Not applicable.

Informed Consent Statement: Not applicable.

Data Availability Statement: The data presented in this study are available in this article.

Acknowledgments: We acknowledge Michael Hanford from Universidad of Chile, for providing the translation and discussion support. We kindly thank Alejandro Riquelme of Centro de Estudios Avanzados en Fruticultura. We also appreciate the valuable contributions of the reviewers of this manuscript, they have been very useful to improve its quality.

Conflicts of Interest: The authors declare no conflict of interest.

References

1. Flowers, T.J.; Colmer, T.D. Salinity tolerance in halophytes. *New Phytol.* **2008**, *179*, 945–963. [CrossRef]
2. Killi, D.; Haworth, M. Diffusive and metabolic constraints to photosynthesis in quinoa during drought and salt stress. *Plants* **2017**, *6*, 49. [CrossRef]
3. Adolf, V.I.; Shabala, S.; Andersen, M.N.; Razzaghi, F.; Jacobsen, S.E. Varietal differences of quinoa's tolerance to saline conditions. *Plant Soil* **2012**, *357*, 117–129. [CrossRef]
4. Rengasamy, P. World salinization with emphasis on Australia. Journal of Experimental Botany. *Plants Salin. Spec. Issue* **2006**, *57*, 1017–1023. [CrossRef]
5. Chen, W.; Hou, Z.; Wu, L.; Liang, Y.; Wei, C. Evaluating salinity distribution in soil irrigated with saline water in arid regions of northwest China. *Agric. Water Manag.* **2010**, *97*, 2001–2008. [CrossRef]
6. Hanson, B. Drip irrigation and salinity. Agricultural Salinity Assessment and Management. Manuals and reports on engineering practice 71 (2 edn). *Am. Soc. Civ. Eng. Rest.* **2012**, *5*, 539–560.
7. Mounzer, O.; Pedrero-Salcedo, F.; Nortes, P.A.; Bayona, J.M.; Nicolás, E.; Alarcón, J.J. Transient soil salinity under the combined effect of reclaimed water and regulated deficit drip irrigation of Mandarin trees. *Agric. Water Manag.* **2013**, *120*, 23–29. [CrossRef]

8. Hasewaga, P.M.; Bressam, R.A.; Zhu, J.K.; Bohnert, H.J. Plant cellular and molecular response to high salinity. *Ann. Rev. Plant Physiol. Plant Biol.* **2000**, *51*, 463–499. [CrossRef] [PubMed]
9. Almeida-Machado, E.M.; Serralheiro, R.P. Soil Salinity: Effect on Vegetable Crop Growth Management Practices to Prevent and Mitigate Soil Salinization. *Horticulturae* **2017**, *3*, 30. [CrossRef]
10. Blumwald, E. Sodium transport and salt tolerance in plants. *Curr. Opin. Cell Biol.* **2000**, *12*, 431–434. [CrossRef]
11. Tsugane, K.; Kobayashi, K.; Niwa, Y.; Ohba, Y.; Wada, K.; Kobayashi, H. A recessive Arabidopsis mutant that grows photoautotrophically under salt stress shows enhanced active oxygen detoxification. *Plant Cell* **1999**, *11*, 1195–1206. [CrossRef]
12. Tsunekawa, K.; Shijuku, T.; Hayashimoto, M.; Kojima, Y.; Onai, K.; Morishita, M.; Ishiura, M.; Kuroda, T.; Nakamura, T.; Kobayashi, H.; et al. Identification and characterization of the Na+/H+ antiporter Nhas3 from the Thylakoid Membrane of Synechocystis sp. PCC 68036803. *J. Biol. Chem.* **2009**, *284*, 16513–16521. [CrossRef]
13. Yamaguchi, T.; Hamamoto, S.; Uozumi, N. Sodium transport system in plant cells. *Front. Plant Sci.* **2013**, *4*, 410. [CrossRef]
14. Rivelli, A.R.; Lovelli, S.; Perniola, M. Effects of salinity on gas exchange, water relations and growth of sunflower (Helianthus annuus). *Funct. Plant Biol.* **2002**, *29*, 1405–1415. [CrossRef] [PubMed]
15. Centritto, M.; Loreto, F.; Chartzoullakis, K. The use of low [CO_2] to estimate diffusional and non-diffusional limitations of photosynthetic capacity of salt-stressed olive saplings. *Plant Cell Environ.* **2003**, *26*, 585–594. [CrossRef]
16. Orsini, F.; Accorsi, M.; Gianquinto, G.; Dinelli, G.; Antognoni, F.; Ruiz-Carrasco, K.B.; Martinez, E.A.; Alnayef, M.; Marotti, I.; Bosi, S.; et al. Beyond the ionic and osmotic response to salinity in *Chenopodium quinoa* functional elements of successful halophytism. *Funct. Plant Biol.* **2011**, *138*, 818–831. [CrossRef] [PubMed]
17. Flexas, J.; Bota, J.; Escalona, J.M.; Sampol, B.; Medrano, H. Effects of drought on photosynthesis in grapevines Ander field conditions: An evaluation of stomatal and mesophyll limitattions. *Funct. Plant Biol.* **2002**, *29*, 461–471. [CrossRef]
18. von Caemmere, S.; Evans, J.R. Determination of the CO_2 pressure in chloroplast from leaves of several C3 plants. *Aust. J. Plant Physiol.* **1991**, *18*, 287–305.
19. Bongi, G.; Loreto, F. Gas—Exchange properties of salt stressed olive (*Olea europea* L.) Leaves. *Plant Physiol.* **1989**, *90*, 1408–1416. [CrossRef]
20. Lopez-Climent, M.F.; Arbona, V.; Pérez-Clemente, M.; Gómez-Cadenas, A. Relationship between salt tolerance and photosynthetic machinery performance in citrus. *Environ. Exp. Bot.* **2008**, *62*, 176–184. [CrossRef]
21. Bota, J.; Medrano, H.; Flexas, J. Is photosynthesis limited by decresed Rubisco activity and RuBP content under progressive water stress? *New Phytol.* **2004**, *126*, 671–681. [CrossRef] [PubMed]
22. Loreto, F.; Centritto, M.; Chartzoullakis, K. Photosynthetic limitations in olive cultivars with differente sensitivity to salt stress. *Plant Cell Environ.* **2003**, *26*, 595–601. [CrossRef]
23. Delfine, S.; Alvino, A.; Concetta, M.; Loreto, F. Restrictions to carbon dioxide conductance and photosynthesis in Spinach leaves recovering from salt stress. *Plant Physiol.* **1999**, *199*, 1101–1106. [CrossRef]
24. Flexas, J.; Medrano, H. Drought-inhibition of Photosynthesis in C3 Plants: Stomatal and Non-stomatal Limitations Revisited. *Ann. Bot.* **2002**, *89*, 183–189. [CrossRef]
25. Orrego, M.; Riquelme, A. *Efecto de la Concentración de Sodio(Na+) Sobre Algunos Parámetros Hídricos y Fenología de la Quinua (Chenopodium quinoa Willd)*; Tesis para optar al Título de Ingeniero de Ejecución Agrícola. Pregrado, Universidad Arturo Prat: Iquique, Chile, 2012.
26. Jacobsen, S.E.; Mujica, A.; Jensen, C.R. The resistance of quinoa (*Chenopodium quinoa* Willd.) to adverse abiotic factors. *Food Rev. Int.* **2003**, *19*, 99–109. [CrossRef]
27. Bosque, H.; Lemeur, P.; Van Damme, P.; Jacobsen, S.E. Ecophysiological Analysis of drought and salinity stress of quinoa (*Chenopodium quinoa* Willd). *Food Rev. Int.* **2003**, *19*, 11–19.
28. Ruffino, A.M.C.; Rosa, M.; Hilal, M.; González, J.A.; Prado, F.E. The role of cotyledon metabolism in the establishment of quinoa (*Chenopodium quinoa*) seedlings growing under salinity. *Plant Soil* **2010**, *326*, 213–224. [CrossRef]
29. Jensen, C.R.; Jacobsen, S.-E.; Andersen, M.N.; Núñez, N.; Andersen, S.D.; Rasmussen, L.; Mogensen, V.O. Leaf gas exchange and water relation characteristics of field quinoa (*Chenopodium quinoa* Willd.) during soil drying. *Eur. J. Agron.* **2000**, *13*, 11–25. [CrossRef]
30. Delatorre-Herrera, J.; Pinto, M. Importance of ionic and osmotic componets of salt stress on the germination of four quinua (*Chenopodium quinoa* Willd.) selections. *Chil. J. Agric. Res.* **2009**, *69*, 477–485. [CrossRef]
31. Hariadi, Y.; Marandon, K.; Tian, Y.; Jacobsen, S.E.; Shabala, S. Ionic and osmotic relations in quinoa (*Chenopodium quinoa* Willd.) plants grown at various salinity levels. *J. Exp. Bot.* **2011**, *62*, 185–193. [CrossRef] [PubMed]
32. Razzaghi, F.; Jacobsen, S.E.; Jensen, C.R.; Andersen, M.N. Ionic and photosynthetic homeostasis in quinoa challenged by salinity and Drought—Mechanisms of tolerance. *Funct. Plant Biol.* **2015**, *42*, 136–148. [CrossRef]
33. Geissler, N.; Mervat, S.H.; El-Far, M.M.; Koyro, H.-W. Elevated atmospheric CO_2 concentration leads to different salt resistance mechanisms in a C3 (*Chenopodium quinoa*) and a C4 (*Atriplex nummularia*) halophyte. *Environ. Exp. Bot.* **2015**, *118*, 67–77. [CrossRef]
34. Maughan, P.J.; Bonifacio, A.; Jellen, E.N.; Stevens, M.R.; Coleman, C.E.; Ricks, M.; Mason, S.L.; Jarvis, D.E.; Gardunia, B.W.; Fairbanks, D.J. A genetic linkage map of quinoa (*Chenopodium quinoa*) based on AFLP, RAPD, and SSR markers. *Appl. Genet.* **2004**, *109*, 1188–1195. [CrossRef]
35. Schlink, G.; Bubenheim, D. Quinoa crop NASA's Controlled Ecological Life Support Systems. In *Progress in New Crops*; ASHS Press: Arlington, VA, USA, 1996; pp. 632–640. ISBN 0961502738.

36. Hunt, J. Dilute hydrochloric acid extraction of plant material for routines cation analysis. *Commun. Soil Sci. Planta Anal.* **1982**, *13*, 49–55. [CrossRef]
37. Bates, L.S.; Waldren, R.P.; Teare, I.D. Rapid determination of free proline for water-stress studies. *Plant Soil* **1973**, *39*, 205–207. [CrossRef]
38. Grieve, C.M.; Grattan, S.R. Rapid assayfor determination of water soluble quaternary ammonium compounds. *Plant Soli* **1983**, *70*, 303–307. [CrossRef]
39. Lambers, H.; Chapin, F.S.; Pons, T.L. *Plant Physiological Ecology*; Springer: New York, NY, USA, 2008; p. 624. ISBN 978-0-387-78340-6. [CrossRef]
40. Farquhar, G.D.; Von Caemmerer, S.; Berry, J.A. A biochemical model of photosynthetic CO_2 assimilation in leaves of C3 species. *Planta* **1980**, *149*, 78–90. [CrossRef]
41. Farquhar, G.D.; Von Caemmerer, S. Modelling of photosynthetic responses to environmental conditions. In *Physiological Plant Ecology II. Water Relations and Carbon Assimilation*; Lange, O.L., Nobel, P.S., Osmond, C.B., Ciegler, H., Eds.; Springer: Berlin, Germany, 1982; pp. 549–587.
42. Harley, P.C.; Loreto, F.; Di Marco, G.; Sharkey, T.D. Theoretical considerations when estimating the mesophyll conductance to CO2 flux by analysis of the response of photosynthesis to CO_2. *Plant Physiol.* **1992**, *98*, 1429–1436. [CrossRef] [PubMed]
43. Maxwell, K.; Johnson, G.N. Chlorophyll fluorescence—A practical guide. *J. Exp. Bot.* **2000**, *51*, 659–668. [CrossRef]
44. Genty, B.; Briantais, J.-M.; Baker, N.R. The relationship between the quantum yield of photosynthetic electron transport and quenching of chlorophyll fluorescence. *Biochim. Biophys. Acta BBA Gen. Subj.* **1989**, *990*, 87–92. [CrossRef]
45. Björkman, O.; Demmig, B. Photon yield of O2 evolution and chlorophyll fluorescence characteristics at 77 K among vascular plants of diverse origins. *Planta* **1987**, *170*, 489–504. [CrossRef] [PubMed]
46. Ruiz-Carrasco, K.; Antognoni, F.; Coulibaly, A.K.; Lizardi, S.; Covarrubias, A.; Martínez, E.A.; Molina-Montenegro, M.A.; Biondi, S.; Zurita-Silva, A. Variation in salinity tolerance of four lowland genotypes of quinoa (*Chenopodium quinoa* Willd.) as assessed by growth, physiological traits, and sodium transporter gene expression. *Plant Physiol. Biochem.* **2011**, *49*, 1333–1341. [CrossRef]
47. Ashraf, M.; Foolad, M.R. Roles of glycine betaine and proline in improving plant abiotic stress resistance. *Environ. Exp. Bot.* **2007**, *59*, 206–216. [CrossRef]
48. Habib, N.; Ashraf, M.; Ali, Q.; Perveen, R. Response of salt stressed okra (*Abelmoschus esculentus* Moench) plants to foliar-applied glycine betaine and glycine betaine containing sugarbeet extract. *S. Afr. J. Bot.* **2012**, *83*, 151–158. [CrossRef]
49. Zhao, Y.; Ma, Y.Q.; Weng, Y.J. Variation of betaine and proline contents in wheat seedlings under salt stress. *J. Plant Physiol. Mol. Biol.* **2005**, *31*, 103–106.
50. Morales, A.J.; Bajgain, P.; Garver, Z.; Maughan, P.J.; Udall, J.A. Physiological responses of *Chenopodium quinoa* to salt stress. *Int. J. Plant Physiol. Biochem.* **2011**, *3*, 219–232.
51. Zhang, L.; Gao, M.; Hu, J.; Zhang, X.; Wang, K.; Ashraf, M. Modulation role of abscisic acid (ABA) on growth, water relations and glycinebetaine metabolism in two maize (*Zea mays* L.) cultivars under drought stress. *Int. J. Mol. Sci.* **2012**, *13*, 3189–3202. [CrossRef] [PubMed]
52. Ruiz, K.B.; Rapparini, F.; Bertazza, G.; Silva, H.; Torrigiani, P.; Biondi, S. Comparing salt-induced responses at the transcript level in a salares and coastal-lowlands landrace of quinoa (Chenopodium quinoa Willd). *Environ. Exp. Bot.* **2017**, *139*, 127–142. [CrossRef]
53. Guarino, F.; Ruiz, K.B.; Castiglione, S.; Cicatelli, A.; Biondi, S. The combined effect of Cr (III) and NaCl determines changes in metal uptake, nutrient content, and gene expression in quinoa (*Chenopodium quinoa* Willd.). *Ecotoxicol. Environ. Saf.* **2000**, *193*, 110345. [CrossRef]
54. Carillo, P.; Mastrolonardo, G.; Nacca, F.; Parisi, D.; Verlotta, A.; Fuggi, A. Nitrogen metabolism in durum wheat under salinity: Accumulation of proline and glycine betaine. *Funct. Plant Biol.* **2008**, *35*, 412–426. [CrossRef]
55. Flowers, T.J.; Troke, P.F.; Yeo, A.R. The mechanism of salt tolerance in halophytes. *Annu. Rev. Plant Physiol.* **1997**, *28*, 89–121. [CrossRef]
56. Annunziata, M.G.; Ciarmiello, L.F.; Woodrow, P.; Dell'Aversana, E.; Carillo, P. Spatial and Temporal Profile of Glycine Betaine Accumulation in Plants Under Abiotic Stresses. *Front. Plant Sci.* **2019**, *10*, 230. [CrossRef] [PubMed]
57. Shabala, S. Signalling by potassium: Another second messenger to add to the list. *J. Exp. Bot.* **2017**, *68*, 4003–4007. [CrossRef]
58. Shabala, S.; Hariadi, Y.; Jacobsen, S.-E. Genotypic difference in salinity tolerance in quinoa is determined by differential control of xylem Na+ loading and stomatal density. *J. Plant Physiol.* **2013**, *170*, 906–914. [CrossRef] [PubMed]
59. Tarczynsk, M.C.; Jensen, R.G.; Bohnert, H.J. Stress Protection of Transgenic Tobacco by Production of the Osmolyte Mannitol. *Science* **1993**, *259*, 508–510. [CrossRef]
60. Munns, R.; Tester, M. Mechanisms of salinity tolerance. *Annu. Rev. Plant Biol.* **2008**, *59*, 651–681. [CrossRef] [PubMed]
61. Munns, R.; Sharp, R. Involvement of absicic acid in controlling plant growth in soils of low water potential. *Aust. J. Plant Physiol.* **1993**, *20*, 425–437.
62. Murguia, J.R.; Belles, J.M.; Serrano, R. A Salt-Sensitive 3′(2′),5′-Bisphosphate Nucleotidase involved in Sulfate activation. *Science* **1995**, *267*, 232–234. [CrossRef]
63. Shabala, L.; Mackay, A.; Tian, Y.; Jacobsen, S.E.; Zhou, D.; Shabala, S. Oxidative stress protection and stomatal patterning as components of salinity tolerance mechanism in quinoa (*Chenopodium quinoa*). *Physiol. Plant.* **2012**, *146*, 26–38. [CrossRef]

64. Maser, P.; Eckelman, B.; Vaidyanathan, R.; Horie, T.; Fairbairn, D.J.; Kubo, M.; Yamagami, M.; Yamaguchi, K.; Nishimura, M.; Uozumi, N.; et al. Altered shoot/root Na+ distribution and bifurcating salt sensitivity in Arabidopsis by genetic disruption of the Na+ transporter AtHKT1. *FEBS Lett.* **2002**, *531*, 157–161. [CrossRef]
65. James, R.A.; Munns, R.; von Caemmerer, S.; Trejo, C.; Miller, C.; Condon, A.G. Photosynthetic capacity is related to the cellular and subcellular partitioning of Na+, K+ and Cl− in salt-affected barley and durum wheat. *Plant Cell Environ.* **2006**, *29*, 2185–2197. [CrossRef] [PubMed]
66. Chen, Z.; Pottosin, I.I.; Cuin, T.A.; Fuglsang, A.T.; Tester, M.; Jha, D.; Zepeda-Jazo, I.; Zhou, M.; Palmgren, M.G.; Newman, I.A.; et al. Root plasma membrane transporters controlling K+/Na+ homeostasis in salt-stressed barley. *Plant Physiol.* **2007**, *145*, 1714–1725. [CrossRef] [PubMed]
67. Wullschleger, S.D. Biochemical Limitations to Carbon Assimilation in C3 Plants. A Retrospective Analysis of the j Curves from 109 Species. *J. Exp. Bot.* **1993**, *44*, 907–920. [CrossRef]
68. Solomon, A.; Beer, S.; Waisel, Y.; Jones, G.P.; Paleg, L.G. Effects of NaCl on the carboxylating activity of Rubisco from *Tamarix jordanis* in the presence and absence of proline-related compatible solutes. *Physiol. Plant* **1994**, *90*, 198–204. [CrossRef]
69. Seemann, J.R.; Critchley, C. Effects of salt stress on the growth, ion content, stomatal behaviour and photosynthetic capacity of a salt-sensitive species, *Phaseolus vulgaris* L. *Planta* **1985**, *164*, 151–162. [CrossRef]
70. Wang, Y.; Nii, N. Changes in chlorophyll, ribulose bisphosphate carboxylase-oxygenase, glycine betaine content, photosynthesis and transpiration in *Amaranthus tricolor* leaves during salt stress. *J. Hortic. Sci. Biotechnol.* **2000**, *75*, 623–627. [CrossRef]
71. Seki, M.; Narusaka, M.; Ishida, J.; Nanjo, T.; Fujita, M.; Oono, Y. Monitoring the expression profiles of 7000 Arabidopsis genes under drought, cold and high-salinity stresses using a full-length cDNA microarray. *Plant J.* **2002**, *31*, 279–292. [CrossRef] [PubMed]
72. Sobhanian, H.; Razavizadeh, R.; Nanjo, Y.; Ehsanpour, A.A.; Jazii, F.R.; Motamed, N. Proteome analysis of soybean leaves, hypocotyls and roots under salt stress. *Proteome Sci.* **2010**, *8*, 19. [CrossRef] [PubMed]
73. Manter, D.K.; Kerrigan, J. A/Ci curve analysis across a range of woody plant species: Influence of regression analysis parameters and mesophyll conductance. *J. Exp. Bot.* **2004**, *55*, 2581–2588. [CrossRef]
74. Allakhverdiev, S.I.; Murata, N. Salt stress inhibits photosystems II and I in cyanobacteria. *Photosynth. Res.* **2008**, *98*, 529–539. [CrossRef]
75. Feng, Z.T.; Deng, Y.Q.; Zhang, S.C.; Liang, X.; Yuan, F.; Hao, J.L.; Zhang, J.-C.; Sun, S.F.; Wang, B.S. K+ accumulation in the cytoplasm and nucleus of the salt gland cells of *Limonium bicolor* accompanies increased rates of salt secretion under NaCl treatment using NanoSIMS. *Plant Sci.* **2015**, *238*, 286–296. [CrossRef]
76. Zhao, Y.Q.; Ma, Y.C.; Duan, H.M.; Liu, R.R.; Song, J. Traits of fatty acid accumulation in dimorphic seeds of the euhalophyte Suaeda salsa in saline conditions. *Plant Biosyst.* **2019**, *153*, 514–520. [CrossRef]
77. Allakhverdiev, S.I.; Sakamoto, A.; Nishiyama, Y.; Inaba, M.; Murata, N. Ionic and osmotic effects of NaCl-induced inactivation of photosystem I and II in *Synechococcus* sp. *Plant Physiol.* **2000**, *123*, 1047–1056. [CrossRef] [PubMed]
78. Song, Y.; Li, J.; Sui, Y.; Han, G.; Zhang, Y.; Guo, S.; Sui, N. The sweet sorghum SbWRKY50 is negatively involved in salt response by regulating ion homeostasis. *Plant Mol. Biol.* **2020**, *102*, 603–614. [CrossRef] [PubMed]
79. Zhang, Q.-Y.; Wang, L.-Y.; Kong, F.-Y.; Deng, Y.-S.; Lia, B.; Meng, Q.-W. Constitutive accumulation of zeaxanthin in tomato alleviates salt stress-induced photoinhibition and photooxidation. *Physiol. Plant.* **2012**, *146*, 363–373. [CrossRef]
80. Sui, N.; Han, G.L. Salt-induced photoinhibition of PSII is alleviated in halophyte *Thellungiella halophila* by increases of unsaturated fatty acids in membrane lipids. *Acta Physiol. Plant* **2014**, *36*, 983–992. [CrossRef]
81. Maeda, M.; Sakuragi, Y.; Bryant, D.A.; DellaPenna, D. Tocopherols Protect *Synechocystis* sp. Strain PCC 6803 from Lipid Peroxidation. *Plant Physiol.* **2005**, *138*, 1422–1435. [CrossRef]
82. Tsao, R. Chemistry and Biochemistry of Dietary Polyphenols. *Nutrients* **2010**, *2*, 1231–1246. [CrossRef]
83. Escribano, J.; Cabane, J.; Jiménez-Atiénzar, M.; Ibañez-Tremolada, M.; Gómez-Pando, R.L.; García-Carmona, F.; Gandía-Herrero, F. Characterization of betalains, saponins and antioxidant power in differently colored quinoa (*Chenopodium quinoa*) varieties. *Food Chem.* **2017**, *234*, 285–294. [CrossRef]
84. Gomez-Caravaca, A.M.; Segura-Carretero, A.; Fernandez-Gutierrez, A.; Caboni, M.F. Simultaneous determination of phenolic compounds and saponins in quinoa (*Chenopodium quinoa* Willd) by a liquid chromatography-diode array detection-electrospray ionizationtime- of-flight mass spectrometry methodology. *J. Agric. Food Chem.* **2011**, *59*, 10815–10825. [CrossRef] [PubMed]
85. Graf, B.L.; Rojas-Silva, O.; Rojo, L.E.; Delatorre-Herrera, J.; Baldeón, M.E.; Raskin, I. Innovations in Health Value and Functional Food Development of Quinoa (*Chenopodium quinoa* Willd.). *Compr. Rev. Food Sci. Food Saf.* **2015**, *14*, 431–445. [CrossRef]
86. Kurepin, L.V.; Ivanov, A.G.; Zaman, M.; Pharis, R.P.; Hurry, V.; Hüner, N.P.A. Interaction of Glycine Betaine and Plant Hormones: Protection of the Photosynthetic Apparatus During Abiotic Stress. In *Photosynthesis: Structures, Mechanisms, and Applications*; Hou, H., Najafpour, M., Moore, G., Allakhverdiev, S., Eds.; Springer: Cham, Switzerland, 2017. [CrossRef]
87. Park, S.-Y.; Yu, J.-W.; Park, J.-S.; Li, J.; Yoo, S.-C.; Lee, N.-Y.; Lee, S.-K.; Jeong, S.-W.; Seo, H.-S.; Koh, H.-J.; et al. The Senescence-Induced Staygreen Protein Regulates Chlorophyll Degradation. *Plant Cell* **2007**, *19*, 1649–1664. [CrossRef] [PubMed]
88. Chen, T.H.; Murata, N. Glycinebetaine protects plants against abiotic stress: Mechanisms and biotechnological applications. *Plant Cell Environ.* **2011**, *34*, 1–20. [CrossRef] [PubMed]

89. Delatorre-Herrera, J.; Rojas-Urrutia, J.; Rojo, L.E.; Graf, B.L. Osmotic stress in Chenopodium quinoa Willd: Variations in osmoprotectants at different phenological stages. *Afr. J. Agric. Res.* **2019**, *14*, 361–368. [CrossRef]
90. Elbein, A.D.; Pan, Y.T.; Pastuszak, I.; Carroll, D. New insights on trehalose: A multifunctional molecule. *Glycobiology* **2003**, *13*, 17R–27R. [CrossRef] [PubMed]

plants

MDPI

Article

Quinoa Productivity and Stability Evaluation through Varietal and Environmental Interaction

Elhadji Thiam *, Asmaa Allaoui and Ouafae Benlhabib *

Département de Production, Protection et Biotechnologies Végétales, Institut Agronomique et Vétérinaire Hassan II, Rabat B.P 6202, Morocco; allaouiasmaa81@gmail.com
* Correspondence: elasthiam0@gmail.com (E.T.); o.benlhabib@gmail.com (O.B.)

Abstract: *Chenopodium quinoa* is a pseudocereal species identified as a potential crop to mitigate world food security. It has the ability to adapt to diverse agro-ecosystems ranging from sea level to over 4000 masl. Its cultivation in Morocco began in 1999, as it is tolerance to drought, salinity, and frost, and it can grow on marginal soils. It has exceptional nutritional value, as it is rich in proteins, essential amino acids, mineral nutrients, trace elements, vitamins, and unsaturated fatty acids. The present study aims to evaluate the adaptation of 14 quinoa varieties and lines from four different origins through fourteen agro-morphological characters. The experimental trials were conducted at five contrasted agro-climatic sites across the central part of Morocco. The data analysis showed high variability among the tested varieties and between sites for all assessed traits. The Meknes (foot-hill plain) site was the most productive; its grain yield reached 78.6 qx/ha. At the Rabat (coastal land) and Berrechid (continental plain) sites, grain production was respectively 56.4 and 45.9 qx/ha. The SW2 Moroccan line produced the highest grain yield that reached 78.3 qx/ha across sites. The Danish variety Titicaca presented the best harvest index (HI = 0.69) as well as the best "thousand kernel weight" (TGW = 3.4 g). As the mildew infection evaluation, the Vikinga and Titicaca varieties ranked the most sensitive to *Peronospora farinosa*. The germination rates of the harvested seeds were prejudiced by the sites' high temperatures and were low in Tinejdad (oases site) and El Kbab (mountain plateau). The best average germination rate across sites was that of the Puno variety (84.5%). According to the Additive Main effects and Multiplicative Interaction analysis (AMMI), 23% of the grain yield variability is due to the genotype, while 32% is due to the site by the variety interaction contribution to the production variability. AMMI analysis also ranked the varieties according to their productivity and stability value. Accordingly, two varieties that have yielded above the overall average (42.7 qx/ha) are considered stable; those are Riobamba and W11, which is a local selected line. Titicaca, ILLPA, Atlas cultivars and the SW2 local line presented the best grain yield in one of the experimental sites but performed not as well on the others.

Keywords: quinoa; agro-morphological traits; genotype × environment interaction; AMMI; yield; stability

Citation: Thiam, E.; Allaoui, A.; Benlhabib, O. Quinoa Productivity and Stability Evaluation through Varietal and Environmental Interaction. *Plants* **2021**, *10*, 714. https://doi.org/10.3390/plants10040714

Academic Editor: Cataldo Pulvento
Received: 22 December 2020
Accepted: 26 January 2021
Published: 7 April 2021

1. Introduction

Global warming in recent decades reached a critical level exposing some populations at risk, particularly in terms of the food supply. Food security is also threatened by population growth and rainfall decrease. Grain crops statistics predict a yield decrease between 3.1 and 7.4% for every 1 °C temperature increase unless breeders create new varieties for warmer environments [1]. Thus, there is a great challenge to overcome by exploring and expanding new crops that have a better adaptation to the harsh climates.

Quinoa is a pseudocereal native to the Andean highland of the South. It gained worldwide attention because of its tolerance to abiotic stresses [2] such as drought, heat, frost, and salinity [3]. It expends progressively in different parts of the world; it is growing in more than 95 countries [4]. The high resistance to the abiotic stresses in quinoa results from its vast genetic diversity and harsh environmental conditions prevailing in its zone of origin [5]. It sustains five ecotypes based on their adaptation ability to specific agro-ecological environments: sea level from Southern Chile, Andean valleys, Yungas from the subtropical rainforest, Salar from the Lake Titicaca highland in Southern Bolivia, and Altiplano from the Andean high plateaus (ca. 4000 m) [6].

Quinoa is also known for its high nutritional value; its seeds are exceptionally rich in proteins and essential amino acids. Temperature and photoperiod are the two main factors that affect significantly quinoa production [7]. Quinoa yield is generally insignificant in regions where the temperatures go higher than 32 °C [4,8]. According to Bertero et al., [9], the genotype's performance depends largely on the cultivar genetic makeup, the environment, and their interaction. In general, quinoa shows high genotype × environment interaction (GEI) under multi-environments trials (MET). Such significant GEI affects breeding efficiency strength [10]. Understanding and assessing genotype and GEI effects are required to enhance the selection efficiency in crop breeding [11]. Furthermore, MET has the advantage of helping identify genotypes that have large adaptations or the ones that adjust to a specific environment [12].

Several statistical analyses are used to evaluate the yield performance of genotypes across the environments such as the regression coefficient of the GEI effects on the environmental [13], the coefficient of variation (CV) [14], the non-parametric stability [15], the harmonic mean of the genotype relative performance value [16], and the AMMI—Additive Main effects and Multiplicative Interaction model [17]. There are different crops that benefit from these approaches, including corn [18], rice [19], bread wheat [20], maize [21], and quinoa [22,23]. AMMI analysis allowed us to analyze the GEI effects in multi-location trials. This method has the advantage of generating outputs that can help easily diagnose the varietal adaptability and yield stability [24].

Quinoa was selected among half a dozen species as elite to enhance crop diversity of the cropping systems in Morocco. The quinoa introduction's main objective was to improve the food security of mountain sustenance farmers in climate change and overgrazing occurrence. Since then, quinoa became the focus of several studies, germplasm productivity, and adaptation evaluation, screening for diseases and abiotic stresses tolerance, water-use efficiency and drought tolerance, genetic and molecular characterization, etc. [25–27]. Consequently, quinoa cultivation has expanded across several regions and environments in Morocco; hence, there is a real need to provide well-adapted cultivars to farmers and producers for different regions. Therefore, the main objectives of the present study are to evaluate the productivity and stability of 14 quinoa genotypes (11 varieties and three Moroccan lines) under field conditions at five contrasting environments and to interpret the genotypes adaptation through the tested cropping conditions and through 14 evaluation descriptors.

2. Material and Methods

We conducted the experiments at five sites during the 2017–2018 cropping seasons. Table 1 summarizes the specifications of the experimental locations, the quinoa genotypes under evaluation, the type of environmental climates (oceanic to semi-desert), and altitude ranges (low to high) of the sites. The field experiment design was a randomized complete block with four replications. The elementary plot size varies according to the available land space for the trial. Six rows represented the elementary plots in Rabat and El Kbab, four rows represented the elementary plots in Berrechid and Tinejdad, and three rows represented the elementary plots in Meknes. The row length was respectively 1.6, 2.0, or 2.5 m with a 0.2, 0.5, or 0.6 m's inter-row spacing. All the cropping tasks (planting, weeding, harvesting, and cleaning seeds) were manual. The fourteen genotypes tested are from different origins: Titicaca, Puno, Vikinga (Quinoa Quality, Regstrup, Denmark), Atlas, Pasto, Riobamba (Wageningen University, Wageningen, The Netherland), ILLPA, Amarilla de Marangani, Altiplano INIA 431, Salcedo INIA, Passankalla INIA 415 (Universidad Nacional Agraria de La Molina, Lima, Peru), and SW2, W11, and W16 (IAV Hassan II,

Rabat, Morocco). We sowed all the genotypes at a rate of 0.66 g/m in February during the cropping season expect El Kbab, which was sowed in April since the snow.

Table 1. Location and description of the experiment agro-climatic sites.

Location	Geographic Position		Alt (m)	Soil Type	Temperature (°C)			Rainfall (mm)	Sowing Date
	Latitude	Longitude			Min	Max	Mean		
Rabat	34°03′31″ N	6°79′10″ W	135	Sandy-Silty	12.6	26.2	19.8	229	20/02/18
Berrechid	33°18′12″ N	7°47′59″ W	309	Clayey-Silty	12	30	20.3	318	16/02/18
Meknès	33°85′42″ N	5°66′17″ W	592	Clayey	11	33	19.3	703	19/02/18
El Kbab	32°71′31″ N	5°55′23″ W	1503	Sandy-Silty	8.2	37.2	19.1	550	06/04/18
Tinejdad	31°54′26″ N	5°19′39″ W	1062	Silty-Clayey	12.1	40.3	23	200	23/02/18

The mildew (*Peronospora farinosa*) assessment took place under the natural infection conditions at the vegetative and flowering plant growth stages. We first collected three leaves from three stem levels, the third-bottom, the middle, and the apical per genotype and block, and scored them according to the leaf area surface covered by the fungus [28].

At harvest, we scored the agro-morphological traits. We pulled up separately four rooted plants per elementary plot. Subsequently, we assessed seven IBPGR traits (descriptors International Board for Plant Genetic Resources) that are good in discriminating traits between genotypes or well correlated to the yield in our former screening tests [29,30]. These gathered plant height, stem diameter, root length, panicle length and width, and yield components.

Grain and dry matter yield per plant measurements took place after threshing and drying the seeds at 35 °C and brushwood at 70 °C immediately. Three samples of one thousand kernels that were weighted were obtained by counting the seeds manually. For the grain diameter, thirty seeds per genotype and per block were measured using a binocular magnifying glass with mm graph paper. Yield and dry matter per hectare were estimated by through the grain yield per plant, plant density, and elementary plot area. Harvest index was calculated as defined below:

$HI = (GW/(BW + GW)) \times 100)$

GW: Grain weight

BW: Brushwood weight

For the germination test, we held three samples of 20 kernels per treatment (genotype × site), washed them in 12% sodium hypochlorite solution, and then placed them at room temperature (20.2 °C) in sterilized Petri dishes. Daily recording germinating rate data recording was for ten days.

The database went first through the descriptive analyses and ANOVA test to assess the variability between the genotypes within and between sites. Genotypes homogeneous groups were according Newman–Keuls posthoc test. The Principal Component Analysis [31] and the Additive Main effects and Multiplicative Interaction (AMMI) model [17] performed the genotype stability and productivity. The version 3.5.1 of R software version 3.5.1 is the one used in the analyses. The AMMI model combines both the ANOVA and the PCA. The principal additive effects divide the effect of the GEI into interaction principal component axes (IPCA) provided by the PCA. The calculation of the AMMI Stability Value (ASV) is according to Purchase [32], where $\frac{\text{Sum of Square}_{IPCA1}}{\text{Sum of Square}_{IPCA2}}$ is the weight given to the IPCA1 value that is equal to the IPCA1 sum of squares divided by the IPCA2 sum of squares. Genotypes with small values of IPCA1, IPCA2, and ASV are more stable across environments. It also means that the genotypes with the ASV close to zero are the most stable ones [32].

3. Results

Analysis of variance showed a high degree of morphologic and agronomical variability among the genotypes. For all the investigated traits, a significant effect exists except for the germination rate and the seed size in Rabat and Tinejdad (Table 2). The plant height fluctuates between sites from 34.85 to 127.35 cm at El Kbab and Tinejdad, respectively. The ANOVA reveals the tall size of the Peruvian varieties; Amarilla de Marangani reached 2.30 m at Rabat. Pasto and Vikinga cultivars produced the smallest size plants. At the Meknes site, the Moroccan lines were taller than some Peruvian varieties. Intermediate sizes over the sites were mainly those of Puno and Titicaca, which are the Danish cultivars.

Table 2. Agro-morphological traits variance analysis of the quinoa genotypes at the five experiment sites.

Sites	Rabat		Berrechid		Meknès		Tinejdad		El Kbab	
Traits	F	Mean	F	Mean	F	Mean	F	Mean	F	Mean
Plant height (cm)	75.623 ***	106.62	23.068 ***	79.33	56.871 ***	97.99	47.938 ***	127.35	3.112 *	34.85
Stem diameter (mm)	13.327 ***	7.67	4.323 ***	8.09	13.543 ***	8.13	19.859 ***	10.97	2.100	4.51
Root length (cm)	2.113 *	15.09	1.249	15.54	11.101 ***	14.77	-	-	1.052	12.56
Panicle length (cm)	14.401 ***	31.73	5.275 ***	24.14	15.831 ***	27.91	-	-	3.765 **	9.15
seed size	NS	1.91	7.639 ***	1.93	28.721 ***	1.99	NS	1.99	33.452 ***	1.72
Panicle width (cm)	7.43 ***	4.74	4.538 ***	4.66	15.831 ***	5.18	-	-	4.373 **	2.03
Dry matter per plant (g)	17.790 ***	18.96	3.688 ***	17.49	9.926 ***	22.29	12.861 ***	53.02	5.935 ***	2.31
Yield per plant (g)	15.711 ***	10.82	5.935 ***	10.14	13.049 ***	14.22	21.652 ***	9.77	7.400 ***	0.81
Harvest Index	59.049 ***	0.4	53.668 ***	0.37	53.136 ***	0.42	26.545 ***	0.22	6.263 ***	0.29
Thousand kernel weight (g)	17.986 ***	2.6	8.323 ***	2.32	1.415	2.85	17.375 ***	2.28	5.682 ***	1.59
Yield (qx/ha)	11.947 ***	56.41	5.541 ***	43.37	3.643 ***	78.64	12.987 ***	9.75	3.310 **	14.92
Dry matter (qx/ha)	15.385 ***	137.91	9.074 ***	132.3	16.796 ***	169.95	20.296 ***	529.39	2.030	383.35
Mildew sensibility (%)	10.196 ***	26.84	-	-	17.540 ***	17.59	-	-	-	-
Germination rate (%)	NS	81.15	NS	93.85	NS	94.62	NS	27.78	NS	29.72

***, **, *: significant at 0.1%, 1%, 5% respectively; NS = Non-significant; Gen = Genotype.

Plant dry matter varied between sites, from an average of 2.31 to 53.02 g in respectively El Kbab and Tinejdad; while plant grain average yield ranged from 0.81 to 14.22 g at El Kbab and Meknes, respectively (Table 2). Therefore, plants produced more biomass at Tinejdad and less at El Kbab and more seeds in Meknes than El Kbab. Furthermore, Atlas had the highest grain yield per plant in Meknes, averaging 50.75 g; it also had a good performance in Tinejdad and Berrechid, where its plant grain yields reached 32.17 and 21.38 g, respectively. In Berrechid, Titicaca ranked first with a plant grain yield of 25.19 g. In Rabat, W11 had the best grain yield per plant of 33.06 g, which was followed by both other lines, W16 and SW2 (27.63 and 25.63 g). The Peruvian varieties were the least and even did not produce any seeds in Tinejdad.

The harvest index fluctuated significantly between sites from 0.22 to 0.42. The highest harvest index of 0.69 belongs to Titicaca at Meknes. Puno presents the best harvest indexes of 0.64 and 0.65 at Rabat and Berrechid, respectively (Table 2).

The powdery mildew (*Peronospora farinosa*) sensibility evaluation took place in Rabat and Meknes sites. The reaction to the disease was more pronounced at Rabat (26.84%) than Meknes (17.59%). Four varieties were the most susceptible, Vikinga with 45.50 and 58.50% respectively at Meknes and Rabat, Titicaca with 51.75 and 34.00%, SW2 with 37.25 and 40.50%, and W16 with 43.00 and 33.25%. Pasto and Puno were the most tolerant with 0.50 and 13.25%, and 8.25 and 12.50%, respectively, at Meknes and Rabat (Table 2).

After their harvest and conditioning, collected seeds were placed in Petri dishes for germination to test their viability. The site effect was significant on the germination rate. Thus, the data reveal Meknes and Berrechid as the best sites for seed production; their germination rates exceeded 93%. Rabat recorded a germination rate close to 80% (Table 2). El Kbab and Tinejdad germination rates were below 30%. Puno had the best germination percentage across sites (84.5%) with 100% at Meknes and 50% at Tinejdad.

To assess the varietal variation, principal component analysis (PCA) was performed by considering simultaneously all the variables. The first two principal components explain 72% of the total variability between the 14 genotypes under study. Plant height, stem diameter, dry matter per plant, and harvest index were the major contributors to PC1. Thus, PC1 is considered as a biomass production indicator. The second principal component (PC2) was more related to grain yield per plant, root weight, and plant density, which partially explains the seed production ability per surface unit. The PC3 contributes 10.24% to the total variance; root length, panicle width, grain yield, and plant density present the largest coefficients. The PC4 axis links mainly to the mildew sensitivity (52.41%) and the thousand kernel weight (19.34%). The three first components explained 85.49% of the variability. Thus, these axes were useful to identify homogeneous genotype clusters.

According to the Pearson correlations matrix, the variables that contribute more to the biomass are plant height, stem diameter, root and stem weight, panicle length and width, and plant dry matter. They all are positively correlated to the dry matter and negatively correlated to the harvest index and thousand kernel weight. Genotypes with high growth vigor did not produce much seed; that was the case of the Peruvian cultivars. In Rabat, Passankalla, Salcedo, ILLPA, and Altiplano, dry matter ranged between 14.6 and 19.3 g/plant, while in the Danish cultivars, it varied from 3.9 to 5.2 g. At the same time, the Danish genotypes harvest index (0.62 to 0.64) was higher than the Peruvian ones (0.04 to 0.17).

Grain yield, grain yield per plant, harvest index, and thousand kernel weight are the main variables that are positively correlated. Since plant grain yield correlates negatively to the dry matter, cultivars with moderate growth vigor and a short growth cycle seem more productive under the experimental conditions if we consider the February sowing date. Susceptibility to mildew does not correlate with the other traits; rather, it depends on the environment's climate, mainly temperature and humidity. A strong correlation exists between the plant height and the stem diameter (0.92).

The PCA biplot gathered the 14 genotypes into four clusters (Figure 1). Amarilla de Marangani forms Cluster I (blue) and has high biomass components (plant height, stem diameter, panicle length, and dry matter) as well as a harvest index close to zero. Cluster II (red) includes SW2, W11, W16, and Atlas that have in common relatively low plant density and high grain yield: 63.7 qx/ha on average over the five sites. Cluster III (green) gathers only European cultivars, Titicaca, Puno, Vikinga, Pasto, and Riobamba. This group had relatively low biomass, 7.23 g/plant on average over the five sites. Their cumulative dry matter/plant (36.08 g) does not surpass that of Altiplano (43.15 g). In addition, their harvest index (HI = 0.5) is much higher compared to the other varieties, as is their grain yield average across the sites (49.0 qx/ha). Titicaca seems to have additional special traits than the other cultivars; it has higher yield (72.6 qx/ha), harvest index, and thousand kernel weight, but it is more susceptible to mildew. Cluster IV (yellow) holds Altiplano, Passankalla, Salcedo, and ILLPA Peruvian cultivars; they produced few seeds, an average of 24.5 qx/ha over the five sites; their harvest index is exceedingly low at 0.09 on average, but they are more tolerant to mildew.

In multi-location trials, the selection for grain yield stability involves an estimation of the interaction between the genotypes and the environments (GEI), which is sometimes highly significant. The AMMI analysis on yield facilitates the identification of the most productive and stable genotypes. The analysis of variance (ANOVA) associated with the AMMI model revealed the highly significant sites, genotypes, and interaction effects (Table 3). The environment explains the largest grain yield variability (45.17%) followed by the interaction (31.77%); this indicates the great influence of the selected environments on the cultivars' behavior.

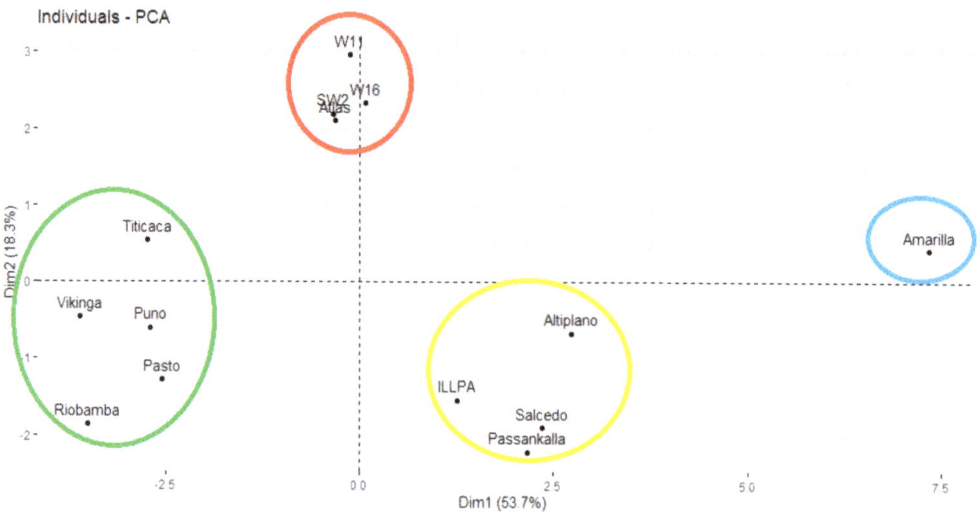

Figure 1. Biplot principal component 1 (PC1) × principal component 2 (PC2) of the 14 quinoa genotypes derived from the average linkage cluster analysis.

Table 3. Additive Main effects and Multiplicative Interaction analysis (AMMI) analysis of variance for grain yield of 14 quinoa genotypes in five locations.

	D	Sum Sq.	Mean Sq.	*F* Value	Pr. (>F)	Var. (%)
Environment	4	822,732	205,683	68.3663 ***	9.01×10^{-10}	45.17
Rep (Env)	15	45,128	3009	2.0224 *	0.01166	-
Genotype	13	419,875	32,298	21.7112 ***	2.20×10^{-16}	23.05
Interactions	52	578,715	11,129	7.4812 ***	2.20×10^{-16}	31.77
IPCA1	16	69,269.584	4329.349	2.91 ***	0.0001	53.4
IPCA2	14	32,940.884	2352.9203	1.58	0.0785	25.4
Residuals	983	1,462,333	1488			

***, *: significant at 0.1%, 5% respectively.

Furthermore, the first interaction component IPCA1 generated by the model explains 53.4% of the total yield variation, while the second axis justifies an additional 25.4%. Therefore, the IPCA1 vs. grain yield biplot fully describes the quinoa genotypes' behavior as confirmed through its significance (p-value = 0.0001).

Figure 2 illustrates the genotypes grain yield across the five sites. Meknes is the most productive site (78.6 qx/ha), while the lowest yielding site is El Kbab (9.5 qx/ha). Meknes, a propitious pluvial zone with an average annual rainfall of 576 mm/year, mean temperature of 19–22 °C in May–June, and relative humidity of 60–78%, gave much more auspicious growing conditions. The coastal site of Rabat presented a quite high yield of 56.4 qx/ha. The growing conditions of the oasis site of Tinejdad affected significantly the quinoa genotypes' performance, especially Peruvian cultivars. As a result of the late sowing at El Kbab, the quinoas' reproduction-phase matches the long photoperiod and high-temperature period (37.2–40 °C), while almost no irrigation is applied. El Kbab conditions impact the 14 quinoa genotypes and more the Peruvian cultivars.

Figure 2 reveals Meknes and Rabat as the most interactive sites with the first IPCA, since they reached the highest IPCA1 scores of 6.88 and −8.54, respectively. These two sites were judged appropriate to discriminate and assess the genotype grain yield variance stability. More than 50% of the genotypes produced over-mean yields; those are W16, Pasto, Riobamba, W11, Puno, Titicaca, Atlas, and SW2. According to Purchase [32], the closer the genotype score is to the center of the IPCA1 by IPCA2 biplot, the more stable is the genotype, and the opposite is true. The highest yield of 78.3 qx/ha was that of SW2. This local accession reveals low stability; it is placed far from the IPCA1–IPCA2 biplot center (Figure 3). Atlas and Titicaca have significant yields of respectively 75.6 and 72.6 qx/ha and are also unstable. The lowest yields were those of Amarilla de Marangani (0 qx/ha), Passankalla (9.2 qx/ha), and Viking (18.6 qx/ha) (Figure 2). Nevertheless, their stability scores were valuable, of 2.46, 1.88, and 3.09, respectively.

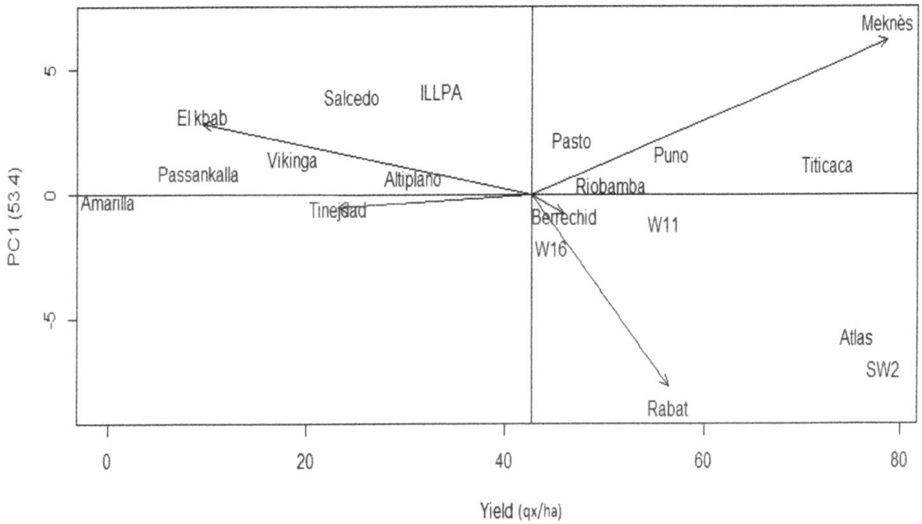

Figure 2. Biplot of the first principal component of the interaction (IPCA1) and the grain yield of the 14 quinoa genotypes.

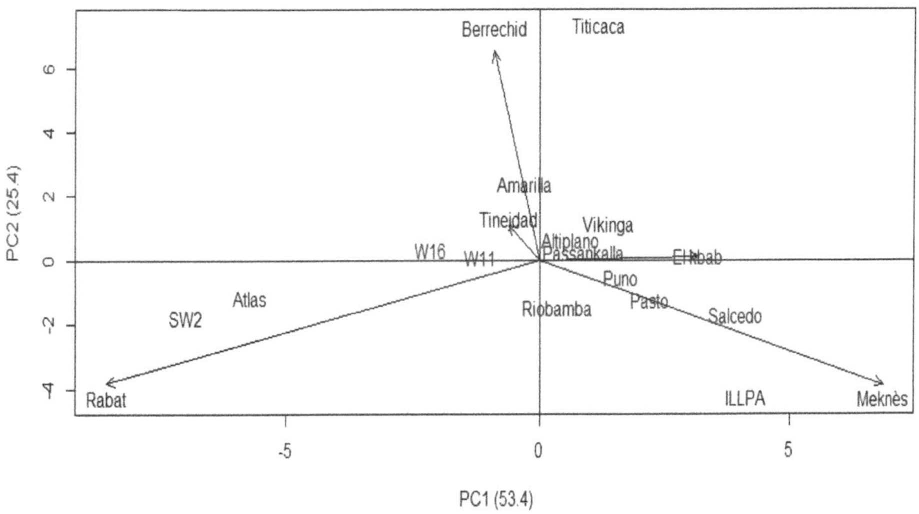

Figure 3. Biplot between the first component (IPCA1) and the second component (IPCA2) of the genotype × environment interaction (GEI) interaction for the grain yield of the 14 quinoa genotypes.

Genotypes projection, aside from the environmental vectors, indicated specific interactions. The Figure 3 biplot displays varietal adaptation degree to a specific site. Titicaca is better suited to Berrechid (143.2 qx/ha); ILLPA interacts positively with Meknes environment (125.4 qx/ha), as Atlas (143.9 qx/h) and SW2 (160.4 qx/ha) interact with Rabat, where they recorded their best grain yields. W11 (56.0 qx/ha) and Riobamba (50.7 qx/ha) genotypes, with a yield over the total mean, presented good yield stability.

In summary, the AMMI analysis provides the genotypes' productivity and the AMMI Stability Values (ASV), as shown in Figure 4. Both parameters allowed gathering the cultivars into three groups.

The first group includes five cultivars, Altiplano, Riobamba, Passankalla, W11, and Amarilla; these are most stable with ASV scores ranging from 1.40 to 2.46. Two of them, Riobamba and W11, are above the average yield (Figure 4).

The second group gathers Vikinga, Puno, W16, and Pasto genotypes; their ASV stability scores range from 3.09 to 4.75; three among them, Puno, W16, and Pasto, overpass the average yield of 42.7 qx/ha.

The third group gathers Titicaca, Salcedo, ILLPA, Atlas, and SW2; their stability scores are sited between 7.75 and 14.78 ASV values. Atlas and SW2 have the highest productivity of 75.7 and 78.3 qx/ha, respectively. Both accessions were ranked unstable according to their ASV scores (12.1 and 14.8, respectively).

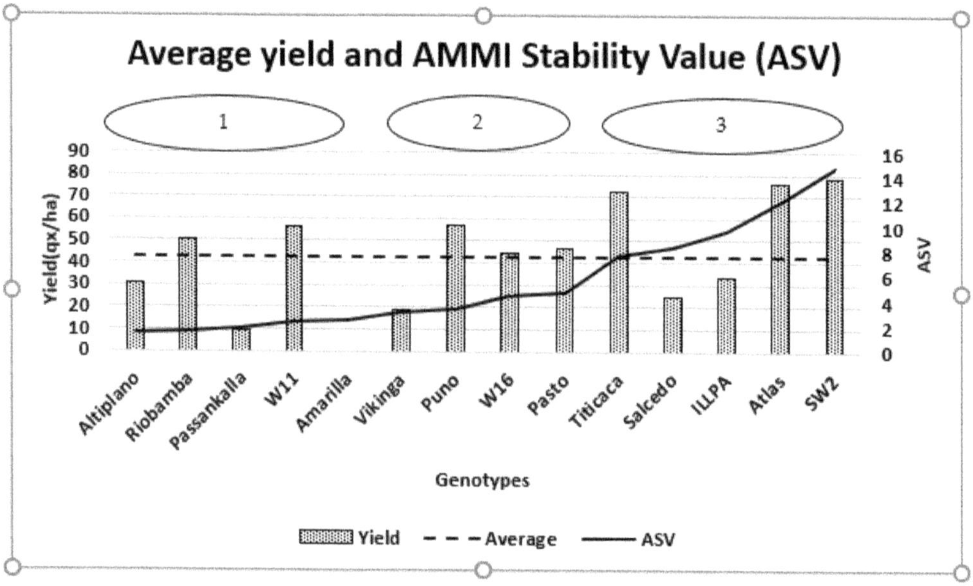

Figure 4. Grain yield and evolution of the AMMI stability values (ASV) of the 14 quinoa genotypes.

4. Discussion

Quinoa is an alternative to global stable food products, such as rice and wheat; it is superior in terms of nutritional value and abiotic stress tolerance [33]. Recently, quinoa cultivars have been widely tested outside the species center of origin in Latin America [4]. Quinoa has great morphological, agronomic, and physiological variability, predisposing it to wide environmental adaptation. This great intraspecific diversity connects to the quinoa five ecotypes of different origins and its allotetraploid status [6]. Hence, there is a real need to consider the effect of environmental factors such as temperature and photoperiod to optimize the introduction of quinoa in new regions [34]. On the other hand, when we select genotypes only for their predicted genotypic productivity, it does not guarantee that they will maintain their performance when cultivated under other environments [35]. Thus, it is difficult to identify consistently superior genotypes across environments when the G $\times$ E interaction is highly significant [20]. Therefore, this present study seeks to assess the stability of the grain yield of 14 quinoa genotypes across five agro-ecological zones in Morocco.

The data analyses reveal an important agro-morphological diversity between the genotypes. Peruvian cultivars plants have a large size and tend to produce much biomass. In general, their height is within the range reported by Jacobsen and Stolen [36]. According to these authors, quinoa plant size varies between 0.5 and 3 m with an average of 1 to 1.5 m. Amarilla de Marangani gave the highest plant height (2.30 m) as expected; it belongs to the Inter-Andean valleys' ecotype [37]. Its origin latitudinal score is 13°3 S. According to Tapia et al. [6], this ecotype grows between 2000 and 4000 m, and it has a high size and late flowering. Generally, plant size is much more variable and depends on the plant branching habit and the inflorescence shape, amaranthiform, or glomeriform [6]. The growth cycle duration explains largely plant size differences between the two groups. European varieties were ready for harvest at least 4 weeks before the Peruvian accessions. Peruvian varieties were still at the flowering stage, even Amarilla de Marangani was at the late vegetative phase when the European cultivars reached their physiological maturity. Amarilla de Marangani is the latest and the tallest among all the varieties.

The growth cycle length seems to have a direct relationship with the photoperiod. According to previous studies conducted by Christiansen et al. [38] and Bertero et al. [7], quinoa is a facultative short-day species rather than qualitative, indicating that regardless of the cultivar adaptation, quinoa can flower under a wide range of day lengths. Thus, genotypes are classified as a short day (12.25 h), long day (14 h), or insensitive, depending on their sensibility to the photoperiod. Bertero [34] reported that long-cycle quinoa cultivars are generally more photoperiod sensitive compared to short-cycle ones. Furthermore, the duration of the phenological phases (emergence to floral initiation, floral initiation to the first anthesis, and first anthesis to physiological maturity) depend on the genotype sensitivity to the photoperiod. Christiansen et al. [38] reveal a significant varietal effect on the leaf and flower nodes initiation when quinoa plants are grown at a long day compared to short-day photoperiods, with Real being the most pronounced reactive cultivar. As Real is a traditional large-seed variety from the Bolivian southern Altiplano, it is late-maturing when compared to the sea level ecotype such as the Danish varieties. Our findings match with these statements. Peruvian late cultivars produce more vegetative organs at the detriment of the seeds. In addition, Bendevis [39], while testing two quinoa cultivars, reported similar adjustments in resource allocation between plants growing under short and long-day photoperiods. They are also in agreement with Dorais and others' [40] conclusions on tomato and sweet pepper crops, stating that an extended photoperiod results in further shoot development and carbohydrate accumulation in leaves over fruit development.

In contrast, when quinoa grows under long photoperiods and high temperatures, it tends to shorten its development cycle. Therefore, plants might reach their physiological maturity even before they complete their development. This should be the case in El Kbab where the plants' size did not exceed 54 cm. It is important to remember that the El Kbab trial was installed 7 weeks later because of the snow, and there was no irrigation stream when the temperature and day length increased. These results confirm the Atkinson and Porter [41] scheme stating that when the growing environment is at risk of high temperature or water stress, plants would hasten their development and form their seeds. Bertero [34] reported that late in the season in the Andes, when water deficit occurs, photoperiod-sensitive cultivars are faster in their growth and seed filling, allowing plants to mature before water or temperature becomes restrictive.

Water supplies may also influence plant growth to some extent. All the genotypes were relatively more vigorous in Tinejdad despite the oases type of climate (average temperature 23 °C, humidity 33.83%). Indeed, the Tinejdad trail received regular irrigation to meet the plant and the atmosphere demand, which contributed to the expansion of the plant size. In fact, excessive irrigation at the seedling stage improves the plant vigor but does not enhance the grain yield [42].

The origin of the variety appears to impact considerably quinoa seed production. Most Peruvian cultivars produced little or no seeds despite their well-established panicles. They were very late when compared to the other genotypes. Amarilla de Marangani did not produce any seeds across the five experiment sites. Several tested genotypes are day-length sensitive and classified as short day. Quinoas are also sensitive to temperature; they require between 15 and 20 °C for optimal growth. Thus, the growing cycle duration depends on the combination of both factors [43]. Bertero [43] reported that far from the equator, short-day plants become more sensitive to temperature and less sensitive to photoperiod. In 2017, Noulas [44] reported that several Latin American cultivars gave no seeds under the central Greece arid environment. Research conducted also in the US stated that temperatures exceeding 35 °C are likely to cause pollen sterility [45]. These facts well justify our current finding on the Peruvian quinoa in the Tinejdad oases. In addition to the high temperatures, a summer-long photoperiod and low air humidity contribute to the seed establishment hindering. On the other hand, the same Peruvian cultivars behave differently at Meknes, where average temperatures are quite mild (19.3 °C). In terms of comparison, ILLPA seed production per plant at Meknes (11.19 g) over-passed Titicaca (8.81 g), Vikinga (6.44 g), Pasto (8.12 g), and Riobamba (6.25 g).

The two most important determinant factors to grain yield are biomass accumulation and its partitioning into the storage organ, which is evaluated through the harvest index. Puno had the best harvest indexes of 0.64 and 0.65 at Rabat and Berrechid, respectively. As for Meknes, the highest harvest index of 0.69 was Titicaca's. These well-known quinoa cultivars by farmers recorded the best yield and harvest index at previous experiments. In 2010–2011, Titicaca had the highest harvest index of 0.47 during a trial carried out in the semi-arid region of Bouchane. It is quite low compared to the 2018 exceptional rainy cropping season. Indeed, the 2018 season was particularly favorable for seed production, especially in Meknes. Overall, the varietal harvest indexes ranked between 0.3 and 0.69 for respectively W11 in Tinejdad and Titicaca in Meknes, except for the Peruvian varieties, which experienced problems seed-setting due to heat. The principal component analysis also disclosed the European cultivars group for their high harvest index. Jacobsen [33] reported that quinoa's perfect variety for seed production in northern Europe is the one that matures early and homogeneously. A growing period shorter than 150 days is advantageous. Quinoa should also have a consistently high seed yield, and it should be short and non-branching to facilitate mechanical harvesting. They used primarily Chilean cultivars that are supposed to be the least sensitive or insensitive to photoperiod for seed filling to achieve these goals [34]. The harvest index represents photosynthetic and translocation seed production capacity. It is also sensitive to agronomic practices and environmental conditions [46]. Rojas et al. [47] reported variation in the quinoa harvest index ranging from 0.06 to 0.87.

Thousand kernel weight (TKW) differences were significant between varieties at Rabat and Berrechid. The TKW means fluctuated from 1.17 g for Salcedo in Berrechid and 3.42 g for Titicaca in Berrechid and W11 in Rabat. The favorable pluvial region of Meknes did not discriminate the varieties through their TKW, as 2018 was an exceptional rainy season, and all the cultivars had reached their best seed weight performance.

Recently, multi-location experiments (MTE) became valuable to assess quinoa genotype adaptation in new regions [9,22,23]. According to Gadisa et al. [20], it is difficult to identify consistently superior genotypes across environments when genotype × environment interactions (GEI) are highly significant. In our present case, the GEI effect explains 31.77% of the total grain yield variation compared to the 23.05% genotype effect. As expected through the AMMI analysis, the environment gathers the largest grain yield variability (45.17%), indicating that grain yield depends highly on the environment. These findings are in line with previous conclusions of other research on quinoa, maize, and bread wheat [22].

AMMI Stability Values (ASVs) of the GEI ranks the studied genotypes into three clusters: stable, moderately stable, and unstable (Figure 4). Among the stable genotypes, two (Riobamba and W11) produced more than the overall mean yield of 42.7 t/ha. The moderate stability group gathers three genotypes that exceed the overall average yield: Puno (56.6 qx/ha), W16 (44.7 qx/ha), and Pasto (46.8 qx/ha). Within the unstable group, Atlas and SW2 genotypes had the highest grain yield of 75.6 and 78.3 qx/ha, respectively. Thus, the grain yield of highly productive genotypes is more likely to be influenced by the growing conditions in different environments. Then, AMMI analysis allowed identifying genotype adaption to a specific site. Titicaca suits better to the Berrechid site, while ILLPA is more productive in Meknes. Atlas and SW2 interact positively with Rabat. According to Mohammadi et al. [48], stability should not be the only parameter for selection; usually, stable genotypes would not necessarily reach the highest yield across the experimental stations. Some authors use the Yield Stability Index as the main selection criterion to rank their germplasm [49,50]. In our case, we used the synthetic graph in Figure 4 to interpret both yield and stability criteria.

For cultivar performance evaluation, we calculated grain yield according to its components. Thus, the estimated yields fit within the range 0.00 qx/ha as a minimum for the Amarilla de Marangani cultivar and 78.3 qx/ha as maximum for the SW2 local advanced line with an overall yield of 42.7 qx/ha. Compared to other countries that have recently introduced quinoa cultivation, grain yields are relatively variable probably because of the large diversity of the cultivars, their origins, and the contrasting environments tested. Dost [51] reported 38.7 qx/ha as the highest yield recorded in Egypt, while in Germany, Präger et al. [52] reached a maximum grain yield of 24.3 qx/ha for the Zeno cultivar and Titicaca produced less in the same experiment. Recently, quinoa grain yield in Egypt revealed a maximum of 30.8 qx/ha under irrigation with the Argentinian Regalona cultivar [53]. The relatively high overall yield of 42.7 qx/ha we obtained through our investigation could have three explanations: the performance of selected cultivars used, the intensive care taken to conduct the trials, and the sampling method based on an individual plant. Moreover, Dost [51] reported a maximum yield of 75.0 qx/ha in Lebanon. Another study investigating the genetic variability of 27 lines originating from different parts of the Andean region and South America conducted in the northern part of India had grain yields varying widely between 4.7 and 60.1 qx/ha [30].

5. Conclusions

The present study allowed detecting the great variability among the 14 quinoa genotypes through the morphological and agronomic traits. It also showed that grain yield is more influenced by the environment and the genotype–environment interaction, and it was able to prove the significance and challenge of evaluating the varietal grain yield stability across the different contrasting sites. According to AMMI analysis, Meknes was the most productive site, followed by Rabat and Berrechid. Riobamba cultivar and W11 local advanced line showed broad adaptation with a grain yield above the overall yield. The Titicaca, ILLPA, Atlas, and SW2 genotypes fitted better in a particular environment. Some Peruvian long-grow cycle cultivars performed less in terms of grain yield because of their sensibility to the photoperiod and high temperatures. Other investigation studies should be held at different sowing dates and agro-climatic zones to be able to select for quinoa growers and farmers adapted cultivars to their local environment and to cover their needs in seeds.

Author Contributions: Conceptualization, O.B.; methodology, O.B.; software, E.T.; validation, O.B., and E.T.; formal analysis, E.T.; investigation, E.T. and O.B.; resources, O.B.; data curation, E.T., and A.A.; writing—original draft preparation, E.T.; writing—review and editing, O.B.; visualization, E.T. and O.B.; supervision, O.B.; project administration, O.B.; funding acquisition, O.B. All authors have read and agreed to the published version of the manuscript.

Funding: This research was funded jointly by the FAO, grant number TCP/MOR/3602/C1 and the IAV quinoa program grant number 2129.

Institutional Review Board Statement: Not applicable.

Informed Consent Statement: Not applicable.

Data Availability Statement: Dataset analyzed in this study is publicly available and can be found here: (https://mega.nz/file/iVc00RxA#Rif61fwxFncPLcNoxz-ciO5sJPxq9cojUlA8dZOX7oE).

Acknowledgments: The authors would like to acknowledge the FAO for the financial support provided for conducting the field trials under the Project TCP/MOR/3602/C1-Quinoa «Quinoa cultivation Promotion in Morocco».

Conflicts of Interest: The authors declare no conflict of interest.

References

1. Zhao, C.; Liu, B.; Piao, S. Temperature increase reduces global yields of major crops in four independent estimates. *Proc. Natl. Acad. Sci. USA* **2017**, *114*, 9326–9331. [CrossRef]
2. Rao, N.K.; Shahid, M. Quinoa-A promising new crop for the Arabian Peninsula. *Am. J. Agric. Environ. Sci.* **2012**, *12*, 1350–1355.
3. Fuentes, F.; Bhargava, A. Morphological analysis of quinoa germplasm grown under lowland desert conditions. *J. Agron. Crop Sci.* **2011**, *197*, 124–134. [CrossRef]
4. Bazile, D.; Jacobsen, S.E.; Verniau, A. The global expansion of quinoa: Trends and limits. *Front. Plant Sci.* **2016**, *7*, 6–22. [CrossRef]
5. Sanchez, H.B.; Lemeur, R.; Damme, P.V.; Jacobsen, S.E. Ecophysiological analysis of drought and salinity stress of quinoa (*Chenopodium quinoa* Willd.). *Food Rev. Int.* **2003**, *19*, 111–119. [CrossRef]
6. Tapia, M.E.; Mujica, A.; Canahua, A. Origen, distribucion geografica y sistemas de produccion de quinoa. In *Primera Reunion sobre Genetica y Fitomejoramiento de la Quinoa*; Universidad Nacional Tecnica del Altiplano: Puno, Peru, 1980; pp. 1–8.
7. Bertero, H.D.; King, R.W.; Hall, A.J. Photoperiod-sensitive development phases in quinoa (*Chenopodium quinoa* Willd.). *Field Crop. Res.* **1999**, *60*, 231–243. [CrossRef]
8. Hinojosa, L.; Matanguihan, J.B.; Murphy, K.M. Effect of high temperature on pollen morphology, plant growth and seed yield in quinoa (*Chenopodium quinoa* Willd.). *J. Agron. Crop Sci.* **2018**, *205*, 34–45. [CrossRef]
9. Bertero, H.D.; de la Vega, A.J.; Correa, G.; Jacobsen, S.E.; Mujica, A. Genotype and genotype-by-environment interaction effects for grain yield and grain size of quinoa (*Chenopodium quinoa* Willd.) as revealed by pattern analysis of international multi-environment trials. *Field Crop. Res.* **2004**, *89*, 299–318. [CrossRef]
10. Mitrovic, B.; Stanisavljevic, D.; Treskic, S.; Stojakovic, M.; Bekavac, G.; Nastasic, A.; Ivanovic, M. GGE biplot analysis of multi-environment trials of NS maize hybrids. *Ratarstvo i Povrtarstvo* **2011**, *48*, 77–82. [CrossRef]
11. Curti, R.N.; de la Vega, A.J.; Andrade, A.J.; Bramardi, S.J.; Bertero, H.D. Multi-environmental evaluation for grain yield and its physiological determinants of quinoa genotypes across Northwest Argentina. *Field Crop. Res.* **2014**, *166*, 46–57. [CrossRef]
12. Van Eeuwijk, F.A.; Malosetti, M.; Yin, X.; Struik, P.C.; Stam, P. Statistical models for genotype by environment data: From conventional ANOVA models to eco-physiological QTL models. *Aust. J. Agric. Res.* **2005**, *56*, 883–894. [CrossRef]
13. Perkins, J.M.; Jinks, J.L. Environmental and genotype-environmental components of variability. *Heredity* **1968**, *23*, 339–356. [CrossRef]
14. Francis, T.R.; Kannenberg, L.W. Yield stability studies in short-season maize. I. A descriptive method for grouping genotypes. *Can. J. Plant Sci.* **1978**, *58*, 1029–1034. [CrossRef]
15. Nassar, R.; Hühn, M. Studies on estimation of phenotypic stability: Tests of significance for nonparametric measures of phenotypic stability. *Biometrics* **1987**, *43*, 45–53. [CrossRef]
16. Resende, M.D.V.d. *Matemática e Estatística na Análise de Experimentos e no Melhoramento Genético*, 1st ed.; Embrapa Florestas: Colombo, Sri Lanka, 2007; 362p.
17. Gauch, H.G. *Statistical Analysis of Regional Yield Trials: AMMI Analysis of Factorial Designs*; Elsevier Science Publishing: Amsterdam, The Netherlands, 1992.
18. De Oliveira, I.J.; Atroch, A.L.; Dias, M.C.; Guimarães, L.J.; de Guimarães, P.E.O. *Selection of Corn Cultivars for Yield, Stability, and Adaptability in the State of Amazonas*; Pesquisa Agropecuária Brasileira: Brasília, Brazil, 2017; Volume 52, pp. 455–463.
19. Filho, J.M.C.; de Resende, M.D.V.; de Morais, O.P.; de Castro, A.P.; Guimarães, É.P.; Pereira, J.A.; Utumi, M.M.; Breseghello, F. Upland rice breeding in Brazil: A simultaneous genotypic evaluation of stability, adaptability and grain yield. *Euphytica* **2013**, *192*, 117–129. [CrossRef]
20. Gadisa, A.W.; Hussein, M.; Dawit, A.; Tesfahun, A. Genotype × Environment Interaction and Yield Stability of Bread Wheat Genotypes in Central Ethiopia. *J. Plant Breed. Genet.* **2019**, *7*, 87–94.
21. Machado, N.G.; Lotufo-Neto, N.; Hongyu, K. Statistical analysis for genotype stability and adaptability in maize yield based on environment and genotype interaction models. *Ciência e Natura* **2019**, *41*, 25. [CrossRef]
22. Ali, M.; Elsadek, A.; Salem, E.M. Stability Parameters and AMMI Analysis of Quinoa (*Chenopodium quinoa* Willd.). *Egypt. J. Agron.* **2018**, *40*, 59–74. [CrossRef]
23. Vasconcelos, E.S.; Echer, M.M.; Kliemann, M.A.; Lang, M.J. Selection and recommend of quinoa (*Chenopodium quinoa*) genotypes based on the yield genotypic adaptability and stability. *Rev. Ceres Viçosa* **2019**, *66*, 117–123. [CrossRef]
24. De Mendiburu, F. Una Herramienta de Analisis Estadistico Para la Investigacion Agricola. Master's Thesis, Universidad Nacional de Ingenieria (UNI-PERU), Rimac, Peru, 2009.

25. Benlhabib, O.; Atifi, M.; Jellen, E.N.; Jacobsen, S.E. The introduction of a new peruvian crop «Quinoa » to a rural community in Morocco. In Proceedings of the VIII European Society of Agronomy Congress, KVL, Copenhagen, Danemark, 11–15 July 2004; pp. 881–882.
26. Mhada, M.; Benlhabib, O.; Jimenez, F.R.; Arano, I.A.; Andrews, D.E.; Maughan, P.J.; Jellen, E.N. Population structure and genetic diversity in quinoa assessed with SNP markers. Molecular screening and diversity of quinoa. Proocedings of the SWUP-MED Project Final International Conference: Sustainable Water Use for Securing Food Production in the Mediterranean Region under Changing Climate, Agadir, Marrocos, 10–15 March 2013; pp. 237–243.
27. Fghire, R.; Anaya, F.; Oudou, I.A.; Benlhabib, O.; Ragab, R.; Wahbi, S. Physiological and photosynthetic response of quinoa to drought stress. *Chil. J. Agric. Res.* **2015**, *75*, 174–183. [CrossRef]
28. Danielsen, S.; Jacobsen, S.E.; Echegaray, J.; Ames, T. *Impact of Downy Mildew on the Yield of Quinoa. Andean Roots and Tubers and Other Crops*; CIP Program Report 1999–2000; The Royal Veterinary and Agricultural University: Copenhagen, Denmark, 2000; pp. 397–401.
29. Benlhabib, O.; Boujartani, N.; Maughan, P.J. Elevated genetic diversity in an F2: 6 population of quinoa (*Chenopodium quinoa*) developed through an inter-ecotype cross. *Front. Plant Sci.* **2016**, *7*, 1222. [CrossRef] [PubMed]
30. Bhargava, A.; Shukla, S.; Rajan, S.; Ohri, D. Genetic diversity for morphological and quality traits in quinoa (*Chenopodium quinoa* Willd.) germplasm. *Genet. Resour. Crop Evol.* **2007**, *54*, 167–173. [CrossRef]
31. Sneath, P.H.A.; Sokal, R.R. *Numerical Taxonomy: The Principles and Practice of Numerical Classification*; Freeman: San Francisco, CA, USA, 1973.
32. Purchase, J.L. *Parametric Analysis to Describe Genotype × Environment Interaction and Yield Stability in Winter Wheat*; University of Free State: Bloemfontein, South Africa, 1997.
33. Jacobsen, S.E. Adaptation and scope for quinoa in northern latitudes of Europe. In *State of the Art Report of Quinoa in the World in 2013*; FAO & CIRAD: Rome, Italy, 2015; pp. 436–445.
34. Bertero, H.D. Response of Developmental Process to Temperature and Photoperiod in Quinoa (*Chenopodium quinoa* Willd.). *Food Rev. Int.* **2003**, *19*, 87–97. [CrossRef]
35. Cruz, C.D.; Carneiro, P.C.S.; Regazzi, A.J. *Modelos Biométricos Aplicados ao Melhoramento Genético*, 3rd ed.; Editora UFV: Viçosa, Brazil, 2014; 668p.
36. Jacobsen, S.E.; Stølen, O. Quinoa: Morphology, phenology and prospects for its production as a new crop in Europe. *Eur. J. Agron.* **1993**, *2*, 19–29. [CrossRef]
37. Bertero, H.D.; King, R.W.; Hall, A.J. Photoperiod and temperature effects on the rate of leaf appearance in quinoa (*Chenopodium quinoa*). *Aust. J. Plant Physiol.* **2000**, *27*, 349–356. [CrossRef]
38. Christiansen, J.L.; Jacobsen, S.E.; Jørgensen, S.T. Photoperiodic effect on flowering and seed development in quinoa (*Chenopodium quinoa* Willd.). *Acta Agric. Scand. B Soil Plant* **2010**, *60*, 539–544.
39. Bendevis, M.A.; Sun, Y.; Rosenqvist, E.; Shabala, S.; Liu, F.; Jacobsen, S.E. Photoperiodic effects on short-pulse 14C assimilation and overall carbon and nitrogen allocation patterns in contrasting quinoa cultivars. *Environ. Exp. Bot.* **2014**, *104*, 9–15. [CrossRef]
40. Dorais, M.; Yelle, S.; Gosselin, A. Influence of extended photoperiod on photosynthates partitioning and export in tomato and pepper plants. *N. Z. J. Crop Hortic.* **1996**, *24*, 29–37. [CrossRef]
41. Atkinson, D.; Porter, J.R. Temperature, plant development, and crop yields. *Trends Plant Sci.* **1996**, *1*, 119–124. [CrossRef]
42. Oelke, E.; Putnam, D.; Teynor, T.; Oplinger, E. *Alternative Field Crops Manual, Quinoa*; University of Wisconsin Cooperative Extension Service, University of Minnesota Extension Service, Centre for Alternative Plant and Animal Products: Madison, WI, USA, 1992; Available online: https://www.hort.purdue.edu/newcrop/afcm/quinoa.html (accessed on 21 April 2016).
43. Bertero, H.D. Effects of Photoperiod, Temperature and Radiation on the rate of leaf appearance in quinoa (*Chenopodium quinoa* Willd.) under field conditions. *Ann. Bot.* **2001**, *87*, 495–502. [CrossRef]
44. Noulas, C.; Tziouvalekas, M.; Vlachostergios, D.; Baxevanos, D.; Karyotis, T.; Iliadis, C. Adaptation, agronomic potential, and current perspectives of quinoa under mediterranean conditions: Case studies from the lowlands of central Greece. *Commun. Soil Sci. Plant Anal.* **2017**, *48*, 2612–2629.
45. AAFRD—Alberta Agriculture, Food and Rural Development. *Quinoa—The Next Cinderella Crop for Alberta*; el Hafid, R., Aitelmaalem, H., Driedger, D., Bandara, M., Stevenson, J., Eds.; Technical Report; Alberta Agriculture, Food and Rural Development (AAFRD): Edmonton, AB, Canada, 2005.
46. Bertero, H.D.; Ruiz, R.A. Reproductive partitioning in sea level quinoa (*Chenopodium quinoa* Willd.) cultivars. *Field Crop. Res.* **2010**, *118*, 94–101. [CrossRef]
47. Rojas, W. Multivariate analysis of genetic diversity of Bolivian quinoa germplasm. *Food Rev. Int.* **2003**, *19*, 9–23. [CrossRef]
48. Mohammadi, R.; Roostaei, M.; Ansari, Y.; Aghaee, M.; Amri, A. Relationships of phenotypic stability measures for genotypes of three cereal crops. *Can. J. Plant Sci.* **2010**, *90*, 819–830. [CrossRef]
49. Farshadfar, E. Incorporation of AMMI Stability Value and Grain Yield in a Single Non-Parametric Index (GSI) in Bread Wheat. *Pak. J. Biol. Sci.* **2008**, *11*, 1791–1796. [CrossRef]
50. Atta, B.M.; Shah, T.M.; Abbas, G.; Haq, M.A. Genotype x environment interaction for seed yield in kabuli chickpea (*Cicer arietinum* L.) genotypes developed through mutation breeding. *Pak. J. Bot.* **2009**, *41*, 1883–1890.
51. Dost, M. Field Evaluation Results across Locations and Identification of Suitable QUINOA Varieties. In *Wrap up Workshop of Regional Quinoa Project (TCP/RAB/3403–FAO)*; Food and Agriculture Organization of the United Nations: Rome, Italy, 2015.

52. Präger, A.; Munz, S.; Nkebiwe, P.M.; Mast, B.; Graeff-Hönninger, S. Yield and Quality Characteristics of Different Quinoa (*Chenopodium quinoa* Willd.) Cultivars Grown under Field Conditions in Southwestern Germany. *Agronomy* **2018**, *8*, 197.
53. Ashraf, E. Multi-environmental evaluation for grain yield and its components of quinoa genotypes across the north western coast of Egypt. *Egypt. J. Desert Res.* **2017**, *67*, 65–82.

plants

MDPI

Review

Seed Dormancy and Preharvest Sprouting in Quinoa (*Chenopodium quinoa* Willd)

Emma M. McGinty [1], Kevin M. Murphy [2] and Amber L. Hauvermale [2,*]

[1] The School of Biological Sciences, Washington State University, P.O. Box 644236, Pullman, WA 99164, USA; emma.mcginty@wsu.edu
[2] Department of Crop and Soil Science, Washington State University, Pullman, WA 99164, USA; kmurphy2@wsu.edu
* Correspondence: ahauvermale@wsu.edu; Tel.:+1-509-335-3661

Abstract: Quinoa (*Chenopodium quinoa* Willd.) is a culturally significant staple food source that has been grown for thousands of years in South America. Due to its natural drought and salinity tolerance, quinoa has emerged as an agronomically important crop for production in marginal soils, in highly variable climates, and as part of diverse crop rotations. Primary areas of quinoa research have focused on improving resistance to abiotic stresses and disease, improving yields, and increasing nutrition. However, an evolving issue impacting quinoa seed end-use quality is preharvest sprouting (PHS), which is when seeds with little to no dormancy experience a rain event prior to harvest and sprout on the panicle. Far less is understood about the mechanisms that regulate quinoa seed dormancy and seed viability. This review will cover topics including seed dormancy, orthodox and unorthodox dormancy programs, desiccation sensitivity, environmental and hormonal mechanisms that regulate seed dormancy, and breeding and non-breeding strategies for enhancing resistance to PHS in quinoa.

Keywords: abscisic acid; desiccation sensitivity; gibberellin; hormone signaling; precocious germination; seed morphology

Citation: McGinty, E.M.; Murphy, K.M.; Hauvermale, A.L. Seed Dormancy and Preharvest Sprouting in Quinoa (*Chenopodium quinoa* Willd). *Plants* **2021**, *10*, 458. https://doi.org/10.3390/plants10030458

Academic Editor: Cataldo Pulvento

Received: 24 December 2020
Accepted: 24 February 2021
Published: 28 February 2021

1. Introduction to Quinoa, Cultivars, Breeding Issues, and Preharvest Sprouting

Quinoa (*Chenopodium quinoa* Willd.) is a pseudocereal originating from the Andes Mountain Range in South America and is a culturally significant staple food source that has been grown for thousands of years [1,2]. Due to its natural drought and salinity tolerance, quinoa has emerged as a favorable crop for production in marginal soils and in highly variable climates [3]). Quinoa's nutrient dense grain is also ideal for supporting human health in diverse global communities, and for this reason, it has agronomic significance in global economies.

There are five distinct quinoa ecotypes that originated from five different localities in South America. The original localities are (1) the valley habitat ranging across Colombia, Ecuador, Peru, and Bolivia, and the ecotype from this region is often tolerant to downy mildew; (2) the altiplano habitat which is near Titicaca Lake on the border of Bolivia and Peru, and the ecotype from this region is tolerant of marginal environments and frost; (3) the salares habitat, ranging across the salt flats of Bolivia and Chile, and the ecotype from this region is tolerant to high salinity; (4) the sea-level habitat, ranging from low-altitude areas of southern and central Chile, and the ecotype from this region is high yielding; (5) the subtropical or yungas habitat, ranging from the low-altitude humid valleys of Bolivia, and this ecotype is known for its late-flowering genotypes [1,3–5]. Until recently, there was limited pedigree information for quinoa, making it more difficult to identify the first quinoa ecotypes cultivated to produce today's modern varieties. However, phenotypic and genotypic clues indicate that modern varieties have little to no seed dormancy, are xerophobic (meaning seeds display desiccation intolerance), and adult plants are salt and drought tolerant. These phenotypes are likely the result, at least in part, of adaptation to the original environments in which the first varieties were cultivated [1,3]. With the recent sequencing of the quinoa genome, it is expected that genetic tools will help to accelerate major breeding efforts for quinoa improvement focused on abiotic stress resistance, yield, and end-use quality [6].

From a production perspective, traits associated with enhanced plant plasticity that help to mitigate abiotic stress are beneficial for integrating quinoa into diverse cropping systems. For this reason, many research efforts have focused on understanding the genetic mechanisms that regulate abiotic stress responses, leading to drought or salt tolerance in adult quinoa plants [3,7–10]. Additionally, from a management perspective, the weak seed dormancy or in some cases, the absence of dormancy observed in quinoa is a desirable characteristic for integration into diverse crop rotations. This is because seeds that germinate readily are less likely to establish stable volunteer seed banks, requiring little to minimal input for management [11]. However, lack of seed dormancy has led to issues with reduced yields due to premature germination prior to harvest and has revealed a critical gap in knowledge about the regulation of seed dormancy in quinoa. Therefore, this review will define seed dormancy and the hormones involved in regulation, the different seed dormancy programs, the differences between orthodox and unorthodox dormancy programs, desiccation sensitivity, environmental mechanisms that regulate seed dormancy, and strategies for enhancing resistance against preharvest sprouting (PHS) in quinoa.

2. Model Systems: A Theoretical Framework for Quinoa Seed Dormancy, Hormone Signaling, and PHS

Seed dormancy is defined as a state in which seeds fail to germinate after receiving favorable environmental cues [12,13]. Dormancy classification is based on several factors including the developmental state of the embryo at the time of seed dispersal, physical characteristics of the seed, and physiological responses of seeds to environmental stimuli [14]. Historically, primary and secondary dormancy are most often used to describe differences in dormancy types for diverse plant species. Primary dormancy is characterized as dormancy induced on the mother plant during embryo maturation. Unlike primary dormancy, secondary dormancy is caused by the environment rather than inherited from the mother plant and occurs after seed maturation [15]. In addition to primary and secondary dormancy, other studies have revealed five additional subcategories of seed dormancy including physiological, morphological, morphophysiological, physical and combinational dormancy (Table 1) [14–18]. These categories, which may also influence primary and secondary dormancy, illustrate the nuanced and complex nature of dormancy. Seed dormancy and dormancy release are not binary processes but instead are regulated by a complex network of molecular, temporal, and physical cues, which ultimately result in germination [19,20]. The same is undoubtedly true for dormancy regulation in quinoa.

Table 1. A summary of recognized seed dormancy categories [15].

Dormancy Category	Description
Primary	Established during embryo maturation by the plant hormone abscisic acid (ABA) and prevent germination.
Secondary	Established in mature seeds by environment stimuli preventing germination.
Physiological	Physiological responses to environmental or hormonal stimuli that prevent germination.
Morphological	Fully differentiated embryos remain physically too small to carry out radicle/cotyledon emergence.
Morphophysiological	Physiological responses and physical limitations that prevent germination.
Physical	Specialized physical features of the seed that prevent germination.
Combinational	Specialized physical features and physiological responses that prevent germination.
Embryo	External or internal physical or biochemical signals that prevent embryo growth and germination.
Seed Coat-imposed	Imposed by a hard, impermeable seed coat requiring physical damage to induce germination.

Primary dormancy is regulated by the plant hormone abscisic acid (ABA) [19,20]. As a seed transitions from dormancy to germination changes in physiology are largely controlled by the plant hormones, ABA, and gibberellin (GA) [21,22]. The hormone balance theory suggests that after seed maturation, as ABA signaling decreases there is a corresponding increase in GA signaling that leads to germination [23–25]. Many dormancy studies using model systems and cereals have established a clear connection between ABA and GA signaling with seed dormancy and dormancy loss and provide a theoretical framework for dormancy regulation in quinoa. Specifically, these studies have demonstrated the following: (1) higher seed dormancy is associated with higher endogenous ABA levels, and increased gene expression of ABA biosynthesis genes 9-cis-epoxycarotenoid dioxygenase 1 and 2 (*NCED1* and *NCED2*), (2) ABA levels and/or sensitivity decline during dormancy loss with a corresponding increase in sensitivity to GA, (3) at physiological maturity, a lack of seed germination in dormant seeds is associated with GA-insensitivity, and germination in nondormant seeds is stimulated by GA, (4) as dormant seeds after-ripen, dormancy is lost in stages reflected by changes in sensitivity to ABA and GA, and (5) with after-ripening seed dormancy loss occurs with decreased ABA hormone levels due to an increased expression of ABA catabolic genes ABA8′-hydroxylase 1 and 2 (*ABA8′OH1* and *ABA8′OH2*) and increased GA signaling [20,23,26–43]. Similar investigations evaluating dormancy release have demonstrated the following: (1) after-ripening decreases ABA sensitivity and increases GA sensitivity through increased GA biosynthesis and hormone accumulation resulting from GA20-oxidase gene expression, (2) decreased expression of GA2-oxidase, a GA catabolism gene, occurs as dormancy declines, (3) the GA GID1 (GA-INSENSITIVE DWARF1) hormone receptor increases with after-ripening and (4) as dormancy is lost ABA hormone accumulation decreases through increased ABA catabolism [19–21,23,37,38,43–46].

Seed dormancy studies in the Amaranthaceae family, that of which quinoa belongs, suggest that different quinoa varieties proceed through a combination of primary and physiological dormancy, or they have no dormancy (Table 1) [15]. Furthermore, close weedy relative of quinoa, *Chenopodium album* (common lambsquarter), and *Chenopodium berlandieri* were previously described as having primary dormancy [47–49]. Although these studies provide important phenotypic clues about the underlying mechanisms of quinoa seed maturation, dormancy, and germination in a broad sense, they fail to evaluate directly if there are different types of seed dormancy across quinoa varieties. For these reasons, more efforts are needed to implement a unified platform for characterizing and cataloging differences in seed dormancy phenotypes across varieties in a similar fashion as was done to link agronomic characteristics with regional ecotype.

PHS is characterized by the germination of mature seeds on the mother plant due to rain or moisture prior to harvest (Figure 1) [50,51]. PHS is most often described through the lens of primary dormancy loss in model species and crops. Domestication and selective breeding for synchronized seedling emergence and stand establishment in many crops has resulted in decreased primary seed dormancy; seed dormancy mechanisms have been tailored to ensure maximum crop performance within the confines of a specific growing season [50,51]. PHS which results from altered primary dormancy is also known to occur through decreased ABA signaling and increased GA signaling [50]. The shift in the balance of hormone signaling leads to an extended seed germination window that is no longer synchronous with harvest [50–52]. It is important to note that both primary dormancy loss and the absence of seed dormancy increase the likelihood of PHS if rain occurs prior to harvest. However, the two physiological states are not synonymous and are likely regulated at the molecular level in different ways. Moreover, dormancy and PHS are complex traits that are regulated by multiple genes [50,51]. Therefore, to understand why quinoa is susceptible to PHS future studies in quinoa will need to evaluate whether PHS susceptibility is due to disruptions that change the architecture of primary dormancy and ABA and GA signaling or because quinoa seeds lack primary dormancy altogether.

Figure 1. Quinoa preharvest sprouting (PHS). Panels A-C show inflorescences severely impacted by PHS in breeding line '3964' from the 2015 and 2019 field seasons. Plants from the 2015 field season (**A**,**C**) were grown in Quilcene, WA on Dharma Ridge Farm. Plants from the 2019 field season (**B**) were grown in Skagit Valley, WA.

3. Orthodox and Unorthodox Seed Types

Seed dormancy is an evolutionary adaptation that ensures the species survival of natural catastrophes within a specific environment [38]. Dormancy type is also associated with seed type and is different for orthodox versus unorthodox seeds (Figure 2). Orthodox seeds often display primary dormancy, which is marked by six chronological phases of development, including the following: (1) embryo growth and differentiation; (2) seed expansion, reserve storage, and vacuole filling; (3) internal desiccation, organellar de-differentiation, and membrane stabilization; (4) metabolic quiescence; (5) imbibition, reserve mobilization, and resumption of metabolic responses to environmental cues; and (6) germination [52]. Orthodox seeds are desiccation tolerant (DT) remaining viable for long periods of time in low moisture conditions [52]. Environmental cues such as light, temperature, and moisture not only impact the depth of primary dormancy, but they also play an important role in modulating secondary dormancy characteristics and the length of time required for complete dormancy release [38].

Seeds that lack one of the six previously described developmental stages associated with primary dormancy are classified as unorthodox seeds [52]. In many plant species, and weeds, unorthodox or discontinuous dormancy ensures germination only in favorable conditions, and it confers environmental plasticity, or the ability to respond to changing biotic or abiotic environmental factors [53]. Desiccation sensitivity (DS) otherwise known as recalcitrance, and vivipary, are two characteristics of unorthodox seeds [52]. Neither DS nor viviparous seeds display primary dormancy and both lack the third step in development necessary for desiccation tolerance. As a result neither seed type survives in low-moisture environments, or through periods of dry storage or freezing [52]. DS seeds are often from tropical environments, and if there is adequate soil moisture, DS seeds germinate immediately after dispersion [52]. Rather than persist in the seedbank similar to DT/orthodox seeds, DS/unorthodox seeds that do not germinate immediately after dispersion die.

Elizabeth Farnsworth first suggested that quinoa produced "recalcitrant" or unorthodox seeds [52]. This characterization is based on the observation that many quinoa varieties that germinate at physiological maturity in wet environments, do not appear to survive as seeds in the soil, or they form stable seed banks. However, other quinoa germination studies have demonstrated that dormancy type and seed characteristics vary depending on quinoa variety, with some behaving as orthodox seeds with primary dormancy, while others do not [15,52]. It has also been hypothesized that seed desiccation status, such as primary and secondary dormancy, is the by-product of a plant's natural environment and selection pressure, which is directly tied to the maternal line [54,55]. Additionally, most studies report that quinoa seeds lose their viability in a short time, especially in conditions of high humidity and temperature [56]. Poor seed viability has been largely attributed to poor storage conditions and a lack of uniform storage conditions. Poor seed viability is itself characteristic of many unorthodox seeds and suggests the possibility that desiccation tolerance or insensitivity in quinoa has largely been under-characterized and is largely not understood. Some research has suggested that there is a correlation between desiccation sensitivity, the generation of reactive oxygen species (ROS), and the occurrence of oxidative damage during dehydration in the seed [57,58]. Furthermore, this research suggests that desiccation tolerance depends on the seeds ability to scavenge ROS compounds by antioxidant defense systems [22,57,59]. Interestingly, heat and drought resistance in adult quinoa plants is thought to occur, at least in part, through the mitigation of ROS, and it is associated with increased peroxisome proliferation [3]. If ROS scavenging pathways are also involved in the desiccation status of quinoa seeds, i.e., tolerant or susceptible, then glyoxysome proliferation may be an important indicator for selecting DT quinoa varieties.

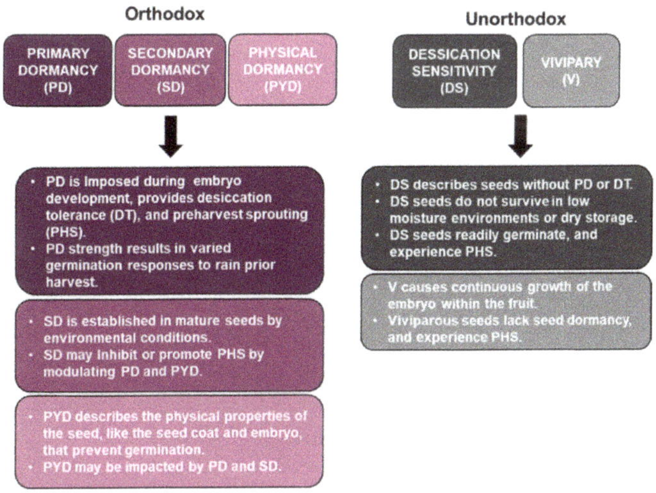

Figure 2. A summary of dormancy types associated with orthodox verses unorthodox seed types. Orthodox seeds display primary dormancy (PD), secondary dormancy (SD), and physical dormancy (PYD). Unorthodox seeds display desiccation sensitivity (DS) and vivipary (V). Specific dormancy programs and changes to these programs have implications for PHS susceptibility or resistance in quinoa.

Viviparous seeds are another group of unorthodox seeds, and they are common in monocot plant families such as Iridaceae (iris family) and Asparagaceae (asparagus family) [58]. Mangroves and corn are also two well-known examples of plants with viviparous seeds [60]. Similar to DS seeds, viviparous seeds have no primary dormancy, have a short viability window, and cannot survive dry storage or freezing [52]. Viviparous seeds also display PHS, although not all seeds that display PHS are viviparous [22,52,56]. Often in the literature, vivipary is used as a synonym for PHS due to the phenotypic similarities between both. However, the molecular architecture of the two seed physiologies is quite different, and to date there have been no studies to evaluate if quinoa PHS results from vivipary, lack of seed dormancy or both [52]. Simple germination screens have routinely been deployed to evaluate characteristic changes in hormone sensitivity associated with primary dormancy and dormancy loss, as well as the underlying causes of PHS physiology in many plant species. Thus, germination screening platforms will be essential for elucidating underlying mechanisms regulating PHS physiology across diverse quinoa germplasm.

4. Physical Dormancy in Quinoa

Physical dormancy is an additional subcategory of orthodox dormancy, involving the embryo, seed coat, or both [61–63]. Embryo dormancy is characterized by an external or internal physical or biochemical block that prevents embryo growth and germination [63]. Seed coat-imposed dormancy often occurs in seeds with a hard, impermeable shell that requires physical perforation or damage to germinate [63]. In some populations, it is possible that both embryo and seed coat-imposed dormancy play a role in quinoa germination programs and need to be carefully described to evaluate if and how each contributes to PHS.

In a recent study, seed coat thickness was measured in two varieties of quinoa, Chadmo and 2-Want, to evaluate if differences in observed dormancy occurred because of seed coat-imposed dormancy and in turn impacted PHS [64]. 2-Want had a thinner seed coat, while the Chadmo had a thicker seed coat. When the seed coats were perforated, both varieties continued to display a basal level of dormancy. Hormone analysis determined that seed coat thickness negatively correlated with the amount of endogenous ABA leached from the seed during development. Varieties with thinner seed coats leached more ABA and were less dormant than varieties with thicker seed coats. However, in both cases seed coat disruption did not completely alleviate dormancy. This finding suggested that in addition to seed coat-imposed dormancy, some quinoa varieties also have embryo dormancy [64]. This result is important because it suggests that at least in some quinoa varieties, vivipary is not a contributing factor in PHS sensitivity.

In addition to seed coat thickness, seed coat color has also been implicated in the regulation of quinoa seed dormancy and may be associated with ABA signaling mechanisms. The connection between seed coat color and dormancy regulation has been documented in other plant species including cereals such as wheat and barley, which are members of the Amaranthaceae family [48,49,65,66]. Comparisons between red versus white wheats indicate that red seed coat color is associated with stronger seed dormancy [50,67]. Additionally, a study evaluating the role of seed coat color in wheat found a quantitative trait locus (QTL) for PHS in the same location as coat color and suggested that coat color is likely to play an important role in PHS sensitivity or tolerance [68]. Interestingly, studies evaluating the link between seed coat color and dormancy depth in close relatives of quinoa, C. *album*, C. *berlandieri*, and C. *bonus-henricus*, observed that darker seed coat color was associated with stronger dormancy [48,61]. Furthermore, studies exploring differences in dormancy associated with heteromorphic populations of *C. album* found that brown seeds had thinner seed coats, and no primary dormancy. Whereas, black-colored seeds had thicker, and sometimes stronger seed coats, which are thought to act additively to enhance primary dormancy [49]. Likewise, seed coat studies evaluating dormancy in quinoa varieties such as Chadmo and Titicaca indicated that dormancy is stronger in seeds with a darker colored coat [64]. Another study, evaluating the *Suaeda salsa* in Amaranthaceae, found similar results [47]. Brown seeds had higher germination rates and absorbed water better than black seeds [47]. However, a key limitation of these studies is that they only compare dormancy associated with very dark (black or brown) and light seed coats (light brown) and do not evaluate the broader range of seed coat colors that exist in quinoa including those that are yellow, red or pink [69–73]. Therefore, future studies will need to examine the genetic connection between seed coat thickness, color, and ABA signaling mechanisms, as well as associated impacts on dormancy programs across a wide collection of quinoa cultivars.

5. Environmental Regulation of Quinoa Seed Dormancy

Secondary dormancy is regulated by the environment and is temporally separated from primary dormancy [38]. Environmental factors reported to enhance secondary dormancy in many species, including quinoa, are photoperiod or day length, temperature, precipitation, and altitude [2,3,6,12,71–75]. Environmental factors that influence the depth of secondary dormancy often do so through increased ABA signaling. Interestingly, previous studies investigating abiotic stress responses, specifically those connected to drought and salinity tolerance in adult quinoa plants also proceed through ABA signaling networks [76,77]. However, there is no research that has investigated whether the ABA signaling networks that contribute to abiotic stress tolerance in adult quinoa plants are also involved in modulating PHS physiology during seed germination. However, what is known is that as the photoperiod increases endogenous, ABA and sugar levels increase in quinoa seeds. Increases in endogenous ABA and sugars are important indicators of embryo dormancy and "ripening" [76–78]. Based on these findings, it was concluded that ABA and sugar signaling are possible mechanisms that regulate the photo-adaptability of adult quinoa cultivars [77]. Long photoperiods may also promote stronger dormancy by increasing ABA levels in seeds [52,78–82].

In addition to photoperiod, temperature plays a role in quinoa development, and it likely impacts seed dormancy and germination. In a study addressing how the environment can affect quinoa, two varieties Chadmo and 2-Want, were exposed to different temperatures and photoperiods to gauge germination and dormancy capacity [78]. The authors discovered that high temperatures and long photoperiod days increased dormancy. It was also determined that the temperature window has increased for quinoa, meaning that modern quinoa varieties have adapted to germinate in colder temperatures than their earlier relatives, and that growing environment is the biggest factor impacting seed dormancy [78].

The cultivation of quinoa across diverse regions, including in hot and dry climates, is likely to have contributed to the requirement for higher germination temperatures. However, understanding the relationship between germination rates and temperature is confounded by the fact that quinoa can germinate in a wide array of varying temperatures. One study reported maximum germination rates at 30 °C [78]. However, other studies have found that the optimum germination temperature is 37 °C [80]. A shift in recent breeding strategies to cultivate quinoa across very diverse environments has likely had a significant impact on plant and seed physiology [76]. The result of changes to breeding practices and environments has resulted in varieties with low dormancy but high adaptability [78]. These results suggest that growth and germination temperatures may have different effects on distinct varieties of quinoa. These results also suggest that these effects may be exacerbated by environments and may have major implications for variations in seed dormancy phenotypes depending on local growing conditions.

Precipitation and altitude also influence quinoa seed dormancy. Its widely known that quinoa is tolerant to drought and mildew, and these desirable characteristics are a direct result of the original growing habitats [3,54]. It is also known that rainfall prior to harvest when seeds are mature may result in PHS. The five original ecotypes come from very diverse environments, with precipitation varying from intense wet mountains to dry sandy regions [3–5,54]. It is important to mention this, because to date, although there have been some studies that have tested how different quinoa varieties respond to varied precipitation and drought treatments, none have directly investigated how precipitation timing near harvest maturity impacts PHS susceptibility.

Seed coat thickness and color, two factors that impact dormancy strength, may also be regulated by the environment. For example, a study evaluating the effect of elevation on seed coat thickness and rates of germination for quinoa's close relative *C. bonus-henricus* found that seeds grown at a lower elevation had thinner seed coats, and increased rates of germination [61]. If seed dormancy mechanisms are conserved in quinoa then it might be expected that varieties grown at higher elevations will have thicker seeds coats, slower rates of germination, and perhaps more seed dormancy than those grown at lower elevations. It is also good to note while looking at color differences, that seeds with thinner seed coats appear to be lighter in color, whereas thicker coats have darker coloration [64].

6. Breeding Strategies to Mitigate PHS in Quinoa

Research studies in cereals such as wheat and barley, and model plants like *Arabidopsis thaliana* has demonstrated a compelling connection between increased seed dormancy and PHS resistance [3,12,45,52,54,78,83,84]. In the cases of wheat and Arabidopsis, several major and minor quantitative trait loci (QTLs) associated with increased PHS resistance map to regions of the genome containing genes previously characterized as regulators of dormancy and seed coat color [50,68,85,86]. Therefore, breeding for stronger seed dormancy in quinoa seems a promising approach for reducing the risk of crop losses due to PHS. Two strategies currently being used to accomplish this task are 1) to make crosses between quinoa varieties displaying different levels of seed dormancy, and 2) to make crosses between quinoa and wild relatives, such as native lambsquarter (*C. berlandieri*) which is a tetraploid similar to quinoa and has more clearly defined seed dormancy. The primary goals of both approaches have been to create hybrid populations that allow for a better understanding of how seed dormancy and PHS are segregating within a population, and to create PHS resistant germplasm.

In the first strategy, breeders have used Titicaca, a cultivar developed in Demark with higher seed dormancy, and 'Chadmo, QQ065-PI 614880′, which is a naturally dormant variety originating from the Chiloe island in Chile, to create quinoa populations with increased seed dormancy and PHS resistance [3,64,78,82]. However, despite these efforts, incidences of PHS in both Titicaca and Chadmo have been reported with adequate rainfall prior to harvest, and across diverse growing regions (breeder listening sessions; International Quinoa Conference, 2020). Additionally, hybrid populations created at the Sustainable Seed Systems Laboratory at Washington State University using Chadmo also displayed frequent PHS in the higher rainfall zones of Western WA, despite initially appearing to be resistant (K. Murphy, personal communication). Taken together these results suggest that while many quinoa varieties appear to be susceptible to PHS, with some displaying some form of seed dormancy, there is not a clear connection between the type or level of seed dormancy with level of PHS susceptibility. These results also suggest that the mechanisms of regulation between dormancy and PHS in quinoa may not be analogous to those in cereals or other model plant species. However, with the recent sequencing of the quinoa genome, many genomics-assisted breeding approaches, including QTL analysis, and the molecular characterization of PHS-specific genes, are now possible [6,50].

The second approach developed to increase PHS resistance in quinoa is to introduce the desired dormancy type and level, i.e., primary and strong, by making a wide cross with a genetically compatible relative. The objective of this approach is to add desirable traits that are currently lacking in the existing genetic pool. Selective breeding approaches have been routinely implemented in other crops, using landraces or weedy relatives to increase physiological plasticity to abiotic stresses, and to increase disease resistance [87].

Although classified as an invasive weed species in the United States, *Chenopodium berlandieri*, also known as pitseed goosefoot, is grown as a seed crop in other parts of the world. *C. berlandieri* has emerged as a possible candidate for increasing PHS resistance in quinoa because (1) it is genetically compatible with quinoa and a cross between the two produces viable offspring, and (2) it displays strong orthodox primary seed dormancy, and this dormancy diminishes in a trackable manner over time that would also allow for simultaneous PHS sensitivity screening. Pitseed goosefoot is comprised of two subspecies, *berlandieri* and *nuttaliae*, and ecotypes can be found growing from southern Mexico and Texas, into southwestern and eastern North America, including along the coasts of the Atlantic and Gulf of Mexico [88]. Since pitseed goosefoot was a source of food for centuries in pre-European indigenous cultures in eastern North America, it is adapted to regions of the U.S. where quinoa struggles to grow. Pitseed goosefoot is a potential donor parent of key agronomic traits in quinoa, including heat tolerance, nutritional value, and resistance to PHS.

However, it is important to mention the possible risks associated with this approach, namely the development of germplasm that will establish a robust volunteer seedbank. From an agronomic perspective, two of the most favorable characteristics of existing quinoa varieties are that they germinate readily and do not survive desiccation, meaning they do not establish weedy seed banks. Hybrid populations generated from wide crosses will serve as essential tools for characterizing dormancy physiology and PHS regulation in quinoa. If used for variety development, future breeding efforts using wide crosses will need to strike a balance between increased seed dormancy and decreased germinability, so as not to trade PHS susceptibility for volunteer seedbank establishment.

In addition to the previously mentioned strategies aimed at increasing PHS resistance through increasing quinoa seed dormancy, another approach is to select for faster maturing varieties. The advantages to this "avoidance strategy" is that neither dormancy status nor PHS sensitivity are factors if harvest time is separated fall rains. However, the challenge with this approach, however, is that it does not address quality issues associated with dormancy, or PHS physiology, relying instead on optimum growing conditions and weather stability. In an age of increasing climate variability, it is unclear how current quinoa production regions might be impacted or how future production systems will need to be tailored to target optimal growing regions.

7. Other Tools for Mitigating PHS in Quinoa

Breeding for increased resistance to PHS in quinoa is a long-term endeavor taking years to produce new varieties. Given that very little information is understood about quinoa seed dormancy structure and PHS physiology, breeding strategies that rely heavily on the conventional wisdom established for PHS in cereals may be confounded by differences in biological and environmental factors that regulate seed maturation, germination, and viability in pseudocereals. Therefore, future research should also incorporate non-breeding, short and intermediate-term PHS mitigation strategies that help alleviate the risk of PHS in quinoa. Quinoa seed physiology may fall outside of an orthodox dormancy regime, with varieties displaying weak dormancy to no primary dormancy, DS, as well as a significant decline in seed viability post-harvest [89]. For these reasons, in addition to trying to increase PHS resistance through breeding for increased seed dormancy, it might also be necessary to minimize PHS risks by transiently modulating seed physiology from DS to DT in the field just prior to harvest.

One way to temporarily change seed physiology is by using growth regulators. Paclobutrazol (PAC) is a GA biosynthesis inhibitor historically used to understand the dynamics of GA and ABA signaling networks in dormant and germinating seeds [90]. Previous studies evaluating PAC treatments in quinoa have done so in adult plants and have focused on increasing drought and salt tolerance as well as yield [3,48,83]. In all cases PAC treatments were efficacious for increasing resistance to abiotic stress and for increasing yields and did so through an indirect increase in ABA signaling resulting from decreased GA signaling. Additional work investing the regulation of DS in the seeds of *Citrus limons*, a species without seed dormancy, found that treating freshly harvested seeds with PAC both slowed the rate of seed germination and extended the seed viability window from weeks to months, which are two phenotypes consistent with orthodox seed dormancy [52].

Expression analysis comparing non-treated and PAC-treated seeds demonstrated that in addition to inhibiting GA biosynthesis and signaling genes, PAC treatment also inhibited other phytohormone pathways associated with cell growth, including auxin and ethylene signaling, as well as causing an increase in both endogenous ABA hormone levels and signaling. Furthermore, studies evaluating the efficacy of ABA treatments for mitigating DS in the seeds of *Acer Saccharinum* indicated that ABA alone was able to rescue a DS phenotype through increased ABA-mediated signaling [90]. Analysis of total protein in ABA-treated seeds found that ABA-treated seeds accumulated higher amounts of Late Embryogenesis Associated (LEA) proteins which are associated with seed dormancy and desiccation tolerance [90]. Although PAC is a powerful research tool for understanding dormancy regulation, due to issues with toxicity, it is not suitable for large-scale production systems. However, unlike PAC, ABA is routinely and safely used to stimulate fruit ripening and enhance fruit color in grape production systems [91]. In a similar way, timed ABA treatments just before seed physiological maturity may prove to be powerful tool for enhancing PHS tolerance and increasing seed viability in quinoa through a transient increase in dormancy.

8. Conclusions

Over the last decade quinoa has emerged as a high-value, nutritious crop to enhance food security; it is artisanal (domestic interest), and it performs well in variable environments and in marginal soils. For these reasons, many research efforts have centered on understanding and improving traits related to abiotic stress tolerance, disease resistance, and yield. In many cases studies have focused on adult plants. Consequently, one largely overlooked area of research has been the study of the mechanisms that regulate quinoa seed dormancy. In the last few growing seasons, across the globe from Rwanda to the Pacific Northwest of the U.S.A., diverse quinoa varieties have been plagued by PHS, including those previously bred for enhanced dormancy. In some instances, untimely rains before harvest have resulted in nearly complete crop losses due to PHS. Therefore, the overall goal of this review paper is to inform breeders and non-breeders alike about the complex physiology leading to PHS in quinoa.

The strategies we believed to be key for understanding the dynamic nature of quinoa seed dormancy and PHS physiology include (1) the development of a high-throughput hormone screening pipeline, to quickly characterize the presence or absence of dormancy, and baseline dormancy strength at physiological maturity, (2) the implementation of a PHS screening platform modeled after those routinely used to evaluate PHS susceptibility in wheat, (3) the development of gene-specific primers to evaluate changes in genes associated with ABA and GA signaling, seed dormancy, PHS tolerance, and desiccation sensitivity/tolerance, and (4) assessing the effects of ABA and PAC treatments on dormancy preservation, seed germination rates, and desiccation tolerance in greenhouse and field trails. These efforts will provide a framework for developing new tools for understanding seed physiology in quinoa and mitigating PHS.

Author Contributions: E.M.M. and A.L.H. conceived and drafted the manuscript. K.M.M. and A.L.H. edited the manuscript, and K.M.M. contributed significantly to additional information added to the final manuscript draft. All authors have read and agreed to the published version of the manuscript.

Funding: This research was funded in part by the Orville A. Vogel Wheat Research Fund.

Acknowledgments: The authors would like to thank attending members of the 2020 International Quinoa Symposium for helpful discussions about issues that impact quinoa seed dormancy.

Conflicts of Interest: The authors declare no conflict of interest.

References

1. Morales, A.; Zurita-Silva, A.; Maldonado, J.; Silva, H. Transcriptional responses of chilean quinoa (*Chenopodium quinoa* Willd.) under water deficit conditions uncovers ABA-independent expression patterns. *Front. Plant Sci.* **2017**, *8*, 1–13. [CrossRef] [PubMed]
2. Albani, D.; Hammond-Kosack, M.C.U.; Smith, C.; Conlan, S.; Colot, V.; Holdsworth, M.; Bevan, M.W. The wheat transcriptional activator SPA: A seed-specific bZIP protein that recognizes the GCN4-like motif in the bifactorial endosperm box of prolamin genes. *Plant Cell* **1997**, *9*, 171–184.
3. Hinojosa, L.; González, J.A.; Barrios-Masias, F.H.; Fuentes, F.; Murphy, K.M. Quinoa abiotic stress responses: A review. *Plants* **2018**, *7*, 106. [CrossRef] [PubMed]
4. Tapia, M. The long journey of quinoa: Who wrote its history. In *State of the Art Report on Quinoa around the World*; Bazile, D., Bertero, H.D., Nieto, C., Eds.; FAO: Santiago, Chile; CIRAD: Montpellier, France, 2013; Volume 1, pp. 1–7. ISBN 978-92-5-108558-5.
5. Risi, J.; Galwey, N.W. The Chenopodium grains of the Andes: Inca crops for modern agriculture. *Adv. Appl. Biol.* **1984**, *10*, 145–216.
6. Jarvis, D.E.; Ho, Y.S.; Lightfoot, D.J.; Schmöckel, S.M.; Li, B.; Borm, T.J.A.; Ohyanagi, H.; Mineta, K.; Michell, C.T.; Saber, N.; et al. The genome of Chenopodium quinoa. *Nature* **2017**, *542*, 307–312. [CrossRef]
7. Sanchez, H.B.; Lemeur, R.; Damme, P.V.; Jacobsen, S.-E. Ecophysiological analysis of drought and salinity stress of quinoa (*Chenopodium Quinoa* willd.). *Food Rev. Int.* **2003**, *19*, 111–119. [CrossRef]

8. Jacobsen, S.-E.; Mujica, A.; Jensen, C.R. The resistance of quinoa (*Chenopodium quinoa* Willd.) to adverse abiotic factors. *Food Rev. Int.* **2003**, *19*, 99–109. [CrossRef]
9. Adolf, V.I.; Jacobsen, S.-E.; Shabala, S. Salt tolerance mechanisms in quinoa (*Chenopodium quinoa* Willd.). *Environ. Exp. Bot.* **2013**, *92*, 43–54. [CrossRef]
10. Ruiz-Carrasco, K.; Antognoni, F.; Coulibaly, A.K.; Lizardi, S.; Covarrubias, A.; Martínez, E.A.; Zurita-Silva, A. Variation in salinity tolerance of four lowland genotypes of quinoa (*Chenopodium quinoa* Willd.) as assessed by growth, physiological traits, and sodium transporter gene expression. *Plant Physiol. Biochem.* **2011**, *49*, 1333–1341. [CrossRef]
11. Walck, J.L.; Baskin, J.M.; Baskin, C.C.; Hidayati, S.N. Defining transient and persistent seed banks in species with pronounced seasonal dormancy and germination patterns. *Seed Sci. Res.* **2005**, *15*, 189–196. [CrossRef]
12. Arc, E.; Sechet, J.; Corbineau, F.; Rajjou, L.; Marion-Poll, A. ABA crosstalk with ethylene and nitric oxide in seed dormancy and germination. *Front. Plant Sci.* **2013**, *4*, 1–19. [CrossRef]
13. Bewley, J.D. Seed germination and dormancy. *Plant Cell* **1997**, *9*, 1055–1066. [CrossRef]
14. Willis, C.G.; Baskin, C.C.; Baskin, J.M.; Auld, J.R.; Venable, D.L.; Cavender-Bares, J.; Donohue Kathleen de Casas, R.R.; The NESCent Germination Working Group. The evolution of seed dormancy: Environmental cues, evolutionary hubs, and diversification of the seed plants. *New Phytolog.* **2014**, *203*, 300–309. [CrossRef] [PubMed]
15. Baskin, J.M.; Baskin, C.C. A classification system for seed dormancy. *Seed Sci. Res.* **2004**, *14*, 1–16. [CrossRef]
16. Nikolaeva, M.G. *Physiology of Deep Dormancy in Seeds*; Leningrad, Russia: Izdateljstvo 'Nauka' (in Russian). [Translated from Russian by Z. Shapiro (1969). National Science Foundation: Washington, DC, USA, 1967; p. 219.
17. Nikolaeva, M.G. On criteria to use in studies of seed evolution. *Seed Sci. Res.* **2004**, *14*, 315–320. [CrossRef]
18. Karssen, C.M. Uptake and effect of abscisic acid during induction and progress of radicle growth in seeds of Chenopodium album. *Physiolog. Plant.* **1976**, *36*, 259–263. [CrossRef]
19. Hauvermale, A.L.; Ariizumi, T.; Steber, C.M. Gibberellin signaling: A theme and variations on DELLA repression. *Plant Physiolog.* **2012**, *160*, 83–92. [CrossRef]
20. Tuttle, K.M.; Martinez, S.A.; Schramm, E.C.; Takebayashi, Y.; Seo, M.; Steber, C.M. Grain dormancy loss is associated with changes in ABA and GA sensitivity and hormone accumulation in bread wheat, *Triticum aestivum* L. *Seed Sci. Res.* **2015**, *25*, 179–193. [CrossRef]
21. Koornneef, M.; Jorna, M.L.; der Brinkhorst-van Swan, D.L.C.; Karssen, C.M. The isolation of abscisic acid (ABA) deficient mutants by selection of induced revertants in non-germinating gibberellin sensitive lines of *Arabidopsis thaliana* L. heynh. *Theor. Appl. Gen.* **1982**, *61*, 385–393. [CrossRef]
22. Finkelstein, R.R.; Reeves, W.; Ariizumi, T.; Steber, C.M. Molecular aspects of seed dormancy. *Annu. Rev. Plant Biol.* **2008**, *59*, 387–415. [CrossRef]
23. Karssen, C.M.; Lacka, E. A revision of the hormone balance theory of seed dormancy: Studies on gibberellin and/or abscisic acid-deficient mutants of Arabidopsis thaliana. *Plant Growth Sub.* **1985**, 315–323.
24. Field, R.J.; Jackson, D.I. A hormone balance theory of apical dominance. In *Bulletin of the Royal Society of New Zealand*; Springer: Berlin, Germany, 1974; pp. 655–657.
25. Luckwill, L. Growth-inhibiting and growth-promoting substances in relation to the dormancy and after-ripening of apple seeds. *J. Hort. Sci.* **1952**, *27*, 53–67. [CrossRef]
26. Walker-Simmons, M. ABA levels and sensitivity in developing wheat embryos of sprouting resistant and susceptible cultivars. *Plant Physiol.* **1987**, *84*, 61–66. [CrossRef]
27. Morris, C.F.; Moffatt, J.M.; Sears, R.G.; Paulsen, G.M. Seed dormancy and responses of caryopses, embryos, and calli to abscisic acid in wheat. *Plant Physiol.* **1989**, *90*, 643–647. [CrossRef] [PubMed]
28. Taylor, I.B.; Burbidge, A.; Thompson, A.J. Control of abscisic acid synthesis. *J. Exper. Bot.* **2000**, *51*, 1563–1574. [CrossRef]
29. Kushiro, T.; Okamoto, M.; Nakabayashi, K.; Yamagishi, K.; Kitamura, S.; Asami, T.; Hirai, N.; Koshiba, T.; Kamiya, Y.; Nambara, E. The Arabidopsis cytochrome P450 CYP707A encodes ABA 8′-hydroylases: Key enzymes in ABA catabolism. *Emb. J.* **2004**, *23*, 1647–1656. [CrossRef]
30. Chono, M.; Honda, I.; Shinoda, S.; Kushiro, T.; Kamiya, Y.; Nambara, E.; Kawakami, N.; Kaneko, S.; Watanabe, Y. Field studies on the regulation of abscisic acid content and germinability during grain development of barley: Molecular and chemical analysis of pre-harvest sprouting. *J. Exper. Bot.* **2006**, *57*, 2421–2434. [CrossRef] [PubMed]
31. Finch-Savage, W.E.; Leubner-Metzger, G. Seed dormancy and the control of germination. *New Phytolog.* **2006**, *171*, 501–523. [CrossRef]
32. Millar, A.A.; Jacobsen, J.V.; Ross, J.J.; Helliwell, C.A.; Poole, A.T.; Scofield, G.; Reid, J.B.; Gubler, F. Seed dormancy and ABA metabolism in arabidopsis and barley: The role of ABA 8′-hydroxylase. *Plant J.* **2006**, *45*, 942–954. [CrossRef]
33. Okamoto, M.; Kuwahara, A.; Seo, M.; Kushiro, T.; Asami, T.; Hirai, N.; Kamiya, Y.; Koshiba, T.; Nambara, E. CYP707A1 and CYP707A2, which encode abscisic acid 8′-hydroxylases, are indispensable for proper control of seed dormancy and germination in Arabidopsis. *Plant Physiol.* **2006**, *141*, 97–107. [CrossRef]
34. Barrero, J.M.; Talbot, M.J.; White, R.G.; Jacobsen, J.V.; Gubler, F. Anatomical and transcriptomic studies of the coleorhiza reveal the importance of this tissue in regulating dormancy in barley. *Plant Physiol.* **2009**, *150*, 1006–1021. [CrossRef]

35. Schramm, E.C.; Abellera, J.C.; Strader, L.C.; Campbell, K.G.; Steber, C.M. Isolation of ABA-responsive mutants in allohexaploid bread wheat (*Triticum aestivum* L.): Drawing connections to grain dormancy, preharvest sprouting, and drought tolerance. *Plant Sci.* **2010**, *179*, 620–629. [CrossRef]
36. Barrero, J.M.; Jacobsen, J.V.; Talbot, M.J.; White, R.G.; Swain, S.M.; Garvin, D.F.; Gubler, F. Grain dormancy and light quality effects on germination in the model grass *Brachypodium distachyon*. *New Phytol.* **2012**, *193*, 376–386. [CrossRef] [PubMed]
37. Ariizumi, T.; Hauvermale, A.L.; Nelson, S.K.; Hanada, A.; Yamaguchi, S.; Steber, C.M. Lifting DELLA repression of Arabidopsis seed germination by nonproteolytic gibberellin signaling. *Plant Physiol.* **2013**, *162*, 2125–2139. [CrossRef] [PubMed]
38. Bewley, J.D.; Bradford, K.J.; Hilhorst, H.W.M.; Nonogaki, H. *SEEDS: Physiology of Development, Germination and Dormancy*; Springer: New York, NY, USA, 2013.
39. Chono, M.; Matsunaka, H.; Seki, M.; Fujita, M.; Kiribuchi-Otobe, C.; Oda, S.; Kojima, H.; Kobayashi, D.; Kawakami, N. Isolation of a wheat (*Triticum aestivum* L.) mutant in ABA 8′-hydroxylase gene: Effect of reduced ABA catabolism on germination inhibition under field condition. *Breed. Sci.* **2013**, *63*, 104–115. [CrossRef] [PubMed]
40. Schramm, E.C.; Nelson, S.K.; Kidwell, K.K.; Steber, C.M. Increased ABA sensitivity results in higher seed dormancy in soft white spring wheat cultivar 'Zak'. *Theor. Appl. Gen.* **2013**, *126*, 791–803. [CrossRef]
41. Barrero, J.M.; Downie, A.B.; Xu, Q.; Gubler, F. A role for barley CRYPTOCHROME1 in light regulation of grain dormancy and germination. *Plant Cell* **2014**, *26*, 1094–1104. [CrossRef]
42. Hauvermale, A.L.; Tuttle, K.M.; Takebayashi, Y.; Seo, M.; Steber, C.M. Loss of *Arabidopsis thaliana* seed dormancy is associated with increased accumulation of the GID1 GA hormone receptors. *Plant Cell Physiol.* **2015**, *56*, 1773–1785. [CrossRef]
43. Koornneef, M.; van der Veen, J.H. Induction and analysis of gibberellin sensitive mutants in *Arabidopsis thaliana* L. heynh. *Theor. Appl. Gen.* **1980**, *58*, 257–263. [CrossRef]
44. Jacobsen, J.V.; Pearce, D.W.; Poole, A.T.; Pharis, R.P.; Mander, L.N. Abscisic acid, phaseic acid and gibberellin contents associated with dormancy and germination in barley. *Physiol. Plant.* **2002**, *115*, 429–441. [CrossRef]
45. Gualano, N.; Carrari, F.; Rodriguez, M.V.; Perez-Flores, L.; Sanchez, R.; Iusem NBenech-Arnold, R.L. Reduced embryo sensitivity to abscisic acid in a sprouting-susceptible sorghum (Sorghum bicolor) variety is associated with altered ABA signaling. *Seed Sci. Res.* **2007**, *17*, 81–90. [CrossRef]
46. Gubler, F.; Hughes, T.; Waterhouse, P.; Jacobsen, J. Regulation of dormancy in barley by blue light and after-ripening: Effects on abscisic acid and gibberellin metabolism. *Plant Physiol.* **2008**, *147*, 886–896. [CrossRef]
47. Li, W.; Liu, X.; Khan, M.A.; Yamaguchi, S. The effect of plant growth regulators, nitric oxide, nitrate, nitrite and light on the germination of dimorphic seeds of *Suaeda salsa* under saline conditions. *J. Plant Res.* **2005**, *118*, 207–214. [CrossRef] [PubMed]
48. Halwas, S.J. Domesticating Chenopodium: Applying Genetic Techniques and Archaeological Data to Understanding Pre-Contact Plant Use. Ph.D. Thesis, Departments of Biological Sciences and Anthropology, University of Manitoba, MB, Canada, 2017.
49. Yao, S.; Lan, H.; Zhang, F. Variation of seed heteromorphism in Chenopodium album and the effect of salinity stress on the descendants. *Annu. Bot.* **2010**, *105*, 1015–1025. [CrossRef]
50. Vetch, J.; Stougaard, R.N.; Martin, J.M.; Giroux, M.J. Review: Revealing the genetic mechanisms of pre-harvest sprouting in hexaploidy wheat (*Triticum aestivum* L.). *Plant Sci.* **2019**, *281*, 180–185. [CrossRef]
51. Benech-Arnold, R.L. Bases of preharvest sprouting resistance in barley: Physiology, molecular biology and environmental control of dormancy in the barley grain. In *Barley Science. Recent Advances from Molecular Biology to Agronomy of Yield and Quality*; Slafer, G.A., Molina-Cano, J.L., Savin, R., Araus, J.L., Romagosa, I., Eds.; Food Product Press: New York, NY, USA, 2001; pp. 481–502.
52. Farnsworth, E. The ecology and physiology of viviparous and recalcitrant seeds. *Annu. Rev. Ecol. Syst.* **2000**, *31*, 107–138. [CrossRef]
53. Baker, H.G. *Characteristics and Modes of Origin of Weeds. In The Genetics of Colonizing Species*; Academic Press: Cambridge, MA, USA, 1965; pp. 147–168.
54. Murphy, K.M.; Matanguihan, J.B.; Fuentes, F.F.; Gómez-Pando, L.R.; Jellen, E.N.; Maughan, P.J.; Jarvis, D.E. Quinoa breeding and genomics. *Plant Breed. Rev.* **2018**, *42*, 257–320.
55. Peroni, P.A. Field and laboratory investigations of seed dormancy in red maple (*Acer rubrum* L.) from the North Carolina piedmont. *For. Sci.* **1995**, *41*, 378–386.
56. Ellis, R.H.; Hong, T.D.; Roberts, E.H. A low-moisture-content to logarithmic relations between seed moisture content and longevity. *Annu. Bot.* **1988**, *61*, 405–408. [CrossRef]
57. Bailly, C. Active oxygen species and antioxidants in seed biology. *Seed Sci. Res.* **2004**, *14*, 93–10761. [CrossRef]
58. Kermode, A.R. Role of abscisic acid in seed dormancy. *J. Plant Growth Regul.* **2005**, *24*, 319–344. [CrossRef]
59. Berjak, P. Unifying perspectives of some mechanisms basic to desiccation tolerance across life forms. *Seed Sci. Res.* **2006**, *16*, 1–15. [CrossRef]
60. Nonogaki, M.; Nonogaki, H. Prevention of preharvest sprouting through hormone engineering and germination recovery by chemical biology. *Front. Plant Sci.* **2017**, *8*, 90. [CrossRef]
61. Dorne, A.J. Variation in seed germination inhibition of Chenopodium bonus-henricus in relation to altitude of plant growth. *Can. J. Bot.* **1981**, *59*, 1893–1901. [CrossRef]
62. Crocker, W.; Barton, L.V. *Physiology of Seeds*; Chronica Bot. Co.: Waltham, MA, USA, 1957; p. 267.
63. Heydecker, W. Glossary of terms. In *Seed Ecology*; Heydecker, W., Ed.; Butterworths: Oxford, UK; Pennsylvania State University Press: University Park, PA, USA, 1973; pp. 553–557.

64. Ceccato, D.; Bertero, D.; Batlla, D.; Galati, B. Structural aspects of dormancy in quinoa (*Chenopodium quinoa*): Importance and possible action mechanisms of the seed coat. *Seed Sci. Res.* **2015**, *25*, 267–275. [CrossRef]
65. Metzger, R.J.; Silbaugh, B.A. Location of genes for seed coat color in hexaploid wheat *Triticum aestivum* L. *Crop Sci.* **1970**, *10*, 495–496. [CrossRef]
66. Himi, E.; Yamashita, Y.; Haruyama, N.; Yanagisawa, T.; Maekawa, M.; Taketa, S. Ant28 gene for proanthocyanidin synthesis encoding the R2R3 MYB domain protein (Hvmyb10) highly affects grain dormancy in barley. *Euphytica* **2012**, *188*, 141–151. [CrossRef]
67. Himi, E.; Mares, D.J.; Yanagiswa, A.; Noda, K. Effect of grain colour gene (R) on grain dormancy and sensitivity of the embryo to abscisic acid (ABA) in wheat. *J. Exp. Bot.* **2002**, *53*, 1569–1574. [CrossRef] [PubMed]
68. Lin, M.; Cai, S.; Wang, S.; Liu, S.; Zhang, G.; Bai, G. Genotyping-by-sequencing (GBS) identified SNP tightly linked to QTL for preharvest sprouting resistance. *Theor. Appl. Gen.* **2015**, *128*, 1385–1395. [CrossRef] [PubMed]
69. Jacques, R. Action de la lumiere par l'intermediaire du phytochrome sur la germination, la croissance et le developpement de *Chenopodium polyspermum* L. *Physiol. Veg.* **1968**, *6*, 137–164.
70. Karssen, C.M. The light promoted germination of the seeds of *Chenopodium album* L. III. Effect of the photoperiod during growth and development of the plants on the dormancy of the produced seeds. *Acta Bot. Neerl.* **1970**, *19*, 81–94. [CrossRef]
71. Pourrat, Y.; Jacques, R. The influence of photoperiodic conditions received by the mother plant on morphological and physiological characteristics of *Chenopodium polyspermum* L. seeds. *Plant Sci. Lett.* **1975**, *4*, 273–279. [CrossRef]
72. Jacobsen, S.E.; Bach, A.P. The influence of temperature on seed germination rate in quinoa (*Chenopodium quinoa* Willd). *Seed Sci. Tech.* **1998**, *26*, 515–523.
73. Bois, J.F.; Winkel, T.; Lhomme, J.P.; Raffaillac, J.P.; Rocheteau, A. Response of some Andean cultivars of quinoa (*Chenopodium quinoa* Willd.) to temperature: Effects on germination, phenology, growth and freezing. *Eur. J. Agron.* **2006**, *25*, 299–308. [CrossRef]
74. Präger, A.; Munz, S.; Nkebiwe, P.M.; Mast, B.; Graeff-Hönninger, S. Yield and quality characteristics of different quinoa (*Chenopodium quinoa* willd.) cultivars grown under field conditions in southwestern Germany. *Agronomy* **2018**, *8*, 197. [CrossRef]
75. Williams, J.T.; Harper, J.L. Seed polymorphism and germination: I. The influence of nitrates and low temperatures on the germination of Chenopodium album. *Weed Res.* **1965**, *5*, 141–150. [CrossRef]
76. Benech-Arnold, R.L.; Fenner, M.; Edwards, P.J. Changes in germinability, ABA content and ABA embryonic sensitivity in developing seeds of *Sorghum bicolor* L. Moench. induced by water stress during grain filling. *New Phytol.* **1991**, *118*, 339–347. [CrossRef]
77. Marques, A.; Nijveen, H.; Somi, C.; Ligterink, W.; Hilhorst, H. Induction of desiccation tolerance in desiccation sensitive Citrus limon seeds. *J. Integr. Plant Biol.* **2019**, *61*, 624–638. [CrossRef]
78. Bendevis, M.A.; Sun, Y.; Shabala, S.; Rosenqvist, E.; Liu, F.; Jacobsen, S.E. Differentiation of photoperiod-induced ABA and soluble sugar responses of two quinoa (*Chenopodium quinoa* Willd.) cultivars. *J. Plant Growth Red.* **2014**, *33*, 562–570. [CrossRef]
79. Ceccato, D.V.; Bertero, H.D.; Batlla, D. Environmental control of dormancy in quinoa (*Chenopodium quinoa*) seeds: Two potential genetic resources for preharvest sprouting tolerance. *Seed Sci. Res.* **2011**, *21*, 133–141. [CrossRef]
80. Gualano NABenech-Arnold, R.L. The effect of water and nitrogen availability during grain filling on the timing of dormancy release in malting barley crops. *Euphytica* **2009**, *168*, 291–301. [CrossRef]
81. Benech-Arnold, R.L. Inception, maintenance and termination of dormancy in grain crops: Physiology, genetics and environmental control. In *Handbook of Seed Physiology: Applications to Agriculture*; Benech-Arnold, R.L., Sanchez, R.A., Eds.; The Haworth Press: New York, NY, 2004; pp. 169–198.
82. Gubler, F.; Millar, A.A.; Jacobsen, J.V. Dormancy release, ABA and preharvest sprouting. *Plant Biol.* **2005**, *8*, 183–187.
83. Gómez, M.B.; Castro, P.A.; Mignone, C.; Bertero, H.D. Can yield potential be increased by manipulation of reproductive partitioning in quinoa (*Chenopodium quinoa*)? Evidence from gibberellic acid synthesis inhibition using Paclobutrazol. *Funct. Plant Biol.* **2011**, *38*, 420–430. [CrossRef]
84. Baskin, C.C.; Baskin, J.M. *Seeds—Ecology, Biogeography, and Evolution of Dormancy and Germination*; Academic Press: San Diego, CA, USA, 1998.
85. Martinez, S.A.; Godoy, J.; Huang, M.; Zhang, Z.; Carter, A.H.; Garland Campbell, K.A.; Steber, C.M. Genome-wide association mapping for tolerance to preharvest sprouting and low falling numbers in wheat. *Front. Plant Sci.* **2018**, *9*, 1–16. [CrossRef] [PubMed]
86. Fofana, B.; Humphreys, D.G.; Rasul, G.; Cloutier, S.; Brûlé-Babel, A.; Woods, S.; Somers, D.J. Mapping quantitative trait loci controlling preharvest sprouting resistance in a red × white seeded spring wheat cross. *Euphytica* **2009**, *165*, 509–521. [CrossRef]
87. Hauvermale, A.L.; Sanad, M.N.M.E. Phenological plasticity of wild and cultivated plants. In *Tech Open. Plant Phenology*; IntechOpen Limited: London, UK, 2019; ISBN 978-953-51-5871-4.
88. Jellen, E.N.; Jarvis, D.E.; Hunt, S.P.; Mangelsen, H.H.; Maughan, P.J. New seed collections of North American pitseed goosefoot (*Chenopdium berlandieri*) and efforts to identify its diploid ancestors through whole-genome sequencing. *Int. J. Agric. Nat. Res.* **2019**, *46*, 187–196.
89. Souza, F.F.D.J.; Devilla, I.A.; de Souza, R.T.; Teixeria, I.R.; Spehar, C.R. Physiological quality of quinoa seeds submitted to different storage conditions. *Afr. J. Agric. Res.* **2016**, *11*, 1299–1308.

90. Beadmoore, T.; Whittle, C.A. Induction of tolerance to desiccation and cryopreservation in silver maple (*Acer saccharinum*) embryonic axes. *Tree Physiol.* **2005**, *25*, 965–972. [CrossRef]
91. Pilati, S.; Bagagli, G.; Sonego, P.; Moretto, M.; Brazzale, D.; Castorina, G.; Simoni, L.; Tonelli, C.; Guella, G.; Engelen, K.; et al. Abscisic acid is a major regulator of grape berry ripening onset: New insights into ABA signaling. *Front. Plant Sci.* **2017**, *21*, 1–16. [CrossRef]

plants

MDPI

Article

A Long Journey of CICA-17 Quinoa Variety to Salinity Conditions in Egypt: Mineral Concentration in the Seeds

Juan A. González [1,*], Leonardo Hinojosa [2], María I. Mercado [3], José-Luis Fernández-Turiel [4], Didier Bazile [5,6], Graciela I. Ponessa [3], Sayed Eisa [7], Daniela A. González [8], Marta Rejas [4], Sayed Hussin [7], Emad H. Abd El-Samad [9], Ahmed Abdel-Ati [10] and Mohamed E. A. Ebrahim [11]

1 Fundación Miguel Lillo, Instituto de Ecología, Comportamiento y Conservación, T4000 Tucumán, Argentina
2 Institute for Biodiversity and Ecosystem Dynamics (IBED), University of Amsterdam, 1012 WX Amsterdam, The Netherlands; l.a.hinojosasanchez2@uva.nl
3 Fundación Miguel Lillo, Instituto de Morfología Vegetal, T4000 Tucumán, Argentina; mimercado@lillo.org.ar (M.I.M.); giponessa@lillo.org.ar (G.I.P.)
4 Geosciences Barcelona, CSIC, 08028 Barcelona, Spain; jlfernandez@geo3bcn.csic.es (J.-L.F.-T.); mrejas@geo3bcn.csic.es (M.R.)
5 CIRAD, UMR SENS, 34398 Montpellier, France; didier.bazile@cirad.fr
6 SENS, CIRAD, IRD, University Paul Valery Montpellier 3, 34090 Montpellier, France
7 Faculty of Agriculture, Ain Shams University, Cairo 11672, Egypt; sayed_eisa@hotmail.com (S.E.); sayed_hussin@hotmail.com (S.H.)
8 Instituto de Bioprospección y Fisiología Vegetal (INBIOFIV), Consejo Nacional de Investigaciones Científicas y Técnicas (CONICET), T4000 Tucumán, Argentina; danigonz37@gmail.com
9 Vegetable Research Department, Agricultural & Biological Research Division, National Research Centre, Giza 12611, Egypt; emadhassanein@hotmail.com
10 Plant Production Department, Ecology and Dry Land Agriculture Division, Desert Research Center, Cairo 11753, Egypt; dr.abdelati@gmail.com
11 General Organization for Agriculture Equalization Fund, Giza 12511, Egypt; mohamedelsawalhy@yahoo.com
* Correspondence: jagonzalez@lillo.org.ar

Citation: González, J.A.; Hinojosa, L.; Mercado, M.I.; Fernández-Turiel, J.-L.; Bazile, D.; Ponessa, G.I.; Eisa, S.; González, D.A.; Rejas, M.; Hussin, S.; et al. A Long Journey of CICA-17 Quinoa Variety to Salinity Conditions in Egypt: Mineral Concentration in the Seeds. *Plants* **2021**, *10*, 407. https://doi.org/10.3390/plants10020407

Academic Editor: Adriano Sofo

Received: 15 January 2021
Accepted: 16 February 2021
Published: 22 February 2021

Abstract: Quinoa may be a promising alternative solution for arid regions, and it is necessary to test yield and mineral accumulation in grains under different soil types. Field experiments with *Chenopodium quinoa* (cv. CICA-17) were performed in Egypt in non-saline (electrical conductivity, 1.9 dS m^{-1}) and saline (20 dS m^{-1}) soils. Thirty-four chemical elements were studied in these crops. Results show different yields and mineral accumulations in the grains. Potassium (K), P, Mg, Ca, Na, Mn, and Fe are the main elements occurring in the quinoa grains, but their concentrations change between both soil types. Besides, soil salinity induced changes in the mineral pattern distribution among the different grain organs. Sodium was detected in the pericarp but not in other tissues. Pericarp structure may be a shield to prevent sodium entry to the underlying tissues but not for chloride, increasing its content in saline conditions. Under saline conditions, yield decreased to near 47%, and grain sizes greater than 1.68 mm were unfavored. Quinoa may serve as a complementary crop in the marginal lands of Egypt. It has an excellent nutrition perspective due to its mineral content and has a high potential to adapt to semi-arid and arid environments.

Keywords: *Chenopodium quinoa* Wild.; salinity; mineral concentration; food; extreme environment

1. Introduction

Climate change is a reality, and we already see today its effects on the physiology, growth, and yield of field crops. For instance, the frequency of heatwaves has increased in large areas of the world, and precipitation changes have become more unpredictable [1]. Besides, the climate change effects and the bad agronomic practices have increased the saline soil areas. Salinity limits crop yields due to a reduction in photosynthesis, respiration, and protein synthesis. Around 7% of all land area in the world (1000 million ha) is affected by soil salinity, and more than 77 million ha from the arable area are affected by high salt contents [2,3].

Nevertheless, the main problem is that the principal crops are using plant species adapted to "old climatic conditions". Hence, it is necessary to look for alternative crops or "new crops" to face the "new climatic conditions". In this sense, it is crucial to consider some species that grew during millennia in mountain regions under extreme environmental conditions. Mountain plants, especially those adapted and cultivated in different altitudinal levels, can be crucial due to the gene pool that allowed these adaptations. In this scenario, quinoa (*Chenopodium quinoa* Willd.), which has grown throughout the Andes in South America for 5000 to 7000 years [4,5], can be considered a good option. During a long period of cultivation by the Aymaras and Inca populations, this crop was grown in different ecological zones, from sea level, in Chilean varieties [6], to 2000 to 4000 m above sea level (a.s.l.) along the Andes. Quinoa presents a C3 photosynthetic pathway according to anatomic and carbon isotope discrimination studies [6], with high photosynthetic assimilation and an intrinsic water use efficiency (iWUE) [7,8].

Several studies confirmed quinoa as an important source of nutritional components such as essential amino acids, fatty acids, minerals, soluble sugars, and bioactive components [9–11]. Furthermore, numerous reports in the field or lab conditions showed that quinoa is a species with high resilience to abiotic stress, including salinity, drought, high temperature, and ultraviolet B (UV-B) radiation [12–14]. Quinoa can tolerate very high salinity concentrations, producing a complete life cycle even at water salinities of 500–750 mM NaCl [15–17]. Thus, it can be grown in very marginal environments, for example, in North Africa, where soil salinization and drought are serious issues. The high nutritional value maintenance under different stresses makes quinoa an excellent crop to grow in the aforementioned marginal environments and face climate change. This quinoa tolerance to edaphic and harsh climatic conditions is related to this crop's high diversity along the Andes. In effect, there are more than 16,000 quinoa accessions stored in different seed banks in 30 countries, most of which are concentrated in Bolivia and Peru [18–20]. These accessions include the five ecotypes classified by Tapia (2015) [21]: (i) Valley quinoa; (ii) Altiplano quinoa; (iii) Salar quinoa; (iv) Sea level quinoa; and (v) Subtropical quinoa. Quinoa accessions of the different ecotypes are considered multipurpose plants: the seeds and leaves can be used as food, the biomass can be used as animal feed or as a cover crop, the colorants and the saponin content can be used in pharmaceutical and agroindustry, and plantings can serve as a phytoremediation tool for environmental cleanup [22–24].

Tapia's classification accepts an implicit fact: each ecotype can thrive in the environment in which it was adapted. However, quinoa has been introduced at higher latitudes as a complementary crop with good adaptation [25]. Currently, quinoa is cultivated and experimented on in almost 130 countries [19], including The United States [26,27], India [28], Italy [29], and Egypt [16], among others. Quinoa adaptation's success is due to its high plasticity to reach places that differ from its original location managing the sowing dates, taking advantage of the environmental offer (basically temperature and light). One of the perfect examples of quinoa plasticity is the CICA-17 variety. It was selected at 3800 m a.s.l. in Cuzco-Peru from the local variety Amarilla de Maranganí at Centro de Investigaciones de Cultivos Andinos (CICA, Universidad Nacional de San Antonio Abad del Cusco, Peru). CICA-17 belongs to the Altiplano ecotype, and it is tolerant to cold temperatures, low precipitation, and high salinity conditions. CICA-17 quinoa was introduced in northwestern Argentina in 1996–1998 from the American and European Test of Quinoa conducted by FAO-CIP [30]. Nowadays, CICA-17 is the variety most used by small producers in Northwest Argentina and especially in arid high mountain valleys (above 2000 m a.s.l.) where the climate is desert type. This variety has been cultivated in Egypt for ten years because of its good adaptation to its marginal places. Egypt has a considerable extension in arid, semi-arid, and marginal lands that constrain classical crop productivity. In this scenario, quinoa is becoming a complementary crop of high nutritional value. CICA-17 has a notable yield (near 2000 kg ha^{-1}) either in mountain valleys at 2000 m a.s.l. as in lowlands at 200 m a.s.l. in northwestern Argentina and Egypt [7,31,32]. Eisa et al. (2018) [33] and Ebrahim et al. [34] showed that the CICA-17 yield varied between 2000 and 3000 kg ha^{-1} in a marginal land at El-Fayoum oasis (Egypt).

CICA-17 quinoa can provide a new complementary crop for dry-saline lands. Still, it makes it necessary to study the mineral concentrations in different grain and seed organs, especially if different soil salinities influence these elements. Often, the mineral study in quinoa was focused on Na and K because of their relationship with the osmotic adjustment mechanisms that halophytes exhibit [35] or on the presence of a few minerals in different quinoa seed organs, and the abrasion effect on the Ca and K pericarp content [36]. All these approaches are essential to understand the physiological behavior of this promising species. It is also relevant to understand how these mineral contents can vary in different soils and climatic conditions in field conditions if quinoa is used as food in marginal lands. Nowadays, it is known that quinoa's seeds and leaves are a significant source of major minerals (calcium, magnesium, potassium, phosphorus, sulfur, and sodium), trace elements (iron, cobalt, zinc, copper, and manganese), and ultratrace elements (chromium, lithium, arsenic, nickel, molybdenum, selenium, tin, and vanadium) and that their content varies according to the genotype and the place where the cultivation is carried out [33,36–38]. However, the detailed mineral composition in different quinoa ecotypes is still scarce and even more so is their spatial distribution. For example, Prado et al. [37] reported 18 minerals present in quinoa grains, while Konishi et al. [36] mentioned only six minerals. Hence, we investigated the grain yield and size, the occurrence and content of minerals and their spatial distribution in the different grain tissues, in the CICA-17 quinoa crop grown in field conditions on soils with different salinity in the marginal lands of the Egypt.

2. Results

2.1. Soil and Irrigation Water Analysis

Electrical conductivity (EC) and organic matter, Cl, Na, Mg, K, Ca, SO_4, and Fe contents are higher in saline soil (Table 1). These parameters increase 43, 24, 17, 9, 2.6, 4, and 1.3%, respectively, compared to the non-saline soil. Table 2 summarizes the water irrigation analysis for both cases.

Table 1. Physical and chemical properties of soil samples collected in two saline and non-saline locations of Egypt.

					Physical Properties					
depth Soil cm	**pH**	**Sp [1] %**	**EC [2] dS m^{-1}**	**DB [3] g cm^{-3}**	**OM [4] %**	**CO_3Ca %**	**Sand %**	**Silt %**	**Clay %**	**Textural class**
0–60 (non-saline)	8.0	66	2.1	1.51	1.25	2.8	30.0	58.0	12.0	silt loam
0–60 (saline)	8.2	75	26.0	1.92	0.58	0.3	33.3	53.2	3.5	silt loam
					Chemical Properties					
	Na	**K**	**Ca**	**Mg**	**Cl**	**$SO_4{}^2$ meq L^{-1}**	**HCO^{-3}**	**$CO_3{}^{-2}$**	**Fe**	**Mn**
0–60 (non-saline)	7.8	0.6	11.7	4.3	6.5	13.5	4.45	0	4.36	5.29
0–60 (saline)	187.5	6.0	47.0	76.8	280.0	35	1.26	0	5.78	2.44

[1] Sp: Saturation percentage; [2] EC: electrical conductivity; [3] DB: bulk density; [4] OM: Organic matter. Each value is the mean of n = 6 samples.

Table 2. Physical and chemical properties of irrigation water of saline and non-saline locations.

Location	EC dS m^{-1}	pH	Na^+	K^+	Ca^{2+}	Mg^{2+}	Cl^- meq L^{-1}	HCO^{-3}	$CO_3{}^{-2}$	$SO_4{}^{-2}$
Non-saline	0.43	7.16	1.04	0.21	1.40	1.22	1.02	0.62	0.00	1.29
Saline	1.70	7.23	6.37	0.35	3.40	4.50	8.50	4.60	0.00	6.12

2.2. Grain Yield

Soil salinity negatively affected grain yield. A reduction of close to 47% concerning the grain yield was obtained in saline conditions compared to the non-saline one (Figure 1).

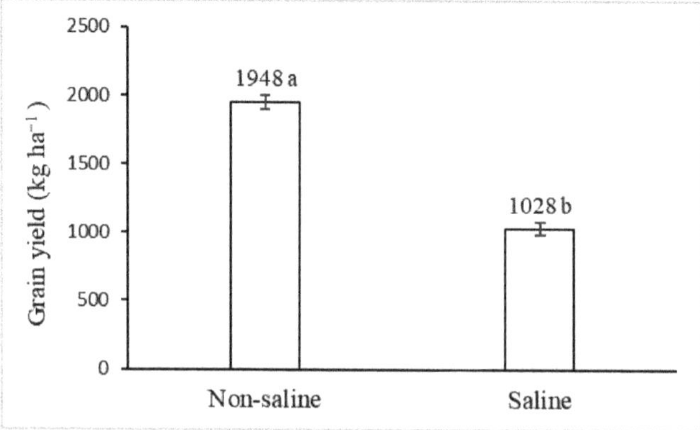

Figure 1. Effect of non-saline and saline soils on grain yield (mean ± SE) on *C. quinoa* cv. CICA-17. Different letters above columns indicate significant differences between means at $p < 0.05$, according to the *t*-test.

2.3. Grain Weight and Sizes

The weight of 1000 seeds of CICA-17 quinoa decreased by 13% in saline conditions compared to non-saline conditions (Figure 2). On the other hand, the grain size distribution (A: <1.41 – >1.0 mm; B: <1.68 – >1.41 mm; C: <2.00 – >1.68 mm; D: ≥2 mm) showed that the two largest grain sizes were unfavored by salinity conditions (Figure 3).

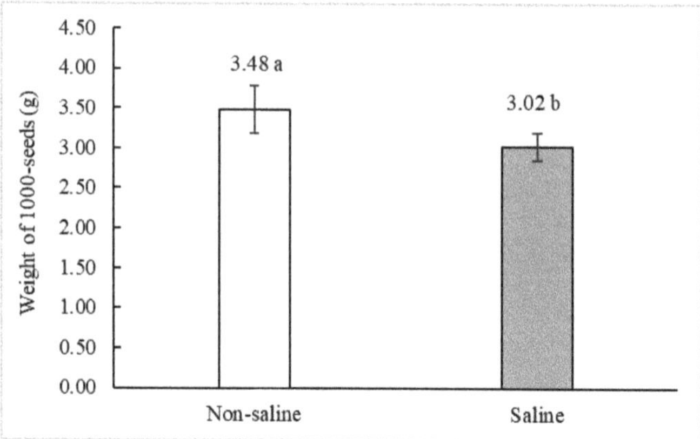

Figure 2. Effects of non-saline and saline soils on the weight (mean ± SE) of 1000-seeds of *C. quinoa* cv. CICA-17 plants. Significant differences between means ($p \leq 0.05$) are indicated by different letters above columns according to the *t*-test.

2.4. Mineral Content in Quinoa Grains

The most important mineral elements with biological activities showed two trends. While phosphorus (P), magnesium (Mg), and sodium (Na) increased their content under saline conditions, silicon (Si), potassium (K), calcium (Ca), and iron (Fe) contents decreased (Table 3). The same increase or decrease were detected for the other elements except for Mn, Cu, and Co, which exhibited the same behavior under non-saline and saline conditions.

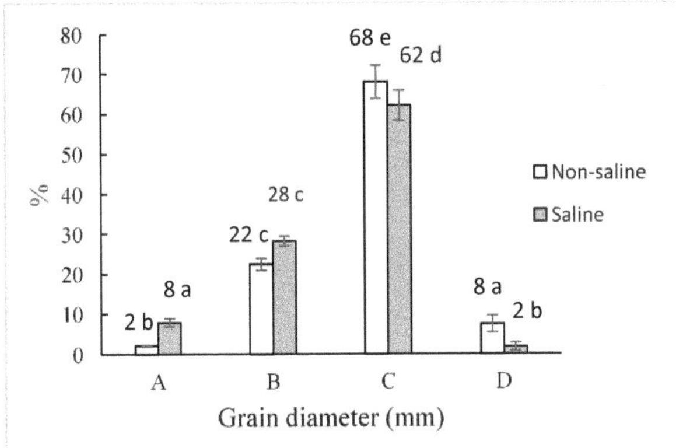

Figure 3. Seeds diameter distribution of *C. quinoa* cv. CICA-17 obtained in non-saline and saline soils. All the differences were significant ($p \leq 0.05$) according to the *t*-test. A: <1.41 – >1.0 mm; B: <1.68 – >1.41 mm; C: <2.00 – >1.68 mm; D: ≥2 mm.

Table 3. Mineral content in quinoa grains (pericarps + seeds) under two saline conditions. Significant differences between means ($p \leq 0.05$) are indicated by different letters behind the values according to the *t*-test.

Element	Non Saline Soil (mg kg^{-1} dry weight)	Saline Soil (mg kg^{-1} dry weight)	Difference (%)	Element	Non Saline Soil (mg kg^{-1} dry weight)	Saline Soil (mg kg^{-1} dry weight)	Difference (%)
	Major elements			Zr *	0.50 a	0.18 b	64.6
K	9707.62 a	8226.40 b	15.3	Ni	0.48 b	0.30 a	37.3
P	3334.57 b	3959.37 a	18.7	Pb *	0.46 a	0.11 b	75.8
Mg	1443.81 b	1690.27 a	17.1	V *	0.29 a	0.11 b	61.1
Ca	678.22 a	447.04 b	34.1	Ce *	0.09 a	0.05 b	41.8
Na	44.17 b	267.00 a	504.5	As	0.06 a	0.03 b	50.5
	Minor or trace elements			Ga *	0.06 a	0.02 b	64.9
Fe	72.82 a	49.92 b	31.4	Sn	0.05 a	0.04 b	20.1
Zn *	27.56 a	8.53 b	69.1	Ge *	0.05 a	0.02 b	58.1
Mn	15.30 a	16.19 a	5.9	La *	0.04 a	0.03 b	40.5
Cu	6.70 a	6.34 a	5.3	Li	0.04 b	0.05 a	19.1
Co	0.06 a	0.06 a	2.8	Nb *	0.04 a	0.03 b	34.2
	Ultratrace elements			Y *	0.04 a	0.02 b	39.0
Si	9968.95 a	7837.06 b	21.4	Nd *	0.04 a	0.03 b	322.0
Al	36.85 a	26.72 b	27.5	Cr	0.29 a	0.16 b	45.3
Ti *	8.61 a	4.18 b	51.5	Pr *	0.01 a	0.01 b	33.5
Sr	3.99 b	7.71 a	92.9	Th *	0.01 a	0.01 b	26.6
Ba *	2.07 a	0.53 b	74.1	Sm *	0.01 a	0.01 b	35.8
Rb *	0.96 b	1.42 a	47.2				

* Element detected for the first time in quinoa.

2.5. Mineral Spatial Distribution on Quinoa Grains

Quinoa seed has different tissues (Figure 4). Using SEM-EDX analysis, we found that only the pericarp accumulated sodium in both non-saline and saline soils. Grains developed under saline conditions showed 6.3 times greater sodium content than those detected under non-saline conditions (Table 4, Figure 5). Chlorine, P, and Br also increased significantly in pericarp in saline conditions. By contrast, soil salinity led to a decrease in Ca, Al, Fe, Cu, and Si. Meanwhile, potassium was located mainly in the pericarp without significant differences due to the soil salinity condition. In the endosperm and the perisperm, the K content increased in saline soil (2.5 and 2.2 times, respectively) while its concentration was more significant (2.7 times) in the embryo in non-saline soil. The other elements studied did not change significantly.

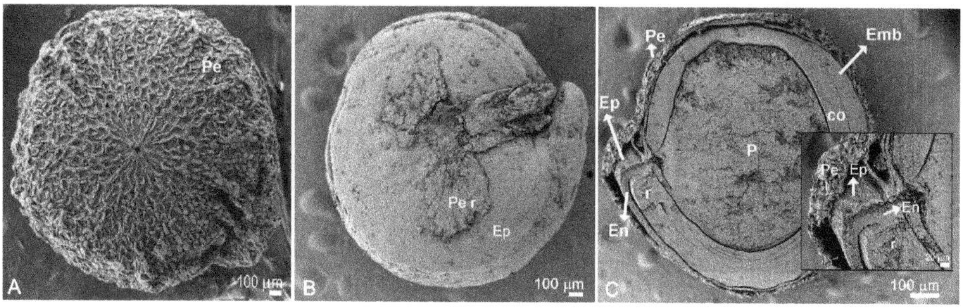

Figure 4. SEM photographs of *C. quinoa* cv. CICA-17. (**A**) grain external view; (**B**) seed with manually removed epicarp; and (**C**) longitudinal medial grain section. Pe, pericarp; Pe r, pericarp partially removed; Ep, episperm; En, endosperm; P, perisperm; Emb, embryo with a radicle-hypocotyl axis (r) and cotyledon (co).

Table 4. SEM-EDX comparative mineral relative percentages on quinoa grain tissues cultivated under non-saline and saline conditions (mean ± SD; n.d., not detected). Significant differences between means ($p \leq 0.05$) are indicated by different letters behind the values according to the *t*-test. Comparisons were made between identical tissues under the two salinity conditions.

Mineral	Pericarp		Embryo Cotyledon Mesophyll		Endosperm		Perisperm	
	Non-Saline	Saline	Non-Saline	Saline	Non-Saline	Saline	Non-Saline	Saline
C	51.89 ± 3.2 a	50.5 ± 3.7 a	61.32 ± 5.70 a	57.37 ± 2.40 a	61.64 ± 1.40 a	66.17 ± 5.3 a	68.60 ± 2.7 a	74.23 ± 7.10 a
O	39.96 ± 1.5 a	41.4 ± 1.00 a	26.51 ± 5.00 a	30.70 ± 1.40 a	25.76 ± 1.40 a	22.41 ± 3.6 a	26.30 ± 2.1 a	23.89 ± 4.90 a
N	n.d.	n.d.	9.75 ± 0.30 a	10.53 ± 0.10 a	11.46 ± 0.05 a	7.31 ± 1.3 b	4.97 ± 0.01 a	2.50 ± 0.70 b
Na	0.08 ± 0.04 b	0.5 ± 0.03 a	n.d.	n.d.	n.d.	N.d.	n.d.	n.d.
Mg	0.23 ± 0.10 a	0.34 ± 0.10 a	0.38 ± 0.07 a	0.39 ± 0.04 a	0.16 ± 0.03 b	0.38 ± 0.04 a	0.01 ± 0.01 a	0.02 ± 0.01 a
Si	0.26 ± 0.10 a	0.21 ± 0.10 b	n.d.	n.d.	n.d.	n.d.	n.d.	n.d.
P	0.06 ± 0.01 b	0.08 ± 0.04 a	1.01 ± 0.07 a	0.65 ± 0.10 b	0.17 ± 0.01 b	1.23 ± 0.05 a	n.d.	n.d.
S	0.19 ± 0.07 a	0.22 ± 0.2 a	0.45 ± 0.07 a	0.15 ± 0.01 b	0.11 ± 0.01 b	0.69 ± 0.1 a	0.02 ± 0.01 a	0.06 ± 0.01 a
Cl	0.60 ± 0.10 b	1.24 ± 0.40 a	n.d.	0.02 ± 0.01 a	0.07 ± 0.02 a	n.d. b	0.08 ± 0.03 b	0.30 ± 0.20 a
K	6.18 ± 10 a	5.46 ± 3.20 a	0.59 ± 0.05 a	0.22 ± 0.01 b	0.52 ± 0.07 b	1.31 ± 0.10 a	0.12 ± 0.10 b	0.26 ± 0.05 a
Ca	0.31 ± 0.20 a	0.16 ± 0.10 b	n.d.	n.d.	0.12 ± 0.01 a	0.04 ± 0.01 b	n.d.	n.d.
Fe	0.08 ± 0.05 a	0.06 ± 0.04 b	n.d.	0.01 ± 0.00 a	n.d.	n.d.	n.d.	n.d.
Br	0.09 ± 0.09 b	0.13 ± 0.10 a	n.d.	n.d.	n.d.	n.d.	n.d.	n.d.
Al	0.05 ± 0.04 a	0.03 ± 0.03 b	n.d.	n.d.	n.d.	n.d.	n.d.	n.d.
Cu	0.05 ± 0.04 a	N.d. b	n.d.	n.d.	n.d.	n.d.	n.d.	n.d.

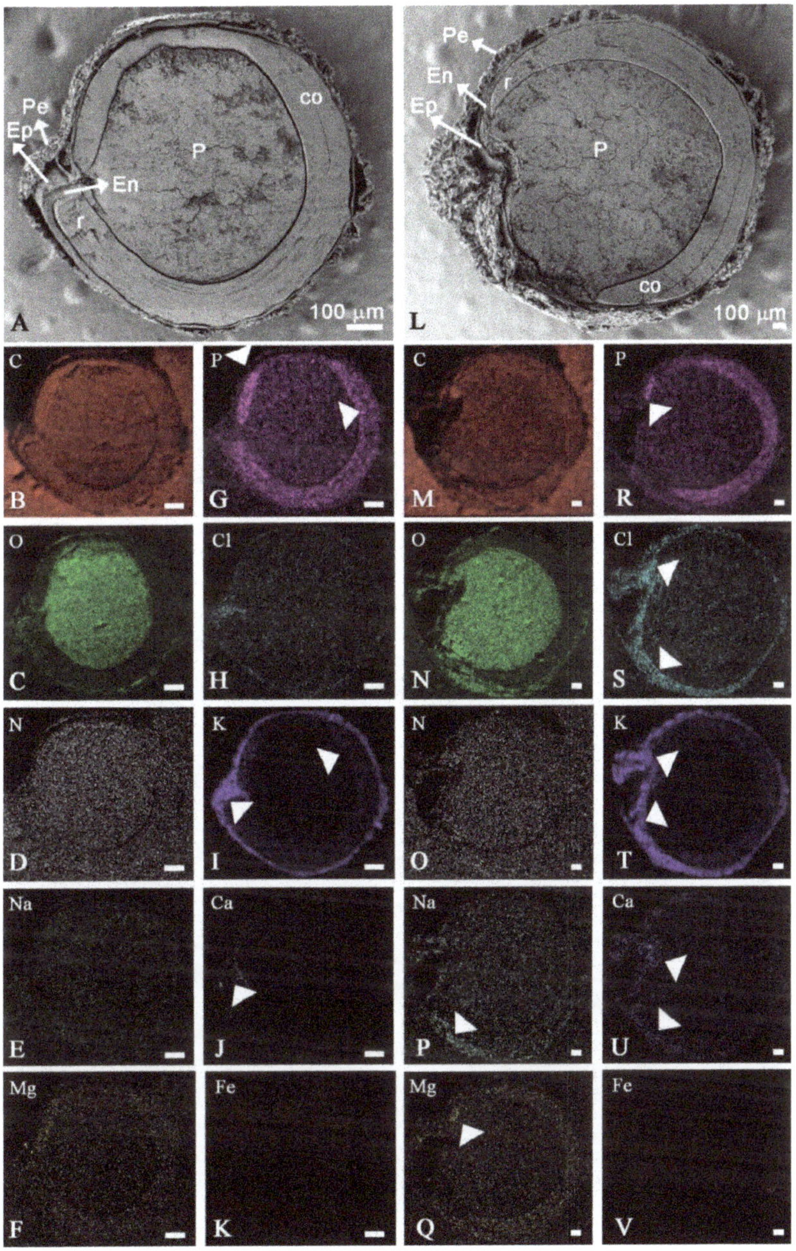

Figure 5. Spatial distribution of elements analyzed by SEM-EDX in grain longitudinal medial sections of *C. quinoa* cv. CICA-17, cultivated under non-saline (**A–K**) and saline (**L–V**) conditions. Ep, episperm; En, endosperm; P, perisperm; Pe, pericarp; Emb, embryo with the radicle-hypocotyl axis (r) and cotyledon (co). Arrowhead indicates the greater abundance of the mapped element.

The Mg content increased significantly in the endosperm under saline soil, while Cl content increased in embryo, and perisperm and decreased in the endosperm. Sulfur content differed between soil salinities. It was stored mainly in the embryo in non-saline conditions, meanwhile, it accumulated more and perisperm in saline soil (6.3 and 3 times, respectively). Nitrogen was detected only in the cotyledons, endosperm, and perisperm, reducing their content in the last two tissues under salinity conditions. Silicon, Fe, Br, Al, and Cu were below the detection limit for embryo, endosperm, and perisperm as for Na, they are detected only in pericarp.

3. Discussion

The high salinity soils of Sah El Tina have not produced harmful effects on the quinoa crop cultivation (cv. CICA-17). As a halophyte, the plant displayed a series of physiological and morphological adaptations that allowed it to complete its cycle [35–39]. Saline soils affect grain yield and seed size, according to previous research [40,41]. In our case, we detected a decrease near 47% compared to the yield get in non-saline conditions. The weight of 1000 seeds decreased by 13% in saline soil, and only smaller grains (< 1.68 mm) were favored. From a commercial point of view, this finding is also essential because markets prefer large grains instead of small ones.

Regardless of the soil's saline conditions where quinoa was grown, the most abundant minerals in the quinoa grains were Si, K, P, Mg, Ca, Na, Mn, Fe, Cu, Al, and Zn. However, an increase in P, Na, and Mg contents was observed due to soil salinity. Sodium, an essential mineral in the cell ionic balance, was detected only in the pericarp and not in other tissues (embryo, endosperm, and perisperm). It is evident that the pericarp structure (with different cell layers) is a shield to prevent the entry of sodium to the underlying tissues, but not for the chlorides that increased in saline conditions. Contradictorily, the mineral content of Mg and Mn was reduced by salinity in saline-sodic soils in Greece [42]. However, the salinity conditions of that field experiment are low (6.5 dS m^{-1}) in comparison to our experiment (26 dS m^{-1}) (Table 1). It is necessary to consider that 26 dS/m is the starting value of the saline soil's electrical conductivity (EC). It probably increases during the life cycle because of the soil and water quality used (with high EC). So we can assume that EC is further increased during cultivation, and the seeds were produced under more significant saline stress conditions than the starting one. This hypothesis must be verified in future studies in the field.

Regarding the occurrence of certain minerals in quinoa, it is necessary to consider whether the analyses are performed on grains (pericarp + embryo) or only in seeds (without pericarp). The desaponification process removes the pericarp and probably all the elements specifically present in this tissue (Table 4) This feature is mentioned in many cases where the analysis was performed on flour, but it is not clear whether the grains used contained the pericarp or not. Our results showed that Na was only in the pericarp, Mg was present in the pericarp, embryo, endosperm, and perisperm. While S was found in all the tissues, P was only absent in the perisperm. All these features should be considered to prepare quinoa-based foods since the product will not have the same mineral composition based on whether or not it is desaponified. It is important to point out that saponins are present in different quinoa organs (leaves, flowers, fruits, and seeds), especially in seed coats (pericarp). Saponins must be removed by different methods (physical or chemical) to avoid conferring a bitter flavor to the quinoa products. In general, saponin concentration ranges from 0.01 to 5% on a dry weight basis [43,44]. There are no data concerning saponin concentration in the quinoa crop of Egypt. However, González et al. [45] observed that saponin content in CICA -17, grown in desert climatic conditions in a high valley in Northwest Argentina, varied from 2.3 to 6.9% according to different nitrogen treatments. Considering that saponins can increase their concentration under saline condition [29], we can conclude that desaponification must be a necessary process before consuming quinoa either as grain or as flour because the maximum acceptable level of saponin in quinoa for human consumption varies from 0.06 to 0.12% [46,47]. Besides, desaponification removes sodium from the grain, avoiding its potential negative effects when consumed.

Except for Sn, Ni, and As, already found by Prado et al. (2014), the other ultratrace mineral elements (Rb, Sn, Th, Nd, Pr, Nb, Sm, Ni, Y, La, Ce, As, Ti, Ge, V, Zr, Ga, Zn, Ba, and

Pb) were first detected in quinoa in both non-saline and saline conditions. The presence of Cs was reported [48], but it was observed in aboveground parts of plants (stem and leaves). For many trace and ultratrace minerals detected in quinoa, the cellular level function is unknown, and their presence is only a passive accumulation from the soil and irrigation water. The role of Cr, Li, Si, Ni in human metabolism was already demonstrated [49,50]. Chromium participates in protein transport and improves diabetes [51], while Li is an essential element for regulating the central nervous system [52]. Following our results, Li has also been found in quinoa and amaranth (*Amaranthus caudatus*) consumed as food in the Northwest of Argentina [53]. Silicon is essential for Ca assimilation, the formation of new cells, and tissue nutrition [54], and Ni is necessary for the proper functioning of the pancreas [55]. Regarding arsenic, which is dangerous in high concentrations, several studies suggest too that it probably plays a physiological role in the metabolism of methionine, acting as an effector of acid metabolism amino sulfur [56]. Aluminum is typically considered a toxic element, but some studies in vitro suggest that this element plays an essential role in different biological systems (e.g., DNA synthesis stimulation or bone formation) [57].

In summary, quinoa can be considered a source of minerals as Cr, Li, Si, Ni, As, and Al concerning nutritious food and health. Besides, considering that quinoa foliage can accumulate some minerals such Ni, Cr, Cu, and Cd [24] and the hyperaccumulation of heavy metals in roots [58], we hypothesize that this species may be a good alternative for the remediation of contaminated soils.

4. Materials and Methods

4.1. Plant Material, Site Description, and Experimental Design

Grains of CICA-17 quinoa cultivar were selected for this study. Field experiments were conducted in 2015/2016 (mid-November to end-March) in two places in Egypt. One was in the Sahl El-Tina plain (named saline in this work), located in the northwestern coast of Sinai Peninsula (31° 02′ N, 32° 35′ E), and other in the Experimental Station (named non-saline) of Ain Shams University, Cairo (30° 03′ N, 31° 14′ E). Both places are arid, with an average annual precipitation of 60 mm yr^{-1} (Sahl El-Tina) and 20 mm yr^{-1} (Cairo), with rainfall concentrated between October and April. The average monthly temperature was 20.5, 12.8, 16.7 °C and 21.7, 12.1, 16.9 °C for maximum, minimum, and mean temperature in saline and non-saline places during the crop cycle, respectively. Representative soil samples in both locations were collected in the center of each plot at 0.60 m depth. Soil samples (6 in total) were obtained with a soil borer. Samples were mixed in the lab, and physical and chemical analyses were performed according to the standard methods published by Page et al. (1982) [59]. Results are listed in Table 1.

Experimental soils were prepared, including the construction of ridges. Compost with a rate of 8 t ha^{-1} and phosphorus at a 120 kg P_2O_5 ha^{-1} was added during the land preparation. Nitrogen was added as side-dressing at 160 kg N ha^{-1} in two equal rates after 30 and 51 days from the sowing date. Potassium was added at a rate of 140 kg K_2O ha^{-1} at the flowering stage. Seeds of quinoa were sterilized with sodium hypochlorite solution (5% active chloride) for 10 min and then washed with distilled water several times and dried with tissue paper before planting. We sowed about ten seeds per hill to ensure germination. A complete randomized block design with six replicates (experimental plots) was used, with an average of 18 m^2 for each plot (6 ridges with 5 m length and 0.6 m width). After four weeks from the sowing date, the seedlings were thinned to two or three seedlings per hill. Local sources provided irrigation water in both places, and Table 2 lists their chemical properties. A detailed description of the procedure is in Eisa et al. (2017) [16].

4.2. Yield Components

Ten quinoa plants in each experimental plot were cut and air-dried for 7–10 days at the harvesting stage. The dried panicles were threshed by hand. Subsequently, we determined the grain yield (kg ha^{-1}), the seed sizes, and the weight of 1000 seeds (g). The percentage of grain sizes was determined with sieves of different meshes.

4.3. Mineral Content

The grain samples were dried in an oven at 65 °C to constant weight. The dried samples were ground into a fine powder, passed through a 60-mesh sieve, and ashed (electric oven at 575 °C for 16 h). After cooling, ash subsamples (0.1 g) were digested with 10 mL of $HF/HClO_4/HNO_3$ (5.0/2.5/2.5, v/v) mixture in a closed teflon vessel (Savillex, Canada) at 90 °C for at least 12 h. Once digestion finished, the remaining acid was evaporated on a hotplate to incipient dryness. Next, 1 mL of HNO_3 was added twice and was evaporated again to incipient dryness. Residual sediment was dissolved in HNO_3 (1 mL) and transferred into a 100 mL volumetric flask. The digestion vessel was rinsed with deionized water several times, and washing water was also transferred to the volumetric flask. Flask volume was made up with deionized water and then used for chemical element analysis. A total of 49 elements, including major (e.g., Ca, K, Mg, Na, and P), minor (e.g., Zn, Fe, Cu. Mn, Co, and Na), and ultratrace mineral elements (e.g., Cr, Li, As, Ni. Mo, Se, Sn, and V) were quantitatively determined by high resolution inductively coupled plasma mass spectrometry (Element XR HR-ICP-MS, Thermo Scientific, Germany) at the labGEOTOP of institute Geosciences Barcelona (Spanish Research Council, CSIC, Barcelona, Spain) [37]. Quality control of element determinations was carried out using internal standards and control samples of known composition for each analyzed element (BDH Chemical, England). Of the 49 elements analyzed, only 34 were listed because the rest (15 elements) were lower than the equipment's detection limit. Element concentration (mean of 3 repetitions) was expressed on a dry weight (dw) basis (Table 1). The accuracy and precision of analytical determinations by HR-ICP-MS were lower than 10%.

4.4. Mineral Spatial Distribution

To determine the spatial distributions and relative abundances of C, O, N, Na, K, Ca, Mg, P, Cl, Cu, Br, Al, Fe, Si, and S, three samples of achenes per saline and non-saline locations were analyzed by scanning electron microscopy (SEM, Supra55VP) coupled to an energy dispersive X-ray analyzer (EDX). SEM-EDX conditions were −10 °C, sample chamber pressure of −50 Pa, accelerating voltage of 20 kV). These analyses were conducted at CIME facilities (Integral Center of Electronic Microscopy of CONICET-UNT, Tucumán, Argentina).

Complete grains (to study pericarps), seeds with manually removed pericarp (to expose the seed coat), and median longitudinal sections of whole grains (cut with a blade) were mounted and fixed with a commercial adhesive on sample plates. Maps of mineral distribution were made in the median longitudinal sections. Semi-quantitative analyses of mineral abundance were obtained from equivalent areas of different tissues or organs (pericarp, episperm, endosperm, perisperm, embryo, and cotyledon mesophyll).

4.5. Statistical Analysis

The mean values between treatments per tissue/organs were compared using the t-test at $p \leq 0.05$ level of probability (statistical package SPSS Inc., version 11.0, Chicago, IL, USA).

5. Conclusions

This study demonstrated the high resilience of the CICA-17 variety of quinoa to two contrasting edaphic situations in Egypt. It was clear that mineral concentration can change when interacting with different soil types, irrigation waters, and other environmental conditions. Soil salinity induces substantial changes in the distribution of minerals in the different grain tissues. Although quinoa is a significant source of major minerals such as K, P, Mg, Ca, and Na, the contents of Si, Ni, Cr, and Li may be important for their healthy behavior in human metabolism. The detection for the first time of 17 elements (Rb, Th, Nd,

Pr, Nb, Sm, Y, La, Ce, Ti, Ge, V, Zr, Ga, Zn, Ba, and Pb) in quinoa seed add a new value to quinoa crops as a potential application for phytoremediation processes. From our results, we can conclude that cv. CICA-17 may be a complementary crop in the marginal lands of high salinity in Egypt and the Mediterranean region, showing high potential concerning healthy food and environmental issues as phytoremediation.

Author Contributions: Conceptualization, J.A.G., S.E., and L.H.; methodology, M.I.M., G.I.P., D.A.G., M.R., M.E.A.E., and J.-L.F.-T.; formal analysis an investigation, J.A.G., S.E., D.B. and S.H.; writing-original draft preparation, J.A.G., S.E.; E.H.A.E.-S.; writing-review and editing, J.A.G., L.H., and D.B.; project administration, E.H.A.E.-S., and A.A.-A.; funding acquisition, J.A.G., and S.E. All authors have read and agreed to the published version of the manuscript.

Funding: The Ministry of Foreign Affairs, International Trade and Worship, Buenos Aires, Argentina, partially supported this research through the South-South cooperation project "Development of quinoa crop in marginal areas of Egypt", FO.AR-6429, and Fundación Miguel Lillo (Miguel Lillo 251, T4000, Tucumán, Argentina) under the Project B-0004-1 ("Ecophysiology of selected varieties of quinoa and others species with nutritional values in lab and field").

Institutional Review Board Statement: Not applicable.

Informed Consent Statement: Not applicable.

Data Availability Statement: The data presented in this study are available in this article.

Acknowledgments: Authors are thankful to Lic. Sebastián Buedo (Fundación Miguel Lillo) for providing laboratory facilities. We acknowledge the analytical assistance of labGEOTOP Geochemistry Laboratory (infrastructure co-funded by ERDF–EU Ref. CSIC08–4E–001).

Conflicts of Interest: The authors declare no conflict of interest.

References

1. IPCC. *Climate Change 2014—Impacts, Adaptation, and Vulnerability: Regional Aspects*; Cambridge University Press: Cambridge, UK; New York, NY, USA, 2014; ISBN 978-1-107-05816-3.
2. Rasool, S.; Hameed, A.; Azooz, M.M.; Muneeb-u-Rehman; Siddiqi, T.O.; Ahmad, P. Salt stress: Causes, types and responses of plants. In *Ecophysiology and Responses of Plants under Salt Stress*; Ahmad, P., Azooz, M.M., Prasad, M.N.V., Eds.; Springer: New York, NY, USA, 2013; pp. 1–24. ISBN 978-1-4614-4746-73.
3. Tester, M.; Davenport, R. Na^+ tolerance and Na^+ transport in higher plants. *Ann. Bot.* **2003**, *91*, 503–527. [CrossRef] [PubMed]
4. Dillehay, T.D.; Rossen, J.; Andres, T.C.; Williams, D.E. Preceramic adoption of peanut, squash, and cotton in Northern Peru. *Science* **2007**, *316*, 1890–1893. [CrossRef]
5. Verena, I.A.; Alanoca, C.; Adriana, A.; Ricardo, A.; Alberto, J.A.; Fabiana, A.; Genaro, A.; Pilar, B.; Andrea, B.; Flora, B.; et al. *State of the Art Report on Quinoa around the World 2013*; Bazile, D., Bertero, H.D., Nieto, C., Eds.; FAO: Santiago, Chile; CIRAD: Montpellier, France, 2015; pp. 1–7.
6. Fuentes, F.; Bazile, D.; Bhargava, A.; Martinez, E.A. Implications of farmers' seed exchanges for on-farm conservation of quinoa, as revealed by its genetic diversity in Chile. *J. Agric. Sci.* **2012**, *150*, 702–716. [CrossRef]
7. González, J.A.; Bruno, M.; Valoy, M.; Prado, F.E. Genotypic variation of gas exchange parameters and leaf stable carbon and nitrogen isotopes in ten quinoa cultivars grown under drought. *J. Agron. Crop Sci.* **2010**, *197*, 81–93. [CrossRef]
8. Hinojosa, L.; Matanguihan, J.B.; Murphy, K.M. Effect of high temperature on pollen morphology, plant growth and seed yield in quinoa (*Chenopodium quinoa* Willd.). *J. Agron. Crop Sci.* **2018**, *205*, 33–45. [CrossRef]
9. González, J.A.; Eisa, S.S.S.; Hussin, S.A.E.S.; Prado, F.E. Quinoa: An incan crop to face global changes in agriculture. In *Quinoa: Improvement and Sustainable Production*; Murphy, K., Matanguihan, J., Eds.; John Wiley & Sons, Inc.: Hoboken, NJ, USA, 2015; pp. 1–18. ISBN 978-1-118-62804-1.
10. Vilcacundo, R.; Hernández-Ledesma, B. Nutritional and biological value of quinoa (*Chenopodium quinoa* Willd.). *Curr. Opin. Food Sci.* **2017**, *14*, 1–6. [CrossRef]
11. Nowak, V.; Du, J.; Charrondière, U.R. Assessment of the nutritional composition of quinoa (*Chenopodium quinoa* Willd.). *Food Chem.* **2016**, *193*, 47–54. [CrossRef]
12. García-Parra, M.A.; Roa-Acosta, D.F.; Stechauner-Rohringer, R.; García-Molano, F.; Bazile, D.; Plazas Leguizamón, N. Effect of temperature on the growth and development of quinoa plants (*Chenopodium quinoa* Willd.): A review on a global scale. *Sylwan* **2020**, *164*, 411–423.
13. González, J.A.; Rosa, M.; Parrado, M.F.; Hilal, M.; Prado, F.E. Morphological and physiological responses of two varieties of a highland species (*Chenopodium quinoa* Willd.) growing under near-ambient and strongly reduced solar UV–B in a lowland location. *J. Photochem. Photobiol. Biol.* **2009**, *96*, 144–151. [CrossRef]

14. Hinojosa, L.; González, J.A.; Barrios-Masias, F.H.; Fuentes, F.; Murphy, K.M. Quinoa abiotic stress responses: A review. *Plants* **2018**, *7*, 106. [CrossRef]
15. Eisa, S.; Hussin, S.; Geissler, N.; Koyro, H.W. Effect of NaCl salinity on water relations, photosynthesis and chemical composition of Quinoa (*Chenopodium quinoa* Willd.) as a potential cash crop halophyte. *Aust. J. Crop Sci.* **2012**, *6*, 357–368.
16. Eisa, S.S.; Eid, M.; El-Samad, E.H.A.; Hussin, S.; Abdel-Ati, A.; El-Bordeny, N.; Ali, S.; Al-Sayed, H.M.; Lotfy, M.; Masoud, A. *Chenopodium quinoa* Willd. A new cash crop halophyte for saline regions of Egypt. *Aust. J. Crop Sci.* **2017**, *11*, 343–351. [CrossRef]
17. Orsini, F.; Accorsi, M.; Gianquinto, G.; Dinelli, G.; Antognoni, F.; Carrasco, K.B.R.; Martinez, E.A.; Alnayef, M.; Marotti, I.; Bosi, S.; et al. Beyond the ionic and osmotic response to salinity *in Chenopodium quinoa*: Functional elements of successful halophytism. *Funct. Plant Biol.* **2011**, *38*, 818–831. [CrossRef] [PubMed]
18. Rojas, W.; Pinto, M.; Alanoca, C.; Gómez Pando, L.; León-Lobos, P.; Alercia, A.; Diulgheroff, S.; Padulosi, S.; Bazile, D. Quinoa genetic resources and ex situ conservation. In *State of the Art Report on Quinoa around the World in 2013*; Bazile, D., Bertero, H.D., Nieto, C., Eds.; FAO & CIRAD: Rome, Italy, 2015; pp. 56–82.
19. Bazile, D.; Jacobsen, S.-E.; Verniau, A. The global expansion of quinoa: Trends and limits. *Front. Plant Sci.* **2016**, *7*, 622. [CrossRef]
20. Ruiz, K.B.; Biondi, S.; Oses, R.; Acuña-Rodríguez, I.S.; Antognoni, F.; Martinez-Mosqueira, E.A.; Coulibaly, A.; Canahua-Murillo, A.; Pinto, M.; Zurita, A.; et al. Quinoa biodiversity and sustainability for food security under climate change. A review. *Agron. Sustain. Dev.* **2013**, *34*, 349–359. [CrossRef]
21. Tapia, M. The Long Journey of Quinoa: Who wrote its history. In *State of the Art Report on Quinoa around the World in 2013*; Bazile, D., Bertero, H.D., Nieto, C., Eds.; FAO and CIRAD: Rome, Italy, 2015; Volume 1, pp. 1–7. ISBN 978-92-5-108558-5.
22. Tang, Y.; Li, X.; Chen, P.X.; Zhang, B.; Hernandez, M.; Zhang, H.; Marcone, M.F.; Liu, R.; Tsao, R. Lipids, tocopherols, and carotenoids in leaves of amaranth and quinoa cultivars and a new approach to overall evaluation of nutritional quality traits. *J. Agric. Food Chem.* **2014**, *62*, 12610–12619. [CrossRef]
23. Stuardo, M.; Martín, R.S. Antifungal properties of quinoa (*Chenopodium quinoa* Willd) alkali treated saponins against botrytis cinerea. *Ind. Crop. Prod.* **2008**, *27*, 296–302. [CrossRef]
24. Bhargava, A.; Shukla, S.; Srivastava, J.; Singh, N.; Ohri, D. Genetic diversity for mineral accumulation in the foliage of *Chenopodium* spp. *Sci. Hortic.* **2008**, *118*, 338–346. [CrossRef]
25. Jacobsen, S.-E. The scope for adaptation of quinoa in northern latitudes of Europe. *J. Agron. Crop Sci.* **2017**, *203*, 603–613. [CrossRef]
26. Peterson, A.; Murphy, K.M. Quinoa cultivation for temperate North America: Considerations and areas for investigation. In *Quinoa: Improvement and Sustainable Production*; Murphy, K., Matanguihan, J., Eds.; John Wiley & Sons, Inc.: Hoboken, NJ, USA, 2015; pp. 173–192. ISBN 978-1-118-62804-1.
27. Hinojosa, L.; Kumar, N.; Gill, K.S.; Murphy, K.M. Spectral reflectance indices and physiological parameters in quinoa under contrasting irrigation regimes. *Crop. Sci.* **2019**, *59*, 1927–1944. [CrossRef]
28. Bhargava, A.; Shukla, S.; Ohri, D. *Chenopodium quinoa*—An Indian perspective. *Ind. Crop. Prod.* **2006**, *23*, 73–87. [CrossRef]
29. Pulvento, C.; Riccardi, M.; Lavini, A.; Iafelice, G.; Marconi, E.; d'Andria, R. Yield and quality characteristics of quinoa grown in open field under different saline and non-saline irrigation regimes. *J. Agron. Crop. Sci.* **2012**, *198*, 254–263. [CrossRef]
30. Mujica, A.; Jacobsen, S.E.; Izquierdo, J.; Marathee, J. Resultados de la Prueba Americana y Europea de la Quinoa. FAO, UNA-Puno, CIP. 2001. Available online: http://www.fao.org/tempref/GI/Reserved/FTP_FaoRlc/old/prior/segalim/prodalim/prodveg/cdrom/contenido/libro03/home03.htm (accessed on 20 November 2020).
31. González, J.A.; Konishi, Y.; Bruno, M.; Valoy, M.; Prado, F.E. Interrelationships among seed yield, total protein and amino acid composition of ten quinoa (*Chenopodium quinoa*) cultivars from two different agroecological regions. *J. Sci. Food Agric.* **2011**, *92*, 1222–1229. [CrossRef] [PubMed]
32. Al-Naggar, A.M.M.; El-Salam, R.M.A.; Badran, A.E.E.; El-Moghazi, M.M.A. Genotype and drought effects on morphological, physiological and yield traits of quinoa (*Chenopodium quinoa* Willd.). *Asian J. Adv. Agric. Res.* **2017**, *3*, 1–15. [CrossRef]
33. Eisa, S.; Abd El-Samad, E.H.; Hussin, S.; Ali, E.A.; Ebrahim, M.; González, J.A.; Ordano, M.; Erazzú, L.E.; El-Bordeny, N.E.; Abdel-Ati, A.A. Quinoa in Egypt-Plant Density Effects on Seed Yield and Nutritional Quality under Marginal Regions. *Middle East J. Appl. Sci.* **2018**, *8*, 512–522.
34. Ebrahim, M.; Abdel-Ati, A.A.; Hussin, S.A.; Ali, S.H.; Eisa, S.S. Evaluation of some *Chenopodium quinoa* cultivars under saline soil conditions in Egypt. *Arab. Univ. J. Agric. Sci.* **2018**, *26*, 337–347. [CrossRef]
35. Adolf, V.I.; Jacobsen, E.; Shabala, S. Salt tolerance mechanisms in quinoa (*Chenopodium quinoa* Willd.). *Environ. Exp. Bot.* **2013**, *92*, 43–54. [CrossRef]
36. Konishi, Y.; Hirano, S.; Tsuboi, H.; Wada, M. Distribution of minerals in quinoa (*Chenopodium quinoa* Willd) seeds. *Biosci. Biotechnol. Biochem.* **2004**, *68*, 231–234. [CrossRef]
37. Prado, F.E.; Fernández-Turiel, J.L.; Tsarouchi, M.; Psaras, G.K.; González, J.A. Variation of seed mineral concentrations in seven quinoa cultivars grown in two agroecological sites. *Cereal Chem. J.* **2014**, *91*, 453–459. [CrossRef]
38. Ruiz, K.B.; Biondi, S.; Martínez, E.A.; Orsini, F.; Antognoni, F.; Jacobsen, S.-E. Quinoa—A model crop for understanding salt-tolerance mechanisms in halophytes. *Plant Biosyst.—Int. J. Deal. Asp. Plant Biol.* **2016**, *150*, 357–371. [CrossRef]
39. Miranda-Apodaca, J.; Agirresarobe, A.; Martínez-Goñi, X.; Yoldi-Achalandabaso, A.; Pérez-López, U. N metabolism performance in *Chenopodium quinoa* subjected to drought or salt stress conditions. *Plant Physiol. Biochem.* **2020**, *155*, 725–734. [CrossRef]

40. Koyro, H.-W.; Eisa, S.S. Effect of salinity on composition, viability and germination of seeds of *Chenopodium quinoa* Willd. *Plant Soil* **2008**, *302*, 79–90. [CrossRef]
41. Hussain, M.I.; Dakheel, A.J.; Reigosa, M.J. Genotypic differences in agro-physiological, biochemical and isotopic responses to salinity stress in quinoa (*Chenopodium quinoa* Willd.) plants: Prospects for salinity tolerance and yield stability. *Plant Physiol. Biochem.* **2018**, *129*, 411–420. [CrossRef]
42. Karyotis, T.; Iliadis, C.; Noulas, C.; Mitsibonas, T. Preliminary research on seed production and nutrient content for certain quinoa varieties in a saline-sodic soil. *J. Agron. Crop Sci.* **2003**, *189*, 402–408. [CrossRef]
43. Kuljanabhagavad, T.; Thongphasuk, P.; Chamulitrat, W.; Wink, M. Triterpene saponins from *Chenopodium quinoa* Willd. *Phytochemistry* **2008**, *69*, 1919–1926. [CrossRef]
44. Ando, H.; Chen, Y.-C.; Tang, H.; Shimizu, M.; Watanabe, K.; Mitsunaga, T. Food components in fractions of Quinoa Seed. *Food Sci. Technol. Res.* **2002**, *8*, 80–84. [CrossRef]
45. González, J.A.; Lizárraga, E.F.; Erazzú, L.E.; Martínez-Calsina, L.; Buedo, S.E.; González, D.A. Goat manure fertilization effect on saponin and protein content in quinoa (*Chenopodium quinoa* Willd) grain of different origin. *Middle East J. Agric. Res.* **2020**, *9*, 434–443. [CrossRef]
46. Bacigalupo, A.; Tapia, M. Potencial agroindustrial de los cultivos andinos subexplotados. In *Cultivos andinos subexplotados y su aporte a la alimentación*; Tapia, M., Ediciones Gegra, S.A., Eds.; FAO: Santiago, Chile, 1990; pp. 136–163.
47. Vega-Gálvez, A.; Miranda, M.; Vergara, J.; Uribe, E.; Puente, L.; Martínez, E.A. Nutrition facts and functional potential of quinoa (*Chenopodium quinoa* Willd.), an ancient Andean grain: A review. *J. Sci. Food Agric.* **2010**, *90*, 2541–2547. [CrossRef]
48. Isobe, K.; Nakajima, E.; Morita, N.; Kawakura, S.; Higo, M. Effects of NaCl on growth and cesium absorption in quinoa (*Chenopodium quinoa* Willd.). *Water Air Soil Pollut.* **2019**, *230*, 66. [CrossRef]
49. Nielsen, F.H. Possibly essential trace elements. Nutrition and Health. In *Clinical Nutrition of the Essential Trace Elements and Minerals: The Guide for Health Professionals*; Bogden, J.D., Klevay, L.M., Eds.; Humana Press: Totowa, NJ, USA, 2000; pp. 11–36. ISBN 978-1-59259-040-7.
50. Nielsen, F. Ultratrace elements. In *Encyclopedia of Human Nutrition*; Sadler, M.J., Strain, J.J., Caballero, B., Eds.; Academic Press: San Diego, CA, USA, 1999; pp. 1884–1887.
51. Cefalu, W.T.; Hu, F.B. Role of chromium in human health and in diabetes. *Diabetes Care* **2004**, *27*, 2741–2751. [CrossRef]
52. Chiu, C.-T.; Chuang, D.-M. Neuroprotective action of lithium in disorders of the central nervous system. *Zhong Nan Da Xue Xue Bao Yi Xue Ban* **2011**, *36*, 461–476. [CrossRef] [PubMed]
53. Coelho, I.; Gueifão, S.; Nascimento, A.C.; Lobo, M.; Gimenez, G.; Samman, N.; Castanheira, I. Proximate and mineral profile of quinoa (*Chenopodium quinoa*), and kiwicha (*Amaranthus caudatus*) consumed in north of Argentina. *Int. Food Data Conf.* **2011**, *9*, 14–17.
54. Chappell, H.F.; Jugdaohsingh, R.; Powell, J.J. Physiological silicon incorporation into bone mineral requires orthosilicic acid metabolism to $SiO_4{}^{4-}$. *J. R. Soc. Interface* **2020**, *17*, 20200145. [CrossRef] [PubMed]
55. Amaral, A.F.S.; Porta, M.; Silverman, D.T.; Milne, R.L.; Kogevinas, M.; Rothman, N.; Cantor, K.P.; Jackson, B.P.; Pumarega, J.A.; López, T.; et al. Pancreatic cancer risk and levels of trace elements. *Gut* **2012**, *61*, 1583–1588. [CrossRef] [PubMed]
56. Uthus, E.O. Arsenic essentiality and factors affecting its importance. In *Arsenic Exposure and Health*; Chappell, W.R., Abernathy, C.O., Cothern, C.R., Eds.; Science and Technology Letters: Northwood, UK, 1994; pp. 199–208.
57. Martin, R.B. The chemistry of aluminum as related to biology and medicine. *Clin. Chem.* **1986**, *32*, 1797–1806. [CrossRef]
58. Thomas, E.C.; Lavkulich, L.M. Community considerations for quinoa production in the urban environment. *Can. J. Plant Sci.* **2014**, *95*, 397–404. [CrossRef]
59. Page, A.L. *Methods of Soil Analysis: Chemical and Microbiological Properties*; American Society of Agronomy: Madison, WI, USA, 1982; ISBN 0-89118-072-9.

plants

Article

Development of Quinoa Value Chain to Improve Food and Nutritional Security in Rural Communities in Rehamna, Morocco: Lessons Learned and Perspectives

Abdelaziz Hirich [1,*], Sifeddine Rafik [2,3], Mohamed Rahmani [3], Amira Fetouab [2], Fatima Azaykou [2], Kaoutar Filali [4], Hayatullah Ahmadzai [5], Younes Jnaoui [2], Aziz Soulaimani [2], Mariam Moussafir [2], Mohamed El Gharous [2], Salwa Karboune [6], Abdelaziz Sbai [4] and Redouane Choukr-Allah [4]

1 African Sustainable Agriculture Research Institute (ASARI), Mohammed VI Polytechnic University (UM6P), Laayoune 70000, Morocco
2 Agricultural Innovation and Technology Transfer Center (AITTC), Mohammed VI Polytechnic University (UM6P), Ben Guerir 43150, Morocco; sifeddine.rafik@um6p.ma (S.R.); amirafetouab@gmail.com (A.F.); azaykoufatima@gmail.com (F.A.); younes.jnaoui@um6p.ma (Y.J.); aziz.soulaimani@um6p.ma (A.S.); mariam.moussafir@um6p.ma (M.M.); mohamed.elgharous@um6p.ma (M.E.G.)
3 Department of Food and Nutritional Sciences, Section of Agricultural and Food Industries, Agronomic and Veterinary Medicine Hassan II Institute, Rabat 10112, Morocco; rahmanimohammed@yahoo.fr
4 Department of Agronomy, Agronomic and Veterinary Medicine Hassan II Institute, Rabat 10112, Morocco; k.filali87@gmail.com (K.F.); asbai2007@gmail.com (A.S.); redouane53@yahoo.fr (R.C.-A.)
5 International Center for Biosaline Agriculture, Directorate of Programs, Dubai 14660, United Arab Emirates; h.ahmadzai@biosaline.org.ae
6 Department of Food Science and Agricultural Chemistry, Macdonald Campus, McGill University, Sainte-Anne-de-Bellevue, Québec, QC H9X 3V9, Canada; salwa.karboune@mcgill.ca
* Correspondence: abdelaziz.hirich@um6p.ma

Citation: Hirich, A.; Rafik, S.; Rahmani, M.; Fetouab, A.; Azaykou, F.; Filali, K.; Ahmadzai, H.; Jnaoui, Y.; Soulaimani, A.; Moussafir, M.; et al. Development of Quinoa Value Chain to Improve Food and Nutritional Security in Rural Communities in Rehamna, Morocco: Lessons Learned and Perspectives. *Plants* **2021**, *10*, 301. https://doi.org/10.3390/plants10020301

Academic Editors: Cataldo Pulvento

Received: 31 December 2020
Accepted: 13 January 2021
Published: 5 February 2021

Abstract: Agricultural production in the Rehamna region, Morocco is limited with various challenges including drought and salinity. Introduction of climate resilient and rustic crops such as quinoa was an optimal solution to increase farmer's income and improve food security. This study summarizes results obtained from a research project aiming to develop quinoa value chain in Morocco. The study tackled several aspects including agronomic traits (yield and growth), transformation, quality (nutritional and antinutritional traits) and economic analysis and, finally, a strength–weaknesses–opportunities–threats analysis, lessons learned and development perspectives were presented. From an agronomic point of view, introduced new quinoa cultivars showed higher performance than locally cultivated seeds and, furthermore, the use of irrigation and organic amendment has tremendously improved seed yield by double and three times, respectively, compared to rainfed conditions. Nutritional analysis revealed that protein and phosphorus content remained stable after seed pearling while most of the micronutrients content decreased after seed pearling. However, saponins content was reduced by 68% using mechanical pearling compared to 57% using both traditional abrasion and washing. The economic analysis showed that production cost of quinoa seeds could be further decreased using mechanized intensive tools along with irrigation and organic amendment supply. This study revealed several lessons learned from the field experience and proposed several development actions for each value chain component that can be implemented within a national quinoa program.

Keywords: production cost; pearling; yield; irrigation; mechanization; harvest

1. Introduction

Today, more than 120 countries around the world cultivate quinoa (*Chenopodium quinoa* Willd.) or try to adapt it to their environmental conditions. The continued expansion of its cultivation in all continents challenges the prejudices of that quinoa is a species, which can only grow in the high plains of the Andes on the shores of Lake Titicaca. After a first boom in quinoa cultivation in the 1990s mainly linked to the demand of vegetarians for products rich in vegetable proteins from organic farming, a second boom in the 2000s was based on the values of fair trade, and, today, we are facing a third quinoa boom at the global level with the production of quinoa in new countries that were not even importing quinoa [1]. Morocco falls into the last category of the country having initiated its cultivation before importing it for its own consumption. These changes on a global scale are such that great transformations in progress in the way quinoa is produced, the networks related to its distribution and in the ways of considering it and incorporating it into various local diets.

The year 2013 has been declared the International Year of Quinoa (IYQ) by the United Nations. This made it possible to recognize the importance of the biodiversity of quinoa and the high nutritional value of its seeds [2]. Within this dynamic, quinoa has been introduced in Morocco since the 1999/2000 season and was considered as an important alternative to traditional crops such as cereals, which are strongly subjected to climate change effects and soil degradation due to salinization making quinoa a judicious solution and potential crop that may contribute to national food security [3]. In this Moroccan context, quinoa is proving to be an interesting solution to limit the risk of agricultural production failure associated with the yield losses observed on traditional cereals cropping systems, which sometimes contribute to soil degradation because of the monoculture practiced in several regions. The fact that quinoa is considered a rustic crop resistant to various abiotic stress makes it a resilient and climate smart crop that could be used for climate change adaptation [4].

In Morocco, quinoa was subjected to several field trials evaluating the performance of introduced cultivars and the effect of various cropping practices on its productivity. First a collection of quinoa accessions was tested for the adaptation goal in the Khenifra region in the year of 2000 resulting in a selection of 14 accessions, which were believed to be tolerant to drought. Then, experiments on quinoa in Morocco were intensified within the SWUP-MED EU funded project (sustainable water use securing food production in dry areas of the Mediterranean region) where quinoa was introduced and tested in several regions including Rehamna, Rabat and Agadir [5]. Secondly, research activities were focusing on testing the effect of several practices on quinoa such as irrigation with saline water [6,7], deficit irrigation [8,9], organic amendment [10], sowing dates [11], use of wastewater for irrigation [12], etc.

At the nutritional level, for some people, quinoa is a new and nutritious food that has recently been found in supermarkets and restaurants and can replace many common grains. Certainly, in many regions of the world, this vision corresponds to reality but it should be known that quinoa was one of the main food crops of the pre-Columbian civilizations of Latin America and remains an important food for the Quechuas and Aymaras settled in rural areas of the Andes, South America. In the Quechua language, quinoa is called chisiya, which means mother grain [13]. Quinoa provides as much energy as foods used in a similar way, such as beans, corn, rice or wheat. It is also an important source of quality protein, dietary fiber, polyunsaturated fatty acids and minerals [14]. Protein content in of quinoa seeds varies between 12% and 20%; however, it is reported as 16% on average [15].

One of the obstacles for quinoa seed valorization is its content in terms of saponins because of their bitter taste and toxic effects, which necessitates their elimination. Several pearling techniques and methods are used to eliminate saponins from the quinoa seeds; the wet technique remains the most used one especially in Morocco combined with preliminary manual abrasion [16].

In this study we provided an evaluation of the agronomic performances of introduced quinoa cultivars grown under different production scenarios and the effect of seed pearling on nutritional and saponin contents. The study also presents a technical and economic analysis of the quinoa production and transformation. Through the conducted investigations the strengths, weaknesses, opportunities, and threats related to the existing quinoa value chain in Morocco were revealed and the lessons learned. Finally, we proposed development perspectives for each value chain component.

2. Results

2.1. New High Yielding Chenopodium Quinoa Cultivars Introduction

Table 1 presents obtained results in terms of plant productivity and growth. The data clearly indicate that introduced ICBA (International Center for Biosaline Agriculture) cultivars performed better than other tested varieties in terms of yield while locally produced bulk seeds showed the lowest performances. Irrigation supply and organic amendment have greatly increased the seed yield for all tested cultivars. Under farm conditions, yield was doubled under irrigation supply and tripled under combined irrigation and amendment application. More or less the same effect has been noticed for plant height. Quinoa dry matter responded very well to the amendment application in the case of ICBA-Q3 cultivar while no significant changes were noticed for local bulk seeds. Results obtained for 1000 seed weight indicate that Titicaca cultivar had the highest seed weight and size while Puno cultivar had the lowest. Organic amendment had a notable effect on 1000 seed weight in the case of local bulk seed while no significant difference was obtained for ICBA-Q3 cultivar.

Table 1. Seed yield, plant height, dry matter and 1000 seed weight of quinoa tested in the Rehamna region under several experimental conditions. Different letters (a, b, ab, c) indicate a significant difference according to the Tukey test ($p < 0.05$).

Trial Conditions	Treatments	Seed Yield ($t \cdot ha^{-1}$) of Tested Varieties							
		ICBA-Q1	ICBA-Q2	ICBA-Q3	ICBA-Q4	ICBA-Q5	Titicaca	Puno	Bulk Seeds
Cultivars performance trial under controlled experimental conditions		1.94 b	3.40 a			3.89 a	1.90 b	1.47 b	1.63 b
Organic amendment trial under controlled experimental conditions	0 T/ha			2.20 a					3.16 a
	5 T/ha compost			2.34 a					2.10 b
	10 T/ha compost			2.43 a					2.05 b
	20 T/ha compost			2.31 a					2.40 b
	10 T/ha manure			2.94 ab					1.87 b
	20 T/ha manure			2.50 a					2.23 b
	40 T/ha manure			4.40 b					2.60 ab
Trials under farm conditions	Rainfed	0.74 a	0.90 a	0.71 a	0.51 a	0.63 a			0.54 a
	Irrigated without manure	2.34 a	2.91 a	1.72 ab	1.55 ab	2.70 a			0.85 b
	Irrigated with manure	3.31 a	3.65 a	1.78 b	1.69 b	3.26 a			1.04 b
		Plant height (cm)							
Cultivars performance trial under controlled experimental conditions		125.73 a	136.38 a			123.18 a	92.40 b	79.92 b	96.28 b
Organic amendment trial under controlled experimental conditions	0 T/ha			97.35 a					75.75 a
	5 T/ha compost			96.64 a					74.72 a
	10 T/ha compost			101.93 a					81.97 a
	20 T/ha compost			104.83 a					84.94 a
	10 T/ha manure			103.51 a					76.61 a
	20 T/ha manure			108.43 a					88.72 a
	40 T/ha manure			117.06 a					86.11 a
Trials under farm conditions	Rainfed	50.83 a	52.16 a	52.83 a	46.05 a	46.77 a			42.83 a
	Irrigated without manure	115.66 ab	127.5 a	107.58 ab	137.5 a	87.66 b			70.85 b
	Irrigated with manure	109.25 bc	131.9 a	121.75 b	122.75 b	92.91 c			79.75 c

Table 1. *Cont.*

Trial Conditions	Treatments	Seed Yield ($t{\cdot}ha^{-1}$) of Tested Varieties							
		ICBA-Q1	ICBA-Q2	ICBA-Q3	ICBA-Q4	ICBA-Q5	Titicaca	Puno	Bulk Seeds
				Dry Matter (g·plant−1)					
Cultivars performance trial under controlled experimental conditions		82.83 b	171.47 a			95.70 b	41.40 b	82.54 b	77.19 b
Organic amendment trial under controlled experimental conditions	0 T/ha			101.32 b					75.62 a
	5 T/ha compost			113.33 ab					58.85 b
	10 T/ha compost			119.65 ab					56.66 b
	20 T/ha compost			100.16					61.24 ab
	10 T/ha manure			133.05 a					55.29 b
	20 T/ha manure			94.61 b					61.48 ab
	40 T/ha manure			137.94 a					43.66 c
				1000 Seed Weight (g)					
Cultivars performance trial under controlled experimental conditions		2.66 ab	2.58 ab			2.76 ab	3.55 a	1.73 b	2.47 ab
Organic amendment trial under controlled experimental conditions	0 T/ha			5.2 a					3.9 b
	5 T/ha compost			6.5 a					6.0 a
	10 T/ha compost			5.5 a					4.6 ab
	20 T/ha compost			5.4 a					6.1 a
	10 T/ha manure			5.1 a					5.2 ab
	20 T/ha manure			5.7 a					4.5 ab
	40 T/ha manure			6.7 a					5.6 ab

2.2. *Quinoa Seed Processing*

2.2.1. Harvest and Postharvest Machines

Several mechanized tools were locally developed or adapted to be used for quinoa harvest and post-harvest operations by a private entrepreneur (BenRim farm) as presented in Table 2. Supporting local private entrepreneurs to manufacture those tools was one of the key outcomes of this project.

Table 2. Harvest and post-harvest machines locally developed.

Machines	Description	Capacity	
Combined harvester	The cereal's combined harvester was used to harvest quinoa with few adaptations at sieves levels to match quinoa seed size	1.5 $ha{\cdot}hr^{-1}$	
Thresher	The machine can be powered by an electric or a diesel engine. It adopts axial-flow roller	200 $kg{\cdot}hr^{-1}$	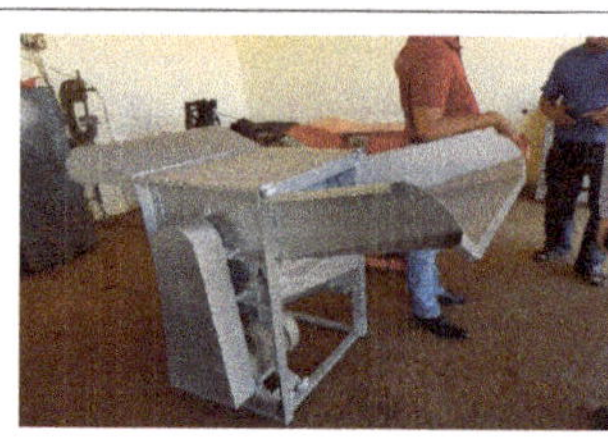

Table 2. *Cont.*

Machines	Description	Capacity	
Winnower	Threshing quinoa panicles results in a mixture of grains, small residues and chaffs. The mechanical winnowing consists of using a winnowing fan that creates wind that blows away the lighter chaff, while the heavier grains fall back down for recovery.	150 kg·hr^{-1}	
Sheller (pearling machine)	It is equipped with two motors, the first one is designed to turn a drum with a rotation speed of 750 rpm. The second one is more powerful (3000 rpm) and designed to extract the fine dust produced during the pearling process. The rotating drum is made of 80 cm long perforated inox steel and has 6 baffles distributed throughout the drum.	200 kg·hr^{-1}	

2.2.2. Quinoa Transformation Pathway

Couscous is a famous Mediterranean dish and widely consumed in Morocco. It is now produced out of quinoa by several women cooperatives in the Rehamna region using both quinoa flour and semolina. The pathway for quinoa-based products processing is described in Figure 1. Quinoa based products described in this study were produced traditionally by women following several steps such as seed pearling, washing and drying to produce processed seeds and milling, grinding and sieving to have quinoa flour and semolina. The key steps of couscous production are manual rolling, forced sieving, precooking and drying. However, 27 steps from raw material reception to final product shipping are required to produce traditional couscous.

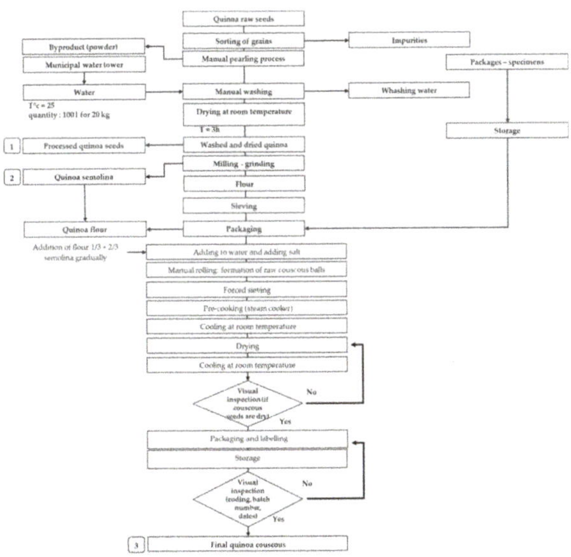

Figure 1. Processing diagram for quinoa based products adapted from the "3rd Millennium" cooperative.

2.2.3. Shelling Impacts on the Nutritional and Antinutritional Aspect of Quinoa Seeds

Table 3 shows the nutrients content of raw, processed seeds and quinoa bran for Puno, Titicaca and ICBA-Q5 cultivars. Data clearly indicate that nutrient content varies from cultivar to another. Titicaca processed seeds presents the highest content in terms of proteins, Mg, P, Ca, Zn and Cu, for Puno cultivar the highest content was observed for C and Fe, while ICBA-Q5 processed seeds presents the highest content in terms of K, Na, Mn and Ash.

Table 3. Nutrients content of raw, processed seeds and quinoa bran for Puno, Titicaca and ICBA-Q5 cultivars.

Cultivar	Products	Protein Content (%)	C (%)	K (%)	Mg (%)	Na (%)	P (%)	Ca (%)	Fe (mg/kg)	Zn (mg/kg)	Cu (mg/kg)	Mn (mg/kg)	Moisture (%)	Ash Content (%)
Puno	Raw seed	14.42	44.70	0.98	0.17	0.09	0.28	0.17	57.35	20.54	5.91	28.22	8	3.74
	Processed seed	15.39	44.35	0.53	0.14	0.06	0.31	0.09	61.55	22.30	4.21	19.73	7	2.30
	Quinoa Bran	14.99	46.48	5.67	0.36	0.11	0.10	0.55	101.64	180.85	13.78	83.78	7	18.00
Titicaca	Raw seed	14.47	43.61	1.33	0.19	0.09	0.35	0.12	67.79	30.30	5.76	32.80	6	3.80
	Processed seed	18.83	43.98	0.72	0.18	0.06	0.43	0.10	51.22	30.94	5.49	23.97	6	2.67
	Quinoa Bran	12.65	42.60	8.87	0.43	0.23	0.19	0.51	521.23	61.98	6.75	109.95	6	17.31
ICBA-Q5	Raw seed	12.18	40.40	1.74	0.20	0.09	0.18	0.10	46.51	26.38	2.48	32.09	7	4.85
	Processed seed	11.07	40.20	1.38	0.16	0.09	0.18	0.05	33.99	27.63	3.80	28.11	7	3.50
	Quinoa Bran	14.61	39.36	6.64	0.77	0.18	0.19	0.54	249.04	68.23	6.02	110.91	7	16.69

Figure 2 presents the saponin content in two different seeds polished using two different methods, bulk seeds polished manually by the women cooperative and Puno seeds polished mechanically using a locally manufactured pearling machine. Obtained results indicate clearly that bulk seeds accumulate more saponins compared to Puno seeds. Pearling using the mechanized tool was shown to be more efficient than manual abrasion as the saponin content was reduced by 68% in the case of Puno and 57% in the case of bulk seeds.

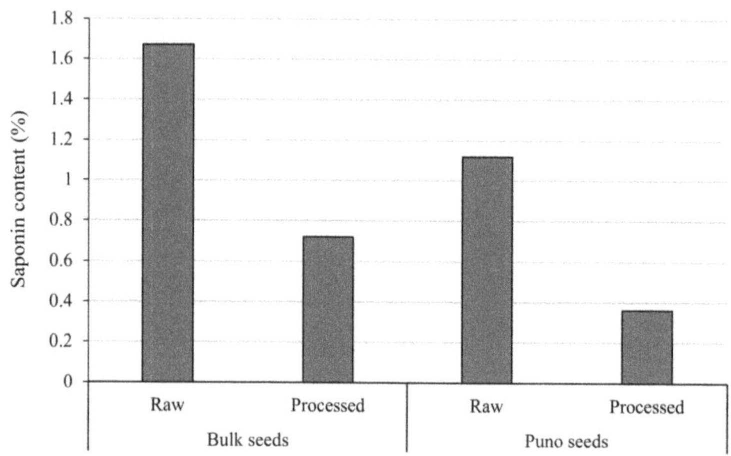

Figure 2. Saponin content (%) of raw and processed bulk and Puno seeds. Bulk seeds were pearled manually and Puno seeds mechanically using quinoa sheller.

2.3. *Quinoa Import*

Import of quinoa in Morocco has known a great evolution since 2015 (first year of record) to reach 84 tons in 2019 with a total value of 2.4 million MAD (Figure 3). The average price per kilogram was greatly decreased. In 2015 the average import price was equal to 80 MAD.kg^{-1} while in 2019 it was equal to 29 MAD·kg^{-1} following the worldwide trend in quinoa price as per the Statista database [17].

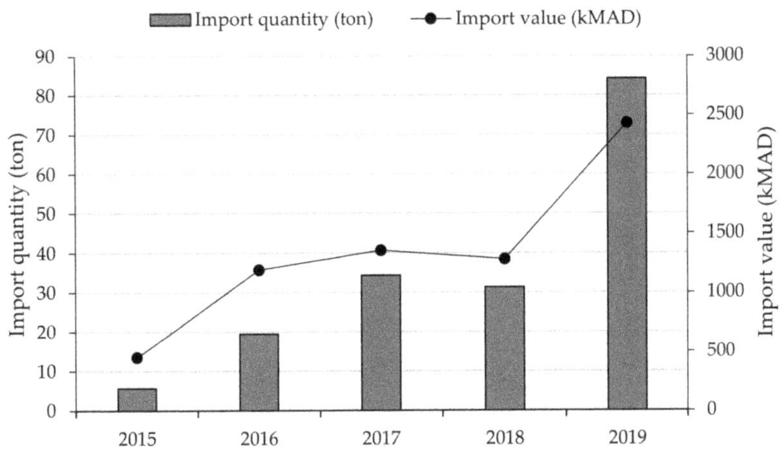

Figure 3. Evolution of quinoa import in Morocco in terms of quantity and value. 1 USD = 9 MAD.

2.4. *Economic Analysis of Quinoa Seed Production*

2.4.1. Quinoa vs. Traditional Cereals: Production Cost and Net Margin

Data presented in Table 4 clearly indicate that for quinoa cultivation, production cost per kilogram is higher under the rainfed and manual production mode compared to the scenario where irrigation is supplied along with fertilizers and organic amendments. Thus, production cost per kilogram decreased from 27 to 11 MAD and this can be explained by the improved yield and reduced cost attributed mainly to irrigation and fertilization and the use of the mechanized tool for quinoa cultivation. Consequently, the net margin was improved significantly due to input supply and the mechanized tools adoption. The presented data also indicate that quinoa is more remunerating than cultivated cereals as it generates five times and twice the net margin generated by barley and wheat, respectively, grown under the rainfed and mechanized scenario.

Table 4. Production cost and net margin of processed quinoa seeds compared to traditional cereals cultivated in the Rehamna region. 1 USD = 9 MAD.

Scenarios	Production System	Yield (kg/ha)	Production Cost (MAD/kg)	Production Cost (MAD/ha)	Net Margin (MAD/ha)
Quinoa Rainfed: Scenario I (manual); Scenario II (mechanized)	Manual	500	26.8	13,400	21,100
	Mechanized	500	19.3	9650	28,850
Quinoa Irrigated with organic amendment and fertilization: Scenario III (manual); Scenario IV (mechanized)	Manual	2000	16.2	32,195	101,805
	Mechanized	2000	11.1	22,445	111,555
Barley under rainfed conditions	Mechanized	2000	1.5	3097	3072
Wheat under rainfed conditions	Mechanized	3000	1.3	3914	6366

2.4.2. Quinoa Production Cost Breakdown

Cost breakdown for quinoa cultivation operations is presented in Table 5. Harvest and post-harvest operations account for the largest part in the overall production cost in the case of rainfed and manual cultivation contributing with 63% in the total production cost. While this percentage was decreased to 50% when mechanized tools for harvest and post-harvest operations were adopted. In the case of the optimized production mode using irrigation and fertilization the part of harvest and post-harvest operations in the total cost was greatly decreased due to high cost attributed to irrigation system depreciation, pumping and fertilizers and was equal to 48 and 26% under manual and mechanized production mode, respectively.

Table 5. Cost breakdown for quinoa cultivation and seed processing operations. 1 USD= 9 MAD.

Operation	Description	Unit	Production Scenario														
			Rainfed × Manual (Objective Yield: 500 kg/ha)			Rainfed × Mechanized (Objective Yield: 500 kg/ha)			Irrigated with Fertilization × Manual (Objective Yield: 2000 kg/ha)			Irrigated with Fertilization × Mechanized (Objective Yield: 2000 kg/ha)		
			Qty	Unit Cost	Total Cost	Qty	Unit Cost	Total Cost	Qty	Unit Cost	Total Cost	Qty	Unit Cost	Total Cost
Ploughing and soil preparation	Deep	Hour	4	200	800	4	200	800	4	200	800	4	200	800
	Shallow	Hour	2	150	300	2	150	300	2	150	300	2	150	300
Fight against bird's attack	Manual	Day	15	100	1500	15	100	1500	15	100	1500	15	100	1500
Sowing	Manual/ seeder	Day /Hour	3	100	300	1	250	250	3	100	300	1	250	250
Irrigation and fertigation (energy, depreciation, and fertilizers)		Ha							1	11795	11,795	1	11,795	11,795
Thinning and weeding	First	Day	10	100	1000	10	100	1000	10	100	1000	10	100	1000
	Second	Day	10	100	1000	10	100	1000	10	100	1000	10	100	1000
Harvest	Manual	Day	30	100	3000				30	100	3000			
Threshing	Manual	Day	20	100	2000				20	100	2000			
Seed cleaning	Manual	Day	15	100	1500				45	100	4500			
Seed washing	Manual	Day	10	100	1000				30	100	3000			
Drying and sieving	Manual	Day	10	100	1000				30	100	3000			
Harvest and threshing (combined)	Mechanized	Ha				1	1800	1800				1	1800	1800
Seed pearling	Mechanized	Kg				500	6	3000				2000	2	4000
Total (MAD)					13,400			9650			32,195			22,445

2.4.3. Sensitivity Analysis of Net Margin vs. Production Cost, Grain Yield and Sale Price

Figure 4 presents the sensitivity analysis elucidating the impacts of +25 and −25% changes of total cost, grain yield and sale price on net profit. Results indicate clearly that net profit is affected greatly by sale price as a reduction or increase of 25% in sale price is likely to have the largest impact on net profit either negatively or positively but with a more pronounced effect on the net profit in the case of price reduction. Holding all other variables at their base-value, a 25% reduction in the output price will reduce net profit by 94.2, 82.8, 74.2 and 67.7 percent under scenario I, II, III and IV, respectively.

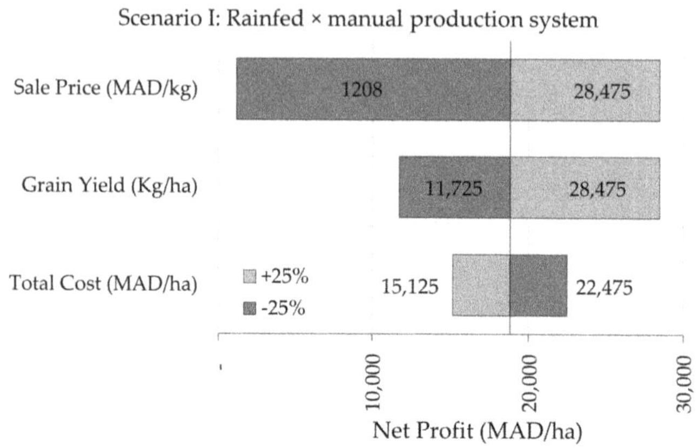

Figure 4. *Cont.*

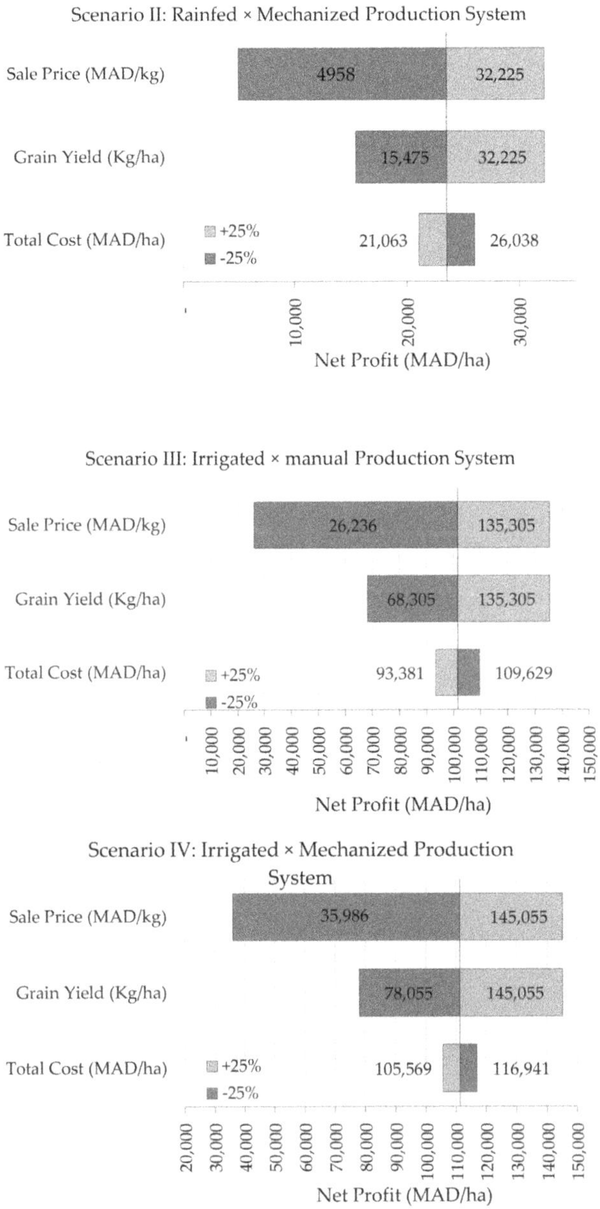

Figure 4. Changes in net profit as responses to ±25% variation in sale price, grain yield and total cost under tested production scenarios. 1 USD = 9 MAD.

2.4.4. Monte Carlo Simulation Analysis (10,000 Iterations)

Monte Carlo simulation performs risk analysis by building models of possible results by substituting a range of values—a probability distribution—for any factor that has inherent uncertainty. It then calculates results over and over, each time using a different set of random values from the probability functions. Depending upon the number of uncertainties and the ranges specified for them, a Monte Carlo simulation could involve thousands or tens of thousands of recalculations before it is complete. In our case the number of iterations made is 10,000 simulating the net profit as affected by changes in market price and yield.

The Monte Carlo simulations presented in Figure 5 as frequencies derived after 10,000 iteration of simulation show that the risk of having a financial loss when producing quinoa is about 2.55%, 0.55%, 0.05% and 0% respectively for scenario I, II, III and IV assuming that changes in yield may occur from 200 to 800 kg·ha^{-1} for scenario I and II and from 1000 to 3000 kg·ha^{-1} for scenario III and IV, and allowing for simultaneous variation in price from a lower bond of 30 MAD·kg^{-1} to an upper bond of 120 MAD·kg^{-1}.

We also carried out a break-even analysis for each of the four cases to identify values of the key parameters that make revenues equal to the cost of production, and as a result the net gains were zero (Table 6). The estimated break-even points were lower than the lower bonds assumed in the simulations (200 kg·ha^{-1} of yield for scenario I and II, 2000 kg·ha^{-1} yield under scenario III and IV and 30 MAD for the price) indicating higher confidence for the main results presented in Table 5 to remain unaffected.

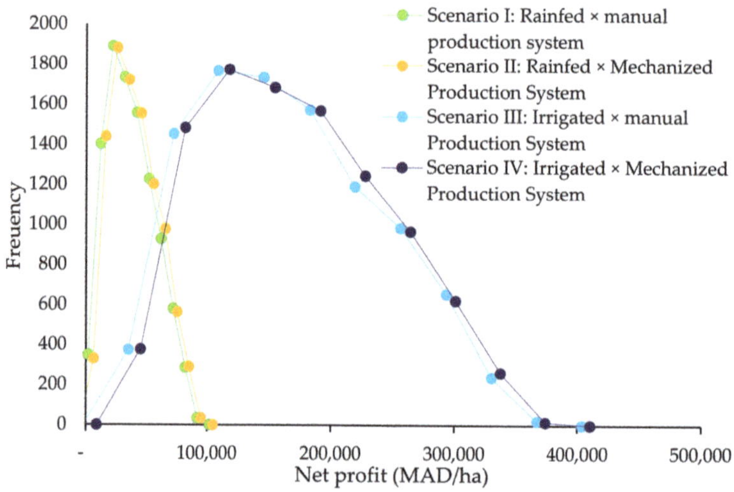

Figure 5. Monte Carlo simulation frequency of net profit as affected by a change in market price and yield. 1 USD = 9 MAD.

Table 6. Break-even analysis. 1 USD = 9 MAD.

Scenario	Yield (kg·ha^{-1})	Price (MAD/kg)	Total Cost(MAD/ha)
Scenario I: Rainfed × manual production system	219.50	29.40	33,500
Scenario II: Rainfed × Mechanized Production System	148.51	19.90	33,500
Scenario III: Irrigated × manual Production System	485.00	16.25	101,505
Scenario IV: Irrigated × Mechanized Production System	339.48	11.37	134,000

2.4.5. Cost Breakdown of Quinoa Based Products

Table 7 presents cost breakdown for quinoa products such as quinoa couscous, processed seeds and quinoa flour in Morocco. Raw material has the largest contribution to the total production cost for all products with almost 80%, 87% and 85% for quinoa couscous, flour and processed seeds, respectively. Labor cost consisting of women working on different steps of production is also important as it represents more than 13% in the case of couscous while it represents only 1.88 and 3.64% of the total production cost for quinoa flour and processed seeds, respectively. Consequently, among all quinoa products, quinoa couscous remains very expensive compared to other cereals couscous (max market price is equal to 20 MAD·kg^{-1}). Obtained data indicate that the women cooperatives producing quinoa products should reduce raw material cost through their own production of quinoa seeds, better planning for quinoa stocks and elaborating the sale contract with quinoa producers for a low price.

Table 7. Cost breakdown for quinoa couscous, flour and processed seeds production. 1 USD= 9 MAD.

Quinoa Product	Inputs	Description	Cost (MAD/kg)	%	
Quinoa Couscous	Raw material	Quinoa semolina	21	28.14%	80.40%
		Quinoa flour	39	52.26%	
		Salt	0.004	0.01%	
	Water	Washing	0.02	0.03%	0.04%
		Processing	0.002	0.00%	
	Women's labor force	Sorting	5	6.70%	13.57%
		Hydration	5	6.70%	
		Packaging	0.12	0.17%	
	Energy	Electricity	0.05	0.08%	0.48%
		Gas	0.3	0.40%	
	Transportation		1	1.34%	1.34%
	Packaging	1 kg package	1.46	1.96%	1.96%
	Depreciation	Dryer	1.64	2.20%	2.22%
		Heat-sealing machine	0.01	0.01%	
	Total		74.62	100%	100%
Quinoa flour	Raw material (Quinoa seeds)		50	87.66%	
	Washing water		0.019	0.03%	
	Energy		0.436	0.77%	
	Labor force		1.073	1.88%	
	Packaging		2.975	5.22%	
	Transport		0.951	1.67%	
	Depreciation		1.580	2.77%	
	Total		57.03	100%	
Quinoa processed seeds 1 kg	Raw material (Quinoa seeds)		50	85.65%	
	Washing water		0.02	0.03%	
	Energy		0.45	0.78%	
	Labor force		2.12	3.64%	
	Packaging		3.12	5.34%	
	Transport		1	1.71%	
	Depreciation		1.65	2.83%	
	Total		58.37	100%	

2.5. SWOT Analysis of the Quinoa Value Chain in Morocco

Table 8 presents analysis of the quinoa value chain in Morocco including its strength. weaknesses. opportunities and threats at different levels.

2.6. Lessons Learned

Several lessons were revealed by the present study as presented in Table 9.

2.7. Perspectives of Development

With several challenges (drought and salinity) facing staple crops such as cereals in many marginal areas in Morocco. quinoa could be a judicious solution to improve food security and increase farmer's income. A national program of quinoa in Morocco is becoming necessary and needs to tackle all value chain components. We suggest several actions to improve the quinoa value chain in Morocco summarized in Table 10.

Table 8. SWOT (strength. weaknesses. opportunities and threats) analysis of the quinoa value chain in Morocco.

Strengths	Weaknesses
In terms of cultivation: • Quinoa is more profitable compared to cereals. • Farmer know-how in terms of cereal production is compatible with quinoa cultivation. • Tolerance of quinoa to various stresses that characterize the region including drought and salinity. • Quinoa byproducts such as leaves. straw and saponin could be potentially valorized. • Low requirement in terms of agricultural inputs (fertilizers. management. pesticides. etc.). *At the gastronomic level:* • High nutritional value compared to cereals. • Quinoa seeds are gluten free with low content in sugar. which make it optimal food for diabetic and celiac consumers. • Fast cooking. • Versatility of quinoa-based recipes. • Easy to integrate into food habits given its resemblance to locally prepared dishes (soup. boiled rice. couscous. bread. etc.).	*At the production level:* • Small production area compared to the potential. • Poor organization of producers among those who have adopted quinoa. • Quinoa is labor intensive with very few mechanized operations (especially at the post-harvest phase). • Problems linked to quinoa establishment at the field level (low germination). • Unavailability of good quality seeds. • Sensitivity to diseases such as downy mildew. • Attacks of birds (during the emergence and maturity stage). • Lodging problem in the case of strong winds. • High post-harvest costs. *At the valorization and marketing level:* • Transformation pathway is not well structured and mastered. • Basic marketing channel. • Lack of promotion and communication around quinoa-based products. • Uncontrolled price formation. • Poor product quality. • High transformation cost.
Opportunities	**Threats**
• The Rehamna region presents edaphic-climatic conditions favorable for rustic crops such as quinoa. • Policies and development program encouraging the introduction and development of alternative crops (e.g. Green Morocco Plan and Generation Green). • Replacement of cactus that was completely devasted by cochineal by quinoa in the Rehamna region. • Growing interest by the national and foreign researchers in the adoption of quinoa by farmers. • Willingness of national and international development agencies such as the OCP group (Office Chérifien des Phosphates). ONCA (Office National du Conseil Agricole). DPA (Direction Provinciale d'Agriculture). Universities. IDRC (International Development Research Center). FAO (UN Food and Agriculture Organization). etc.. to promote and accelerate the process of adopting quinoa in the area. • Availability of national and international agriculture fairs (Salon International d'Agriculture de Meknes) for quinoa products showcase. • Increased interest for healthy food consumption by individual consumers and restaurants. • Growing quinoa international market. • Availability of labours in the rural areas. • Remunerative price. • Possibility for quinoa product export to the European market.	• Competitiveness of local quinoa products compared to imported ones. • Substitute products are numerous. • High cost and slow process of organic certification. • Climatic variability and negative effects of drought and heat waves on quinoa production. • Loss of varietal purity and genetic performance due to the use of harvested seeds for many years. • Quinoa products supply exceed the demand.

Table 9. Lessons learned from the quinoa value chain in Morocco.

Lesson Learned	Action Taken/Needed
At the Farm Level	
First. more awareness should be raised among farmers. relevant government entities. private sector and general public about the economic benefits of quinoa and its potential as an alternative crop tolerant to stress and soil–climate conditions of the Rehamna region.	Several training sessions were organized for farmers. women cooperatives and extension agents about quinoa cultivation and its virtues.
In a process of introducing new crop such as quinoa. more technical and economic information is needed to dispel the hesitations of some farmers who are faithful to their usual practices.	Several extension material and brochures about quinoa including a farmer practical guide about quinoa cultivation were produced and shared with farmers and extension agents during organized trainings and workshops.
At the level of production techniques: In spite of the important peasant know-how. quinoa remains a new crop and therefore obeys a logic of adoption. which means categories of progressive farmers (willing to take risk). neutrals (those who see no objection to the introduction of quinoa) and recalcitrant (who are unwilling to question their crop rotation plan). In the first two categories. even a light training in production techniques is necessary.	Field trials conducted by students can serve as a demonstration platform before generalization. The agricultural advisers (institutional partners of the project. ONCA) should act as a link between the results of the experiments and the introduction of quinoa into the farms.
One of the constraints limiting quinoa production is the labor costs. which are estimated by farmers to be excessive compared to the margin generated by the sale of quinoa.	Mechanization of cropping practices is necessary to reduce labor cost including sowing. harvest. threshing. seed pearling and even packaging. Individual small farmers cannot afford those tools. thus. farmer's organization in cooperatives or associations is considered a judicious option to acquire mechanized tools to be used collectively by the adhered farmers. Several mechanized tools including threshing. winnowing. pearling and seed washing have been locally manufactured and provided to several cooperatives.
Farmers in the Rehamna region usually use the harvested seeds to be sown in the next season for several years. which led to a loss of genetic performance of the initially introduced lines and therefore low germination rate and low performance are usually occurred.	Five introduced varieties (ICBA Q1–Q5) have been registered in the national germplasm catalogue and transferred to a local seed production farm (Benrim Farm) to sustain the production of high quality and homogeneous variety seeds.
In most of farms. quinoa is produced under an organic mode (without application of chemicals) but without certification. Therefore. organic certification is a good option to better valorize quinoa seeds in Morocco and to target international market that require such as certified products (e.g., European market).	The first group of organic quinoa producers has been created in Rehamna in 2018 formed in a first stage by 5 farmers and received organic certification in 2019.
At the Valorization Level (Women Cooperatives)	
Seed pearling and saponin removal remains the most critical post-harvest operation as the final quinoa seed quality depends on this step. In most of the cases seed pearling is performed manually. which increase the cost without reaching the saponin content threshold (0.12%) recommended by the CODEX [18].	A pearling machine or sheller (described in the Harvest and Postharvest machines section) has been locally manufactured and preliminary results show good performance of the machine in removing saponin. Several shellers have been distributed to several farmer's and women's cooperatives.
The breakdown of quinoa-based product processing costs shows that raw materials account for more than 80% of the total costs. followed by labor.	• The reduction in the cost of raw materials can be achieved by a combination of several practices: ○ The own production of quinoa. which can also improve the tracking of production. ○ The purchase of quinoa grains. in large quantities. during the harvest period. ○ The improvement of quinoa yields per hectare using intensive production system (irrigation. fertilization. etc.).
Weak organization of the women cooperatives linked to several administrative issues mainly due to poor management of the unit. poor distribution of responsibilities. lack of operation's records. weak coordination. opportunism. decision making. etc.	Training about best practices for cooperative governance and management has been delivered to several women cooperative members to build their managerial and leadership capacity.

Table 9. *Cont.*

Lesson Learned	Action Taken/Needed
At the Market Level	
The application of high prices of imported quinoa showcased in supermarkets with small quantities to the locally produced quinoa seeds constitutes a bottleneck in the quinoa marketing in Morocco. Pricing strategy should consider the price of similar products based on other cereal (e.g. couscous). import price. the production cost and the willingness price of the Moroccan consumer. The current price of locally produced quinoa seeds remains high as perceived by the Moroccan consumer.	A marketing study was performed in order to determine the psychological price of quinoa in Morocco [19]. The study indicates that the Moroccan consumer is willing to pay 4–5 USD for one kilogram of processed quinoa seeds.
Promotion and communication around quinoa products and virtues in Morocco still needs to be further developed.	Several promotion and communication activities were organized to showcase quinoa products including: • The first quinoa promotion workshop organized in November 2018 inviting a celebrity Moroccan chef to lead a cooking session. • Showcase of quinoa products made by women cooperatives in the international fair of agriculture in Meknes in April 2019 (SIAM 2019). • Tasting sessions organized in several events including SIAM 2019. • Participation in the African Fair of the Social and Solidarity Economy held between 29 October to 5 November 2019 in Senegal by the 3 millennium women cooperative. • Inviting national TV and press to elaborate stories about quinoa in Morocco. • Project video capsules regularly published in YouTube and other social media channels.
There is a need to develop a specific and unified packaging labeled "Quinoa Rehamna" for all beneficiaries involved in the project including the "gluten free" and "organic" label.	A branding and visual identity document about "Quinoa Rehamna" brand has been elaborated and shared with relevant stakeholders.
For the moment. quinoa producers only commercialize their production at local fairs. weekly rural markets. cereal markets. healthy and organic food shops and individual clients (consumers. restaurants. etc.). Quinoa producers in Morocco could not market their products in the local supermarkets due to a lack of a sanitary certificate for quinoa seed processing and transformation.	• Quinoa producers especially women cooperative should first improve their valorization unit to meet the ONSSA (National Food Safety Office) requirements to get the sanitary certificate. which is required by supermarkets. • Quinoa producers should also develop partnership with food industries and bakeries to produce quinoa products adapted to the Moroccan context such as couscous. noodles and bakeries and to be marketed at large scale benefiting of their own distribution networks. • The organic quinoa producer group recently created should explore export opportunities towards the European market.

Table 10. Proposed development actions and involved actors for each value chain components.

Value Chain Component	Development Actions	Involved Actors
Agricultural inputs supply	• Registration of quinoa performing varieties • Improve the availability of high-quality seeds (seed production systems). • Improve the availability of agricultural inputs such as fertilizers. pesticides. etc., especially in remote area (e.g., South of Morocco).	• Agricultural input suppliers. • Seed production organization and private companies (e.g., SONACOS seed company). • Agriculture Ministry departments.
Quinoa production	• Adoption of best cropping practices such as: ○ Optimal sowing date. ○ Irrigation supply (deficit or supplemental irrigation). ○ Application of amendments (manure. compost. etc.). ○ Application of fertilizers. ○ Increase sowing density to reduce plant lodging. ○ Plant thinning and hilling. • Plant protection: ○ Protection from bird attacks (Maghreb lark) at the early stage (emergence and seedling). ○ Protection against caterpillars at early stage. ○ Treatment against downy mildew. ○ Use of a wind-break to protect quinoa from strong wind and reduce lodging. • Harvest: ○ Plant harvest at optimal stage to avoid grain loss. ○ Use of mechanized tool for harvest (e.g., adapted combined harvester). • Continuous training for farmers about best cropping practices.	• Farmers. • Farmer's cooperatives and associations. • Agriculture Ministry departments. • Research and development institutions.
Quinoa transformation and valorization	• Use of a mechanical sheller for saponin removal. • Respect of the transformation line and separation in time and space for gluten free products. • Improve women cooperative's management skills. • Improve packaging and storage conditions. • Build the technical capacity of women in the best quinoa transformation practices.	• Women cooperatives. • Agriculture Ministry departments. • Research and development institutions. • Food industries.

Table 10. *Cont.*

Value Chain Component	Development Actions	Involved Actors
Marketing and distribution	• Elaborating marketing study for specific quinoa products. • Use of social media and influencer to promote quinoa products. • Develop the "quinoa Rehamna" label. • Develop a sale contract with super and hypermarkets. • Target groceries and weekly markets. • Target E-commerce platforms. • Develop contracts with food industries. • Target direct clients such as restaurants. hotels. individuals. etc. • Participation in food and agricultural fairs.	• Farmers. • Farmer's cooperatives and associations. • Women cooperatives. • Agriculture Ministry departments. • Research and development institutions. • Supermarkets. • Retailers.
Consumption	• Conduct promotion activities including: ○ Video capsules in TVs. radios and social media. ○ Organize tasting sessions in food fairs and supermarkets. • Develop promotion materials such as brochures. posters. etc.	• Women cooperatives. • Research and development institutions. • Supermarkets. • Retailers.
Research and development	Further research and development activities are needed in the following subjects: • Quinoa breeding for new adapted and stable varieties. • Optimization of quinoa fertilization. • Mechanized sowing. • Harvest and post-harvest mechanized tools development. • Development of saponin removal methods and techniques. • Valorization of quinoa byproducts (saponins). • Market assessment.	• Farmers. • Farmer's cooperatives and associations. • Women cooperatives. • Agriculture Ministry departments. • Research and development institutions.

3. Discussion

Since its introduction to Morocco in the 2000s. quinoa was seen as a rustic and stress tolerant crop with several potentialities to replace cereals and other traditional crops in the marginal environment of Morocco. Therefore. it was subjected to various trials at the field and pot level in several regions to evaluate its productivity and responses to various stresses. For instance. the finding of this study indicates that quinoa yield was tripled for most of the tested cultivars under full irrigation conditions compared to rainfed. which support the results obtained by Fghire et al. [8] who found that the yield of Puno cultivar (one of the tested cultivars in the present study) conducted in the same study area (Tnin bouchane) was increased by 236% under full irrigation (100% ETp) compared to rainfed irrigation. Our results are in agreement with Geerts et al. [20] who reported that full irrigation increased quinoa yield with 27% compared to rainfed conditions. Contrarily to our case. this low increase percentage is mainly explained by the high amount of rain received (330 mm) for rainfed treatment compared to irrigated treatment (245 mm irrigation

+ 330 mm rainfall). While in the present study quinoa under farm conditions received only 43 mm of rain during its growing season while the irrigated treatment received 200 mm of irrigation. which obviously explain the tremendous increase of yield and other growth parameters as a response to irrigation.

It is well known that organic matter amendment has a positive effect on crop growth and productivity. In the case of quinoa. very few studies are available evaluating the effect of organic amendment on yield; nevertheless. other crops were cultivated under the organic amendment and showed a positive response [21–23]. Our results suggest that the ICBA-Q3 cultivar grown under controlled experimental conditions significantly ($p < 0.01$) responded to the organic amendment only after applying 40 T/ha of manure while no significant difference was observed under a lower dose. While under farm conditions. quinoa yield and plant height were significantly improved under organic amendment application. Our results are in agreement with Hirich et al. [10] who found that that organic amendment of 10 t ha^{-1} and 5 t ha^{-1} significantly increased seed yield by 13% and 3%. respectively. under full irrigation. The yield improvement under organic amendment is explained by a soil content increase in terms of nutrients after mineralization of the organic matter; therefore. the nutrients uptake will be increased. which will result in high plant growth and productivity [24].

Our finding in terms of seed nutrient content indicates that most of the micronutrients content was reduced in the processed seeds due to seed pearling. which means that they are mostly concentrated in the pericarp (bran) as suggested by Konishi et al. [25] and D'Amico et al. [26] who reported that minerals are accumulated in the pericarp (seed out layer) and proteins are mostly accumulated in the endosperm tissues. In terms of mineral content. Konishi et al. [25] reported that phosphorus is mainly localized in the embryonic tissues. which explain why P content in the present study for all tested varieties was higher in the processed seeds and lower in bran. However. in terms of magnesium and potassium our finding indicates high content in the seed bran compared to processed seeds. which disagrees with Konishi et al. [25] who reported that both magnesium and potassium are located in the embryonic tissues.

One of the limiting factors for quinoa valorization and transformation is its content in terms of saponins. which are mainly concentrated in the pericarp or bran and need to be removed before use [27]. The cultivars we tested in this study are classified as bitter [28]. and the saponins removal level by either mechanical abrasion or manual polishing was not enough to classify the quinoa as sweet. since the saponin content threshold for human consumption is equal to 0.12% according to CODEX [18]. Our results indicate that Puno seeds pearling using mechanical abrasion resulted in 68% reduction in terms of saponin content. which confirm the finding of Hirano and Konishi [29] who reported that quinoa seed pericarps contain 67.6% of the total saponin content in the whole grain while the rest is remaining in the seed endosperm and other internal layer. Our results are in agreement with Mhada et al. [30] who reported that mechanical abrasion allowed the reduction of saponins level from 1.4 to 0.51% for the Puno cultivar. a reduction of 64% of the initial saponin level. Our findings are also in line with Gómez-Caravaca et al. [31] who reported that an abrasion degree of 20% allowed reducing the saponin levels in pearled quinoa (129.8 mg/100 g d.w.) more than 50% comparing with the initial saponin content in whole quinoa (244.3 mg/100 g d.w.).

In Morocco the quinoa market is very limited and still a niche market. In order for the market to expand. huge effort is required to promote for quinoa and a rise in awareness among consumers about its health benefits. Like other countries. such as Turkey. quinoa consumption is limited to those with knowledge of health foods for specific health benefits. including its gluten free status. Quinoa is not a product "consumed by the masses". but rather one "discovered" by educated. health-conscious consumers [32]. The economic analysis showed that under rainfed conditions production cost at harvest per hectare varies from 6650 to 7900 MAD (739–878 USD) for the mechanized and manual production mode. respectively. While under irrigated conditions the production cost increased to 18.445 and 19.695 MAD·ha^{-1} (2049 and 2188 USD·ha^{-1}) for the mechanized and manual production mode. respectively. due to depreciation of the irrigation system. energy and fertilizers input costs. The same trend was reported by Yazar et al. [32] in Turkey who found that the production cost of quinoa was equal to 728 and 1650 USD·ha^{-1} under rainfed and irrigated conditions. respectively. Our results are also in agreement with Mercado and Ubillus [33] who reported that the production cost of quinoa in Peru varies from 676 to 2604 USD·ha^{-1} for traditional rainfed and conventional production system. respectively. with a profitability that varies from 100 to 200% and market price varies from 1.7 to 2 USD·kg^{-1}. However. in Morocco the profitability could vary from 150 to 500% due to high quinoa price and the use of intensive production systems (irrigation and mechanized tools).

This study presents a SWOT analysis of the quinoa value chain in Morocco. which revealed that one of the main weaknesses limiting quinoa market expansion in Morocco is the traditional production and valorization and the lack of using intensive production tools. Thus. quinoa price in Morocco remains relatively high above middle class consumer's purchasing power even at farmgate and only rich people can afford it. This way the local quinoa products with a high price and relatively lower quality could never compete with an imported one. which have usually good quality. The trendy nature of the market for quinoa in Morocco has had both positive and negative aspects. Certainly. growers have benefited from the rising prices that the crop commands. though the various intermediaries may reap more of the profits than the small growers [34]. Another bottleneck in the quinoa value chain is a lack of promotion around quinoa benefits using public channels and social media. which is considered a key point for any new product development [19].

4. Materials and Methods

4.1. Study Area

The province of Rehamna is geographically located between Marrakech (South). Settat and El Jadida (North). El Kelaa des Sraghna (East) and Sidi Bennour. Youssoufia to the west (Figure 6). This region is characterized by an average rainfall of 177 mm with an intra and interannual variation (Figure 7). Temperatures are relatively homogenous throughout the zone. with temperatures ranging from 4 to 46 °C. Prevailing winds are from the North-East in winter and from the West in summer. Warm winds (Chergui) are frequent and blow from the East and South. The total agricultural area of the region of Rehamna is 591.125 Ha with an arable land area of 342.500 Ha (35.425 Ha of irrigated area and 307.075 Ha of rainfed area) (DPA Rehamna 2018).

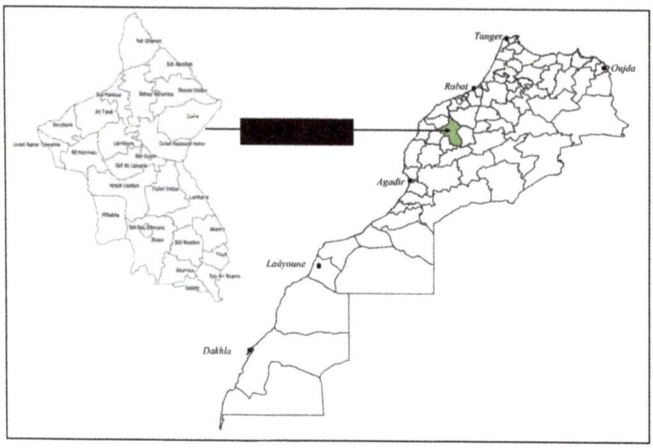

Figure 6. Rehamna province localization (study area).

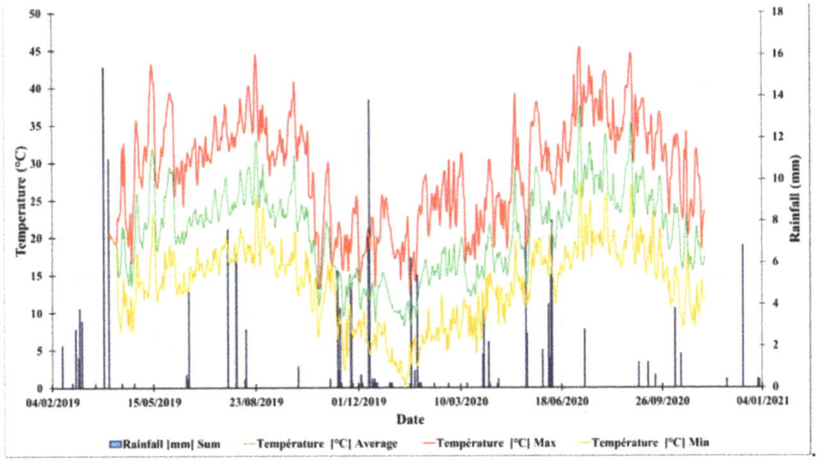

Figure 7. Rehamna climatic conditions taken from the UM6P Experimental Farm meteorological station.

4.2. Soil and Water Analysis

Table 11 presents the physicochemical analysis of soils at the UM6P experimental station and farm level. Soil texture is clay loam at the UM6P experimental station and sandy loam at the farm level. Both soils are considered poor in terms of organic matter and rich in terms of potassium.

Table 11. Physicochemical analysis of soil at the UM6P experimental station and farm level.

Parameter		Unit	At UM6P Experimental Station Level	At Farm Level	Analysis Method
Granulometric composition	Clay	%	32.15	6.03	NF X 31-107
	Silt	%	26.45	23.54	NF X 31-108
	Sand	%	42.34	70.46	NF X 31-109
pH-water			8.38	7.9	NF ISO 10390
Electric Conductivity (EC) 1/5 at 25 °C		$mS \cdot cm^{-1}$	0.19	0.1	NF ISO 11265
Total limestone ($CaCO_3$)		%	4.83	0.2	NF EN ISO 10693
Organic matter		%	1.07	0.8	NF ISO 14235
Phosphorus (P_2O_5)		$Mg \cdot kg^{-1}$	17.15	43.45	NF ISO 11263
Potassium (K_2O)		$Mg \cdot kg^{-1}$	455.45	216	NFX 31-108
Ammonium (NH_4^+)		$Mg \cdot kg^{-1}$	6.45	10.25	NFX 31-109
Nitrate (NO_3^-)		$Mg \cdot kg^{-1}$	7.91	12.50	NFX 31-110

According to Table 12. both irrigation waters were slightly saline with more salinity and mineral content obtained for irrigation water at the UM6P experimental station.

4.3. Tested Cultivars

In this study ICBA (International Center for Biosaline Agriculture) quinoa cultivars were introduced due to their high adaptation to MENA and Morocco conditions and resistance to drought and salinity [4]. The origins of those cultivars are: low land. Bolivia for ICBA-Q1. Q2 and Q3 and coast. Chile for ICBA-Q4 and Q5. Those cultivars were already introduced and tested in the south of Morocco (Laayoune area) within a previous R&D project and showed high performance under salinity conditions with an average seed yield exceeding 2 t/ha. In addition to ICBA cultivars. two Danish public varieties were tested. Titicaca and Puno. two short cycle varieties widely cultivated in Morocco and showed a good adaptation with higher yield compared to other quinoa accessions [4,5,30,35].

Table 12. Physicochemical analysis of irrigation water at the UM6P experimental station and farm level.

Parameter	Unit	At UM6P Experimental Station Level	At Farm Level	Method
pH		9.28	7.91	NM ISO 10523
Electric conductivity at 25 °C	mS·cm^{-1}	2.50	1.44	IM ISO 7888
Dry extract	g.l^{-1}	1.60	0.92	NM 03.7.027
		Cations		
Potassium (K^+)	meq.l^{-1}	0.22	0.01	Continuous Flow Analysis (CFA)
Sodium (Na^+)		14.67	5.40	Continuous Flow Analysis (CFA)
Calcium (Ca^{2+})		3.50	2.31	Continuous Flow Analysis (CFA)
Magnesium (Mg^{2+})		6.80	6.45	Continuous Flow Analysis (CFA)
		Anions		
Chloride (Cl^-)	meq.l^{-1}	17.44	7.17	Continuous Flow Analysis (CFA)
Carbonate ($CO_3{}^{2-}$)		2.40	0.17	NM ISO 9963-1
Bicarbonate ($HCO_3{}^-$)		2.60	0.77	NM ISO 9963-2

4.4. Trial Installation

4.4.1. At the Farm Level

An on-farm trial was conducted at the farm level (Tnin Bouchane. 32°14.6267′ N. 8°19.8181′ W. 280 m + MSL (mean sea level)) testing ICBA cultivars (ICBA Q1-Q5) compared to locally cultivated bulk seeds (mixture of L119 and L143 accession) under rainfed. irrigation and irrigation with cow manure amendment (40 T/ha) conditions in a split plot design with 4 replications (plot size was equal to 10 m^2). Organic amendment was applied along with soil preparation before sowing. Irrigation practices were performed according to farmer usual practices with an irrigation supply of about 200 mm (2000 m^3/ha) for the whole cropping period. using drip irrigation following evapotranspiration demand according to Allen et al. [36]. Quinoa seeds were sown using a plant density of 8 plants/m^2 (50 cm between lines and 25 cm between plants). The trials were carried out between 18 February and 30 June 2018.

4.4.2. At the UM6P Experimental Farm

Another trial was carried out in the UM6P experimental farm (Ben Guerir. 32°13.08″ N. 7°53.23′ W. 468 m + MSL (mean sea level)) to investigate the performance of six quinoa cultivars including ICBA-Q1. ICBA-Q2. ICBA-Q5. Titicaca. Puno and locally cultivated bulk seeds under Rehamna conditions. The objective of this trial was to assess the productivity of tested cultivars and their adaptation to Rehamna agroclimatic conditions. The trial was conducted in a completely randomized block design with four replications. Plot size was equal to 100 m^2. Irrigation was applied following the evapotranspiration method according to Allen et al. [36] using parameters from the existing weather station. Irrigation volume supplied was equal to 300 mm (3000 m^3/ha). Trials were carried out between 21 February and 25 June 2018.

4.5. Agronomic Practices and Seed Yield Determination

All trials were subjected to commercial agronomic practices such as soil preparation. preirrigation. weeding (3 times during the growing period). phytosanitary treatments (application of insecticide treatment against caterpillar in the seedling stage) and plant thinning (keeping only one or two plants per sowing hole). Quinoa seeds were sown using a plant density of 8 plants/m^2 (50 cm between lines and 25 cm between plants).

Seed yield for all trials was determined after maturity. Quinoa panicles were harvested first and dried in open air. Seeds were extracted using manual threshing followed by seed polishing and cleaning.

4.6. Seed Pearling

The saponin elimination process remains as a critical operation in seed processing. Recently. many appropriate technologies have been developed to remove saponins to an acceptable threshold without affecting the nutritional properties of the seed.

Puno seeds processed mechanically by Benrim farm were polished using a pearling machine that was locally manufactured for a duration of two minutes. The machine operates on a semi-industrial scale with a transformation capacity of 120 $kg \cdot hr^{-1}$. it is equipped with two motors. the first one is designed to turn a drum with a rotation speed of 750 rpm. The second one is more powerful (3000 rpm) and designed to extract the fine dust produced during the pearling process. The rotating drum is made of 80 cm long perforated stainless steel and has 6 baffles distributed throughout the drum. During seed processing. the speed of rotation and friction (seed-seed. seed-drum and seed-baffles) gradually increased the temperature of the seeds to 35 °C and decreased the moisture from 13 to 10%.

At the level of the women's cooperatives in Morocco. saponins elimination from quinoa seeds was performed manually with traditional equipment. The majority of valorization units (women's cooperatives. startups. etc.) used combined operations starting with a dry method (manual abrasion) and finishing by a wet one (washing using water). First. a manual abrasion using a glove against a rough surface (e.g., rubber. sieve) is carried out to eliminate the external coat of the episperm (bran). This manual abrasion operation is time consuming and requires effort; it takes one hour to dehull 6 kg of quinoa seeds. Women's cooperatives are using a partial manual abrasion in order to avoid losing the embryo and preserve the seed morphological aspect. However, the residual saponin still remains above the CODEX [36] threshold (0.12% of dry matter) and the bitterness perception is still present. Hence. they add a washing step as a supplementary operation to totally eliminate the bitterness. For the washing operation. seeds soaked with water allowing saponins to dissolve. They use a water quantity of 20 L per 5 kg of quinoa seeds for 10–15 min. This quantity is used three times soaking a total of 15 kg of polished quinoa seeds. Finally, the processed quinoa is dried for 5 h.

4.7. Chemical Analysis

4.7.1. Nutrient Analysis

Raw and mechanically processed seeds and resulted bran of the Puno variety were used for nutrient content determination following the steps below:

- After weighing. harvested samples were ground to a fine powder using the FOSS CT 293 Cyclotec grinder (Fisher Scientific, Canada).
- The moisture content was measured by drying 100 g of sample at 105 °C for 48 h
- Crude protein was determined using Kjeldahl (Buchi, Switzerland, AACC 46–10) method with a conversion factor of 6.25. Micronutrients were determined after sample mineralization.
- Representative samples (0.25 g) were digested with 7.5 mL of HNO_3 acid in the DigiPrep System (SCP SCIENCE, France) during two hours at 100 °C.
- After digestion. the solutions were filtered through 45 µm filters. and the filtrates were diluted to 50 mL with deionized water and acidified (2% HNO_3) in order to undergo the analysis by ICP-OES using Agilent technologies 5110 ICP-OES (Agilent, United States of America) for the elements P. K. Mg. Ca. Cu. Mn. Fe. Zn and B.

4.7.2. Saponin Analysis

In addition to Puno seeds (raw and mechanically processed seeds) used for nutrient content determination. the extraction and quantification of saponin content was performed on locally produced and manually processed by the Bouchane cooperative.

Saponin extraction was performed grinding one gram of dried sample to a fine powder and dissolving in 20 mL of 20% isopropanol. The blend was heated to 86 °C for 20 min for saponin extraction by a microwave-assisted method and filtered (Whatman filter paper) for further quantification. Saponin concentrations were measured by spectrophotometric methods as described by Gianna et al. [37] with minor modifications. The Liebermann-Burchards (LB) reagent was used to quantify saponins. as it is capable of producing a light brown coloration if these compounds are present in a sample. The LB reagent was a 1:5 mixture of acetic acid and sulphuric acid. respectively. Following mixing 1 mL sample solution with 3.5 mL LB reagent. the absorbance at wavelength 580 nm was measured in all samples after 10 min. A calibration curve based on pure quinoa saponins was used for determining the final saponin concentration (mg/mL) in each solution on the basis of absorbance measurements (absorbance = 4.5725 × saponin concentration + 0.0164). The percentage of saponin content was determined on the basis of fresh weight [38]. For nutrient content and saponin determination. three replications have been analyzed. Pictures of grains have been taken using optical microscopy Nikon Eclipse Lv100nd-motorized microscope (Nikon, France) with episcopic/diascopic illumination that enables control of objectives and light intensity from the camera control unit and automatically detects the observation method.

4.8. *Farmer's Survey*

Quinoa field production cost was determined using face-to-face interviews with farmers cultivating quinoa bulk seeds under different cropping systems (rainfed. irrigated. organic amendment and mechanized). Three farmers from each cropping system were selected and interviewed. The following questions related to production cost breakdown were included in the survey:

- Field operation costs;
- Plowing: deep. superficial;
- Organic amendment: quantity. application;
- Irrigation system: purchase. installation;
- Seeds: quantity. price;
- Sowing: manual. seeder;
- Irrigation: workforce. energy;
- Fertilization: manual. fertigation;
- Weeding: manual. chemical;
- Phytosanitary treatment: insecticide. fungicide;
- Harvest: manual. mechanical;
- Other operations;
- Post-harvest operation costs;
- Yield;
- Panicle drying;
- Threshing: Mechanical. Manual;
- Cleaning: Mechanical. Manual;
- Washing: Mechanical. Manual;
- Seed drying;
- Weighing and packaging;
- Labeling;
- Other operations.

4.9. *Sensitivity Analysis and Monte Carlo Simulations*

Quinoa is not yet a well-established crop in the local production systems. A lack of farmers' experience and the possibility of a shortage of inputs. especially planting materials. could possibly result in greater yield variability among farmers. The economic performance of quinoa will depend critically on the actual yield performance of different varieties and households' characteristics and their management practices. which greatly varies among farmers. Beyond production. the markets for quinoa are not well developed. Imperfect and non-competitive markets may fail to clear at competitively determined prices. Poorly functioning markets may therefore pose price risks to the local producers. Hence. such variabilities in yield and potential price volatility may alter the results presented under the base-case scenarios (Table 4). We run simulations of the base-case results to factor in for potential production and price uncertainty.

We first conduct deterministic sensitivity analysis by changing a single parameter. whilst holding all other parameters of the model at their baseline values. In our case. the deterministic analysis was carried out by allowing for a 25% change in yields. prices and total costs to model and assess the sensitivity of net gains for each scenario. While the conventional one-way sensitivity helps determine the scale of impact of a single parameter and its limitation is that it does not proved insights into the probability of such a change (e.g., it does not explain how likely it is for the parameter of interest to take a specific value). Moreover. the deterministic approach fails to take into account the correlation between the values taken by the parameter of interest and other parameters in the model that are held constant [39].

Unlike the deterministic case. simulations allow for stochastic and simultaneous variations and shocks in multiple parameters using the principals of inferential statistics. To evaluate the contemporaneous impact of variations in yields and prices. we then construct a dynamic variant of the model to estimate all possible outcomes given a probabilistic distribution in yields and prices. Using a Monte-Carlo simulation method. we assign multiple values to yields and prices by generating random numbers that follow a symmetric triangular distribution and uniform distribution with lower and upper bonds. respectively. Note that the symmetric triangular distribution is a probability distribution with a probability density function (PDF) shaped like a triangle allowing for central tendency towards the "most-likely or the base-case value". It therefore gives due weightage to the mean value in the yield with frequent outcomes clustered around the most-likely value. Uniform or triangular distribution assumed for the price variable. on the other hand, allow for the randomly generated number to take any value between the specified upper and lower bonds based on a constant probability. Hence. any value in the specified interval is just as likely and probable [40].

4.10. *Valorization Cooperative's Survey*

A technical and financial assessment was carried out conducting a diagnostic of the 3rd Millennium" cooperative (a quinoa valorization unit in Rehamna region) in order to assess the technical pathway of quinoa transformation and determine production costs of processed products. The cooperative used Puno variety. which is the most common used by cooperatives and it is provided by the Benrim farm in the Berrechid area where the quinoa price is the most affordable.

4.11. *Quinoa Import Data*

Quinoa import data in terms of quantity and value have been extracted from the change office database [41]. Data were first downloaded searching for quinoa as a keyword and processed using Excel software.

4.12. *Statistical Analysis*

Differences in response variables to applied treatments were assessed using a general linear model with StatSoft STATISTICA 8.0.550 software (StatSoft Inc. Tulsa, OK, USA). Statistical differences were all significant at $\alpha = 0.05$ or less. The means comparison was based on a one-way ANOVA analysis.

5. Conclusions and Recommendations

In the light of obtained results quinoa was shown to be a potential and resilient crop that could be an alternative to traditional cereals in the marginal area such as the Rehamna region where traditional cereals are not performing well at both the agronomic and economic level. Furthermore. quinoa offers better remuneration and yield under both rainfed and irrigated cultivation. It is also recommended for farmers to adopt mechanized tools for quinoa cultivation and seed processing to reduce production cost and improve their income. The access to those mechanized tools will be easier if farmers are gathered in cooperatives or associations. Quinoa price structuration remains a bottleneck in its value chain in Morocco as quinoa still have a niche market and demand on quinoa products still does not meet the producer's expectations. Furthermore. production cost of the quinoa-based product such as couscous remains very high due to a high cost of raw material and involved labor force. Therefore. it is recommended for women's cooperatives valorizing quinoa to have their own quinoa production.

Several scenarios of cost–benefit analysis were conducted to assess the economic viability of quinoa production in Morocco. The results across multiple scenarios consistently indicated that quinoa is highly profitable. yielding a net margin ranging from 21.100 to 111.555 MAD depending on the scenario (e.g., irrigated vs. rainfed and manual vs. mechanized systems). Further sensitivity analysis and simulations were undertaken to analyze the potential impacts of uncertainty in key variables and assumptions. particularly taking into account variability in yield performance. market prices and production costs. The sensitivity analysis showed that output price has the largest and significant impact on the quinoa profitability. However. as is indicated by the results of the simulation. the likelihood of net profit to be negative is neglected with a probability ranging from 0.5 to 2.55% depending on the scenario.

This study revealed several lessons learned from the field experience and proposed several development actions for each value chain component that can be implemented within a national quinoa program. which may be funded within the new Moroccan agricultural development plan called the "green generation".

The Rehamna Quinoa upscaling project has identified the suitable varieties. and the best production and management practices to maximize yields. In addition. the nutritional analysis of the genotypes with the highest potential was carried out. Seed multiplication of the most promising genotypes has been developed with a private local company to ensure enough seeds are available in the market for the scaling up production. On a global scale. the impact of the project on small farmers and the gender issues has been positive in general. as it secures a minimum revenue for the farmers even in dry years. Therefore. it will be judicious to pursue implementing cutting edge research to collect. screen and identify the quinoa genotypes that have the best potential for wide scale adoption in different agroecological zones and marginal environments.

We recommend then to set up support for all players in the sector through coordination of the quinoa interprofession in Morocco supported by the structures of the Department of Agriculture. Additionally. we need much better organization of the quinoa sector. in order to have more visibility to the consumers. This starts with focusing on more development of quinoa on public awareness and promotion-marketing. More research is needed in reducing saponin content in the different quinoa product. which represent one of the current weak points in the sector and it is necessary to remove this bottleneck as it does affect. today. the quality of the product and the entire value chain. Morocco's situation within the Maghreb region places it in an advance position in the development of quinoa value chain. This should trigger a promoting cooperation within the Maghreb countries. It is also possible to create a Mediterranean or African network on quinoa to extend ideas and research results.

Author Contributions: A.H. supervised agronomic trials, conducted economic assessment, analyzed the data and wrote the manuscript; S.R. conducted chemical analysis and conducted valorization and transformation assessment; M.R. analyzed quinoa transformation pathway; A.F. carried out agronomic trial at the UM6P experimental station about organic amendments; F.A. carried out agronomic trial at the UM6P experimental station about new cultivars performance; K.F. carried out trials at farm levels; H.A. conducted the cost analysis; Y.J. contributed to field trials running; A.S.(Aziz Soulaimani) conducted chemical analysis; M.M. supervised agronomic trial at the UM6P experimental station; M.E.G. supervised all agronomic trials; S.K. conducted saponin analysis; A.S.(Abdelaziz Sbai) and R.C.-A. elaborated SWOT analysis, lessons learned and development perspectives. All authors have read and agreed to the published version of the manuscript.

Funding: This study was carried out within an IDRC (International Development Research Center-Canada) funded project entitled: Scaling up quinoa value chain to improve food and nutritional security in poor rural communities of Morocco under the agreement number 108637-001.

Institutional Review Board Statement: Not applicable.

Informed Consent Statement: Not applicable.

Data Availability Statement: The data generated within this work are open access and available to be shared with interested persons according to signed agreement with IDRC.

Acknowledgments: First. we thank our sponsor the International Development Research Center (IDRC)—Canada for providing fund to conduct the project activities. Besides. we appreciate the efforts deployed by the technical team of UM6P- Agricultural Innovation and Technology Transfer Center (AITTC) and the technical support of the International Center of Biosaline Agriculture (ICBA). Special thanks to members of the 3rd Millennium Cooperative for couscous production and Benrim Farm (Haj Mohamed Ibrahimi) for their cooperation. Our gratitude goes to our local partners and facilitators (farmers. women cooperatives. startups. OCP group. DRA Marrakech-Safi. DPA Rehamna. ONCA Rehamna. ONSSA Rehamna. etc.).

Conflicts of Interest: The authors declare that they have no conflicts of interest.

References

1. Bazile, D.; Jacobsen, S.E.; Verniau, A. The global expansion of quinoa: Trends and limits. *Front. Plant Sci.* **2016**, *7*, 622.
2. Bazile, D.; Bertero., H.D.; Nieto., C. (Eds.) *Estado del Arte de la Quinua en el Mundo en 2013*; FAO: Rome, Italy; CIRAD: Montpellier, France, 2014; 724p, Available online: http://www.fao.org/3/a-i4042s/index.html (accessed on 16 November 2020).
3. Lavini, A.; Pulvento, C.; d'Andria, R.; Riccardi, M.; Choukr-Allah, R.; Belhabib, O.; Yazar, A.; İncekaya, Ç.; Metin Sezen, S.; Qadir, M.; et al. Quinoa's potential in the Mediterranean region. *J. Agron. Crop Sci.* **2014**, *200*, 344–360.
4. Choukr-Allah, R.; Nanduri, K.R.; Hirich, A.; Shahid, A.; Alshankiti, A.; Toderich, K.; Gill., S.; Rahman, B. Quinoa for Marginal Environments: Toward Future Food and Nutritional Security in MENA and Central Asia Regions. *Front. Plant Sci.* **2016**, *7*, 346. [CrossRef]
5. Benlhabib, O.; Jacobsen, S.E.; Jellen, R.; Maughan, J.; Choukrallah, R.; Oussible, M. Status of quinoa production and research in Morocco. In *State of the Art Report on Quinoa around the World in 2013*; Bazile, D., Bertero, H.D., Nieto, C., Eds.; FAO. CIRAD: Santiago, Chile, 2015; pp. 478–491.
6. Brakez, M.; El Brik, K.; Daoud, S.; Harrouni, M.C. Performance of chenopodium quinoa under salt stress. In Developments in Soil Salinity Assessment and Reclamation. Springer: Dordrecht, The Netherlands, 2013; pp. 463–478.
7. Hirich, A.; Jelloul, A.; Choukr-Allah, R.; Jacobsen, S.E. Saline water irrigation of quinoa and chickpea: Seedling rate. stomatal conductance and yield responses. *J. Agron. Crop Sci.* **2014**, *200*, 378–389.
8. Fghire, R.; Wahbi, S.; Anaya, F.; Issa Ali, O.; Benlhabib, O.; Ragab, R. Response of quinoa to different water management strategies: Field experiments and SALTMED model application results. *Irrig. Drain.* **2015**, *64*, 29–40.
9. Filal, K.; Hirich, A.; Benlhabib, O.; Choukr-Allah, R.; Ragab, R. Yield and dry matter simulation using the SALTMED model for five quinoa (Chenopodium quinoa) accessions under deficit irrigation in South Morocco. *Irrig. Drain.* **2017**, *66*, 340–350.
10. Hirich, A.; Choukr-Allah, R.; Jacobsen, S.E. Deficit irrigation and organic compost improve growth and yield of quinoa and pea. *J. Agron. Crop Sci.* **2014**, *200*, 390–398.
11. Hirich, A.; Choukr-Allah, R.; Jacobsen, S.E. Quinoa in Morocco–effect of sowing dates on development and yield. *J. Agron. Crop Sci.* **2014**, *200*, 371–377.
12. Hirich, A.; Choukr-Allah, R.; Jacobsen, S.E. The combined effect of deficit irrigation by treated wastewater and organic amendment on quinoa (Chenopodium quinoa Willd.) productivity. *Desal. Water Treat* **2014**, *52*.
13. Abugoch James., L.E. Quinoa (Chenopodium quinoa Willd.): Composition. chemistry. nutritional and functional properties. *Adv. Food Nutr. Res.* **2009**, *58*, 1–31.
14. Graf, B.L.; Rojas-Silva, P.; Rojo, L.E.; Delatorre-Herrera, J.; Baldeón, M.E.; Raskin, I. Innovations in health value and functional food development of quinoa (Chenopodium quinoa Willd.). *Compr. Rev. Food Sci. Food Saf.* **2015**, *14*, 431–445.
15. Sezgin, A.C.; Sanlier, N. A new generation plant for the conventional cuisine: Quinoa (Chenopodium quinoa Willd.). *Trends Food Sci. Technol.* **2019**, *86*, 51–58.
16. El Hazzam, K.; Hafsa, J.; Sobeh, M.; Mhada, M.; Taourirte, M.; El Kacimi, K.; Yasri, A. An Insight into Saponins from Quinoa (Chenopodium quinoa Willd): A Review. *Molecules* **2020**, *25*, 1059.

17. Statista. Average Price of Quinoa Worldwide from 2010 to 2019 (in U.S. Dollars per Kilogram). Statista Databas. 2020. Available online: https://www.statista.com/statistics/520974/average-price-of-quinoa-worldwide/#:~{}:text=Average%20global%20price%20of%20quinoa%202010%2D2019&text=The%20global%20average%20price%20of.U.S.%20dollars%20on%20average%20worldwide (accessed on 20 November 2020).
18. Codex Alimentarius. Codex Standard for Quinoa—CXS 333—2019. FAO: Roma, Italy, 2019; pp. 2–4.
19. Rafik, S.; Chaoui, M.; Assabban, Y.; Jazi, S.; Choukr-Allah, R.; El Gharouss, M.; Hirich, A. Quinoa value chain. adoption. and market assessment in Morocco. *Environ. Sci. Poll. Res.* **2020**, 1–12.
20. Geerts, S.; Dirk, R.; Magali, G.; Vacher, J.; Mamani, R.; Mendoza, J.; Huanca, R.; Morales, B.; Miranda, R.; Cusicanqui, J.; et al. Introducing deficit irrigation to stabilize yields of quinoa (Chenopodium quinoa Willd.). *Eur. J. Agron.* **2008**, *28*, 427–436.
21. Barzegar, A.R.; Yousefi, A.; Daryashenas, A. The effect of addition of different amounts and types of organic materials on soil physical properties and yield of wheat. *Plant Soil* **2002**, *247*, 295–301.
22. Chung, R.S.; Wang, C.H.; Wang, C.W.; Wang, Y.P. Influence of organic matter and inorganic fertilizer on the growth and nitrogen accumulation of corn plants. *J. Plant Nutr.* **2000**, *23*, 297–311.
23. Quiroga, A.; Funaro, D.; Noellemeyer, E.; Peinemann, N. Barley yield response to soil organic matter and texture in the Pampas of Argentina. *Soil Till. Res.* **2006**, *90*, 63–68.
24. Timsina, J. Can organic sources of nutrients increase crop yields to meet global food demand? *Agronomy* **2018**, *8*, 214.
25. Konishi, Y.; Hirano, S.; Tsuboi, H.; Wada, M. Distribution of minerals in quinoa (Chenopodium quinoa Willd.) seeds. *Biosci. Biotechnol. Biochem.* **2004**, *68*, 231–234.
26. D'Amico, S.; Jungkunz, S.; Balasz, G.; Foeste, M.; Jekle, M.; Tömösközi, S.; Schoenlechner, R. Abrasive milling of quinoa: Study on the distribution of selected nutrients and proteins within the quinoa seed kernel. *J. Cereal Sci.* **2019**, *86*, 132–138.
27. Varriano-Marston, E.; DeFrancisco, A. Ultrastructure of quinoa fruit (Chenopodium quinoa Willd). *Food Struct.* **1984**, *3*, 9.
28. Präger, A.; Munz, S.; Nkebiwe, P.M.; Mast, B.; Graeff-Hönninger, S. Yield and quality characteristics of different Quinoa (Chenopodium quinoa Willd.) cultivars grown under field conditions in southwestern Germany. *Agronomy* **2018**, *8*, 197.
29. Hirano, S.; Konishi, Y. Nutritional characteristics within structural part of quinoa seeds. *J. Jpn. Soc. Nutr. Food Sci.* **2003**, *56*, 283–289.
30. Mhada, M.; Metougui, M.L.; El Hazzam, K.; El Kacimi, K.; Yasri, A. Variations of Saponins. Minerals and Total Phenolic Compounds Due to Processing and Cooking of Quinoa (Chenopodium quinoa Willd.) Seeds. *Foods* **2020**, *9*, 660. [CrossRef]
31. Gómez-Caravaca, A.M.; Iafelice, G.; Verardo, V.; Marconi, E.; Caboni, M.F. Influence of pearling process on phenolic and saponin content in quinoa (Chenopodium quinoa Willd). *Food Chem.* **2014**, *157*, 157–174.
32. Yazar, A.; Incekaya, Ç.; Sezen, S.S.; Tkin, T. Quinoa experimentation and production in Turkey. 586 Chapter 6.1.4. In *State of the Art Report on Quinoa around the World in 2013*; Bazile, D., Bertero, H.D., Nieto, C., Eds.; FAO. CIRAD: Santiago, Chile, 2013; pp. 466–477.
33. Mercado, W.; Ubillus, K. Characterization of producers and quinoa supply chains in the Peruvian regions of Puno and Junin. *Sci. Agropecu.* **2017**, *8*, 251–265.
34. Parker-Gibson. N. Quinoa: Catalyst or Catastrophe? *J. Agric. Food Inf.* **2015**, *16*, 113–122.
35. Ali, O.I.; Fghire, R.; Anaya, F.; Benlhabib, O.; Wahbi, S. Physiological and morphological responses of two quinoa cultivars (Chenopodium quinoa Willd.) to drought stress. *Gesunde Pflanz.* **2019**, *71*, 123–133. [CrossRef]
36. Allen, R.G.; Pereira, L.S.; Raes, D.; Smith, M. Crop evapotranspiration-Guidelines for computing crop water requirements. In *FAO Irrigation and Drainage Paper 56; United Nations Food and Agriculture Organization*; United Nations Food and Agriculture Organization: Rome, Italy, 1998; Volume 300, p. D05109.
37. Gianna, V.; Montes, J.; Calandri, E.; Guzman, C. Impact of Several Variables on the Microwave Extraction of Chenopodium quinoa Willd Saponins. *Int. J. Food Sci. Technol.* **2012**, *47*, 1593–1597. [CrossRef]
38. Fiallos-Jurado, J.; Pollier, J.; Moses, T.; Arendt, P.; Barriga-Medina, N.; Morillo, E.; Arahana, V.; Torres, M.; Goossens, A.; Leon-Reyes, A. Saponin Determination. Expression Analysis and Functional Characterization of Saponin Biosynthetic Genes in Chenopodium quinoa Leaves. *Plant Sci.* **2016**, *250*, 188–197. [CrossRef]
39. McCabe, C.; Paulden, M.; Awotwe, I.; Sutton, A.; Hall, P. One-Way Sensitivity Analysis for Probabilistic Cost-Effectiveness Analysis: Conditional Expected Incremental Net Benefit. *Pharm. Econ.* **2020**, *38*, 135–141.
40. Fairchild, K.W.; Misra, L.; Shi, Y. Using Triangular Distribution for Business and Finance Simulations in Excel. *J. Financ. Educ.* **2016**, *42*, 313–336.
41. Office de Change. Morocco Import Statistic. Available online: https://services.oc.gov.ma/DataBase/CommerceExterieur/requete.htm (accessed on 15 November 2020).

plants

MDPI

Review

Quinoa in Ecuador: Recent Advances under Global Expansion

Leonardo Hinojosa [1,*], Alex Leguizamo [2], Carlos Carpio [3], Diego Muñoz [4], Camilo Mestanza [5], José Ochoa [6,7], Carmen Castillo [6], Angel Murillo [6], Elena Villacréz [6], Carlos Monar [8], Nicolas Pichazaca [9] and Kevin Murphy [10]

1 Institute for Biodiversity and Ecosystem Dynamics (IBED), University of Amsterdam, 1098XH Amsterdam, The Netherlands
2 Comité Europeo Para la Formación y la Agricultura (CEFA), Guayas 22-46 y Venezuela, Lago Agrio EC210105, Ecuador; a.leguizamo@cefaonlus.it
3 Facultad de Recursos Naturales, Escuela Superior Politécnica de Chimborazo, Panamericana Sur km 1 1/2, Riobamba EC060155, Ecuador; ccarpio@espoch.edu.ec
4 Grupo de Desarrollo de Tecnologías para la Reducción y Racionalización de Agroquímicos, Riobamba EC060155, Ecuador; diegom91_@hotmail.com
5 Facultad de Ciencias Pecuarias, Carrera de Ingeniería Agropecuaria, Universidad Técnica Estatal de Quevedo-(UTEQ), km 7 1/2 vía Quevedo–El Empalme, Mocache, Los Ríos EC120509, Ecuador; cmestanza@uteq.edu.ec
6 Instituto Nacional de Investigaciones Agropecuarias Estación Experimental Santa Catalina, Panamericana Sur Km 1, Quito EC171107, Ecuador; jose.ochoa@iniap.gob.ec or jbochoa@uce.edu.ec (J.O.); carmen.castillo@iniap.gob.ec (C.C.); angel.murillo@iniap.gob.ec (A.M.); elena.villacres@iniap.gob.ec (E.V.)
7 Facultad de Ciencias Agrícolas, Universidad Central del Ecuador, Jerónimo Leiton s/n y Av. La Gasca, Quito EC170521, Ecuador
8 Facultad de Ciencias Agropecuarias, Recursos Naturales y del Ambiente, Campus Laguacoto II, Universidad Estatal de Bolívar, Vía Guaranda a San Simón, Guaranda EC020150, Ecuador; cmonar@ueb.edu.ec
9 Asociación de Productores de Semilla y Alimentos Nutricionales Andinos-Mushuk Yuyay (APROSANAMY), Cañar EC030304, Ecuador; maycela@hotmail.com
10 Sustainable Seed Systems Lab, Department of Crop and Soil Sciences, College of Agricultural, Human, and Natural Resource Sciences, Washington State University, Pullman, WA 99164-6420, USA; kmuphy2@wsu.edu
* Correspondence: l.a.hinojosasanchez2@uva.nl

Citation: Hinojosa, L.; Leguizamo, A.; Carpio, C.; Muñoz, D.; Mestanza, C.; Ochoa, J.; Castillo, C.; Murillo, A.; Villacréz, E.; Monar, C.; et al. Quinoa in Ecuador: Recent Advances under Global Expansion. *Plants* **2021**, *10*, 298. https://doi.org/10.3390/plants10020298

Academic Editors: Cataldo Pulvento and Veronica De Micco

Received: 1 January 2021
Accepted: 1 February 2021
Published: 4 February 2021

Abstract: Quinoa is a highly diverse crop domesticated in the Andean region of South America with broad adaptation to a wide range of marginal environments. Quinoa has garnered interest worldwide due to its nutritional and health benefits. Over the last decade, quinoa production has expanded outside of the Andean region, prompting multiple studies investigating the potential for quinoa cultivation in novel environments. Currently, quinoa is grown in countries spanning five continents, including North America, Europe, Asia, Africa, and Oceania. Here, we update the advances of quinoa research in Ecuador across different topics, including (a) current quinoa production situation with a focus on breeding progress, (b) traditional seed production, and (c) the impact of the work of the nongovernment organization "European Committee for Training and Agriculture" with quinoa farmers in Chimborazo province. Additionally, we discuss genetic diversity, primary pests and diseases, actions for adapting quinoa to tropical areas, and recent innovations in quinoa processing in Ecuador. Finally, we report a case study describing a participatory breeding project between Washington State University and the Association of Andean Seed and Nutritional Food Producers Mushuk Yuyay in the province of Cañar.

Keywords: *Chenopodium quinoa*; quinoa production; participatory plant breeding; quinoa processing

1. Introduction

Quinoa (*Chenopodium quinoa* Willd.) is an annual crop from South America. Originally thought to be domesticated in the highland region of the Andes Mountains, a recent report suggests that quinoa domestication from its wild relative *Chenopodium hircinum*

Schard. occurred independently in the Andean highlands and on the Chilean coast [1]. Five ecotypes have been used to classify quinoas: (1) valley = grown at 2000 to 3500 m above sea level (m a.s.l.); (2) altiplano = grown at high altitudes of more than 3500 m a.s.l.; (3) salares = grown in the salt flats of Bolivia and Chile; (4) sea-level = grown at low altitude; (5) subtropical or yungas = grown in humid valleys of Bolivia [2].

Quinoa is adapted to a wide range of marginal environments [3]. It possesses high salinity tolerance and is considered as a facultative halophyte crop [4,5]. Quinoa can tolerate drought [6] and has been shown to grow well in controlled high temperature environments [7–9]; however, the combination of drought and heat can considerably reduce seed yield [9,10]. Quinoa seeds are highly nutritious, containing high contents of fiber, vitamins such as ascorbic acid, alpha-tocopherol, thiamin, and riboflavin, and minerals. Quinoa protein quality is high as it contains the nine essential amino acids in significant amounts, though not always in sufficient quantities to be a complete protein source to consumers across all age groups [11]. Moreover, quinoa possesses phytochemicals such as phenolics and bioactive peptides; these components have demonstrated beneficial effects on metabolic, cardiovascular, and gastrointestinal health [12–14].

Due to the characteristics described above, quinoa production has been expanding rapidly outside of the Andean region of South America over the last decade [15,16]. Peru and Bolivia are the major quinoa producers in the world, responsible for 86,011 and 70,763 tonnes (t) in 2018, respectively [17]. Ecuador was considered the third largest quinoa producer globally [17]; however, the extent of recent quinoa expansion worldwide is difficult to measure and it is a challenge to accurately rank Ecuador in quinoa production compared to other countries. Ecuador, one of the 17 most megadiverse countries in the world, is a small country in South America, divided into three continental regions: Amazon, mountains and coast, and one maritime region, the Galapagos Islands. [18,19].

The aim of this article is to summarize the advances in quinoa research in Ecuador. We discuss diverse aspects of the current quinoa production situation, and focus on breeding progress, traditional seed production in the province of Bolívar, and the impact of the quinoa work of the nongovernmental organization (NGO) "European Committee for Training and Agriculture" (CEFA) with farmers in the province of Chimborazo. Potential quinoa pests and diseases, genetic diversity, strategies for adapting quinoa to tropical areas, and innovations in quinoa processing are reported. Moreover, we include a case study highlighting a participatory breeding project between Washington State University (WSU) and the Association of Andean Seed and Nutritional Food Producers Mushuk Yuyay (Asociación de Productores de Semillas y Alimentos Nutricionales Andinos Mushuk Yuyay (APROSANAMY)) in the province of Cañar.

2. Quinoa Production in Ecuador

Quinoa has been cultivated from southern Colombia to the coast of southcentral Chile, including parts of northwest Argentina and some subtropical lowlands in Bolivia [20]. It was one of the main crops and food sources in Incan society and later was partially displaced by cereals such as barley and wheat that were introduced during the Spanish conquest [2]. According to historians, Ecuadorian quinoa was cultivated before and after the Spanish conquest and has been one of the most important food sources in the indigenous communities in the Ecuadorian highlands. Traditionally, quinoa is cultivated in Ecuador between 2400 to 3400 m a.s.l. in different production systems [21].

Traditionally, Ecuadorian quinoa production was relatively constant and did not exceed 1000 t prior to 2010. In 2001, international partners started to work with Ecuadorian quinoa on different projects such as establishing producers associations and initiating participatory plant breeding projects [21–23]. These efforts resulted in slight production increases beginning in 2010. In 2015, quinoa production achieved a record high, reaching 12,000 t (Figure 1A). This increase was primarily due to government policies that encouraged quinoa production. The Ecuadorian Ministry of Agriculture created the project "Promotion of Quinoa Production in the Ecuadorian Sierra", which delivered production kits that included seeds, fertilizers, and pesticides to hundreds of farmers. In 2015, the total production area reached 7800 hectares, partially fulfilling the goals of the project. However, the increased quinoa grain production did not have a readymade market; thus, a large amount of quinoa seed was stored for the following two years, and the fall in quinoa prices explains the drop in production in 2017 [16] (Figure 1A). International demand was high in 2015 and 2016, and quinoa production from Peru and Bolivia combined reached 180,000 t in 2015 [24].

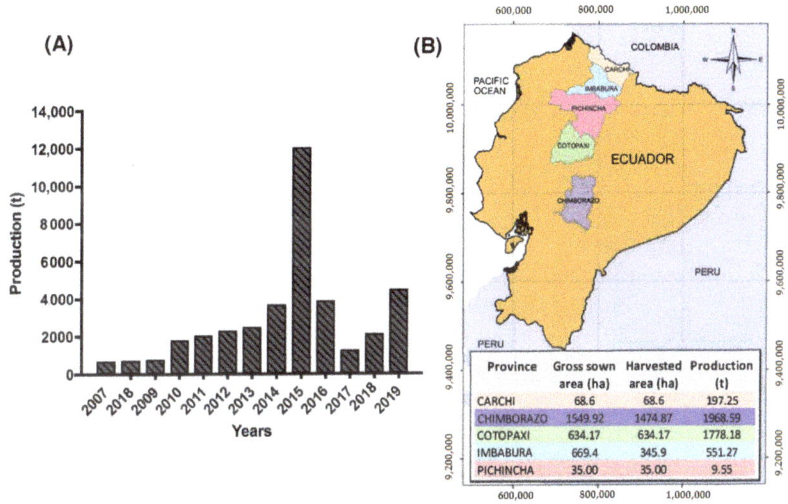

Figure 1. Quinoa production in Ecuador. (**A**) In the period from 2007 to 2019; (**B**) map including gross sown and harvested areas by province in 2019. Source: FAO and INEC-ESPAC [17,25].

In 2019, quinoa production increased again (4504 t) since many farmers' associations, mainly in the provinces of Chimborazo and Cotopaxi, have stable international markets. NGOs such as Fundación Mujer y Familia Andina (Andean Woman and Family Foundation), European Committee for Training and Agriculture (CEFA), and the McKnight Foundation encouraged organic quinoa production in these two provinces [21,26]. The total area sown in 2019 was 2057 ha, led by Chimborazo province (1549 ha and 1968 t of production) and followed by Cotopaxi, Imbabura, Carchi, and Pichincha (Figure 1B).

2.1. *Quinoa Breeding in Ecuador*

In the 1980s, the national germplasm bank of quinoa was established at Ecuador's National Institute of Agricultural Research (Instituto Nacional de Investigaciones Agropecuarias (INIAP)) with support from international cooperators and in consultation with Bolivian scientists. The first quinoa varieties, "INIAP Cochasqui" and "INIAP Imbaya", were selected from the germplasm accessions, and in 1992 two low saponin content quinoa varieties, "INIAP Ingapirca" and "INIAP Tunkahuan", were released [21]. Currently, "INIAP Tunkahuan" is the most widely planted variety among farmers, comprising 66% of total quinoa area with an average productivity of 1300 kg ha^{-1}. In 2007, "INIAP Pata de Venado" was released due to its excellent adaptation to and production in high elevation environments (3000 to 3600 m a.s.l.) [21,24].

Quinoa hybridization started in Ecuador in 2008 at the INIAP Santa Catalina Experimental Station (Estación Experimental Santa Catalina (EESC)) at 3050 m a.s.l. using "INIAP Tunkahuan" and "INIAP Pata de Venado" in the crossing block. The primary breeding objective for this hybridization was to obtain early maturing lines, with large seed diameters and low saponin contents. In 2009, crosses were made using the Bolivian variety "Jacha" in order to obtain larger grain size (>2.2 mm). Evaluation of this population from the F_2 to F_4 was carried out at EESC using the pedigree method. Later, homogenous lines (F_5–F_8) were evaluated using participatory selection methodology in farmers' fields in the provinces of Imbabura, Cotopaxi, Chimborazo, Cañar, and Bolívar. In 2019, the F_{10} lines LQEP4 and LQEP8, derived from the cross "INIAP Tunkahuan" × "INIAP Pata de Venado", and lines EQ26 and EQ28, derived from the cross "INIAP Tunkahuan" × "Jacha" were selected. All four selected lines reached maturity in less than 170 days, had plant heights < 150 cm, were resistant to downy mildew, and yielded >1500 kg ha^{-1} (Table 1). Researchers at INIAP plan to release one or two new quinoa varieties from the four advanced lines in 2021 [27].

Table 1. Average yield, plant height, days to harvest, and downy mildew severity of four promising quinoa lines evaluated during 2015–2019 in different locations in Ecuador.

Quinoa Line	Yield (kg ha^{-1})	Plant Height (cm)	Days to Harvest	Mildew Severity [1]
LQEP4	1559	130	162	3
LQEP8	1708	125	160	3
EQ26	1522	121	161	5
EQ28	1508	120	160	4
INIAP Tunkahuan	1330	152	180	4

[1] Severity scale of mildew caused by *Peronospora variabilis*: 1 to 3, first lower third of the plant (35%); 4 to 6, second middle third (35%); 7 to 9, last upper third of the plant (30%).

Additional hybridizations using accessions from the national germplasm bank of quinoa were made in 2016. The crosses ECU255 × EQ28 and ECU-248 × ECU-6205 produced F_5 lines Q1 and Q2, characterized by high yield performances, early maturity, and short statured plants. In 2018, new crosses were made to obtain lines with low saponin contents, high yields, and different colors of seed such as black, red, and yellow.

2.2. Participatory Research

Participatory research has been conducted in Ecuador for many years; for example, participatory quinoa breeding involved farmers from Cotopaxi, Chimborazo, and Cañar in order to select promising breeding lines and future varieties [22,23]. Participatory research has been carried out in the province of Bolívar. Here, we synthesize the main results from this participatory research.

Bolívar province is located in the central region of Ecuador. It has a population of 200,000, of whom 60% live in rural areas, and agriculture is the main economic activity. The primary crops are corn (*Zea mays* L.), barley (*Hordeum vulgare* L.), bean (*Phaseolus vulgaris* L.), potato (*Solanum tuberosum* L.), faba bean (*Vicia faba* L.), Andean grains including Andean lupin "chocho" or "tarwi" (*Lupinus mutabilis* Sweet), and quinoa.

Through a cooperative sustainability strategy with the Validation, Technology Transfer and Training Unit of Bolívar (Unidad de Validación, Transferencia de Tecnología y Capacitación-Bolívar (UVTT/C-B)) in Bolívar province, INIAP established local strategic alliances with the Ecuadorian Populorum Progressio Fund (Fondo Ecuatoriano Populorum Progressio (FEPP)), Ministry of Agriculture and Livestock (MAG), Instituto Técnico Superior Agropecuario "Tres de Marzo", State University of Bolívar (Universidad Estatal de Bolívar (UEB)), and Alto Guanujo producers' organizations. To initiate participatory research, the UVTT/C-B, selected two agronomic zones: (i) El Alto Guanujo (2900 to 3600 m a.s.l), its production system consisting of 98% potato and pasture, and (ii) Middle Zone (2500 to 2850 m a.s.l.), where the base crop is a monocropping of white soft corn associated with beans and to a lesser extent, wheat, barley, quinoa, and other legumes such as peas, lentil, and chocho.

Different people were involved in each zone because of diverse weather, social, culture and economic conditions. For instance, farmers from eight communities actively participated with researchers in a farmer field school in Alto Guanujo, the Comités de

Investigación Agrícola Local (CIAL) "Progressio a la Vida". In the Middle Zone, the main participants were students, professors, and researchers from UEB as well as INIAP technicians, and farmers. The result of this participatory research was the validation and technology transfer of new climate-resilient varieties of quinoa, potato, barley, faba bean, chocho, and pasture in the Alto Guanujo and quinoa, potato, bean, corn, wheat, barley, peas, lentils, and chocho in the Middle Zone. Moreover, it trained farmers to diversify their production systems by including these new varieties. Here, we summarize the main results for quinoa (Table 2).

Table 2. Results of the participatory research on quinoa at two agronomic zones in the province of Bolivar, Ecuador.

Zone	Period	Variety	Selection Criteria	Comments
Alto Guanujo	2005–2016	"INIAP Pata de Venado"	Early, adapted to high altitude, resistant to freezing, wind, and mildew. Good flavor in soup. Used in rotation with potatoes.	Variety named "Pata de Venado" (Deer leg) (because deer are light, fast, and can resist cold, wind, and drought)
Middle Zone	2005–2019	"INIAP Tunkahuan Line ECU-6717"	Moderately early, tolerant to mildew, and resistant to lodging. The grain is white and sweet. Excellent for soups, flour, cookies.	Variety was grown in an intercrop system with corn "INIAP 111", local corn varieties, chocho, and lentils

2.3. Traditional Seed Production

When a variety has been generated, validated, and selected by the beneficiaries, it is essential for its dissemination to be designated "Quality Seed", which must comply with specified physical, genetic, physiological, and health attributes [28]. Through a strategic alliance with MAG-Bolívar, UEB designed and implemented a small-scale seed research and production program to accelerate seed distribution to farmers. It was accredited as an official seed producer in November 2017 by the Ministry of Agriculture and Livestock under the rules and regulations of the Ecuadorian seed law. In collaboration with INIAP, MAG-Bolívar, and organizations of seed producers, UEB produces seed both in the conventional ("Certified" seed class) and artisanal models (Selected seed), whereas INIAP produces pure seed in "Basic" and "Registered" classes.

To obtain high quality quinoa seeds, negative selection was applied in field seed production; thus, atypical plants, wild quinoas, and inferior plants were eliminated. In addition, UEB designed and implemented a participatory strategy to produce seed in the local area. Specifically, seeds were delivered to trained organizations of producers and agronomy students who planted the seeds and harvested the crop. After only one growing season, they returned to UEB twice the number of seed they originally received, thereby simultaneously increasing the volume of available seed and advancing its diffusion. The success of this seed program is linked to clearly determining quinoa seed variables of production, quality, quantity, continuity, and price.

2.4. The Nongovernmental Organization (NGO) "European Committee for Training and Agriculture" (CEFA) in Chimborazo

CEFA is a nonprofit, nonpolitical, and nondenominational NGO founded in Italy in 1972 and legally recognized in Ecuador through a technical cooperation framework agreement in 2009. In 2017, CEFA created the Inclusive and Sustainable Value Chains Program (Programa Cadenas de Valor Inclusivas y Sostenibles (PCV-IS)), funded by the European Union and the Italian Agency for Development Cooperation—AICS—whose objective is to create the quinoa chain in the province of Chimborazo, and within the framework of the MAG strategy, the popular and solidarity economy, and fair trade. The PCV-IS has three main components: (a) socio-organizational strengthening; (b) productive development; (c) associative marketing. In addition, implementing adaptation measures to climate change and including young people in the value chain are two transversa axes of the PCV-IS.

2.4.1. Quinoa Chain in Chimborazo

Chimborazo province is located in the central area of the Ecuadorian highlands (Figure 1B). It owes its name to the Chimborazo volcano that rises to 6268 m a.s.l. and is located 150 km south of the city of Quito in the Western Cordillera of the Andes. Chimborazo has a population of 524,004 inhabitants and is the ninth most populous province of Ecuador [29].

Cultivation of quinoa in the province of Chimborazo was promoted by the Fundación Escuelas Radiofónicas Popular del Ecuador (Popular Radio Schools of Ecuador Foundation (ERPE)) with the support of international cooperators. They led to the first export of quinoa with organic certification in 1998. Prior to 1998, quinoa was cultivated primarily for self-consumption. In 2003, the Corporación de Productores y Comercializadores Orgánicos Bio Taita Chimborazo (Bio Taita Chimborazo Organic Producers and Marketers Corporation (COPROBICH)) was formed, an organization that brought together producer families to comply with ERPE's commercial agreements. Sumak Life Cía. Ltd. a joint venture between ERPE and COPROBICH, was formed in 2006. COPROBICH commercially dissociated itself from Sumak Life Cía. Ltd. in 2012 to enter the markets of France, Belgium, and Canada with certified organic quinoa and the Small Producers seal. In the same year, the Maquita Foundation began exporting certified organic quinoa with the fairtrade certification of the World Fair Trade Organization (WFTO).

Quinoa in Chimborazo is produced under the organic certification regulations of the EU and the USA. In economic terms, according to estimates by the PCV-IS based on information contained in the certification companies' internal control systems, it represents a gross income of approximately USD 2,000,000 each year for 1500 producer families located in the cantons of Colta, Riobamba, Guamote, and Guano. According to the General Coordination of National Agricultural Information of the MAG, in 2019 the sale of quinoa to exporters was the main source of income for 60% of these families [30].

Sumak Life Cía. Ltd., COPROBICH, and Maquita Foundation together exported around seven hundred metric tons of quinoa grain every year, mainly to the European Union and North American markets, which corresponds to around 95% of the total production in Chimborazo province. The difference is divided between self-consumption and sale to local markets. Producers generally rotate quinoa with barley, wheat, faba bean, or a mix of vetches (*Vicia* spp.) and oats (*Avena sativa*) that are used as forage for their livestock.

2.4.2. Main Actions of the PCV-IS in the Province of Chimborazo

Creation of a Technical Committee of Quinoa in Chimborazo

The objective of this technical committee is to analyze problems in production, transformation, commercialization, and consumption links of the quinoa chain in order to prioritize and articulate the actions of the direct and indirect actors. The direct actors in the quinoa chain are represented by Sumak Life Cía. Ltd., COPROBICH, Maquita Foundation, Sumak Tarpuy Corporation, and the Asociación de Producción y Comercialización de Productos Alimenticios Emprendedores Nutriandina (Association for the Production and Marketing of Food Products Entrepreneurs Nutriandina). Additionally, the indirect actors are represented by those who have support, regulation, investigation, or control functions in the chain, such as INIAP, MAG, Ecuadorian Agency for Quality Assurance in Agriculture (AGROCALIDAD), and the Polytechnic Higher School of Chimborazo (ESPOCH).

Promotion of Participatory Research and Links with the academy to Improve the Dynamics of Innovation in the Value Chain

Participatory research, led by INIAP, occurred in 18 field schools that were established in representative areas of the quinoa sector. Several studies were carried out on fertilization, intercropping systems, participatory selection of Chimborazo quinoa ecotypes, validation of the adaptation of quinoa advanced lines, and planting systems. In addition, ESPOCH implemented the project *"Diseño e Implementación del Proyecto de Producción, Transformación, Comercialización y Promoción de Consumo de la Quinua y sus Derivados"* (Design and Implementation of the Project for Production, Transformation, Commercialization and Promotion of Consumption of Quinoa and its Derivatives). This project involved seven departments and 60 researchers from different areas. Several undergraduate research and community linking projects were conducted as part of the project, engaging more than 260 students.

Identification and Validation of Three Species for Intercropping with Quinoa

Common vetch (*Vicia sativa*) was sown in an intercropped system at the time of quinoa hilling. After the quinoa was harvested, the vetch remained a remnant for the grazing of breeding animals. A faba bean variety, "INIAP-441 Serrana", was sown in the system comprised of one row of faba bean x three rows of quinoa, for areas located above 3200 m a.s.l. in loamy to clay loam soils. Faba bean contributes nitrogen to the quinoa crop and is used for family consumption and occasionally for the marketing of surpluses. Finally, a chocho variety, "INIAP-450 Andino", was planted in a system comprised of one row of chocho × three rows of quinoa for farms below 3200 m a.s.l. in sandy and sandy loam soils (Figure 2).

Figure 2. Intercrop system of chocho (*Lupinus mutabilis*) with quinoa. Comunidad Achullay, Guamote. Juan Yuquilema (farmer), Fausto Yumisaca (INIAP Chimborazo), Rodrigo Aucancela, and Galo Morocho (CEFA). Photo: A. Leguizamo.

Use of Local Organic Matter Source for Soil Improvement

ESPOCH, through the Faculty of Sciences and the local government of Riobamba, will begin to recover the vegetable residues from local markets, compost them, and then supply the compost to quinoa farmers. Approximately 1000 t of compost per year are expected to be recovered from this initiative.

Development and Implementation of Adaptation Measures to Climate Change

Quinoa soils are vulnerable to water erosion during intense rains, especially fields with newly planted crops. Therefore, a transition from the traditional tillage system to one of minimum tillage, including cover crops such as vetch and direct seeding of quinoa, is required.

3. Pests and Diseases

3.1. Quinoa Pests in Ecuador

A formal study of insect pests affecting quinoa in Ecuador was conducted 30 years ago [31]. Species identification was performed by international experts at that time.

Seven lepidopterans from the Noctuidae family were identified affecting quinoa. Cutworms damaging plantlets in the field such as *Agrotis deprivata* Walker and *A. ipsilon* (Hufnagel) were reported. The following leaf or grain chewers were also reported: *Peridroma saucia* (Hübner), *Dargida grammivora* Walker, *Spodoptera* sp., and two nonidentified species of *Copitarsia*. Two other lepidopterans were identified—*Scrobipalpula* sp. in the Gelechiidae family affected the leaves and an *Ephestia* sp. of the Pyralidae family damaged the stored quinoa seed. Two tachinid flies were found: *Incamyia* sp. was identified parasitizing both *Agrotis* species and *Elfia* sp. was a parasitoid of *Scrobipalpula* sp. Two species within the Coleoptera order were identified: the curculionid *Naupactus* sp., which eats quinoa leaves and grains, and *Oryzaephilus surinamensis* (L.) (Cucujoidea), which affects stored quinoa flour. *Liriomyza* sp. (Diptera: Agromyzidae) is a leaf miner reported in quinoa. The sap feeders *Paratanus yust* Young (Hemiptera: Cicadellidae) and *Proba sallei* (Stal) (Hemiptera: Miridae) were identified as affecting quinoa plants in the field. The same study reported the presence of a few natural enemies of quinoa pests—two tachinid flies were found: *Incamyia* sp. was identified as parasitizing both *Agrotis* species and *Elfia* sp. as a parasitoid of *Scrobipalpula* sp. [31].

Primary pests of quinoa are only briefly mentioned in crop management documents. One publication reports cutworms (*Agrotis* sp.) in young plants and birds are the most common pests [32]. Recently, between June and August 2020, an insect inventory was carried out to survey entomofauna associated with organic quinoa fields at six locations in the province of Chimborazo. The most abundant insects collected in this inventory were from the following orders: Hemiptera (one family), Coleoptera (three families), Lepidoptera (two families); parasitoid insects from the order Hymenoptera (one family), and predatory insects from the Neuroptera (one family) (Table 3). The main phytophagous insects observed on the plants belonged to the family Aphididae (Hemiptera), which attack the leaves, and Gelechiidae (Lepidoptera), which feeds on the inflorescence (Table 3). It is also important to point out the presence of parasitoids the of Braconidae family (Hymenoptera).

Table 3. Families of insects collected in destructive sampling of quinoa plants. The families are ordered in relation to abundance (from highest to lowest).

Order/Family	Functional Role
Hemiptera: Aphididae [1]	Phytophagous
Coleoptera: Curculionidae	Phytophagous
Lepidoptera: Gelechiidae	Phytophagous
Coleoptera: Elateridae	Phytophagous
Coleoptera: Latriididae	Phytophagous
Lepidoptera: Arctiidae	Phytophagous
Hymenoptera: Braconidae	Parasitoid
Hemiptera: Aphididae [1]	Phytophagous
Neuroptera: Chrysopidae	Predator
Lepidoptera: Noctuidae	Phytophagous

[1] Different morphospecies.

Insects of the family Cucurlionidae were the most abundant phytophagous insects found in the soil around the roots (Table 4). Cucurlionidae fed on the external parts of roots and did not bore into the roots. Additionally, a greater quantity of Annelidae (earth worms) was found in the plots from the lower elevation locations.

Insects from the following orders were collected at the lower elevation locations using yellow bowl pan traps placed 40 cm above the soil: Coleoptera (two families), Diptera (16 families), Hemiptera (four families), Hymenoptera (15 families), Lepidoptera (one family), and Thysanoptera (one family). In the traps of the plots in the location at the highest altitude, insects from the following orders were collected: Coleoptera (one family), Diptera (21 families), Hemiptera (three families), Hymenoptera with 10 families (Table 5), Neuroptera (one family), and Thysanoptera (one family).

Table 4. Insect families collected in soil blocks of quinoa plants removed from plots. The families are ordered in relation to abundance (from highest to lowest).

Order/Family	Functional Role
Coleoptera: Curculionidae [1]	Phytophagous
Coleoptera: Sthaphylinidae	Predator
Coleoptera: Tenebrionidae	Detritivore
Lepidoptera: Noctuidae	Phytophagous
Himeptera: Aphidae	Phytophagous
Dermaptera: Anisolabididae	Detritivore, predator
Coleoptera: Curculionidae [1]	Phytophagous

[1] Different morphospecies.

Table 5. Families of Hymenoptera collected in yellow bowl pan traps in the province of Chimborazo. The families are ordered in relation to abundance (from highest to lowest).

Family	Functional Role
Halictidae	Pollinator
Braconidae	Parasitoid
Chalcididae	Parasitoid
Ichneumunidae	Parasitoid
Crabronidae	Predator
Pteromalidae	Parasitoid
Bethylidae	Parasitoid, predator
Megaspilidae	Parasitoid
Diapriidae	Parasitoid
Figitidae	Parasitoid

The richness of individuals collected of Hymenoptera may indicate the good health of the plots sampled, since we know that individuals of this order are sensitive to the application of insecticides. In the plots from the lower altitude locations, most of the Hymenoptera belonged to Halictidae, Braconidae, Chalcididae, and Ichneuminidae, whereas most of the Hymenoptera in the plots from the highest altitude locations belonged to Crabronidae, Pteromalidae, Bethylidae, and Megaspilidae (Table 5).

3.2. *Quinoa Diseases in Ecuador*

In most cultivated plants, diseases are major biotic constraints usually involving a wide range of pathogens. Unlike most crops, quinoa is in general a marginal host for plant pathogens. In Ecuador, neither viruses nor bacteria have been reported to infect quinoa [33], though this could be due to a lack of pathology research. *Cercospora* sp. and *Phoma* sp., the most common foliar fungi [34], are low damaging pathogens. In Ecuador, the most economically important quinoa disease is downy mildew (DM), caused by the oomycete *Peronospora variabilis*.

This low range of pathogens infecting quinoa in Ecuador agrees to some extent with the worldwide quinoa disease review reported by Danielsen et al. [35]. However, *Sclerotium rolfsii* and *Pythium zingiberum*, causing seedling diseases, and *Ascochyta hyalospora* and *Phoma exigua* var. *foveata*, causing ascochyta leaf spot and brown stalk rot, respectively, have not been reported in Ecuador. *Rhizoctonia solani* and *Fusarium* sp. causing damping off have been observed sporadically but are not considered threats for quinoa in Ecuador.

3.2.1. Downy Mildew

Downy mildew (DM) is the most damaging disease that threatens quinoa production worldwide [35]. In Ecuador, this disease is especially important in commercial monoculture of uniform varieties. The Department of Plant Pathology and the Quinoa Breeding Program of INIAP have been conducting research to identify and introgress durable resistance (DR) into quinoa DM, which is highlighted in this report.

Peronospora variabilis, previously classified as *P. farinosa* f. sp. *chenopodii*, was considered specific to quinoa; however, the pathogen has also been found infecting *Chenopodium album* and *C. murale* [36,37]. In Ecuador, isolates infecting *C. album* belong to the same population that infects quinoa and since *C. album* is a common weed of quinoa, the secondary host status of *C. album* is an important aspect to consider for DM management as well as in breeding for DR.

P. variabilis reproduces by heterothallic sexual reproduction [38], and two mating types appear to occur frequently, since oospores are massively produced on the leaves and carried by the seed [39,40]. In addition, oospores transmitted by seed are the primary pathogen inocula, and seed transmission is the main dispersion strategy for the pathogen to reach regions outside the Andes.

Studies of pathogenic diversity have shown that *P. varibilis* is a variable and adaptable pathogen [41]. Four virulence groups (Vs) of *P. varibilis* have been identified with respect to R-3 factors (R). V1 infects only susceptible lines, V2 infects susceptible lines and lines carrying R1, V3 infects susceptible lines and lines carrying R1 and R2, and V4 infects susceptible lines and lines carrying R1, R2, and R3. Pathogen evolution that has taken place in Ecuador is associated with single successive step mutations, which is in turn associated with the breeding initiatives carried out during the 1980s and early 1990s, which exploited the three R-factors, and R2 and R3 were released in varieties INIAP Ingapirca and INIAP Imbaya, respectively. Levels of resistance to DM at time of variety release were lower than 4 on a 0–9 scale, values located in the resistance fraction of the scale [21].

P. varibilis evolution is associated with mutation events. Although genetic recombination has not yet been evident, heterothallism is a potential mechanism by which the pathogen evolves beyond R-factor adaptation to include improved adaptation to environmental conditions and increased aggressiveness. Efficient seed transmission of the pathogen facilitates this potential mode of pathogen evolution. These epidemiological aspects are primarily associated with modern quinoa cultivation, in which cultivating genetically uniform varieties drives higher selection pressures on the pathogen.

Development of strategies to search for DR to DM quinoa should be an important objective of modern breeding programs. Since R-factors are often ephemeral, screening for new sources of R-factors is not advisable. Therefore, non-race-specific resistance appears to be the most effective strategy to attain DR in the quinoa/downy mildew pathosystem. In Ecuador, "INIAP Tunkahuan" carries this type of resistance and has remained effective since variety release in 1992. Resistance of "INIAP Tunkahuan" can be considered durable since it aligns with the requirements for durable resistance as proposed by Johnson [42]. Thus, "INIAP Tunkahuan" has been hegemonically cultivated in commercial monoculture in a relatively large area, in conditions conducive to the disease, and for a long period of time (more than 25 years).

Levels of non-race-specific resistance were studied in four locations in Ecuador conducive to DM [22]. Resistance varied from low to high in lines without R-factors and in lines with defeated R-factors (residual resistance). Furthermore, lines carrying non-race specific resistance scored area under disease progress curve (AUDPC) values as low as lines carrying efficient R-factors, indicating that non-race-specific resistance can provide the same high level of protection as provided by R-factors. This resistance of non-race-specific type is quantitative in nature, and in other downy mildews, as in lettuce, this resistance is polygenic, with major and minor effects [43,44]. The high variation of AUDPC in lines with non-race-specific resistance suggests that this type of resistance is polygenic. In this study, lines were found with similar or higher levels of resistance than "INIAP Tunkahuan", suggesting that sources of non-race-specific resistance are available in quinoa germplasm of the Quinoa Breeding Program at INIAP.

The early Ecuadorian experience in selection for DM resistance showed that breeders are interested in high levels of resistance, and by selecting lines with low disease severities, they are unconsciously favoring R-factors. However, they eventually also select high levels of non-race-specific resistance when lines carry similar disease severities to R-factors, which is the case of "INIAP-Tunkahuan". Since R-factors resistance appears ephemeral, non-race-specific resistance is a more durable resistance strategy. Discarding the most susceptible and most resistant lines, as proposed by Parlevliet (1993) appears a practical strategy to favor selection of non-specific resistance and to select for DR to DM [45]. However, by discarding resistant lines, other promising traits and residual resistance can be inadvertently discarded. Therefore, to select non-race-specific resistance more efficiently, discard the most susceptible lines with the assistance of R-factors postulation to assess residual resistance in the field. This strategy is presently being implemented by the Quinoa Breeding Program of INIAP.

3.2.2. Minor Diseases

Quinoa leaf spot caused by *Cercospora* sp is considered a minor disease in Ecuador [34]. Symptoms regularly develop late in the season on leaves of the upper part of the plant. Symptoms are characterized by round to oval and brown to grey lesions with diameters of less than 1 cm with darker brown reddish margins. Similar symptoms are also found in Ecuador on *Chenopodium album* [34]. This pathogen was also detected in the USA on quinoa, and the fungus was phylogenetically similar to *Passalora dubia*, which infects *C. album*; therefore, *P. dubia* is presently the taxonomical status of the causal agent of the quinoa leaf spot, known in the USA as *Passalora* leaf spot [46]. The incubation period of *P. dubia* of around 30 days [46] is significantly longer than that of *Cercospora* sp., which might explain its limited pathogenicity and why epidemics appear late in the quinoa growing season.

Eye-like stalk spot caused by *Phoma* sp. is also a disease that affect quinoa late in the cropping season. Symptoms are gray spots with diameters of 1–3 cm surrounded by brown to dark margins. Dark pycnidia of the pathogen become regularly visible in the gray background of the lesion [34,35]. In conducive conditions, lesions can join and cover a considerable area of stems. Lesions can also be produced by mechanical injuries such as those caused by hailstorms.

4. Genetic Diversity

Ecuadorian quinoa belongs to the valley ecotype. This quinoa ecotype is characterized by tall, late-maturing plants with high seed yield potentials and biomass accumulation [21]. Seed yield is negatively affected when the photoperiods exceeds 14 h in high temperature environment [2,47]. Currently, the quinoa germplasm at INIAP contains 608 accessions—283 from Ecuador (46%) and the rest from other countries, primarily Peru and Bolivia [48].

In recent decades, significant genetic erosion of quinoa landraces has been observed in the field. For instance, 30 years ago, variation in quinoa grain color was easy to find, including colors such as white, cream, brown, yellow, red, pink, and black. Currently, just white- or cream-colored quinoa grains are grown by most farmers, due to market demand and eating patterns in production areas [49]. A core quinoa collection using 469 quinoa accessions from the Ecuadorian quinoa germplasm bank was evaluated, and discriminating traits such as downy mildew reaction, seed color, days to flowering, and seed yield differentiated four genetic groups [50].

Traditionally, Ecuadorian quinoa is positioned to have limited genetic diversity, with Altiplano (Peru–Bolivia) quinoa indicated as the most probable point of introduction for Ecuadorian accessions [20]. This evidence came from the molecular and genetic diversity studies where just two Ecuadorian quinoa accessions were included [51,52]. Recently, a molecular characterization of diversity of Ecuadorian quinoa was conducted using 15 species-specific SSR markers on 84 quinoa accessions [53]. The results showed that Ecuadorian quinoa was highly diverse—196 alleles were detected, and the genetic heterozygosity (H_E) was 0.71. Three subgroups were determined by a phenetic analysis—one group corresponded to the "INIAP Tunkahuan" variety that has been exposed to intense inbreeding and selection. The three groups included quinoa from the seven provinces where seed was collected, and indicated that ancestral landrace populations had been disseminated throughout Ecuador mainly due to farmers' seed interchange.

Wild Quinoa Relatives

Wild quinoa is known as "ashpa quinoa" in the Quichua language or "ajara" in the Aymara language. The wild quinoa in Ecuador has been poorly studied; however, there is a report that a free-living (wild, weedy, spontaneous) quinoa population named as *C. quinoa* ssp. *milleanum* Aellen or var. *melanospermum* is distributed mainly in the mountains of Ecuador [54]. There is a hypothesis that *C. quinoa* ssp. *milleanum* could be closely related to the North American tetraploid species *C. berlandieri* subs. *zschachkei* and the South American tetraploid species *C. hircinum* ssp. *catamarcense* [54,55].

A phylogenetic study focused on the amplification and sequences of gene rpoB in 32 wild quinoa accessions collected from four provinces including Carchi, Imbabura, Cotopaxi, and Chimborazo showed high inter- and intrapopulation genetic diversity across the accessions [56]. Accessions from Carchi and Imbabura had 77.4% of the variability and genotypes from Cotopaxi and Chimborazo had 70.2%. In this study, all the wild quinoas were designated as *C. quinoa*; this description resulted in a misunderstanding between the cultivated quinoa and the wild quinoa in Ecuador. Thus, it is necessary to conduct more molecular studies including *C. berlandieri*, *C. hircinum*, and quinoa varieties as controls to understand the evolutionary origin of Ecuadorian quinoa.

A collection of wild quinoas in Carchi and Chimborazo showed a significant difference in seed morphology (unpublished data). The seed gathered from farmers' fields in Carchi showed a pitted surface (Figure 3A). It is likely that these plants are *C. berlandieri*, *C. hircinum*, or the *C. quinoa* ssp. *milleanum* [57]. On the other hand, the seed collected in Chimborazo had a smooth surface (Figure 3B). It is possible that this species belongs to the *C. album* complex, which has a cosmopolitan distribution around the world [58]. Molecular studies are necessary to confirm this hypothesis.

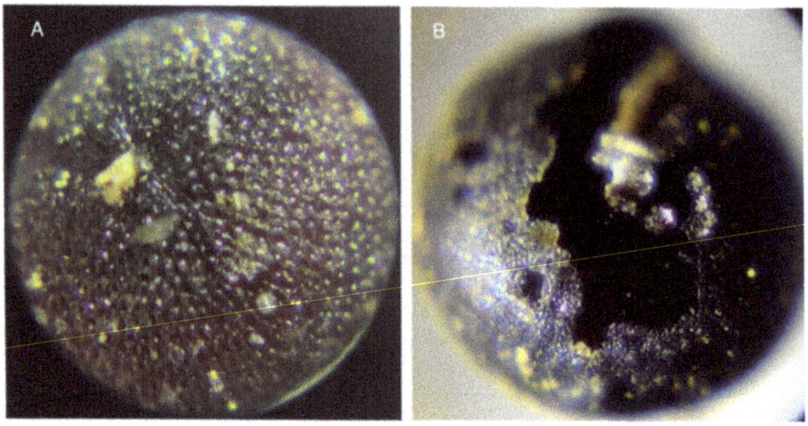

Figure 3. Wild quinoa relatives' seed surfaces. (**A**) Probably *C. quinoa* ssp. *milleanum*, or *C. berlandieri*, or *C. hircinium*, with a pitted surface. (**B**) Probably *C. album*, with a smooth surface. Magnitude: 60×. Photos: L. Hinojosa.

Interspecific hybridizations in nature between *C. quinoa* and *C. berlandierie*, *C. hircinum*, and *C quinoa* ssp. *milleanum* have been reported [54,55,59]. Plant breeders have intentionally created interspecific crosses (simple and reciprocal) in order to recombine favorable traits (e.g., heat tolerance) present in different species and to concentrate them in the selected offspring [60]. Studies are required to identify if wild quinoas in Ecuador can cross with cultivated quinoas; this feature could be beneficial to breeding programs.

5. Quinoa Adaptation to Tropical Areas

Of the five quinoa ecotypes described above, the valley ecotype is predominantly produced and consumed in Ecuador [61]. Little information exists on the adaptation of quinoa to coastal or tropical areas in Ecuador. In 2014, Peralta et al. evaluated the adaptability of 269 quinoa accessions from the quinoa germplasm existing at INIAP, under the agro-ecological conditions of the Santa Elena Peninsula; however, all accessions evaluated proved to be poorly adapted to the area mainly due to high temperature [62]. The interest in producing quinoa on the coast of Ecuador arises from the increase in consumption worldwide, high international prices [63], and the experiences of coastal production in Chile [64] and Peru [65]. In addition, knowing quinoa's great capacity for adaptation [3] makes the northern area of the province of Los Rios (Quevedo) a suitable niche for the production of rainfed quinoa (Figure 4).

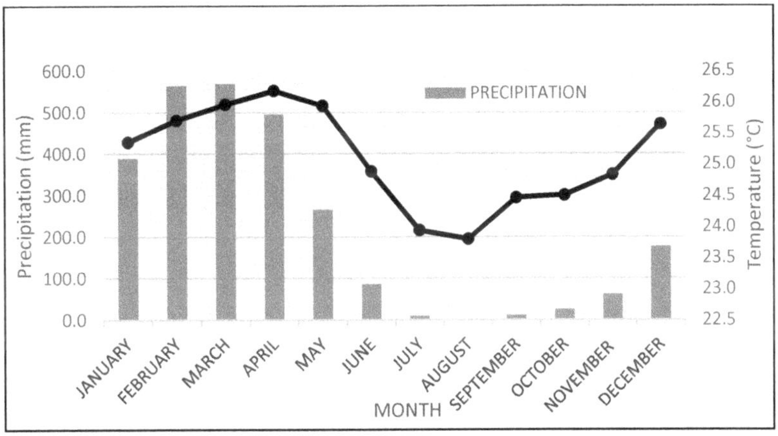

Figure 4. Average monthly precipitation and temperature for the last ten years (2010–2019) in the Quevedo area.

In collaboration with other researchers, multiple trials were established in 2017 to evaluate quinoa genotypes from the coast and south of Chile, Argentina, and Ecuador. In the first trial, 21 accessions were sourced from Chile and Argentina, and grown from August to December 2017 at the research station "La Maria" of the State Technical University of Quevedo located in Mocache, Quevedo [66]. The results demonstrated the feasibility of quinoa production in this area. Days to maturity ranged from 90 to 143 days after sowing and yield ranged from 346 to 470 g m^{-2}. Several colors in the grain such as yellow, white, red, and black were described.

In the second trial at Mocache, Quevedo, genotypes from Ecuador, Chile and Argentina were surface-seeded and grown from June to September 2019 [67]. Although this trial was affected by poor germination, the genotype O-10 (red leaf color) performed well, with a crop cycle of 110 days, a grain yield of 3723 kg ha^{-1}, a harvest index of 0.41, and plant height of 90 cm.

Follow-up investigations were carried out (data not yet published), which demonstrate that the best genotype (O-10) under these conditions, needs 1300 accumulated degree days to finish its cultivation cycle in 110 days. In addition, based on the efficient use of nitrogen in rainfed conditions, it is recommended to apply 62 kg N ha^{-1} if the objective is to achieve maximum N use efficiency. However, if the objective is to achieve maximum grain performance, it is recommended to apply 110 kg N ha^{-1} to achieve a grain yield of 4700 kg ha^{-1}. This research also showed that an increase in the nitrogen rate from 0 to 200 kg ha^{-1} results in an increase in protein in the grain from 13% to 17%.

Currently, at the State Technical University of Quevedo (Universidad Técnica Estatal de Quevedo), research continues on adapting the red O-5 genotype and performing hybridizations with the cream-white O-3 genotype to obtain quinoa populations with adaptation criteria of the O-3, but with certain characteristics of O-5. Finally, it should be noted that our small collection of quinoas has shown the feasibility of being able to carry out two cultivation cycles a year, the first cycle from May to August and the second cycle from September to December. January to April is reserved for maize production because none of our quinoa genotypes will tolerate the excessive humidity that is generated during the first months of the year (Figure 4).

6. Quinoa Processing and Agroindustry

Quinoa was widely used in South America during pre-Hispanic times; however, after the Spanish conquest, its cultivation and consumption were reduced to small areas scattered in mountainous areas of the Andes [32]. Since then, the grain was scarcely known or commercialized; however, the "re-discovery" of quinoa resulted in an explosion of its consumption, especially in Europe and the United States, in recent decades. This growth is largely a result of the high nutritional value of its gluten-free seeds and leaves, which have a moderate protein content (>15%), a good balance of all essential amino acids, dietary fiber, lipids, carbohydrates, vitamins, and minerals, and low glycemic index when consumed [13,68]. Functional compounds in quinoa such as polyphenols, phytosterols, and flavonoids have also been found to have important nutraceutical benefits [13,69].

Quinoa Transformation

The Ecuadorian agroindustry processes innovative quinoa products for culinary, pharmaceutical, and cosmetic uses. Farmers and agribusiness enterprises that consider only grain production without the application of transformation technologies downgrade the commercial value of quinoa [21]. Several public and private institutions, universities, and research centers in Ecuador are currently developing improved varieties, including thoroughly characterizing the physico-chemical and nutritional traits, protein and amino acid contents, fatty acids, vitamins and minerals, phytohormones, antioxidants, phytosterols, and dietary fiber, as well as the identification and use of saponins [70–73]. Additionally, applications of thermal processes and bioprocesses have been studied in order to increase mineral bioavailability, the level of acceptability, and the nutraceutical value of quinoa grain [74].

The range of quinoa products that are produced and marketed in Ecuador include: whole grain, quinoa flour, quinoa flour mixtures with oats or amaranth, baby food porridge, granola, energy bars, soft drinks, quinoa expanded as breakfast cereal, biscuits, alfajores (a traditional dessert), and quinoa bread with substitution percentages reaching 30% [61,75]. On a smaller scale, and only in warehouses of products produced by farmers (fair trade), noodles are sold with some percentages of substitution. The Food World Programme (WFP), PANN-2000 (Programa de Alimentos), and "Aliméntate Ecuador" ("Feed Ecuador") established as a mandate that both porridge and drinks for children must include quinoa as a raw material in their formulas [21]. Currently, solid fermentation processes are being tested to obtain vegetable meat and technologies are being optimized to produce milk- and yogurt-type drinks, products desired by children, adolescents, and adults, with national and international market potential among subjects with gluten allergy problems [75]. Additionally, precooked quinoa facilitates quinoa preparation and consumption in schools and assistance centers. The product can be seasoned with vegetables and can be introduced without much difficulty in various markets in Europe and the United States.

Quinoa leaves can be used as a nutritious vegetable, as they contain carotenoids, iron, and other micronutrients. The phytochemical profile shows that the leaves (60–80 days of cultivation) have low saponin contents. Their nutritional profile surpasses the grain, and they possess interesting potential in nutrition and food. Quinoa leaves can be used in the preparation of salads, soups, main dishes, and as raw material to enter the food industry, which wants to offer a continuous movement of new products in order to remain in the market, thereby contributing to the change of the production matrix [76,77].

It is worth highlighting the collaborations that private industry establishes with farmers capable of supplying large volumes of grain. Manufacturers in the food sector establish contracts with farmers. Contract farming is opening a very interesting way to consolidate the sustainability of the value chain [21]. Producers in the province of Chimborazo focus their attention on the production of organic fair-trade quinoa for traditional food applications in those areas less appropriate for intensification. However, Ecuadorian farmers face competition from other countries, as new varieties and models pave the way for large-scale production under different conditions.

7. Participatory Plant Breeding Project between WSU and APROSANAMY (Case Study)

The Association of Producers of Seeds and Nutritious Andean Foods Mushuk Yuyay (APROSANAMY) was formed in 1994, with the main objective to focus on seed production of Andean crops such as barley, quinoa, faba bean, amaranth, and peas. In 2004, APROSANAMY started to add value to their production—for example, in the processing of barley and quinoa flour. In 2018, a new processing and storage facility was built to allow for the expansion of production. Currently, APROSANAMY has 69 producers from 18 communities from the cantons Cañar, Suscal, and El Tambo in Cañar province [78]. The province of Cañar, with a population of 281,396 inhabitants, is located between the provinces of Chimborazo and Azuay, in southcentral Ecuador [29].

Quinoa researchers at Washington State University (WSU) were first connected with APROSANAMY through Alan Adams, a former Peace Corps volunteer who worked in Quilloac, Cañar province, from 1967 to 1969. In May 2013, Alan received an email message from Nicolas Pichazaca requesting that he join in the work of APROSANAMY. Alan began researching, writing grants, and reaching out to potential collaborators. Based on a short article on quinoa in National Geographic that his wife Paulette read about WSU quinoa research, Alan contacted researcher Kevin Murphy and his Ecuadorian PhD student, Leonardo Hinojosa [79]. This began a formal arrangement between APROSANAMY and WSU to collaborate on quinoa and barley research. After initial discussions between Murphy and Hinojosa of WSU and Nicolas Pichazaca, Jose Luis Pichazaca, and Antonio Guaman of APROSANAMY regarding collaboration strategies with quinoa breeding, a two-pronged approach was implanted, including (a) evolutionary participatory breeding (EPB), and (b) evaluations of homogeneous cultivars and advanced breeding lines.

The first strategy (EPB) included sharing a diverse breeding population to stimulate a participatory selection approach [80]. This approach involved selection from diverse bi-parental populations by APROSANAMY farmers and researchers, as well as natural selection in the different environments of Cañar, utilizing an evolutionary participatory breeding (EPB) approach [81,82]. EPB combines evolutionary breeding through natural selection with participatory selection by farmers and breeders to develop populations adapted to locally prevalent diseases and abiotic pressures [83,84]. These populations retain a significant amount of genetic variation, allowing the population to rapidly evolve in response to climate change and the resulting unpredictable fluctuations in environmental conditions [85,86]. Recent shifts in rainfall patterns and climate instability are key concerns voiced by Kichwa Kañari farmers.

Population development proceeded in two distinct phases. Phase I of the EPB approach focused on population development that incorporated traits of importance to Kichwa Kañari farmers. Based on multiple discussions between WSU and APROSANAMY, traits such as (a) early maturity to mitigate problems with preharvest sprouting due to increasingly unpredictable late season rains, (b) large seed size, (c) seed yield, and (d) red leaf color emerged as particularly important. Leonardo Hinojosa crossed cultivars such as "Titicaca" and "Puno", and the red-leafed breeding line "3964" (developed collaboratively by researchers at Brigham Young University and WSU) with more regionally adapted varieties such as "Pasankalla", a cultivar widely grown in Peru, and "CICA", an Altiplano cultivar from Peru that was shared kindly by Juan Antonio Gonzales from Argentina. Additionally, the "Titicaca", "Puno", "3964", "QQ74", "Colorado 407D", "Cahuil", and "Kaslaea" varieties were grown as control cultivars next to the populations (Table 6).

Table 6. Phase I and Phase II WSU evolutionary participatory breeding (EPB) populations grown by APROSANAMY in Cañar province, Ecuador.

Phase I		Phase II	
Pedigree		**Pedigree**	
Female	**Male**	**Female**	**Male**
Colorado 407D	3964	Pison	3964
Titicaca	3964	Japanese Strain	3964
Titicaca	Pasankalla	QQ74	3964
Colorado 407D	Pasankalla	23TES	3964
Titicaca	Cica	Titicaca	MisaMisa
Puno	Pasankalla	QQ74	MisaMisa
		Puno	Pasankalla
		Colorado 407D	Pasankalla
		Kaslaea	INIAP Tunkahuan
		Titicaca	Cica

These Phase I populations were planted by an APROSANAMY-WSU team in 2016 (Figure 5C) and have been replanted each year since then. By March 2018, Nicolas Pichazaca had selected the "Puno/Pasankalla" population and the "Kaslaea", "QQ74", "Colorado 407D", and "3964" cultivars (Figure 5D), which were the most promising genotypes mainly due to early maturity, large seed size, and leaf color. These will continue to be grown in the future.

Phase II crosses continued to introduce new genetic variations designed to introgress traits such as heat tolerance ("Pison" and "QQ74") [8], early maturity (Japanese Strain, "Titicaca") red leaf color ("3964" and "MisaMisa"), and seed yield ("Titicaca", "Kaslaea") into adapted varieties such as "MisaMisa", "Pasankalla", and "INIAP Tunkahuan" (Figure 5E–G and Table 6).

The second strategy focused on testing pureline varieties and breeding lines. Fifty lines from the F_6 population ("Cahuil"/"PI 510534"), created by former WSU graduate student Adam Peterson, were planted in Cañar in February 2019. The parents from this population include "Cahuil", a variety that originated from Chile with good adaptation to Washington State, and "PI 510534", an accession from National Plant Germplasm System of USDA that originated from the valley of Peru. Unfortunately, the seed germination from the lines was very low and drought and frost conditions caused the loss of the trial; however, new plantings are planned in the future with the same lines.

In October 2019, Kevin Murphy, Nicolas Pichazaca, Cristina Ocaña Gallegos (a new graduate student at WSU) planted 37 quinoa lines from the previous year's selections. The trial was affected by high pressure of downy mildew and lines derived from "Puno"/"Pasankalla" showed high resistance to this pathogen.

In Table 7, we detail the timeline of the collaboration between APROSANAMY and WSU from 2015 to 2020. It is necessary to emphasize that in addition to the work on quinoa, WSU and APROSANAMY have been working in parallel with barley—including malting barley and hulless food barley. Currently, there are five promising lines under evaluation. Lines 13WA146-7, 13WA-146.3, and 13WA-146.1 were selected for their elongated grain shapes suitable for barley flour, "machika", and pleasant smells during the manufacturing-grinding process; lines 13WA-146-10 and 13WA-126.6 were selected for "Arroz de Cebada" ("Barley Rice"), a traditional dish in the Ecuadorian highlands.

Figure 5. Asociación de Productores de Semillas y Alimentos Nutricionales Andinos Mushuk Yuyay (APROSANAMY) and Washington State University (WSU) activities. (**A**) Kañari planting ceremony; (**B**) field preparation, plowing (Macy Hagler, WSU student); (**C**) planting quinoa and barley with WSU students; (**D**) Line 3964 (red) and Colorado 407D (yellow); (**E**–**G**) introduction of new genetic variation (red color leaves). Photos by: L. Hinojosa, N. Pichazaca, and J. Kellogg.

The next step in the quinoa EPB is to plant the selected quinoa lines at different locations and altitudes in the province of Cañar to evaluate yield performance and disease resistance. Moreover, APROSANAMY is interested in introducing the color panicle quinoa lines such as red, yellow, and orange close to archaeological sites where tourists are keen to see quinoa diversity.

Table 7. Timeline of APROSANAMY and WSU collaboration, including dates, locations, and activities.

Planting	Date	Location	Elevation (m a.s.l.)	Activity
	2015	Cañar Comunidad San Rafael (2°32′55.2″ S 78°57′07.0″ W)	3073	- Initiation of APROSANAMY and WSU collaboration
Phase I	23 May 2016	Molobog Chico (2°36′27.252″ S 78°52′48.432″ W)	3203	- Kañari planting ceremony (Figure 5A) - Field preparation, plowing (Figure 5B) - Planting of barley populations with WSU students
		La Posta (2°36′27.252″ S 78°52′48.432″ W)	2956	- Planting of quinoa and barley populations with WSU students (Figure 5C)
	18 March 2017	La Posta	2956	- Leonardo Hinojosa and Dan Packer (WSU) visited second cycle of planting populations
Phase II	22 May 2018	La Posta	2956	- Nicolas Pichazaca planted 10 F_2 populations (Table 6)
Phase I, II and Pure Line Selection strategy	2 February 2019	La Posta	2956	- Leonardo Hinojosa and Nicolas Pichazaca increased seed from Puno/Pasankalla (Phase I) - They planted F_3 population (Titicaca/CICA), (Titicaca/MisaMisa) from Phase II - They planted 60 F_6 lines of (Cahuil/PI 510534) (Phase III)
	15 October 2019	La Posta	2956	- Kevin Murphy, Nicolas Pichazaca, and Cristina Ocaña Gallegos planted 37 quinoa lines

8. Conclusions

Ecuador was considered the third largest quinoa producer globally, but with more recent expansion of this crop in other countries it is hard to assume that this statement is still true. The current quinoa production in Ecuador is around 4500 t per year and the Central provinces (Chimborazo and Cotopaxi) are the major producers. Participatory research has been carried out on quinoa for many years, mainly in the province of Bolívar. The National Agricultural Research Institute of Ecuador (INIAP) has released all the Ecuadorian quinoa varieties and new varieties are expected to be released soon from its hybridization program. The European Committee for Training and Agriculture (CEFA) has been working in the province of Chimborazo to support quinoa production; its main input was to develop a quinoa technical committee. On the other hand, new insects have been reported in quinoa fields, many of which are predators and parasitoids. Downy mildew (*Peronospora variabilis*) is the only damaging quinoa disease in Ecuador and durable resistance is the main objective of the quinoa breeding program at INIAP.

Ecuadorian quinoa is highly diverse contrary to previously thought. Wild quinoa has been poorly studied in Ecuador and there are no strong molecular studies showing its origin and identification; thus, new molecular works are necessary and must include wild quinoa and more landraces. On the other hand, new initiatives have begun to adapt quinoa to tropical coastal areas. Great progress is occurring in the province of Los Rios with the start of a new quinoa hybridization program at the State Technical University of Quevedo. In addition, Ecuador has developed different quinoa transformation strategies to link farmers and agribusiness enterprises, and Ecuadorian universities have conducted several research studies on nutrition and food sciences in the last five years.

A great effort has been made in the last five years to increase quinoa diversity in the Cañar province through an evolutionary participatory breeding project between Association of Producers of Seeds and Nutritious Andean Foods Mushuk Yuyay (APROSANAMY) and Washington State University (WSU). New quinoa colors, bigger seed size, and different panicle shapes have been introduced as well as early maturing quinoa. These new traits will help Cañar quinoa producers to achieve higher grain production and avoid late season rains and unpredictable droughts.

The current information presented in this article suggests that quinoa research and production in Ecuador is on the right track; however, it will be difficult to compete with the big technologies of countries such as China, Australia, and Spain where quinoa production has been recently rebounding. As a country, Ecuador needs to reinforce local consumption and internal demand for quinoa.

Author Contributions: L.H. led the manuscript planning, reported information about quinoa genetic diversity and contributed in the Case Study; A.L. contributed with CEFA information and results; C.C. (Carlos Carpio), C.C. (Carmen Castillo), and J.O. provided significant information about quinoa pests and disease; D.M. collected the pests in quinoa fields; C.M. (Camilo Mestanza) contributed with information about quinoa adaptation in tropical areas; A.M. provided the quinoa breeding strategies from INIAP; E.V. contributed with information about processing quinoa; C.M. (Carlos Monar) synthesized the information on participatory research in Bolívar; N.P. conducted the quinoa trials with APROSANAMY; K.M. edited the manuscript and provided significant additional information. All authors have read and agreed to the published version of the manuscript.

Funding: This research was funded by IBED University of Amsterdam (accession no. R.2323.0106). The WSU-APROSANAMY research was supported in part by USDA NIFA Hatch funds (accession no. 1014754). The insect inventory was supported by CEFA.

Acknowledgments: We acknowledge Jorge Cajarmarca and Victor Lema for providing the updated quinoa production data and for assisting us with Figure 1. We kindly thank all the Kichwa Kañari farmers that helped to plant and maintain the trials in the case study with WSU and APOSANAMY. In addition, we thank to Annie Pollard and Adam Alans for their comments and suggestions to our manuscript.

Conflicts of Interest: The authors declare no conflict of interest. The funders had no role in the design of the study; in the collection, analyses, or interpretation of data; in the writing of the manuscript, or in the decision to publish the results.

References

1. Jarvis, D.E.; Ho, Y.S.; Lightfoot, D.J.; Schmöckel, S.M.; Li, B.; Borm, T.J.A.; Ohyanagi, H.; Mineta, K.; Michell, C.T.; Saber, N.; et al. The Genome of *Chenopodium quinoa*. *Nature* **2017**, *542*, 307–312. [CrossRef]
2. Tapia, M. The Long Journey of Quinoa: Who Wrote Its History. In *State of the Art Report on Quinoa Around the World 2013*; Bazile, D., Bertero, H.D., Nieto, C., Eds.; FAO: Rome, Italy; CIRAD: Paris, France, 2015; Volume 1, pp. 1–7. ISBN 978-92-5-108558-5.
3. Hinojosa, L.; González, J.A.; Barrios-Masias, F.H.; Fuentes, F.; Murphy, K.M. Quinoa Abiotic Stress Responses: A Review. *Plants* **2018**, *7*, 106. [CrossRef]
4. Adolf, V.I.; Jacobsen, S.-E.; Shabala, S. Salt Tolerance Mechanisms in Quinoa (*Chenopodium quinoa* Willd.). *Environ. Exp. Bot.* **2013**, *92*, 43–54. [CrossRef]
5. Ruiz, K.B.; Biondi, S.; Martínez, E.A.; Orsini, F.; Antognoni, F.; Jacobsen, S.-E. Quinoa—A Model Crop for Understanding Salt-Tolerance Mechanisms in Halophytes. *Plant Biosyst. Int. J. Deal. Asp. Plant Biol.* **2016**, *150*, 357–371. [CrossRef]
6. Bosque Sanchez, H.; Lemeur, R.; van Damme, P.; Jacobsen, S.-E. Ecophysiological Analysis of Drought and Salinity Stress of Quinoa (*Chenopodium quinoa* Willd.). *Food Rev. Int.* **2003**, *19*, 111–119. [CrossRef]
7. Hinojosa, L.; Matanguihan, J.B.; Murphy, K.M. Effect of High Temperature on Pollen Morphology, Plant Growth and Seed Yield in Quinoa (*Chenopodium quinoa* Willd.). *J. Agron. Crop Sci.* **2019**, *205*, 33–45. [CrossRef]
8. Hinojosa, L.; Kumar, N.; Gill, K.S.; Murphy, K.M. Spectral Reflectance Indices and Physiological Parameters in Quinoa under Contrasting Irrigation Regimes. *Crop Sci.* **2019**, *59*, 1927–1944. [CrossRef]
9. Hinojosa, L.; Sanad, M.N.M.E.; Jarvis, D.E.; Steel, P.; Murphy, K.; Smertenko, A. Impact of Heat and Drought Stress on Peroxisome Proliferation in Quinoa. *Plant J.* **2019**, *99*, 1144–1158. [CrossRef] [PubMed]
10. Yang, A.; Akhtar, S.S.; Amjad, M.; Iqbal, S.; Jacobsen, S.-E. Growth and Physiological Responses of Quinoa to Drought and Temperature Stress. *J. Agron. Crop Sci.* **2016**, *202*, 445–453. [CrossRef]
11. Craine, E.B.; Murphy, K.M. Seed Composition and Amino Acid Profiles for Quinoa Grown in Washington State. *Front. Nutr.* **2020**, *7*, 26. [CrossRef]

12. Wu, G. Nutritional Properties of Quinoa. In *Quinoa: Improvement and Sustainable Production*; Murphy, K., Matanguihan, J., Eds.; John Wiley & Sons, Inc.: Hoboken, NJ, USA, 2015; pp. 193–210. ISBN 978-1-118-62804-1.
13. Vilcacundo, R.; Hernández-Ledesma, B. Nutritional and Biological Value of Quinoa (*Chenopodium quinoa* Willd.). *Curr. Opin. Food Sci.* **2017**, *14*, 1–6. [CrossRef]
14. Noratto, G.D.; Murphy, K.; Chew, B.P. Quinoa Intake Reduces Plasma and Liver Cholesterol, Lessens Obesity-Associated Inflammation, and Helps to Prevent Hepatic Steatosis in Obese Db/Db Mouse. *Food Chem.* **2019**, *287*, 107–114. [CrossRef]
15. Bazile, D.; Jacobsen, S.-E.; Verniau, A. The Global Expansion of Quinoa: Trends and Limits. *Front. Plant Sci.* **2016**, *7*, 622. [CrossRef]
16. Alandia, G.; Rodriguez, J.P.; Jacobsen, S.-E.; Bazile, D.; Condori, B. Global Expansion of Quinoa and Challenges for the Andean Region. *Glob. Food Secur.* **2020**, *26*, 100429. [CrossRef]
17. FAO. FAOSTAT: Food and Agriculture Data. Crop Statistics. Available online: http://www.fao.org/faostat (accessed on 14 August 2020).
18. Penafiel, D.; Vanhove, W.; Espinel, R.L.; Van Damme, P. Food Biodiversity Includes Both Locally Cultivated and Wild Food Species in Guasaganda, Central Ecuador. *J. Ethn. Foods* **2019**, *6*, 25. [CrossRef]
19. Mestanza-Ramón, C.; Henkanaththegedara, S.M.; Vásconez Duchicela, P.; Vargas Tierras, Y.; Sánchez Capa, M.; Constante Mejía, D.; Jimenez Gutierrez, M.; Charco Guamán, M.; Mestanza Ramón, P. In-Situ and Ex-Situ Biodiversity Conservation in Ecuador: A Review of Policies, Actions and Challenges. *Diversity* **2020**, *12*, 315. [CrossRef]
20. Zurita-Silva, A.; Fuentes, F.; Zamora, P.; Jacobsen, S.-E.; Schwember, A.R. Breeding Quinoa (*Chenopodium quinoa* Willd.): Potential and Perspectives. *Mol. Breed.* **2014**, *34*, 13–30. [CrossRef]
21. Peralta, I.E.; Mazon, N. Quinoa in Ecuador. In *State of the Art Report of Quinoa in the World in 2013*; Bazile, D., Bertero, D., Nieto, C., Eds.; FAO: Rome, Italy; CIRAD: Paris, France, 2015; ISBN 978-92-5-108558-5.
22. McElhinny, E.; Peralta, E.; Mazón, N.; Danial, D.L.; Thiele, G.; Lindhout, P. Aspects of Participatory Plant Breeding for Quinoa in Marginal Areas of Ecuador. *Euphytica* **2007**, *153*, 373–384. [CrossRef]
23. Mazón, N.; Peralta, E.; Murillo, Á.; Rivera, M.; Guzmán, A.; Pichazaca, N.; Nicklin, C. It's Not Just the Technology, Its the Surrounding System: How Researchers in Ecuador Found Ways to Make Themselves Useful to Farmers through Quinoa and Lupin Seed Systems. *Exp. Agric.* **2019**, *55*, 107–124. [CrossRef]
24. Basantes-Morales, E.R.; Alconada, M.M.; Pantoja, J.L. Quinoa (*Chenopodium quinoa* Willd) Production in the Andean Region: Challenges and Potentials. *J. Exp. Agric. Int.* **2019**, 1–18. [CrossRef]
25. INEC (Instituto Nacional de Estadísticas y Censos), ESPAC (Encuesta de Superficie y Producción Agropecuaria Continua). Available online: https://www.ecuadorencifras.gob.ec/estadisticas-agropecuarias-2 (accessed on 12 September 2020).
26. Cárdenas Oleas, S.E. Case Study: A Sustainable Livelihoods Analysis of a Non-Governmental Organization Working with Quinoa Producers in the Province of Chimborazo, Ecuador. Master's Thesis, Iowa State University, Ames, IA, USA, 2015; p. 7896981.
27. Murillo, A.; Mazon, N.; Rodriguez, L.; Vega, L. Mejoramiento Genético de Quinua (*Chenopodium quinoa* Willd.) en Ecuador. In Proceedings of the 7th Congreso Mundial de la Quinua y Otros Granos Andinos, Santiago, Chile, 25–28 March 2019.
28. Monar, C.; Silva, D. *Informe Anual de Actividades. Programa de Investigación y Producción de Semillas*; Universidad Estatal de Bolívar, Facultad de Ciencias Agropecuarias Recursos Naturales y Del Ambiente: Guaranda, Ecuador, 2014.
29. Sistema Nacional de Información. Available online: https://sni.gob.ec/inicio (accessed on 16 November 2020).
30. Informe de Rendimientos Objetivos Quinua. 2019. Available online: http://sipa.agricultura.gob.ec/index.php/informe-de-rendimientos-objetivos/rendimiento-de-quinua (accessed on 16 November 2020).
31. Ruales, C.; Fiallo, E. Plagas. In *Quinua: Hacia su Cultivo Comercial*; Wahli, C., Ed.; Latinreco S.A: Quito, Ecuador, 1990; p. 71, ISBN 978-9978-9901-3-1.
32. Jacobsen, S.-E. *Cultivo de Granos Andinos en Ecuador: Informe Sobre Los Rubros Quinua, Chocho y Amaranto*; Editorial Abya Yala: Quito, Ecuador, 2002; ISBN 978-9978-22-258-4.
33. Ministerio de Agricultura y Ganadería del Ecuador. Programa Nacional de Protección Vegetal. In *Inventario de Plagas, Enfermedades y Malezas Del Ecuador: Quinoa*; Ministerio de Agricultura y Ganadería Quito: Quito, Ecuador, 1986; p. 152.
34. Orellana, H.; Peralta, E.; Castillo, R. Enfermedades de Quinoa. In Proceedings of the Curso de Quinoa, Quito, Ecuador, 16–18 October 1985; pp. 1–5.
35. Danielsen, S.; Bonifacio, A.; Ames, T. Diseases of Quinoa (*Chenopodium quinoa*). *Food Rev. Int.* **2003**, *19*, 43–59. [CrossRef]
36. Choi, Y.-J.; Danielsen, S.; Lübeck, M.; Hong, S.-B.; Delhey, R.; Shin, H.-D. Morphological and Molecular Characterization of the Causal Agent of Downy Mildew on Quinoa (*Chenopodium quinoa*). *Mycopathologia* **2010**, *169*, 403–412. [CrossRef]
37. Baiswar, P.; Chandra, S.; Kumar, R.; Ngachan, S.V. *Peronospora variabilis* on *Chenopodium murale* in India. *Australas. Plant Dis. Notes* **2010**, *5*, 45–47. [CrossRef]
38. Danielsen, S. Heterothallism in *Peronospora farinosa* f.sp. chenopodii, the Causal Agent of Downy Mildew of Quinoa (*Chenopodium quinoa*). *J. Basic Microbiol.* **2001**, *41*, 305–309. [CrossRef]
39. Danielsen, S.; Mercado, V.H.; Ames, T.; Munk, L. Seed Transmission of Downy Mildew (*Peronospora farinosa* f. sp. chenopodii) in Quinoa and Effect of Relative Humidity on Seedling Infection. *Seed Sci. Technol.* **2004**, *32*, 91–98. [CrossRef]
40. Testen, A.L.; del Mar Jiménez-Gasco, M.; Ochoa, J.B.; Backman, P.A. Molecular Detection of *Peronospora variabilis* in Quinoa Seed and Phylogeny of the Quinoa Downy Mildew Pathogen in South America and the United States. *Phytopathology* **2013**, *104*, 379–386. [CrossRef] [PubMed]

41. Jacobs, T. Postulation of Virulence Groups and Resistance Factors in the Quinoa/Downy Mildew Pathosystem Using Material from Ecuador. *Plant Pathol.* **1999**, *48*, 425–430. [CrossRef]
42. Johnson, R. A Critical Analysis of Durable Resistance. *Annu. Rev. Phytopathol.* **1984**, *22*, 309–330. [CrossRef]
43. Norwood, J.M.; Crute, I.R.; Johnson, A.G.; Gordon, P.L. A Demonstration of the Inheritance of Field Resistance to Lettuce Downy Mildew (*Bremia lactucae* Regel.) in Progeny Derived from Cv. Grand Rapids. *Euphytica* **1983**, *32*, 161–170. [CrossRef]
44. Grube, R.C.; Ochoa, O.E. Comparative Genetic Analysis of Field Resistance to Downy Mildew in the Lettuce Cultivars "Grand Rapids" and "Iceberg". *Euphytica* **2005**, *142*, 205–215. [CrossRef]
45. Parlevliet, J.E. What is Durable Resistance, A General Outline. In *Durability of Disease Resistance. Current Plant Science and Biotechnology in Agriculture*; Jacobs, T., Parlevliet, J.E., Eds.; Springer: Dordrecht, The Netherlands, 1993; pp. 23–39. ISBN 978-94-011-2004-3.
46. Testen, A.L.; McKemy, J.M.; Backman, P.A. First Report of Passalora Leaf Spot of Quinoa Caused by *Passalora dubia* in the United States. *Plant Dis.* **2012**, *97*, 139. [CrossRef] [PubMed]
47. Bertero, H.D.; King, R.W.; Hall, A.J. Photoperiod-Sensitive Development Phases in Quinoa (*Chenopodium quinoa* Willd.). *Field Crops Res.* **1999**, *60*, 231–243. [CrossRef]
48. Mazón, N.; Rivera, M.M.; Peralta, I.E.; Estrella Engelmann, J.; Tapia, B.C. *Catálogo del Banco de Germoplasma de Quinua (Chenoodium Quinoa Willd.) de INIAP-Ecuador*; INIAP: Quito, Ecuador, 2001.
49. Delgado, A.; Tapia, B.C.; Naranjo, E.; Borja, E.; Martin Clemente, J.P. Variación de la Agrobiodiversidad en Variedades Tradicionales de Quinua (*Chenopodium Quinoa* Willd) de la Región Andina de Ecuador. In Proceedings of the 7th Congreso Mundial de la Quinua y otros Granos Andinos, Iquique, Chile, 25–28 March 2017; p. 41.
50. Tapia, B.C.; Peralta, E.; Mazón, N. Colecciones Núcleo de Quinua (*Chenopodium Quinoa* Willd.) del Banco de Germoplasma del INIAP, Ecuador. *Axioma* **2014**, *2*, 5–9.
51. Jarvis, D.E.; Kopp, O.R.; Jellen, E.N.; Mallory, M.A.; Pattee, J.; Bonifacio, A.; Coleman, C.E.; Stevens, M.R.; Fairbanks, D.J.; Maughan, P.J. Simple Sequence Repeat Marker Development and Genetic Mapping in Quinoa (*Chenopodium quinoa* Willd.). *J. Genet.* **2008**, *87*, 39–51. [CrossRef]
52. Christensen, S.A.; Pratt, D.B.; Pratt, C.; Nelson, P.T.; Stevens, M.R.; Jellen, E.N.; Coleman, C.E.; Fairbanks, D.J.; Bonifacio, A.; Maughan, P.J. Assessment of Genetic Diversity in the USDA and CIP-FAO International Nursery Collections of Quinoa (*Chenopodium quinoa* Willd.) Using Microsatellite Markers. *Plant Genet. Resour.* **2007**, *5*, 82. [CrossRef]
53. Salazar, J.; de Lourdes Torres, M.; Gutierrez, B.; Torres, A.F. Molecular Characterization of Ecuadorian Quinoa (*Chenopodium quinoa* Willd.) Diversity: Implications for Conservation and Breeding. *Euphytica* **2019**, *215*, 60. [CrossRef]
54. Wilson, H.D. Quinua Biosystematics II: Free-Living Populations. *Econ. Bot.* **1988**, *42*, 478–494. [CrossRef]
55. Jellen, E.N.; Kolano, B.A.; Sederberg, M.C.; Bonifacio, A.; Maughan, P.J. Chenopodium. In *Wild Crop Relatives: Genomic and Breeding Resources*; Kole, C., Ed.; Springer: Berlin/Heidelberg, Germany, 2011; pp. 35–61. ISBN 978-3-642-14386-1.
56. Navarrete, S.; Suasnavas, B. *Análisis Filogenético de Quinua Silvestres (Chenopodium Quinoa Willd.) Utilizando el Gen rpoB en Muestras Provenientes de las Provincias de Carchi, Imabura, Cotopaxi y Chimborazo*; Biotecnología de los Recursos Naturales, Salesian Polytechnic University: Quito, Ecuador, 2015.
57. Jellen, E.N.; Jarvis, D.E.; Hunt, S.P.; Mangelsen, H.H.; Maughan, P.J. New Seed Collections of North American Pitseed Goosefoot (*Chenopodium berlandieri*) and Efforts to Identify Its Diploid Ancestors through Whole-Genome Sequencing. *Cienc. Investig. Agrar.* **2019**, *46*, 187–196. [CrossRef]
58. Yao, S.; Lan, H.; Zhang, F. Variation of Seed Heteromorphism in *Chenopodium album* and the Effect of Salinity Stress on the Descendants. *Ann. Bot.* **2010**, *105*, 1015–1025. [CrossRef]
59. Wilson, H.; Manhart, J. Crop/Weed Gene Flow: *Chenopodium quinoa* Willd. and *C. berlandieri* Moq. *Theor. Appl. Genet.* **1993**, *86*, 642–648. [CrossRef]
60. Murphy, K.M.; Matanguihan, J.B.; Fuentes, F.F.; Gómez-Pando, L.R.; Jellen, E.N.; Maughan, P.J.; Jarvis, D.E. Quinoa Breeding and Genomics. In *Plant Breeding Reviews*; John Wiley & Sons, Ltd.: Hoboken, NJ, USA, 2018; pp. 257–320. ISBN 978-1-119-52135-8.
61. Hernández Medina, P.; Hernández Maqueda, R.; Gavilanes, P.; Marin Quevedo, K.P. Culture and Healthy Lifestyle: Factors Influencing the Decision to Buy Quinoa in the City of Latacunga in Cotopaxi Province, Ecuador. *J. Food Prod. Mark.* **2020**, *26*, 440–455. [CrossRef]
62. Peralta, E.; Murillo, A.; Mazón , N.; Rodriguez Ortega, D.G.; Minchala, L.; Dominguez Olmedo, D.E.; Mina Chala, D.F. Evaluación de 239 Accesiones y 30 Líneas de Quinua (Chenopodium Quinoa Willd.) en Las Condiciones Agroecológicas de la Península de Santa Elena, Ecuador. In Proceedings of the 5th Congreso Mundial de la Quinua, 2nd Simposio Internacional de Granos Andinos, San Salvador de Jujuy, Argentina, 27–30 May 2015; FAO: Rome, Italy; p. 13.
63. Angeli, V.; Miguel Silva, P.; Crispim Massuela, D.; Khan, M.W.; Hamar, A.; Khajehei, F.; Graeff-Hönninger, S.; Piatti, C. Quinoa (*Chenopodium quinoa* Willd.): An Overview of the Potentials of the "Golden Grain" and Socio-Economic and Environmental Aspects of Its Cultivation and Marketization. *Foods* **2020**, *9*, 216. [CrossRef]
64. Madrid, D.; Salgado, E.; Verdugo, G.; Olguín, P.; Bilalis, D.; Fuentes, F. Morphological Traits Defining Breeding Criteria for Coastal Quinoa in Chile. *Not. Bot. Horti Agrobot. Cluj-Napoca* **2018**, *46*, 190–196. [CrossRef]
65. Apaza, V.; Cáceres, G.; Estrada, R.; Pinedo, R. *Catalogue of Commercial Varieties of Quinoa in Peru: A Future Planted Thousands of Years a Ago*; FAO: Rome, Italy, 2015; ISBN 978-92-5-108765-7.

66. Mestanza Uquillas, C.A.; Zambrano Calderón, K.; Pinargote Alava, J.; Veliz Zamora, D.; Vásconez Montufar, G.; Fernández-García, N.; Olmos, E. Evaluación agronómica de genotipos de quinua (*Chenopodium quinoa* Willd.) En condiciones agroclimáticas en la zona de mocache. *Cienc. Tecnol.* **2019**, *12*, 19–30. [CrossRef]
67. Mestanza Uquillas, C.; Santana Chávez, J.; Veliz Zamora, D.; Vasconez, G. Rendimiento de grano de genotipos de quinua (*Chenopodium quinoa* Wild) sembrado a chorro continua, en el campus "La Maria.". *Univ. Cienc. Tecnol.* **2020**, *1*, 53–60.
68. Tang, Y.; Li, X.; Chen, P.X.; Zhang, B.; Hernandez, M.; Zhang, H.; Marcone, M.F.; Liu, R.; Tsao, R. Lipids, Tocopherols, and Carotenoids in Leaves of Amaranth and Quinoa Cultivars and a New Approach to Overall Evaluation of Nutritional Quality Traits. *J. Agric. Food Chem.* **2014**, *62*, 12610–12619. [CrossRef]
69. Schoenlechner, R.; Drausinger, J.; Ottenschlaeger, V.; Jurackova, K.; Berghofer, E. Functional Properties of Gluten-Free Pasta Produced from Amaranth, Quinoa and Buckwheat. *Plant Foods Hum. Nutr. Dordr. Neth.* **2010**, *65*, 339–349. [CrossRef]
70. Altuna, J.L.; Silva, M.; Álvarez, M.; Quinteros, M.F.; Morales, D.; Carrillo, W. Ecuadorian Quinoa (*Chenopodium quinoa* Willd) Fatty Acids Profile. *Asian J. Pharm. Clin. Res.* **2018**, *11*, 209. [CrossRef]
71. Vilcacundo, R.; Miralles, B.; Carrillo, W.; Hernández-Ledesma, B. In Vitro Chemopreventive Properties of Peptides Released from Quinoa (*Chenopodium quinoa* Willd.) Protein under Simulated Gastrointestinal Digestion. *Food Res. Int.* **2018**, *105*, 403–411. [CrossRef]
72. Vilcacundo, R.; Martínez-Villaluenga, C.; Hernández-Ledesma, B. Release of Dipeptidyl Peptidase IV, α-Amylase and α-Glucosidase Inhibitory Peptides from Quinoa (*Chenopodium quinoa* Willd.) during in Vitro Simulated Gastrointestinal Digestion. *J. Funct. Foods* **2017**, *35*, 531–539. [CrossRef]
73. Fiallos-Jurado, J.; Pollier, J.; Moses, T.; Arendt, P.; Barriga-Medina, N.; Morillo, E.; Arahana, V.; de Lourdes Torres, M.; Goossens, A.; Leon-Reyes, A. Saponin Determination, Expression Analysis and Functional Characterization of Saponin Biosynthetic Genes in *Chenopodium quinoa* Leaves. *Plant Sci.* **2016**, *250*, 188–197. [CrossRef]
74. Quelal, M. Análisis de la Cadena Agroproductiva de la Quinua (*Chenopodium Quinoa* Willd), en Las Provincias de Chimborazo e Imbabura. Bachelor's Thesis, Ingeniera Agroindustrial, Escuela Politecnica Nacional, Quito, Ecuador, 2009.
75. Villacrés, E.; Peralta, E.; Egas, L.; Mazon, N. *Potencial Agroindustrial de la Quinua*; Instituto Nacional Autónomo de Investigaciones Agropecuarias: Quito Ecuador, 2011; p. 32.
76. Villacrés, E. *La Riqueza Oculta de las Hojas de Quinua (Chenopodium quionoa Willd)*; INIAP, Estación Experimental Santa Catalina, Departamento de Nutrición y Calidad: Quito, Ecuador, 2016.
77. Vázquez-Luna, A.; Pimentel, V.C.; Carmona, F.F.; Sobac, R.D. Quinoa Leaf as a Nutritional Alternative. *Int. J. Agric. Nat. Resour.* **2019**, *46*, 137–143. [CrossRef]
78. Pichazaca, N.; Quishpilema, L.D.; Pichazaca, J.L. Innovación de la agricultura familiar kañari a través de la cadena de valor de granos para sostener la seguridad alimentaria y conservar la agro-biodiversidad local. *Mem. Univ. Azuay* **2018**, 183–198.
79. Adams, A. Lino's Dream. Available online: https://www.esperanzaproject.com/2019/native-american-culture/linos-dream-kanari-future-traditions-ecuadorean-andes-2/ (accessed on 3 November 2020).
80. Murphy, K.M.; Bazile, D.; Kellogg, J.; Rahmanian, M. Development of a Worldwide Consortium on Evolutionary Participatory Breeding in Quinoa. *Front. Plant Sci.* **2016**, *7*, 608. [CrossRef]
81. Döring, T.F.; Knapp, S.; Kovacs, G.; Murphy, K.; Wolfe, M.S. Evolutionary Plant Breeding in Cereals—Into a New Era. *Sustainability* **2011**, *3*, 1944–1971. [CrossRef]
82. Kellogg, J.; Murphy, K. Evolutionary Participatory Quinoa Breeding for Organic Agro-Ecosystems in the US Pacific Northwest. In *Farmers and Plant Breeding Current Approaches and Perspectives*; Westengen, O.T., Winge, T., Eds.; Routlege Taylor & Francis Group: New York, NY, USA, 2019; p. 24, ISBN 978-1-138-58942-8.
83. Murphy, K.; Lammer, D.; Lyon, S.; Carter, B.; Jones, S.S. Breeding for Organic and Low-Input Farming Systems: An Evolutionary–Participatory Breeding Method for Inbred Cereal Grains. *Renew. Agric. Food Syst.* **2005**, *20*, 48–55. [CrossRef]
84. Suneson, C.A. An Evolutionary Plant Breeding Method. *Agron. J.* **1956**, *48*, 188–191. [CrossRef]
85. Murphy, K.M.; Carter, A.H.; Jones, S.S. Evolutionary Breeding and Climate Change. In *Genomics and Breeding for Climate-Resilient Crops: Volume 1 Concepts and Strategies*; Kole, C., Ed.; Springer: Berlin/Heidelberg, Germany, 2013; pp. 377–389, ISBN 978-3-642-37044-1.
86. Merrick, L.F.; Lyon, S.R.; Balow, K.A.; Murphy, K.M.; Jones, S.S.; Carter, A.H. Utilization of Evolutionary Plant Breeding Increases Stability and Adaptation of Winter Wheat across Diverse Precipitation Zones. *Sustainability* **2020**, *12*, 9728. [CrossRef]

MDPI
St. Alban-Anlage 66
4052 Basel
Switzerland
Tel. +41 61 683 77 34
Fax +41 61 302 89 18
www.mdpi.com

Plants Editorial Office
E-mail: plants@mdpi.com
www.mdpi.com/journal/plants

www.ingramcontent.com/pod-product-compliance
Lightning Source LLC
Chambersburg PA
CBHW061453220926
51850CB00004B/84
9783036570433